Gravity and Magnetic Exploration

Principles, Practices, and Applications

This combined study and reference text provides a comprehensive account of the principles, practices, and application of gravity and magnetic methods for exploring the subsurface using surface, subsurface, marine, airborne, and satellite measurements. Key current topics and techniques are described, including high-resolution magnetic investigations, time-variation gravity analysis from surface and satellite gravity measurements, absolute and gradient gravimetry, and the role of GPS in mapping gravity and magnetic fields. The book also describes the physical properties of rocks and other Earth materials that are critical to the effective design, implementation, and interpretation of surveys, and presents an overview of digital data analysis methods used to process and interpret anomalies for subsurface information.

Each chapter starts with a general overview and concludes with a list of key concepts that help readers review what they have learned. An appendix provides a grounding on basic data analysis using simple and accessible mathematical notation. Study questions and problem sets on an accompanying website, together with computer-based exercises available online, give readers hands-on experience of processing, modeling, and interpreting gravity and magnetic anomaly data. A comprehensive suite of full-color case histories on the book's website illustrates the practical utility of modern gravity and magnetic surveys in energy, mineral, environmental, archaeological, and engineering exploration and lithospheric studies, as well as their potential limitations.

This book is an ideal text for advanced undergraduate and graduate courses but also serves as a reference for research academics, professional geophysicists, and managers of exploration programs that use gravity and magnetic methods. It is a valuable resource for all those interested in petroleum, engineering, mineral, environmental, geological, and archaeological exploration of the lithosphere.

WILLIAM J. HINZE is Emeritus Professor of Geophysics at Purdue University, and has taught exploration geophysics for 40 years. His extensive teaching and research experience is complemented by industrial consulting experience in the geological and engineering applications of gravity and magnetic fields. He has authored or co-authored more than 130 journal publications, and four books, and has served as Associate Editor of *Geophysics* and Senior Editor of *The Journal of Geophysical Research – Solid Earth*. Professor Hinze has been a member of numerous government and scientific panels dealing with key issues from the geophysics and geology of continents to nuclear waste disposal and geophysical data. He is a member of the Society of Exploration Geophysicists, the American Geophysical Union, and the Geological Society of America.

RALPH R. B. VON FRESE is Professor of Earth Sciences at The Ohio State University, where he has taught undergraduate and graduate courses across geophysics and Earth systems since 1982. His research has focused mostly on archaeological and planetary applications of gravity and magnetic fields, and he has authored or co-authored more than 100 journal publications and served on several government and scientific panels. Professor von Frese is the founding co-chair of the Antarctic Digital Magnetic Anomaly Project (ADMAP), an international collaboration of the Scientific Committee for Antarctic Research (SCAR) and the International Association of Geomagnetism and Aeronomy (IAGA). He is a member of the Society of Exploration Geophysicists, the American Geophysical Union, and the Geological Society of America.

AFIF H. SAAD is a Geophysical Consultant to the oil and gas exploration industry, specializing in integrated interpretation, modeling, magnetic depth estimation, software development, and training. He has over 40 years of broad experience in the theory and practice of gravity and magnetic methods in exploration, including 25 years in the oil industry, and has provided training at several schools and workshops in basic, intermediate, and advanced gravity and magnetics in the United States and overseas. Dr Saad has authored or co-authored several publications, has peer-reviewed numerous papers and textbooks in gravity and magnetics for *Geophysics* and the *Geophysical Journal International*, and has served as Associate Editor for *Magnetic Exploration Methods for Geophysics*. He is a member of the Society of Exploration Geophysicists (SEG), the SEG Gravity and Magnetics Committee, the SEG Global Affairs Committee, and the Geophysical Society of Houston.

"Written by three leading researchers, this is a comprehensive textbook that takes its readers from the fundamentals of potential fields through modern data acquisition, processing, modeling and inversion to practical interpretation. The theory and mathematical derivations are suitable for both beginners and experienced geophysicists. Well-organized and nicely illustrated, it is both informative and clearly written."

– Professor Dr Alan Green

Institute of Geophysics, ETH-Swiss Federal Institute of Technology

"This extensive work is much more than just a textbook: it includes detailed discussions, such as the philosophy of modeling and the nature of errors, which are critical to properly interpreting gravity and magnetic data, but are often glossed over. A very useful addition to practitioners' reference shelves and an excellent textbook for advanced students."

– Roger C. Searle, Professor Emeritus

Department of Earth Sciences, Durham University

"The geophysical applications of gravity and magnetic techniques have advanced a great deal in the twenty-first century. Thus, this rigorous book covering the physical basis, analysis, interpretation, and applications of these techniques is a timely and important contribution. It is designed to serve both the student and practitioner and is enhanced by an innovative website."

– Professor G. Randy Keller

Geology and Geophysics, University of Oklahoma;
Director of Oklahoma Geological Survey

Gravity and Magnetic Exploration

Principles, Practices, and Applications

William J. Hinze
Purdue University

Ralph R. B. von Frese
The Ohio State University

Afif H. Saad
Saad GeoConsulting

CAMBRIDGE
UNIVERSITY PRESS

CAMBRIDGE
UNIVERSITY PRESS

University Printing House, Cambridge CB2 8BS, United Kingdom

One Liberty Plaza, 20th Floor, New York, NY 10006, USA

477 Williamstown Road, Port Melbourne, VIC 3207, Australia

314-321, 3rd Floor, Plot 3, Splendor Forum, Jasola District Centre, New Delhi - 110025, India

79 Anson Road, #06-04/06, Singapore 079906

Cambridge University Press is part of the University of Cambridge.

It furthers the University's mission by disseminating knowledge in the pursuit of education, learning and research at the highest international levels of excellence.

www.cambridge.org
Information on this title: www.cambridge.org/9780521871013

First published 2013
Reprinted 2017

A catalogue record for this publication is available from the British Library

ISBN 978-0-521-87101-3 Hardback

Additional resources for this publication at www.cambridge.org/gravmag

To our wives, Marilyn Ann Hinze, Janet Shaw-von Frese, and Linda C Saad, whose understanding and encouragement enabled us to complete this book.

Contents

Preface *page* xi
Objectives of this book xi
Related books xii
Organization of this book xii
Study questions and exercises xiii
Units xiii
Acknowledgements xiv

1 Introduction 1
1.1 Overview 1
1.2 The Earth and its planetary force fields 2
1.3 Basis of the gravity and magnetic methods 3
1.4 Foundations of geophysical methods 5
1.5 Geophysical practices 7
1.6 Nature of geophysical data 11
1.7 Key concepts 14

Part I Gravity exploration 17

2 The gravity method 19
2.1 Overview 19
2.2 Role of the gravity method 20
2.3 The Earth's gravity field 21
2.4 History of the gravity method 27
2.5 Implementing the gravity method 32
2.6 Key concepts 36

3 Gravity potential theory 38
3.1 Overview 38
3.2 Introduction 38
3.3 Gravity effects of a point mass 39
3.4 Gravity effects of an extended body 41
3.5 Idealized source gravity modeling 48
3.6 General source gravity modeling 53
3.7 Gauss' law 59
3.8 Gravity anomaly ambiguity 60
3.9 Poisson's theorem 61
3.10 Pseudoanomalies 61
3.11 Key concepts 62

4 Density of Earth materials 64
4.1 *Overview* 64
4.2 *Introduction* 64
4.3 *Types of densities* 65
4.4 *Density of the Earth's interior* 66
4.5 *Rock densities* 68
4.6 *Density measurements* 75
4.7 *Density tabulations* 85
4.8 *Key concepts* 86

5 Gravity data acquisition 88
5.1 *Overview* 88
5.2 *Introduction* 89
5.3 *Measuring gravity* 89
5.4 *Gravity surveying* 107
5.5 *Gravity measurements from space* 113
5.6 *Key concepts* 120

6 Gravity data processing 122
6.1 *Overview* 122
6.2 *Introduction* 122
6.3 *Extraneous gravity variations* 123
6.4 *Gravity anomalies* 143
6.5 *Anomaly isolation and enhancement* 155
6.6 *Key concepts* 173

7 Gravity anomaly interpretation 175
7.1 *Overview* 175
7.2 *Introduction* 175
7.3 *Interpretation parameters* 181
7.4 *Simplified interpretation techniques* 189
7.5 *Modeling anomaly sources* 200
7.6 *Key concepts* 211

Part II Magnetic exploration 213

8 The magnetic method 215
8.1 *Overview* 215
8.2 *Role of the magnetic method* 215
8.3 *The Earth's magnetic field* 216
8.4 *History of the magnetic method in exploration* 229
8.5 *Implementing the magnetic method* 231
8.6 *Key concepts* 233

9 Magnetic potential theory 235
9.1 *Overview* 235
9.2 *Introduction* 235
9.3 *Magnetic potential of a point dipole* 236
9.4 *Magnetic effects of a point dipole* 236
9.5 *Magnetic effects of an extended body* 238
9.6 *Idealized source magnetic modeling* 243
9.7 *General source magnetic modeling* 244
9.8 *Total magnetic moment* 250

9.9 Magnetic source ambiguity 250
9.10 Combined magnetic and gravity potentials 250
9.11 Key concepts 251

10 Magnetization of Earth materials 252
10.1 Overview 252
10.2 Introduction 252
10.3 Magnetism of Earth materials 253
10.4 Mineral magnetism 257
10.5 Magnetic susceptibility 259
10.6 Magnetization of rocks and soils 261
10.7 Magnetic property measurements 272
10.8 Magnetic property tabulations 273
10.9 Key concepts 274

11 Magnetic data acquisition 276
11.1 Overview 276
11.2 Introduction 276
11.3 Instrumentation 277
11.4 Survey design and procedures 284
11.5 Magnetic measurements from space 291
11.6 Key concepts 297

12 Magnetic data processing 300
12.1 Overview 300
12.2 Introduction 301
12.3 Extraneous magnetic variations 302
12.4 Anomaly isolation and enhancement 314
12.5 Key concepts 336

13 Magnetic anomaly interpretation 338
13.1 Overview 338
13.2 Introduction 339
13.3 Interpretation constraints 342
13.4 Interpretation techniques 355
13.5 Modeling anomaly sources 394
13.6 Key concepts 411

Part III Applications 413

14 Applications of the gravity and magnetic methods 415
14.1 Introduction 415
14.2 General view of applications 415
14.3 Near-surface studies 416
14.4 Energy resource applications 417
14.5 Mineral resource exploration 418
14.6 Lithospheric investigations 420

Appendix A Data systems processing 422
A.1 Overview 422
A.2 Introduction 422
A.3 Data bases and standards 423
A.4 Mathematical methods 424

A.5 Anomaly analysis 444
A.6 Data graphics 463
A.7 Key concepts 475
References 477
Index 502

Preface

Objectives of this book

Investigations of terrestrial gravity and magnetic fields are among the oldest methods for determining the nature and processes of the Earth. Despite the development of an increasing number of additional complementary investigative methods, some of which have better subsurface resolution, the gravity and magnetic methods continue to have an important, often decisive role, in a wide variety of terrestrial investigations. In contrast to the several available texts on specific topics in gravity and magnetics, this book provides an overall, modern resource on the principles, practices, and applications of both the gravity and magnetic methods to exploring the Earth. Although these aspects of the gravity and magnetic methods are well grounded in widely described and accepted principles, they are continually undergoing practical improvements and expanded understanding. The continued improvements result from enhanced technology for acquiring, processing, and interpreting data made possible largely by increasing computational power and new techniques. Special emphasis in gravity and magnetic exploration is being placed on high sensitivity mapping of anomalies at the extremities of their spectra, both shorter and longer wavelengths, increasing the vertical and horizontal resolution of individual anomaly sources, and investigating the temporal variations in the gravity and magnetic fields of the Earth which permit new insights into Earth processes. It is hoped that this book which sets a benchmark in our current knowledge of gravity and magnetic exploration will encourage further developments in these methods and applications.

Both the theory and practice of gravity and magnetic methods applied to subsurface investigations are described in this book. The book considers the methods from the planning and organization of surveys through the analysis and interpretation of observations by digital computations incorporating both physical and geological principles. Case histories in supplemental chapters, which are available on an accompanying website, illustrate the advantages and limitations of the methods in a variety of applications including near-surface engineering and archaeological studies, mineral and energy resource investigations, and planetary crustal and subcrustal studies. A final chapter of this book summarizes these applications.

This presentation of gravity and magnetic methods differs substantially from existing books. For example, it takes into account the rapidly accelerating availability of subsurface, terrestrial, marine, airborne, and satellite measurements that the current information age is mapping in prodigious volumes. Further, it develops strategies on the combined use of these anomaly fields for solving subsurface problems. It also describes the newest standards and methods for reducing gravity and magnetic data to a usable form, as well as modern instrumentation for observing both absolute and relative total field and vectorial components.

The book emphasizes practical procedures and applications, and the physical properties and geological principles that constrain subsurface analyses of gravity and magnetic fields. It describes all major modern topics of gravity and magnetic methods that are likely to be of general interest. More general descriptive material and procedures are highlighted, and reference is made to alternate procedures or approaches. However, necessarily numerous computational and interpretational procedures described in the geophysical literature which may have a more limited application are not included because of space limitations. Wherever possible, examples of these methods are briefly described and pertinent literature is cited to guide the interested reader to additional information. Derivations of equations are restricted to those of general interest and details of many procedures are not included in the descriptions. Therefore the reader of this text will need to make abundant use of the many references cited for more detailed and comprehensive treatment of specialized topics.

The book can serve university courses in exploration and general Earth geophysics, gravity exploration, and magnetic exploration at the advanced undergraduate and beginning graduate levels. It uses mathematics up to and including basic differential equations, but develops the methods in simple digital array operations with minimal use of arcane and complex notations from integral and differential calculus. Thus, this book is much more computationally oriented than most previous works because the reader can readily implement and explore analytical results with electronic computing. Consistent abbreviations and terminology are used throughout the book.

The book is not only a textbook, but also serves as a reference for practioners, that is for professional geophysicists, geologists, engineers, and other scientists who have limited experience in the application of the gravity and magnetic methods to subsurface problems. Especially useful in this regard are the supplemental case histories, which have extensive reference lists. The explanation of the gravity and magnetic methods along with the case histories will also facilitate the efforts of project managers to develop optimal strategies for implementing geophysical methods in solving subsurface problems. Thus, the book serves a wide audience ranging from advanced undergraduate and beginning graduate students in the Earth and planetary sciences and engineering to professional geoscientists and engineers in academia, government, and industry.

Related books

This comprehensive book is unique in the breadth of the content and approach. However, a number of books pertaining to the gravity and magnetic exploration methods and their interpretation are available that the reader may find to be useful supplements to this book. A selected list of these books includes *Introduction to Potential Theory* (SIGL, 1985), *Interpretation of Filtered Gravity Maps* (STEINER and ZILAHI-SEBESS, 1988), *Geophysical Data Analysis* (MENKE, 1989), *Geophysical Inverse Theory* (PARKER, 1994), *Potential Theory in Gravity and Magnetic Applications* (BLAKELY, 1995), *Geologic Applications of Gravity and Magnetics: Case Histories* (GIBSON and MILLEGAN, 1998), *Geophysical Inverse Theory and Regularization Problems* (ZHDANOV, 2004), *Principles of the Gravitational Method* (KAUFMAN and HANSEN, 2008), *Principles of the Magnetic Methods in Geophysics* (KAUFMAN *et al.*, 2008), *Gravity and Magnetic Interpretation in Exploration Geophysics* (MURTHY, 2010), *Gravity and Magnetic Methods for Geological Studies* (MISHRA, 2011), *Field Geophysics* (MILSON and ERIKSEN, 2011), Fundamentals of Gravity Exploration (LAFEHR and NABIGHIAN, 2012), and Acquisition and Analysis of Terrestrial Gravity Data (LONG and KAUFMANN, 2013).

Organization of this book

This book consists of 14 chapters organized into two principal parts, dealing with gravity and magnetic exploration, plus an introductory chapter and a final chapter dealing with applications. The initial chapter describes the basis and foundations of the gravity and magnetic methods, the general components of the geophysical process, and the nature of geophysical data. The final chapter considers the application of gravity and magnetic methods to near-surface studies including engineering, environmental, and archaeological investigations, energy and mineral resource exploration, and geologic studies of the lithosphere.

Each of the two principal parts consists of six chapters involving an introduction to the method, germane potential theory, the physical property involved in the method, and three final chapters which describe the acquisition, processing, and interpretation of the data. Finally, an appendix deals with data systems processing principles which are important to the processing and interpretation of gravity and magnetic methods. Included in the appendix are a discussion of gravity and magnetic data bases and standards, mathematical methods widely employed in gravity and magnetic methods, anomaly analysis, and data graphics. It serves as important background to both the gravity and magnetic chapters.

The gravity and magnetic parts each contain six chapters that may be used independently in separate courses or study programs. Overlap between the parts is minimal, but some methods and principles are more fully developed in one part than in the other part. For example, some depth determination methods used in both gravity and magnetic studies are more comprehensively described in the magnetics chapters because they are more widely used and successful in this application. The text explains the need for the reader to study fuller explanations elsewhere to achieve a comprehensive understanding of the methods.

As a study aid to the reader, each chapter is introduced with an overview and concludes with a summary section listing the key concepts of the chapter. The overview is not an abstract, but rather provides the reader with a broad, generalized summary of the chapter that is a useful guide in reading and studying the chapter. The summary key concepts draw the reader's attention to the more important concepts that are presented. They are not a listing of what should be known upon reading the chapter, but rather guide the reader in reviewing what has been emphasized in the chapter.

Important elements of the book are the illustrations and tables, which aid in understanding principles and provide useful data tabulations and examples of gravity and magnetic methods and their analysis. They are derived both from the literature and from original composition for the book. Numerous references guide the reader to the source of specialized information and additional details on a topic. Readers seeking additional definitions of the geoscience terms in geophysics and geology used in this book should consult the *Encyclopedic Dictionary of Exploration Geophysics* (SHERIFF, 2002) and the *Glossary of Geology* (NEUENDORF *et al.*, 2005), respectively.

A website, www.cambridge.org/gravmag, accompanying the book provides useful supplemental material for the user of the book. Several black and white illustrations in the book are reproduced in color on the website to enhance their utility to the reader. Additionally, a few complex figures are shown in expanded size for improved legibility. Four additional chapters are provided on the website that describe the applications of both the gravity and magnetic methods to near-surface investigations, energy resources, mineral resources, and lithospheric investigations. The application chapters review the use of the methods and cite examples. The examples demonstrate the scale and magnitude of the anomalies, their breadth of application, and the limitations of the methods. The selection of examples is necessarily limited, but we have attempted to reach a balance in the selection process of including traditional applications plus unique examples to illustrate the range of applications. Placing these chapters on the website has permitted greater coverage, extensive use of color in the illustrations, and provision for updating examples with improvements in technology and broadening the range of problems addressed by the gravity and magnetic methods.

Study questions and exercises

In addition to the supplemental material, the website contains study questions for the first 13 chapters of the book. These questions foster further understanding of the topics of these chapters and serve as a study guide for the reader and a resource for instructors. Exercises, problem sets, and practical examples of application of the gravity and magnetic methods are also presented on the website. Exercises have been developed with the cooperation of Geosoft, Inc. utilizing special versions of their Oasis montaj software. The reader of this book can readily access the Geosoft Inc. software through the link provided on the website to process and interpret the data supplied for the exercises. The exercises are keyed to specific sections of chapters of the book where the user can gain knowledge of the basis, background, and application of the methodologies employed in the programs. Efforts will be made to keep the software available and the exercises up to date with current practice in the gravity and magnetic exploration profession. Updates of the content are identified on the website.

Some of the exercises on the website involve the forward calculation of anomalies, both gravity and magnetic, and discussion of the results of the modeling. These calculations can be performed using the special Oasis montaj versions for this book. Alternatively, the software for modeling, inversion, and filtering of profile gravity and magnetic anomaly data prepared by Professor Gordon R. J. Cooper, University of Witwatersrand can be used. The latter software can be obtained on the website http://www.wits.ac.za/academic/science/geosciences/research/geophysics/gordoncooper/6511/software.html. Descriptions of these programs and their use are presented with the software.

Units

The units used in geophysical methods are diverse, depending largely on the application of the methods. In this book the normal practice is to use SI units. SI is the abbreviation for Le System International d'Unites, which is a system of units that is broadly accepted internationally by governmental agencies and professional societies. This system has similarities to the metric system of units, but is not identical to it. The base and supplementary units of the SI system together with their combinations, called derived units, that are common to geophysical studies in this book are listed in Table 1.2 of Chapter 1. In this book the exception to the use of SI units is in the description of certain case histories. The original units used in reporting the results of the study, which may not be SI units, are used for consistency with the original information.

Acknowledgements

We wish to acknowledge our colleagues and students who have encouraged us to write a modern book on the gravity and magnetic methods and supported us in this effort. We particularly recognize the students, faculty, and administration of Purdue University (WJH) and The Ohio State University (RvF) and the support of the Chevron Oil Company (AHS) in the development of interpretational procedures described in this book. We acknowledge the contribution of the following colleagues who have reviewed early drafts of one or more sections of the book: Mohammad F. Asgharzadeh, Lawrence W. Braile, Val W. Chandler, David A. Chapin, John D. Corbett, Hyung Rae Kim, Dmitry Koryakin, Xiong Li, Neil M. Coleman, Dhananjay Ravat, Michal Ruder, Richard W. Saltus, Patrick T. Taylor, and Daniel Winester. Their advice has been invaluable to us, but they are not responsible for errors of omission or commission in the final manuscript. We also acknowledge the cooperation of Colin Reeves in making available his useful web-based publication entitled "Aeromagnetic Surveys."

Special thanks are extended to Ian MacLeod, Elizabeth Baranyi, and Gerry Connard of Geosoft Inc. for their continuing cooperation in providing software and assistance in organizing and implementing geophysical processing and interpretation exercises for this book. We are particularly grateful to Elizabeth Baranyi of Geosoft for preparing the geological model and its anomaly fields shown on the cover of this book and assisting the authors in the preparation of the exercises on the website that employ Geosoft Inc. software. We acknowledge with gratitude the cooperation of Professor Gordon R. J. Cooper of the University of Witwatersrand, Johannesburg, South Africa in providing the users of this book with ready access to his computational software. Yuriy Yeremenko has provided invaluable assistance in preparing the figures. For this we express our deep appreciation.

We also want to acknowledge the continuing assistance from our Cambridge University Press editor, Laura Clark, and also her patience in dealing with three authors with a wide variety of commitments that complicated the completion of the book.

1 Introduction

1.1 Overview

Geophysics, the science of the physics of the Earth from its magnetosphere to the deep interior, is useful in characterizing the subsurface Earth. Solid-Earth geophysics employs techniques involving the measurement of force fields to study subsurface features and the processes that act upon them. Thus, geophysical studies serve a broad variety of geologic, natural resource, engineering, and environmental purposes. Gravity and magnetic methods, which measure very small spatial and temporal changes in the terrestrial gravity and magnetic force fields, have a wide range of uses from submeter to global scales. Although these methods in most cases fail to match the resolution and precision of direct observations, they are rapid, cost-effective, and non-invasive procedures of studying the inaccessible Earth and optimizing the location of drill holes for direct studies and other remote sensing studies which have higher resolution capabilities.

The application of gravity and magnetic methods generally involves a common approach consisting of planning, data acquisition, data processing, interpretation, and reporting phases. During the planning phase the appropriate method(s) are selected for meeting the objective of the study, and procedures for data acquisition, processing, and interpretation are established. These decisions are reached on the basis of experience, model studies, or test surveys. Special care is taken to determine an error or noise budget for the survey and to consider the propagation of errors, both random and systematic, through the data acquisition and processing chain. Selection of the distribution of observations in the survey region includes consideration of the objective of the study, the geologic, topographic, vegetative cover, and cultural features of the area, access over the region, and financial considerations. The geophysical observations are subject to numerous analytical processing steps to minimize effects from non-germane sources. Interpretation of these processed data involves not only determining the distribution of anomalous masses in the subsurface, but the nature of these masses. The latter commonly requires the translation of properties directly measured by the geophysical method into secondary properties, such as lithology, porosity, and strength, which are more directly related to the survey objective. Interpretation is achieved by transforming the survey data to quantitative models of the subsurface that satisfy the data. However, all interpretations are subject to ambiguities that to a degree depend on the implemented method and procedures and the integration of the results with collateral geological and geophysical information.

1.2 The Earth and its planetary force fields

Geophysics is an interdisciplinary science that integrates the observations, hypotheses, and laws of geology with the techniques and principles of physics to understand the composition, nature, structure, and processes of the Earth. Geophysics involves measuring and interpreting phenomena related to the physical nature of the Earth, from its center some 6,371 km beneath the surface to the outer limits of its magnetosphere at altitudes many times the

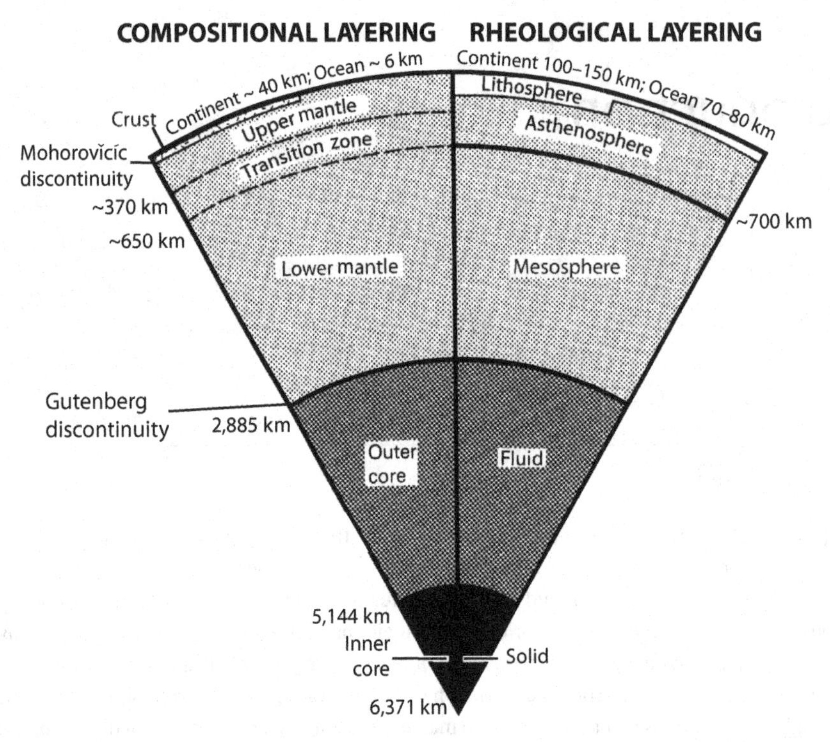

FIGURE 1.1 Cross-section of a segment of the Earth showing major first-order internal subdivisions in composition and mechanical or rheological properties. Table 1.1 lists the mean densities and magnetizations of the Earth's major structural components. Adapted from KEARY and VINE (1990).

radius of the Earth. Thus, it incorporates investigations of the subsurface, hydrosphere, atmosphere, ionosphere, and magnetosphere. In this book, the focus is on solid-Earth geophysics, considering the properties and processes of the Earth primarily within the crust and uppermost mantle (lithosphere) as reflected in the spatial and temporal variations in gravity and magnetic force fields. We are all very aware of these planetary fields. The gravity field is the source of the force which causes all objects to be attracted toward the Earth, and the geomagnetic field controls the compass which is useful in determining geographic directions. These force fields have been and continue to be an important part of the science of geophysics.

Applications of gravity and magnetic methods include micro-scale surveys to map the physical property variations of the upper meter or two of the subsurface, or conducted within drill holes to establish the physical properties of the adjacent rocks. Larger-scale applications include regional to global surveys designed to image the deeper variations of the Earth's crust, mantle, and core (Figure 1.1). The crust is the outermost surface rind that consists of surface-like rocks extending to depths as great as 70 km. The crust overlies the mantle, made up of higher density and velocity but generally non-magnetic rocks and

TABLE 1.1 The average densities $<\sigma>$ and magnetizations $<J>$ in kg/m^3 and A/m, respectively, of the Earth's major structural elements shown in Figure 1.1.

Structure	$<\sigma>$	$<J>$
Upper crust	2,200–2,900	0–5
Lower crust	2,800–3,100	2–10
Upper mantle	3,300	0
Asthenosphere	3,300–4,000	0
Lower mantle	4.400–5,500	0
Outer core	9,900–12,200	0*
Inner core	12,800–13,100	0

*If the terrestrial field were caused by magnetization in the Earth's outer core, its effective magnetization would be $\sim 1.7 \times 10^3$A/m.

extending to a depth of roughly 2,900 km, which in turn lies directly on the roughly spherical, dense, largely metallic core of the Earth in which the main terrestrial magnetic field originates. The lithosphere is the outermost semi-rigid shell consisting of the crust and uppermost mantle. It normally has a thickness of roughly 150 km beneath the continents and less in oceanic regions, and is the source of most of the variations in the gravity and magnetic fields of the Earth.

The crust exhibits highly complex structural and compositional properties that reflect the effects of erosion, sedimentation, metamorphism, tectonics, and igneous activity, and the plastic movement of the mobile asthenosphere underlying the lithosphere that have occurred over the Earth's 4,600-Myr history. These processes have led to the differentiation of chemical elements, deposition of a variety of sediments, vertical and horizontal movements, zones of crustal weakness, and the focusing of geological processes, such as volcanism, in limited regions of the Earth. These variations in the nature of the lithosphere and specifically the crust that solid-Earth geophysicists map and investigate are of societal interest because they control the formation and distribution of the Earth's resources, and volcanic, earthquake, and other natural hazards. Geophysics is an efficient and effective method of conducting these investigations, avoiding the problems of direct sampling of the hidden Earth. Nonetheless, these studies come at a cost, because the results of their interpretation are to varying degrees ambiguous and lack the accuracy of direct measurements.

1.3 Basis of the gravity and magnetic methods

Gravity and magnetic methods are commonly referred to as potential field methods because the measurements involve a function of the potential of the observed field of force, either the terrestrial gravity or magnetic field, at the observation site. These methods are widely used at a variety of scales to investigate the Earth because in comparison to most other geophysical methods the acquisition of data is inexpensive and rapid, and for many applications the reduction and interpretation of the observations are relatively simple. Furthermore, gravity and magnetic methods always provide information about the subsurface. In addition, there is a large reservoir of these data covering the entire Earth in varying detail that are publicly available at minimal cost to the user.

1.3.1 Gravity

The gravity method involves measurement of very small variations in the Earth's gravitational field, of the order of a few parts per million or lower, caused by lateral variations in density. Most observations are made with highly specialized weighing devices, called gravimeters or gravity meters, which measure the acceleration of gravity. Less frequently, they are made with instrumentation which measures the gradient or vector components of the gravity field. Gravity variations useful for studying the solid Earth are observed on land, in surface and subsurface water vessels, in drill holes, in the air, and from satellites orbiting the Earth. The variations they measure are dependent on Newton's universal law of gravitation, which takes into account the differential mass and the distance between the source and observation point. Because density is a universal property of matter, gravity is ever-present, but only where the density of the Earth varies laterally will gravity variations be noted that can be related to changes in the nature and structure of the Earth.

Observed gravity variations called anomalies are the differences between the observed and the theoretical field based on planetary considerations and the assumption of radial symmetry of the Earth layers. The anomalies may be either positive or negative depending on the presence of mass excesses or deficiencies, as illustrated in Figure 1.2. Their interpretation is subject to uncertainties in the observation, reduction to anomaly form, and processing and limitations resulting from the inherent ambiguity in their interpretation. However, meaningful interpretations can be obtained with proper use of constraining, collateral geologic, and geophysical information.

A wide range of densities occurs within the crust, from the essentially zero density of air-filled voids in near-surface formations, to densities of unconsolidated sediments with their interstitial openings filled with either air or water, to the highest densities related to iron/magnesium-rich crystalline rocks and metallic ores. Even higher densities are associated with the radial shells that make up the mantle and the core of the Earth. The potentially broad range of contrasting densities in the near-surface, in the crust, and in subcrustal rock materials leads to the wide range of applications of the gravity method.

1.3.2 Magnetics

The magnetic method (commonly referred to as magnetics) is similar to the gravity method in that variations in a planetary field are measured, in this case the magnetic field of the Earth. Observations are readily made to a high precision with portable electronic magnetometers on land, drill holes, sea, and air including measurements from planet-orbiting satellites. Most land areas of the Earth have now been measured by airborne observations and much of the ocean area has been observed either by airborne or shipborne measurements, at least by widely spaced observations. Variations, or anomalies, in the magnetic field of the Earth obey Coulomb's law, which is comparable to Newton's law of gravitation, but takes into account the

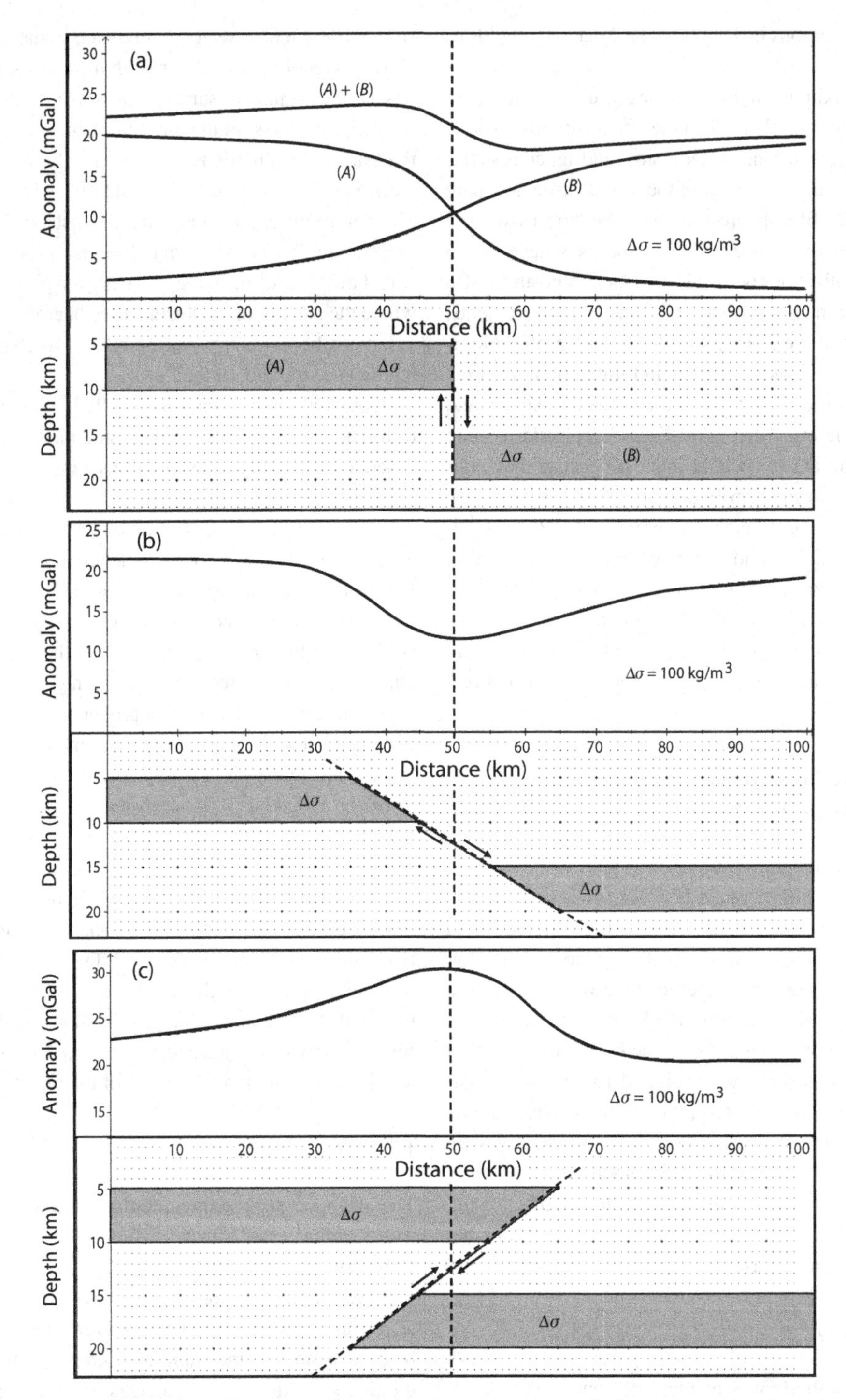

FIGURE 1.2 Examples of vertical gravity effects derived from faulting that offsets a layer of higher density within a formation of lower density. Panel (a) shows the layers vertically offset with the profiles of the gravity effects for the individual layer components (A) and (B), as well as for their total or superimposed effects (A + B). Panels (b) and (c) show the total gravity effect profiles for normal and reverse faulting of the horizontal layers. The illustrated gravity effect is the vertical acceleration of gravity given in units of milligals (mGal) where 1 mGal is equal to 10^{-3} cm/s^2 or 10^{-5} m/s^2.

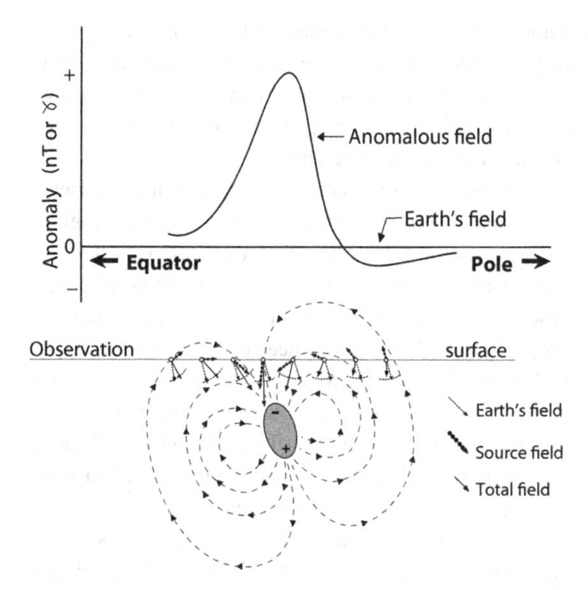

FIGURE 1.3 Example of the total magnetic field variation (anomaly) caused by the magnetic effect of a buried magnetic object (shaded area) magnetized by the Earth's field. The sum of the magnetic field of the anomalous source and the Earth's field produce the observed total magnetic intensity anomaly profile at the top of the figure. Note the asymmetry of the anomaly owing to the inclined magnetization and the negative component of the anomalous field caused by the positive pole of the magnetic source.

magnetic polarization variations of the Earth rather than its mass. Magnetic polarization is dependent on the magnetic susceptibility and the remanent or permanent magnetization of Earth materials. Magnetic susceptibility is a measure of the ease with which a material can be magnetized in the current magnetic field of the Earth; remanent magnetization is the permanent magnetization previously acquired and retained by the material.

All magnetic materials, including the Earth, have two poles, north and south or positive and negative, and thus are called dipolar. Objects of high magnetic susceptibility become polarized or magnetized when they are present in the Earth's dipolar geomagnetic field. Magnetic field observations taken over a buried magnetized object will measure both the positive and negative fields associated with the dipolar magnetization of the object. The resulting anomaly from a high magnetic susceptibility object will combine the fields of both poles as, for example, illustrated in Figure 1.3, but will be dominated by the pole nearest to the observation. The magnetic field of the Earth will induce in the northern geomagnetic hemisphere a negative pole near the top of the anomalous source and a positive pole near the base of the source. The negative pole, being in closer proximity to the observations, will produce a greater attraction on a north-seeking pole (+) than the repulsion from the negative pole. Accordingly, the magnetic field over the anomalous source will be dominated by an increase in the magnetic field over the Earth's field as shown in Figure 1.3. An inverse anomaly dominated by a decrease in the field would be observed over an object with lower magnetization than the surrounding Earth materials. The magnetization of an object magnetized in the Earth's magnetic field will align with the Earth's field. As a result the anomalous field will vary with location on the Earth's surface owing to the dipolar nature of the main field which roughly aligns with the axis of rotation and the resulting changes in the main field over the Earth's surface.

Unlike crustal rock densities, which generally vary by less than a half-order of magnitude and are directionally independent, magnetic polarization commonly varies over several orders of magnitude, giving rise to large property contrasts, and is directionally variable. The directional attribute of magnetization complicates the interpretation of measurements of the magnetic field, as do the presence of both positive (attractive) and negative (repulsive) poles within all magnetic materials as illustrated in Figure 1.3.

An advantage of the magnetic method over the gravity method is that the field varies inversely one power faster with distance to the source than does the gravity field from the same source. As a result, the magnetic method is more sensitive to the source depth, which is commonly an important objective in interpretation of the observations. Furthermore, the resolving power of the method to distinguish independent sources is greater than that of the gravity method. Magnetic field variations are derived from only a few minerals, and these occur only as accessory minerals in most rocks. However, the measured variations are several parts in a hundred thousand or greater; thus magnetic variations are easier and less costly to map than are gravity anomalies from similar sources. Also, magnetic measurements can readily be made from simple, mobile platforms increasing the surveying rate, making them cost-efficient. As a result the magnetic method is widely applied as a reconnaissance tool in geophysical studies and has several specialized applications in shallow subsurface and crustal studies.

1.4 Foundations of geophysical methods

The foundations of geophysics were developed in the last few centuries through scientific studies of surface geological features by pioneering geologists and the study by early physicists of natural force fields of the Earth. From the seventeenth century onwards, geologists such as Steno,

Smith, Werner, Hutton, Playfair, and Lyell established the basic laws of geology, and explained the formation of rocks. These and other principles are explained in introductory geological texts. The reader unfamiliar with the basics of geology is encouraged to learn the key concepts presented in these books because they are fundamental to the understanding and application of geophysical methods in general, and the gravity and magnetic methods in particular.

Contemporaneous to the early geologic studies, physicists investigated a variety of terrestrial force fields and developed theories and laws to explain their observations. Beginning in the sixteenth century, such prominent scientists in the history of physics as Newton, Galilei, Gilbert, Gauss, Coulomb, Volta, Oersted, Ampere, Bouguer, Faraday, Fresnel, and Maxwell contributed greatly to the science of geophysics. By the mid-nineteenth century, they and their peers had essentially established the foundations of gravitational, magnetic, and electrical fields of the Earth and the basic theory upon which are based current studies of these fields in geophysics. At about that time, instrumentation was becoming available for field geophysical measurements and the potential for subsurface studies with these measurements was being identified.

Building upon improved instrumentation and interpretational techniques developed in the succeeding decades of the twentieth century, particularly after World War I, great progress was made in the use of gravity and magnetic methods in the search for Earth resources. Technological developments, primarily in electronics, during World War II made instrumentation improvements that led to the broad use of computers, electronic magnetometers, accelerometers, ground-penetrating radar, digital recording, and other advanced instrumentation of geophysics. Post-World War II geophysical investigations in geomagnetism, seismology, paleomagnetism, and isotopic age dating of rocks led to the development of the paradigms of seafloor spreading and plate tectonics by Vine, Matthews, Morley, Morgan, Sykes, Runcorn, Oliver, Wilson, Heirtzler, Dalrymple, and others. These paradigms explain the slow movement of crustal units over the Earth's surface as well as the destruction of existing crust and the construction of new crust by interaction with the Earth's mantle. These concepts are essential to understanding the evolution and the geological and physical processes of the Earth, and thus to the application of gravity and magnetic methods.

The latter half of the twentieth century saw technological improvements that resulted in more precise, portable, and inexpensive instrumentation and faster computations. The continuing improvement in computers has been fundamental to all of geophysics. Not only have computers made it possible to collect and store huge amounts of data, but they are the keystones to current data processing and presentation technology responsible for today's broad success and acceptance of geophysics.

Progress in geophysics has been driven by societal needs and economic factors as well as by technological advances. In the 1920s and 1930s, the worldwide surge in the number and use of automobiles, with their gasoline-powered internal combustion engines, increased the need for petroleum products. This need could not be met solely with production from petroleum traps located by surface geologic information and wildcat drilling. Geophysics stepped into this void by greatly increasing the chances of discovery. The growth in petroleum exploration geophysics was accelerated by the ever-increasing demands of the post-war surge in the world's economy. The societal and economic pressure caused a revolution in petroleum exploration geophysics that continues today. In a similar way, post-World War II industrial developments and the depletion of mineral resources during the global war forced the broadening of mineral exploration to the geophysical search for new mineral districts and ore deposits which have little or no surface indication.

Petroleum geophysical exploration began with instrumental developments in the early part of the twentieth century that permitted gravity to be measured with a precision necessary to study subsurface geologic structures. These developments led to the first geophysical discovery of petroleum in the United States, which was by the gravity method, in the early 1920s. The use of gravity in petroleum exploration reached a peak shortly after World War II, but its relative role decreased as the reflection seismic method was improved, largely as a result of the computational power of computers and related theoretical developments. Nonetheless, the gravity method has a significant niche and is especially valuable used in concert with the reflection seismic method to constrain possible interpretations. The improvements in the gravity method for hydrocarbon exploration have given impetus to its use not only for this application, but also for shallow zone and regional exploration of the Earth. The successful development of techniques for measuring gravity to a precision useful for exploration using airborne and satellite platforms has given the method a new range of applications.

The magnetic method has been used since the seventeenth century in mineral exploration, especially for iron ore prospecting, but with the advent of airborne magnetic observations after World War II, it has been used on a broad basis for regional geological studies in petroleum

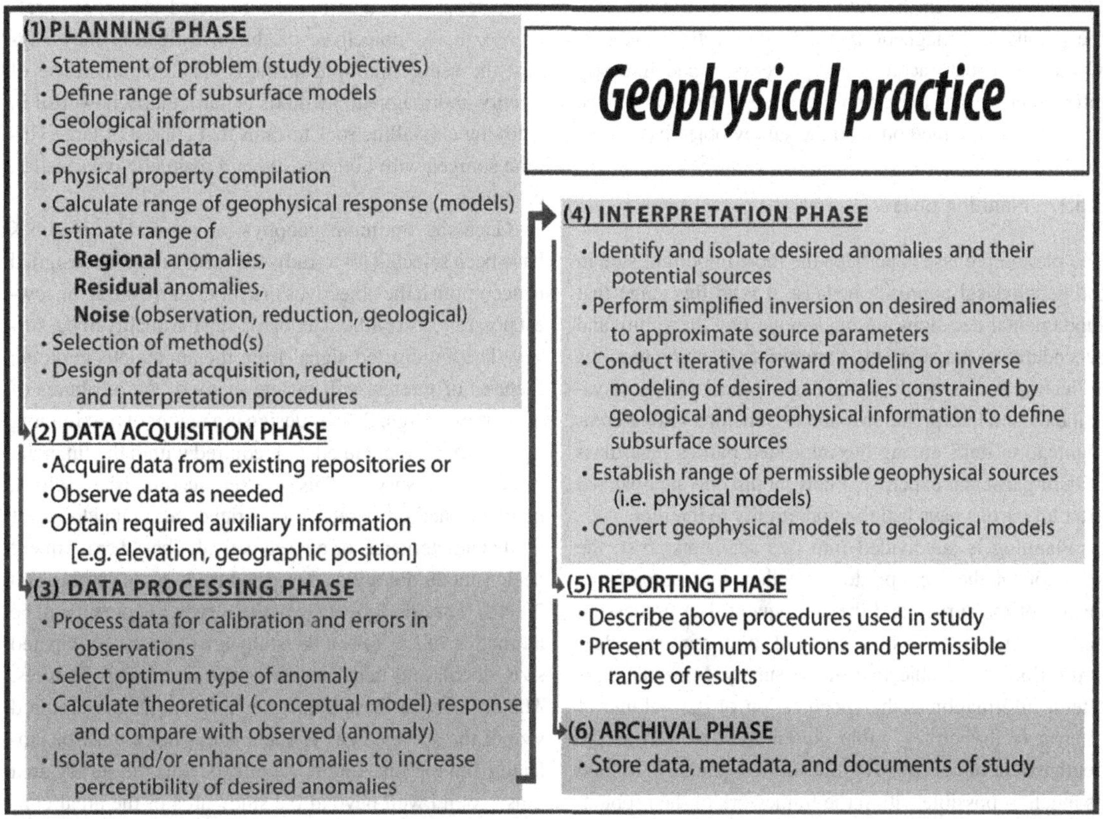

Geophysical practice

(1) PLANNING PHASE
- Statement of problem (study objectives)
- Define range of subsurface models
- Geological information
- Geophysical data
- Physical property compilation
- Calculate range of geophysical response (models)
- Estimate range of
  Regional anomalies,
  Residual anomalies,
  Noise (observation, reduction, geological)
- Selection of method(s)
- Design of data acquisition, reduction, and interpretation procedures

(2) DATA ACQUISITION PHASE
- Acquire data from existing repositories or
- Observe data as needed
- Obtain required auxiliary information [e.g. elevation, geographic position]

(3) DATA PROCESSING PHASE
- Process data for calibration and errors in observations
- Select optimum type of anomaly
- Calculate theoretical (conceptual model) response and compare with observed (anomaly)
- Isolate and/or enhance anomalies to increase perceptibility of desired anomalies

(4) INTERPRETATION PHASE
- Identify and isolate desired anomalies and their potential sources
- Perform simplified inversion on desired anomalies to approximate source parameters
- Conduct iterative forward modeling or inverse modeling of desired anomalies constrained by geological and geophysical information to define subsurface sources
- Establish range of permissible geophysical sources (i.e. physical models)
- Convert geophysical models to geological models

(5) REPORTING PHASE
- Describe above procedures used in study
- Present optimum solutions and permissible range of results

(6) ARCHIVAL PHASE
- Store data, metadata, and documents of study

FIGURE 1.4 The geophysical practice for implementing the gravity and magnetic methods involves a sequence of six phases.

exploration, both detailed and regional studies in mineral exploration, and in geologic mapping of crystalline rock terrains where the rock units have varying magnetic polarization. It has also proved useful in identifying ferrometallic objects in the near surface, such as buried well casings, storage containers, and unexploded ordnance and the study of archaeological sites.

The development of satellites and other space age technologies since the 1960s has greatly advanced regional exploration of the surfaces and deep interiors of the Earth and other planetary bodies. Political considerations do not limit satellite operations so that essentially any region is available to satellite remote sensing and geophysical mapping efforts. Unprecedented timing and positioning data from the constellation of Global Positioning Satellites (GPS) greatly expedite modern geophysical survey efforts, while the communication capabilities of satellites allow access to geophysical experiments literally worlds away from our offices and laboratories. Satellite gravity and magnetic observations, in particular, are yielding important new insights on the nature, architecture, and dynamics of the Earth and other terrestrial planets.

1.5 Geophysical practices

The gravity and magnetic methods are described in individual chapters that follow. However, in addition to the fundamentals specific to the individual methods, there are general principles and practices that are used in geophysical exploration programs. They are sufficiently general that a description of them serves as an introduction to the use of both gravity and magnetic methods. Whether they deal with the selection of the geophysical method, the design of a data acquisition and processing program, the reporting of an investigation or any one of the numerous components that mark a successful geophysical campaign, they are for the most part nothing more than the application of appropriate scientific methodology (Figure 1.4). The following description assumes that the program involves all phases from planning to report preparation and archiving the data and results. Programs may also focus on previously surveyed data that already have been reduced to anomaly form, where only the latter phases are applicable. Nonetheless, considering the factors for the phases described below will help to determine the usefulness of existing data. In the individual sections dealing with the gravity and magnetic methods, descriptions are provided of the data acquisition, processing, and interpretation phases, but the planning phase is focused on the principles of the methods and the survey objectives.

1.5.1 Planning phase

The planning phase is perhaps the most important step in the geophysical approach because it is in this stage that fundamental decisions are made regarding the nature and procedures of the program. Appropriate planning requires collecting and using all available geological and geophysical data and interpretations, and establishing strong communication links among the interested parties regardless of their particular expertise. Plans should only be finalized after all parties have had the opportunity to interact.

Planning is subdivided into two segments: first, the selection of the appropriate method(s) and, second, the design of the survey and the subsequent data processing and interpretation. To be successful, both require a clear exposition of the objective of the survey. Important collateral information is the specification of the volume of interest to the survey – that is, the areal as well as the depth extent of interest. This subsurface volume is limited as much as possible within the framework of the problem because the areal extent of the survey is a major factor in determining the cost of the survey. In addition, the survey procedures are tuned to the depth of interest as dictated by the survey objectives.

The most important consideration in the selection of the method for a study is to determine if the target sources will produce an observable anomaly even in the presence of extraneous signals. This requires estimation of the anticipated source volume, depth, and physical property contrast as well as evaluation of potential geologic, observational, and processing noise and errors. Information on the physical properties of the Earth materials in the subsurface volume being investigated is important to all phases of the application of geophysical methods, but particularly in planning studies when target anomalies are being estimated. Rock property data are obtained from *in situ* measurements on the site, sample measurements, and general tabulations. The character of target anomalies may come from experience in related situations, forward modeling of both anticipated anomalies and potential anomaly noise, or test surveys. The latter are particularly useful where information needed for modeling of sources and estimating noise and errors is lacking. The parameters of the source targets commonly cover a range of values necessitating the study of a distribution of anomaly characteristics. Evaluation of anticipated anomalies in reference to the objectives of the investigation may suggest the use of multiple methods. The combined use of gravity and magnetic methods is particularly powerful in studying crystalline rock terrains that consist of large volume sources with both density and magnetic polarization contrasts.

Once the optimum geophysical method or methods have been selected for a study, the survey must be designed to accomplish the objectives in a minimum time at the lowest possible cost without jeopardizing the quality of the survey. The anticipated signal from the anomalous geologic features of interest will dictate many of the attributes of the survey design. Survey design is a matter of maximizing the information obtained and required within the financial limits of the survey. This is often accomplished with a heuristic method based on experience and knowledge of field characteristics, or on a statistically based experiment design methodology as described by Curtis (2004a and 2004b). The areal coverage of the survey, of course, will be a function of the size of the study area and the anticipated size, depth, and depth extent of the anomalous features. The greater are these parameters, the larger the required size of the survey area. The anomalies often must be isolated from regional and noise effects, thus the survey area must extend well beyond the study area or the areal configuration of the anomalous feature. This is well illustrated for both the gravity and magnetic methods in Figures 1.2 and 1.3, respectively, where the anomaly needed for identification and analysis of the subsurface feature (that is, the fault in the gravity anomaly illustration and the ferrous source in the magnetic anomaly illustration) extends well beyond the immediate region of the anomalous feature.

Critical concerns in planning surveys are selection of the data density and precision. These are determined by the objectives of the survey and characteristics of the anticipated signals. For most objectives the anomalous signal including the maximum gradients must be fully measured, not simply the maximum amplitude of the anomalous signal. In gravity and magnetic surveys it is necessary to map the gradients of the anomalies to effect a useful interpretation. This requirement necessitates closely spaced and high-precision observations. Forward modeling of the range of anticipated anomalies, including their size, properties, and position, provides a basis for selecting the required data density and precision.

In general, sampling theory specifies that the station interval should be no greater than half the length of the smallest dimension that needs to be mapped in the survey. This interval or spacing is referred to as the Nyquist wavelength and its inverse as the Nyquist frequency. It is the maximum spacing that should be used between measurement points, although in practice the sampling interval should be considerably less than the Nyquist sample interval if the gradients of the measurements are of interest or the higher-frequency noise components must be mapped and isolated from the desired signal.

In gravity and magnetic methods, the separation between observations is directly proportional to the depth of investigation – that is, the greater the target depth, the greater is the permissible station spacing. Often a separation approximately equal to half of the depth to anticipated sources is used in surveys. However, generalizations regarding separation of measurement points are of limited value because of the need to consider the specific attributes of the survey. As a result, it may be desirable to determine quantitatively the probability that a specific anomaly will be detected utilizing the sampling theorem, with stations located either randomly through a region or on a regular grid (e.g. SAMBUELLI and STROBBA, 2002). Wherever possible, it is desirable to conduct test surveys over a limited, representative portion of the survey area or noise tests to select the optimum survey layouts.

1.5.2 Data acquisition phase

In the data acquisition phase, the actual field and necessary related data are measured and recorded. Auxiliary observations that are made in addition to the primary geophysical measurements include essential data for the reduction of the measurements to an interpretable form (e.g. station elevation or flight altitude, surrounding topographic relief, and water depth). Instrumentation must be selected to meet the precision requirements of the survey as established in the planning phase as efficiently as possible. Actual field procedures are dictated by the survey objectives, sources of noise, surface and weather conditions, instrumentation, and access within the survey area. For example, access may limit the survey to discrete observations along roads rather than a grid pattern more useful in interpretation. Observations are commonly made along traverses which are oriented perpendicular to the prevailing strike direction of the anomalies, separated at greater distances than the observations along the traverse. The distance between traverses is determined by estimating the length of the continuity of the character of the anomalies along their strike direction.

Gravity and magnetic methods are particularly effective in a reconnaissance or regional study mode because they are fast and efficient; while other methods may provide better detail and resolution, they are likely to be more costly and time-consuming. Use of gravity and magnetic methods early in an exploration program sequence can delimit an area for detailed investigations with other methods and improve the survey design to obtain maximum information from measurements. For example, magnetic measurements which often can be taken quickly and inexpensively from an airborne platform may be used to delineate likely faulted areas. In this way limited sectors of a large region can be isolated for study and evaluation in much greater detail by slower and more costly methods, like the seismic reflection method. Similarly, regional gravity surveys may be used to determine the strike direction of prevailing geological features within an area which can guide the selection of the direction of detailed traverses along which gravity or other geophysical fields or forces are measured.

1.5.3 Data processing phase

The nature and role of the data processing phase may vary considerably between gravity and magnetic methods and with the survey objectives. In general, the data processing requirements of the magnetic method are considerably simpler than for the gravity method largely because of the intense magnetic polarization contrasts in the crust of the Earth. For example, in some magnetic studies to locate near-surface ferrometallic bodies, the amplitudes of anomalies are sufficiently large that no data processing is needed. However, this is the uncommon situation, particularly with the increasing demands for precision in the results of geophysical studies. Accordingly, most survey objectives and methods lead to data in which the signal to be used in interpretation is significantly distorted by extraneous effects. Data processing is used to remove these extraneous effects and enhance the desired signal for interpretational purposes. Generally, data processing is performed subsequent to acquisition of the raw or observed data, but field processing may be used to minimize unwanted signals. For example, the stabilization of a gravimeter in field procedures may minimize wind-driven accelerations acting on the meter, and field processing by digital filtering can be used to supplement the effect of the field procedure.

Data processing may include several steps. The first is to prepare the data for interpretation by removing the effects of instrumentation as calibration adjustments and correction for instrument instability. These data are then reduced for known or predictable effects upon the observed data by calculating the theoretical value of the observation at a specific site using all known variables, such as elevation and planetary effects, and subtracting this predicted value from the observed measurement to obtain the anomalous value or anomaly. Data available in publicly available data banks commonly are at this level of processing. The next step is some form of digital filtering to remove wavelengths smaller (noise) and larger (regional) than the anomalies of interest. The purpose of this stage is to enhance particular attributes of the anomaly or signal that will increase its perceptibility and to isolate anomalies of interest for interpretational purposes. Although these procedures are highly automated in most data processing schemes, human interaction is required to establish optimum procedures and parameters.

1.5.4 Interpretation phase

The procedures involved in interpretation of the force field measurements into the nature and distribution of subsurface materials or associated processes relevant to the objectives of the investigation are highly varied depending on the goals and scope of the survey and the experience and skills of the interpreter. Successful interpretation commonly involves intangible qualities of the interpreter such as experience, observational powers, ability to visualize in three dimensions, and the intellectual capacity to organize and integrate a variety of often disparate types of information. As such, it is viewed in some quarters as an art, but most interpretation follows an orderly logical process, often called scientific methodology, using the methods of deduction or induction to proceed from processed data to a successful conclusion.

For simple survey problems, the interpretation phase is largely qualitative and is essentially terminated with the successful identification and isolation of anomalies. For example, if the location of bedrock highs is the objective of a gravity survey, the interpretation is completed with the isolation of gravity anomaly highs associated with the greater density bedrock contrasting with the overburden of unconsolidated sediments. In this and similar cases, the distinction between the data processing and interpretation phases becomes blurred. If the goal of the gravity survey is not simply to isolate bedrock highs, but is to determine the bedrock configuration, a more quantitative interpretation procedure must be applied.

Quantitative interpretation uses inversion to quantify possible geometric and physical property parameters of the subsurface that can satisfy the observed data. The essential element of inversion is the forward model that produces a synthetic set of estimates or effects for comparison with the actual observations. Acceptable models of the subsurface are typically judged by how well the modeled effects match the observed data in amplitude and shape. However, an acceptable model cannot guarantee a unique solution for the data because the forward model is always a mathematical simplification of the subsurface conditions, and it and the observed data always contain errors. In addition, the non-uniqueness of gravity and magnetic inversions is further exacerbated by the inherent source ambiguity of any anomaly solution. Thus, ancillary geological and geophysical constraints are commonly invoked to limit the range of acceptable models from an inversion.

Multiple approaches are available for solving inversion problems. A common methodology involves trial-and-error comparisons of observed geophysical signals with the effects from a presumed subsurface model. Through an iterative process the parameters of the presumed subsurface model are modified until a close match is obtained between the observed and estimated values. This so-called forward or direct modeling approach is especially appropriate when dealing with a relatively small data set and a simple subsurface model where only a few unknown parameters must be evaluated.

For more complex inverse problems involving greater numbers of observations and model unknowns, so-called inverse modeling approaches are desirable. These approaches commonly assume the forward model of the relevant volume of the Earth from the measurements, their distribution, and boundary conditions imposed by the geological setting. Modern inversions typically invoke the linear forward model as a series of simultaneous equations where the unknown geometric or physical property parameters can be estimated by fast matrix inversion methods.

The nonlinear forward model has more limited use because the computational labor of implementing the related inversion is much greater than for linear inversion. The nonlinear inversion requires the investigator to explore solution space generated typically by a large number of simulations where values of the unknown parameters have been randomly selected. These simulations, for example Monte Carlo simulations, are graphically or numerically processed for solution maxima or minima that may mark acceptable solutions.

A final step in the interpretation phase is to transform the quantitative model obtained by inversion into appropriate geological parameters. That is, the geometric and physical property estimates that satisfy the anomalous field must be converted into an effective geological context. For example, the geological significance of the physical properties or property contrasts interpreted from the observed data is best appreciated when related to the lithology and secondary physical characteristics of the formations. This is illustrated in Figure 1.5 where an observed gravity anomaly is shown that is closely matched by the effect of a cylindrical source of a positive density contrast.

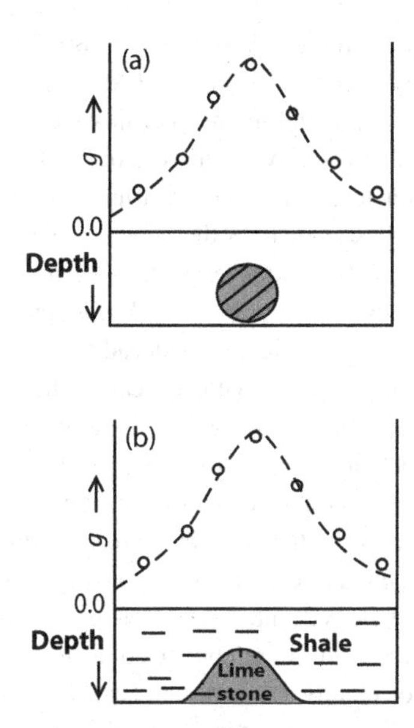

FIGURE 1.5 (a) Schematic illustration of the match of observed gravity anomalies (open circles) with the theoretical gravity effects (dashes) from a cylindrical source. (b) Translation of the horizontal cylindrical source in (a) into the geologic context of a limestone bed folded into an overlying lower-density shale. The gravitational effects of the idealized cylindrical source and the folded limestone are essentially equivalent.

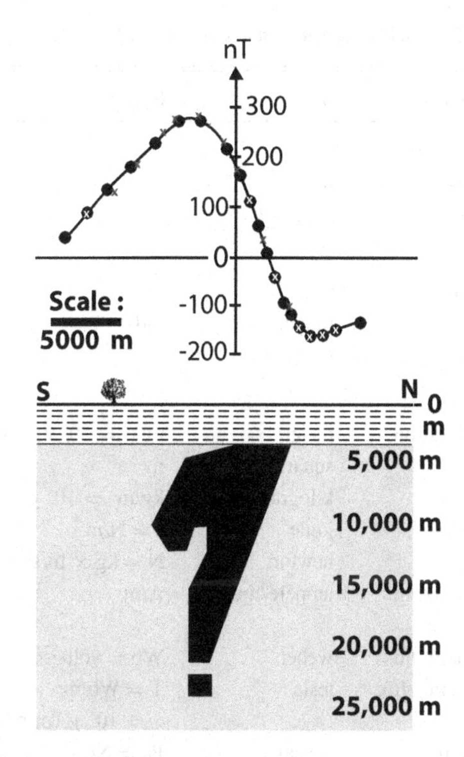

FIGURE 1.6 Comparison of the observed and calculated magnetic values across a subsurface source near Lausanne, Switzerland, with the configuration of the magnetic source used in the calculation shown in the underlying geological cross-section. The absurdity of the shape of the source denies the credibility of the interpretation. The observed data (dots) were modeled (crosses) with the black 2D body of volume magnetic susceptibility 0.00345 CGSu subjected to a polarizing field intensity of 46,450 nT with an inclination of 62°. Adapted from Meyer de Stadelhofen and Juillard (1987).

In geological terms, the cylindrical source effect can be taken as the effect of an anticline which brings limestone into juxtaposition with overlying lower-density shale.

Consideration of the stages in the interpretational process clearly shows that the process is subject to ambiguity, where multiple solutions are equally compatible with the available geophysical information. This is an inherent property of potential field methods regardless of the accuracy of the measurements and processing and the sophistication of the interpretational procedures.

A striking example of the ambiguity of geophysical data is presented in the tongue-in-cheek illustration of Figure 1.6. This figure shows the close correspondence between the observed field and calculated magnetic values along a profile across a magnetic anomaly near Lausanne, Switzerland, of unknown origin. The close correspondence of the field and calculated values gives a false sense of credibility to the forward model interpretation. Clearly the question-mark shape indicates the absurdity of the source configuration.

Fortunately, the interpreter can decrease the potential ambiguity in interpretation, although it can never be completely eliminated. For example, ambiguity can be minimized by the integrated interpretation of two or more geophysical signals derived from common sources. This takes advantage of the different physical responses of fields and their varying sources and degrees of ambiguity. In a similar manner, geological and physical property information can be extremely valuable in establishing boundary conditions for possible interpretations. Because interpretations are fundamentally non-unique, however, it is appropriate to conduct several analyses based on different assumptions to find the range of possible interpretations. This is referred to as a sensitivity analysis.

1.6 Nature of geophysical data

Data in this book are presented in SI units (SIu), except in specific situations where alternative units are more meaningful to the user of the data or the interpretation. SI is the abbreviation for Le System International d'Unites, which is a system of units that is broadly accepted internationally by governmental agencies and professional societies. This system has similarities to the metric system of units, but

TABLE 1.2 SI units (SIu) used in this book.

Quantity	SIu	Symbol
Base units		
Length	meter	m
Mass	kilogram	kg
Time	second	s
Supplementary units		
Plane angle	radian	rad
Solid angle	steradian	sr
Derived units		
Acceleration	meter/s^2	m/s^2 = 10^5 mGal
Area	square meter	m^2
Density	kilogram/m^3	kg/m^3 = 10^3 g/cm^3
Energy	joule	J = N/m
Force	newton	N = kg × m/s^2
Magnetizing force	ampere/meter	A/m
Magnetic flux	weber	Wb = volts × s
Magnetic flux density	tesla	T = Wb/m^2 = 10^9 γ (or nT)
Pressure	pascal	Pa = N/m^2
Viscosity (dynamic)	pascal second	Pa × s
Work	joule meter	J × m

is not identical to it. The base and supplementary units of the SI system together with their combinations, called derived units, that are common to geophysical studies in this book are listed in Table 1.2. Note that SIu is used as the abbreviation for SI units throughout this book.

It is important that only significant figures be retained in the data: that is, no figures should be kept in the data stream beyond the first doubtful one. In rounding off numbers to the nearest significant figure, the number should be increased to the next highest digit if the following number is 5 or more. Zeros are significant only if they are preceded by digits or are necessary to establish the position of the decimal point.

1.6.1 Data documentation

Data measured in geophysical surveys, whether consisting of a single observation at a single site or repetitive observations at multiple sites, require annotations to describe the survey, the specific site of the measurement, instrument characteristics, processing applied, and in many cases the environmental conditions of the measurement. These annotations are commonly referred to as metadata. The specific annotation and the manner in which the annotations are recorded vary with the type of measurement.

The adequate annotation of measured data is only one example of "best" practices that are important in quality assurance. Quality assurance has long been important to geophysical investigators, but as a formal process it has been recognized only in recent decades. It has become particularly important in geophysical studies related to engineering problems such as the siting of critical structures. The potential impact of failure of these structures as a result of incorrect conclusions drawn from faulty investigations, with the resulting effect on humans and the environment, has encouraged regulatory and licensing agencies of the government as well as private industry to insure that the studies are conducted at the highest possible level of quality. As a result, quality assurance has become a required element in the acquisition and processing of geophysical data in many types of both commercial and governmental geophysical surveys. This is intended to insure the integrity of the studies and the quality of the data and resulting data processing and interpretation.

1.6.2 Data errors

Measurements of forces and fields are subject to uncertainty as a result of a variety of errors. Errors or noise are the difference between the truth and the actual measurement, and thus are analogous to anomalies which are at the heart of most geophysical studies. Errors may originate from the instrumentation system and the observer, the reduction and processing of data, and geophysical interpretation. Errors or noise also may be caused by mistakes by the geophysical analyst, but not all errors are mistakes. In fact most errors in geophysical studies are not caused by humans, and in many cases the sources of noise are unknown and unavoidable.

Errors are of two basic types, systematic and random. Systematic errors are consistent deviations within a measurement system. They may be constant or vary in either a linear or nonlinear manner with some attribute of the system or its environment, such as the amplitude of the measurement or the temperature of the system. They are caused, for example, by incorrect instrumentation calibration, poor design of apparatus, incorrect identification of baseline values, and some personal errors. The latter may originate from the tendency of an observer to consistently misread a galvanometer that is used in the measurement system or by a consistent bias in interpretation by an analyst. The conclusions from studies which have only systematic errors may be consistent in themselves and therefore be precise, but inaccurate on an absolute basis. Systematic errors or noise can be difficult to detect because they will not show up in repetitive measurements by the same system. They can only be identified by making the same measurement with a different measurement system involving changes in the observer/analyst, measurement instrumentation, data processing scheme, and interpretation procedures.

Random errors are deviations from the truth that occur by chance, and thus are unpredictable and subject to the laws of probability. They are the deviations or errors we observe in repetitive observations. They exhibit no correlation with attributes of the measurement system or source of the field measured and are unrelated to other measurements made by the system. They arise from inconsistencies in the sources, instrumentation instabilities, and observer/analyst non-systematic errors. Random errors take on a normal distribution, that is a large number of repetitive observations or results will assume a normal (i.e. bell-shaped or Gaussian) distribution around a central value that is the arithmetic mean of a set of numbers. Statistical tests can be used to determine if a data set reasonably approximates a normal distribution. Individual numbers of the set of normally distributed values are equally likely to be positive or negative relative to the average, but extreme variations are less likely to occur. The frequency of occurrence is much greater for those values that are nearer to the average value.

The arithmetic mean is usually taken as the most probable value of the quantity. However, the arithmetic value is not the true value of the quantity because the mean depends on the number of measurements used in the calculation. The accuracy will increase as the number of values in a set of data increases. Because the mean is not the exact value of the quantity, it is common practice to estimate the accuracy of the calculated mean value. One method is to calculate the standard deviation which takes into account the number of observations. The standard deviation σ of a set of n observations $x_1, x_2, \ldots, x_n$ with mean $\bar{x}$ is

$$\sigma = \sqrt{\frac{1}{n-1}\sum_{i=1}^{n}(x_i - \bar{x})^2}. \tag{1.1}$$

There is a 68.26% chance that the true value is within one standard deviation of the mean value and a 95.46% chance that it is within two standard deviations.

In a set of observations it is not uncommon to experience an outlier value that departs widely from the others. The question thus arises as to whether the outlier is due to a blunder and therefore should be rejected or should be considered in determining the statistics of the data set. If no obvious mistake is evident, a statistical test can be employed to test the validity of the measurement as a member of the set. One test, which is based on probability, states that if a single observation departs from the mean by more than three times the standard deviation, the observation should be discarded because the chances are 400 : 1 that the observation is due to a random error or blunder.

Most geophysical measurements require some mathematical manipulation or data processing before they are used for interpretational purposes. Commonly several mathematical steps are involved with multiple measurements and parameters, each with their own error expressed as a standard deviation. As a result, errors will propagate through the mathematical steps. Of course, it is possible for positive errors to be offset by negative errors, but because it is impossible to determine that this is the case, it is the norm to be conservative. Thus, the rules for calculating the net error for each mathematical manipulation assume the maximum error. The rule for both addition and subtraction is to add the standard deviations. Thus, the sum of two sets of numbers with means N_1 and N_2 and respective standard deviations σ_1 and σ_2 is

$$(N_1 \pm \sigma_1) + (N_2 \pm \sigma_2) = (N_1 + N_2) \pm (\sigma_1 + \sigma_2), \tag{1.2}$$

and their difference is

$$(N_1 \pm \sigma_1) - (N_2 \pm \sigma_2) = (N_1 - N_2) \pm (\sigma_1 + \sigma_2). \tag{1.3}$$

In multiplication, the rule is

$$(N_1 \pm \sigma_1)(N_2 \pm \sigma_2) = N_1 N_2 \left[1 \pm \left(\frac{\sigma_1}{N_1} + \frac{\sigma_2}{N_2}\right)\right], \tag{1.4}$$

whereas for division, it is

$$\frac{(N_1 \pm \sigma_1)}{(N_2 \pm \sigma_2)} = \frac{N_1}{N_2}\left[1 \pm \left(\frac{\sigma_1}{N_1} + \frac{\sigma_2}{N_2}\right)\right]. \tag{1.5}$$

The standard deviations also may be considered in determining the number of significant figures. As discussed above, in addition and subtraction, the standard deviation is considered directly, and in multiplication and division the percentages of the standard deviations are added to determine the number of significant figures.

Random errors or deviations in geophysical observations as described above may originate from a variety of sources including effects from geologic heterogeneities. These effects can be minimized by adding together a series of observations in which measurements include a coherent signal with random errors or signals superimposed. This procedure, commonly referred to as stacking, attenuates the random errors or noise by a factor of $\sqrt{n}$ with respect to the coherent signal, where n is the number of elements in the series. This procedure is used widely in geophysics to minimize random errors or noise.

1.7 Key concepts

- Geophysics involves the application of physical principles to the study of the Earth from its magnetosphere to its central core. Solid-Earth geophysics evolved from the study of surface geological units and observations of the terrestrial force fields, and the development of related laws and principles of geology and physics.
- Common characteristics of geophysical methods are their strong base in physical principles, the need for geologic information in all phases of conducting the methods, and the intensity of computational aspects necessary to make force-field observations geologically significant.
- Geophysical methods are advantageous because they can investigate portions of the Earth unavailable to direct exploration, but they are limited in their resolution, ability to provide unambiguous results, and direct information on the physical properties of the Earth.
- The gravity and magnetic methods have a rich tradition of investigating the Earth for over a century. They find useful applications in engineering and environmental studies, resource investigations, and general studies of the nature and processes of the Earth.
- Gravity and magnetic methods are employed in Earth studies at a variety of scales ranging from submeter to several hundreds of kilometers. Observations are made in drill holes within the Earth, on the Earth's surface, marine areas, and from altitudes of a few meters to hundreds of kilometers.
- Progress in geophysics has been driven by societal needs and economic factors and is made possible by technological advances leading to rapid, efficient data acquisition, processing, presentation, and interpretation.
- Geophysical methods are applied in a process that involves the key elements of planning, field measurement, data processing, and interpretation.
- The critical planning phase requires strong communications among the geophysicist, project engineer and scientist, and end-user. Essential steps during this phase are selection of the appropriate geophysical methods and procedures for data acquisition, processing, and interpretation. These decisions require an understanding of geophysical methods, the geological nature of the site, and the objective of the survey that is reached on the basis of experience, model studies, or test surveys.
- Data acquired during geophysical surveys are subject to a wide range of both random and systematic errors or noise. These need to be understood and minimized by field and data processing procedures to levels appropriate to the precision requirements of the objective. Error budgets are made and consideration is given to the propagation of the errors through the data stream from acquisition to interpretation.
- Field procedures must be adjusted to the objective of the survey, the local surface and cultural features, and the resources available. Planning of these procedures should be based on consideration of the cost/benefit ratio. The distribution of observations over the survey area is tuned to the objective of the survey and detail required by the range of anticipated anomalous signals.
- The physical properties of the Earth are important in all phases of applying geophysical methods. Knowledge of the properties at a specific site can be obtained from local *in situ* measurements, sample study, and generalized tabulations. However, care must be exercised in the use of physical property tables because of the commonly strong effect of local conditions and the environment on properties. In many near-surface studies, the objective is to determine properties of the Earth that are not directly measured, but are of a secondary nature, being related to primary properties measured by the method using empirical relationships which are subject to error.
- After the observed data are processed to isolate the desired signal from extraneous effects, interpretation proceeds by data inversion using inverse or iterative forward modeling of the subsurface until the processed data are matched. A wide variety of qualitative and quantitative procedures are used in the interpretation process, depending on the objectives of the survey and the character of the measured data. The result of the interpretation is a physical model that must be transformed into a geological model to achieve the objectives of the survey.
- Geophysical data are subject to a wide variety of errors, both systematic and random, despite protocols for quality assurance. Evaluation of these errors and their impact on the results is an important element of high-quality studies.

- The two main parts of this book provide comprehensive discussions of the respective gravity and magnetic exploration methods. The material is presented using current data analysis methods and jargon that are described in greater detail in Appendix A. Readers who are not familiar with the fundamentals of digital data analysis may wish to review Appendix A before proceeding to the gravity and magnetic exploration parts of the book.

Supplementary material and Study Questions are available on the website www.cambridge.org/gravmag.

Part I
Gravity exploration

2 The gravity method

2.1 Overview

The gravity method of geophysical exploration is based on the measurement of variations in the gravity field caused by horizontal variations of density within the subsurface. It is an important technique for many problems that involve subsurface mapping, and it is the principal method in a number of specific types of geological studies. The method, which has its roots in geodetic studies from the seventeenth to the twentieth centuries, has developed rapidly over the past century as a result of significant technological advances: primarily high-accuracy gravity measuring instruments for use on land, sea, and air, and in space, as well as the increasing computational and graphics power of digital computers. These advances are backed up by continuing improvements in data processing, interpretational schemes, practical experience, and surveying methodologies.

Traditionally the gravity method has been used primarily in regional characterization of the Earth for determining the architecture of the crust, identifying potentially favorable regions for resource exploration, and developing conceptual exploration models. This is made possible by the millions of gravity observations now available worldwide in both public and commercial data sets. The accuracy now available in gravity observations and the relative ease of measurement have made the method viable for exploration objectives such as assessing subsurface changes over time and combined interpretation with seismic reflection mapping.

The method is founded on measuring and analyzing perturbations in the terrestrial gravity field. The main gravity field of the Earth is a function of the planet's mass, size, and rotational characteristics. The field is described by the universal law of gravitation which relates the force of attraction between objects to the product of their masses and inversely to the square of the distance between them. Spatial variations in gravity over the surface of the Earth are caused by mass heterogeneities within the Earth and latitudinal effects related to the change in the radius of the Earth and centrifugal force of the rotating Earth from the equator to the poles. Temporal variations arise from tidal effects due to extra-terrestrial bodies of the solar system as well as from fluctuations in the fluid and gas content of the Earth.

Most gravity surveying presumes to map the radial component of the gravity force per unit mass, that is the acceleration of gravity in the vertical direction. This component is measured in most exploration surveys to the precision of one part in one hundred million or better of the Earth's gravity field. In most exploration surveys of relative gravity, this high sensitivity is achieved by spring balances in which the change in length of the spring caused by a change in gravitational attraction, calibrated in units of gravitational acceleration, is observed. Specialized instrumentation and procedures are used for determining absolute gravity and in measuring gravity and its vectorial components in dynamic environments.

Successful application of the gravity method requires care in planning, acquiring, and processing of data to isolate anomalous variations in gravity associated with the targets of investigation. The effects of these targets are commonly only of the order of a millionth of the total gravity, which is the cumulative effect of all mass sources within the Earth. Thus, considerable effort is used to extract the signal derived from the target residual sources from those of deeper regional sources and shallower noise sources that are not germane to the problem of interest. The resulting anomaly is a function of the mass differential, shape, and depth of the anomalous source. Interpretation of residual anomalies usually begins with estimations of the source parameters based on experience and approximation techniques largely derived from simplification of theoretical gravity responses by making assumptions about the geometry of the sources. Further interpretation is based on anomaly inversion, commonly using either trial-and-error forward modeling approaches or the more direct linear inversion of the anomaly field for the parameters of a postulated mathematical model of the sources. The trial-and-error approach involves the computation of an anomaly field from an assumed, mathematical model of the subsurface with iterative re-computation of the field based on plausible alterations of the characteristics of the model until the anomaly field is replicated. Neither this approach nor direct inversion leads to unambiguous results without invoking other constraining geologic and geophysical information.

2.2 Role of the gravity method

The gravity method has been used for a wide variety of purposes over the past century. Although the method lacks the resolution, depth control, and subsurface imaging capabilities of several other non-intrusive procedures, notably the seismic reflection and ground-penetrating radar methods, it has a significant role in the early stages of an exploration campaign and in specialized local problems. It is particularly successful when integrated with other geophysical data and available geologic information, thus minimizing the inherent ambiguity of the method.

The gravity method is an important approach to many problems that involve subsurface mapping. It is the principal method in a number of specific geological studies, as for example in mapping near-surface voids, quantitative studies of metallic ore bodies, characterizing salt structures, and monitoring changes of fluid/gas content in volcanoes, but the method seldom provides a comprehensive answer on its own. The gravity method has also been used in regional characterization of the Earth for determining the architecture of the crust, identifying potentially favorable regions for resource exploration, and developing conceptual exploration models. Achievement of these goals is made possible by the millions of gravity observations from extensive observation campaigns by both industry and government agencies and now available in both public and commercial data sets as a result of development of portable, accurate gravimeters (Figure 2.1) in the 1930s and 1940s.

FIGURE 2.1 Scintrex CG-5 gravity meter set up for measuring gravity at a land surface station. Courtesy of Scintrex, Ltd. A color version of this figure is available at www.cambridge.org/gravmag.

Additionally, increasingly available marine, airborne, and satellite gravity observations have mapped regional-to-global density variations related to larger-scale compositional, structural, hydrological, and thermal variations and evolution of the Earth's surface, crust, mantle, and core. Satellite gravity surveys also have provided regional insights on the subsurface properties and evolution of the Moon and extraterrestrial planets. In addition to the importance of gravity to regional investigations, the accuracy now available in gravity observations and the relative ease of making these measurements have made the method viable to attacking local subsurface problems which require a high accuracy and data density. Engineering and environmental problems such as siting of critical structures, determination of the location and configuration of metallic ore deposits, and temporal mass variations in the subsurface are a few examples of local applications of the gravity method.

The gravity method is implemented in drillhole, terrestrial, marine, airborne, and satellite survey environments. It is based on measurement of perturbations in the Earth's gravitational field caused by lateral variations in density of subsurface units. The component of gravity which is derived from these density variations is extracted from the measured signal and interpreted in geologic terms by inversion. The long history of development has produced an effective process for using the gravity method in subsurface exploration.

2.3 The Earth's gravity field

The gravity method is a passive exploration method in that it is based on a natural and ever-present gravitational force field. It requires no active energy source, as is necessary with seismic and most electrical exploration methods. This is an obvious advantage, but also a disadvantage because the field cannot be modified to suit the particular application. And although the Earth's gravity field is ever-present, it varies both spatially and temporally to distort the changes in the anomaly fields caused by local subsurface conditions of interest in gravity surveying. Thus, to produce interpretable measurements, these variations must be removed from the observed data.

2.3.1 Gravitational force

The gravity force field is caused by the fundamental phenomenon of gravitation that attracts bodies toward each other. Specifically, a mass in the presence of another body, like the Earth, has energy due to the gravitational attraction which is called gravitational potential. This energy causes objects to accelerate towards each other if they are free to move. The force of gravity between objects on the surface of the Earth is not observed because their mass is so much smaller than the Earth's, and thus their attractive effect on each other is negligible.

It is the mass of an object, a property determined by its volume, atomic content, and the packing of the atoms, that is the source of the gravitational field. In geophysical considerations, mass is replaced by the product of density (in other words the mass per unit volume) and the volume of the object. Thus, density is the operative material physical property for the gravity method. Gravitational force is proportional to the product of the masses of the attracting bodies (Figure 2.2). It is not affected by the presence of material between the masses, temperature, physical state, or other environmental conditions. In the case of the

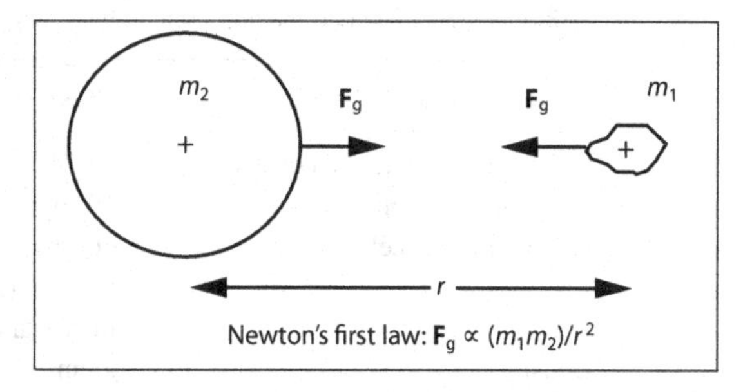

FIGURE 2.2 Illustration of two bodies attracted toward each other with a gravitational force $\mathbf{F}_g$ proportional to the product of their masses m_1 and m_2 and inversely proportional to the square of the distance r between the centers of the masses.

Earth, the gravity force field is dependent on the attraction between the mass of the Earth and the mass of an object. The weight of an object, or the gravitational force on it, is the attraction between the Earth and the object. Thus, the mass of an object is invariant, but its weight is dependent on the attractive body (here, the Earth). On the surface of the Moon, an object has the same mass as on the Earth, but its weight is one-sixth of that on the Earth because the mass of the Moon is only about 1.25% of that of the Earth and its radius is about a quarter of the Earth's radius.

In the seventeenth century, Newton showed that gravity force is inversely dependent on the distance between objects. In the case of sources whose dimensions are small with respect to the separation, e.g. the planets revolving around the Sun, the force is inversely proportional to the square of the distance between the center of the objects. Thus, if the distance between bodies is doubled, the force on them is reduced to a quarter. Newton showed from Kepler's observations of the motion of the planets around the Sun that the radial attractive force of the Sun on the planets is

$$\mathbf{F}_g = \frac{4\pi^2}{\mathbf{r}^2} m, \tag{2.1}$$

where the vector $\mathbf{F}_g$ is the attractive force, the scalar m is the mass of the planet, and $\mathbf{r}$ is the displacement vector describing the distance between the center of the Sun and the planet. Furthermore, because action and reaction are equal and opposite by Newton's third law, the planets must exert a similar force on the Sun. Thus,

$$\mathbf{F}_g = G\frac{m_1 m_2}{\mathbf{r}^2}, \tag{2.2}$$

which is Newton's universal law of gravitation where, in SIu, $\mathbf{F}$ is the force of attraction in newtons (N) between masses m_1 and m_2, in kilograms, that are separated by a distance $\mathbf{r}$ in meters. The universal gravitational constant G has the value $(6.674\,215 \pm 0.000\,092) \times 10^{-11}$

TABLE 2.1 Parameters commonly used in the gravity method in equivalent CGS and SI systems of units.

Gravity parameter	CGSu	SIu
Universal gravitational constant	6.674×10^{-8} cm^3/g s^2	6.674×10^{-11} m^3/kg s^2
Force of attraction	10^5 dynes	newton (N)
Gravitational acceleration	cm/s^2	10^{-2} m/s^2
	milligal (mGal)	10^{-5} m/s^2
	microgal (μGal)	10^{-8} m/s^2
Density	g/cm^3	10^3 kg/m^3 (metric tonne)

($m^3 \times kg^{-1} \times s^{-2}$) or ($N \times m^2/kg^2$) in SIu (GUNDLACH and MERKOWITZ, 2000). This constant can be determined by a variety of field and laboratory methods, with most recent precise determinations made in the laboratory by measuring the force between two masses. An additional result of Newton's studies is that the attraction on the surface of the Earth is equivalent to the situation where the mass of the Earth is concentrated at its center.

The gravitational force of the Earth cannot be specified or measured independently of mass, so the acceleration of a mass falling in response to the gravity field is used to describe the gravitational force. The acceleration **a** of a freely falling mass in the Earth's gravitational field is related to the gravitational force $\mathbf{F}_g$ through Newton's second law

$$\mathbf{F}_g = m_1 \mathbf{a}, \tag{2.3}$$

which gives the force acting on the mass m_1 due to the presence of another mass m_2. In the Earth's gravitational field, the force on the body m_1 is exactly the same as if it were being accelerated at the rate

$$\mathbf{a} = \mathbf{F}_g / m_1 = G\frac{m_2}{\mathbf{r}^2}, \tag{2.4}$$

where m_2 in the terrestrial situation is the mass of the Earth. Therefore, the attraction of the Earth can be considered the force per unit mass and is equivalent to the acceleration caused by the free fall of the mass in the gravitational field of the Earth. The gravitational acceleration **a** is the quantity measured in geophysical exploration where it is commonly defined by the special symbol **g** or

$$\mathbf{g} = \mathbf{F}_g / m_1 = G\frac{m_2}{\mathbf{r}^2}. \tag{2.5}$$

This equation holds only for homogeneous or radially stratified spherical bodies and equivalent compact sources. The relationship changes from this simple inverse square law when the acceleration is integrated over more complex sources.

The force of gravitational attraction is always positive, that is all objects are attracted towards each other, but variations in gravity may be negative in geophysical exploration as a result of a lower than normal mass due to horizontal variations in density within the Earth. These negative values are not a result of repulsion between objects, but simply a lesser attraction than normal. This concept emphasizes that what is measured and analyzed in the gravity method of geophysical exploration is the relative spatial difference in gravity.

2.3.2 Gravity units

Gravitational acceleration or simply gravity is measured in N/kg or m/s^2 in SIu. Table 2.1 presents the equivalent units of common parameters used in the gravity method in both CGSu and SIu. In geophysical exploration, the gal is used for gravitational acceleration and is equivalent to 0.01 m/s^2 in SIu or 1 cm/s^2 in CGSu. The gal honors Galileo Galilei who in the late sixteenth and early seventeenth centuries pioneered the investigation of the motion of the planets around the Sun and the nature of gravitation. However, the gal is a large unit compared with the changes in the gravitational acceleration caused by variations in the Earth's subsurface masses. As a result, the unit milligal (mGal) or one-thousandth of a gal is used in geophysical exploration, while in engineering geophysics and other high-sensitivity studies the microgal (μGal = 0.001 mGal) is frequently used as the unit because of the very small magnitude of many of the significant anomalies. Another unit sometimes used in gravity exploration in the petroleum industry is the gravity unit (g.u.), which is equivalent to 0.1 mGal. The use of the g.u. should be avoided because of possible confusion with the unit milligal commonly used in gravity exploration.

2.3.3 Basis of the gravity method

The gravity method is focused on measuring and analyzing perturbations or anomalies in the terrestrial field caused by lateral variations in density within the subsurface. The

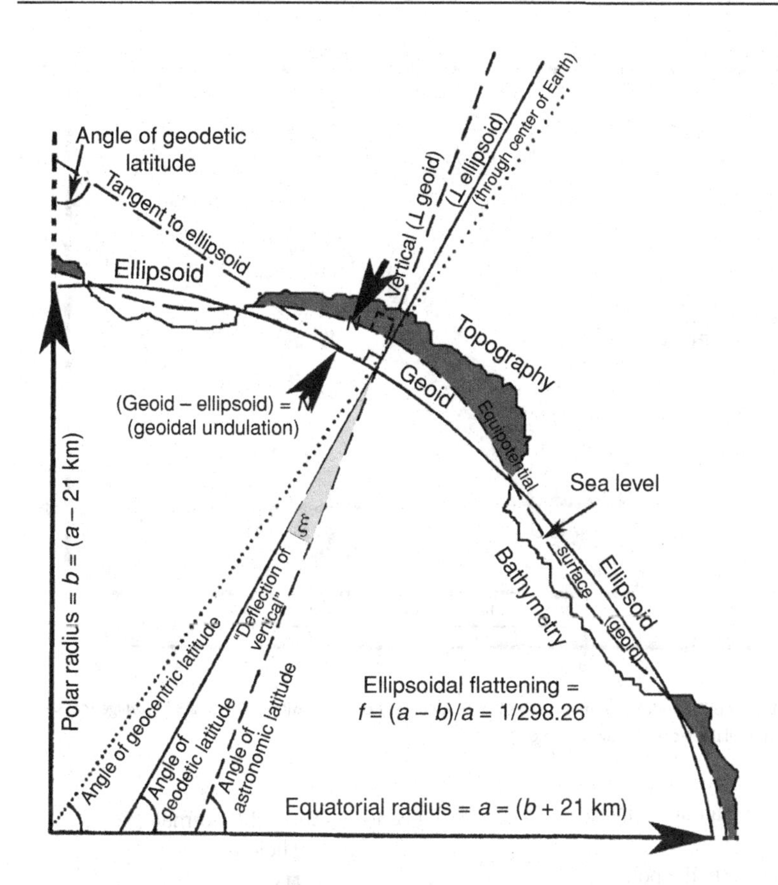

FIGURE 2.3 Gravimeters measure vertical gravity (g) relative to the geoid (dashed line), which is the equipotential surface represented by mean sea level in the oceans. Normal gravity (g_N) is defined perpendicular to the oblate ellipsoid or spheroid of revolution (full line) that is a mathematical approximation of the geoid. Geocentric latitude, which is also called co-latitude, is referenced to the center of the Earth, whereas astronomic latitude is measured relative to the true local vertical defined by the geoid. The deflection of vertical is the angular difference between the verticals observed on the geoid and defined by the ellipsoid, whereas height differences (N) between geoid and ellipsoid are geoidal undulations. Adapted from SHERIFF (2002) with surfaces not to scale.

theoretical and physical property principles underlying the measurements are explained in Chapters 3 and 4, respectively, whereas the mapping and data reduction and analysis procedures are described in Chapters 5–7.

The Earth's gravity field includes spatial variations due to the size, shape, and rotational properties of the Earth and temporal variations related to the differential gravity effects of the Moon and Sun on the Earth. Every gravity observation at or near the Earth's surface includes these normal gravity effects, which in the case of spatial variations and some temporal variations are large compared with many geological gravity effects. They must be removed from gravity measurements to isolate the gravity effects of subsurface targets for analysis. The next two subsections provide general views of the terrestrial gravity field and its principal variations over the Earth; these are developed in greater detail in later chapters as needed to address the acquisition, processing, and interpretation of gravity data for Earth exploration purposes.

2.3.4 Spatial variations

The normal gravitational acceleration g_N of the Earth is modeled in terms of a mathematical oblate ellipsoid that best fits the geoid (Figure 2.3). The geoid is defined as a hypothetical surface from which topographic heights and ocean depths are measured. It is represented by sea level in the oceans and its continuation into the continents. It is parallel to the surface defined by a spirit level, and the net acceleration on the surface of the Earth is directed perpendicular to it. It is often described as the gravitational equipotential surface which most closely corresponds to the shape of the Earth. That is, the summation of the potential of the gravitational attraction and centrifugal force of the Earth's rotation is constant on this surface which most closely corresponds to the ocean's surface. The gravitational accelerations due to these forces are not constant on this surface because the force is a function of the spatial rate of change of potential (e.g. Equation 3.1).

The maximum deviation of the geoid from the reference ellipsoid is minor compared with the radius of the Earth, about ± 100 m, as illustrated by the geoidal undulation map of Figure 2.4. This is only roughly one part in 64,000 of the average radius of the Earth. As a result the deflection of the true vertical (the angle ξ in Figure 2.3), which is the angular difference between perpendiculars to the geoid and the reference ellipsoid, is extremely small,

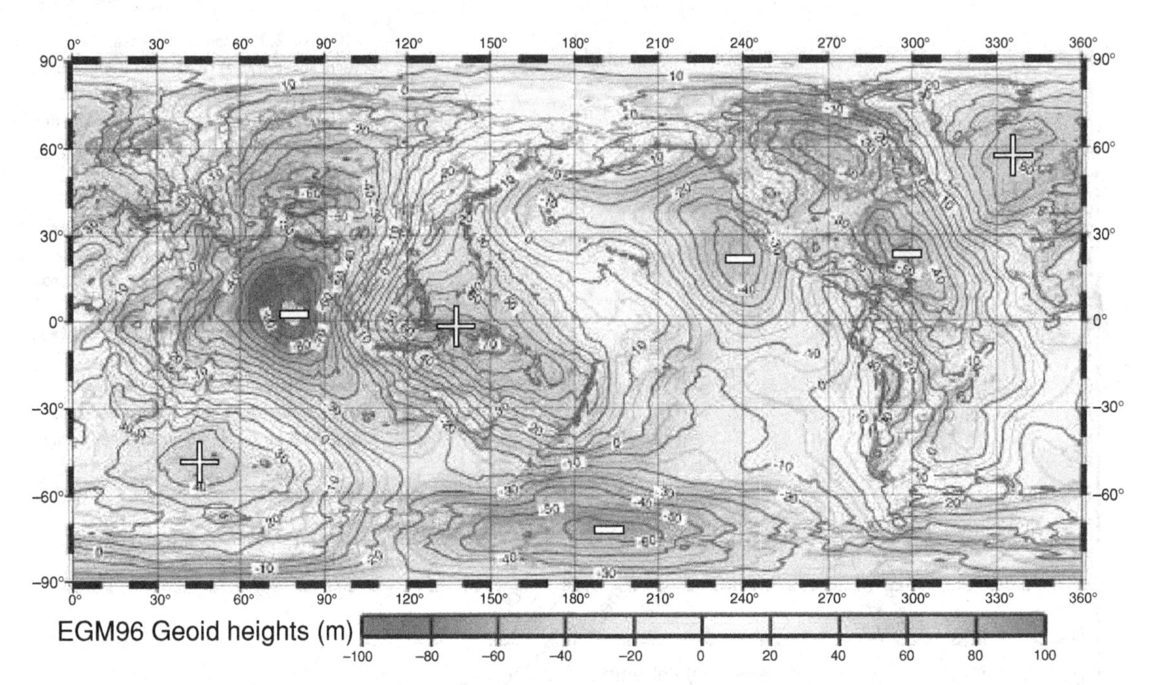

FIGURE 2.4 Earth Gravity Model-08 (EGM08) geoidal undulation map of the Earth. Courtesy of the National Geospatial-Intelligence Agency (NGA). A color version of this map is available at www.cambridge.org/gravmag.

reaching maximum values of little more than one minute of arc.

Normal gravity varies from the equator to the poles of rotation by about 0.5%, changing by slightly more than 5 gal from approximately 978 gal at the equator to 983 gal at the poles. The source of this planetary variation is two-fold. First, and more important because it accounts for more than 3 of the 5 gal, is the change in the centrifugal force over the Earth's surface caused by the Earth's rotation around its axis as shown in Figure 2.5. This is a pseudo-force that results from the rotation of the coordinate system that is fixed to the Earth and whose origin is the center of mass of the Earth. Note that Equation 2.5 is for a non-rotating body. The centrifugal force imparts an acceleration to a body that is equal to the product of the square of the angular velocity of the Earth and the perpendicular distance between the axis of rotation and the body (the radius of gyration). This acceleration is inseparable from the gravitational acceleration of the Earth and is vectorially added to the gravitational acceleration. The centrifugal acceleration effect is maximum at the equator because it is directly opposite to the gravitational acceleration and the radius of gyration is a maximum. In contrast, the centrifugal effect is zero at the poles because the radius of gyration is zero.

The second cause of the planetary change in gravity over the Earth's surface is the change in the radius from

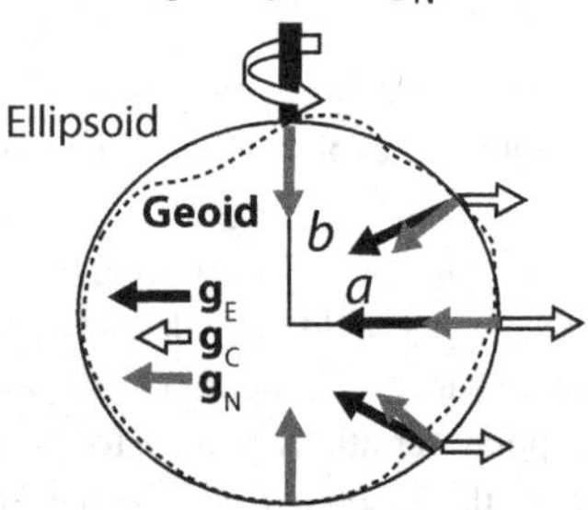

FIGURE 2.5 The globally best-fitting ellipsoid (solid line) to the geoid (dashed line) with equatorial and polar axes a and b, respectively, is the reference surface for defining the rotating Earth's normal gravity g_N, which is the vector sum of the mass effects of the Earth g_E and the centrifugal force g_C due to the Earth's rotation about its polar axis – i.e. g_N is the gravity effect of a homogeneous non-rotating Earth over the surface of the ellipsoid.

the equator to the poles. The gradual change in the radius, amounting to a maximum of 21 km, results in an ellipsoidal cross-section as shown in Figures 2.3 and 2.5. In particular, the Earth's shape is that of an oblate spheroid in which the equatorial radius is greater than the polar radius. This is the shape transcribed by the revolution of the elliptical cross-section of the Earth. The shape of the Earth is described by the geometrical (or polar) flattening (f) given by the ratio of the difference in the equatorial and polar radii to the equatorial radius which is approximately 1/298 (Figure 2.3). This minor flattening is verified by viewing the apparent circular cross-section of the Earth from space. The flattening of the Earth is an equilibrium condition between gravitational forces attempting to make the body spherical and rotational forces that are trying to flatten it out. The flattening of 1/298 reflects the increasing density of the Earth with depth as a result of compositional, temperature, and pressure variations, as well as lateral (non-radial) variations in the density of the Earth. The fact that the Earth's shape can be produced by an ellipsoid shows that there is no planetary longitudinal variation in gravity. As a result, the planetary field can be described by an equation that only considers a latitudinal function.

The absolute value of gravity on the surface of the Earth has been the subject of considerable interest, but unfortunately measurements were not very accurate prior to recent improvements in electronics and in the accuracy of the measurement of time and length, necessary parameters in the determination of gravity. The theoretical or normal gravity accounting for the mass, shape, and rotation of the Earth on the best-fitting terrestrial ellipsoidal surface is the 1980 Geodetic Reference System of the International Union of Geodesy and Geophysics (Moritz, 1980b). Additional details on this standard and its use in geophysical studies are given in Chapter 6.

2.3.5 Temporal variations

The Earth's gravity field varies over a broad range of periods and amplitudes from a variety of internal and external sources. Fortunately for the exploration of the Earth, most periods and amplitudes of these changes do not seriously conflict with gravity observations. Fluctuations generally do not exceed amplitudes of 1 mGal (most are only a small percentage of that), and their periods are long with respect to gravity observations. Time variations of gravity originate from planetary sources, such as wobbling of the axis of rotation of the Earth, which affects centrifugal accelerations; from local and regional atmospheric pressure perturbations that vary the attraction of the atmosphere; from natural and anthropogenic changes in the subsurface mass associated with transport of fluid and gas in subsurface pore and fracture space, and movement of magma related to igneous events and volcanic activity; and from tidal effects due to varying positions of the Sun and Moon relative to a location on the surface of the Earth. Furthermore, the tides can lead to changes in surface elevation that will also affect surface gravity measurements. The principal temporal variation in gravity in terms of interference with exploration gravity measurements is the tidal effect of the Sun and Moon and related solid-Earth tides which change the radius of the Earth. The maximum tidal effect is only roughly $3 \times 10^{-5}\%$ of the gravity attraction on the surface of the Earth or 0.33 mGal over an approximately 1-day period. The combined effect of the gravitational attraction of the Sun and the Moon and the associated change in surface elevation can be calculated from standardized equations. Chapter 6 considers gravity temporal variations more fully in terms of their sources, characteristics, and elimination from observations during exploration surveying.

2.3.6 Measurement

The two observable quantities in gravity are the so-called "big G" and "little g," representing respectively the gravitational constant and the acceleration of gravity or gravity force per unit mass. The measurement of little g has exercised both geophysicists and geodesists for a long time. Geodesists measure gravity to determine the shape of the Earth, and geophysicists use the measurements to predict changes in the subsurface Earth and study geophysical phenomena. The requirements in the accuracy of gravity measurements are generally in the milligal range, but an accuracy of 1 μGal or better has been the target of an increasing number of measurements for near-surface studies, geodetic measurements, and drillhole gravity surveys. Obtaining these levels of precision is a remarkable achievement considering that the Earth's gravity field is roughly 1000 gal, which means the measurement of one part in a billion of the total field is obtainable.

Measurements of gravity may be either absolute or relative. Absolute measurements have been made only on a limited basis because they are unnecessary for exploration purposes and until recently have been more difficult and time-consuming to make and have had a lower accuracy than relative measurements. However, with absolute gravimeters that use laser interferometry to measure distance and atomic clocks to measure the time for an object to free fall or rise and fall (ballistic instruments), it is possible to obtain absolute gravity in the microgal range within a few minutes with an instrument roughly comparable in portability to relative gravimeters. Absolute gravity measurements provide the actual value of gravity without reference to a station that previously has been observed with an absolute gravimeter. Absolute values are useful in geodetic work, in establishing gravity benchmarks for use with and calibration of relative gravimeters, and in measurements for temporal studies of the Earth's field because they are free of inherent drift effects.

For most purposes, relative gravity observations are quite satisfactory. They can be made with gravimeters that use the same principle as simple kitchen scales: measuring the weight of an object by the change in the length of a spring to which the mass is attached. In the gravimeter a constant mass is attached to the spring, and the change in gravity from site to site is determined by the change in the length of the spring calibrated in units of gravity acceleration. A variety of these instruments have been developed, incorporating a wide range of schemes for magnifying the change in the spring length to achieve the required precision. Although improvements in gravimeters continue to be made, the fundamentals and practice of gravimeter construction are mature and well documented. Most relative gravimeters today are based on the principle of the inclined zero-length spring sensor (Section 5.3.1) in which the lever arm provides mechanical advantage to amplify small relative displacements of the mass. The actual measurement can be based on the positional change in the mass or the change in tension on the spring upon applying a torque to the spring to bring the mass to a null position. Either can be readily calibrated into units of gravity acceleration.

The gravimeter measurement is referenced by the use of spirit levels to the local horizontal or geoid to which the direction of observed gravity is everywhere perpendicular. The measurement assumes that the Earth's normal $\mathbf{g}_N$ and anomaly $\Delta\mathbf{g}_z$ gravity vectors are co-linear (i.e. parallel), and thus the scalar magnitude of the total gravity vector is simply

$$g_T = g_N + \Delta g_z \ \ni \ \Delta g_z = g_T - g_N. \tag{2.6}$$

However, $\mathbf{g}_N$ and $\Delta\mathbf{g}_z$ are not exactly co-linear because $\mathbf{g}_N$ is taken perpendicular to the ellipsoid, which is only a mathematical approximation of the geoid that the gravimeter's spirit levels observe (Figure 2.3).

Thus, for the general gravity anomaly $\Delta\mathbf{g} = \Delta\mathbf{g}_z + \Delta\mathbf{g}_t$ that also includes a transverse component $\Delta\mathbf{g}_t$ (Figure 2.6), the use of the anomaly estimate in Equation 2.6 involves the error

$$\begin{aligned} e_{\Delta g_z} &= g_T - (g_N + \Delta g_z) \\ &= \sqrt{(g_N + \Delta g_z)^2 + \Delta g_t^2} - (g_N + \Delta g_z). \end{aligned} \tag{2.7}$$

However, expanding the square-root term by the binomial theorem and ignoring terms greater than second order shows that the maximum error in the anomaly estimate is

$$e_{\Delta g_z}(\max) \approx \frac{1}{2}\left(\frac{\Delta g_t^2}{g_N + \Delta g_z}\right), \tag{2.8}$$

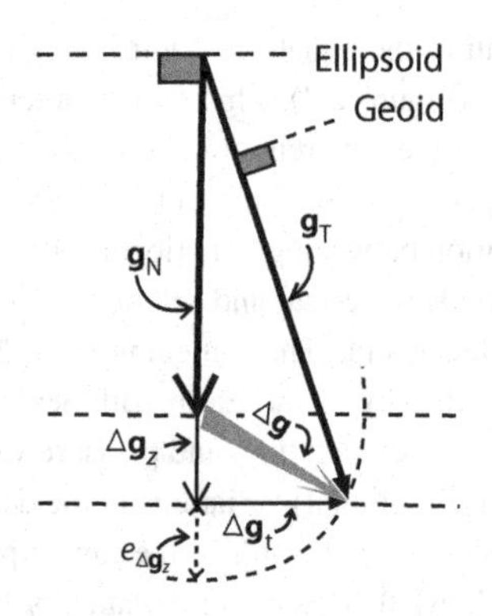

FIGURE 2.6 Assessing the error magnitude Δg_z in using the simple scalar difference in Equation 2.6 for estimating the intensity of the anomalous vertical gravity component Δg_z.

which is negligible on the Earth and other planetary bodies where $g_N \gg \Delta g_z$ and Δg_t. As an example, for $g_N = 10^6$ mGal and $\Delta g_z = \Delta g_t = 10$ mGal, the maximum error $e_{\Delta g_z}(\max) \approx 5 \times 10^{-5}$ mGal is smaller than is measured by available gravimeters.

Gravimeters have been adapted for making measurements in the dynamic environments of marine vessels and both fixed- and rotary-winged aircraft using gyrostabilized platforms to minimize unwanted accelerations. The precision of these measurements is sufficient to warrant their use in extensive exploration programs including investigations for both energy and mineral resources. New developments in gravimeters, employing paired mechanical accelerometers, now make it possible to measure both the gravity field and the full gravity tensor in a variety of dynamic environments for exploration purposes.

Earth-orbiting satellites have significantly increased information on the size and shape of the Earth and its gravity field. Although regional in scope these measurements are being extensively used in exploration of marine regions of the Earth. One system employs satellite altimetry that uses pulse-limited radar systems and precision tracking of the satellite to determine the height of the ocean surface. Subtracting these heights from the elevation of the reference ellipsoid can be used to map the geoidal surface (sea level) to a precision useful in geological studies of ocean basins. The geoidal variations are converted to grids of vertical deflection from which the vertical gradient is calculated as gravity anomaly estimates. The precision of these determinations is of the order of several milligals, but the anomalies must have a wavelength exceeding roughly 15 km to be mapped by the satellite.

In addition, accurate tracking of satellites carrying GPS receivers allows the satellite orbits to be compared to standard Earth orbits to identify orbital deviations that can be analyzed for the Earth's gravity field variations. Radars on two or more satellites measuring the distances between them as they orbit the Earth yield range rate data that also translate into gravity field variations. Most recently, accelerometers have been deployed in orbit to measure the full gravity tensor. The accuracy of these various satellite systems is on the order of several milligals, but the anomaly resolution is limited by the orbital altitudes above ground surface which typically range between 300 and 550 km. These measurements are very useful, however, for mapping large-scale features of the lithosphere that are difficult to study by airborne and shipborne surveying.

Relative gravity observations have been made over much of the accessible land surface of the Earth with gravimeters and over vast areas of the oceans with a combination of satellite-derived measurements and gravimeters mounted in surface ships. Many of these data are available from a wide variety of geophysical data centers. However, for the most part the station interval of the data in these centers limits their use to studies of anomalies covering several kilometers or more. Additional details concerning the measurement of gravity observations are described in Chapter 5.

2.4 History of the gravity method

The effects of gravity have been recognized for millennia and began to be formalized in the third century before the common era with Aristotle's incorrect assertion that heavier objects fall faster than lighter objects. However, he did correctly deduce the basic spherical shape of the Earth from lunar eclipses and other data. In the second century before the common era, the Librarian of Alexandria and father of geodesy, Eratosthenes, accurately measured the Earth's circumference, and hence its radius and the volume necessary for deducing the density of the Earth from gravity observations. The validity of Aristotle's assertion was questioned by the experiments of John Philoponus in the fifth century of the common era and conclusively rejected by Galileo Galilei's investigations in the late sixteenth century. Since the Renaissance there has been an accelerating understanding of gravity and expanded application of it to the study of the interior of the Earth.

The classical history of the gravity method in imaging the Earth can be divided into four broad and overlapping periods that largely were initiated by technological developments. The first period (Table 2.2) extended from the beginning of the seventeenth century, near the end of the Renaissance, until the beginning of the twentieth century. During these three centuries, studies of gravity were related to geodetic investigations of the size, shape, mass, and density of the Earth using the newly invented pendulum for measuring gravity. It was during this period that much of the basic theory of gravity and the terrestrial gravity field was described.

At the beginning of the twentieth century, a new period (Table 2.3) of gravity studies was initiated with recognition of the potential of gravity for mapping of the subsurface, especially oil- and gas-bearing structures. The incentive for geophysical exploration of oil and gas drove the development of high-accuracy, portable instruments for measuring relative gravity into the 1950s. Many of the basic concepts of geological mapping with gravity and related interpretational methods were founded in this period.

The third period (Table 2.4), extending from the 1950s through the 1980s, was marked by the rapid increase in computational power and speed available in digital computers, and the improved processing and modeling methodologies made possible by this increased computing power. Also during this period, advances were made in instrumentation for marine and drillhole measurements of gravity capable of an accuracy useful for geologic studies. Since the 1990s and continuing to the present (Table 2.5), the development of accurate surveying and observation of gravity by satellites especially in marine regions, the availability of exploration-grade airborne gravity measurements, and improved methods of conducting independent and joint inversion with other geophysical data of gravity measurements have led to the recent renaissance in gravity methods.

Numerous publications have described the history of the gravity method of exploring the Earth (e.g. Eckhardt, 1948; Jakosky, 1950; Sazhina and Grushinsky, 1971; Torge, 1989). A particularly comprehensive review of its modern development is presented by Nabighian *et al.* (2005a). The description of the gravity method given below from the Renaissance period to the present and detailed in Tables 2.2, 2.3, 2.4 and 2.5 is largely derived from these references and serves to place the method in a historical context.

As outlined in Table 2.2, current knowledge of gravity began with the work of Galileo Galilei who in roughly 1590 showed that the speeds of objects falling in the Earth's gravitational field are independent of their mass, that they accelerate the longer they fall, and that the distance an object falls is proportional to the square of the elapsed time. Although his observations are often related to experiments conducted from the famed leaning tower of Pisa, Italy, they were actually made from observing the movement of objects sliding on an inclined plane, which permitted greater accuracy in the measurement of time than in a free fall. He used the pulse rate of his heart

TABLE 2.2 Classical gravity milestones I. Development of gravitational theory and geodetic measurements from the seventeenth to early twentieth centuries.

Date	Event
1590	*Galileo Galilei* establishes the basic principles of the gravity field
1605	*J. Kepler* deduces the laws of movement of the planets around the Sun based on *Brahe's* observations of planetary orbital motion
1657	*C. Huygens* develops pendulum clock and mathematical theory of the pendulum which showed that gravity acceleration is related to the pendulum's period
1672	*J. Richer* deduces that the change in period of pendulum clock in different places is due to variation in gravity
1687	*I. Newton* formulates the universal law of gravitation and that the source of gravity of the Earth is the result of the Earth's mass and rotation
1735–1743	The first relative measurements of gravity; *P. Bouguer* develops the gravity correction for the height of an observation relative to sea level
1743	*A. C. Clairaut* publishes a theorem that permits the determination of geometric parameters of the Earth's figure from gravity data and vice versa
1797	*H. Cavendish* measures the gravitational constant "big *G*" with high accuracy
1775–1850	*J. L. Lagrange, P. S. Laplace, A. M. Legendre, C. F. Gauss, S. D. Poisson*, and *G.Green* develop the classical theory of potential fields
1854	*G. Stokes* describes a theorem that shows there is no unique solution to the inverse problem of potential fields and the geoid can be determined from global gravity
1855	*G. B. Airy* and *J. H. Pratt* give geological models for how regional terrestrial topographic features are in hydrostatic equilibrium in accordance with *G. Everest's* measurements of the deflection of the vertical in India in the 1850s
1889	*C. E. Dutton* coins the term *isostasy* to describe the hydrostatic equilibrium condition of mountains, basins, and other crustal features with considerable relief

and those of his assistants to establish the travel-times of moving objects in his experiments. Galilei also discovered that the period of a pendulum is a function only of its length and probably made the first measurement of gravity.

Building on Galilei's work and Johannes Kepler's observations of the movements of the planets around the Sun, Isaac Newton in the latter part of the seventeenth century formulated the law of universal gravitation, the famed inverse square law. Newton described his studies in a 1687 publication which showed that gravity is produced by the attraction of the Earth's mass and the centrifugal force due to the Earth's rotation, and that weight is the effect resulting from gravity. His theory is applicable across the universe at all scales. He also suggested that the equatorial bulge due to the Earth's rotation accounted for latitudinal differences in gravity as measured by pendulums and that gravity increases from the equator to the poles as the square of the sine of latitude.

A fundamental aspect of Newton's formulation is that the force of attraction between masses, such as the Sun and its planets, is a function of a proportionality constant, the gravitational constant. Although Newton had to approximate this value, Henry Cavendish roughly a century later (around 1797) used torsion balance measurements to determine the constant to within 1% of the currently accepted value. With big G determined, gravity measurements or little g could be evaluated for the mass of the Earth via the extension of Newton's second law in Equation 2.5.

Dividing the Earth's mass by its volume as obtained from Eratosthenes' observations showed that the mean density of the Earth is about 5500 kg/m^3, which is roughly twice the average density of the rocks at the surface. Thus, the density clearly must increase with depth, which is also consistent with a mass-differentiated Earth involving crustal, mantle, and core components of increasing densities (Figure 1.1).

Building upon Galilei's studies, Christian Huygens in the seventeenth century developed the mathematical theory of the pendulum and built the first pendulum clock which led to the discovery that gravity varies over the Earth's surface. Jean Richer in the 1670s, upon moving a pendulum clock from Paris, France to French Guinea in South America, found that the clock had lost about 2.5 minutes per day. Richer correctly deduced that this change was due to variation in gravity as predicted by the pendulum law of Galilei and resulted from the change in latitude from Paris to near the equator.

The first relative gravity measurements were made during the French expedition to Peru from 1735 to 1743 to determine the shape of the Earth. It was this study together with the results of a complementary expedition to Lapland that showed a degree of latitude decreases in length from the equator to the poles, proving that the shape is that of an oblate spheroid, that is, the equatorial radius of the Earth is roughly 21 km greater than the polar radius. It was during this expedition that Pierre Bouguer developed the gravity correction for the mass effect of the ground defined by the height of the observation relative to sea level.

Continuing gravity observations were made into the nineteenth century primarily for the purposes of determining the gravitational constant and the mass and mean density of the Earth. The reversible pendulum for making relatively accurate gravity measurements was developed by H. Kater in 1819, and it was improved with the development of the invariable pendulum which measures the relative gravity change from a base station. The pendulum with various modifications to minimize errors and improve the timing of the pendulum swings was used for relative gravity measurements until around 1930 and for absolute measurements until the 1970s when free-fall instrumentation displaced it.

In the mid-eighteenth century, A. C. Clairaut published his now widely used theorem that permits the determination of geometric parameters of the Earth's figure from gravity data and vice versa. During the latter part of the eighteenth and into the early nineteenth century, classical mathematical studies on potential fields by J. L. Lagrange, P. S. Laplace, A. M. Legendre, S. D. Poisson, C. F. Gauss, and G. Green led to the theoretical foundations of both the gravity and magnetic methods. In the mid-nineteenth century, G. Stokes showed that there was no unique solution to the inverse problem of potential fields, and that the flattening of the Earth's ellipsoid and the deviation of the best fitting equipotential surface, the geoid, can be determined from the global gravity. His theorem, which relates the shape of the Earth and gravity, led to campaigns to measure gravity over the Earth's surface.

In the railroad surveys of India, G. Everest observed increasing closure errors as the surveys approached the Himalaya Mountains. He deduced that the mountains involved mass deficits which affected the plumb bobs in the surveying instruments to cause the erroneous deflections of the vertical. This confirmed the speculation by Leonardo da Vinci that terrestrial topographic features tend to be in hydrostatic equilibrium. C. F. Dutton in 1889 coined the term isostasy for this effect that describes the hydrostatic equilibrium of the Earth at some level by changes in subsurface masses despite the variation in the elevation of the Earth's surface. G. B. Airy and J. H. Pratt in 1855 described quite different, but equally plausible explanations for this phenomenon. Airy proposed that the mountains were of constant density but had roots. Pratt proposed that the mountains were less dense the higher they were. This prompted a variety of geophysical studies to determine which theory was valid. During the early part of the twentieth century, J. F. Hayford, W. Bowie and others made extensive gravity observations to investigate the relationship between surface topography and the isostatic state of the Earth. This initiated the use of gravity for geological purposes.

Despite the continued use of pendulums for measuring gravity through the nineteenth century, increasing attention was devoted to develop more efficient, portable, and accurate methods of measuring gravity in the latter part of the nineteenth century and into the twentieth century (Table 2.3). The work of R. von Eötvös at the turn of the century in developing the portable double-beam torsion balance which measures gradients of gravity is notable. Studies in the first decade of the twentieth century showed that gravity measurements could be used to study subsurface features. In 1915, J. Fekete and D. Pékar conducted the first practical torsion balance survey over the Egbell oilfield in Czechoslovakia and in 1917, H. V. Boeckh explained why geological structures such as salt domes and anticlines have associated gravity anomalies. At the end of World War I, H. Shweydar performed a torsion balance survey for petroleum exploration at the site of a salt dome in northern Germany. He also developed schemes for calculating and removing topographic effects from gravity observations. These surveys continued in northern Germany. As part of post World War I reparations, in 1922 Everett Lee De Golyer brought the torsion balance to the Gulf Coast region of the United States where a positive anomaly was found associated with the cap rock of the Spindletop dome in Texas. Drilling in 1924 on an anomaly mapped by the torsion balance that is associated with the Nash dome in Texas discovered the first petroleum in America by geophysical methods. The torsion balance led to several other discoveries, but by 1940 it was entirely replaced by gravimeters.

During the 1920s, there was intense interest in improving the accuracy and ease of gravity measurements, driven by the potential of the gravity method in petroleum exploration. This led to the development and greatly expanded use of gravimeters of a variety of designs, but generally based on variation in the length of a spring. It was also during this time that the gal was first used in Germany as a unit of gravity acceleration and subsequently accepted internationally (1930). Standard data reduction procedures for pendulum and gravimeter observations were discussed by M. K. Hubbert and F. A. Melton in 1928 and were

TABLE 2.3 Classical gravity milestones II. Development of gravity for mapping the subsurface from the early twentieth century into the 1950s.

Date	Event
1900	Development of gravimeters for measuring gravity field commences
1902	*R. von Eötvös* develops the double-beam torsion balance for measuring gravity gradients and conducts first survey in Hungary
1915	*J. Fekete* and *D. Pékar* conduct torsion balance survey over Egbell oilfield
1917	*H. V. Boeckh* explains why geological structures in sedimentary basins have associated gravity anomalies
1922	*H. Schweydar* performs torsion balance survey over a salt dome and develops schemes for removing topographic effects from gravity observations
1922	Torsion balance brought to USA where a positive gravity anomaly is mapped, associated with the cap rock of the Spindletop dome in Texas
1920–1930s	Intense interest in improving the ease and accuracy of gravity measurements leading to development and greatly expanded use of gravimeters based on various designs, but generally on the measurement of the length of a spring holding a constant mass. Gravimeters replace torsion balances for measuring gravity by 1940
1920s	*F. A. Vening Meinesz* modifies the pendulum for obtaining geologically usable gravity observations at sea
1924	Drilling on an anomaly mapped by the torsion balance associated with the Nash Dome in Texas produces probably the first petroleum in the USA discovered by geophysical methods
1928	First use of gal for the CGS-unit of gravity acceleration, with international acceptance of unit in 1930. Standard data reductions for pendulum and gravimeter data discussed by *M. K. Hubbert* and *F. A. Melton*, and expanded by *S. Hammer*, *L. L. Nettleton*, and others
1930–1935	Portable pendulum used in exploration for measuring relative gravity
1934	*L. J. B. LaCoste* develops the zero-length spring principle which obtains the highest accuracy measurements in any relative gravimeters
1940s	*S. Worden* incorporates universal temperature compensation into quartz spring gravimeters which achieve high accuracy without external power supply
1940–1950s	Gravimeter in wide use in oil and gas exploration and introduced in mineral exploration
1949	*L. L. Peters* publishes methodology for enhancing potential field data useful in interpretation

expanded upon by S. Hammer, L. L. Nettleton, and others. It was also during the 1920s that F. A. Vening Meinesz modified the pendulum to achieve geologically usable gravity observations at sea, although O. Hecker had made a few hundred gravity observations in the major oceans prior to this, during the period from 1901 to 1909, that showed the oceans to be in isostatic equilibrium except for deep-sea trenches of the Pacific Ocean.

In 1934, L. J. B. LaCoste developed the principle of the zero-length spring which opened the way for greater accuracy in relative gravity measurements. Subsequent development of gravimeters based on this principle has led to the majority of gravimeters used today for land, sea, and air observations. L. M. and F. W. Mott-Smith were among the first to take advantage of the excellent mechanical and thermal properties of quartz in building a gravimeter. By 1945, gravity was being used extensively in petroleum exploration as evidenced by the peak of 170 crews observing gravity in the USA alone. The use of quartz elements in gravimeters was expanded after World War II by S. Worden to include temperature compensation which permitted high-accuracy measurements without an external power supply. The resulting meter and its derivatives are very portable, and thus widely used in many gravity applications today. G. P. Woollard showed that this meter could be used for making accurate observations over a wide range of values, and thus could be used for geodetic as well as exploration purposes. Additionally, these meters and gravimeters capable of 0.001 mGal accuracy have found significant uses in mineral exploration, engineering, and environmental studies. Woollard also was responsible for popularizing gravity as a useful method of studying regional geological features during and after World War II. L. J. Peters in 1949 described methods of enhancing magnetic anomaly data in the space domain which can also be used for gravity anomaly data.

Although gravity measurements were made at sea after World War II using modified land gravimeters set on the sea floor or on surface vessels using a variety of specially designed instruments, it was the introduction during the 1960s of the gyrostabilized platform that led to extensive measurements at sea. Modified land gravimeters are placed on this platform, which eliminates longer-wavelength (period) accelerations associated with the movement of the meter from gravity observations. Similar instruments were successfully used from the early 1980s in aircraft for exploration purposes. The availability of the Global Positioning System for positioning of gravity measurements on land, sea, and air since the 1990s has greatly improved the accuracy and efficiency of gravity measurements.

The ready availability of digital computers and their continuing improvements since the 1950s have led to many computational improvements in the processing and interpretation of gravity anomaly data and, during roughly the past decade, in the presentation of anomaly data (Table 2.4). Use of the Fourier transform in potential field analysis by DEAN (1958) and others, the computation of gravity effects from arbitrarily shaped two- and three-dimensional sources by TALWANI *et al.* (1959) and TALWANI and EWING (1960) and inversion techniques of interpretation by BOTT (1960), CORBATO (1965), OLDENBURG (1974), LI and OLDENBURG (1998a) and others brought notable improvements in gravity data applications. In addition, during this period the development of Geographic Information Systems (GIS) enabled interpreters to view and analyze multiple data sets in a synergistic manner.

The global coverage of gravity measurements has markedly increased with the use of artificial Earth satellites (Table 2.5). The observation and analysis of satellite orbits as early as 1958 were used to establish long-wavelength components of the Earth's field to a greater accuracy than given by surface observations. The number and sophistication of satellites, some designed especially for potential field studies, increased the accuracy of the results and the spectrum of gravity anomalies that can be mapped from satellites. Of particular note are the satellite altimetry studies of the ocean surface beginning in the late 1990s by SANDWELL and SMITH (1997) with observations from the Geosat and ERS-1 and ERS-2 satellites that are capable of mapping marine gravity anomalies greater than roughly 15 km in minimum dimension. The GRACE satellite is particularly noteworthy in measuring temporal variations in gravity such as those caused by changes in the water stored in surface water basins and ice in the great ice sheets of Greenland and Antarctica.

Further significant developments in gravity over the past decade are the use of inertial accelerometers for measuring the gravity tensor which is finding increased use in gravity interpretation. Airborne gravity gradiometers are being used routinely in exploration, and the GOCE satellite gradiometer observations are providing significant new constraints on the regional mass variations of

TABLE 2.4 Classical gravity milestones III. Development of improved computer data processing and modeling capabilities from the 1950s through the 1980s.

Date	Event
1958	*W. C. Dean* publishes filter theory for potential fields in frequency domain
1959	*M. Talwani* and others develop algorithms for calculating the gravity effect of 2D bodies of arbitrary shape and in 1960 *M. Talwani and M. Ewing* publish modeling algorithms for the gravity effects of 3*D* bodies of arbitrary shape
1960	*M. H. P. Bott* publishes a method for interpretation of gravity data by inversion
1960s	Development of the gyrostabilized gravity meter make possible exploration grade gravity measurements from a moving platform
1965	*C. E. Corbató* publishes a method of linearizing nonlinear inversion of 2D bodies of arbitrary shape
1970s	Geographic information system (GIS) technology applied to potential field analysis
1972	*R. W. Parker* uses Fourier transforms to calculate gravity anomalies from complex bodies
1972	First gravity measurements on the surface of the Moon
1974	*D. W. Oldenburg* modifies *Parker's* Fourier transform method of forward modeling to inversion of gravity data
1977	Airborne gravity measurements used in oil and gas exploration
1978	Free-air gravity anomalies of the global oceans mapped from Seasat and subsequent satellite altimetry missions

TABLE 2.5 Classical gravity milestones IV. Gravity advances due to improved navigation capabilities, mobile platforms, and inversion procedures from the 1990s to the present.

Date	Event
1990s	Oil and gas reservoir monitoring with gravity measurements initiated
1990s	Development of portable falling body absolute gravity measuring instrument with an accuracy approaching 1 μGal
1997	*D. T. Sandwell* and *W .H. F. Smith* map global marine gravity anomalies from Geosat and the ERS altimetry satellite missions
2000	US military discontinues degradation of GPS signals, allowing high-accuracy navigation of gravity surveys
2000	CHAMP geophysical research satellite launched to map the global gravity and magnetic fields at 300–450 km altitude
2002	GRACE satellites launched to map temporal gravity variations due to water and glacial mass dynamics of the Earth at about 400 km altitude
2000s	Improvements continue in development of independent and joint inversion of gravity data
2000s	Rotating disk gravity gradiometers developed for military uses, but declassified in the mid-1990s, are adapted to exploration purposes and used increasingly in both oil and gas exploration and mineral exploration
2009	GOCE satellite mission launched to map the global gravity gradients of the Earth at roughly 250 km altitude

the Earth's crust, mantle, and core. Finally, much of the global data is maintained in public and commercial data repositories and can be readily accessed by users.

2.5 Implementing the gravity method

Whatever the objective, applying the gravity method to imaging the subsurface follows the common sequence of geophysical practice which is described in Section 1.5. The implementation sequence involves the planning, data acquisition, data processing, and interpretation phases outlined in Figure 2.7. This section broadly introduces the topic, with more objective-specific details described in Chapters 5, 6, and 7.

2.5.1 Planning phase

Once it is deemed advisable to apply the gravity method to a particular problem, the planning phase is initiated. In this phase the areal size of the survey, data density, and the accuracy of the measurements and their processing are all of interest. The first step in this phase is to develop a model of the subsurface sources of interest. This means that the general location of the sources and their geometries, sizes, depths, and density contrasts with the horizontally adjacent rock units must be ascertained or estimated. Commonly the available subsurface information is inadequate to define these parameters very accurately or there is a range of sources of interest so that a spectrum of models of sources of interest covering the range of uncertainty in the definitive parameters is defined.

All parameters of the sources are of interest, but depth is a particularly significant factor because of the ubiquitous role of the inverse distance function in gravity theory (e.g. Equation (3.26)). The density differential is commonly difficult to estimate because of a paucity of direct density data, particularly on the formations adjacent to the sources of interest. These country rocks generally are poorly studied, and their potential heterogeneity complicates the estimation of their density. The lack of direct data typically leads to estimating the densities from generalized tables based on lithologies and a consideration of the controlling characteristics specified in Section 4.5.

In the planning phase one must estimate the minimal characteristics of the anomalies anticipated in the study. This can be based on experience in similar projects, but more often involves the approximation of the anticipated sources of anomalies with simplified geometries and the calculation of their anomalies based on estimated densities of the sources and their adjacent formations. The result is a suite of anomaly dimensions and amplitudes that are likely to be encountered.

In addition, the nature of anticipated noise in the gravity measurements and anomalies is estimated, again either from experience or by calculation of effects from anticipated sources. A common source of noise, that is anomalies that have wavelengths approximating or less than the wavelengths of sources of interest, results from errors in measuring and processing the gravity field to the anomalies. Occasionally these sources also may produce longer-wavelength anomalies due, for example, to slowly varying calibrations of instrumentation. Other important sources of noise are derived from local geological conditions. These latter effects cannot be eliminated, while the former can be minimized through improved measurement and reduction procedures if they are found to be a significant factor

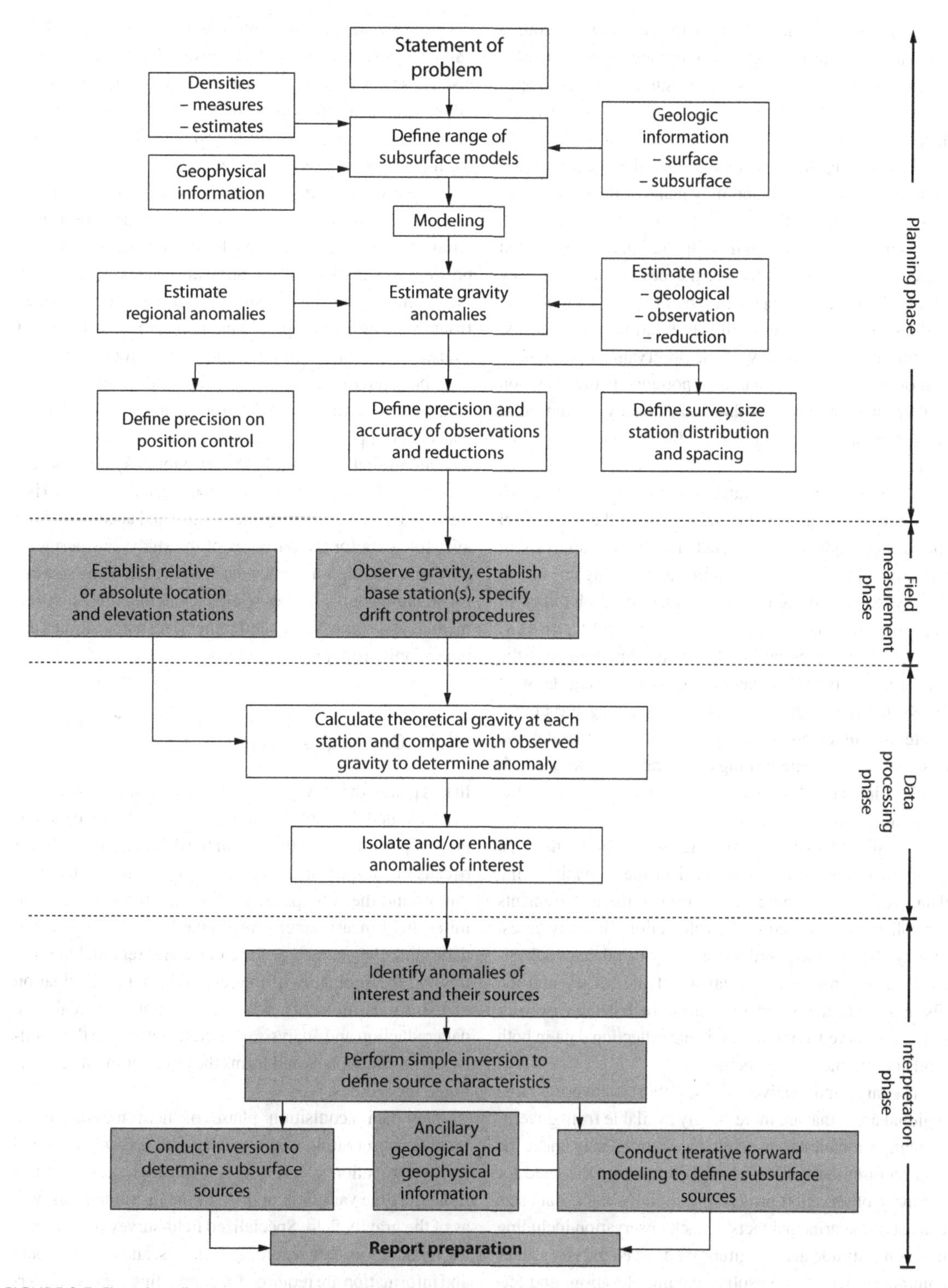

FIGURE 2.7 Flow chart illustrating the principal steps in conducting the gravity method.

in analyzing anomalies. The anticipated noise and source anomaly characteristics are used in selecting data density and accuracy requirements of the survey. On the opposite end of the spectrum, anomalies due to broader and deeper features than the sources of interest, resulting in longer wavelengths or regional anomalies, generally are estimated from regional anomaly maps of the study area.

Once models of the anticipated sources are specified and their anomalies determined in relation to the expected noise and regional anomalies, the design of the survey begins. This requires specification of the size and location of the survey, the layout of the observations, data density, and the required accuracy of the observations and reduction of the data. Of particular importance is the selection of data density, which is dictated not only by the characteristics of sources but also by the intended use of the data.

A survey may be conducted simply to detect potentially interesting anomalous values above the noise level for further study or to conduct quantitative analysis of the anomalies. The latter requires a much higher station density. Survey design may consider the probability of detecting an anomaly (e.g. SAMBUELLI and STROBBA, 2002) and may incorporate the fractal dimensions of the source anomaly and the network of stations (e.g. DIMRI, 1998). Design of the survey means selecting the appropriate instrumentation and procedures for making the observations and determining collateral information, such as elevation and surrounding topographic relief, for the reduction of the observations.

It is also necessary at this stage to determine the type of anomaly(ies) that will be used in the analysis of the data. The type of anomaly determines the requirements for collateral data used in the calculation. In many cases, gravity data are acquired secondarily to other geophysical data which control the layout of the survey and the distribution of the observation sites. Such is the case with gravity observed along with seismic reflection data in both land and marine environments.

The large and relatively dense data sets covering large regional areas that are increasingly available from governmental, commercial, or academic sources may meet the data requirements of the survey, eliminating the need for a gravity observation program. Generally, these data sets consist of the principal facts of each observation including position (latitude and longitude), observed gravity values commonly tied to an absolute datum, elevation, and terrain correction that permit the reduction of gravity data to either free-air anomaly or Bouguer anomaly based on an assumption of the mass between the station and the elevation datum. In many areas several data sets are available covering the same region, but these may not be of equal value because of unequal data density. Care must be exercised to select a data set that has erroneous values removed either through a comprehensive quality assurance program or filtering of high-wavenumber components and also has the highest-quality terrain corrections.

In many cases the data sets have (in addition to the principal facts) the anomaly values at the observation sites, eliminating the need for any additional processing before posting and analysis of the anomaly values. Commonly, the anomaly data sets consist of anomaly values gridded from the point values, providing a uniform distribution of values for ease in presentation and further processing. Care must be taken to evaluate the grid interval compared to the distribution of the original observations so that aliasing concerns (Appendix A.4.3) are minimized. Where gravity data are available, actual field surveying may be unnecessary. However, care must be taken in deciding to use existing data by first evaluating the density and accuracy of the available data for the purposes of the study. Furthermore, the dense and high-accuracy data required in many surveys for shallow-source studies related to engineering, environmental, and some energy and mineral resource studies are unavailable from public data centers.

2.5.2 Data acquisition phase

In this phase, the survey as designed in the planning stage is implemented. It is not uncommon that the plans are incomplete or preliminary at the start of the data acquisition phase because of a lack of information on the subsurface conditions and the anticipated gravity signatures. As a result initial field measurements may take the form of a test to determine the character of the noise and regional anomalies and the anomalies of interest in the study. Evaluation of these measurements, which is facilitated by real-time data reduction and analysis, is used to determine the validity of the initial plans and forms the basis for modifications in the procedures.

The data acquisition phase of field measurements involves observation of the gravity data and, where called for, reoccupation of previously measured sites to determine the time variation, or drift, of the instruments as well as of the gravity field. Specialized field-survey procedures, documentation, data recording facilities, and ancillary data and information are required for conducting surveys underwater, underground, and in ships and aircraft (e.g. TORGE, 1989). In these surveys, as well as land-surface surveys, the final step in this phase is to record and store the data in standardized formats in digital form.

The timing of the gravity survey is important where gravity surveying will be deployed along with other geophysical techniques. It should be done early in the process because the method generally is one of the least costly of the geophysical methods on a per-unit-area or line-distance basis. Thus, it can provide useful information for targeting the positioning of the other more expensive and time-consuming geophysical procedures.

2.5.3 Data processing phase

This phase in conducting gravity surveys consists of two basic steps: data reduction, and isolation and enhancement, leading to data that are suitable for interpretation. The field measurements, which commonly consist of the turns on a screw or some variation thereof that are required to bring the mass within a gravimeter to a null position, are converted to gravity units with the calibration unique to the instrument. These observations are then subjected to a broad range of adjustments for the time variation of gravity and to relate them to gravity datums. The resulting values are compared with the theoretical or modeled gravity to produce the gravity anomaly, the difference between the observed gravity and the modeled value at a site.

The process of calculating the gravity anomaly is referred to as reduction, but this should not be construed to consist of reduction of the observations to a datum. This step involves determining the theoretical gravity at the site, taking into account planetary as well as local effects. A variety of types of anomalies have been developed for specific interpretational purposes. Each takes into account certain effects which may include predicted geological effects. The purpose of the reduction procedure is to convert the field measurements into interpretable form by eliminating all predictable gravity effects leaving only the effects of unknown subsurface mass variations.

After the observed data have been reduced to anomaly form, they are subjected to a process that isolates or enhances the anomalies of interest in the particular survey. The reduced anomaly data consist of a spectrum of anomalies derived from sources covering a range of depths, volumes, geometries, and density contrasts. The cumulative effect of the multiplicity of anomalies commonly leads to distortion of the anomalies of interest and difficulties in identifying them.

Isolation and enhancement procedures effectively serve as filters to minimize the effect of short-wavelength noise and long-wavelength regional anomalies upon the anomalies of interest in the survey. This filtering commonly is conducted by mathematical digital methods, but in the case of simple extraneous effects the process can be achieved by graphical procedures. Isolation techniques attempt to delineate the interesting anomalies with a minimum of distortion so as to maintain the integrity of the anomalies for interpretational purposes, whereas enhancement procedures seek to accentuate the anomalies of interest by emphasizing particular attributes of the anomaly such as amplitudes, gradients, and strike directions.

The procedures used in data reduction generally are standardized, with the interpreter only making limited decisions about the parameters involved and the types of factors to be considered. As a result, anomalies calculated by different experienced analysts generally are essentially equivalent. The results of the isolation and enhancement process are somewhat different. Experienced analysts often apply specialized filters to anomaly data to isolate or enhance the anomalies of interest, possibly producing quite different gravity maps or profiles for interpretation. The interpreter's view of the geology of the region and the anomaly sources becomes an important factor in selecting procedures and defining the range of anomalies to isolate or enhance.

2.5.4 Interpretation phase

The interpretational phase of the gravity method consists of making inferences about the subsurface from anomaly values derived from the data processing phase. The process may be quite involved, using intensive mathematical computations, or may simply consist of a qualitative appraisal of the anomalies. The actual process employed is determined by the complexity of the anomaly pattern, the objective of the survey, the experience of the interpreter, and the available resources.

Interpretation of gravity data usually involves either iterative forward modeling of geologically reasonable subsurface sources until an approximate match is obtained to the observed anomalies, or inversion wherein the attributes of the sources are determined to some prescribed quantitative level of error directly from the anomalies. Both procedures can take on various degrees of complexity, but the results of even the most sophisticated methods are not unique. All gravity interpretations are subject to ambiguity, the degree of which is determined by the control that can be exerted on the interpretation from the results of other geophysical methods and related geological information.

The initial interpretation of gravity anomaly data is a physical model or a range of possible models of a specified density contrast with the surrounding Earth materials. In the final stage, this physical model or models must be transcribed into a geological model that will answer the objective of the survey. As a result a strong element of geological expertise is required in gravity interpretation.

2.5.5 Reporting phase

The process is completed with a report on the various phases used in the application. Of particular importance is a clear specification of the assumptions made in the various phases, the error budget, and the optimum interpretation of the gravity measurements and the robustness of this interpretation in view of the limitations of the gravity data and the constraints imposed by collateral information. Computer graphics are very effective and routinely used in promoting gravity data analyses and interpretations.

2.6 Key concepts

- The gravity method is based on measurement of perturbations or anomalies in the Earth's gravitational field caused by horizontal variations in the density of Earth materials. Generally the vertical component of gravity or, in more limited exploration investigations, the gradient component of the total gravity field, is measured. The measured gravity component is the vectorial summation of all sources. As a result, the measured fields are compared to model fields based on planetary and known geological effects to obtain the anomalies from the subsurface. Additionally, the computed anomalies can be subject to a variety of filtering processes to isolate or enhance the anomaly associated only with the specific targeted geologic source.
- The gravity field of the Earth is caused by the fundamental phenomenon of gravitation which causes bodies to be attracted toward each other. The gravitational force, as defined by the universal law of gravitation, is proportional to the product of the masses of the bodies and inversely proportional to the square of their separation. This force cannot be measured independent of mass, so the force per unit mass is measured, which in the terrestrial situation is the acceleration of a body caused by the free fall of the mass in the gravitational field of the Earth.
- The common unit of measurement in gravity is the gal in CGSu which is equivalent to 1 cm/s^2, and the m/s^2 (= 100 gal) in SIu. Relative measurements of the acceleration of gravity generally are made in the study of the subsurface with specialized spring balances in which the change in length of the spring calibrated into units of acceleration is measured as a function of the changing gravity. Typically the precision of these measurements is of the order of one part in a hundred million or better of the Earth's gravity field, as required by the small variations in density and resulting mass changes in the subsurface. Increasingly, tensor and gradient components of the acceleration of gravity as well as absolute gravity are measured with specialized instrumentation in exploration studies.
- Gravity changes over the surface of the Earth by about 0.5% as a result of the decreasing radius and centrifugal acceleration of the Earth from the equator to the poles. Minor temporal variations arise from tidal effects due to the varying attraction of extraterrestrial bodies of the solar system.
- The classical history of the gravity method in imaging the Earth can be divided into four broad and overlapping periods that largely were initiated by technological developments. The *first period* extended from the beginning of the seventeenth century, near the end of the Renaissance, until the beginning of the twentieth century. During these three centuries, studies of gravity were related to geodesy, investigation of the size, shape, mass, and density of the Earth using the newly invented pendulum for measuring gravity. It was during this period that much of the basic theory of gravity and the terrestrial gravity field was described.

  At the beginning of the twentieth century, the *second period* of gravity studies was initiated with recognition of the potential of gravity for mapping of the subsurface, especially oil- and gas-bearing structures. The incentive of geophysical exploration for oil and gas drove the development of high-accuracy, portable, relative gravity measuring instruments into the 1950s. Many of the basic concepts of geological mapping with gravity and related interpretational methods were founded in this period.

  The *third period*, extending from the 1950s into the 1990s, was marked by the rapid increase in computational power and speed available in digital computers, and improved processing and modeling methodologies made possible by this increased computing power. Also during this period, there were advances in instrumentation for marine and drillhole measurements of gravity capable of an accuracy useful for geologic studies.

  In the *fourth period*, since the 1990s and continuing to the present, the development of accurate surveying and observation of gravity by satellites especially in marine regions, availability of exploration-grade airborne gravity measurements, and improved methods of conducting independent and joint inversion with other geophysical data of gravity measurements have led to the recent renaissance in gravity methods.
- Regardless of the objective, applying the gravity method to imaging the subsurface follows common geophysical

practice, involving a logical sequence of steps that consists of planning, data acquisition, data processing, and interpretation.

- The planning phase involves definition of the range of anomaly characteristics which are anticipated from the sources of interest and the nature of the expected anomalous noise that interferes with the identification and analysis of the anomalies of interest. Decisions regarding the pattern, density of observations, instrumentation, processing and interpretational procedures etc. are based on these anomaly and noise characteristics, taking into account cost/benefit considerations.
- Data acquisition follows the procedures outlined in the planning stage so that observations are made with appropriate accuracy and coverage density. However, the increasing amount of data now compiled in governmental and commercial data repositories often meets survey requirements, eliminating the need for acquiring new data, and greatly decreasing the costs of the gravity program.
- Processing of the acquired data generally involves two steps. First is the reduction of the data into an interpretable form by eliminating all predictable gravity effects leaving only the gravity acceleration of the unknown subsurface sources. This is accomplished by comparing the observed data with a theoretical model of the gravity acceleration at each observation site. The second involves the isolation or enhancement of the anomalies of interest so that they can be identified and analyzed. The former step is usually unnecessary for gravity observations in data repositories. The latter is usually oriented to a specific set of target anomalies, and thus is required as part of the processing program.
- The next stage is the interpretation of the anomalies of interest. This may simply be a matter of identifying the location of the specific sources, but more commonly includes quantified interpretation involving inversion of the selected anomalies for their sources. The inversion process commonly estimates not only the location and configuration of the source, but also its depth and property contrast with the surrounding rocks. This inversion process is inherently ambiguous, but the range of possible sources generally is limited by incorporating other geological and geophysical data into the analysis.
- Finally, the physical model of the inversion is converted into geological terms in the reporting phase of the program based on the geology of the region and constraints imposed by the laws and theorems of geology. An error analysis with a clear specification of the error budget is an important element of reporting on a gravity study.

Supplementary material and Study Questions are available on the website www.cambridge.org/gravmag.

3 Gravity potential theory

3.1 Overview

The gravity method is commonly referred to as a potential field method because it involves measurements that are a function of the potential of the observed gravitational field of force of the Earth. "Potential" is defined by the amount of work done in moving a particle from one position to another in the presence of a force field acting upon the particle. Thus, potential is a function of force-field space such that its rate of change is the component of force in that direction, which is related to the acceleration of gravity measured in the gravity method. This concept leads to a number of fundamental laws and theorems, such as Laplace's and Poisson's equations, Gauss' law, and Poisson's theorem, which are useful in understanding the properties of the gravity field and its analysis. These results extend to an arbitrary distribution of source particles by summing at the observation point the potential effects of all the particles of the body.

Forward models with closed form analytical expressions involve idealized sources with simple symmetric shapes (e.g. spherical, cylindrical, prismatic) that can be volume-integrated in closed form at the observation point. Forward models without closed form expressions involve general sources with irregular shapes that must be numerically integrated by filling out the volumes with idealized sources and summing the idealized source effects at the observation point. The numerical integration can be carried out with least-squares accuracy by distributing idealized sources throughout the general source according to the Gauss–Legendre quadrature decomposition of the irregular volume. The gravity effects of all conceivable distributions of mass can be modeled to interpret the significance of gravity anomalies.

3.2 Introduction

The concepts of the potential and the potential field (e.g. KELLOGG, 1953; RAMSEY, 1961; SIGL, 1985) are fundamental to understanding the gravity method. Consideration of the potential is useful because it simplifies the analysis of the force field and aids in its understanding. In particular, potential is a scalar quantity, and thus is independent of direction, in contrast to field forces and their derivatives, which are vectors. Thus, potential is used wherever possible to avoid complexities in mathematical operations imposed by the directional attributes of forces.

Potential is described in terms of relative values based on the potential difference which is defined as the amount of work done in moving a particle from one position to another within a force field. Thus, potential (work) is defined as a function of force-field space, such that its rate of change (derivative) is the component of force in that direction. In the Cartesian (x, y, z)-coordinate system, the force $\mathbf{F}$ may be expressed as

$$\mathbf{F} = +\frac{\partial \Phi}{\partial \mathbf{r}} \text{ or } \mathbf{F} = -\frac{\partial \Phi}{\partial \mathbf{r}} \tag{3.1}$$

respectively opposite to or in the direction of the 3D displacement vector $\mathbf{r} = (\Delta x)\hat{\mathbf{e}}_x + (\Delta y)\hat{\mathbf{e}}_y + (\Delta z)\hat{\mathbf{e}}_z$ with unit vectors $\hat{\mathbf{e}}_x$, $\hat{\mathbf{e}}_y$, and $\hat{\mathbf{e}}_z$ in the respective principal x-, y-, and z-directions, and Φ is the potential at the observation point (x, y, z) at the distance

$$|\mathbf{r}| = r = \sqrt{(x-x')^2 + (y-y')^2 + (z-z')^2}$$
$$= \sqrt{\Delta x^2 + \Delta y^2 + \Delta z^2} \tag{3.2}$$

from the source point (x', y', z').

The literature considers Equation 3.1 in both sign conventions because the use of either sign convention yields the same theoretical results. In this book, however, the more commonly used conventions are adopted for analyzing gravity and magnetic potentials (e.g. KELLOGG, 1953; BLAKELY, 1995). Specifically, the positive gradient of the positive potential U (i.e. $\mathbf{F} = +\partial U/\partial \mathbf{r}$) is invoked for gravity fields where particles of like sign attract each other and the potential U equals the work done by the field or the negative of the particle's potential. For magnetic fields, on the other hand, the positive gradient of the negative potential V (i.e. $\mathbf{F} = -\partial V/\partial \mathbf{r}$) is invoked where particles of like sign repel each other and the potential equals the work done against the field by the particle or the potential of the particle.

The force vector $\mathbf{F}$ has the orthogonal vector components $\mathbf{F}_x$, $\mathbf{F}_y$, and $\mathbf{F}_z$ given by

$$\mathbf{F} = \pm\nabla\Phi = \pm\frac{\partial\Phi}{\partial x}\hat{\mathbf{e}}_x \pm \frac{\partial\Phi}{\partial y}\hat{\mathbf{e}}_y \pm \frac{\partial\Phi}{\partial z}\hat{\mathbf{e}}_z$$
$$= \pm F_x\hat{\mathbf{e}}_x \pm F_y\hat{\mathbf{e}}_y \pm F_z\hat{\mathbf{e}}_z = \mathbf{F}_x + \mathbf{F}_y + \mathbf{F}_z. \tag{3.3}$$

The gradient of the potential, $\nabla\Phi$, is a vector representing the maximum space rate of change of Φ which is in a direction normal to the equipotential surface.

Taking the spatial derivatives of the vector components, one obtains the gradient tensors of $\mathbf{F}$ given by

$$F_{xx} = \frac{\partial F_x}{\partial x},\ F_{xy} = \frac{\partial F_x}{\partial y},\ F_{xz} = \frac{\partial F_x}{\partial z},$$
$$F_{yx} = \frac{\partial F_y}{\partial x},\ F_{yy} = \frac{\partial F_y}{\partial y},\ F_{yz} = \frac{\partial F_y}{\partial z},$$
$$F_{zx} = \frac{\partial F_z}{\partial x},\ F_{zy} = \frac{\partial F_z}{\partial y},\ F_{zz} = \frac{\partial F_z}{\partial z}. \tag{3.4}$$

However, $\mathbf{F}$ is a conservative vector field so that its curl given by the vector cross product $\nabla \times \mathbf{F}$ is zero, i.e.

$$\nabla \times \mathbf{F} = \begin{vmatrix} \hat{\mathbf{e}}_x & \hat{\mathbf{e}}_y & \hat{\mathbf{e}}_z \\ \frac{\partial}{\partial x} & \frac{\partial}{\partial y} & \frac{\partial}{\partial z} \\ F_x & F_y & F_z \end{vmatrix}$$
$$= (F_{zy} - F_{yz})\hat{\mathbf{e}}_x + (F_{xz} - F_{zx})\hat{\mathbf{e}}_y + (F_{yx} - F_{xy})\hat{\mathbf{e}}_z$$
$$= 0. \tag{3.5}$$

For Equation 3.5 to be true, the tensors in each set of parentheses must be equal, and thus three of these six tensors are redundant. Furthermore, the divergence of $\mathbf{F}$ given by the vector dot product $\nabla\cdot\mathbf{F}$ in free space is Laplace's equation where

$$\nabla\cdot\mathbf{F} = \nabla^2 U = F_{xx} + F_{yy} + F_{zz} = 0, \tag{3.6}$$

so that one of these three tensors is also redundant. Accordingly, the nine tensors in Equation 3.4 can be expressed as the symmetric matrix

$$\begin{bmatrix} F_{xx} & F_{xy} & F_{xz} \\ F_{yx} & F_{yy} & F_{yz} \\ F_{zx} & F_{zy} & F_{zz} \end{bmatrix} = \begin{bmatrix} F_{xx} & F_{xy} & F_{xz} \\ F_{yx} & F_{yy} & F_{yz} \\ F_{zx} & F_{zy} & (-F_{xx} - F_{yy}) \end{bmatrix}$$
$$= \begin{bmatrix} \frac{\partial}{\partial x} \\ \frac{\partial}{\partial y} \\ \frac{\partial}{\partial z} \end{bmatrix} \begin{bmatrix} F_x \\ F_y \\ F_z \end{bmatrix}^t, \tag{3.7}$$

where only five of them are unique and t is the matrix transpose (Appendix A.4.1).

The above relations are fundamental to potential fields because they show that the force can be obtained by taking the space derivative of the scalar potential solution to a problem. More specifically, they reveal that the amount of work done in going from r_1 to r_2 ($= \int_{r_1}^{r_2} \mathbf{F}\cdot\partial\mathbf{r}$) or from r_2 to r_1 ($= -\int_{r_2}^{r_1} \mathbf{F}\cdot\partial\mathbf{r}$) is independent of any path joining these two points in the field. Accordingly, around any closed path $\oint \mathbf{F}\cdot\partial\mathbf{r} = 0$, where the scalar potential can be estimated from

$$\Phi = \pm\int \mathbf{F}\cdot\partial\mathbf{r} = \pm\int (F_x\partial x + F_y\partial y + F_z\partial z). \tag{3.8}$$

In principle, all lower- and higher-order properties of the potential can be derived from each other because the derivatives of the argument ($\frac{1}{r}$) are linearly related as

$$\frac{\partial^n}{\partial r^n}\left(\frac{1}{r^k}\right) = \frac{(-1)^n}{r^{k+n}} \times \frac{(k+n-1)!}{(k-1)!}. \tag{3.9}$$

To model the potential field effects for an extended body, the above effects for the point source (i.e. the integrands or kernels) are summed (integrated or superimposed) at the observation point for all point sources making up the body. The next two sections further detail the above relations for the gravity point mass and the extended bodies that are filled up with these point sources.

3.3 Gravity effects of a point mass

Figure 3.1 illustrates the concept of the gravitational potential, U, in the presence of gravitational force field $\mathbf{F}_g$ ($\equiv \mathbf{F}$) due to a mass m. The potential can be evaluated by integrating the gravitational force between two points at

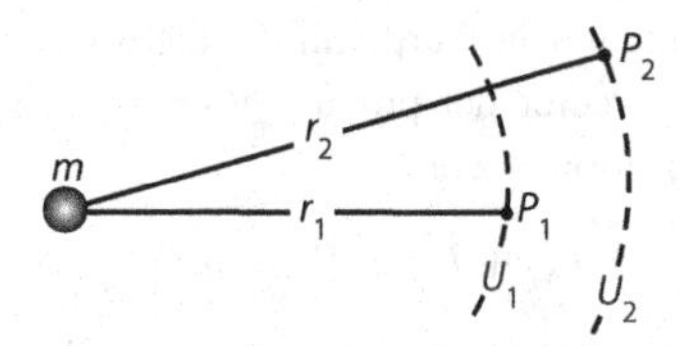

FIGURE 3.1 Spatial arrangement of points P_1 and P_2 at distances r_1 and r_2 from a mass m used in the calculation of potentials U_1 and U_2, respectively. The dashed curves are portions of cross-sections of equipotential surfaces surrounding the mass.

distances r_1 and r_2 using the finite form of Equation 3.8 given by

$$U = \int_{r_1}^{r_2} \mathbf{F}_g \cdot \partial \mathbf{r}. \tag{3.10}$$

The potential is not completely determined, because of the constant of integration of this integral. However, by assuming the potential is zero at infinity, the potential at a point can be completely defined. In moving a unit mass from r_2 to r_1, the gravitational force $\mathbf{F}_g$ due to m acting on a unit mass does work $W(r_2, r_1)$ given by

$$W(r_1, r_2) = \int_{r_1}^{r_2} \mathbf{F}_g \cdot \partial \mathbf{r}. \tag{3.11}$$

Now the force on a unit mass due to a concentrated mass m at the distance $\mathbf{r}$ is

$$\mathbf{F}_g = G\frac{m}{\mathbf{r}^2}, \tag{3.12}$$

where G is the universal gravitational constant, so that by Newton's universal law of gravitation

$$W(r_1, r_2) = G\int_{r_1}^{r_2} \frac{m}{r^2}\partial r = Gm\left(\frac{1}{r_1} - \frac{1}{r_2}\right) \tag{3.13}$$

and the potential difference is

$$\Delta U = W(r_1, r_2) = [U(r_1) - U(r_2)] = Gm\left(\frac{1}{r_1} - \frac{1}{r_2}\right). \tag{3.14}$$

Assuming r_2 is removed to infinity where $U(r_2) = 0$ results in the simplification

$$U(r_1) = G\frac{m}{r_1} = \int_{r_1}^{r_2} \mathbf{F}_g \cdot \partial \mathbf{r}, \tag{3.15}$$

or more generally,

$$U(r) = G\frac{m}{r} = G\int \frac{\partial m}{r}. \tag{3.16}$$

In SIu, the gravity potential or work per unit mass is given in m^2/s^2 which is equivalent to gal × m.

From the above results, the gravitational acceleration or force at the distance $\mathbf{r}$ from a point source with mass m is

$$\mathbf{F}_g = \frac{\partial U}{\partial \mathbf{r}} = \frac{\partial U}{\partial r}\left(\frac{\mathbf{r}}{r}\right) = -G\frac{m}{r^2}\left(\frac{\mathbf{r}}{r}\right) = F_x\hat{\mathbf{e}}_x + F_y\hat{\mathbf{e}}_y + F_z\hat{\mathbf{e}}_z, \tag{3.17}$$

where the vector component magnitudes are

$$F_x = \frac{\partial U}{\partial x} = -G\frac{m\Delta x}{r^3}, \quad F_y = \frac{\partial U}{\partial y} = -G\frac{m\Delta y}{r^3},$$
$$F_z = \frac{\partial U}{\partial z} = -G\frac{m\Delta z}{r^3} \equiv g. \tag{3.18}$$

The principal component of gravity measured by gravimeters and used in most gravity analyses is the intensity F_z of the force per unit mass in the vertical direction, which in the literature is often indicated by the special symbol g. It also is commonly called little g, in contrast to big G which refers to the universal constant of gravity.

Before g became routinely accessible with the advent of modern gravimeters in the mid 1940s, subsurface exploration invoked mostly torsion balance measurements of gravity vector and tensor components. These laborious, exactingly oriented surveys were sufficiently accurate, however, to locate, for example, salt domes trapping highly profitable oil and gas deposits in the sedimentary rocks of the USA Gulf Coast. Recent advances in gravity measuring instrumentation and procedures have renewed interest in the geological utility of the vector components (e.g. FEATHERSTONE *et al.*, 2000) and their gradient tensors (e.g. ZHDANOV *et al.*, 2004). Of the nine gradient tensors for the point mass, the unique ones include any two of the diagonal elements of the matrix in Equation 3.7 given by

$$F_{xx} = Gm\left(\frac{3\Delta x^2}{r^5} - \frac{1}{r^3}\right),$$
$$F_{yy} = Gm\left(\frac{3\Delta y^2}{r^5} - \frac{1}{r^3}\right),$$
$$F_{zz} = Gm\left(\frac{3\Delta z^2}{r^5} - \frac{1}{r^3}\right), \tag{3.19}$$

and the off-diagonal tensors

$$F_{xy} = Gm\left(\frac{3\Delta x\Delta y}{r^5}\right) = F_{yx},$$
$$F_{yz} = Gm\left(\frac{3\Delta y\Delta z}{r^5}\right) = F_{zy},$$
$$F_{xz} = Gm\left(\frac{3\Delta x\Delta z}{r^5}\right) = F_{zx}. \tag{3.20}$$

To facilitate quantitative implementations of the gravity point mass effects, an example of the potential is given in Figure 3.2 for a point source with mass $m = 10^{12}$ kg

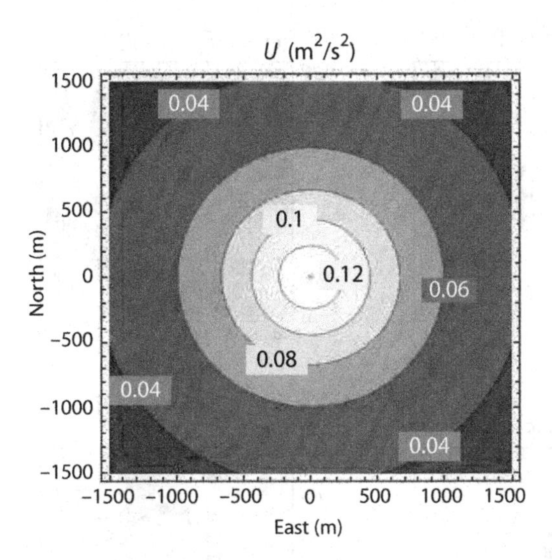

FIGURE 3.2 Gravitational potential U (Equation 3.16) for a point source with mass $m = 10^{12}$ kg at depth $z = 500$ m. The point mass has the same geometric parameters as the point dipole in Figure 9.2.

at depth $z = 500$ m below the observations. In addition, Figure 3.3 shows the corresponding vector and gradient tensor components modeled from Equations 3.18–3.20. The gradient tensor components are useful in improving the identification and resolution of the anomalous source.

3.4 Gravity effects of an extended body

An extended mass body made up of many point poles has gravity effects that can be evaluated at an observation point from the superimposed effects of the differential point masses. However, the above equations are for the gravity effects of the point mass only, whereas in geophysical practice the operative physical property is the mass per unit volume (in kg/m^3) or its density σ of the source so that

$$m = \iint_v\int \partial m = \sigma \times \iint_v\int \partial x'\partial y'\partial z' = \sigma \times v, \quad (3.21)$$

where $\partial m = \sigma \times \partial x'\partial y'\partial z'$ and v is the source's volume (in m^3). Thus, taking density into account and integrating Equation 3.16 over the lower a- and upper b-limits in each spatial dimension of the 3D source gives its gravity potential (in joules/kg)

$$U = U(x, y, z)$$
$$= \int_{z'_a}^{z'_b}\int_{y'_a}^{y'_b}\int_{x'_a}^{x'_b}\left\{U' = G\left(\frac{1}{r}\right)\sigma\right\}\partial x'\partial y'\partial z', \quad (3.22)$$

where the gravitational constant $G = 6.67 \times 10^{-11}\,\mathrm{m^3\,kg^{-1}s^{-2}}$. Note that the kernel U' is the potential per unit volume or point mass effect in Equation 3.16.

The kernel can also be generalized in terms of the product of the density functional and the inverse displacement distance functional for the generalized gravity potential

$$U = U(x, y, z)$$
$$= G\int_{z'_a}^{z'_b}\int_{y'_a}^{y'_b}\int_{x'_a}^{x'_b}\sigma\left\{\frac{1}{r}\right\}\partial x'\partial y'\partial z'. \quad (3.23)$$

Indeed, all gravity forward modeling equations in all coordinate systems have kernels that can be generalized as the product of a physical property functional times a geometry functional or Green's function.

The gravity vector components of the 3D body from integrating Equation 3.18 over the body's finite volume limits are

$$F_x = \frac{\partial U}{\partial x}$$
$$= \int_{z'_a}^{z'_b}\int_{y'_a}^{y'_b}\int_{x'_a}^{x'_b}\left\{F'_x = -G\left(\frac{\Delta x}{r^3}\right)\sigma\right\}\partial x'\partial y'\partial z', \quad (3.24)$$

$$F_y = \frac{\partial U}{\partial y}$$
$$= \int_{z'_a}^{z'_b}\int_{y'_a}^{y'_b}\int_{x'_a}^{x'_b}\left\{F'_y = -G\left(\frac{\Delta y}{r^3}\right)\sigma\right\}\partial x'\partial y'\partial z', \quad (3.25)$$

and

$$F_z = \frac{\partial U}{\partial z}$$
$$= \int_{z'_a}^{z'_b}\int_{y'_a}^{y'_b}\int_{x'_a}^{x'_b}\left\{F'_z = -G\left(\frac{\Delta z}{r^3}\right)\sigma\right\}\partial x'\partial y'\partial z'$$
$$\equiv g. \quad (3.26)$$

These vector components make up the generalized gravity vector of the 3D body given by

$$\mathbf{F}_g = \mathbf{F}_g(x, y, z)$$
$$= -G\int_{z'_a}^{z'_b}\int_{y'_a}^{y'_b}\int_{x'_a}^{x'_b}\sigma\left\{\nabla\left(\frac{1}{r}\right)\right\}\partial x'\partial y'\partial z'. \quad (3.27)$$

In addition, integrating Equation 3.19 over the body's 3D volume limits results in its three diagonal tensors that

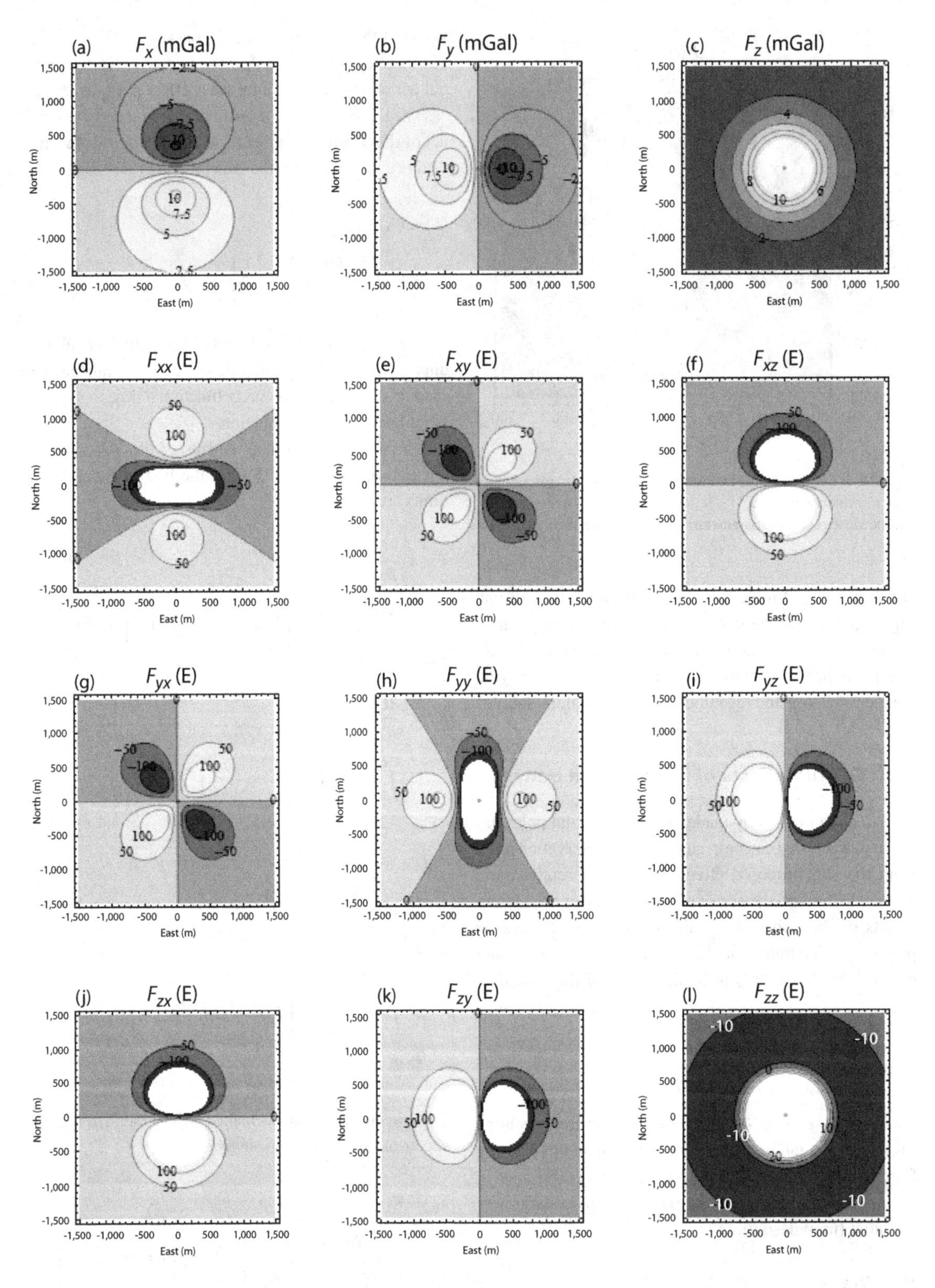

FIGURE 3.3 Examples of the gravity vector components F_x, F_y, and F_z (Equation 3.18) and gradient tensors (Equations 3.19 and 3.20) in Eötvös units ($1\ \mathrm{E} = 0.1\ \mathrm{mGal/km}$) for the point mass of Figure 3.2.

satisfy Laplace's Equation 3.6 given by

$$F_{xx} = \frac{\partial^2 U}{\partial x \partial x} = \int_{z'_a}^{z'_b}\int_{y'_a}^{y'_b}\int_{x'_a}^{x'_b}\left\{F'_{xx} = G\left(\frac{3\Delta x^2}{r^5} - \frac{1}{r^3}\right)\sigma\right\}\partial x'\partial y'\partial z' = -(F_{yy} + F_{zz}), \quad (3.28)$$

$$F_{yy} = \frac{\partial^2 U}{\partial y \partial y} = \int_{z'_a}^{z'_b}\int_{y'_a}^{y'_b}\int_{x'_a}^{x'_b}\left\{F'_{yy} = G\left(\frac{3\Delta y^2}{r^5} - \frac{1}{r^3}\right)\sigma\right\}\partial x'\partial y'\partial z' = -(F_{xx} + F_{zz}), \quad (3.29)$$

and

$$F_{zz} = \frac{\partial^2 U}{\partial z \partial z} = \int_{z'_a}^{z'_b}\int_{y'_a}^{y'_b}\int_{x'_a}^{x'_b}\left\{F'_{zz} = G\left(\frac{3\Delta z^2}{r^5} - \frac{1}{r^3}\right)\sigma\right\}\partial x'\partial y'\partial z' = -(F_{xx} + F_{yy}). \quad (3.30)$$

Further integrating Equation 3.20 yields the body's three unique off-diagonal tensors

$$F_{xy} = \frac{\partial^2 U}{\partial x \partial y} = \int_{z'_a}^{z'_b}\int_{y'_a}^{y'_b}\int_{x'_a}^{x'_b}\left\{F'_{xy} = G\left(\frac{3\Delta x\Delta y}{r^5}\right)\sigma\right\}\partial x'\partial y'\partial z' = F_{yx}, \quad (3.31)$$

$$F_{yz} = \frac{\partial^2 U}{\partial y \partial z} = \int_{z'_a}^{z'_b}\int_{y'_a}^{y'_b}\int_{x'_a}^{x'_b}\left\{F'_{yz} = G\left(\frac{3\Delta y\Delta z}{r^5}\right)\sigma\right\}\partial x'\partial y'\partial z' = F_{zy}, \quad (3.32)$$

and

$$F_{xz} = \frac{\partial^2 U}{\partial x \partial z} = \int_{z'_a}^{z'_b}\int_{y'_a}^{y'_b}\int_{x'_a}^{x'_b}\left\{F'_{xz} = G\left(\frac{3\Delta x\Delta z}{r^5}\right)\sigma\right\}\partial x'\partial y'\partial z' = F_{zx}. \quad (3.33)$$

These nine gradient tensors make up the generalized gradient tensor of the 3D body given by

$$\nabla \mathbf{F}_g = \nabla \mathbf{F}_g(x, y, z) = G\int_{z'_a}^{z'_b}\int_{y'_a}^{y'_b}\int_{x'_a}^{x'_b}\sigma\left\{\nabla\left[\nabla\left(\frac{1}{r}\right)\right]\right\}\partial x'\partial y'\partial z'. \quad (3.34)$$

For 2D modeling of a profile on the x-axis across an elliptical or narrow elongated anomaly with essentially infinite strike length along the y-axis and the vertical z-axis positive downwards, the gravitational potential is

$$\mathcal{U}(x, z) = \int_{z'_a}^{z'_b}\int_{x'_a}^{x'_b}\partial x'\partial z'\left\{\int_{-\infty}^{\infty} G\left(\frac{1}{r}\right)\sigma\partial y'\right\} = \int_{z'_a}^{z'_b}\int_{x'_a}^{x'_b}\left\{\mathcal{U}' = 2G\log\left(\frac{1}{r}\right)\sigma\right\}\partial x'\partial z'. \quad (3.35)$$

For the above logarithmic potential, the displacement vector is 2D in the (x, z)-plane with magnitude

$$r = \sqrt{(x - x')^2 + (z - z')^2} = \sqrt{\Delta x^2 + \Delta z^2}, \quad (3.36)$$

and the operative physical property is the body's mass per unit length or surface density (in kg/m^2)

$$\frac{m}{y'} = \int\int_S \frac{\partial m}{y'} = \sigma \times \int\int_S \partial x'\partial z' = \sigma \times S, \quad (3.37)$$

where $(\partial m/y') = \sigma \times \partial x'\partial z'$ and S is the source's cross-sectional area or surface (in m^2). Thus, the gravitational field components of the 2D source are

$$\mathcal{F}_x = \frac{\partial \mathcal{U}}{\partial x} = \int_{z'_a}^{z'_b}\int_{x'_a}^{x'_b}\left\{\mathcal{F}'_x = \frac{\partial \mathcal{U}'}{\partial x} = -2G\left(\frac{\Delta x}{r^2}\right)\sigma\right\}\partial x'\partial z', \quad (3.38)$$

and

$$\mathcal{F}_z = \frac{\partial \mathcal{U}}{\partial z} = \int_{z'_a}^{z'_b}\int_{x'_a}^{x'_b}\left\{\mathcal{F}'_z = \frac{\partial \mathcal{U}'}{\partial z} = -2G\left(\frac{\Delta z}{r^2}\right)\sigma\right\}\partial x'\partial z' \equiv \mathcal{G}. \quad (3.39)$$

In addition, the two unique gravity tensors of the 2D source are

$$\mathcal{F}_{zz} = \frac{\partial^2 U}{\partial z \partial z} = \int_{z'_a}^{z'_b}\int_{x'_a}^{x'_b}\left\{\mathcal{F}'_{zz} = 2G\left(\frac{2\Delta z^2 - r^2}{r^4}\right)\sigma\right\}\partial x'\partial z' = -\mathcal{F}_{xx}, \quad (3.40)$$

and

$$\mathcal{F}_{xz} = \frac{\partial^2 U}{\partial x \partial z} = \int_{z'_a}^{z'_b}\int_{x'_a}^{x'_b}\left\{\mathcal{F}'_{xz} = 4G\left(\frac{\Delta x\Delta z}{r^4}\right)\sigma\right\}\partial x'\partial z' = \mathcal{F}_{zx}. \quad (3.41)$$

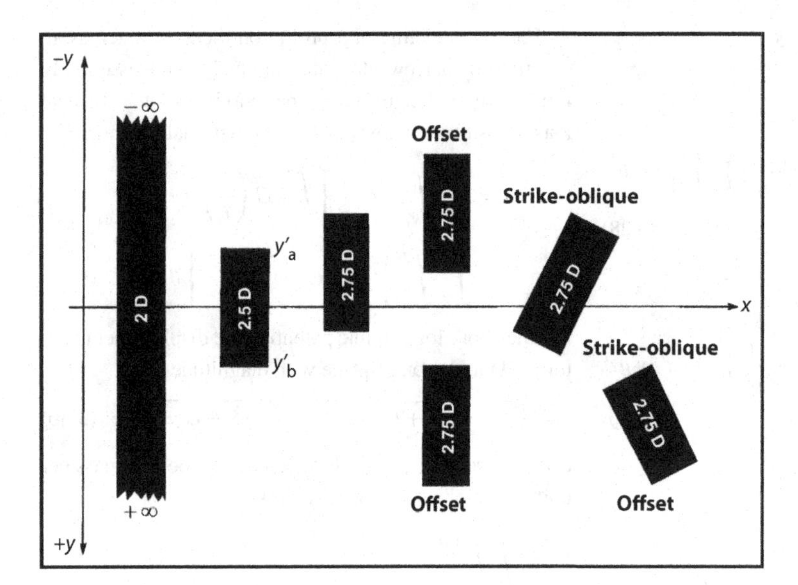

FIGURE 3.4 Geometric attributes of elongated bodies with polygonal cross-sections in the (x, y)-plane used in GAMMA for the symmetric 2D and 2.5D and asymmetric 2.75D cases relative to the principal profile along the x-axis. Adapted from SAAD and BISHOP (1989).

For an elongated source that is finite with ends at distances $\Delta y'_a$ and $\Delta y'_b$ from the profile, the 2D effects can be corrected based on their ratios to the related 3D effects. For example, using Equations 3.26 and 3.39, the 3D gravity effect g of an elongated finite mass can be expressed as the mass's infinitely elongated 2D effect $\mathcal{G}$ multiplied by the factor

$$\frac{g}{\mathcal{G}} = \frac{1}{2}\int_{y'_a}^{y'_b}\left(\frac{1}{r}\right)\partial y'. \tag{3.42}$$

TELFORD *et al.* (1990) note that this correction factor for an elongated source with circular cross-section (i.e., a horizontal cylinder, HC) may be approximated by

$$\frac{g(\mathrm{HC})}{\mathcal{G}(\mathrm{HC})} \simeq \frac{1}{2}\left[\frac{y'_a}{\left(r_c{}^2 + y'_a{}^2\right)^{\frac{1}{2}}} + \frac{y'_b}{\left(r_c{}^2 + \Delta y'_b{}^2\right)^{\frac{1}{2}}}\right], \tag{3.43}$$

where r_c is the distance from the station to the central axis of the source, g(HC) is the 3D effect of the finite y-length horizontal cylinder, and $\mathcal{G}$(HC) is the 2D effect of the infinite y-length horizontal cylinder. The so-called 2.5D effect is obtained when $|y'_a| = |y'_b| < \infty$, whereas for $|y'_a| \neq |y'_b|$, the corrected result is called the 2.75D effect. Note that the terms 2.5D and 2.75D are misnomers because they refer in reality to 3D bodies with 3D gravity potentials, and vector and gradient tensor components given by Equations 3.23, 3.27, and 3.34, respectively. In the exploration literature, however, they commonly refer to end corrections developed for elongated sources with arbitarily shaped cross-sections that can be represented by polygons (e.g. RASMUSSEN and PEDERSEN, 1979; CADY, 1980).

In general, the end-corrected, 2D polygonal source finds extensive use for interactive, forward modeling interpretation of both gravity and magnetic anomalies. A particularly comprehensive suite of algorithms for 2D, 2.5D, and 2.75D Gravity and Magnetic Modeling Applications (GAMMA) developed by SAAD (1991, 1992, 1993) facilitates efficient interactive modeling of combined gravity and magnetic anomalies. The GAMMA algorithms for gravity and gravity gradients are outlined below, whereas the corresponding magnetic algorithms are described in Section 9.7.1. The magnetic equations are the same as those describing the gravity gradient tensor components after re-scaling.

The GAMMA algorithms are based on the geometric attributes of an elongated source with a polygonal cross-section perpendicular to the strike that is invariant along the strike direction (Figures 3.4 and 3.5). The strike length along the y-axis can be either infinite (the 2D case), finite and symmetrical with respect to the profile or x-axis (the 2.5D case), or finite and asymmetrical (the 2.75D case). The most general 2.75D case includes not only bodies that are crossed by the profile, but also bodies that are completely offset from the profile, as well as bodies striking obliquely at any arbitrary angle from the profile (Figures 3.4 and 3.5).

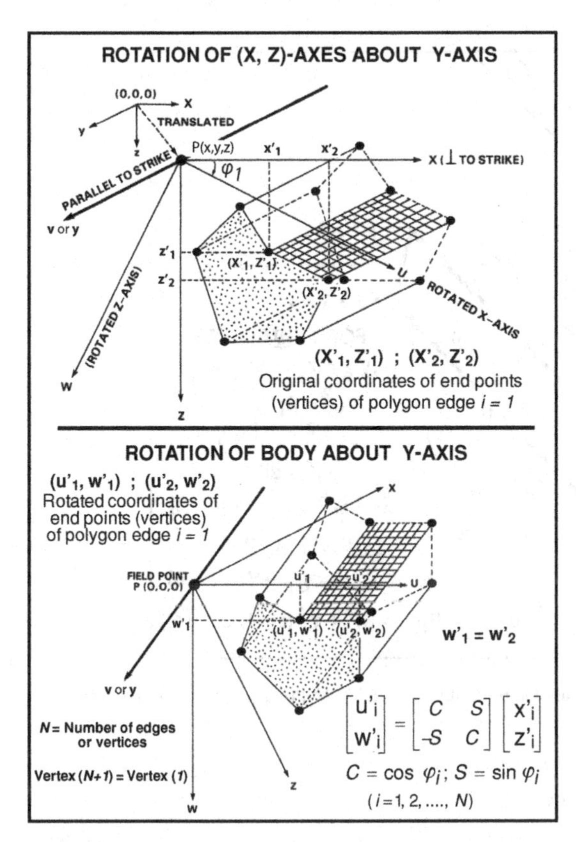

FIGURE 3.5 Coordinate transformation used in GAMMA to shift the origin to the observation point and rotate the (x, z)-axes or body about the y-axis. Adapted from SAAD and BISHOP (1989).

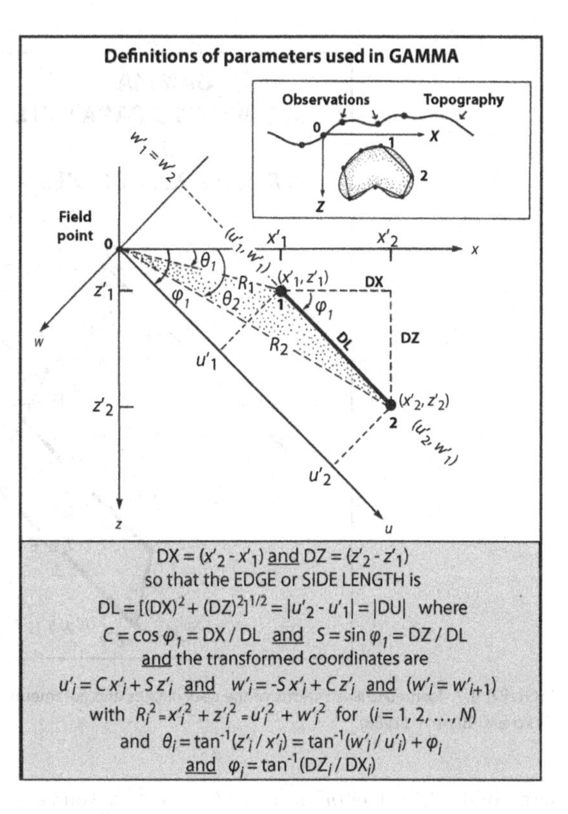

FIGURE 3.6 Definitions of edge parameters used by the GAMMA algorithms in the (x, z)- or rotated (u, w)-planes. Adapted from SAAD and BISHOP (1989).

The derivations of the GAMMA algorithms for gravity are developed from the basic equations of the Newtonian gravitational potential for 3D sources (Equation 3.23) and the logarithmic potential for 2D sources (Equation 3.35) and consider the 2.5D source as a special case of the 3D source. The volume integrals for the 3D sources, and surface integrals for the 2D sources are reduced to surface and line integrals, respectively, using Gauss' divergence theorem (e.g. Equation 3.99). To evaluate the surface integrals for 2.5D bodies and line integrals for 2D bodies, a method similar to that described by OKABE (1979) or RASMUSSEN and PEDERSEN (1979) is used. The method involves a coordinate transformation to shift the origin to the observation point followed by rotating the (x, y, z)-coordinate system about the y-axis to the (u, v, w)-coordinate system (Figure 3.5). In the rotated system, the u-axis is parallel to the polygonal edge or facet for which the integrals are to be evaluated, the w-axis is perpendicular to the facet, and the v-axis coincides with the y-axis. The (x, z)-axes are rotated clockwise about the y-axis by the dip angle φ of the polygonal edge. Mathematically, the rotation is achieved according to the matrix equation at the bottom of Figure 3.5 or the equations in Figure 3.6.

The geometric parameters used in GAMMA for each polygonal edge are illustrated in Figures 3.6 and 3.7. Notice that all the parameters and functions used by the GAMMA algorithms are expressed fully in terms of the (x', z')-coordinates of the polygonal vertices and the lower a- and upper b-limits of the y'-coordinates of the end faces in the 2.5D and 2.75D cases.

The geometric functions described in Table 3.1 for the 2D, 2.5D and 2.75D cases apply to both gravity and magnetic algorithms. These functions are expressed for a given polygonal edge as differential natural logarithmic (ln) and arctangent ($\tan^{-1}$) functions – i.e. the difference in values at two successive vertices (1, 2) which define the edge. For the 2D case, the functions are designated DLR (differential logarithm of radial distances), and DAT (differential arctangent). Both of these functions are fully defined in

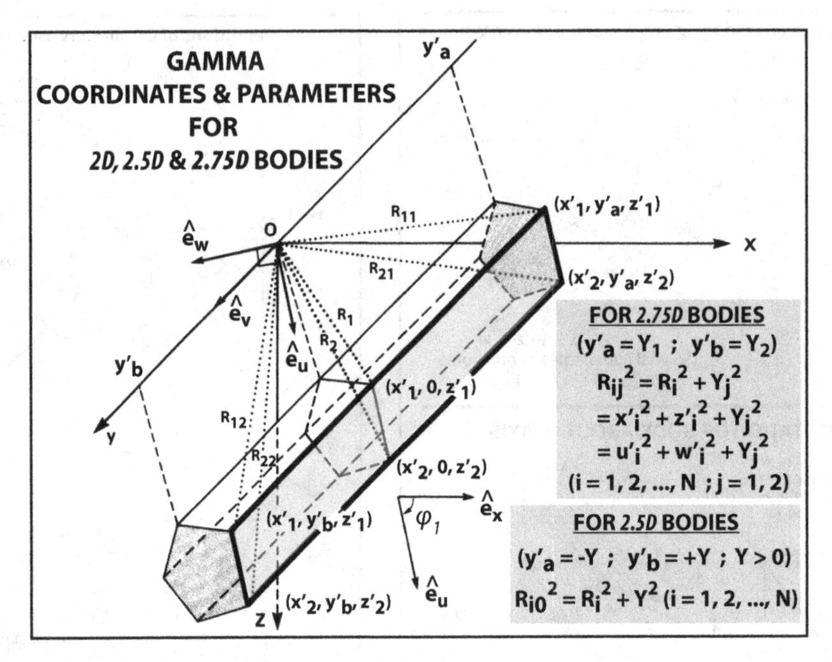

FIGURE 3.7 Isometric illustrations of the coordinates and parameters used in the 2.5D and 2.75D GAMMA algorithms. Adapted from SAAD and BISHOP (1989).

terms of the (x', z')- or (u', w')-coordinates of the vertices and invariant under this coordinate transformation. Notice that DAT is the angular difference $(\theta_1 - \theta_2)$ shown in Figure 3.6. For the 2.5D and 2.75D cases, the logarithmic and arctangent functions depend on the (x', z') or (u', w')-coordinates of the vertices, as well as on the half-strike length $Y = 0.5(y'_b - y'_a)$ in the symmetric 2.5D case or on the y'-coordinates of the end faces y'_a and y'_b of the strike length in the asymmetric 2.75D cases as shown in Figure 3.7 and Table 3.1.

The GAMMA algorithms for the 2D, 2.5D, and 2.75D gravity vector components can be expressed compactly in terms of the above-defined parameters and functions. In the symmetric 2D case, for example,

$$\mathcal{F}_x = 2G\sigma \sum_{i=1}^{N} w'_1[C \times \text{DLR} + S \times \text{DAT}] \tag{3.44}$$

and

$$\mathcal{F}_z \equiv \mathcal{G} = 2G\sigma \sum_{i=1}^{N} w'_1[S \times \text{DLR} - C \times \text{DAT}], \tag{3.45}$$

where N is the total number of polygonal sides or vertices, σ is the density (or density contrast of the polygon), and G is the gravitational constant. For the symmetric 2.5D case, on the other hand, the gravity vector components are

$$F_{(2.5)x} = 2G\sigma \sum_{i=1}^{N} w'_1[C \times \text{DLR}Y + S \times \text{DAT}Y] - S[Y \times \text{DLR}U], \tag{3.46}$$

$$F_{(2.5)y} = 0 \quad \text{(because of symmetry in y)}, \tag{3.47}$$

and

$$F_{(2.5)z} \equiv g = 2G\sigma \sum_{i=1}^{N} w'_1[S \times \text{DLR}Y - C \times \text{DAT}Y] + C[Y \times \text{DLR}U]. \tag{3.48}$$

For the asymmetric 2.75D case, the gravity vector components are

$$F_{(2.75)x} = G\sigma \sum_{i=1}^{N} w'_1[C \times \text{DLR}Y21 + S \times \text{DAT}Y21] - S[y'_b \times \text{DLR}U2 - y'_a \times \text{DLR}U1], \tag{3.49}$$

$$F_{(2.75)y} = G\sigma \sum_{i=1}^{N} w'_1[\text{DLR}U21] + [y'_b \times \text{DAT}Y2 - y'_a \times \text{DAT}Y1], \tag{3.50}$$

TABLE 3.1 Definitions of functions applied in both gravity and magnetic GAMMA algorithms as illustrated in Figures 3.4–3.7.

2D – symmetric

$$\mathrm{DLR} = \ln(R_1) - \ln(R_2) = \ln(R_1/R_2),\quad \text{where } \ln \equiv \log_e$$

$$\mathrm{DAT} = \tan^{-1}(z_1'/x_1') - \tan^{-1}(z_2'/x_2') = \tan^{-1}(w_1'/u_1') - \tan^{-1}(w_2'/u_2');\quad w_2' = w_1'$$

2.5D – symmetric

$$\mathrm{DLR}Y = \ln\left[\frac{(R_{20}+Y)}{(R_{10}+Y)} \times \frac{R_1}{R_2}\right] = \mathrm{DLR} - \ln\left[\frac{(R_{10}+Y)}{(R_{20}+Y)}\right]$$

$$\mathrm{DLR}U = \ln\left[\frac{(R_{20}+u_2')}{(R_{10}+u_1')}\right]$$

$$\mathrm{DAT}Y = \tan^{-1}\left(\frac{u_2' \times Y}{w_2' \times R_{20}}\right) - \tan^{-1}\left(\frac{u_1' \times Y}{w_1' \times R_{10}}\right);\quad w_2' = w_1'$$

2.75D – asymmetric

$$\mathrm{DLR}Y21 = \ln\left[\frac{(R_{22}+y_b')}{(R_{12}+y_b')} \times \frac{(R_{11}+y_a')}{(R_{21}+y_a')}\right]$$

$$\mathrm{DLR}U1 = \ln\left[\frac{(R_{21}+u_2')}{(R_{11}+u_1')}\right];\quad \mathrm{DLR}U2 = \ln\left[\frac{(R_{22}+u_2')}{(R_{12}+u_1')}\right]$$

$$\mathrm{DLR}U21 = \ln\left[\frac{(R_{22}+u_2')}{(R_{12}+u_1')} \times \frac{(R_{11}+u_1')}{(R_{21}+u_2')}\right] = \mathrm{DLR}U2 - \mathrm{DLR}U1$$

$$\mathrm{DAT}Y2 = \tan^{-1}\left(\frac{u_2' \times y_b'}{w_2' \times R_{22}}\right) - \tan^{-1}\left(\frac{u_1' \times y_b'}{w_1' \times R_{12}}\right);\quad w_2' = w_1'$$

$$\mathrm{DAT}Y1 = \tan^{-1}\left(\frac{u_2' \times y_a'}{w_2' \times R_{21}}\right) - \tan^{-1}\left(\frac{u_1' \times y_a'}{w_1' \times R_{11}}\right);\quad w_2' = w_1'$$

$$\mathrm{DAT}Y21 = \mathrm{DAT}Y2 - \mathrm{DAT}Y1$$

Saad and Bishop, 1989.

and

$$F_{(2.75)z} \equiv g = G\sigma \sum_{i=1}^{N} w_1'[S \times \mathrm{DLR}Y21 - C \times \mathrm{DAT}Y21] + C[y_b' \times \mathrm{DLR}U2 - y_a' \times \mathrm{DLR}U1]. \quad (3.51)$$

It should be noted that polygonal 2D, 2.5D, and 2.75D bodies with arbitrary strike directions that are not necessarily perpendicular to the profile (x-axis) can be accommodated within GAMMA by using another simple coordinate transformation or rotation of the polygon's vertices about the vertical axis.

To facilitate modeling magnetic vector components of polygonal bodies (Section 9.7.1) via Poisson's theorem (Section 3.9), the GAMMA algorithms developed by Saad in 1989 express the gravity gradient tensor components in terms of the normalized second derivative of the gravitational potential or Green's function $D_{ij} = F_{ij}/(G \times \sigma)$ (Saad and Bishop, 1989). Thus, the gravity gradient tensor components for the symmetric 2D case can be expressed as the product of the scalar physical property functional $(G \times \sigma)$ and the D-matrix of geometric functionals or Green's functions given by

$$\begin{pmatrix} \mathcal{F}_{xx} & \mathcal{F}_{xz} \\ \mathcal{F}_{zx} & \mathcal{F}_{zz} \end{pmatrix} = \begin{pmatrix} \mathcal{D}_{xx} & \mathcal{D}_{xz} \\ \mathcal{D}_{zx} & \mathcal{D}_{zz} \end{pmatrix} (G \times \sigma), \quad (3.52)$$

where

$$\mathcal{D}_{xx} = -\mathcal{D}_{zz} = 2\sum_{i=1}^{N} S[C \times \mathrm{DLR} + S \times \mathrm{DAT}] = 2\sum_{i=1}^{N} C[S \times \mathrm{DLR} - C \times \mathrm{DAT}], \quad (3.53)$$

and

$$\mathcal{D}_{xz} = \mathcal{D}_{zx} = 2\sum_{i=1}^{N} S[S \times \mathrm{DLR} - C \times \mathrm{DAT}] = -2\sum_{i=1}^{N} C[C \times \mathrm{DLR} + S \times \mathrm{DAT}]. \quad (3.54)$$

For the symmetric 2.5D case, on the other hand, the gravity tensor components are

$$\begin{pmatrix} F_{(2.5)xx} & F_{(2.5)xy} & F_{(2.5)xz} \\ F_{(2.5)yx} & F_{(2.5)yy} & F_{(2.5)yz} \\ F_{(2.5)zx} & F_{(2.5)zy} & F_{(2.5)zz} \end{pmatrix} = \begin{pmatrix} D_{(2.5)xx} & 0 & D_{(2.5)xz} \\ 0 & D_{(2.5)yy} & 0 \\ D_{(2.5)zx} & 0 & D_{(2.5)zz} \end{pmatrix} (G \times \sigma), \quad (3.55)$$

where the y-symmetry makes $F_{(2.5)xy} = F_{(2.5)yx} = F_{(2.5)yz} = F_{(2.5)zy} = 0$,

$$D_{(2.5)xx} = -[D_{(2.5)yy} + D_{(2.5)zz}] = 2\sum_{i=1}^{N} S[C \times \mathrm{DLR}Y + S \times \mathrm{DAT}Y], \quad (3.56)$$

$$D_{(2.5)xz} = D_{(2.5)zx} = 2\sum_{i=1}^{N} S[S \times \mathrm{DLR}Y - C \times \mathrm{DAT}Y], \quad (3.57)$$

$$D_{(2.5)yy} = -[D_{(2.5)xx} + D_{(2.5)zz}] = -2\sum_{i=1}^{N} \mathrm{DAT}Y, \quad (3.58)$$

$$D_{(2.5)zx} = D_{(2.5)xz} = -2\sum_{i=1}^{N} C[C \times \mathrm{DLR}Y + S \times \mathrm{DAT}Y], \quad (3.59)$$

and

$$D_{(2.5)zz} = -[D_{(2.5)xx} + D_{(2.5)yy}] = -2\sum_{i=1}^{N} C[S \times \mathrm{DLR}Y - C \times \mathrm{DAT}Y]. \quad (3.60)$$

In addition, the gravity tensor components for the asymmetric 2.75D case are given by

$$\begin{pmatrix} F_{(2.75)xx} & F_{(2.75)xy} & F_{(2.75)xz} \\ F_{(2.75)yx} & F_{(2.75)yy} & F_{(2.75)yz} \\ F_{(2.75)zx} & F_{(2.75)zy} & F_{(2.75)zz} \end{pmatrix} = \begin{pmatrix} D_{(2.75)xx} & D_{(2.75)xy} & D_{(2.75)xz} \\ D_{(2.75)yx} & D_{(2.75)yy} & D_{(2.75)yz} \\ D_{(2.75)zx} & D_{(2.75)zy} & D_{(2.75)zz} \end{pmatrix} (G \times \sigma), \quad (3.61)$$

where

$$D_{(2.75)xx} = -[D_{(2.75)yy} + D_{(2.75)zz}] = \sum_{i=1}^{N} S[C \times \mathrm{DLR}Y21 + S \times \mathrm{DAT}Y21], \quad (3.62)$$

$$D_{(2.75)xy} = D_{(2.75)yx} = \sum_{i=1}^{N} [S \times \mathrm{DLR}U21], \quad (3.63)$$

$$D_{(2.75)xz} = D_{(2.75)zx} = \sum_{i=1}^{N} S[S \times \mathrm{DLR}Y21 - C \times \mathrm{DAT}Y21], \quad (3.64)$$

$$D_{(2.75)yy} = -[D_{(2.75)xx} + D_{(2.75)zz}] = -\sum_{i=1}^{N} \mathrm{DAT}Y21, \quad (3.65)$$

$$D_{(2.75)yz} = D_{(2.75)zy} = -\sum_{i=1}^{N} [C \times \mathrm{DLR}U21], \quad (3.66)$$

$$D_{(2.75)zx} = D_{(2.75)xz} = -\sum_{i=1}^{N} C[C \times \mathrm{DLR}Y21 + S \times \mathrm{DAT}Y21], \quad (3.67)$$

and

$$D_{(2.75)zz} = -[D_{(2.75)xx} + D_{(2.75)yy}] = -\sum_{i=1}^{N} C[S \times \mathrm{DLR}Y21 - C \times \mathrm{DAT}Y21]. \quad (3.68)$$

In summary, the extended body integrals described in this section are readily adapted for modeling the gravity effects of any spatial distribution of mass. The extended body integrals can be analytically evaluated for idealized gravity sources with simple symmetric shapes like spheres, cylinders, and prisms, whereas numerical integration is necessary for more complicated general gravity sources with arbitrary shapes. One of the more interesting sources from the standpoint of the Earth is the gravitational attraction of a uniform thin spherical shell. Outside the shell, the gravitational attraction of the shell is equivalent to that of the mass of the shell concentrated at the center of the sphere which is also the center of the shell, but inside the shell the gravitational attraction is readily shown to be zero: the effect of any segment of the shell is canceled by an equal and opposite effect of the antipodal segment. This is the case for any number of concentric thin spherical shells. The gravitational potential inside the shell is constant because the attraction anywhere inside the shell is zero. The gravity effects of idealized and general sources are detailed further in the sections below.

3.5 Idealized source gravity modeling

The gravity effects of idealized 3D and 2D bodies are modeled by setting the appropriate limits and executing the relevant integrals in Equations 3.22–3.33 and 3.35–3.41, respectively. Tables 3.2 and 3.3 list some of the more useful and widely used idealized source equations in Cartesian coordinates from the gravity modeling literature (e.g. SHAW, 1932; NETTLETON, 1942; TELFORD *et al.*, 1990). These equations are remarkably robust for making gravity anomaly calculations with errors due to the limiting assumptions generally small over a large range of dimensions (e.g. HAMMER, 1974).

Figures 3.8 and 3.9 depict the geometric parameters of the sources in Tables 3.2 and 3.3, respectively, beneath

TABLE 3.2 Gravity effects of the point, spherical, line, and cylindrical sources.

Source	Gravity effect in spatial domain
(1) Point mass	$g_{\#1} = G \times (m = \Delta\sigma \times \text{volume}) \times (z/r^3)$
(2) Sphere	$g_{\#2} = (4\pi G\Delta\sigma R^3)/(3z^2[1 + (x^2/z^2)]^{3/2})$
(3) ∞ H-line mass	$g_{\#3} = 2G \times (m = \Delta\sigma \times \text{area}) \times (z/r^2)$
(4) ∞ H-cylinder	$g_{\#4} = (2\pi G\Delta\sigma R^2 z)/[x^2 + z^2]$
(A) 2.5D	$g_{\#4A} = g_{\#4} \times \{[1 + (r/Y)^2]^{-\frac{1}{2}}\}$
(B) 2.75D	$g_{\#4B} = g_{\#4} \times 0.5\{[1 + (r/y'_a)^2]^{-\frac{1}{2}} + [1 + (r/y'_b)^2]^{-\frac{1}{2}}\}$
(5) ∞ V-line mass	$g_{\#5} = (\pi G\Delta\sigma R^2)/[x^2 + z^2]^{1/2}$
(6) ∞ V-cylinder	$g_{\#6} = (K_A)ga_{\#6A} + (K_B)g_{\#6B} + (K_C)g_{\#6C}$
(A) $K_A = 0$ & 1 $\forall(x = 0)$	$ga_{\#6A} = 2\pi G\Delta\sigma[(z^2 + R^2)^{1/2} - z]$ and
(B) $K_B = 0$ & 1 $\forall(0 < x \le R)$	$g_{\#6B} = 2\pi G\Delta\sigma R\{\frac{1}{2}(\frac{R}{\sqrt{x^2+z^2}}) - \frac{1}{16}(\frac{R}{\sqrt{x^2+z^2}})^3(\frac{2z^2-x^2}{x^2+z^2}) + \frac{1}{128}(\frac{R}{\sqrt{x^2+z^2}})^5(\frac{35z^4}{(x^2+z^2)^2} - \frac{30z^2}{x^2+z^2} + 3) + \cdots\}$ and
(C) $K_C = 0$ & 1 $\forall(x > R)$	$g_{\#6C} = 2\pi G\Delta\sigma R\{1 - (\frac{z}{R}) + \frac{1}{4}(\frac{2z^2-x^2}{R^2}) - \frac{1}{8}(\frac{8z^4-24z^2x^2+3x^4}{R^4}) + \cdots\}$
(7) Finite V-cylinder	$g_{\#7} = g_{\#6}(@\ z_1) - g_{\#6}(@\ z_2)$
(A) $\forall(z > x = 0)$	$ga_{\#7A} = (2\pi G\Delta\sigma)[r_1 + \delta z - r_2]$ $\forall r_i = [R^2 + z_i^2]^{1/2}$ & $i = 1, 2$
(B) $\forall(z = x = 0)$	$ga_{\#7B} = (2\pi G\Delta\sigma)[\delta z + R - (\delta z^2 + R^2)^{1/2}]$
(C) for $(R \longrightarrow \infty)$	$ga_{\#7C} = 2\pi G\Delta\sigma\delta z$
(8) V-cylinder segment	$ga_{\#8} = \theta G\Delta\sigma[r_1 - r_2 + r_3 - r_4]$ $\forall r_i = [R_j^2 + z_i^2]$ & $i = 1, 2, 3, 4$ & $j = 1, 2$

The symbol $\forall$ means "for all" and the abbreviations include H for horizontal and V for vertical. Here, g is the vertical gravity anomaly in gal ($\text{m/s}^2 = 10^5$ mGal) calculated at the orgin (0, 0, 0) due to a density contrast $\Delta\sigma$ in g/cm^3 at distances x and z; ga is the vertical gravity anomaly on the axis of the source; G is the gravitational constant ($= 6.674 \times 10^{-11}$ m^3 kg^{-1} s^{-2}), r is the distance from the calculation point to the center, centerline, axis, or edge of the source; $Y = 0.5(y'_b - y'_a)$ is the uniform distance to each end of the cylinder from the bisecting profile; $\Delta y'_a$ and $\Delta y'_b$ are the two not necessarily equal distances from the profile to each end of the cylinder; z_1 and z_2 are respectively the vertical distances from the calculation point to the upper and lower surfaces of the source so that $\delta z = (z_2 - z_1)$; R is the radius of the sphere or cylinder; and ϕ is the angle from the horizontal to the appropriate point of the source in radians. All distances are in centimeters (meters) with parentheses containing SIu. The geometric parameters of these sources are illustrated in Figure 3.8.

the principal profile along the x-axis. For these effects, the principal profile is centered on the central anomaly value CA$[g]_{\#}$ and trends orthogonally across the strike of the source. In addition, the equations implement the density contrast ($\Delta\sigma$) because the relative sign of the anomaly always corresponds to the sign of the density contrast of the source.

The gravity effect of the spherical source, for example, is $g_{\#2} = g_z = g \times \cos\theta = g(z/r) = Gmz/r^3$ so that converting the mass into the density contrast over the volume [$= (4/3)\pi R_{\#3}^3$] of the spherical body gives the result in Table 3.2. NETTLETON (1942) recommended dividing the gravity effect by z^3 so that it may be expressed as the product

$$g_{\#2} = \left\{\left(\frac{4\pi R^3}{3}\right)\frac{G\Delta\sigma}{z^2} = \text{CA}[g]_{\#2}\right\} \times \left\{f_g\left[\frac{x}{z}\right]_{\#2} = \left[1 + \left(\frac{x^2}{z^2}\right)\right]^{-\frac{3}{2}}\right\}. \quad (3.69)$$

TABLE 3.3 Gravity effects of tabular and sheet sources.

Source	Gravity effect in spatial domain
(9) Bouguer H-slab	$g_{\#9} = 2\pi G \Delta\sigma \delta z = ga_{\#7C}$
(10) Thin α-inclined sheet (2D)	$g_{\#10} = 2G\Delta\sigma t\left[\sin\alpha \ln\left(\frac{r_2}{r_1}\right) - (\theta_2 + \theta_1)\cos\alpha\right]$
(11) Thin V-sheet (2D)	$g_{\#11} = (2G\Delta\sigma t)\ln\left[\left(z_2^2 + x^2\right)/\left(z_1^2 + x^2\right)\right]$
(12) Thin H-sheet (2D)	$g_{\#12} = 2G\Delta\sigma\delta z\Delta\phi$
(13) Thin S-∞ H-slab (fault)	$g_{\#13} = 2G\Delta\sigma\delta z[(\pi/2) - \tan^{-1}(x/z)]$
(A) $\forall (x = 0)$	$ga_{\#13A} = \pi G\Delta\sigma\delta z$
(14) Thick S-∞ H-slab (fault)	$g_{\#14} = 2G\Delta\sigma[x\ln(r_2/r_1) + \pi t - z_2\phi_2 + z_1\phi_1]$
(15) Oblique faced S-∞ H-slab	$g_{\#15} = 2G\Delta\sigma\left[z_2\phi_2 - z_1\phi_1 + \left[\frac{x_2 z_1 - x_1 z_2}{\Delta x^2 + \delta z^2}\right]\left[\delta z \ln\left(\frac{r_2}{r_1}\right) + \Delta x\Delta\phi\right]\right] \forall \Delta x = (x_2 - x_1)\ \&\ \Delta\phi = (\phi_1 - \phi_2)$

Abbreviations and variables are the same as listed for Table 3.2, but also includes S for semi. The geometric parameters of these sources are illustrated in Figures 3.8 and 3.9.

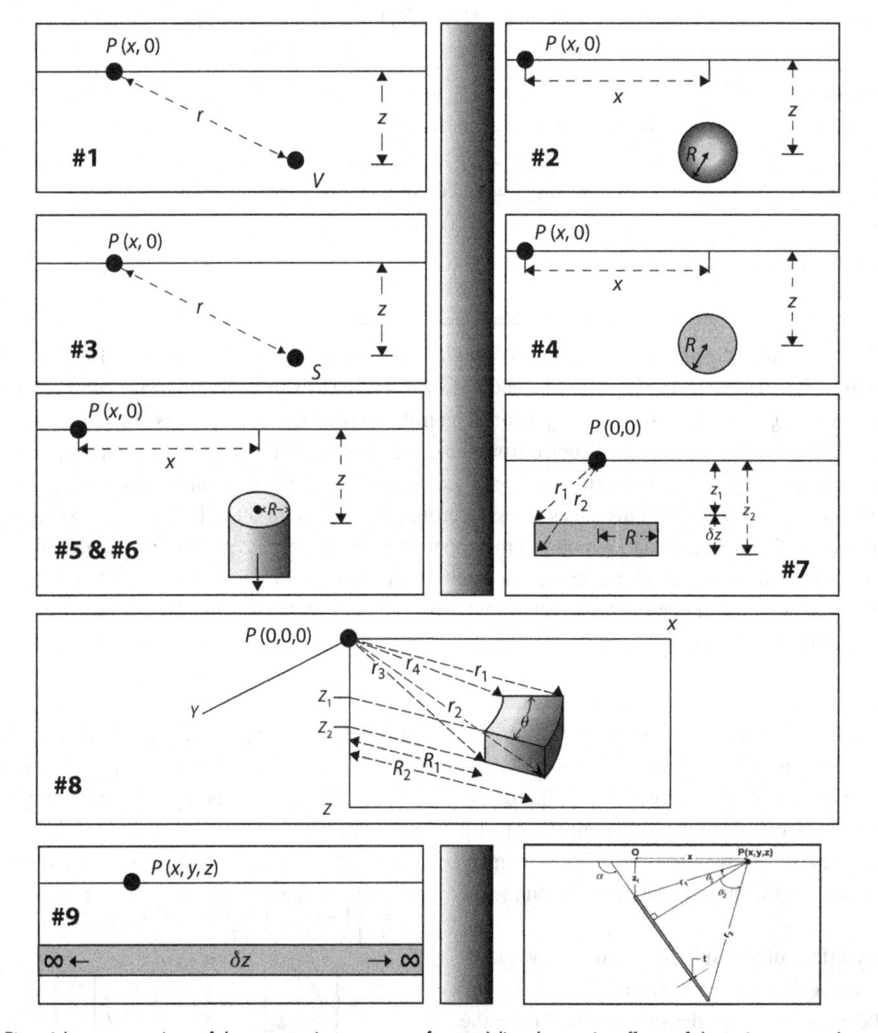

FIGURE 3.8 Pictorial representations of the geometric parameters for modeling the gravity effects of the point mass, sphere, line mass, and cylinder identified in Table 3.2 and the horizontal and inclined slabs identified in Table 3.3.

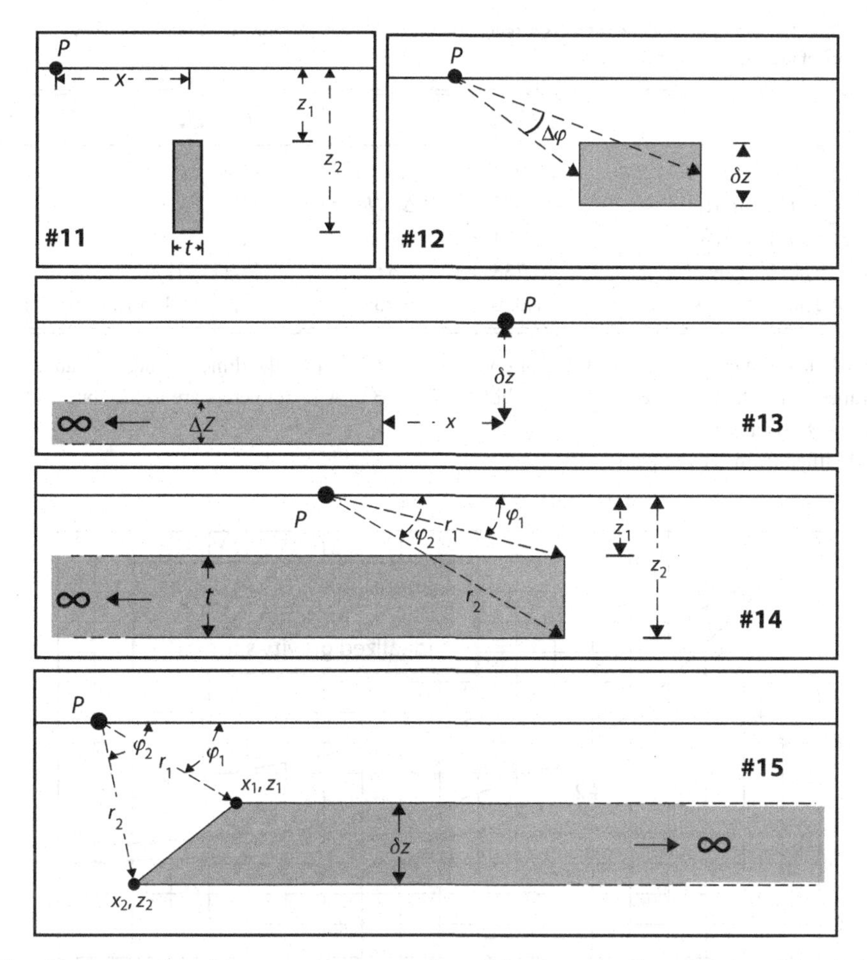

FIGURE 3.9 Pictorial representations of the geometric parameters for modeling the gravity effects of tabular and sheet bodies identified in Table 3.3.

In this more simplified format, $\mathrm{CA}[g]_{\#2}$ is the central anomaly maximum or minimum that reflects the sign of the density contrast, and varies with depth $z_c^{(n=2)}$ where the decay rate n is one power lower than the number of effective volume dimensions of the source. Figure 3.10 illustrates the dimensionless curve $f_g[x/z]_{\#2}$ that describes the decay of $\mathrm{CA}[g]_{\#2}$ with offset x.

Table 3.4 lists the more easily applied formats for several of the idealized body equations that NETTLETON (1942) developed, and Figure 3.10 illustrates the related dimensionless curves that modulate the $\mathrm{CA}[g]_{\#}$ with offset from the central value. These results are useful starting points for the interpretation of simple anomalies and for developing preliminary conceptual models. Furthermore, the plot of the normalized distance function, which gives the shape of the anomaly from sources irrespective of amplitude, is helpful in comparing anomalies to differentiate sources.

In contrast to the 3D effect of the sphere, the infinite horizontal cylinder equation (#4) exemplifies the 2D effect. It is applicable for a body with horizontal length ($L_{\#4}$) that is ten or more times greater than its radius ($R_{\#4}$), which in turn is less than or not much greater than half the depth (z_c) to the central axis of the source (i.e. $L_{\#4} \geq 10 \times (R_{\#4} \leq 0.5z_c)$). The effective mass here, however, is the mass per unit length of the source given by $m_e(\#4) = \Delta\sigma \times [\pi R_{\#4}^2 = A(\#4)]$, where now the effective volume is the cross-sectional area of the source $A(\#4)$. In other words, the effective mass for the 2D source is the mass per unit length given by the product of its density contrast and cross-sectional area so that $\mathrm{CA}[g]_{\#4}$ varies with $z_c^{(n=1)}$.

The effect of finite horizontal cylinder length can be accommodated by relatively simple end corrections. For example, Equation $g_{\#4.A}$ in Table 3.2 gives the so-called 2.5D effect for the finite cylinder where the principal

TABLE 3.4 Examples of dimensionless attenuation factors ($f_g[x/z]_\#$) for calculating the gravity effect of idealized geometric sources.

Source	CA[g]#	$f_g[x/z]_\#$
(2) Sphere	$27.979 \times 10^{-6}\Delta\sigma_v(R^3/z^2)$	$[1+(x/z)^2]^{-\frac{3}{2}}$
(4) ∞ H-cylinder	$41.936 \times 10^{-6}\Delta\sigma(R^2/z)$	$[1+(x/z)^2]^{-1}$
(5) ∞ V-line mass	$20.984 \times 10^{-6}\Delta\sigma(R^2/z)$	$[1+(x/z)^2]^{-\frac{1}{2}}$
(11) Thin V-sheet (2D)	$30.738 \times 10^{-6}\Delta\sigma t$	$\log[1+(z/x)^2]^{-\frac{1}{2}}$
(14) Thin S-∞ H-slab (fault)	$41.936 \times 10^{-6}\Delta\sigma\delta z$	$(1/2)+(1/\pi)\tan^{-1}(x/z)$

CA[g]$_\#$ is the center anomaly value or constant and $f_g[x/z]_\#$ is the dimensionless distance function for the numbered sources in Tables 3.2 and 3.3. All parameters are in SIu, but the CA[g]$_\#$ are in mGal.
Modified from NETTLETON (1942).

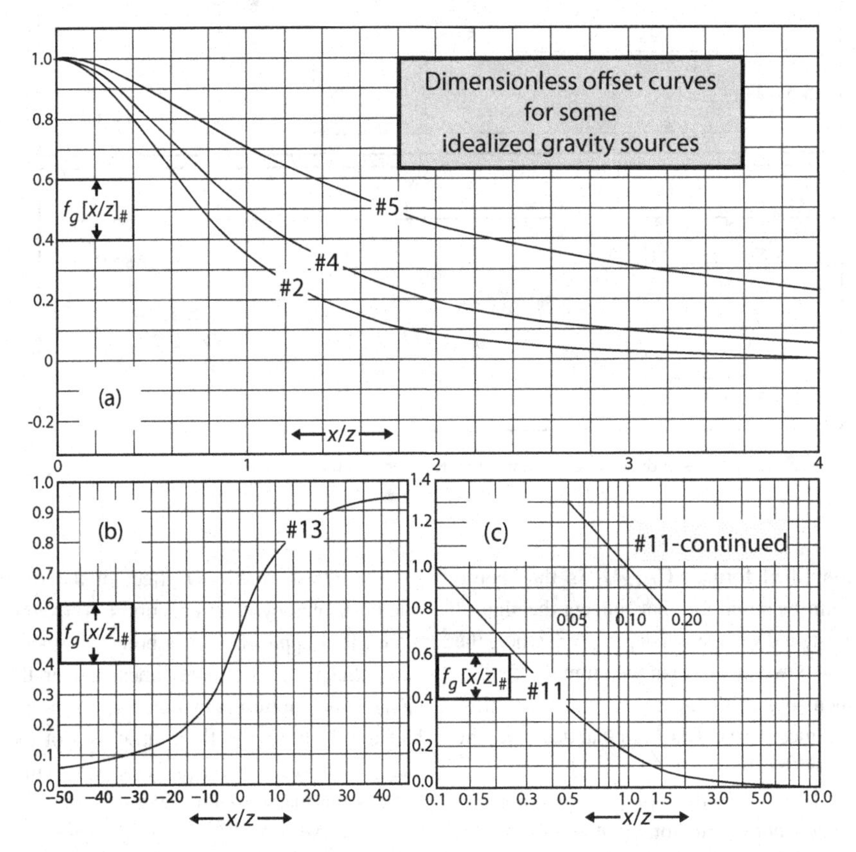

FIGURE 3.10 Distance functions from Table 3.4 used in calculating the gravity anomaly of (a) the sphere (source #2), infinite horizontal cylinder (source #4), and infinite vertical line mass (source #5); (b) the thin, horizontal semi-infinite slab (fault) (source #13); and (c) the thin vertical sheet (source #11).

profile is an equal distance Y from both ends of the cylinder. Equation $g_{\#4.\mathrm{B}}$ gives the 2.75D effect where the distances y'_a and y'_b to the ends of the finite horizontal cylinder from the profile are not equal.

For a 1D source like the Bouguer slab, the effective mass is simply $m(\#9) = 2\pi\Delta\sigma$, where 2π is the solid angle that the infinite slab subtends at any observation point (e.g. NETTLETON, 1942). Thus, any tabular source with horizontal dimensions that are large relative to the elevation of the observation point has a gravity effect that varies with $z^{(n=0)}$.

A disk with large radius (i.e. $R \longrightarrow \infty$) compared with its depth z_t, for example, has the same effect (Equation $g_{\#7.C}$) as the Bouguer slab. Hence, a disk with a smaller radius will have an effect $\omega/2\pi$ as large, or $g = \omega G \Delta\sigma \delta z$, where ω is the solid angle that the disk subtends at the observation point. This approximation holds to within about 5% or less for disk thicknesses $\delta z \leq 0.5 z_t$ (e.g. DOBRIN and SAVIT, 1988), and is especially useful for numerically modeling the effect of an arbitrarily shaped geological body. The calculation involves filling in the volume of the body with disks and summing or integrating the effects of the disks at the observation point.

In computing these gravity effects, effective ω estimates typically are visually selected from charts because of the difficulty in calculating solid angles, even for simple geometric bodies (e.g. NETTLETON, 1942; DOBRIN and SAVIT, 1988). However, TELFORD *et al.* (1990) presented the infinite vertical cylinder effect in terms of Legendre polynomials $P_{ni}(\cos\theta = z/r)$ of degree ni, which offers a less tedious modeling approach. For efficient digital computation, Equation $g_{\#6}$ gives these effects with the Legendre polynomials expanded through degree $ni = 4$ by the recurrence formula

$$\left[(ni+1)P_{(ni+1)} = (2ni+1)\left(\frac{z}{r}\right)P_{ni} - (ni)P_{(ni-1)}\right]$$
$$\forall\, ni \geq 0 \;\&\; P_0 \equiv 1. \qquad (3.70)$$

These results can provide the gravity effect of a disk with top and bottom surfaces at z_1 and z_2, respectively, by subtracting the effect of a cylinder with top at z_2 from the effect of a cylinder with top at z_1, as described in Equation $g_{\#7}$ of Table 3.2.

3.6 General source gravity modeling

In contrast to idealized body effects with closed analytical solutions, general sources refer to arbitrarily shaped 3D and 2D geometric models with gravity effects that must be numerically evaluated from Equations 3.22–3.33 and 3.35–3.41, respectively. These effects can be obtained from the integrated effects of idealized sources that effectively fill out or occupy the volume of the general body. Figure 3.11 gives an example of modeling the gravity anomaly over the Minden salt dome in Louisiana by several equivalent disks with varying density contrasts that differentiate the higher-density cap-rock from the underlying lower-density salt units (NETTLETON, 1943). Here,

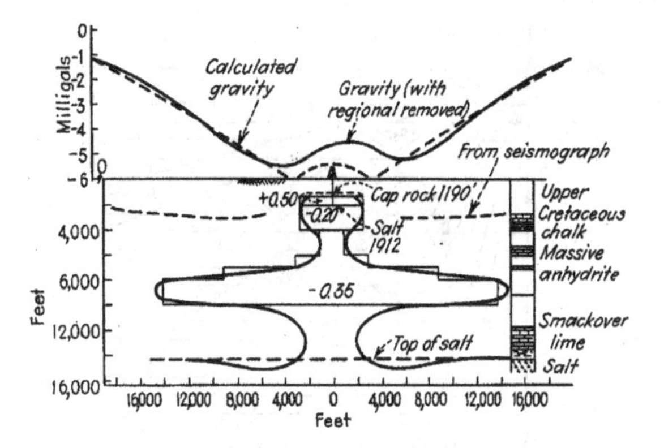

FIGURE 3.11 Gravity modeling of the Minden salt dome, Louisiana, with an equivalent stack of disks of varying density contrasts constrained by available seismic and drillhole data. Adapted from NETTLETON (1943).

the forward gravity model was developed to be consistent with available seismic and drillhole data.

A variety of schemes are available for gravity modeling depending on the geometric attributes of the source (e.g. SILVA *et al.*, 2001). For a 2D source of constant cross-section along one horizontal dimension that is four or more times longer than the other horizontal dimension, the gravity effect can be established in terms of the integrated analytical gravity effects of the (x', z')-line elements that follow the boundary of the cross-section. For a 3D source that varies also in the vertical z'-dimension, the gravity effect can be determined by the more complicated effort of establishing the varying (x', y')-boundaries of horizontal cross-sectional slabs that fill up the body. In addition, numerical least-squares gravity effects for 3D and 2D sources can be determined from idealized bodies distributed throughout each source according to the Gauss–Legendre quadrature decomposition of its volume and cross-section, respectively. Details on these modeling approaches are described in the sections below.

3.6.1 Generic 2D modeling procedures

In general, the gravity effects of complex shapes are commonly modeled by subdividing the body into cells and summing the effects of all the cells at each observation point. HUBBERT (1948) developed a classical graphical approach for modeling the gravity effect of an elongated 2D geological feature with arbitrarily shaped, constant cross-section perpendicular to the anomaly profile. Hubbert showed that the gravity effect of the cross-section is equivalent to the line integral effect around the periphery of the cross-section. This result follows Green's theorem, which equates the double integral over the

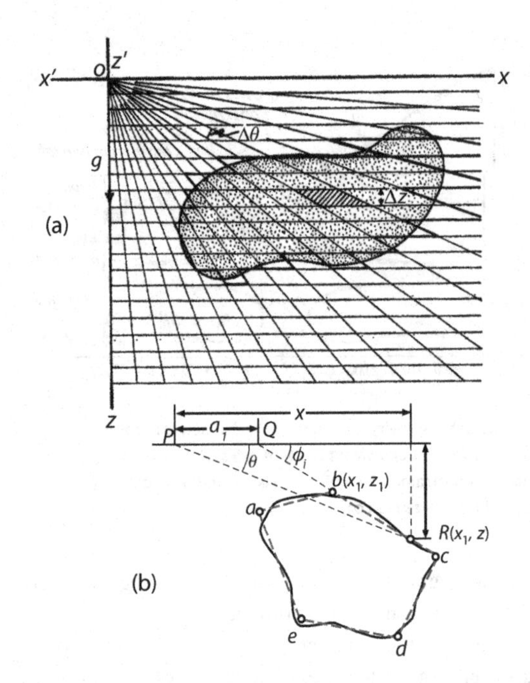

FIGURE 3.12 (a) Template for evaluating the gravity effect of 2D bodies of irregular cross-section. Adapted from HUBBERT (1948). (b) Polygon approximation of the irregular cross-section for a 2D body with line segment endpoints a, b, c, d, and e in the (x, z)-plane. Adapted from TALWANI *et al.* (1959).

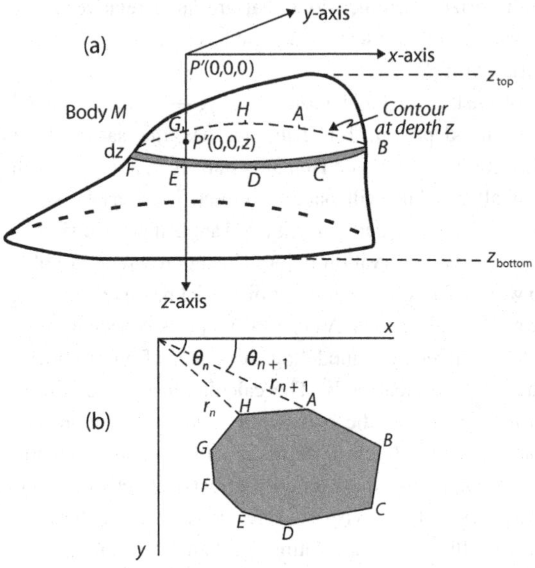

FIGURE 3.13 (a) Determining gravity at point $P'(0,0,0)$ from a 3D body of irregular shape that can be represented by a series of horizontal laminas each bounded by a connected sequence of straight line segments. (b) Polygon approximation of the irregular cross-section for a 3D body with line segment endpoints A, B, C, D, E, F, G, and H in the (x, y)-plane. Adapted from TALWANI and EWING (1960).

cross-section's area to the line integral around the boundary (e.g. KELLOGG, 1953).

To evaluate the line integral, a template or graticule is used as shown in Figure 3.12(a). It is scaled to the representation of the geologic cross-section and subdivides the subsurface by a pattern of radial lines directed from the station location and arcuate lines centered on the station or a series of equally spaced horizontal lines. The intersecting lines of the template define closed areas of equal gravity attraction at the station. The attraction of these compartments depends on their size, the scale represented, and the density differential between the 2D body and adjacent rock formations. The gravity effects of the areas and portions of areas intersecting the geological body's cross-section are summed for the gravity effect at the station location. This process is repeated for all stations on the gravity profile. The effect of finite length for the mass of any compartment of the template can be taken into account by the end correction in Equation 3.42, and variations in density of the conceptual model can be considered as well by applying the appropriate density to the area or portion thereof on the template.

Digital computing methods have essentially displaced the graphical procedure, which is effective and straightforward, but lengthy and tedious to implement. The widely used digital modeling approach due to TALWANI *et al.* (1959) approximates the boundary of the cross-section by straight line segments as shown in Figure 3.12(b). Here, the line integral effect used in HUBBERT (1948) is approximated by the summed gravity effects of connected line segments at the observation point where all geometric elements of the integral are expressed completely in terms of the 2D coordinates (x', z') of the end points or vertices of the line segments.

3.6.2 Generic 3D modeling procedures

The line integration method was also adapted by TALWANI and EWING (1960) to model the gravity effects of an irregular 3D source represented by a vertical stack of ni infinitely thin, but disconnected horizontal laminas. As illustrated in Figure 3.13, the ith lamina is approximated by a polygon with straight line segments. The gravity effect is $G(\Delta\sigma)S(z_i')$ with the line integral around the polygon $S(z_i')$ evaluated at the observation point completely in terms of the 3D coordinates (x', y', z_i') of the end points of the line segments.

In contrast to the analytical or closed-form gravity effect of the polygon, vertical integration of the 3D source must be carried out numerically by summing at the observation point the effects of the *ni* polygons that approximate the elevation contours of the arbitrarily shaped body. The integration can be effectively implemented using a relatively large number of polygons for the top third of the body, and two-thirds and one-third of this number of polygons for the middle third and bottom third portions of the body, respectively. The thicknesses of the laminas usually are restricted to less than one-tenth its depth. Another common approach uses a conventional scheme like Simpson's rule that assumes equal spacing of the laminas over the z'-limits of integration. However, the integrand is analytical, and thus the numerical integration can be carried out more efficiently and with least squares accuracy by the Gaussian quadrature method (e.g. STROUD and SECREST, 1966; KU, 1977; VON FRESE *et al.*, 1981b).

3.6.3 Least-squares 3D modeling

In the generic modeling application of the previous section, interpolation polynomials approximate the integrand by a summation formula like

$$\int_{z'_a}^{z'_b} S(z')\partial z' \simeq \sum_{i=1}^{ni} A_i S(z'_i), \tag{3.71}$$

where the *ni* values A_i are the weights to be given to the *ni* functional values $S(z'_i)$ evaluated at the interpolation coordinates, z'_i. However, because the integrand is analytical, Gaussian quadrature formulae can be developed to yield selected values of the interpolation z'_i and coefficients A_i so that the sum in Equation 3.71 gives the integral exactly when $S(z')$ is a polynomial of degree $2ni$ or less (e.g. STROUD and SECREST, 1966; CARNAHAN *et al.*, 1969).

The Gaussian coefficients A_i are obtained from a polynomial of order *ni* which is orthogonal over the interval of integration such that the *ni* points of interpolation z'_i are the zeros of the polynomial. Orthogonal polynomials commonly used to develop Gaussian quadrature formulae include Legendre, Laguerre, Chebyshev, and Hermite polynomials. Here, however, only the prototype Gaussian method involving Legendre polynomials is considered.

Legendre polynomials $P_{ni}(\bar{z})$ of order *ni* that are orthogonal over the interval $(-1 \le \bar{z} \le 1)$ are given by

$$P_{ni}(\bar{z}) = \left(\frac{1}{2^{ni} ni!}\right)\left(\frac{d^{ni}}{d\bar{z}^{ni}}[\bar{z}^2 - 1]^{ni}\right) \;\forall\; P_0(\bar{z}) \equiv 1, \tag{3.72}$$

and abide by the recurrence relations expressed in Equation 3.70. Thus, the standard Gauss–Legendre quadrature (GLQ) integration over the interval $(-1, 1)$ is

$$\int_{-1}^{1} S(\bar{z})\partial\bar{z} \simeq \sum_{i=1}^{ni} A_i S(\bar{z}_i), \tag{3.73}$$

where the interpolation points $\bar{z}_i$ at which the integrand is evaluated are the zeros of Equation 3.72 and the Gaussian coefficients are

$$A_i = \frac{2(1-\bar{z}_i^2)}{ni^2[P_{ni-1}(\bar{z}_i)]^2}. \tag{3.74}$$

Now, for arbitrary limits of integration such as in Equation 3.71 it is necessary to map the standard interval $(-1 \le \bar{z}_i \le 1)$ into the interval of integration $(z'_a \le z'_i \le z'_b)$ by the transformation

$$z'_i = \frac{\bar{z}_i(z'_b - z'_a) + (z'_b + z'_a)}{2}. \tag{3.75}$$

Thus, the integral in Equation 3.71 can be approximated as

$$\begin{aligned}\int_{z'_a}^{z'_b} [S(z')]\partial z' &= \int_{-1}^{1}\left[S\left\{\frac{\bar{z}_i(z'_b - z'_a) + (z'_b + z'_a)}{2}\right\}\right]\partial z' \\ &\simeq \frac{(z'_b - z'_a)}{2}\sum_{i=1}^{ni} A_i S(z'_i),\end{aligned} \tag{3.76}$$

so that the GLQ expression for the 3D gravity effect of an irregularly shaped body is

$$g = G\Delta\sigma \int_{z'_a}^{z'_b} S_i(z')\partial z' \simeq G\Delta\sigma\left[\frac{(z'_b - z'_a)}{2}\right]\sum_{i=1}^{ni} A_i S(z'_i). \tag{3.77}$$

A large value of the order *ni* insures a very accurate solution, but meaningful improvements in solution accuracy are difficult to achieve in practice once the nodal spacing (i.e. the spacing between the roots or zeros of Equation 3.72) is smaller than the depth to the source (e.g. KU, 1977; VON FRESE *et al.*, 1981b). Thus, choosing the lowest order *ni* so that $(\Delta z'_i < z'_b)$ facilitates efficient implementation of Equation 3.77.

The Gaussian coefficients A_i and corresponding nodes $\bar{z}_i$ are tabulated to 30 significant figures for Legendre polynomials of orders $ni = 2$ to 512 in STROUD and SECREST (1966). However, algorithms for digitally computing these values are also available in books on numerical methods (e.g. CARNAHAN *et al.*, 1969; PRESS *et al.*, 2007). Experience suggests that these values to 10 digit accuracy for orders up to $ni = 16$ are sufficient for most geological applications.

Equation 3.77 models the anomalous gravity effect of an irregular 3D body by summing at each observation point the gravity effects of ni polygons located at source coordinates given by Equation 3.75 and weighted by GLQ coefficients A_i. Relative to more conventional numerical integrations, GLQ integration minimizes the number ni of laminas for accurate least-squares modeling because the integrand (i.e. the gravity effect of the polygon) is analytical, and thus can be taken at the nodes or roots $z'_i(\bar{z}_i)$ of the Legendre polynomial selected to span the integration interval (z'_a, z'_b).

A further simplification extends the GLQ integration to evaluate the line integral around the polygon $S(z'_i)$ using the versatile analytical point pole or mass gravity expressions at the nodes of Legendre polynomials of orders nj and nk that respectively span the polygon's x' and y' limits. This extension transforms the 3D Equation 3.77 into Equation 3.26 with kernel F'_z and least-squares numerical solution

$$g \equiv F_z \simeq \left(\frac{y'_{kb} - y'_{ka}}{2}\right) \times \sum_{k=1}^{nk} \left\{ \left(\frac{x'_{jb} - x'_{ja}}{2}\right) \times \sum_{j=1}^{nj} \left\{ \left(\frac{z'_{ib} - z'_{ia}}{2}\right) \sum_{i=1}^{ni} [F'_z] A_i \right\} A_j \right\} A_k. \quad (3.78)$$

Here the $ni \times nj \times nk$ point pole gravity effects per unit volume F'_z are evaluated at the coordinates within the body given by

$$z'_i = 0.5[\bar{z}_i(z'_{ib} - z'_{ia}) + z'_{ib} + z'_{ia}],$$
$$x'_j = 0.5[\bar{x}_j(x'_{jb} - x'_{ja}) + x'_{jb} + x'_{ja}],$$
$$\text{and } y'_k = 0.5[\bar{y}_k(y'_{kb} - y'_{ka}) + y'_{kb} + y'_{ka}], \quad (3.79)$$

where $\bar{z}_i$, $\bar{x}_j$, and $\bar{y}_k$ are the respective ni, nj, and nk Gaussian nodes in the standard $(-1, 1)$ interval.

For a uniformly dimensioned body like the prism, the integration limits for evaluating Equation 3.78 are easy to specify. In this case, for example, $(x'_{ja}, x'_{jb}) = (x'_a, x'_b)$, $(y'_{ka}, y'_{kb}) = (y'_a, y'_b)$, and $(z'_{ia}, z'_{ib}) = (z'_a, z'_b)$. However, for the irregular body, the differential integration limits in each of the three dimensions must be interpolated at every node from a set of body point coordinates that provide an approximation of the surface envelope of the body (e.g. Ku, 1977; von Frese *et al.*, 1981b). Typically, the body is considered for its longest dimension, which for the sake of argument might be the x' dimension. Then the a-lower and b-upper x'-limits of the body (x'_{ja}, x'_{jb}) are established from which the nj Gauss–Legendre nodes x'_j are determined. Next, interpolations of the body point coordinates are performed at each x'_j to determine the (y'_{ka}, y'_{kb})-limits of the body for the nk nodes y'_k. Similarly, the vertical coordinates of the body points are interpolated at each horizontal coordinate (x'_j, y'_k) to appropriate vertical (z'_{ia}, z'_{ib})-limits of integration for the ni nodes z'_i. In practice, maximum accuracy effectively results when the spacing between the equivalent point source nodes is smaller than the depth to the top of the source (e.g. Ku, 1977; von Frese *et al.*, 1981b).

The 3D body accordingly is represented by a distribution of $ni \times nj \times nk$ point poles from which all gravity effects of the body can be least-squares estimated by

$$[U; F_x; F_y; F_z \equiv g; F_{xx}; F_{yy}; F_{zz}; F_{xy}; F_{xz}; F_{yz}] \simeq \left(\frac{y'_{kb} - y'_{ka}}{2}\right) \sum_{k=1}^{nk} \left\{ \left(\frac{x'_{jb} - x'_{ja}}{2}\right) \times \sum_{j=1}^{nj} \left\{ \left(\frac{z'_{ib} - z'_{ia}}{2}\right) \sum_{i=1}^{ni} [U'; F'_x; F'_y; F'_z; F'_{xx}; F'_{yy}; F'_{zz}; F'_{xy}; F'_{xz}; F'_{yz}] A_i \right\} A_j \right\} A_k. \quad (3.80)$$

Here, the unprimed variables in the left portion of the equation are the complete gravity effects of the extended 3D body in Equations 3.22–3.33 as derived from the weighted triple sum of the primed variables in the right portion which are the corresponding integrands of the equations. However, for the integration to hold, these kernels must be evaluated at the strategic (x'_j, y'_k, z'_i)-coordinates within the body given by Equation 3.79. A 3D example giving the details of estimating the gravity effect ($F_z \equiv g$) of a rectangular prism by GLQ integration is shown in Figure 3.14.

3.6.4 Least-squares 2D modeling

For the 2D body, the gravity effects are similarly obtained by fitting the x'- and z'-limits of the cross-section with Legendre polynomials of orders nj and ni, respectively (e.g. Ku, 1977). The body accordingly is represented by a 2D distribution of $nj \times ni$ point poles that estimate the source's least-squares gravity effects by

$$[\mathcal{U}; \mathcal{F}_x; \mathcal{F}_z \equiv \mathcal{G}; \mathcal{F}_{zz} = -\mathcal{F}_{xx}; \mathcal{F}_{xy}] \simeq \left(\frac{x'_{jb} - x'_{ja}}{2}\right) \sum_{j=1}^{nj} \left\{ \left(\frac{z'_{ib} - z'_{ia}}{2}\right) \sum_{i=1}^{ni} [\mathcal{U}'; \mathcal{F}'_x; \mathcal{F}'_z; \mathcal{F}'_{zz} = -\mathcal{F}'_{xx}; \mathcal{F}'_{xz}] A_i \right\} A_j. \quad (3.81)$$

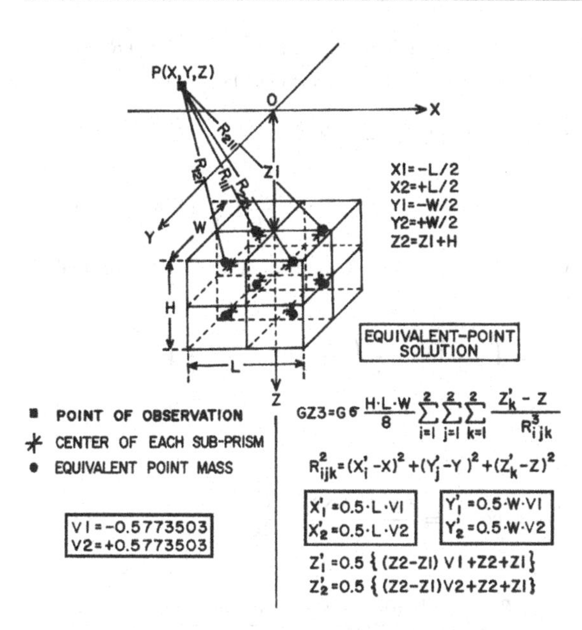

FIGURE 3.14 Details of the $2 \times 2 \times 2$ Gauss–Legendre quadrature formula for estimating the vertical gravity anomaly $GZ3 = F_z \equiv g$ of a uniform density, 3D prism. Adapted from Ku (1977). Here, the Gaussian coefficients $V1$ and $V2$ in all spatial dimensions are equal in magnitude, but opposite in sign because the height H, width W, and length L of the cube are equal. The GLQ formula for the prism's magnetic total field effect is illustrated in Figure 9.8.

Here again, the unprimed variables in the left portion of the equation are the complete gravity effects of the 2D body in Equations 3.35–3.41 as derived from the weighted double sum of the primed variables in the right portion which are the corresponding integrands of the equations evaluated at the strategic (x'_j, z'_i)-coordinates within the cross-section according to Equation 3.79.

3.6.5 Least-squares modeling accuracy

In 2D and 3D cell applications, the minimum numbers of equivalent point sources required for GLQ gravity modeling are $nj \times ni = 2 \times 2 = 4$ and $nk \times nj \times ni = 2 \times 2 \times 2 = 8$, respectively. As illustrated in Ku (1977), the modeling assumes that the cell is subdivided into quadrants where mass is concentrated at points displaced towards the corners relative to the centers of the quadratures. If the estimates are made too close to the body, the gravity effect is dominated by the individual point source effects, whereas adding sources so as to make the source spacing smaller than the elevation of the estimates integrates the point effects to a closer least-squares estimate of the body's gravity effect.

Thus, the practical trade-off between accuracy and the speed and computational labor of GLQ integration is controlled by the number of point sources. The integration must be computed at a sufficiently great distance from the body that the individual point pole effects coalesce into an acceptable approximation of the analytical solution. In practice, this can be achieved either by subdividing the body into smaller blocks or by simply increasing the number of point sources until the distance between them is smaller than the depth to the top of the body (e.g. Ku, 1977; von Frese *et al.*, 1981b). For example, Gaussian nodes and weights tabulated to 30 significant figures for Legendre polynomial orders 2 to 512 are readily available (e.g. Stroud and Secrest, 1966). The tabulated values could be used to model the gravity effects of the prism by a $(512 \times 512 \times 512 = 134,217,728)$-point GLQ formula, but it would match the analytical solution in many more significant figures than is applicable in practice.

In general, changes in the significant figures of the GLQ estimate decrease as the number of nodes increases. Thus, the trade-off in accuracy and computational effort is effectively optimized by the smallest number of point source nodes beyond which changes in the estimate's least significant figure are negligible.

3.6.6 Least-squares modeling in spherical coordinates

Regional gravity anomalies registered in spherical coordinates are becoming increasingly available for geological studies of large areas of the Earth and other planetary bodies of the solar system as the result of satellite measurements and continental-scale and larger compilations of terrestrial, marine, and airborne surveys. The above modeling results can be adapted by determining the equivalent point source (EPS) effects in spherical coordinates. This conversion requires the Cartesian-to-spherical coordinate transformations given by

$$x = r\cos(\theta)\cos(\varphi),\ y = r\cos(\theta)\sin(\varphi),\ \text{and}$$
$$z = r\sin(\theta), \tag{3.82}$$

where the body-fixed spherical coordinates include geocentric r-distance, θ-latitude (i.e. the co-latitude), and φ-longitude so that

$$r = z\cos(\theta) + K\sin(\theta),\ \theta = -z\sin(\theta) + K\cos(\theta),$$
$$\text{and } \varphi = y\cos(\varphi) - x\sin(\varphi) \tag{3.83}$$

with $K = [x\cos(\varphi) + y\sin(\varphi)]$ (Figure 3.15).

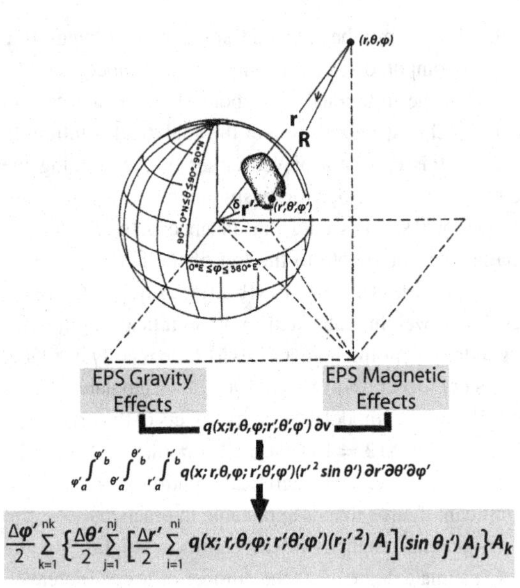

$$\int_{\varphi'_a}^{\varphi'_b}\int_{\theta'_a}^{\theta'_b}\int_{r'_a}^{r'_b} q(x; r,\theta,\varphi; r',\theta',\varphi')(r'^2 \sin\theta')\,\partial r'\partial\theta'\partial\varphi'$$

$$\frac{\Delta\varphi'}{2}\sum_{k=1}^{nk}\left\{\frac{\Delta\theta'}{2}\sum_{j=1}^{nj}\left[\frac{\Delta r'}{2}\sum_{i=1}^{ni} q(x; r,\theta,\varphi; r',\theta',\varphi')(r_i'^2)A_i\right](\sin\theta_j')A_j\right\}A_k$$

FIGURE 3.15 Gravity and magnetic anomaly modeling in spherical coordinates by Gauss–Legendre quadrature integration of a body of any shape and uniform physical property x in density and magnetization, respectively. Adapted from VON FRESE et al. (1981b). The equivalent point source (EPS) effects refer to potential, vector, and gradient tensor effects of the gravity point pole and magnetic point dipole that are described in full analytical detail in ASGHARZADEH et al. (2007) and ASGHARZADEH et al. (2008), respectively. The EPS effects generalize as a generic integrand or q-function over a differential volume ∂v that can be analytically or numerically volume integrated for the body's effect by the respective triple integral and quadrature series expressions at the base of the equation tree. The GLQ estimate includes $\Delta\varphi' = (\varphi'_{kb} - \varphi'_{ka})$, $\Delta\theta' = (\theta'_{jb} - \theta'_{ja})$, and $\Delta r' = (r'_{ib} - r'_{ia})$.

By Equation 3.82, the Cartesian gravity point pole potential (Equation 3.16), for example, transforms into the spherical coordinate EPS expression given by

$$U(r,\theta,\varphi) = \frac{G\times m}{|\mathbf{r}-\mathbf{r}'|} = \frac{G\times m}{R}, \tag{3.84}$$

where G is the universal gravitational constant, m is the mass (or mass contrast) of the point pole, $R = \sqrt{r^2 + r'^2 - 2rr'\cos(\delta)}$, $\mathbf{r}$ is the radius vector directed from the Earth's center to the observation point (r,θ,φ), $\mathbf{r}'$ is the radius vector directed from the Earth's center to the source point (r',θ',φ'), δ is the angle between $\mathbf{r}$ and $\mathbf{r}'$ such that $\cos(\delta) = \cos(\theta)\cos(\theta') + \sin(\theta)\sin(\theta')\cos(\varphi - \varphi')$, (θ,θ') are the geocentric latitude (i.e. co-latitude) coordinates of the observation and source points, respectively, and (φ,φ') are the longitude coordinates of the respective observation and source points.

Thus, the gravitational EPS force field $\mathbf{F}_g(R)$ at the distance R in spherical coordinates is

$$\mathbf{F}_g(R) = \nabla U(R) = \frac{\partial}{\partial R}[U(R)]\nabla R = (G\times m)\left[\frac{\partial}{\partial R}\left(\frac{1}{R}\right)\nabla R\right], \tag{3.85}$$

where the gradient of R in observation point coordinates is

$$\nabla R = \left(\frac{\partial R}{\partial r}\right)\hat{\mathbf{e}}_r + \frac{1}{r}\left(\frac{\partial R}{\partial\theta}\right)\hat{\mathbf{e}}_\theta + \frac{1}{r\sin\theta}\left(\frac{\partial R}{\partial\varphi}\right)\hat{\mathbf{e}}_\varphi \tag{3.86}$$

with

$$\frac{\partial R}{\partial r} = \cos\psi = \frac{r - r'\cos\delta}{R} = \frac{C}{R}, \tag{3.87}$$

$$\frac{1}{r}\left(\frac{\partial R}{\partial\theta}\right) = \frac{r'[\sin\theta\cos\theta' - \cos\theta\sin\theta'\cos(\varphi-\varphi')]}{R} = \frac{E}{R}, \tag{3.88}$$

$$\frac{1}{r\sin\theta}\left(\frac{\partial R}{\partial\varphi}\right) = \frac{r'\sin\theta'\sin(\varphi-\varphi')}{R} = \frac{H}{R}, \tag{3.89}$$

and $\hat{\mathbf{e}}_r$, $\hat{\mathbf{e}}_\theta$, and $\hat{\mathbf{e}}_\varphi$ are the spherically orthogonal unit basis vectors at the observation point. With the results from the above three equations, Equation 3.85 becomes

$$\mathbf{F}_g(R) = (G\times m)\left\{\frac{\partial}{\partial R}\left(\frac{1}{R}\right)\left[\frac{C}{R}\hat{\mathbf{e}}_r + \frac{E}{R}\hat{\mathbf{e}}_\theta + \frac{H}{R}\hat{\mathbf{e}}_\varphi\right]\right\}, \tag{3.90}$$

where the radial scalar component is

$$F_r = (G\times m)\left\{\frac{\partial}{\partial R}\left(\frac{1}{R}\right)\frac{\partial R}{\partial r}\right\} = \left\{\frac{-G\times m}{R^3}\right\}C \equiv g, \tag{3.91}$$

the horizontal θ-co-latitude scalar component is

$$F_\theta = (G\times m)\left\{\frac{\partial}{\partial R}\left(\frac{1}{R}\right)\left(\frac{1}{r}\frac{\partial R}{\partial\theta}\right)\right\} = \left\{\frac{-G\times m}{R^3}\right\}E, \tag{3.92}$$

and the horizontal φ-longitude scalar component is

$$F_\varphi = (G\times m)\left\{\frac{\partial}{\partial R}\left(\frac{1}{R}\right)\left(\frac{1}{r\sin\theta}\frac{\partial R}{\partial\varphi}\right)\right\} = \left\{\frac{-G\times m}{R^3}\right\}H. \tag{3.93}$$

Additional details on these spherical gravity components, the associated gradient tensors, and their implementation for modeling the gravity effects of extended bodies are described by ASGHARZADEH *et al.* (2007).

The equation tree at the bottom of Figure 3.15, for example, shows the adaption of Equation 3.80 to spherical coordinates where (r, r') are the respective observation and source point radial distances from the Earth's center, (θ, θ') are the respective observation and source point co-latitudes, and (φ, φ') are the longitude coordinates of the respective observation and source points. However, to convert density to mass in spherical coordinate applications, the standard Cartesian unit volume $\partial v(x', y', z') = \partial x' \partial y' \partial z'$ is replaced by $\partial v(r', \theta', \varphi') = (r'^2 \sin\theta')\partial r' \partial\theta' \partial\varphi'$ as a result of the Cartesian-to-spherical coordinate transformations (Equation 3.82).

The EPS gravity effects listed at the top of the equation tree in Figure 3.15 include the spherical coordinate point pole potential U', and vector $F'_r \equiv g'$, F'_θ, and F'_φ components from Equations 3.84, 3.91, 3.92 and 3.93, respectively. The effects also apply to the gradient tensors F'_{rr}, $F'_{\theta\theta}$, $F'_{\varphi\varphi}$, $F'_{r\theta} = F'_{\theta r}$, $F'_{r\varphi} = F'_{\varphi r}$ and $F'_{\theta\varphi} = F'_{\varphi\theta}$ that are analytically detailed in ASGHARZADEH *et al.* (2007). Each of the listed point pole effects can be generalized as $q(x \equiv \sigma G; r, \theta, \varphi; r', \theta', \varphi') \times \partial v(r', \theta', \varphi')$ where the generic q-function is made up of the product of the functional involving density and big G times the geometric functional involving only the coordinates of the observation point (r, θ, φ) and the source point (r', θ', φ'). Hence, the gravity effects of an extended body are given by the triple integral, which for an idealized body can be evaluated in closed form, whereas for an irregularly shaped body the triple series at the bottom of the equation tree provides a least-squares numerical solution.

3.7 Gauss' law

In theory, the mass of a subsurface volume can be uniquely evaluated from its gravity effects independent of its geometry. This result can be useful in estimating tonnages of mineral ores, hazardous waste, and other subsurface sources of gravity variations. It is a consequence of Gauss' law that relates the total mass enclosed within a region to the normal component of its gravitational attraction integrated over the surface of the region.

In particular, the gravitational flux, ∂N, through an elemental surface area, ∂S, is

$$\partial N = (\hat{\mathbf{e}}_n \cdot \mathbf{g}) \times \partial S = g(\partial S)\cos(\theta), \quad (3.94)$$

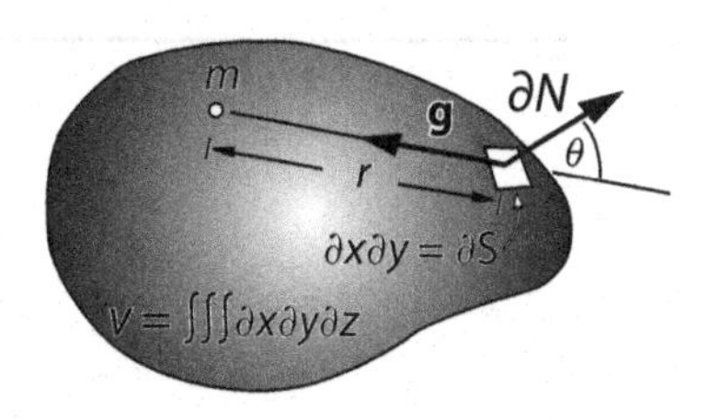

FIGURE 3.16 Gauss' law (Equation 3.98) involves the spatial arrangement of the gravitational flux element ∂N, normal through a unit surface element ∂S, due to a mass m that causes a gravitational attraction g at the center point of ∂S, on the surface S enclosing the volume v.

where the dot product $(\hat{\mathbf{e}}_n \cdot \mathbf{g})$ is the gravitational component normal to the surface with the unit normal vector $\hat{\mathbf{e}}_n$. Thus, according to Equation 3.1 and Figure 3.16, the flux through the element of the surface ∂S from a small mass m placed within the surface S is

$$\partial N = g(\partial S)\cos(\theta) = -Gm(\partial S)\cos(\theta)/r^2. \quad (3.95)$$

However, the element of solid angle $\partial\Omega$ subtended at the location of m by the elemental surface ∂S is

$$\partial\Omega = (\partial S)\cos(\theta)/r^2, \quad (3.96)$$

which is the area swept out on a sphere of unit radius by a radius from m that sweeps over the surface element ∂S. Substituting Equation 3.96 into 3.95 gives the flux through the surface element ∂S as

$$\partial N = Gm(\partial\Omega), \quad (3.97)$$

so that the total gravitational flux through the closed surface is

$$\int_S\int \partial N = N = -4\pi Gm. \quad (3.98)$$

If the mass is located exterior to the surface, the flux is zero because the fluxes of the paired surfaces subtended by the solid angles at the exterior mass are equal and opposite.

Gauss' law (Equation 3.98) can also be derived using Gauss' divergence law of vector calculus which relates the vector field over a volume to the normal component of the field over the surface enclosing the volume. Specifically, for any vector point function $\mathbf{F}(x, y, z)$, Gauss' law gives

$$\int\int_v\int (\nabla \cdot \mathbf{F}_g)\partial v = \int_S\int (\hat{\mathbf{e}}_n \cdot \mathbf{F}_g)\partial S, \quad (3.99)$$

where v is the volume enclosed within the surface S, and $\hat{\mathbf{e}}_n$ is the unit vector perpendicular to the surface S pointing out from the volume at each point of S (Figure 3.16). By this theorem, $(\hat{\mathbf{e}}_n \cdot \mathbf{F}_g)$ is the component of $\mathbf{F}_g$ normal to S so that the total amount of $(\hat{\mathbf{e}}_n \cdot \mathbf{F}_g)$ inside v is equal to the total flux of $\mathbf{F}_g$ outward through the surface S.

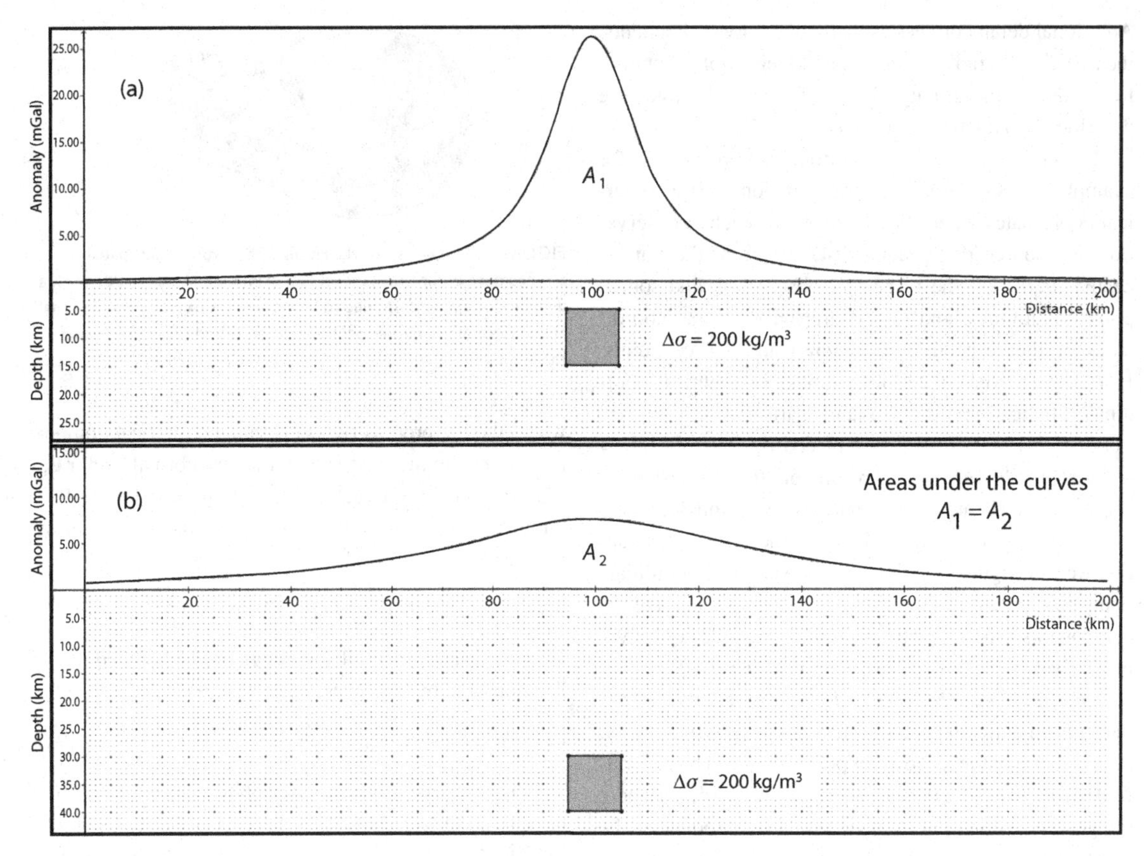

FIGURE 3.17 Gravity anomaly profile due to a 2D (strike-infinite) excess mass located near the surface (a) and deeper (b) within the Earth. The equal mass anomalies of (a) and (b) result in equal cross-sectional anomaly areas A_1 and A_2 beneath the profiles.

Returning to geophysical applications of Gauss' law, Equation 3.98 can be rearranged using Equation 3.94 so that integrating the total flux over the surface S gives

$$\int_S\int g\partial S = 4\pi Gm, \tag{3.100}$$

because gravity measurements relate to the normal component of gravitational attraction that is taken positively downwards (i.e. towards the Earth). For gravity measurements on the surface of the Earth, the summation of the gravity anomaly over its entire spatial extent is equal to $2\pi Gm$ or one-half of the right-hand side of Equation 3.100 because the anomaly is observed over only half of the surface enclosing the anomalous mass, m. Thus,

$$\sum_i g_i \times \partial S_i = 2\pi Gm, \tag{3.101}$$

where g_i is the vertical component of gravity (i.e. $g_z(i)$) over the horizontal surface element ∂S_i with $\sum_i \partial S_i = S/2$.

This equation is a useful means of uniquely determining the anomalous mass causing a gravity anomaly. The anomalous mass, m, is calculated irrespective of assumptions regarding the location and geometry of the source by numerical integration of the gravity anomaly. The only assumption required is that the numerical integration has captured the vast majority of the total energy of the anomaly (anomalous gravity value times area).

Gauss' law also shows that the total gravity anomaly energy is the same for equal masses or mass differentials regardless of the shape and depth of the anomalous mass. As a result the anomaly from a shallow mass will be sharper with a higher amplitude than the anomaly from the same mass at a greater depth (Figure 3.17). The utility of Gauss' law in practice is limited by the ability to evaluate the energy in the tails or extreme edges of the anomalies where they asymptotically merge into the non-anomalous gravity field (e.g. HAMMER, 1945; LAFEHR, 1965).

3.8 Gravity anomaly ambiguity

Gauss' law uniquely relates the integral of the source's gravity effect to its mass, but not to how this mass is distributed. Interpretational ambiguity occurs because the gravity potential is a function of $\frac{1}{r}$ with derivatives and integrals that are linearly related, and thus the potential can be evaluated from any component of the field without knowing the actual distribution of the causative mass contrast or source. Of course, the anomaly component is mapped in practice to evaluate its integral and derivative properties for insight on how the source is spatially distributed. However, for any mass variation whose effects match the observed anomaly component, a wide range of source variations can always be found with matching effects as shown below.

Gauss' law shows that the flux N is independent of the position and configuration of m within the surface S. Thus, for a continuous distribution of mass within S, Equation 3.98 becomes

$$\int_S\int(\hat{\mathbf{e}}_n\cdot\mathbf{g})\partial S = -\int\int_v\int(4\pi G\sigma)\partial v. \tag{3.102}$$

Applying Gauss' divergence theorem (Equation 3.99) to the left side of the above equation gives

$$\int_S\int(\hat{\mathbf{e}}_n\cdot\mathbf{g})\partial S = \int\int_v\int(\nabla\cdot\mathbf{g})\partial v. \tag{3.103}$$

Thus, subtracting Equation 3.102 from 3.103 yields

$$\int\int_v\int(\nabla\cdot\mathbf{g}+4\pi G\sigma)\partial v = 0, \tag{3.104}$$

which holds for any volume v only if the integrand vanishes so that

$$\nabla\cdot\mathbf{g} = \frac{\partial g_x}{\partial x}+\frac{\partial g_y}{\partial y}+\frac{\partial g_z}{\partial z} = -4\pi G\sigma(x,y,z). \tag{3.105}$$

Substituting Equation 3.1 into the above equation yields Poisson's equation

$$\nabla^2 U = \frac{\partial^2 U}{\partial x^2}+\frac{\partial^2 U}{\partial y^2}+\frac{\partial^2 U}{\partial z^2} = -4\pi G\sigma, \tag{3.106}$$

and Laplace's equation (Equation 3.6) in free space where $\sigma = 0$. The equations of Poisson and Laplace lead to the inescapable conclusion that a wide distribution of equivalent sources can produce an equivalent gravity anomaly. Thus the gravity anomaly is said to have inherent source ambiguity because any spatial distribution of mass that satisfies it as a solution is not unique.

3.9 Poisson's theorem

A point mass that is also magnetized has point magnetic dipolar potential $V(r)$ given by Equation 9.12 which is developed in detail in Section 9.3. It has exactly the same displacement distance r that holds for the gravity point mass potential $U(r)$ in Equation 3.16. Thus, the gravitational and magnetic potentials of a point source, as well as its extensions to anomalous bodies, can be directly compared. Poisson's theorem (Poisson, 1826) formalized this connection by which all properties of the magnetic field due to a homogeneous body are derivable from its gravity field and *vice versa*. A pseudoanomaly refers to an anomaly of one type (i.e. gravity or magnetic) that has been transformed from the equivalent anomaly of the other type (i.e. magnetic or gravity) via Poisson's theorem.

At an observation point distance r from the source of constant density σ and magnetization with intensity J and direction i, Poisson's theorem connects the gravity $U(r)$ and magnetic $V(r)$ potentials by

$$V(r) = \frac{J}{G\sigma}\left[\frac{\partial U(r)}{\partial i}\right]. \tag{3.107}$$

Thus, the magnetic potential and first derivative of the gravitational potential in the direction of magnetization are linearly related by the scalar proportionality $(J/G\sigma)$. Furthermore, because the derivatives of the potentials are linear transformations of each other as shown in Equation 3.9, Poisson's theorem can be readily extended to relate the (n)th derivative magnetic effect to the relevant $(n+1)$th derivative gravity effect.

The Poisson relation (Equation 3.107) can be obtained by generalizing the magnetic potential in Equation 9.25 as

$$V(r) = \mathbf{J}\cdot\nabla\int\int_v\int\left(\frac{1}{r}\right)\partial v = J\frac{\partial}{\partial i}\int\int_v\int\left(\frac{1}{r}\right)\partial v, \tag{3.108}$$

where $\mathbf{J}\cdot\nabla = J(\partial/\partial i)$ is taken because the magnetization throughout the volume v is constant. However, consideration of the corresponding gravitational potential (Equation 3.22) shows that

$$\int\int_v\int\left(\frac{1}{r}\right)\partial v = \frac{U(r)}{G\sigma}, \tag{3.109}$$

which when substituted into the magnetic potential yields the Poisson relation.

3.10 Pseudoanomalies

Poisson's theorem also clearly holds for the gravity and magnetic vector and tensor components of a common source because they are spatial derivatives of the potentials. For induced magnetization at the geomagnetic field poles where $i = z$, for example, the vertical magnetic field component B_z can be related to the gravity tensor F_{zz}

as

$$B_z = -\frac{\partial V(r)}{\partial z} = \frac{-J}{G\sigma}\left[\frac{\partial}{\partial z}\left(\frac{\partial U(r)}{\partial z} = F_z \equiv g\right)\right]$$
$$= \frac{-J}{G\sigma}[F_{zz}]. \qquad (3.110)$$

Comparable relationships can be established for any of the potential field components of the source with uniform density and magnetization.

The magnetic effect (e.g. B_z) derived from observed gravity anomaly values (e.g. F_{zz}) is called the pseudomagnetic effect, and the pseudogravity effect is the gravity effect (e.g. F_{zz}) obtained from magnetic anomaly values (e.g. B_z). Comparing pseudomagnetic or gravity effects against the respectively surveyed magnetic or gravity effects provides an important test for relating the effects to a common source and reducing interpretational ambiguities. Pseudomagnetic determinations are described further in Sections 6.5.3 and 7.4.6, and additional descriptions of pseudogravity calculations are offered in Sections 12.4.2 and 13.4.2.

The validity of Poisson's theorem is independent of the shape or depth of the source, with the minor exception of sources that have significant demagnetization effects. The theorem permits the transformation of the source's magnetic anomaly into its equivalent pseudogravity effect and its gravity anomaly into the equivalent pseudomagnetic effect. Thus, the theorem provides the quantitative basis for comparing gravity and magnetic anomalies to help reduce ambiguities in interpretation, quantify anomaly correlations, and establish other anomaly attributes related to the distribution and physical properties of anomaly sources. These applications are considered further in the gravity and magnetic interpretation Sections 7.4.6 and 13.4.2, respectively.

3.11 Key concepts

- The gravity method is described as a potential field method because it measures a function of the potential of the gravity field of force. Potential is a scalar quantity of force-field space such that its rate of change is the component of the force in that direction.
- The gravity effect of an extended body is derived by summing at each observation point the effects of point sources that fill out the body. The operative physical property for gravity modeling is the mass per unit volume or density.
- For an idealized source with simple symmetric shape such as a sphere, cylinder, or prism, the gravity effect can be integrated by a closed-form analytical expression. Tables 3.2 and 3.3 list the gravity effects for 16 idealized sources, ranging from the 3D effects of the point mass and sphere to the 2D effects of vertical and horizontal cylinders, and the 1D effects of thick and thin slabs dipping horizontally to vertically with oblique-to-vertical edges.
- A general source with arbitrary shape, on the other hand, must be evaluated by numerical integration from the superimposed effects of idealized sources that fill up the arbitrary volume. General 2D and 3D sources have been widely modeled using, respectively, a vertical lamina and a vertical stack of horizontal laminas through the irregularly shaped body. In both cases, the lamina is approximated by a polygon with straight line edges whose gravity effect involves geometric elements completely described by the end point coordinates of the line elements.
- The elongated finite-length source with constant polygonal cross-section perpendicular to its strike is widely used for interactive modeling of gravity vector and gradient tensor components. The modeling is said to be 2.5D where the principal profile bisects the source perpendicular to strike and 2.75D otherwise, even though both terms in reality refer to 3D modeling. By Gauss' divergence theorem, the volume integral is reduced to a surface integral which in turn is reduced to line integrals that are evaluated along the line segments making up the edges of the polygonal cross-section and the rectangles that connect the edges of the polygon at the ends of the body. Poisson's relation can extend the gravity modeling to magnetic modeling efforts.
- Idealized sources have analytical effects, and thus can also be positioned throughout the general source according to the Gauss-Legendre quadrature (GLQ) decomposition of its effective volume/area/length so that the numerical integration is obtained with least-squares accuracy. The analytical simplicity of the point mass effect makes it particularly effective for least-squares gravity modeling. The 2D and 3D gravity effects are accurately computed by summing at each observation point the anomalous effects of equivalent point sources located within the source, where each point source effect is weighted by appropriate GLQ coefficients and the coordinate limits of the body's cross-sectional area and volume, respectively.
- For any mass distribution, the gravity effect (potential, geoidal undulation, vector anomaly components, spatial derivatives of all orders, etc.) involves the related gravity effect of the point mass in the integrand of the

modeling equation. Thus, the gravity effects are linearly related so that an observed gravity anomaly can be linearly transformed into any other gravity component to investigate its subsurface significance.
- Fundamental laws and theorems of potential fields, such as Gauss' law and the equations of Laplace and Poisson, are useful in understanding the nature of fields and in their analysis. They lead to the inescapable conclusion that potential fields can be produced by a wide distribution of equivalent sources. Ambiguities of interpretation can be reduced only by the imposition of ancillary geological, geophysical, and other constraints.

Supplementary material and Study Questions are available on the website www.cambridge.org/gravmag.

4 Density of Earth materials

4.1 Overview

The gravity method measures horizontal spatial changes in the gravitational field that result from mass differentials which in turn are controlled by the volume and contrasting densities of anomalous masses. As a result, an understanding of the density of Earth materials is essential in planning surveys as well as in interpreting gravity anomalies.

Density is a function of the mineral composition of Earth materials as well as their void volume and the material filling the voids. As a result, densities of rocks can be estimated by considering their origin and the processes that subsequently have acted upon them. However, it is advisable to measure densities either directly or indirectly wherever possible, preferably *in situ*, because of the difficulties in obtaining samples that are representative of the actual geological setting. *In situ* measurements may be obtained from the relationship of gravity anomalies to topography or determined indirectly from correlative measurements such as seismic wave velocity or attenuation of gamma rays.

4.2 Introduction

Knowledge of germane Earth material densities within a study region is required for effective planning and implementation of gravity surveys. Accordingly, as an introduction to the gravity method, the sections below describe the fundamentals of this property and the controls on it, together with methods of determining density. Representative values are presented of the density of a variety of Earth materials including igneous, metamorphic, and sedimentary rocks, sediments, and soils to aid the explorationist in the use of the gravity method. In actual practice it is not the density that is the controlling property of gravity anomaly fields, but rather the density contrast between the anomaly source and the laterally adjacent formations, often termed the country rocks, which are assumed to be constant and the norm. As a result, knowledge of densities within a survey region involves both the anomalous mass and the country rock. The sign of the gravity anomaly is positive if the anomalous mass exceeds that of the country rock and is negative for the reverse.

Densities of rocks and other Earth materials are largely dependent on their major mineral composition and void space and typically have a limited range from 1000 to 3000 kg/m^3 in SIu, or 1 to 3 g/cm^3 in CGSu. Studies of well-logs, gravity anomaly data, and physical samples indicate that density is scale-independent within the crust (e.g. Pilkington and Todoeschuck, 1995, 2004; Todoeschuck *et al.*, 1994) and generally varies within Earth formations. As a rule, density is not clearly diagnostic of most common rock types because of overlapping ranges, and it may be variable at a range of scales within a formational unit. As a result, accurately specifying properties by measurements of limited samples requires significant care.

A common decision to be made early in the geophysical process is whether to use general tabulations and extrapolations from them or to make measurements in the survey region. The latter alternative is often desirable because of limitations in existing tables of values and the inability to predict the properties of the site materials accurately enough from available information. Of course, the decision regarding site-specific measurements depends on the availability of samples and *in situ* measurement opportunities and resources. If measurements are to be made, follow-up decisions are whether observations should be made on samples or *in situ*, and which specific procedures should be used. The answers to these questions depend on local conditions. The sections below on densities and their measurements aim to help in reaching decisions on these questions.

Density, a scalar property, is the simplest of the geophysical properties and most readily measured to first-order accuracy without sophisticated instrumentation. However, measurements of the density of many Earth materials are not easy to make accurately. As a result, there is a limited base of high-quality measurements reported in sufficient detail to permit their evaluation and general use. Most tabulations of measurements are based on observations of a limited number of samples and are largely restricted to crystalline and well-lithified sedimentary rocks that have been subjected to only minor secondary processes that will alter their density.

Densities of crystalline and well-lithified sedimentary rocks, being largely dependent on their major component minerals, are relatively uniform to the extent that the rock mass is homogeneous. These densities can be predicted rather accurately because the major minerals commonly can be estimated knowing the origin of the rock and the tectonic and crystallization history of igneous and metamorphic rocks, or depositional environment of sediments and sedimentary rocks. Secondary processes, both physical and chemical, may alter the density of rocks by causing fracture, solution, or interstitial void space with varying void fillings. Alteration of the more chemically susceptible minerals may further modify the density of the rock mass because secondary minerals are typically lower in density than the primary minerals of the rock. As a result, available density tabulations are only useful for generalizations and starting points for further estimations.

Typically, crystalline and other well-lithified rocks have minimal void spaces, generally less than 1%. This greatly simplifies the measurement and increases the accuracy of estimating their densities. In contrast, materials of the near-subsurface may have large void space with varying degrees of void filling. Their *in situ* densities are difficult to measure accurately, and many of the materials are highly heterogeneous at a range of scales. Thus, these Earth materials are poorly reported and documented in the literature.

The available literature shows that there is considerable overlap in the densities of subsurface materials. In addition, the distribution of densities within most specific rock units or lithologies is broad and commonly skewed towards lower densities as a result of secondary processes.

TABLE 4.1 Types of densities used in gravity methods.

Type	Symbol	Definition
True density	σ_t	Mass of unit volume of solid material without voids
Grain density	σ_g	Mass of unit volume of mineral grains, equivalent to σ_t
Specific gravity		Ratio of the mass of a unit volume of material to the mass of the same volume of chilled gas-free distilled water; excludes voids
Bulk density	σ_B	Mass per unit volume of a dry rock; includes both solid material and void space
Natural density	σ_n	Mass per unit volume of a water-saturated Earth material; includes both solid material and void space filled with water
Saturated bulk density		Equivalent to σ_n

4.3 Types of densities

Several types of densities have been identified in the literature and are regularly used in reporting on measurements (see Table 4.1). The various types recognize that all terrestrial materials consist of three phases to a greater or lesser degree: the solid mineral grains, the void spaces, and the void filling material, either liquids or gases or both. In the saturated zone of the Earth beneath the water table, all void space is assumed to be filled with liquids, but locally gases may be entrapped or occur in their movement from generation within the Earth to the surface. In the vadose zone above the water table, the assumption is made that void space is filled with air, but at least partial water saturation occurs owing to the delay and entrapment of downward-infiltrating surface waters or upward movement of water by capillary action. Void spaces within rocks, regardless of their origin, tend to close up under increasing lithostatic pressure. As a result, at depths of several kilometers, the void space generally is assumed to be negligible.

4.3.1 True density

The mass of a unit volume of solid material where the volume excludes the voids in the rock is the true density, σ_t. It is equivalent to the grain density, σ_g, of a rock which is the mass of a unit volume of grains and is calculated from

$$\sigma_t = \sigma_g = \sum \sigma_i v_i, \tag{4.1}$$

where σ_i and v_i are respectively the grain density and volume percentage of each ith mineral component within the rock. True specific gravity is the ratio of the mass of a unit volume of material to the mass of the same volume of gas-free distilled water at a temperature of 4 °C, in which the volume is the impermeable part of the solid material. Specific gravity is numerically equivalent to density, but this term is seldom used in geophysical exploration.

4.3.2 Bulk density

The bulk density, σ_B, which is sometimes referred to as dry bulk density, is the density of a thoroughly dry rock including both the solid material and the void space: thus

$$\sigma_B = \sigma_t - \frac{v_p \sigma_t}{100}, \tag{4.2}$$

where v_p is the percentage of the volume that is total void or pore space and the density of the gas filling the void space is negligible. The total pore space is the sum of the effective porosity which permits significant fluid flow, plus diffusion porosity, which only allows fluid movement by diffusion because of the limited size of the interconnection between the pores. In many rocks, such as unfractured igneous and metamorphic rocks and clay-rich sediments and sedimentary rocks, the void space is primarily diffusion porosity, and thus the total porosity is markedly different than the effective porosity. The openings in rocks occur at a range of scales with cross-section dimensions measured in micrometers to tens of meters and lengths up to kilometers in solution cavities and shear zones (e.g. MANGER, 1963; EMERSON, 1990).

4.3.3 Natural density

The natural density, σ_n, which is sometimes also called the saturated or wet bulk density, is the density of a rock with all the pore space, both flow and diffusion, filled with water. This is assumed to be the normal subsurface situation, but is not the case within the unsaturated zone at the Earth's surface and in pore volumes where gases are entrapped. The equation for calculating natural density is

$$\sigma_n = \sigma_B + \frac{v_f \sigma_f}{100} = \sigma_t - \frac{v_p \sigma_t}{100} + \frac{v_f \sigma_f}{100}, \tag{4.3}$$

where v_f is the percentage of the volume that is void space filled with fluid of density σ_f. In Earth materials that have all the void space filled with fluid, $v_p = 0$. The fluid density of chilled, gas-free distilled water is 1,000 kg/m^3, whereas sea water has a density of 1,030 kg/m^3, subsurface brines have densities of the order of 1,100 kg/m^3, and gases presumably have negligible densities.

The density unit traditionally has been given in CGSu, i.e. in g/cm^3, where density and specific gravity are numerically equivalent. The scientific community, however, is increasingly reporting densities in SIu, i.e. in kg/m^3, which is equivalent to 10^3 g/cm^3. To avoid the large numbers of the SIu, densities are sometimes reported in Mg/m^3, which is numerically equivalent to g/cm^3.

4.4 Density of the Earth's interior

One of the first clues to the nature of the interior of the Earth was obtained from determinations of its density. The density of surface rocks is of the order of 2,500 to 3,000 kg/m^3, but the calculated density of the entire Earth is approximately 5,520 kg/m^3, which provides unquestioned evidence that the Earth is not homogeneous, but increases markedly in density with depth. Calculation of the Earth's density was first made possible by measurement of the surface gravitational acceleration, and the gravitational constant which was measured rather precisely by Henry Cavendish in 1798. When these quantities are substituted into Newton's universal law of gravitation, the mass of the Earth is determined to be 5.97×10^{24} kg. By considering the average radius of the Earth, the density can be calculated. The mass can also be calculated independently by measuring the period of rotation of a satellite around the Earth.

But how does the density vary with depth? The distribution of mass within the Earth controls its dynamic behavior and its moment of inertia. The moment of inertia is a measure of the resistance to change in the angular acceleration of a rotating body upon the application of a torque. It is dependent on the shape and size of a body as well as on its mass and how the mass is distributed. The moment of inertia of the Earth is 8.07 kg m^2, which is only 83% of the value it would have if it were homogeneous. This confirms that the mass is concentrated toward the center, and thus the Earth's density increases with depth.

Detailed estimates of the change in density with depth have been established from compressional and shear seismic wave velocity variations with depth, obtained by inversions of observed seismic travel times, using a self-compression model of the Earth based on the Adams–Williamson equation. The Earth's overall density and moment of inertia are also used to constrain the determinations. The Adams–Williamson method is applicable to a region of uniform composition which is in a state of hydrostatic stress.

The change in density with depth in a homogeneous body may be determined from the change in compressional and shear wave velocities, V_p and V_s, respectively, where

$$V_p = \left[\left(k + \frac{4}{3}\mu\right) \Big/ \sigma\right]^{\frac{1}{2}}, \tag{4.4}$$

and

$$V_s = [\mu/\sigma]^{\frac{1}{2}}, \tag{4.5}$$

and k is the bulk modulus, μ is the shear modulus, and σ is the density. Assuming hydrostatic conditions then gives

$$\partial p/\partial r = -\sigma \times g, \tag{4.6}$$

where ∂p is the increase in pressure p due to change in the radius ∂r, and σ and g are the density and gravitational attraction at radius r. From Equations 4.4, 4.5, and 4.6, the change in density with depth is

$$\partial \sigma/\partial r = -\sigma \times (g/\phi), \tag{4.7}$$

where the seismic parameter ϕ is

$$\phi = k/\sigma = V_p^2 - \frac{4}{3}V_s^2. \tag{4.8}$$

The calculation of the change in density as a function of depth with the Adams–Williamson Equation 4.7 requires several simplifying assumptions that lead to uncertainties in the results. The equation only provides the change in density with depth, so the density of the top of each homogeneous unit must be estimated. Making the assumption that the mantle and core are independent homogeneous units, the density of the top of each of these shells is estimated from laboratory studies of velocities of rocks that are believed to make up these units. In addition, the Adams–Williamson equation does not consider the effects of increasing temperature with depth, which will cause the density to decrease with depth as a result of volumetric thermal expansion and contraction effects of self-compression. The temperature effect is likely to be minor except where there are strong thermal gradients, as might exist in the lower mantle from the heat derived from the core. Other limitations include the existence of compositional and phase changes within the mantle and core, especially in zones marked by steep velocity gradients.

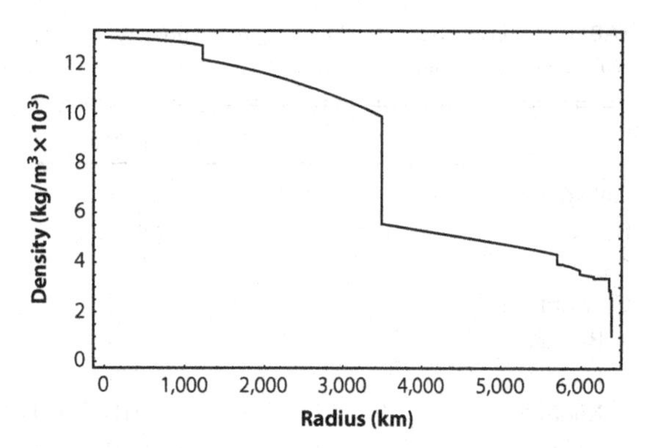

FIGURE 4.1 Preliminary Reference Earth Model (PREM) of the density of the Earth. Adapted from DZIEWONSKI and ANDERSON (1981).

Several other approaches have been used for estimating density with depth within the Earth. Prominent among these is the use of the period and distribution of standing waves or free oscillations which depend upon the variation of elastic constants and density with depth. The Earth, being elastic, will oscillate as a result of a large earthquake. These oscillations may continue for hours or even days after the occurrence of a large earthquake. Seismic surface waves of very long wavelength are reinforced at specific periods to produce standing waves which sample the elastic constants and density deep within the Earth. Inversion determines an Earth density model that best fits the periods of the observed free oscillations. Unfortunately, this modeling is not unique and must be combined with seismic velocity, moment of inertia, and mass of the Earth constraints for effective determination of the density variation with depth. PRESS (1970) used a Monte Carlo inversion method to limit the range of uncertainty in the models. His results suggest significantly greater uncertainty in the density of the inner core than in the mantle and outer core.

A model for the variation in a number of parameters as a function of depth within the Earth, including density, is the Preliminary Reference Earth Model (PREM) (DZIEWONSKI and ANDERSON, 1981). The density profile of this model is presented in Figure 4.1. The properties of the interior of the Earth remain an active research area, so refinements are being made in the density profile, but no significant variations from the PREM density profile are anticipated except in the high-gradient regions of the lowermost mantle and the inner core.

TABLE 4.2 Examples of densities in kg/m^3 of primary sedimentary rock minerals.

Name	Density
Anhydrite	2,960
Barite	4,480
Calcite	2,710
Clay minerals	
Bentonite	3,600
Dichite	2,620
Kaolinite	2,610–2,680
Glauconite	2,300
Montmorillonite	2,610
Vermiculite	2,300
Coal	
Anthracite	1,300–1,500
Bituminous	1,100–1,300
Dolomite	2,870
Halite	2,160
Gypsum	2,310
Sylvite	1,990

Adapted primarily from JOHNSON and OLHOEFT (1984).

TABLE 4.3 Examples of densities in kg/m^3 of igneous and metamorphic rock minerals.

Name	Density
Augite	3,300
Biotite	3,360
Ca Al pyroxene	3,360
Chlorite	2,800
Diamond	3,520
Feldspars	
Albite to anorthosite	2,620–2,760
Microcline	2,560
Orthoclase	2,570
Sanadine	2,560
High-grade metamorphic minerals	
Sillimanite	3,250
Kyanite	to
Garnet, etc.	4,300
Hornblende	3,080
Muscovite	2,560
Olivine	
Fosterite to fayalite	3,210–4,390
Quartz	2,650
Serpentine	2,600
Talc	2,780

Adapted primarily from JOHNSON and OLHOEFT (1984).

4.5 Rock densities

The primary controls on the density of subsurface materials are mineral composition and void space, which are largely dependent on the lithology (rock type) and the chemical and physical effects of secondary processes including rock fracturing, solutioning, and chemical alteration of minerals. The density of minerals varies from 1,990 kg/m^3 for sylvite, the potassium salt, to about 20,000 kg/m^3 for gold, but the vast majority of commonly occurring minerals range from 2,500 to 3,500 kg/m^3, although ore minerals of metals are in the 4,000–6,000 kg/m^3 range. Examples of mineral densities are given in Tables 4.2, 4.3, and 4.4, where several generalities can be noted. For example, the density of minerals tends to decrease with an increase in SiO_2 content, and rock densities decrease with an increase in H_2O content. Furthermore, minerals formed under high-pressure conditions in metamorphic rocks tend to have higher densities, and non-metalliferous resources (e.g. coal, salts, clays) have lower than average densities, while metalliferous ores have densities exceeding those of common rock-forming minerals. In addition, densities are affected by lithostatic pressure and temperature, both of which are primarily functions of depth within the Earth.

4.5.1 Lithology

For purposes of considering density, Earth materials are conveniently classified into crystalline rocks, sedimentary rocks, and unconsolidated sediments. Crystalline rocks include igneous rocks, both plutonic and volcanic, that originate from magma that has solidified, respectively, within the Earth and at the surface. In addition, they include metamorphic rocks, derived from both igneous and sedimentary rocks that have been altered deep in the crust by increased lithostatic pressure, tectonic stress, and enhanced temperatures. Unconsolidated sediments are made up of the fragments derived from erosion of pre-existing rocks that are commonly deposited in water or less commonly in air and by chemical precipitants. Sedimentary rocks consist of sediments that have been lithified by lithostatic pressure and chemical precipitants. The component minerals and the origin and nature of the void space in these three classes generally occur within specified ranges, facilitating consideration of their densities.

TABLE 4.4 Examples of densities in kg/m^3 of ore minerals and metals, and terrestrial waters.

Name	Density
Ore minerals and metals	
Barite	4,480
Cinnabar	8,187
Chalcopyrite	4,200
Chalcocite	5,793
Copper	8,934
Corundum	3,987
Galena	7,600
Gold	19,282
Halite	2,163
Hematite	5,275
Iron	7,875
Kaolinite	2,594
Lead	11,343
Limonite	4,880
Magnetite	5,200
Malachite	4,031
Pyrite	5,010 (4,950–5,030)
Pyrrhotite	4,610
Sphalerite	4,089
Uraninite	10,970
Water	
Fresh (at 4 °C)	1,000
Ice	890–910
Brine	1,125
Sea water	1,030

Adapted primarily from JOHNSON and OLHOEFT (1984).

Crystalline rocks

Unaltered plutonic igneous rocks that are formed deep within the crust or upper mantle characteristically have minimal void space, generally less than 1%, and values seldom exceed 3%. A minor portion of this volume is originally due to intergrain voids, but the vast majority of voids are caused by chemical and physical weathering within the upper few hundred meters of the surface and by fracturing and faulting within the brittle crust plus cooling cracks. Generally, voids occurring within fractures such as rock joints and in faults largely, but not completely, close up under the effect of lithostatic pressure. As a result, both densities and associated seismic velocities approach constant values with increasing pressure at values of the order of 600 MPa (1 Pa = 10^{-8} kilobars [kb]) which is reached at depths within the Earth of the order of 15–20 km. The decrease in density in fault zones is not only a result of the increase in volume during fracturing, but also the result of physiochemical alteration of the rocks to lower-density clay minerals.

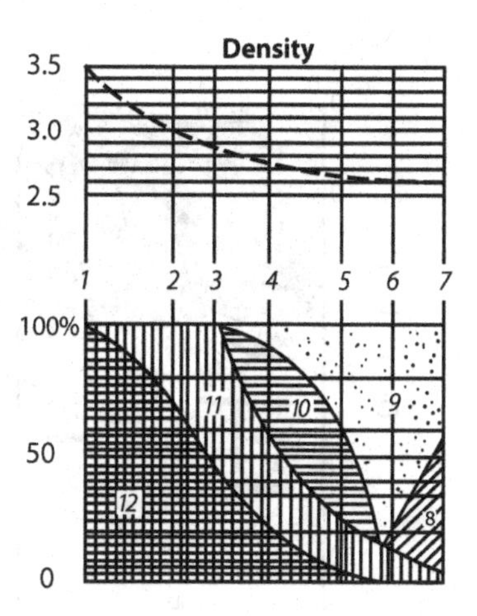

FIGURE 4.2 Schematic illustration showing the dependence of the density of plutonic igneous rocks upon their mineralogical composition. The lower panel shows the proportional content of the major minerals of igneous rocks, whereas the upper panel gives the range of densities, in kg/m^3 × 10^3, of the corresponding rock types. 1 – peridotite and pyroxenite; 2 – gabbro; 3 – diorite; 4 – granodiorite; 5 – granite; 6 – syenite; 7 – nepheline syenite; 8 – nepheline (σ = 2,600 kg/m^3); 9 – potassium feldspar (σ = 2,500 kg/m^3); 10 – quartz (σ = 2,600 kg/m^3); 11 – plagioclase (σ = 2,600 to 4,800 kg/m^3); 12 – iron-magnesium minerals (σ = 3,100 to 3,500 kg/m^3). Adapted from PICK *et al.* (1973).

Given the minor void space in plutonic rocks, the density of these rocks is primarily the result of the densities and constituent proportions of the minerals present. The lowest-density minerals (see Table 4.3) commonly present in plutonic rocks are quartz and orthoclase, which have densities around 2,600 kg/m^3. Rocks with relatively high proportions of these minerals are called felsic (or acidic) and have the lowest densities among plutonic igneous rocks. In contrast, rocks containing significant proportions of plagioclase feldspars, biotite, and hornblende, and thus rich in calcium, magnesium, and iron, have greater densities. These are informally termed mafic (or basic) rocks. The variation in the density of plutonic rocks with varying mineralogical composition is generalized in Figure 4.2.

The crystalline rocks of the continental crust in general become more mafic with depth, and thus densities increase with depth over and above the effect of increasing lithostatic pressure. However, the continental crust is not layered in the simple view of an upper granitic layer underlain by a basaltic layer. Rather, all evidence points to a crust, as illustrated in Figure 4.3, that varies significantly

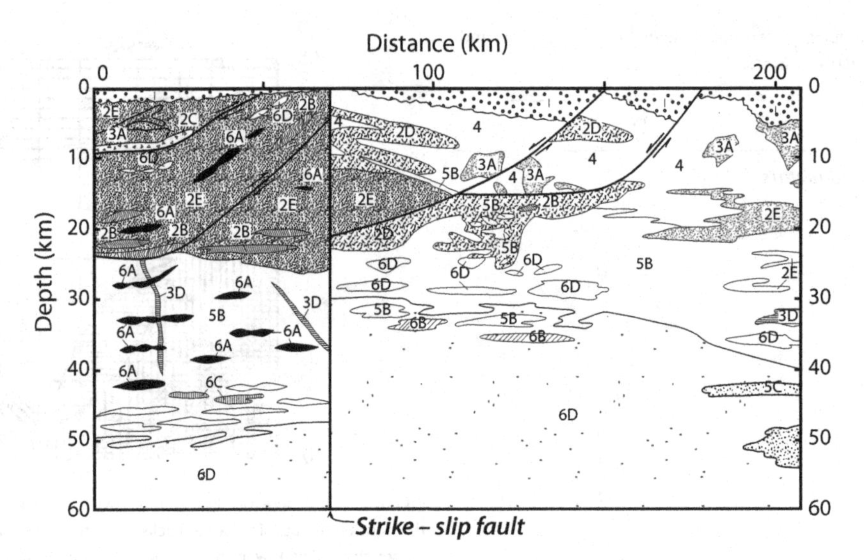

FIGURE 4.3 Hypothetical cross-section of the crystalline crust and upper mantle. (1) Sedimentary rocks; (2) metamorphosed supracrustal rocks; (3) plutonic rocks; (4) gneiss and migmatite; (5) mafic rocks; (6) ultramafic rocks. The Moho is the line separating mafic gneiss (5B) from the peridotite (6D) of the upper mantle. Adapted from FOUNTAIN and CHRISTENSEN (1989).

in both composition and density over short lateral and vertical dimensions (FOUNTAIN and CHRISTENSEN, 1989). In a general sense, increasing metamorphic grade is evident with increasing depth, supporting the view of increasing density with depth. The densities of surficial rocks vary greatly because of the presence of materials ranging from unconsolidated sediments to mafic crystalline rocks. However, the overall average of 2,670 kg/m^3 reflects their general granitic (felsic) nature (HINZE, 2003).

The increase in density with depth to values around 3,100 kg/m^3 in the lower continental crust results in an average crustal density of roughly 2,830 kg/m^3 (CHRISTENSEN and MOONEY, 1995). The lower crust has an average chemical composition equivalent to gabbro, but garnet becomes more abundant with depth. At the base of the crust, mafic garnet granulite is most abundant. Granulite is a relatively coarse-grained, high-grade metamorphic rock that is formed at the high temperatures of the lower crust. Despite the widespread lateral density variations in the crust, it is useful to consider an average density profile through the continental crust, illustrated in Figure 4.4 as a starting point for investigating heterogeneities. However, it is important to remember that this is an average profile when developing density profiles of specific regions of the crust. For example, velocity studies of the Archean Pilbara Craton of northwest Australia show two layers with a transition zone of only a few vertical kilometers between them that defines a density contrast of the order of 50 kg/m^3 at depths of about 10 to 15 km (DRUMMOND, 1982).

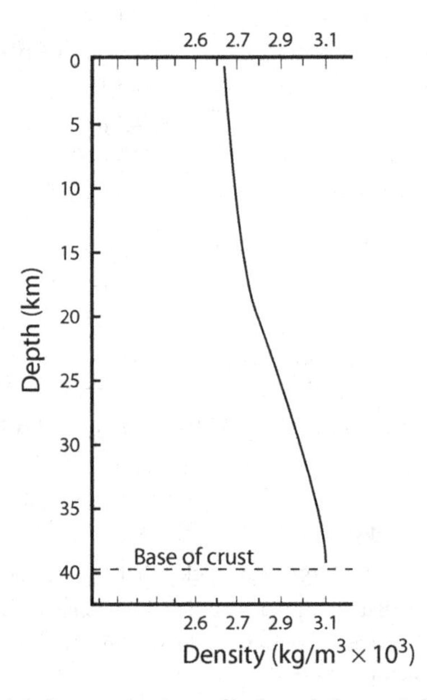

FIGURE 4.4 Average density profile through the crust. Adapted from CHRISTENSEN and MOONEY (1995).

The oceanic crust is apparently more laterally homogeneous in both composition and density than the continental crust and consists primarily of mafic extrusive and intrusive rocks beneath a layer of sediments and sedimentary rocks. However, there are lateral differences associated with varying age of the crust, alteration (or metamorphic grade), tectonic province, and heat flow (e.g. CHRISTENSEN and SALISBURY, 1975; CARLSON and HERRICK, 1990). Carlson and Herrick's analysis of laboratory and downhole logging of oceanic crystalline materials and seismic investigations of the oceanic crust shows that the upper crystalline layer ("layer 2"), which consists largely of lavas extruded on the sea floor, has a density ranging from 2,620 to 2,690 kg/m^3, and the lowermost layer of the oceanic crust ("layer 3"), consisting of dikes and mafic magma chambers, has an estimated average density of 2,920 to 2,970 kg/m^3. They estimate the average density of the entire crystalline oceanic crust to be 2,860 $\pm$30 kg/m^3, very similar to Christensen and Mooney's estimate of the density of the continental crust. Thus, gravity anomalies at the continent/ocean interface are primarily due to changes in the thickness of the crust and sedimentary rock accumulations.

The ultramafic upper mantle has a density of approximately 3,300 (3,270–3,320) kg/m^3, but varies with rock composition, temperature, pressure effects, and the degree of partial melting. Peridotite is the principal lithology of the upper mantle based on geophysical studies and direct observation of mantle rocks exposed on the Earth's surface. These rocks are rich in olivine, but another important lithology of continental upper-mantle is eclogite containing garnet. These latter rocks, which are proposed to originate from tectonic processes involving remnants of subducted crust into the mantle, may have densities of 200 kg/m^3 greater than peridotite. Upper-mantle continental rocks generally have undergone periods of multiple melting leading to depletion in Ca, Fe, Al, and Ti. These depleted rocks are rich in olivine and Mg content and have a lower density. Clearly, the upper mantle, particularly in the continents, is subject to variations leading to lateral changes in density. For example, KABAN and MOONEY (2001) describe a $\pm$3% variation ($\sim$100 kg/m^3) in the density of the upper mantle of the southwestern USA, which they ascribe largely to compositional differences rather than temperature, on the basis of seismic wave velocities. In analyzing the gravity anomalies associated with the Kenya rift in East Africa, RAVAT *et al.* (1999) find on the basis of seismic velocities and gravity modeling that the mantle plume beneath the rift at a depth of about 100 km has a density contrast of approximately -30 kg/m^3 with the surrounding mantle.

In addition, the density of the upper mantle incorporated into subducted slabs increases as the slab undergoes metamorphic reactions with increasing depth of subduction (e.g. GROW and BOWIN, 1975; TESSARA *et al.*, 2006). A thermal–petrologic–seismological model of subduction zones (HACKER *et al.*, 2003, 2004) is available which includes provision for calculating the density of subduction zone rocks at high pressures and temperature. The asthenosphere, which is recognized as the low-velocity zone at the base of the lithosphere in the upper mantle, is believed to have a small (<50 kg/m^3) negative density contrast with the surrounding mantle, and ŚWIECZAK *et al.* (2009) suggest the density of the asthenosphere beneath Europe is in the range 3,100–3,340 kg/m^3.

The density contrast across the Moho (crust/mantle interface) has been assigned a range of values from of the order of 200 to 450 kg/m^3. Based on a density of the lowermost crust of 3,100 (Figure 4.4) and a mean density of the upper mantle of 3,300 kg/m^3, the density contrast at the Moho is 200 kg/m^3. But where the crust/mantle interface has a relief that reaches several kilometers or more, the density contrast is greater. In fact, SIMPSON *et al.* (1986) find that a density difference across the Moho in the conterminous USA of 350 kg/m^3 provides Moho depths consistent with seismic refraction measurements. However, CHAPIN (1996) arrives at a value of 450 kg/m^3 for the crust/mantle density contrast across the South American continent, based on analysis of observed continental gravity values, and KABAN and MOONEY (2001), on the basis of seismic velocity data, find a density contrast of 420 kg/m^3 across this interface in the southwestern USA. Clearly the crust/mantle interface density contrast is a non-linear function with relief as a result of the changing density with depth and is likely to vary depending on the geologic terrane.

Igneous rocks also form close to or on the Earth's surface producing volcanic or extrusive igneous rocks. The rapid crystallization of igneous melts in this environment leads to a fine-grained texture in contrast to the coarse texture characteristic of plutonic rocks. The fine-grained texture tends to lower the density despite the rock having a similar composition, but the effect generally is less than 10%. The occurrence of non-crystalline glasses in these rocks such as obsidian and vitrophyre, for example, leads to even lower densities. Commonly the density of volcanic rocks is lowered by the presence of voids resulting from gas cavities frozen into the rock during its rapid crystallization. This situation is particularly prevalent in the upper portion of volcanic flows where gas rising through the flow accumulates. In extreme situations, the densities of these rocks, such as pumice and scoria, will be lowered to less than that of water by the cavities filled with entrapped gas.

Volcanic ash deposits typically are highly porous, and thus are significantly lower in density than their plutonic or lava flow equivalents, but pore space is readily decreased when ash is welded during the depositional process by its internal heat or compacted by burial. For example, felsic

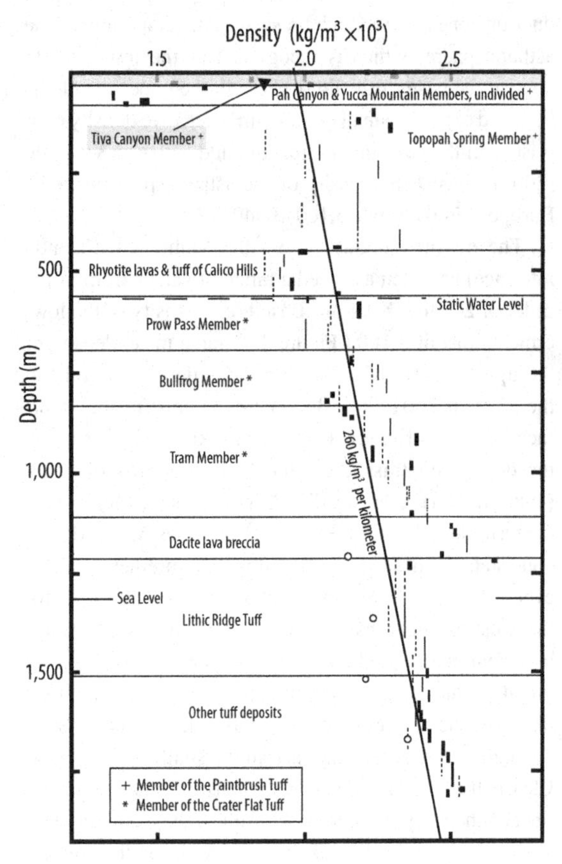

FIGURE 4.5 Density versus depth plot for Tertiary tuff at Yucca Mountain, Nevada. Solid rectangles indicate drillhole gravity measurements, depth range, and error range; dashed lines are gamma–gamma density measurements (see Section 4.6.3); circles are density measurements on unsaturated core from saturated zone at Yucca Mountain. Solid line is least-squares fit to the data. Adapted from SNYDER and CARR (1984).

Tertiary tuffs of Southern Nevada vary in density from about 1,700 to 2,500 kg/m^3, with respective porosities from a few percent to 50%. A density profile from a drill hole into these rocks, obtained from a variety of measurement methods, shows a generally linear increase in density as illustrated in Figure 4.5. In this case the increase in density is linear, but in other situations the density increases exponentially with the largest increase near the surface. BLAKELY *et al.* (1999) report that in the volcanic basins of the Basin and Range Province, volcanic rocks increase in density by 200 kg/m^3 from the surface to 1.2 km. The density of volcanic flows may be lowered by fracturing associated with movement of the crystallizing melt and by the effect of chemical alteration of the flows leading to less dense minerals.

The density of metamorphic rocks that make up the majority of the continental crust depends primarily on the original mineral composition of the rocks, but is also strongly influenced by the degree and type of metamorphism which largely reflects the temperature and pressure to which the rocks have been subjected. In general, densities increase with chemical compositions of rocks higher in iron, magnesium, and calcium. Numerous studies show the broad range of densities of exposed metamorphic rocks of Precambrian shields (e.g. WOOLLARD, 1962; GIBB, 1968; SMITHSON, 1971; SUBRAHMANYAM and VERMA, 1981; KORHONEN *et al.*, 1993). Typically, mean densities of metamorphic rocks occur in the range 2,700 to 2,800 kg/m^3, but many metamorphic rocks derived from felsic igneous rocks and metasedimentary rocks, such as granite gneiss, have densities in the range of 2,600 to 2,700 kg/m^3. In contrast, intermediate to mafic metavolcanic rocks have significantly higher densities. GUPTA and GRANT (1985) report a mean density for these rocks of the order of 2,850 kg/m^3 in the Sudbury–Cobalt area of Canada, and GIBB (1968) has found similar values for these rocks in the Precambrian shield of Manitoba, Canada. The highest-density metamorphic rocks are those that have been metamorphosed to the eclogite grade in the lower crustal environment where garnet has replaced plagioclase. These rocks typically have densities in the range 3,000 to 3,300 kg/m^3.

Sedimentary rocks and unconsolidated sediments

The densities of unconsolidated sediments and their lithified equivalents, sedimentary rocks, are primarily controlled by their void space. Their mineral components are rather limited, and those that are present do not vary greatly in density. The most common constituent is quartz derived from the chemical and physical breakdown of igneous rocks and their metamorphic equivalents. Clay minerals originating by chemical modification of feldspars and other silicate minerals which dominate in clays and shales are another important component. Sediments and sedimentary rocks formed by chemical precipitation are largely monomineralic, and thus are of constant density except where post-depositional processes primarily cause dissolution that produces secondary porosity. Limestone (calcite) and dolomite are the principal chemical sediments. However, salt deposits generally consisting of halite are important locally in producing density contrasts within sedimentary basins. All of these mineral constituents, quartz, clays, and calcite, have densities in the range 2,600 to 2,700 kg/m^3, whereas dolomite has a density 2,870 kg/m^3 and salt deposits have densities in the

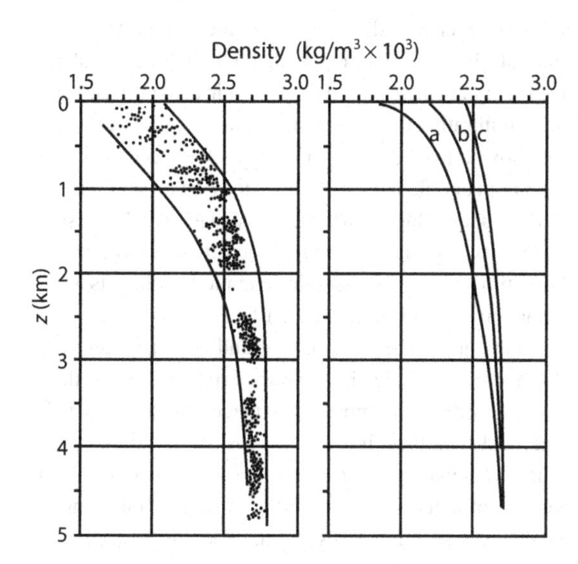

FIGURE 4.6 Density of sedimentary rocks vs. depth (z) for the North German–Polish basin. The left panel includes sandstones and siltstones; the right panel gives the mean values for the (a) Quaternary, Cretaceous, Jurassic, (b) Bunter sandstone (Lower Triassic), and (c) Permian stratigraphic units. Adapted from SCHÖN (1996).

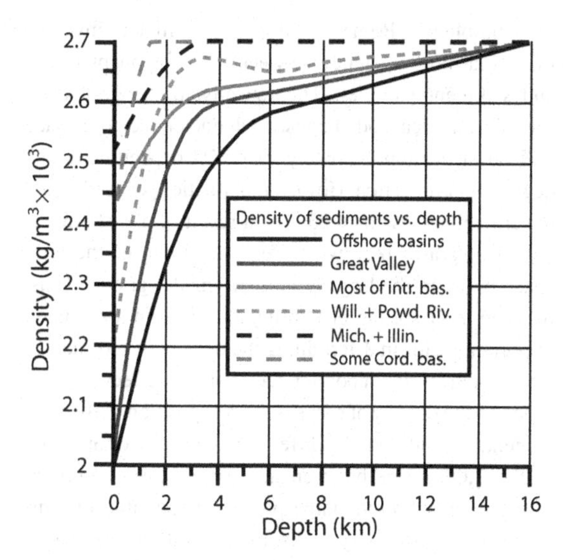

FIGURE 4.7 Smoothed density of sedimentary rocks vs. depth for several sedimentary basins on the North American continent including some Cordilleran (Cord.) basins; Michigan and Illinois basins (Mich. + Illin.); Williston and Powder River basins (Will. + Powd. Riv.); most intracratonic basins; Great Valley of California; and offshore basins. Adapted from MOONEY and KABAN (2010).

range 2,000 to 2,200 kg/m^3. Methane hydrate, which may occur in deep oceanic sediments, has a density of the order of 900 kg/m^3.

Densities of sedimentary rocks increase with depth rapidly at shallow depths and less rapidly at progressively greater depths owing to lithostatic pressure as well as lithification. ATHY (1930) and others have shown that this variation can be approximated by an exponential function with a negative exponent. Densities also generally increase with the age of the sedimentary rocks. Older rocks tend to occur at greater depth or have at one time been at greater depth, and thus have undergone greater compaction and lithification which leads to increased densities. Examples of increasing density with depth and age are shown in Figures 4.6 and 4.7. Note that in the right panel of Figure 4.6, the older rocks of (b) and (c) do not increase in density as rapidly at shallower depths as younger rocks, probably as a result of the lower compaction possible in the older rocks. Figure 4.7 was assembled by MOONEY and KABAN (2010) from drillhole logs in the upper 2 to 3 km and extended to greater depths with seismic constraints. In this figure, the density variation with depth in the Michigan and Illinois basins, which contain significant quantities of carbonate rocks, is much less than in the offshore basins which consist largely of poorly lithified relatively young sediments subject to compaction with depth.

4.5.2 Lithostatic pressure and void space

Lithostatic pressure derived from the weight of overlying rocks has a profound effect on the density of subsurface materials, particularly sediments and sedimentary rocks. This is primarily due to the decrease in void space with increasing pressure. As explained above, both primary and secondary void space of crystalline rocks are minimized with increasing pressure leading to higher densities, with the process being especially effective in igneous ash flows and lava flows and near the surface.

Numerous relationships have been developed to describe the changing density with depth and lithostatic pressure. One form of this is

$$\sigma_{P2} = \sigma_{P1}\{1 + [(P2 - P1)/K]\}, \tag{4.9}$$

where σ_{P1} and σ_{P2} are the respective densities of rocks in common units at pressures $P1$ and $P2$ in pascals (Pa) and K is the bulk modulus of the rock in Pa. The assumed constant K for the upper crust is roughly 52 GPa, whereas the assumed values for the lower crust and upper mantle are approximately 75 and 130 GPa, respectively (DZIEWONSKI and ANDERSON, 1981).

Void space in sediments and sedimentary rocks has a variety of origins and tends to be highly variable, accounting for the observed wide range of densities (e.g. HALL and HAJNAL, 1962; EATON and WATKINS, 1970) of these lithologies. Primary void space in sediments involves interstitial pore openings between the component grains and sediment particles. Theoretically, if the grains are roughly spherical and of constant diameter, the pore space will vary from approximately 50 to 25% depending on the degree of compaction. However, porosities generally tend toward values less than 25% as a result of a wide distribution of grain sizes, which leads to filling of the pore space between the larger grains by smaller grains. Thus, the degree of sorting or distribution of grain sizes is an important factor in determining the density of sediments.

Of course, an important factor in the effect of void space on the density of rocks is the composition of the pore content. Considering the difference in density of approximately 1,000 kg/m^3 between atmospheric gases and pore-filling liquids, and the large pore space possible in sediments and sedimentary rocks, the pore filling may cause significant density contrasts within the Earth. For example, a rock with 30% void volume has a difference in density of 300 kg/m^3 depending on whether the pores are filled with air or water. Pore space within the vadose zone above the water table are generally filled with air, but the water content may become important where there are perched water zones and an extensive capillary fringe. In contrast, the assumption of complete water saturation below the water table is incorrect where gas is entrapped within the sediments.

The porosity of sedimentary rocks is significantly less than sediments in most geological situations as a result of the lithification process, specifically the compaction of granular materials and cementation of the grains decreasing the interstitial pore space. This is particularly evident in the conversion of surface clays with their platy texture to shale rocks: clays at the surface which have densities of the order of 1,500 kg/m^3 are transformed into shales that exponentially reach the approximate density of clay minerals (~2,600–2,700 kg/m^3) at depths of the order of several kilometers.

Compaction has less effect on sandstones and chemical sedimentary rocks than on clay, with relatively consistent values reached at shallower depths than in the case of shale. The exponential change of density with depth of alluvial sediments in the basins of the Basin and Range Province of the southwestern United States as a result of compaction is reported by BLAKELY *et al.* (1999). They found densities of roughly 2,050 kg/m^3 between the surface and 200 m, 2,150 kg/m^3 between 200 and 600 m, 2,350 kg/m^3 between 600 and 1,200 m, and 2,450 kg/m^3 at greater depths.

Secondary void space also exists in sedimentary rocks as a result of fracturing (faulting and jointing), bedding planes, and dissolution. Dissolution is especially important in the chemical sedimentary rocks limestone and dolomite, because of the relatively high solubility of calcite and dolomite when exposed to carbonic acid derived from water infiltrating these Earth materials and picking up carbon dioxide from the atmosphere and from decomposing organic material. This void space may be several tens of percent, or even higher in some areas, markedly decreasing the density, especially in the high-density dolomitic rocks. The ultimate effect of dissolving chemical sediments is the formation of caves in the near-surface.

Faults, joints, bedding planes, and cooling cracks are also a source of voids in Earth materials. However, their effect is normally at most a few percent of the rock volume, and thus their impact on densities is minimal. Of course, this may not be true where shear zones are broad because of extensive reactivation, or where fracturing is locally extensive. Fractures within rocks can be accompanied by chemical alteration, either from infiltrating ground water or from hydrothermal solutions driven by buoyancy from deeper geologic units. This alteration usually leads to formation of lower-density minerals, but under appropriate chemical conditions, fluid movement may result in mineral deposition within the voids and increased density of the rocks.

4.5.3 Temperature

Temperature has only a minor role in controlling the density of Earth materials, except in abnormal situations, because of the low volume thermal coefficient of expansion of rocks. Most Earth materials have volume thermal coefficients of expansion of roughly 20 to 40 $\times 10^{-6}/{}^\circ$C; thus a temperature differential of approximately +100 °C is required to produce a density decrease of 100 kg/m^3. RAVAT *et al.* (1999) have modified the lithostatic pressure Equation 4.9 to incorporate the effects of volume change with temperature as

$$\sigma_{P2} = \sigma_{P1}\{1 + [(P2 - P1)/K] - \alpha\Delta T\}, \tag{4.10}$$

where α is the coefficient of thermal expansion (set at $25 \times 10^{-6}/{}^\circ$C in their study) and ΔT is the change in temperature in °C.

Density decrease is also associated with partial rock melts where a magma chamber or a geologic unit consists of a mix of rock crystals and melted rock. For example, the density of basaltic magma at the Earth's surface is approximately 10% less than solidified basalt. As a result, basalt containing 20% melt by volume will be less dense by approximately 50 kg/m^3.

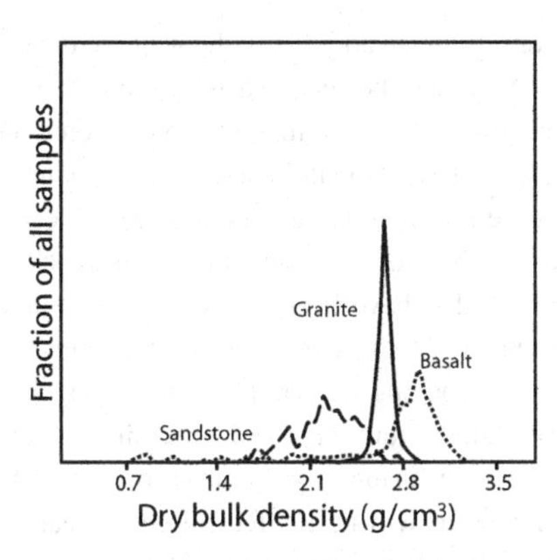

FIGURE 4.8 Comparison of frequency distributions of dry bulk densities of granite, basalt, and sandstone samples. Adapted from JOHNSON and OLHOEFT (1984).

4.5.4 Summary

Several of the factors controlling the density of rocks are illustrated by the histogram of bulk density measurements of granite, basalt, and sandstone compiled by JOHNSON and OLHOEFT (1984) that is reproduced in Figure 4.8. The distribution of the densities of granite is largely restricted to a tight normal distribution around an average value of roughly 2,650 kg/m^3. Variations around this central value probably reflect slight differences in the relative proportion of quartz and feldspars and, primarily, the minor proportion of mafic minerals. A few lower densities extending from the central peak are probably associated with alteration of the feldspars and mafic minerals to less dense minerals, and perhaps with intergrain voids. In contrast, the histogram for basalt peaks at roughly 2,900 kg/m^3 with a broader range of values reflecting variations in the mineral content, largely the types of feldspars and mafic minerals, and even the presence of limited quantities of metallic oxides or sulfides, and, particularly at lower densities, the existence of voids, probably largely gas cavities.

The third histogram in this figure is for sandstone which shows a broad, irregular distribution between 1,800 and 2,600 kg/m^3. The broad distribution primarily reflects the variation in the interstitial pore volume, although minor effects may be due to mineral content. Clearly, the tighter the distribution of the measurements, which is indicative of greater homogeneity, the more likely the density of the rock can be estimated correctly from tables or from fewer measurements. Based on the above analysis, the natural density (saturated bulk density) of granite likely will not change significantly from the bulk densities shown in Figure 4.8, but because of water filling of the voids, all the values of the sandstone will increase, as will the lower values of the basalt extending from the major grouping of measurements.

4.6 Density measurements

In many gravity investigations, rock and sediment densities for planning, reduction, and interpretation of surveys are estimated from tables of densities or extrapolated from tables using knowledge of the geology of the area and the effects of various geological processes on densities. This is a necessity in many surveys because of lack of access to samples or instrumentation for measurements or because of limitations in time and resources. However, where information on densities is sparse within a survey region, it is desirable to determine the local formation and rock densities. Furthermore, surveys commonly require an accuracy in densities that cannot be achieved from available tabulations. This is particularly true of surveys that involve near-surface materials including unconsolidated sediments, poorly compacted and highly fractured rocks, and weathered materials that are inadequately represented in most compilations. Thus, measuring rock densities in a survey area is highly desirable if not absolutely necessary.

In principle, determining rock density is simple, but in practice these measurements can be complex and subject to sometimes significant errors. Problems arise particularly from the heterogeneous nature of rocks and difficulties in making accurate measurements on friable samples and those with irregularly distributed, high void space. Densities are measured in three general ways: (1) laboratory measurements on samples; (2) gravity measurements; and (3) measurement of correlative properties. Another method is to calculate densities knowing the mineralogical or chemical composition of generic rock types (e.g. SOBOLEV and BABEYKO, 1994), but this is rarely used in geophysical exploration because the composition of rocks is seldom known accurately enough. Each of these methods has its relative advantages and disadvantages that must be considered in selecting the appropriate method for a particular survey problem.

Laboratory measurements are the simplest and least expensive when representative samples of the rocks of the area can be collected from the site and taken to the laboratory for investigation. Gravity measurements are particularly useful because they provide *in situ* values of relatively large volumes of rocks, but they require specialized instrumentation and surveys either in drill holes or on the Earth's surface. Correlative measurements, especially seismic velocity and gamma-ray attenuation, can provide more volume-restricted *in situ* density determinations of the subsurface provided that these measurements are available by virtue of other studies or can readily be made in drill holes.

4.6.1 Laboratory measurements

With reasonable care, laboratory measurements of most consolidated rocks of low void space closely approximate *in situ* rock densities (McCulloh, 1965). However, difficulties are encountered in selecting representative samples, obtaining unweathered samples from outcrops, making measurements on friable, unconsolidated samples, and measuring the density of rocks with high porosity. As a result, *in situ* measurements of density should be preferred where these difficulties are anticipated. As this is not always possible, care should be exercised to avoid problems associated with these difficulties. Representative samples need to be acquired in proportion to the volume representation of the variable attributes of the rock units. Generally, a minimum of 30 samples is desirable where significant heterogeneity is anticipated in the units. Ideally, unweathered samples should be collected at appropriate sites and stripped of any surficial effects. Unconsolidated formations require specialized procedures involving weighing in air a pre-set sample volume. In rocks which have significant porosity, generally over a few percent, it is advisable to determine the porosity of the rock independently and take the measured value of porosity into account in determining the *in situ* density.

The measurement of density in the laboratory (Emerson, 1990) involves determination of the mass (weight) and volume of the sample. In low-porosity rocks or if the matrix or grain volume or density is desired, the volume can be determined by the loss of weight of the sample in water using Archimedes' principle, which states that a mass in water is buoyed up by a force (weight) equal to the volume of the displaced fluid. The bulk density is calculated from

$$\sigma_B = \frac{W_a \times \sigma_f}{(W_a - W_f)}, \tag{4.11}$$

where W_a and W_f are the weights of the sample in air and fluid, respectively, and σ_f is the density of the fluid. It is advisable to use pure water in making these measurements, and the density of the water should be normalized to a constant temperature.

The total or bulk volume can be determined in porous rocks by sealing the sample from infiltrating water with a thin layer of paraffin or similar material before inserting the sample in water for measurement. In this way the natural density can be calculated. Care must be exercised to avoid air bubbles on samples by using warm water, agitating the samples, or putting an additive in the water to decrease its viscosity. Water can be replaced with other fluids if the rocks contain water-soluble minerals, but account must be taken of the density of the fluid. Typically, samples of 200 to 1,000 g are used, but larger samples are subject to less error and may be more representative of the rock formation. Samples that have been allowed to dry out before measurement need to have water drawn into their voids by vacuum or by soaking the samples for at least 24 hours to achieve water saturation for determining either grain or natural density using appropriate procedures. Simple soaking, however, will be ineffective when dealing with micropores which have low permeabilities.

A wide variety of methods have been developed for measuring density and pore volume using two of either void volume, bulk volume, or grain density, depending on the nature and size of the samples and the accuracy desired. They are discussed and evaluated by Johnson and Olhoeft (1984).

Commonly, density measurements have to be made on previously obtained rock samples and cores that have been allowed to dry out. The saturated bulk density or natural density, σ_n, can be obtained by resaturating the sample and performing volume and mass measurements or by determining the dry bulk density, σ_B, and calculating the natural density from this measurement. The latter requires estimation of the grain density, σ_g, and the density of the fluid occupying the void volume, σ_f. The natural density is calculated from

$$\sigma_n = \sigma_B \left(1 - \frac{\sigma_f}{\sigma_g}\right) + \sigma_f. \tag{4.12}$$

If the dried-out samples have undergone shrinkage, the calculated densities will be decreased by a few percent.

It may be useful to make measurements on cuttings from drill holes or other small, irregularly shaped specimens. The density of these rock fragments can be determined by using a small glass flask of precisely determined volume, or pycnometer, which has a glass stopper extending into a fine capillary tube that provides an overflow for excess water so that the flask is filled with a constant volume of water. The density of the specimen is the quotient of the product of the mass of the specimen in air and the density of the water in the pycnometer divided by the sum of the mass of the specimen and the pycnometer filled with water minus the mass of the pycnometer. Gas pycnometry, in which the volume of specimens is determined by the gas displacement method, is widely used to determine the density of small, irregular specimens. The sample is sealed in a chamber of known volume, then an inert gas such as helium is admitted into the chamber, and the gas expanded into a connecting chamber of precise volume. The measured pressure change upon expansion into the secondary chamber is proportional to the volume of the specimen. This volume together with the mass of the specimen can be used to calculate the bulk density. Detailed protocols for conducting density measurements of small sample fragments and soils are given in publications of the American Society for Testing Materials.

4.6.2 Gravity measurements

Gravity measurements are directly related to the mass of the adjacent materials, and thus can be used with specialized techniques to determine the *in situ* density of subsurface materials. These gravity observations can be made over the surface to determine the density of the material included over the elevation range of the measurements, or in mines and drill holes to measure the density of the rock formations between vertically separated observations. These methods have the advantage of determining density under natural conditions and over a large volume. They are used not only for determining density for use in planning, analyzing, and interpreting gravity data, but increasingly in remotely sensing density, and thus investigating numerous significant characteristics of the nearby Earth.

Surface measurements

The density of surface materials which often are of interest to near-surface studies can be measured by making a series of gravity observations over a topographic feature using the so-called Nettleton density profile method (NETTLETON, 1939). The density of the material included within the topography can be determined by finding the density that will produce the minimum correlation between the Bouguer gravity anomaly and the topography. In determining the Bouguer gravity anomaly at an observation site the gravitational attraction of the mass of the Earth between the datum of the survey and the site is calculated. If the correct density has been used to determine the gravitational attraction, the observed effect and thus the relationship between the anomaly and the topography will be cancelled in the anomaly calculation. As a result the density of the surface material can be estimated from the density used in the mass calculation which leads to the minimum correlation between the Bouguer gravity anomaly and the topographic relief. The method is normally implemented by taking a series of closely spaced observations over an erosional feature that is not correlated with subsurface anomaly sources and primarily consists of a single geologic formation (PARASNIS, 1952). Gravity anomalies

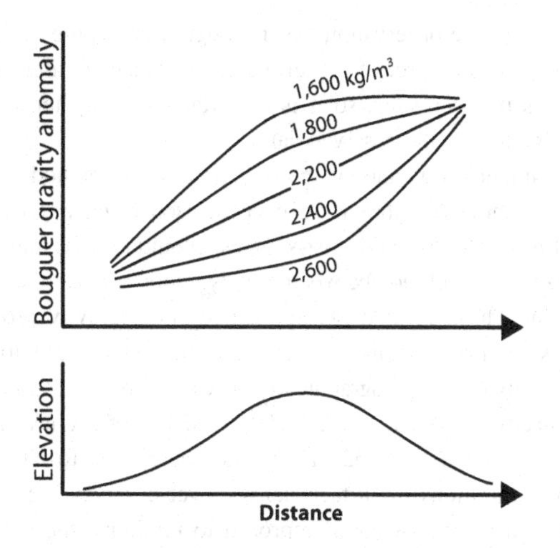

FIGURE 4.9 Illustration of the use of a profile of gravity observations over a topographic feature to determine the included density by minimizing the correlation between the elevation of the stations and the gravity anomaly calculated assuming a range of densities. The density of the material making up the hill in this illustration is 2,200 kg/m^3.

are calculated taking into account spatial and elevation effects using a variety of densities to determine the effect of the mass between the lowest observation level and the elevation of the stations. The density which shows minimum correlation between the calculated gravity anomaly and the topography is the density of the surface material making up the topography.

This method is illustrated in Figure 4.9, where 2,200 kg/m^3 is shown to be the density of the surface material by the lack of correlation between the anomaly, calculated using this density in the mass correction, and the topography. Anomalies calculated with densities higher than 2,200 kg/m^3 have an inverse relationship to the topography because they remove too much mass from the observed gravity, and lower densities produce the converse. The relationships between too great or too low a density are reversed for a topographic depression. Several stations are placed outside of the topographic feature in order to establish the gradient of the gravity anomaly upon which the local topographic anomaly is superimposed.

The required elevation range of the feature, ΔE in meters, can be determined by considering the accuracy of the gravity anomaly, Δg in milligal, and the desired accuracy of the density, $\Delta\sigma$ in kg/m^3, using the relationship

$$\Delta E = (24 \times 10^3)\frac{\Delta g}{\Delta \sigma}. \tag{4.13}$$

Thus, if the density needs to be determined to an accuracy of 10 kg/m^3 and the gravity anomaly has an accuracy of 0.02 mGal, the elevation range must be 48 m. By the very nature of the observations over topographic features, the gravitational effect of the terrain can produce significant errors in the results, so it is important to apply terrain corrections to the observed data.

Numerous variations on this graphical method of determining densities have been devised using analytical methods based on least-squares techniques that minimize the correlation between gravity anomaly and topography. These methods as applied to either gravity profiles or maps assume that the deviation of the gravity anomaly due to topographic effects can be treated as random errors (e.g. LEGGE, 1944; PARASNIS, 1952; GRANT and ELSAHARTY, 1962). The method can be used to determine variations of surface density over a survey region using a moving-window approach to minimize the correlation between topography and gravity on a region-by-region basis. SIEGERT (1942) has suggested an alternative method to minimize the subjectivity of the density determination, referred to as the triplet method. It is based on estimating the elevation correction factor, and thus the density of the surface material, from three successive gravity observations along a survey line. The difference (h_i) between the elevation of the intermediate station and the weighted mean of the elevations of the adjacent stations is compared with the difference (g_i) between the observed gravity reading and the weighted mean of those of the adjacent stations to determine the correction factor. Weighting is normally inversely proportional to the distance between the stations. The correction factor is given by

$$k = -\frac{\sum_i h_i g_i}{\sum_i h_i}, \tag{4.14}$$

where the summation is over the triplets in the gravity survey line. Siegert also suggested a methodology for estimating the error in the determination of the correction factor.

Another method to determine the density of the topography from gravity measurements takes advantage of the fact that, in general, topography is scale-independent or self-similar, and thus can be studied using fractal methods (e.g. TURCOTTE, 1997). Although various techniques have been used for this (e.g. THORARINSSON and MAGNUSSSON, 1990; CHAPIN, 1996; HISARLI and ORBAY, 2002), the basic approach is to determine the density that reduces the fractal nature of the Bouguer gravity anomaly, that is the surface density which has the minimum relationship to topography. CHAPIN (1996) achieves this by determining the fractal dimension corresponding to the straight line slopes in log–log plots of the radially averaged power spectra of the Bouguer gravity anomaly calculated over a range of densities. Removal of the linear effect, caused by the scale-dependence term of the Bouguer correction, from a plot of fractal dimensions versus density values is used to identify the minimal fractal dimension, corresponding to the density which has the least relationship to topography.

Underground measurements

Underground gravity measurements also are used to determine the density between two co-located but vertically displaced observations. Originally, these measurements were made with surface gravimeters with observations on the surface and in mines and tunnels (e.g. ALGERMISSEN, 1961; HAMMER, 1950; MCCULLOH, 1965). Studies of this nature continue (e.g. CAPUANO *et al.*, 1998), but with the advent of drillhole gravimeters of sufficient stability and precision, measurements are now routinely made in drill holes (e.g. MCCULLOH, 1966; LAFEHR, 1983).

The standard method assumes that the rock layers between observations are horizontal and of constant density, σ_L, and that no local anomalies are present that will disturb the normal vertical gradient of gravity, $(\Delta g)/(\Delta h)$, which is measured over a vertical range Δh. The observed gravity difference, Δg, between the two stations must be corrected for temporal variations due to instrumental drift and natural changes in gravity in the observations, and the effect of surface terrain features. The difference between the two vertically separated observations is the vertical gradient of gravity due to the mass of the Earth minus twice the effect of the mass of the plate of material between the observations; once for the increased gravity at the upper observation due to the mass of the plate between the two observations, and again at the lower site for the upward attraction of the slab. The density, σ_L in kg/m^3, of the material between the two vertically separated stations is

$$\sigma_L = \left[3.6816 - 128.5761\left(\frac{\Delta g - \Delta T}{\Delta h}\right)\right] \times 10^3, \tag{4.15}$$

where ΔT is the change in the surface terrain effect between the two observations.

In many petroleum applications, the terrain effect is negligible, and thus disregarded because the depth of the measurements below the surface is large in comparison to the vertical interval over which the measurements are made. However, terrain effects can be important depending on the local terrain where the observations are relatively shallow (< 1 km). It should be noted that terrain effects for surface stations are always added to the observed gravity, but may be either added or subtracted for underground measurements depending on the variation of the surface topography from the elevation of the surface vertically above the measurements.

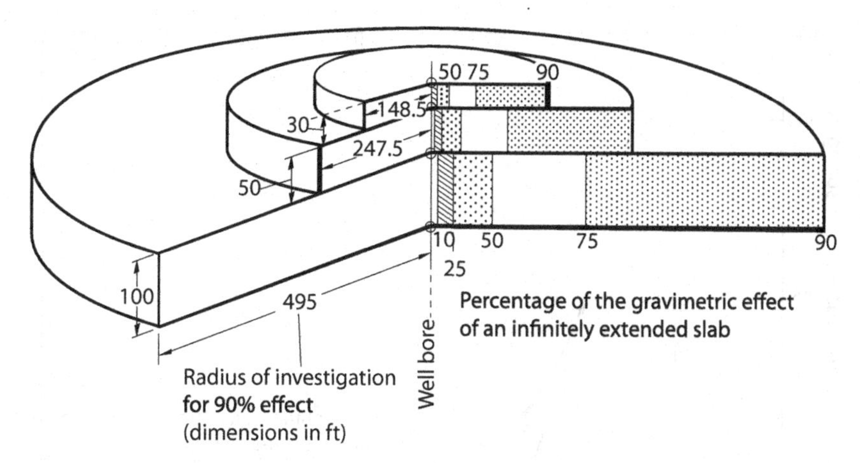

FIGURE 4.10 Percentage gravity effect as a function of distance of an infinitely extending slab on the vertical gravity gradient measured along a drill hole through the slab. Adapted from HEARST (1977b), from a figure originally by McCULLOH (1966).

In Equation 4.15, the assumption is made that the vertical gradient of gravity is 0.3086 mGal/m. The accuracy of this gradient can be improved by considering second-order effects of elevation and the latitude of the observations. However, the major variations from the assumed normal gradient of gravity are local anomalous effects unrelated to the density of the rocks between the two vertically separated stations. The presence of these anomalous effects can be determined by considering the local variations in the gravity anomalies surrounding the drill hole. The vertical gradient can be calculated from these local observations and projected to the elevation of the drillhole observations. Alternatively, measurements can be directly made of the vertical gradient of gravity with measurements on the surface and in a tower placed over the drill hole.

Drillhole gravity measurements are made at specified depths in drill holes with specialized instruments that employ a modified zero-length spring sensor (Chapter 5). The standard instrument used for this purpose can operate in holes that deviate up to 14° from the vertical and have diameters exceeding 10 cm. They are operative in environments that do not exceed a temperature of 260 °C. Measurement precisions of 3 μGal are approached where multiple individual measurements are made. A slim hole gravimeter is also available that can be used at depths of less than 3 km in holes that deviate from the vertical up to 60° and that have diameters over about 5.75 cm. Great care is also used to achieve precision in the depth measurements and to locate observations at formation boundaries identified by other logging techniques.

The density calculated in Equation 4.15 is given numerous names in the geophysical literature to identify it as a calculated density from drillhole gravity observations. The most widely used term is apparent density (LAFEHR, 1983) which recognizes that the calculation is based on the assumption that the slab of rock between the observations is horizontal, infinite, and of constant thickness and density. This density is an approximation to the saturated bulk density of the rock material. The calculation of density from gravity measurements is complicated in regions of dipping formations (e.g. HEARST, 1977a; BROWN and LAUTZENHISER, 1982). However, as an approximation, bed dips of less than 7° from the horizontal can be analyzed with Equation 4.15.

A consideration in drillhole gravity studies is the sensitivity of the measurements to the effects of adjacent formations at distances which are considerably greater than in typical (≤15 cm) formation density logging instrumentation. This is one of the significant advantages of the method, but it does have very definite limits imposed by the magnitude of the mass differentials, the distance from the observations, and the accuracy of the gravity and depth measurements. An illustration first shown in the context of drillhole gravimetry by McCULLOH (1966) displays the percentage of the gravimetric effect of an infinitely extending slab (Figure 4.10). This figure shows that approximately 90% of the gravitational effect is derived from the mass of the rock within approximately five times the vertical distance Δh between the stations. Unfortunately, this figure has been erroneously interpreted to suggest that with increasing distance between observations, the range of the density measurements can be increased. This is not the case, rather the illustration simply shows that for a slab of homogeneous density, the outer part of the slab has far less effect than the inner part. Several articles provide further details on the effective range of investigation of drillhole

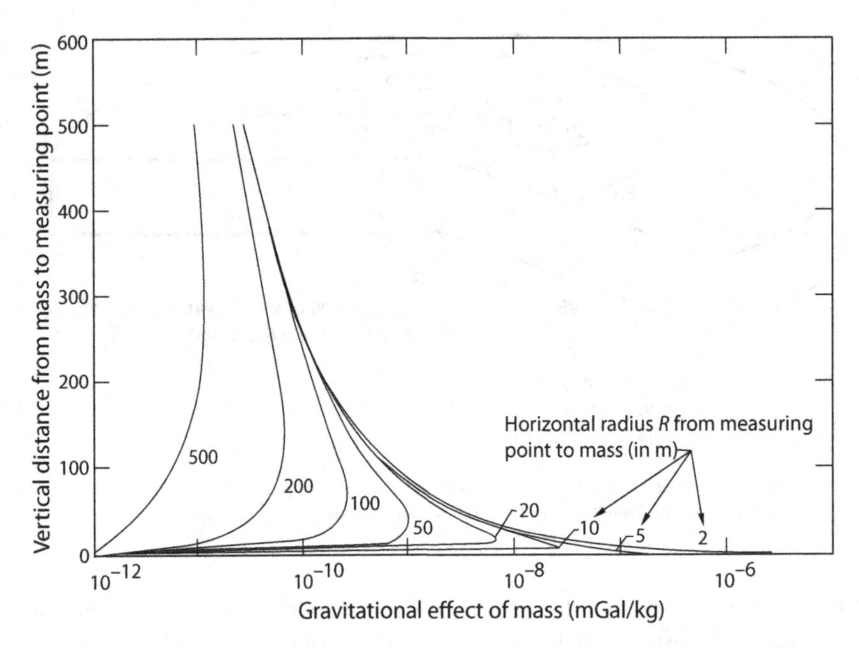

FIGURE 4.11 Point mass gravity effects versus horizontal distances from the drill hole. Adapted from HEARST (1977b).

gravity measurements (e.g. HEARST, 1977b; LAFEHR, 1983).

The only reliable method of judging whether remotely sensed density variations can be identified in a drillhole gravity survey is to make simplifying assumptions regarding the body, its location, size, and density contrast, and calculate the gravitational effect of this simplified model. As a guide, Figure 4.11 shows the gravitational effect of a point source at a range of vertical and horizontal distances from the drill hole. At vertical distances below the center of gravity of the anomalous mass, the gravitational effects (not shown in the figure) will have the opposite sign and the curves will be mirror images of those shown. This figure illustrates the rapid decrease in gravity effect with distance and the increase in the vertical distance to the maximum anomalous effect as the distance to the source increases. For example, a point mass located 500 m from a drill hole will reach a maximum (or minimum) value of roughly 10^{-11} mGal/kg at several hundred meters from the depth to the point mass, while a source at 50 m will reach a maximum value of 10^{-9} mGal/kg at a vertical distance of approximately 30 m from this depth.

Drillhole gravity measurements are useful for determining densities of formations for gravity interpretation, but can also be used as a remote sensing tool determining the occurrence of density anomalies outside the range of more conventional logging techniques. Density is directly related not only to rock composition, but also to porosity, whether air- or fluid-filled, and indirectly to the strength of rocks. Thus, gamma–gamma density logging (Section 4.6.3) finds several applications, particularly in hydrocarbon exploration, and potentially the redistribution of fluid/gas content associated with reservoir development could be monitored with repeat borehole gravity measurements (SCHULTZ, 1989). Nonetheless, drillhole gravimetry is restricted in its applications because of the costs involved with the required specialized equipment, availability of structurally stable drill holes with diameters greater than 10 cm, and limitations on the minimum distances between observations that will permit the desired accuracies in measured density: if a precision in the measurement of the gravity difference is 0.01 mGal and an accuracy in density of 10 kg/m^3 is desired, then the separation between observations must be around 12 m or more. There are additional complications in drillhole gravimeter studies including obtaining precise gravity and depth measurements, washouts of formations adjacent to the drill hole, and anomalous vertical gradients. As a result, great care is exercised in making and interpreting drillhole surveys.

An example of the results of a drillhole gravity survey made at contacts between major formation units from the surface into the basement Precambrian sedimentary rocks over a range of 4 km in the center of the Michigan basin is shown in Figure 4.12. The mean density of the Phanerozoic sedimentary rock column was determined to

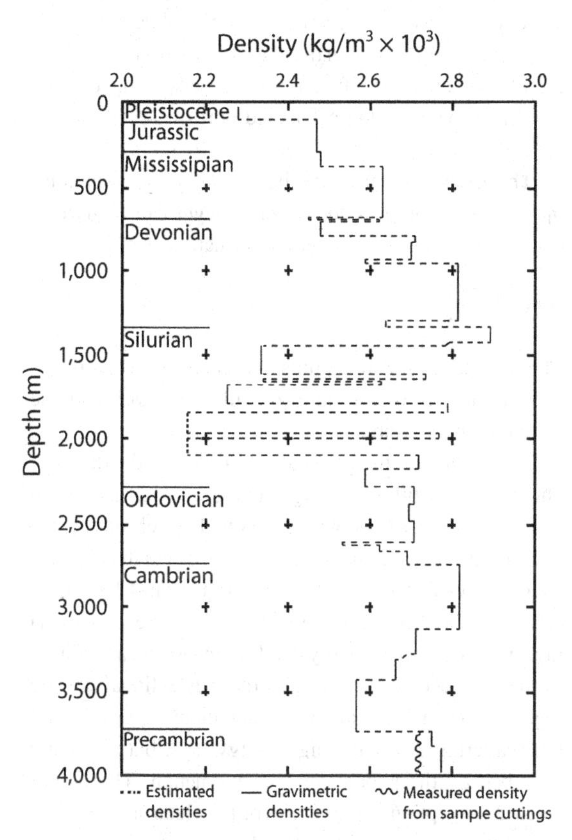

FIGURE 4.12 Profile of gravimetric densities obtained by making gravity measurements at major rock unit boundaries in a drill hole in the center of the Michigan basin. Note the discrepancy between laboratory and gravimetric densities in the Precambrian rocks underlying the basin, believed to be due to abnormal gravity gradients caused by a deeply buried high-density mass. Adapted from HINZE *et al.* (1978).

be 2,610 kg/m^3 after correcting the vertical gradient for the anomalous gradient caused by deeply buried Precambrian mafic rocks. Several of the densities of evaporite and shale formations were determined from gamma–gamma density logging because of washouts of these formations which prevented precise density calculations from the gravity measurements. The discrepancy between measured densities of the Precambrian sedimentary rocks and the drillhole density calculations in Figure 4.12 was interpreted as the possible anomalous effect on the vertical gradient due to Precambrian mafic rocks beneath 3.5 km which were not encountered in the drilling.

4.6.3 Correlative property measurements

Density is a controlling factor in several rock properties. As a result it is possible to measure these properties and through appropriate calibration relate them directly to density. This indirect method of determining densities is useful because the correlative measurements may be made more easily or measured for other purposes, and thus may be readily available. Also, the correlative measurements are commonly made of the material in place, and thus densities can be determined without the difficulties normally encountered in determining *in situ* densities.

Gamma–gamma density log

The gamma–gamma log, or the density log as it is commonly called, is useful in directly determining the *in situ* density of drilled formations. This logging tool was originally developed for measuring densities for use in gravity interpretation, but now finds its major use in determining formation properties, particularly porosity in the hydrocarbon exploration industry. The method primarily is used in non-cased holes which may be either dry or fluid-filled. The method measures photon radiation from a gamma radiation source (e.g. ^{137}Cs or ^{60}Co) located at a fixed distance from a radiation detector, usually less than 0.5 m. Both source and detector are shielded within the logging sonde except for windows directed toward the formation wall. Most gamma–gamma logging devices employ two detectors to correct for drillhole diameter effects and changes in the mudcake lining holes which have been drilled with mud to bring the drill cuttings to the surface. The radiation measured by the detector closest to the source is more affected by local variations in the hole, and thus the ratio of the radiation measured at the near and far detectors can be used to eliminate drillhole effects. This procedure can be performed within the instrument system, in which case the log is referred to as compensated. A schematic cross-section of a gamma–gamma logging sonde is shown in Figure 4.13.

The basic principle of the gamma–gamma log is that the gamma radiation from the source is attenuated in proportion to the electron density of the material making up the walls of the borehole, and thus the saturated bulk density of the formation. Considering their energy levels, gamma rays from the source interact with the formation primarily by Compton scattering which results in the ejection of a Compton recoil electron and a scattered gamma ray (photon) of somewhat lower energy than the incident ray. As the gamma rays diffuse through the formation, the backscattered photons will continue to lose energy and eventually be captured. This attenuation is directly proportional to the number of electrons per unit volume or the formation electron density. As the electron density increases, the count rate of the detector will decrease. The

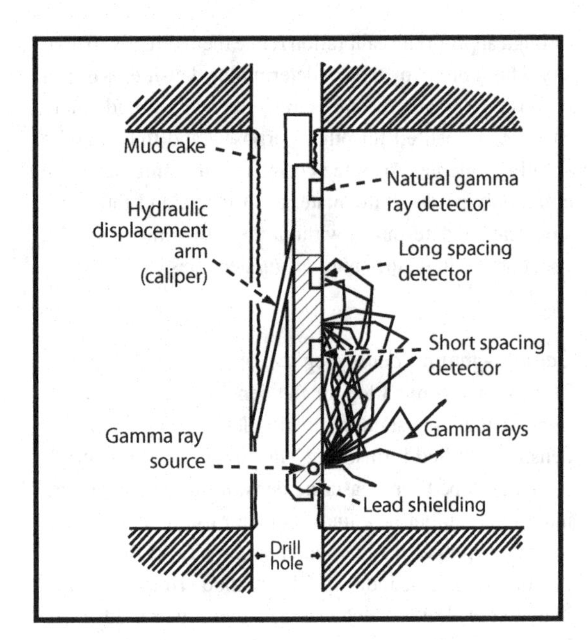

FIGURE 4.13 Schematic cross-section of a gamma–gamma density (compensated) logging sonde using a double detector to eliminate the effect of the drillhole mudcake on the measurements and a separated natural gamma ray detector to permit stripping of natural gamma rays.

bulk density, σ_B, is related to electron density, σ_e, by the expression

$$\sigma_B = \sigma_e \times N \times \left(\frac{Z}{A}\right), \tag{4.16}$$

where N is Avogadro's number, Z is the atomic number of the material, and A is the atomic weight of the material.

The electron density within a range of roughly 15 cm of the source is inversely proportional to the count rate. This is valid even in the presence of natural radiation because the wavelength of the gamma rays produced by the source is different than that of natural radiation. Thus, the bulk density can be calculated if the ratio (Z/A) for the material is known. For most common sediments and sedimentary materials, the ratio is relatively consistent at roughly 0.5. Thus, a calibration of electron density to bulk density is widely applicable, but corrections are necessary in formations consisting of elements which have a ratio that departs from those used as the basis of calibration. As a result, density logs which are commonly calibrated for fresh water-saturated limestone give erroneous bulk densities without a correction in the case of formations of coal, halite, gypsum, anhydrite, and many crystalline rocks. The accuracy of saturated bulk density determinations from gamma–gamma logs approaches $\pm 10\,\mathrm{kg/m^3}$ in limestone, dolomite, and sandstone, but is significantly poorer ($\pm 50\,\mathrm{kg/m^3}$) where the measurements are made through the casing. A typical compensated density log is shown in Figure 4.14 with the gamma radiation and caliper log illustrated on the left, and the compensation, compensated bulk density, and calculated porosity, ϕ, shown on the right.

The measurement of the bulk density of formations, σ_B, from the compensated density logger can be used to determine the porosity, ϕ, of the formations from

$$\phi = \frac{\sigma_g - \sigma_B}{\sigma_g - \sigma_f}. \tag{4.17}$$

The porosity estimate requires that the matrix density, σ_g, and the density of the fluid, σ_f, filling the voids of the formation be available.

A variation on the gamma–gamma log is the density–photoelectron lithology log or the litho-density tool (LABO, 1986) which was developed to eliminate the need to assume the matrix density in calculating porosity of the drilled formations from Equation 4.17. In the density–photoelectron lithology tool, low-energy gamma rays within a narrow energy band are measured in addition to the high-energy rays used in the conventional density log. The low-energy radiation is not attenuated by Compton scattering as are the high-energy rays, but are rather absorbed by the photoelectric phenomenon. The photoelectric absorption is a function of the electrons per atom (photoelectric cross-section) or the average atomic number, Z, of the sampled material. As a result, the absorption is dependent on the chemical composition, and thus is an indicator of the lithology of the material. The log is particularly effective in distinguishing changes in density that result from the presence of gas in rock formations.

The principles employed in gamma–gamma logging also are applied to surface density measurements, where the gamma-ray source and the detector of the Compton scatter radiation are placed on the surface at a fixed separation. The relationship between electron density and bulk density has also been used to measure surface densities by measuring the radiation received at a surface detector from a radiation source inserted by a probe into unconsolidated near-surface materials to a fixed depth of 0.5 to 1 m. In this latter case, the measured radiation is inversely proportional to the bulk density (e.g. HOMILIUS and LORCH, 1957; BLAKE and HARTGE, 1986).

Density from seismic methods

The velocity of propagation of seismic waves, particularly compressional or longitudinal waves, through Earth materials is extensively measured both in the field and in the laboratory, and the strong correlation between seismic compression wave velocity and density is commonly used to estimate densities. Theoretically, the velocity of

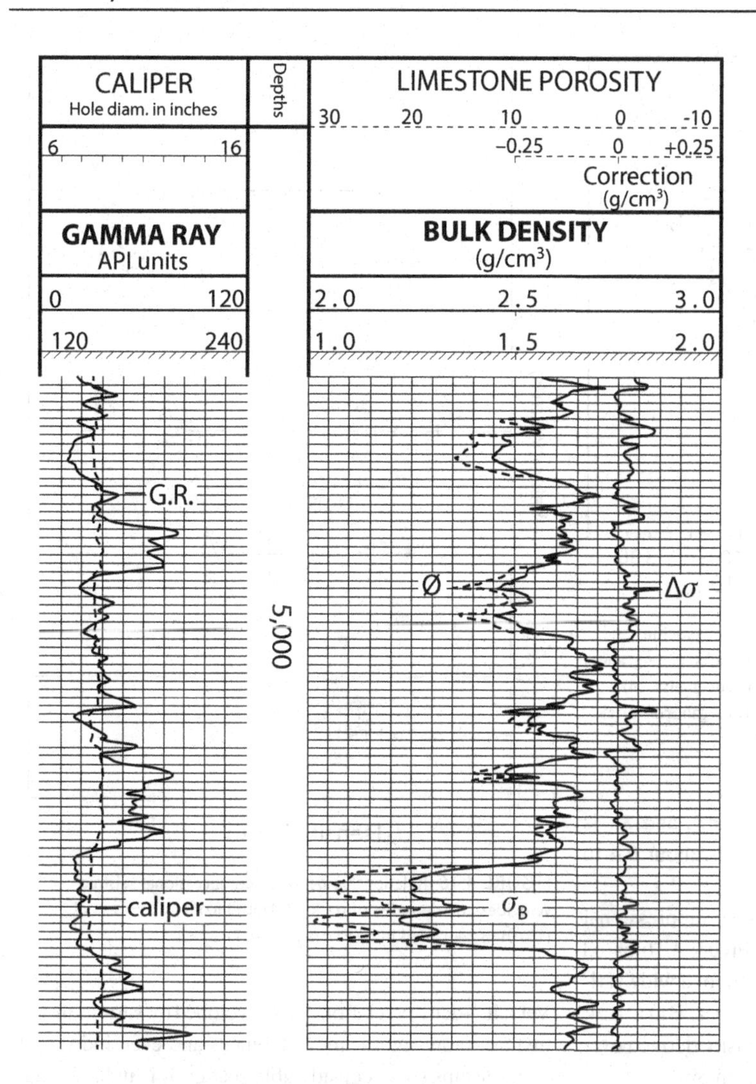

FIGURE 4.14 Example of a compensated gamma–gamma density log. The left panel shows the natural gamma ray (G.R.) activity and the hole diameter. The right panel shows the compensation for the drillhole mudcake ($\Delta\sigma$), the bulk density (σ_B), and the density-derived porosity (Ø). Units and scale are shown in the headers. Adapted from LABO (1986).

propagation of compressional waves is inversely proportional to density, but in actuality the relationship is direct because of the dominance of the elastic properties on velocity and the direct relationship between elastic properties and density of Earth materials. The relationship has been the subject of intensive study by laboratory workers who measure seismic velocity under compression to duplicate subsurface conditions where lithostatic pressure minimizes effects due to microfractures.

Early comprehensive studies of the relationship between velocity and density of crystalline rocks by BIRCH (1961) showed that velocity is approximately linearly related to density for materials that have a common mean atomic weight up to 10 kbar (1 GPa). The relationship, which is referred to as Birch's law, is given by

$$\sigma = a(\bar{M}) + bV_p, \tag{4.18}$$

where $\bar{M}$ is the mean atomic weight of the elements making up the rock, ranging roughly from 21 for felsic rocks to 22 for mafic rocks, and V_p is the compressional velocity in km/s.

Numerous workers have built upon Birch's pioneering studies in identifying the composition of the Earth from its radial distribution of velocity and density. CHRISTENSEN and MOONEY (1995), for example, have compiled extensive information on the velocities and densities of crustal rocks (except for volcanic and monomineralic rocks). They found the best-fitting linear regression line is

$$\sigma = 541 + 360V_p. \tag{4.19}$$

DORTMAN (1976) also assembled useful information showing the correlation of compressional wave velocity and density for several suites of igneous and metamorphic rocks (Figure 4.15). His best-fitting curve to the data is represented by the equation (SCHÖN, 1996)

$$V_p = 5.45 \exp[0.5(|\sigma| - 2.6)] \pm K_T, \tag{4.20}$$

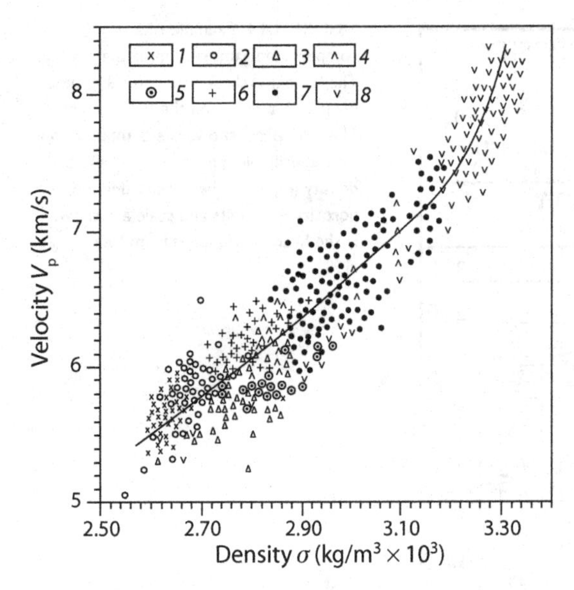

FIGURE 4.15 Comparison of compressional wave velocities and densities for suites of igneous and metamorphic rocks. (1) – granite, (2) – biotitic/amphibolitic gneiss, (3) – garnet/biotitic gneiss, (4) – amphibole gneiss, (5) – granulite, (6) – diorite, (7) – gabbro/norite, (8) – ultrabasite. Adapted from SCHÖN (1996).

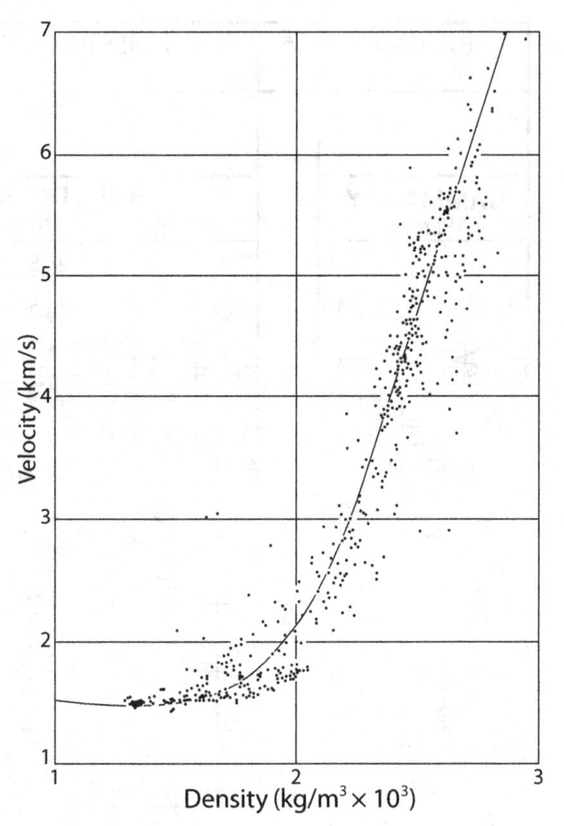

FIGURE 4.16 Relationship between compressional seismic velocities and densities of rocks. Adapted from LUDWIG *et al.* (1970) as reprinted in BARTON (1986).

where $|\sigma|$ is the density magnitude in g/cm^3, and K_T is mainly controlled by texture effects and is about 0.4 to 0.5 km/s.

A commonly used relationship between compressional wave velocity and density is shown in Figure 4.16 based on data obtained by LUDWIG *et al.* (1970) for a wide variety of sedimentary and crystalline rock types. BROCHER (2005) developed an equation for the density derived over a velocity range from 1.5 to 8.5 km/s given by

$$\sigma = 1661.2V_p - 472.1V_p^2 + 67.1V_p^3 - 4.3V_p^4 + 0.19V_p^5, \tag{4.21}$$

where the density is in g/cm^3 and the velocity in km/s.

In dealing with sedimentary rocks (Figure 4.17), the velocity–density relationship obtained by GARDNER *et al.* (1974), which is given by

$$\sigma = 310V_p^{0.25}, \tag{4.22}$$

is useful. For use with dolomite, however, 110 kg/m^3 must be added to the right-hand side of the equation.

The ease of converting velocities to densities and the wealth of available seismic velocity data fosters the use of this method of determining densities over direct measurements. However, the procedure is not without pitfalls, as has been pointed out by BARTON (1986) and others. Velocity–density relationships developed by various investigators show significant differences and even individual relationships show considerable scatter. It is unlikely that these differences and the dispersion in observed measurements are a result of errors in measurements; instead, they probably reflect different suites of rocks and variations within a rock type. The differences can be significant. For example, in the relationship given by LUDWIG *et al.* (1970) the density for a velocity of 5.5 km/s is 2,660 kg/m^3, but values range from a minimum of 2,390 to 2,780 kg/m^3 (BARTON, 1986). This range of values can have a profound impact on gravity interpretation. Thus, care must be used in converting velocities to densities and using these relationships in gravity studies.

DEBSKI and TARANTOLA (1995) have shown that it is possible to obtain subsurface densities from the amplitude of reflected waves measured at varying incidence angles. Although the use of amplitude-variation-with-offset (AVO) to determine density contrasts is difficult and tends to lead to unstable results, BEHURA *et al.* (2010) show that it is possible to extract density from AVO

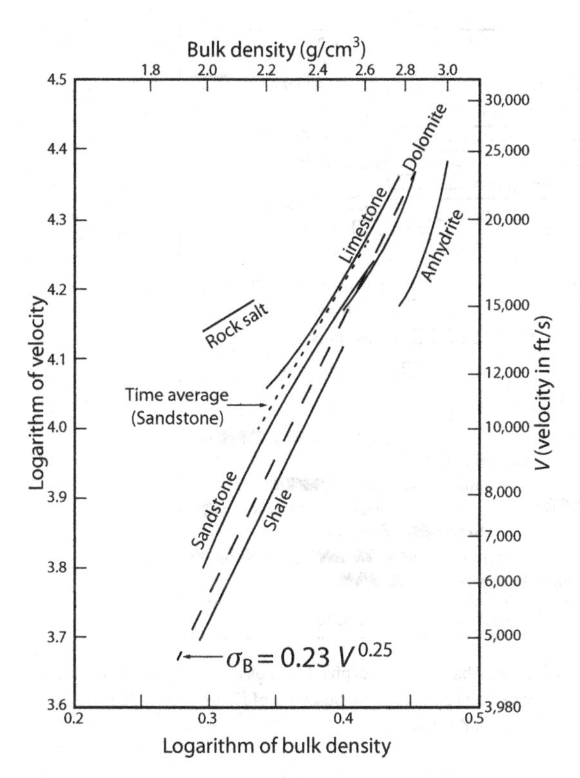

FIGURE 4.17 Compressional wave velocity and density relationships for various sedimentary rocks. Adapted from GARDNER *et al.* (1974).

inversion if the density contrast is the major contributor to the reflection at the interface.

4.7 Density tabulations

Tables of typical values of rock densities are presented in most geophysical exploration texts. Tabulations of densities of Earth materials including some sediments are also given in special publications dealing with the physical properties of rocks and minerals (e.g. DALY *et al.*, 1966; PARASNIS, 1971; JUDD and SHAKOOR, 1981; JOHNSON and OLHOEFT, 1984). In addition, there are numerous publications that report on the densities of rocks within a specific geographic region (e.g. GIBB, 1968; SOBCZAK *et al.*, 1970; CHANDLER and LIVELY, 2003; BROCHER, 2005; GRAUCH *et al.*, 2009). Most of the compilations deal with crystalline and sedimentary rocks which have only a minimal void space. Tabulations of the densities of soils, alluvial material, and organic materials which have high-void volumes and are of special importance to near-surface studies are limited (e.g. HALL and HAJNAL, 1962; MCGINNIS *et al.*, 1963; KLEMETTI and KEYS, 1983).

Typical values of rocks and some unconsolidated materials can be obtained from these compilations, but care is needed because of the lack of specificity regarding what was measured, related statistics, and the differences in the density of rocks of similar composition from different tectonic and geologic terranes.

Figure 4.18 gives representative values for saturated bulk or natural densities for consolidated and unconsolidated Earth materials. The values are useful when no other information is available on the densities of materials of a specific region and as a starting point for the estimation of densities. When choosing a value within the range given in the table, a default selection will be the central value. This assumes that the distribution of values is normal, but as illustrated in Figure 4.8, the frequency distributions of densities of rocks are different depending on the controlling factors and their relative importance. Thus, it is advisable to consider each individual type of material and the local conditions in selecting a specific density from this figure.

Numerous generalizations can be made from the densities presented in Figure 4.18, but only a few of these are mentioned here. All rocks have a range of natural densities, even monomineralic rocks, because of the occurrence of voids and impurities. Most rock types have a range of mineral composition so that even where the effect of voids is negligible, the densities cover a range. Those types which have the widest diversity of compositions have the broadest range of densities except where the effect of void space plays a major role. Neglecting the effect of gas cavities and other voids in volcanic rocks, the density of igneous rocks decreases with increasing felsic mineral content, but overlaps exist in the ranges. The densities of metamorphic rocks have broad ranges which overlap the densities of most igneous rocks. The broad ranges of density in these rock types are caused primarily by a wide variety of compositions from felsic to mafic. Metamorphic rock classification is based more on texture reflecting the origin of the rock rather than upon the mineral composition as in the case of igneous rocks. The density of metamorphic rocks increases with metamorphic grade – that is, increasing temperature and pressure lead to the highest densities where the minerals have been modified to high-temperature silicates of iron, calcium, and magnesium.

The chemically precipitated sedimentary rocks, limestone and dolomite, overlap the ranges of the more felsic igneous and metamorphic rocks, while sedimentary rocks derived from clastic sediments are less dense than most igneous and metamorphic rocks. The density ranges of shales, siltstones, and sandstones overlap over most of their

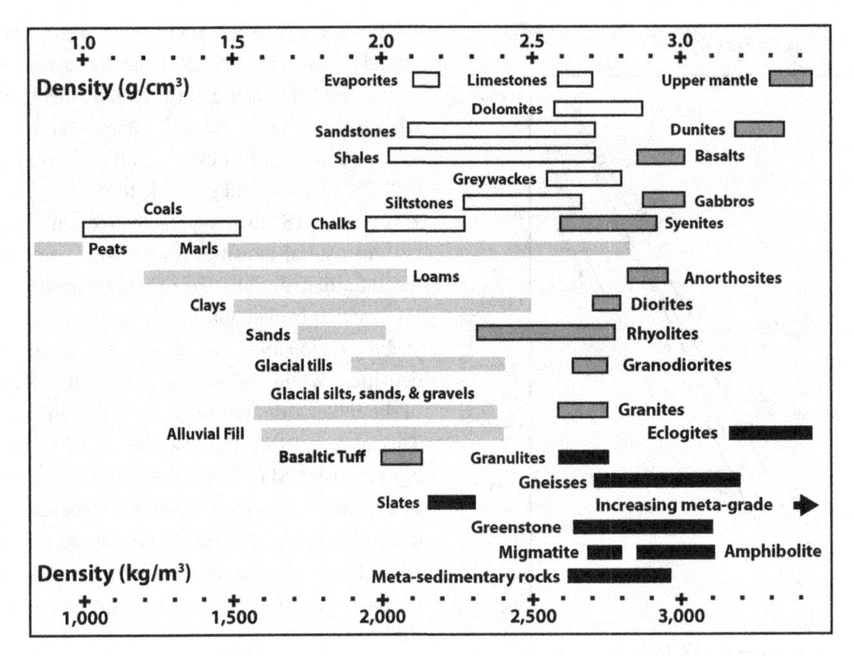

FIGURE 4.18 Natural densities (saturated bulk density) of igneous (bordered gray bars), metamorphic (solid black bars), and sedimentary rock types (bordered white bars), and sediments and soils (solid gray bars). Based on rock density tabulations of DALY *et al.* (1966); PARASNIS (1971); JOHNSON and OLHOEFT (1984); EATON and WATKINS (1970), and others.

total range. Although the density of sediments is less than that of related sedimentary rocks, the ranges do overlap. Densities of less than 2,000 kg/m^3 generally only occur in rocks with interstitial pore space that have undergone only limited compaction or lithification.

The broad and overlapping ranges of density mean that care must be used in determining if there is a significant difference between the average values of the densities of rock units based on the measured densities. Differences may simply be due to variability within a rock unit resulting in no significant difference that will produce a mappable density contrast by the gravity method. Assuming that the distribution of densities is normal, which is generally permissible for many rock types, it is possible to evaluate the statistical significance of the difference in the average values by calculating the standard error of the difference, ϵ, from

$$\epsilon = \left[\frac{s_1^2}{n_1} + \frac{s_2^2}{n_2}\right] \tag{4.23}$$

for two groups of n_1 and n_2 sample measurements with standard deviations s_1 and s_2, respectively. If the observed difference in the average values is greater than 3ϵ, the probability is less than 0.5% that the groups of samples are from the same population, and thus the differences in the averages can be considered significant, being from two different density units. Consider, for example, two groups of samples with density means and standard deviations of 2,520 and 100 and 2,620 and 100 kg/m^3, respectively, based on the measurement of 10 and 20 samples. The standard error $\epsilon = 38.73$ fails to meet the requirement that the measured values are from two statistically different populations.

Clearly, the use of *in situ*, laboratory, and correlative property measurements is fraught with pitfalls for estimating possible subsurface density variations. For example, the variability in the relationship between rock density and magnetic susceptibility measurements (e.g. HENKEL, 1976; SUBRAHMANYAM and VERMA, 1981) is commonly so great that only first-order density estimates are possible from the correlations. However, even in this more extreme case, the relationship between density and magnetic susceptibility can exclude large ranges of physical property values and related subsurface conditions that otherwise might have been included in modeling coincident gravity and magnetic anomalies.

4.8 Key concepts

- Density, or in practice the density contrast between the anomalous source and the laterally adjacent country rocks, is the controlling physical property in the gravity method. It is a function of the mineral content of the Earth material as well as the void volume and the filling material of the voids. The natural density, or the saturated bulk density, of Earth materials, is the density with all the void space filled with water, a condition approximated in many subsurface formations. This is the density commonly reported in the literature. Tables of density values show that there is considerable overlap in the densities of different lithologies and the distribution of densities within most units or lithologies is broad with distributions particularly skewed towards lower densities as a result of secondary processes which increase pore volume and alter minerals to less-dense forms.

- Densities play an important role in the planning, reduction, and interpretation of gravity surveys. In principle their measurement is simple, but in practice these measurements can be complex and subject to numerous errors. Accurate measurements are particularly difficult to obtain in place and in friable or high-void volume materials. Three general measurement techniques are used: (1) laboratory measurements on samples extracted from the Earth; (2) gravity measurements, both surface and underground; and (3) measurement of correlative properties, such as seismic velocity and gamma ray attenuation, using empirical relationships.
- Plutonic igneous rocks normally have minimal void space and densities ranging from roughly 2,650 to 3,000 kg/m^3. Densities increase from felsic to mafic rocks. Igneous rocks formed near the surface as a result of volcanic activity have similar mineral composition to their plutonic equivalents, but they generally have a higher pore volume resulting in lower densities.
- Sediments and sedimentary rocks primarily consist of minerals whose densities range from 2,600 to 2,700 kg/m^3, but their densities are generally controlled by void space and the density of the void filling. Void space is dictated by the origin of the sediments and sedimentary rocks and their subsequent geological environment. Compaction by lithostatic pressure is a primary control on the density of these materials with density increasing exponentially with depth. Sedimentary rocks formed by chemical precipitation are normally higher in density than other sedimentary rocks except where they have been partly dissolved leading to increasing void space.
- Metamorphic rocks generally are equal to or higher in density than the rocks from which they have formed. Higher densities are caused by the transformation of minerals to a high-density form as a result of a high pressure/temperature environment and a decrease in the void volume due to pressure effects.

Supplementary material and Study Questions are available on the website www.cambridge.org/gravmag.

5 Gravity data acquisition

5.1 Overview

Although simple in principle, the measurement of gravity for geologic purposes to an accuracy of the order of 10^{-8} to 10^{-9} of the Earth's gravitational field requires highly sophisticated instrumentation and rigorous survey procedures. Fundamentally because of the nature of the measurement, all gravity instrumentation must be mechanical. Most land gravity measurements are made with relative-measuring instruments using the zero-length spring principle to achieve high sensitivity in portable instrumentation. Although extensive governmental and commercial databases containing millions of observations exist, additional gravity surveying continues to achieve greater detail and accuracy in the data. Random errors are minimized in modern instrumentation by performing the observations automatically. However, residual errors remain, owing both to inherent instrumentation problems and non-geologic acceleration components that need to be considered while conducting surveys.

Increasingly accurate observations and improved anomaly resolution are being achieved especially in marine and airborne observations of gravity. A variety of instrumentation is used for measurements on mobile platforms, including modified zero-length spring gravimeters and electromagnetic accelerometers mounted on gyrostabilized platforms to minimize short-period horizontal and vertical accelerations due to movements of the ship or aircraft. In addition, highly accurate absolute gravity measurements are being made with stable, portable free-fall instruments, and rotating-disk gravity gradiometers are being used to observe the gravitational tensor components. These improvements are useful not only in mapping components of the gravity field related to variations in subsurface geology, but also in monitoring time-variable processes within the Earth associated with mass changes.

The accuracy of gravity measurements has been markedly improved with the use of GPS for both horizontal and vertical position control. Despite ever-improving terrestrial topographic and bathymetric models, a major limitation to the accuracy of anomalies obtained from many gravity surveys, especially land surface surveys, is the assessment of the effect of local and regional terrain. The usefulness of gravity measurements is greatly improved by surveying and measurement procedures designed for the specific objectives and constraints of the survey.

Gravity measurements have been made on a global basis from Earth-orbiting satellites for several decades, but since the 1960s new satellite missions have markedly improved the synoptic view of the terrestrial gravity as well as the gravity field of other planetary bodies. Active measurements involving radar topographical mapping of the Earth's ocean surface from satellites have provided new insight into the undulations of the Earth's geoid, and thus the gravity in marine regions, at accuracies of a few milligals with a minimum full-wavelength resolution of about 15 km. Several recent satellites dedicated to terrestrial gravity observations have been used in passive ways to map regional gravity anomalies on a worldwide basis. Variations in this gravity signal with time are being used to determine temporal mass variations caused by, for example, changing surface ice and water distribution on the continents.

5.2 Introduction

The acquisition of gravity data, which is relatively simple in principle, is based on determining the gravity force field acting on a test mass within the instrument. This can be accomplished in a number of different ways, from dropping the test mass in free fall, to pendulums, to the downward force of the test mass measured by a spring, to electromagnetic accelerometers. The accuracy and horizontal spatial resolution requirements of gravity used in the study of the Earth are high, and thus care must be exercised in selection of survey instrumentation, design of the investigation, and the procedures used in data acquisition. Most geological studies require an accuracy of 10^{-6} to 10^{-8} of the surface gravity (1 to 0.01 mGal), while measurements for many shallow-zone investigations and the mapping of time variations in the Earth's gravity field associated with subsurface migration of fluids and elevation changes seek accuracies at the 10^{-9} (0.001 mGal or 1 μGal) level. These sensitivity requirements are among the most stringent encountered in scientific studies, making gravity instrumentation among the most sensitive mechanical devices humans have produced. As a result of the specialized instrumentation, great care must be employed in making the measurements and adjusting them for use in subsurface interpretation. This is especially true in high-resolution surveys where the station interval is measured in meters or tens of meters, and sensitivities at the microgal level are required.

In addition to the accuracy of measurements, horizontal spatial resolution, which is the minimum separation that permits recognition of adjacent individual sources, is an important consideration because it determines the smallest features that can be identified in a survey. Resolution in gravity data is commonly expressed in terms of the minimum half-wavelength that can be readily mapped in a survey provided the amplitude of the signal exceeds the accuracy limits of the observations and reductions. Resolution is a function of the measurement interval, either real or virtual. In land-based measurements it is specified by the actual data interval which can be adjusted to the objectives of the survey. However, for measurements made on a moving platform such as marine vessels, aircraft, and satellites, the resolution is not necessarily determined by the measurement interval, but rather by a virtual interval established by the requirements for filtering over a series of consecutive observations to eliminate undesirable noise components due to motion of the platform. As a result there is a non-linear relationship between accuracy and spatial resolution leading to a decrease in accuracy as the spatial resolution or the minimum observable wavelength of the survey decreases. Thus, in gravity data acquisition, it is necessary to obtain the accuracy and resolution dictated by the objectives of a survey or the converse will be true – that is, the objectives of a study will be determined by the accuracy and resolution of the data acquisition.

The accuracy and spatial resolution of a particular measurement system for a specific survey are complicated, especially in dynamic survey systems. The noise envelope of the measurements which affects both accuracy and resolution depends not only on the instrumentation, but also on the environment of the measurements including the speed and accelerations of the platform and the non-geologic sources of the gravity force field, and on the schemes for noise rejection and filtering of the observations. However, "best possible" results can be compared based on experience with the individual systems. FAIRHEAD and ODEGARD (2002) have shown such a comparison with a time-trend plot of "best possible" gravity accuracy versus resolution (Figure 5.1). The arrows show the change in these parameters over time, with the arrowhead being the current estimate at the time of the 2002 publication. Although somewhat dated, this figure provides order of magnitude results that are generally applicable today. All the systems have improved markedly with time, and that trend continues today. For example, the use of a lighter-than-air airship for the platform of an airborne gradiometry survey has led to improvements in both accuracy and resolution because of the stability of the airborne system and its relatively low speed (HATCH *et al.*, 2007; HATCH and PITTS, 2010).

5.3 Measuring gravity

The gravity method uses the measurement of spatial and temporal variations in the intensity or gradients of the planetary force field. The actual measurements are made indirectly using observations of parameters such as time, linear or rotational displacement, spring tension, or an electrical component that can be related through fundamental relationships to the value of planetary gravity or its spatial rate of change. Traditionally in exploration, the spatial and temporal change in gravity or its gradients are measured rather than the absolute value of gravity intensity. This is due to the high sensitivity requirement of absolute measurements, which is of the order of 10^{-8} for 0.01 mGal resolution, compared with a significantly lower sensitivity requirement for relative instruments.

Although a wide variety of gravimeters (gravity meters) has been developed for measuring relative gravity and used since the first one was developed early in the nineteenth century (CHAPIN, 1998), most are spring-type in which

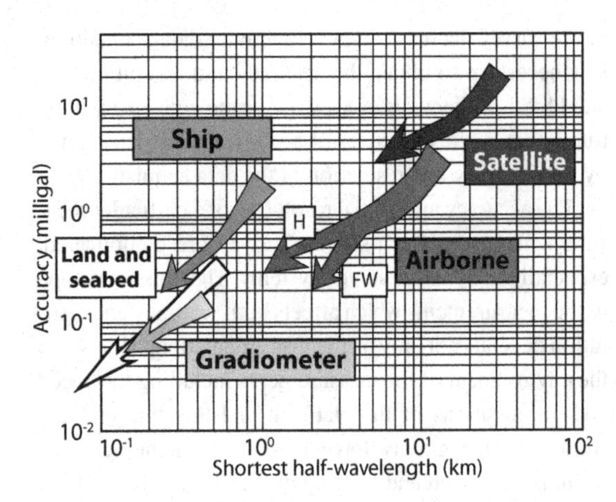

FIGURE 5.1 Relative accuracies and wavelength resolution of gravity surveying methodologies. The log–log time-trend plot gives inferred optimal gravity resolutions of survey systems where the arrow points represent current claims. Here FW = fixed wing and H = helicopter aircraft. In addition, drillhole gravimeters have an accuracy of a few microgals and a resolution of approximately 10 m, and absolute gravity measurements have an accuracy of approximately one microgal. Adapted from FAIRHEAD and ODEGARD (2002).

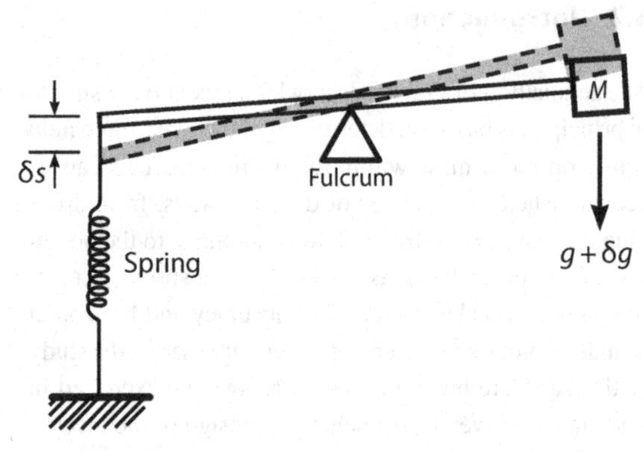

FIGURE 5.2 Principle of the gravimeter showing how a change in gravity, δg, causes a displacement of the lever arm, δs, which can be used to measure the change in gravity.

the relative gravity is measured by the change of strain (length) of a spring which accompanies a change in g (Figure 5.2). This is the same principle used in a sensitive weighing device. A critical problem in this type of gravimeter is to achieve sufficient amplification of the change in length associated with the required sensitivity. Without amplification a spring would have to have an absurdly long length of tens of meters or have an extremely heavy test mass to achieve a sensitivity equivalent to modern gravimeters.

Several methods of optical and electrical amplification have been employed in gravimeters, but the most successful and most widely used is mechanical magnification. This is achieved through astaticization in which by construction a mechanical force reinforces the gravity force and opposes the restoring force of the spring. In this manner, a small change in gravity produces a large, measurable displacement of the test mass. Astaticization causes the system to operate very near a point of instability. In this position, the period of oscillation becomes very long because of the near equivalence of the restoring force and the gravitational force. The sensitivity is proportional to the square of the period of oscillation of a system. To avoid undesirable non-linear response of an astaticized system, the measurement of gravity in land meters is made by bringing the system to a reference position by a force from a calibrated nulling screw. The null position is placed near the horizontal to minimize effects of misleveling.

Several principles other than spring-type balances are the basis for construction of gravimeters (e.g. CHAPIN, 1998). One of these that is useful in studies requiring very high sensitivity is the so-called superconducting gravimeter in which the vertical position of a niobium sphere suspended between superconducting coils is electrically monitored (e.g. CROSSLEY, 1994; GOODKIND, 1999). This instrument is purported to have sensitivities of the order of 1 part in 10^{12} of the Earth's surface gravity, sufficient to detect a change in elevation equal to the thickness of a piece of writing paper, opening up new opportunities for studying internal changes within the Earth as well as Earth dynamics and rotation with temporal variations in gravity using continuous gravity observations.

Specialized gravimeters have been constructed and placed on gyroscopically controlled stable platforms for making measurements on moving ships and aircraft (LACOSTE, 1967). These measurements have lower sensitivity and resolution than land surface measurements, and therefore are generally not suitable for detailed surveys requiring an accuracy higher than 0.1 mGal. However, the development of highly accurate airborne gravity gradiometers has broadened the use of airborne measurements for exploration purposes, even for mineral resource exploration where the targets are small and the amplitudes of the anomalies limited.

Advances continue to be made in the operational system of gravimeters utilizing new technology to increase their accuracy by minimizing environmental and human effects, and to increase the efficiency of the meters by making them more automatic. CHAPIN *et al.* (1999) have found, in considering the accuracy, convenience, and speed of measurement of several different static gravimeters in surveying stations separated by several tens of meters, that all of the gravimeters have relative advantages and limitations. Their conclusions reinforce the need for careful selection of a specific gravimeter based on its design and operational characteristics, and on the objectives and nature of the survey. Recent advances in gravimetry for exploration purposes are focused primarily on improving the accuracy and resolution of measurements obtained from mobile platforms and space satellites, and notable progress has been made in the past decade (e.g. FAIRHEAD and ODEGARD, 2002; HERRING and HALL, 2006; LANE, 2010). Significant advances have also been made in absolute gravity measurements (e.g. NIEBAUER *et al.*, 1995) that are finding new uses in gravity exploration especially related to temporal variations and in the use of the Global Positioning System (GPS) to improve the accuracy of airborne, marine, and land gravity positioning, and thus the processing of observations to anomalies.

5.3.1 Land surface measurements

Most land surface gravity observations are made with the gravimeter placed on a tripod set firmly on the ground surface. Virtually all modern exploration gravimeters used in this mode are based on the zero-length spring principle originally developed by LACOSTE (1934) for increasing the sensitivity of vertical long-period seismographs. The zero-length spring is wound in a pre-stressed condition like the spring of an ordinary screen door. A specific force is required to open the spring so that if it were capable of collapsing to zero length, the force would be zero. The advantage of the zero-length spring concept is that it involves a key principle in measuring the gravity effect on the displacement of the test mass. That principle is that the spring behaves in a known, reliable way. Prior to zero-length springs, spring gravimeters underwent extensive laboratory testing to determine the property of the spring in each gravimeter, resulting in gravimeters of the same model behaving either reliably or erratically over different ranges of measurement. With the zero-length spring, the only calibration needed within the gravimeter was the mechanism to stretch the spring (commonly with a micrometer screw).

Gravimeters based on the principle of a zero-length spring are commonly classified as either metal or quartz zero-length spring meters. An inventory of gravimeters found that about 4,000 instruments based on the zero-length principle had been constructed up to 1998 with roughly 3,000 of them made in equal numbers with metal springs by LaCoste & Romberg Gravity Meters, Inc., and with quartz springs by the Worden Gravity Meter Co. (CHAPIN, 1998). Today, the only manufacturer of new relative gravity meters is LaCoste & Romberg-Scintrex (LRS), Inc., which makes an automated zero-length spring quartz instrument (currently the CG-5). Nevertheless, many older gravity instruments, if properly maintained, are satisfactory for most surveying purposes. Indeed, some professionals prefer the older instruments because most of the post-manufacture instrument drift is no longer present.

These gravimeters normally are employed in a static mode on land and in underground workings where the instrument is mounted on a tripod and leveled to align the sensor sensitivity axis in the local vertical direction. However, they have been modified to include automatic leveling, nulling, and readouts for use in specialized housings in drill holes and underwater, or when deployed on the surface from hovering helicopters, or simply to minimize human errors in land observations. Typically sea-bottom surveys have been limited to depths of a few hundred meters with deployment and observations made by cables reaching from the remotely operated instrument to a surface vessel. However, recent interest in monitoring sub-sea petroleum reservoirs with time-lapse gravity measurements has led to development of remotely operated vehicle-deployed deep ocean systems (SASAGAWA *et al.*, 2003; ZUMBERGE *et al.*, 2008). The measurements are made automatically and remotely using a standard quartz zero-length spring gravimeter in a waterproof housing which is deployed by a remotely operated vehicle onto the ocean bottom. Recent systems are reported to have a repeatability of 3 μGal in gravity and 5 mm in station depth (ZUMBERGE *et al.*, 2008). This measurement accuracy is desirable because of the small size (generally less than 100 μGal) of the variations anticipated in time-lapse gravity to monitor petroleum of the reservoirs.

Zero-length spring concept

A critical breakthrough in modern-day gravimeters occurred when LACOSTE (1934) described a method of achieving great sensitivity in a vertical seismograph which measured the acceleration of the ground motions of the Earth upon the passage of a seismic wave. This type of instrument, including both metal and quartz spring meters, is capable of measuring the change in g when seismic vibrations are not prominent. Astaticization, and thus high sensitivity, is achieved through the principle of the zero-length spring without a long spring. For example, a 10 cm spring will experience a 10^{-7} m change with a 10 μGal variation in gravity (CHAPIN, 1998). This change in length is readily measured with current technology. As a result, gravimeters based on the zero-length spring concept can be made relatively small for ease of portability and the required consistency of the environment within the sensor housing.

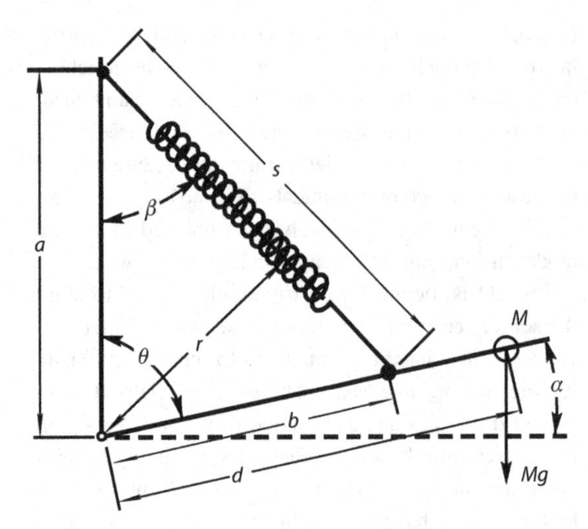

FIGURE 5.3 Diagram of an inclined zero-length spring assembly used in modern gravimeters showing the parameters that describe the location of a reference mass M in the Earth's gravity field on a lever attached to the spring. Courtesy of LaCoste & Romberg-Scintrex (LRS), Inc.

The zero-length spring, which counteracts the force of gravity on a test mass in the gravimeter, is constructed so that if the mass were removed, the force would be zero and the spring would have zero length. This condition is attained by pre-stressing the spring while it is being wound so that it will be in compression before its coils are separated by stretching. In quartz instruments, this is achieved by constructing the quartz spring so that the spring coils spiral inward such that it would literally become a nested flat zero-length spring if allowed to be free of the mechanism.

Figure 5.3 from LACOSTE and ROMBERG (2004) illustrates the principle of the zero-length spring gravimeter, whereby the restoring force of the helical spring counteracts the force of gravity acting on the test mass M, bringing the mass arm to a neutral, equilibrium position. A change in g causes the mass arm to deflect to a new position, raising it for a decrease in gravity and lowering it for an increase in gravity.

The principle can be appreciated by considering the clockwise torque τ_c acting on the lever arm b in Figure 5.3 due to the force Mg, which is

$$\tau_c = (Mgd)\sin\theta = (Mgd)\cos\alpha. \tag{5.1}$$

The counterclockwise torque τ_{cc} due to the restoring force of the spring, on the other hand, is

$$\tau_{cc} = kSr \tag{5.2}$$

because the spring with spring constant k is constructed (i.e. prestressed) so that its length S is zero when no force is applied to it. Furthermore,

$$S = (b)\frac{\cos\alpha}{\sin\beta} \quad \text{and} \quad r = a\sin\beta, \tag{5.3}$$

so that

$$\tau_{cc} = (kba)\cos\alpha. \tag{5.4}$$

At equilibrium $\tau_c = \tau_{cc}$ where $(Mgd)\cos\alpha = (kba)\cos\alpha$ or

$$Mgd = kba. \tag{5.5}$$

Therefore, the gravimeter is insensitive to the angles θ, α, and β. It will be in equilibrium over a small range of angles and there is no restoring force if displaced from an equilibrium point by only a small amount. There is a basic relationship between a system's free-period of oscillation and its sensitivity: that is, the sensitivity is proportional to the square of the period. Thus, theoretically the zero-length spring instrument can be constructed to have an infinite natural period and an infinite sensitivity. The only limitation is the quality and length of the micrometer screw that nulls it. In modern instruments, the action of nulling the spring using the micrometer screw is achieved through an electronic nulling system. In modern gravimeters, the full range of the instrument can be nulled in this manner, eliminating the screw altogether. In metal spring devices, large ranges can be achieved electronically, and the screw is only used to re-range the device so that electronic nulling is possible.

Zero-length gravimeters became commercially available immediately after World War II with several manufacturers producing instruments based on this principle. In theory, these instruments should be capable of achieving sensitivities of less than 1 μGal, but in practice most gravimeters are built with a sensitivity of 10 μGal. Gravimeters built for microgravity surveys and measurement of time variations in gravity at a site typically have sensitivities approaching 1 μGal, but have limited gravity ranges without resetting of the instrument's mass arm. Both LaCoste & Romberg type gravimeters using a metal, inclined zero-length spring and the Worden type using a quartz vertical spring are manufactured in models that automatically measure and record the observations corrected for tidal effects and take successive observations to achieve greater precision through averaging. These fully portable instruments automatically level themselves and have sensor-housing heaters powered by internal batteries. They are easier and faster to use than previous meters and potentially more accurate because of elimination of

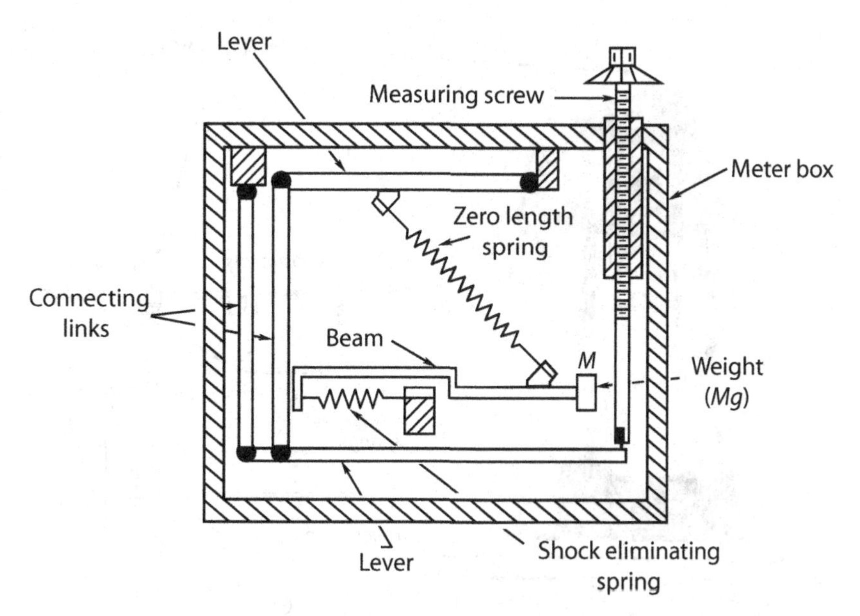

FIGURE 5.4 Simplified diagram of the LaCoste & Romberg gravimeter using the inclined zero-length spring. Courtesy of LaCoste & Romberg-Scintrex (LRS), Inc.

the possibility of human errors. A special model has a reported sensitivity of 0.1 μGal for microgravity studies. Specialized zero-length meters of somewhat less accuracy which incorporate automatic leveling and reading have been designed for operation on ice and underwater, using increased damping of the sensor element and filtering to remove short-period oscillations.

Metal zero-length spring gravimeters

The LaCoste & Romberg type gravimeter measures the change in gravity between observation sites utilizing a metal zero-length spring to achieve high sensitivity. The major components of the instrument are shown in the simplified diagram of Figure 5.4. The sensor itself is roughly a 5 cm cube. The test mass located at the extremity of the pivoted beam is held in place by a metal, inclined zero-length spring. To achieve the necessary sensitivity, the mass is relatively large and is made of dense tungsten in some models and gold in others. As a result, the beam must be arrested when the meter is moved to avoid the possibility of mechanical damage to the sensor. Metal is used for the spring because of its strength and the low coefficient of thermal expansion that is attained by selected metallic alloys. To achieve constant thermal behavior of the sensor, the instrument is heated to a constant temperature.

When the sensor experiences a change in gravity, the pivoted beam moves to a new equilibrium position. The position of the beam is monitored either optically or electrically, and the meter is returned to a reference null position by changing the torque on a screw system that modifies the position of the upper connection of the zero-length spring. The movement of the beam is linear near the null point, allowing a large range over which precise measurements can be made. The torque is calibrated in units of gravity so that the change in gravity causing the displacement of the beam can be determined. Calibration is performed in the laboratory by placing a standardized weight on the test mass and measuring the displacement of the beam with the nulling screw. Alternatively calibration can be achieved by tilting the meter by a known angle and comparing the observation against the calculated change in gravity due to the tilt. These methods are backed up by checking the calibration in the field with observations between sites of known gravity values.

The LaCoste & Romberg type gravimeters consist of two basic models for land surface measurements. The common variety is the Model G meter which has a worldwide range (7,000 mGal), a repeatability of 0.01 mGal and an accuracy better than 0.04 mGal. With very careful observations and field procedures, some investigators claim a precision of the order of 0.005 mGal for this model. The Model D meter is similar to the Model G, but has two nulling screws, one for the worldwide range of gravity and the other for the limited measuring range (slightly more than 200 mGal). This model has a sensitivity of 0.001 mGal (1 μGal) over its fine-adjustment measuring screw, making it useful for microgravity surveys.

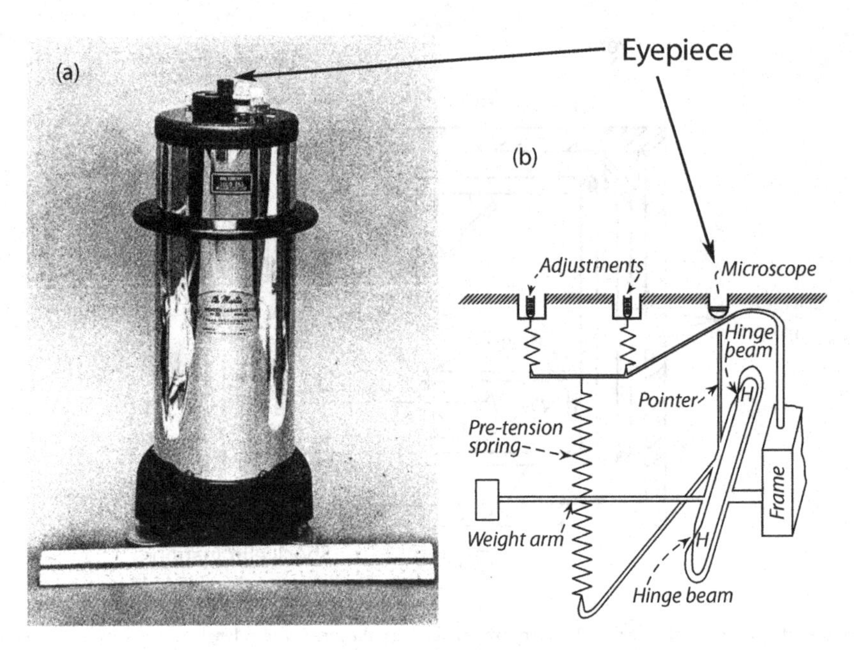

FIGURE 5.5 The Worden gravimeter pictured in (a) uses the vertical, quartz zero-length spring assembly shown in (b). Courtesy of Worden Gravity Meter, Inc.

A specialized gimbal-mounted metal zero-length type gravimeter is used in drill holes. These instruments are leveled and read remotely achieving a practical noise level of roughly 2 μGal. Drillhole gravimeters used in petroleum exploration have a minimum diameter of roughly 10 cm. They operate in holes that are inclined up to 14° from the vertical, and at temperatures up to 115 °C, or more using special housings for the meter.

Variations in the temperature of the gravimeter sensor are potentially a principal source of error in gravity readings. These variations will modify the geometry of the mechanical system and the density of the air, and thus its buoyancy. As a result the zero-length spring is made of metals with low thermal coefficient of expansion and is operated as near as possible to its zero thermal coefficient. Nonetheless, the system requires a near-constant temperature to achieve high precision. The meter is placed in a thermally insulated housing and is maintained at a near-constant temperature by a thermostated heating system powered by a battery. Typically, temperatures are maintained within the sensor housing to a thousandth of a degree centigrade or less, because instruments have a temperature coefficient of roughly 0.0003 °C/μGal.

Air pressure variations within a gravimeter will change the buoyancy of the air, and thus affect the geometry of the sensor system. This effect is minimized by sealing the sensor against air pressure variations and placing a barometric compensator on the pivot beam. Finally, the metal components of the sensor are subject to magnetic effects which are controlled under normal conditions by demagnetization of the metal parts and placing a magnetic shield around the sensor.

Quartz zero-length spring gravimeters

Another gravimeter using the zero-length spring concept to attain great sensitivity is the Worden type instrument (Figure 5.5(a)) which became commercially available in the mid-1940s. This meter, because of its sensitivity, portability, and robustness, has been widely used in geophysical exploration. The original Worden meter has been reproduced with modifications by several other instrument manufacturers.

Although similar to the LaCoste & Romberg type gravimeter, there are notable differences which bear upon its operating characteristics, advantages, and limitations. Instead of a metal, inclined zero-length spring, the Worden type meter uses a vertical quartz spring (Figure 5.5(b)). In fact the majority of the sensor system is constructed from fused quartz. Quartz is well suited to use in gravimeter sensors because of its excellent elastic properties, limited thermal response, and, unlike metals, it has limited fatigue or memory effects and is essentially non-magnetic.

A schematic illustration of the design of the Worden type meter sensor is presented in Figure 5.6. The sensor is housed in an evacuated cylindrical flask to minimize ambient air temperature variations and to isolate the

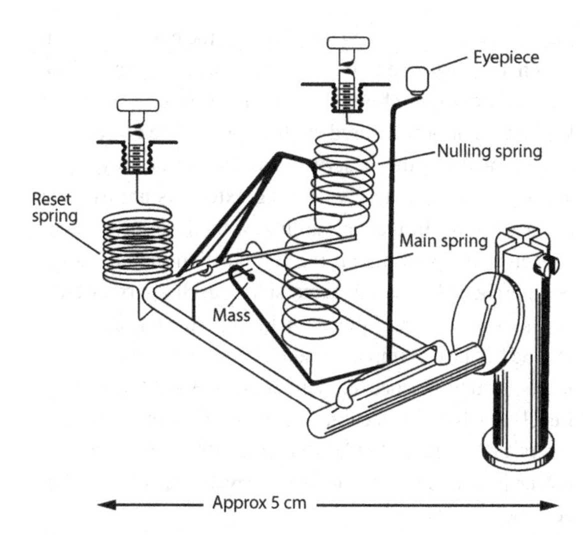

FIGURE 5.6 Schematic diagram of the Worden gravimeter sensor showing the major components. Courtesy of Worden Gravity Meter, Inc.

sensor from the effects of varying barometric pressure. The sensor fits into a roughly 5 cm cube, and the meter housing is about 18 cm in diameter and 36 cm in height. The weight is approximately 3.5 kg. The mass beam is connected to the bottom of the zero-length spring and also to the indicator arm through a quartz torsion hinge. The spring is commonly adjusted to have a natural period of 5 to 8 seconds which permits a precision of the order of 0.01 mGal. The mass is allowed to move freely between two stops; the low weight of the test mass and the weight arm prevent destructive oscillations caused by moving the meter. But this also is a disadvantage because the low test mass ultimately limits sensitivity.

Temperature compensation is achieved by a metallic fiber which is connected to the upper and lower ends of the arm that is attached to the zero-length spring. The difference in the temperature coefficients of these two components causes a rotation of the arm which compensates for the change in length of the spring with temperature. A universal compensation covering the full range of surface gravity and a broad range of temperature variations can be implemented with construction of a non-linear compensator. To further minimize rapid and large fluctuations in temperature, some Worden type instruments include a thermostatically controlled heating element which maintains the temperature within a limited range. But it does not have to be heated to the high temperature required for operating metal mechanisms. Therefore, a small battery is sufficient which has a long lifetime in the field.

Worden type gravimeters are null-reading instruments in which a lever arm connected to the mass arm is returned to a null position by adjustable springs. This procedure is used to make the readings linear. Nulling is achieved by applying a torque to one of the two springs. A coarse or geodetic spring which covers most of the surface gravity range of the Earth is used to bring the measuring spring within range by adjusting for the latitudinal variation of gravity. The measuring screw often referred to as the "small dial" has a range of approximately 100 mGal and a sensitivity of approximately 0.01 mGal. Modern instruments are fully capable of measuring over the entire range of gravity on the surface of the Earth. In the more modern versions of this instrument the position of the lever is monitored and moved to a null position with a capacitance indicator system and a force feedback system as in many LaCoste & Romberg type gravimeters. The torque applied to the spring is calibrated into gravity units by tilting the instrument in the laboratory to prescribed angles and verifying the results against field occupation of sites of established gravity values.

A fully portable automated version of the quartz spring instrument has been manufactured by Scintrex which repetitively samples the observations and corrects them for tilt errors, temperature variations, and tidal effects, improving the productivity and reliability of the measurements and minimizing operator error. Models with sensitivities of 10 and 1 μGal are available. More recently, automated land gravimeters such as the Scintrex CG-5 have become available which are fully automated and self-leveling (Figure 5.7). The meter levels itself with servo-driven, telescoping legs for a true, in-level gravity reading. The repeatability of this instrument under field conditions is purported to be 0.003 mGal and mature meters have an inherent drift of less than 0.5 mGal per month. Meters are now available with an adjustable level of accuracy. An optional filter is available for removing microseismic activity on the observations.

In addition, a slim drillhole gravimeter using a quartz spring is available for shallow-zone applications, less than 3,000 m, which has a sensitivity of roughly 5 μGal over a 7,000 mGal range and can operate successfully in holes that are greater than 5.72 cm in diameter and are deflected from the vertical less than 60°.

Instrumental error sources

The high sensitivity required in many gravity surveys, $1{:}10^8$ in exploration surveys and $1{:}10^9$ in microgravity studies, leads to the need to consider a broad variety of potential error sources inherent to gravity instrumentation as well as those derived from external environmental variations (e.g. Torge, 1989; Rymer, 1989; Ander *et al.*, 1999). For example, Butler and Llopis (1991) found

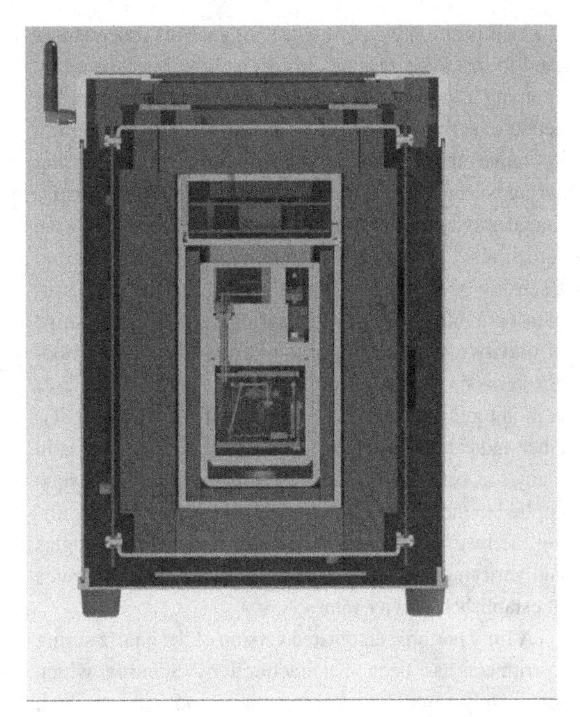

FIGURE 5.7 Schematic illustration of the cross-section of the Scintrex CG-5 gravimeter showing its major components. The sensor unit is housed in the lower center rectangular opening. The meter is roughly 31 cm high and 27 cm wide. Courtesy of Scintrex, Ltd.

in the mechanically noisy conditions of an urban environment that 50% of the observations repeat at less than 10 μGal and 90% at less than 30 μGal, but with more exacting procedures 60% of the stations will vary by less than 5 μGal and 95% by less than 10 μGal. The impact of these errors on the accuracy of the measurements can be diminished to a significant degree by the design and construction of the instruments. The more recently designed instruments, especially gravimeters which automatically level and take repeated observations at a station, are less prone to errors, particularly random errors. Nonetheless, residual error sources remain in gravity observations that should be evaluated when selecting a gravimeter for a particular survey and designing a data acquisition program.

Instrumental errors include effects from reading and leveling errors, elastic hysteresis, variable power supplies, and calibration. Reading errors are particularly a problem in older instruments still in active use today which require visual observation of the null position of the lever arm of the sensor through the meter's optical system. These random errors are decreased by repeated observations and by use of an automated scheme for monitoring and nulling the instrument. Random errors also may result from above-the-surface differences in the height of the base plate upon which the instrument is set during measurements. The instrument height above the ground surface needs to be kept constant or measured so that appropriate corrections can be made to the observations. This error may reach 10 μGal or much more where leg extenders are used on the base plate. In manual nulling of the lever arm, backlash errors may arise; thus the null position should always be approached from the same side. Leveling errors arise because of the lack of precision in centering the spirit levels and because of improper adjustment of the levels. Spirit levels need to be checked daily and adjusted if necessary. Levels that have an accuracy of 10 seconds generally have errors of less than 2 μGal (Torge, 1989). Some of the newer instruments use electronic levels that have higher accuracy.

Springs are subject to elastic hysteresis effects that occur upon resetting the measurement range of a meter and unclamping a sensor. They cause rapid changes in the gravimeter reading or tares, which generally decrease to the microgal level several minutes after resetting the range. A tare also can occur when the instrument is shocked during field transport causing sudden mechanical slippage within the instrument mechanism. In this case, the tare is essentially a break in the trend of the gravity reading, which is a reset in relative value.

The instruments should always be allowed time to settle thermally, especially when first turned on. If a voltage regulator is not used on the power supply which keeps the temperature of the sensor constant, unstable voltages may occur, affecting the temperature of the system and leading to errors in the gravity observations. Problems with voltages especially occur immediately after reconnecting a battery to the system and after a strong discharge of the battery.

Periodic calibration errors, also referred to as circular errors, may result from non-linear effects caused by variations in the pitch and shape of the nulling screws and imperfections in the springs or levers of the nulling system. These errors are particularly troublesome when high accuracy is required over a large measurement range. Careful construction can minimize these errors which can be identified by making measurements at a series of stations convering a range of observed gravity. Also, electronic nulling systems help to minimize the problem by eliminating use of the screw except to re-range the device. Calibration shifts may occur in LaCoste & Romberg type gravimeters over periods of several months. Carbone and Rymer (1999) have observed such calibration changes of up to 0.1% which is comparable with the calibration drift of new quartz type instruments. Ukawa *et al.* (2010) also have observed calibration variations in the Scintrex CG-3M gravimeter of the order of 10 ppm/year even several years after construction of the meter. Clearly, it is advisable to periodically check the calibration over an established series of stations whose values cover the normal reading range of the instrument or at the start and conclusion of high-sensitivity surveys.

Numerous environmental factors may have a disturbing effect upon gravity observations. Of particular importance are changes in temperature and atmospheric pressure, but the effects of mechanical shocks and magnetic fields can also be troublesome. The design and construction of instruments are focused on minimizing these effects. Instruments are thermally insulated and maintained at a near-constant temperature and/or thermally compensated. Generally, instruments that rely on thermal compensation, such as the older Worden type instruments, have measurable temperature effects depending on the magnitude and character of the temperature change. To ameliorate these it is advisable to maintain the instrument at as constant an external temperature as possible. This is true even for instruments that are temperature-controlled, because of temperature gradients that are developed across the sensor housing when the temperatures change markedly. The external temperature variations are decreased and delayed in reaching the sensor by its surrounding insulation and the vacuum maintained in the housing. A thermal tare is likely to occur in the readings when the power supply is cut off from temperature controlled meters. Even if the loss of power is only for a few minutes, the recovery period from the tare may take several hours (Rymer, 1989).

Atmospheric pressure changes due to either elevation or meteorological changes will affect gravimeter readings largely because of the buoyancy effect on the sensor, given by Archimedes' principle. For this reason gravity sensors are housed in evacuated chambers and/or have pressure compensating devices built into the sensor. Older Worden type sensors are placed in a glass flask vacuum chamber. The glass flask sets a limit to the robustness of the gravimeter. As long as the hermetic seals of the housings are maintained, the effects of atmospheric pressure are neglected in gravity measurements.

Magnetic effects may cause measurable effects in metal gravimeter sensors if the metal is not demagnetized or if magnetic shielding is insufficient. Normally, magnetic fields are neglected without deleterious consequences, but in regions of abnormally high and variable fields, it may be necessary to make the measurements with a consistent instrument orientation with respect to the ambient field. However, this is seldom a problem with most modern instruments.

Perhaps the most vexing of the external environment effects stem from mechanical motion or shocks to the gravimeter sensor, particularly during transport between observation sites, even in systems which are arrested during transport (Hamilton and Brulé, 1967). The effect depends on the type and magnitude of the shock and on the construction of the instrument. Shock sustained during transport depends on the mode of transport and the instrumental shock absorption that is employed. Vibrations will cause a continuous change in the time rate of change of the instrument, whereas sudden discrete mechanical shocks may cause rapid changes in the gravimeter reading (tares). The isolated nature of the tares makes them difficult to recognize and thus particularly troublesome. One type of shock parallel to the axis of the screw may cause the calibration screw to slip a thread, leading to instrument calibration errors.

Both continuous and discrete changes due to exterior mechanical effects decrease with time in a manner that is difficult to predict. Even clamped systems are subject to increased time variations when subjected to vibrations that excite the natural periods of the sensor. Mechanical vibrations caused by vehicular traffic, operating machines, seismic waves, microseismic effects, wind, loose ground, etc. also may cause difficulty in nulling the gravimeter. Thus, care needs to be taken to avoid these disturbances by proper selection of the observation site or time. Microseismic events near coastal regions due to ocean waves breaking on shore are eliminated by filtering in some gravimeters. However, the passage of seismic waves, particularly seismic surface waves, associated with either local earthquakes or major seismic events anywhere on the Earth may cause disturbances exceeding 1 mGal that cannot be eliminated under most conditions, so gravimeter observations must be halted during these periods. The effects of major seismic events may last for several hours or more.

Manufacturers of gravimeters discuss potential sources of error and methods of minimizing errors in their instruction manuals. Careful study of these manuals is a necessity in understanding and dealing with individual models of instruments. Error budgets can be assembled for the various model gravimeters from the information provided in these manuals and tests on individual instruments. It is necessary to perform individual error tests because of idiosyncrasies of particular instruments and the specific effects on the readings of the manner in which the gravimeter is handled and the procedures used in making observations. Individualized errors are much less a factor in the error budget of automated instruments which remove most but not all the operator considerations in the measurements.

A common method of ascertaining the error of a particular instrument is to make repeated measurements for several days at the same location (usually indoors where the meter can be left without being disturbed). The instrument should be able to track diurnal Earth tides during the period of observation. If not, distortion from the predicted tides gives information on the drift characteristics of the meter. If the difference between the observed and predicted tides is pronounced, there may be a problem with the instrument.

TORGE (1989) has presented examples of error budgets for LaCoste & Romberg gravimeters using normal as well as specialized procedures that would be employed in microgravity surveying. These budgets include a wide variety of potential errors, both random and systematic, as well as total errors that include temporal gravity changes. The main sources of error are temperature variations, mechanical shocks, calibration errors, and gravimetric tides. The total error in the observations using standard procedures is around 0.02 to 0.03 mGal, and with precision procedures this can be reduced to approximately 0.01 mGal. Similar accuracies have been quoted for quartz spring gravimeters. In general, the errors of the Model D LaCoste & Romberg gravimeter are slightly less than those of the Model G.

Errors due to random events can be reduced by making repeated observations at a site. Non-linear and periodic calibration effects can be minimized or at least identified by occupying a station with more than one gravimeter. The main sources of errors and problems in establishing the drift of the gravimeter are reduced notably in microgravity surveys where the time between observations is short and their gravity differences are less than a few milligals. Under these conditions the overall error budget is less than 10 μGal. Where the meter is not moved between observations and the sensitivity of the meter permits, precision of the order of a few microgals or less may be obtained. These types of observations are useful in measuring temporal variations in gravity due to mass transfer at depth and changes in surface elevation.

ANDER *et al.* (1999) in reviewing the potential sources of errors in the LaCoste & Romberg type gravimeters conclude that there are no fundamental instrumental limitations which prevent the meter from making observations to the 0.1 μGal level. Importantly, they find the test mass of this instrument is subject to temporal variations, even over the length of a survey, which increase the drift and subject the instrument to tares. These changes are believed to be caused by contamination of the mass by dust, oil, and other volatiles within the sensor unit which can be largely eliminated by vacuum cleaning for several days.

Vibrating string gravimeter

Another type of gravimeter that has been used for measuring land and drillhole gravity, as well as gravity on the surface of the Moon (TALWANI, 2003), is the vibrating string gravimeter. Although not widely used, it has the advantages of being small and having a large dynamic range. The meter is developed on the principle that the resonant frequency of a mass-loaded vertical string is proportional to gravity (GILBERT, 1949; HOWELL *et al.*, 1966). The basic instrument consists of a flat, thin, conducting ribbon that is suspended vertically and weighted with a mass. The resonant frequency of this system is

$$f = \frac{1}{2L}\sqrt{\frac{mg}{\sigma_l}}, \tag{5.6}$$

where L is the length of the string, m is the mass suspended from the string, g is the gravity acceleration, and σ_l is the density of the string per unit length. Differentiating the above expression relates the relative frequency change to gravity change by

$$\frac{\partial f}{f} = \frac{1}{2}\left(\frac{\partial g}{g}\right). \tag{5.7}$$

The ribbon or string is caused to vibrate at its resonant frequency in a magnetic field generating a recorded oscillating voltage. The system has been constructed using a variety of configurations of ribbons and masses. The vibrating spring meter used for the measurements on the Moon was a double string mounted vertically and separated by a weak spring with the strings constrained at both ends. The difference in the resonant frequencies of the two strings is a function of the gravity.

5.3.2 Moving platform measurements

Gravity observations made on a moving platform, either a ship or aircraft, have a significant role in exploration studies despite their generally lower accuracy and lower spatial resolution compared with land surface measurements. The advantage of measurements on a moving platform is that surveys can be made rapidly and efficiently in marine regions and in land areas that are not easily accessible to land surveying. Shipborne measurements are generally made as auxiliary observations on vessels conducting surveys for other purposes, while airborne measurements are made either from helicopters or fixed-winged aircraft flown specifically for gravity surveying. The principal challenge is the dynamic environment, with the associated problems of keeping the gravimeter vertical and sorting out inertial accelerations of the moving platform from the gravity effect of subsurface sources.

Useful general descriptions of gravity instrumentation for moving platforms have been published by DEHLINGER (1978), TORGE (1989), HERRING and HALL (2006), and DIFRANCESCO *et al.* (2008).

Although several gravity measuring instruments have been used in the marine environment, including pendulums and gravimeters, it was the use of an overdamped version of the land LaCoste & Romberg gravimeter on surface ships that led to broad commercial marine gravity exploration. A key to the success of measuring gravity on moving platforms beginning in the mid-1960s was the understanding that no matter what the state of the sea, the effects of wave motions on the sensor, the inertial accelerations, average out to zero over time. This means that a gyrostabilized platform can compensate for sea states. The gyrostabilized platform upon which the gravimeter is mounted counteracts extraneous motion by minimizing horizontal accelerations and the tilt of the platform. Servomotors actuated by accelerometers keep the platform horizontal, and thus the gravimeter vertical, within a few seconds of arc. The widely used LaCoste & Romberg sea gravimeter beam is heavily damped to match the behavior of the platform so that the sensor does not overreact to each motion. As a result of the heavy damping, observations of the displacement of the beam are not practical. Accordingly, the rate of change of the beam motion which is a measure of the gravity acting on the test mass is observed.

Extraneous vertical accelerations of the sensor are reduced by applying low-pass filters that match the sea state, thus isolating the longer period components of the gravity measurements associated with subsurface sources. This low-pass filtering can be accomplished physically by mechanical, magnetic, or hydraulic damping of the sensor mass, electronically by analog or digital filtering of the recorded signal, or computationally. Ocean wave effects that have wavelengths longer than a minute generally have amplitudes smaller than one milligal, minimizing their interference with Earth gravitational effects (HERRING and HALL, 2006). This is in contrast to the short wavelength (5–10 s) ocean wave vertical fields that can reach amplitudes of from 10,000 to 50,000 mGal. An alternative to the gyrostabilized platform is to use the gravimeter in a strapped-down mode and correct the off-level error effects in the gravity measurement with positioning data obtained by highly accurate GPS surveying (e.g. VERDUN and KLINGELÉ, 2005).

Despite the actions of the stable platform in moving measurements, some residual horizontal accelerations persist. These accelerations produce fictitious vertical accelerations in the inclined zero-length spring of the LaCoste & Romberg gravimeter, and resulting errors in the gravity measurements. These cross-coupling effects can reach values of several tens of milligals owing to circular motions of the transport vessel, and thus must be removed from the measurements used in subsurface studies. LACOSTE *et al.* (1967) refer to these as inherent cross-coupling, to differentiate them from cross-coupling due to imperfections in the gravimeter which can be greater than the inherent cross-coupling. The residual horizontal accelerations cause a torque on the sensor beam which is proportional to this acceleration and to the angular relationship between the beam and the horizontal, which varies with the frequency of the vessel's motion and the phase difference between the horizontal and vertical accelerations.

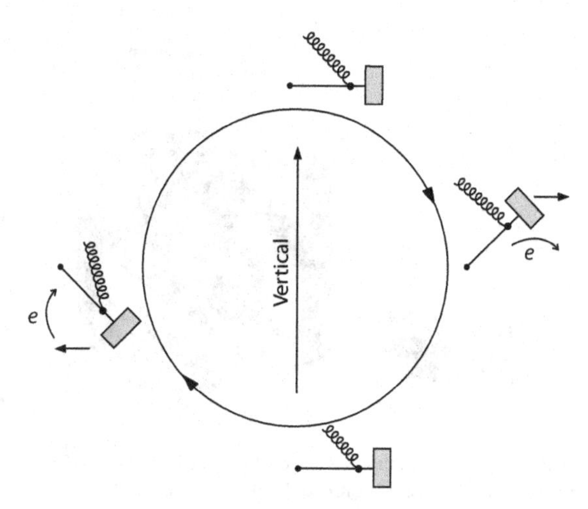

FIGURE 5.8 Schematic illustration of the inherent cross-coupling of a highly damped inclined-spring gravimeter during a wave cycle. The vectors in the 3-o'clock and 9-o'clock positions show the direction of the force due to the motion of the meter during the wave cycle, and the resulting shift of the sensor beam causing an error (e) in the gravity measurement from the true gravity indicated by the sensor beam at the 12-o'clock and 6-o'clock positions. Adapted from LACOSTE *et al.* (1967).

The vertical movements of a highly damped sensor beam resulting from residual accelerations during a wave cycle are shown schematically in Figure 5.8. At the 12-o'clock and 6-o'clock positions the rates of the beam displacement give a true gravity measurement, but at the 3-o'clock and 9-o'clock positions a spurious measure is given by the beam. During the down part of the wave cycle the apparent measurement is too low and for the up part of the cycle the converse is the case. The errors in the up and down part of the cycle usually do not counteract each other. The cross-coupling error can be minimized by construction and appropriate operation of the meter, and also removed by calculations based on measured

FIGURE 5.9 The LaCoste & Romberg System 6 Dynamic (Air/Sea) gravity meter. Courtesy of LaCoste & Romberg-Scintrex (LRS), Inc.

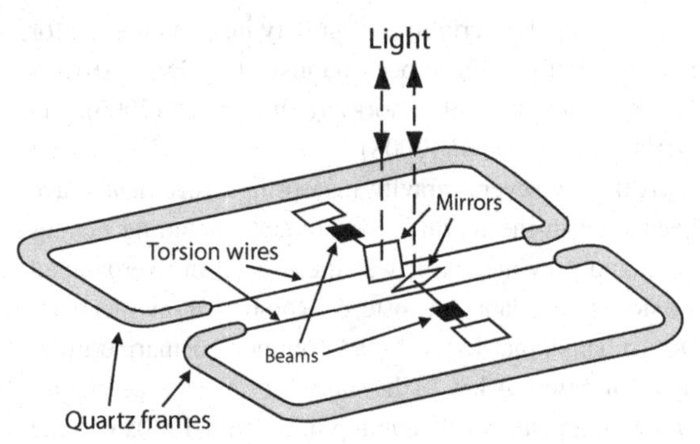

FIGURE 5.10 Schematic diagram of the Chekan-AM gravimeter quartz double-beam sensor. Adapted from STELKENS-KOBSCH (2005).

horizontal accelerations and deflection of the mass beam from the horizontal.

A specialized gravimeter was designed and constructed by LACOSTE (1983) to eliminate cross-coupling effects in measurements made on a moving platform, the so-called LaCoste & Romberg S-gravimeter. The sensor of this meter employed a modified suspension of the test mass in which the movement is vertically in a straight line rather than in an arc. In the past decade, Zero Length Spring, Inc. developed the ZLS Dynamic Meter™ Sensor based on this principle. However, most marine and airborne gravity measurements have been made with highly damped beam-type LaCoste & Romberg sensors with a metal zero-length spring placed on a gyrostabilized platform (LACOSTE *et al.*, 1967; VALLIANT, 1992). The LaCoste & Romberg System 6 Dynamic (Air/Sea) gravity meter is illustrated in Figure 5.9, but several modifications of this meter are widely used (SAGE™, ZLS UltraSys, AirSea System II, TAGS Air III, etc). The modifications are mostly related to upgraded electronics and software of the stabilized platform and gravity meter control as well as data logging systems. Important technical improvements since the original design included changing the beam position indicator from optical to capacitive, and transitioning from two single-axis gyros to a single three-axis gyro pack and finally to modern fiber-optic gyros (FOG). The latter gyros have no moving parts, and thus are advantageous because they are insensitive to mechanical effects. FOGs sense the platform movement photometrically by way of an interference pattern of a light beam which is injected in opposite directions of a very long strand of fiber optic cable wound on a coil.

Several other gravimeters designed for moving measurements using stabilized platforms have or are being used in exploration. For example, the updated Askania gravimeter developed by the Bodenseewerk employs a vertical spring with an attached mass which is largely unaffected by cross-coupling errors. Another beam-type marine gravity meter, GMN-K, similar in principle to the highly damped metal spring LaCoste & Romberg gravimeters, was developed in Russia in the late 1970s and used in marine surveys in the 1980s and 1990s. This gravity meter used the quartz astatic-spring beam-type sensor system with very high liquid damping, mounted on a passive gyrostabilizer. A somewhat similar gravimeter used in marine as well as airborne applications is the Russian Chekan-AM meter. This gravity meter has a quartz double-beam sensor (Figure 5.10) which consists of two identical torsion frames holding pre-stressed quartz fibers to which horizontal beams are attached. The beams are oriented in opposite directions. As a result, when the outputs of the beams are summed, the cross-coupling effects of the beams cancel each other to the extent that the properties of the two systems are identical. Both beam systems are placed in the same housing with highly viscous liquid for initial physical damping of vertical accelerations, and the entire unit is placed on a gyrostabilized platform. Further descriptions of the meter are provided by STELKENS-KOBSCH (2005) and KREYE *et al.* (2006).

Alternatively, gravity measurement systems originally developed by the US military and declassified in the 1990s, using electromagnetic accelerometers rather than spring systems, have been used increasingly because they have a purported lower noise level and higher resolution than previous systems. The BGM-3 marine gravimeter as described by BELL and WATTS (1986) uses this technology, consisting of a test mass which is limited to movement in the vertical direction between two permanent magnets (Figure 5.11). The mass is kept in a balanced

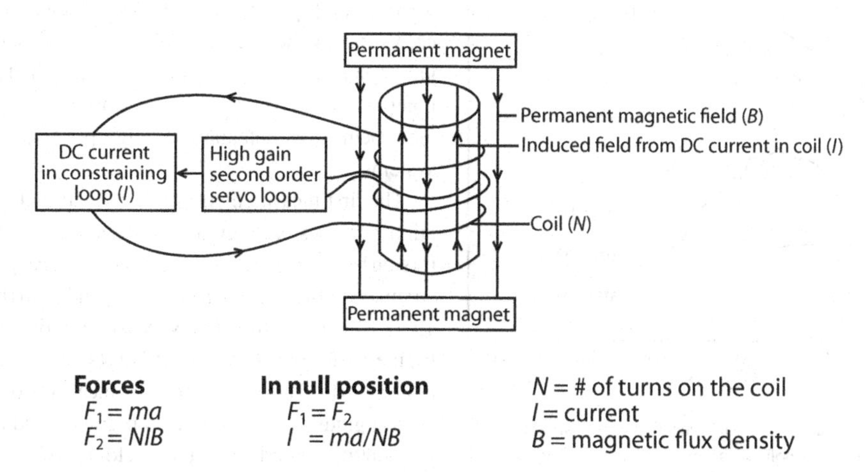

FIGURE 5.11 Schematic diagram of the BGM-3 accelerometer configuration with the test mass wrapped in a coil between two permanent magnets. The mass is kept in a null position balanced between the local gravitational force and the electromagnetic force induced by the d.c. current flowing in the coil. With a change in the gravitational force and movement of the mass, a current is induced in a servo-loop which causes a change in the coil current to bring the mass back to the null position. The change in this current, which is measured, is proportional to the change in the gravity field. Adapted from BELL and WATTS (1986).

position between the magnets by the gravitational force acting upon the mass and the electromagnetic force induced in the mass by a direct current passing through a coil which is wound around the mass. Upon experiencing a change in the gravitational field and a resulting movement of the mass, a field is generated in a secondary servo-loop around the mass which modifies the direct current in the main coil to bring the mass back to the null position. The change in the coil current is proportional to the change in gravity. The sensor is mounted on a stable platform to minimize the effect of inertial accelerations due to the changing horizontal and vertical movements of the gravimeter system.

Over the past few decades, several other gravimeters have been developed primarily for making observations in either the marine or airborne environment. All are mounted on high-precision gyrostabilized platforms to minimize inertial accelerations, but they use a variety of sensors. For example, the Russian-developed GT-1A meter has a test mass on a spring suspension which minimizes cross-coupling (GABELL *et al.*, 2004). The Sander Geophysical, Inc. AIRGrav system, which was built specifically for airborne applications (SANDER *et al.*, 2004; STUDINGER *et al.*, 2008), uses a three-axis stabilized inertial platform with three orthogonal accelerometers and two two-degrees-of-freedom gyroscopes. GPS-derived accelerations of the aircraft which are synchronized with gravity measurements are subtracted from the measured accelerations and filtered to remove additional acceleration noise. The platform keeps the vertical accelerometer oriented for measurement of changes in the vertical gravity field, but the combination of the output of the high precision accelerometers and gyroscopes can be used to determine the vector gravity components.

The accuracy and resolution of these various gravimeters are dependent on the gravity meter system, the speed and direction of the vessel, and the sea state or aircraft turbulence. Thus, it is difficult to give precise values for these parameters or error budgets for the systems as used in either marine or air observations, but accuracies of slightly better than 0.1 mGal and a half-wavelength resolution of 300 m have been reported.

A significant complication to moving gravity measurements is the effect of the changing Coriolis acceleration, the centrifugal force due to the horizontal movement of the observation platform over the rotating Earth's surface. This so-called Eötvös effect causes an increase in the gravity readings during east-to-west movement and a decrease for movement in the opposite direction. This effect must be removed from the moving observations to equate the measurements to stationary observations and is calculated from the velocity and direction of the platform. The calculation and application of Eötvös effects are considered further in Section 6.3.8. Improvements in the accuracy of determining the location of observations and the velocity of the platform from differential

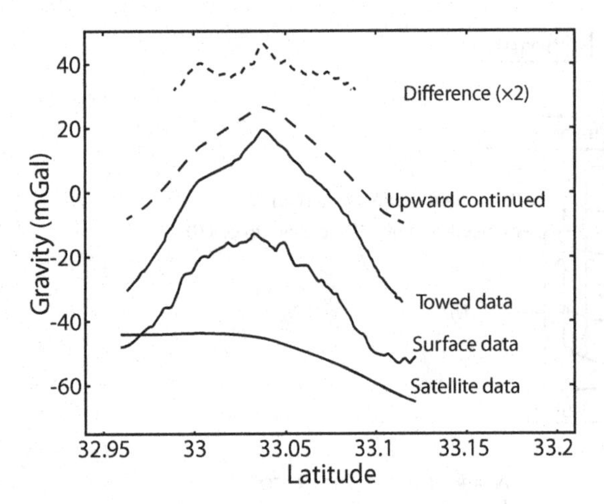

FIGURE 5.12 Comparison of near-coincident gravity anomaly profiles over the Emery Knoll, offshore California at the ocean surface and at a depth of 610 m obtained during a deep-tow gravity survey. The profiles are arbitrarily shifted to separate the anomalies from each other. The deep-tow profile is expanded by a factor of 2 over the other profiles to make the 5 mGal anomaly observed over the center of the Knoll more perceptible. Adapted from Ridgway and Zumberge (2002).

GPS positioning and Doppler shift of the GPS carrier phase signal have had a profound effect on increasing the accuracy of the Eötvös effect and improving gravity measurements from moving platforms (Herring and Hall, 2006). Accordingly, modern data are likely to be significantly more accurate than those observed prior to enhanced GPS position control.

To improve the resolution of marine measurements which is roughly limited to the order of the depth of the water, the LaCoste & Romberg marine gravimeter has been placed in a submerged robot pressure-vessel and towed behind a marine vessel along a track only a few tens of meters above the ocean bottom (Zumberge *et al.*, 1997). This system, which is credited with a repeatability of less than 0.4 mGal, is used to increase the resolution of marine gravity measurements while still making observations rapidly and continuously. The depth and location of the meter, monitored by conventional methods, is used for reducing the observations to anomaly form and eliminating the Eötvös effect from the measurements. The enhanced resolution of deep-tow gravity measurements over ocean surface observations is illustrated in Figure 5.12 from a survey over the Emery Knoll offshore California (Ridgway and Zumberge, 2002). The deep-tow measurements show the presence of a local 5 mGal anomaly over the center of the Knoll not observed in the surface measurements.

5.3.3 Gravity gradiometry

Recent developments in gravimetry include the marked expansion of the use of shipborne and airborne gravity gradiometers. Gradients or tensors of the gravity field have inherent advantages in interpretation of gravity data and lend themselves potentially to high precision in onboard systems.

The first measurements of the gravity field for exploration purposes were gradients observed with the Eötvös torsion balance, which remained the principal gravity measurement until replaced in part with pendulums by the early 1930s and almost completely by astatic spring gravimeters by the mid-1930s (Bell and Hansen, 1998; Speake *et al.*, 2001). The Eötvös balance measures both the horizontal gradient of the horizontal gravity and the vertical gradient of the vertical gravity field with an accuracy of a few Eötvös units ($1\,\mathrm{E} = 10^{-9}/\mathrm{s}^2 = 0.1\,\mathrm{mGal/km}$). These gradient measurements, despite their relatively high resolution, were displaced because they were time-consuming (requiring several hours for a measurement) and subject to errors from non-geological gradients, primarily terrain effects (Bell and Hansen, 1998). Subsequently the measurement of gravity gradients has been used for only specialized objectives, largely related to near-surface compact geological features of large amplitude gradients (e.g. Agar and Liard, 1982). However, gravity gradients are valuable in interpretation, particularly in increasing the perceptibility of shallower anomalies and in mapping the margins of anomalous sources. As a result, gradients of the vertical force of gravity have been approximated from the spatial relationship among gravity anomaly observations, either in the space or in the frequency domain.

Recently, there has been increasing interest in measuring gravity gradients for exploration purposes. This stems from improvements in gradiometers developed since the 1960s for military-related objectives, and the availability of digital terrain models and computational power to readily and accurately calculate the effect of terrain-related gradients, which are a primary source of perturbing non-instrumental gradients on land surveys. The rotating-disk military gradiometer, which is a full tensor gradiometer (FTG) that measures the five independent tensors of gravity (Chapter 3), has been modified for exploration surveying. The tensor measurements largely eliminate the regional gravity effects from deep sources which have low gradient anomalies, and focus instead on high gradient anomalies from shallow sources, resulting in an interpretational advantage (e.g. Pedersen and Rasmussen, 1990; Pawlowski, 1998; Saad, 2006; Mims and Mataragio, 2010).

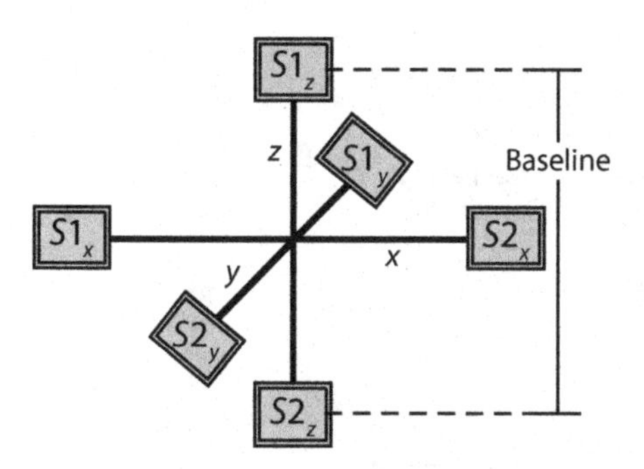

FIGURE 5.13 Schematic illustration of the principle of the full tensor gravity gradiometer. Individual accelerometers are labeled S. The difference in the measurements of the acceleration divided by the baseline separation distance is the gradient of gravity in each of the three orthogonal directions x, y, and z. Combinations of these approximations give nine different tensor elements of gravity, of which only five are unique. Adapted from CHAPIN (1998).

Gradiometer measurements are capable of high accuracy in the dynamic environment of shipborne and airborne systems because they are less susceptible to large translational accelerations that affect conventional gravimeters. The FTG instrumentation consists of 12 paired force-rebalance accelerometers mounted on three rotating disks (Figure 5.13). The effect of extraneous accelerations is minimized because the gradients are measured between two closely spaced sensors that are subject to essentially the same accelerations resulting from the motion of the platform. By subtracting the observations from each other to obtain the gradient, these extraneous effects are eliminated as long as the response of the sensors is identical and the distance between sensors is small enough that the gradient is essentially linear.

With the release of the rotating-disk FTG technology to the public in 1994, the gradiometer system has been modified and redesigned to meet exploration objectives, to be easily portable, and to accommodate the high dynamic environment of airborne surveys. Although other systems for gravity gradiometers are being investigated, including the superconducting instrument (MOODY *et al.*, 1986), at this time the vast majority of airborne gradiometer surveys are being conducted with derivatives of the military FTG technology. The airborne FTG (MIMS and MATARAGIO, 2010) has been modified directly from the military system and takes advantage of the full gravity tensors in subsurface interpretation (STASINOWSKY, 2010). The other system, the Falcon technology, was developed from FTG technology for airborne surveys to detect and map small, low-amplitude anomalies encountered in mineral exploration surveys. This latter system measures only the vertical gravity and its vertical gradient (LEE, 2001). Both of these systems lead to accuracies after processing of the order of 3 to 7 Eötvös units and resolution of the order of 0.3 to 1 km. Improved characteristics are anticipated with further developments in noise reduction.

5.3.4 Absolute gravity

Relative gravity values measured with gravimeters are the norm in geophysical exploration in both static and dynamic environments because the measurements are highly sensitive and rapidly made, and the instrumentation is relatively portable and robust. Nonetheless, absolute gravity values are of interest and have special uses in exploration. They are used to establish reference points to tie together individual surveys, to reference surveys to national and international datums, to establish gravity benchmarks for calibration of gravimeters, and for specialized studies involving high-precision time variations in gravity free from the drift of relative gravimeters (NIEBAUER, 2007). In addition, absolute gravity is useful in metrology, for example, in calibration of mechanical force standards and for Earth parameters such as the wobble in the Earth's rotational axis.

Historically, absolute gravity measurements were made by observing the oscillation time (i.e. period of oscillation) of a pendulum, which is related to the acceleration of gravity through the physical characteristics of the pendulum. The simple pendulum, for example, which has a weightless shaft of length l connecting a mass to the fulcrum of the pendulum, oscillates with a period T equal to $2\pi\sqrt{l/g}$. In practice, the measurement of gravity used a compound pendulum which relates the period of oscillation to gravity through the moment of inertia of the pendulum. Unfortunately, multiple sources of noise and constraints on determining the physical parameters of the pendulum limit the accuracy of the most sensitive pendulums to roughly 0.1 mGal.

As a result, most absolute observations are now made by measuring the time t for the free fall of an optical mass over a vertical distance d within an evacuated chamber using a variety of instruments (e.g. TORGE, 1989). Specifically, the time t for a mass to travel the distance d in the gravitational field of acceleration g is $\sqrt{2d/g}$. Thus, direct measurements of the distance and time parameters give the acceleration of the mass due to the gravitational field. Recent improvements in instrumentation, especially in measuring time with a calibrated laser standard coupled with an atomic clock, have made it possible to obtain

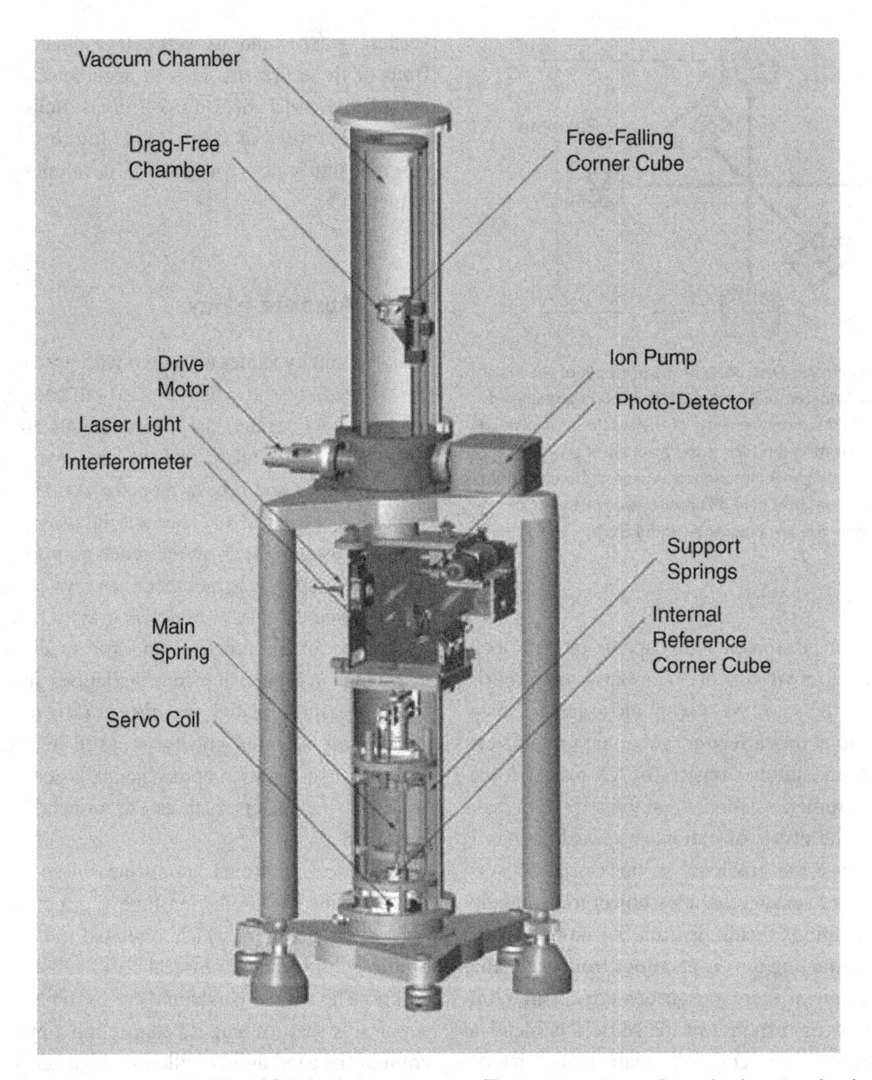

FIGURE 5.14 Schematic cross-section of the FG-5 absolute gravity meter. The upper section encloses the dropping chamber, the middle section is the interferometer for measuring the position of the falling mass, and the lower section contains the springs which isolate the meter from Earth vibrations. Courtesy of LaCoste & Romberg-Scintrex (LRS), Inc.

accuracies of the order of 1 μGal and to reduce observation times to an hour or less. The accuracy of the instrument is enhanced by averaging the results over repeat measurements.

The modern absolute gravity measuring instrument consists of three major components (Figure 5.14); an upper chamber in which the mass element drops at roughly 100 cycles/minute, a central interferometer measurement system, and a lower long-period seismometer (Super Spring) which isolates the free-fall chamber from Earth vibrations which can seriously downgrade the precision of the measurements by changing the frame of reference for the measurement (NIEBAUER *et al.*, 1995). This instrument has undergone numerous improvements (MARSON and FALLER, 1986) in recent decades, including decreasing the length of the free-fall chamber and thus making the instrument smaller and more portable. In addition, it has been made more robust, making it feasible to use in field situations, and measurement times have been decreased.

The FG-5 free-fall instrument shown schematically in Figure 5.14 has a reported absolute accuracy of ±2 μGal, a measurement precision of ±1 μGal, and an integration time of about 1 hour for 1 μGal measurement precision. A more fieldworthy version of this instrument, the FGL Absolute Gravimeter, has an instrument set up, observation, and tear-down time measured in tens of minutes. The absolute accuracy of this instrument is reported to be

±10 μGal with a measurement precision of ±5 μGal. The high accuracy and precision of these instruments, together with the lack of inherent drift and tares in measurements and the elimination of calibration errors, make these instruments attractive for measuring small changes in gravity occurring over long time periods. These might be associated with, for example, crustal deformation, volcanic activity, movement of fluids and gases within Earth reservoirs, and tidal activity (Brown *et al.*, 1999). Recently, on an experimental basis, absolute measurements have been made from an airborne platform over the Swiss Alps, with a reported uncertainty of 6.9 mGal for wavelengths of 12 km based on comparison with upward-continued gravity values from the surface to an altitude of 2,500 m (Baumann *et al.*, 2011).

5.3.5 Navigation control

Establishing horizontal and vertical position is extremely important in gravity surveying. Since the deployment of the multiple Earth-orbiting satellites known as the Global Positioning System (GPS) in 1985, considerable advances have occurred in accurately locating gravity observations and reducing the costs and errors in survey navigation.

Horizontal positioning

Horizontal control is important in accurately locating gravity stations for determining the Eötvös effect of measurements made on moving platforms, mapping purposes, and in reducing gravity observations for the planetary latitudinal gravitational variation. The latter, as explained in Section 6.3.8, are due to changes in the radius of the Earth from the equator to the poles and the change in the centrifugal force due to the Earth's rotation. The latitudinal variation, which is a non-linear function of north–south distance, is zero at the poles and the equator, and maximum at 45° north or south latitude with a value of approximately 0.8 μGal/m. The required accuracy of the horizontal location is a function of the sum of the effect of the latitudinal effect and the gradient of the anomalies in the survey area.

The gradient of anomalies primarily is a function of the geometry and depth of the source. The maximum gradient will be observed over a concentrated source (e.g. a sphere) and will be located at a distance from the anomaly maximum equal to one-half of the depth to the center of the source. It is useful to calculate the maximum gradient anticipated in a survey in order to estimate the required precision in the horizontal position control. The maximum anomalous horizontal gradient of a spherical source in μGal/m is

$$\left(\frac{\partial g}{\partial x}\right)_{\max} \simeq 2.4 \times 10^{-2}\sigma\left(\frac{R^3}{z^3}\right), \tag{5.8}$$

where σ is the density contrast in kg/m^3, R is the radius of the sphere in meters, and z is the depth to the center of the sphere in meters.

For example, consider a survey designed to search for air-filled cavities in limestone that have a radius of 5 m and whose center depth is 10 m. Assuming a density contrast of 2,700 kg/m^3, the maximum gravity anomaly is roughly 95 μGal so that by the above equation, the maximum horizontal gravity gradient due to the assumed cavities is 8 μGal/m. Thus, if the error due to the mislocation of the station is to be kept to less than approximately 1%, the stations must be located to an accuracy of roughly 0.1 m. Of course, the latitudinal gravity effect also must be considered in evaluating the potential error from mislocation.

The horizontal locations of gravity stations are determined in a variety of ways. Until a few decades ago most stations in surface surveys were located by traditional optical surveying using tachymetry, a geometric technique in which the vertical distance between cross-hairs observed on a stadia rod through the surveying instrument is a measure of the distance between the instrument and the rod. Alternatively, electromagnetic distance measurement (EDM) techniques have been used, employing the timing of the travel time of microwaves for long distances and lasers or infrared radiation for short distances to determine the distance between the transmitter and a reflector which returns the signal to the transmitter/receiver. Reflectorless laser rangefinders are now available for measuring distances without the need to place a reflector at the survey location. When outfitted with inclinometers and internal fluxgate compasses, these instruments provide a complete surveying system, measuring not only distance, but direction and vertical angle as well. These units are particularly useful in mapping the location of local cultural and geographic features with respect to gravity stations and in specifying nearby topography for gravity terrain corrections. In more regional surveys, stations have been located on maps or aerial photographs using identifiable natural or cultural landmarks. This situation has changed with the availability of the GPS system for defining locations both in a static and in a dynamic environment (e.g. Fairhead *et al.*, 2003). The use of GPS in position control is so widespread that many newer gravimeters have built-in GPS receivers.

GPS, a USA system, and its counterparts from other nations which make up the Global Navigation Satellite

System (GNSS), have had a profound effect on surveying the location and elevation of gravity observation sites by increasing the efficiency of the surveying and improving the accuracy of the measurements over all but the most precise optical surveying procedures (AIKEN *et al.*, 1998). Optimum techniques lead to accuracies in horizontal positions of a centimeter or better depending on the system and methodology. The method is based on locating the position of a site while simultaneously receiving microwave signals from four of roughly 30 special satellites orbiting the Earth at an elevation of about 20,000 km. The time that it takes for the microwave signal to travel from a satellite to the antenna of the GPS receiver located at the gravity station is converted to distance, and thus, knowing the exact position of the satellite and repeating this for three additional satellite signals, it is possible by triangulation to locate the site of the GPS receiver antenna.

Each satellite transmits two carrier frequencies modulated by its own unique random binary code of on/off cycles. A number of options are available for measuring the three-dimensional position of GPS receiver sites depending on the required accuracy, observation spacing, survey time, and required productivity. It is desirable to select the method based on the requirements and resources of the survey.

The two basic procedures involve either code phase or carrier phase measurements. Code phase measurements are based on cross-correlation of the unique random code modulating the carrier frequency to determine the time delay of the signal in reaching the receiver from the transmitting satellite, and thus determine the distance between them. The receivers are tuned to several channels (currently up to 20) simultaneously for locating the receiver by triangulation. These signals are readily amplified permitting the use of small receiver antennas. Carrier phase measurements, which are inherently more accurate but more difficult to make, determine the phase difference between the incoming and receiver-generated high-frequency carrier waves. Most methodologies today are based on differential GPS (DGPS) using the measurement of codes or carrier phase. Both are commonly referred to as DGPS. This technique was essential for high-precision surveying before the intentional degradation of GPS for civilian applications was lifted by the US government on 1 May 2000, but it is still useful for eliminating inherent error associated with atmospheric effects on the transmission of microwave radiation, multipath and clock errors, receiver noise, and orbital uncertainties.

In the DGPS method, the measurements of a roving GPS receiver are compared with observations at a nearby receiver located at a site whose position is known with great accuracy (Figure 5.15). Additional precision can be obtained by using dual-frequency receivers to minimize ionospheric and tropospheric effects and by using receivers capable of receiving signals from all of the GNSS satellites. Corrections determined at the base site and received by radio at the roving receiver are used to adjust observations at the roving receiver. Processing of GPS observations can be done after the survey or on a real-time basis. The base receiver may be part of a stationary system of receivers or established only for the purposes of the specific survey. Currently, real-time kinematic (RTK) phase-difference GPS receivers used in a differential mode yield horizontal position accuracies of the order of a centimeter and somewhat lower accuracies in the vertical. However, the accuracy and efficiency of GPS mapping and navigating are improving rapidly so that RTK studies, especially useful for positioning of measurements in transit, are capable of achieving this accuracy at second or sub-second rates.

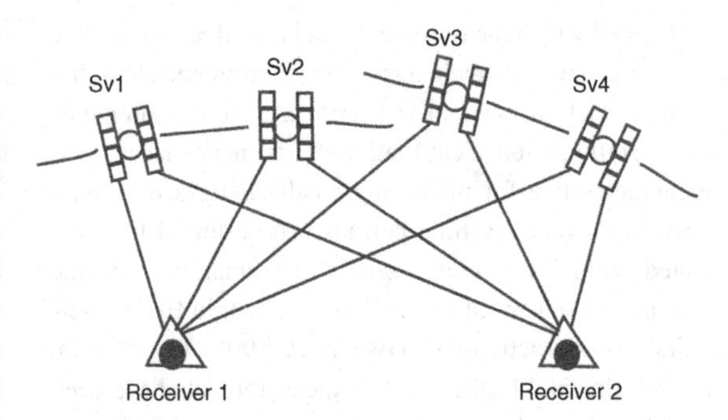

FIGURE 5.15 The principle of differential and relative GPS. A base receiver located at a known position and a roving receiver observe the signals from the same four satellites simultaneously to reduce common errors and yield accurate coordinate differences. Adapted from FEATHERSTONE (1995).

GPS measurements have the added advantage of providing information on the uncertainties in the observations and calculations. Furthermore they can be incorporated into the digital recording of the observed gravimeter measurements for quality control of the measurements, instantaneous recognition of observation problems, and potentially real-time reduction of the observations to anomaly form (e.g. CHEN *et al.*, 2005).

Vertical position

Reduction of gravity observations to account for station elevation variations is typically the most important control on the accuracy of computed gravity anomalies. Thus, determination of the elevation of the gravity stations is of paramount concern in planning and conducting a survey.

The combined effect of the vertical gradient of gravity and the attractive effect of the included mass between the elevation datum and the observation site is

$$g_z = 4.193\sigma \times 10^{-5}\ h \tag{5.9}$$

where σ is the density of the included mass in kg/m^3 and h is the height of the station in meters. Assuming the density of surface material is of the order of 2,000 kg/m^3, the vertical gradient of gravity is approximately 84 μGal/m. As a result the surface elevation must be known to an accuracy of approximately 1 cm to achieve a 1 μGal accuracy.

The effect of the anomalous vertical gradient is not significant considering the elevation accuracy required for most surface surveys. The maximum vertical gradient of gravity in μGal/m caused by an anomalous source in the shape of a sphere is

$$\left(\frac{\partial g}{\partial z}\right)_{\max} \simeq 5.6 \times 10^{-2}\sigma\left(\frac{R^3}{z^3}\right), \tag{5.10}$$

where the variables are the same as given for Equation 5.8. Thus, for the earlier example of an air-filled cavity in limestone with a radius of 5 m at the depth-to-center of 10 m, the maximum anomalous vertical gradient that occurs directly over the center of the sphere is roughly 19 μGal/m, which is negligible where the elevation accuracy is of the order of 1 cm.

Vertical control for gravity surveying can be achieved by measuring differences in atmospheric pressure, determining vertical angles and distances, spirit leveling, and most recently by GPS surveying. The first method is seldom used because of potential errors in using aneroid barometers due to difficulties in monitoring local transient atmospheric variations. However, the elevations of many stations in national data bases where the observations were made prior to the availability of GPS are based on barometric altimeter elevation. The most accurate methodology is spirit leveling, but this surveying is slow even with self-leveling instruments and requires line-of-sight access. GPS methods, especially those employing carrier phase measurements, are being used for essentially all gravity position surveying including marine and airborne surveys. Elevations determined by GPS are inherently less accurate than GPS-determined horizontal positions by a factor of roughly 1.5 (Aiken *et al.*, 1998). As a result, accuracies of GPS elevations using rapid methods are of the order of a few centimeters. Subcentimeter GPS accuracies currently use static methods which typically require occupation times of an hour or more. This long period of time invalidates a major advantage of GPS, its speed, and thus its cost-saving attribute.

Elevations determined with GPS methods are referenced to a mathematical model of the Earth's shape – a smooth reference ellipsoid (World Geodetic System of 1984, WGS84) or local datum – rather than the conventional geoid which is what the Earth's shape really is at sea level. If GPS elevations are to be combined with sea level elevations, a correction must be made to the GPS measurements to account for this difference from the mathematical model of the shape of the Earth. This is done by interpolating of the geoid height at the observation location from a digital grid of geoid heights and adjusting the observed elevation accordingly. Geoid digital models are included in some GPS receivers so that this can be done automatically. In regions of great topographic variability such as mountains, the local variations in the geoid roughly parallel the gravity observations corrected for the vertical gradient of gravity or the topography. In contrast, in regions of subdued topography, the geoid primarily reflects the gravity associated with subsurface density variations.

As a rule of thumb, a 3 mGal gravity variation results in a geoidal variation (N) of approximately 1 cm. Regional geoidal undulations are available on a global basis, but more detailed, higher frequency geoidal fluctuations can be readily computed from Bruns' formula given by Equation 5.13 below (See Section 5.5.2).

5.4 Gravity surveying

Gravity field mapping is implemented by ground, marine, airborne, and satellite surveys. This section considers the design, measurement, and other factors that influence the effectiveness of surface surveys for mapping the gravity field, whereas Section 5.5 below describes gravity surveying from space by satellites. General guides to this topic for land surface surveys are available from Seigel (1995) and Milson (2003) and in instrument manuals.

5.4.1 Survey design

A critical step in gravity surveys, as in all geophysical studies, is the design of the layout of stations to be observed in the study. The factors to be considered include the purpose of the survey, the minimal amplitudes and the areal size and configuration of the anomalies of interest, terrain of and access to the survey area, nature of noise in the gravity field, size of the area, available instrumentation, and the time and financial resources for the survey.

Land surface surveys

Land surface surveys performed for regional studies generally involving anomalies derived from sources within the

Earth's crust or for reconnaissance of the types of anomalies that are present in a region typically have stations distributed in as uniform a grid as possible considering the limitations imposed by access, and at station intervals measured in kilometers. In many land areas of the Earth, regional surveys are not required because of the availability of data from governmental, academic, or commercial data banks. Generally these data are available as discrete stations as well as in gridded formats of anomaly values at an interval appropriate to the spacing of the original data. In contrast, most gravity studies for near-surface and upper crustal mapping purposes require detailed surveys of rather limited areas. Station intervals for these studies are measured in meters to hundreds of meters. These types of data are not available in repositories for most regions, and thus need to be acquired in modern surveys.

Stations are distributed in a regular grid pattern over the entire area of interest where equidimensional anomalies are of interest, but if reconnaissance surveying or geological information suggests that the anticipated anomalies are much longer in one predictable direction than the other, surveying commonly is performed along straight, parallel profiles perpendicular to the longer dimension. The direction of the profiles may be modified somewhat from this idealized situation by terrain and access considerations. The distance between profiles is usually kept to a maximum of five times the distance between observations along the profile. Where the only purpose of the survey is to detect anomalies, statistical principles can be used to determine the probability of encountering a specific anomaly size for a particular layout of stations (e.g. Agocs, 1955).

Selection of the size of the area covered by a survey and the length of profiles is based on two primary considerations: (1) the maximum depth of the anticipated sources, and (2) the areal size and amplitude of the regional anomalies in the area. As described in the next chapter, the anomalies of interest in a survey must be isolated from the effects of deeper, broader, so-called regional anomalies, which produce gravity gradients that can distort the anomalies of interest. These regional anomalies must be mapped by gravity surveying extending beyond the area of immediate interest or determined from pre-existing gravity data. A common rule of thumb is that the gravity surveying should extend a minimum of three times the maximum depth of the sources of interest beyond the limits of the area of interest, as the anomaly from a concentrated mass will decrease to 10% or less of the maximum amplitude at this distance. The distance derived from the "three times the depth" rule should be considered a minimum, especially where the depth extent of the source exceeds one-tenth of the depth.

In selecting the station spacing it is important to consider the trade-offs between numerous factors. Critical concerns include the amplitude and areal size of the anomalies of interest, the objective of the study, the gravity noise, surface terrain and access, and the available resources. Given a minimal amplitude and size of anomalies of interest, a principal criterion in selection of station spacing is the objective of the survey. Surveys may be used to detect anomalies or map the anomalies for quantitative analysis and modeling as well as for calculation of gradients and other short-wavelength components of the gravity field from the observed measurements. The optimum station spacing for these different objectives is not the same.

In the simplest of cases for detection of anomalies based on amplitude, it is necessary to have only three stations per anomaly, or, as stated in the Nyquist sampling theorem in Appendix A.4.3, the minimum number of equally spaced observations that will define a specified wavelength is three located at a separation equal to one-half the wavelength. If there are not three observations per wavelength, the anomaly will be aliased or folded into observations of longer wavelengths.

Where significant gravity anomaly noise is present, the selection of station spacing may become more complex depending on the wavelength and amplitude of the noise. In the exreme case, it may be necessary to choose a spacing that will map the noise so that it can be identified and extracted from the observations. However, it is commonly difficult to ascertain the full noise characteristics prior to a survey, so it is desirable to minimize sources of noise, such as terrain effects which cannot be precisely estimated, by proper siting of the station locations and other appropriate survey procedures. A general guideline for detecting anomalies is that the distance between stations should not exceed the depth to the center of the anticipated concentrated sources in the region. This assures a minimum of one station within the area in which the anomaly from a concentrated source exceeds 70% of its maximum amplitude.

This criterion is inappropriate for mapping anomalies and performing other quantitative calculations on the anomaly, including the calculation of gradients from the discrete observations. Detection is based on absolute amplitudes, but the success of modeling and other quantitative calculations is very much a function of accurately mapping spatial changes in amplitude, that is their gradients. It is best, then, to map anomalies for interpretation purposes at an interval over which it is possible to approximate the gradient with a linear function. A reasonable linear approximation can be achieved for a concentrated source, which has the highest gradient anomaly among the various idealized anomaly sources, with an interval of one-quarter the depth to the source.

Marine surveys

Most marine gravity surveys are conducted as an add-on to seismic reflection surveys. This minimizes the cost of the surveys and generally allows the use of larger vessels with greater stability for the gravity measurements along a grid of survey lines. In this situation the survey specifications are largely dictated by the requirements of the seismic study. This was not always the case, because prior to the 1980s many marine gravity surveys operated from small dedicated ships with survey specifications designed specifically for the objectives of the survey or employed underwater surveys where the gravimeters were placed on the sea bottom for observations. Stand-alone marine gravity surveys are still performed for specific purposes. For such surveys, a stable and easily steered survey vessel is desirable, enhancing the accuracy of the gravity observations. The tracks of these surveys are supplemented by orthogonal tie lines to provide data on intersection misties that are used in leveling of the survey observations. The surveys should always include onboard data quality control processing for near real-time quality evaluation.

Prior to 1984 when the first commercial, surface marine gravity survey was conducted, navigation was a primary limitation on the accuracy of marine surveys because of the Eötvös effect. Thus, significant attention was directed toward accurately determining the location of the ship, but this changed with the global availability of highly accurate positioning using GPS. The position accuracies, and thus the velocities of the ship, obtained with GPS are especially important in surveys conducted on seismic survey ships. This is because the heavy cable towed by the seismic survey ship results in frequent course and velocity changes that affect the Eötvös correction (HERRING and HALL, 2006).

Marine gravity observations that are made continuously are determined by averaging the values over a period of time, commonly 30 to 60 seconds to minimize short-period accelerations of the ship that cannot be eliminated by data processing or instrument construction. As a result there is a trade-off between virtual station interval and accuracy of the observations – the longer the averaging of observations, the greater the accuracy, but the lower the resolution or the greater the effective data interval. Thus, marine surveys require decisions on the averaging period that is commensurate with the objectives of the survey, the velocity of the ship, and the noise parameters of the observations. In addition, water depth is an important parameter in the processing of the gravity data, and thus high-resolution bathymetry is a standard component of marine surveys. However, errors in bathymetry may be as large as 10% of the water depth.

Airborne surveys

In contrast to marine gravity surveys, essentially all airborne surveys are designed and conducted based on the anticipated gravity anomalies of the study area and the objectives of the survey. These are dedicated surveys optimized for gravity measurement. Selection of the appropriate aircraft is an important initial decision based on the survey objectives, terrain of the study region, and access to the region from air terminals with appropriate maintenance facilities. Generally, the aircraft is either a single- or twin-engine aircraft capable of relatively slow speeds (~50 to 100 m/s) to achieve the desired accuracy to resolution ratio. In regions of steep topography, cultural features, and continuous cloud cover, safety considerations may suggest that the fixed-winged aircraft be replaced with a helicopter. Topography and access determine the specifications of the aircraft with regard to rate of climb, descent, and range. Commonly, onboard magnetic sensors and recording equipment supplement the gravity measurements, but specifications are dictated by the gravity measurements.

Gravity observations are usually made along parallel lines perpendicular to the anticipated strike-direction of the gravity anomalies with orthogonal tie lines at a spacing of 5 to 10 times the line spacing. These tie lines are used for quality control of the survey and as an aid in data processing. The basic flight line spacing is dictated by the accuracy and resolution required for the objectives of the survey. Higher accuracy and shorter resolutions require closer line spacings. Thus, in high-resolution surveys for near-surface source mapping, a closer line spacing is used. Closer line spacing also significantly improves digital elevation models that are mapped by altimeters on the aircraft and can be used in calculating terrain effects. Flight line spacings are between 50 and 3,000 m or more for most regional objectives. Generally, line spacing is oversampled for the purposes of the survey to minimize noise and is less than the length of the filter applied to the measured data to eliminate short-period inertial accelerations caused by aircraft motions.

Closely associated with the selection of the line spacing is the decision on flight height or altitude above the ground. To enhance resolution and increase the perceptibility of the anomalies the flight height should be as low as possible taking into account aircraft safety, governmental regulations, and near-surface turbulence which may decrease the accuracy of the gravity measurements. Consideration must be made of longer filters applied to the data to achieve necessary accuracy in the survey resolution for identifying individual anomalies. Typically, survey altitudes are a few hundred meters. Surveys are commonly flown at a constant altitude, but in regions of marked topographic variations, it is desirable to drape the survey over the terrain, taking into account safe climb and descent rates of the aircraft, to avoid poor anomaly resolution in lower elevation regions. The position of the drape may be established from regional elevation models and aircraft characteristics and entered directly in the onboard GPS for controlling the flight path of the aircraft. Draped surveys more closely mimic land gravity surveys (which follow the topography). Thus, in attempting to merge airborne surveys with ground surveys it may be desirable to drape the airborne survey.

Decisions on the anomaly accuracy and resolution based on the objectives of the survey and anticipated anomalies must be considered with the allowable errors in altitude and position of the aircraft that will not interfere with anomaly accuracy. These allowable errors dictate navigational requirements of the survey. As in the case of other gravity surveys, real-time quality control of the data acquisition and processing is important to achieving the goals of the survey.

5.4.2 Survey procedures

Successful acquisition of gravity data for exploration purposes depends on the selection of instrumentation, survey design, and measurement procedures appropriate for the objectives of the study, as well as the methodology that will be used for reduction and interpretation. To insure that the process meets the intended goals requires a comprehensive quality control procedure at each stage in the process and a thorough knowledge of the principles of gravity and their application to exploration. Most gravity surveys other than land surveys are conducted by industrial contractors, research institutions, and governmental agencies which generally have defined procedures for the acquisition of data for specific objectives. Thus, this section is largely devoted to survey procedures for land surveying, but some of the procedural considerations described are applicable to other survey types. Where standardized procedures are used by contract crews, it is advisable for the user of the data to be well acquainted with these and restrictions they may place on the data.

Gravimeter considerations

The efficient measurement of gravity to one part in 10^8 or better requires instrumentation that is subject to numerous potential errors. It is necessary to select the most appropriate gravimeter, to maintain it in the optimum working condition, and to fully understand the characteristics of the particular gravimeter under the survey conditions. In marine and airborne surveys, significant criteria are the range and accuracy of the instrument's measurements on a moving platform. Depending on the objectives of the survey, a meter that measures either gravity or its vector or tensor components may be the best choice. For land surface surveys, critical attributes are the sensitivity and range of the instrument and its portability. Both the LaCoste & Romberg and Worden type gravimeters are constructed in highly portable versions with a large range and a sensitivity of approximately 0.01 mGal or, in specialized meters, 1 μGal.

It is advisable to use the microgal meter only where the precision is required, because of increased meter costs and reading time and the more limited operational range of these meters without reset. Resetting requires time and may induce tares in the observations. Where portability is of concern, it is desirable to use instruments that require a minimum battery supply to keep the meter at a constant temperature. Batteries add to the weight and require recharging, which can be a problem where observations are made over many hours in remote locations where electrical power is unavailable for recharging.

Gravimeters that are kept at a constant temperature, which is the norm today, should be brought to the specified temperature a few days or at least several hours prior to the start of the survey to remove temperature gradients within the meter, and the meter temperatures should be kept constant throughout the survey including overnight periods. The meter temperature should be set well above the anticipated maximum ambient temperature and low enough to minimize the power requirements in cold environments.

Worden-type meters that are not temperature-controlled are highly portable, but subject to generally high drift rates following a rapid, large change in temperature. As a result these gravimeters should be protected from temperature transients while maintaining as constant an ambient temperature as possible. All meters should be kept out of direct sunlight because of possible thermal effects as well as differential thermal expansion which may alter the level of the instrument.

Gravimeters that have a limited operational range should be reset only when absolutely necessary because of possible tares in the observations that may accompany readjustment of the coarse spring which keeps the lever arm carrying the mass within observational range. Adjustment of the meter to an appropriate operational range should be done at the start of a survey with due consideration of the anticipated gravity changes from the elevation and latitudinal ranges over the survey area. Immediately following resetting of the range it is advisable to monitor the gravity changes in a fixed location for several tens of minutes to be certain that high drift rates are not present.

KAUFMANN and DOLL (1998) have shown that the LaCoste & Romberg microgal (Model D) gravimeter is subject to calibration errors in the fine-nulling screw when the coarse (ranging) screw is moved to different positions. The resulting so-called circular errors may be significant to microgravity applications, so the position of the coarse dial should not be changed during the course of a survey without checking the calibration of the fine screw. This is especially true where a small percentage change in calibration may have a large impact, as in those surveys where gravity varies greatly between stations.

Each type of gravimeter has specific drift characteristics that may vary somewhat among individual instruments. Thus, it is advisable to conduct drift tests on a gravimeter by frequent (several minute) reoccupations of the same station over a period of several hours under the normal conditions of the survey. This will establish the anticipated drift rate of the meter, which can be used to determine the period of time allowable between reoccupation of stations for a specified accuracy in the drift control. Observed drift rates that exceed the normal drift range given by the manufacturer indicate that a problem is present, either in the meter itself or the manner in which the meter is handled between and during the observations. In the latter case, trial and error procedures will isolate the problem and suggest changes that need to be made. A common source of high drift rates, especially among meters that do not require arresting of the mass lever arm during moving of the meter, is inadequate protection from vibrations and shocks during transit. Recommendations from the manufacturers can be useful in eliminating this problem. In any event, it is advisable to maintain the instrument in a near-vertical position at all times. Tipping of meters, especially quartz-spring meters, beyond 45° from the vertical should be avoided because of the possibility of kinking or tangling of the springs, necessitating repair by the manufacturer. Additional details on sources of instrumental drift and corrective procedures are given in Section 6.3.1.

Measurement procedures

Gravimeters are capable of great precision and accuracy, but under field conditions this requires exacting measurement procedures. Of particular importance is an in-depth knowledge of the type of gravimeter being used and its inherent characteristics. The source and magnitude of errors of the particular meter should be well understood. Ways in which they can be minimized must be considered if high accuracy is required. In many surveys the success of the measurement is determined by the manner in which the drift of the meter due to tidal effects, spring relaxation, and temperature variations is monitored and removed from the observations. Also, procedures need to identify and remove tares in the observations. Measurement procedures must be particularly well controlled in dealing with microgal sensitivity surveys, as illustrated by ZUMBERGE *et al.* (2008) for ocean bottom monitoring of petroleum reservoirs by time-lapse gravity measurements, and by JACOB *et al.* (2010) for a land survey. CHEN *et al.* (2005) make a compelling case for real-time data acquisition and analysis of both gravity and position data to detect errors during the surveying which may necessitate unplanned measurements.

Measurement procedures vary between static land surface and moving platform observations. Moving platform measurements are made essentially continuously during the survey with averaging of values over a specified period to minimize high-frequency, non-geological accelerations in the data. However, observations made during turns of the vessel or aircraft to take a new track are usually removed from the processing. Additionally, some straight sections after a turn may need to be removed to allow time for the meter observations to settle down. This is especially true of airborne surveys, where turns tend to be more abrupt. To further minimize extraneous accelerations the instrumentation is set up near the center of gravity of the marine vessel or aircraft.

In detailed surface surveys where the station interval is measured in meters or tens of meters, the exact position of the station is commonly prescribed, but in more regional surveys, the position of the station may be varied somewhat to minimize noise factors and to increase the accuracy of the observation. Wherever possible, stations should be located to decrease the effect of local terrain on the gravity measurement. Moving a station even a few meters may decrease the terrain effects to negligible levels. Making estimates of the gravity effect of local terrain requires considerable experience, so charts showing the gravity effect of simplified topographic features are useful to the instrument operator in estimating these effects.

The impact of terrain corrections is very much dependent on local, nearby topographic features because the inverse square law may cause the effect of a relatively small volume feature close to the gravity station to be much greater than that derived from a much larger volume located at a distance. As a result, judicious location of

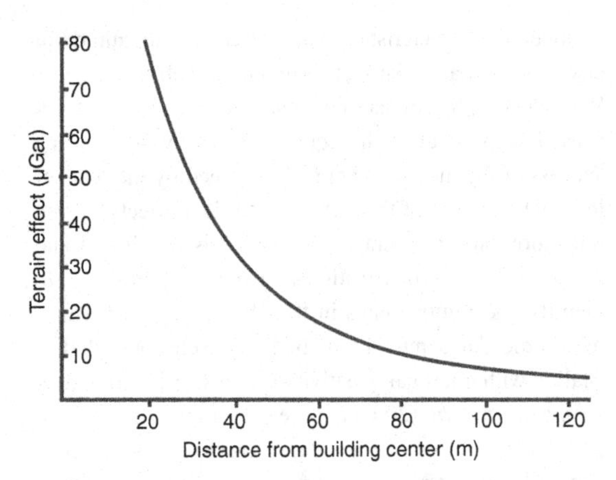

FIGURE 5.16 The terrain effect of a building, 27 m square, 100 m high, 25×10^6 kg mass and average density of 340 kg/m^3, as a function of distance from the center of the building as calculated using Hammer's template method. If the building was constructed over a basement 6.9 m deep, the terrain effect would have to be increased by 50%. Adapted from NOWELL (1999).

stations relative to local terrain is extremely important: for example, it is undesirable to site a station at the edge of a cliff or over a culvert. The 1-in-20 approximation rule is useful here: the terrain correction may be constrained to a few μGal if the station is at a distance of roughly 20 or more times the vertical distance to the topographic feature from the elevation of the station.

In addition, the spatial gradient of the terrain effects is much greater for local features than for distant topography, making local terrain effects much more subject to error. Local terrain effects may be a result not only of natural topography, but man-made features as well such as buildings (Figure 5.16) and basements whose effects may reach upwards of 100 μGal at nearby distances, and meter-deep drainage ditches which cause terrain effects of up to 50 μGal.

The increasing requirements for high accuracy in exploration surveys have directed attention to the need for improved terrain corrections. The advent of precise, rapid methods of topographic surveying at modest cost by GPS methods has greatly reduced position errors in gravity surveying which have long been the major factor limiting the accuracy of surveys, and improved gravimeters have decreased reading errors. As a result, terrain corrections are now commonly a major source of error in accurate land-surface gravity surveying (FILS *et al.*, 1998). These errors have been offset by the increased availability of digital elevation models (DEM) which provide topographic elevations on 30 to 750 m grids for continental-sized areas. These grids are very useful in calculating regional terrain corrections. In high-accuracy surveys, especially where gradients are measured, nearby elevations may require greater detail and accuracy than available from DEMs. These additional elevations are readily surveyed with reflectorless laser ranging devices combined with inclinometers and compasses. Some of these methods, particularly the reflectorless laser ranging devices, have allowed gravity surveyors to site stations in locations hitherto forbidden because of terrain errors that now can be accurately accounted for. Finally, there has been a marked improvement in computer computation methodologies for dealing with both nearby and regional terrain effects and in terrain grids. Global topography and bathymetry grids are now available at roughly 30 arcsecond (0.66 km at 45° latitude) and 2 arcminute (2.6 km at 45° latitude) grids (e.g. http:www.gina.alaska.edu/page.xml?group=data&page=griddata). Additional details concerning terrain gravity effects are described in Section 6.3.7.

Another factor affecting the choice of location is the stability of the surface material. Leveling is one of the more time-consuming aspects of making accurate surface gravity observations. An error in the level at right angles to the mass balance arm produces a variation in gravity proportional to the cosine of the angle of error. The error is roughly 1 μGal per second of arc. Automatic leveling which is available on many land surface gravimeters greatly simplifies the problem, but at some sites, the surface material is unable to support the gravimeter in a level position for a measurement.

Earth and wind vibrations have a deleterious effect on the accuracy of gravity observations by causing the mass balance arm to oscillate. Sometimes these vibrations can be minimized or eliminated by choice of station location or time at which the observations are made, but in many cases they cannot be totally eliminated. The effect of the wind can be limited by firmly emplacing the meter's tripod into the surface material and shielding the meter from the wind. Current meters which average values of several repetitive observations at a station minimize this effect.

Earth vibrations are more difficult to avoid, and they cover a wide range of sources and periods. Vibrations may come from both natural and man-made sources. Traffic and machinery vibrations are particularly troublesome in urban and industrial regions. Largely, they are avoided by making observations during quieter periods of the day. Many natural vibrations with periods of a second to an hour are derived from ocean noise and the passage of seismic waves. Microseismic effects derived largely from the ocean (waves and longer period events) commonly have amplitudes of the order of 10 μGal. Seismic waves derived from large, global earthquakes may make it impossible

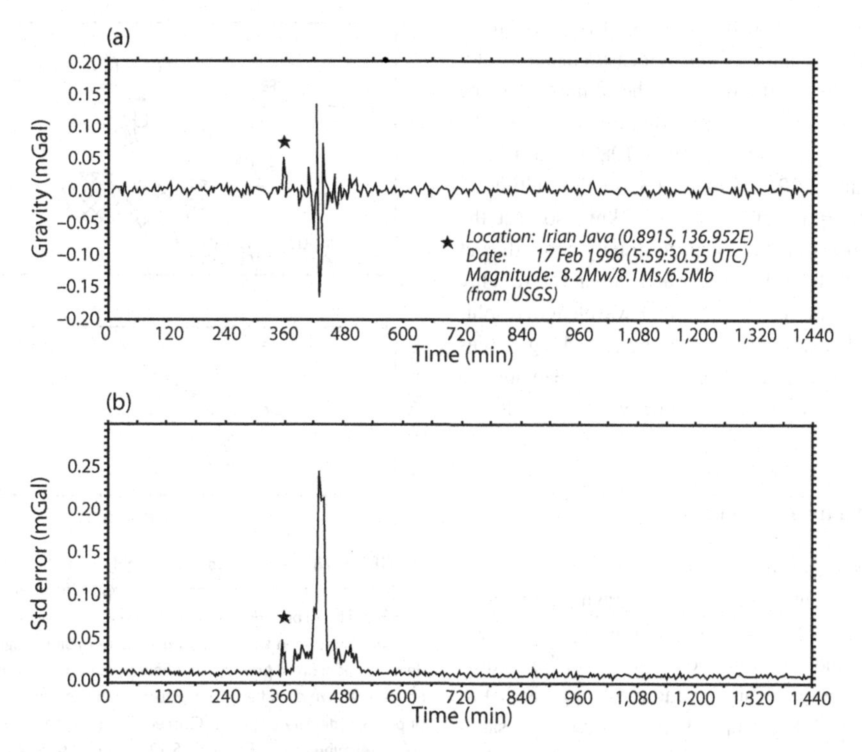

FIGURE 5.17 Response of a gravimeter to the passage of seismic waves associated with a teleseismic event of magnitude 8.2 Mw. Residual gravity response (a) was obtained by removal of Earth tides and linear gravimeter drift. The corresponding standard error in the gravity measurements is shown in (b). Note that the errors in measurements due to the passage of the seismic waves persist for several hours and reach amplitudes of 0.25 mGal. Adapted from Bonvalot *et al.* (1998).

to make precise observations for several hours or more (Figure 5.17).

All relative gravimeter observations are measured in reference to a specific station. In regional surveys this station commonly has been observed with an absolute measuring instrument or tied to the national gravity network by gravimeter observations. In this way the regional survey is converted from relative to absolute gravity units and tied to international gravity measurements. The precise location, elevation, absolute gravity value, and description of these stations which are distributed across many countries generally are available from world, national, or governmental agency geophysical data centers. It is advantageous to tie into these national bases because then the measurements can be related in an absolute way to surrounding surveys. Absolute gravity base stations are often established at harbors and airports to facilitate tying marine and airborne gravimeter setups to absolute values with portable gravimeters.

To minimize non-geologic accelerations in marine surveys, measurements are made on stabilized platforms as described previously and, as much as possible, during low sea states. Similarly, airborne gravimeter observations are commonly made during the early morning or evening periods when the air turbulence is a minimum.

5.5 Gravity measurements from space

Satellites have mapped the gravity fields of the Earth, Moon, Mars, and Venus for fundamental constraints on planetary properties and evolution. Satellites map the gravity field in either passive or active survey modes. Passive measurements include tracking a satellite's orbit relative to the ground or to other satellites, and satellite-hosted gravimeter observations, whereas active measurements use satellite-generated altimetry (i.e. electromagnetic or laser pulses) to image the geoid in oceans. Variations in both sets of measurements can be related to gravity field anomalies.

Gravity anomalies from geological features that measure less than several hundred kilometers are highly attenuated at satellite elevations. As a result they are not observed in passive measurements. Active measurements, on the other hand, provide higher-resolution data over the sea surface, but these data are restricted to off-shore marine environments in contrast to the broader coverage including over the land surface obtained by passive measurements. The resolution of active measurements is dictated primarily by the footprint of the altimeter on the ocean surface that is a very close approximation of the geoid. For example, Geosat (1985–1989) and ERS-1 (1991–1996) and ERS-2 (1995–2003) satellite altimetry data have a footprint of about $7 \times 7\,\text{km}^2$, so that the marine altimetry data have mapped gravity anomalies at wavelengths of about 15–20 km and larger for roughly 70% of the Earth's surface. The shorter-wavelength resolution limit largely results from waveform retracking studies which have improved the extraction of the ocean surface signals in the altimetry data (e.g. ANDERSEN *et al.*, 2010).

5.5.1 Passive measurements

Tracking the satellite in its orbit can determine the basal harmonic components of the planetary gravity field. Satellite orbits are mostly a function of the lower-order gravity harmonics and the effects of solar wind, lunar tidal forces, and other perturbing parameters (e.g. KAULA, 1966; REIGBER, 1989). Depending on altitude, the satellite is primarily affected by the gravity harmonics through degree 20 or 30 with the power of the higher-order harmonics tapering off to near zero through degree 70.

A model for predicting a satellite's location (C) may be calculated from *a priori* estimates of the above-mentioned parameters (P_i) by

$$C = \sum_{i=1}^{n} P_i. \tag{5.11}$$

As shown in Figure 5.18, differencing the calculated orbital location with the actual observed location (O) of the satellite yields a residual that is a function of both the errors in the model parameters (ΔP_i) and the observations (e_{obs}). Transformations between the Earth-centered and satellite-centered frames of reference must be determined, as must the relationships between the orbital path and the various parameters used to predict the orbital path.

Knowing these relationships, the difference between the actual location (as determined from observation) and the predicted location may be expressed by a well-established observation equation (e.g. KAULA, 1966; TORGE, 1989; REIGBER, 1989) according to the general formula

$$(O - C) = \sum_{i=1}^{n} \Delta P_i + e_{\text{obs}}. \tag{5.12}$$

An *a priori* model is assumed to be close to the ideal parameters, so that the desired corrections are small and

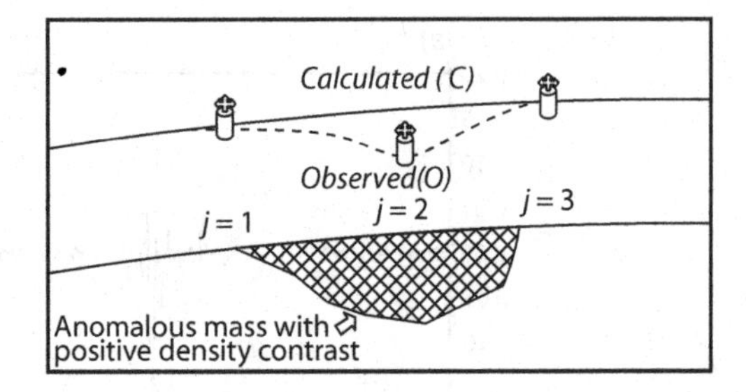

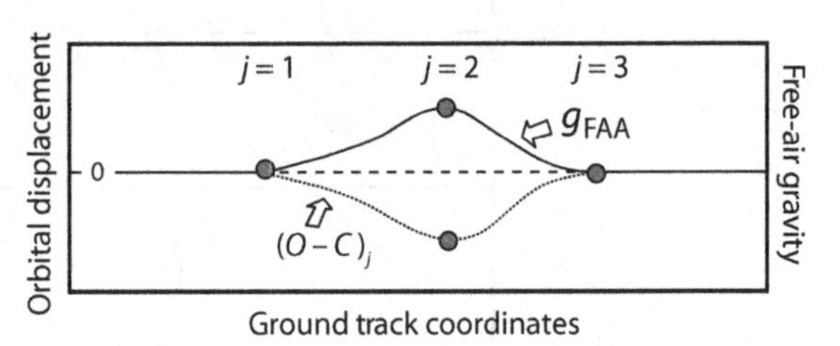

FIGURE 5.18 Generalized gravity field measurement by ground-based tracking of a satellite. Adapted from VON FRESE *et al.* (1999c). The orbital displacement $(O - C)_j$ at location j is the difference between the calculated satellite orbital height (C) determined from *a priori* gravity field parameters and its observed height (O), which reflects the actual Earth with an anomalous mass of positive density contrast. Corrections (ΔP_i) to the *a priori* values are determined using Equation 5.12 to generate *a posteriori* parameters, which are used to estimate the free-air gravity anomalies (g_{FAA}).

may be linearized. With multiple (j) observations, this result becomes a linear system of equations that may be solved by least squares for modifications (ΔP_i) to the parameters. This approach, however, involves tedious satellite tracking calculations, and will also be limited over regions where satellite tracking data are sparse.

Additional refinements to this method have been offered (e.g. LERCH, 1991; ANDERSON and CAZENAVE, 1986; LEMOINE *et al.*, 1998) that include schemes for weighting Equation 5.12 and adapting it for the gravity inversion of satellite-to-satellite tracking data. Satellite-to-satellite data in High–Low and Low–Low (e.g. NASA's GRACE mission) tracking configurations (Figure 5.19) provide more continuous coverage for determining the gravity field. In addition, the use of GPS for tracking satellites, such as in the TOPEX/Poseidon (1992–2006) and GRACE (2002–) missions, yields exceptionally good orbit determinations. However, these data still limit the solution of the gravity field to the lower-order components that predominate at satellite altitudes owing to attenuation of the signal.

The accuracy of the passive satellite gravity field estimates also is greatest for the lowest-order components and decreases with increasing order. The GRACE mission,

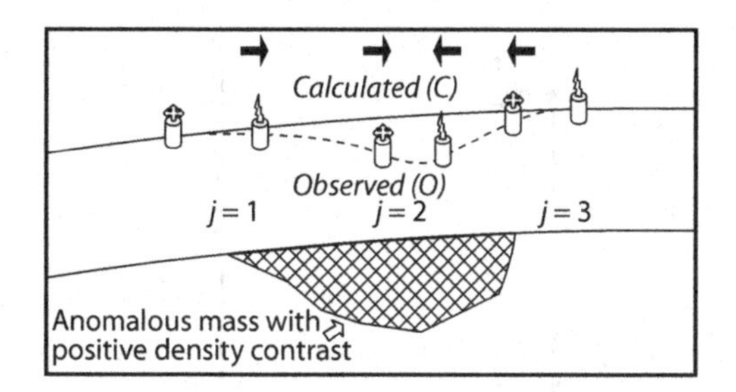

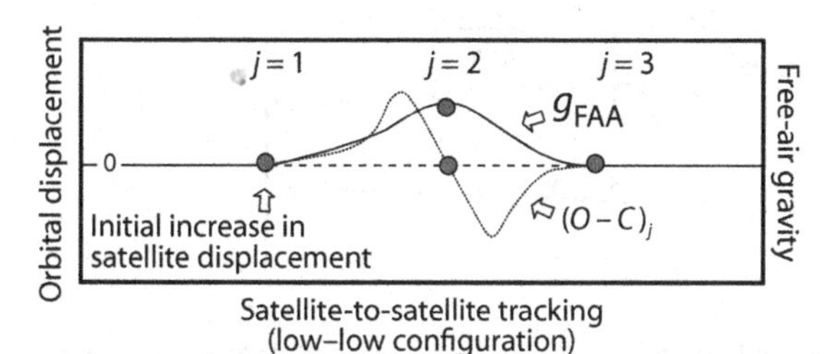

FIGURE 5.19 Generalized gravity field measurement by satellite-to-satellite tracking in the Low–Low configuration. Adapted from VON FRESE *et al.* (1999c). The fundamental observation at location j is the orbital displacement $(O - C)_j$ between the satellites in the orbital plane. The calculated distance (C) is the difference between satellite positions in the absence of an anomalous mass as determined from *a priori* gravity field parameters. An anomalous mass with positive density contrast will cause the observed displacement (O) to increase at $j = 1$ as the leading satellite is pulled forward, then decrease at $j = 2$ as the trailing satellite surges forward while the leading satellite decelerates, and finally increase back to the initial displacement at $j = 3$ as the satellite pair moves beyond the anomalous mass effect. These changes in the displacement may be used in Equation 5.12 to refine the gravity field parameters and thereby estimate the free-air gravity anomalies (g_{FAA}).

operating at altitudes of roughly 450–500 km, for example, has produced the most precise set of gravity field observations to date. However, the repeatability of the shortest 500 km wavelength components is no better than about 5 mGal, whereas the wavelength components in excess of several thousand kilometers are coherent or repeatable to within several microgals.

The GOCE mission (2009–) hosts the first and to date only spaceborne gravity gradiometer. The gradiometer consists of six accelerometers arranged to form three gradiometer arms mounted on an ultra-stable diamond structure or cage. The gradiometer measures the application of electrostatic forces that maintain a levitated proof mass at the center of the cage. These results give the acceleration difference between each pair of identical accelerometers along each arm that is about 50 cm in length. The three arms are orthogonal to each other with one aligned along the satellite's trajectory, another perpendicular to the trajectory, and the third one pointed to the Earth's center. The differential accelerations are combined with the star tracker and GPS data to estimate the gravity gradient components and perturbing angular accelerations. A primary objective of the GOCE mission is to use the gravity gradient tensor observations in combination with the satellite tracking data to resolve the Earth's gravity field components over wavelengths from a few thousand to about 200 km with an accuracy of 1 mGal (e.g. DRINKWATER *et al.*, 2003).

5.5.2 Active measurements

Satellite altimeters map significantly higher-order components of the Earth's anomalous gravity field over the oceans than observed at satellite altitudes. They are truly measuring gravity at the Earth's surface. The altimetry determines the distance from the satellite to the surface of the ocean (Figure 5.20) which serves as a proxy indicator of the Earth's geoid undulation. The geoid represents the effect of the disturbing potential at the Earth's surface, and therefore includes the higher-order gravity components. By remotely measuring this surface with radar, remarkably detailed and accurate gravity information has been retrieved for exploration of the Earth's oceanic crust.

The accuracy and resolution of altimetry-derived gravity anomalies are significantly limited by errors in determining the satellite orbits. Geodetic satellites commonly operate in inclined polar orbits where half of the ground tracks ascend from south to north across the Earth to intersect the other half that descend from north to south. The orbital mismatches at the cross-over points can be used to reduce satellite orbit errors significantly for optimal altimetry estimates of the gravity field (e.g. KIM, 1996; SCHARROO and VISSER, 1998). The areal density of cross-over points increases markedly towards the poles where the dense track coverage results in enhanced spatial resolution of the height variation of the sea surface. For example, Geosat data, which cover the southern oceans to 72° S, have an along-track data interval of roughly 7 km, whereas the spacing between tracks at 60° S is only about 2–3 km.

The observed altimetry data (ρ_{obs}) must first be screened and corrected for errors caused by the wet and dry troposphere, atmospheric pressure, the ionosphere, and other factors (Figure 5.20). The corrected altimetry data (ρ) can then be differenced with orbital elevations (H) determined from tracking and orbital models to generate sea surface height (SSH) variations. Corrected SSHs are generated by applying models for other surficial factors such as the static sea surface topography (SSST) and

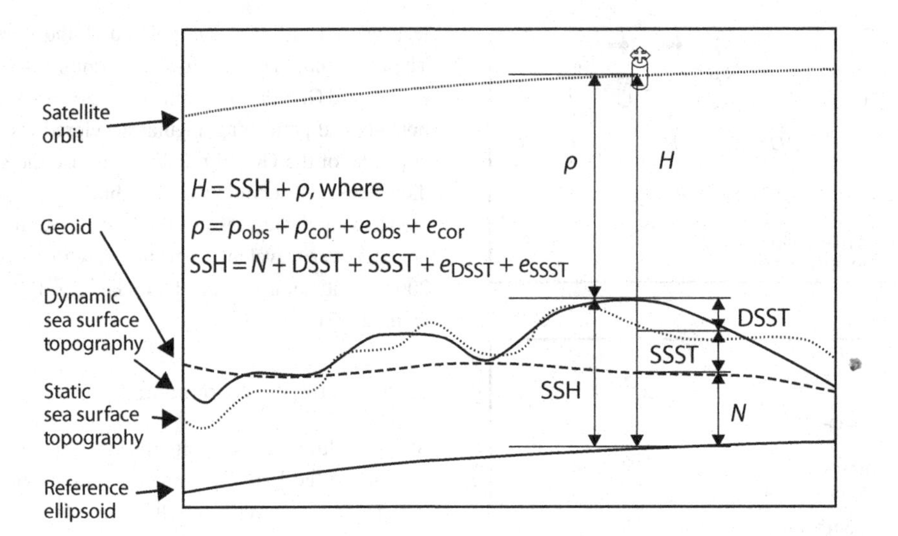

FIGURE 5.20 Generalized gravity field measurement by satellite altimetry. Adapted from VON FRESE *et al.* (1999c). Altimetry observations (ρ_{obs}) with errors e_{obs} are corrected for various atmospheric effects (ρ_{cor}) with errors e_{cor}. The corrected altimetry data (ρ) then are differenced with the orbital heights (H) to generate sea surface heights (SSH). Removal of the dynamic sea surface topography (DSST) and static sea surface topography (SSST) with errors e_{DSST} and e_{SSST} leaves the desired geoid undulations (N) that are converted into free-air gravity anomalies (g_{FAA}) by the fundamental equation of geodesy (Equation 5.14).

performing a cross-over adjustment to reduce the dynamic Sea Surface Topography (DSST) and orbit errors arising in the determination of H.

Models for correcting the ρ_{obs} are generally available with the altimetry data (e.g. KIM, 1996), so that the geoidal components can be readily recovered. However, the corrected SSHs contain residual errors, and thus do not accurately reflect the geoid undulations at this level of processing.

The common approach for extracting gravity anomalies from the corrected SSHs is to determine along-track vertical deflections from the descending and ascending orbits at the cross-over points. These vertical deflections can be converted directly into gravity anomalies without the necessity of computing the geoid (e.g. HAXBY *et al.*, 1983; SANDWELL, 1992; MCADOO and MARKS, 1992; KIM, 1996; SANDWELL and SMITH, 1997). However, the effort to evaluate geoid undulations from the corrected SSHs facilitates the imposition of strong geological constraints such as the depths of possible sources and enhanced reduction of residual orbital errors in the altimetry-derived gravity estimates (e.g. KIM and VON FRESE, 1993; VON FRESE *et al.*, 1999c).

In this geoid-mapping approach, the two sets of subparallel ascending and descending tracking orbits are processed for separate geoid estimates. Lithospheric signals that are larger than the spacing between the ground tracks of the subparallel orbits are coherent between two neighboring tracks, and hence may be extracted with spectral correlation filters. Inversely transforming the correlative wavenumber components from each data track yields geoid undulation estimates where non-correlative features are suppressed, involving presumably crustal effects that are small compared with track spacing and non-lithospheric effects from temporal and spatial variations of ocean currents, measurement and data reduction errors, etc.

Separate geoids then can be produced from gridding the correlation filtered ascending and descending data tracks. Typically, the geoid estimates exhibit strong washboard effects that parallel the orbital tracks and reflect errors in orbit determination, along-track data processing, and other non-geologic effects (e.g. KIM, 1996; KIM *et al.*, 1998a). In the spectrum of each map, the track-line noise is restricted predominantly to two of the quadrants with the other two being relatively uncontaminated. The pair of clean quadrants in one map is orthogonal to the clean pair in the other map because the track orientations of the ascending and descending data sets are different. Hence, a new spectrum can be constructed by combining the two pairs of clean quadrants to yield geoid undulation estimates where track-line noise is severely suppressed.

The resultant grid of geoid undulations (N) can be used to estimate free-air gravity anomalies (g_{FAA}) because it is a function of the disturbing potential (T) as expressed in Bruns' formula (HEISKANEN and MORITZ, 1967)

$$T = N \times g_N, \tag{5.13}$$

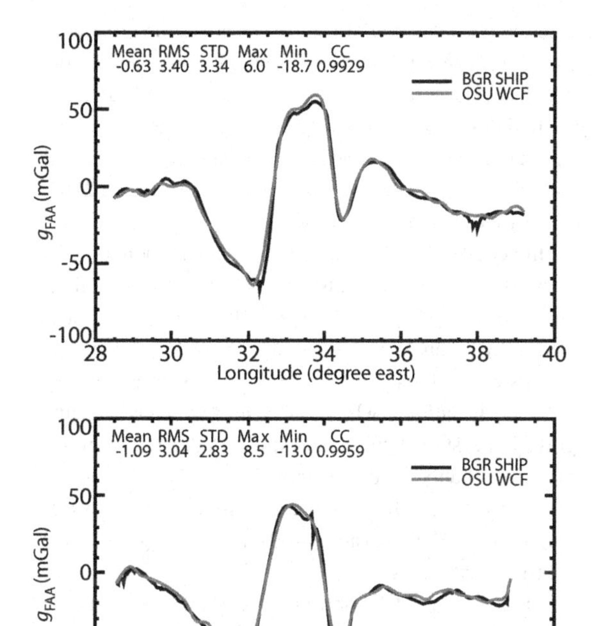

FIGURE 5.21 Free-air gravity anomaly (g_{FAA}) comparisons between ship measurements (BGR SHIP) and wavenumber correlation filtered Geosat altimetry-derived estimates (OSU WCF) along west–east running tracks that cross the Gunnerus Ridge in the Larsen Sea, Antarctica, near 65°S (top) and 67°S (bottom). Adapted from VON FRESE *et al.* (1999c). The statistics of their differences, which are given in the upper left corner of each profile comparison, include the mean, root-mean-square (RMS), standard deviation (STD), maximum (Max), and minimum (Min) values. The coefficient of correlation (CC) between the observed and estimated anomaly profiles is also given.

where g_N is normal gravity given by the International Gravity Formula for a standard Earth of homogeneous mass. Typically, geoid undulations are used to predict g_{FAA} through the fundamental equation of geodesy

$$g_{FAA} = -\frac{\partial T}{\partial r} - \frac{2T}{a_e} = -\frac{\partial (N \times g_N)}{\partial r} - \frac{2(N \times g_N)}{a_e}, \tag{5.14}$$

where a_e is the Earth's mean radius, and the radial (r) gradient of N is estimated by the spectral filter.

Figure 5.21 gives examples of the remarkable capacity of satellite altimetry to estimate marine gravity anomalies (VON FRESE *et al.*, 1999c). Both west–east running profiles compare Geosat altimetry-derived free-air gravity anomaly estimates with ship observations across the Gunnerus Ridge in the Larsen Sea, off Japan's Syowa Station on the coast of Queen Maud Land, East Antarctica. The Geosat estimates accurately map detailed features of the marine gravity field, even for regions such as this where the sea surface is relatively stormy and contaminated by floating ice. These results are quite close to the theoretical accuracy limit of 3 mGal for the Geosat-derived g_{FAA} found by a coherency analysis of hundreds of exact repeat orbits (SAILOR and DRISCOLL, 1993). In practice, however, errors in estimating satellite altimetry-derived g_{FAA} range more typically around 4–6 mGal, owing to the effects of water depth, sea state, ocean currents, sea ice, and other factors on local sea surface conditions (e.g. ANDERSEN *et al.*, 2010).

5.5.3 Satellite gravity mapping progress

Gravity observations at satellite altitudes provide important and relatively newly available boundary conditions for interpreting variations in the near-surface gravity field. In practice, a model satisfying just a single altitude slice of the gravity field effectively accounts for field behavior only to vertical distances within roughly a station interval or so of the observation surface because of measurement errors. However, the model that jointly satisfies the sets of satellite-altitude and near-surface observations offers an enhanced picture of how the gravity field may vary through the unsurveyed altitudes between the boundary conditions. Thus, the jointly constrained model provides more insight into the altitude behavior of the field for interpreting anomalous subsurface mass variations than the single altitude model can, with or without implementing the classical Poisson or far-field constraint of zero effect at infinite distance (e.g. Appendix A.4.3).

Satellite-altitude gravity surveys of the Earth have been carried out to date by geodesists for mapping the geoidal undulations and the temporal variations of the very low-order gravity components that may constrain temporal changes of the crustal waters and ice sheets. Current state-of-the-art satellite gravity data are being collected by the GRACE and GOCE missions at orbital altitudes of about 450–500 km and 270 km, respectively. The satellites operate in near-Sun-synchronous orbits in the dawn–dusk meridian plane of the Earth to maximize solar charging of batteries. Each orbit takes roughly 96 minutes to complete, by which time the Earth has rotated through an equatorial distance of about 2,667 km.

The near-polar orbits are inclined relative to the Earth's equatorial plane at 89° for GRACE, and at 96.7° for GOCE. Thus, the spacing between ground tracks decreases greatly towards the poles where they become tangential to the margins of the pole-centered holes or coverage gaps of diameters 2° and 13.4° for the respective GRACE and GOCE missions. In addition, the ground tracks for the GRACE and GOCE satellites roughly repeat every 61 revolutions/4 days and 467 revolutions/29 days, respectively. Thus, the track coverage from the GRACE and GOCE missions varies from track intervals of essentially zero kilometers at the margins of the polar holes to maximum intervals at the equator of about 44 and 6 km, respectively.

The orbital r-radial, θ-colatitude, and φ-longitude coordinates of each satellite altitude gravity observation are accurately located by on-board GPS receivers with on-board star cameras giving the attitude of the observation for interpreting the measured gravity component. The GRACE gravity data are inferred from dual one-way range changes between the two satellites measured by a K-band microwave ranging system with a precision of about 1 μm per second, whereas the GOCE gradiometry data are collected at intervals of about 47 km along the mission tracks. Because the observations have essentially no sensitivity for gravity wavelengths smaller than the orbital altitudes above the Earth's surface, the along-track gravity data are averaged in bins that are one or two degrees on a side and Δr km thick. The bin averaging ignores the effects of the altitude variations Δr in the observations that may range from a few kilometers to several tens of kilometers. It consists simply of computing an initial average and standard deviation, deleting all outliers with values equal to or greater than the mean plus or minus three standard deviations. From the remaining values, a new bin average and standard deviation divided by the square root of the number of data points are computed for the final gravity estimate and its standard error, respectively.

The binning process yields a grid of gravity estimates at the mean altitude of the bins that is transformed for analysis and interpretation into a set of spherical harmonic coefficients. These coefficients give the amplitudes of orthogonal sine and cosine functions that span the Earth's spherical surface at the average radius a_e ($\approx$ 6,378.1 km) which closely approximate the bin-averaged gravity values when weighted for the altitude difference of the grid relative to a_e and summed at the grid points.

The spherical harmonic model is based on the Cartesian-to-spherical coordinate transformation in Equation 3.82. Because Laplace's equation (Equation 3.6) in spherical coordinates can be solved by separation of variables, the gravitational potential may be written as

$$U(r,\theta,\varphi) = \frac{\xi}{r}\sum_{n=0}^{\infty}\left(\frac{a_e}{r}\right)^n \sum_{m=0}^{l} P_{n,m}[\sin(\theta)] \times [C_{n,m}\cos(m\varphi) + S_{n,m}\sin(m\varphi)], \quad (5.15)$$

where ξ ($\approx 3.986\,004\,415 \times 10^{14}$ m^3/s^2) is the product of big G and the Earth's mass, and the second summation, which is independent of r, involves the spherical harmonic coefficients $C_{n,m}$, $S_{n,m}$ of degree n and order m, and the associated Legendre functions $P_{n,m}[\sin(\theta)]$ of degree n and order m (e.g. Pick *et al.*, 1973; Press *et al.*, 2007; Blakely, 1995).

In general, the gravitational spherical harmonic model consists of a set of constants that specify ξ and a_e and the $C_{n,m}$, $S_{n,m}$ coefficients. However, the modeled gravitational variations have wavelengths of the Earth's circumference divided by m in longitude and by (n − m) in latitude where $-n \le m \le n$. Thus, satellite gravity measurements at 400 km altitude, for example, may be reliably modeled only for orders $m \le 100$ (= $[2\pi \times a_e \approx 40{,}000]/400$) because they observe wavelengths no smaller than about 400 km, assuming negligible measurement errors.

In 2008, the US National Geospatial Intelligence Agency (NGA) released the Earth Gravitational Model EGM2008 (Figure 5.22). It is the most comprehensive model to date of the Earth's satellite, airborne, marine, and terrestrial gravity observations (http://earth-info.nga.mil/GandG/wgs84/ gravitymod/egm2008). This gravitational model is complete to spherical harmonic degree and order 2,159 and contains additional coefficients extending to degree 2,190 and order 2,159. Thus, the model accounts for anomaly wavelengths of about 18–20 km and longer, but only where the constraining gravity observations provided commensurate or higher-wavelength resolution. Over Antarctica, for example, the gravity field is basically constrained by satellite altitude observations so that the EGM08 coefficients are most reliable for modeling the Antarctic gravity field only at satellite altitudes. Thus, care is required in applying these coefficients for subsurface investigations, owing to the uneven spatial coverage and wavelength resolution of the gravity surveys used in producing the spherical harmonic model.

Spherical harmonic models of satellite altitude gravity observations for the Moon, Mars, and Venus that fundamentally constrain the subsurface properties of these mass-differentiated planetary bodies have also been constructed. The gravity observations are based on the tracking of the satellites by NASA's Deep Space Network (DSN), which is an international network of antennas supporting interplanetary spacecraft missions and radio and radar astronomy observations for the exploration of the solar system and the universe. The DSN currently consists of three deep-space communications facilities placed about 120° apart around the world, at Goldstone in California's Mojave Desert and near Madrid, Spain, and Canberra, Australia. This configuration of antennas permits constant observation of

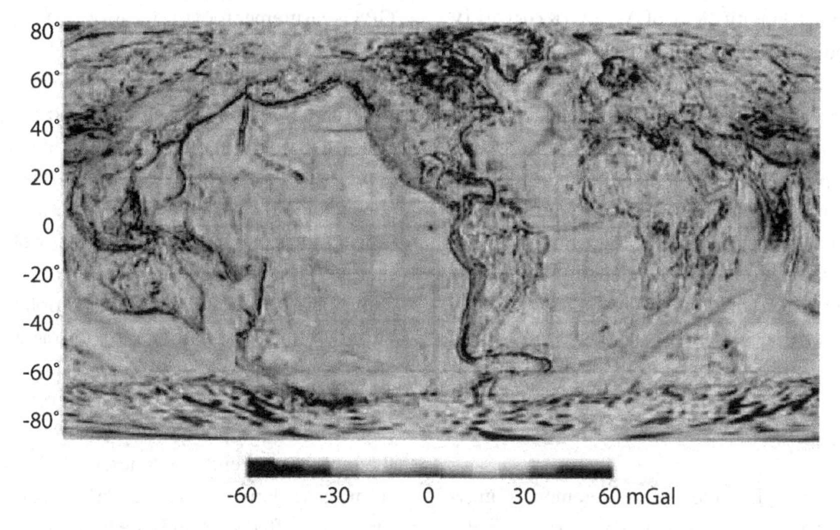

FIGURE 5.22 Free-air gravity anomalies evaluated from the 2008 spherical harmonic Earth gravity model (EGM08) averaged over 5 arcminute by 5 arcminute cells on the Earth's surface. A color version of this figure is available at www.cambridge.org/gravmag. Adapted from the National Geospatial Intelligence Agency (NGA).

spacecraft as the Earth rotates, and helps to make the DSN the largest and most sensitive scientific telecommunications system in the world.

Lunar gravity exploration began with the tracking of the Apollo satellites (1963–72) that discovered a gravity field pockmarked with circular large-amplitude positive gravity anomalies marking regions of concentrated mass or "mascons", centered on giant impact basins with diameters of 300 km or more. The inverse correlation of the mascon and impact basin topography has been interpreted as indicating impacted crust that had recoiled to raise a plug of underlying mantle material. The mascon is produced presumably because the raised plug of excess mantle material is being held in the crust against isostasy by the strength of the lithosphere.

The USA Clementine (1994) and Lunar Prospector (1998–99) polar orbiting missions mapped greatly improved details of the lunar gravity field. In addition, Clementine obtained laser altimetry of the Moon's surface that yielded the spherical harmonic lunar topographic model GLMT 2 to degree and order 72 with total radial error of about 100 m (Smith *et al.*, 1997). The Lunar Prospector (LP) satellite was operated in near-circular orbits at altitudes of about 100 km and 30 km above the mean lunar surface radius of a_{m} =1,738 km. The LP data were combined with Apollo and Clementine data to produce the comprehensive LP165P model of the lunar gravity field to degree and order 165 (Konopliv *et al.*, 2001).

The Moon is locked into an orbit where only its near side is visible to the line-of-sight observations of the DSN. The DSN cannot observe the far side where less certain gravity estimates resulted from numerically integrating the equations of spacecraft motion accounting for all forces including the gravity field. However, significantly improved mapping of the farside gravity field was established by Japan's polar orbiting SELENE mission (2007–09). This mission consisted of the main satellite, Kaguya, in a circular 100 km orbit, and the relay sub-satellite Okina and the very long baseline interferometry (VLBI) sub-satellite Ouna. The four-way Doppler measurement of Kaguya via Okina obtained the world's first direct observations of the lunar farside gravity field to degree and order 100 (Namiki *et al.*, 2009).

Mapping of the gravity field of Mars is based on the DSN's radiometric tracking of the Mariner-9 (1971–1972) and Viking 1 and 2 orbiters (1975–80), and the Mars Global Surveyor (1996–2006). The polar orbiting Mars Global Surveyor (MGS) collected the most extensive set of gravity observations at an average altitude of 378 km above the mean martian surface radius $a_{\mathrm{M}} = 3{,}339$ km, along with laser altimetry of the martian surface. The MGS gravity observations have been converted into the degree and order 80 Goddard Mars Model 2B, which reveals numerous mascons, with the largest free-air gravity anomaly overlying the 20 km high Olympus Mons which is the largest known volcano of the solar system. The MGS carried the Mars Orbiter Laser Altimeter (MOLA) which yielded a 360 degree and order spherical harmonic model of the martian topography (Smith *et al.*, 1999).

Radiometric tracking data from NASA's Pioneer Venus (1978–1992) and the Magellan (1989–1994) missions have been combined to produce a 180 degree and order spherical harmonic gravity field of Venus (Konopliv *et al.*, 1999). The polar orbiting Magellan mission obtained gravity data over altitudes of about 200 km above the mean surface radius $a_V = 6{,}052$ km with complementary synthetic aperture radar (SAR) data that provided a spherical harmonic topography model to degree and order 360 (Rappaport and Plant, 1994). The correlation spectrum between venusian gravity and topography reveals strong positive correlation at all wavelengths, suggesting that both the surface tectonics and gravity field are strongly related to interior mantle dynamics and convection.

5.6 Key concepts

- Although simple in principle, the measurement of gravity for geologic purposes to an accuracy of 10^{-8} or better requires highly sophisticated instrumentation and rigorous survey procedures. Current instrumentation makes it possible to measure relative changes in gravity for exploration purposes on land, in a variety of marine vessels and aircraft, in space, in drill holes, and at the bottom of and within oceans and lakes.
- Numerous principles have been used to measure gravity, but land surface gravity measurements generally are made of the relative vertical gravity with gravimeters that use a lever arm held up by a zero-length spring to amplify the change in the position of a test mass with varying gravity. This is accomplished through the process of astaticization in which by construction, a mechanical force reinforces the gravity force acting upon a test mass and opposes the restoring force of the spring. To avoid non-linearity issues the mass is brought to a null position by a force calibrated in gravity units. Zero-length spring meters, either the inclined metal spring type or the vertical quartz spring type, are highly portable and capable of sensitivities of up to 1 μGal.
- To minimize the effects of temperature and atmospheric pressure on gravity observations, gravimeters are either temperature- and pressure-controlled or compensated. Despite this, observations are subject to time variations or drift, generally less than 0.1 mGal/hr, owing to relaxation of the meter's spring and environmental effects. The surveying procedure is designed to determine the gravimeter's drift, which is meter dependent, through reoccupations of previous measurements. Using normal surveying procedures, the total error in standard land measurements is of the order of 0.02 to 0.03 mGal, but this can be reduced with rigorous procedures to approximately 0.01 mGal.
- GPS measurements of the location and elevation of gravity stations has greatly increased the efficiency and accuracy of gravity surveys. As a result, the error in the locations and elevations of stations is no longer the major concern in observing gravity; rather in land surveying it is the estimation of the gravitational effect of the terrain surrounding the gravity observation site.
- Gravity measurements from moving platforms, both marine vessels and aircraft, have become the norm, using a broad variety of instrumentation including variations of the zero-length type of meter and vertical accelerometers.
- The significant challenge of measurements on a moving platform is to keep the gravimeter vertical and sort out the inertial accelerations inherent to the moving vessel from those derived from subsurface sources. To minimize these effects, the gravimeter is placed on a gyrostabilized platform. The platform counteracts its motion and that of the gravimeter by minimizing horizontal accelerations and the tilt of the platform. Servomotors actuated by accelerometers keep the platform horizontal, and thus the gravimeter vertical within a few seconds of arc. High-frequency vertical accelerations are minimized by low-pass filtering and averaging the observations over a period of time. The combination of filtering, whether physical, electronic, or mathematical, applied to the recorded gravity signal, sampling time intervals, and the velocity of the gravimeter over the Earth are major controls on the spatial resolution of the measurements.
- Measurements on a moving platform are subject to errors due to the change in the vertical component of the Coriolis effect caused by motion over the Earth's surface. This so-called Eötvös effect causes an increase in the measured gravity for an east-to-west movement and a decrease for movement in the opposite direction. This effect, which must be removed from the moving observations to equate the measurements to stationary observations, is calculated from the velocity and direction of the platform. The use of GPS for position control has greatly improved the accuracy of these corrections.
- In addition to measuring the vertical component of gravity for geological studies, improvements in instrumentation allow precision measurements of the five independent tensors of gravity which are proving useful in increasing the perceptibility of shallow-source anomalies and in mapping margins of anomalous sources. Highly accurate rotating-disk gradiometer measurements are readily made in the dynamic environment of shipborne and airborne systems because they are less

susceptible to large inertial accelerations that affect conventional gravimeters.

- Significant technological advances have led to the development of accurate measurements of the absolute gravity with highly portable, efficient instruments based on the measurement of the time for a mass to fall in the Earth's gravity field. The accuracy of the measurement depends on the integration time for repeated falls of the mass element. Accuracy is purported to reach a few microgals for measurement times of approximately an hour or 10 microgals for times measured in tens of minutes. An attractive attribute of these measurements is that they are not prone to the time variations (drift) of relative gravimeters.
- Survey design is an important element in data acquisition particularly in land and airborne surveys. Marine gravity surveys are largely performed as add-ons to seismic reflection surveys, and in these cases the survey design is not usually a consideration. The factors to be considered in survey design include the purpose of the survey, the minimal amplitudes and the areal size and configuration of the anomalies of interest, the terrain of and access to the survey area, the nature of the noise in the gravity field, the size of the area, the available instrumentation, and the time and financial resources.
- The spacing between gravity observations is an important parameter of the design of a survey. In the selection of the station spacing it is important to consider the trade-offs between numerous factors listed on the previous bullet point. The data interval of marine surveys is determined by the period of time over which the observations are averaged to minimize the short-period accelerations to the ship. Thus, marine surveys require decisions on the averaging period that is commensurate with the objectives of the survey, the velocity of the ship, and the noise parameters of the observations.
- Airborne surveys consist of parallel flight tracks with tie-lines flown at a spacing of 5 to 10 times the line spacing. Typically, survey heights are a few hundred meters. Surveys are generally flown at a constant altitude, but in regions of marked elevation variations it is desirable to drape the survey over the terrain, taking into account safe climb and descent rates of the aircraft to avoid poor anomaly resolution in lower-elevation regions. As in marine surveys the virtual data spacing is determined by averaging measurements over a period of time to minimize the effects of extraneous, non-geological vertical accelerations. The length of this period is determined by the desired resolution and accuracy of the observations and the noise envelope of the survey.
- Satellite gravity surveys provide important boundary conditions for interpreting regional sources of anomalies. Gravity estimates can be obtained from tracking the deviations that satellites make from standard orbits about a homogenous Earth. These passive satellite observations resolve gravity variations to wavelengths of the satellite altitude and longer with accuracies ranging from several microgals at the very lowest frequencies to several milligals at the shortest wavelengths. Typically they are described by spherical harmonic models for analysis and interpretation.
- Over the oceans, satellite altimetry maps geoid undulations which can be spatially differentiated to determine the gravity field. These active satellite measurements resolve the marine gravity field to 15–20 km wavelengths and longer with accuracies of several milligals. Currently available satellite altimetry data have a $7 \times 7\,\text{km}^2$ footprint on the ocean surface that limits the wavelength resolution of the gravity anomaly estimates.

Supplementary material and Study Questions are available on the website www.cambridge.org/gravmag.

6 Gravity data processing

6.1 Overview

Gravity observations include the combined effects of instrumental, surface, terrain, and planetary sources in addition to the subsurface mass variations that are the objective of an exploration gravity survey. To isolate the effects of subsurface sources, extraneous effects which include both temporal and spatial variations are removed from the data using theoretical considerations, geological information, and empirical observations. Some are considered universally, but others only in specific geological, surface, and observational conditions. There is an increasing need to eliminate a broader range of extraneous effects more precisely as the objectives of gravity surveying require higher precision and accuracy and are focused on both long- and short-wavelength anomalies. Unwanted effects are removed by calculating the gravity anomaly, which is the arithmetic difference between the observed vertical acceleration of gravity and the predicted or theoretical acceleration at the observation site. Theoretical gravity is based on a conceptual model of the sources of gravity variations. This model varies depending on the intended use of the gravity anomaly and the conditions of the survey.

Three classes of anomalies are recognized. The primary class, planetary anomalies, incorporates only analytically determined planetary considerations in the theoretical model, e.g. the Bouguer gravity anomaly. A second type, geological anomalies, applies additional effects from known or postulated subsurface geological conditions in the model, e.g. the isostatic residual gravity anomaly. The third type, filtered anomalies, is calculated by removal of arbitrary gravity effects caused by unknown sources, empirically or analytically determined by filtering that enhances particular attributes of the spatial pattern of the anomalies, e.g. wavelength-filtered gravity anomalies. The latter type seeks to isolate or enhance those of interest at the expense of other anomalies.

6.2 Introduction

Raw or field gravity measurements require processing to prepare the data for interpretation of subsurface variations in density. The low amplitude of subsurface gravity signals necessitates removal of predictable or measurable unwanted gravity variations, both temporal and spatial, from the raw data. The result is the calculation of a gravity anomaly at the observation site. The anomaly is the difference in either relative or absolute terms between the theoretical value of gravity at the gravity station and the observed value. The theoretical value is calculated assuming a conceptual planetary and a geological model that describes the gravity sources at the observation site. A variety of anomaly types are used in gravity interpretation depending on its intended use. Each type is produced from a specific model incorporating a different set of extraneous gravity effects.

To interpret anomaly data, specific variations in the cumulative gravity signal that relate to subsurface sources of interest must be determined. Generally, this is accomplished by identifying particular spatial spectral characteristics of certain anomalies. This requires the observation of gravity over a broad enough area and with ample detail and accuracy to map and isolate these anomalies. Unfortunately, the spectral characteristics of a wide variety of subsurface sources tend to overlap. Thus a significant part of the gravity data processing procedure is to isolate the specific anomalies of interest. This is achieved by a filtering process based on the characteristics of the anomaly and their pattern that either isolates these anomalies or enhances them at the expense of other anomalies that are unwanted in the interpretation. A variety of methodologies have been developed to achieve either isolation or enhancement. Particularly useful in this regard is spectral filtering as described in Appendix A.5.

This chapter describes and evaluates the sources of gravity variations that may be considered and removed from observed values to calculate the anomalies useful for exploring the Earth. It also considers the computation of a variety of gravity anomalies together with their potential applications to the study of the subsurface. Additionally, the relative merits of isolating and enhancing gravity anomalies are described.

6.3 Extraneous gravity variations

Raw or field gravity measurements are influenced by a wide variety of terrain, surface, geological, instrumental, and planetary sources. These sources have different amplitudes, periods in the case of temporal variations, and wavelengths in spatial variations that generally mask or at a minimum distort anomalies that are the objective of the gravity study. As a result gravity observations must be modified to eliminate these unwanted, geologically extraneous effects. The corrections are obtained by a combination of theoretical, geologic, and empirical methods. Some are applied universally to gravity data used in geological exploration while others are used only in specific geological, surface, and observational conditions. This process is commonly referred to as the correction or reduction of gravity data. Correction does not suggest that the observed data are in error, nor does reduction imply that the data of multiple observations are reduced or continued to a common vertical datum. Rather, both terms refer to the conversion of raw gravity observations to a usable form – that is, an anomaly which is the difference between the raw observed gravity and the modeled or predicted value of gravity at the station.

The significance of unwanted gravity variations varies with their attributes and the characteristics of the anomalies of interest, which in turn depend on the objective of the survey. In many surveys, the effects described below are negligible for achieving study objectives. However, special care is warranted in surveying and correcting gravity data obtained in mountainous regions (e.g. STEINHAUSER *et al.*, 1990) because of the profound effect of elevation on the sources of many gravity variations. Furthermore, special consideration must be given to corrections applied to surveys studying the long- and short-wavelength extremes of the anomaly spectrum because of their overlap with extraneous gravity effects, as well as to high-accuracy surveys designed to investigate temporal variations in subsurface conditions, or small sources. The precision and accuracy requirements of the survey are important factors in selecting the corrections, the methods used in their determination, and their accuracies.

In the following sections the predictable or measurable sources of variations of gravity are explained, as are the methods of correcting for them. The description is not comprehensive, but is enough for most gravity survey objectives. Ample references are provided to direct the reader to more extensive descriptions. The descriptions are directed to data processing for geophysical problems rather than geodetic problems, which are in most situations quite different.

6.3.1 Temporal variations

Observations of the gravitational field as a function of time at a station show that there are temporal variations or drift in the measurements with a wide range of periods. These changes, which vary depending on geographic location, are the result of a combination of effects from instrumental drift, planetary variations and tidal effects from extraterrestrial bodies, atmospheric pressure (mass) changes, both natural and anthropogenic-derived redistribution of mass in the subsurface, and elevation changes (e.g. VAN DAM and OLIVIER, 1998). In general, temporal variations due to changes in rotation of the Earth, oscillatory motions in the fluid core of the Earth, and tectonic movements (e.g. CROSSLEY, 1994) need not be considered because their amplitudes or periods do not overlap anomalies of interest. Shorter-period temporal effects due to Earth tides (e.g. LONGUEVERGNE *et al.*, 2009) and instrumental drift, on the other hand, can be important when dealing with surveys with long periods of time between repeat readings or when high accuracy is needed because of the low amplitude of the effect of geological sources of interest.

Inherent instrumental drift

Modern absolute gravity measuring instruments are not subject to instrumental drift or the inherent change of the sensor's response in observing gravity at a site with time. This is not the case with gravimeters that measure the relative change in gravity. Every gravimeter is unique in its mechanical properties, and thus has a distinctive instrumental drift curve describing how its measurement of gravity changes over time. Because of this uniqueness, the drift curve of one gravimeter cannot be used to correct the observations of another. Indeed, to correct for these temporal effects, there is no recourse but to reoccupy a previously observed station at appropriate time intervals to estimate the drift of the meter observations.

(A) Sources Gravimeters are subject to temporal variations (drift) due to changes in the environment and the mechanical and electrical systems of the instrumentation. Gravity measuring instruments that are based on mechanical systems are particularly susceptible to drift. Springs, whether constructed of metal or fused quartz, are subject to slow relaxation and strain imposed by stresses altering the mechanical system which in turn affects the gravity observations. Creep of the components is usually slow and relatively consistent; however, it can be accelerated by a mechanical or thermal shock. The result is that the meter will suffer a rapid, step-function-like variation, referred to as a tare (Section 5.3.1). A tare is essentially a break in the trend of gravity readings that represents a reset in relative value. Usually, these are less than 0.05 mGal. Tares are part of the total time variation in the gravity measurements, but their effect is rapid compared with the normal slowly varying effects of elastic relaxation. Especially significant tares may occur if a metal-spring meter, which requires clamping of the mass arm prior to disturbing the meter, is moved without arresting the mass. Typically, these tares are the result of the micrometer null adjustment screw slipping a thread when mechanically shocked in an unclamped state.

Gravimeters are subject to temperature and atmospheric pressure effects acting on the mechanical balance of the sensors. Temperature is especially critical. Instruments that are maintained at a constant temperature and have springs made of metal have significantly less effect from temperature than instruments that rely solely on compensation of the mechanical system for temperature variations and are constructed of quartz. The latter may have drift rates of 0.1 mGal/hour or more. The actual response of a gravimeter to sudden temperature changes can be complex, including changes in the sign of the gradient of the drift. Regardless of the instrument, drift is minimized by maintaining the instrument at a consistent temperature. Meters that lose temperature-maintaining power for even a few minutes may undergo increased drift, as indicated by monitoring the change in the readings.

(B) Correction procedures Numerous methods have been used to determine the instrumental drift of gravity measurements, involving real or virtual repeated observations at a site with the same meter. Repeat observations are required with the actual survey meter because the responses vary among meters. Thus, dedicating a gravimeter to taking measurements during a survey at a base station is not useful for determining instrumental drift.

Modern gravimeters have incorporated automated drift corrections into the instruments based on the assumption that the drift is caused by slow changes in the sensor system and including a temperature-measuring circuit which is used to minimize the effect of very small temperature changes. The drift from these sources is assumed to be linear with time as determined by repeated measurement of gravity (minus the tidal effects) with that particular instrument. Owing to minor non-linearity this drift function is generally updated at intervals of roughly a few months. To supplement this correction or to take its place where gravimeters without automated drift correction are used, a program of repeat readings at a base station must be established.

The simplest and one of the most accurate methods to establish drift control is the base check method wherein the gravimeter is returned to the survey base station at specified intervals to determine the drift of the meter, both inherent and environmental. The assumption is made that the observed gravity at the base remains constant at an accuracy prescribed by the survey over the period of the observations. Generally, the period of base return should be shorter than the diurnal Earth tides (typically every 4–6 hours), so that tides can be accounted for correctly. An example is illustrated in Figure 6.1, which shows the variation in instrumental effects (drift) at a gravity base as a function of time. The correction for the drift at each survey station is specified by the interpolation between these readings at the time of the measurement. Generally, the interpolation between base station readings can be and is assumed to be linear provided that the time interval between observations is no more than a few hours. When the interval exceeds this, a low-degree polynomial function taking into account the gradients of the theoretical tidal accelerations is appropriate. The polynomial can be determined by least squares, but the results need to be checked to determine that the errors of the approximation do not exceed an acceptable value for the specific survey. The base check method is one of the commonest of the drift check methods for exploration surveys because of its potential high accuracy. It is particularly useful where surveys are conducted within a limited area, centered around a common base station, and where there are no travel

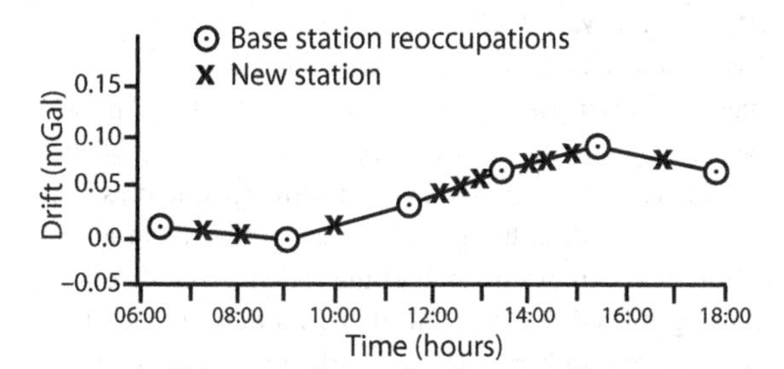

FIGURE 6.1 Graph showing the drift of a gravimeter as established by reoccupation of a base station. The base stations, indicated by the circles, are connected together by linear interpolation. The gravimeter drift correction for a survey station, indicated by the cross, is shown at the time of the measurement at this station.

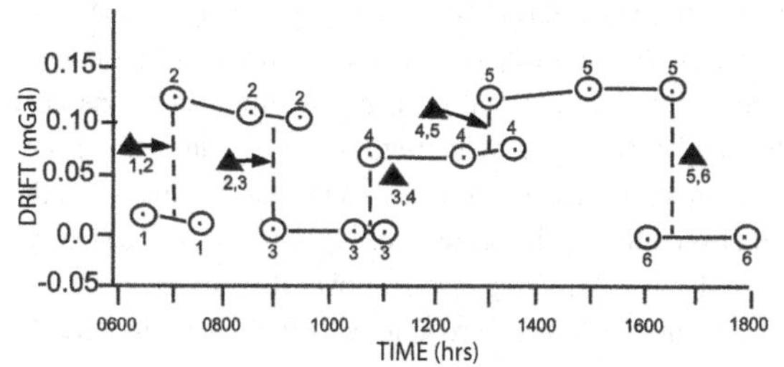

FIGURE 6.2 Graph of an example of the drift of a gravimeter as determined by the looping method. At each integer-labelled station, the gravity observation is shown by the dot within a circle. The difference between gravity observations of successively reoccupied stations in the looping procedure is indicated by a vertical dashed line with an associated delta symbol pointing to it. The numbers assigned to each delta (e.g. 2,3) indicate the stations being tied to each other by the looping procedure (e.g. stations 2 and 3).

restrictions preventing ready access between the survey and base stations.

The time interval between base check reoccupations depends on the accuracy requirements, characteristics of the instrument, environmental conditions (particularly the temperature variations), travel restrictions, and handling of the instrument (e.g. Chapter 5). It is desirable at the start of the survey to test the characteristics of the instrument, and the effect of the temperature variations and the handling of the instrument on its drift, by reoccupying, under the same environmental conditions, the same station at time intervals approximately equivalent to those anticipated during the survey. This information provides a basis for determining the base reoccupation interval for the required accuracy. Generally, a reoccupation interval of 1 to 2 hours is desirable for exploration surveys. Errors caused by the assumption of linear drift over this interval are generally negligible. However, the validity of this assumption for a specific instrument and environmental conditions can be tested by comparing the results with a drift curve based on more frequent reoccupations of the base station.

A potentially more precise method of determining drift is the looping method, which is based on the average difference between reoccupation drift curves prescribing the gravity difference between successive stations. For example, the observational sequence of consecutively numbered stations would be 1, 2, 1, 2, 3, 2, 3, 4, 3, 4, 5, . . . , 1 (Figure 6.2). Closed sequences of the individual loops permit the accuracy to be checked and errors distributed throughout the looping procedure. This method is useful in high-accuracy surveys and in setting secondary bases to which subsequent gravity surveying can be tied. The secondary bases are closely tied to the original base station, and thus ties to them are virtual base-station reoccupations.

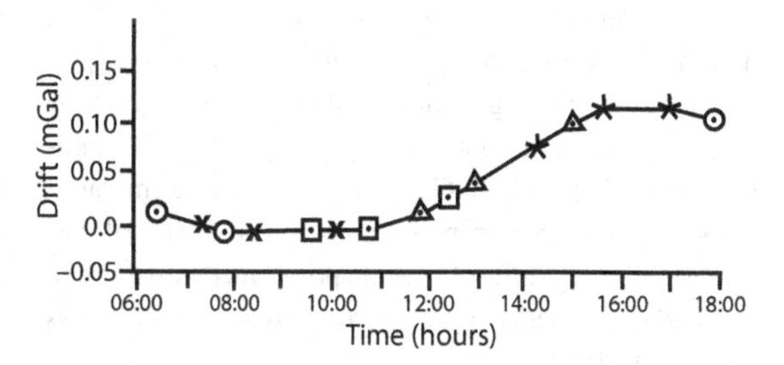

FIGURE 6.3 Graph showing the drift of a gravimeter as established by the cross-check method. Each symbol represents reoccupations of a specific station. The circles represent observations at the primary base. The other symbols are secondary bases which are positioned on the curve relative to the drift established by prior and subsequently reoccupied stations. The established drift curve is used to correct the stations of the survey for the gravimeter drift during the observational period.

The method, although capable of high accuracy, requires a considerable amount of travel time.

A drift check method which is useful in surveys of long traverses where travel to and from the primary base station is difficult and/or time consuming is the cross-check or traveling base-station method. This method decreases the travel time in making reoccupations, and thus speeds up the survey. In it, the secondary bases are established during the course of the traverse measurements. For successive stations numbered in sequence, the measurement pattern is as follows: 1, 2, 3, . . . , 9, 10, 1, 10, 11, 12, . . . , 19, 20, 10, 20, 21, 22, . . . , 1. As illustrated in Figure 6.3, the secondary base is established as part of the observational sequence, which makes it unnecessary to return to the primary base station (1) after completion of the first loop until the sequence is closed out at the end of the survey period. The advantage of this method is that travel time is decreased for a survey involving a linear traverse of stations, whereas the disadvantage is that errors may build up in the course of the measurements. The latter can be overcome by taking special care in making the reoccupation measurements, closing out the survey with an observation at the primary base at the end of the survey period and distributing the error over the survey.

Other drift check procedures, often involving more complex tying together of the reoccupied stations, have been designed to minimize the error in the drift prediction. For example, GETTINGS *et al.* (2008) used a drift correction procedure, the so-called staircase drift function procedure, based on least-squares inversion of multiple reoccupations of a base over the course of the survey. The staircase drift function in their study resulted in zero residuals at each repeat observation with linear drift between reoccupations.

Instrumental drift rates fundamentally control the design of drift monitoring procedures. For example, pre-1990s quartz-spring gravimeters of the Worden type have inherent short-term, non-linear drift measured in multiple hours of up to 0.1 mGal/hour, and longer-term variations of several days or more can reach 0.1–1.0 mGal/day. The short-term drift of metal-spring meters is up to 5 μGal/hour, whereas longer-term drift is generally less than 10 μGal/day.

Tares or instantaneous changes in the readings of a gravimeter will not be correctly monitored by the normal drift check methods because they assume that the change in readings is linear, or at least nearly linear, between reoccupations. As a result, tares will lead to errors in all the stations occurring in the base reoccupation interval during which the tare occurred. The only way to control these is to identify situations that may lead to tares in the gravimeter readings. For example, if the meter is mechanically shocked, handled roughly, or is subject to a rapid temperature change, it should be returned immediately to the last station that was observed and the change in the reading noted. If the drift is significantly changed and a tare is identified, the meter should be read semi-continuously at the last station until the tare dissipates or the meter readings regain the normal drift gradient of the instrument.

Modified procedures are required in marine and airborne gravity surveys because of the impossibility of reoccupying stations. However, instrumental drift can be largely eliminated if tie lines crossing the survey lines are observed and the differences in the anomaly values at the intersection points are minimized by a network adjustment procedure. Essentially, the line intersections serve as locally reoccupied base stations.

Atmospheric variations

Local atmospheric pressure variations enter into gravimeter observations in several ways. As described previously, they may cause a change in the buoyancy of the gravimeter sensor, and, in addition, atmospheric pressure changes will affect the gravitational attraction of the atmosphere that opposes the pull of the solid Earth, as well as causing related deformation of the Earth's surface. The effect of atmospheric pressure variations upon the mass of the atmosphere and the resulting effect on gravity measurements are normally incorporated into corrections for inherent instrumental drift. However, in high-accuracy surveys, it may be necessary to determine the gravitational effect of atmospheric pressure variations, especially when the period of the variations is small compared with the time between base station reoccupations. Surveys in mountainous areas are particularly vulnerable to short-term variations in pressure because of high, variable-velocity winds.

The period and amplitude of atmospheric variations cover a broad spectrum. For example, atmospheric tides, which have periods measured in hours, generally have pressure variations of < 1 hPa (1 hPa = 1 mbar), whereas cyclonic events that have periods of several days and affect areas measured in 100s or 1,000s of kilometers have a maximum range of ± 60 hPa, and seasonal variations are of the order of ± 10 hPa. The effect of these pressure variations on gravity observations is complex, but simplifying the atmosphere to a horizontal slab, a valid assumption in all but the most accurate surveys, shows that the microgal change in gravity Δg_{atm} due to local atmospheric pressure variation (within a radius of about 50 m) Δp in hPa is

$$\Delta g_{\mathrm{atm}} = -0.43 \Delta p, \tag{6.1}$$

where the negative sign of the correction reflects the inverse relationship between pressure and observed gravity (MERRIAM, 1992).

The change in gravity per millibar of pressure is not a constant but varies with the size and configuration of the pressure cell causing the mass change and its distribution relative to the site. MERRIAM (1992) reports that the value used in the above relationship is for large cells of radii of several hundred kilometers or more, and the value varies to about ± 0.35 μGal/mbar for cells of several tens of kilometers radius. As a result, TORGE (1989) estimates that short-term pressure variations cause an effect of only a few tenths of a microgal, with a maximum effect of $\pm 20 - 30$ μGal over several days, and roughly ± 2 μGal over a season. Corrections for measured atmospheric pressure changes are generally assumed to reduce the error due to these changes to less than a microgal. NEUMEYER *et al.* (2004) describe the relative merits of procedures for

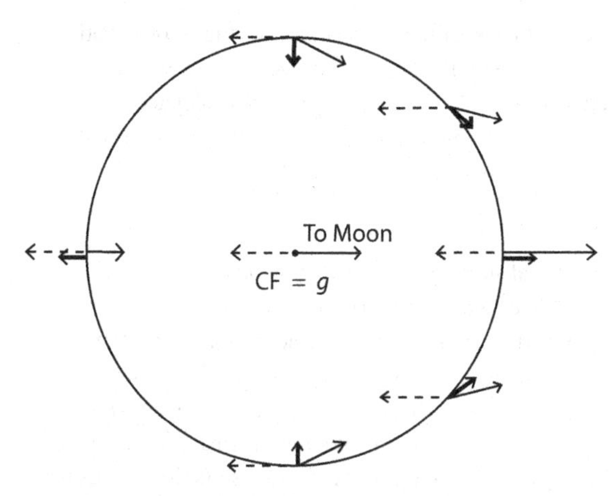

FIGURE 6.4 Schematic illustration of the tidal effect of the Moon upon the Earth (wide solid vectors) due to the varying gravitational effect (g) of the Moon (solid line vectors) upon the constant centrifugal acceleration (CF) caused by the orbital motion of the Earth and Moon around a common center of gravity (dashed vectors).

modeling the gravitational effect from atmospheric pressure changes. They recommend that the surface deformation can be satisfactorily modeled with 2D surface atmospheric pressure data, but that the attraction effect should be modeled with 3D pressure data.

Earth tides

A primary source of temporal variations in gravity observations is the tidal effect caused by gravitational forces of the Moon and Sun, which result in variations of up to several tenths of a milligal over time periods measured in hours. Gravitational tides (accelerations) originate in the difference between the gravitational attraction of an extraterrestrial body at a specific site and centrifugal acceleration resulting from the rotation of the system of celestial bodies around a common center of mass. The motion of the bodies around each other is translational rather than rotational resulting in a constant centrifugal force over the Earth, but the gravitational attraction from the extraterrestrial body differs over the surface of the Earth as a result of the varying distance between the surface site and the extraterrestrial body. The difference is the tidal effect which can be calculated from the position and masses of the bodies (Figure 6.4). Despite the great mass of the Sun, its tidal effect is only about 46% of the effect of the Moon because the Moon is so much closer to the Earth. Other extraterrestrial bodies have negligible effect on the surface of the Earth.

The cumulative effect of the tidal attraction of the Moon and Sun on the Earth varies in a periodic manner, with a common dominant period of roughly 12 hours, depending on the relative positions of these bodies as they orbit around the common center of mass. The maximum effect of the Moon is approximately ± 0.11 mGal, and the Sun causes a ± 0.05 mGal maximum variation. The maximum total variation within 24 hours is roughly 0.33 mGal and, during maximum variation, the effect has a period of about 25 hours with the period decreasing with the effect. The maximum rate of change is 0.05 mGal/hour.

The tidal effect also manifests itself as an elastic deformation of the Earth (body or solid-Earth tide), causing a periodic change in the radius of the Earth which commonly attains ± 20 cm, but can infrequently exceed 30 cm. To account for this effect, the values of the gravitational tide given in the previous paragraph include a 16% amplification. The body tide does not occur instantaneously, so there is a phase lag of the observed tidal effect. The elastic deformation is likely to change with location owing to variations in elasticity, but the variation of the effect with the elasticity of the nearby Earth is unlikely to be observed in surveys for geological investigations. As a result, an average elasticity of the Earth is commonly assumed, but more precise methods of determining the body tides can be used (e.g. TAMURA, 1987; HARTMANN and WENZEL, 1995).

The nature of tides is complicated near oceans because of the changing water mass as a result of ocean tides and the deflection of the surface due to the changing water mass. These may be out of phase with the Earth tides. The effect of ocean loading is generally less than 3 μGal, but can be as large as 20 μGal (TORGE, 1989), while the effects of the ocean tides themselves are highly variable, but may be as large as several milligals. An example of the combined lunar and solar tides upon a specific location on the Earth is shown in Figure 6.5.

Tidal accelerations are commonly calculated using the closed-form equations presented by LONGMAN (1959) employing an amplification factor of 1.16 which assumes an average elasticity of the Earth. The input to these equations is the latitude and longitude of the station and the Coordinated Universal Time/Greenwich Mean Time (UTC/GMT) of the observation. These equations are used in software incorporated into modern gravimeters or correction programs that remove the tidal effect from the raw observations. Unfortunately, theoretical calculations may show a discrepancy in phase and/or amplitude with the tidal effects observed by a gravimeter. These differences arise because the amplification of anelastic tidal calculations varies with latitude, with the elasticity of the nearby Earth, and with changes in space and time of the coastal effect associated with ocean tides. TRACEY (2006), in a

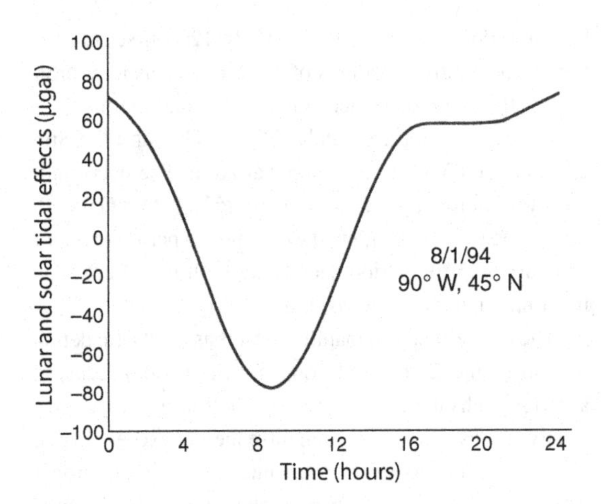

FIGURE 6.5 An illustration of the combined lunar and solar gravitational tides on the Earth assuming an amplification factor of 1.2 due to the elasticity of the Earth.

study of the accuracy of Earth tide programs, concludes that Longman's program (LONGMAN, 1959) satisfactorily predicts Earth tidal gravity for most exploration investigations, but that for studies requiring accuracies of the order of 10 microgals and where the observations are within about 500 km of the ocean coastline, it is necessary to use a tidal prediction program that incorporates the effect of the ocean load tide (SCHWIDERSKI, 1980).

When theoretical tides are being considered for correcting gravimeter measurements in a region, it is advisable to test their validity against semi-continuous observations over the course of at least a day, usually at one location. Instead of a theoretical formula, a continuous-reading base station gravimeter (Earth tide meter) can be used (e.g. MACQUEEN, 2010). If a continuous base station is used, the inherent instrumental drift of the Earth tide meter must also be evaluated. The use of a continuous-recording Earth tide meter is generally desirable for extremely high-accuracy work. Errors between the actual Earth tide measurements and the theoretical tide formula estimates usually reflect a slight phase shift because local elasticity effects are not accounted for in the theoretical formula.

Generally, tidal accelerations for exploration surveys are removed by normal reoccupation procedures, but in some regional surveys it is not feasible to reoccupy a base station at intervals of less than half a day. Under these circumstances, theoretical tides can be useful in predicting drift between the widely dispersed reoccupation of gravity bases. It is also occasionally desirable to compare predicted tidal variations against observed gravimeter drift to determine the nature of the short- and long-term inherent drift of the meter. Marine and airborne gravity surveys may not have sufficient accuracy, owing to observational and location uncertainties, to warrant concern with tidal effects, but tidal effects are readily calculated from the time and location of the observation from theoretical formulations such as presented by TORGE (1989).

Temporal subsurface mass variations

Subsurface mass may vary with time over a wide spectrum of periods from minutes to decades as a result of redistribution of magma, fluids, and gases in the subsurface. In some cases these mass changes are the objective of the survey, as is the case in studying magma movements, but in many others they cause extraneous gravity signals that may need to be considered in high-accuracy gravity surveying. Of special importance in exploration surveys are complex variations in gravity which have been noted due to groundwater changes as a result of rainfall, snowmelt, and flooding. These processes lead to increases in mass of the subsurface and gravity. A gravity decrease has also been noted from lowering of levels of standing bodies of water and direct changes in water levels due to groundwater withdrawal. Seasonal changes in water levels are generally no more than the order of 10 μGal, but may reach several tens of microgals in some situations such as tropical regions of high seasonal rainfall. A 1 m change in regional water levels with an effective porosity of roughly 20% will cause about a 10 μGal gravity change. Groundwater recharge from rainfall or snowmelt infiltrating through the unsaturated zone will cause similar gravity fluctuations. Significant variations are associated with the surface depressions that develop in some urban centers (e.g. Mexico City, Mexico; Houston, TX; and Phoenix, AZ) around groundwater cones of depressions caused by intense withdrawal of ground water and limited recharge. These gravity changes, where they interfere with the required accuracy of the survey, can be largely avoided by judicious location of gravity stations. A further description and illustration of gravity effects from fluid withdrawal are provided in the chapter on near-surface investigations in the website accompanying this book.

Subsurface mass variations and their accompanying gravity effects have also been related to viscoelastic rebound associated with recent deglaciation in southeast Alaska. SUN *et al.* (2010) measured a −3.5 to −5.6 μGal/year change over a high-accuracy gravity network that cannot be accounted for by crustal uplift. They therefore related the change to subsurface mass transfer associated with isostatic movements accompanying the deloading of the crust following glacial retreat and other tectonic sources.

6.3.2 Datums

All gravity surveys are tied to geographic, vertical, and gravity datums even if they default to arbitrary, local, datums, which is the case for many surveys conducted for limited objectives over small areas. However, it is advantageous in gravity surveying to relate all position, height, and gravity values of stations to established datums. That way, different gravity surveys can be used together. There are several choices in datums, but efforts are being made to establish global standards so that survey data can be readily joined without conversion or boundary discontinuities. In this section the optimum datums are specified, and methods are described for adjusting gravity station values to a datum.

Geographic and vertical datums

Globally, use of local geographic datums, based on different Earth ellipsoids, may lead to errors in position of up to 1 km. Accordingly, the North American Gravity Data Committee (NAGDC, 2005) has recommended that all geographic positions (latitude and longitude) of gravity stations be provided in decimal degrees to seven decimal places using the International Terrestrial Reference Frame (ITRF), with the 1980 Geodetic Reference System (GRS80) ellipsoid. Comparison of the 1984 World Geodetic System (WGS84) shows negligible differences with the ITRF. Precise WGS84 coordinates, used in the global positioning system (GPS), agree with internationally accepted ITRF coordinates to within 10 cm. These datums are subject to updating, so care must be exercised to use the updated datums or insure that results from these cause no significant discrepancy with older datums.

Traditionally, the vertical datum used in gravity surveying is the geoid or sea level, the same datum used in essentially all surface elevations. However, there is a global difference of approximately ± 100 m in the height between the geoid and best-fitting ellipsoidal surface of the Earth (Figures 2.3 and 2.4), the GRS80 ellipsoid, which is the datum for theoretical gravity over the Earth and GPS positions. As a result of this discrepancy there is a variable long-wavelength error in modeling gravity at a station.

To eliminate such problems, the revised standard vertical datum for the North American Gravity Database is the GRS80 ellipsoid (NAGDC, 2005). This approach is also being used for the Australian National Gravity Database (TRACEY *et al.*, 2007) and in some local surveys (e.g. WEBRING *et al.*, 2004). A generalized methodology for calculating gravity anomalies using the ellipsoid as the datum is described by KELLER *et al.* (2006). To differentiate anomalies calculated this way, the anomaly is preceded by the term "ellipsoidal." This revision in the vertical datum eliminates the need for the "geophysical" indirect effect described below which is used to adjust the modeled gravity at a station for the difference in height between the ellipsoid and datum. The adjective "geophysical" is used to differentiate this indirect effect from that used in geoidal studies. Using the ellipsoid as a vertical datum eliminates boundary inconsistencies in gravity anomaly values between areas using different vertical datums or geoids. Furthermore, the ellipsoid is the reference vertical datum for GPS that is now widely used in determining the height of gravity observations (FEATHERSTONE, 1993). Computer programs are available for conversion of elevations referenced to the geoid-to-ellipsoidal heights taking into account the most recent internationally accepted geoid (currently GEOID03 in the USA).

Gravity datum

Although gravity is commonly presented as an absolute value based on measurements that incorporate the mass of the Earth with its rotational properties, gravity measurements are generally made with an accelerometer which only measures the change in gravity acceleration, rather than its absolute value. Furthermore, absolute measurements of gravity have historically not been very accurate. A measurement of absolute gravity at Potsdam, Germany, in 1909 was the basis of gravity measurements for nearly 60 years and the absolute standard used in establishing the 1930 International Gravity Formula (IGF). All gravity measurements were tied to this observation by way of relative measurements established globally largely by gravimeters. However, by the 1960s it was apparent that the Potsdam measurement was considerably in error. Comprehensive studies identified that absolute gravity needed to be decreased by about 14 mGal. Further improvements in measuring absolute gravity have been incorporated into the currently accepted 1980 IGF (MORITZ, 1980b). National datums are continually being improved as a result of more accurate absolute measurements with modern portable absolute meters (e.g. the new Australian gravity datum; TRACEY *et al.*, 2007) which have an estimated accuracy of the order of 0.01 mGal.

Absolute gravity values are obtained by tying gravimeter observations to a gravity benchmark or base station that in turn is tied to the International Gravity Standardization Net 1971 (IGSN71) (MORELLI, 1974). Published IGSN71 values include a correction, the Honkasalo term (HONKASALO, 1964), which removes the average part of the Earth gravity tide. The use of this term is now no longer recommended by the International Association of

Geodesy (Uotila, 1980) because of errors it imposes on the calculation of the geoid from gravity measurements. The correction is removed from the IGSN71 values by adding the following gravity value:

$$\Delta g_{hon} = 0.0371(1 - 3\sin^2\theta), \tag{6.2}$$

where θ is the latitude (north or south) of the station. This correction varies from ± 0.04 mGal at the equator to ± 0.02 mGal at $\pm 45°$ and to -0.07 mGal at the poles. Accordingly, the effect is only of concern to accurate long-wavelength objective surveys.

Gravity datum adjustment

As described above, gravimeter observations are non-absolute, being measured only with reference to the other stations of the survey. To eliminate time-to-time changes in the gravitational acceleration as measured by the gravimeter, all daily observations are tied to a single base station or to one of a group of secondary bases which previously have been tied together and to a primary base. The process of tying the observations together is performed by reoccupation of stations during the course of a survey. In the case of marine surveys, observations made with a shipborne meter can be tied with a portable land gravimeter to absolute bases at ports of call. Datum adjustment is generally not a concern in airborne measurements, but can be approximated by overlapping the airborne survey onto a well-mapped surface gravity survey and tying the airborne survey into the upward-continued results from the land survey.

In most cases it is desirable to have the observed gravity tied to an absolute gravity benchmark, but this is not essential in surveys covering limited areas for restricted objectives. Absolute ties permit putting the gravity survey into context with other data, particularly observations available from national or regional data bases which may be useful in determining the nature of the regional gravity anomaly. Also, an approximation to the mean regional anomaly of an area can be estimated by calculating the gravitational effect of the mass between sea level and the regional elevation of the area of the survey. This assumes that the area is in isostatic equilibrium (Section 6.3.8), with the excess gravitational effect of the mass above sea level being compensated by an equivalent deficiency of mass at depth, and that the mass between the datum and the station is accounted for in the anomaly calculation as it is in the Bouguer gravity anomaly. The gravity effect of this deficiency in mass is the mean regional anomaly of the area.

6.3.3 Latitude variation

The gravitational attraction of the Earth varies from the equator to the poles as a result of the decrease in radius of about 21 km and a change in centrifugal acceleration due to the difference in the radius of gyration from the equatorial radius to zero at the poles. This total difference is roughly 5,000 mGal and is called the latitude correction. It is represented by the International Gravity Formula described below and is the first-order correction performed in nearly all gravity measurements.

As a result of the Earth's rotation, the variations in centrifugal force

$$\mathrm{CF} = \omega^2 r \tag{6.3}$$

with angular velocity $\omega = 7.292\,11 \times 10^{-5}$ rad/s and radius of gyration r account for approximately 3/5 of the total change in the roughly 5.2 gal variation between the equator and the poles. The variation in normal gravity, g_θ, as a function of latitude (θ) either north or south on the ellipsoidal surface best fitting the shape of the Earth commonly is given by the second-order series expansion:

$$g_\theta = g_e[1 + A\sin^2(\theta) - B\sin^2(2\theta)], \tag{6.4}$$

where g_e is the statistically determined value of gravity at the equator on the ellipsoid, A is the gravitational flattening given by the difference between the gravity on the ellipsoid at the equator and poles divided by the equatorial gravity, and B is a function of the flattening (f) of the Earth given by the difference between the semimajor and semiminor axes of the ellipsoid divided by the semimajor axis, and the ratio of the equatorial centrifugal force and the equatorial gravity (Figure 2.3).

According to the most recently accepted international standard, GRS80 (Moritz, 1980b), the numerical values for the components of Equation 6.4 are $g_e = 978.0327$ gal, $A = 0.005\,302\,4$, $B = 0.000\,005\,8$, and $f = (1/298.257)$. In closed form, the normal gravity (in gal) in latitude (Somigliana, 1930) based on the GRS80 is

$$g_\theta = 978.032\,677\,15\left[\frac{1 + 0.001\,913\,185\,135\,3\sin^2(\theta)}{\sqrt{1 - 0.006\,694\,380\,022\,9\sin^2(\theta)}}\right]. \tag{6.5}$$

The GRS84 ellipsoid has no significant differences from the GRS80 ellipsoid in geophysical uses of gravity data.

As noted from Equation 6.5, the normal gravity of the Earth increases by roughly 5.186 gal from the equator to the poles, but decreases as a function of latitude from both the poles and equator to a latitude of $\pm 45°$ by approximately 0.006 gal because of the $\sin^2(2\theta)$ term.

Equation 6.5 is commonly referred to as the International Gravity Formula 1980 (IGF) and g_θ as the theoretical or normal gravity.

The Geodetic Reference System 1980 differs very little from the previous Geodetic Reference System 1967 (GRS67) (MORELLI, 1974), which has been used to calculate gravity anomalies in many databases. In the 1980 formula, g_e has been increased by 0.9 mGal and the value of the $\sin^2(2\theta)$ term at its maximum at 45° latitude has decreased by 0.1 mGal.

In many exploration surveys covering limited areas, the station locations are not referenced to global latitude. As a result, the latitudinal effects on the stations of the survey are referenced to a single station (usually the base station) and determined by assuming a linear gradient of the normal gravity as a function of north–south distance. This gradient in mGal/m can be determined by differentiating Equation 6.5 with respect to north–south distance x_{N-S} so that

$$\frac{\partial g_\theta}{\partial x_{N-S}} = 0.8144 \times 10^{-3} \sin(2\theta), \tag{6.6}$$

assuming a mean radius for the Earth of 6,367.45 km and the spherical coordinate transformation of x given in Equation 3.82. The gradient is zero at the equator and poles, and maximum (0.8144 mGal/km) at 45° latitude, but at 45° the gradient is more nearly linear. Normal practice is to calculate the gradient for the mid-latitude of the survey area and to check this value against the gradient at the latitude extremes to insure linearity over the area within the range of acceptable error in the survey. If the observed station is north of the reference station in the northern hemisphere, the gradient multiplied by the north–south horizontal distance between the two stations is added to the normal gravity (calculated or assumed) at the reference station to determine the normal gravity at the observed station. This is reversed for surveys in the southern hemisphere.

6.3.4 Atmospheric mass correction

As discussed above, the temporal deformation of the gravity observation site elevation and gravitational attraction due to varying atmospheric pressure may be considered in high-accuracy gravity surveys. In addition, there is an atmospheric correction considering the elevation of the observation site, because the normal gravity on the Earth's ellipsoid includes the mass of the total atmosphere and assumes that this mass is within the Earth's ellipsoid. This assumption is invalid when considering gravity measurements on the topographic surface of the continents because of the change in thickness of the actual atmosphere. As a result, an atmospheric correction may be applied to the normal gravity on the ellipsoid given by the IGF 1980 that takes into account the elevation of the station in high-accuracy surveys, especially those that cover a wide range of elevations. This correction is subtracted from the theoretical or normal gravity at a station. Consideration of this correction increases a gravity anomaly by approximately 0.86 mGal at 100 m and by roughly 0.77 mGal at 1,000 m.

Numerous methods have been described to calculate the effect of the atmospheric mass. A method based on a model atmosphere using an analytical expression described by ECKER and MITTERMAYER (1969) is reprinted in MORITZ (1980b), or the effect can be calculated to the nearest hundredth milligal for the atmosphere up to a height of 10 km using

$$g_{\text{atm}} = 0.874 - 9.9 \times 10^{-5}h + 3.56 \times 10^{-9}h^2, \tag{6.7}$$

where h is the height of the observation in meters (WENZEL, 1985) and the correction is in milligals. This formula was used by LI *et al.* (2006) to calculate the gravity effect of the atmosphere over North America (Figure 6.6). The variation is about 0.3 mGal over the continent (0.88 to 0.58 mGal) from sea level to the highest elevations. There are seasonal and diurnal changes in the mass of the atmosphere as a function of latitude due to variations in the Earth's rotational and orbital characteristics and in radiation received from the Sun that complicate the calculation based on the simplifying assumption of homogeneous shells of the atmosphere. However, MERRIAM (1992) has pointed out that these variations cause changes only in the submilligal range because 90% of the atmospheric signal is produced by the atmosphere within a 50 km radius of the gravity station. In addition, NIEBAUER (1988), MIKUSKA *et al.* (2008), and others have noted the effect of topographic variations on the simple assumptions used in calculating the gravity effect of the overlying atmosphere. However, the effect is minimal considering that the density of the atmosphere is only of the order of 0.04% of the density of surface material densities.

6.3.5 Height effect

The height correction, which is also commonly called the free-air and sometimes Faye's correction, accounts for the decrease in gravity with increasing height. This is because the IGF (Equation 6.5) is only valid on the best fitting ellipsoid to the surface of the Earth. In other words, the IGF is being corrected for the fact that the observation is not on the Earth's ellipsoid. However, this correction does not account for the mass of material between the datum and the station, which explains the origin of the term free-air correction. The vertical gradient of gravity at

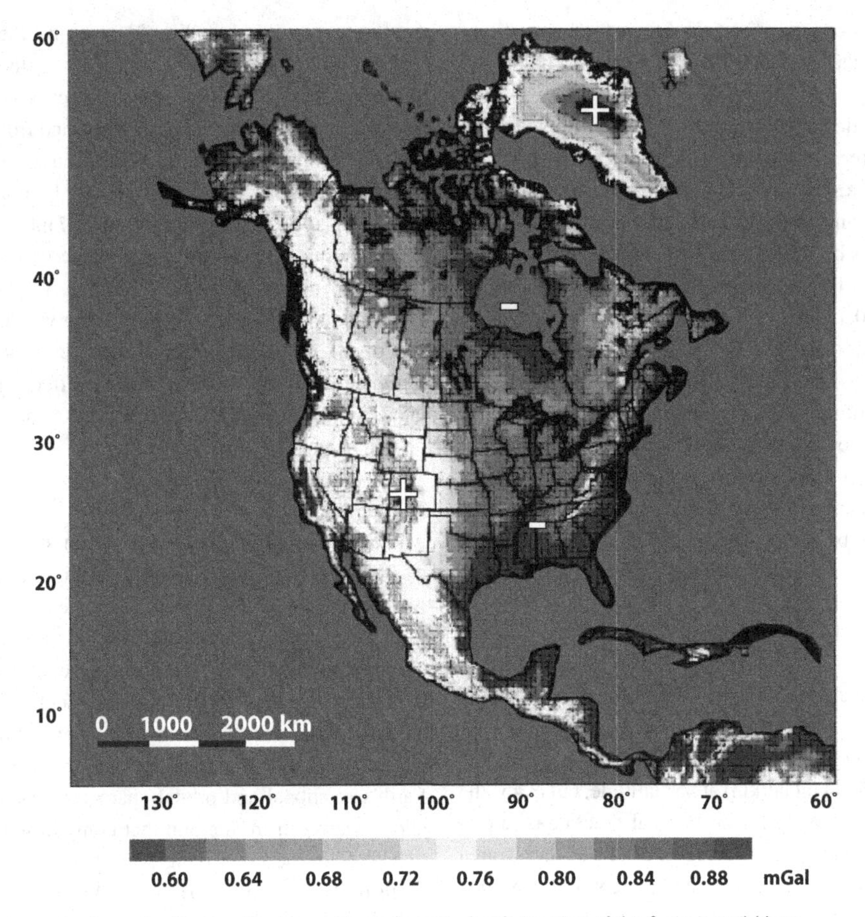

FIGURE 6.6 Atmospheric gravity effect in milligals over North America. A color version of this figure is available at www.cambridge.org/gravmag. Adapted from Li *et al.* (2006).

the surface of the Earth can be determined by considering the mass of the Earth concentrated at its center with the potential in Equation 3.16. Thus, differentiating the point mass potential in the r-radial direction gives the radial component gravity effect

$$F_r = \frac{\partial U}{\partial r} = -\frac{GM_e}{r^2} \equiv g(r), \tag{6.8}$$

where M_e is the mass of the Earth. The gravity effect at a small distance h from r can be estimated from the Taylor's series expansion

$$g(r+h) = g(r) + h(\frac{\partial g(r)}{\partial r}) + \cdots, \tag{6.9}$$

so that the first-order difference is

$$g(r+h) - g(r) \approx h\left(\frac{2g(r)}{r}\right), \tag{6.10}$$

and thus the free-air correction is positive for $h < 0$ and negative for $h > 0$.

Evaluating Equation 6.10 at the Earth's spherical surface with radius $r = a_e = 6{,}378{,}137$ m, where $g(a_e) = g_e = 978{,}032.67715$ mGal, gives the free-air or height correction in mGal

$$g_h = 0.3086 \times h. \tag{6.11}$$

Equation 6.11 represents the change in normal gravity at the station with the height difference h in meters relative to the datum height. The datum may be an arbitrary level, the level of the reference station, the ellipsoid, or sea level depending on the information available and the purpose of the survey. This is the most commonly used formula for the free-air correction.

The above consideration of the vertical gradient of gravity does not take into account the gravitational or shape ellipticity of the Earth or the non-linear components of the gradient with increasing elevation from the surface. These effects are commonly neglected because the gradient changes by less than 0.2% from the equator to the pole and less than 0.02% with height from sea level to 10 km. Taking into account the effect of latitude (θ) and height above sea level (h), a commonly used expanded version of the height correction in mGal based on the GRS80 ellipsoid is

$$g_{h\theta} = \frac{-2g_e}{a}\left[1 + f + m + \left(-3f + \frac{5m}{2}\right)\sin^2\theta\right]h + \frac{3g_e h^2}{a^2}, \quad (6.12)$$

where the GRS80 ellipsoid has the parameter values $a =$ 6,378,137 m for the semimajor axis, $b =$ 6,356,752 m for the semiminor axis, $f = 0.003\,352\,810\,681$ for the flattening, $g_e = 978{,}032.67715$ mGal, and $m = \omega^2 a^2 b^2 / GM_e = 0.003\,449\,786\,003\,08$ for the angular velocity $\omega = 7{,}292{,}115 \times 10^{-11}$ rad/s, and the geocentric gravitational constant $GM_e = 3{,}986{,}005 \times 10^8$ m^3/s^2.

For the GRS80 ellipsoid, this second-order formula becomes

$$g_{h\theta} = [0.308\,769\,1 - 4.398 \times 10^{-4}\sin^2\theta]h + 7.2125 \times 10^{-8}h^2, \quad (6.13)$$

in mGal. This second-order formula is recommended for surveys of high accuracy where station elevations vary by more than several hundred meters. Abnormal vertical gradients result from local terrain features and subsurface mass variations; however, the former are removed through proper reduction of the observed data and the latter are the subject of the investigation. In local surveys requiring high accuracy, some investigators (e.g. GEHMAN *et al.*, 2009) measure the vertical gradient over the limited area of study with observations at different vertical levels and use the resulting gradient to correct for the varying elevations of the stations. This removes the major subsurface gravity effect. However, great care is needed in making observations for determining the vertical gradient, and measurements are needed across the survey area to validate the use of the averaged gradient.

6.3.6 Mass effect

The mass or Bouguer correction takes into account the gravitational attraction of the Earth material that occurs between the station and the vertical datum of the survey. The correction is equal to the gravity effect of a horizontal slab of the material of thickness h and infinite radius. This effect is given by the Bouguer slab Equation 7.11 with $\Delta z' = h$ so that the mass correction in milligals is

$$g_m = 2\pi G\sigma h = 4.193 \times 10^{-5}\sigma h, \quad (6.14)$$

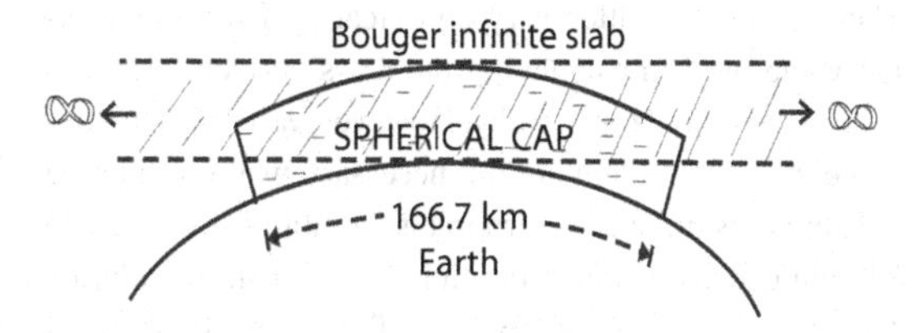

FIGURE 6.7 Schematic illustration (not to scale) comparing the cross-section of the infinite slab used in computing the mass correction, the spherical cap with a radius of 166.7 km that approximates the gravitational effect of the infinite slab, and the spherical shell of the Earth. The effect of the curvature of the Earth is considered in a special correction, the curvature or Bullard B correction.

where the gravitational constant $G = 6.674 \pm 0.0001 \times 10^{-11}$ m^3/(kg s^2), and σ is the density in kg/m^3 of the material between the datum and the station at height h in meters.

This approximation closely approaches the gravitational effect of a spherical cap of the Earth centered on the station that has a radius of 166.7 km (Figure 6.7). It is one-half that of the attraction of a complete spherical shell ($4\pi G\sigma h$) of the Earth. However, the 2π term is used for exploration surveys because surface masses beyond 167 km produce negligible gravitational effects on the station. The mass correction is opposite in sign to the height correction owing to the vertical gradient of gravity and typically is roughly one-third of this effect. The two corrections, height and mass, commonly are combined into a single elevation correction factor because both are a function of the height difference between the datum and station. Typically, a crustal density of 2,670 kg/m^3 is assumed in gravity surveys resulting in a mass correction factor in milligals

$$g_m(2670) = 0.1119 \times h \quad (6.15)$$

with the combined height in meters and mass (i.e. elevation) correction factor given by

$$g_{elev}(2670) = 0.1976 \times h. \quad (6.16)$$

Although a density of 2,670 kg/m^3 is an appropriate estimate of the average density of the continents above sea level for local surveys in many terrains (HINZE, 2003), this factor is commonly modified to accommodate the generally lower densities of actual surface materials. Even in regional surveys (e.g. HUANG *et al.*, 2001; KUHN, 2003), if information on surface densities is available on a station by station basis, the correction for mass is more accurate using variable densities, although estimating densities within several percent of their actual value is seldom possible. The use of variable densities is especially important where actual densities vary significantly from the average value used in calculating the mass correction and in regions of rugged topography. However, variable densities in the mass correction are not necessary where modeling of anomalies directly incorporates variable densities. The following densities are traditionally used in calculating the mass effect: sea water $\sigma_{sw} \approx 1{,}027\,\mathrm{kg/m^3}$, fresh water $\sigma_{fw} \approx 1{,}000\,\mathrm{kg/m^3}$, and ice $\sigma_{ice} \approx 917\,\mathrm{kg/m^3}$.

In actual survey conditions, the surface of the Earth may not be flat enough to warrant the assumption of the horizontal infinite slab used in the mass correction. Accordingly, corrections may be necessary for the deviant terrain effects as described in the next section, and for the curvature of the Earth's surface.

Mass corrections in the marine environment require special procedures. Water bottom surveys, for example, include a correction for the mass of the water column above the observation and the difference in elevation between the sea surface (sea level) and the site of the measurement. The mass of the water column actually is considered twice, once for removal of the water column effect on the normal gravity and again for its effect on the actual measurement on the bottom surface. In addition, in calculating Bouguer gravity anomalies, the water column is replaced by a horizontal slab of appropriate density so that

$$g_m = 2\pi G(\sigma_r - \sigma_{sw}) \times d, \tag{6.17}$$

where σ_r is the density of the rock replacing the water and d is the water depth.

In marine surface surveys, the water column is simply replaced by an appropriate rock column using Equation 6.17 to adjust the anomaly for the low density of water. Terrain corrections may be required for the water depth correction in shallow water regions of rugged bottom relief because the correction assumes a horizontal water column slab of uniform thickness. In contrast to land terrain corrections, which are always the same sign regardless of the relief, the terrain correction is positive for a rise in the sea floor and negative for increased water depths. These corrections are readily adapted for use in lakes taking into account the elevation of the stations above sea level. In airborne surveys, the mass effect is calculated in the normal manner using the elevation of the ground surface directly below the airborne observation and the appropriate density for the local geology. The altitude of the aircraft at operating levels is not considered because the Bouguer slab effect is independent of the distance above it. However, this assumes that the ground surface is flat. In regions of rugged topography it is important to consider the 3D terrain effects (e.g. TZIAVOS *et al.*, 1988; MACQUEEN and HARRISON, 1997; DRANSFIELD and ZENG, 2009). This is particularly true of gravity gradiometer measurements which are notably affected by terrain effects.

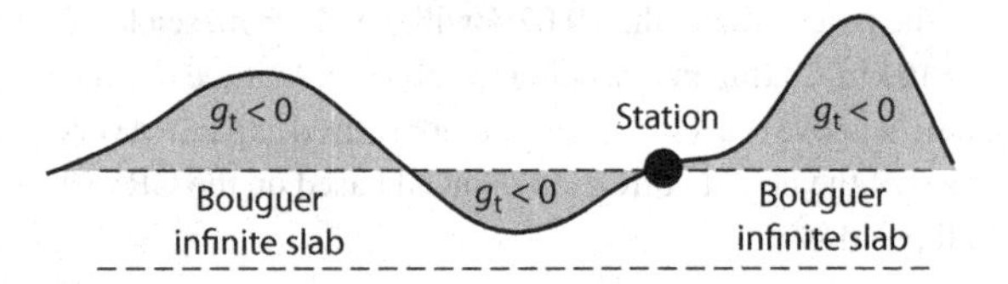

FIGURE 6.8 Schematic illustration showing the gravitational terrain effect of both positive and negative changes in the surface topography relative to the elevation of a gravity station as a result of the horizontal, infinite slab assumed in the correction for the mass of material between the station and the datum. Note that both positive and negative changes cause a decrease in the gravitational effect of the assumed slab.

6.3.7 Terrain effect

Terrain corrections are applied to gravity observations to adjust the Bouguer correction for the effects of variations in thickness of the horizontal, infinite slab used in calculating the mass effect at a station. The gravity effect of this correction reflects the deviation of the Earth's surface from the Bouguer slab, hence the term terrain correction, although some geophysicists prefer the term topographical correction. The effect of surface relief on a surface station whether above or below the station is always to decrease the mass correction (and thus the gravity anomaly) for surface measurements. The gravitational attractive force of topography above the observation site is away from the Earth, thus decreasing the gravitational effect of the slab. The effect of topography below the station on the gravitational acceleration caused by the slab is also negative because of the lack of material, and thus a deficiency in attraction (Figure 6.8).

The importance of this effect is very much dependent on local, nearby topography because of the inverse square law of gravity. However, it also depends on the volume of the terrain deviation from the slab. Thus, for generally flat regions where the elevation range is measured in at most tens of meters, terrain corrections typically range from 0.1 to 1 mGal. In hilly regions where relief ranges up to a few hundred meters, the corrections are of the order of 1 to 10 mGal; and in mountainous regions the corrections range upwards of several tens of milligals. These values are only crude ranges, but they do illustrate the importance of the correction and the need to make accurate estimates in the calculation of anomalies. Bathymetric corrections are equivalent to terrain corrections, where the gravity effect is determined for the change in water-bottom relief in either ocean or lake environments.

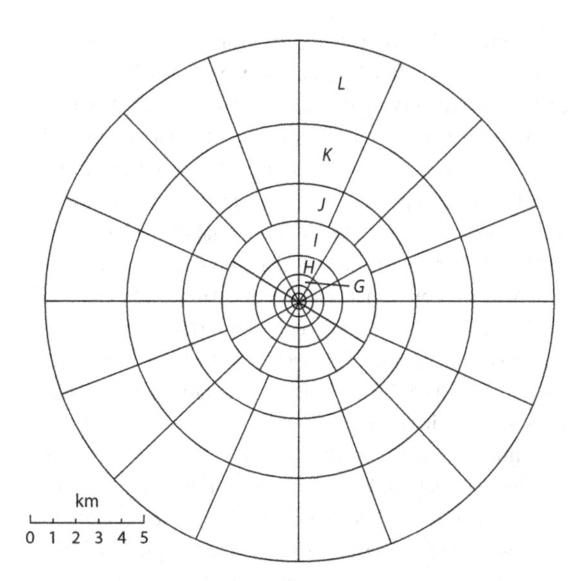

FIGURE 6.9 Portion of template used in determining average elevation within the sectors of the annular rings (zones) specified by HAMMER (1939) to calculate terrain corrections for exploration gravity surveys in hilly country. Note scale. HAMMER (1939) tabulated the radii and terrain corrections for the annular rings A to M from the station to 21.94 km.

In considering terrain corrections, it is important to differentiate between absolute or total corrections and relative terrain effects within the survey area. In dealing with surveys of small geographical extent for specific geologic objectives, it is unnecessary to have absolute anomaly values because terrain effects that are nearly constant or change in a regular manner over the area of the survey need not be calculated. Effects of this nature commonly originate from distant major topographic features. Their effect on the survey stations can be removed from the anomalies if necessary during the regional elimination phase of the data processing. This may greatly simplify the terrain correction calculations. In other words, terrain effects closest to the station are more important to remove but those from distant terrain may be ignored, depending on the objectives and size of the survey.

The classical manual methodology for calculating terrain corrections for exploration surveys was developed by HAMMER (1939). This approach has been largely superseded by more efficient, automated methods using digital elevation models (DEMs). However, examining Hammer's method is instructive in understanding terrain corrections and their calculation because it is emulated in the automated approach.

Hammer's method is based on determining the gravity effect of a series of segmented annular rings, called zones, centered around the station as shown in Figure 6.9. The system was designed to produce approximately equidimensional sectors in each ring by relating the radius of the ring to the number of sectors, thereby improving the accuracy of the corrections. Hammer's terrain charts were designed for an accuracy of roughly 0.1 mGal when used in hilly country. The method and related look-up tables for rapidly determining corrections have been expanded to achieve greater accuracy and to be applicable to areas of higher relief (e.g. BIBLE, 1962; DOUGLAS and PRAHL, 1972; HAMMER, 1982) by decreasing the radii of the inner rings (or zones) and the interval of the elevation of the sectors and increasing the total range of the elevations. Additional modifications in the method have been suggested by replacing the flat-topped sectors based upon the average height with complex but more realistic sloping surfaces (e.g. CAMPBELL, 1980; OLIVIER and SIMARD, 1981; ZHOU *et al.*, 1990).

Hammer's method is implemented by preparing a transparent template at the scale of the available topographic map of the survey region, overlaying the template of the segmented annular zones on the topographic map with its center at the location of the gravity station, determining the difference in elevation between the station and the average elevation of each sector from the topographic contours, and using a look-up table as provided by HAMMER (1939) or subsequently prepared modifications to determine the terrain effect of each sector. The effects of the sectors are summed and the result modified for the appropriate density of the materials making up the terrain. Care needs to be exercised in averaging elevations within a sector that are near the elevation of the station because heights above and below the station elevation do not counteract each other. As a result, the average of the elevation differences must be determined and not the difference between the average elevation of the sector and the station elevation.

Typically no correction is applied to Hammer's *A*-zone, from 0 to roughly 2 m (6.56 ft), unless the slope in the region exceeds approximately 30°. This is best compensated by proper placement of the observation station in generally flat areas away (by a few meters) from local mounds or gullies. Most topographic maps and DEMs do not provide sufficient detail to accurately specify the elevations within the sectors of the *B*-, *C*-, and *D*-zones, which have bounding radii of roughly 2 to 16.6 m, 16.6 to 53.3 m, and 53.3 to 170.1 m, respectively, or of these zones divided into sub-zones (e.g. HAMMER, 1982). Instead, the elevations in the sectors of these zones are estimated visually with the aid of optical or electronic devices such as reflectorless laser rangefinders or from high precision and accuracy contour maps at scales larger than 1:24,000. Even with judicious location of stations, the effects of terrain in these inner zones can be a major contributor to the total correction, especially in areas of moderate-to-high surface relief, and thus must be carefully considered. Even low relief can have a significant effect on terrain corrections (LEAMAN, 1998). For example, ditches or ridges of only a meter or two in depth or height can cause terrain effects of several hundredths of a milligal if the observation is made at the edge or on the local topographic feature.

The manual calculation of terrain effects by Hammer's method is time-consuming, tedious, and subject to error as a result of visually estimating the average elevation of the sectors of the template. As a result, manual calculations have been largely superseded by automated calculations using digital elevations available in a gridded data set, although the elevations of the available data grid may have to be interpolated to a finer grid to include sufficient data points to account for the effect of rugged relief. Numerous methodologies have been developed for implementing computer calculation of terrain effects. For example, Hammer's template method can be automated by calculating the average value of the gridded elevations occurring within a sector of the template centered on the station and determining the correction by reference to a digital look-up table, which gives the terrain effect corrected for the appropriate density of the surface material of each sector of each ring as a function of the vertical distance between the station elevation and the averaged difference between the station and the elevation of the sector.

In contrast, the terrain effect can be numerically estimated by calculating the gravitational effect of a rectangular vertical prism centered on each gridded height point which has a height equal to the difference between the station and the grid height and summing the effects over the regional area (e.g. BOTT, 1959; KANE, 1962; PLOUFF, 1976; BLIZKOVSKY, 1979; COGBILL, 1990). This can be simplified by approximating the mass of the prism with a vertical line element where the distance to the prism is 10 or more times the grid interval (e.g. MA and WATTS, 1994). Section 3.5 describes additional idealized geometric forms with closed-form analytical solutions that are useful for approximating topography, buildings, excavations, and other features in estimating terrain corrections.

The terrain correction process can be speeded up with spectral analysis methods (e.g. SCHWARZ *et al.*, 1990; LI and SIDERIS, 1994; PARKER, 1995, 1996) by taking advantage of the regular grid of elevations available in DEMs and the fact that the terrain effect equation is a spatial convolution of the DEM and an appropriately weighted inverse distance function (e.g. SIDERIS, 1985; FORSBERG, 1985). The method employs the 2D fast Fourier transform (FFT) to evaluate the convolution as the correction over a regional area exclusive of the effects in the immediate vicinity of the station. This method has been used to calculate regional terrain effects over continental-sized regions (e.g. KIRBY and FEATHERSTONE, 1999), which, when supplemented with terrain corrections of nearby relief interior to the grid interval of the DEM, can be used to rapidly determine the complete terrain effect. It should be noted that the use of 30 m DEMs can lead to significant errors in gravitational terrain effects when used to calculate nearby topographic features (e.g. COGBILL, 1990). The inner terrain zones require more detailed consideration in these situations.

Another popular methodology is the so-called 3D Bouguer correction. Instead of computing a Bouguer slab and then correcting that slab separately for terrain effects, some workers now simply compute the complete Bouguer correction directly as a complex DEM surface at each observation. Many marine and airborne datasets are computed in this manner.

A universally recommended methodology for calculating terrain corrections is difficult to prescribe because of the numerous variables to be considered in an actual survey and the processing of its data. These variables include the required accuracy of the corrections, the height and nature of the relief both near and far, the availability of analog or digital heights, the variability of the density of the surface material in the survey and adjacent regions, the availability of computer algorithms, and the resources considering the size of the survey. Realizing the potential complexity of the problem, it is useful to consider a few critical concerns.

First, a computational procedure needs to be decided upon, taking into account the variables just listed. All methods are based on assumptions, and thus have certain limitations. The importance of these limitations depends on the characteristics of the survey and the topography and relief of the region, so generalizations should be treated cautiously. A useful review of methods is presented by NOWELL (1999). Wherever possible, it is best to use digital methods of computation because of their speed, reproducibility, and consistency. These attributes relieve the analyst of the tedious, and inherently erroneous, visual/manual evaluation of elevations from topographic maps and make it easier to include a larger region in greater detail in the analysis. Spectral methods of determining terrain corrections using the FFT are particularly rapid, but generally do not adequately consider the nearby topography. Results of regional studies of terrain effects utilizing this methodology may be available, and thus can be incorporated into local surveys. In this manner regional terrain corrections can readily be included with little additional effort.

Computation of terrain corrections is generally divided into two segments which independently consider the near and far topography. The near topography is commonly limited to the maximum radius of Hammer's zone D (170.1 m) or in some cases distances as great as 1 or 2 km. The distance separating near and far corrections is not only based on the relief of the area, but commonly reflects the grid interval of the available elevation data set, with the terrain within the nearest grid interval being treated independently of the grid values. In surveys of limited areas, terrain corrections are limited to the near region because the effect of more distant topography is constant over the survey area or is linear enough not to affect the interpretation of the survey. To achieve adequate accuracy in terrain correction calculations in high-relief areas, it is necessary to subdivide Hammer's zones or to modify the surfaces of the sectors of the rings from flat to sloping. The terrain correction g_{ter} derived from a constant slope over zone A is the gravitational effect at the station calculated from

$$g_{ter} = 2\pi G\sigma R(1 - \cos\theta), \tag{6.18}$$

where R is the radius of the zone and θ is the vertical angle between the center of the zone and the maximum elevation at R. In essentially all surveys, it is highly desirable to calculate the effect of the far topography from the most detailed DEM that is available, using vertical, rectangular prisms where the distance to the grid point is less than 10 times the grid interval and using the simplified equation of the vertical line mass for greater distances. Terrain corrections for shipborne, underwater, underground, and airborne gravity surveys require specialized treatments (e.g. Nowell, 1999).

The second critical concern in arriving at a reasonable terrain effect methodology is the importance of considering the source and nature of the elevation data. Surface elevations may be obtained from digital elevation models, topographic contour maps, or discrete elevations obtained in the near vicinity of the stations by optical or electronic methods at critical locations at the extremes of elevation variance from the station elevation. Wherever possible, it is desirable to use a digital elevation model for regional areas, understanding that the values are in error by an amount depending on the quality of the original map from which they were derived. The error generally is of the order of magnitude of one-tenth of the map's contour interval. Cogbill (1990) suggests that use of the 30 m DEMs prepared by the US Geological Survey in calculating terrain corrections leads to accuracies of roughly 5 μGal for topography more than 250 m from the station. Topography located within 250 m of the station causes errors of 50 μGal or less if the total terrain effect is less than 300 μGal. In the case of terrain corrections exceeding 300 μGal, it is advisable to have supplemental elevations to help decrease the errors.

In any case, it is useful to determine the elevations within the B, C, and D zones of Hammer's template if the 1-in-20 approximation rule (Section 5.4.2) is not met. The effect of errors in elevations on terrain corrections varies in a complex, non-linear manner with distance to the station and the elevation difference between a sector of a zone and the station. Thus, in evaluating potential errors it is advisable to consider the accuracy requirements of a survey and the nearby relief.

The third critical concern in choosing a methodology is that consideration needs to be given to the extent of the region included in the terrain correction. This is particularly true in manual determination of terrain corrections because of the significant increase in the effort needed to obtain corrections at increasing distances. It is much less of a problem in digital computations provided that the DEMs cover a large enough region. With the available global elevation data sets, it is possible to calculate the effects of topography over the entire Earth at any point on the surface. Mikuska *et al.* (2006) have calculated the gravity effect of topography and bathymetry from relief beyond an angular distance of approximately 1.5° (roughly 167 km) based on a spherical Earth approximation. These global values change slowly with distance, and thus do not have to be considered in surveys of local areas with only moderate relief, but the effect can be significant in continental-wide surveys and in mountainous regions. The horizontal gradient due to distant relief seldom exceeds 0.03 mGal/km and the vertical gradient never exceeds 3 mGal/km.

Generally, the greater the relief in and adjacent to the area of the survey, the greater is the region that should be considered in terrain corrections. In most regional surveys, the maximum extent of the terrain correction calculations is 166.7 km (1.5°), but in exploration surveys the maximum radius often is limited to Hammer's zone M (22 km). LaFehr (1991b) makes a case for calculating terrain effects to 166.7 km and even beyond. He points out that the absolute effect of topography beyond the 166.7 km distance may amount to several milligals in mountainous areas. This curvature effect is called the Bullard-B correction, which is described below in Section 6.3.8. However, Talwani (1998) recognizes that the correction from terrain beyond 166.7 km generally varies slowly over the area of a survey. Thus, where absolute values are not of importance, the Bullard-B effects can be neglected for surveys of a few kilometers or less in survey dimension.

Beyond a distance of roughly 22 km, the curvature of the Earth should be considered in the calculations. In many local surveys the limit of the region considered in terrain corrections is much less than this. BUTLER (1985) suggests that in microgravity surveys, terrain effects generally need to be corrected on a station-by-station basis for the topography that occurs only within a distance from the survey margin equal to the maximum dimension of the survey area. Clearly the decision on the size of the region considered needs to be evaluated on a case-by-case basis and determined by calculations showing the minimum-sized area that will provide the accuracy required in the results.

A final concern in calculating terrain corrections is the treatment of densities of the topographic features. Commonly, terrain correction tables and charts are based on a density of the surface material of 2,000 kg/m^3, and actual corrections are modified proportionally for the correct density. The densities may be modified, taking into account a constant density that closely approximates the true density or a variable density based upon geologic information. Typically, an approach based on a constant density is used, and variations in density of the surface material are considered in the interpretation process (LAFEHR, 1991a; LEAMAN, 1998). However, the use of variable densities may be desirable in local surveys where the density of topographic features, especially those through Hammer's D zone, are variable, but known or can be readily estimated.

6.3.8 Miscellaneous effects

The gravity effects discussed above are the most routine, and their corrections are generally incorporated into the calculation of gravity anomalies used in exploration investigations. But under specialized conditions, it may be useful or even necessary to incorporate additional corrections described below which enhance the accuracy of anomalies or make them more readily interpretable.

Geological effect

The geological correction is an arbitrary correction that is applied to observed gravity anomaly data in an attempt to remove the effect of known geological or manmade features. Additionally, the effect may be based on models constructed assuming geological hypotheses, such as isostatic equilibrium within the Earth. This correction has been used for many years to decrease the complexity of gravity anomaly data, making these data more readily amenable to interpretation. Its use commonly is based on eliminating the effect of near-surface geology, which is better known than effects from deeper geological sources. An example of the method, the gravity stripping method, has been described by HAMMER (1963). Stripping off the gravitational effects of near-surface geology can be accomplished during the basic reduction of the observed data or later in the analysis procedure during the isolation and enhancement of anomalies.

Geological corrections are a useful method of improving the interpretability of gravity data, but can be readily abused leading to distortion of gravity anomalies. To avoid distortions, this method should only be applied to data where the sources whose effects are being removed from the data are well established by direct information on their location, size, geometry, and physical properties. Furthermore, the procedures used to approximate the gravity effect of subsurface sources (e.g. sources of idealized geometric configuration) should be sufficiently precise that distortions are not imposed on the resulting geologically corrected data.

Isostatic effect

The isostatic effect is a specialized geological correction in which the assumption is made that, at some level within the Earth, there is a condition of gravimetric (or hydrostatic) equilibrium. This is brought about by a change in mass so as to compensate for the effect of the varying mass of surface relief. When a region is not subject to vertical buoyancy forces, the region is considered to be in isostatic equilibrium. Two end-member theories, the Airy–Heiskanen and Pratt–Hayford hypotheses, have been proposed to describe the manner in which the mass change at depth accounts for the support of surface topography. Both of these are operative within the Earth. Based on these theories, the gravity effect of global topography (and bathymetry) and the associated compensation of mass at depth can be calculated. This is the isostatic correction.

The Airy–Heiskanen theory of isostasy is based on the assumption that crustal and mantle densities are laterally constant and the variation in surface relief is compensated by the change in the thickness of the crust or depth to the Moho discontinuity which marks the boundary between the crust and the underlying mantle. In contrast, the Pratt–Hayford theory assumes that all compensation is due to varying lateral density in the crust at a compensation depth. To a first order, the Airy–Heiskanen theory closely matches modern plate tectonic theory, and thus remains the most popular method used to explain isostasy. It is appropriate for the marked contrast in relief between the continents and the oceans, and the observed direct relationship between surface elevation and depth to the Moho. This is evident in that the average crustal thickness is of the order of 35 km in the continents and 10 km in the oceans. It is further supported by the thickening of the crust beneath mountain ranges to depths of 60 km and more. However, geological and geophysical studies show that the crust and mantle are not laterally homogeneous, and therefore not at the same density as the Airy–Heiskanen theory requires. In some cases the forces derived from these mass variations exceed the elastic limit of the lithosphere, leading to localized isostatic compensation as suggested by the Pratt–Hayford hypothesis. Alternatively, the isostatic compensation may be achieved regionally as explained by the Vening–Meinesz regional-compensation theory. This theory is a hybrid between the previous two that tends to minimize the flaws of each. Globally, observations suggest that isostasy is achieved with a variable depth of compensation in a complex manner based on changes in thickness and density of the crust, mantle variations in density, and regional accommodation taking into account the viscosity of the crust and upper mantle which are controlled to a significant degree by the ambient temperature. In many cases, by looking at deviations from the simple isostatic models, these variations may become the objective of the study. Fortunately, the gravity effect from large-scale horizontal variations in density, such as those related to isostatic compensation, are nearly insensitive to depth. As a result the manner of compensation, either by the Pratt–Hayford or Airy–Heiskanen theories, does not affect the magnitude of the isostatic effect.

The isostatic gravity effect is calculated from the gravity effect of the deviation in subsurface mass which compensates for, and thus is equal to, the variation in mass due to the surface elevation such that the lithostatic pressure at some level within the Earth is constant. On the continents this involves the actual surface elevation, while in the oceans the bathymetry is considered together with the water column. In most calculations of the correction, the assumption is made that compensation is achieved by the Airy–Heiskanen theory on a local basis where the lithostatic pressure of a unit area of the surface mass above sea level is equal to the change in thickness of the crust. That is,

$$\sigma h = \Delta\sigma(T_h - T_{SL}) \tag{6.19}$$

where T_{SL} and T_h are respectively the thickness of the crust where the surface elevation is sea level and elevation h, the density of the crust at the surface is σ, and $\Delta\sigma$ is the difference in density between the crust and upper mantle. Knowing the surface elevation and determining or making assumptions regarding the crustal density and thickness at sea level, the thickness of the crust at elevation h can be determined from

$$T_h = T_{SL} + \frac{\sigma h}{\Delta\sigma} \tag{6.20}$$

using consistent units.

Classically, isostatic effects have been determined by means of a template method similar to that used in Hammer's terrain correction (Hammer, 1939). With the template centered on the gravity station, the elevation is estimated for each sector of the annular rings from global elevations. The compensating effect for the topography is then determined from the average elevation assuming isostatic equilibrium has been achieved in a specific manner. Finally, the gravimetric effect of the average elevation of each sector and its compensating mass is determined from look-up tables as in the manual determination of terrain corrections. The gravitational effect is computed using the formulation for the effect of a mass element on a spherical Earth (e.g. Heiskanen and Vening Meinesz, 1958; Mikuska *et al.*, 2006). Recently, isostatic corrections have been computed more efficiently in the spectral domain using global digital elevation (and bathymetry) models and the fast Fourier transform in the same manner as they have been used to calculate terrain corrections (e.g. Simpson *et al.*, 1983, 1986; Blakely, 1995).

In the calculation of the isostatic effect, the three parameters of the right-hand side of Equation 6.20 must be specified; the surface density of the Earth material, the thickness of the crust at sea level, and the density contrast between the crust and the material. For over a century, typically the density of the surface has been assumed to be 2,670 kg/m^3 (Hayford, 1909). Hinze (2003) considered the origin of this value and its appropriateness and found that it was a reasonable approximation. However, considering that roughly 75% of the continental surface is made up of sedimentary rock, a density of the order of 2,600 kg/m^3 is probably more appropriate. This is the value obtained by Chapin (1996) for the surface density of the South American continent. The surface density is an important parameter because, together with elevation, it determines the mass that is compensated at depth. Thus, it is important to decrease the uncertainty in this parameter. Using a global average density simplifies the calculation, but it is likely, as in the determination of regional terrain corrections, that future studies will increasingly incorporate variable densities taking into account the appropriate specific lithologies of the surface rocks.

The thickness of the crust at sea level has been estimated for purposes of calculating the isostatic effect at between 30 and 40 km. Although variations in this value are anticipated across the globe, the value is likely to be closer to 30 km on average than 40 km. The value of the density contrast across the crust–mantle boundary generally has been assigned values ranging from 200 to 450 kg/m^3. Variation in this value is expected because of heterogeneities in the composition and temperature of both the crust and mantle as well as the relief on the Moho (Section 4.3). As cited in Chapter 4, comprehensive investigations indicate a density contrast between the crust and

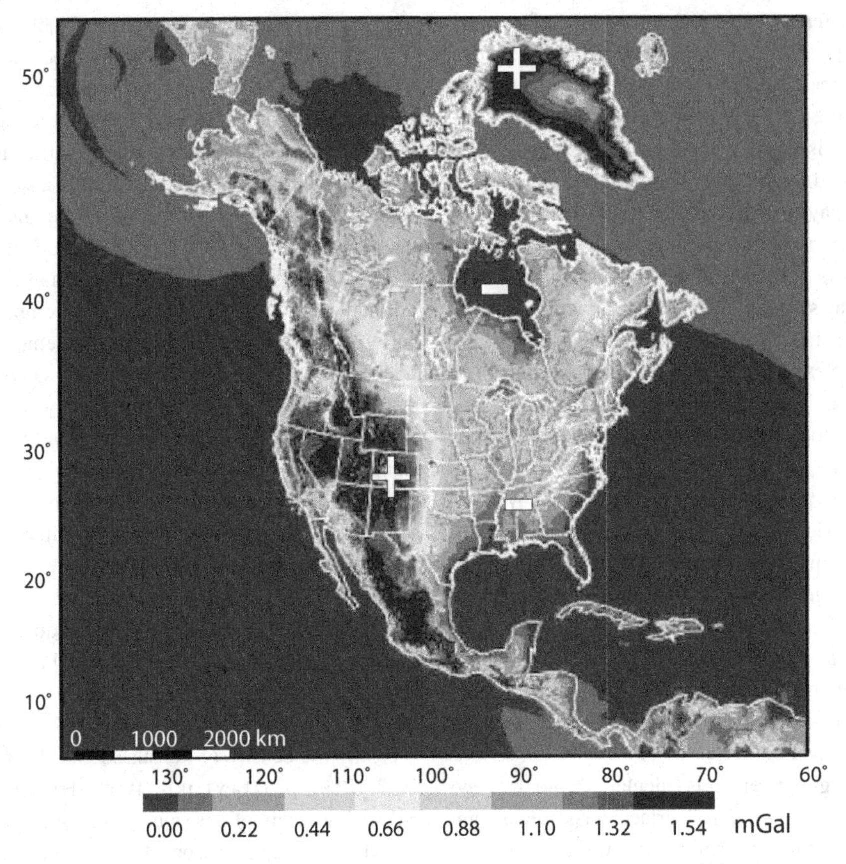

FIGURE 6.10 Curvature correction in milligal over North America. A color version of this figure is available at www.cambridge.org/gravmag. After LI *et al.* (2006).

mantle varying from 350 to 450 kg/m^3. Anticipated uncertainties in the assumptions of T_{SL} and $\Delta\sigma$ will have only minor impact on the results because errors in $\Delta\sigma$ will be compensated by changing $(T_h - T_{SL})$ values and the depth insensitivity of the gravity computation.

Curvature effect

In calculating the mass effect of Earth material between the station and the elevation datum using a horizontal disk as in the generally used Bouguer (mass) effect calculation, no consideration is given to the curvature of the Earth (see Figure 6.7). However, the differential effect between a flat and a curved disk (spherical cap) may become significant for high-accuracy surveys and for surveys that cover a broad range of elevations. The effect of the curvature of the Earth is referred to as the curvature or Bullard-B correction. It is essentially a correction connected to the terrain effect discussed previously. This correction is largest at higher elevations.

The correction considers two effects, the curvature of the Earth and the termination of the spherical cap. The first involves an increase in the gravitational attraction at a station due to the downward curvature of the cap as compared with the horizontal slab, and the second incorporates the decrease in the gravitational effect of the cap due to its termination at 166.7 km (Figure 6.7). The former dominates up to an elevation of approximately 4,150 m where the effect of the truncation and curvature of the cap are equal. The total effect attains a positive value of roughly 1.5 mGal at approximately 2,100 m and continues to decrease at elevations of greater than 4,150 m, attaining a value of roughly −1.5 mGal at 5,000 m. The curvature correction map of North America is shown in Figure 6.10. The correction attains values of the order of +1.5 mGal along the Cordillera from Mexico to Alaska. Note the decrease in the correction from roughly 1.5 mGal at the highest elevations in this mountain chain and in the interior of Greenland. LAFEHR (1991b) derived the following closed-form formula for the spherical cap of radius 166.7 km for land observations

$$g_{curv} = 2\pi\sigma G(\mu h - \lambda R), \quad (6.21)$$

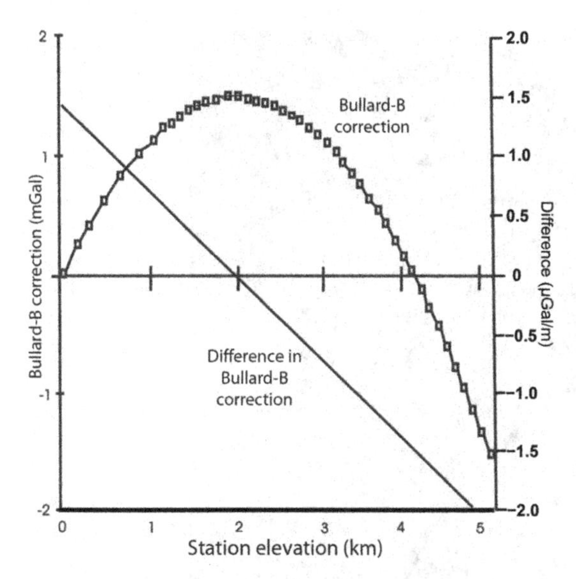

FIGURE 6.11 Curvature correction and the vertical gradient of this correction as a function of station elevation. Calculations based on an Earth radius of 6,371 km and a density of 2,670 kg/m^3 for a spherical cap of radius of 166.7 km. Adapted from LAFEHR (1991a).

where μ and λ are dimensionless coefficients defined by LaFehr, $R = (a_e + h)$ is the radius of the Earth at the station with height h above the ellipsoid of mean radius a_e, and σ is the density of the material making up the spherical cap. A numerical estimate of the curvature correction in milligals obtained from the power series expansion of Equation 6.21 is

$$g_{curv} = 1.464\,139 \times 10^{-3}h - 3.533\,047 \times 10^{-7}h^2 + 1.002\,709 \times 10^{-13}h^3 + 3.002\,407 \times 10^{-18}h^4, \quad (6.22)$$

where the elevation h of the station is in meters assuming a density for the surface material of 2,670 kg/m^3 and the radius of the Earth $a_e = 6{,}371$ km.

The correction and its vertical gradient are presented graphically in Figure 6.11. Note that near sea level, the gradient of the correction is roughly 1.4 μGal/m and that it decreases with elevation. Thus, in high-accuracy microgravity surveys that have an elevation range of more than a few meters, this correction can be significant. The difference in the Earth's radius at the poles and equator from the mean radius a_e changes the curvature correction from Equation 6.22 by only a few microgals in total up to 1 km elevation (LAFEHR, 1991a). The role and significance of the curvature correction are discussed in detail by TALWANI (1998) and LAFEHR (1998).

Extensions of LaFehr's equation are presented by ARGAST *et al.* (2009) for marine and airborne gravity observations taking into account the non-linear relationship between the inhomogeneous gravity effects of multiple layers of different materials (density) as encountered in the computation of the spherical cap in marine and airborne surveys.

Indirect effect

The indirect reduction or correction, which is applicable to the calculation of all gravity anomalies, is the gravitational effect resulting from the use of different vertical datums for establishing the height and theoretical gravity at a station's position (LI and GÖTZE, 2001). It represents the difference between the geoidally obtained gravity measurements and the ellipsoidally referenced mapping system. In gravity exploration this is sometimes referred to as the "geophysical indirect effect," to differentiate it from the geodetic indirect effect which is the correction applied to the compensated geoid (or cogeoid) derived from gravity anomalies downward-continued to the geoid (HACKNEY and FEATHERSTONE, 2003). Conventionally in gravity reductions for exploration purposes, the elevation (or orthometric height) datum is the geoid, whereas the gravity datum is the internationally accepted ellipsoid of the Earth. The difference between ellipsoid (geometric) height and elevation relative to the sea level or the geoid has a range of ±100 m globally and 80 m over North America. Geoidal heights relative to the ellipsoid of the Earth are shown in Figure 2.4. Conventionally, elevations are used in calculating gravity anomalies because heights typically are given with respect to sea level rather than the ellipsoid, owing to historical limitations in mapping the geoid.

The indirect effect combines the gravitational effect of the difference in height between the geoid and ellipsoid at the station and the mass effect of the included material. Assuming a horizontal layer of thickness equal to the difference in height (the geoidal height or undulation) and a density of 2,670 kg/m^3, the effect is

$$g_{ind} = (0.3806 - 0.1119)N = 0.1976N \quad (6.23)$$

where N is the geoidal height in meters and g_{ind} is the indirect reduction in milligals. Thus, globally the indirect effect attains a maximum value of the order of ±20 mGal. A similar correction of 0.2655N mGal/m is required of marine gravity observations (CHAPMAN and BODINE, 1979). If the geoid is above the ellipsoid, the indirect reduction has the same sign as the elevation correction and *vice versa*. The indirect effect primarily interferes with the accuracy of long-wavelength anomalies. Geoidal heights change slowly with distance, less than 10 cm for

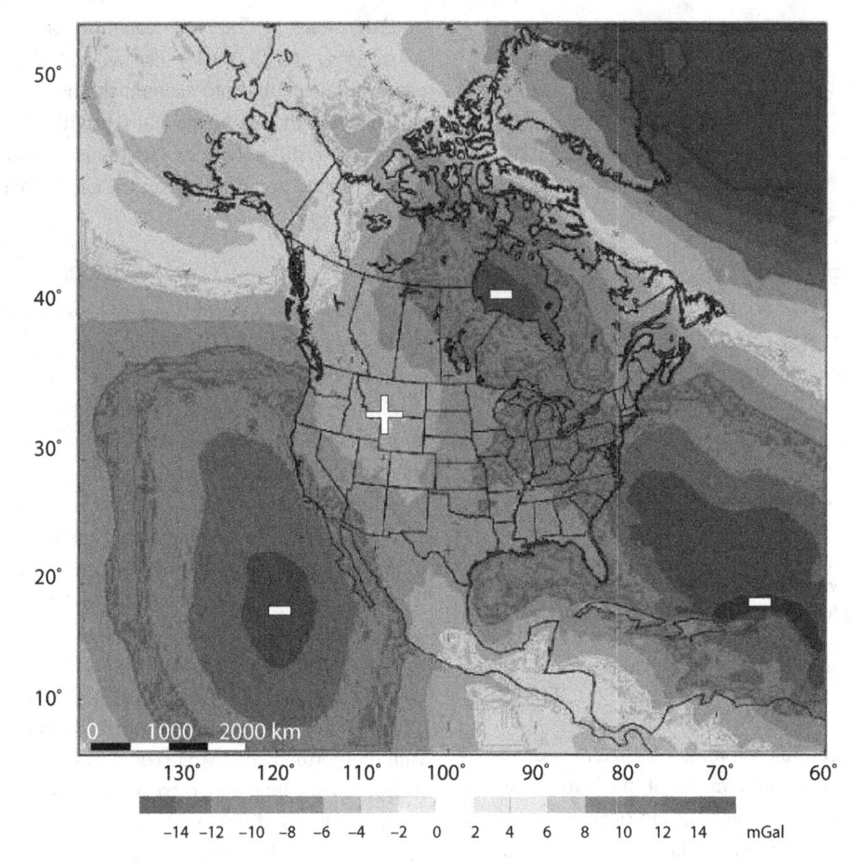

FIGURE 6.12 Geophysical indirect gravity effect in milligals over North America. A color version of this figure is available at www.cambridge.org/gravmag. Adapted from LI *et al.* (2006).

distances shorter than 10 km, and thus this correction has little or no impact on the relative value of gravity over a limited survey area. The correction is neglected in these situations or removed with the regional anomalies. However, in mountainous areas and other areas of strong tectonic disturbance, the spatial variation in geoidal heights may reach several meters per 100 km, which may necessitate an indirect reduction correction (TALWANI, 1998). Figure 6.12 shows the indirect effect over North America, which ranges over roughly ±16 mGal.

Eötvös effect

Previous corrections assumed that the gravity observation is stationary. However, gravity measurements made from a moving platform, such as a ship or aircraft, are altered by the vertical components of centrifugal and Coriolis accelerations acting upon the gravity instrument owing to its motion over the Earth's surface. The correction for this alteration is the Eötvös correction, named after Baron Roland von Eötvös who first recognized the effect in gravity measurements of the early twentieth century. This effect is important because it can be much larger than most of the anomalies of interest. Shipborne observations are subject to Eötvös effects measured in tens of milligals. Those made on airborne platforms are commonly in the hundreds of milligals because of their increased velocities.

The centrifugal force or acceleration, a, acting upon the surface of the Earth at radius R and a latitude θ rotating with an angular velocity ω is

$$a = R\omega^2 \cos\theta \tag{6.24}$$

and the vertical component, parallel to the gravitational component, a_v is

$$a_v = R\omega^2 \cos^2\theta \tag{6.25}$$

as shown in Figure 6.13. The change in this component due to a change in the angular velocity is

$$\Delta a_v/\Delta\omega = 2R\omega \cos^2\theta \tag{6.26}$$

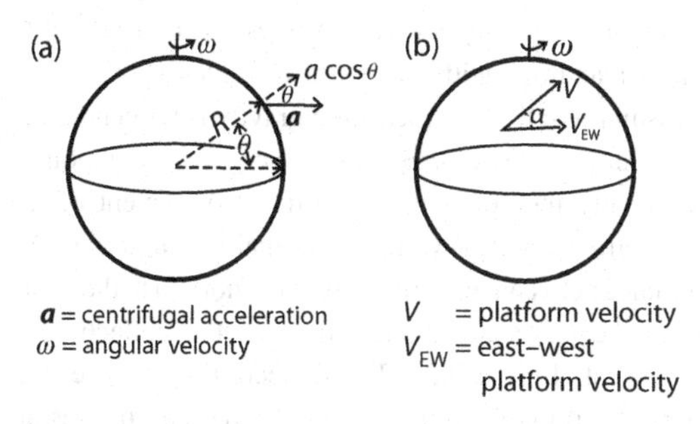

FIGURE 6.13 (a) Vector diagram of Eötvös effect. (b) Platform direction of motion used in evaluating the Eötvös effect (Equation 6.30).

and the velocity of the gravity measurement platform, V, in the east–west direction is

$$V_{ew} = V \cos\alpha, \tag{6.27}$$

where α is the heading of the platform with respect to east (Figure 6.13).

The change in angular velocity due to this motion is

$$\Delta\omega = V \cos\alpha / R \cos\theta. \tag{6.28}$$

As a result, the change in the vertical acceleration of the platform is

$$\Delta a_v = 2R\omega \cos^2\theta (V \cos\alpha)/R \cos\theta = 2V\omega \cos\theta \cos\alpha. \tag{6.29}$$

In addition there is a centrifugal acceleration, V^2/R, in the vertical direction due to the outward acceleration of the platform as it moves over a curved surface which is independent of the Earth's rotation and the direction of motion. Thus, the total Eötvös acceleration is

$$g_{etv} = 2V\omega \cos\theta \cos\alpha + (V^2/R). \tag{6.30}$$

Using an average radius of the Earth and neglecting the elevation of the platform, and expressing the velocity in km/hr and the gravity correction in milligals, the Eötvös correction is

$$g_{etv} = 4.040V \cos\theta \cos\alpha + 0.00121V^2, \tag{6.31}$$

of if V is in knots,

$$g_{etv} = 7.487V \cos\theta \cos\alpha + 0.004V^2. \tag{6.32}$$

This correction is subtracted from measurements on a westward-directed platform and added to measurements on an eastward-moving platform. The second term on the right of Equation 6.31 is generally neglected for marine measurements because it is negligible in comparison with other sources of measurement errors, but it is evaluated in airborne measurements where this term is large because of the greater velocity of airborne measurements. However, airborne gravity gradiometer observations are not subject to this effect because the differential accelerometer measurements effectively cancel out linear as well as rotational acceleration gradients associated with the motion of the platform.

A related correction that may be necessary if the observation platform is subject to strong horizontal accelerations is the Browne correction (e.g. BROWNE, 1937; LACOSTE *et al.*, 1967). This correction that seeks to eliminate the effect of the horizontal acceleration by multiplying the observed gravity g_{obs} by the cosine of the angle of deflection ξ (Figure 2.3) between true g_N and observed gravity is

$$g_{Bro} = g_{obs} \cos(\xi) \approx g_{obs} - (g_N \xi^2/2). \tag{6.33}$$

The increased accuracy of navigation achieved with GPS has markedly decreased errors due to velocity and heading determination of moving platforms. One promising use is to utilize GPS velocities to compute the Eötvös correction directly, rather than differentiating the standard GPS position information over time. Thus, GPS has greatly improved the corrections and accuracy of gravity measurements from mobile platforms.

6.4 Gravity anomalies

As described in Section 6.3, gravity measurements are subject to a wide range of effects apart from the subsurface conditions that are the objectives of gravity surveys. The sources of these variations and methods of determining corrections for them have been detailed in the previous section. The application of these corrections leading to the calculation of a variety of the gravity anomalies is the subject of the following discussion.

6.4.1 Fundamental elements

There are two fundamental components in the calculation of the gravity anomaly, namely the observed and theoretical gravity at the observation point. The gravity anomaly is defined as the difference between the observed and the theoretical or predicted vertical acceleration of gravity. The observed gravity is determined by converting the measurement of the change in position of a mass in a gravimeter between a base location and the observation station into units of gravitational acceleration through a calibration constant. This measured value is modified for the relative change in gravity from the base to the station due to the time variation in gravity as measured by the gravimeter during the time interval between the base and station measurements, the drift of the meter, and the Eötvös effect for measurements made on a moving platform. Finally, the drift-corrected measurements related to the base station are adjusted to a common base or to a particular gravity datum.

The result of these computations is the observed gravity at a station, relative to an arbitrary datum, or an absolute value tied to an internationally recognized base station that has an established absolute gravity value.

The other component of the gravity anomaly is the theoretical or modeled gravity at the station. This is the gravity at the station taking into account variations over the surface of the Earth due to planetary and terrain factors and in specialized conditions additional corrections. Several gravity anomalies as described below have been defined on the basis of an array of corrections considered in calculating the theoretical gravity value (e.g. Heiskanen and Vening Meinesz, 1958; Scheibe and Howard, 1964). All anomalies have specific uses, but the most useful anomaly for general exploration purposes is the complete Bouguer gravity anomaly or some variation of it. Some anomalies, such as the Helmart and Faye gravity anomalies which take into account the latitudinal, height, and terrain effects of the Earth in the modeled gravity, are used exclusively by geodesists in the study of the size and shape of the Earth.

Unfortunately there is confusion regarding gravity anomalies because it is not unusual for different nomenclature to be used for the same anomaly. For example, the term Bouguer gravity anomaly may be used without the adjectives complete or simple. A complete Bouguer gravity anomaly has terrain corrections applied; a simple Bouguer gravity anomaly does not. Furthermore, geophysicists and geodesists differ in their use of the term "anomaly." Geodesists use the term gravity disturbance rather than anomaly when the vertical datum for reduction purposes is the ellipsoid, and only use the term gravity anomaly when the modeled gravity at a station is based on the geoid as the vertical datum. In contrast, in exploration geophysics, the term anomaly is used regardless of the vertical datum. The term anomaly is tightly woven into the geophysical culture and semantics so that it is unlikely to change. To geophysicists, the anomaly refers to anything left over in the observation after removing the known gravity effects. As a result, the North American Gravity Data Committee recommends that the term ellipsoid precede the anomaly when the vertical datum is the ellipsoid in geophysical studies (NAGDC, 2005).

Generally, gravity anomaly values are understood to be in scalar form without regard to direction, based on the assumption that the measured gravity is perpendicular to the Earth's ellipsoid. However, the gravity component traditionally measured is the vertical component of the Earth's gravity vector which is normal to the geoid. The difference between the directions of the normal to the geoid and to the ellipsoid is referred to as the deflection of the vertical (Figure 2.3). The deflection of the vertical can be determined from the geoid model and the resulting change in the gravity measurement calculated due to its projection onto the normal to the ellipsoid. In actuality, the difference between the scalar value measured in gravity and the value normal to the ellipsoid is negligible because the deflection of the vertical is typically a few arcseconds, reaching as much as an arcminute at the extreme. The projection results in a relative change of up to $1 - \cos(1') = 4.23 \times 10^{-8}$. As a result, considering the approximate absolute value of gravity on the Earth's surface, 980 Gal, the maximum error from the assumption that the measurement is perpendicular to the Earth's ellipsoid is around 0.04 mGal over distances measured in hundreds of kilometers.

The definition of the gravity anomaly as the difference between the observed and theoretical or modeled gravity at an observation site should be kept in mind in calculating and interpreting anomalies. If the anomaly is calculated by subtracting the theoretical model as determined by the summation of planetary, terrain, and geological effects from the observed gravity, there is less chance for an error in determining the sign of the gravity effect than by the arithmetically equivalent method of correcting (reducing) the observed gravity value to a level datum, normally sea level. Furthermore, the computed gravity anomaly is the value at the precise site of the station, not at the datum level. There is no standard downward-continuation reduction procedure that corrects an anomalous gravity value to a height datum. This is particularly important in dealing with surveys in rugged terrain where shallow anomalous masses are of interest. In this case, the anomaly values calculated from the surface measurements will change rapidly and significantly over the elevation range, owing to the proximity of the sources.

Airborne gravity gradiometer observations consist of differential measurements of accelerometers which effectively cancel linear as well as rotational gradients related to motion of the platform and any regional and planetary gradients. As a result, processing of the observations into an interpretable form is greatly simplified once the instrumentation and platform issues have been effectively handled. However, these observations are subject to

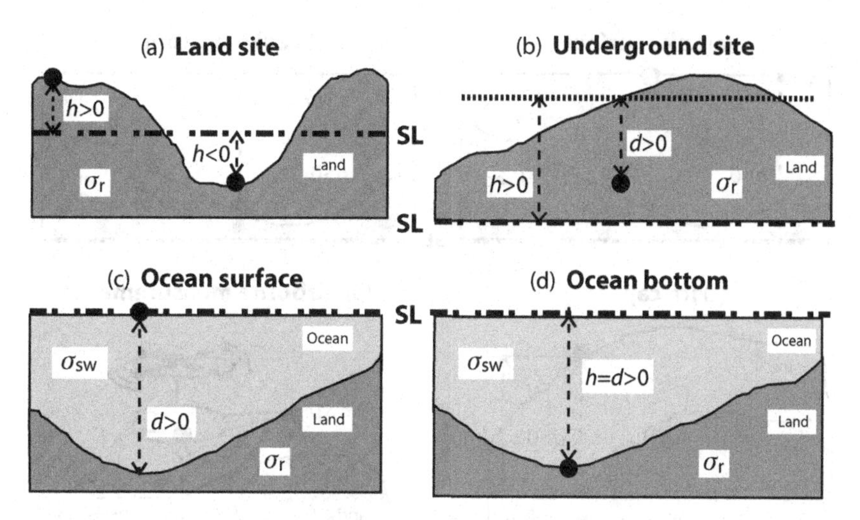

FIGURE 6.14 Parameters for calculating gravity anomalies at (a) land surface (Eqs. 6.35 and 6.38), (b) underground (Eqs. 6.42 and 6.43), (c) ocean surface (Eqs. 6.44 and 6.45), and (d) ocean bottom (Eqs. 6.46 and 6.47) stations. The location of the station site is indicated by the filled circle. SL is sea level and the remaining symbols are defined in the text.

significant error from terrain which must be removed from the observations.

6.4.2 Classes of gravity anomalies

Gravity anomalies of interest to exploration of the Earth can be divided into three classes, based on the models assumed in the calculation of the theoretical gravity, which is subtracted from the observed gravity to produce the anomaly. The primary class incorporates only analytically determined planetary considerations into the model, so they are called planetary anomalies. The second type applies additional effects from known or postulated subsurface geological conditions into the model. These anomalies are referred to as geological anomalies. The third type consists of filtered anomalies which result from the removal of arbitrary gravity effects caused largely by unknown sources that are empirically or analytically determined and by filtering to enhance particular attributes of the spatial pattern of the gravity anomalies. The anomalies become more arbitrary moving from planetary to geological to filtered anomalies, but they become more useful in identifying and resolving gravity effects of interest to explorationists. In the following sections these anomalies and their uses are described.

Planetary gravity anomalies

Free-air and Bouguer gravity anomalies are planetary gravity anomalies based on theoretical models of the gravity at a station that incorporate effects caused by varying gravity over the surface of the Earth due to its rotation and changing radius, and the topography of the Earth and the related effect of the atmosphere. The Bouguer gravity anomaly is particularly valuable because it includes the gravitational effect of the mass within the included topography which greatly enhances the identification and study of subsurface mass variations.

Examples of scenarios for calculating free-air and Bouguer gravity anomalies and their relevant equations are presented in the following descriptions and in Figures 6.14 and 6.15.

(A) Free-air gravity anomaly The free-air gravity anomaly takes into account the latitudinal change in gravity on the Earth's best-fitting ellipsoid represented by GRS80 and the vertical change in gravity between the reference datum and the observation height assuming that the gravity station is located in free air, hence the name free-air anomaly. In this anomaly, the intervening space between the observation and the height datum is assumed to have no mass and no gravitational effect. This is the most basic of anomalies used in geologic studies because unlike other anomalies no assumptions are made about the Earth's masses. For this reason WOOLLARD (1966) referred to this anomaly as the natural anomaly. It has also been referred to in the past as the total gravity anomaly. The free-air anomaly does not take into account the effect of terrain departures from the height of the observation, but in some geodetic applications it is incorporated into the calculation. The resulting anomaly is sometimes called the

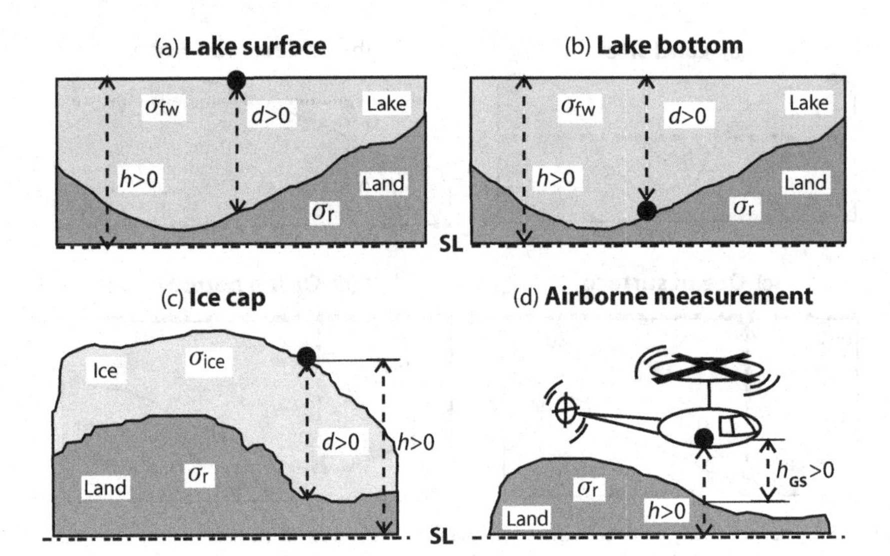

FIGURE 6.15 Parameters for calculating gravity anomalies at (a) lake surface (Eqs. 6.48 and 6.49), (b) lake bottom (Eqs. 6.50 and 6.51), (c) ice cap (Eqs. 6.52 and 6.53), and (d) airborne (Eqs. 6.54 and 6.55) stations. The location of the station site is indicated by the filled circle. SL is sea level and the remaining symbols are defined in the text.

Faye anomaly. The free-air gravity anomaly g_{FAA} is calculated from

$$g_{FAA} = g_{obs} - g_{mod}, \tag{6.34}$$

where g_{obs} is the observed gravity and g_{mod} is the theoretical gravity model at the observation site taking into account the change in gravity with latitude, g_θ, and with height, g_h, relative to the vertical datum (Equation 6.11). When the free-air anomalies are referenced to sea level (geoid) as a datum, the height (h) is appropriately referred to as elevation, rather than the generic term height which refers to the vertical distance between a point and a horizontal datum. Thus, because $g_{mod} = g_\theta - g_h$, the free-air gravity anomaly is

$$g_{FAA} = g_{obs} - g_\theta + 0.3086 \times h. \tag{6.35}$$

The free-air anomaly is unsuitable for most terrestrial geological problems except under special circumstances because it contains the gravitational effect of uncompensated topographic masses which generally obscure the effect of subsurface masses in exploration surveys. In marine applications, it is sometimes useful where bathymetric effects are negligible or where the sea bottom is very far from the sensor.

The free-air anomaly is the difference between actual gravity and the theoretical gravity on the ellipsoid that best fits the physical surface of the Earth. As such the g_{FAA} is a measure of the mass of the subjacent Earth including the topographic masses. The term subjacent acknowledges that the gravity effect of equivalent masses increases with their proximity to the observation site. If there were perfect isostatic equilibrium regardless of the size of the topographic feature, this anomaly would be zero everywhere when averaged over an area. In fact, when the continent-wide values of the free-air gravity anomaly are plotted against the station height, the values are distributed around the zero anomaly value regardless of height, confirming the isostatic equilibrium hypothesis on a regional basis.

Regional free-air anomalies may exist where the stress field of the Earth maintains the geological structure in a non-isostatic condition. Examples include basins, ore bodies, and other small geologic structures that may be held in the crust in a non-isostatic condition by the strength of the lithosphere. Subduction zones also are characterized by prominent local free-air anomalies because the inherent strength of the Earth's lithosphere is sufficient to maintain the local mass imbalance. In this case, the Earth does not deform elastically in a vertical sense. The island of Hawaii is another example where the load of the volcanic pile that makes up the island deforms the surrounding ocean crust in a series of concentric rings, detectable as free-air anomalies. The free-air anomalies here tend to detect geologic features out of vertical isostatic balance. The mass that can be supported without deformation depends on the nature of the rocks involved and their temperature.

The size of surface and crustal features that are in isostatic equilibrium varies between continents and oceans and within them as well (Watts, 2001). Topographic features over the North American continent of the order of 100 km and larger are generally in isostatic equilibrium (e.g. WOOLLARD, 1966; DORMAN and LEWIS, 1972). However, this size varies considerably. For example, significantly smaller areas are in isostatic equilibrium in the western USA where the crust is thinner and the lithosphere hotter than in the more stable cratonic areas of the continent.

In contrast to the plot of regional values of the free-air anomalies as a function of station elevation, the graph of local values of g_{FAA} versus station elevation, assuming negligible regional anomalies and terrain effects, has a positive slope which is equal to the quantity $2\pi G\sigma$ in the mass correction Equation 6.14, where σ is the density of the surface topography. This provides a simple means of approximating the density of near-surface materials of a region where the assumptions are appropriate.

Although the free-air anomaly is usually not helpful for the interpretation of local gravity studies, it is useful in regional, continental-scale studies. It can be used on a regional basis as a simplified isostatic anomaly and for interpretation of airborne and satellite survey results that are observed at high altitudes compared with the terrain relief. In both cases, the effect of local mass variations due to topographic mass irregularities is averaged out of the results. In addition, free-air anomalies are widely used in interpretation of surveys in marine areas where computational substitution of rock for water will lead to large changes in gravity anomaly level that may distort the results regionally, or where anomaly distortions are likely to be associated with local bathymetry as a result of a mismatch between the density of the substitute rock in the mass correction and the density of the bathymetric features. This is why global or continental gravity surveys that include both continents and adjacent oceans typically use Bouguer gravity anomalies over land areas and free-air anomalies in marine areas. Where the values are adjusted to a sea level datum, the Bouguer and free-air anomalies are equivalent at the ocean shoreline so that the contouring of anomaly values is continuous across this boundary. However, the anomalies will differ slightly at the shoreline if the Bouguer gravity anomalies incorporate terrain effects, as in the complete Bouguer gravity anomaly described below, or if the reduction of the anomalies is referenced to the Earth's ellipsoid.

(B) Bouguer gravity anomaly The Bouguer gravity anomaly is the most frequently used of the gravity anomalies in surveys of continental and near-shore marine areas. It differs from the free-air gravity anomaly by including the gravitational acceleration of the mass between the site of the observation and the datum level of the survey. This attraction is calculated using the horizontal (or Bouguer) slab equation (Equation 6.14) and modifying it for the deviation of the surface relief from the horizontal slab.

Bouguer gravity anomalies usually do not correlate with local topography unless these features are related to structural or stratigraphic variations that cause density variations below the datum level. On the contrary, the relationship of these anomalies to broad topographic variations that are isostatically compensated is negative. The negative linear relationship has been used to study isostatic relationships within the Earth (e.g. SIMPSON *et al.*, 1986; CHAPIN, 1996). In high mountains, the value of Bouguer anomalies may reach minus 400 mGal or less, in deep oceans the average value is positive 300 mGal or more, and in continental areas averaging an elevation of around 300 m, the Bouguer gravity anomalies are typically around minus 30 mGal. The trend of the relationship between Bouguer gravity anomalies and the height of the stations for continental-scale data sets can be predicted based on the Bouguer slab equation. However, the mean density of the slab specified by this trend is less than the density contrast between the surface and the atmosphere, which is typically assumed to be 2,670 kg/m^3. As pointed out by CHAPIN (1996), the slope of the trend is the mean Bouguer slab effect using the density difference between the surface–atmosphere and the crust–mantle density contrasts because the gravity effects from both density contrasts are height-dependent.

(1) Simple and complete Bouguer gravity anomalies. The theoretical gravity used in the calculation of the Bouguer gravity anomaly for surface land stations is

$$g_{mod} = g_\theta - g_{atm} - g_h + g_{ind} + g_m - g_{ter} + g_{curv}. \qquad (6.36)$$

The respective atmospheric (g_{atm}), curvature (g_{curv}), and indirect (g_{ind}) effects generally are applied only under specialized conditions, and thus may or may not be included in calculating the Bouguer gravity anomaly with parameters shown for the examples in Figures 6.14 and 6.15.

When the terrain correction and the specialized terms of the theoretical gravity equation are not used, the simple Bouguer gravity anomaly, g_{SBA}, is calculated for land surface stations (Figure 6.14a) by

$$g_{SBA} = g_{obs} - g_{mod} = g_{obs} - g_\theta + 0.3086 \times h - 2\pi G\sigma_r \times h. \qquad (6.37)$$

When the theoretical gravity also takes into account the terrain correction, the complete Bouguer gravity anomaly,

g_{CBA}, is obtained from

$$g_{CBA} = g_{obs} - g_\theta + 0.3086 \times h - (2\pi G\sigma_r \times h - g_{ter} \equiv g_{tef}), \quad (6.38)$$

where the terrain effect g_{tef} is modeled from the combined mass and terrain corrections. The term complete Bouguer gravity anomaly is not limited to those gravity anomalies that consider the terrain correction to any particular distance. Even when the terrain is considered only in the near vicinity of the station, the assumption is made that the terrain effect from the far-distance region is negligible over the survey area, and thus the term complete Bouguer gravity anomaly is used.

The description of the reduction procedure accompanying a gravity map or profile should clearly state the corrections that have been employed. It should also include collateral information such as the maximum distance from the station used in the terrain correction calculations, and the density that was assumed in the mass and terrain corrections.

The term Bouguer gravity anomaly is normally used for anomalies that are based on sea level as a reference datum, but in many gravity surveys of limited areal extent, the height datum is established at an arbitrary, convenient level. In the latter case, the datum is assumed to parallel the ellipsoid upon which the normal gravity is calculated as in either Equation 6.4 or 6.5, or the normal gradient of gravity with north–south distance from an arbitrary reference point is calculated using Equation 6.6. As a result, the Bouguer gravity anomaly can be either absolute or relative.

Absolute anomalies have the observed gravity given in absolute values achieved by tying the observations to an absolute gravity base and elevations referenced to sea level (or the ellipsoid). In contrast, relative anomalies have the observed gravity and height referenced to arbitrary levels. Relative Bouguer gravity anomalies generally are simpler to calculate when the survey extent is limited to several kilometers or less and when gravity base stations are not readily available. Further, they are equally useful for interpretation. As a result, they are the norm in areally limited gravity surveys. The nature of the datums used in the calculations should be clearly described in the text accompanying reports on gravity surveys.

(2) Example calculations. The following example calculations illustrate the computation of relative and absolute simple and complete Bouguer gravity anomalies for surface land gravity stations. They also help demonstrate the physical significance of the sign of the corrections that must be applied to the anomaly determinations.

The first example illustrates the computation of the absolute complete Bouguer gravity anomaly for the key site parameters shown in Figure 6.14(a). Consider a station with observed gravity 5.382 mGal greater than the gravity observed at the base which is 980.465 322 Gal, where the sea level (or ellipsoid) gravity at the latitude θ of the station is $g_\theta = 980.502\,130$ Gal, the elevation of the station above sea level is $h = 100$ m, the density of Earth material between the datum and the station is $\sigma = 2{,}000$ kg/m^3, and $g_{ter} = 1.023$ mGal is the terrain correction out to 166.7 km. Accordingly, the absolute complete Bouguer gravity anomaly in milligals is

$$\begin{aligned} g_{CBA} &= g_{obs} - g_\theta + (0.3086 - 2\pi G\sigma) \times h + g_{ter} \\ &= (980{,}465.322 + 5.382) - 980{,}502.130 \\ &\quad + [0.3086 - (4.193 \times 10^{-5} \times 2000)]100 \\ &\quad + 1.023 = -7.923 \text{ mGal}. \end{aligned} \quad (6.39)$$

As another example, consider the computation of the relative complete Bouguer gravity anomaly in the northern hemisphere for $g_{obs} = 3.025$ mGal at the station relative to the base, the elevation of the station above the base $h = 30.02$ m, the location of the station is 300 m north of the base, the latitude (normal) correction $g_\theta = 0.8130 \times 10^{-3}$ mGal/m, the density of Earth material between the base and station $\sigma = 1{,}900$ kg/m^3, and the terrain correction out to 22 km is $t_{ter} = 0.890$ mGal. Accordingly, the relative complete Bouguer gravity anomaly in milligals is

$$\begin{aligned} g_{CBA} &= g_{obs} - g_\theta + 0.3086 \times h - 2\pi G\sigma \times h + g_{ter} \\ &= 3.025 - (300 \times 0.810\,30 \times 10^{-3}) \\ &\quad + [0.3086 - (4.193 \times 10^{-5} \times 1{,}900)] \\ &\quad \times 30.02 + 0.890 = 10.545 \text{ mGal}. \end{aligned} \quad (6.40)$$

As a further example, consider the computation of the relative simple Bouguer gravity anomaly in the northern hemisphere for $g_{obs} = -3.025$ mGal at the station relative to the base. The station is 16.78 m below the base and located 200 m south of the base where the latitude (normal) correction is 0.8130×10^{-3} mGal/m, and the density of surface material is $\sigma = 1{,}900$ kg/m^3. The terrain correction is negligible and is not included. Accordingly, the relative simple Bouguer gravity anomaly in milligals is

$$\begin{aligned} g_{\mathrm{SBA}} &= g_{\mathrm{obs}} - g_{\theta} + 0.3086 \times h - 2\pi G\sigma \times h \\ &= -3.025 - (-200 \times 0.810\,30 \times 10^{-3}) \\ &\quad + [0.3086 \times (-16.78)] - [4.193 \times 10^{-5} \\ &\quad \times 1900 \times (-16.78)] = -6.704 \text{ mGal}. \end{aligned} \tag{6.41}$$

(3) Anomaly equations. Gravity measurements are made in a number of scenarios in addition to the surface land observations described in previous sections. Measurements are observed on the surface of oceans, lakes, and ice caps, and underground as well as underwater in oceans and lakes. The calculation of anomalies at these measurement sites requires the development of appropriate equations that are the difference between the observed gravity, g_{obs}, and the theoretical or modeled gravity, g_{mod}, at the site. The process is simplified by calculating the modeled gravity at the site taking into account the known parameters of the site and adjacent natural conditions, as in the derivations of the Bouguer gravity anomalies described below.

In the following equations, the atmospheric, curvature, and indirect effects are not considered; elevations, h, are given relative to sea level with positive values above sea level, and depths, d, are given relative to the surface directly above the station and positive below the surface. If, as has been suggested for the revised gravity data standards, the ellipsoid is used as the elevation datum, the anomaly is referred to as the ellipsoidal anomaly. The vertical gradient of gravity in free air, g_h, is as given in Equation 6.11 and the mass effect, g_{m}, is the Bouguer slab effect, $2\pi G\sigma$ times the height or depth (Equation 6.14). The symbol for density of the slab, σ, is specified for fresh water, σ_{fw}, sea water, σ_{sw}, ice, σ_{ice}, and land, σ_{r}.

The equations are for the complete Bouguer gravity anomaly, but the method of calculation of the terrain effect is not specified because it is dependent on the requirements of the survey and the nature of the topography as described previously. The terrain effect for surface land stations is always the same sign: regardless of a positive or negative topographic feature relative to the height of the station, the modeled gravity is always less as a result of the terrain. This is not the case for observations made underground or from airborne platforms. For these situations, the sign of the topography must be taken into account as well as the distance above or below the surface height. In oceanic and lake environments, changes in bathymetry may also be included in the equations, as are terrain effects. The sign of the bathymetry effect is variable as is the terrain effect depending on the position of the measurement site at the surface or at the bottom of the body of water.

Figure 6.14(b) illustrates key site parameters for calculating gravity anomalies underground. The respective free-air anomaly (g_{FAA}) and complete Bouguer gravity anomaly (g_{CBA}) at a measurement site beneath the surface of the Earth, but above sea level are

$$\begin{aligned} g_{\mathrm{FAA}} &= g_{\mathrm{obs}} - g_{\mathrm{mod}} \\ &= g_{\mathrm{obs}} + 2\pi G\sigma_{\mathrm{r}} d - [g_{\theta} - 0.3086(h-d) - 2\pi G\sigma_{\mathrm{r}} d] \\ &= g_{\mathrm{obs}} - g_{\theta} + 0.3086(h-d) + 4\pi G\sigma_{\mathrm{r}} d \end{aligned} \tag{6.42}$$

and

$$\begin{aligned} g_{\mathrm{CBA}} &= g_{\mathrm{obs}} - g_{\mathrm{mod}} \\ &= g_{\mathrm{obs}} - [g_{\theta} - 0.3086(h-d) - 2\pi G\sigma_{\mathrm{r}} d \\ &\quad + 2\pi G\sigma_{\mathrm{r}}(h-d) \pm g_{\mathrm{ter}}] \\ &= g_{\mathrm{obs}} - g_{\theta} + 0.3086(h-d) + 4\pi G\sigma_{\mathrm{r}} d \\ &\quad - 2\pi G\sigma_{\mathrm{r}} h \mp g_{\mathrm{ter}}. \end{aligned} \tag{6.43}$$

The gravitational effect of the mass above the station is added to the observed gravity. The terrain effect, g_{ter}, is positive for terrain above the level of the surface directly above the observation (Figure 6.14(a)) and vice versa. This formulation excludes the corrections for specialized gravity effects such as the indirect, atmospheric, and curvature effects, as do the following anomaly equations.

Figure 6.14(c) illustrates the principal site parameters for evaluating gravity anomalies at the ocean surface. The respective free-air and simple Bouguer gravity anomaly equations at a measurement site on the surface of the ocean assuming no nearby land terrain effects and no significant bathymetric changes, are

$$\begin{aligned} g_{\mathrm{FAA}} &= g_{\mathrm{obs}} \pm g_{\mathrm{etv}} - g_{\mathrm{mod}} \\ &= g_{\mathrm{obs}} \pm g_{\mathrm{etv}} - g_{\theta}, \end{aligned} \tag{6.44}$$

and

$$\begin{aligned} g_{\mathrm{SBA}} &= g_{\mathrm{obs}} \pm g_{\mathrm{etv}} - g_{\mathrm{mod}} \\ &= g_{\mathrm{obs}} \pm g_{\mathrm{etv}} - (g_{\theta} + 2\pi G\sigma_{\mathrm{sw}} d - 2\pi G\sigma_{\mathrm{r}} d) \\ &= g_{\mathrm{obs}} \pm g_{\mathrm{etv}} - g_{\theta} + 2\pi G d(\sigma_{\mathrm{r}} - \sigma_{\mathrm{sw}}). \end{aligned} \tag{6.45}$$

If land surfaces adjacent to the ocean produce significant gravitational effects, the terrain effects are subtracted from the overall modeled gravity and added to the final equation. If gravitational effects of bathymetry are significant, the effect of changes in depth of the ocean are taken into account as in surface land measurements, except that the signs differ, depending on the sign of the deviation of the adjacent sea bottom from the depth at the observation site. For a deepening of the ocean the modeled gravity at the observation site will be decreased and for a rise in the sea bottom the modeled gravity will be increased. The density differential is the difference between land and sea water density. This formulation also includes the correction of the observed gravity for the Eötvös effect due to the motion of the ocean surface vessel over the surface of the Earth during the measurement process. It does not include consideration of vertical tidal motion of the ocean surface which may require attention in special situations.

Figure 6.14(d) shows the key site parameters for computing gravity anomalies on the ocean bottom. Here the respective free-air and simple Bouguer gravity anomaly equations, assuming no gravitationally significant bathymetric changes, are

$$\begin{aligned} g_{\mathrm{FAA}} &= g_{\mathrm{obs}} - g_{\mathrm{mod}} \\ &= g_{\mathrm{obs}} + 2\pi G\sigma_{\mathrm{sw}}d - (g_\theta + 0.3086d - 2\pi G\sigma_{\mathrm{sw}}d) \\ &= g_{\mathrm{obs}} - g_\theta - 0.3086d + 4\pi G\sigma_{\mathrm{sw}}d, \end{aligned} \quad (6.46)$$

and

$$\begin{aligned} g_{\mathrm{SBA}} &= g_{\mathrm{obs}} - g_{\mathrm{mod}} \\ &= g_{\mathrm{obs}} + 2\pi G\sigma_{\mathrm{sw}}d - (g_\theta + 0.3086d - 2\pi G\sigma_r d) \\ &= g_{\mathrm{obs}} - g_\theta - 0.3086d + 2\pi G(\sigma_r + \sigma_{\mathrm{sw}})d. \end{aligned} \quad (6.47)$$

As in the case of subsurface measurements, the mass of the water above the measurement is added to the observed gravity, and as in the case of ocean surface measurements, ocean-bottom gravity observations may be affected by adjacent land surfaces and corrected in a similar manner. If bathymetry changes are not gravimetrically negligible, they are corrected as in surface land gravity measurements for terrain, but the density differential is between sea water and land density.

Figure 6.15(a) shows the site parameters for calculating gravity anomalies on a lake surface. The respective free-air and simple Bouguer gravity anomaly equations where the lake is above sea level are

$$\begin{aligned} g_{\mathrm{FAA}} &= g_{\mathrm{obs}} \pm g_{\mathrm{etv}} - g_{\mathrm{mod}} \\ &= g_{\mathrm{obs}} \pm g_{\mathrm{etv}} - (g_\theta - 0.3086h) \\ &= g_{\mathrm{obs}} \pm g_{\mathrm{etv}} - g_\theta + 0.3086h, \end{aligned} \quad (6.48)$$

and

$$\begin{aligned} g_{\mathrm{SBA}} &= g_{\mathrm{obs}} \pm g_{\mathrm{etv}} - g_{\mathrm{mod}} \\ &= g_{\mathrm{obs}} \pm g_{\mathrm{etv}} - [g_\theta - 0.3086h + 2\pi G\sigma_{\mathrm{r}}(h-d) \\ &\quad + 2\pi G\sigma_{\mathrm{fw}}d] \\ &= g_{\mathrm{obs}} \pm g_{\mathrm{etv}} - g_\theta + 0.3086h - 2\pi G[\sigma_{\mathrm{r}}(h-d) \\ &\quad + \sigma_{\mathrm{fw}}d], \end{aligned} \quad (6.49)$$

where d is the depth of the lake at the observation site. The same comments made in considering the terrain and bathymetry of surface ocean measurements are applicable to lake surface measurements.

Figure 6.15(b) illustrates site parameters for a lake bottom. For cases where the lake is above sea level, the respective free-air and simple Bouguer gravity anomaly equations are

$$\begin{aligned} g_{\mathrm{FAA}} &= g_{\mathrm{obs}} - g_{\mathrm{mod}} \\ &= g_{\mathrm{obs}} + 2\pi G\sigma_{\mathrm{fw}}d - [g_\theta - 0.3086(h-d) - 2\pi G\sigma_{\mathrm{fw}}d] \\ &= g_{\mathrm{obs}} - g_\theta + 0.3086(h-d) + 4\pi G\sigma_{\mathrm{fw}}d, \end{aligned} \quad (6.50)$$

and

$$\begin{aligned} g_{\mathrm{SBA}} &= g_{\mathrm{obs}} - g_{\mathrm{mod}} \\ &= g_{\mathrm{obs}} - [g_\theta - 0.3086(h-d) - 2\pi G\sigma_{\mathrm{fw}}d \\ &\quad + 2\pi G\sigma_r(h-d)] \\ &= g_{\mathrm{obs}} - g_\theta + 0.3086(h-d) + 4\pi G\sigma_{\mathrm{fw}}d \\ &\quad - 2\pi G(\sigma_{\mathrm{fw}}d + \sigma_r h), \end{aligned} \quad (6.51)$$

where d is the depth of the lake at the observation site. As in the case of subsurface land measurements, the mass of the water above the measurement is added to the observed gravity, and like for ocean bottom measurements, lake bottom gravity observations may be affected by adjacent land surfaces and corrected in a similar manner. If bathymetry changes are not gravimetrically negligible, they are corrected as in surface land gravity measurements for terrain, but the density differential is between sea water and land density.

Figure 6.15(c) illustrates the site parameters for determining gravity anomalies on an ice cap. The respective free-air and complete Bouguer gravity anomaly equations for this site where the bottom of the ice is above sea level are

$$\begin{aligned} g_{\mathrm{FAA}} &= g_{\mathrm{obs}} - g_{\mathrm{mod}} \\ &= g_{\mathrm{obs}} - (g_\theta - 0.3086h) \\ &= g_{\mathrm{obs}} - g_\theta + 0.3086h, \end{aligned} \quad (6.52)$$

and

$$
\begin{aligned}
g_{\mathrm{CBA}} &= g_{\mathrm{obs}} - g_{\mathrm{mod}} \\
&= g_{\mathrm{obs}} - [g_\theta - 0.3086h + 2\pi G\sigma_{\mathrm{ice}}h \\
&\quad + 2\pi G\sigma_{\mathrm{r}}(h-d) - g_{\mathrm{ter}}] \\
&= g_{\mathrm{obs}} - g_\theta + 0.3086h - 2\pi G[\sigma_{\mathrm{ice}}d + \sigma_{\mathrm{r}}(h-d)] \\
&\quad + g_{\mathrm{ter}}, \qquad (6.53)
\end{aligned}
$$

where d is the depth of the ice.

Figure 6.15(d) illustrates the site parameters for obtaining airborne gravity anomalies. Both free-air and Bouguer gravity anomalies are widely used in airborne measurements. The free-air gravity anomaly is

$$
\begin{aligned}
g_{\mathrm{FAA}} &= g_{\mathrm{obs}} - g_{\mathrm{mod}} \\
&= g_{\mathrm{obs}} - [g_\theta - 0.3086h] \pm g_{\mathrm{etv}} \\
&= g_{\mathrm{obs}} - g_\theta + 0.3086h \pm g_{\mathrm{etv}}, \qquad (6.54)
\end{aligned}
$$

where h is the elevation of the airborne observation above sea level.

The complete Bouguer gravity anomaly equation for airborne measurements is

$$
\begin{aligned}
g_{\mathrm{CBA}} &= g_{\mathrm{obs}} - g_{\mathrm{mod}} \\
&= g_{\mathrm{obs}} - [g_\theta - 0.3086h + 2\pi G\sigma_{\mathrm{r}}(h - h_{\mathrm{GS}}) \pm g_{\mathrm{ter}}] \\
&\quad \pm g_{\mathrm{etv}} \\
&= g_{\mathrm{obs}} - g_\theta + 0.3086h - 2\pi G\sigma_{\mathrm{r}}(h - h_{\mathrm{GS}}) \mp g_{\mathrm{ter}} \\
&\quad \pm g_{\mathrm{etv}}, \qquad (6.55)
\end{aligned}
$$

where h_{GS} is the height of the instrument above the ground surface directly beneath the observation. The terrain correction is positive for heights above the surface height directly beneath the airborne measurement and *vice versa.*

There are numerous other scenarios that are involved in gravity observations, but the equations for them can be determined on the basis of principles used in deriving the above equations. The appropriate constants in the equations above are specified in Section 6.3.

(C) Bouguer gravity anomaly limitations Despite the common use of Bouguer gravity anomalies in gravity interpretation, the anomaly does have several limitations that may affect their utility in specific geologic situations. First, in areas of rugged relief and near oceanic coastlines, gravity anomalies can be distorted in a regional sense by topographic features not considered in terrain corrections or not accurately evaluated, or due to isostasy. These effects are of consideration in interpreting deep mass variations in the Earth, and thus are usually not of concern in local gravity surveys of limited areal extent.

Second, the Bouguer gravity anomaly will be in error if the density of the material between the gravity station and the datum used in the Bouguer correction is incorrect (Figure 6.16). The magnitude of the error is 0.004193 mGal/m for each 100 kg/m^3 error in density. As a result, care is necessary in selecting densities used in establishing the mass reduction, especially in surveys of areas that have considerable surface relief. Overestimating or underestimating the terrain density results in erroneous anomalies that respectively correlate positively or negatively with the topography. Wherever possible, *in situ* densities should be used because of the inherent difficulties in estimating or measuring in the laboratory the density of the unconsolidated materials often found at the surface of the Earth. The density of 2,670 kg/m^3 normally used is generally erroneous for most surveys in areas of unconsolidated surface materials and clastic sedimentary rocks.

Third, special problems are encountered in surveys that are conducted over areas where the surface formations have a varying density. Questions arise about the use of a varying density or an average density, and how the mass correction should be calculated if a varying density is used. Unless specific information is available bearing on the variation in density of the surface material over a survey area, an average density is preferable (VAJK, 1956), but care must be taken to evaluate the magnitude of the error anticipated from the variation in density and the amount of surface relief.

If a variable density is used in the mass correction, it is advisable to use a variable density factor down only to the lowest surface level and a constant density from this level to sea level because of the lack of information on the densities of the subsurface formations. Variations in density below the lowest surface level can be accounted for in the interpretation process. Corrections can be modified by taking into account the changes in density of the formations adjacent to the station in a manner similar to terrain correction calculations by evaluating the gravity effect of annular rings and zones around a station. In practice this is seldom done because of the lack of detailed information and the fact that roughly 75% of the mass effect is derived within a radius equal to the elevation of the station with respect to the datum. These remarks emphasize the need for caution in using Bouguer gravity anomaly values from data banks where the mass correction is calculated using a value for density that might be acceptable on a regional basis, but not for accurate surveys of localized regions.

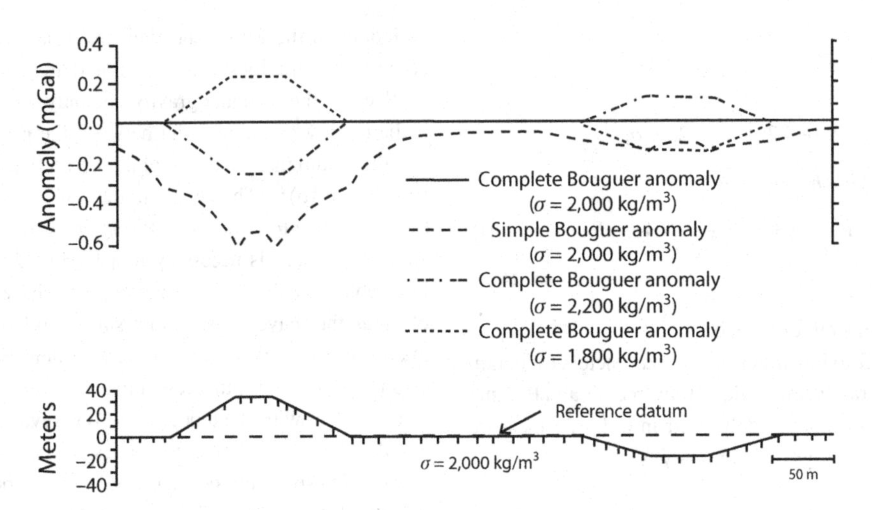

FIGURE 6.16 Gravity effect in milligals of surface relief (assuming 2D topographic features) and a range of densities used in the Bouguer (mass) correction on the Bouguer gravity anomaly. The correct Bouguer gravity anomalies use a reduction density of 2,000kg/m^3, the density of the surface material.

A final concern in the use of Bouguer gravity anomalies is the invalid assumption, often made in the interpretation of these anomalies, that the bedrock surface is a horizontal plane. Variations in the depth to bedrock produce large horizontal density differentials which give rise to anomalies that distort and confuse subsurface gravity anomalies. The effect of bedrock relief is particularly troublesome because the density contrast between the overburden and the bedrock commonly exceeds the geologically interesting density contrasts. The problem is complicated because often major bedrock relief is controlled by the lithology of the bedrock, and thus the gravity variations due to lithology and bedrock relief are superimposed. The effect of bedrock topography is dependent not only on the amplitude of the relief and the density contrast, but also on the overall depth of the bedrock, with the effect of the bedrock relief amplified by near-surface bedrock. Auxiliary geologic or geophysical data are needed to take this effect into account in most geological situations.

(D) Error assessment The assessment of errors in the gravity anomaly is an important final step in the calculation of the anomaly. The cumulative error in the calculation bears on the mode of presentation of the anomaly and its interpretation. For example, estimated errors need to be considered in the contour level of gravity anomaly maps, the vertical scale of anomaly profiles, and the level of interpretation, which should not exceed the overall accuracy of the anomalies. Interpretation carried beyond this level is misleading and can be overly time-consuming.

Standard procedures for evaluating the propagation of errors through the anomaly calculation should be used. The cumulative error is equal to the square root of the sum of the squares of the estimated errors of the individual components of the calculation. Typically the estimated errors are the computed standard deviations of the components. Errors occur in both fundamental components of the anomaly calculation – that is, in the observed gravity at a station, and the predicted theoretical gravity at the site. Sources of errors in the observed gravity were described at length in Chapters 5 and Section 6.3. The principal source of error in theoretical calculations based on data acquired before the availability of GPS elevations is the accuracy in station height. With the advent of rapid methods of accurately determining heights on a global basis, this is no longer true. However, data bases that include stations established prior to roughly 2000 are subject to errors in heights where they have been obtained with aneroid altimeters or from contour maps. Typically these errors in heights are of the order of 3 m, although greater errors are not uncommon.

The principal sources of error in post-2000 observations are in the accuracy of calculating the mass or Bouguer correction, owing to errors in prescribing the density of the surface materials in the mass correction and in terrain corrections. Both of these sources of error are amplified in areas of high surface relief. Assuming a gravity observation at 45° N latitude and at a height of 200 m, an error in the observed gravity of ±0.05 mGal, an error in north–south distance of ±50 m, an error in height of ±0.05 m, a terrain correction error of ±0.1 mGal, and an error of roughly ±10% in density, the total error in the complete Bouguer gravity anomaly is dominated by the density error and is of the order of ±1.7 mGal. Thus, despite great care to obtain accurate observations and position and height control, the total error is relatively large for high-accuracy surveys if the *in situ* density is not known accurately for calculating the mass effect.

Geological gravity anomalies

To better isolate unknown sources of anomalies, gravity anomalies can be altered to eliminate known features or effects based on geological hypotheses. The resulting anomalies, geological gravity anomalies, have a variety of forms including both specialized and more broadly used anomalies. An example is the anomaly obtained by stripping off the gravity effects of near-surface geological features that are known from either direct (e.g. drilling and physical property measurements in the hole or drill core) or indirect (e.g. seismic reflection studies) methods (HAMMER, 1963) or obtained from inversion of regional data (LI and OLDENBURG, 1998a). The resulting geological anomaly can be a powerful tool in interpretation free from marked density contrasts in the near surface. In a similar manner, it may be possible to determine the broad gravity anomalies associated with deep sources, and thus isolate near-surface gravity effects. For example, KABAN and MOONEY (2001) have removed the effect of variations in the thickness of the crust as determined by seismic methods from the regional gravity field of the southwestern USA, and IBRAHIM and HINZE (1972) have illustrated how regional gravity anomalies can be removed to isolate near-surface gravity anomalies associated with changes in depth to bedrock in the glaciated northern midcontinent of the USA that are underlain by thick sedimentary rocks. The most basic form of the geological gravity anomaly is the Bouguer gravity anomaly described in the previous section, where the mass of the topography is taken into account in land observations and the ocean water in marine measurements is replaced by typical continental rocks.

(A) Isostatic residual gravity anomaly The best known and most widely used geological gravity anomaly is the isostatic residual gravity anomaly. Hayford and Bowie published an isostatic gravity anomaly map of the conterminous USA in 1912. This map based on only 89 observations, was the first in a long series of maps prepared for the purpose of determining the isostatic state of the Earth. Until relatively recently these maps were referred to as simple isostatic gravity anomaly maps, but SIMPSON *et al.* (1986) suggested that the more meaningful term isostatic residual anomaly map be used in its place. It is more meaningful because the anomaly is not the gravitational effect of the deep mass variations which cause the isostatic state of the Earth, but rather the Bouguer gravity anomaly with this isostatic gravitational effect removed, hence the residual isostatic effect. The isostatic regional gravity field, caused by isostatic effects derived from topographic and bathymetric variations, is illustrated in Figure 6.17. The effects are negative in the continents and positive in the oceans, and are inverse to the topography on the continents. The highest gradients of the regional isostatic effect are over the continental slopes within the oceans.

The Bouguer gravity anomaly, if properly calculated, is free of effects of local topographic relief, but, as discussed in the previous section, it is profoundly affected by regional changes in elevation which originate in subsurface mass variations related to the state of isostatic balance within the Earth. The Earth seeks to reach a condition of equal mass at some level by changing the surface elevation or by readjusting the subsurface masses. If the Earth were homogeneous, its shape would be a perfect ellipsoid without regional surface relief, but in the actual situation the surface is raised in regions of lower mass and *vice versa*. Thus, in regions where the lighter crust is thickened, the surface rises to form mountains or plateaus; where it is thinned, as in the oceans, the regional surface is depressed. In the computation of the Bouguer gravity anomaly, the effects of varying surface relief are accounted for, but they do not model the gravity effects of the anticipated subsurface changes in mass related to this relief. The isostatic anomaly takes these subsurface changes in mass into account in calculating the theoretical gravity anomaly anticipated at a gravity observation site.

The isostatic residual anomaly g_{IRA} is calculated by subtracting the gravitational effect of global topography and the related subsurface masses from the theoretical gravity used in the calculation of the complete Bouguer gravity anomaly. The isostatic effect does not include the mass effect of the topography included in the mass or Bouguer effect. The results are quite insensitive to the mode of isostatic compensation used in calculating the correction; usually, however, the Airy–Heiskanen model is used which assumes a fixed crustal density of varying thickness as described in the previous section on isostatic gravity effects. This is the principal mode of compensation at the ocean–continent topographic level and in mountainous regions. The appropriate equation used in calculation of the isostatic anomaly is

$$g_{IRA} = g_{CBA} + g_{IE}, \quad (6.56)$$

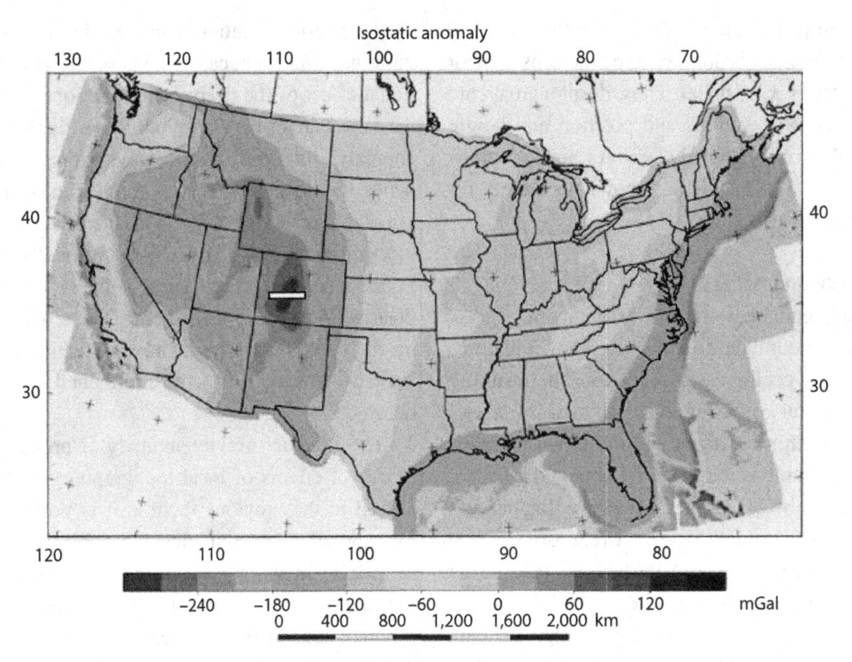

FIGURE 6.17 Isostatic regional gravity anomaly field (in milligals) of the conterminous USA. A color version of this figure is available at www.cambridge.org/gravmag. Adapted from SIMPSON *et al.* (1986).

where g_{IE} is the effect of global terrain and compensation for the topography at the location of the station.

In effect, the isostatic anomaly is a long-wavelength cut filter that eliminates the effect of subsurface mass variations associated with regional changes in elevation. These masses are broad and deep, generally at the base of the crust or within the upper mantle, and thus their effect slowly varies over a region, affecting the long-wavelength components of the gravity field. Under ideal isostatic conditions, all regional topography would be exactly counterbalanced by the assumed subsurface masses and the isostatic residual anomaly would be zero. Exceptions to this situation are caused by the regional spread of the anomaly derived from the compensating mass located at depth and slow-acting subsurface processes that have not had time to adjust to surface masses. The latter exists, for example, in regions of the Earth where subsurface adjustment for the removal of the mass of the Pleistocene continental glaciers has not reached equilibrium since the melting of the glaciers. In addition, the lithosphere may be strong enough to withstand mass variations without adjusting for them.

The principal use of the isostatic residual gravity anomaly in exploration surveys is to eliminate the change in Bouguer gravity anomaly due to subsurface adjustments for regional topography. In the conterminous USA, the relationship between Bouguer gravity anomalies and topographic height is approximately −106 mGal/km (HINZE and BRAILE, 1988). As a result, the range of anomalies is drastically decreased and anomalies are focused upon mass variations occurring in the upper and middle portions of the crust (e.g. SIMPSON *et al.*, 1986; WOOLLARD, 1966). The isostatic gravity anomaly is particularly useful in studying crustal sources that occur near marked changes in topography or bathymetry where regional gradients in the Bouguer gravity anomaly are high.

(B) Decompensative and mantle gravity anomalies Specialized geological gravity anomalies such as the decompensative gravity anomaly have found use in geological studies for very specific objectives. The decompensative gravity anomaly is akin to the isostatic residual gravity anomaly in that it attempts to account for the gravitational effects of compensating masses. Decompensative anomalies attempt to account for mass variations in the crust that are unrelated to topography. The process is somewhat complex, and thus has not been broadly used in geological interpretation. The gravity effect at depth within the mantle of opposite sign to the interpreted mass variations in the crust is calculated and upward-continued to the interpreted level of the crustal mass, and the effects of the crustal source and the mantle compensating source are combined algebraically. CORDELL *et al.* (1991) show that the result of subtracting this anomaly, calculated in the wavenumber domain, from the Bouguer gravity anomaly focuses the resulting anomaly upon the gravity effects of subcrustal sources largely free from the effects of shallower sources. A limitation of this anomaly is the necessary assumption that the subsurface mass variations are sufficiently large and old that they are in isostatic equilibrium.

Another anomaly that has been used especially in the interpretation of marine gravity over mid-oceanic ridges is the mantle Bouguer gravity anomaly or simply the mantle gravity anomaly (Kuo and Forsyth, 1988). In marine environments the free-air anomaly is modified for the gravity effect of the varying seafloor depth and mass variations within the crust. Calculation of this anomaly is based on the assumption that the crust has a constant density, and the thickness changes according to a specific model of the crust that incorporates the effect of cooling of the aging oceanic crust as it moves away from its origin at the center of mid-oceanic ridges. Variations from this assumed mantle are reflected in the resultant mantle Bouguer gravity anomaly. This anomaly has also been used in continental areas (e.g. Kaban *et al.*, 2003) for tectonic analysis. In this case, it is the free-air anomaly minus the crustal gravity effect including topography, based on independent knowledge of the thickness and density of the crust and the surface topography.

Filtered gravity anomalies

Gravity anomaly data bases typically consist of numerous anomalies of varying amplitudes, dimensions, and gradients derived from a wide variety of subsurface sources, but generally those of interest have a limited range of characteristics. By identifying anomalies within this range, it is possible to focus upon those important to interpretation. Although anomalies can be pinpointed using a variety of characteristics, in practice they are usually localized on the basis of their wavelength characteristics. Components that are of shorter wavelengths than those of interest to interpretation are considered noise and removed with a high-frequency cut filter, but generally it is more critical to remove longer-wavelength anomalies to isolate or highlight anomalies of interest. These longer-wavelength anomalies, related to deeper and larger sources, typically have greater amplitudes than the local anomalies of interest, and thus distort the target anomalies (Skeels, 1967). A variety of analytical and graphical methods have been developed for isolating or enhancing the desired wavelength gravity components (e.g. Blakely, 1995; Nabighian *et al.*, 2005a). These methods eliminate components of an arbitrary nature, unlike the removal of prescribed planetary or geological gravity effects in the types of anomalies described above. Additional details on the use of filtered gravity anomalies are described in the next section.

6.5 Anomaly isolation and enhancement

Appendix A on data systems processing includes a mathematical treatment of analytical methods of processing data by filtering to remove or emphasize particular wavelength components of potential fields and describes the fundamentals of the processing methodologies for the isolation and enhancement of particular anomalies. In this section, these methods and others are explained in terms of their potential usefulness to gravity anomaly analysis.

6.5.1 Residual and regional gravity anomalies

Corrected observed gravity data are presented in either map or profile form for analysis and interpretation depending on the objectives and coverage of the survey and the attributes of the observed anomalies. The scales and contour intervals of these presentations are a function of the magnitude and areal size of the significant anomalies of interest in the study, the accuracy of the data, and the range of the anomaly values. These data are the summation of all anomaly sources regardless of their depth or distance from the observation point and any errors in the processing of the data. The errors are assumed to be negligible in comparison to the magnitude of the anomalies of interest. As a result of the inverse distance function of gravity anomalies, which describes the exponential attenuation of anomalies with increasing distance, the gravity anomaly is dominated by the effect of nearby sources and large-volume geological features at depth. The gravitational effects of other horizontal variations contribute less significantly to the overall observations.

The cumulative effect of gravity anomalies is illustrated in Figure 6.18. The schematic gravity effect of four geological sources of excess mass (density) are portrayed over the simplified geological cross-section showing the anomaly sources. Sources 1 and 2 are equivalent sources located near to the surface which are the mass anomalies of interest in the survey. These sources produce anomalies that are relatively high amplitude and sharp with resulting high anomaly gradients. Source 3 is equivalent to sources 1 and 2, but is barely discernible in the profile because it is buried at a greater depth, and thus is lower in amplitude, and the anomaly is broader with low gradients. This anomaly source is not a viable target for a gravity survey. Source 4 is deep, as is source 3, but its volume is large, and thus produces a large-amplitude anomaly. As a result, it is

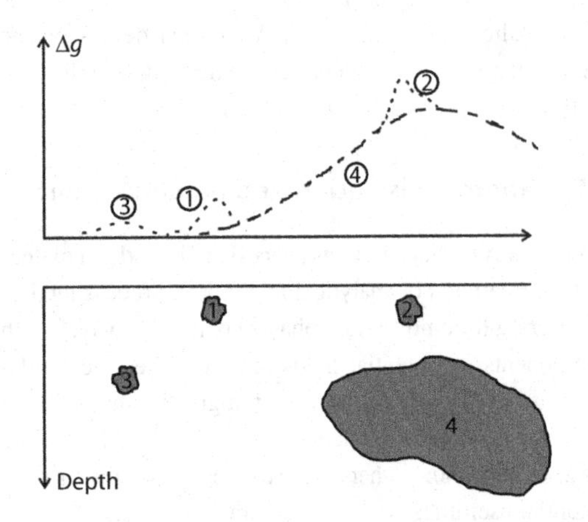

FIGURE 6.18 Schematic gravity anomaly profile over four positive density differential sources and their cumulative effect. This profile illustrates the concept of regional and residual anomalies, and the need for isolation and enhancement of residual gravity anomalies in order to identify and analyze them. The profiles are identified with the number of the anomalous source.

readily observed despite the low gradient of the anomaly. This figure is independent of scale in terms of both areal size and amplitude of anomalies. That is, the superposition of anomalies and their cumulative effect is valid regardless of the size and depth of the anomaly sources.

In most gravity surveys, especially those for exploration purposes, the anomalies of interest are derived from relatively shallow sources, such as sources 1 and 2 in Figure 6.18. These are called the residual anomalies. The definition of residual anomaly is not very precise (e.g. NETTLETON, 1954; SKEELS, 1967), but generally it is defined as the anomaly derived from a source of interest.

In exploration surveys, the residual anomaly is the one arising from, for example, a cave, ore body, or buried anticline that is the target of the survey. The anomaly derived from source 4 is called the regional anomaly, which is defined as the anomaly or anomaly gradient that is areally larger than the anomaly of interest, that is the residual anomaly. The amplitude is not necessarily a defining quality of the regional anomaly, although the obvious regional gravity anomaly is usually of greater amplitude than the residual anomaly. In Figure 6.18 the anomaly of source 3 is a regional anomaly or part of the sum of the regional anomalies, but is insignificant because of its low amplitude.

To illustrate the scale independence of the figure and the residual–regional concept, let source 4 be the target of interest in the survey, for example a dense mass in the bedrock. In this case the anomaly from source 4 would be the residual anomaly, and deeper sources of large volume would cause the regional anomaly of the survey. Thus, the specification of a residual anomaly depends on the survey objectives. A specific anomaly may be the residual in one survey and the regional in still another.

In the case where the target of the survey is source 4, the anomalies of sources 1 and 2 constitute areally smaller variations which are considered noise in the survey. Noise can be described as variations in gravity smaller in size than the residual anomaly. Noise may originate from shallow geological features which are restricted in horizontal dimensions, or from errors in the surveying and processing of the data. However, shallow features and errors also may produce anomalies of equivalent or greater size than the residual anomaly.

The gravity anomaly (A) resulting from the reduction of observed gravity data thus consists of three components; the residual anomaly (A_r) which is the target of the survey, the regional anomaly (A_R) which is spatially larger than the residual anomaly, and the noise (N) which is spatially smaller than the residual. That is,

$$A = A_r + A_R + N. \tag{6.57}$$

Generally, the noise, N, is considered to be negligible and disregarded, in which case the anomaly of interest, the residual gravity anomaly, is simply

$$A_r = A - A_R. \tag{6.58}$$

In actual practice, the subtraction is directly performed graphically or arithmetically, or indirectly performed by emphasizing the residual anomaly at the expense of the regional anomaly.

Elimination of the regional gravity anomaly from the observed anomaly is a critical step in the identification and analysis of gravity anomalies. Improper definition of A_R can cause errors in the attributes of the residual anomaly, its amplitude, gradients, trend, etc. Interpretation, both qualitative and quantitative, is particularly sensitive to errors in amplitudes and gradients. However, the definition process can be problematic and especially difficult where the character of the regional anomaly approximates that of the residual anomaly. This is illustrated diagrammatically in Figure 6.19 in which the fields of the amplitude and maximum gradient of the regional and residual gravity anomalies, as well as the noise, are mapped as a function of the areal dimension of the anomaly.

In the cases illustrated in Figure 6.19(a) and (c), the characteristics of the residual anomaly A_r are well separated from the noise and regional gravity anomaly. In these cases, the separation of the residual anomaly is likely to be accomplished rather accurately, but in the cases

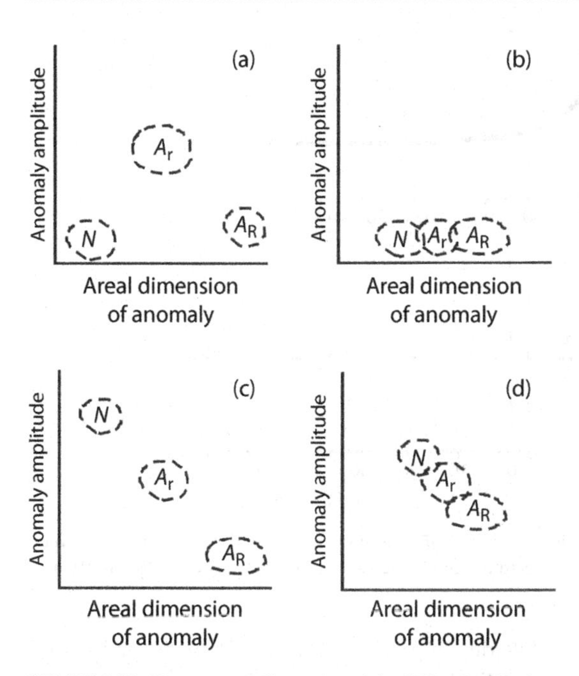

FIGURE 6.19 Diagrammatic illustration of the fields of residual gravity anomalies A_r, regional gravity anomalies A_R, and gravity anomaly noise N plotted as functions of their amplitudes and spatial dimensions. In the cases mapped in panels (a) and (c), the residual anomaly is likely to be easily distinguished from the noise and regional. But in the cases of panels (b) and (d), it is more difficult to identify the residual anomaly because the anomaly amplitudes and spatial dimensions overlap each other.

illustrated in Figure 6.19(b) and (d), the fields overlap, making the process more difficult.

In any event, there is no unique solution to identifying the residual anomaly from the observed anomaly. The process is indeterminate because two parameters must be solved from a single equation, Equation 6.58. Although there are an indefinite number of solutions to this equation, in practice the bounding characteristics of regional anomalies and geologic considerations combined with experience and judgment of the analyst limit the number of effective solutions.

The problem of separating the regional gravity anomaly from the residual anomaly is not only a matter of isolating the regional anomaly, but also of obtaining sufficient survey coverage to identify it. By definition the regional anomaly is broader, more extensive than the residual anomaly. As a result, gravity surveys must be extended beyond the immediate area of interest to delineate the regional gravity anomaly. This is clearly illustrated in Figure 6.18. The residual gravity anomaly could not be determined without observations extending beyond the location of these anomalies. Extending the surveys beyond the area of immediate interest increases the cost of surveys and is often made difficult because of surface physical features and access problems. In some circumstances, observations from regional gravity data sets may be used to supplement survey data, and thus minimize the areal extent of the survey. A useful rule of thumb is to have spatial control in excess of twice the anticipated size of the objective anomalies. For example, if residual anomalies of exploration interest have a 1 kilometer wavelength, it would be appropriate to obtain survey data that extends to 2 kilometers or more.

Numerous schemes have been developed for eliminating regional anomalies (e.g. Zurflueh, 1967; Ku *et al.*, 1971; Meyer, 1974; Naidu and Mathew, 1998). An aura of mystery has developed around these methods because of the indeterminate nature of the problem and lack of understanding of the advantages and limitations of the methods. In addition, industry rumors pertaining to secret or patented techniques are not uncommon. As a result, the residual anomalies derived from the residual–regional separation process are sometimes viewed with skepticism by the end user of the gravity analysis.

To minimize this problem, it is important that each residual anomaly map or profile be clearly annotated with the attributes of the method used to determine the residual anomaly. In addition, the analyst should be careful to consider not only the residual anomaly, but also the regional anomaly and its viability in view of the geology of the site. One of the significant advantages of the current ease in performing calculations and presenting the results on maps and profiles is that a number of methods of residual separation can be tested and compared to obtain optimum results and a range of possible solutions.

6.5.2 Fundamental principles

There are two broad approaches to the residual–regional anomaly separation problem that can be classified as either isolation or enhancement techniques (Hinze, 1990). Isolation techniques attempt to eliminate from the observed anomaly field all anomalies that do not have a certain set of specified characteristics as defined by the objective of the survey. Thus, the anomaly that is isolated, the residual anomaly, is the gravity expression of the significant source. The fundamental premise of the isolation approach is that the geologically significant anomaly is minimally modified by the regional gravity field and by the process. As a result, the residual anomaly is amenable to quantitative analysis, inversion, and modeling.

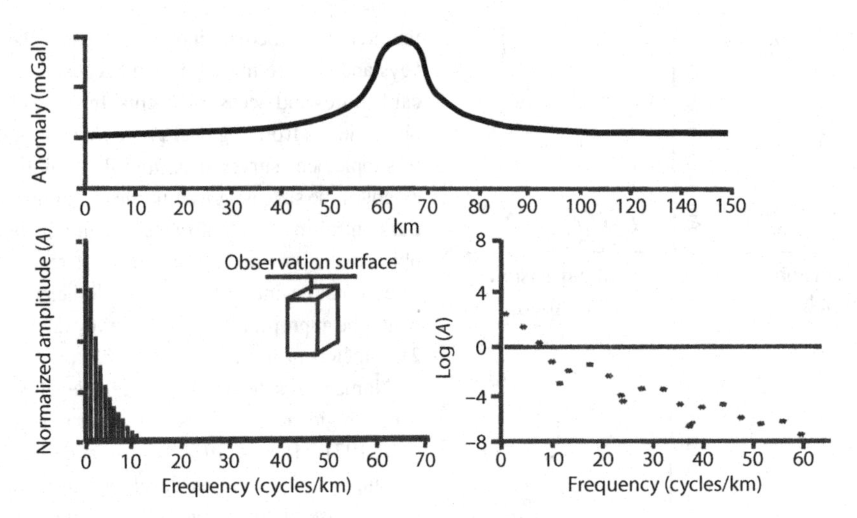

FIGURE 6.20 Gravity anomaly in milligals and frequency spectra of a prism source with dimensions in kilometers. The source of the gravity anomaly is a 10×10 km vertical prism that is located at a depth of 1 km beneath the observation surface and extends vertically for 50 km.

Enhancement techniques, in contrast, are a broad group of methods which accentuate a particular characteristic or group of attributes that is definitive of anomalies that have significance to the objective of the analysis. In the process of enhancing these characteristics, the anomalies are accentuated to increase their perceptibility. As a result, the anomalies are distorted and may no longer be generally useful for quantitative analysis or inversion. Enhanced anomaly maps and profiles are widely used in qualitative visual inspection analysis and interpretation, although they do have some specialized uses in quantitative analysis.

The residual–regional separation process is comparable to the use of filters to pass/reject only the desired anomaly frequencies (or wavelengths), as described in Appendix A.5. The ideal residual–regional separation method will pass only those anomalies that are significant to the investigation without distortion and reject all other components of the observed anomaly, both larger and smaller. Properly designed methods have very useful filtering characteristics, but the cutoffs of these filters usually are not sharp, and they can lead to both amplitude and phase distortion.

This problem is complicated by the wide spectrum of wavenumbers that may be present in a gravity anomaly, as illustrated for the anomaly derived from a prism source in Figure 6.20. In filtering, that is in residual–regional separation, it is impossible to retain the entire spectrum of the residual anomaly without also passing components of the regional anomaly through the filter. As a result, the emphasis in the process is to minimize the distortion of the anomaly associated with a target source.

Filtering of gravity anomalies can be on the basis of the amplitude, spatial dimensions, sharpness (gradients), and directional characteristics of the anomaly. All of these attributes are useful for interpretation, but as the characteristics of residual anomalies become similar to the regional anomaly or noise, they become increasingly difficult to separate. Such is the case in situations illustrated by the plotted anomaly attributes in Figure 6.19(b) and (d). In these and similar cases, mathematical filters can be helpful for mapping residual anomalies based on one or more of the anomaly characteristics.

6.5.3 Isolation and enhancement methods

As described in Appendix A.5, numerous methods of varying complexity have been used to isolate and enhance gravity methods for interpretation. They are referenced in a variety of geophysical journals and books (e.g. Gupta and Ramani, 1980; Roach *et al.*, 1993; Blakely, 1995; Naidu and Mathew, 1998; Hearst and Morris, 2001; Nabighian *et al.*, 2005a). Most are briefly described here in the context of their use in dealing with gravity anomalies. Some of the methods, particularly those that are of a more subjective nature, have been largely superseded by more objective analytical techniques that depend heavily upon the ready availability of massive computing power, like the spectral techniques. However, more subjective methods are also included because occasionally they have a role in gravity processing. Numerous subjective methods are documented and many gravity surveys have been interpreted based on these methods. Thus, it is advantageous to understand them in comparison with other more current methods of anomaly isolation and enhancement.

Geological methods

One of the more successful methods of isolating gravity anomalies is the elimination of the gravity effects of known or hypothesized geological sources in the calculation of the geological gravity anomalies as described previously. The isostatic residual gravity anomaly is only one of these anomalies that are widely used in gravity interpretation. It seeks to eliminate upper mantle sources of anomalies by removing those related to supporting regional topographic changes. The isostatic residuals are essentially derived by subtracting a 3D mathematical model of the crust based on topography. On a much smaller scale, a similar approach can be used to remove the gravitational effects of, for example, plutons or salt domes that are known from supplemental geologic and geophysical studies so as to isolate residual gravity anomalies associated with much smaller features or more subtle sources. In contrast to automated, numerical methods of residual–regional separation which are relatively rapid, geological methods can be time-consuming. Nonetheless, it is advantageous to eliminate all known sources of gravity anomalies from an observed data set prior to interpretation. For example, ROACH *et al.* (1993) have compared the geological (or forward modeling in their nomenclature) method in isolating residual anomalies of Tasmania, Australia with those of trend surface, upward continuation, and spectral filtering. After removal of known or hypothesized gravity effects due to the differences in the oceanic and continental crust, the depth to the base of the crust, and oceanic water surrounding the island, they find that the geological method provides a superior residual gravity anomaly for quantitative analysis.

A specialized form of this method is the gravity stripping method (HAMMER, 1963). In this method the gravitational effect of near-surface geologic features is determined by computing their gravity response from information obtained, for example, from shallow drill holes. The method is illustrated in Figure 6.21, where the gravity effect of a geological cross-section as determined by drill holes and geologic inference is calculated to a depth datum which marks the lower limit of the information. The sum of the gravity response of the known geology, which in effect is a part of the regional anomaly, is shown as the total correction in the figure. The gravity effects, as explained by Hammer, commonly are calculated assuming the geology specified in the drill holes can be approximated by horizontal slabs. This greatly simplifies the calculation procedure, but more complicated shapes involving either two- or three-dimensional sources can be readily calculated as well. The observed gravity anomaly minus the total correction for the known geology is the geologically corrected anomaly which can be used directly as the residual gravity for interpretation or can be subject to further residual–regional separation procedures if required.

This method assumes there are many constraints to the understanding of subsurface geology. If any of these assumptions are incorrect, they can severely distort the results in a way that is unpredictable and confusing. However, if the subsurface is understood at least to some degree, then it can produce highly desirable results that are essentially the unknown aspects of the subsurface.

Graphical methods

As described in Appendix A, a variety of graphical methods are available to separate the residual from regional gravity anomalies. The methods are appropriately considered non-linear, in the sense that it is unlikely that results from multiple analysts will duplicate each other. However, they are simple and easily applied and are flexible, permitting the use of analyst's experience and auxiliary data in their application. The success of the methods is dependent on the experience of the analyst, especially in the use of gravity for a specific geological objective, the simplicity of the regional anomaly, and the perceptibility of the residual anomaly. The methods were used extensively before computers were commonly available to perform analytical, linear residual–regional separations rapidly and inexpensively. Generally, the methods are currently only applied to surveys of limited extent with simple regionals and relatively obvious residual anomalies. In this situation the residual anomaly can be isolated relatively free of distortion. When graphical methods are used, the geological reasonability of the regional should be ascertained as fully as possible as a means of validating the viability of the separation process.

Graphical methods can be used on gravity anomaly data in either profile or map format. In profile data the regional anomaly is visually established as a smooth curve through the gravity anomaly values that excludes the anomalous portion of the profile that includes the residual anomaly. The process is illustrated in Figure 6.22(a), where the residual and regional can be easily distinguished from one another. Successful application of the method requires gravity data well beyond the limits of the target area. The regional anomaly is subtracted from the observed anomaly to determine the residual anomaly. In a situation where the regional is more complex and the residual less identifiable, as in Figure 6.22(b), the separation of regional from

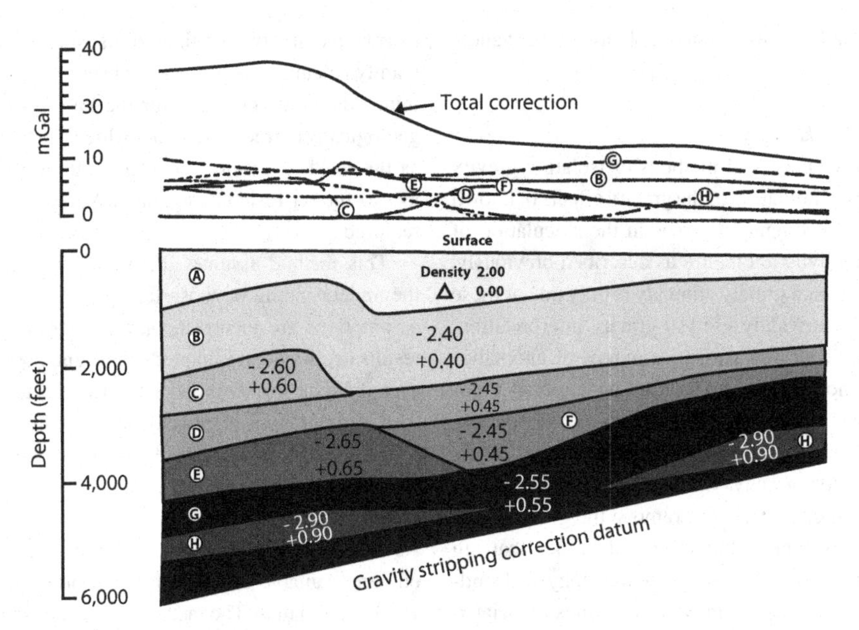

FIGURE 6.21 Gravity stripping calculations of individual geologic units and their total effect in milligals over the hypothetical geological profile with the density of the geologic units given in g/cm^3 and the density contrast from the density of the surface unit (2.00 g/cm^3). The sum of the individual gravity anomaly effects is shown as the total correction. Adapted from HAMMER (1977).

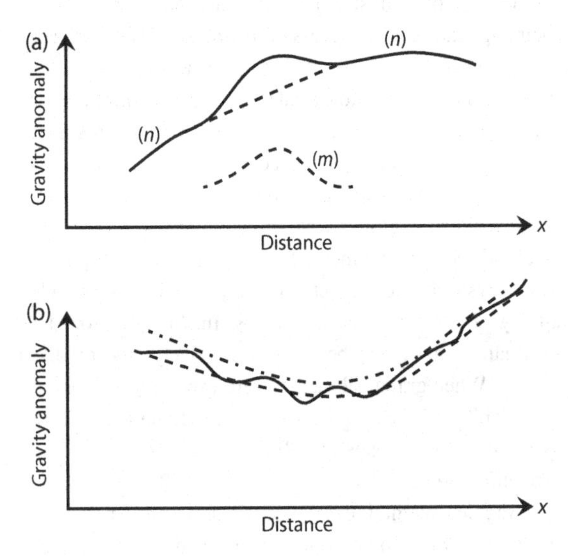

FIGURE 6.22 Schematic gravity anomaly profiles illustrating regional–residual separation by graphical methods. (a) The residual anomaly, m, is easily separated from the regional, n, by interpolation of the regional anomaly across the segment of the observed anomaly perturbed by the residual. (b) Complex observed anomaly consisting of the superimposed effects of several sources of both positive and negative density differentials. The regional (indicated by dashed lines) is not obvious, which leads to a non-reproducible regional–residual separation.

residual by graphical methods is difficult and likely to be non-reproducible.

Where gravity data are available in map form, residual anomalies are localized by subtracting smoothed contours from the observed anomaly map or from smoothed right-angle profiles prepared from the mapped data. The smoothed values at intersections of the profiles are required to have the same value, which constrains the assumed regionals. Trend surface analysis is an analytical approach to residual–regional separation by graphical methods that represents the residual–regional component anomalies as least-squares polynomial approximations (e.g. AGOCS, 1951; COONS *et al.*, 1967; THURSTON and BROWN, 1992). It is particularly successful in mapping obvious residual anomalies where the regional trend computation excludes the region of the residual anomaly (SKEELS, 1967). The method is used on either profiles or maps where the regional anomalies are complex and the differences between the regional and residual anomalies are subtle. It produces objective, automatic computations that aid the rapid exploration of observed gravity anomalies for residual components. However, the application and interpretation of the trend surface method is still subject to concerns regarding subjectivity and non-uniqueness of the graphical methods (e.g. BELTRÃO *et al.*, 1991) and to problems with linearity and phase distortion (e.g. THURSTON and BROWN, 1992). As a result, the method does not have broad usage, but it is used to precondition gravity anomaly data by removing low-order polynomial components before subjecting the residual gravity anomaly to other isolation and enhancement techniques.

Analytical grid methods
A large body of geophysical literature starting over a half century ago deals with filtering gravity anomalies in the spatial domain using convolution operations. These methods determine the regional value at a grid point based on anomaly values within a template centered on the point. The procedure is repeated for each grid point in a map or on a profile until the entire grid of values has been transformed into the desired analytical output. Depending on the weighting of the sampled values surrounding the central grid point, the output can be high- or low-wavenumber filtered results, upward- and downward-continuation anomalies, and derivatives and integrals of the observed anomalies. These methods are appropriately termed analytical grid methods because they operate on an equidimensional grid and involve an analytical expression in the calculation. The filtering characteristics of the various analytical grid templates have been extensively investigated by empirical studies (e.g. GRIFFIN, 1949; SWARTZ, 1954; HAMMER, 1977), but their wavenumber responses, which can be obtained by taking the Fourier transform of the filtering function or weights (FULLER, 1967), are particularly useful in evaluating the methods.

Spectral filtering methods
The spatial domain approach to isolation and enhancement of gravity anomalies has been largely replaced in modern processing by spectral methods. As described in Appendix A, digital data analysis in the space domain is used to a lesser extent because spectral analysis is easier to implement, has broader application, is more efficient for larger data sets, and suffers less edge effect. The spectral approach is especially effective where there is a distinct separation in the dimensions of the regional and residual components. However, gravity observations represent the superimposed or overlapping effects of the components so that error in isolating or enhancing either component necessarily distorts both components. Thus, in practice, both components merit evaluation by ancillary geological and geophysical constraints if either is to be used for interpretation and quantitative analysis. Spectral methods are used to produce a variety of filtered anomaly maps.

(A) Wavelength filters Wavelength filters pass or reject wavenumber components of the gravity anomalies based on their wavelength (or wavenumber) properties. Low-pass/high-cut filters are used to remove undesirable high-wavenumber components from small, shallow sources and to reduce errors in observations and reduction of the gravity anomalies. High-pass/low-cut filters, on the other hand, remove the longer-wavelength components associated with large, deep sources that are usually regarded as the regional anomaly. In addition, band-pass/reject filters can help to suppress/enhance the noise and regional characteristics in the gravity data.

To illustrate the application of band- and high-pass filters, consider the gravity anomaly map in Figure 6.23, which was constructed by compositing several residual and regional anomalies with superimposed low-amplitude noise (XINZHU and HINZE, 1983). Figure 6.24 is a residual gravity map based on the same data (Figure 6.23) prepared using a narrow band-pass filter in which the anomalies between 2 and 4 km were passed. The low-pass component of the filter removed the noise and the residual anomalies with horizontal dimensions in the range of 1 km, whereas the high-pass portion eliminated the long wavelength components of the regional anomalies. The prevailing left-to-right gradient, which is derived from density variations within the basement rocks of the model, was dampened but not eliminated, and the amplitude and gradients of the residual anomalies were significantly dampened. The locations of the residual anomalies are clearly indicated on the figure, but the residual anomalies are not faithfully reproduced, limiting the use of the map for quantitative analysis. A more restricted range of wavelengths (300 to 1,000 m) of the anomaly map is shown in Figure 6.25. This map has many of the attributes of Figure 6.24, but the local anomalies are even more isolated and the regional more attenuated. Both of these maps as well as several succeeding maps show marked edge effects which are blocked out by the gray shading on the maps. These anomalies are not derived from subsurface features, but rather are artifacts of the filtering process and are disregarded in the interpretation process.

Figure 6.26 shows the anomalies with wavelengths of less than 1 km that were obtained by high-pass filtering the gravity map in Figure 6.23. This map shows the location of the smaller horizontal dimension anomalies including those with discernible dimensions of less than 1 km that were excluded from the previously discussed band-pass filtered map (Figure 6.24). Again this map is useful for isolating residual anomalies, but limited for quantitative computations.

An objective of wavelength filtering is to focus attention on anomalies derived from a specified depth range. Although depth is a major factor in determining the spectrum of an anomaly, it is impossible to reach this objective

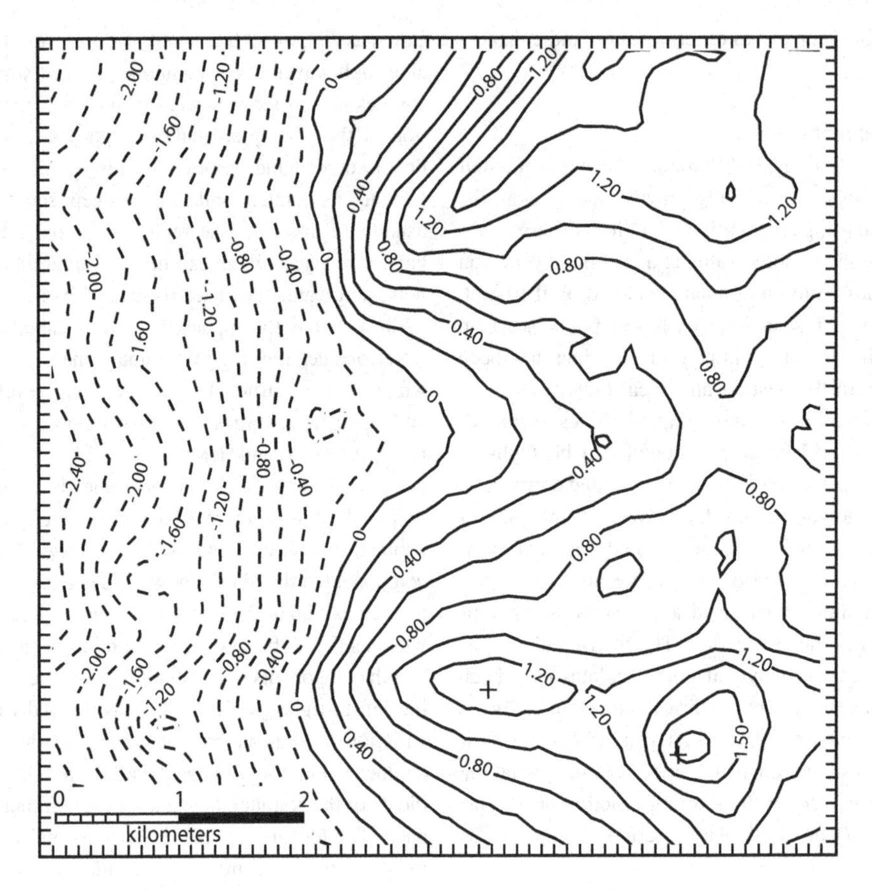

FIGURE 6.23 Observed gravity anomaly map (with mean removed), which is a composite of several residual and regional anomalies plus noise. Contour interval is 0.1 mGal. Adapted from XINZHU and HINZE (1983).

because of the breadth of the anomaly spectrum and the effect of geometry and size of sources on the spectrum. Nonetheless, a rule of thumb in gravity analysis is that $Z_c \approx \lambda/6$, where λ is the limiting wavelength for a source whose depth is Z_c.

This rule assumes that residual sources are concentrated masses, an assumption only valid in some investigations. It is based on the fact that the anomaly wavelength of a concentrated, isolated source is approximately equal to $8X_{1/2}$ where $X_{1/2}$ is half the width of the anomaly at half its amplitude. The maximum depth to the center of a concentrated mass Z_c is $\leq 1.3X_{1/2}$. Thus, $Z_c = 1.3(\lambda/8)$ or $Z_c \approx \lambda/6$. For example, to perform wavelength filtering to obtain the residual anomalies from concentrated sources (e.g. roughly equidimensional caves in a carbonate bedrock) that are assumed to be at a depth of less than 40 m, a high-pass filter passing wavelengths of less than approximately 240 m is applied to the data. Similarly, to focus on concentrated sources within the crust of the Earth whose thickness is estimated at 40 km, a filter passing only wavelengths less than 240 km is used.

(B) Matched filters Matched filters are designed to match the spectra of the anomalies that are desired in the residual anomaly patterns. Accordingly, the spectra of the type of target anomalies that are of interest in a data set can be calculated using simplifying, idealized sources. The spectra can then be used to design a filter that will pass anomalies primarily with the desired spectral characteristics (e.g. SPECTOR and GRANT, 1970; SYBERG, 1972; COWAN and COWAN, 1993). The filter is commonly derived by spatial or spectral cross-correlation analysis or from the correlation spectrum of the desired and residual anomaly patterns (Appendix A.5.1).

Matched filters are potentially of important use, but the results also include the wavelength components of other anomalies that overlap into the range of the desired anomalies, thus decreasing the resolution of the filter and its usefulness in isolating anomalies for analytical interpretation. This is especially the case in dealing with gravity anomalies because of the extensive overlap in gravity anomaly spectra derived from sources at different depths.

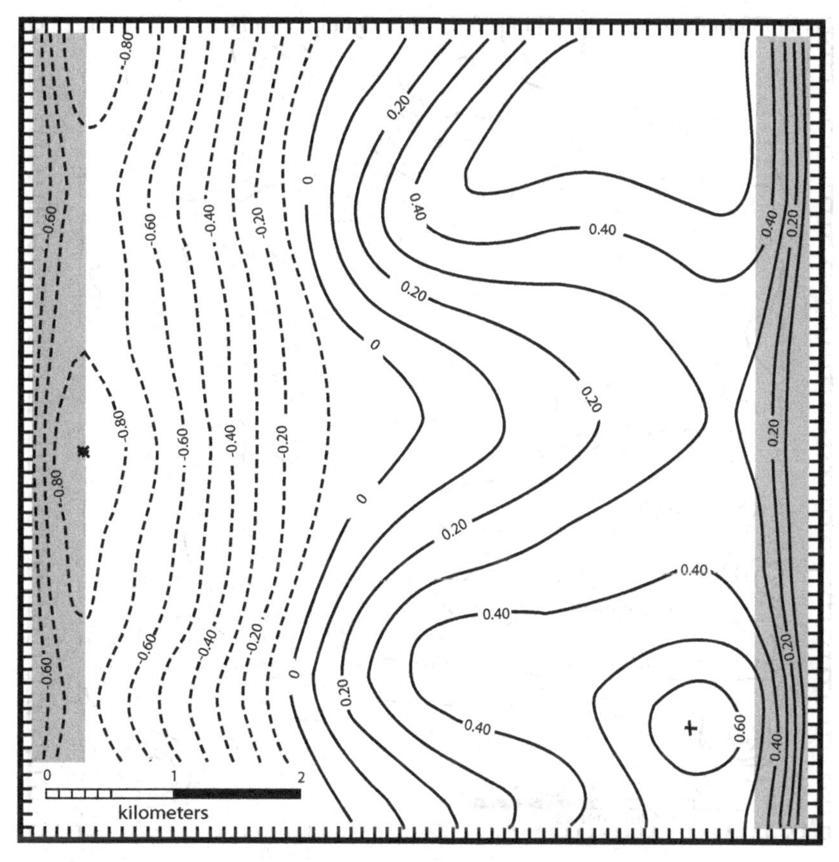

FIGURE 6.24 Band-pass filtered gravity anomaly map ($2 < \lambda < 4$ km) of the gravity anomaly map shown in Figure 6.23. Contour interval is 0.1 mGal. The gray-shaded margins include edge effects from the filtering that are typically avoided in anomaly interpretation. Adapted from XINZHU and HINZE (1983).

(C) Correlation filters The correlation filter passes or rejects wavenumber components between co-registered data sets based on the correlation coefficient between the common wavenumber components as given by the cosine of their phase difference (Appendix A.5.1). Thus, correlation filtering aids in quantifying gravity anomaly associations with geologic, topographic, photometric, heat flow, magnetic, and other geophysical data to help minimize interpretational ambiguities. In addition, anomalies with sources at depths greater than the line or track spacing of marine, airborne, and satellite surveys are coherent between two or more neighboring tracks, and thus can be extracted from these tracks by correlation filtering.

For example, in processing satellite sea surface altimetry for lithospheric gravity anomalies, correlation filtering facilitates separating the spatially and temporally static variations of lithospheric sources from the dynamic variations of non-lithospheric effects (Section 5.5.2). As an illustration, Figure 6.27 shows five track-pairs of geoidal undulations from ascending orbits of the US Navy's Geosat-GM altimetry mission over a region east of the Gunerus Ridge which extends offshore of East Antarctica's Queen Maud Land. The tracks for each of these pairs are separated by mean distances of about 2 to 5 km. Figure 6.28 gives the geologically coherent undulations obtained by track-to-track correlation filters designed to pass all wavenumber components satisfying $r(k) > 0.80$. Comparing the mean correlation coefficients of the track pairs in Figures 6.27 and 6.28 shows an improvement of nearly 11% in the positive correlations between track undulations, which reflects an enhancement of about 36% in the ratio of lithospheric signal to non-lithospheric noise (Equation A.72).

In any filtering application, it is always prudent to check the rejected components for any inconsistencies with the underlying assumptions. Accordingly, Figure 6.28 was subtracted from Figure 6.27 for Figure 6.29 to reveal the features suppressed by the wavenumber correlation filtering. These broadband non-correlative features presumably

FIGURE 6.25 Band-pass filtered gravity anomaly map ($300 < \lambda < 1{,}000$ m) of the gravity anomaly map shown in Figure 6.23. Contour interval is 0.05 mGal. The gray-shaded margins include edge effects from the filtering that are typically avoided in anomaly interpretation. Adapted from XINZHU and HINZE (1983).

include crustal signals that are smaller than the track spacing and the dynamic signals from temporal and spatial variations of the ocean currents, waves and ice, measurement and data reduction errors, and other non-lithospheric effects.

Correlation filtering also provides insights on regional gravity anomaly correlations with topography, which are commonly studied to assess the isostatic attributes of the lithosphere (e.g. DORMAN and LEWIS, 1970, 1972; FORSYTH, 1985; WATTS, 2001; LEFTWICH *et al.*, 2005). For example, the correlation between free-air anomalies and the gravity effects of the topography can reflect the isostatic compensation of the topography as illustrated in Figure 6.30. For the 100% compensated crustal column (panel (f)), the difference between the root and topographic effects yields essentially a zero free-air anomaly that minimally correlates with the other effects. The correlation coefficient at the crustal surface (i.e. $r \approx 0.58$) is dominated by topographic edge effects (panel (e)) that cannot be fully compensated by the effects of the more distant crustal root, whereas at high altitude (i.e. 20 km) compared with the topographic relief, the gradients of these effects are substantially attenuated and the correlation coefficient is reduced significantly (i.e. $r \approx 0.12$). However, where the crust is too thin and overcompensated by the mass of the mantle, the correlation coefficient between the gravity effects (e.g. panels (g) and (h)) increases dramatically to nearly its maximum value. In this case, the isostatic anomaly is the topographically correlated free-air anomaly which must be removed from the complete Bouguer anomaly to minimize isostatic errors in crustal thickness modeling.

The sign of the correlations is reversed where crustal thickness is too great and undercompensated by the mantle. The nature of the compensation given by the sign of the correlation coefficient between free-air and terrain gravity effects is lost, however, in coherency analysis which is based on the squared correlation coefficient (i.e. r^2). Thus, coherency analysis is implemented over low-resolution wavenumber bands (Appendix A.5.1) that

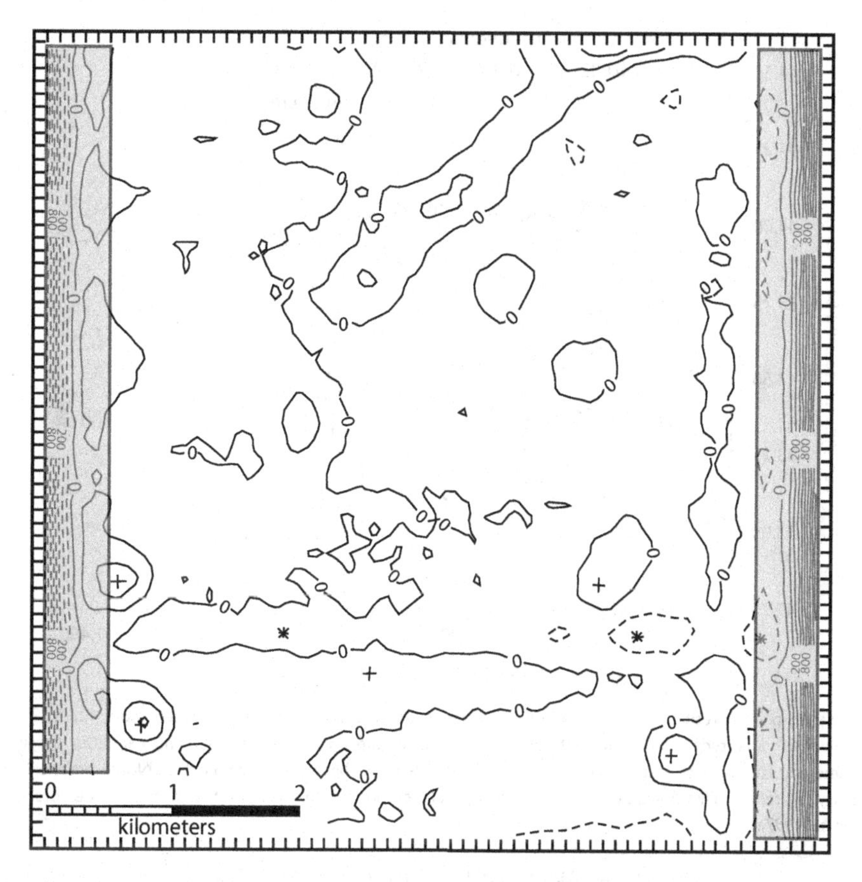

FIGURE 6.26 High-pass filtered gravity anomaly map ($\lambda < 1$ km) of the gravity anomaly map shown in Figure 6.23. Contour interval is 0.1 mGal. The gray-shaded margins include the high-gradient edge effects from the filtering that are typically ignored in anomaly interpretation. Adapted from XINZHU and HINZE (1983).

broadly compare the band-averaged spectral properties of the crustal relief and gravity for insights on the response of the lithosphere to regional tectonic loads (e.g. WATTS, 2001). In general, crustal applications of spectral correlation theory and filtering are growing with the increasing availability of regional and global digital gravity and terrain data (e.g. VON FRESE *et al.*, 1997a,b,c, 1999c; LEFTWICH *et al.*, 1999; KIM *et al.*, 2000; WATTS, 2001; LEFTWICH *et al.*, 2005). Additional examples of spectral correlation filtering are considered in Section 12.4.2 and Appendix A.5.1.

(D) Derivative and integral filters Derivative filters are powerful in enhancing the higher-wavenumber, shorter-wavelength components of a gravity anomaly field and identifying source configuration. They are valuable in locating subtle changes in the gravity field. Both vertical and horizontal derivatives are used with vertical derivatives useful in increasing the perceptibility of anomalies that are derived from shallow sources, whereas horizontal derivatives are primarily used for locating the edges of broad anomaly sources (e.g. SIMPSON *et al.*, 1986). Both the first and second vertical derivatives of gravity are used, whereas horizontal derivatives are normally restricted to first derivatives. The first derivative is the gradient of the anomaly in either the vertical or horizontal direction, and the second derivative is the gradient of the gradient.

Disadvantages of derivatives are that closely spaced, high-accuracy observations are required to implement them satisfactorily, the derivatives do not look like the anomalies or geologic sources from which they are derived, and the anomaly amplitudes, which are a critical parameter, can be greatly distorted in the derivative calculation. Quantitative analysis of derivative anomalies is recommended when they are directly measured, but otherwise their calculations are approximations that may be poorly suited for quantitative applications.

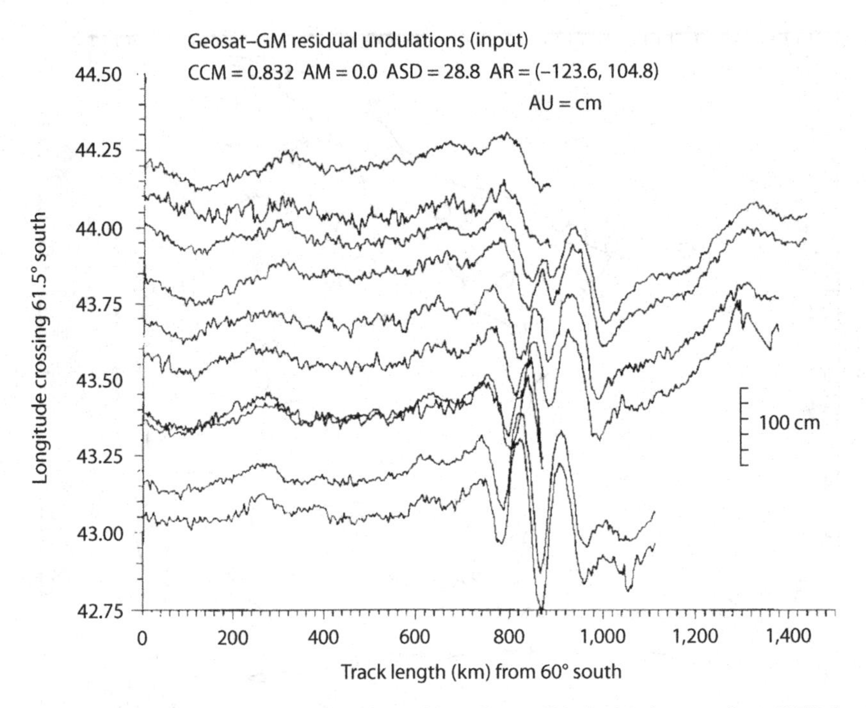

FIGURE 6.27 Five track-pairs of residual geoid undulations obtained from Geosat-GM altimetry between about 42.75° E and 44.50° E longitude and 60° S and 70° S latitude. Additional graphical attributes include the amplitude units (AU) and a scale for the undulation amplitudes in centimeters. The mean correlation coefficient (CCM) is about 0.83 between the data orbits. Non-lithospheric features include the effects of spatially and temporally varying ocean currents, as well as measurement uncertainties, data reduction errors, etc. Adapted from VON FRESE *et al.* (1997a).

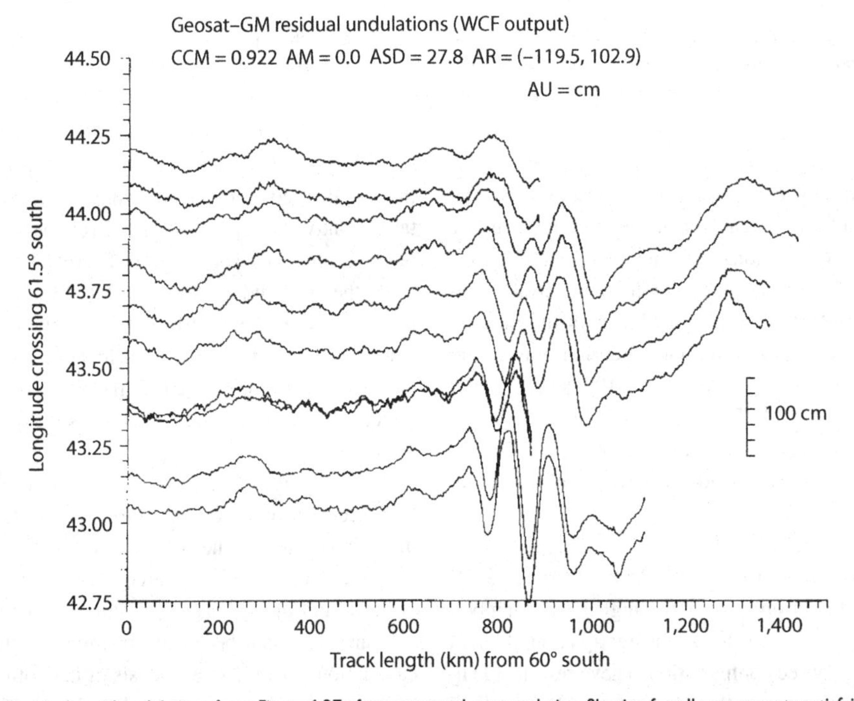

FIGURE 6.28 Residual geoid undulations from Figure 6.27 after wavenumber correlation filtering for all components satisfying $r(k) > 0.80$. The mean correlation coefficient (CCM) between the tracks has been enhanced to 0.92 as would be expected when the anomalies are increasingly dominated by lithospheric effects which are larger than the track spacing. Adapted from VON FRESE *et al.* (1997a).

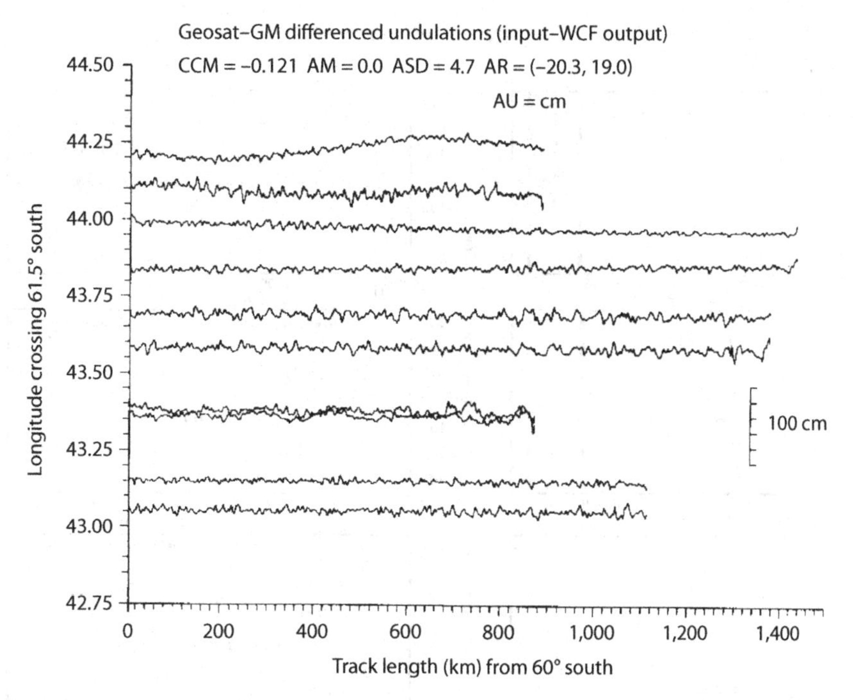

FIGURE 6.29 Non-correlative residual geoid undulations suppressed by the wavenumber correlation filtering. Adapted from VON FRESE *et al.* (1997a).

Figure 6.31 illustrates the characteristics of derivatives for gravity anomaly profiles from a linear concentrated excess mass where the vertical and second derivative anomalies consist of a high with adjacent lows. The second derivative high is narrower and the total amplitude range is greater than for the vertical gradient, reflecting the higher pass characteristics of the second derivative filter.

In Figure 6.32, the horizontal gradient of the vertical contact anomaly as shown is a maximum over the inflection point of the anomaly which marks the approximate position of the vertical contact between units of differing densities. This illustrates the utility of mapping density boundaries in the subsurface with maximum horizontal gradients. The accuracy of boundary mapping is determined by the data control and errors, the variation of the dip of the boundary from vertical, and the depth and depth extent of the body.

Gravity gradients are useful in outlining shapes of subsurface features. For instance, NETTLETON (1976) found that all the key oil-producing anticlines in the Los Angeles basin can be imaged by mapping the zero contour of the second vertical derivative of gravity. These gradients are also useful for exploiting Poisson's relation for possible pseudomagnetic effects of subsurface mass variations.

Spectral integration is obtained by simply applying the relevant inverse derivative coefficients to the spectral model of the gravity anomalies. This approach is useful for estimating gravity anomalies from gradiometry data, geoid variations from free-air observations, and other integral properties of gravity measurements.

(E) Continuation filters Continuation filters are important isolation and enhancement methods that project the observed gravity anomaly to other surfaces either above (upward continuation) or below (downward continuation) the original observation surface (e.g. PAWLOWSKI, 1995). Upward continuation is a smooth, low-wavenumber pass filter which emphasizes the anomalies from the broader, deeper sources at the expense of the shallow-sourced anomalies and noise in the anomaly field. Because it is a slowly varying, smooth filter, it preserves the character of the broader anomalies making it possible to quantitatively analyze the upward-continued anomalies. However, adjacent anomaly sources will produce increasingly overlapping anomalies as the height of the continuation is increased. This decreases the perceptibility of individual anomalies, and highly attenuates the anomalies of short wavelength that are considered noise because they distort the anomalies of interest.

The smoothing effect of upward continuation is evident in Figure 6.33, which is the anomaly pattern of Figure 6.23 upward-continued 300 m. Comparison of the

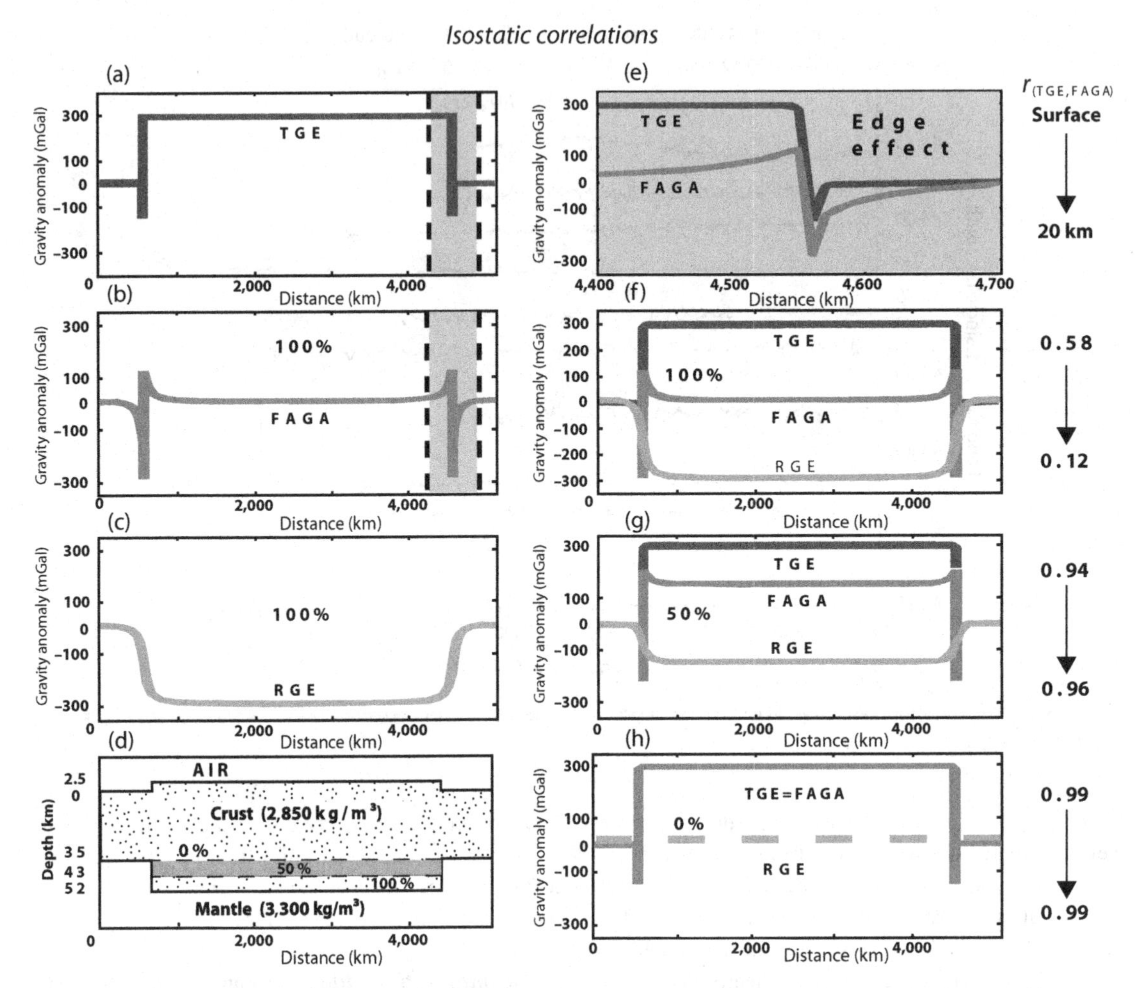

FIGURE 6.30 Gravity anomaly variations associated with varying degrees of isostatic compensation. The left column gives the (a) gravity effects of the terrain (TGE), (b) free-air gravity anomaly (FAGA), and (c) its Airy root (RGE) at station intervals of 20 km for the regional crustal and mantle mass distribution in panel (d). Panel (e) compares the panel (a) and (b) edge effects within the gray-shaded region that dominate the signal correlation at the terrain surface. The middle column shows these effects at 100% (f), 50% (g), and 0% (h) isostatic compensations as modeled from the respective maximum (solid line), middle (gray-shaded area), and zero (upper dashed line) extensions of the crustal root in panel (d). The right column gives the corresponding correlation coefficients between FAGA and TGE at the surface and 20 km altitude.

two maps shows that gravity anomaly noise and anomalies from the smaller, shallower sources are strongly suppressed by the upward-continuation process. The result of upward continuation of a data set has been used as an approximation to the regional field which is subtracted from the observed anomaly field to determine the residual anomalies (Jacobsen, 1987). The height of the upward continuation to achieve an appropriate regional will vary depending on the nature of the anomalies. The decision on the appropriate height is arbitrary and can best be determined by comparing results from a range of heights.

In contrast to upward continuation, downward continuation is a high-pass filter which emphasizes the smaller, shallower anomalies with steep gradients at the expense of those from broader, deeper sources, in other words from the regional anomalies. In that sense, downward continuation is an enhancement technique. However, even though the anomaly reflects the decreased distance between the source and the level of the anomaly field, the properties of the anomaly are preserved on downward continuation, and thus the anomaly may be isolated for quantitative interpretation. The restriction is that the original anomaly field must be free from significant noise because the

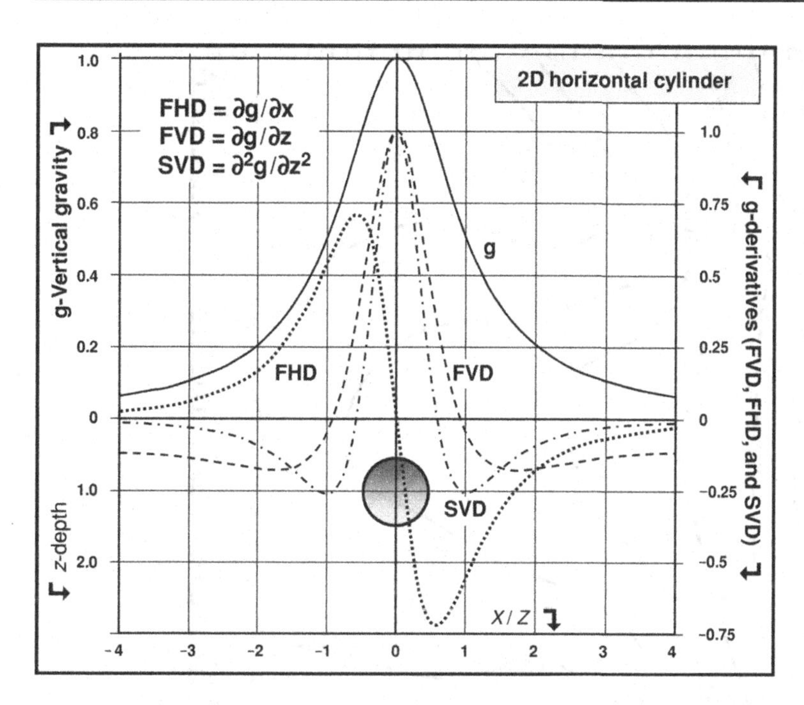

FIGURE 6.31 Illustrations of the vertical gravity (g), first vertical derivative (FVD, dashed curve), the first horizontal derivative (FHD, dotted curve), and second vertical derivative (SVD, dot-dashed curve) gravity anomaly profiles over an infinitely long 2D horizontal cylinder striking perpendicular to the profile using normalized vertical and horizontal scales.

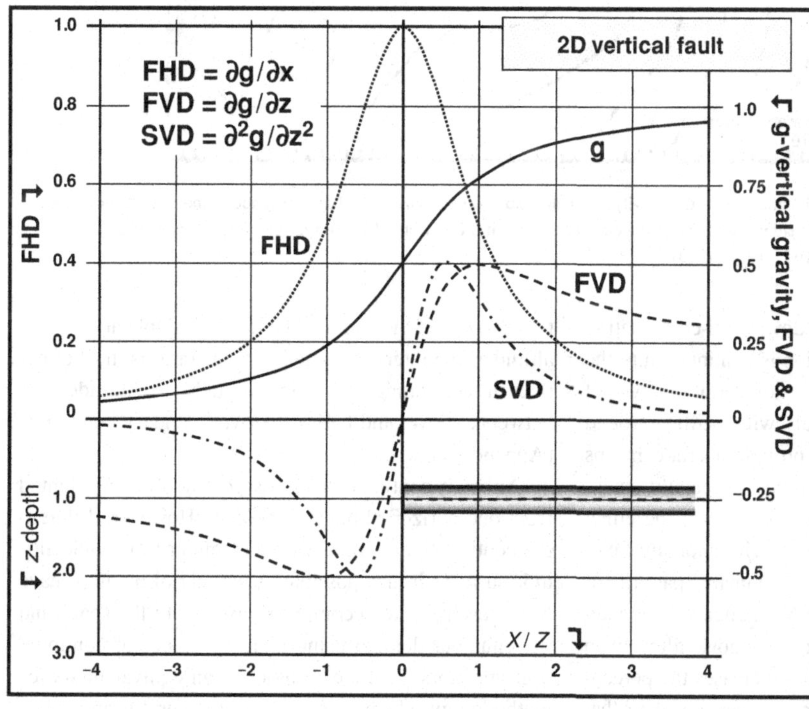

FIGURE 6.32 Illustrations of the vertical gravity (g), first vertical derivative (FVD), the first horizontal derivative (FHD), and second vertical derivative (SVD) gravity anomaly profiles over an infinitely long vertical contact between geological units of differing densities which also represents the edge of a broad anomaly source. Line styles as in Figure 6.31.

downward-continuation filter will emphasize all short-wavelength components. This restriction is sometimes overcome by subjecting the anomaly field to a high-cut noise filter before or after downward continuation of the data.

The downward-continuation process, as well as upward continuation, assumes that all the observed data are observed on a common horizontal plane. In the case of downward continuation, the assumption also is made that there are no sources of anomalies between the

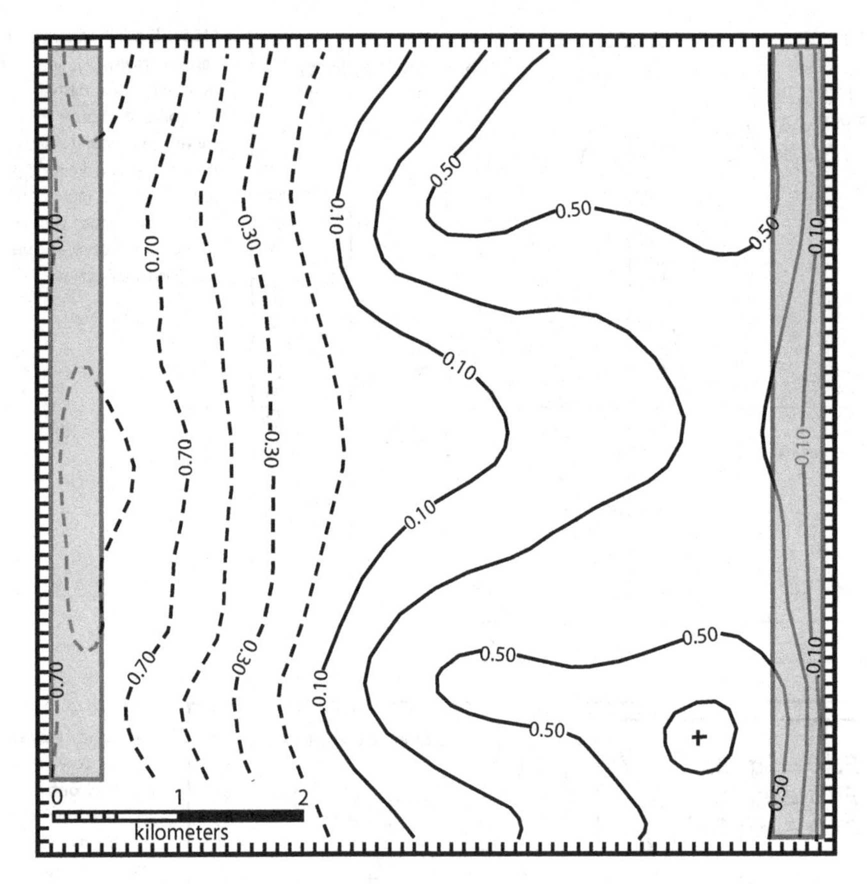

FIGURE 6.33 Upward-continued gravity anomaly of the gravity anomaly map shown in Figure 6.23 at 300 m above the observation level. Contour interval is 0.2 mGal. The gray-shaded margins include edge effects from the filtering that are typically ignored in anomaly interpretation. Adapted from XINZHU and HINZE (1983).

original anomaly level and the downward-continuation level. If the downward continuation is continued into the level of the sources, the amplitude of the anomaly will markedly increase and the anomaly will oscillate around the background level. This can provide a crude means of assessing the depth of the sources, especially small, shallow-sourced anomalies. As an example of the effect of downward continuation on the gravity anomaly field, Figure 6.34 shows the gravity anomaly pattern of Figure 6.23 downward-continued 50 m after the anomaly pattern has been subjected to a high-cut noise filter. Even this limited downward continuation increases the perceptibility of the small, shallow-sourced anomalies over that of the original map.

Upward and downward continuations are routinely used to compare data at different datums. For instance, a land gravity survey can be upward-continued for comparison with an airborne magnetic survey; or a marine gravity survey can be downward-continued to merge it with a water bottom survey. On the other hand, where two surveys of the anomaly field are available at different altitudes, the interval continuation operator is most effective in continuing the anomaly field to altitudes in-between, above, and below the two boundary conditions (Appendix A.5.1E).

Anomaly continuation is most commonly implemented from one horizontal plane to another. However, differential continuation of anomaly data from or to variable altitude surfaces is also possible. A number of methods have been developed to accomplish this in both the space and wavenumber domain using either a variety of continuation operators or, more commonly, an equivalent source methodology which is described in general terms below (e.g. BHATTACHARYYA and CHAN, 1977; HANSEN and MIYAZAKI, 1984; PILKINGTON and URQUHART, 1990; CORDELL, 1985; HUSSENOEDER *et al.*, 1995; PILKINGTON, 1998; FEDI *et al.*, 1999; NAKATSUKA and OKUMA, 2006). Variable upward and downward continuations are useful for mathematically moving observed anomaly data from a topographically draped configuration to a constant

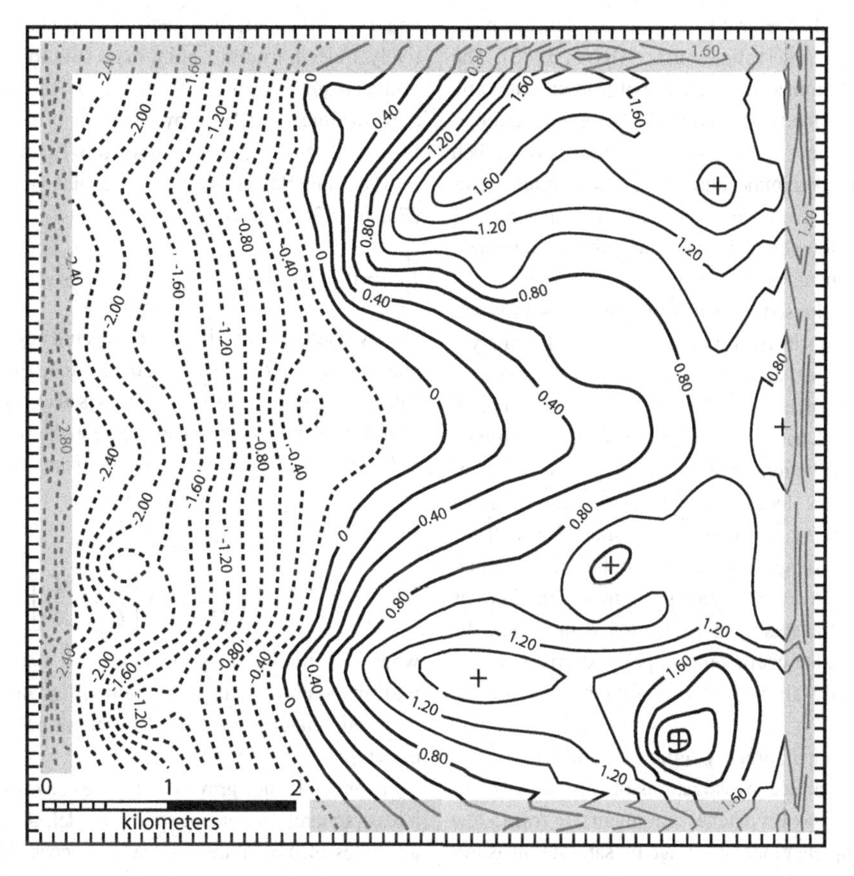

FIGURE 6.34 Downward-continued gravity anomaly of the gravity anomaly map in Figure 6.23 at a level of 50 m below the observation level. Contour interval is 0.2 mGal. The gray-shaded margins include edge effects from the filtering that are typically ignored in anomaly interpretation. Adapted from XINZHU and HINZE (1983).

altitude datum, such as may be needed to compare a land gravity survey to an aerogravity survey.

(F) Pseudomagnetic filters Pseudomagnetic effects can be calculated from observed gravity anomalies for sources of correlative density and magnetization variations via Poisson's theorem (Equation 12.7). For example, the vertical magnetic anomaly component obtained by Equation 3.110 forms the Fourier transform pair

$$B_z = -\frac{\partial V}{\partial z} = \frac{-J}{G\sigma}\frac{\partial g}{\partial z} \Longleftrightarrow \frac{-J}{G\sigma}\left(2\pi\sqrt{k_x^2 + k_y^2}\right)\mathcal{G} = \mathcal{B}_z, \tag{6.59}$$

where $\mathcal{G}$ and $\mathcal{B}_z$ are the Fourier transforms of the gravity anomaly g and the vertical component magnetic anomaly B_z, respectively, and $(\partial/\partial z) \Longleftrightarrow 2\pi\sqrt{k_x^2 + k_y^2}$ is the Fourier transform pair for the vertical derivative operator as shown by Equation A.81.

Inversely transforming $\mathcal{B}_z$ obtains the vertical pseudomagnetic anomaly B_z in the spatial domain from the wavenumber spectrum $\mathcal{G}$ of the gravity anomaly g due to a source of uniform density σ and magnetization with intensity J in direction z. However, this approach clearly can be used to transform any derivative and integral components of the gravity potential into any desired derivative and integral component of the pseudomagnetic potential. Examples of pseudomagnetic anomaly applications are consider further in Section 8.4.

Equivalent source methods

An alternative method of calculating upward- or downward-continued as well as derivative and integral gravity anomalies from an observed gravity anomaly data set is to calculate these anomalies from equivalent source effects. Equivalent source methods are based on the principle that observed potential fields can be duplicated by a distribution such as a grid of appropriate sources positioned below the observation surface (Roy, 1962). The sources are determined from the gravity observations by inversion as described in Appendix A. The gravity anomalies calculated from these equivalent source solutions then can be used to determine anomalies at other levels, or to calculate derivatives, integrals, and other attributes of the equivalent sources to isolate and enhance residual–regional components (Dampney, 1969).

Equivalent point source (EPS) analysis is a very effective alternative to spherical harmonic analysis for analyzing gravity data registered in spherical coordinates over finite spherical patches of the Earth and other planetary bodies. Spherical harmonic coefficients are determined from global distributions of data, and thus have very limited sensitivity for local data coverage details and the various subsurface components in the anomalies that the local drilling, geology, and other geophysical information may constrain. Furthermore, local gravity predictions from these coefficients reflect not only data within the study area, but also coefficient-propagated errors due to coverage gaps and other errors in the data from outside the study area.

A more computationally efficient and responsive approach to subsurface investigations of spherically composited gravity data over patches ranging in size from a few degrees to complete global coverage by satellite measurements is to relate the data (i.e. $b_{i,1} \in \mathbf{B}$) to the masses (i.e. $x_{1,j} \in \mathbf{X}$) of a specified spherical distribution ($a_{i,j} \in \mathbf{A}$) of gravity point poles by least-squares matrix inversion (i.e. $\mathbf{X} = (\mathbf{A}^t\mathbf{A})^{-1}\mathbf{A}^t\mathbf{B}$) so that $\mathbf{AX} \simeq \mathbf{B}$ (Appendix A.4.2). To achieve a desired data-processing objective such as data gridding, continuation, differentiation, integration, or other field transformations, the initial design matrix $\mathbf{A}$ in the forward model $\mathbf{AX}$ is exchanged for a new design matrix $\mathbf{A}'$ so that $\mathbf{A}'\mathbf{X}$ models the desired processing objective.

The gravity component g is most commonly mapped for regional investigations of the internal properties of the Earth and other planetary bodies. Thus, the basis for processing of these data over spherical patches of coverage by EPS inversion is Equation 3.91. By prescribing the distribution of gravity point poles, the analyst fixes the curly bracketed values as the coefficients of the design matrix so that the unknown values of the point masses can be determined by least-squares matrix inversion. With the point masses thus determined, the analyst can modify the curly bracketed values to satisfy, for example, desired data interpolation or continuation objectives by prescribing new coordinates of prediction. Or to estimate the horizontal F_θ and F_ϕ gravity components from g, the analyst simply exchanges the C-coefficient in Equation 3.91 for the E and H coefficients in Equations 3.92 and 3.93, respectively. Indeed, from the point mass solution to g, all other affiliated gravity components can be estimated because they are linearly related by the spatial derivatives of the inverse distance function as shown in Equation 3.9.

For example, the vertical or radial r-derivative of g (Equation 3.91) given by

$$\frac{\partial g}{\partial r} = (G \times m)\left[\frac{1}{R^3} - \frac{3C^2}{R^5}\right] = g\left[\frac{1}{C} - \frac{3C}{R^2}\right] \quad (6.60)$$

clearly demonstrates the linear dependency that allows g and its spatial derivatives to be determined from each other. Thus, where the gravity point mass is also magnetized, the magnetic component corresponding to the vertical or radial r-direction of magnetization by Poisson's theorem is

$$B_r(I' = 90^o) = \left(\frac{j}{G \times m}\right)\frac{\partial g}{\partial r} = g\left(\frac{j}{G \times m}\right)\left[\frac{1}{C} - \frac{3C}{R^2}\right], \quad (6.61)$$

where j is the intensity of the dipolar magnetic moment with inclination I'. In other words, Equation 6.61 gives the pseudomagnetic effect of the point pole with gravity effect g.

In general, once gravity anomalies have been related to a least-squares equivalent source model, the least-squares attributes of the anomalies are also accessible, including the affiliated potentials, geoidal undulations, vector and tensor components, continuations, and other geophysically interesting integral and derivative components (e.g. von Frese *et al.*, 1981a). Additional examples illustrating the use of these EPS results for modeling gravity effects of extended bodies in spherical coordinates were considered in Section 3.6.5.

6.5.4 Summary

A wide variety of methods are available to separate residual anomalies from regional anomalies in observed gravity anomaly data. The choice of method is based on the objectives of the analysis, but is also dependent on the experience of the analyst, the time and resources available, the complexity of the gravity field, the geological information available, the data coverage, and the quality of the data. The rules for choosing a procedure are not rigorous, and it is often desirable to use more than one method and compare their results. This comparison may lead to a range of viable solutions. In the presence of complex anomaly fields, analytical solutions are preferred, but in simple situations where geological information is available, graphical methods may be useful because of the ease of imparting geological constraints.

In gravity analysis, residual–regional separation commonly is focused on separation of anomalies on the basis of the source depth. Unfortunately, this problem is indeterminate without making simplifying and often indefensible assumptions. Many of the methods must really be considered as simply enhancing the residual anomalies of interest so as to increase their perceptibility, because the results distort the anomalies to a degree that prohibits quantitative analyses. Finally, in evaluating the validity of the residual–regional separation process, always consider whether the anomalies seem reasonable, both residual and regional, given both the attributes of gravity fields and the geologic sources of density differences within the local geological context.

6.6 Key concepts

- Gravity observations include the effects of a wide range of temporal and spatial variations that must be removed to condition the data for interpretation into subsurface variations in density. These extraneous variations result from instrumental, terrain, surface conditions, geological, and planetary sources.
- Temporal variations in the gravity measurements are the combination of instrumental drift, tidal effects from extraterrestrial bodies, atmospheric pressure (mass) changes, and both natural and anthropogenic-derived redistribution of mass in the subsurface and associated elevation changes. Typically drift rates are less than 0.1 mGal/hour and are monitored by reoccupation of previously observed stations at intervals of an hour to several hours because instrumental drift is peculiar to a specific instrument. Where reoccupation of stations is not possible, a major source of daily time variations in gravity due to tidal variations is approximated by theoretical calculations. Generally, other temporal variations have a much longer period and need not be considered.
- Traditionally, the vertical datum for gravity observations is the geoid or sea level, and the gravity datum is the 1980 Geodetic Reference System ellipsoid that best fits the shape of the Earth. To avoid inconsistencies due to the deviation of the geoid of ± 100 m from this ellipsoid over the surface of the Earth, a correction for the vertical distance and mass of the material between the geoid and the ellipsoid, the indirect effect, can be evaluated and considered in modeling the theoretical gravity at the measurement site. But the low gradient of this effect only interferes with long-wavelength gravity anomalies.
- In modeling the theoretical gravity at an observation site, consideration is given to either the absolute or relative change in position on the Earth's surface and the elevation of the observation site. It should also take into account the mass of the Earth material between the vertical datum and the site. This is usually based on the gravity effect of a horizontal plate of Earth material of a thickness equivalent to the elevation of the surface station. Deviations of the surface of the Earth from this plate, the curvature of the Earth, and the effect of the varying overlying atmosphere also may be taken into consideration in modeling the gravity. In observations from platforms moving over the Earth's surface, evaluation of the model gravity must consider the change in the vertical component of the centrifugal force and the Coriolis accelerations acting upon the gravity instrument because of its motion over the surface, relative to the theoretical gravity, which assumes the platform is stationary.
- The difference between the observed gravity and the conceptual model of gravity at the observation site is referred to as the gravity anomaly. This is the value used in interpreting the location and nature of subsurface mass variations. This process is often referred to as a reduction or correction procedure. However, the gravity is not being reduced to a common vertical datum nor being corrected for errors, but rather the observed data are being compared to a conceptual model of gravity assuming that the Earth is laterally homogeneous in density.
- Numerous types of anomalies have been established by taking into account a variety of sources of gravity variations. These are organized into three classes of anomalies: planetary, geological, and filtered. Planetary anomalies employ only planetary effects such as location and elevation in determining the theoretical model of gravity at an observation site. For example, the free-air gravity anomaly only takes into consideration the location of the station and the elevation of the station without considering the gravitational effect of the material between the station and the observation site, hence the name free-air. It is unsuitable for most exploration problems except under special conditions because it contains the gravitational effect of isostatically uncompensated topographic masses which commonly obscure the effect of subsurface density variations. Gravity anomalies can also be difficult or impossible to identify because of high-wavenumber elevation-related effects. Nonetheless, it has uses in regional gravity surveys, in airborne and satellite surveys where the altitude of the survey is great compared with the surface relief, and in deep-water marine gravity surveys.
- The most widely used gravity anomaly in exploration is also a planetary anomaly, the Bouguer gravity anomaly, which is determined by including the mass of material

between the elevation datum and the site of the observation in the free-air gravity anomaly in the theoretical model. When the mass calculation is modified for the local terrain (the terrain correction), the anomaly is referred to as the complete Bouguer gravity anomaly. If the terrain is not considered, the term used is simple Bouguer gravity anomaly. Bouguer anomalies generally do not correlate with local topography unless these features are related to structural or stratigraphic variations in density below the elevation datum level. They do correlate inversely with regional topography that is isostatically compensated. The simple Bouguer gravity anomaly is equivalent to the free-air anomaly at the ocean shoreline. Bouguer gravity anomalies usually are referenced to absolute values of gravity in regional surveys, but sometimes are only based on a local arbitrary value in local surveys.

- A density of 2,670 kg/m^3 is used as the default value for computation of the mass effect in the Bouguer gravity anomaly. This is considered a reasonable average value for continental regions, but the range of surface densities varies considerably from this value. This can lead to significant errors in regions of high topographic relief where the density deviates from the 2,670 kg/m^3 value. This error and errors in terrain corrections are the principal sources of errors in modern gravity surveys.
- A powerful tool in isolating unknown subsurface sources of gravity anomalies is to include the gravitational effect of known or hypothesized geological sources in the theoretical model. The resulting anomaly, the geological gravity anomaly, includes only the effects of unknown subsurface density variations. A common example is the isostatic residual gravity anomaly. The assumption is made in this anomaly that surface topography is isostatically compensated at depth so that the Earth is in hydrostatic equilibrium, and this effect is included in the calculation of the theoretical gravity at the observation site. Its principal use in exploration is to minimize the deep regional effects due to compensation for topography by changes in density, and thus concentrate the anomaly on sources located within the crust of the Earth.
- A third class of gravity anomalies is the filtered gravity anomaly. A wide variety of analytical and graphical methods have been developed for isolating and enhancing the desired components of the anomalous gravity. These methods, unlike planetary and geological gravity anomalies, may eliminate arbitrary components of the anomaly field within a prescribed range of attributes. The resulting anomaly is commonly referred to as the residual anomaly field that remains after removal of longer-wavelength components, the regional field, and shorter-wavelength anomalies, or noise. Filtering may be based on amplitudes, gradients, anomaly trends, or wavelengths, depending on the nature of the anomaly field and the desired residual anomaly.
- There are two general approaches to separation of the residual gravity anomaly from the observed anomaly field. Isolation techniques attempt to eliminate all anomalies of the observed field that do not have a prescribed set of characteristics as defined by the objective of the survey. The fundamental premise of this technique is that the geologically significant anomaly is minimally modified in the process of removing the other components of the field. Thus, the isolation technique can produce results that may be readily interpreted in terms of the attributes of the anomaly source. In contrast, other methods are based on enhancement of the desired anomaly at the expense of other anomalies. The residual anomaly is distorted in the process, but may be more readily identified in the anomaly field.
- A wide variety of methods is available for separating the residual anomaly from the observed field. The most common approach is to filter the data in the spectral domain based on anomaly wavelength characteristics. However, there is no rigorous rule for selecting the optimum procedure. The choice is based largely upon the nature of the anomaly field, the objective of the survey, the quality and coverage of the data, and the experience of the analyst. Analytical methods are readily applied to mass calculations on complex anomaly fields and are generally preferred because they minimize personal bias. It is often best to use more than one procedure permitting comparison of results that may aid the interpretation.

Supplementary material and Study Questions are available on the website www.cambridge.org/gravmag.

7 Gravity anomaly interpretation

7.1 Overview

Raw gravity observations are reduced for their non-geologic effects to one of a variety of anomalies which in turn are processed by isolation and enhancement procedures into residual anomalies that map the gravity effects of interest in interpretation. Anomaly interpretation, which models the gravity anomalies for the nature and processes of the subsurface, is a relatively straightforward process compared with the measurement, reduction, and residual–regional separation phases of the gravity method. However, anomaly interpretation is never unique, owing to the ubiquitous presence of data errors and the inherent source ambiguity of the gravity potential. Thus, ancillary geological, geophysical, and other constraints on the subsurface are essential to help limit the ambiguity.

Gravity interpretations can be qualitative where the analysis objectives are satisfied by the mere presence or absence of an anomaly. Interpretation also can be highly quantitative with comprehensive modeling of the geometric and physical properties of the anomaly sources. Effective interpretation requires knowledge of the key geological variables that influence the anomaly's amplitude and geometry. It also requires an understanding of the key geophysical variables that control the inverse problem of estimating source parameters from the anomaly.

Gravity interpretation generally is initiated using simplified techniques to estimate preliminary source depths, depth extents, margins, density contrast, and mass. These estimates are often enhanced by more comprehensive analyses that include forward modeling using trial-and-error inversion methods if only a few unknown modeling parameters are involved. For more unknowns or where they must be estimated by least squares or some other error norm, inverse modeling is commonly implemented by matrix inversion. Both inversion methods compare the predicted anomaly from an assumed forward model with the observed gravity anomaly.

7.2 Introduction

The general approach to the interpretation of geophysical measurements was discussed in Section 1.5.4. In this chapter, the generalized interpretation approach is focused specifically on the gravity method. Gravity anomaly interpretation covers many techniques depending on the quality of the data and the objectives of the analysis. It ranges from qualitative reviews of the data in various presentation forms to highly quantitative modeling.

The stages in the interpretation of gravity anomaly data have been discussed by a number of authors with varying procedures being described (e.g. Simpson and Jachens, 1989; Chapin, 1998). Differences in the procedures reflect personal experience, the available data, computational facilities and software, and the objectives of the interpretation. However, the common view is to tune the interpretation techniques to the quality and coverage of the available data, and to integrate the interpretation with all pertinent subsurface information. Most importantly it

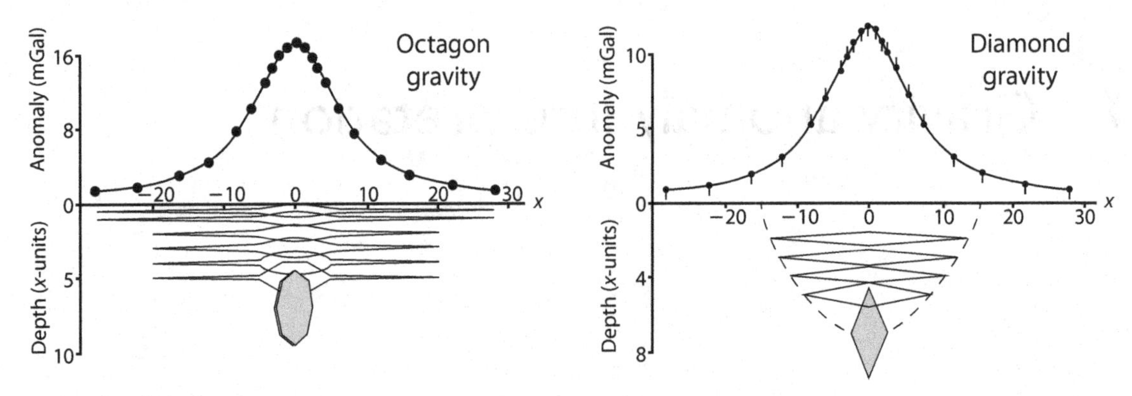

FIGURE 7.1 Examples illustrating the equivalence of the Bouguer gravity anomaly derived from a variety of sources include octagonal and diamond distributions of mass at depth (gray shaded) that spread out with decreasing depth (unshaded). The depth and horizontal distance x are in arbitrary units. Adapted from JOHNSON and VAN KLINKEN (1979).

needs to incorporate known geological information from the surface and subsurface. This must be done as early as possible and not left to the final stages of interpretation. The use of auxiliary geophysical and geological data is important in defining subsurface structure and formation lithologies. This is especially critical in estimating the densities of the formations. The amplitude of all gravity anomalies and geologically derived gravity noise which may perturb interpretations is directly related to the contrast in densities. Wherever possible, these densities should be determined directly from measurements made on samples obtained in the region or from *in situ* observations using methodologies described in Section 4.3.5. However, the lack of samples or *in situ* measurements may require evaluation of local densities on the basis of general tables, modified by consideration of local factors and lithological attributes as discussed in Section 4.3.6.

7.2.1 Ambiguity in gravity interpretation

An important reason for using auxiliary geophysical and geological data is to decrease the ambiguity of gravity interpretation (SKEELS, 1947). Figure 7.1 illustrates a number of different subsurface bodies that all produce equivalent gravity anomalies (JOHNSON and VAN KLINKEN, 1979). Non-uniqueness is an overriding concern in all gravity interpretation, although ambiguity is present to varying degrees in all geophysical interpretations. Of course, the availability of appropriate constraints can limit these ambiguities.

Ambiguity in gravity analysis arises because of limitations in data coverage and quality, and the inherent nature of gravity fields. Limitations of data sampling may cause aliasing where station density is insufficient to provide a minimum of two station intervals or three observations within each desired anomaly wavelength. Errors of the observation, anomaly calculation, and residual anomaly isolation procedures will distort the anomaly. Although these sources of ambiguity can be controlled to a significant degree, they cannot be completely eliminated, primarily as a result of difficulties in isolating the desired residual anomaly.

The inherent ambiguity in gravity interpretation, even when the data are reliable, is a more fundamental problem. This cannot be eliminated without auxiliary information on the sources of the anomalies. The inherent ambiguity of potential fields, such as gravity anomalies, is discussed in Sections 3.4 and 3.6 in dealing with Laplace's and Poisson's equations, as well as in Appendix A.4.2 where it was shown that any inverse problem in practice leads to non-unique solutions due to errors in the observations and the assumed forward problem. Figure 7.2, from HUTCHINSON *et al.* (1983), gives a further example of the inherent ambiguity. This figure shows the gravity effects of three conceptual crustal models constrained by deep seismic studies that compare favorably with the Appalachian Mountain paired gravity anomaly found in profiles from New Jersey to Georgia. Even though all three models match the seismic refraction and gravity data, they are entirely different because there are not enough constraints available to distinguish geologically between them.

7.2.2 Two- versus three-dimensional interpretation

A decision that needs to be made early in the interpretation of gravity data is whether the interpretation will be made using 2D or 3D approaches. Although these are seldom

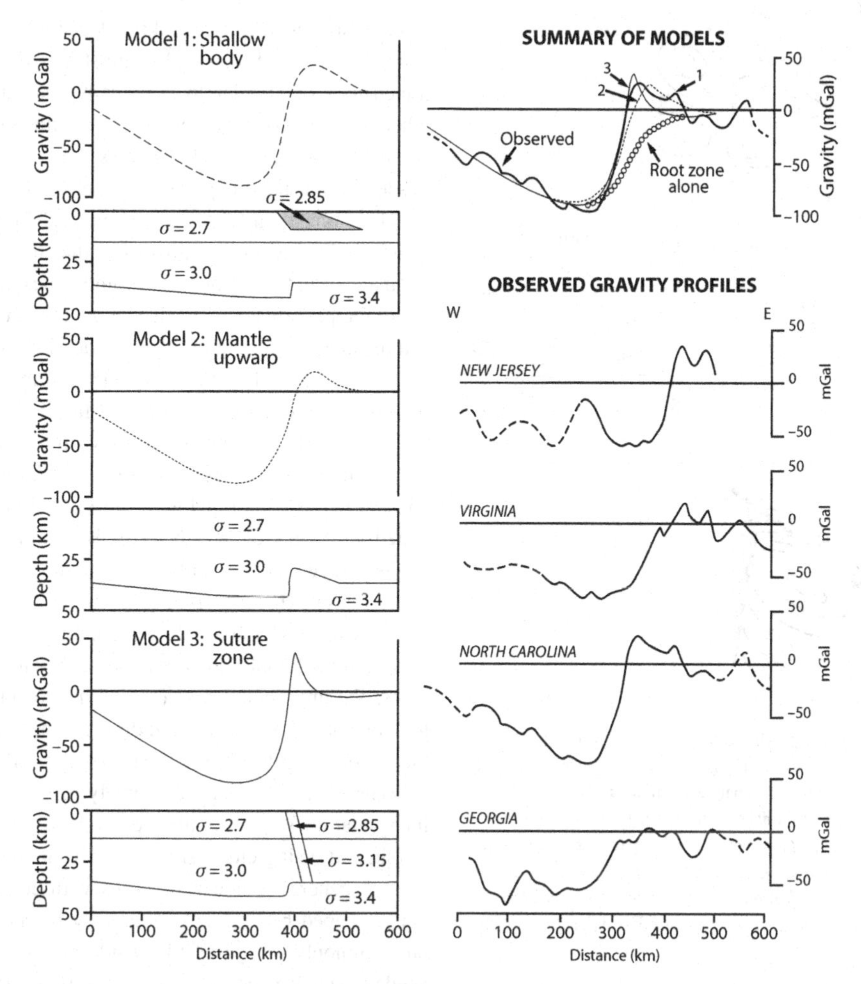

FIGURE 7.2 Ambiguity of the gravity anomaly interpretation illustrated by three different subsurface models that are consistent with seismic constraints and the observed gravity anomaly profiles. Densities σ are given in g/cm^3. Adapted from HUTCHINSON *et al.* (1983).

mutually exclusive, in most interpretations the emphasis is either on the interpretation of profiles or maps.

Profile interpretation normally is performed on the principal profile that is positioned through the central maximum or minimum amplitude of the anomaly and perpendicular to the anomaly contours. This method is commonly used for two reasons. First, there may already be a seismic or geologic cross-section over the same location. Second, generally speaking, a single profile interpretation is relatively fast and efficient to perform. Profile interpretation is commonly described as two-dimensional (2D), based on the assumption that the geological cross-section extends to infinity without change into or out of the profile. When the strike length of the source is finite, two-and-a-half dimensional (2.5D) interpretations are invoked where the strike lengths from the profile to each end of the source are the same. When the ends of the source are at unequal distances from the profile, or the source strikes at an angle different from perpendicular to the profile, the interpretations are sometimes called two-and-three-quarters dimensional (2.75D).

Map interpretation assumes a finite extent of sources within the boundaries of the map. Gravity data are acquired along multiple traverses directed perpendicular to the anticipated strike direction of the sources of interest, or more commonly on a more or less regular grid pattern with one of the grid directions perpendicular to the prevailing geological strike of the region. The data of the grid or series of traverses are combined into a contour map and subjected to an evaluation of the source of the anomalies.

Circular or elliptical anomalies indicate sources of limited strike length (Figure 7.3(a)) which need to be analyzed on a 3D or at a minimum of a 2.5D or 2.75D basis. In contrast, elongate anomalies (Figure 7.3(b)) may be

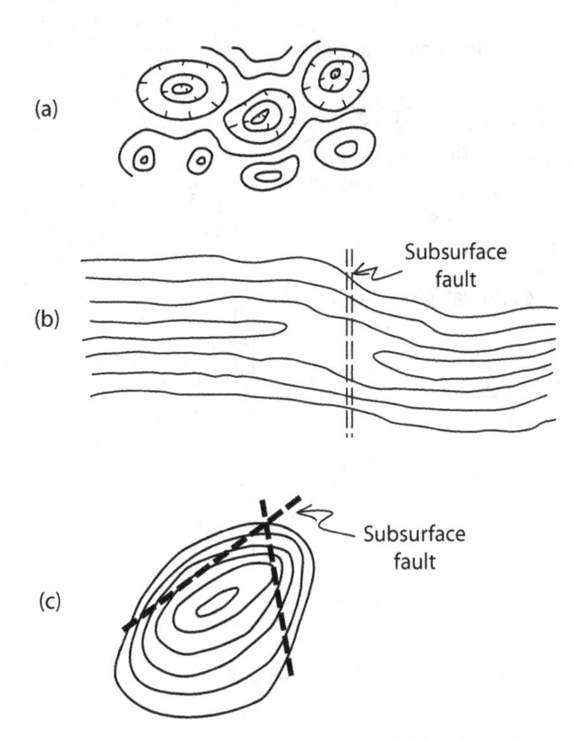

FIGURE 7.3 Gravity anomaly contour maps. (a) Roughly equidimensional anomalies requiring analysis by 3D interpretational methods. (b) Linear gravity anomaly that can be analyzed by 2D methods. The right-lateral offset of the anomaly shows the position of a subsurface fault. (c) Complex anomaly pattern with different gradients indicating high-angle subsurface faults which offset the source of the anomaly. An alternative interpretation is that the variable gradients are due to a source which has dipping margins.

considered on a 2D or 2.5D basis. NETTLETON (1940) showed that the error in gravity calculations assuming an infinite strike length for the source (i.e. 2D source) will be less than 10% where the strike length of the source is a minimum of four times the depth to the center of the source on either side of the profile. This is a commonly employed approximation to determine whether 2D analysis can be used. However, Nettleton's evaluation is based on the gravity effect of a horizontal rod whose radius is small in comparison with the depth to its center. As the source width increases, the required strike length needed to maintain two-dimensionality increases, but very slowly. The length must increase significantly to maintain two-dimensionality as the depth extent or depth to the top of the source increases. The maximum horizontal dimensions of the source can be approximated to first order from the horizontal distance between the inflection points of the anomaly gradients.

Figure 7.3 shows schematic examples of gravity anomaly contour maps illustrating the patterns of anomalies that are considered in making decisions regarding 2D and 3D analysis. Figure 7.3(a) illustrates the near-equal dimensions of circular to elliptically shaped anomalies where the low gradients relative to the dimensions of the sources suggest deep depths to the sources. These anomalies are best analyzed by 3D methodologies.

The elongate anomalies of Figure 7.3(b) can be analyzed by 2D techniques providing that the ends of the anomalies and the segment of the anomalies near the flexure in the pattern are avoided. The flexure in the anomaly can be interpreted as fault offset of the linear source.

Figure 7.3(c) shows the complication in the anomaly pattern where offsets along vertical faults have disturbed the source of the anomaly. Faults may either cause a flexure in the anomaly pattern as shown in Figure 7.3(b) or marked changes in gradients as in Figure 7.3(c). The anomaly associated with the fault which parallels the long dimension of the source can be readily incorporated into the profile interpretation. However, profiles drawn across the elongate anomaly in the vicinity of the cross fault will be disturbed, preventing a 2D interpretation approach. This illustration shows the importance of evaluating the contour map pattern of anomalies before deciding on the interpretation methodology and selecting the profiles for 2D analysis.

Where the nature of the anomaly pattern suggests 3D interpretation, sources of limited horizontal extent must be considered, using either simple idealized source geometries or general source geometries with complex, irregularly shaped bodies. The latter may be more realistic, but commonly is unjustified considering the effort that is needed to analyze complex sources. Often 2.75D analysis can be used to approximate a 3D body as a starting point in the interpretation.

The 2D, or 2.5D, or 2.75D interpretation techniques are commonly preferred over 3D methods because they are faster, simpler to implement, and provide greater detail than all but the most exhaustive 3D analyses. As long as the observation traverses are effectively perpendicular within a few tens of degrees to the strike of the anomaly pattern, it is advantageous to use the actual anomaly values in constructing the profiles for analysis. Profiles extracted for analysis from contour maps so as to achieve an orthogonal relationship to the anomaly strike direction are always high-cut filtered as a result of the gridding and contouring process. The effect may be detrimental to the analysis depending on the nature of the data.

7.2.3 The interpretation process

The above discussion makes it clear that there is no standard template for gravity interpretation. There are

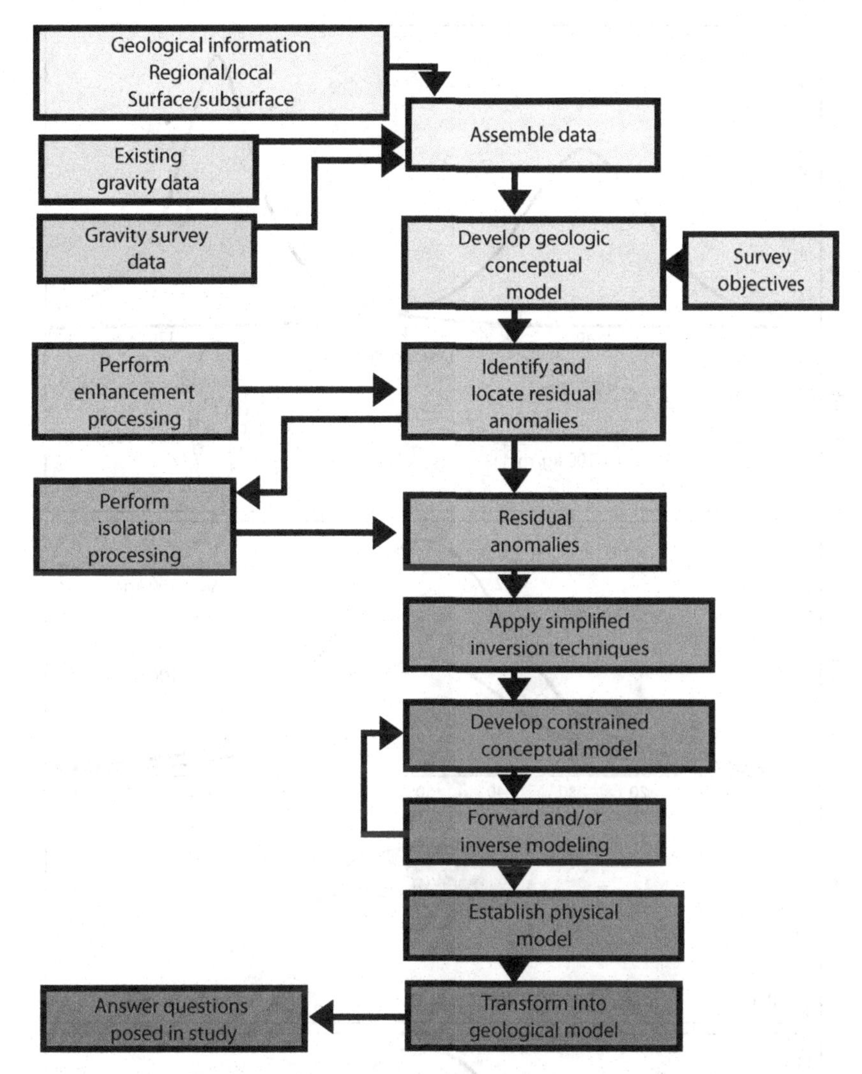

FIGURE 7.4 Flow chart showing the components involved in the interpretation of gravity data and their inter-relationships.

too many variables that enter into the process. Evaluation of these variables is a matter of knowledge and experience, which vary among interpreters. Nonetheless, there are certain common steps (Figure 7.4) that should be considered in the interpretation process after the gravity data are assembled and conceptual forward models of the anomaly sources are developed or inversion is performed on the basis of the objectives of the survey and the auxiliary surface and subsurface information.

For example, after the survey data have been reduced to anomaly values and the ancillary data are assembled, enhancement techniques may be applied to the gravity anomaly data if the residual anomalies are not readily apparent in the anomaly data. Enhancement techniques (Section 6.5.3) may distort the anomaly pattern by emphasizing certain characteristics of the sought-after anomalies. As a result, these techniques help to identify and locate residual anomalies.

In qualitative interpretation, the process terminates with the evaluation of the residual gravity anomaly and the derived or enhanced anomalies, either in map or profile form. However, it is very desirable for map views to supplement profile analyses.

In quantitative interpretation, on the other hand, isolation techniques are applied to gravity anomaly data to help extract anomalies of interest in the survey from other anomalies of either shorter or longer wavelength. This stage is necessary because anomalies located by enhancement techniques commonly may be distorted to a degree

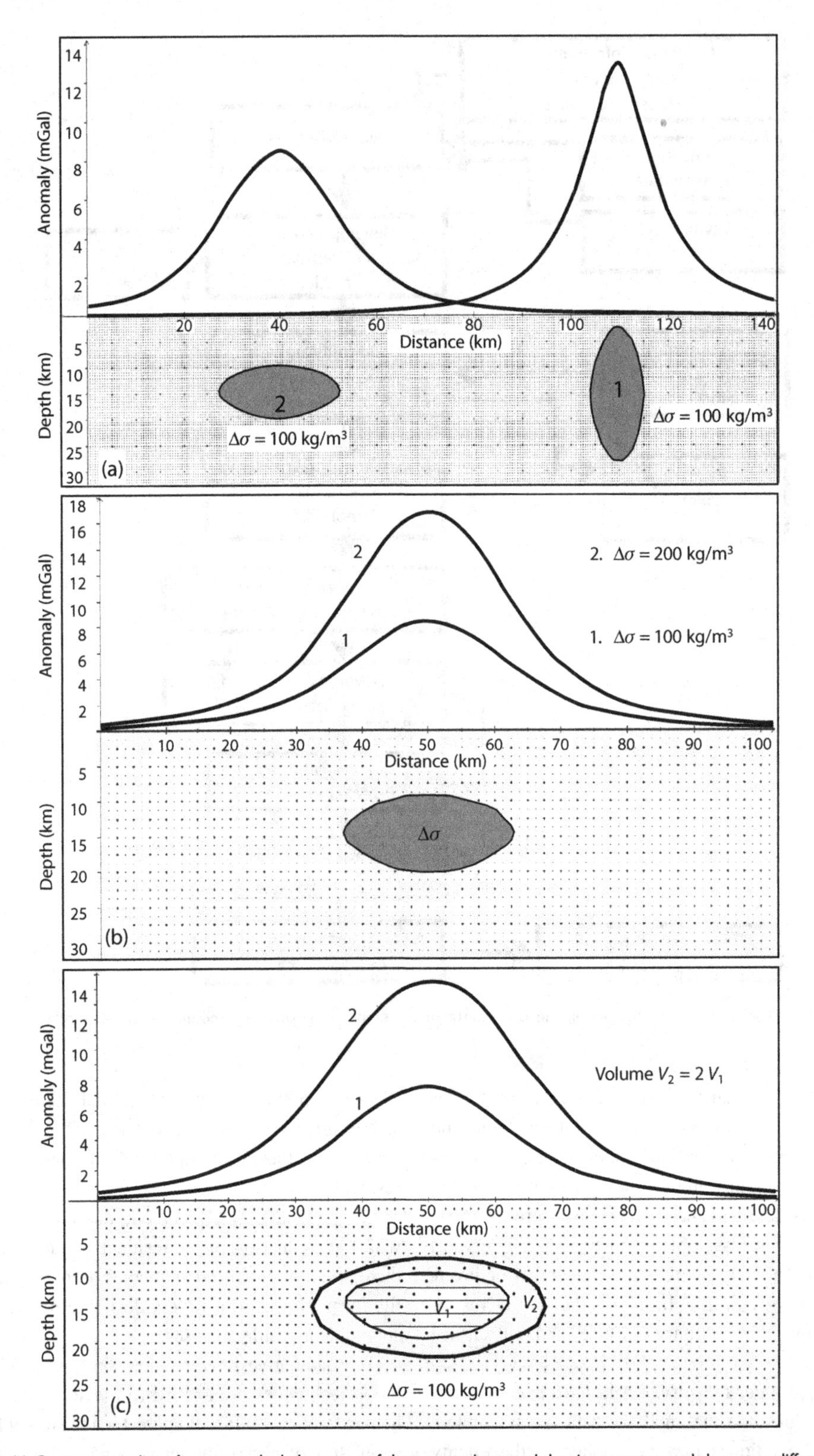

FIGURE 7.5 (a) Gravity anomalies of two gray-shaded sources of the same volume and density contrast, and thus mass differential, and central source depth. The two anomalies, although of quite different amplitude and shape, have equivalent total integrated anomaly energy. (b) Doubling the density contrast of one of the sources shown in (a) doubles the amplitude of the anomaly without changing its shape. (c) Doubling the volume of one of the sources shown in (a) changes the amplitude and shape of the anomaly.

that prevents them from being directly useful in modeling and other quantitative interpretation schemes.

Simplified inversion techniques may then be applied to the isolated residual anomalies to aid in constraining the characteristics of the sources for use with other geophysical and geological information. For example, procedures may be used to determine the depth to the source from the gradients of the anomalies and identify the subsurface geological contacts from the anomalies themselves.

More comprehensive inversions can establish additional source characteristics that better explain the observed anomalies in the context of constraining auxiliary information. They also produce error analyses which can help resolve the range of effective parameter values that may characterize the subsurface sources. Finally, the source characteristics determined in the previous stages are put into geologic terms that answer the questions posed in defining the objectives and parameters of the gravity survey.

7.3 Interpretation parameters

In dealing with the gravity method there are two basic questions that need to be considered. First, can a particular problem be solved with gravity, and second, what can be determined from gravity anomaly values? The procedures used in answering these questions, which are central to the interpretation process, involve a number of factors which ROMBERG (1958) called the key variables of gravity. The first question is based on key geologic variables and the second on key geophysical variables. Interpretation of gravity measurements is dependent on effective knowledge and use of both sets of variables.

7.3.1 Key geological variables

The amplitude and character of the gravity anomaly signatures derived from subsurface variations in mass are dependent on several key geologic variables. The anomalies vary with source geometry including volume and shape, source depth, isolation from other sources, and the density contrast between the source and the horizontally adjacent formations. Consideration must be given to sources of disturbing gravity effects and the error budget of the gravity anomalies as discussed previously.

The integrated effect or energy of the gravity anomaly – that is, the spatial summation of the products of the amplitude and the associated area over the total anomaly – is a function of the mass differential caused by the anomalous source. As such, the volume of the anomalous mass and its density contrast with the horizontally adjacent rocks are key variables. This is illustrated in Figure 7.5(a) in which two sources of the same volume, density contrast, and central source depth produce quite dissimilar anomaly shapes, but have equivalent total integrated anomalies. Doubling the mass differential by doubling either the volume or density contrast will double the total integrated anomaly. Doubling the density contrast of either volume will double the amplitude of the anomaly without changing its shape (Figure 7.5(b)), but doubling the volume will change the shape of the source or its depth, and thus will change the shape of the anomaly (Figure 7.5(c)).

The effect of anomalous source shape on the shape and amplitude of its gravity anomaly is reflected in its inverse distance function, and thus the change in anomaly with depth to the source. The inverse distance function, $1/r^N$, represents the decay or fall-off in amplitude in the analytical function for the gravity anomaly due to a source of prescribed shape. The quantity r is the distance between the source and the observation point, and N is the decay rate or structural index (Table 7.2), which changes from $N = 2$ for a concentrated 3D source to $N = 0$ for a long, wide horizontal source, with intermediate shaped sources having intermediate real values of the decay rate (i.e. $0 \leq N \leq 2$). As a result, concentrated sources change in amplitude and shape rapidly with variations in depth, while the amplitude of anomalies derived from broad horizontal sources such as faulted margins of flat sedimentary strata will not be notably affected by varying depth (Figure 7.6).

As a general rule, linear changes in density are expressed as linear changes in only the amplitude of the anomaly. Linear changes in size or volume, on the other hand, are cubic functions in both the magnitude and gradient of the anomaly. Linear changes in depth of concentrated sources, however, are squared functions that also affect both the anomaly's amplitude and wavelength. Thus, the anomaly's shape and amplitude are most sensitive to variations in size first, then depth, and lastly density.

The shape of the anomalous source is a key variable profoundly affecting anomaly shape and amplitude. A concentrated, compact differential mass will produce the same integrated anomaly as a horizontally distributed source with the same mass differential, but the shape and maximum amplitude of the anomalies will be quite different. A compact source will produce an anomaly of higher amplitude, but narrower width than a horizontally distributed source. The result is that a compact source is more readily discerned in a gravity field as long as the observations are dense enough to map it, and the anomaly is more easily extracted from low-gradient regional gravity anomalies.

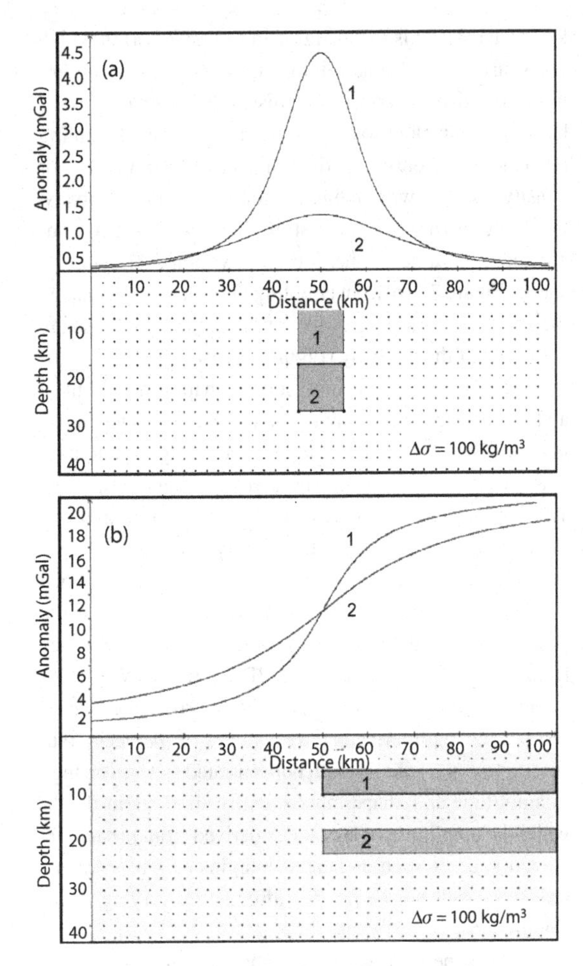

FIGURE 7.6 Gravity anomalies of two different anomalous source shapes at two different depths.

The isolation of a source from other anomalous sources is also a key factor in determining whether the gravity method is a suitable approach to solving a particular problem. This results from the potential overlap of anomalies and the resulting distortion of the individual anomalies. The extreme of this distortion is the superposition of effects to the point where the existence of multiple source bodies is not discernible.

The resolution, which is the minimum horizontal distance between equivalent sources that still permits recognition of the individual sources, is a function of the sharpness of the anomaly, controlled by the shape and depth of the source. A general rule is that identical individual sources located at the same depth must have a horizontal separation of at least twice the depth to the sources. However, in optimal cases, resolution for concentrated sources can be achieved at separation approximately equal to the depth.

This key variable is a major concern in many engineering and environmental as well as mineral exploration studies where the resolution of multiple sources is of particular concern. An example of the effect of overlapping anomalies due to equivalent sources is illustrated in Figures 7.7(a) and (b). The concerns with isolation are not just limited to the superposition of the effects of sources at the same depth, but also include multiple sources at different depths. This is especially problematic where the sources are not separated by a vertical distance that exceeds several times the depth to the shallowest source. As illustrated in Figure 7.7(c), the individual sources lying above one another cannot be distinguished until their vertical separation is several times the depth to the upper source.

The vertical resolution problem becomes more critical as the geometric shape of the source causes their anomalies to broaden out. Figure 1.2, for example, illustrated the problems in identifying anomalous sources by the gravity method caused by vertical superposition of anomalies due to a faulted, high-density layer within a lower-density medium. In Figure 1.2(a), the individual anomalies of the two high-density units are shown together with their summed anomaly where the fault is vertical. The upper high-density A-layer causes a positive anomaly to the left and the lower B-layer causes a positive anomaly to the right. These effects destructively interfere, but because the sources are at different depths, the summation $(A + B)$-anomaly shows a positive to the left and a negative to the right, and the typical fault anomaly is no longer present. Further complications in the summed effects occur where the fault is dipping as shown in Figures 1.2(b) and 1.2(c), with the inflection point of the summation anomaly no longer at the midpoint of the fault.

The effect of various geologic variables on gravity anomalies is illustrated in Figures 7.8 to 7.11. Gravity anomaly profiles in Figures 7.8 and 7.9 illustrate a long subsurface void with an approximately 12 m by 12 m cross-section and a density contrast of 2,700 kg/m^3. The gravity anomalies decrease rapidly with increasing depth to the top of the void from roughly 1.5 to 30 m (Figure 7.8). However, the amplitudes of the gravity anomalies do not change significantly as the strike length of the void whose upper surface is at 3 m changes from 6 m to infinity on either side of the cross-section shown in the figure.

The anomalies of Figure 7.8 support the use of 2D methods of gravity interpretation. The gravity anomalies from equivalent mass differences used in Figures 7.8 and 7.9 are shown in Figures 7.10 and 7.11, except the void shape has been changed by halving its vertical extent and doubling its width. The amplitudes of the resulting

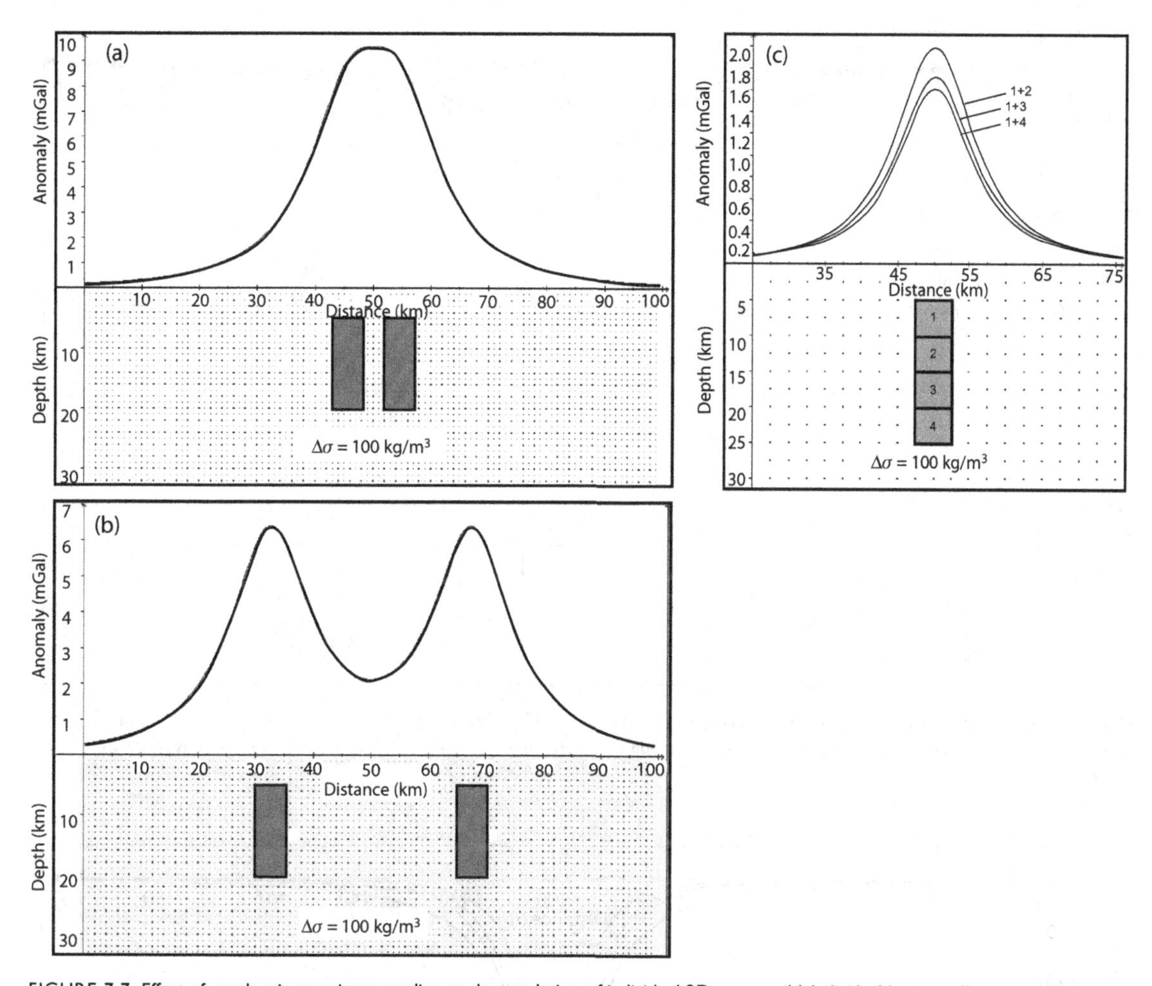

FIGURE 7.7 Effect of overlapping gravity anomalies on the resolution of individual 2D sources. (a) Individual horizontally separated sources are too close together to be recognized. (b) Individual sources of (a) are separated sufficiently to recognize the existence of two anomalies. (c) Summation effects of anomalies caused by prismatic sources centered on the same vertical line. Note the difficulty in isolating the individual sources owing to lack of vertical resolution.

anomalies have not been modified greatly, but the widths of the anomalies have increased in response to the greater widths of the sources.

7.3.2 Key geophysical variables

Key geophysical variables address the inverse problem of estimating the geological parameters. Here, the problem is the interpretation of the nature of the subsurface from the gravity anomalies. Key geophysical variables used in interpretation are the amplitude, sharpness, shape, and perceptibility of the gravity anomaly, as well as its correlation with other geophysical and geological features.

Amplitude is the most important of the geophysical variables. It is the most obvious thing observed on a gravity anomaly map or profile, and portions of the survey area are defined in terms of their relative amplitude. The mass differential, a product of the volume and density contrast of the source, directly controls the amplitude of the anomaly as explained in Section 3.4.2. In fact the mass differential can be determined uniquely by Gauss' law (Equation 3.101), which relates this quantity to the numerical integration of the energy of the anomaly.

As pointed out above, the anomaly amplitude depends on the depth to the source, but the sensitivity to depth changes with the shape of the causative mass. The amplitude of an anomaly from a concentrated mass such as an ore body or a solution cavity in a limestone formation will be most sensitive to depth, where the inverse distance function, $1/r^N$, varies with $N = 2$. The anomaly from a long, concentrated source such as a buried bedrock ridge or a tubular solution cavity in limestone, on the other hand, will

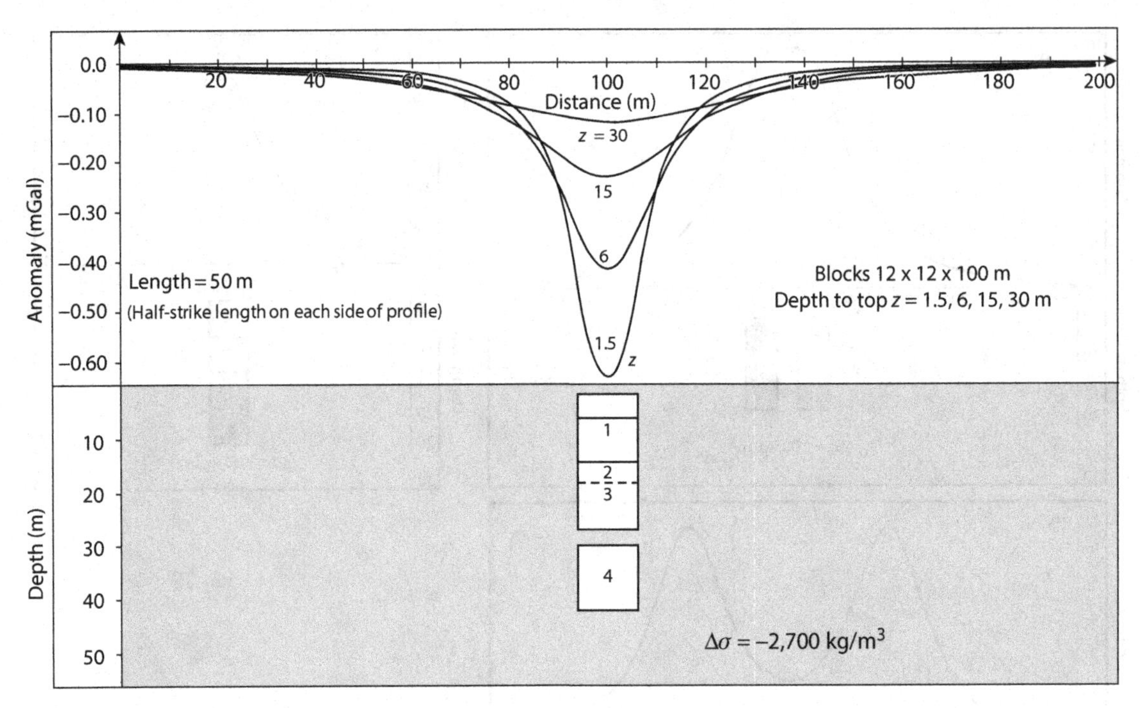

FIGURE 7.8 Gravity anomaly profiles across a rectangular void $12 \times 12 \times 100$ cubic meters filled with air in a rock unit of density $2,700\,\mathrm{kg/m^3}$. The length of the void perpendicular to either side of the profile is 50 m. The depth to the top of the void is varied from 1.5 to 30 m.

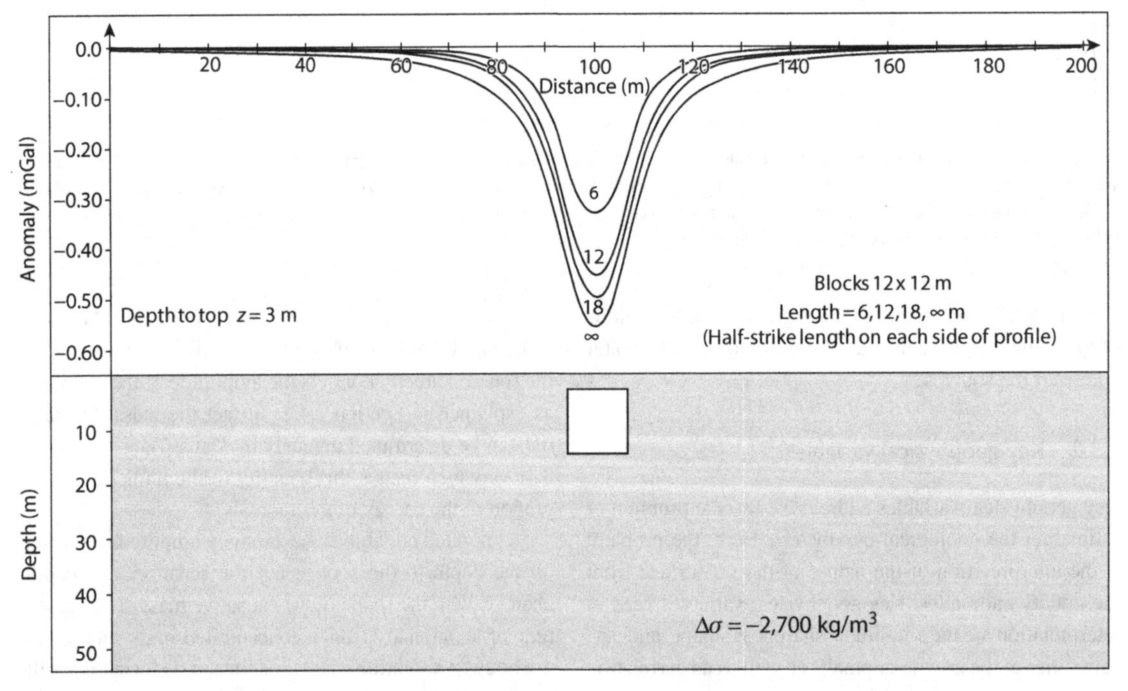

FIGURE 7.9 Gravity anomaly profiles across a rectangular void 12×12 square meters in cross-section filled with air in a rock unit of density $2,700\,\mathrm{kg/m^3}$. The depth to the top of the void is 3 m and the length of the void perpendicular to either side of the cross-section is varied from 6 m to infinity.

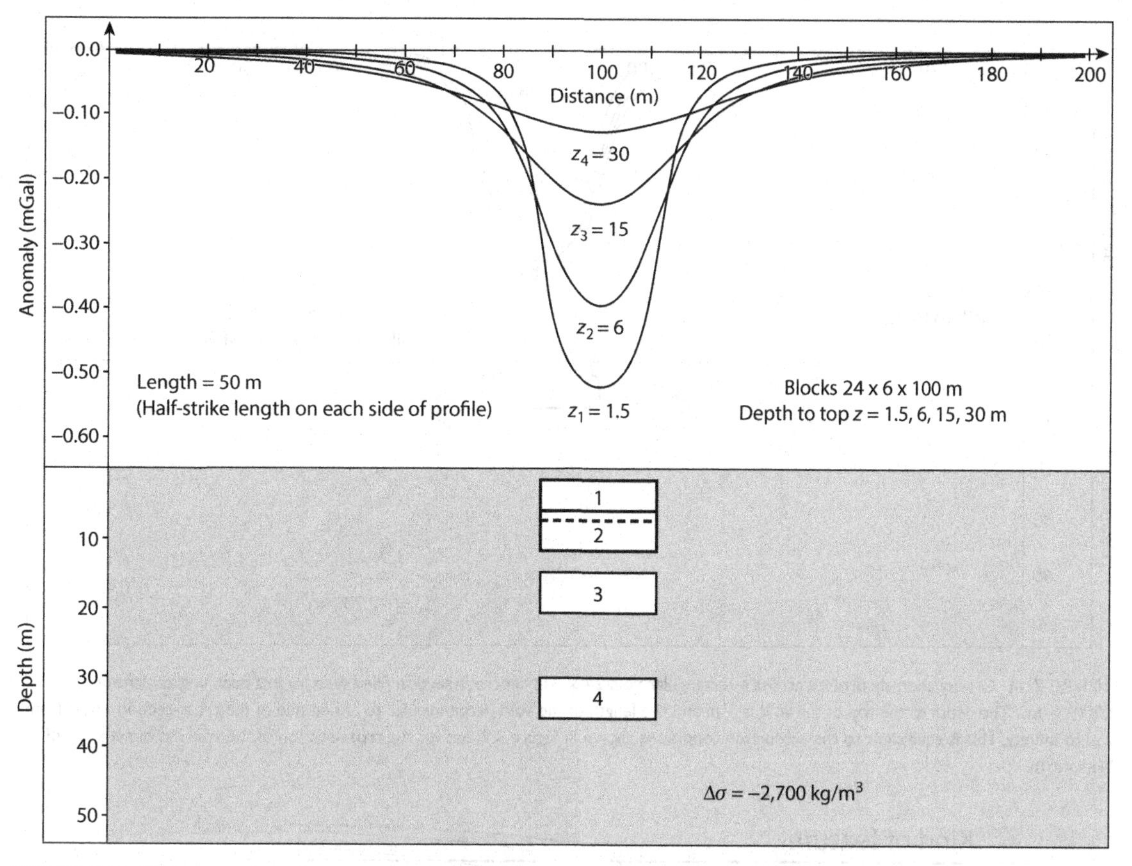

FIGURE 7.10 Gravity anomaly profiles across a rectangular void $24 \times 6 \times 100$ cubic meters filled with air in a rock unit of density $2{,}700\,\text{kg/m}^3$. The length of the void perpendicular to either side of the cross-section is 50 m. The depth to the top of the void is varied from 1.5 to 30 m. This is equivalent to the subsurface conditions shown in Figure 7.8, except the cross-section of the void has dimensions of $24\,\text{m} \times 6\,\text{m}$.

vary with depth according to $1/r$ where $N = 1$. In addition, the amplitude of the anomaly derived from a long, wide mass such as a buried bedrock plateau or a horizontal variation in lithology within a series of flat sedimentary formations varies as a function of $1/r^0$, where $N = 0$ results in no change in amplitude with depth. This is identical to the gravity effect of the slab described above. The effect of depth on the amplitude of anomalies derived from these three geometrically shaped sources is illustrated in Figure 7.12.

The sensitivity of the amplitude to depth to the source and the varying degree of sensitivity has an important effect upon the application and interpretation of gravity anomalies. For example, the sharp drop-off in amplitude of concentrated sources with depth results in increasing difficulty in recognizing anomalies as these sources increase in depth. Also, the lack of a depth effect on the amplitude of anomalies from long, wide sources makes it possible to estimate the product of the thickness and density differential of a source of this nature from the anomaly amplitude without concern for the depth to the source.

The next most important geophysical variable is the sharpness of the anomaly. Sharpness is defined by the spatial rate of change in the anomaly. Sharp anomalies are easily distinguished in either map or profile, whereas broad anomalies have low gradients making them difficult to observe where other anomalies are present. Sharpness can be observed on contour maps by the distance between the contours or, if the map is colored with a linear scale, by rapid color changes. High-pass filtering of gravity anomalies, such as second derivative filtering, emphasizes the sharpness of anomalies.

The sharpness of an anomaly is a function of the depth and the geometric configuration of the source, decreasing one power faster than the amplitude – i.e. the inverse distance function for sharpness is $1/r^{N+1}$. This has significant implications for the interpretation of gravity anomalies. For example, although the amplitude of the anomaly of a

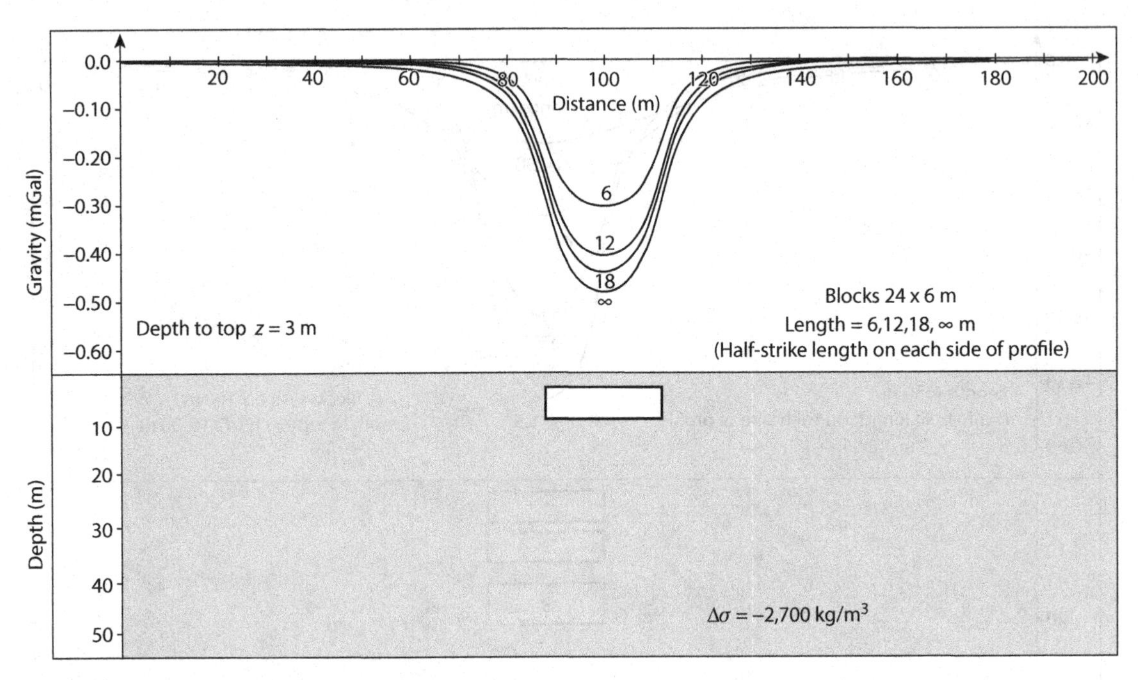

FIGURE 7.11 Gravity anomaly profiles across a rectangular void $24 \times 6\,m^2$ in cross-section filled with air in a rock unit of density $2,700\,kg/m^3$. The depth to the top of the void is 3 m and the length of the void perpendicular to either side of the cross-section varies from 6 m to infinity. This is equivalent to the subsurface conditions shown in Figure 7.9, except the cross-section of the void has dimensions of $24\,m \times 6\,m$.

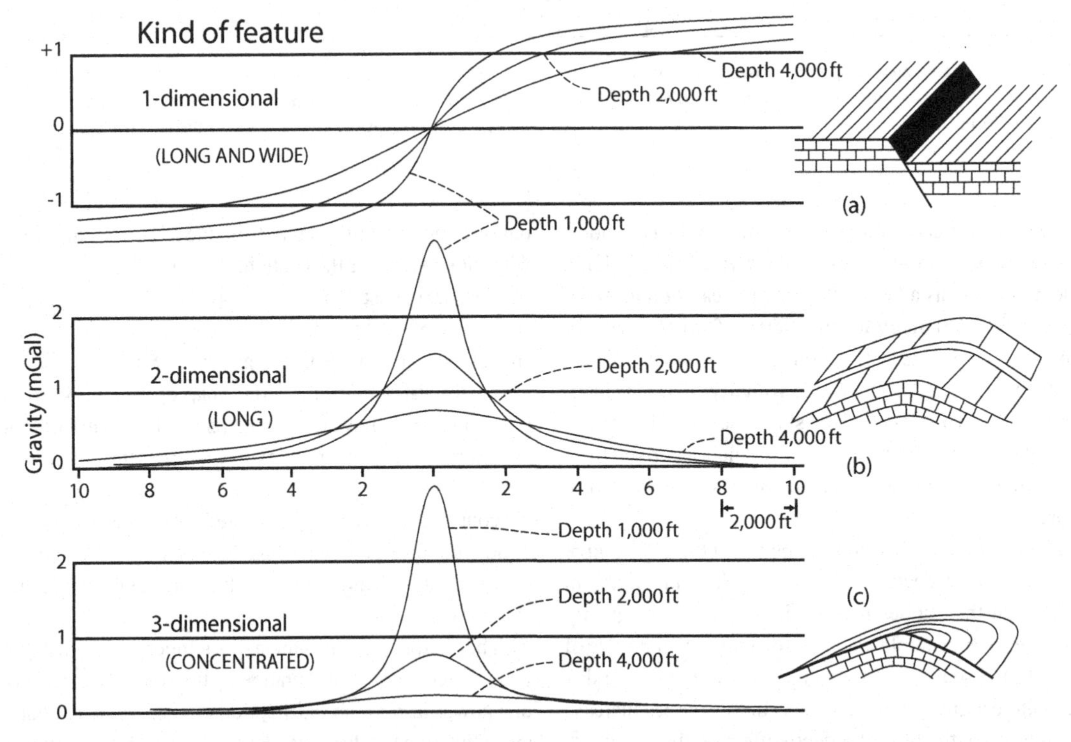

FIGURE 7.12 Gravity anomalies (left) derived from depths of 1,000, 2,000, and 4,000 feet for (a) concentrated spherical, (b) long horizontal line, and (c) long, wide tabular sources illustrated on the right. Adapted from ROMBERG (1958).

long, wide source does not vary with depth, the marginal gradients of the anomalies change with depth as illustrated in Figure 7.12(c). Unlike amplitude, sharpness does not depend on the mass differential of the source. Accordingly, it is much more useful in determining the depth of an anomaly source. A variety of methodologies have been developed for relating anomaly depth to its sharpness, such as the half-width depth rules discussed in the next section.

The shape of a gravity anomaly, as determined by its areal distribution on maps or continuity between profiles, is an important geophysical variable. Elongation of a gravity anomaly is indisputable evidence of the strike length of the source, but this is only the case where the length of the source is of the order of twice the depth to the source. Flexures in the elongation indicate a change in strike direction as a result of, for example, folding or an offset of the source due to faulting.

Unfortunately, the inverse is not true. The ellipticity of an anomaly is only clearly visible where the depth-to-length ratio of the source is relatively small, approximately less than one. The broadening of a gravity anomaly with increasing distance from the source attenuates the shape of the source. In the same way that increasing snow depth over a surface structure will mask its shape, the shape of a subsurface feature will be smoothed out in the gravity anomaly observed at an increasing distance from the source.

The tendency of gravity anomalies to hide the elongation or eccentricity of the plan view shape of the source with increasing source-depth is a serious limitation to gravity interpretation. The decreasing resolution of individual sources with increasing depth also is caused by this same effect that broadens anomalies relative to the sizes of the anomalous sources as source depths are increased. As an approximation, elongation in the source is seldom recognized in gravity anomalies unless the source length is at least twice the depth to the source.

The shape of a gravity anomaly is also affected by the vertical cross-sectional shape of the source. As the configuration of the source changes with depth or is displaced from the upper surface by the dip of the source, the shape of the anomaly will be modified, the marginal gradients of the anomalies will differ, and the anomaly peaks will be displaced.

Figure 7.13 illustrates the effect of width dip, vertical extent, and strike length of an anomaly source on a gravity profile observed perpendicular to the source. The effect of dip is to displace the axis of the anomaly and to cause a shallower anomaly gradient in the down-dip direction of the source. The asymmetry in the gradients decreases as the depth extent of the source decreases and its dip increases. The modest effect on the anomaly of increasing strike length is evident in the profiles. If the depth extent of the dipping sources decreases, then the maxima of the anomalies move toward the center of the axis of the top surface of the body, the symmetry of the anomaly increases, and the percentage effect of increasing strike length decreases.

The importance of subtle variations in gradient in the interpretation of gravity anomalies shown in Figure 7.13 is emphasized in the anomaly contour map of Figure 7.14. This figure shows the contoured anomaly values from a prism which is three depth units square in plan view, but dips at 30° with a vertical extent of two depth units. The upper surface of the prism (thick solid line) and the lower surface (thick dashed line) are included in the figure.

Another important geophysical variable is anomaly perceptibility. Perceptibility concerns the ease by which the anomaly can be separated from other anomalies. It is a combined effect that depends on the amplitude, sharpness, and strike of the anomaly and the nature of the superimposed gravity anomaly noise and regional anomalies. As the characteristics of the noise or regionals and the anomalies of interest approach each other, perceptibility decreases. The role of the isolation and enhancement processing of the gravity data is to increase the perceptibility by either enhancing the anomalies of interest at the expense of the gravity noise and regional gravity anomalies or isolating them from the noise and regional effects.

Perceptibility is a subjectively determined variable which may well vary across the extent of the gravity survey. Its influence is illustrated in Figure 7.15 which shows the change in perceptibility of the same anomaly in a noise-free segment of the profile and a noisy segment subject to a strong-gradient regional anomaly. The anomaly can barely be recognized in the noisy portion of the profile where there is a regional.

Perceptibility is also enhanced by the spatial correlation of the gravity anomaly with other constraints on the subsurface source, such as may be inferred from topography, photographic and other remote sensing imagery, geology, and magnetic, thermal, electrical and electromagnetic, and other geophysical anomalies. The latter geophysical anomalies just listed are all based on $(1/r)$-potentials, and thus can be transformed into pseudogravity anomalies (Section 3.6.2) for comparison with the observed gravity anomaly to help minimize interpretation ambiguities. The correlation, of course, cannot guarantee a common source for the responsible physical property variations because of the effects of anomaly superposition and the source ambiguity which is inherent to the analysis of potential

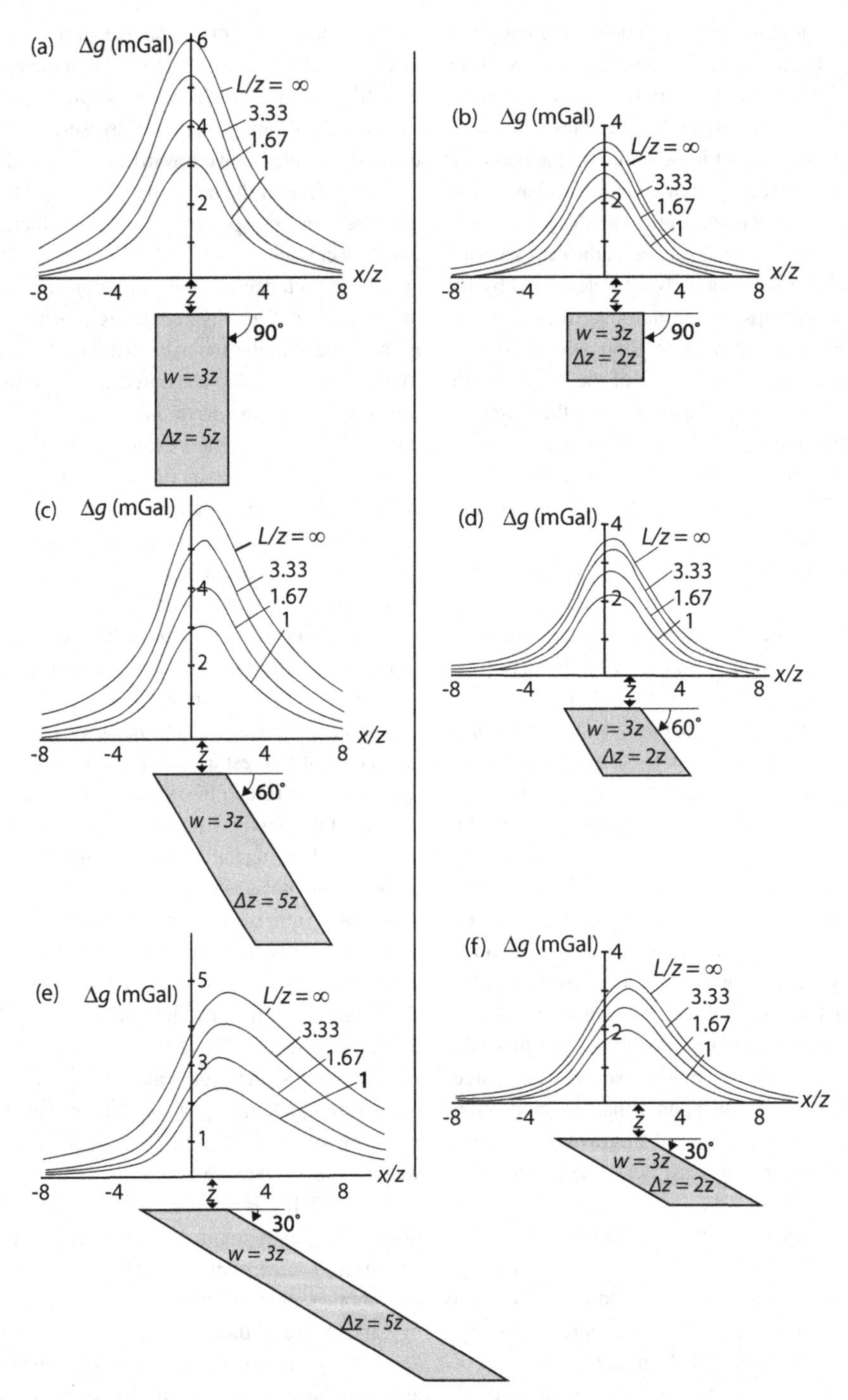

FIGURE 7.13 Gravity anomaly profiles over dipping anomalous prisms showing the effect of strike length L relative to the width (w), dip, and depth extent (Δz) of the prisms on the shape and amplitude of the anomalies. The width of the prisms is $3z$ and their vertical extent, Δz, is $5z$ in (a), (c), and (e), and $2z$ in (b), (d), and (f). The density contrast is $1{,}000\,\text{kg/m}^3$. Adapted from HJELT (1974).

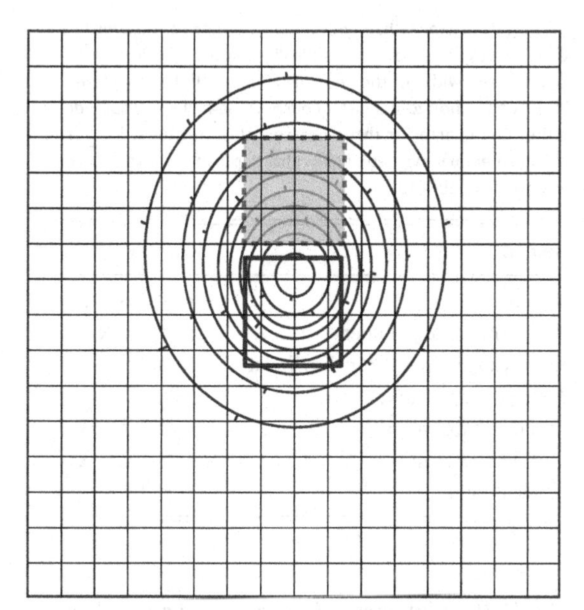

FIGURE 7.14 Gravity anomaly contour map of a prism dipping at an angle of 30° towards the top of the figure. The prism's upper surface (thick solid line) is at a depth of three grid units and separated from its lower gray-shaded surface (thick dashed line) by one depth unit. The density contrast of the prism is 1,000 kg/m^3 and the contour interval is 0.2 mGal. Adapted from Hjelt (1974).

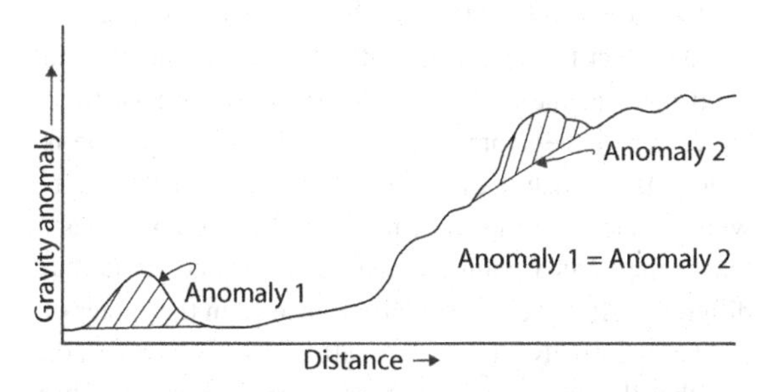

FIGURE 7.15 Gravity profile which has the same residual anomaly in both the left and right portions. Note the difficulty in recognizing this anomaly in the right-hand portion which contains strong gravity noise and a large regional gradient.

fields. However, the correlations, either direct or inverse, or even the lack of correlation can be useful in limiting the possible geological interpretation of specific anomalies.

7.4 Simplified interpretation techniques

After residual anomalies have been identified and located, the interpretational process commonly involves the use of simplified inversion techniques to ascertain important characteristics of the source of the anomalies (Figure 7.4). These techniques are approximate or rule-of-thumb methods of analysis that in some cases are empirical in origin. But most rule-of-thumb methods are based on simplification of theory by making appropriate assumptions regarding the geometry of the source.

Their advantage is that they can be used rapidly and simply to reach approximate answers regarding the nature of the sources of the anomalies. As such, they are often used in preliminary, back of the envelope analysis of data as a guide to developing conceptual models for more comprehensive modeling. However, in some studies they meet the objectives of the analysis, and thus are the termination point in the interpretational process. The successful application of these techniques by the interpreter requires a thorough understanding of their basis, theoretical or empirical, and the assumptions used in their development.

7.4.1 Depth

Source depth is one of the critical characteristics that is commonly approximated by simplified interpretational techniques. It is determined by one of several measures of the sharpness of the anomaly. The techniques generally give the maximum depth either to the top or the center of the source, depending on the source being invoked. Their accuracy is less than that of magnetic depths because of the greater sensitivity of magnetics to depth to the source and the greater overlap of the anomaly spectra of gravity sources.

Graphical techniques

Depth determination generally starts with identification of the approximate source geometry from the anomaly's spatial pattern and gradients, and geological information derived from other geophysical data or collateral geological data. For example, a roughly equidimensional anomaly is likely to be derived from a concentrated source which can be approximated by the anomaly of a sphere or from a source which approximates a vertical cylinder. The vertical cylinder source will result in an anomaly with lower gradients than a sphere. A sphere might geologically represent an isolated ore body, whereas a vertical cylinder might be the geological equivalent of an intrusive pipe. The choice between the two different sources may be suggested by ancillary information or comparison of the observed anomaly profile with the theoretical anomalies derived from the two sources. If a definitive choice is not possible, the depth determinations based on the two anomaly source geometries will provide a depth range for the source.

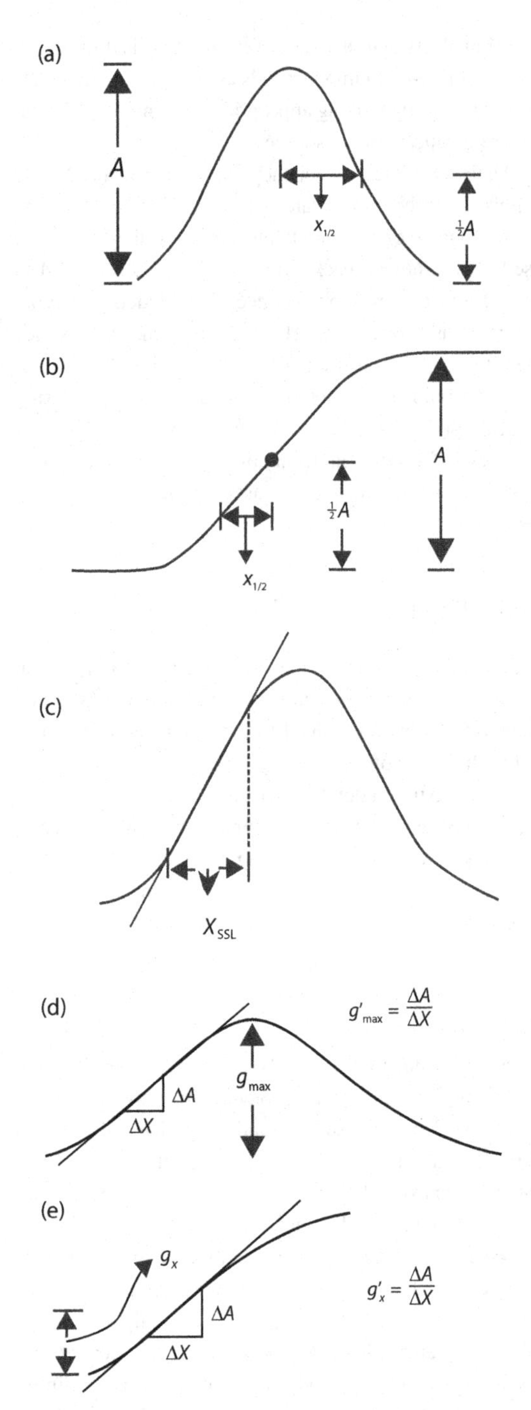

FIGURE 7.16 Measurements used in estimating source depth from gravity anomalies. (a) Half-width method based on a spherical source. (b) Half-width method applied to an anomaly derived from a fault. (c) Straight-slope method. (d) Smith method based on the entire anomaly. (e) Smith method based on a partial anomaly. In these estimates, A is the maximum amplitude of the anomaly, X is horizontal distance, and $\Delta A/\Delta X$ is the change in amplitude (ΔA) over the horizontal distance change (ΔX). See text for further descriptions.

TABLE 7.1 Gravity half-width depths either to the tops (z_t) or the centers (z_c) of various idealized sources in terms of one-half the width of the anomaly at one-half the amplitude, $X_{1/2}$, or the horizontal distance from one-half the amplitude to either one-quarter or three-quarters the amplitude, $X^*_{1/2}$. The complementary magnetic half-width depths for these sources are given in Table 13.1.

Source	Gravity depth
Sphere	$z_c \leq 1.3 \times X_{1/2}$
Thin horizontal cylinder	$z_c \leq 1.0 \times X_{1/2}$
Deeply extending vertical cylinder	$z_t \leq 0.58 \times X_{1/2}$
Narrow vertical dike	
if depth extent $\approx Z_t$	$z_t \approx 0.7 \times X_{1/2}$
if depth extent $\gg Z_t$	$z_t \approx 1.0 \times X_{1/2}$
Vertical fault	$z_c \leq 1.0 \times X^*_{1/2}$

(A) Half-width method The half-width method of depth determination (Nettleton, 1940, 1942) has been widely used in gravity interpretation. It is based on equating the gravity effect of an idealized geometric source to half of its amplitude and solving for the source's depth in terms of the horizontal distance between the anomaly peak and one-half of the peak amplitude.

The method is based on simplification of the theoretical gravity effect from an idealized geometry assumed in the application of the method. The horizontal distance from the center of the anomaly to one-half of its amplitude is called the anomaly half-width (Figure 7.16(a)). The half-width distance for the vertical edge of the horizontal slab, that is the vertical fault anomaly, is measured somewhat differently. It is the horizontal distance from the center of the fault anomaly that is one-half of its total amplitude to either the one-quarter or three-quarters anomaly value (Figure 7.16(b)). For various idealized sources with simple geometric forms, Table 7.1 lists the relationships between the anomaly half-width distance, $X_{1/2}$, and the depth either to the top, z_t, or the center, z_c, of the source.

(B) Straight-slope method Anomaly source depths also may be estimated from the horizontal distance over which the maximum gradient of the anomaly remains essentially constant. The method is sometimes easier to perform on the computed vertical derivative of the gravity anomaly, rather than the gravity anomaly itself where the flat spots at the peaks and troughs are measured. This is the so-called straight-slope distance method because it is based on the distance over which the slope at the inflection point of the anomaly profile remains straight or the gradient is constant. This method is not based on theoretical formulations, but rather on empirical evidence from case histories or anomalies calculated from idealized sources. DAMPNEY (1977) found that the relationship $z_c \approx 2 \times X_{SSL}$ is useful in determining the depth z_c to the center line of a vertical fault from the straight slope length, X_{SSL} (Figure 7.16(c)). RAM BABU *et al.* (1987) found a similar relationship and developed other straight-slope length relations for additional geometric forms including $z_c \approx 2 \times X_{SSL}$ for spheres and horizontal cylinders, and $z_c \approx 1.22 \times X_{SSL}$ for thin horizontal plates.

(C) Smith rules SMITH and BOTT (1958) and SMITH (1959, 1960) developed several depth determination rules based on horizontal derivatives of gravity. These rules, commonly referred to as the Smith Rules, are independent of source geometry, and thus are potentially useful where the geometry of the source is unknown or cannot be approximated with a simple shape. Where the entire anomaly is isolated, the approximate depth, z_t, to the top of the source is

$$z_t \leq K \times (g_{max}/g'_{max}), \tag{7.1}$$

where $K = 0.65$ for the 2D source and 0.86 for the 3D source, g_{max} is the peak or maximum anomaly value, and g'_{max} is the absolute maximum horizontal derivative (i.e. slope) of the gravity anomaly in gravity anomaly unit per depth unit (Figure 7.16(d)). Where only a portion of the anomaly is mapped and isolated (Figure 7.16(e)), the depth approximation is

$$z_t \leq K' \times (g_x/g'_x), \tag{7.2}$$

where $K' = 1.0$ for the 2D source and 1.5 for the 3D source, and g_x is the gravity anomaly value where the absolute horizontal derivative (slope) is g'_x.

If the absolute maximum density contrast $\Delta\sigma_{max}$ occurring within the source can be specified as well as the absolute maximum second horizontal derivative of the gravity anomaly g''_{max}, then the depth to the top of the source is

$$z_t \leq 5.4 \times G(\Delta\sigma_{max}/g''_{max}), \tag{7.3}$$

where G is the gravitational constant and all the variables must be consistent units. If the depth to the source can be estimated from other sources, Equation 7.3 can be inverted to determine the maximum density contrast. The results can also be improved if the density contrast within the source is positive throughout. In this case, $\Delta\sigma_{max}$ is replaced by $\Delta\sigma_{max}/2$.

Semi-automated approaches

Several techniques have been developed that evaluate possible source depths from the anomaly data localized in data patches or windows which are maneuvered across the data set. These depth estimates can vary widely as the size of the window is changed, and thus the interpreter's experience, with the method and knowledge of the geologic framework of the application are critical for selecting the optimum depths from these multiple results.

All depth determination techniques are approximations and are potentially seriously erroneous depending upon the quality of the measurements, the isolation of the anomaly, and the degree to which the simplifying assumptions are met by the actual geologic conditions. Furthermore, most methods of depth determination provide only a maximum depth for the source.

(A) Euler deconvolution To help discriminate anomaly sources, the decay rate N, which is also called the structural index (THOMPSON, 1982), can be investigated using Euler's homogeneity equation

$$\Delta x \frac{\partial g(x,y,z)}{\partial x} + \Delta y \frac{\partial g(x,y,z)}{\partial y} + \Delta z \frac{\partial g(x,y,z)}{\partial z} = -N \times g(x,y,z), \tag{7.4}$$

where $\Delta x = (x - x')$, $\Delta y = (y - y')$, and $\Delta z = (z - z')$. Euler's equation can be solved in either 2- or 3D form for the respective (x', z')- or (x', y', z')-position of the source provided that the value of the attenuation rate or structural index N is known or assumed and that the gravity values and derivatives are measured or can be calculated accurately from the observed gravity values. Solution is readily achieved through matrix inversion techniques applied to a series of anomaly values within a selected window of the data.

This methodology was originally applied to magnetic interpretation, and thus is more fully described in the magnetics interpretation (Section 13.4.1). However, Euler deconvolution also has been effectively adapted for gravity interpretation, generally using gravity values occurring within overlapping windows (e.g. MARSON and KLINGELÉ, 1993; STAVREV, 1997; ROY *et al.*, 2000; REID *et al.*, 2003). The method is particularly useful where the anomalies are derived from concentrated, idealized sources, and is difficult to interpret for more complex shapes where the attenuation rate varies with distance from the source (e.g. RAVAT, 1994). Table 7.2 summarizes the structural indices N for the gravity effects of some simple mass models.

TABLE 7.2 Structural indices N for the gravity anomaly (GA), first derivative (FD), and second derivative (SD) gravity anomalies of some simple mass models.

Model	GA	FD	SD
Sphere (point mass)	2	3	4
Horizontal cylinder (horizontal line mass)	1	2	3
Vertical pipe (vertical line mass)	1	2	3
Faulted thin-bed (small-throw double-sided fault)	1	2	3
One-sided fault (small-throw semi-infinite horizontal sheet)	0	1	2
Contact/edge	−1?	0	1

Euler deconvolution can be used for gravity as well as gravity gradient interpretation. Figures 7.17, 7.18, and 7.19 show examples of the Euler interpretation of the vertical gravity, and first and second vertical derivative anomalies due to a 2D horizontal cylinder (line of mass). In these figures, at least five structural indices, $N = 0.5, 1.0, 1.5, 2.0$, and 3.0 are used and the depth solutions are plotted with plot symbols of 1, 2, 3, 4, and 5, respectively. A tolerance TOL is used in each case to minimize the scatter in the number of solutions. In addition, two or three passes are used with increasing window sizes. These figures show that when the structural index N used for the particular model and input field is correct, the solutions are accurate and well-clustered.

(B) Werner deconvolution This method was also originally developed for magnetic interpretation, and thus is more fully described in the magnetics interpretation Section 13.4.1. However, Werner deconvolution has also been used for gravity interpretation although rarely (e.g. Kilty, 1983). The method is particularly useful when the profile anomaly of interest can be expressed as a rational function of the form

$$\frac{A\Delta x + B\Delta z}{\Delta x^2 + \Delta z^2}, \tag{7.5}$$

which is similar to the magnetic expression of the field due to a thin sheet or the derivative of a magnetic interface. As such, the method can be extended to gravity models such as a horizontal cylinder whose vertical gravity component is given by

$$\frac{-G(\Delta M)\Delta z}{\Delta x^2 + \Delta z^2}, \tag{7.6}$$

where ΔM is excess mass of the cylinder per unit length and G is the gravitational constant.

Another gravity expression of the proper form for Werner deconvolution is the second horizontal derivative of gravity over a horizontal fault of small throw T and density contrast $\Delta\sigma$ (Kilty, 1983), which is given by

$$\frac{4G(\Delta\sigma)T\Delta x\Delta z}{\Delta x^2 + \Delta z^2}. \tag{7.7}$$

An example of the application of Werner deconvolution to gravity data from Kilty (1983) is shown in Figure 7.20.

(C) Statistical spectral techniques Depth estimation of gravity sources can be obtained from the slope of the logarithmic power spectrum of the gravity profile data or from the slope of the logarithmic radially averaged power spectrum of the gravity gridded data as explained in the magnetic interpretation Section 13.4 (Saad, 1977b). However, because the vertical gradient of gravity is equivalent to a pole-reduced pseudomagnetic field, a simple correction to the gravity power spectrum is required prior to making the slope-depth estimation. The correction simply consists of adding $[2\ln(k)]$ to ln (PS) where $k = 2\pi f$ is the wavenumber and PS the power spectrum (i.e. PS is multiplied by k^2). The correction has the effect of reducing the gravity data in the space domain to gravity gradient or, equivalently, to pseudomagnetic data. The spectral slope method can be applied directly to observed gravity gradient data or transformed to pseudomagnetic data. Figure 7.21 shows an example of spectral analysis of gravity profile data from a 2D prismatic model. The power spectrum (right panel) is computed by the maximum entropy method (MESA) using a prediction error operatar of length LPEO $= 30$, in contrast to LOPT $= 17$ which is the optimum LPEO according to the final prediction criterion for the prediction error filter (Saad, 1978). The sampling interval used is $\Delta x = 1$ kft yielding a Nyquist wavenumber of $k_n = 2\pi f_n = \pi$ rad/kft; other information and model parameters are given in Figure 7.21.

To obtain accurate depth information from the spectral slope of gravity, it must be corrected by adding $[2\ln(k)]$ to ln (PS). The original and corrected MESA spectra are compared in the right panel of Figure 7.21. The depth $z = 12$ obtained from the slope of the original spectrum

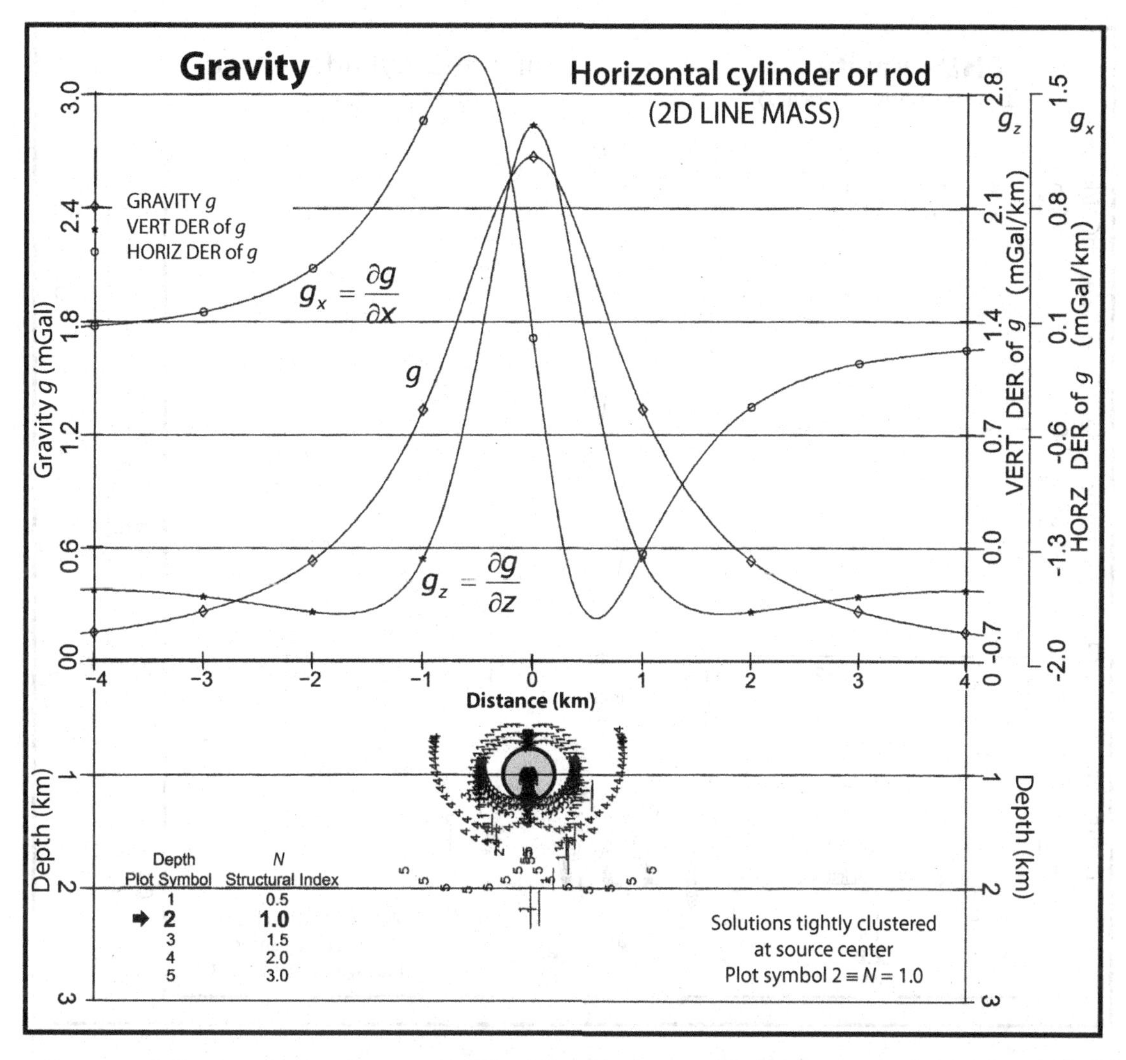

FIGURE 7.17 Euler depth estimation of the gravity anomaly due to a line mass or 2D horizontal cylinder with structural index (SI) of $N = 1$. Solutions are plotted for $N = 0.5$, 1.0, 1.5, 2.0, and 3.0 using plot symbols 1, 2, 3, 4, and 5, respectively. In this example, three passes over the profile data are used with increasing window sizes, and a tolerance TOL is used to minimize solution scatter in each case. Source location is indicated by the circle on the depth plot. Notice that the solutions for the correct SI of $N = 1$ are exact and well-clustered, whereas other solutions are scattered both horizontally and vertically, being generally shallower for the smaller SIs and deeper for the larger SIs.

is highly overestimated with respect to the actual depth of 7.5, whereas the corrected spectrum gives the more accurate depth $z = 8.0$.

7.4.2 Depth extent

Several methods of determining the depth extent of a gravity anomaly source have been developed for specialized situations. The methods involving the use of families of curves (e.g. Grant and West, 1965) generally have limited application, and thus are not widely used. However, a simple approximation based on the gravity formula for a long, wide anomalous mass, such as a vertical fault, has broad application and is especially useful for sources which are areally extensive compared with their depth. This method is independent of depth as long as the mass is much broader than its depth. Based on the equation for the gravity effect of a vertical fault (or a horizontal slab) juxtaposing units of differing density, the thickness of material of density contrast 1,000 kg/m^3 required to produce a 1.0 mGal anomaly is 23.85 m. This result relates the vertical thickness or depth extent of material, $\Delta z'$, in meters

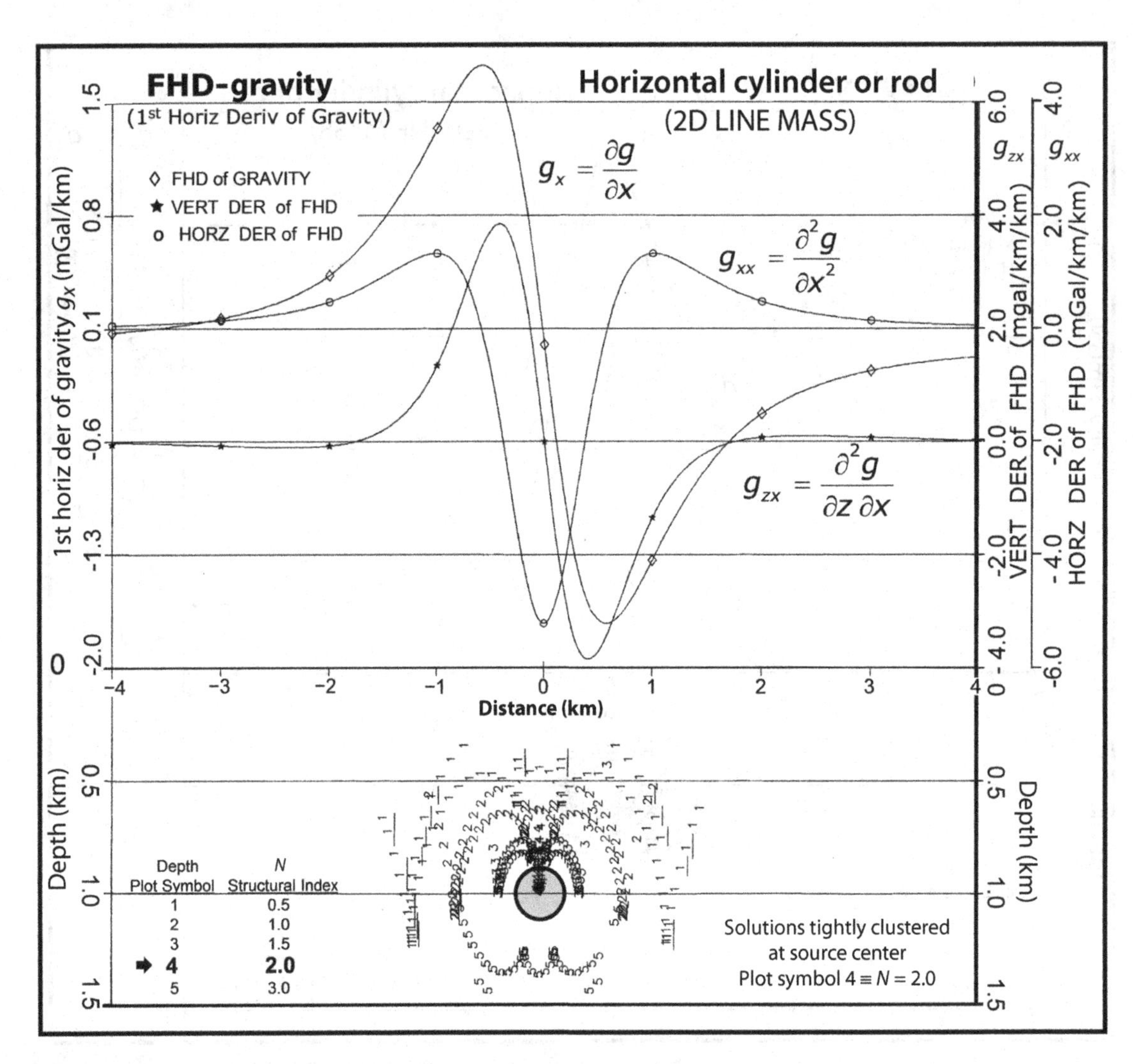

FIGURE 7.18 Euler depth estimation of the first horizontal derivative (FHD) of gravity anomaly due to a line mass or 2D horizontal cylinder with structural index (SI) of $N = 2$. Solutions are plotted for five SIs as in Figure 7.17. In this example, however, two passes are used with different window sizes, and a tolerance TOL is used to minimize solution scatter. Source location is indicated by the circle on the depth plot. Notice that the solutions for the correct SI of $N = 2$ are exact and well-clustered, whereas other solutions are scattered both horizontally and vertically, being generally shallower for the smaller SIs and deeper for the larger SIs.

to the anomaly amplitude, Δg, in mGal and the density contrast, $\Delta\sigma$, in kg/m^3 by

$$\frac{\Delta z'}{\Delta g} = \frac{23.85 \times 10^3}{\Delta\sigma}. \tag{7.8}$$

More generally, this equation permits estimating any one of the parameters when the other two are known.

This approximation technique is theoretically limited to long, wide sources, but it is in error by less than 5% where a 2D slab subtends an angle of greater than 110° on the principal profile. The results will be in error by less than 10% for the horizontal circular disk-shaped body whose diameter is five times or more the depth to the disk. Accordingly, this approximation has broad application as illustrated in Figure 7.22.

7.4.3 Ideal body depths

SKEELS (1963) developed characteristic curves that permit estimating the maximum depth of an anomaly source assuming a maximum density differential for the source approximated by either a vertical-sided prism or cylinder.

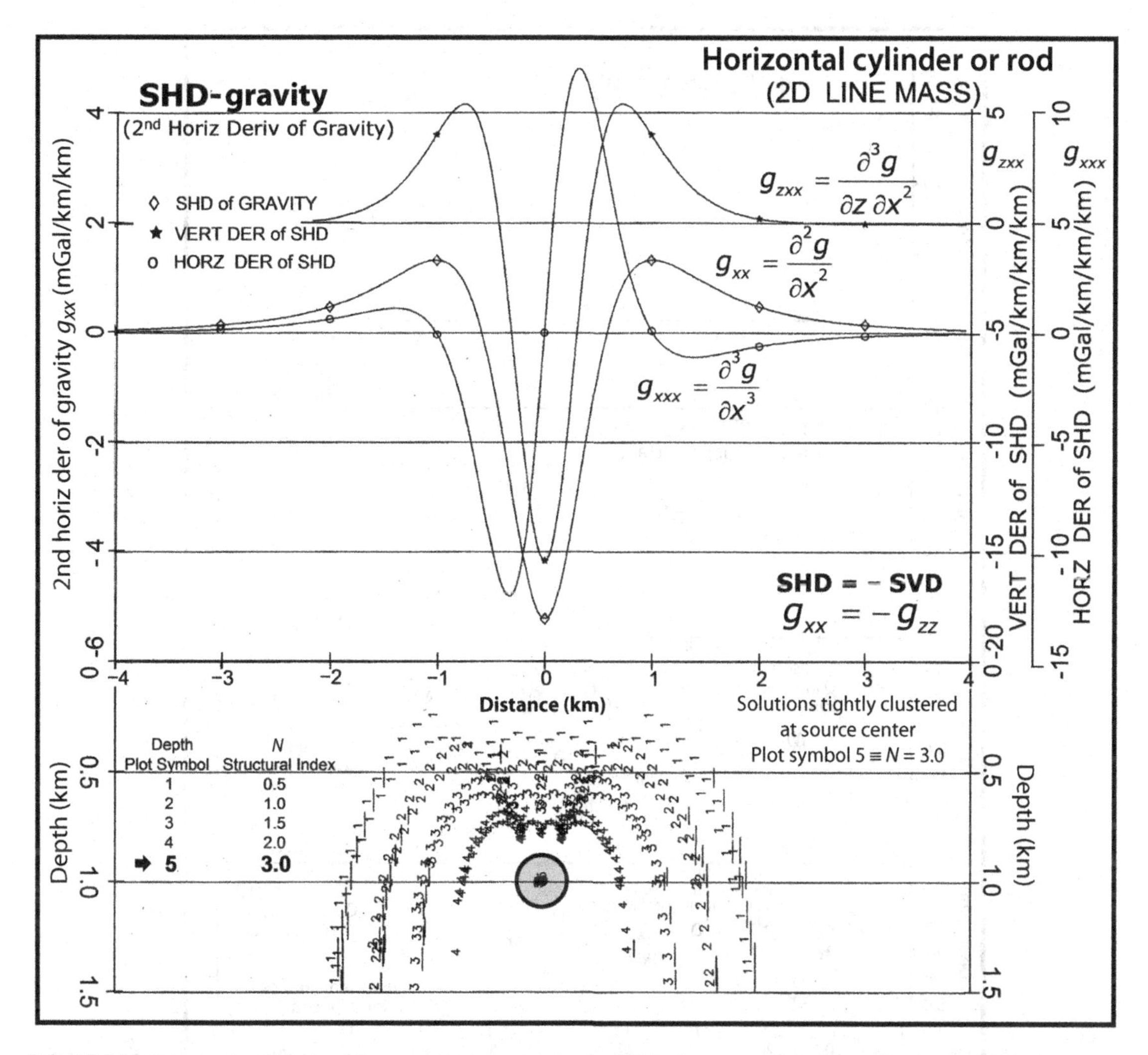

FIGURE 7.19 Euler depth estimation of the second horizontal derivative (SHD) of gravity anomaly due to a line mass or 2D horizontal cylinder with structural index (SI) of $N = 3$. Solutions are plotted for five SIs as in Figure 7.17. In this example, however, two passes are used with different window sizes, and a tolerance TOL is used to minimize solution scatter. Source location is indicated by the circle on the depth plot. Notice that the solutions for the correct SI of $N = 3$ are exact and well-clustered, whereas all other solutions for the smaller SIs are shallower and scattered both horizontally and vertically.

PARKER (1975) suggested a more general approach to the problem based on ideal body theory that is not restricted to specific source geometries. In this method, the ideal body has the smallest density contrast for a source that will provide a match with the residual anomaly. There are infinitely more sources of higher density that will satisfy the anomaly and none of lower density. The depth to the top of the ideal body source of the anomaly is the maximum depth for the estimated maximum density contrast. A plot of the minimum densities versus depth-to-top defines a field of permissible ideal models for the anomaly. The maximum possible depth of the source based on an estimation of the maximum density differential from collateral geological data is at the intersection of this density with the mathematical bound defined by the ideal body minimum densities as shown in Figure 7.23 from HILDENBRAND and RAVAT (1997). BLAKELY (1995) describes methods developed by PARKER (1975) and others for analytical solutions for ideal bodies.

ANDER and HUESTIS (1987) took Parker's approach a further step by focusing on the extremal bounds. These are end-member depth limits that describe whether certain solutions might exist. For instance, the question might be whether basement could be as shallow as a prescribed

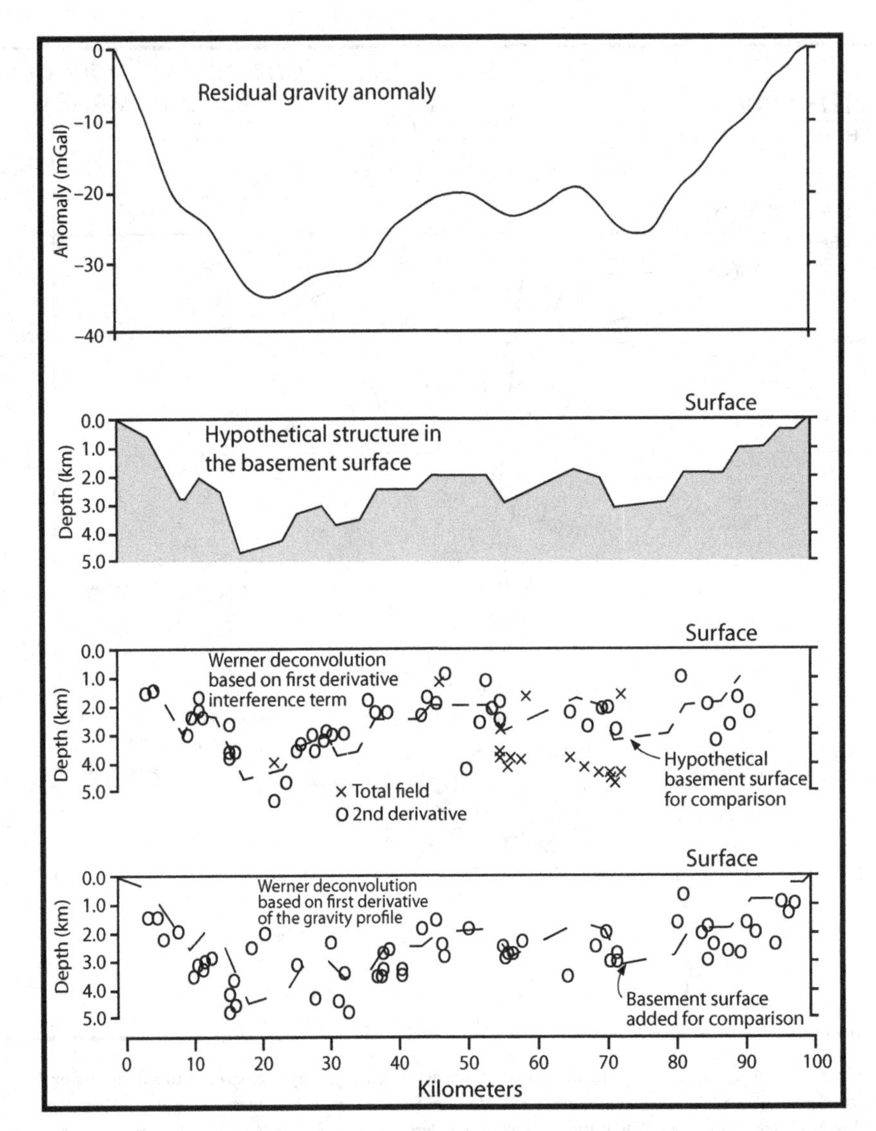

FIGURE 7.20 Werner deconvolution of the gravity effect of hypothetical basement structure at four times vertical exaggeration. Estimated depths have considerable scatter so that an exact depth estimate is not possible. However, the maximum basement depth may be estimated because the deepest estimates lie near or below the basement surface. Additionally, estimates cluster at specific locations (e.g. 12, 16, 38, 56, and 70 km) that can be correlated between profiles to facilitate studies of basement structural trends. Adapted from KILTY (1983).

depth given normal Earth materials. Extremal bounds would either conclusively accept or reject the possibility with no ambiguity, even though any acceptable solution or ideal body is not unique.

7.4.4 Formation boundaries

A critical step in interpretation and developing conceptual models for analysis is the specification of formation boundaries of anomalous subsurface masses from their gravity anomalies. The most widely used technique for accomplishing this is to relate the boundaries to the inflection point of the marginal gradients of the anomalies. Inflection points are mapped by either the zero second vertical derivative (VACQUIER *et al.*, 1951) or the equivalent maximum in the horizontal gradient (BLAKELY and SIMPSON, 1986). In map form, they are represented as the zero contour of the second vertical derivatives or the axis of the maximum of the horizontal gradient.

The correlation between the inflection point and the formation boundary assumes the boundary is vertical and the width of the anomaly source is large with respect to

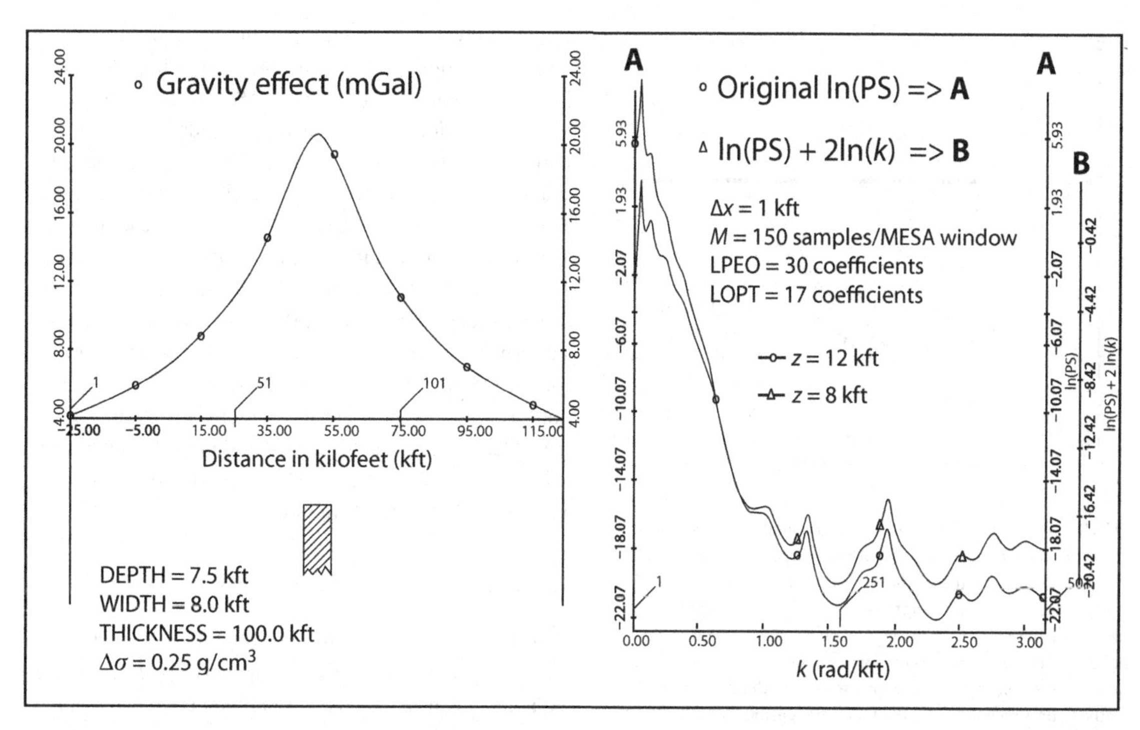

FIGURE 7.21 Maximum entropy spectral analysis (MESA) of a gravity profile data due to a 2D prismatic model (left panel). The power spectrum (right panel) is computed by the maximum entropy method using a prediction error operator of length LPEO = 30. Depths-to-top of source estimates z are obtained from the slope of the logarithmic power spectrum with and without the correction $2\ln(k)$ added to $\ln(\mathrm{PS})$. Applying the correction is equivalent to transforming gravity data to gravity-gradient or pseudomagnetic data.

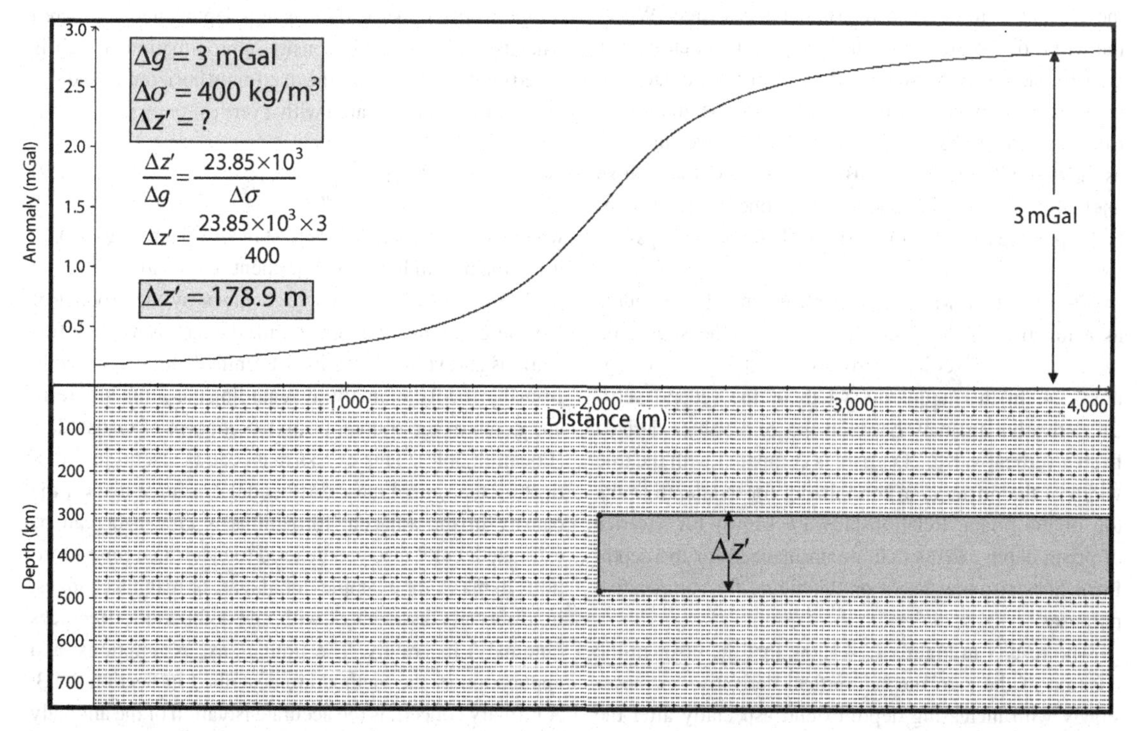

FIGURE 7.22 Example of estimating the thickness of a long, wide tabular source from its gravity anomaly. Note that no assumption must be made regarding the depth of the anomalous source.

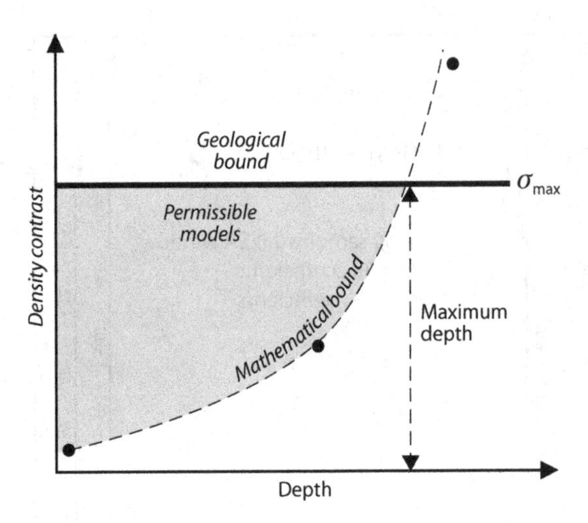

FIGURE 7.23 Plot of the mathematical bound of lowest density contrast versus depth to the top of the source, employing ideal body analysis. The area to the left of the mathematical bound defined by the calculated minimum density contrast for a given depth to the ideal body (heavy dots) is the field of permissible models. The horizontal line is the maximum density contrast of the body based on geological information. The intersection of this line and the mathematical bound defines the maximum permissible depth to the source. Adapted from HILDENBRAND and RAVAT (1997).

its depth. The effect of dip is minimal where the dip of the contact is toward the interior of the source. Where the width of the gravity anomaly source is much greater than the depth to the source, the dip of the contact can be estimated by comparing the ratio of the maximum to minimum of the second vertical derivative anomaly associated with the contact (BOTT, 1962). For a positive anomaly, the ratio will be greater than one where the contact dip is greater than 90° and less than one for dips less than 90°.

The effect of width and depth extent of the source as a function of the depth to the top of the source on the zero second vertical derivative is illustrated in Figure 7.24. In this figure the width of the second vertical derivative anomaly between the zero values derived from a vertical-sided, 2D prism is compared with the true width of the prism after normalizing for the depth to the top of the prism. Relationships are shown for a series of prism depth extents that are normalized for the depth. Error lines also are shown on this figure for comparative purposes.

The results in Figure 7.24 show that the error in the position of the zero second vertical derivative increases slowly with increasing depth extent, especially after the ratio of depth extent to depth exceeds a value of roughly 10. The error in location of formational boundaries by this method is high where the ratio of width to depth is small, exceeding 50% where the ratio is less than one. The error is less than 10%, or 5% for each boundary, where the width of the source exceeds 4.5 times the depth to the source. As a result, the inflection points of gravity anomalies derived from widely spaced, high-angle faults are particularly useful in defining the location of the fault by way of the inflection point of the fault anomaly. Figure 7.24 can be used not only for determining the conditions under which this method can be applied with minimum error, but also in correcting the observed position of the zero vertical derivative or maximum horizontal gradient for the actual geological conditions. However, a number of factors can lead to errors in defining the boundaries with the horizontal gradient maximum or the zero second vertical derivative (e.g. GRAUCH and CORDELL, 1987).

7.4.5 Density contrast and mass

The density contrast of the horizontally extended gravity anomaly source may be approximated by using Equation 7.8 if the anomaly amplitude is known and the thickness of the source can be estimated from independent information. In addition, the density contrast can be estimated from the maximum horizontal gradient of an anomaly derived from a horizontal slab-like or tabular source whose upper surface is at a depth small compared with its thickness. This density approximation, which is based on the equation for the maximum horizontal gradient of the anomaly associated with a vertical fault, is given by

$$\Delta\sigma \approx 3.55 \times 10^4 \left(\frac{dg}{dx}\right)_{\max}, \tag{7.9}$$

where $\Delta\sigma$ is the density contrast in kg/m^3 and $(dg/dx)_{\max}$ is the maximum horizontal gradient in mGal/m.

The excess mass of a source, given by the product of the source's density contrast with the enclosing rock formations and volume, can be determined uniquely directly from the gravity anomaly, disregarding the geometric form and depth of the source, using the concept of Gauss' law (Equation 3.98). This excess mass can be useful in investigating the source mass differential or source mass if the density of the adjacent formations is known or can be estimated, but it also can be used to determine time variations in mass, for example liquid, gas, or magma changes, in reservoirs associated with the temporal differences (anomaly) in gravity. The limitations, assumptions, and application of the method are described in Section 3.4.3. A primary concern is the accurate isolation of the anomaly including its asymptotic approach to zero anomaly values at the flanks. This is also a concern in the interpretation

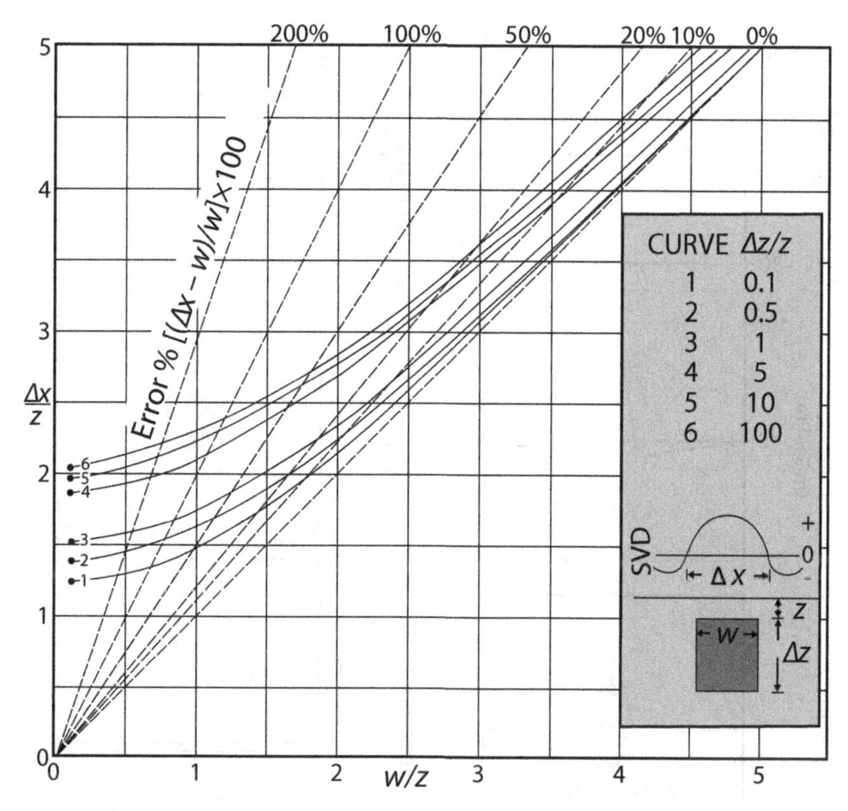

FIGURE 7.24 Normalized distance between zero second vertical derivative (SVD) gravity values versus normalized width of anomalous 2D vertical-sided prisms for various depth extents of the prisms. Dotted lines are percentages of errors for source widths estimated from the distances between zero second vertical derivative values. Adapted from data in SPURGAT (1971).

of mass in the spectral domain (REGAN and HINZE, 1976, 1978).

7.4.6 Pseudomagnetic anomalies

Pseudomagnetic anomalies (Section 3.6.2), determined from gravity anomalies by filtering or by calculation of magnetic anomalies from equivalent point sources obtained by inversion (Section 8.5.3), can be useful in some interpretation problems. For example, Poisson's theorem (Section 3.6.1) has been applied to compare calculated pseudomagnetic and observed magnetic anomalies, to aid in identification of common anomaly sources and their property contrasts (e.g. GARLAND, 1951). In another study, CORDELL and TAYLOR (1971) used pseudoanomalies to determine the physical properties of a North Atlantic Ocean seamount. Furthermore, CHANDLER *et al.* (1981) developed a method for combined analysis of gravity and magnetic anomalies using a moving-window approach to the data, illustrated for a synthetic example in Figure 7.25. In this example, the correlation coefficients between the first vertical derivative of gravity and the reduced-to-pole magnetic anomalies at each window position are compared with the ratios of the magnetization-to-density contrasts obtained from the slopes of the linear regressions of the windowed point values, and with the related intercepts on the magnetic axis that measure the isolation of a common anomaly source within the window. Supplemental descriptions of pseudogravity anomalies and their uses are given in Section 13.4.2.

CHANDLER and MALEK (1991) illustrate the use of the combined gravity and magnetic analysis to the study of basement crystalline rock terranes in Minnesota, which provides useful constraining information for quantitative modeling. Similarly, CHANDLER *et al.* (1981) provide an illustration of the results of the method applied to a coincident magnetic and gravity anomaly in Lake County, Michigan (Figure 7.26). The first vertical derivative of gravity (i.e. the pseudomagnetic) map (c) derived from the Bouguer gravity anomaly map (a) compares well with the reduced-to-pole magnetic map (b), as indicated by the relatively high correlation coefficients in the central region of the map (e). The $(\Delta J/\Delta\sigma)$-values in the central portion of

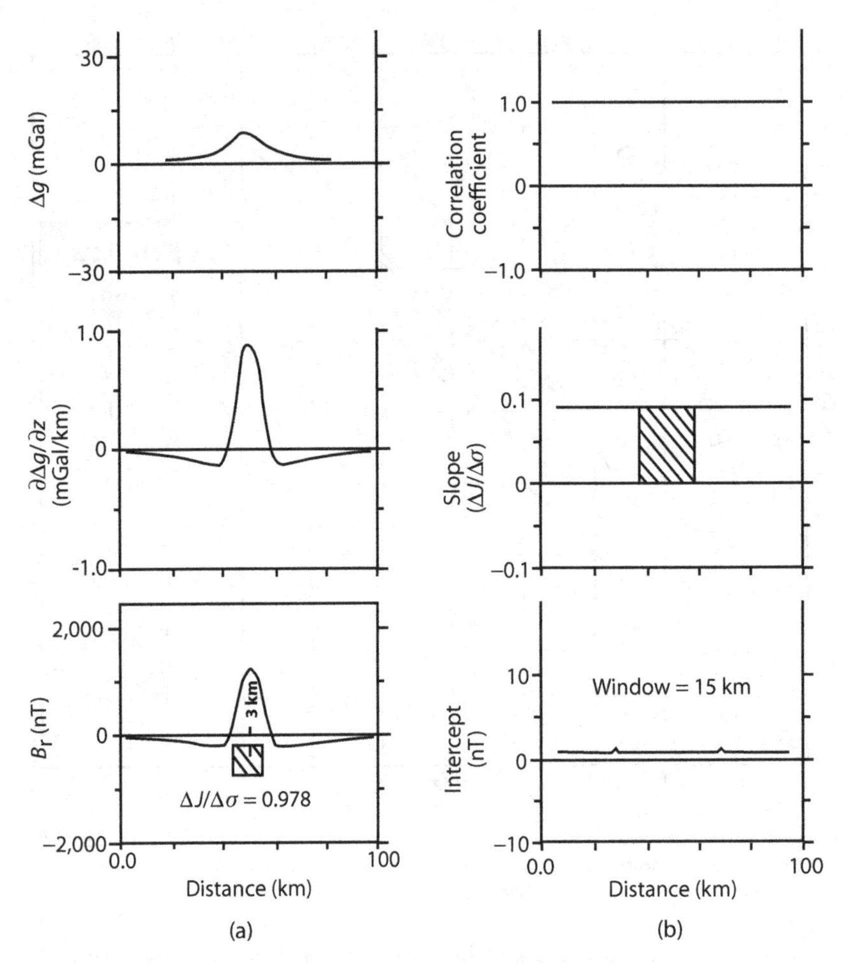

FIGURE 7.25 Analysis of a profile of an isolated gravity and magnetic source using the moving-window approach to analysis of coincident gravity and magnetic anomalies based on Poisson's theorem. (a) Coincidence between the gravity anomaly (top), vertical gradient of gravity (pseudomagnetic) anomaly (middle), and reduced-to-pole magnetic anomaly (bottom). (b) The results of the moving-window analysis that include a coefficient of correlation between the magnetic and pseudomagnetic anomalies showing the direct correlation of the anomalies, the slope (magnetization/density contrasts) of the regression of the point values of these anomalies within a moving window, and the intercept of the regression line on the magnetic anomaly axis indicating no interfering anomalies. The value of calculated magnetization to density contrasts is the same as that of the modeled source. Adapted from CHANDLER *et al.* (1981).

map (d) compare favorably with the CGSu-value of 0.013 previously estimated for the anomaly source. The intercept map (f) suggests that the isolation of the anomalies has not been complete or that the source is not completely homogeneous.

7.5 Modeling anomaly sources

Quantitative interpretation of the geologic source is the inversion of observed and isolated gravity anomalies (residual anomalies) using a conceptual geologic model predicated by collateral geologic and geophysical information, qualitative assessment of the observed gravity signature, and the results of simplified analysis of the anomalies as discussed in the previous section. As outlined in Appendix A.5 and Figure 7.4, modern gravity inversions are most commonly achieved either through trial-and-error or least-squares modeling (BHATTACHARYYA, 1978). These respective non-linear and linear inversion methods are rooted in simplification of the probable geologic sources of the anomaly into an analytically tractable forward model.

In gravity applications, establishing a solution by trial-and-error adjustments of the unknown parameters until the predictions of the forward model reproduce the observed data within a specified margin of error is called forward modeling, whereas determining the unknown parameters directly from the observed data using an optimization

FIGURE 7.26 Anomaly and regression coefficient maps of the Lake County, Michigan anomaly using the moving window method of applying Poisson's theorem. Map (a) is the Bouguer gravity anomaly with contour interval CI $= 2\,\mathrm{mGal}$; map (b) is total magnetic intensity anomaly reduced-to-the-pole with CI $= 100\,\mathrm{nT}$; and map (c) is the first vertical derivative of gravity (pseudomagnetic map) with CI $= 1\,\mathrm{mGal/km}$. All anomaly data have been upward-continued 1 km to minimize high-wavenumber components. Map (d) is the $(\Delta J/\Delta\sigma)$-map with CI $= 0.005$ CGSu; map (e) is the correlation coefficient map with CI $= 0.4$; and map (f) is the magnetic intercept of the regression line map with CI $= 100$ nT. After CHANDLER *et al.* (1981).

process comparing the observed and predicted data is called inverse modeling.

7.5.1 Forward modeling

Trial-and-error inversion is the most widely used and successful approach for interpreting gravity anomalies. This forward modeling method involves computing the effects from a simplified mathematical model (i.e. the forward model) of the presumed subsurface conditions (i.e. the conceptual model) with iterative re-computation of the effects based on alterations to the parameters of the forward model until an "acceptable" correlation with the residual anomaly is achieved. The procedure is used for either profile or map analysis. Map calculations employ 3D sources which are more difficult and time-consuming to input and modify to achieve an acceptable match. Thus, most forward modeling uses profile analysis with limits placed on the two-dimensionality of the individual modeled bodies based on the map view of the associated anomalies.

The acceptance of the match between the residual anomaly and calculated effect is highly subjective and likely to vary with interpreter, amount of geologic and geophysical subsurface control, objective, and resources. Simple error estimates such as the root mean square deviations should be augmented with estimates of the correlation coefficients to insure an effective match in the horizontal gradients between the residual anomalies and calculated effects. Indeed, the most crucial control on the match of anomalies is the similarity of horizontal gradients. This can be checked by calculating and comparing the observed and modeled horizontal derivatives across the modeled profile. The amplitude of the modeled anomaly can be matched at the final stage by fine-tuning the density contrasts. In general, the forward model is modified to obtain an improved correlation of the observed and calculated anomalies. These modifications are guided by the difference in amplitudes and horizontal gradients over the extent of the anomalies based either on plotted or visual observation of differences.

The most critical aspect of forward modeling is to match the horizontal derivatives or phase properties of the observed data. These properties are controlled by the geometric parameters of the problem that are isolated in the curly-bracketed functionals of the kernels in the generalized gravity forward modeling Equations 3.23, 3.27, and 3.34. These functionals typically include the nonlinear effects of depth, thickness, and shape of the source that are modified by trial-and-error adjustments until the phase properties of the observed data are effectively matched.

The final modification adjusts the linear functional in the density σ until the predicted amplitudes match the observed amplitudes. This functional is linear because the proportional change in the physical property value leads to the same proportional change in the predicted amplitude. Thus, the final amplitude adjustment is the most straightforward modification to make in anomaly modeling.

Of course, no gravity inversion can guarantee a unique solution no matter how closely it matches the computed effects with the observed anomaly. The ambiguity inherent in gravity interpretation indicates that the solution is only one of a family of possible interpretations. Thus, a sensitivity study on the solution is commonly performed by modifying critical attributes of the forward model to determine their impact on the gravity anomaly. The range of source parameters that are compatible with the anomaly can be determined by taking into account the error envelope of the observed anomaly. This does not limit the ambiguity of the interpretation, but it does indicate the possible range of sources within the context of the conceptual model.

Forward modeling offers several advantages for gravity data inversion. For example, it requires only a moderate amount of experience to implement. In addition, the forward modeling exercise quickly provides insights on the range of data that the given model satisfies, as well as the sensitivity of the model's parameters to the data. Furthermore, applications of trial-and-error inversion often provide moderately experienced investigators with considerable practical insights into the subsurface implications of gravity data.

However, difficulties with the approach include the absence of reliable error statistics for the final solution. Convergence to the final solution by trial-and-error inversion also can become intolerably laborious and slow for models involving more than a few unknown parameters. Hence, for more complicated problems with larger numbers of unknowns, inverse modeling is commonly implemented.

7.5.2 Inverse modeling

Trial-and-error forward modeling is most effective where the number of unknown parameters of the forward model to be estimated is relatively small, whereas least-squares inversion is commonly implemented for larger-scale inversions. As described in Appendix A, both forward and inverse modeling can be generalized as the inversion of observed anomaly values in $\mathbf{B}$ for the unknown parameters in $\mathbf{X}$ of a presumed forward model $\mathbf{AX}$ with known coefficients in $\mathbf{A}$ so that $\mathbf{AX} = \mathbf{B}$ (Equation A.2). Least-squares inverse modeling determines the parameters ($\mathbf{X}$) of the linear forward model (**AX**) so that the sum of the squares of the residuals between the observed anomalies (**B**) and the determined model's computed effects or predictions ($\hat{\mathbf{B}} = \mathbf{AX} \simeq \mathbf{B}$) is minimum. The universal least-squares solution is $\mathbf{X} = (\mathbf{A}^t\mathbf{A})^{-1}\mathbf{A}^t\mathbf{B}$ (Equation A.16), where the coefficients of the design matrix, **A**, result from calculating the forward model with the unknown parameters set or initialized to unity.

Since the advent of electronic computing, considerable progress has been made in the application of inverse modeling for assessing the source attributes of potential field anomalies (e.g. OLDENBURG, 1974; PARKER, 1974; LINES and TREITEL, 1984; LINES and LEVIN, 1988). Consideration of the generalized forward modeling Equations 3.23, 3.27, and 3.34 shows that the gravity effects are quantified as the sum of products of the density functional in σ and the geometry functional in the inverse of the source-to-observation point displacement r. Solving for values of the physical property functional is the classical linear inversion problem described by Equations A.2, A.16, and A.18. However, estimating parameters of the source (shape, thickness, depth, etc.) in the geometric functional is a linear inversion problem only if the parameters are part of a closed form solution for the integral. These solutions occur for idealized sources that have simple symmetric shapes (spherical, cylindrical, prismatic, etc.) with volumes given by closed form equations in the fundamental shape parameters (thickness, length, width, height, radius, etc.).

As an example, consider the multi-sourced Equation A.8 where the closed form gravity effects of the infinite horizontal cylinder are given in the first term that is superimposed on a regional anomaly C in the second term. This forward modeling equation was processed for the coefficients in Equation A.10 of the design matrix **A** to obtain matrix inversion estimates of the cylinder's density ($\Delta\sigma$) and the regional anomaly (C), assuming that its radius (R) and central axis depth (z) are known. In other words, the modeling algorithm computed the coefficients of **A** simply by evaluating Equation A.8 with $|\Delta\sigma| \equiv 1 \equiv |C|$ and the known (or assumed) values of the other parameters.

Equation A.8, however, is also linear in the source parameter R^2 which can be similarly estimated with C using the design matrix coefficients

$$a_{n1} = [41.93(\Delta\sigma/z)]/[(d_n^2/z^2) + 1],\ a_{n2} = 1.0, \tag{7.10}$$

with presumed or known values for the above listed parameters. Thus, Equation A.8 provides the interpreter with the capability to obtain least-squares estimates of the regional anomaly C and one of the source parameters $\Delta\sigma$ or R in the context of presumed values of the other variables. Of course, where the parameters C, $\Delta\sigma$, and R have been effectively established, the depth z to the center of the cylinder can be estimated from any value of the gravity effect $g_z(d_n)$.

It is incumbent on the interpreter to identify and mathematically articulate effective forward models for analyzing anomaly data. By the source ambiguity principle of potential fields, the anomaly can be modeled by an indefinite number of equivalent sources. Thus, the interpreter must take care to select a model that is computationally efficient for achieving the analysis objectives, especially because forward modeling is the most computationally laborious and time-consuming element of inversion.

The gravity anomaly profile at the top of Figure 7.27 can be modeled, for example, by the effects of a single infinite horizontal cylinder (Model 1), a series of horizontal cylinders (Model 2), and a cross-sectional array of variable-density 2D prisms (Model 3). The choice of which of these equivalent sources to use for anomaly analysis depends on the objective of the analysis. To estimate some continuation, interpolation, or other integral or derivative property of the anomaly, for example, the single horizontal cylinder solution may be the most efficient option. The cylinder equation would be solved for a least-squares estimate of say $\Delta\sigma$ in the context of presumed values for R and z, and the solution subjected to a new set of model coefficients that takes the desired anomaly transformation as suggested in Equation A.65.

However, unacceptable residuals ($\mathbf{B} - \hat{\mathbf{B}}$) can result where errors in the presumed values of R and z are significant. These residuals can be reduced by trying out new presumed values or simply updating the modeling by fitting the residuals with additional cylinders. In the latter case, the collection of all cylinder solutions is integrated at the observation point for the desired integral or derivative transformation of the anomaly. Alternatively, the single cylinder density solution could be replaced with a variable density solution for a horizontal array of uniformly dimensioned cylinders (Model 2). This model can offer better sensitivity for the spatial details of the anomaly, but at significantly increased computational labor because of the greater number of unknown densities $\Delta\sigma(i)$ that must be determined.

In general, the density solution for estimating the free space properties of the anomaly does not have to be physically realistic. It simply provides a number or set of numbers that allows the effects of the cylinder model with presumed parameters to closely match the anomaly observations. However, a sensitivity analysis is always recommended to establish the utility of the model for estimating acceptable transformations of the anomaly (Appendix A.4).

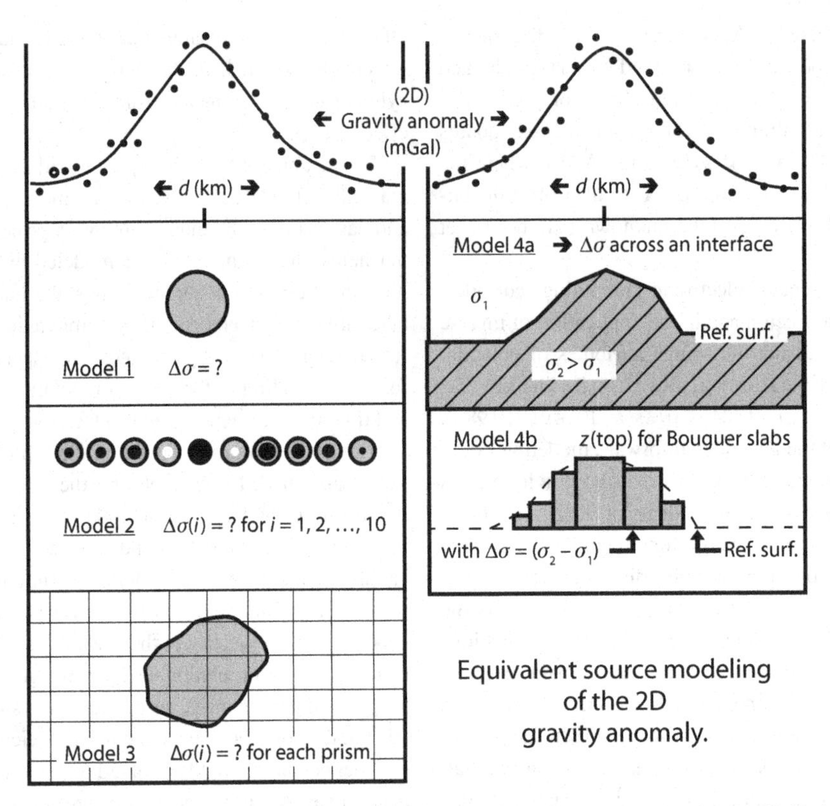

FIGURE 7.27 Observed (dots) and theoretical (line) gravity anomalies (top profiles) that can be modeled in the 2D (d, z)-plane by the effects of an infinite horizontal cylinder with unknown density $\Delta\sigma$ (Model 1), a lateral distribution of horizontal cylinders with unknown densities $\Delta\sigma(i)$ (Model 2), a cross-sectional distribution of horizontal prisms with unknown densities $\Delta\sigma(i)$ (Model 3), and the horizontal density contrast $\Delta\sigma$ (Model 4*a*) approximated by horizontal prisms with unknown top elevations (z(top)) and bottoms on the mean boundary or reference surface (Ref. Surf.) separating the two layers (Model 4*b*).

In contrast to evaluating anomaly transformations, the interpreter may want to drill or directly investigate the source-solution, in which case a more detailed partitioning of solution space is desired such as shown in Model 3. Here, the subsurface has been modeled for the densities of a (d, z)-array of small 2D rectangles. The density variations enhance the resolution of the spatial details of the possible source as suggested by the density contour, although the solution is not unique owing to errors in the observations (**B**), the design matrix (**A**), and the inherent source ambiguity of potential fields. This solution involves determining a large number of unknown densities $\Delta\sigma(i)$, and thus is computationally laborious. Clearly, the enhanced effort is not warranted for obtaining the anomaly transformations that the simpler cylinder Models 1 or 2 can provide.

In addition to idealized sources with closed analytical effects, general sources have arbitrary shapes that must be numerically integrated for their gravity effects. These effects are modeled from the integrated effects of idealized bodies that effectively fill out the arbitrarily shaped volume of the general source. A non-linear modeling strategy uses idealized bodies with geometric parameters that can be iteratively adjusted against successive residuals $(\mathbf{B} - \hat{\mathbf{B}})$ until the updates start oscillating about the final ith solution where $(\mathbf{B} - \hat{\mathbf{B}}_{i+1}) \approx (\mathbf{B} - \hat{\mathbf{B}}_i)$. Improving the residual beyond this convergence point requires a new parameterization of the inversion – that is, a new set of idealized bodies.

As an example, the interpreter may want to model the residual gravity anomaly with a 2D density contrast across an arbitrarily shaped interface like in Model 4a for which no closed form solution is available. However, the interpreter knows the mean thickness and density σ_1 of the top horizontal unit and the higher density σ_2 of the underlying unit, perhaps from drilling, seismic, geological, or other ancillary information. Thus, the interpreter envisions here an anomaly source due to the lateral density contrast $\Delta\sigma = \sigma_2 - \sigma_1$ across the boundary where the overlying material is thinned locally by the underlying material. In other words, the inverse modeling objective is to estimate the top surface of the contrasting density source indicated by the region in Model 4a that extends above the mean horizontal boundary or reference surface (Ref. surf.) separating the two layers.

A relatively efficient approach is to estimate the thickness variations of the source region using iterative applications of the Bouguer slab (BS) equation

$$g_{\mathrm{BS}} \equiv F_z = -\int_{z'_a}^{z'_b}\int_{-\infty}^{\infty}\int_{-\infty}^{\infty}\left\{G\left(\frac{\Delta z}{r^3}\right)\sigma\right\}\partial x'\partial y'\partial z'$$
$$= 2\pi G\sigma(z'_b - z'_a) = 2\pi G\sigma\Delta z', \quad (7.11)$$

which is Equation 3.26 integrated over infinite horizontal limits. At the anomaly observation b_j, this equation is evaluated for $\Delta z'_j$ assuming $\sigma = \Delta\sigma$ with $\Delta z'_j$ added to the mean bottom of the top unit (i.e. the reference surface, Ref. Surf.) for an estimate of the top surface of the source. The 2D gravity effects $\hat{B}_j$ of the collection of $\Delta x' \times \Delta z'$ prisms are then calculated from Equation 3.39 over the finite x'- and z'-limits of the prisms.

The residual $(b_j - \hat{b}_j)$ is evaluated next and if positive or negative is modeled via Equation 7.11 for a thickness correction that respectively lowers or raises the previous estimate of the upper surface of the source. The effects of the updated prisms are again evaluated by Equation 3.39 for a new set of residuals. Usually, after several iterations, the updated residuals start oscillating about the optimal solution (e.g. Model 4b) for the given parameterization of the problem.

Another effective approach to non-linear inverse modeling is to take the first-order Taylor series expansion of the non-linear forward model (e.g. Bevington, 1969; Blakely, 1995; Abdeslem, 2000; Abdelrahman *et al.*, 2001; Tlas *et al.*, 2005). The expansion linearizes the forward model in terms of adjustments that can be applied to initial estimates or guesses of the unknown parameters to minimize the squared sum of residuals between the predicted and original anomalies. Iterative estimation of the adjustments on successive sets of decreasing residuals leads to a minimum about which the residuals oscillate that produces the optimum solution.

Exploring solution space for the optimum solution is another common strategy for solving the non-linear inverse problem (e.g, Bevington, 1969; Harbaugh and Bonham-Carter, 1970; Koch and Link, 1971). Specifically, the behavior of an objective function describing the sum-squared residuals between the predicted and original data is investigated over ranges of the unknown parameter values that the investigator presumes appropriate for the problem. The objective function can be set up in terms of the root-mean-squared differences or the chi-squared (χ^2) statistic or some other measure of the prediction residuals to map out the minimum where the sought-for parameter values are optimum for solving the problem (Figure 7.28). Care must be taken, however, to insure that the global minimum was identified and not one of possibly numerous local minima within the solution space. The problem can be handled by choosing enough different starting points that the exploration strategy discovers all of the minima within the presumed solution space.

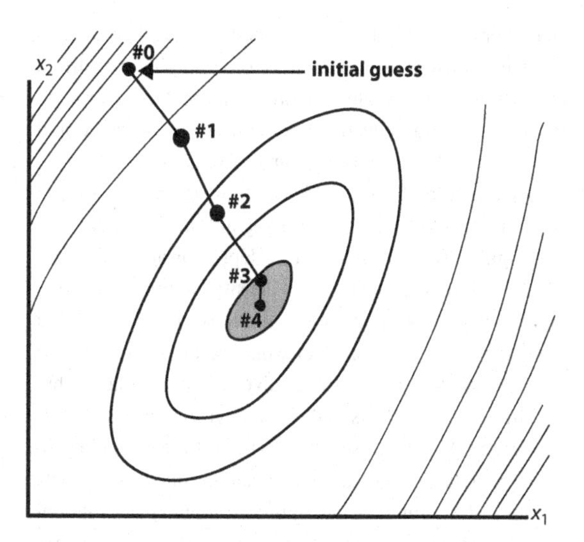

FIGURE 7.28 Example of an objective function for a non-linear inverse model involving the unknowns x_1 and x_2. The optimum values of the unknowns are indicated where the contours of the objective function are minimum. They were established from an initial guess and four iterations of an exploration strategy with enhanced sensitivity for paths of maximum negative gradient or steepest descent in solution space.

In general, the anomalies of any conceivable density distribution can be iteratively modeled using the closed form effects of idealized bodies that fill out the distribution. The generic modeling Equations 3.23, 3.27, and 3.34 govern all gravity modeling applications.

7.5.3 Idealized body interpretation

Generally, idealized body interpretation is used for simple anomalies that can be approximated with a single source, but it may be used for more complex anomalies by summing the effects of multiple simple sources. The approximate geometric form is estimated from the anomaly characteristics (e.g. Figures 3.8 and 3.9), collateral subsurface information, and results from applying simplified interpretation procedures. Gravity anomalies computed from the resultant equations are iteratively compared and modified until the residual anomaly is matched.

As an example, suppose that the exploration targets for a high-resolution gravity survey are limestone caverns of roughly 20 m in diameter at depths of about 200 m. The caverns constitute negative density contrasts with horizontal dimensions that are small relative to the projected depths. Hence, the interpreter would scrutinize the survey data for radially symmetric negative gravity anomalies that are similar to the effects of the spherical source (source #2) in Figure 3.8. Of course, identifying such an anomaly in the survey data cannot guarantee that its source is a cavern because of the fundamental source ambiguity of potential fields. However, where the anomaly is due to a cavern, the gravity effects of the sphere (source #2) in Table 3.2 can be fit to principal profiles across the minimum of the anomaly to help resolve effective, but non-unique quantitative estimates of the cavern's boundary, depth, and density attributes.

Highly elliptical gravity anomalies, on the other hand, mark the presence of buried anticlines, synclines, lava tubes, mines, bedrock valleys, and other subsurface sources of horizontally elongated density contrasts. Thus, gravity surveying helps to map these features in the subsurface where the density contrasts are sufficient to cause recognizable anomalies. Again, the relative sign of the gravity anomaly indicates the sign of the density contrast in the subsurface. Furthermore, when the long axis of the anomaly is significantly greater than the orthogonal short axis, the infinite horizontal cylinder equation (source #4) in Table 3.2 can be used for the inversion of anomaly values from the principal profile to obtain effective, but non-unique quantitative estimates of the boundary, depth, and density attributes of the source.

The elegant analytical properties of the inverse distance function, which is central to potential field theory, offer significant advantages for using the idealized body equations in anomaly interpretation. This interpretational power comes about mainly because the inverse distance function linearly relates the anomalous effect to its spatial derivatives. Thus, the half-width rules for estimating source depths in Table 7.1 were developed by assessing the idealized body equations at the anomaly half-widths ($X_{1/2}$).

Another approach for estimating these source depths compares the maximum central anomaly value CA to its vertical derivative $\mathrm{CA}' = \partial(\mathrm{CA})/\partial z$. In the case of the spherical source, for example, the relationship becomes

$$\frac{\partial(\mathrm{CA}[\#2])}{\partial z} = -2\left(\frac{\mathrm{CA}[\#2]}{z}\right) \ni z_c = -2\left(\frac{\mathrm{CA}[\#2]}{\mathrm{CA}'[\#2]}\right), \tag{7.12}$$

where z_c is the depth to the center of the source. In similar fashion, the depths to the central axis (z_c) and top (z_t) of the respective infinite horizontal and vertical cylinder bodies can be estimated simply from the negative ratios of their gravity-to-first vertical derivative anomalies.

For these applications, the vertical derivative CA′ can be established by direct measurement or numerically estimated from the residual anomaly by convolution or Fourier transform derivative operators (Appendix A.4.3). Extending these numerical procedures to second vertical derivative estimates of the residual anomaly can help delineate the lateral subsurface boundaries of the source in terms of the zero second vertical derivative contour centered on CA.

With the geometrical properties of the source established from the residual anomaly and its vertical derivatives, an appropriate density estimate for the source can be evaluated from the idealized source equation. This estimate can be significant for constraining the source's geological context and evolution, and mechanical and other physical attributes.

Compared with interpreting idealized bodies in the spatial domain, there are interpretational advantages to using the spectral domain. Numerous studies have defined the Fourier transforms of idealized bodies and their use in interpreting the size, depth, and mass of the specified sources (e.g. ODEGARD and BERG, 1965; SHARMA *et al.*, 1970; REGAN and HINZE, 1976, 1977, 1978). These transforms are summarized in Table 7.3, but their uses are limited by the need to isolate the anomaly and problems in transforming the spatial anomaly into the spectral domain (e.g. DAVIS, 1971; REGAN, 1973).

The idealized body equations are, in general, remarkably robust in providing useful results (e.g. HAMMER, 1974). However, errors result where the subsurface conditions deviate from the assumptions underlying these equations. Thus, care must be taken to insure that the assumed idealized source geometries do not invalidate the use of the equations.

7.5.4 Gridded anomaly modeling

Geological mass variations are being increasingly represented in gridded formats where the unit element is the prism with fixed horizontal coordinates and variable

TABLE 7.3 Fourier transforms of the gravity effects of some of the idealized sources in Tables 3.2 and 3.3. Additional variables used here include ω for the angular frequency, and k_0 and k_1 for the respective zero- and first-order modified Bessel functions of the second kind.

Source	Gravity effect in frequency domain
(2) Sphere	$\mathcal{B}(g_{\#2}) = (4/3)GR(\Delta\sigma)\omega K_1(\omega z)$
(4) ∞ H-cylinder	$\mathcal{B}(g_{\#4}) = \pi G(R\Delta\sigma)e^{-z\lvert\omega\rvert}$
(7) Finite V-line mass	$\mathcal{B}(g_{\#7}) = (G\Delta\sigma/\pi)[k_0(z_1\omega) - k_0(z_2\omega)]$
(11) Thin V-sheet (2D)	$\mathcal{B}(g_{\#11}) = (2G\Delta\sigma/\omega^2)\sin(\omega t/2)[e^{-z_1\omega} - e^{-z_2\omega}]$
(14) Thick S-∞ H-slab (fault)	$\mathcal{B}(g_{\#14}) = (G\Delta\sigma)[\pi t\delta(\omega)/2 - j(\omega^{-2}e^{-z_1\omega} - \omega^{-2}e^{-(z_1+t)\omega})]$

After REGAN and HINZE (1976).

vertical coordinates. Gridding in profile, map, and volume dimensions is a relatively routine process (Appendix A.6.2), as is the construction of 3D distributions of prisms to represent geological features by interactive graphical computing (e.g. PIGNATELLI *et al.*, 2011). Additionally, as described in Section 3.6.3, Gauss–Legendre quadrature (GLQ) modeling of the gravity effects of the prism is almost trivial because the integration limits are all fixed by the specified dimensions of the prism and no body point interpolations are required.

The accuracy of the quadrature formulation can be readily controlled by adjusting the number, and hence spacing, of the Gauss–Legendre nodes relative to the elevation of the calculation point. These adjustments can be made in practice by subdividing the body into smaller bodies, or increasing the number of equivalent point sources, or increasing the distance between the calculation point and the body. The latter consideration suggests that the GLQ formulation is well suited for satellite altitude gravity modeling because of the high altitude of the measurements. On the other hand, accurate gravity modeling on the surface of the body is also possible because the nodes are displaced away from the surface into the interior of the body (e.g. Figure 3.14). Thus, the lack of modeling singularities at the surface allows Equations 3.80 and 3.81 to compute accurate gravity effects anywhere on or outside the body's surface. Furthermore, these estimates are always least-squares estimates for any selected number (≥ 2) of point sources (STROUD and SECREST, 1966).

In general, the gravity effect of the prism is finding increasing use for relating gridded geological mass variations to gravity observations. For the Cartesian prism, the analytical gravity effect is described by NAGY (1966) and PLOUFF (1976), whereas the laminar method of TALWANI and EWING (1960) and the GLQ method of KU (1977) are effective for numerically modeling the gravity effect. The GLQ method also readily calculates the body's gravity potential, vector, and tensor components because these components of the point mass are easily derived (Section 3.4.2) and included in the integration. The fundamental simplicity of point source gravity analysis also facilitates implementing different coordinate systems in GLQ gravity modeling.

Spherical coordinate GLQ adaptations, for example, yield the complete gravity components of the spherical prism in support of regional gravity applications which are becoming increasingly available at the Earth's surface, airborne, and satellite altitudes (e.g. VON FRESE *et al.*, 1981b; ASGHARZADEH *et al.*, 2007). These anomalies, complemented by comparably gridded regional and global terrain elevation data, provide fundamental constraints on the geological properties and history of the crust, mantle, and core. In addition, critical insights into the internal mass properties of the Moon, Mars, Venus, and other planetary bodies are resulting from satellite measurements of the related gravity and topographic data. Spherical coordinate adaptations of Equation 3.80 facilitate computing theoretical gravity effects of geologic features for assessing the geological significance of the growing volumes of regional and global gravity and terrain elevation data (e.g. VON FRESE *et al.*, 1981b, 1982, 1997b, 1999a, 2009; LEFTWICH *et al.*, 1999, 2005; POTTS and VON FRESE, 2003).

STROUD and SECREST (1966) provide additional details on the theory and errors of GLQ integration. They also present GLQ formulae for a variety of idealized geometries that can be useful for geophysical applications. The wedge formula, for example, can efficiently represent regional mass variations of the crust in satellite gravity studies at altitudes that are large compared to crustal thickness. The tetrahedron formula, on the other hand, facilitates modeling the gravity effects of mass variations registered in a triangular irregular network (TIN). In general, the point source gravity effects can be readily incorporated into GLQ formulae for modeling the gravity effects of any conceivable mass distribution.

7.5.5 Forward modeling strategies

Most quantitative gravity interpretation is based on comparing gravity anomalies with the gravity effects of conceptual geological models. These models are constrained by available geological and geophysical information or by simplified inversion of the anomalies themselves. The models are modified in a trial-and-error procedure (forward modeling) until a reasonably close match is obtained between the observed and calculated anomalies. Strategies for implementing this procedure are highly diverse depending on the objectives, nature of the observed anomalies, resources, experience of the interpreter, and the software available for the computations and presentation of the results. Software is particularly important because essentially all modeling requires intensive computational power and graphics capabilities, and significant differences exist in available algorithms.

Modeling as a methodology for interpreting gravity anomalies may take several forms. It may be as simple as comparing the observed gravity anomaly or its spectrum with the theoretical response of idealized source configurations that are readily computed analytically. However, this procedure seldom permits the consideration of more than one or at most a few anomalies. Anomaly sources can be constructed from volumes made up of multiple sources, generally of the same configuration. Similarly, 3D modeling of arbitrarily shaped sources is usually limited to no more than a few sources. Even in these cases the 3D modeling process is time-consuming because computational programs for these bodies generally have limited interactive capabilities, preventing relatively rapid modifications of the sources to adjust the model for discrepancies between the observed and calculated anomalies. As a result, most forward modeling for interpretational purposes uses 2D profile analysis that can incorporate limited lateral extent of the sources from the principal profile being modeled while maintaining the same cross-section. Accordingly, the modeling is modified as needed from the infinite extent perpendicular to the profile (2D) to restricted lengths as suggested by off-profile map views of the anomaly. These lengths may be equivalent on either side of the profile (i.e. 2.5D) or different on either side (i.e. 2.75D).

The following description of modeling procedures is focused on trial-and-error forward modeling of gravity anomaly profiles, but many of the concepts pertain to modeling of idealized sources and 3D modeling as well. These guidelines are necessarily described in a general manner because of the varying nature and complexity of the modeling programs that are available. Computational programs differ in a number of ways. For example, they vary in the ease of describing the shapes and properties of the conceived sources, establishing body parameters that cannot be changed in the iterative process, the anomalies and their various derivatives that are calculated, the treatment of surface topography, and the graphic display of the anomalies and the changing cross-section of the geologic profile.

Although iterative forward modeling using available software is relatively simple to implement, its successful use is dependent on a thorough knowledge of the principles and practices of the gravity method as described in this and preceding chapters, and of the density of Earth formations and the factors that control this property as described in Chapter 4. There is not a unique geological model solution to the gravity anomaly that in principle can be duplicated by the effects of numerous possible geological models. These solutions often can be narrowed down to only a few appropriate models given the geologic constraints. This requires adequate knowledge of both the geophysical data and available geological data. However, even here there are likely to be numerous acceptable solutions. As a result it is desirable to conduct sensitivity studies on likely solutions by varying the parameters of the model to determine the range of parameters that will permit acceptable matches to the observed gravity anomaly.

Selecting the anomaly

In Chapter 6 several types of gravity anomalies have been defined and their relative advantages and disadvantages considered. Of these the Bouguer gravity anomaly, preferably the complete anomaly incorporating terrain and bathymetry corrections, is the anomaly of choice for modeling ground surface or airborne gravity measurements. The Bouguer gravity anomaly, using either a constant density or where sufficient information is available variable densities, estimates the gravity effect of the Earth minus the effect of the mass between the observation site and the datum elevation of the survey. This datum generally is sea level, but it can be set at any elevation that suits the survey objectives. During the modeling process the density of the material between the stations and the datum is set to zero because the Bouguer gravity anomaly has included the effect of this material.

Optimally, the Bouguer gravity anomaly values should be those observed at stations along the profile. This makes it possible to establish the elevation of the station, which is the position used in calculating the gravity anomaly from the conceptual model. However, in many situations gravity modeling is performed using gridded Bouguer gravity anomaly values or those interpolated from gridded data where observations are not located optimally for constructing profiles for analysis. In this case the elevations of observations cannot be prescribed to specific locations along the profile. One option is to assume that the observations are at a constant elevation over the length of the profile. This results in errors depending on the actual range of profile elevations and the depths of the sources. An alternative option is to assume that the elevations can be estimated as a mean value or specific values along the profile from digital elevation models. The error resulting from making either of these assumptions can be evaluated by comparing the anomaly values of the model at the highest and lowest elevations along the profile as indicated by elevation models.

In some cases it may be desirable to select a residual gravity anomaly derived from filtering the Bouguer gravity anomaly or to use a geologically corrected anomaly. In both cases, great care must be used to preserve the gradients and amplitudes of the isolated anomalies. It is difficult to insure that this is the case, but there is less likelihood of significant alteration of the anomalies if the filter is not severe and does not greatly change the anomaly. Geologically corrected gravity anomalies, such as isostatic residuals, are potential candidates for use in modeling, but many of these anomalies are based on generalized assumptions that may not be universally valid. As a result, if sufficient information is available for making a geological correction, it is desirable to make this correction an actual part of the modeling procedure.

In marine areas it is standard procedure to use the free-air anomaly for modeling. In this case the bathymetry is used to establish the gravity effect of the water column which is incorporated into the gravity effect of the conceptual model. The gravitational effect of the water column is not considered in modeling if the Bouguer gravity anomaly is used for analysis because bathymetric effects have been incorporated into the calculation of the Bouguer gravity anomaly. However, in the case of free-air gravity anomaly modeling, the interpreter needs to include bathymetric effects.

Constructing the model

The results of trial-and-error forward modeling significantly depend on the starting point of the conceptual model. The validity of this initial model is governed by *a priori* information that constrains the model, by how this information is incorporated into the model, and by knowledge of the geologic and tectonic history of the region. The model should be as simple as possible with a minimum number of sources. Complexities should only be incorporated based on available supplemental geologic and geophysical information and qualitative and simplified quantitative interpretation of the gravity anomalies along the profile. The model needs to be tuned to the objectives of the interpretation including such factors as depth of sources and the detail desired.

The geometric configuration of the sources needs to be synchronized with the source densities. For instance, sedimentary rocks occurring as relatively parallel and low-dip strata generally will have a density contrast range of 100 to 200 kg/m^3, while basement crystalline rocks have a range of density contrast of 100 to 300 kg/m^3, although contrasts that fall outside these ranges are not uncommon. Densities assigned to sources are best obtained from *in situ* methods such as correlation with measured seismic velocity, but may need to be taken from general density tables associated with specific lithologies. In the case of gravity modeling used to validate the interpretation of seismic reflection profiling, the configuration of the sources can be taken directly from the interpretation of geometry of the seismic reflectors. Formation densities are assigned from determined seismic velocities and interpreted lithologies suggested by their interpreted structural attributes.

Either actual densities or density contrasts from a selected background density can be used in calculating the gravity anomaly from the model. However, when actual densities are used, which intuitively is the most direct approach, the density of the geologic unit that forms the boundary of the model must be used to calculate the gravitational effect of the bounding formation using the Bouguer slab formula. This effect must be subtracted from all the calculated anomaly values so that the resulting anomaly is based on density contrasts. Gravity anomalies reflect the difference in mass between the geologic unit and the surrounding formations, not the total mass of the unit. The density contrast is normally assumed to be a constant value for a specific source, but this is not always the case as in the contrast between sedimentary basins and the enclosing basement rocks. Although there are notable exceptions, typically the densities of the basement rocks are appropriately assumed to be a constant. In contrast, the density of sedimentary rocks often increases in a non-linear fashion with depth and varies with horizontal location in basins. Numerous techniques have been proposed to calculate the gravity anomaly associated with the less dense sedimentary rocks of the basin, taking into account the variable density contrast over and within the basin. These may involve either 2D forward modeling in the space (e.g. Cordell, 1973) or wavenumber domain (e.g. Chai and Hinze, 1988), or 3D inversion (e.g. MARTINS *et al.*, 2009) taking into consideration the non-linear decrease in density with depth in sedimentary basins using idealized geometric sources (e.g. RAO, 1990) or arbitrarily shaped 2D sources (e.g. ZHANG *et al.*, 2001; ZHOU, 2010).

The length of the profile involved in the model is a function of the geological objective. The profile must involve all the anomalies of interest in the interpretation, but also capture the depth extent of the source bodies. Generally, the anomaly of an isolated source extends for a length of eight to ten times the maximum depth of the source. This is a minimum length for the profile because the profile length needs to extend for a distance of four to five times the maximum depth on either side of the region of interest. However, this is the very minimum length, because additional coverage on the margins is necessary to isolate the anomalies from the regional background anomaly, suggesting that the length of the profile on the margins of the region of interest should be of the order of magnitude of 10 times the maximum depth. Furthermore, the model should extend for distances of the order of 20 times the maximum depth of the model when actual densities rather than density contrasts are used in the calculation to avoid edge effects of the termination of the model. Seldom is the curvature of the Earth taken into account regardless of the length of the profile. Assuming a smooth surface for a spherical Earth, the curvature will cause a vertical deviation of roughly 125 m for a 1,000 km long profile. The effect of this deviation will change the distance between the buried sources and the surface of the Earth, causing a smoothly varying amplification of the gravity anomalies with increasing distance, but the effect is negligible in comparison with other sources of error in most modeling efforts.

The depth extent of the model is dictated by the objectives of the survey. Modeling of sedimentary rocks is generally limited to the maximum depth to basement, although in models that incorporate basement relief, the models may extend to a depth which includes the lowest level of the basement. If the depths extend to the basement, modeling needs to incorporate the sedimentary rock/basement effect, which can reach density contrasts of a few hundred kilograms per cubic meter. When modeling crustal anomaly sources with models of several hundreds of kilometers in length, the models generally extend to sub-Moho depths. The density contrast across the Moho has a range of 200 to 450 kg/m^3 (Chapter 4). Assuming a contrast of 300 kg/m^3, each kilometer of change in Moho depth results in a gravity anomaly of roughly 12.5 mGal. Where the Moho can be assumed to be flat over the length of the profile, and making the common assumption that the lower crust does not have sufficiently sharp density contrasts to produce identifiable gravity anomalies at the Earth's surface, the depth extents of the models terminate at the Conrad discontinuity, which is assumed to be the base of the upper crust at a depth of about 15 km. In summary, the depth extent of the model is generally considered to be the deepest depth to where the geology starts to approximate horizontal slabs relative to the profile's lateral extent. In dealing with models that extend into the mantle, an approximate check can be made of their validity by calculating the total mass of a series of geologic columns over the model. If the region is in isostatic equilibrium, the mass of the columns should be roughly equal.

The upper surface of the model is appropriately established at the elevation of the gravity stations, and the gravity anomalies are calculated from this surface. This is particularly important in dealing with topographic relief of a few tens of meters or more where source bodies might be near the surface. If the gravity anomaly values are obtained from gridded data that have no assigned elevations, it is necessary to evaluate the effect on the calculated anomaly of the range of elevations existing in the region of the profile while considering the depths of the anomaly sources. This can be achieved by comparing the gravity effect at the minimum elevation with that of the maximum elevation over the length of the profile. If the difference in these anomaly profiles is significant, it might be necessary to consider interpolating elevations along the profile from digital elevation models. In some models of deep crustal sources, gridded data are used and the elevation of the surface is not considered as significant to the geologic problem. Rather, the surface elevation is assumed to be at sea level.

Matching the observed and modeled gravity anomalies

The gravity model is computed for effects to compare with the observed anomaly. The difference in the gradient and amplitude of the anomalies can be used to modify the shapes of the modeled geological sources. It is important to consider the gradients in initial modification of the sources. If the calculated anomaly has a shallower gradient than the observed anomaly the sources are moved deeper until the gradients match. Some interpreters account for the shallower gradient by decreasing the dip of the margins of the sources. This has the same effect as deepening the source. Although this change in configuration of the source may be justifiable in terrains where dipping bodies are indicated by collateral but independent evidence, decreasing the dip is generally not advisable. Using more vertical dips will lead to

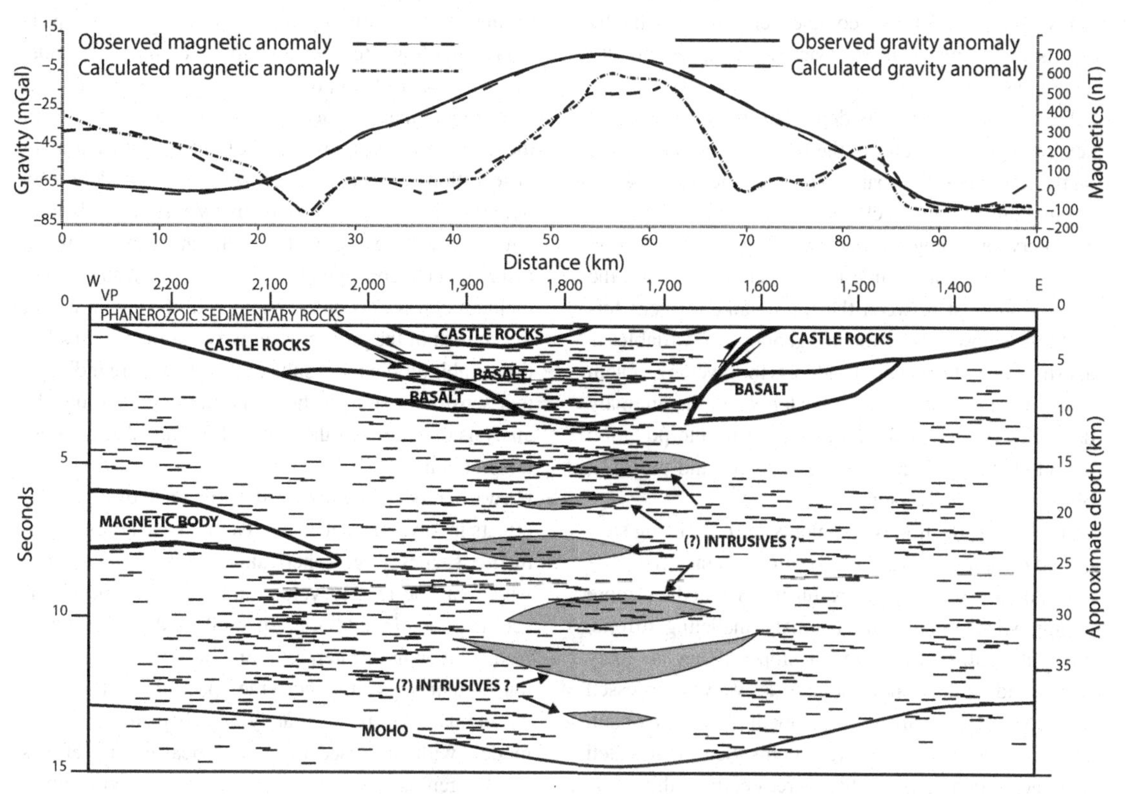

FIGURE 7.29 Geological model of the Midcontinent Rift System in eastern Kansas, based on seismic reflection and combined gravity and magnetic analysis. Adapted from WOELK and HINZE (1991).

an interpretation that has the sources at a maximum depth which is a useful constraint in evaluating the matched model.

With gradients matched, the amplitudes of the anomalies can be matched by the linear process of adjusting the density contrasts of the sources relative to the background density. The change in density can be estimated for wide bodies, that is for bodies that subtend an angle of 90° or more at the observation or calculation point, by rearranging the Bouguer slab equation to calculate the density change needed to produce the observed anomaly difference. This approximation is possible because the amplitude of the computed effect of the Bouguer slab is independent of depth. Care must be taken to insure that the modeled anomaly curve does not intersect the background level, but only approaches it asymptotically. Further refinements in the model are often possible if the sources in the model have a magnetization contrast, thus producing magnetic anomalies that can also be matched. This joint modeling is useful in constraining the possible geometries and developing increased confidence in the results. Numerous examples of the use of combined gravity and magnetic modeling have been given (e.g. WOELK and HINZE, 1991; BROCHER *et al.*, 1998; BLAICH *et al.*, 2011). Figure 7.29 shows an example of combined gravity and magnetic modeling based on the results of seismic reflection profiling across the Midcontinent Rift System in Kansas. Similarly, 3D interactive modeling of gravity and magnetic data has proved to be an important aid in interpretation of potential-field data (e.g. GÖTZE and LAHMEYER, 1988; PEDREIRA *et al.*, 2007).

7.6 Key concepts

- Gravity interpretation alone is never unique, owing to the ubiquitous presence of data errors and the theoretical source ambiguity of any gravity field variation. Subsurface constraints from drilling, geology, geophysical, and other observations can help limit interpretational ambiguities of gravity analysis, but cannot eliminate them completely.
- Gravity interpretations range from qualitative analysis, where objectives are satisfied by the presence or absence of an anomaly, to comprehensive quantitative modeling of the anomaly data. Quantitative interpretations include 3D modeling efforts over anomaly

maps that are relatively complex compared with 2D efforts that apply to profiles across elongated gravity anomalies.

- Effective interpretation is dependent on knowledge of key geological variables of the source body as a control on the amplitude and spatial characteristics of the gravity anomaly. Anomaly characteristics vary with source geometry including volume and shape, source depth, the isolation of the source from other sources, and the density contrast between the source and adjacent formations. Knowledge of the key geophysical variables is also necessary to effectively address the inverse problem of estimating geological parameters from the anomaly data. Key geophysical variables used in interpretation include the amplitude, sharpness, shape, and perceptibility of the gravity anomaly.
- Gravity interpretation is commonly initiated using simplified techniques to ascertain critical characteristics of the anomaly source that are further detailed by more comprehensive follow-up anomaly modeling efforts. Anomaly half-widths, straight slopes along anomaly flanks, and the horizontal anomaly derivatives expressed by the Smith Rules are graphical methods used to obtain first-order estimates of source depths. Semi-automated approaches consider possible source depths within moving windows of data by the Euler and Werner deconvolution methods, as well as depths from the slopes identified in the power spectrum of the anomaly data. Initial depth extents or source thickness estimates can be derived from the Bouguer slab approximation, whereas the inflection points on the flanking gradients of anomalies help map out the lateral margins of gravity sources. The anomalous mass of a source, which is the product of the source's density and volume, can be determined uniquely from the source's anomaly by Gauss' law, assuming that the anomaly has been effectively isolated.
- Gravity modeling proceeds either through forward modeling using trial-and-error methods where only a few unknown parameters of a model must be determined, or by inverse modeling using usually least-squares inversion methods where a greater number of unknowns must be estimated. Both approaches require specifying a forward model that produces predictions to compare with the gradients and amplitudes of the gravity anomaly observations. Forward modeling, which is widely used in gravity interpretation, is relatively easy to implement with available software. However, care is necessary to insure that the conceptual model that is used adequately includes constraints on the geology of the region within the context of the physics of gravity anomalies. In addition, ambiguity in the results of this procedure indicates that sensitivity studies should be incorporated into the analysis to determine the range of permissable parameters in the model.
- Forward models with closed form analytical gravity effects involve idealized sources that have simple symmetric shapes (e.g. spherical, cylindrical, prismatic) which can be readily volume-integrated at the observation point. In addition, spectral analyses of idealized sources have interpretational advantage over spatial domain analysis where the spectrum of the gravity anomaly can be accurately approximated. Forward models without closed form analytical gravity effects involve general sources with irregularly shaped volumes which must be numerically integrated at the observation point with considerably greater care and computational effort.
- Both forward and inverse modeling are inversions that estimate unknown parameters of the assumed forward model from the model's known parameters and anomaly observations. However, because the known or assumed parameters and observations contain errors, the inversion solution is not unique, but only one of a range of solutions that the error bars accommodate. In addition, the gravity solution suffers from inherent source ambiguity because many source distributions satisfy it. Thus, constraints from ancillary geological, geophysical, and other observations are essential to help resolve effective solutions for a gravity investigation.

Supplementary material and Study Questions are available on the website www.cambridge.org/gravmag.

Part II
Magnetic exploration

8 The magnetic method

8.1 Overview

The magnetic method is the oldest and one of the most widely used geophysical techniques for exploring the Earth's subsurface. It is a relatively easy and inexpensive tool to apply to a wide variety of subsurface exploration problems involving horizontal magnetic property variations from near the base of the Earth's crust to within the uppermost meter of soil. These variations cause anomalies in the Earth's normal magnetic field that are mapped by the magnetic method.

The geomagnetic field is caused by electrical currents associated with convective movements in the electrically conducting outer core of the Earth. Additional significant components of the field are derived from variations in the magnetic properties of the lithosphere and electrical currents in the ionosphere as well as in the Earth. The main field roughly coincides with the Earth's axis of rotation and is dipolar in nature. The force of attraction between opposite magnetic poles, or repulsion between like poles, is proportional to the product of the strength of the poles and inversely proportional to the square of the distance between the poles in a manner analogous to the gravitational attraction between masses. For most exploration purposes the magnetic force per unit pole, or magnetic field strength, is measured either in a directional mode where the sensor is spatially oriented or in a non-directional mode where the sensor obtains total field measurements that are presumed to be collinear with the Earth's magnetic field. Modern magnetic measurements are obtained electronically by measuring the precession of atomic particles which is proportional to the ambient magnetic field. Typically measurements to map variations in the magnetic properties of the lithosphere are made to a precision of roughly one part per 500,000 of the terrestrial field.

Successful applications of the magnetic method require an in-depth understanding of its basic principles, and careful data collection, reduction, and interpretation. Interpretations may be limited to qualitative approaches which simply map the spatial location of anomalous subsurface conditions, but under favorable circumstances the technological status of the method will permit more quantitative interpretations involving specification of the nature of the anomalous sources. No other geophysical method provides critical input to such a wide variety of problems. However, seldom does the magnetic method yield an unambiguous answer to an investigation problem. As a result, it is generally used in concert with other geophysical and geological data to limit its interpretation ambiguities.

8.2 Role of the magnetic method

The magnetic method has been employed for a wide variety of problems dealing with mapping of the subsurface since it was first used in the seventeenth century to locate buried, highly magnetic, iron ore deposits. With the advent of precise airborne magnetometers during World War II, the use of the method changed from a focus on localized studies, largely for mineral exploration purposes, to a regional tool for mapping the thickness of sedimentary basins and the geologic structure and nature of buried crystalline rocks. The subsequent development of atomic (resonance) magnetometers capable of a precision of a few orders of magnitude greater than previous instrumentation and requiring no orientation has broadened and simplified the use of magnetic surveying.

Modern magnetometers are deployed in drill holes, mines, submarines and ships, ground surveys and observatories, rotary- and fixed-winged aircraft, balloons, space shuttles, and satellites. In geological practice, geomagnetic surveys are mostly conducted on the ground and sea surface, at altitudes up to a few kilometers by aircraft, and at roughly 350–650 km by satellites. These various platforms are capable of directly mapping the geomagnetic field originating from the relatively shallow subsurface. They are used for a wide variety of applications ranging from archaeological and engineering site investigations, to exploration for economic resources hidden within the Earth, and studies of crustal tectonic features and processes.

The results of magnetic investigations are seldom sufficiently diagnostic to provide the complete answer to problems concerning the subsurface and, as is the case of other potential field techniques such as the gravity method, the interpretations are not unique. Accordingly, the magnetic method is normally employed in conjunction with other geophysical methods and direct subsurface information. The relative ease and simplicity of applying magnetics make it a desirable first-choice methodology wherever it can solve or contribute to the solution of subsurface studies.

Like gravity (Chapter 2.3), the magnetic method is a passive exploration method in that it is based on mapping the natural or normal magnetic force field of the Earth. The geomagnetic field is ever-present, but it varies both spatially and temporally. These variations distort the anomalous changes in the field (i.e. anomalies) caused by local subsurface conditions of interest in magnetic surveying, and thus must be removed from the observed data to produce interpretable measurements. The next section gives an overview of the Earth's magnetic field and its principal variations, which are developed in greater detail in the subsequent chapters on magnetic data acquisition, processing, and interpretation for subsurface exploration.

8.3 The Earth's magnetic field

The terrestrial magnetic field is the summation of several magnetic components, derived from both within and outside the Earth, which vary spatially and temporally over and above the Earth. The main field generated by electromagnetic currents in the outer core of the Earth accounts for about 98% or more of the geomagnetic field. The highly dynamic external fields due to the interaction of solar plasmas with the core field contribute most of the remaining field. Superimposed on the main and external fields are the relatively minor static effects from subsurface magnetization contrasts that are of interest in exploration studies used to help determine the compositional, structural, and thermal properties, and thus the history of the Earth's crust and uppermost mantle.

Relatively simple procedures have been developed for measuring and separating the components of the geomagnetic field so that it has become perhaps the most measured geophysical attribute of the Earth from both airborne and satellite surveys. Additional details are given below concerning the fundamental principles underlying the behavior of the geomagnetic field in space and time, the units in which the variations of the field are measured, the broad spatial and temporal characteristics of the field, and the principal measurements used to assess the properties of the field.

8.3.1 Magnetic force

Magnetism is associated with the motion of electric charges. At the most fundamental level, the magnetic properties of matter originate in a dipole moment caused by spin and orbital motion of electrons around the nucleus of atoms, and coupling of spins between particular adjacent atoms. The connection between electrical and magnetic fields was first studied by Ørsted and Ampere in the early nineteenth century, and later in the century Maxwell achieved a unified mathematical theory of the field.

Magnetism involves fields which exert a force on other magnetic bodies and electrically conducting materials. Unlike the monopolar phenomenon of gravitation, magnetism is dipolar with each magnetic component consisting of two poles that attract each other. Magnetic poles occur near free surfaces of a magnetized body that are not parallel to the internal magnetic field. Thus, for example, the magnetization of a linear body magnetized along its length (e.g. a bar magnet) is represented by two magnetic poles where the magnetization is concentrated near the opposite ends of the body. By convention these fictitious points, which are analogous to point masses or electric charges in classical physics, are referred to as north or positive poles if when free to move they tend to align with the north geographic pole of the Earth. It is the pole strength of an object, a property determined by the dipole magnetic moment of the constituent atoms and molecules, that is the source of the magnetic field.

In geophysical considerations, pole strength is replaced by magnetization, which is a function of the pole strength (and thus the magnetic moment) and the volume of the object. Magnetization or the magnetic moment per unit volume is the operative physical property for the magnetic method of geophysical exploration. It is the vector sum of an induced magnetization given by the product of magnetic susceptibility (the ease with which an object is magnetized in the ambient magnetic field) and the intensity of the terrestrial field, as well any permanent or remanent magnetizations imprinted on the object by previous magnetic fields.

The terrestrial magnetic field as measured on the surface of the Earth has numerous sources, but the primary component, which is the largest under most circumstances and without which the other components would not exist, is the planetary field derived from deep within the Earth, commonly referred to as the main or core magnetic field. Although numerous suggestions have been made for the origin of this field (e.g. MERRILL and MCELHINNY, 1983), the available evidence strongly supports an origin by electrical currents associated with convective movements in the electrically conducting outer core of the Earth. The convective movements probably are driven by the transport towards the surface of the Earth of heat derived from the solidification of the liquid outer core of the Earth onto the solid inner core and possibly by thermal sources, perhaps associated with the decay of radioactive isotopes. These movements are also affected by rotation of the Earth, and perhaps precession of the core, which in turn may play a significant role in the generation and nature of the field.

The approximate alignment of the dipolar field and the axis of rotation suggests a connection to the forces derived from the Earth's spin. The exchange of the mechanical energy of mass movement to magnetic energy is achieved by a dynamo. As applied to the generation of the Earth's magnetic field, the internal generation of the field is referred to as the geomagnetic dynamo. All planets of our solar system including the Moon appear to have or have had dynamo-driven magnetic fields. The dynamo, and thus the field, is self-sustaining in the Earth by virtue of the radial motion and rotation of the fluid core, but is subject to instabilities causing the magnetic field to reverse polarity at irregular intervals throughout geologic history, most recently about 700,000 years ago. The detailed nature of the core's electrical currents and how they lead to reversal of the field remains a matter of continuing research.

The magnetic force between two magnetic poles is proportional to the product of their strengths (Figure 8.1). If the poles are of opposite sign, the force is attractive, and thus operating to move them towards each other, whereas if the poles are the same sign, the force is repulsive and operating to push them apart. The relationship between forces and the poles of magnets was first investigated experimentally by Coulomb in the late eighteenth century and shortly thereafter was put into mathematical terms by Poisson. This empirical law, referred to as Coulomb's law, is similar in structure to Newton's law of gravity force and states that

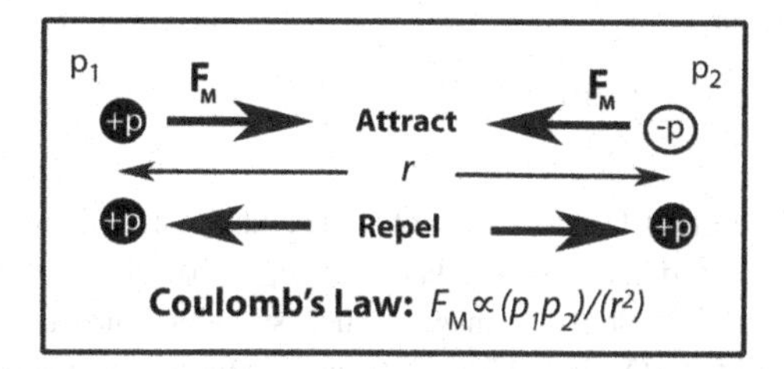

FIGURE 8.1 Schematic illustration of the attractive and repulsive magnetic forces ($\mathbf{F}_M$) generated between two magnetic poles by Coulomb's law. The unit magnetic dipole (top) consists of two fictitious point poles of equal strengths (p), but opposite signs and separated by an infinitesimal distance (r).

$$\mathbf{F}_{\mathrm{M}} = C_{\mathrm{m}} \frac{p_1 p_2}{\mathbf{r}^2}, \tag{8.1}$$

where p_1 and p_2 are point magnetic poles of strength p and r is the distance between them. In CGSu, a unit strength magnetic pole will create a force of one dyne when separated from an equivalent pole by one centimeter. C_{m} is a magnetic constant depending on the units used in the formulation. In classical studies employing CGSu, C_{m} is $1/\mu$ where μ is the magnetic permeability of the medium surrounding the poles. Magnetic permeability is a constant of a material and is a measure of the ease with which the magnetic field is passed through the material. It has a value of one in free space and is essentially the same in air or non-magnetic Earth materials, thus C_{m} is essentially unity in CGSu. However, magnetic calculations are generally performed in SIu which are required by most scientific publications. In SIu the force $\mathbf{F}_{\mathrm{m}}$ is in newtons (N), pole strength is in amperes times meters (A $\times$ m,), and the separation, r, is in meters (m). The constant C_{m} in these units is $(\mu_o/4\pi)$ where μ_o is the magnetic permeability of free space defined as $4\pi \times 10^{-7}$ henry per meter (H/m) or newton per square meter (N/m^2). In this book, the constant C_{m} is included in equations for calculating magnetic fields and usually takes a value of $(\mu_o/4\pi)$ or 10^{-7} to accommodate SIu. Where the constant does not appear in the equations, it is assumed the units are CGSu.

As in the case of gravitational fields, the magnetic force cannot be determined independently of the fundamental property of pole strength. Thus, a more useful quantity is the magnetic field, $\mathbf{B}$, which is the force on a unit pole, or

$$\mathbf{B} = \frac{\mathbf{F}_\mathrm{M}}{p_1} = \frac{C_\mathrm{m} p_2}{\mathbf{r}^2}, \qquad (8.2)$$

where p_1 is a fictitious unit pole at a point in space where **B** is specified. It is assumed that $p_2 \gg p_1$ so that p_1 does not disturb the field **B**. The magnetic field strength is measured in oersteds (Oe) in CGSu or EMu, and in amperes per meter (A/m) in SIu, where 1 A/m $= 4\pi \times 10^{-3}$ Oe $= 0.0126$ Oe.

Magnetic force has rotational and translational components causing magnets to mechanically interact at a distance. Thus, a magnet moving through the field of a second magnet in free space experiences, for example, a turning force which seeks to align the axis of its poles with the field lines of the second magnet. Translational forces of attraction and repulsion also operate owing to the interaction of the magnet's poles with the respective opposite and like poles of the second magnet. Single magnetic poles have yet to be observed in practice, although some theories of modern physics predict their existence. As a result, all observed positive magnetizations have not only a positive response, but also a negative one from the complementary pole and *vice versa.* As a result, a negative magnetic field anomaly may originate from a lower than normal magnetic polarization due to lateral variations in magnetic polarization or from the effect of a complementary pole associated with an increase in magnetic polarization.

The properties of a magnetic field are measured in terms of the mechanical work it takes to move a magnetic dipole in the field. The magnetic pole of strength p webers experiences a force of intensity

$$F_\mathrm{M} = pB \qquad (8.3)$$

in a field of intensity B, whereas the magnetic moment of the dipole is defined as

$$\mathbf{j} = (p \times l)\mathbf{r}_l, \qquad (8.4)$$

where p is the magnitude of either pole separated by the infinitesimal distance l with unit vector $\mathbf{r}_l$ and the direction of **j** is from the negative to positive pole.

Thus, placing a dipole in a uniform magnetic field **B** where the dipole moment **j** makes an angle Θ with **B** (Figure 8.2) produces a zero net force because the forces on the two poles are clearly equal and opposite by Equation 8.3. However, the magnitude of the torque τ acting on a pole p with respect to the center of the dipole at O with infinitesimal length l is by definition $[F_\mathrm{M}(l/2)\sin(\Theta)]$ so that the net torque on the dipole about its center is

$$\tau = 2F_\mathrm{M}(l/2)\sin(\Theta) = jB\sin(\Theta) = \mathbf{j} \times \mathbf{B}, \qquad (8.5)$$

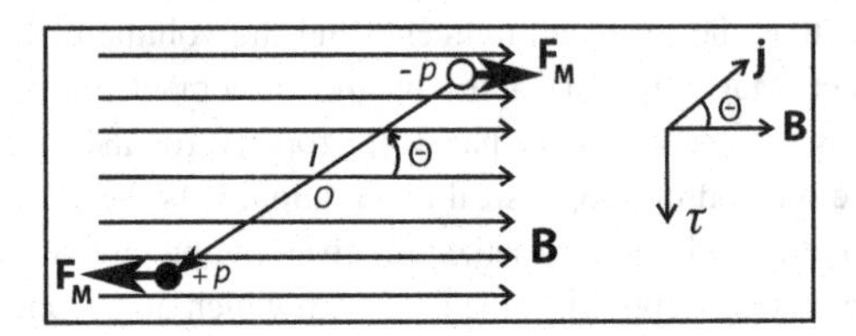

FIGURE 8.2 A uniform magnetic field **B** subjects a magnetic dipole with infinitesimal length l and moment **j** to a torque τ through an angle Θ.

where the vector notation refers to the cross-product of **j** and **B** with τ perpendicular to their plane in a right-handed system (Figure 8.2). In addition, the work or change in potential (V) required to turn the dipole's axis from say $\Theta_o = 90°$ to an arbitrary angle Θ is just

$$V = \int_{\Theta_o}^{\Theta} jB\sin(\Theta)d\Theta = -jB\cos(\Theta) = -\mathbf{j}\cdot\mathbf{B}, \qquad (8.6)$$

where the vector notation signifies the dot product of **j** and **B**. Thus, spatial changes or gradients in the scalar potential V give the magnetic force field τ to within a constant of integration.

The product of **j** and **B** governs not only the turning movements of a dipole in a magnetic field, but also its translational action in a spatially variable field, such as the repulsive and attractive motions a magnet makes in the vicinity of another magnet. It also illustrates the mechanical principle underlying magnetic measurements such as the changes of (1) magnetic moment **j** per unit volume or magnetization under a known **B** or (2) the field **B** under a known magnetization (**j**/volume). The first condition is the basis for measuring the susceptibility and other magnetization properties of materials. The second condition is the basis for magnetometers that measure magnetic field variations in time and space. Additional details on the design and operational principles of magnetic property measurements and magnetometers are considered in Sections 10.7 and 11.3, respectively.

8.3.2 Units

The units of magnetic field strength of oersted (Oe) in electromagnetic units (EMu) and tesla (T) in the SI system of units (SIu) are inconveniently large for most magnetic geophysical exploration usage (Table 8.1). The magnetic field commonly is presented in EMu of induction or flux density. The unit of B in EMu is the gauss, or in geophysical usage the gamma (γ), which is equal to 10^{-5} gauss. The unit of B in SIu is the tesla (T) which is equivalent to 10^4 gauss, but in geophysics the unit used is the

TABLE 8.1 Magnetic parameters.

Parameters commonly used in the magnetic method in equivalent CGSu *or* EMu *and* SIu. *Magnetic susceptibility is also given sometimes in electromagnetic units (*EMu*) of (*EMu/cm^3*) or (*EMu/g*) to distinguish between volume and mass (or specific) susceptibility, but the quantity is dimensionless.*

Magnetics parameter	CGSu or EMu	SIu
Force of attraction and repulsion	10^5 dynes	newton (N)
Magnetic field strength or intensity	oersted (Oe)	10^{-3} ampere (A)/m
	$\gamma = 10^{-5}$ gauss	nanotesla (nT) $= 10^{-9}$ tesla
Magnetic induction	10^4 gauss	tesla (T)
Magnetic permeability of free space (μ_o)	gauss/oersted (gauss = oersted)	$4\pi \times 10^{-7}$ henry (H)/m or $4\pi \times 10^{-7}$ N/A^2
Magnetization	EMu/cm^3	10^3 A/m
Magnetic susceptibility	dimensionless	4π (dimensionless)

nanotesla (nT) or 10^{-9} tesla, which is numerically equivalent to the gamma. Although the gamma is widely found in the geophysical literature, the more appropriate unit is the nanotesla based on the modern international acceptance of the SIu.

8.3.3 Basis of the magnetic method

The magnetic method is based on the measurement and analysis of perturbations or anomalies, in the terrestrial magnetic field caused by lateral variations in the magnetization of subsurface materials. Anomaly mapping together with the magnetic potential field theory and magnetization properties respectively described in Chapters 9 and 10 below, provides a foundation for applying the magnetic method in subsurface exploration. Table 8.2 gives the conversions between CGSu and SIu for the parameters commonly used in magnetic investigations.

Magnetic exploration is a potential field method where the magnitude and direction of the effect depends on the relative positions of the source of the magnetization and the observation point, and the contrast of the source magnetization magnitude and direction with the laterally adjacent materials. In many ways the magnetic method parallels the gravity method in that both are potential field methods employing perturbations in a planetary field. However, the magnetic method offers several advantages over the gravity method. First, most measurements are made of the absolute total field to a high-degree of precision from a non-oriented sensor, which greatly simplifies and speeds up field observations. Second, the dynamic range of magnetization contrasts is much greater than density contrasts, which increases the amplitude of anomalous variations. For example, in many applications the magnetizations of investigated sources are six orders of magnitude greater than adjacent terrestrial materials. Third, the magnetization of materials is dipolar in contrast to the monopolar (attractive) properties of mass. The effects of this are profound. For example, the inverse distance function varies with distance one power faster than for the gravity force for the same source configuration. The result is an increase in the sensitivity of the magnetic method to the distance to the source and in resolving individual sources.

8.3.4 Spatial variations

Efforts to determine the spatial and temporal properties of the geomagnetic field began with Gilbert's classical studies in the late sixteenth century. The importance of geomagnetism to navigation led to the development of a global network of terrestrial geomagnetic observatories to better determine and monitor the behavior of the Earth's magnetic field. Observatory data in combination with geomagnetic observations from satellites since the early 1960s reveal that the Earth's main dipole magnetic field is distorted into the comet-like shape in Figure 8.3 called the magnetosphere. The flow of corpuscular radiation or plasma from the Sun, called the solar wind, causes the main field to be compressed towards the Earth in the direction of the Sun and to extend out into space in the antipodal position.

The solar winds, which may be toxic to life, are deflected from the Earth by the Earth's dipolar magnetic field. When the main or planetary magnetic field shuts down, as it did on Mars some 4 Ga ago, the solar winds strip the planetary surface of the light elements and volatiles that are essential for known life on Earth. Thus, life on Earth could not exist without the protection of the main field.

TABLE 8.2 Conversion of magnetic parameters from CGSu or EMu to SIu.

Parameter (symbol)	CGS units (or emu)	SI units	Conversion CGS ➡ SI	Conversion SI ➡ CGS
Magnetic induction (**B**) (Magnetic flux density)	gauss (G)	tesla (T) (Wb/m^2) (H A/m^2)	1 G = 10^{-4} T 1 G = 10^5 gamma (γ) = 10^5 nanotesla (nT)	1 T = 10^4 G = $10^9\gamma$ = 10^9 nT 1 nT = 1 gamma (γ)
Magnetic field intensity (**H**)	oersted (Oe)	ampere/meter (A/m)	1 Oe = $10^3/(4\pi)$ A/m	1 A/m = $4\pi \times 10^{-3}$ Oe $\approx 1257\gamma$
Magnetic polarization (**J**) (magnetization)	emu/cm^3 $\mathbf{B} = \mathbf{H} + 4\pi\mathbf{J}$ $\mathbf{J} = \mathbf{M}$	tesla (T) $\mathbf{B} = \mu_0\,\mathbf{H} + \mathbf{J}$ $\mathbf{J} = \mu_0\,\mathbf{M}$	1 emu/cm^3 = $4\pi \times 10^{-4}$ T	1 T = $10^4/(4\pi)$ emu/cm^3 (or gauss G)
Magnetization (**M**) (magnetic momen/unit vol)	emu/cm^3 (or gauss G)	A/m	1 emu/cm^3 = 10^3 A/m	1 A/m = 10^{-3} emu/cm^3 (or gauss G)
Magnetic pole (p)	unit pole	A m	1 unit pole = 10^{-8} A m	1 A m = 10^8 unit pole
Magnetic dipole (**m** = p**b**) (dipole moment)	pole cm (or emu) (or G cm^3)	A m^2	1 pole cm = 10^{-10} A m^2 1 G cm^3 = 10^{-3} A m^2	1 A m^2 = 10^{10} pole cm = 10^3 G cm^3 (or emu)
Magnetic volume susceptibility (k)	CGSu (or emu/cm^3 (dimensionless) $\mathbf{M} = \mathbf{J} = k\mathbf{H}$	SIu (dimensionless) $\mathbf{M} = \mathbf{J}/\mu_0 = k\mathbf{H}$	1 CGSu = 4π SIu	1 SIu = $1/(4\pi)$ CGSu
Magnetic mass (specific) susceptibility (k_m)	CGSu cm^3/g ($k_m = k/\sigma$)	SIu m^3/kg ($k_m = k/\sigma$)	1 CGSu cm^3/g = $4\pi \times 10^{-3}$ SIu m^3/kg	1 SIu m^3/kg = $10^3/(4\pi)$ CGSu cm^3/g
Koenigsberger ratio (Q)	dimensionless	dimensionless	1 emu = 1 SIu	1 SIu = 1 emu
Magnetic flux (Φ)	maxwell (Mx)	Wb (V s)	1 Mx = 10^{-8} Wb	1 Wb = 10^8 Mx
Magnetic potential (V) $\mathbf{B} = -\text{grad } V$	gilbert (Gi)	tesla m (T m) (or Wb/m)	1 Gi = 10^{-6} T m = $10/(4\pi)$ A	1 T m = 10^6 Gi
Magnetic force (**F**)	dyne (dyn)	newton (N)	1 dyn = 10^{-5} N	1 N = 10^5 dyn

Notes: G = gauss; γ = gamma; Oe = oersted; T = tesla; nT = nanotesla; Wb = weber; H = henry; A = ampere; Gi = gilbert; Mx = maxwell; g = gram; σ = density (g/cm^3 = 10^{-3} kg/m^3); N = newton; Wb = H A = V s; $\mu_0 = 4\pi \times 10^{-7}$ N/A^2 (or H/m)

Out to about 800 km above the Earth's surface, the main magnetic field is roughly equivalent to the field that would exist if a bar magnet with a dipole moment of 7.95×10^{22} A m^2 was located near the center of the Earth, but tilted at an angle of approximately 11° in a meridian passing through the eastern margin of the United States and Canada. This dipole field, which accounts for roughly 98% of the observed field on the surface, changes in amplitude with time, having decreased by about 10% in the past century and a half. The electromagnetic interactions of the solar plasma with the core field introduce additional temporal variations with significant amplitudes ranging over periods of seconds to a month or more. In general, every magnetic observation at or near the Earth's surface includes these normal magnetic effects which are large compared with geological magnetic effects of the subsurface. Thus, they must be removed from magnetic measurements to help isolate the magnetic effects of the subsurface targets for analysis. Additional description of the spatial and temporal variations of the geomagnetic field and the removal of these effects to prepare magnetic anomaly maps for geologic interpretation are presented in Chapter 12.

The Earth's main field is commonly portrayed by field lines or lines of force as illustrated in Figure 8.4(a). Although shown here in two dimensions, lines of force are curvilinear lines in three dimensions that are everywhere tangential to the ambient field, thus showing the

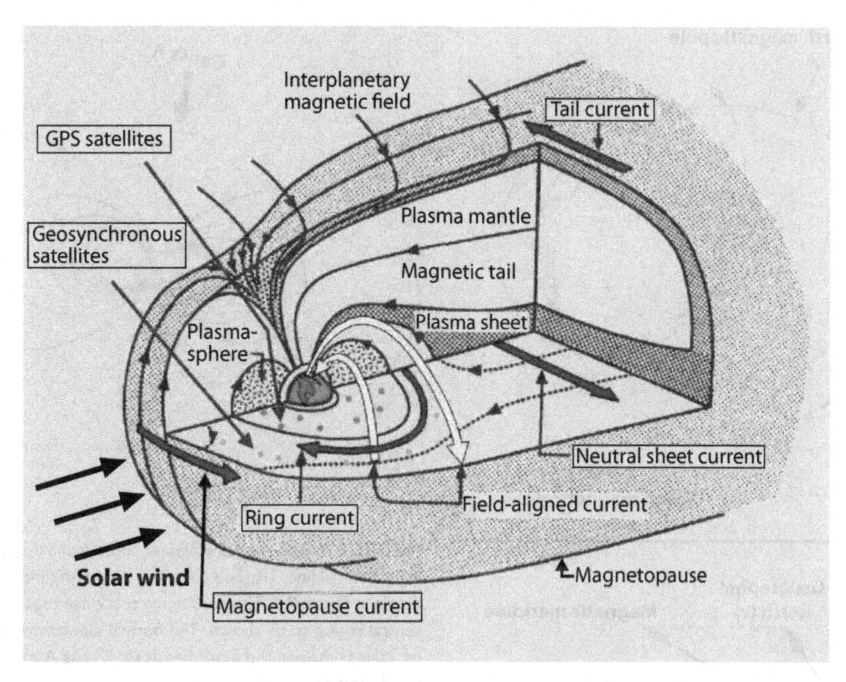

FIGURE 8.3 Schematic illustration of the magnetosphere surrounding the Earth. The Earth's magnetic field is confined by the solar wind with arrows indicating the electrical current directions. The magnetopause is the boundary between the solar wind and the Earth's magnetic field. The plasmas are electrically neutral containing both positively and negatively charged particles. The geomagnetic dipolar field properties shown in Figure 8.4 dominate the outer part of the Earth up to about 800 km altitude. Courtsey of the US National Oceanic and Atmospheric Administration (NOAA).

direction of the field. The number of lines of force passing through a unit area perpendicular to their direction (i.e. the flux in maxwells) is a relative measure of the amplitude or strength or intensity of the field. Figure 8.4(a) indicates that the Earth's field is vertical at the magnetic poles, horizontal at the magnetic equator, and has an amplitude at the poles of roughly twice that at the equator. By definition the pole attracted toward the north geographic pole is called the north-seeking (or positive) pole. Because opposites attract, this means that the north geographic pole is the south (or negative) magnetic pole and *vice versa* for the south geographic pole.

By convention, the main field of the Earth is inclined downwards from the horizontal in the northern geomagnetic hemisphere and upwards in the southern hemisphere with inclinations of the field increasing toward the poles from the equator. The geomagnetic poles are defined as the position where the line passing through the best-fitting dipole at the center of the Earth intersects the surface. These differ in a minor way from the actual magnetic poles where the field is directed vertically owing to non-dipolar components which are the residual of the difference between the best-fitting geocentric dipole field and the actual field of the Earth. The geomagnetic equator is defined by the plane perpendicular to the axis of the best-fitting dipole that passes through the center of the Earth. At the magnetic equator, which only roughly coincides with the geographic equator because of the tilt of the main field, the magnetic field is directed horizontally.

The principal components of the Earth's magnetic field at mid-latitudes are illustrated in Figure 8.4(b). The field $\mathbf{B}_\mathrm{N}$, commonly referred to as the Earth's normal field or main field, is resolved into horizontal $\mathbf{B}_{\mathrm{N}(H)}$ and vertical $\mathbf{B}_{\mathrm{N}(Z)}$ components, with positive z-direction downward along the local vertical. $\mathbf{B}_{\mathrm{N}(H)}$ in turn is resolved into the geographic north $\mathbf{B}_{\mathrm{N}(X)}$ and east $\mathbf{B}_{\mathrm{N}(Y)}$ components. The angular relationship between $\mathbf{B}_{\mathrm{N}(H)}$ and $\mathbf{B}_{\mathrm{N}(X)}$ is the declination D and between $\mathbf{B}_{\mathrm{N}(H)}$ and $\mathbf{B}_\mathrm{N}$ is the inclination I of the field. At the magnetic poles the components diagram collapses to a vertical line, and at the magnetic equator to a horizontal line in the direction of the magnetic meridian.

Thus, the strength or intensity of the normal field vector, $\mathbf{B}_\mathrm{N}$, in its geomagnetic components is

$$B_\mathrm{N} = \sqrt{B^2_{\mathrm{N}(H)} + B^2_{\mathrm{N}(Z)}} = \sqrt{B^2_{\mathrm{N}(X)} + B^2_{\mathrm{N}(Y)} + B^2_{\mathrm{N}(Z)}}, \tag{8.7}$$

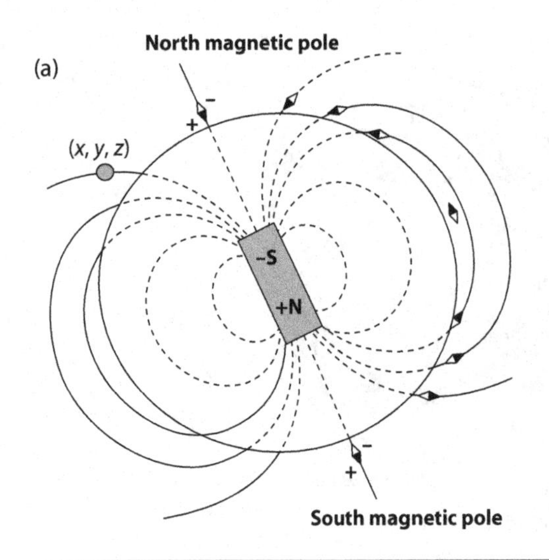

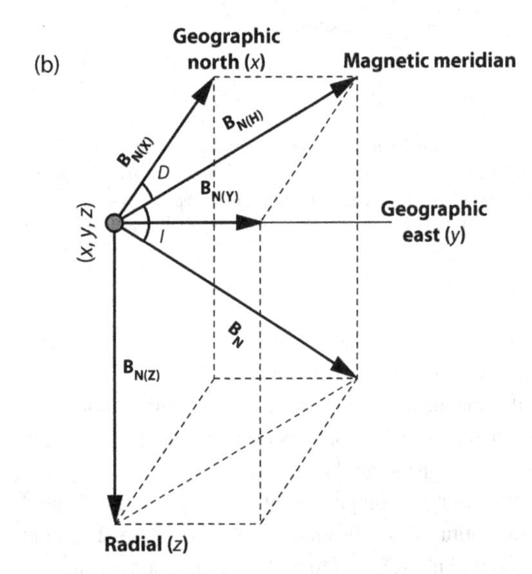

FIGURE 8.4 (a) The Earth's normal magnetic field $\mathbf{B}_N$ represented by lines of force from a dipole located at the center of the Earth, but inclined 10.9° from the Earth's rotation axis. The positive or north (N) pole of the compass needle points to the negative or south (S) pole of the Earth's central dipole and *vice versa*. (b) At the observation point (x, y, z), the normal field has components as described in the text.

where

$$B_H = (B_N)\cos I = \sqrt{B_{N(X)}^2 + B_{N(Y)}^2}, \tag{8.8}$$

$$B_{N(Z)} = (B_N)\sin I = B_{N(H)}\tan I, \tag{8.9}$$

$$B_{N(X)} = B_{N(H)}\cos D, \tag{8.10}$$

and

$$B_{N(Y)} = B_{N(H)}\sin D. \tag{8.11}$$

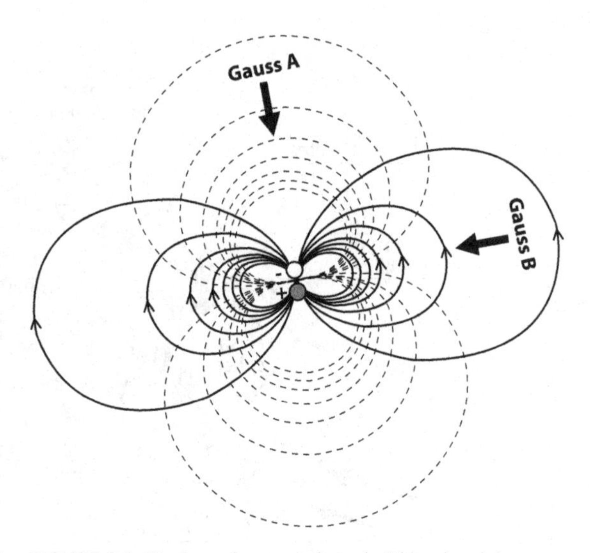

FIGURE 8.5 The lines of magnetic force (solid lines) and the equipotential lines (dashed lines) for the Earth's inclined central dipole in Figure 8.4. The lines come too close together in the central region to be shown. The normal dipolar magnetic field takes on values between the extremes at the Gauss A and B positions (bold arrows).

The geomagnetic field can be considered to first approximation as originating from a small dipole centered in the Earth. For any plane intersecting both poles of the dipole, Equation 8.6 (or 9.6) gives the magnetic field components along the axial or r-direction and the co-latitudinal or Θ-direction normal to r (Figure 8.4). Thus, taking the gradient of the potential V in spherical coordinates yields the components of the $\mathbf{B}_N$-field given by

$$B_{N(r)} = -\frac{\partial V}{\partial r} = (2j)\frac{\cos\Theta}{r^3}, \tag{8.12}$$

and

$$B_{N(\Theta)} = -\left(\frac{1}{r}\right)\frac{\partial V}{\partial \Theta} = (j)\frac{\sin\Theta}{r^3}. \tag{8.13}$$

The two extremes of the magnetic field of the dipole occur for $\Theta = 0$ at the so-called Gauss A or end-on position, and for $\Theta = \pi/2$ at the Gauss B or side-on position (Figure 8.5). At $\Theta = 0$, $B_\Theta = 0$, and

$$B_{N(r)} = 2\left(\frac{j}{r^3}\right), \tag{8.14}$$

whereas at $\Theta = \pi/2$, $B_{N(r)} = 0$ so that

$$B_{N(\Theta)} = \frac{j}{r^3}. \tag{8.15}$$

Thus, at the magnetic poles, the Earth's magnetic field is radial and twice that at the magnetic equator where the field is horizontal with no radial (vertical) component.

The total resultant magnitude of the field in terms of its components in Equations 8.12 and 8.13 is

$$B_N = \frac{j}{r^3}[4\cos^2\Theta + \sin^2\Theta]^{\frac{1}{2}}, \quad (8.16)$$

where Θ ($= \text{latitude} - \pi/2$) is the co-latitude of the Earth, and the direction α of the total field with respect to the radial can be obtained from

$$\tan\alpha = 2\cot\Theta. \quad (8.17)$$

Thus, by knowing the latitude θ ($= \pi/2 + \Theta$) on the Earth, the inclination I ($= \pi/2 - \alpha$) of the total field can be estimated from

$$\tan I = 2\cot\Theta. \quad (8.18)$$

These results describe the quantitative effects of the simple Earth-centered bar magnet that accounts for most of the spatial variations of the normal geomagnetic field $\mathbf{B}_N$ near the Earth's surface. Accordingly, the magnitude of the field $|\mathbf{B}_N|$ along any ray from the Earth-centered dipole decreases with the cube of the inverse distance from the dipole and is twice as large at a specific distance along the axial pole as it is at the same distance on a geomagnetic equatorial radius.

In addition to the dipolar field, the magnetic field at the surface of the Earth has components from the interior of the Earth as well as from external sources. The component which is of particular interest in geophysical exploration is derived from horizontal variations in magnetic polarization or magnetization of the lithosphere, the roughly 150 km mechanically strong outer shell of the Earth which consists of the crust and the very outermost portions of the mantle (Figure 1.1).

The crust is the primary source of this component because the rocks of the mantle are characteristically low in magnetic minerals. In any event, the lowest depth of magnetization is determined by temperature within the Earth: deeper in the Earth, in regions above its Curie temperature, a mineral is no longer able to retain its intense magnetism. The Curie temperature of most commonly occurring magnetic minerals is less than roughly 600 °C, a temperature reached under normal conditions on the continents at roughly the bottom of the crust and in the oceans slightly below this boundary. The magnetic components derived from this depth range are for all practical purposes of a permanent nature in terms of human spans of time. At the surface, they have dimensions measured in distances that are generally less than a few hundred kilometers.

8.3.5 Temporal variations

The geomagnetic field also exhibits a non-dipolar main field component originating within the outer core of the Earth that is superimposed on the dipolar field. The dimensions of these fields are measured in thousands of kilometers, and they have magnitudes of the order of 10% of the main field. The most dominant portion of the non-dipolar field drifts westward at a rate of roughly 0.18°/yr. This drift is readily observable on isomagnetic charts, which are contour maps of equal magnitude or direction of the Earth's field. These maps are prepared at intervals of 5 years, a span of time referred to as an epoch in geomagnetic studies. Over a human life span, long-term variations in both magnitude and direction are readily observed in both dipole and non-dipole fields of the Earth. These are termed secular variations. Their rate of change is of the order of 80 nT/yr, but varies greatly with time and position. Their internal origin and rapid variation with respect to geologic time argues for a source within the liquid outer core of the Earth associated with the principal as well as secondary fluid currents. Models designed to predict future secular variations have proved to be imprecise.

The secular variation in both the angular relationships and amplitude of the main magnetic field components has necessitated the periodic updating of the charts representing the Earth's magnetic field. These charts have a variety of purposes including navigation and definition of regional fields for use in geophysical exploration. For these purposes it is advantageous to have an internationally agreed approximation to the field. Each half decade since 1965 an International Geomagnetic Reference Field (IGRF) has been computed and internationally adopted, based on magnetic observatory and survey measurements plus, in recent years, observations of the Earth's field from satellites (e.g. Langel and Hinze, 1998; Constable, 2007; Olsen *et al.*, 2007).

The IGRF represents a best fitting approximation to the field by a series of spherical harmonics of increasing degree. Harmonics of up to degree 13, approximately 3,100 km and longer at the Earth's surface, are dominated by the core field, whereas higher degrees are related to lithospheric magnetization anomalies. The IGRF accommodates the flattening of the Earth's shape and considers elevation relative to sea level, so that all components of the Earth's magnetic field can be computed from the equation at any altitude. It also incorporates a predictive term for the secular variation of the field based on extrapolation of previous rates. Unfortunately, this cannot be done precisely. Thus, every 5 years a Definitive Geomagnetic Reference Field (DGRF) is calculated taking into account

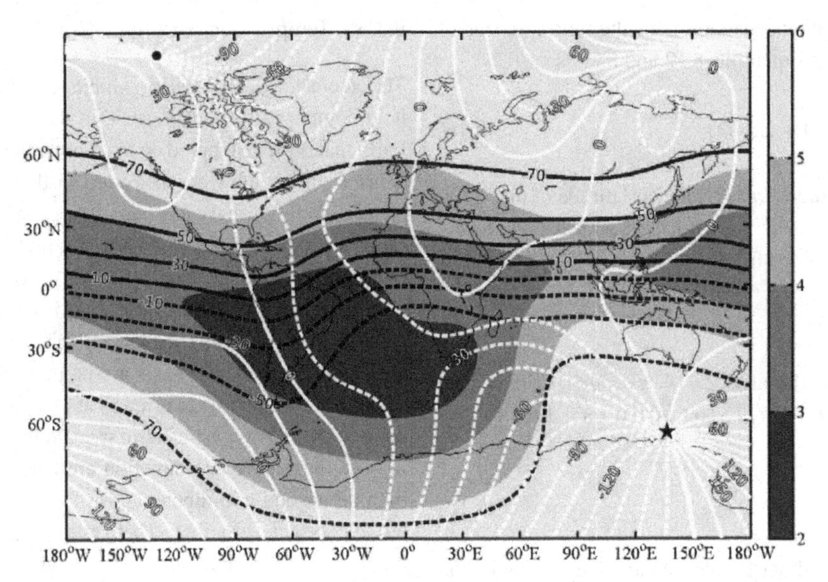

FIGURE 8.6 The total magnetic field of the Earth at 1 January 2011 (= 2011.0) as defined by the IGRF-11. The gray shading gives the intensities in 10^4 nT, whereas the black and white contours give the inclinations and declinations, respectively, in degrees. Dashed contours mark negative values. A large black dot and a star mark the respective positions of the north and south geomagnetic poles. Additional details are available at ngdc.noaa.gov/geomag/WMA.

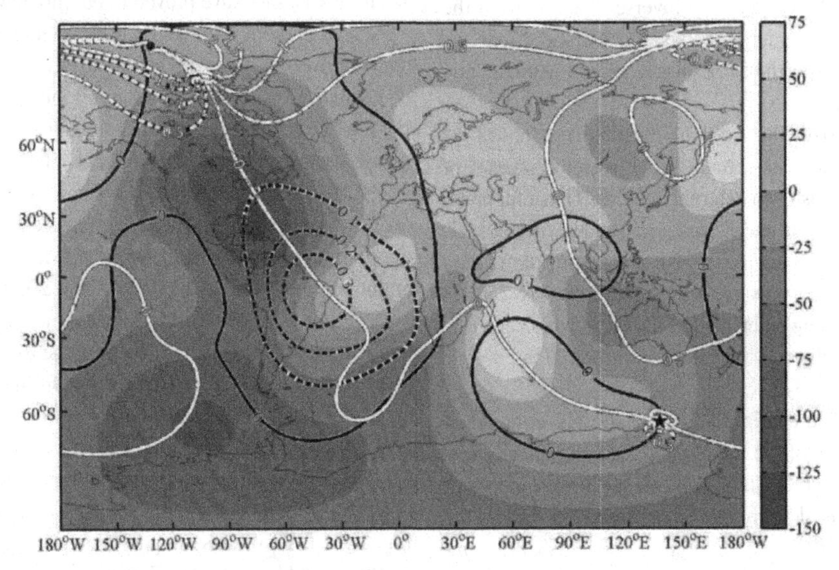

FIGURE 8.7 Annual secular variations for the period 2010.5–2011.5 as predicted by the IGRF-11. The gray shading gives the annual variations of intensity in nT/yr, whereas the black and white contours give the annual variations of inclination and declination, respectively, in degrees/yr. Dashed contours mark negative values. A large black dot and a star mark the respective positions of the north and south geomagnetic poles. Additional details are available at ngdc.noaa.gov/geomag/WMA.

an approximation to the actual secular variation that has been observed over the previous epoch.

The IGRF model for the beginning of 2011 (IGRF11) in Figure 8.6 shows how intensity changes from roughly 60,000 nT at the poles to 30,000 nT at the equator, along with the superimposed variations in inclination and declination. Figure 8.7 indicates annual changes of the geomagnetic field between 2010.5 and 2011.5 that include a maximum decrease in intensity of about 120 nT over Cuba and the eastern and central United Stastes and a

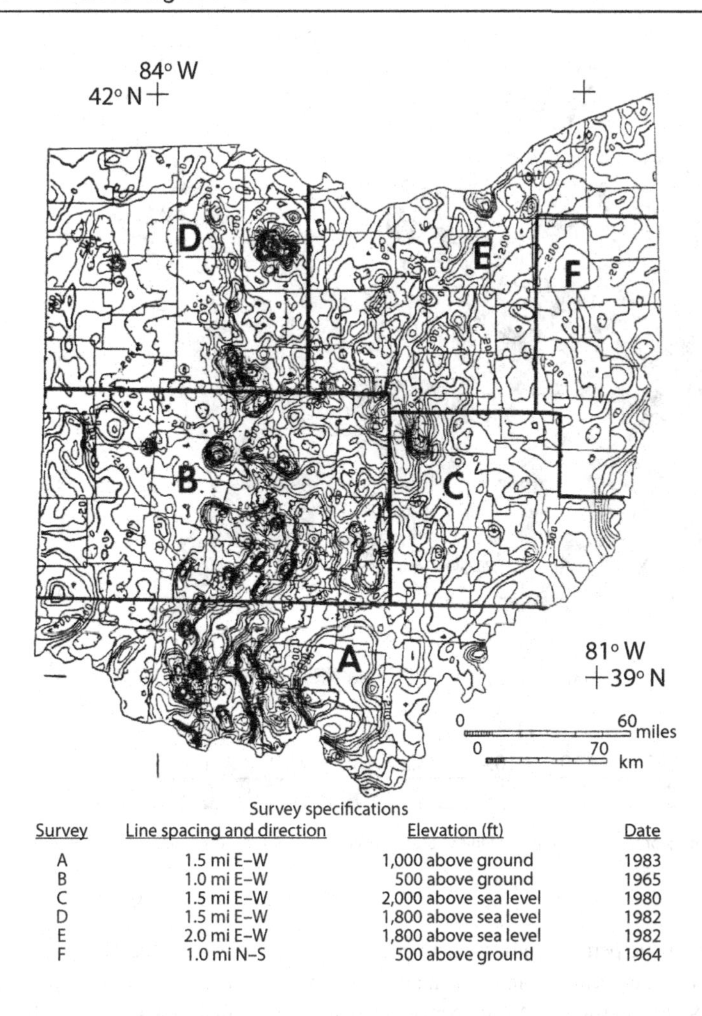

Survey specifications

Survey	Line spacing and direction	Elevation (ft)	Date
A	1.5 mi E–W	1,000 above ground	1983
B	1.0 mi E–W	500 above ground	1965
C	1.5 mi E–W	2,000 above sea level	1980
D	1.5 mi E–W	1,800 above sea level	1982
E	2.0 mi E–W	1,800 above sea level	1982
F	1.0 mi N–S	500 above ground	1964

FIGURE 8.8 Total magnetic anomaly map of Ohio with specifications for the six component airborne surveys. Amplitudes range mostly over $\pm 1{,}000$ nT and are contoured at intervals of 100 nT.

comparable magnitude increase over the Indian Ocean between Madagascar and Australia.

Secular variations as shown in Figure 8.7 are important for constraining geomagnetic dynamo studies and combining magnetic survey data taken during different years. As an example of the latter, Figure 8.8 shows the composite total field aeromagnetic anomaly map of Ohio along with the specifications for the six component surveys collected between 1964 and 1983. Extrapolation of the annual secular variation over the 19 year period of the surveys suggests that the main field intensity of Ohio decreased by about 1,500 nT. Removing the appropriate main field secular variation for each survey accounts for this change in main field intensity which otherwise would severely distort the smaller magnetic anomalies of the subsurface geology.

The daily revolution of the Earth causes a variation in the magnetic field at the surface due to asymmetry of the main field which is related to the relative position with respect to the Sun. Additional variations in the surface magnetic field are caused by external sources, which generally make up only about 1 or 2% of the total field. The source of these externally derived fields is in the ionospheric dynamo, its interaction with solar and corpuscular radiation from the Sun, and the tidal effects of the Sun and the Moon upon the ionosphere.

The Earth's ionosphere, beginning at about 50 km above the surface and extending above 1,000 km, is a shell of ionized gas or plasma formed primarily by sunlight stripping electrons from neutral atoms. Movements of the ions, caused for example by solar heating, result in electrical currents that cause a dynamo action which in turn produces magnetic fields that modify the internally derived field. During periods of solar storms when the Earth is subjected to a strong flux of charged particles spewed out from the Sun, there is intense interaction between the plasma from the Sun and the Earth's magnetic field resulting in increased magnetic fields that may cause disturbances of hundreds of nanoteslas or more. These varying magnetic fields induce secondary currents in the

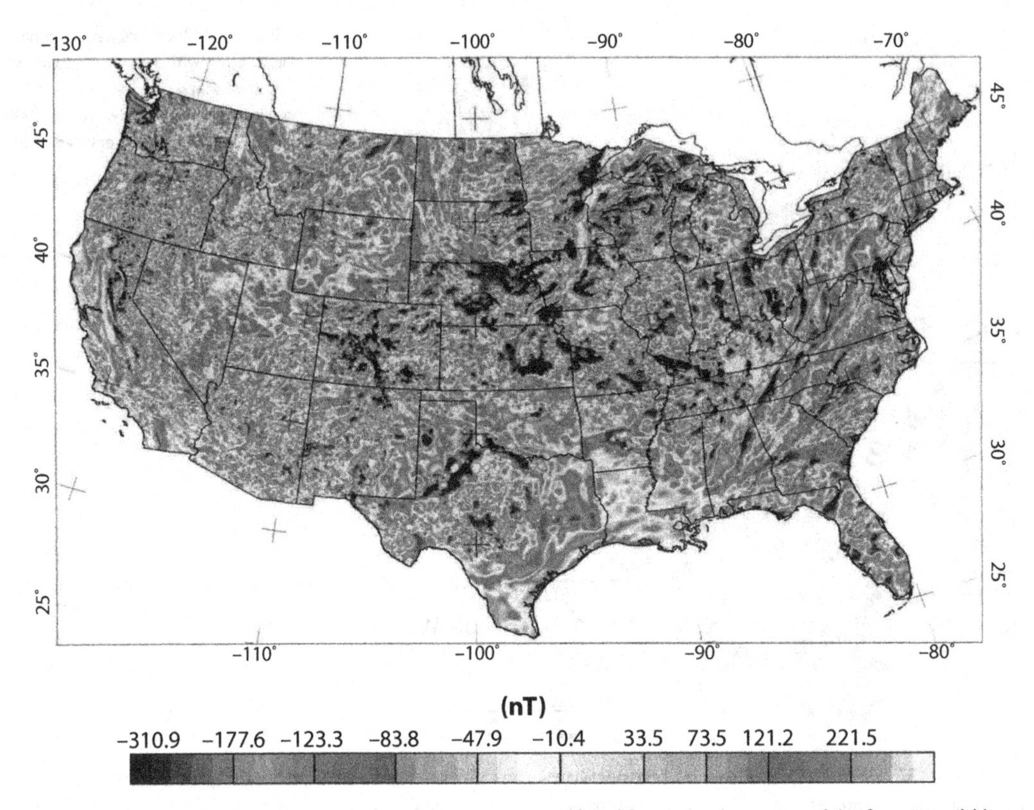

FIGURE 8.9 Reduced-to-pole magnetic anomaly map of the conterminous United States. A color version of this figure is available at www.cambridge.org/gravmage. Adapted from RAVAT *et al.* (2009).

conducting Earth, resulting in additional magnetic perturbations that have been intense enough to knock out power grids in Canada and the northern United States and will cause problems in isolating the lithospheric component in magnetic surveys.

A more comprehensive approach to isolating lithospheric magnetic anomalies from both extraneous internally and externally derived components of the terrestrial field, suggested by LANGEL *et al.* (1996), incorporates a spatially and temporally continuous model of the magnetic field. This Comprehensive Magnetic Field (CMF) model includes not only the main field and its secular variations but also the quiet-time long-wavelength ionospheric and magnetospheric fields together with internally induced fields derived from external currents. Current versions of this CMF model (SABAKA *et al.*, 2004) include data from worldwide magnetic observatories and measurements from POGO, Magsat, Ørsted, and CHAMP satellites. The utility of this model in isolating the lithospheric magnetic component has been shown in many surveys and in the merging of near-surface surveys (e.g. RAVAT *et al.*, 2003). The CMF model has been used by RAVAT *et al.* (2009) to eliminate discontinuities at survey boundaries existing in previous versions of the US magnetic anomaly map. The short-wavelength components of the North American magnetic anomaly data set have been incorporated with the long-wavelength components of the US data set processed with the CMF model to prepare a new magnetic anomaly map for the conterminous United States, shown in Figure 8.9.

Externally derived magnetic variations are commonly referred to as diurnal variations, but they have a wide variety of periods ranging from fractions of a second to several days or more, and magnitudes which generally increase with their period. Their magnitudes are sufficiently great that in many magnetic exploration surveys their effect must be monitored and removed from the observed signal for interpretational purposes. They cannot be predicted either by theory or empirical evidence to the precision required for most surveys except under the least variable of magnetic conditions. Magnitudes over a 24 hour period are generally less than 50 nT (Figure 8.10) except during periods of strong corpuscular bombardment of the Earth by the Sun during magnetic storms (Figure 8.11). Tidal effects upon the ionosphere contribute roughly one-tenth of the normal daily variation. Micropulsations over periods of

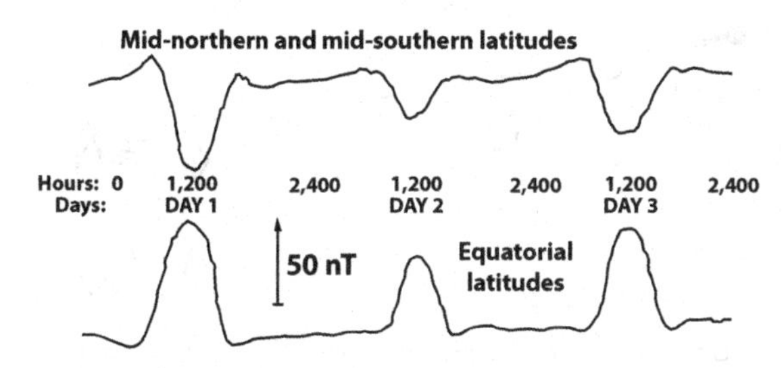

FIGURE 8.10 Typical temporal variations in total magnetic intensity for the middle (upper record) and equatorial (lower record) geomagnetic latitudes. Adapted from BREINER (1973).

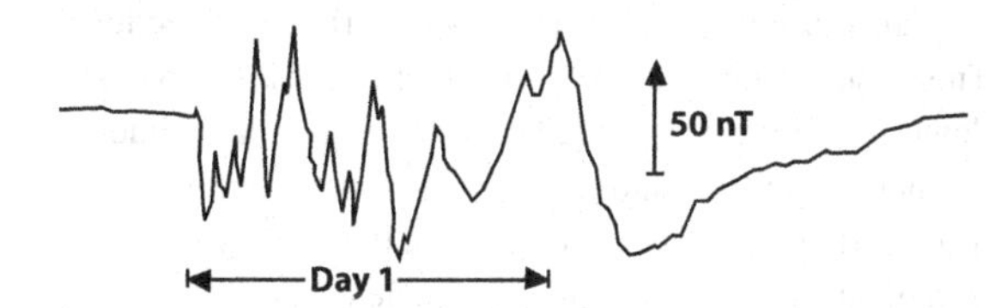

FIGURE 8.11 Typical magnetic storm record in total magnetic intensity. Adapted from BREINER (1973).

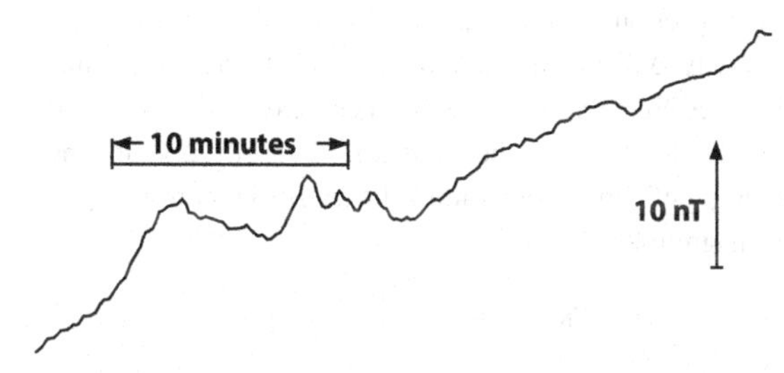

FIGURE 8.12 Typical record of micropulsations in total magnetic intensity. Adapted from BREINER (1973).

tens of minutes (Figure 8.12) are an important consideration for resolving shorter-wavelength, low-amplitude magnetic anomalies such as from soil features in archaeological, engineering, and other high-resolution applications.

8.3.6 Measurement

Numerous techniques have been used to measure the angular relationships and magnitude of the various components of the Earth's magnetic field since the seventeenth century. The simplest procedures used for exploration applications until the mid 1940s used a magnetized element and involved the observation of the angular deviation from the magnetic meridian, its oscillation period, or the interaction between the element and a mechanical torque caused by the magnetic field. Modern observations are principally made electronically using either resonance (atomic) or flux-gate magnetometers. These two types of magnetometers complement each other. The resonance magnetometers are scalar in nature measuring the absolute total intensity (B_T) of the magnetic field without being oriented. In contrast, flux-gate magnetometers are vector magnetometers, measuring the relative magnitude of the field in the direction of the orientation of the sensing element. Resonance magnetometers, including proton-precession, alkali-vapor, and Overhauser devices, measure the precession of atomic particles which is proportional to the ambient magnetic field. The flux-gate magnetometer, which can be oriented to measure specified components of the Earth's field, consists of a transformer in which the saturation of the core is biased by the magnetic field in the direction of the core. The magnitude of the field is measured by this effect on the saturation of the core. Observations can be made to a precision of 0.1 nT or better. Additional details on modern magnetometers including gradiometers which are primarily used to increase the resolution of the observations are given in Section 11.3.

Most exploration magnetometers measure only the scalar magnitude or intensity of the geomagnetic field. However, when resources and effort are expended to orient the sensors, vectorial components of the magnetic field can be observed and used to enhance interpretation. Vector components are most commonly measured by geomagnetic observatories and with satellite systems which use star cameras to orient the orbiting magnetometers (LANGEL and HINZE, 1998). Most ground, marine, and airborne magnetic surveys, however, observe the simpler total field intensity (B_T) using unoriented sensors. Interpreting the total field observation (B_T) for the anomalous intensity (ΔB_T) from a magnetization variation assumes that the anomaly $\Delta \mathbf{B}_T$ perturbs the Earth's main field $\mathbf{B}_N$ only in its principal direction, so that the simple scalar relationship

$$\Delta B_T = B_T - B_N \tag{8.19}$$

holds. This assumption is analogous to the gravity case in Equation 2.6 which relates the difference between the scalar intensity of the total gravity component measured by the gravimeter and the Earth's normal gravity field to anomalous gravity in the normal component.

In general, the total field ($\mathbf{B}_T$) is the vector sum of the main field ($\mathbf{B}_N$) and the anomaly ($\Delta\mathbf{B}_T$) so that the measured scalar total field intensity is

$$B_T = \sqrt{B_N^2 + \Delta B_T^2}. \tag{8.20}$$

However, when the vectors ($\mathbf{B}_N$) and ($\Delta\mathbf{B}_T$) are assumed to be co-linear as shown in Figure 8.13, the scalar total

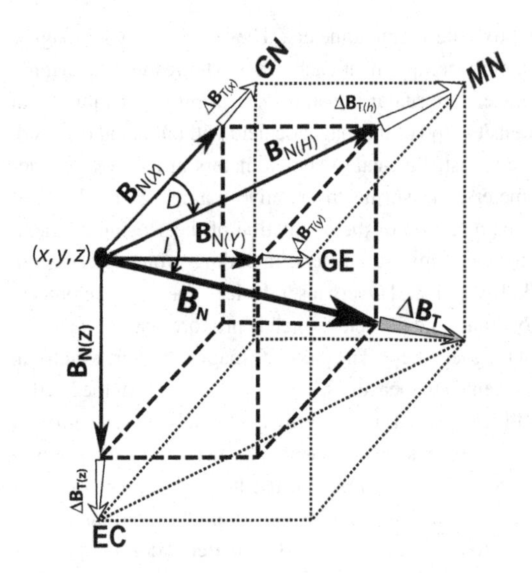

FIGURE 8.13 Geometric interpretation of the total field vector $\mathbf{B}_T = \mathbf{B}_N + \Delta\mathbf{B}_T$ for co-linear main field ($\mathbf{B}_N$) and anomaly ($\Delta\mathbf{B}_T$) vectors at an observation point (x, y, z), where GN is geographic north, MN is magnetic north, GE is geographic east, and EC is towards the Earth's center. Equations 8.7–8.11 describe the components of normal field ($\mathbf{B}_N$) vector and Equations 8.21–8.22 the anomaly ($\Delta\mathbf{B}_T$) vector components. D is the declination of the magnetic field from true north and I is its inclination below the horizontal.

field intensity anomaly reduces to the simple form given in Equation 8.19. Expressing this simple result in terms of the parameters in Figure 8.13 gives

$$\Delta B_T = (B_T - B_N) = \sqrt{\Delta B^2_{T(x)} + \Delta B^2_{T(y)} + \Delta B^2_{T(z)}}, \tag{8.21}$$

with the

$$\begin{aligned} \text{GN-component} &\Longrightarrow \Delta B_{T(x)} = [\Delta B_T] \cos I \cos D, \\ \text{GE-component} &\Longrightarrow \Delta B_{T(y)} = [\Delta B_T] \cos I \sin D, \\ \text{MM-component} &\Longrightarrow \Delta B_{T(h)} = [\Delta B_{T(x)}] \cos D \\ &\qquad + [\Delta B_T(y)] \sin D \\ &\qquad = [\Delta B_T] \cos I, \text{ and} \\ \text{EC-component} &\Longrightarrow \Delta B_{T(z)} = [\Delta B_T] \sin I. \end{aligned} \tag{8.22}$$

Here, GN and GE are the respective geographic north and east components, MM is the magnetic meridian component directed along magnetic declination towards the geomagnetic pole, and the Earth-centered component EC is directed towards the Earth's center.

The interpretation is valid as long as the magnetic anomaly is not large enough to perturb the main field significantly in any direction other than its principal direction. This assumption holds in most applications, but breaks down for anomaly amplitudes with absolute magnitudes of about 10,000 nT and larger, such as may occur in the near-field of large steel objects, or in the vicinity of some iron ore deposits and certain ultramafic rocks.

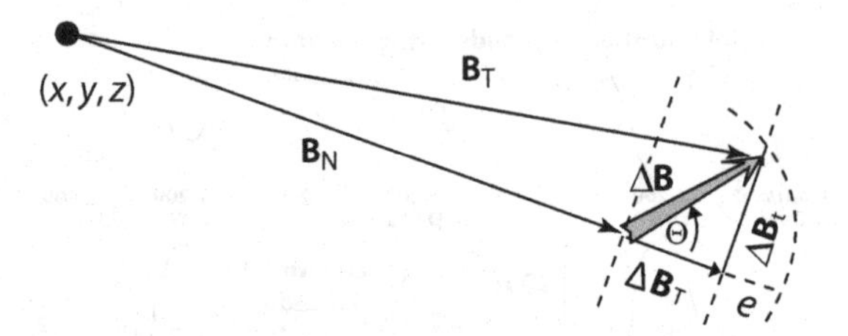

FIGURE 8.14 Geometric interpretation of the error ($e = e_{\Delta B_T}$) caused by assuming that $\Delta\mathbf{B}_T = \Delta\mathbf{B}$, and thus ignoring the transverse $\Delta\mathbf{B}_t$ component by ascribing the simple difference $(B_T - B_N)$ to the scalar total intensity anomaly ΔB_T.

The error inherent to this assumption was described by Klontis and Young (1964) who considered the more general case shown in Figure 8.14, where $\Delta\mathbf{B}_T$ is the projection onto the main field vector $\mathbf{B}_N$ of an arbitrary anomaly $\Delta\mathbf{B}$ directed at some angle Θ to $\mathbf{B}_N$ which thus is not co-linear with it. In the general case, the commonly held assumption that the anomalous intensity ΔB_T can be simply defined by Equation 8.19 leads to an error ($e_{\Delta B_T}$) of magnitude

$$\begin{aligned} e_{\Delta B_T} &= B_T - (B_N + \Delta B_T) \\ &= \sqrt{(B_N + \Delta B_T)^2 + \Delta B_t^2} - (B_N + \Delta B_T), \end{aligned} \tag{8.23}$$

where $\Delta\mathbf{B}_t$ is the anomaly component transverse to the main field (Figure 8.14).

To estimate this error, the square-root term is expanded using the binomial theorem to obtain

$$\begin{aligned} e_{\Delta B_T} &\simeq (B_N + \Delta B_T)(\sqrt{1 + [\Delta B_t/(B_N + \Delta B_T)]^2} - 1) \\ &\simeq (B_N + \Delta B_T)(1 + \{\Delta B_t/[2(B_N + \Delta B_T)]\}^2 - 1) \\ &\simeq \frac{1}{2}\left(\frac{\Delta B_t^2}{B_N + \Delta B_T}\right). \end{aligned} \tag{8.24}$$

Clearly, the error is a minimum (i.e. $e_{\Delta B_T}(\min) = 0$) when $\Delta\mathbf{B} = \Delta\mathbf{B}_T$ and thus lies along $\mathbf{B}_N$. The worst case is for $\Theta = 90°$ where $e_{\Delta B_T}(\max) = (\Delta B_T^2/2B_N)$. However, this error is negligible as long as $B_N \gg \Delta B$, in which case the approximation in Equation 8.19 is reasonably valid. Table 8.3 gives some examples of how $e_{\Delta B_T}(\max)$ varies with intensity variations of the main field ($\mathbf{B}_N$) and anomaly ($\Delta\mathbf{B}$).

Extensive areas of the continents and oceans have been surveyed with magnetometers measuring the total

TABLE 8.3 Examples showing how the maximum error [$e_{\Delta B_T}$(max)] varies with main field [B_N] and anomaly [ΔB] intensities for the total field assumption that $\Delta B_T = \Delta B$.

B_N	ΔB	$e_{\Delta B_T}$(max)
30,000 nT	500 nT	4 nT
30,000 nT	10,000 nT	1,667 nT
70,000 nT	10,000 nT	714 nT

magnetic field from mobile platforms, especially from aircraft and satellites. These data are available from governmental data repositories as well as commercial data centers, either in the form of observed or gridded data. Governmental repositories generally contain regional data sets with observations spaced at several kilometers or more, whereas commercial data centers focus on detailed surveys of more limited regions.

8.4 History of the magnetic method in exploration

There is little doubt that the Chinese were the first to observe and employ the directional properties of magnetic objects in navigation. They may have invented and used the compass as early as 2600 before the Common Era, but more likely it was as late as 1100 in the Common Era before the compass was used as a navigational instrument. The compass was first noted in Europe around 1200, but the reference by an English monk suggests that the compass had previously been known for an extended period of time. In 1269, P. de Maricourt (P. Peregrinus) investigated the properties of a spherical lodestone. He studied the dipolar nature of magnetic objects and named the poles after the geographic directions that they pointed to (i.e. north and south). He also identified the forces of attraction between unlike poles and the repulsion between like poles.

Although the matter was possibly observed earlier by the Chinese and also by Christopher Columbus in his voyages of discovery of the Caribbean Islands in 1492, N. Hartmann in 1510 is credited as the first European to record the declination of the geomagnetic field. In 1544, he was also the first to observe the inclination of the field. Towards the end of the sixteenth century, R. Norman constructed the first inclinometer with the dipping magnetic needle located on a horizontal pivot. With it he was able to show that the inclination of the field corresponds to that on a spherical lodestone, indicating that the cause of the directional component of the compass was within the Earth. This led in 1600 to the publication of William Gilbert's famous book *De Magnete*, which postulated that the Earth is a giant magnet.

C. Coulomb markedly improved the sensitivity of the compass in the 1770s. With this instrument he was able to study daily variations in the field that had been discovered earlier in the century by G. Gellibrand and G. Graham. From his observations, he discovered the inverse square law of magnetics that bears his name and is equivalent to Newton's law of gravitation. In 1820, H. C. Ørsted observed the effect on a magnet of a nearby current-carrying wire, thus for the first time relating electricity and magnetism. In that same year, A. Ampere explained these observations in terms of forces between electrical fields, showing that the origin of magnetism lies in electrical fields. The actual origin of the geomagnetic field has long been the subject of speculation. Early theories focused on the presence of permanently magnetized material distributed in various ways within the Earth. However, in 1919, J. Larmor proposed an origin by an internal self-maintaining dynamo such as has been suggested for the origin of strong magnetic fields observed on the Sun. In the late 1940s, W. Elasser and E. Bullard contributed to the theory that the origin lies in motions of electrically conducting material in the fluid outer core of the Earth. These studies on the origin and reversal of the geomagnetic field continue today.

In the early nineteenth century, Baron von Humboldt made magnetic intensity measurements in widely separated locales, observing that the intensity of the field varies over the surface of the Earth. He also conducted the first magnetic survey associated with local geology. However, local variations in the direction of the magnetic field had been mapped as early as 1640 in Sweden in the search for buried magnetic iron ore deposits. Declination measurements on a local basis were made with a Sun compass, where the shadow cast by the Sun is used to determine geographic north which is then compared with the direction of magnetic north indicated by the compass heading. These measurements were brought to New York and New Jersey to locate hidden iron ore deposits early in the eighteenth century, and similar measurements in Michigan in 1844 were used to discover the iron-rich rock formations of the Lake Superior region.

In the early part of the nineteenth century, C. F. Gauss developed instrumentation and procedures for measuring the absolute intensity of the geomagnetic field, in contrast to the relative observations made of declination and inclination. Gauss also published papers between 1832 and 1840 giving the framework for our current theories of terrestrial magnetism. His associate W. E. Weber introduced the Earth inductor, which measures the intensity of the field by measuring the induced current in an oriented wire coil that rotates in the geomagnetic field. In 1842, H. Lloyd made a significant contribution by constructing a counterbalanced magnetic needle. Observations with this instrument can be used to determine both dip and intensity of the magnetic field. A variant of this instrument, the dip needle, a hand-held counterbalanced needle that oscillates in a vertical plane, was used extensively in mineral exploration from the early 1900s to after World War II. A more sensitive variation of this instrument is the Hotchkiss superdip, in which the counterbalance is moved off the axis of the magnetic needle; this was developed in 1915 and used throughout the Lake Superior iron ore district and other mining areas. However, the most widely used Lloyd-type instrument was developed by A. Schmidt in 1915. This instrument was capable of the precision required to map geological formations containing minor magnetic minerals, not just high-magnetic polarization units. It was the standard for geological mapping of both the vertical and horizontal intensity of the geomagnetic field for 40 years. In 1930, A. S. Eve and D. A. Keys reported on the use of this instrument in mapping an iron-copper sulfide ore body and dikes in the Sudbury mining district of Canada. A successful magnetic gradiometer was constructed and used in geological mapping by I. Roman and T. C. Sermon in the early 1930s, using two Earth inductors at a fixed distance from each other.

Airborne measurements of adequate sensitivity for geological studies were initiated shortly after World War II, with a flux-gate magnetometer incorporating the basic sensor developed by J. D. C. Hare and further developed and patented in 1940 by V. Vacquier. This was made into a practical instrument during the war for use in detecting submerged submarines. The flux-gate magnetometer, which measures the relative total intensity of the geomagnetic field, has been largely replaced by the resonance magnetometer which measures the absolute scalar magnetic intensity without an orientation requirement. The first full-scale airborne magnetic survey for geological purposes was conducted in 1945 in the Naval Petroleum Reserve No. 4 in northwestern Alaska by the US Geological Survey. The proton-precession magnetometer, developed by R. Varian, became available for geomagnetic exploration in the 1950s and was rapidly converted to use in marine and airborne measurements as well as satellite magnetic studies and gradiometer studies. Although other types of magnetometers have been developed through the years following World War II, variations of the resonance magnetometer are the principal instrument used in magnetic observations today. It is these magnetometers that have permitted the global mapping of the geomagnetic field at a range of scales.

Observations of the near-Earth magnetic field have been made from Earth-orbiting satellites since Sputnik 3 was launched in 1958. Improved measurements covering larger portions of the globe were made using subsequent satellites, notably Cosmos 49, the polar orbiting geophysical observatories (POGO) of the late 1960s, and Magsat which orbited the Earth for 7 months starting in November 1979. The latter mission was dedicated to magnetic observations, providing useful information on the nature of the lithosphere as well as global temporal and spatial variations in the geomagnetic field. More recently, the satellite missions Ørsted (1999) and CHAMP (2000) have given new insight into the geomagnetic field and its use in lithospheric studies. Magnetometers deployed on the lunar Apollo (1960s), Clementine (1992), and Lunar Prospector (1997) missions mapped the weak magnetic field of the Moon, whereas the Viking mission (1979) obtained magnetic field observations of Venus. The Mars Global Surveyor (1996) mission has also mapped the martian magnetic field. These magnetic observations from space have provided new insight into the origin and tectonic evolution of the rocky planetary bodies, as well as a better understanding of the Earth's magnetic field and its interpretation.

No history of geomagnetism would be complete without mention of paleomagnetism, which is the study of the permanent magnetization of rocks, and its impact on exploring the Earth. Intense permanent magnetization was recognized in rocks other than lodestones in the late eighteenth century, but it was largely believed to be the result of lightning strikes. It was the observation by Delasse in the mid-nineteenth century that permanent magnetization in some rocks parallels the Earth's magnetic field that opened up opportunities to use paleomagnetism to study the Earth and its processes. P. David and B. Brunhes reported in the early twentieth century that some rocks are reversely magnetized to the current geomagnetic field. In 1926, P. Mercanton observed that rocks from widely scattered locations across the Earth are reversely as well as normally magnetized, suggesting that the reversal of the permanent magnetization is a result of ancient reversals of the magnetic field. He suggested that paleomagnetism could be used to evaluate hypotheses of polar wandering and continental drift because of the approximate correlation of the axes of rotation and the geomagnetic field. In the ensuing few decades, secular variation and intensity of the Earth's magnetic field were studied in rocks, baked hearths, and pottery. In 1952, P. M. S. Blackett reported on an astro-magnetism experiment in which improved techniques were developed for measuring the moment and direction of the magnetic field of materials. Studies of the paleomagnetism of rocks throughout the world by J. Graham, P. M. S. Blackett, K. Runcorn, E. Irvine and others led to the support of the continental drift hypothesis.

By accurately dating rocks whose magnetization had been determined, A. Cox and R. R. Dalrymple in 1960 established the first magnetic field reversal chronology over the past 3.6×10^6 years. Subsequently, this chronology has been greatly expanded and its resolution improved. The history of reversals of the geomagnetic field led F. J. Vine and D. H. Matthews in 1963 to explain the origin of the symmetric magnetic anomalies on either side of the oceanic ridges. They proposed that the oceanic crust splits along the ridge, allowing intrusion of basalt, which, upon cooling, takes on a remanent magnetization parallel to the geomagnetic field. The alternating sign of the magnetic anomalies reflects the reversing magnetization of the oceanic crust and its movement away from the oceanic ridge with time as new material is intruded into the central ridge. Thus, the production of oceanic crust results in a magnetic tape-recording that forms the basis of the theory of seafloor spreading and plate tectonics. Observations of the magnetic field at sea, and an understanding of the reversal and timing of the reversal of the geomagnetic field as identified in rocks, led to the hypothesis of plate tectonics that dominates modern Earth sciences.

Additional details on the history of geomagnetism and its application to the study of the Earth can be found in Jakosky (1950), Parkinson (1983), Merrill and McElhinny (1983), Malin (1987), Multhauf and Good (1987), and Hanna (1990).

8.5 Implementing the magnetic method

The magnetic method is applied to subsurface studies in much the same way as the gravity method described in Section 6.5 and, more specifically, follows the same logical sequence of steps of planning, field measurement, data processing, interpretation, and documentation shown in Figure 2.7. Accordingly, this section considers only those variations that are specific to the magnetic method.

8.5.1 Planning phase

The planning phase of the magnetic method is similar to that of the gravity method, but significant differences in the methods must be considered. For example, in calculating the magnetic effects of possible sources in a survey, the greater sensitivity of the magnetic anomaly to source depth and width, as compared with a gravity anomaly from the same source, needs to be recognized. Depths to sources must be approximated as closely as possible, and the data observation interval must be decreased.

Additionally, the large dynamic range of magnetizations may make its estimation for calculation of anticipated anomalies difficult. The problem is eased somewhat by the bimodal nature of terrestrial magnetizations, which often makes it possible to assume that the magnetization of host materials to the anomalous sources is effectively nil. A further complication is that magnetizations are difficult to estimate from lithology. The range of typical magnetizations for a specific crystalline rock type is likely to be wide and overlapping with other lithologies, and can be complicated by remanent magnetizations with magnitudes and directions that can be difficult to estimate.

The noise and regional magnetic field in a survey area must also be considered. Noise can be anticipated from cultural effects, both ferrous objects and electrical fields, but this is difficult to estimate quantitatively. Regional effects normally are not a major concern in the magnetic method because of the relatively low amplitude of deeply buried, regional sources.

8.5.2 Data acquisition phase

After the anomaly responses to the anticipated sources have been determined and related to possible noise and regional effects, the design of the survey begins with specification of the layout of the magnetic observations and the required precision of the measurements. In planning ground surveys, the effects of elevation and terrain are generally assumed to be negligible. The vertical gradient of the geomagnetic field is only about 0.03 nT/m at the poles where its effect is greatest. As a result, the elevation of observations generally is unnecessary and station locations are established by use of GPS technology. Terrain effects are not required where surface materials are essentially non-magnetic and are difficult to determine precisely where highly magnetic rocks, such as volcanic rocks, are at the surface, because of their heterogeneity of magnetization and their proximity to the observation sites.

An important decision for near-surface surveys is whether surface magnetic surveying is a viable approach or whether airborne surveying techniques should be used. Airborne surveying has the advantages of being rapid, able to cover areas otherwise difficult to access, and subject to minimal noise from cultural effects. However, the greater elevation above the surface decreases the resolution of the observations, and thus poses serious difficulties to mapping concentrated, near-surface sources. Furthermore, it may be impractical for covering limited areas because of the cost of mobilization and demobilization of the aircraft carrying the instruments. Although most airborne surveying is with fixed-wing aircraft, helicopter-borne surveying may be necessary where fixed-wing aircraft cannot maintain a desired fixed elevation above rugged surface terrain, or provide enough anomaly resolution and accuracy, or where the survey altitude is too low. This leads to additional surveying costs. Marine magnetic surveying may be necessary where water-covered regions need to be surveyed at high resolution or where the area is too limited to warrant mobilizing an airborne survey.

Another option for surveying is to conduct gradient measurements. The advantage of vertical gradient surveying is that the resolution of the observations is increased and temporal magnetic variations are largely eliminated. Horizontal gradient measurements are employed in special situations, but offer comparable advantages and can be used in the identification of off-track sources. Vectorial measurements are not commonly used in near-surface magnetic surveying, but may under special conditions be useful, for example in localizing the presence of a magnetic source from drillhole measurements. Vector magnetic observations are common in satellite surveys using GPS and star cameras that identify the star patterns to orient the measurements.

Care is needed in all magnetic surveying to isolate the sensor from local magnetic variations caused by either electrical currents or magnetic materials. In satellite surveying, for example, booms extend the magnetometers several meters from the satellite. In airborne installations, this is done by placing the sensor at an appropriate distance from the aircraft and/or compensating the magnetometer for inherent magnetic effects of the aircraft. Similar concerns arise in marine magnetic observations. Ground magnetic observations may be affected by magnetic belt buckles, steel shanks in boots, screws in eyeglasses, and similar magnetic materials worn by the operating personnel when they are close to the measurement. Thus, caution is required to eliminate all possible magnetic noise effects on instrument observers and to avoid nearby sources.

8.5.3 Data processing phase

Data processing in the near-surface magnetic method is simpler than in the gravity method. The principal correction is to eliminate the temporal variations caused primarily by ionospheric electric currents (Section 8.3.4). These effects are ever-present, and thus cannot be avoided. Generally, their broad characteristics are consistent over local regions and their magnitudes low enough that they can be eliminated by simple monitoring procedures. However, during magnetic storms, their magnitude becomes large and their variability increases, making this a difficult problem. It is therefore advisable to avoid making observations during these storm periods. Corrections in the reduction of the observed data also are commonly made for adjustment of observations to a datum and elimination of the planetary geomagnetic field.

Satellite magnetic data include contributions of roughly 98% from the core field, some 0.1% from the lithosphere, and slightly less than 2% from external fields. Correlation filtering (Appendix A.5.1) of neighboring orbital data tracks and anomaly maps at different local magnetic times can be applied to separate the predominantly uncorrelated effects of the spatially and temporally dynamic external field components from the correlated signals of the static internal components of the core and crust. The satellite-altitude core and crustal magnetic fields also have significantly overlapping wavelength characteristics so that using the International Geomagnetic Reference Field (IGRF) to separate them is not as routine as it is for near-surface surveys (Section 12.2).

Isolation and enhancement procedures are used in the magnetic method as in the gravity method, but the emphasis is on enhancement of the reduced data to increase perceptibility rather than on the isolation of magnetic anomalies. Isolation is less of a problem than in gravity because of the lower-amplitude regional magnetic anomalies. However, there are exceptions to this generality.

8.5.4 Interpretation phase

Interpretation of magnetic data follows much the same procedures as in the gravity method with the use of inversion. Because of the nature of magnetic fields and the magnetic characteristics of terrestrial materials, however, the emphasis is on determining depths to the sources and structural mapping rather than on lithology. Both forward and inverse modeling approaches are widely applied, but the outcomes of course are never unique. Thus, access to and expertise in using ancillary geological, geophysical, and other subsurface constraints are critical for effective magnetic anomaly interpretation.

8.5.5 Reporting phase

As in the gravity method, the process is completed with a report on the objectives of the survey, field procedures, the assumptions that were made in the various phases of the survey, the error budget, and the optimum interpretation of the magnetic measurements and its robustness considering the constraints imposed by collateral information and sensitivity studies. Modern reports include the liberal use of computer graphics to promote understanding of the magnetic survey design, measurement, reduction, processing, and interpretation elements. Archiving of metadata regarding the data, their acquisition, and processing is an important follow-on activity.

8.6 Key concepts

- The magnetic method is based on measurement of perturbations or anomalies in the Earth's magnetic field caused by lateral variations in the magnetic moment per unit volume (magnetization) of Earth materials. Generally the total intensity component B_T, or in more limited exploration investigations one of the gradient components of the magnetic field, is measured. The normal or main field intensity is subtracted from the measured total intensity to define the intensity of the anomaly component along the direction of the normal field. This total field anomaly is well estimated by unoriented magnetometer sensors as long as the absolute magnitude of the anomaly is smaller than about 10,000 nT.
- The measured magnetic component is the vectorial summation of all sources of magnetization. As a result, the measured fields are compared to model fields based on planetary and known geological effects to obtain the anomalies from the subsurface. Additionally, the computed anomalies are filtered to isolate or enhance the anomaly associated only with the specific targeted geologic source.
- The magnetic method is a potential field method because a component of the field is measured that is a function of the work performed on a magnetic pole. The magnitude of the magnetic effect depends on the distance between the observation point and the anomalous magnetic pole.
- The theoretical basis of the method derives from Coulomb's law, which states that the attraction between magnetic poles is a direct function of the product of the pole strengths and inversely proportional to the squared distance between their centers. The force of attraction between poles cannot be measured independently of pole strength, so the magnetic field is measured or the force per unit pole.
- It is the magnetic dipole rather than the single pole that is the smallest subdivision of magnetic material. The dipole consists of two equal-strength poles of opposite signs separated by an infinitesimally small distance. The magnetic moment of the dipole is a vector directed from the negative to the positive pole with intensity given by the pole strength times the distance between poles. The dot product of the dipole's moment and gradient of the inverse displacement is the dipole's scalar magnetic potential, which also is a potential field because the magnetic effects of magnetic poles superpose at an observation point.
- The geomagnetic field originates in electrical currents associated with fluid movements in the outer core of the Earth. This dipolar field, which is roughly aligned with the axis of rotation of the Earth, is modified by long-term secular variations originating in the outer core, and temporal variations covering a broad range of periods due to ionospheric electrical currents caused by solar electromagnetic and corpuscular radiation and tidal effects. In addition, lateral variations in the magnetic polarization of the materials largely in the crust of the Earth cause spatial changes in the Earth's magnetic field. The latter are the components that are identified and interpreted in magnetic exploration.
- Analogous to the gravitational attraction of bodies due to their mass, the magnetization of bodies causes a magnetic force proportional to the product of their pole strengths and inversely proportional to the square of the distance between the poles. Magnetic fields are dipolar, and thus the field may be attractive where the poles are of different sign or repulsive where they have the same sign.
- Typically measurements of the magnetic field strength, the force per unit pole, are made to a precision of one part in five hundred thousand of the Earth's field for purposes of mapping the magnetization of the subsurface. The units are gammas in EMu ($1\gamma = 10^{-5}$ oersted or gauss) or preferably nanoteslas in SIu ($1\,\text{nT} = 1\gamma$).
- Magnetic observations in exploration geophysics are primarily made with resonance magnetometers (these measure the precession of atomic particles, which is proportional to the ambient magnetic field) or with directionally sensitive flux-gate magnetometers (which measure the field by determining its effect on the saturation of the core of a transformer). Observations made by resonance magnetometers are absolute, while those of the flux-gate magnetometer are relative.
- Geomagnetism was perhaps first exploited by the invention and use of the compass nearly three millennia ago. With the compass and the discovery of electromagnetism, the inclination, declination, and intensity variations of the Earth's magnetic field were being mapped by the early nineteenth century. The first magnetic exploration applications initiated in Sweden in the mid-seventeenth century used compasses to search for buried iron ore deposits. The oriented magnetic needle was extensively used to map the vertical and horizontal intensity of the geomagnetic field in mineral

exploration from the early to mid-twentieth century. These devices were supplanted by flux-gate and resonance magnetometers, developed at the end of World War II, that measure the total field intensity with largely unoriented sensors. The resonance magnetometer is the principal instrument used today in ground, marine, airborne, and satellite magnetic surveys as well as gradiometer studies.

- Applying the magnetic method to subsurface exploration follows common geophysical practice regardless of the objective. The logical sequence of steps consists of planning, data acquisition, data processing, interpretation, and documentation.
- The planning phase of a magnetic study involves definition of the range of anomaly characteristics which are anticipated from the sources of interest and the nature of the expected anomalous noise that interferes with the identification and analysis of the anomalies of interest. All decisions regarding the pattern, density of observations, instrumentation, processing, and interpretational procedures, etc., are based on these anomaly and noise characteristics taking into account the cost/benefit considerations of the survey.
- Data acquisition follows the procedures outlined in the planning stage so that observations are made with appropriate precision and coverage density to resolve the anomalies of interest. However, the increasing amount of data now compiled in governmental and commercial data repositories often meet survey requirements, eliminating the need for the acquisition of new data and greatly decreasing the costs of the program.
- Processing of the acquired data generally involves two steps. First is the reduction of the data into an interpretable form by eliminating all predictable magnetic effects, leaving only the residual magnetic effects of the unknown subsurface sources. This is done by comparing the observed data with a theoretical model of the geomagnetic field at each observation site and removing temporal variations in the Earth's field. The second involves the isolation or enhancement of the anomalies of interest so that they can be identified and analyzed. The former step is usually unnecessary for magnetic observations in data repositories, but the latter is usually oriented to a specific set of target anomalies, and thus is required as part of the processing program.
- The next stage is the interpretation of the anomalies of interest. This may simply be a matter of identifying the location of the specific sources, but more commonly includes quantified interpretation involving inversion of the selected anomalies for their sources. The inversion process generally involves not only describing the location and configuration of the source, but also its depth and property contrast with the surrounding rocks. This inversion process is inherently ambiguous, but the range of possible sources generally is limited by incorporating other geological and geophysical data into the analysis.
- Finally, the physical model of the inversion is converted into geological terms in the reporting phase of the program based on the geology of the region and constraints imposed by the laws and theorems of geology. An error analysis with a clear specification of the error budget is an important element of reporting on a magnetic study.

Supplementary material and Study Questions are available on the website www.cambridge.org/gravmag.

9 Magnetic potential theory

9.1 Overview

Like the gravity method described in Chapter 3, the magnetic method is a potential field method. Magnetic potential is defined by the amount of work done in moving a point dipole from one position to another in the presence of a magnetic force field acting upon the point dipole. The rate of change of the potential is the component of force in that direction which is related to the magnetic field strength measured in the magnetic method. As in the gravity method, Laplace's and Poisson's equations, Gauss' law, and Poisson's theorem are useful in understanding the properties of the magnetic field and in its analysis.

The magnetic effects of an arbitrary distribution of magnetization are evaluated by the summed effects of the point dipoles that fill out the volume. Magnetic effects of idealized bodies with simple symmetric shapes can be modeled with closed-form analytical expressions. For a general source with irregular shape, however, the magnetic effect must be numerically integrated by filling out the volume with idealized sources and summing the idealized source effects at the observation point. The numerical integration can be carried out with least-squares accuracy by distributing the idealized sources throughout the irregular body according to its Gauss–Legendre quadrature decomposition. Thus, the magnetic effects of all conceivable distributions of magnetization can be modeled to investigate the significance of magnetic anomalies.

9.2 Introduction

As in the gravity method (Chapter 3), the concepts of the potential and the potential field are fundamental to understanding the magnetic method. Specifically, consideration of the magnetic potential is useful because it simplifies the analysis of the magnetic force field and aids in its understanding. The potential is a scalar quantity, and thus is independent of direction, in contrast to the magnetic field force and its derivatives which are vectors. Thus, the magnetic potential is used wherever possible to avoid complexities in mathematical operations imposed by the directional attributes of magnetic forces.

Magnetic potential fields are continuous functions in space, as are their corresponding derivatives. They often are described in terms of their characteristics, but in its simplest form, the magnetic potential is the energy at a point in space due to a force field. This is the kinetic energy used by the force field in moving a unit or point dipole between points in the force field space and is defined by the amount of work (= force × distance) used in moving the dipole. The energy consumed is independent of the path taken by the dipole in moving between points, and thus the magnetic potential field is termed a conservative field.

The next section gives an overview of the analytical properties of the magnetic potential of the point dipole. Subsequent sections specifically detail these properties for the magnetic effects of extended bodies of dipoles.

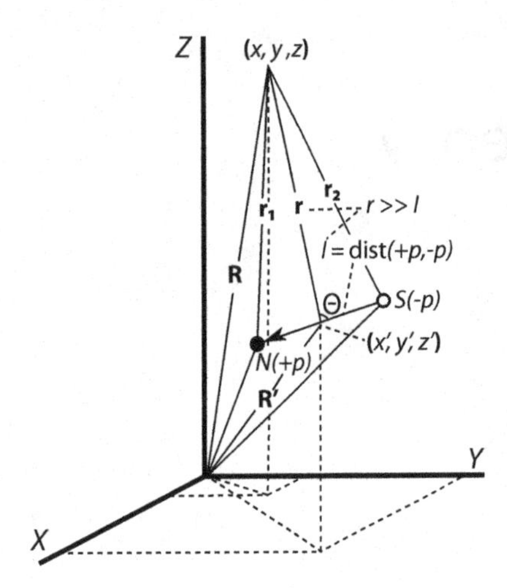

FIGURE 9.1 Spatial arrangement of north (N) or positive $(+p)$ and south (S) or negative $(-p)$ point magnetic poles separated by the infinitesimal distance l defining a magnetic point dipole. The dipole is located at the source point (x', y', z') at the distance $\mathbf{r}$ from the observation point (x, y, z) such that $\mathbf{r} = \mathbf{R} - \mathbf{R}'$ in the Earth-centered Cartesian coordinate system.

9.3 Magnetic potential of a point dipole

The potential field relationships for gravity fields also hold true for the magnetic fields where the force between bodies is both attractive and repulsive and depends on the distance between the bodies. The magnetic dipole in Figure 9.1, for example, which consists of two opposite poles of strength p separated by an infinitesimal distance l, has potential V at distance r given by

$$V(r) = p\left(\frac{1}{r_1} - \frac{1}{r_2}\right) C_{\mathrm{m}}, \quad (9.1)$$

where r_1 and r_2 are the distances to r from the positive (north) and negative (south) poles, respectively, and C_{m} is the magnetic constant described in Section 8.3.1.

To make the magnetic potential more tractable analytically, Equation 9.1 is generalized in terms of r rather than r_1 and r_2. Accordingly, Figure 9.1 shows that

$$r_1^2 = r^2 + \left(\frac{l}{2}\right)^2 - (l \times r)\cos\Theta, \quad (9.2)$$

so that dividing through by r^2 gives

$$\left(\frac{r_1}{r}\right)^2 = 1 + \left(\frac{l}{2r}\right)^2 - \left(\frac{l}{r}\right)\cos\Theta. \quad (9.3)$$

Solving the above for (r_1/r) and expanding with the binomial theorem gives

$$\begin{aligned}\frac{r_1}{r} &= \left(1 + \left[\left(\frac{l}{2r}\right)^2 - \left(\frac{l}{r}\right)\cos\Theta\right]\right)^{\frac{1}{2}} \\ &= 1 - \frac{1}{2}\left[\left(\frac{l}{2r}\right)^2 - \left(\frac{l}{r}\right)\cos\Theta\right] \\ &\quad + \frac{(-\frac{1}{2})(-\frac{3}{2})}{2!}\left[\left(\frac{l}{2r}\right)^2 - \left(\frac{l}{r}\right)\cos\Theta\right]^2 + \cdots \\ &\simeq 1 + \left(\frac{l}{2r}\right)\cos\Theta + \left(\frac{l}{2r}\right)^2\left(\frac{3\cos\Theta - 1}{2}\right), \end{aligned} \quad (9.4)$$

where terms of order higher than $(l/r)^2$ have been neglected because $l \ll r$. Similarly,

$$\frac{r_2}{r} \simeq 1 - \left(\frac{l}{2r}\right)\cos\Theta + \left(\frac{l}{2r}\right)^2\left(\frac{3\cos\Theta - 1}{2}\right), \quad (9.5)$$

so that the magnetic scalar potential of the dipole becomes

$$V(r) \simeq (p \times l)\frac{\cos\Theta}{r^2}C_{\mathrm{m}} = \frac{\mathbf{j}\cdot\hat{\mathbf{r}}}{r^2}C_{\mathrm{m}} = \mathbf{j}\cdot\nabla\left(\frac{1}{r}\right)C_{\mathrm{m}}, \quad (9.6)$$

where the vector dot product involves the dipole moment $\mathbf{j}$ directed from $-p$ to $+p$ with magnitude $j = (p \times l)$ in A × m^2, $\hat{\mathbf{r}}$ is the unit vector such that $\mathbf{r} = r \times \hat{\mathbf{r}}$ and C_{m} is a constant depending on the units (Section 8.3.1). The unit vector $\hat{\mathbf{r}}$ has components $\hat{\mathbf{r}}_{\mathbf{x}} = (r \times \cos\alpha)\hat{\mathbf{e}}_{\mathbf{x}}$, $\hat{\mathbf{r}}_{\mathbf{y}} = (r \times \cos\beta)\hat{\mathbf{e}}_{\mathbf{y}}$, and $\hat{\mathbf{r}}_{\mathbf{z}} = (r \times \cos\gamma)\hat{\mathbf{e}}_{\mathbf{z}}$, where the direction cosines of $\mathbf{r}$ satisfy the condition $\cos^2\alpha + \cos^2\beta + \cos^2\gamma = 1$ so that $|\hat{\mathbf{r}}| = |\hat{\mathbf{e}}| = 1$.

9.4 Magnetic effects of a point dipole

To ascertain the potential field attributes of the magnetic point dipole it is desirable to consider the relevant vectors explicitly in terms of the respective source and observation point coordinates (x', y', z') and (x, y, z). Thus, the displacement vector $\mathbf{r}$ in Figure 9.1 has magnitude given by Equation 3.2 where its gradients at the source and observation points are

$$\nabla'(r) = -\frac{\mathbf{r}}{r} = -\nabla(r), \quad (9.7)$$

and the gradients of its inverse are

$$\nabla'\left(\frac{1}{r}\right) = \frac{\mathbf{r}}{r^3} = -\nabla\left(\frac{1}{r}\right) = \left(\frac{1}{r^2}\right)\nabla(r). \quad (9.8)$$

Now, all magnetic vectors can be expressed directly in terms of their directional attributes (e.g. Figure 8.4). Thus, writing the magnetic moment of the point dipole, or the dipole magnetic moment, in its inclination I' and declination D' gives

$$\mathbf{j} = \mathbf{j_x} + \mathbf{j_y} + \mathbf{j_z}, \tag{9.9}$$

where

$$\mathbf{j_x} = j \times (\cos I' \cos D')\hat{\mathbf{e}}_\mathbf{x}, \quad \mathbf{j_y} = j \times (\cos I' \sin D')\hat{\mathbf{e}}_\mathbf{y},$$
$$\mathbf{j_z} = j \times (\sin I')\hat{\mathbf{e}}_\mathbf{z}, \tag{9.10}$$

so that

$$\mathbf{j} = j \times [(\cos I' \cos D')\hat{\mathbf{e}}_\mathbf{x} + (\cos I' \sin D')\hat{\mathbf{e}}_\mathbf{y} + (\sin I')\hat{\mathbf{e}}_\mathbf{z}]$$
$$= j \times \hat{\mathbf{u}}' \tag{9.11}$$

with the unit vector $\hat{\mathbf{u}}'$ in source point coordinates.

Substituting Equations 9.8 and 9.11 into Equation 9.6 yields the scalar magnetic potential in source point coordinates given by

$$V(r) \simeq \mathbf{j} \cdot \nabla\left(\frac{1}{r}\right) C_\mathrm{m} = -j \times \hat{\mathbf{u}}' \cdot \nabla'\left(\frac{1}{r}\right) C_\mathrm{m}. \tag{9.12}$$

The SIu of magnetic potential or work per unit pole strength is the ampere (A) which is equivalent to nT×m. In addition, the vector magnetic field **B** at **r** is

$$\mathbf{B}(r) = -\nabla V(r) = \nabla\left[\hat{\mathbf{u}}' \cdot \nabla'\left(\frac{1}{r}\right)\right](C_\mathrm{m} \times j)$$
$$= \mathbf{B_x} + \mathbf{B_y} + \mathbf{B_z} \tag{9.13}$$

with the vector component magnitudes given by

$$B_x = \frac{3C_\mathrm{m}}{r^5}\left[j_x\left(\Delta x^2 - \frac{r^2}{3}\right) + j_y(\Delta y \Delta x) + j_z(\Delta z \Delta x)\right], \tag{9.14}$$

$$B_y = \frac{3C_\mathrm{m}}{r^5}\left[j_x(\Delta x \Delta y) + j_y\left(\Delta y^2 - \frac{r^2}{3}\right) + j_z(\Delta z \Delta y)\right], \tag{9.15}$$

and

$$B_z = \frac{3C_\mathrm{m}}{r^5}\left[j_x(\Delta x \Delta z) + j_y(\Delta z \Delta y) + j_z\left(\Delta z^2 - \frac{r^2}{3}\right)\right]. \tag{9.16}$$

The gradient tensors of the magnetic field ($\nabla\mathbf{B}(r)$) involve nine components that form a symmetric matrix (e.g. Equation 3.7) where the three diagonal tensors are

$$B_{xx} = \frac{3C_\mathrm{m}}{r^5}\left[j_x\left(3\Delta x - \frac{5\Delta x^3}{r^2}\right) + j_y\left(\Delta y - \frac{5\Delta x^2 \Delta y}{r^2}\right) + j_z\left(\Delta z - \frac{5\Delta x^2 \Delta z}{r^2}\right)\right], \tag{9.17}$$

$$B_{yy} = \frac{3C_\mathrm{m}}{r^5}\left[j_x\left(\Delta x - \frac{5\Delta y^2 \Delta x}{r^2}\right) + j_y\left(3\Delta y - \frac{5\Delta y^3}{r^2}\right) + j_z\left(\Delta z - \frac{5\Delta y^2 \Delta z}{r^2}\right)\right], \tag{9.18}$$

and

$$B_{zz} = \frac{3C_\mathrm{m}}{r^5}\left[j_x\left(\Delta x - \frac{5\Delta z^2 \Delta x}{r^2}\right) + j_y\left(\Delta y - \frac{5\Delta z^2 \Delta y}{r^2}\right) + j_z\left(3\Delta z - \frac{5\Delta z^3}{r^2}\right)\right], \tag{9.19}$$

which satisfy Laplace's equation (3.6). The three unique off-diagonal tensors, on the other hand, are

$$B_{xy} = \frac{3C_\mathrm{m}}{r^5}\left[j_x\left(\Delta y - \frac{5\Delta x^2 \Delta y}{r^2}\right) + j_y\left(\Delta x - \frac{5\Delta y^2 \Delta x}{r^2}\right) - j_z\left(\frac{5\Delta x \Delta y \Delta z}{r^2}\right)\right] = B_{yx}, \tag{9.20}$$

$$B_{xz} = \frac{3C_\mathrm{m}}{r^5}\left[j_x\left(\Delta z - \frac{5\Delta x^2 \Delta z}{r^2}\right) - j_y\left(\frac{5\Delta x \Delta y \Delta z}{r^2}\right) + j_z\left(\Delta x - \frac{5\Delta z^2 \Delta x}{r^2}\right)\right] = B_{zx}, \tag{9.21}$$

and

$$B_{yz} = \frac{3C_\mathrm{m}}{r^5}\left[-j_x\left(\frac{5\Delta x \Delta y \Delta z}{r^2}\right) + j_y\left(\Delta z - \frac{5\Delta y^2 \Delta z}{r^2}\right) + j_z\left(\Delta y - \frac{5\Delta z^2 \Delta y}{r^2}\right)\right] = B_{zy}. \tag{9.22}$$

To facilitate quantitative implementation of the magnetic point dipole effects, an example of the potential is given in Figure 9.2 for a point source with dipole moment $j = 10^9$ A×m^2 (= 10^{12} gauss × cm^3), inclination $I' = 75°$ N and declination $D' = 45°$ E at depth $z = 500$ m below the observations. The corresponding vector (Equations 9.14–9.16) and gradient tensor components modeled from Equations 9.17–9.22 are shown in Figure 9.3. As in the case of gravity tensors, this illustration shows the increased identification and resolution of the source attainable from tensors over that of the simple field value.

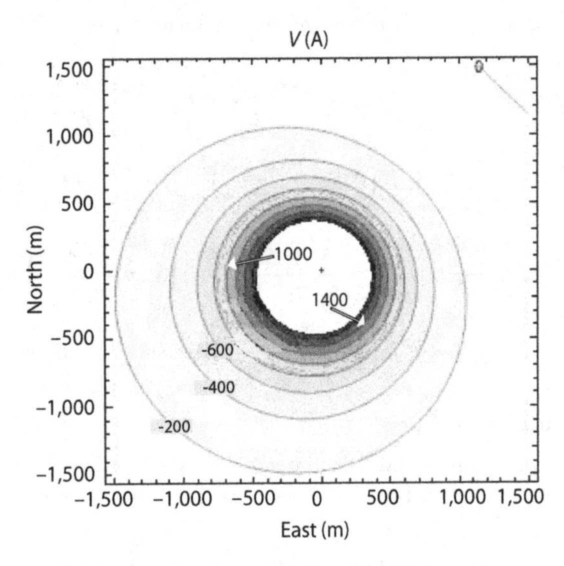

FIGURE 9.2 Magnetic potential V (Equation 9.12) for a point dipole the depth $z = 500\,\mathrm{m}$ with $j = 10^9\,\mathrm{A} \times \mathrm{m}^2$, $I' = 75°\,\mathrm{N}$ and $D' = 45°\,\mathrm{E}$. The point dipole has the same geometric parameters as the point mass in Figure 3.2.

The results for the vector field $\mathbf{B}$ (Equations 9.13–9.16) and its gradient tensors (Equations 9.17–9.22) hold only for oriented sensor measurements of the magnetic field. Oriented measurements are obtained by surface and airborne vertical gradient sensors for archaeological and engineering site investigations and mineral exploration, special airborne and marine surveys that use high-accuracy navigation systems to constrain sensor orientation, and most satellite magnetic surveys. Oriented measurements were also common to terrestrial surveys until the late 1940s when they were largely supplanted by the advent of total field measurement instrumentation and procedures. The ease with which total field measurements were obtained with minimal consideration of sensor orientation revolutionized efforts to map and understand the geomagnetic field in terms of geological sources.

As described in Section 8.3.5, the scalar magnitude of the total field B_{T} is interpreted as the projection of magnetic field $\mathbf{B}$ along the Earth's main or core field $\mathbf{B}_{\mathrm{N}}$ with inclination I and declination D at the observation point. Thus, the total field effect of the point dipole is given by the vector dot product

$$B_T = \hat{\mathbf{u}} \cdot \mathbf{B}(r) = (\cos I \cos D)B_x + (\cos I \sin D)B_y + (\sin I)B_z, \tag{9.23}$$

where $\hat{\mathbf{u}} = (\cos I \cos D)\hat{\mathbf{e}}_{\mathbf{x}} + (\cos I \sin D)\hat{\mathbf{e}}_{\mathbf{y}} + (\sin I)\hat{\mathbf{e}}_{\mathbf{z}}$ is the unit geomagnetic field vector at the observation point.

Examples of the dipole's total field anomalies at various declinations are illustrated in Figure 9.4. The basic method of combining the components was originally established by HUGHES and PONDROM (1947) to obtain the then more commonly measured vertical component from aeromagnetic flux-gate magnetometer measurements of the total field.

9.5 Magnetic effects of an extended body

The extended magnetic body has magnetic effects that can be evaluated at an observation point from the superimposed effects of the differential point dipoles that fill up the source. However, the above equations are for the magnetic effects of the point dipole only, whereas in geophysical practice the operative physical property is the magnetic moment per unit volume (in A/m) of the source or its magnetization $\mathbf{J}$ so that

$$\mathbf{j} = \int\!\!\int\!\!\int_v \partial\mathbf{j} = \mathbf{J} \times \int\!\!\int\!\!\int_v \partial x'\partial y'\partial z' = \mathbf{J} \times v, \tag{9.24}$$

where $\partial\mathbf{j} = \mathbf{J} \times \partial x'\partial y'\partial z'$ and v is the source's volume (in m^3). Thus, taking magnetization into account and integrating Equation 9.12 over the lower a- and upper b-limits in each spatial dimension of the 3D source gives its scalar magnetic potential

$$V(x, y, z) = \int_{z'_a}^{z'_b}\int_{y'_a}^{y'_b}\int_{x'_a}^{x'_b}\left\{V' = \frac{\mathbf{J}\cdot\mathbf{r}}{r^3}C_{\mathrm{m}} = \frac{J_x\Delta x + J_y\Delta y + J_z\Delta z}{(\Delta x^2 + \Delta y^2 + \Delta z^2)^{3/2}}C_{\mathrm{m}}\right\} \times \partial x'\partial y'\partial z', \tag{9.25}$$

where $\mathbf{J} = \mathbf{J}_{\mathrm{ind}} + \mathbf{J}_{\mathrm{rem}}$ is the polarization vector of induced and remanent magnetizations $\mathbf{J}_{\mathrm{ind}}$ and $\mathbf{J}_{\mathrm{rem}}$, respectively. Additionally, the scalar magnetization components are $J_x = J(\cos I' \cos D')$, $J_y = J(\cos I' \sin D')$, and $J_z = J(\sin I')$ for polarization inclination I' and declination D'. Of course, in the absence of remanence, the magnetization is induced only by the geomagnetic field with scalar intensity B_{N}, and inclination and declination I' and D', respectively, so that $J = \Delta k \times B_{\mathrm{N}}$, where Δk is the volume magnetic susceptibility (or susceptibility contrast) of the source. Note that the kernel V' is the potential per unit volume or the point dipole effect in Equation 9.12. The 3D magnetic potential can also be generalized as the product of functionals of the magnetization intensity and geometry (i.e. Green's

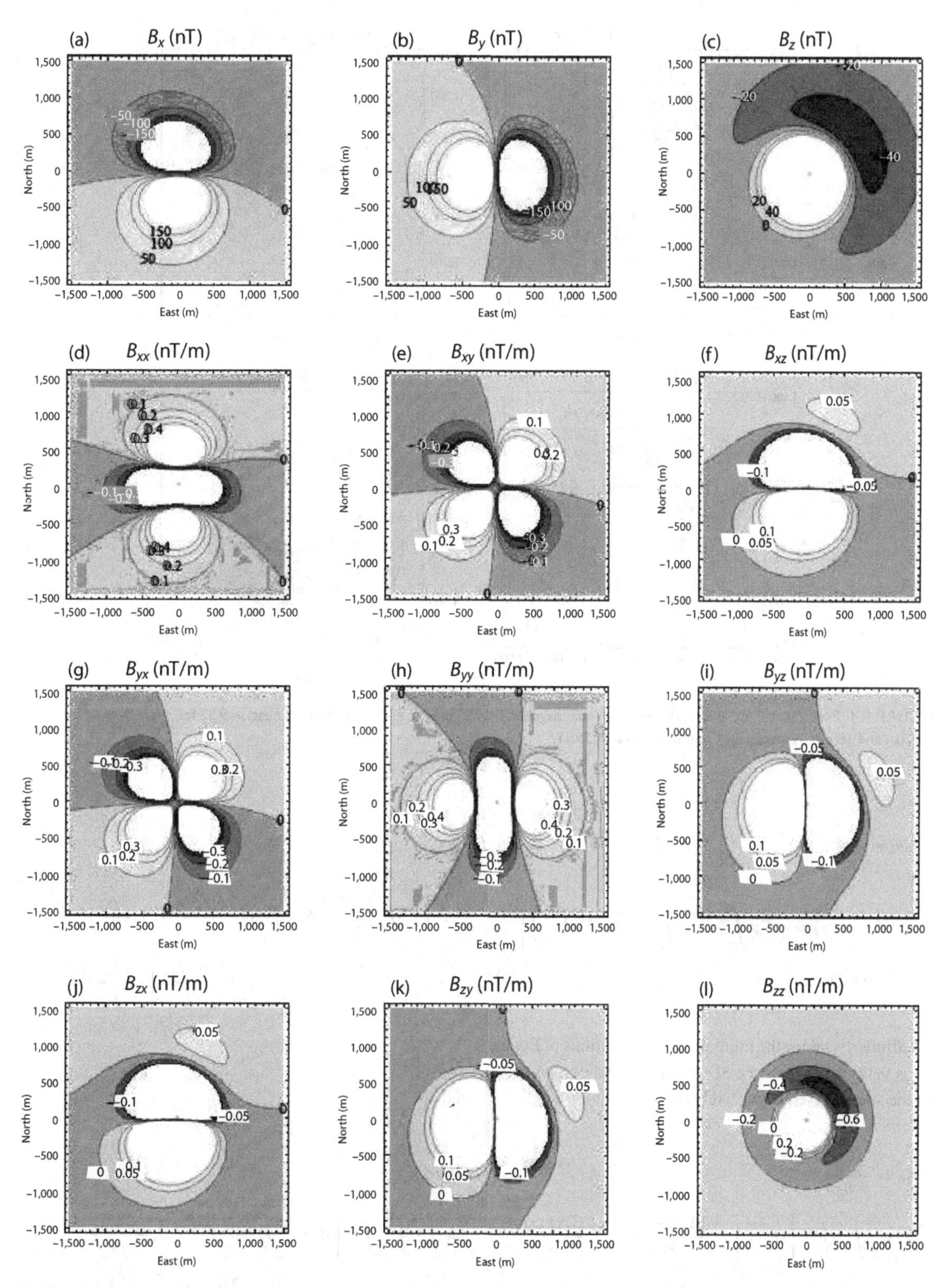

FIGURE 9.3 Examples of the magnetic vector components B_x, B_y, and B_z (Equations 9.14–9.16) and gradient tensors (Equations 9.17–9.22) for the point dipole in Figure 9.2.

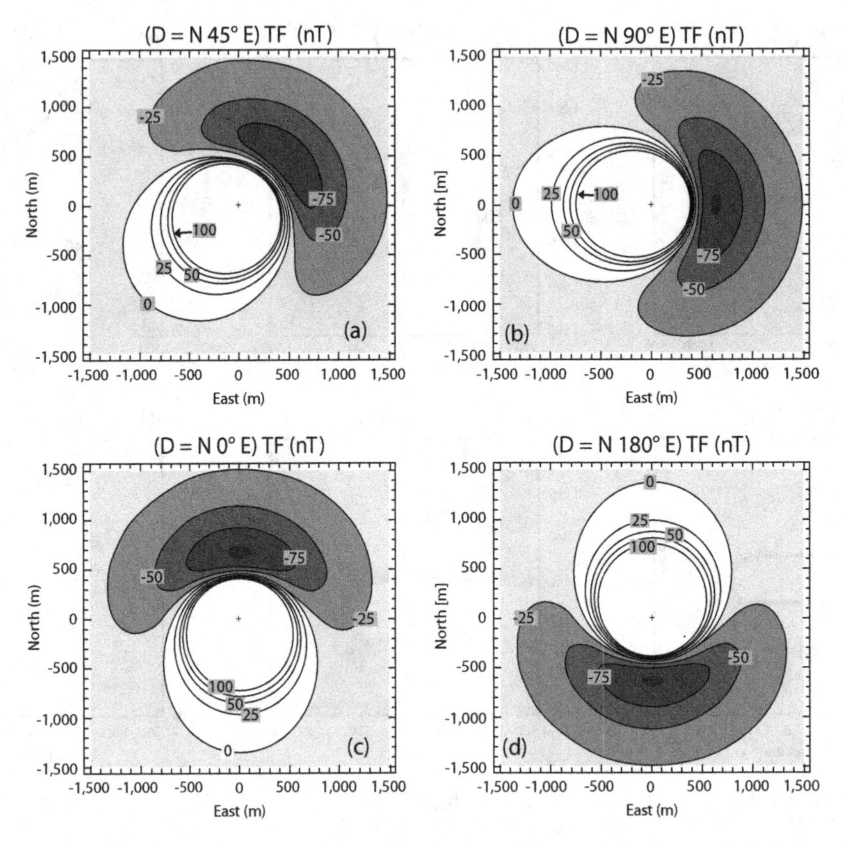

FIGURE 9.4 Maps (a), (b), (c), and (d) present the total magnetic field anomalies (TF) described in Equation 9.23 for $D = D' = 45°$, $90°$, $0°$, and $180°$ E, respectively. Contour interval = 25 nT.

function) by

$$V = V(x, y, z) = -C_m \int_{z_a'}^{z_b'} \int_{y_a'}^{y_b'} \int_{x_a'}^{x_b'} J \left\{ \hat{\mathbf{u}}' \cdot \nabla' \left(\frac{1}{r} \right) \right\} \times \partial x' \partial y' \partial z'. \tag{9.26}$$

Similarly, taking the finite volume integrations of Equations 9.14–9.16 yields the 3D magnetic field components of the source given respectively by

$$B_x = -\frac{\partial V}{\partial x} = \int_{z_a'}^{z_b'} \int_{y_a'}^{y_b'} \int_{x_a'}^{x_b'} \left\{ B_x' = \left[J_x \frac{3\Delta x^2 - r^2}{r^5} + J_y \frac{3\Delta x \Delta y}{r^5} + J_z \frac{3\Delta x \Delta z}{r^5} \right] C_m \right\} \partial x' \partial y' \partial z', \tag{9.27}$$

$$B_y = -\frac{\partial V}{\partial y} = \int_{z_a'}^{z_b'} \int_{y_a'}^{y_b'} \int_{x_a'}^{x_b'} \left\{ B_y' = \left[J_x \frac{3\Delta y \Delta x}{r^5} + J_y \frac{3\Delta y^2 - r^2}{r^5} + J_z \frac{3\Delta y \Delta z}{r^5} \right] C_m \right\} \partial x' \partial y' \partial z', \tag{9.28}$$

and

$$B_z = -\frac{\partial V}{\partial z} = \int_{z_a'}^{z_b'} \int_{y_a'}^{y_b'} \int_{x_a'}^{x_b'} \left\{ B_z' = \left[J_x \frac{3\Delta z \Delta x}{r^5} + J_y \frac{3\Delta z \Delta y}{r^5} + J_z \frac{3\Delta z^2 - r^2}{r^5} \right] C_m \right\} \partial x' \partial y' \partial z'. \tag{9.29}$$

Furthermore, the total magnetic field anomaly of the 3D source, by Equation 9.23, is

$$B_{\mathrm{T}}(x,y,z) = B_x(\cos I \cos D) + B_y(\cos I \sin D) + B_z(\sin I), \tag{9.30}$$

for geomagnetic inclination I and declination D at the observation point. Thus, the generalized 3D magnetic vector is given by

$$\mathbf{B} = \mathbf{B}(x,y,z) = C_{\mathrm{m}} \int_{z'_a}^{z'_b} \int_{y'_a}^{y'_b} \int_{x'_a}^{x'_b} J \left\{ \nabla \left[\hat{\mathbf{u}}' \cdot \nabla' \left(\frac{1}{r} \right) \right] \right\} \partial x' \partial y' \partial z', \tag{9.31}$$

whereas the generalized total field anomaly is

$$B_T = B_T(x,y,z) = C_{\mathrm{m}} \int_{z'_a}^{z'_b} \int_{y'_a}^{y'_b} \int_{x'_a}^{x'_b} J \left\{ \hat{\mathbf{u}} \cdot \nabla \left[\hat{\mathbf{u}}' \cdot \nabla' \left(\frac{1}{r} \right) \right] \right\} \partial x' \partial y' \partial z'. \tag{9.32}$$

Additionally, the three diagonal tensors that satisfy Laplace's equation (3.6) from the finite volume integrations of Equations 9.17–9.19 are

$$B_{xx} = \int_{z'_a}^{z'_b} \int_{y'_a}^{y'_b} \int_{x'_a}^{x'_b} \left\{ B'_{xx} = \frac{3C_{\mathrm{m}}}{r^5} \left[J_x \left(3\Delta x - \frac{5\Delta x^3}{r^2} \right) + J_y \left(\Delta y - \frac{5\Delta x^2 \Delta y}{r^2} \right) + J_z \left(\Delta z - \frac{5\Delta x^2 \Delta z}{r^2} \right) \right] \right\} \times \partial x' \partial y' \partial z' = -(B_{yy} + B_{zz}), \tag{9.33}$$

$$B_{yy} = \int_{z'_a}^{z'_b} \int_{y'_a}^{y'_b} \int_{x'_a}^{x'_b} \left\{ B'_{yy} = \frac{3C_{\mathrm{m}}}{r^5} \left[J_x \left(\Delta x - \frac{5\Delta y^2 \Delta x}{r^2} \right) + J_y \left(3\Delta y - \frac{5\Delta y^3}{r^2} \right) + J_z \left(\Delta z - \frac{5\Delta y^2 \Delta z}{r^2} \right) \right] \right\} \times \partial x' \partial y' \partial z' = -(B_{xx} + B_{zz}), \tag{9.34}$$

and

$$B_{zz} = \int_{z'_a}^{z'_b} \int_{y'_a}^{y'_b} \int_{x'_a}^{x'_b} \left\{ B'_{zz} = \frac{3C_{\mathrm{m}}}{r^5} \left[J_x \left(\Delta x - \frac{5\Delta z^2 \Delta x}{r^2} \right) + J_y \left(\Delta y - \frac{5\Delta z^2 \Delta y}{r^2} \right) - J_z \left(3\Delta z - \frac{5\Delta z^3}{r^2} \right) \right] \right\} \times \partial x' \partial y' \partial z' = -(B_{xx} + B_{yy}). \tag{9.35}$$

The three unique off-diagonal tensors, on the other hand, from respectively integrating Equations 9.20–9.22 are

$$B_{xy} = \int_{z'_a}^{z'_b} \int_{y'_a}^{y'_b} \int_{x'_a}^{x'_b} \left\{ B'_{xy} = \frac{3C_{\mathrm{m}}}{r^5} \left[J_x \left(\Delta y - \frac{5\Delta x^2 \Delta y}{r^2} \right) + J_y \left(\Delta x - \frac{5\Delta y^2 \Delta x}{r^2} \right) - J_z \left(\frac{5\Delta x \Delta y \Delta z}{r^2} \right) \right] \right\} \times \partial x' \partial y' \partial z' = B_{yx}, \tag{9.36}$$

$$B_{xz} = \int_{z'_a}^{z'_b} \int_{y'_a}^{y'_b} \int_{x'_a}^{x'_b} \left\{ B'_{xz} = \frac{3C_{\mathrm{m}}}{r^5} \left[J_x \left(\Delta z - \frac{5\Delta x^2 \Delta z}{r^2} \right) - J_y \left(\frac{5\Delta x \Delta y \Delta z}{r^2} \right) - J_z \left(\Delta x - \frac{5\Delta z^2 \Delta x}{r^2} \right) \right] \right\} \times \partial x' \partial y' \partial z' = B_{zx}, \tag{9.37}$$

and

$$B_{yz} = \int_{z'_a}^{z'_b} \int_{y'_a}^{y'_b} \int_{x'_a}^{x'_b} \left\{ B'_{yz} = \frac{3C_{\mathrm{m}}}{r^5} \left[-J_x \left(\frac{5\Delta x \Delta y \Delta z}{r^2} \right) + J_y \left(\Delta z - \frac{5\Delta y^2 \Delta z}{r^2} \right) + J_z \left(\Delta y - \frac{5\Delta z^2 \Delta y}{r^2} \right) \right] \right\} \times \partial x' \partial y' \partial z' = B_{zy}. \tag{9.38}$$

Thus, the nine magnetic gradient tensors make up the generalized gradient tensor of the 3D body given by

$$\nabla\mathbf{B} = \nabla\mathbf{B}(x,y,z) = C_{\mathrm{m}} \int_{z'_a}^{z'_b} \int_{y'_a}^{y'_b} \int_{x'_a}^{x'_b} J \left\{ \nabla \left(\nabla \left[\hat{\mathbf{u}}' \cdot \nabla' \left(\frac{1}{r} \right) \right] \right) \right\} \times \partial x' \partial y' \partial z'. \tag{9.39}$$

Elliptical or narrow elongated anomalies are also commonly encountered. This may be modeled as a 2D magnetic anomaly with infinite strike length. Taking the x-axis perpendicular to the strike of the 2D body, the y-axis parallel to the strike, and the z-axis positive downwards, the 2D magnetic effects can be modeled from the potential

$$\mathcal{V}(x,z) = C_{\mathrm{m}} \int_{z'_a}^{z'_b} \int_{x'_a}^{x'_b} \partial x' \partial z' \left\{ \int_{-\infty}^{\infty} \frac{J_x \Delta x + J_y \Delta y + J_z \Delta z}{(\Delta x^2 + \Delta y^2 + \Delta z^2)^{3/2}} \partial y' \right\} = \int_{z'_a}^{z'_b} \int_{x'_a}^{x'_b} \left\{ \mathcal{V}' = 2C_{\mathrm{m}} \frac{J_x \Delta x + J_z \Delta z}{\Delta x^2 + \Delta z^2} \right\} \partial x' \partial z', \tag{9.40}$$

where the 2D displacement vector $\mathbf{r}$ in the (x, z)-plane has the magnitude given by Equation 3.36. The operative physical property of the 2D source is its magnetic moment per unit length or surface magnetization (in A/m^2)

$$\frac{\mathbf{j}}{y'} = \int\int_S \frac{\partial \mathbf{j}}{y'} = \mathbf{J} \times \int\int_S \partial x' \partial z' = \mathbf{J} \times S, \tag{9.41}$$

where $(\partial \mathbf{j}/y') = \mathbf{J} \times \partial x'\partial z'$ and S is the source's cross-sectional area or surface (in m^2). In addition, the 2D magnetization components are $J_x = J[\cos I' \cos(\alpha - D')]$ and $J_z = J[\sin I']$, where α is the strike of the profile across the 2D body – i.e. the angle that the x-axis makes measured positive clockwise from geographic north.

Accordingly, the magnetic field components of the 2D source are

$$\mathcal{B}_x = -\frac{\partial \mathcal{V}}{\partial x} = \int_{z'_a}^{z'_b}\int_{x'_a}^{x'_b}\left\{\mathcal{B}'_x = C_{\mathrm{m}}\frac{J_x(\Delta x^2 - \Delta z^2) + 2J_z\Delta x\Delta z}{(\Delta x^2 + \Delta z^2)^2}\right\} \times \partial x'\partial z', \tag{9.42}$$

$$\mathcal{B}_y = -\frac{\partial \mathcal{V}}{\partial y} = 0 \tag{9.43}$$

because there is no horizontal field strength in the y-direction, and

$$\mathcal{B}_z = -\frac{\partial \mathcal{V}}{\partial z} = \int_{z'_a}^{z'_b}\int_{x'_a}^{x'_b}\left\{\mathcal{B}'_z = C_{\mathrm{m}}\frac{2J_x\Delta x\Delta z - J_z(\Delta x^2 - \Delta z^2)}{(\Delta x^2 + \Delta z^2)^2}\right\} \times \partial x'\partial z'. \tag{9.44}$$

In addition, the total magnetic field anomaly is

$$\mathcal{B}_T(x, z) = \mathcal{B}_x[\cos I \cos(\alpha - D)] + \mathcal{B}_z[\sin I], \tag{9.45}$$

and the two unique tensors for the 2D source are

$$\mathcal{B}_{xx} = \int_{z'_a}^{z'_b}\int_{x'_a}^{x'_b}\left\{\mathcal{B}'_{xx} = 4C_{\mathrm{m}}\left[\frac{J_x(3\Delta x\Delta z^2 - \Delta x^3) - J_z(3\Delta x^2\Delta z - \Delta z^3)}{(\Delta x^2 + \Delta z^2)^3}\right]\right\} \times \partial x'\partial z' = -\mathcal{B}_{zz}, \tag{9.46}$$

and

$$\mathcal{B}_{xz} = \int_{z'_a}^{z'_b}\int_{x'_a}^{x'_b}\left\{\mathcal{B}'_{xz} = 4C_{\mathrm{m}}\left[\frac{J_x(\Delta z^3 - 3\Delta x^2\Delta z) + J_z(\Delta x^3 - 3\Delta x\Delta z^2)}{(\Delta x^2 + \Delta z^2)^3}\right]\right\} \times \partial x'\partial z' = \mathcal{B}_{zx}. \tag{9.47}$$

To better account for the magnetic effects of the finite elongated source, end corrections are applied to the 2D magnetic calculations, like the gravity corrections described in Section 3.4. The corrected effect is called a 2.5D effect where the profile is equidistant from the ends of the source, and a 2.75D effect when the distances are not equal. Magnetic end corrections are particularly useful for the finite elongated source with constant polygonal cross-section which is extensively used for interactive forward modeling anomaly interpretation (e.g. SHUEY and PASQUALE, 1973; RASMUSSEN and PEDERSEN, 1979; CADY, 1980; SAAD, 1991).

The algorithms for Gravity and Magnetic Modeling Applications (GAMMA) outlined in Section 3.4 from SAAD and BISHOP (1989), and SAAD (1991, 1992, 1993) include magnetic expressions derived directly from the gravity expressions using Poisson's relation (e.g. Section 3.9). Accordingly, the magnetic 2D, 2.5D, and 2.75D effects may be expressed in terms of the same geometric parameters and functions that are illustrated in Figures 3.4, 3.5, 3.6, and 3.7, and summarized in Table 3.1. By Poisson's theorem, gravity gradient tensors can express the magnetic vector components within a proportionality constant involving the magnetization intensity-to-density ratio of the body. Thus, the magnetic vector components for the symmetric 2D case can be obtained from

$$\begin{pmatrix}\mathcal{B}_x\\ \mathcal{B}_z\end{pmatrix} = C_{\mathrm{m}}\begin{pmatrix}\mathcal{D}_{xx} & \mathcal{D}_{xz}\\ \mathcal{D}_{zx} & \mathcal{D}_{zz}\end{pmatrix}\begin{pmatrix}J_x\\ J_z\end{pmatrix}, \tag{9.48}$$

where the geometric functionals or Green's functions that make up the elements of the $\mathcal{D}$-matrix are given in Equations 3.53–3.54. These elements are the second derivatives of the 2D logarithmic potential of gravity normalized by the physical property functional $(G \times \sigma)$. Thus, the equations for the magnetic vector components are exactly the same as for gravity gradients except that the density functional $(G \times \sigma)$ is replaced by the vector components of the magnetization functional.

Similarly, for the symmetric 2.5D case, the magnetic vector components are available from

$$\begin{pmatrix}B_{(2.5)x}\\ B_{(2.5)y}\\ B_{(2.5)z}\end{pmatrix} = C_{\mathrm{m}}\begin{pmatrix}D_{(2.5)xx} & 0 & D_{(2.5)xz}\\ 0 & D_{(2.5)yy} & 0\\ D_{(2.5)zx} & 0 & D_{(2.5)zz}\end{pmatrix}\begin{pmatrix}J_x\\ J_y\\ J_z\end{pmatrix}, \tag{9.49}$$

with the $\mathbf{D}_{(2.5)}$-matrix elements given by Equations 3.56–3.60. In addition, the magnetic vector components for the asymmetric 2.75D case are given by

$$\begin{pmatrix}B_{(2.75)x}\\ B_{(2.75)y}\\ B_{(2.75)z}\end{pmatrix} = C_{\mathrm{m}}\begin{pmatrix}D_{(2.75)xx} & D_{(2.75)xy} & D_{(2.75)xz}\\ D_{(2.75)yx} & D_{(2.75)yy} & D_{(2.75)yz}\\ D_{(2.75)zx} & D_{(2.75)zy} & D_{(2.75)zz}\end{pmatrix}\begin{pmatrix}J_x\\ J_y\\ J_z\end{pmatrix}, \tag{9.50}$$

where Equations 3.62–3.68 give the normalized elements of the $\mathbf{D}_{(2.75)}$-matrix.

TABLE 9.1 Equations for computing vertical magnetic anomalies B_z in nT of idealized sources with vertical polarization intensity J. *If induced, $J = \Delta k \times B_N$ where Δk is the CGSu magnetic susceptibility contrast over the volume Vol, and B_N is the vertical geomagnetic field intensity in nT. In addition, r is the distance between observation point and center (sphere) or axis (cylinders) or centerline (sheets); x is the horizontal distance along the principal profile from the center of the source to the observation point; z_c is the depth to center (sphere) or central horizontal axis (H-cylinder) or top (z_t) or bottom (z_b) of sheet or centerline (V-fault); t is thickness; R is radius; E is depth extent; θ is angle from horizontal; S-∞ is semi-infinite; V is vertical; and H is horizontal. All distances are measured in common linear units (centimeters, meters, kilometers, inches, feet, yards, kilofeet, miles, etc.). The numbers listed for each source are keyed to the gravity sources in Tables 3.2 and 3.3 that are schematically illustrated in Figures 3.8 and 3.9, respectively.*

Source	Vertical magnetic anomaly
General	$B_z = (J \times \text{Vol})/r^n$, where $J = \Delta k \times B_N$ if induced
(2) Sphere	$B_z = \left(8\pi \times J \times R^3/3z_c^3\right)\left(1 - x^2/2z_c^2\right)/\left(1 + x^2/z_c^2\right)^{5/2}$
(4) H-cylinder	$B_z = (2\pi \times J \times R^2\left[z_c^2 - x^2\right])/\left(x^2 + z_c^2\right)^2$
(5) V-cylinder	$B_z = (\pi \times J \times R^2 \times z_t)/\left(x^2 - z_t^2\right)^{3/2}, \quad E \gg z_t$
(11) Thin V-sheet (2D)	$B_z = (2J \times t \times x)\left(\left[z_t/\left(x^2 + z_t^2\right)\right] - \left[z_b/\left(x^2 + z_b^2\right)\right]\right)$
(12) Thin H-sheet (2D)	$B_z = (2J)(\theta_1 - \theta_2), \quad E \ll z_t$
(13) Thin S-∞ H-slab (V-fault)	$B_z = (J \times t \times x)/\left(x^2 + z_c^2\right)$

After NETTLETON (1942).

Thus, the GAMMA algorithms compute gravity, magnetic, and gravity gradient components separately or simultaneously. Their striking analytical similarities allow common functions and terms to be computed only once so that CPU time for simultaneous interactive gravity, magnetic, and gravity gradient modeling and integration is almost as fast as for a single field application. SAAD (1993) gives further examples illustrating the utilities of GAMMA in interactive gravity and magnetic data modeling.

In general, the extended body integrals presented in this section are readily adapted for modeling the magnetic effects of any spatial distribution of magnetization. The integrals can be analytically evaluated for idealized magnetic sources with simple symmetric shapes like spheres, cylinders, and prisms, whereas numerical integration is necessary for more complicated general magnetic sources with arbitrary shapes. The next sections further detail the magnetic effects of idealized and general sources, respectively.

9.6 Idealized source magnetic modeling

The magnetic effects of idealized 3D and 2D bodies are modeled by setting the appropriate limits and executing the desired integrals in Equations 9.25–9.30 and 9.40–9.45, respectively. NETTLETON (1942) developed a set of idealized body magnetic equations that have found wide application in magnetic anomaly interpretation. However, the most widely used idealized body is the 2D dike, which is developed in greater detail in Section 13.5.3.

Using Poisson's theorem (Section 3.9), the derivatives of the idealized body gravity effects (Tables 3.2 and 3.3) taken in the magnetization direction give the related idealized magnetic effects. The derivatives are generally taken in the vertical direction for the reduced-to-pole or vertically polarized magnetic effects of the idealized bodies (e.g. NETTLETON, 1942), which is equivalent to evaluating the magnetic integral Equations 9.25–9.45 with magnetic inclinations $I = I' = 90°$. Table 9.1 gives some of the more widely used of these simplified magnetic effects, where the numbers listed for each source are keyed to the idealized gravity sources in Tables 3.2 and 3.3 that are schematically illustrated in Figures 3.8 and 3.9, respectively.

The simplified magnetic effects are for principal profiles crossing the centers of the idealized bodies with assumed vertical magnetizations with intensity J. Where the magnetization is induced by a geomagnetic field of intensity B_N, $J = \Delta k \times B_N$ for the susceptibility contrast Δk of the source so that the central magnetic anomaly value at $x = 0$ will have the same polarity as Δk.

A remarkable number of subsurface conditions are accommodated by these idealized source effects. The spherical source (2), for example, is applicable for the 3D dipolar magnetic effects of small ore bodies, archaeological refuse pits, and other magnetic objects with horizontal dimensions that are substantially less than the depth. The horizontal cylinder source (4) is appropriate for the 2D elliptical linear dipolar anomaly effects of folded rocks,

TABLE 9.2 Examples of generalized dimensionless equations for calculating the magnetic effect of idealized geometric sources. *$CA[B_Z]_\#$ is the center magnetic anomaly value or constant and $f_{B_z}[x/z]_\#$ is the attenuation factor for the numbered sources in Table 9.1. The vertical $CA(B_Z)_\#$ and magnetizations (J) are in CGSu, and all spatial dimensions are in common linear units (meters, feet, etc.). Modified from* NETTLETON *(1942).*

Source	$CA(B_Z)_\#$	$f_{B_z}[x/z]_\#$
(2) Sphere	$(8.38 \times 10^5)J(R/z)^3$	$[1 - x^2/(2z^2)][1 + (x/z)^2]^{-\frac{5}{2}}$
(4) ∞ H-cylinder	$(6.28 \times 10^5)J(R/z)^2$	$[1 + (x/z)^2][1 + (x/z)^2]^{-2}$
(5) ∞ V-Line mass	$(3.14 \times 10^5)J(R/z)^2$	$[1 + (x/z)^2]^{-\frac{3}{2}}$
(11) Thin V-sheet (2D)	$(2 \times 10^5)J(t/z)$	$[1 + (x/z)^2]^{-1}$
(13) Thin S-∞ H-slab (V-fault)	$(2 \times 10^5)J(t/z)$	$(x/z)[1 + (x/z)^2]^{-1}$

volcanic tubes, river channels, archaeological trenches, tunnels, and other objects with one horizontal dimension four or more times longer than the orthogonal horizontal dimension. The vertical cylinder source (5), on the other hand, yields an effective monopolar anomaly because its bottom end is too far away to have much effect. It accommodates the circular 3D magnetic effects of volcanic pipes, igneous plugs, mine shafts, archaeological wells, and other vertical pipe-like features for which the horizontal dimension is considerably less than the depth extent.

The narrow vertical sheet expression (11) is useful for approximating the elliptical 2D monopole effects of dikes and other thin vertical bodies whose thickness is small compared with the other spatial dimensions and the depth-to-top. The thin slab expression (12) is appropriate for the elliptical 2D sheet dipole effects of suprabasement sources, sills, and other thin flat-lying bodies of finite width and much greater relative length. The vertical fault expression (13) approximates the 2D magnetic slope effect at the edge of thin sheet sources that are thin relative to the depth-to-top. However, this approximation is quite good even for bodies of considerable thickness: for example, the error is less than 7% for the slab with thickness half its mean depth (NETTLETON, 1942).

NETTLETON (1942) suggested a further simplification of these idealized body expressions that reduces their geometric components to dimensionless terms in the ratio (x/z), where x is the distance from the effective center of the body to the observation point on the principal profile. In this format, the effect may be expressed as the product of the magnetic central amplitude $CA(B_z)_\#$ at $x = 0$ times the dimensionless magnetic fall-off factor $f_{B_z}[x/z]_\#$. The amplitude $CA(B_z)_\#$ varies with depth (z^n) at decay rate (n), and reflects the sign of the susceptibility contrast.

Table 9.2 lists several of the idealized body equations from NETTLETON (1942), and Figure 9.5 illustrates the related dimensionless $f_{B_z}[x/z]_\#$ curves as a function of the offset distance x. Comparisons of the related idealized body gravity effects in Table 3.4 show that the anomaly fall-off with distance x along the principal profile for the magnetic vertical cylinder is the same as that for the gravity of a sphere ($f_{B_z}[\frac{x}{z}]_{\#5} = f_g[\frac{x}{z}]_{\#2}$), whereas the narrow vertical magnetic sheet is equivalent to the fall-off for the horizontal gravity cylinder ($f_{B_z}[\frac{x}{z}]_{\#11} = f_g[\frac{x}{z}]_{\#5}$).

9.7 General source magnetic modeling

A general source involves an arbitrarily shaped body with magnetic effects that are evaluated numerically, in contrast to the idealized source with magnetic effects given by closed-form analytical expressions. The numerical evaluation consists of filling in the arbitrarily shaped body with a collection of idealized bodies and integrating at each observation point the analytical magnetic effects of the idealized bodies.

9.7.1 Generic 2D modeling procedures

The gravity modeling procedures described in Section 3.6.1 have been adapted for modeling the magnetic effects of general sources. TALWANI and HEIRTZLER (1964), for example, developed a widely used procedure that models the magnetic effects of a uniformly magnetized 2D body with constant polygonal cross-section along the y'-axis in the (x', z')-plane of the principal profile by summing the magnetic effects of the bounding line elements at the observation point (Figure 9.6). Like the complementary 2D gravity modeling procedure due to TALWANI *et al.* (1959), all geometric elements of the magnetic line

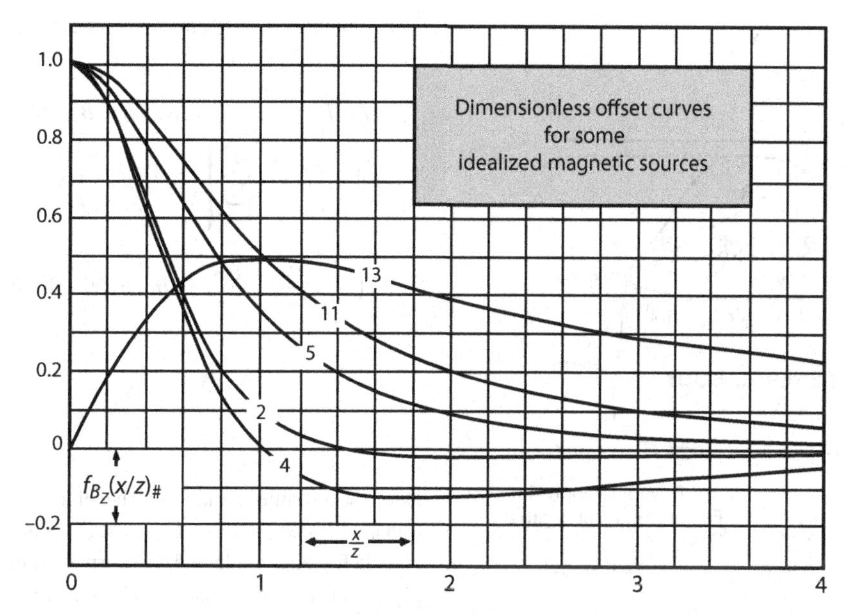

FIGURE 9.5 Distance functions from Table 9.2 used in calculating the magnetic anomaly of the sphere (source 2), infinite horizontal cylinder (source 4), infinite vertical line mass (source 5), thin vertical sheet (source 11), and thin, horizontal semi-infinite slab (fault) (source 13).

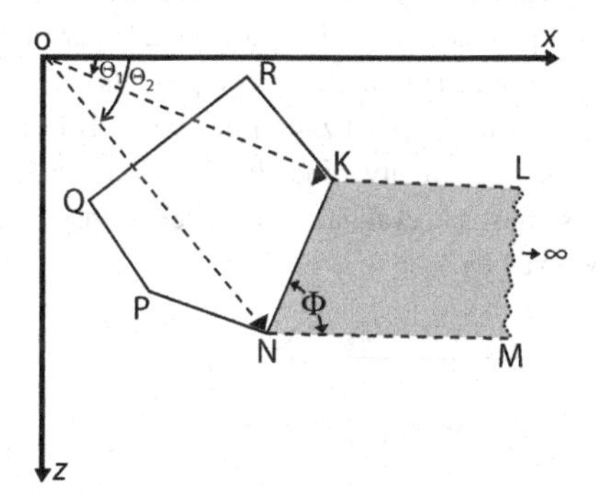

FIGURE 9.6 Geometric conventions for evaluating the magnetic effects of a 2D body with uniform polygonal section $KNPQRK$ from the integrated effects of semi-infinite prisms like $KLMN$. The magnetic profile is calculated by moving the origin successively to various points along the x-axis. Adapted from TALWANI and HEIRTZLER (1964).

integral at the observation point are expressed completely in terms of the 2D coordinates (x', z') of the end points or vertices of the bounding line segments. Accordingly, the magnetic vector and total field components are obtained by setting the origin of the coordinate system at the observation point and proceeding clockwise around the vertical polygonal section.

The 2D magnetic calculations are widely applied and popular, but the source in reality never has infinite strike. Thus, as for 2D gravity calculations (e.g. Equation 3.42), end corrections have been developed for enhanced magnetic modeling of elongated, finite-length magnetic sources (e.g. SHUEY and PASQUALE, 1973; RASMUSSEN and PEDERSEN, 1979; CADY, 1980; SAAD, 1992).

9.7.2 Generic 3D modeling procedures

The line integration method was also adapted by TALWANI (1965) for modeling the magnetic effects of a homogeneously magnetized, irregular 3D source represented by a vertical stack of thin horizontal laminae. As illustrated in Figure 9.7, the procedure involves determining the magnetic effect for each lamina and summing the effects of all the laminas at the observation point. With the origin of the coordinate system set at the observation point and using a polygon to represent the lamina bounded by the body's isopach or thickness contour, the magnetic effect of the lamina can be completely expressed in the coordinates of the corners of the polygon. In contrast to the closed-form magnetic solution for each polygon, the vertical integration combining the magnetic effects of all the polygons must be performed numerically at the observation point.

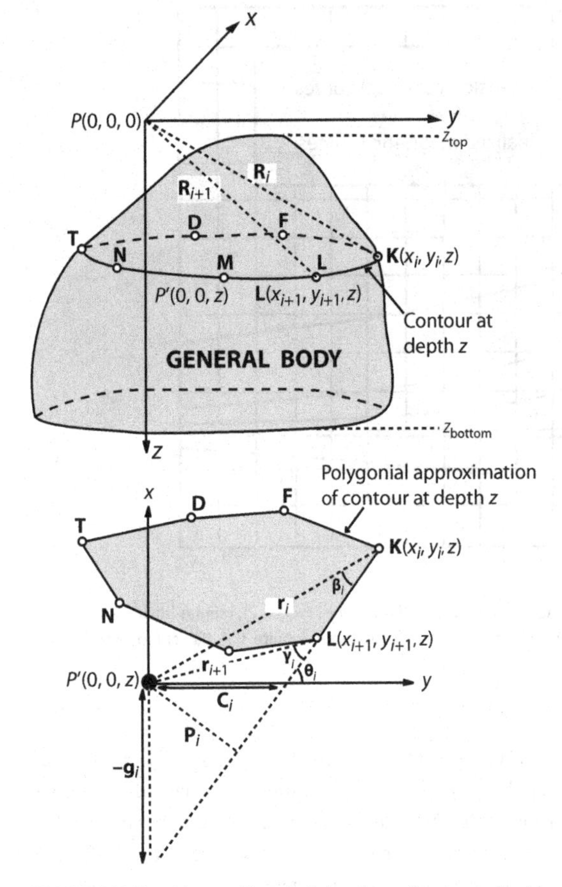

FIGURE 9.7 Variables used in calculating the surface integrals of polygons representing isopach contours of the body so that the magnetic effect of the body is the sum of the polygon effects at the observation point. Adapted from TALWANI (1965).

9.7.3 Least-squares 3D modeling

As described for gravity modeling in Section 3.6.3, the vertical integration can be efficiently performed with least-squares accuracy by the Gauss–Legendre quadrature (GLQ) method. However, rather than dealing with cumbersome magnetic expressions of laminas, a more analytically versatile approach is to implement the GLQ integration using the point dipole effects in Equations 9.12–9.23. This adaptation in effect fits the top and bottom vertical limits of the body with a Legendre polynomial of order ni to determine the vertical z'-coordinates of the roots or zero nodes of the Legendre polynomial for locating the polygons. The x'- and y'-limits of each polygon, in turn, are fitted with Legendre polynomials of orders nj and nk, respectively.

The GLQ decomposition of the body's volume produces a 3D distribution of $ni \times nj \times nk$ point dipoles from which the least-squares magnetic effects of the body are estimated by

$$[V; B_x; B_y; B_z; B_T; B_{xx}; B_{yy}; B_{zz}; B_{xy}; B_{xz}; B_{yz}] \simeq \left(\frac{y'_{kb} - y'_{ka}}{2}\right) \sum_{k=1}^{nk} \left\{ \left(\frac{x'_{jb} - x'_{ja}}{2}\right) \sum_{j=1}^{nj} \times \left\{ \left(\frac{z'_{ib} - z'_{ia}}{2}\right) \sum_{i=1}^{ni} [V'; B'_x; B'_y; B'_z; B'_T; B'_{xx}; B'_{yy}; B'_{zz}; B'_{xy}; B'_{xz}; B'_{yz}] A_i \right\} A_j \right\} A_k. \tag{9.51}$$

Here, the unprimed variables in the left portion of the equation are the complete magnetic effects of the extended 3D body in Equations 9.25–9.38, as derived from the weighted triple sum of the primed variables in the right portion, which are the corresponding integrands of the equations. The z', x', and y' sums are respectively weighted by the Gaussian quadrature coefficients A_i, A_j, and A_k obtained from the relevant Legendre polynomials.

The integrands are simply the magnetic effects of the point dipole (i.e. Equations 9.12–9.23) per unit volume with the magnetic moment components replaced by the corresponding magnetization elements. However, as described in the gravity modeling Section 3.6.3, these integrands must be evaluated at the coordinates within the body given by

$$x'_j = \frac{\bar{x}_j(x'_b - x'_a) + (x'_b - x'_a)}{2}, \quad y'_k = \frac{\bar{y}_k(y'_b - y'_a) + (y'_b - y'_a)}{2}, \quad \text{and } z'_i = \frac{\bar{z}_i(z'_b - z'_a) + (z'_b - z'_a)}{2}, \tag{9.52}$$

where $\bar{x}_j$, $\bar{y}_k$, $\bar{z}_i$ are the Gaussian nodes or coordinates in the standard interval $(-1, 1)$ of the roots for the Legendre polynomials of respective orders nj, nk, and ni that the interpreter chooses to span the lower a- to upper b-limits of the body's spatial dimensions. Thus, Equation 9.52 transforms the integration weighted by Gaussian weights or coefficients A_i, A_j, A_k from the standard $(-1, 1)$-interval to the actual (a, b)-intervals of the 3D body. In application, the Gaussian coefficients and corresponding nodes are commonly implemented from tables (e.g. STROUD and SECREST, 1966) or computed (e.g. Section 3.6.3).

The vertical magnetic vector component B_z of the extended body, for example, is obtained by running the

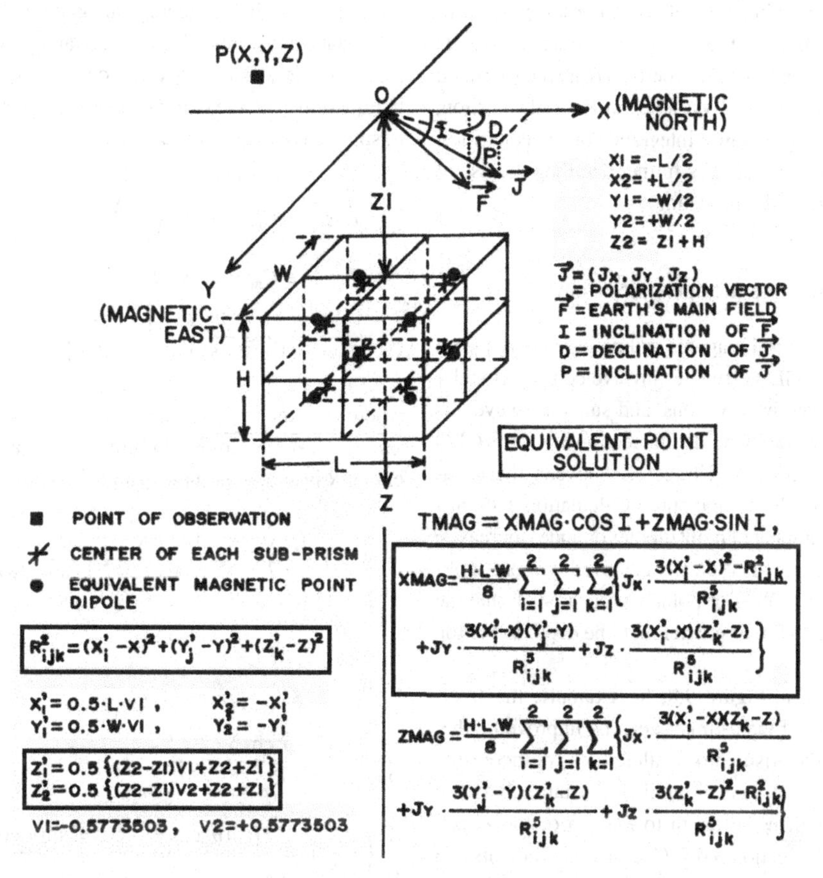

FIGURE 9.8 Details of the $2 \times 2 \times 2$ Gauss–Legendre quadrature formula for estimating the magnetic total field anomaly TMAG $= B_T$ in terms of the horizontal XMAG $= B_x$ and vertical ZMAG $= B_z$ anomaly components of a uniformly magnetized 3D rectangular prism. Adapted from Ku (1977). The notation for the variables is the same as in the text except that $\vec{F}$, D, and P in the figure correspond to $\mathbf{B}_N$, D', and I' in the text. In addition, the Gaussian coefficients $V1$ and $V2$ in all spatial dimensions are equal in magnitude, but opposite in sign because the height H, width W, and length L of the cube are equal. The complementary GLQ formula for the prism's gravity effect is illustrated in Figure 3.14.

weighted triple sum over the vertical point dipole effect per unit volume B'_z that is the integrand of Equation 9.29. The other magnetic effects of the extended body are similarly obtained by repeating this process using the related unit volume point dipole effects evaluated at the appropriate body coordinates x'_j, y'_k, z'_i.

Figure 9.8 gives a further example comparing the details of estimating the vertical B_z, horizontal $B_h = \sqrt{B_x^2 + B_y^2}$, and total field B_T magnetic effects of a prism with uniform dimensions. For GLQ integration, the limits are specified by $(x'_{ja}, x'_{jb}) = (x'_a, x'_b)$, $(y'_{ka}, y'_{kb}) = (y'_a, y'_b)$, and $(z'_{ia}, z'_{ib}) = (z'_a, z'_b)$. For a more general irregular body, the integration limits in each of the body's three dimensions are also readily determined at every node as described in Section 3.6.3.

9.7.4 Least-squares 2D modeling

For the 2D body, the magnetic effects are similarly obtained by fitting the x'- and z'-limits of the cross-section with Legendre polynomials of orders nj and ni, respectively, so that the body is represented by a 2D distribution of $nj \times ni$ point dipoles that estimate the source's least-squares magnetic effects by

$$[\mathcal{V}; B_x; B_z; B_T; B_{xx} = -B_{zz}; B_{xy}] \simeq \left(\frac{x'_{jb} - x'_{ja}}{2}\right) \times \sum_{j=1}^{nj}\left\{\left(\frac{z'_{ib} - z'_{ia}}{2}\right)\sum_{i=1}^{ni}[\mathcal{V}'; B'_x; B'_z; B'_T; B'_{xx} = -B'_{zz}; B'_{xz}]A_i\right\} A_j. \quad (9.53)$$

Here again, the unprimed variables in the left portion of the equation are the complete magnetic effects of the 2D body in Equations 9.40–9.47, as derived from the weighted double sum of the primed variables in the right portion, which are the corresponding integrands of the equations. KU (1977) provides examples further detailing the use of GLQ integration for 2D modeling.

9.7.5 Least-squares modeling accuracy

In contrast to analytical magnetic solutions such as shown in Figure 9.8, the GLQ estimates involve computationally simple multiplications, divisions, and sums. However, as pointed out in the gravity modeling Section 3.6.5, the GLQ estimate always involves a trade-off between the accuracy and speed of the computation. Calculation accuracy increases as the number of point dipoles or nodes increases, but effective accuracy is typically achieved in most applications when the number of point sources is such that the distance between them is smaller than the depth to the top of the source.

For the prism in Figure 9.8, for example, the least-squares accuracy of the estimates can be improved either by subdividing the prism into smaller blocks or increasing the number of point sources until the distance between them is smaller than the depth to the top of the prism. As pointed out in Section 3.6.5, Gaussian coefficients and corresponding nodes have been tabulated to 30 significant figures for Legendre polynomials of orders 2 to 512 (e.g. STROUD and SECREST, 1966) so that GLQ modeling can be accurately computed to many more significant figures than can be measured in practice. Furthermore, no modeling singularities occur at the surface of the prism because the point sources are always displaced away from the surface into the interior of the body. Thus, GLQ integration can compute the prism's magnetic effects to essentially any desired degree of accuracy anywhere on or outside the surface of the prism.

9.7.6 Least-squares modeling in spherical coordinates

As in gravity studies at regional and global scales (Section 3.6.6), equivalent point source (EPS) inversion is a very effective alternate to spherical harmonic modeling for analyzing magnetic anomaly data registered in spherical coordinates over finite spherical patches of the Earth and other planetary bodies. EPS modeling in general offers greater sensitivity for the local data coverage details, and the locally available geological, geophysical, and other subsurface constraints for anomaly analysis.

By the Cartesian to spherical coordinate transformation in Equation 3.82, the scalar magnetic point dipole potential in spherical coordinates is

$$\begin{aligned} V(r,\theta,\phi) &= C_{\mathrm{m}} \times \mathbf{j}\cdot\nabla'\left(\frac{1}{R}\right) \\ &= C_{\mathrm{m}} \times \mathbf{j}\cdot\nabla' R \\ &\quad \times \left\{\frac{\partial}{\partial R}\left(\frac{1}{R}\right) = \frac{-1}{R^2}\right\}, \end{aligned} \tag{9.54}$$

where the displacement distance $R = \sqrt{r^2 + r'^2 - 2rr'\cos\delta}$ (Figure 3.15), and the gradient of R in source point coordinates is

$$\nabla' R = \left(\frac{\partial R}{\partial r'}\right)\hat{\mathbf{e}}_{\mathbf{r}'} + \frac{1}{r'}\left(\frac{\partial R}{\partial \theta'}\right)\hat{\mathbf{e}}_{\theta'} + \frac{1}{r'\sin\theta'}\left(\frac{\partial R}{\partial \phi'}\right)\hat{\mathbf{e}}_{\phi'} \tag{9.55}$$

with

$$\frac{\partial R}{\partial r'} = \frac{r' - r\cos\delta}{R} = \frac{C'}{R}, \tag{9.56}$$

$$\begin{aligned} \frac{1}{r'}\left(\frac{\partial R}{\partial \theta'}\right) &= \frac{r[\sin\theta'\cos\theta - \cos\theta'\sin\theta\cos(\phi-\phi')]}{R} \\ &= \frac{E'}{R}, \end{aligned} \tag{9.57}$$

$$\frac{1}{r'\sin\theta'}\left(\frac{\partial R}{\partial \phi'}\right) = \frac{-r\sin\theta\sin(\phi-\phi')}{R} = \frac{H'}{R}, \tag{9.58}$$

and $\hat{\mathbf{e}}_{\mathbf{r}'}$, $\hat{\mathbf{e}}_{\theta'}$, and $\hat{\mathbf{e}}_{\phi'}$ are the spherically orthogonal unit basis vectors at the source point.

In addition, the $(j_{r'}, j_{\theta'}, j_{\phi'})$-components of the dipolar magnetic moment (or moment contrast) vector $\mathbf{j}$ can be expressed in terms of its inclination I' and declination D' (Figure 8.4) by

$$\begin{aligned} \mathbf{j} &= j\{\hat{\mathbf{u}}'\} \\ &= j\{[\sin(I')]\hat{\mathbf{e}}_{\mathbf{r}'} + [\cos(I')\cos(D')]\hat{\mathbf{e}}_{\theta'} \\ &\quad + [\cos(I')\sin(D')]\hat{\mathbf{e}}_{\phi'}\} \\ &= (j_{r'})\hat{\mathbf{e}}_{\mathbf{r}'} + (j_{\theta'})\hat{\mathbf{e}}_{\theta'} + (j_{\phi'})\hat{\mathbf{e}}_{\phi'} = \mathbf{j}_{r'} + \mathbf{j}_{\theta'} + \mathbf{j}_{\phi'}, \end{aligned} \tag{9.59}$$

where $\hat{\mathbf{u}}'$ is the unit magnetic moment vector. Note that if the moment is induced only by the normal field

$\mathbf{B}_{\mathrm{N}} = B_{\mathrm{N}}\{\hat{\mathbf{u}}'\}$ at the source point, then

$$\mathbf{j} = k \times \mathbf{B}_{\mathrm{N}} = k \times B_{\mathrm{N}}\{\hat{\mathbf{u}}'\} = j\{\hat{\mathbf{u}}'\}, \tag{9.60}$$

where k is the magnetic susceptibility (or susceptibility contrast).

Thus, using the results from Equations 9.56–9.59, the dipolar magnetic potential in Equation 9.54 becomes

$$\begin{aligned} V(r,\theta,\phi) = & -\left(\frac{C_{\mathrm{m}} \times j}{R^3}\right) \\ & \times [C' \sin I' + E' \cos I' \cos D' \\ & + H' \cos I' \sin D'] \\ = & (-C_{\mathrm{m}}/R^3) \times [C'(j_{r'}) + E'(j_{\theta'}) + H'(j_{\phi'})]. \end{aligned} \tag{9.61}$$

The magnetic field is given by $\mathbf{B} = -\nabla V$ with components

$$\begin{aligned} B_r = & \frac{-\partial V}{\partial r} = -\left(\frac{-C_{\mathrm{m}} \times j}{R^3 \times r}\right)\{(r' - C')\sin I' \\ & + E' \cos I' \cos D' - H' \cos I' \sin D' - \left(\frac{3C \times r}{R^2}\right) \\ & \times [C' \sin I' + E' \cos I' \cos D' + H' \cos I' \sin D']\} \end{aligned} \tag{9.62}$$

for $C = r - r' \cos \delta$ (Equation 3.87),

$$\begin{aligned} B_\theta = & \frac{-\partial V}{r\partial\theta} = \left(\frac{-C_{\mathrm{m}} \times j}{R^3}\right)\{(-K')\sin I' \\ & + L' \cos I' \cos D' - M' \cos I' \sin D' - \left(\frac{3K' \times r'}{R^2}\right) \\ & \times [C' \sin I' + E' \cos I' \cos D' + H' \cos I' \sin D']\} \end{aligned} \tag{9.63}$$

for $K' = -\cos\theta' \sin\theta + \sin\theta' \cos\theta \cos(\phi - \phi')$, $L' = \sin\theta' \cos\theta - \cos\theta' \sin\theta \cos(\phi - \phi')$, and $M' = \cos\theta \sin(\phi - \phi')$, and

$$\begin{aligned} B_\phi = & \frac{-\partial V}{r \sin(\theta)\partial\phi} = \left(\frac{-C_{\mathrm{m}} \times j}{R^3}\right)\{(-N')\sin I' \\ & + P' \cos I' \cos D' - Q' \cos I' \sin D' - \left(\frac{3N' \times r'}{R^2}\right) \\ & \times [C' \sin I' + E' \cos I' \cos D' + H' \cos I' \sin D']\} \end{aligned} \tag{9.64}$$

for $N' = -\sin\theta' \sin(\phi - \phi')$, $P' = \cos\theta' \sin(\phi - \phi')$, and $Q' = \cos(\phi - \phi')$. In addition, the magnetic total field of the point dipole is

$$\begin{aligned} B_{\mathrm{T}} = & \sqrt{{B_r}^2 + {B_\theta}^2 + {B_\phi}^2} \simeq (\hat{\mathbf{u}} \cdot \mathbf{B}) \\ = & C_{\mathrm{m}} \times j\left\{-\hat{\mathbf{u}} \cdot \nabla\left[\hat{\mathbf{u}}' \cdot \nabla'\left(\frac{1}{R}\right)\right]\right\} \\ = & B_r \sin I + B_\theta \cos I \sin D + B_\phi \cos I \sin D, \end{aligned} \tag{9.65}$$

where $\hat{\mathbf{u}} = (\sin I)\hat{\mathbf{e}}_r + (\cos I \cos D)\hat{\mathbf{e}}_\theta + (\cos I \sin D)\hat{\mathbf{e}}_\phi$ is the unit vector of the normal field $\mathbf{B}_{\mathrm{N}}$ with inclination I and declination D and spherically orthonormal basis vectors $\hat{\mathbf{e}}_{\mathbf{r}}$, $\hat{\mathbf{e}}_\theta$, and $\hat{\mathbf{e}}_\phi$ at the observation point.

Although the magnetic field components in Equations 9.62–9.64 appear algebraically complicated, they are easily evaluated by electronic computing because they depend only on the spherical (r', θ', ϕ') coordinates and the magnetic moment inclination (I'), declination (D'), and intensity (j) values at the source point, and the (r, θ, ϕ)-coordinates of the observation point. For total magnetic field computations (Equation 9.65), the inclination (I) and declination (D) of the main field at the observation point are also involved. ASGHARZADEH *et al.* (2008) gives additional details on these spherical magnetic field components, associated gradient tensors, and their implementation for modeling the magnetic effects of extended bodies.

Figure 3.15, for example, shows the spherical coordinate adaptation of Equation 9.51. Note, however, that to convert magnetization to magnetic moment in spherical coordinate applications, the standard Cartesian unit volume $\partial v(x', y', z') = \partial x' \partial y' \partial z'$ must be replaced by $\partial v(r', \theta', \phi') = (r'^2 \sin\theta')\partial r' \partial\theta' \partial\phi'$ as a result of the Cartesian-to-spherical coordinate transformations (Equation 3.82).

The primed EPS magnetic variables listed at the top of the equation tree in Figure 3.15 include the potential V', and vector B_r', B_θ', B_ϕ', and B_T components of the point dipole in spherical coordinates from Equations 9.61, 9.62, 9.63, 9.64, and 9.65, respectively. The list also includes the gradient tensors B_{rr}', $B_{\theta\theta}'$, $B_{\phi\phi}'$, $B_{r\theta}' = B_{\theta r}'$, $B_{r\phi}' = B_{\phi r}'$ and $B_{\theta\phi}' = B_{\phi\theta}'$ that are analytically detailed in ASGHARZADEH *et al.* (2008). Each of the listed point pole effects can be generalized as $q(x \equiv C_{\mathrm{m}}\mathbf{J}(J, I', D'); r, \theta, \phi; r', \theta', \phi') \times \partial v(r', \theta', \phi')$ where the generic q-function is made up of the product of the functional involving magnetization, times the geometric functional involving only the coordinates of the observation point (r, θ, ϕ) and the source point (r', θ', ϕ'). For calculations of the total magnetic field B_{T}, the generic x-variable also includes the inclination I and

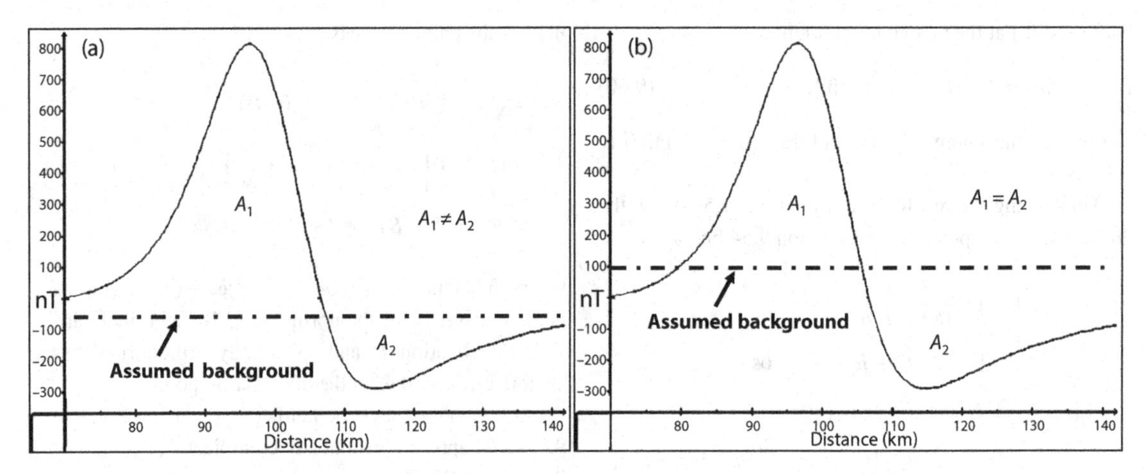

FIGURE 9.9 Magnetic anomaly, ΔB, due to a buried positive magnetization contrast between the source and the surrounding Earth. In (a) the assumed background or regional is too low, as shown by the inequality of A_1 (positive anomaly) and A_2 (negative anomaly). In (b) the assumed background is a closer approximation, as indicated by the equal positive and negative anomaly areas so that $(A_1 - A_2) = 0$.

declination D of the normal field B_N at the observation point. Hence, the magnetic effect of an extended body is given by the triple integral, which for an idealized body can be evaluated in closed analytical form, whereas for an irregularly shaped body the triple series at the bottom of the equation tree provides a least-squares numerical solution.

9.8 Total magnetic moment

Sources of induced magnetic anomalies have equal attractive as well as repulsive magnetic characteristics so that $\nabla \cdot \mathbf{B}(r) = 0$ and the application of Gauss' law to the magnetic anomaly field results in

$$\int\int_v\int \nabla \cdot \mathbf{B}(r)\partial v = \int_S\int (\hat{\mathbf{e}}_n \cdot \mathbf{B})\partial S = 0, \qquad (9.66)$$

where ∂S is a unit element of the surface S enclosing a magnetic body with volume v (e.g. Figure 3.16). Thus, the total energy from an anomalous source regardless of its characteristics equals zero because the positive areas (or volumes) are cancelled out by the negative areas (or volumes), assuming that the vast majority of the surface area or profile in the case of a strike-infinite source is mapped and isolated.

This result is especially useful in determining the quality of the procedure used in isolating a magnetic anomaly because the total energy (anomaly volume or area for a profile of an infinitely long source) will be zero (Figure 9.9). Specifically, areally extensive magnetic surveys should contain essentially equal energy in the positive and negative anomalies if the proper regional effect has been removed from the survey data. Otherwise, the data still reflect the presence of the regional effect.

9.9 Magnetic source ambiguity

The magnetic field satisfies Gauss' law (Equation 9.66) and Laplace's equation in both source-filled and -free space. This means that for any magnetization, there are in principle an infinite number of other magnetizations with variations that can equally satisfy the equations of Poisson and Laplace. The magnetic anomaly is thus said to have inherent ambiguity because any spatial distribution of magnetic moment that satisfies it as a solution is not unique. The lack of uniqueness of magnetic solutions is completely analogous with the interpretational ambiguity of gravity fields (Section 3.8) and will be taken up further in Section 13.2.1.

9.10 Combined magnetic and gravity potentials

For a point dipole which also has mass differential, the magnetic and gravity potentials have exactly the same displacement distance r. Thus, the magnetic and gravitational potentials of a point source, as well as its extensions to anomalous bodies, can be directly compared. As pointed out in Section 3.9, this result is formalized in Poisson's theorem, which shows that all properties of the magnetic field due to a homogeneous body are derivable from its gravity field and *vice versa* (POISSON, 1826). The validity of Poisson's theorem is independent of the shape or depth of the source, with the minor exception of sources that have significant demagnetization effects. As described in Section 3.10, it permits the transformation of the source's magnetic anomaly into its equivalent pseudogravity effect and its gravity anomaly into the equivalent pseudomagnetic effect. Thus, Poisson's theorem provides a quantitative basis for comparing magnetic and gravity anomalies to help reduce ambiguities in interpretation, quantify anomaly correlations, and establish other anomaly attributes related to the distribution and physical properties of anomaly sources. Applications of these results for magnetic anomaly interpretation are considered further in Sections 13.4.2 and 7.4.6.

9.11 Key concepts

- The magnetic method is a potential field method because it measures a function of the potential of the magnetic field of force. The magnetic potential is a scalar quantity of magnetic field space such that its rate of change is the component of the force in that direction.
- The magnetic effects of an extended body are derived by summing at each observation point the effects of the point dipoles that fill out the body. The operative physical property for magnetic modeling is the magnetic moment per unit volume or magnetization.
- The magnetic effect can be integrated in closed-form analytical expression for a sphere, cylinder, prism, and other idealized sources with simple symmetric shapes. A general source with arbitrary shape, on the other hand, must be evaluated by numerical integration from the superimposed effects of idealized sources that fill up the arbitrary volume.
- Idealized sources which have analytical effects can be distributed throughout a general source according to the Gauss–Legendre quadrature decomposition of its effective volume/area/length so that the numerical integration is obtained with least-squares accuracy.
- The infinite-length 2D source, with constant polygonal cross-section perpendicular to its strike, and related finite-length so-called 2.5D and 2.75D sources are widely used in interactive modeling of magnetic anomaly components. The magnetic modeling can be derived from the gravity equations of these sources by Poisson's relation, which also facilitates efforts of combined gravity and magnetic modeling.
- Fundamental laws and theorems of potential fields, such as Gauss' law and the equations of Laplace and Poisson, are useful in understanding the nature of fields and in their analysis. They lead to the inescapable conclusion that magnetic potential fields can be produced by a wide distribution of equivalent sources. Thus, the results of the magnetic method have inherent ambiguities of interpretation that can be reduced only by the imposition of ancillary geological, geophysical, and other constraints.
- Poisson's theorem relates the magnetic and gravity effects from a common source, and thus can help reduce interpretational ambiguities where applicable. Pseudomagnetic effects are derived from the source's gravity effects, whereas pseudogravity effects are derived from its magnetic effects. In addition to limiting ambiguities of interpretation, pseudoanomaly transformations help quantify anomaly correlations and other anomaly attributes related to the distribution and physical properties of anomaly sources.

Supplementary material and Study Questions are available on the website www.cambridge.org/gravmag.

10 Magnetization of Earth materials

10.1 Overview

The magnetic method is based on variations in the magnetic field derived from lateral differences in the magnetization of the subsurface. As a result, an understanding of the magnetization of Earth materials, and the physical and geologic factors that control it, is essential in planning surveys as well as interpreting magnetic anomalies.

Magnetization consists of the vectorial addition of induced and remanent components. Induced magnetization depends on the magnetic susceptibility of the material and the magnitude and direction of the ambient magnetic field, while remanent magnetization reflects the past magnetic history of the material. This makes the prediction of the magnetization of Earth materials difficult in many geological situations. This problem is amplified because, unlike rock densities which vary by only a few orders of magnitude, magnetizations commonly have a range of 10^3 or more. The resulting uncertainty in estimating magnetization is made greater by the fact it is controlled by a few minerals that occur only as accessory constituents in essentially all Earth materials. As a result, material types do not have diagnostic magnetic properties, but useful generalizations can be made based on an understanding of the nature of the constituent magnetic minerals and the thermal and magnetic history of a specific geologic formation. Measurements of magnetic susceptibilities generally are made on samples using an induction balance, and remanent magnetism is determined by measuring the total effect on a magnetic sensor of rotating an oriented sample around three perpendicular axes.

10.2 Introduction

Knowledge of germane Earth material magnetizations within a study region is required for effective planning and implementation of magnetic surveys. Accordingly, as an important component to the magnetic method, this chapter describes the fundamentals of this property and the controls on it, together with methods of measuring it. Representative values are presented of the magnetization properties of a variety of Earth materials including igneous, metamorphic, and sedimentary rocks, sediments, and soils to aid the explorationist in the use of the magnetic method.

Magnetization is directionally dependent, consisting of the vectorial addition of induced and remanent components. Induced magnetization is a function of the magnetic susceptibility of the materials (Section 10.5) and the magnitude and direction of the ambient magnetic field, whereas remanent magnetization reflects the history of the material. In practice, it is not the magnetization that is the controlling property of magnetic anomaly fields, but rather the magnetization contrast between the anomalous source and the laterally adjacent formations, the so-called country rocks, that are assumed to be constant and the norm. As a result, magnetization within a survey region involves both the anomalous volume and the country rock. The magnetization contrast of the anomaly source is said to be positive if the magnetization of the anomalous body exceeds that of the country rock and negative for the reverse.

Induced and remanent magnetization are difficult to estimate by visual inspection or by rock type identification because magnetization is controlled by the previous thermal, chemical, and magnetic history of the material which may be poorly known at best. In addition, the magnetization of most rocks and sediments results from only a few minerals that occur as accessory constituents rather than as major minerals that are used to categorize rocks. Although there are numerous tabulations of magnetic properties, they are primarily focused on rocks that are of interest in mineral resource exploration, which are atypical of most Earth materials, and on measurements of remanent magnetization made for paleomagnetism studies. The latter are used to study the previous magnetic field of the Earth and indirectly a number of features of geological interest, but for the most part the magnitudes of measurements made for paleomagnetism purposes are below the threshold of interest in exploration applications.

The magnetic polarization of rocks and other Earth materials largely comes from the accessory mineral magnetite and varies in common rocks over a range of 10^3 or more. This is considerably larger than the range of densities in Earth materials. As a result, magnetic anomalies have a much larger dynamic range than gravity anomalies. Studies of well-logs, magnetic anomaly data, and geologic samples indicate that magnetization is scale-independent within the crust (e.g. Pilkington and Todoeschuck, 1995, 2004; Todoeschuck *et al.*, 1994) and show that there is a great deal of variation within formations. Remanent magnetization tends to be highly variable depending on the primary and secondary geological processes involved over the history of the rock. Generally, values are of the order of 0.1 to 1.0 times the induced magnetization, but much larger values are observed, especially in young, mafic volcanic rocks and some iron and steel objects. As a general rule, magnetization is not clearly diagnostic of most common rock types because of overlapping ranges and may be variable at a range of scales within a formation. This is especially true because magnetic properties are particularly prone to modification by secondary processes and the heterogeneous distribution of magnetite. As a result, accurately specifying properties by measurements on samples requires care, and use of magnetic property tabulations is problematic. In view of the latter concern, measurements of magnetic properties of samples from the local region are preferable in magnetic surveying, particularly if they are *in situ*. Unfortunately these are not easily made. Measurements of both magnetic susceptibility (which controls induced magnetization) and remanent magnetization require specialized instrumentation (see Section 10.7) and access to multiple samples. As a result, *in situ* measurements are not available for most investigations.

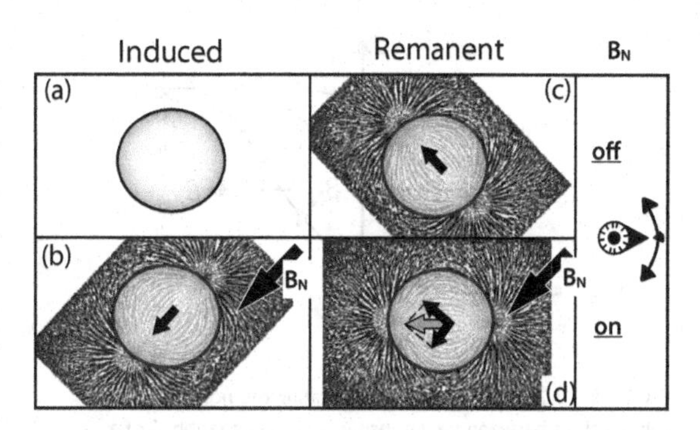

FIGURE 10.1 Matrix illustrating the magnetization set up when a magnetized body (a) is subjected to (b) an externally applied magnetic field $\mathbf{B}_N$. Note that the gray vector in (d) is the resultant magnetization of the induced (b) and remanent (c) components. Simulated iron filings illustrate the patterns of the related force fields.

10.3 Magnetism of Earth materials

Magnetic fields originate in the movement of electrical charges according to the laws of electromagnetism. Atoms, which consist of a positively charged nucleus encircled by shells of orbiting and spinning electrons, are thus a potential source of magnetism. If the individual magnetic moments of neighboring atoms are aligned with one another within a material, they may form a region of uniform magnetization known as a magnetic domain. The regions separating domains with different directions of magnetization are called domain walls, and the movement and blocking of these walls is fundamental to the magnetic properties of rocks. Grains containing just one magnetic domain throughout have strong and stable magnetic moments, whereas multi-domain grains have mobile domain walls with significantly lower and less stable magnetic properties. However, some large grains, pseudo-single-domains, do occur that take on some of the properties of single domains, complicating the simple view of the magnetic properties of materials based on small and large domains (Dunlop, 1995).

Figure 10.1 illustrates the relationships between the induced ($\mathbf{J}_{ind}$) and remanent ($\mathbf{J}_{rem}$) magnetization components and an applied external magnetic field ($\mathbf{B}_N$) for a magnetized body. The induced component of magnetization does not exist in the absence of an external field because the magnetic moments of adjacent atoms due to orbital or spin motions of the electrons are randomly oriented by thermal motions (Figure 10.1(a)). However, in the presence of an external field such as the Earth's magnetic field, the magnetic moments are aligned with the external field resulting in a net magnetic moment or magnetization

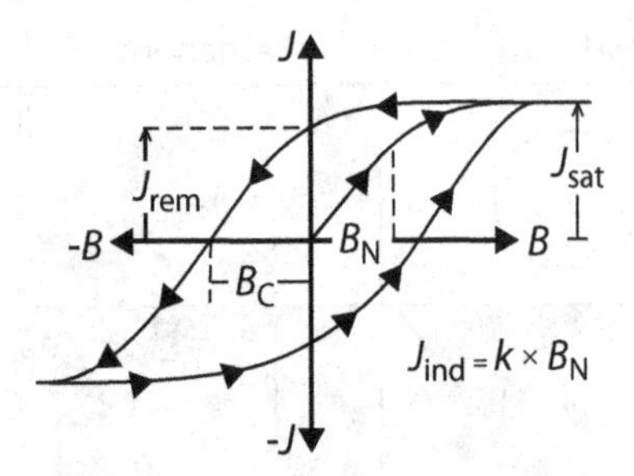

FIGURE 10.2 Hysteresis loop illustrating the non-linear relationships between the intensities of magnetization J of a magnetic body and the varying magnetic field B in which it is placed. J_{sat} and J_{rem} are the respective saturation and remanent magnetization intensities, B_C is the coercive force (field) intensity, and J_{ind} is the intensity of the induced magnetization which is the product of the normal magnetic field strength (B_N) and the magnetic susceptibility (k).

$\mathbf{J}_{ind}$ (Figure 10.1(b)). Any remanent component of the body's magnetization is present, of course, in the absence of an external field (Figure 10.1(c)) and adds vectorially to the induced component to produce the total magnetization $\mathbf{J}_{tot}$ of the body (Figure 10.1(d)).

Figure 10.2 generalizes the non-linear and hysteretic behavior of magnetization **J** for a magnetic body placed in a varying applied magnetic field **B**. Subjecting an unmagnetized sample to the inducing magnetic field with increasing intensity causes the sample's magnetization to increase along the curve until its saturation magnetization J_{sat} is reached. Upon decreasing the applied magnetic field to zero, the sample's magnetization does not fall to zero along the same curve but decreases to the remanent magnetization $\mathbf{J}_{rem}$. Reversing the polarity of the inducing field and increasing its intensity until the remanence is zero gives the coercive force strength B_C of the sample's magnetization. Further increase of the intensity of the reversed applied field results in the magnetization attaining its negative saturation point that is antipodal to the positive saturation point. Again reversing the applied field and decreasing its intensity to zero and then increasing its intensity causes the sample's magnetization to follow the lower curve in Figure 10.2 to its positive saturation level again. Varying the inducing field in a smaller cycle describes a smaller hysteresis loop for the sample's magnetization.

In the weak terrestrial magnetic field, the slope of the magnetization curve in Figure 10.2 defines the sample's magnetic susceptibility, k, or the ease with which a substance is magnetized by the external field. Magnetic susceptibility is a dimensionless quantity given by

$$k = \frac{J_{ind}}{B}. \tag{10.1}$$

TABLE 10.1 Types of magnetization, their sources, and magnetic susceptibility ranges in CGSu.

Susceptibility	Type	Source
$k < 0$	Diamagnetism	Replusive force due to the Larmor precession of orbits of electrons about an applied magnetic field.
$k \equiv 0$	Vacuum	
$0 < k < 10^{-6}$	Paramagnetism	Attractive force due to alignment of electron spin moments.
$10^{-6} < k < 1$	Ferrimagnetism	Adjacent magnetic domains occur in opposition, but with unequal magnetic moments resulting in a net magnetic moment in one direction.
$1 < k < 10^{6}$	Ferromagnetism	Quantum-mechanical exchange forces among atoms causing adjacent magnetic moments to orient parallel to each other forming magnetic domains.

Thus, in the Earth's weak magnetic field $\mathbf{B}_N$, the intensity of induced magnetization is $J_{ind} = k \times B_N$, where k must be measured in a weak inducing field of the order of magnitude of the Earth's field (i.e. 40 A/m, or 0.5 Oe, or 50,000 nT). Equation 10.1 is the same in CGSu or SIu, but the susceptibility in CGSu is 4π times the value in SIu.

The interaction of the ambient (applied) magnetic field with the atoms of Earth materials leads to several types of responses which are identified as types of magnetization (Table 10.1). An understanding of these types of magnetization is important to understanding how magnetic property variations in Earth materials produce magnetic anomalies.

10.3.1 Diamagnetism

Diamagnetism occurs where each filled shell of electrons orbiting the nucleus of an atom has an even number of electrons with one half orbiting in one direction and the other half in the opposite direction. Similarly, the spin directions of electrons are equally divided in opposite directions. This internal symmetry results in a net zero magnetic moment, and thus no external field. However, if the atom is placed in a magnetic field, the electrons will be subjected to a force causing a precession of the orbits about the direction of the external field. The precession results in an additional angular momentum in the direction of the field and a magnetic moment opposite to the applied field. The net opposing field is termed diamagnetism. This magnetism, which is universal to all atoms because the orbits of all electrons experience the precessional effect, opposes the applied magnetic field, the Earth's field in the case of terrestrial applications. The magnitudes of these opposing fields are very small and limited to the case where the electron shells are all filled so that the net magnetic moment is zero in the absence of an ambient field.

Several significant minerals are diamagnetic, including quartz, feldspars, and halite (rock salt). Their susceptibilities are of the order of 10^{-5} SIu. These minerals will produce a field that counteracts the Earth's field, but the magnitude of the field is negligible in comparison to the fields derived from other rock magnetization components, and thus the diamagnetic effect is observed only in special geologic conditions, such as in the presence of massive salt deposits that contrast with the positive magnetic susceptibility of the adjacent sedimentary rocks containing detrital magnetic grains.

10.3.2 Paramagnetism

Paramagnetism is caused by electron spins in atoms that are not offset or compensated by opposing spins of electrons within a shell of orbiting electrons. The uncompensated spins produce a magnetic field external to the atom. Such is the case for transitional elements like iron, titanium, chromium, and nickel. The spinning electrons produce a magnetic dipole. In the absence of an external field the net magnetic moment of these unpaired electrons in atoms is zero because of the disorganizing effect of thermal motions. Thus, the magnitude of paramagnetism is inversely proportional to temperature, a relationship referred to as Curie's law. However, in the presence of a magnetic field, the magnetic moments will favor a parallel alignment to the direction of the field. This results in a weak magnetization called paramagnetism which has a positive magnetic susceptibility, generally in the range of 10^{-3} to 10^{-5} SIu, and is inversely dependent on the temperature. Strong carriers of paramagnetism like the ions Fe^{2+}, Fe^{3+}, and Mn^{2+} cause many common rock-forming minerals such as biotite, pyroxene, amphibole, olivine, and garnet to be paramagnetic.

10.3.3 Ferromagnetism

Ferromagnetism and its various subclasses, which are caused by interactions among neighboring atoms, results from groups of atoms, referred to as domains, aligning their magnetic moments parallel to each other. These magnetic domains, which have dimensions of the order of 10^{-6} m, expand in size under the influence of an external field by alignment of magnetic moments of neighboring domains in the direction of the field or, less commonly, by rotation of the domains into the direction of the external field. The latter situation may result because defects in the atomic arrangement impede growth, or can arise where the grain size is so small that growth potential is limited. The alignment of the domains in ferromagnetic materials leads to a magnetism that far exceeds that of para- or diamagnetism. This is the magnetism of ferrous metals like iron. A prominent characteristic of these materials is that they retain a magnetic moment in the direction of the inducing field after the field is removed. This hysteresis results from the blockage of the domains in their current alignment, effectively retaining a memory of the past field. Above a mineral-specific temperature these ferromagnetic properties are lost. At this temperature, the Curie temperature, ferromagnetic materials take on the properties of a paramagnetic material without the intense magnetism and the magnetic field memory (Figure 10.3).

The exchange forces between atoms in ferromagnetic materials cause adjacent atomic magnetic moments to be oriented in parallel, but other Earth materials exist in subclasses of ferromagnetism in which the adjacent atomic magnets are in opposition. These so-called antiferromagnetic materials behave much like paramagnetic materials, but they exhibit hysteresis, and their magnetic susceptibility increases with temperature up to a mineral-specific value at which the exchange forces disappear and the material behaves as a paramagnetic substance.

10.3.4 Ferrimagnetism

Ferrimagnetism is the most common form of magnetism causing magnetic anomalies. This is similar to antiferromagnetism in that the adjacent magnetic moments are in opposition, but the moments in the two allowable directions are unequal, resulting in a net magnetic moment parallel to the ambient magnetic field. Ferrimagnetic materials, possessing the properties of ferromagnetic substances, are the source of essentially all the magnetization in Earth materials. Magnetite (ferrite having the chemical

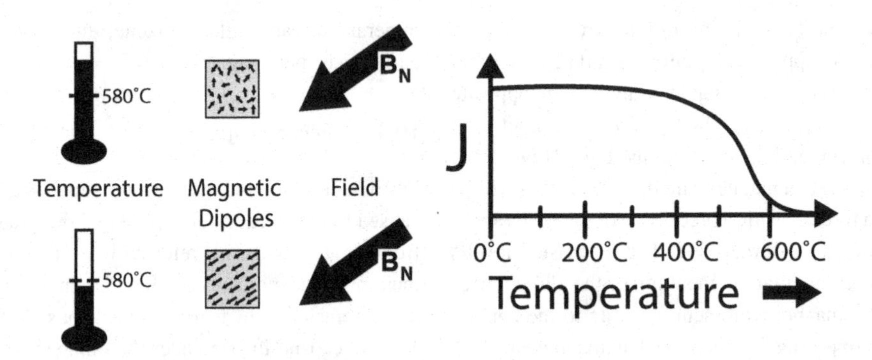

FIGURE 10.3 Ferromagnetic materials lose their magnetism above the Curie temperature because the thermal energy is sufficient to maintain a random alignment of the magnetic moments of the iron minerals. Materials acquire magnetism within roughly a few tens of °C of the Curie point as they cool through the Curie temperature to lower temperatures.

composition $FeO–Fe_2O_3$) is the principal naturally occurring ferrimagnetic component in the Earth. It has a much greater magnetic susceptibility than paramagnetic substances because of the interactions among adjacent atoms and the net magnetic moment in the direction of the external applied magnetic field.

10.3.5 Remanent or permanent magnetism

Remanent magnetization or magnetic remanence refers to the magnetization of ferrimagnetic materials that is taken on and retained from a prior magnetic environment. Unlike induced magnetization, remanent magnetization does not immediately disappear on termination of the ambient magnetic field. It is this form of magnetization that is the basis of paleomagnetic studies and an important source of magnetic anomalies. The relative importance of induced and remanent magnetization in Earth materials is presented in the form of the ratio of remanent to induced magnetization, known as the Koenigsberger ratio or Q. This ratio is highly variable among rock types: it can be 10 or higher in fine-grained rocks, such as basalt, which acquire an intense, stable remanent magnetization, whereas it seldom reaches values of 1 in coarse-grained plutonic rocks.

A wide variety of remanent magnetizations can be acquired by Earth materials during their formation and subsequent geological history. The summation of these magnetizations, both primary and secondary, is natural remanent magnetization (NRM). Primary magnetizations are acquired at the time of the formation of rocks and sediments. The most intense and stable is thermal remanent magnetization or thermoremanent magnetization (TRM) which is imposed upon a rock as it cools through the Curie temperature of the contained ferrimagnetic minerals (Figure 10.3). Most of the magnetization is obtained within a few tens of degrees of the Curie temperature, which is of the order of 560 °C for magnetite. Partial thermoremanent magnetization (PTRM) is the magnetization acquired over a specified interval of cooling temperature. TRM is the summation of a complete spectrum of individual PTRMs. Under rare conditions, which involve the crystallization of a particular mineral, ilmenohematite, the magnetization acquired is reversed to the ambient field and is called reversed thermoremanent magnetization (RTRM).

Detrital or depositional remanent magnetization (DRM) occurs in sedimentary rocks and sediments as a result of the rotation of interstitial grains into the preferred orientation of the ambient field during deposition. Its intensity is generally much less than that of TRM, and thus is not normally important in magnetic mapping.

Chemical remanent magnetization (CRM) is acquired during growth of magnetic minerals in an analogous manner to TRM, but magnetization occurs as the grain size increases while the temperature remains constant. The relationship between grain size and magnetization is complex, but for most minerals the remanent properties are "frozen in" at sizes well below 1 μm and remain constant or decrease with increasing size. An example of CRM is the formation of magnetite grains during serpentinization of ultramafic rocks (e.g. SAAD, 1969a,b).

Magnetizations are also imposed upon Earth materials subsequent to their formation. Anhysteretic remanent magnetization (ARM) is an intense form of magnetization which occurs over small surface areas that have been subjected to alternating fields associated with lightning strikes superimposed upon the steady main magnetic field. A prominent secondary magnetization that is found to a greater or lesser degree in all Earth materials is viscous remanent magnetization (VRM). Acquired over extended periods of time, VRM parallels the ambient field and can reach magnitudes approaching the magnetization induced by the terrestrial field. The time for domains to overcome internal energy barriers, which inhibit the rotation of the magnetic moments into the ambient field, is called the relaxation time; it is directly proportional to the volume of the grains and inversely proportional to the ambient temperature. It increases with the base-10 logarithm (i.e. the log) of the period of time that the material is subjected to the field; that is, the change from time t to $10t$ is the same as the change from $10t$ to $100t$, etc. Thus, over extended periods of time, VRM may become a significant component of the NRM. In a similar manner the magnetization acquired in previous field directions is lost as a function of the log of time. Relaxation times of rocks have a range of values, but the time for the acquisition or decay of VRM may be very long, particularly in fine-grained rocks. This permits VRM, once acquired, to exist for geologic periods measured in millions of years.

10.4 Mineral magnetism

The vast majority of the minerals making up rocks, sediments, and other Earth materials are either diamagnetic or paramagnetic, so they have little impact on the magnetic character of these materials and their role in magnetic mapping is restricted to special situations. However, there are several accessory minerals in rocks that are ferrimagnetic, making them important in geological mapping with the magnetic method. They occur in the titanomagnetite, titanohematite, and iron sulfide series of solid solutions.

10.4.1 Titanomagnetite series

The magnetization of Earth materials is primarily due to magnetic minerals of the ternary system FeO–Fe_2O_3–TiO_2, with minor additional minerals of the Fe–Ni–S system, and ferrous metal alloys. The latter include native metals, especially iron, which occurs rarely in nature, and anthropogenic ferrous metals. The ternary system, shown in Figure 10.4, includes three solid-solution series involving most of the minerals that control magnetism within the Earth, although not all components of the series contribute to the magnetic properties. The most magnetically significant components occur in the titanomagnetite series which joins ulvöspinel (Fe_2TiO_4) and magnetite (FeO–Fe_2O_3). As indicated in the diagram, the series generally is not complete because, under most geological conditions, cooling of the series from high temperature causes exsolution to the end components, essentially pure magnetite and either ilmenite ($FeTiO_3$) or ulvöspinel at room temperatures. Under oxidizing conditions, a common state in nature, ulvöspinel will be converted to an intergrowth of ilmenite and magnetite.

The magnetic properties of titanomagnetites vary as the ratio of ulvöspinel to magnetite changes. The Curie temperature and saturation magnetization increases with decreasing titanium content in a complicated manner depending on the arrangement of the metal ions in the lattice structure. The saturation magnetization is the maximum value of magnetization that can be obtained by the mineral, although the terrestrial field is not strong enough to cause saturation. Magnetic susceptibility and the intensity of TRM are not strongly dependent on composition, because of the overriding effect of grain size (CLARK and EMERSON, 1991). Fine-grained titanomagnetites of the order of 1 μm in size are lower in magnetic susceptibility and higher in specific intensity of the TRM than are larger-grained components. However, the susceptibility is effectively zero for ulvöspinel contents greater than about 70%. Ulvöspinel is paramagnetic at normal temperatures.

Magnetite, the most important and most common magnetic mineral, is ferrimagnetic and has a Curie temperature of about 560 °C, a temperature not exceeded in crust with normal geothermal gradients. The presence of increasing titanium in the solid solution series decreases the Curie temperature. As a result, a Curie temperature of 500° to 550 °C is more realistic for much of the magnetite in the crust because of its titanium content. Magnetite has a strong saturation magnetization and an intense volume magnetic susceptibility. DUNLOP and ÖZDEMIR (2007) give the volume magnetic susceptibility of multidomain magnetite as 3 SIu, whereas CLARK and EMERSON (1991) give volume susceptibilities of single-domain grains of dispersed magnetite from approximately 1 to 6 SIu with susceptibilities of massive, coarse-grained magnetite ranging upward from these single-grain values to 1 or more SIu. These values will be decreased by internal demagnetization as a result of grain-shape anisotropy. The magnetic susceptibility decreases as the Curie temperature is approached, becoming paramagnetic at that temperature.

10.4.2 Titanohematite series

The solid solution between ilmenite ($FeTiO_3$) and hematite (Fe_2O_3), referred to as the titanohematite or ilmenohematite series (Figure 10.4), is second in importance to the titanomagnetite series in Earth materials. The magnetic characteristics of the series are complex and strongly dependent on composition. Upon slow cooling from a high temperature, the components of this series exsolve into the end members, hematite, which is antiferromagnetic but carries a weak magnetism due to a parasitic ferromagnetism in the basal plane of the crystal structure, and

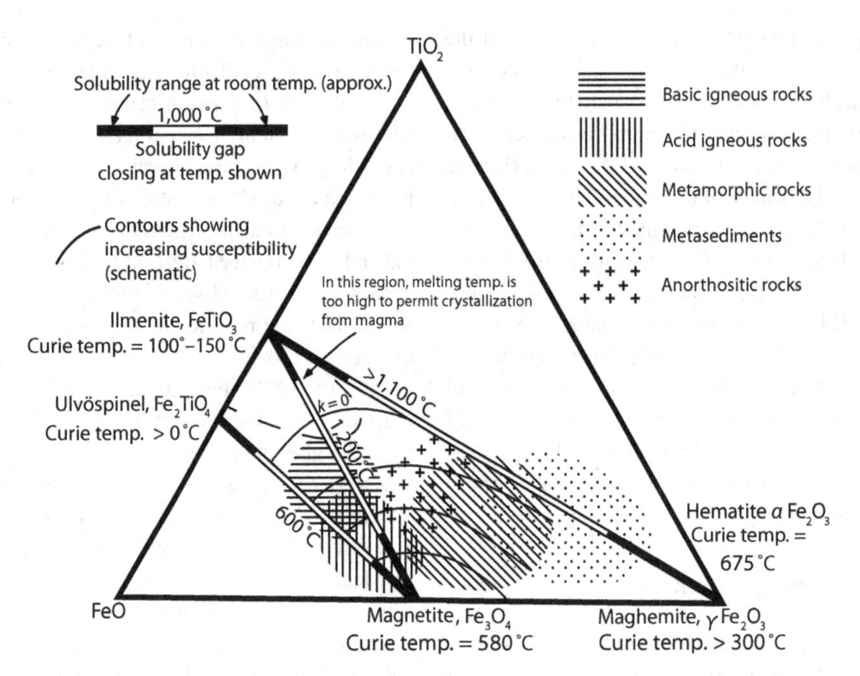

FIGURE 10.4 Review of mineralogical and magnetic data pertaining to the $FeO–TiO_2–Fe_2O_3$ system. After GRANT (1984a,b).

ilmenite, which is paramagnetic at normal temperatures. In the range of this series where ilmenite makes up between 45 and 90% of the total composition, the product is ferrimagnetic and, under special conditions of rapid cooling and specific grain sizes, the products undergo spontaneous self-reversal from the ambient magnetic field. However, in the normal situation involving slow cooling of the rock, the series exsolves to its end members, ilmenite and hematite.

A ferrimagnetic form of Fe_2O_3, maghemite or gamma-hematite (γFe_2O_3), forms under low temperature ($< 200\,^\circ C$) oxidation. Maghemite forms a solid solution series with magnetite that has overall properties similar to magnetite. The other components of the $FeO–TiO_2–Fe_2O_3$ ternary system are paramagnetic, and thus are not an important part of the magnetism of terrestrial materials for exploration purposes.

10.4.3 Iron sulfides

Iron sulfides (FeS_{1+x}) with a range of compositions and magnetic properties occur in a variety of Earth materials. The common form, pyrite (FeS), is paramagnetic, and thus is not an important magnetic constituent. However, monoclinic pyrrhotite (Fe_7S_8) and greigite (Fe_3S_4) are ferrimagnetic. The Curie temperature of pyrrhotite is roughly 320 °C, and the magnetic susceptibility is an order of magnitude less than magnetite, but it is a function of grain size with higher susceptibilities associated with coarser-grained pyrrhotite. As a result, pyrrhotite can be an important component of the magnetic character of certain Earth materials such as deep-seated igneous rocks and amphibolites.

For example, pyrrhotite is the dominant carrier of magnetization at depths greater than about 300 m in the 9 km deep drill hole in Bohemia (BOSUM *et al.*, 1997; BERCKHEMER *et al.*, 1997). The pyrrhotite has a Koenigsberger ratio generally greater than 1, owing to a soft chemical remanent magnetization. Pyrrhotite also occurs as a secondary mineral in strongly reducing environments in recent sulfide-rich sediments and sedimentary rocks which interact with hydrocarbon seepage.

Greigite, which is the sulfide analog of magnetite, has a similar high magnetic susceptibility. Apparently, it is chemically unstable in most sedimentary environments where it is likely to form. Like pyrrhotite, it can be produced by sulfate-reducing bacteria. Native iron, which is strongly ferromagnetic, also can contribute to the magnetization of rocks, but it occurs only rarely in nature because of the ease with which it is oxidized.

10.4.4 Fluids and gases

Fluids and gases within the Earth have very minor influence on the magnetic properties of rocks and are diamagnetic. The susceptibilities of both water and oil are of the

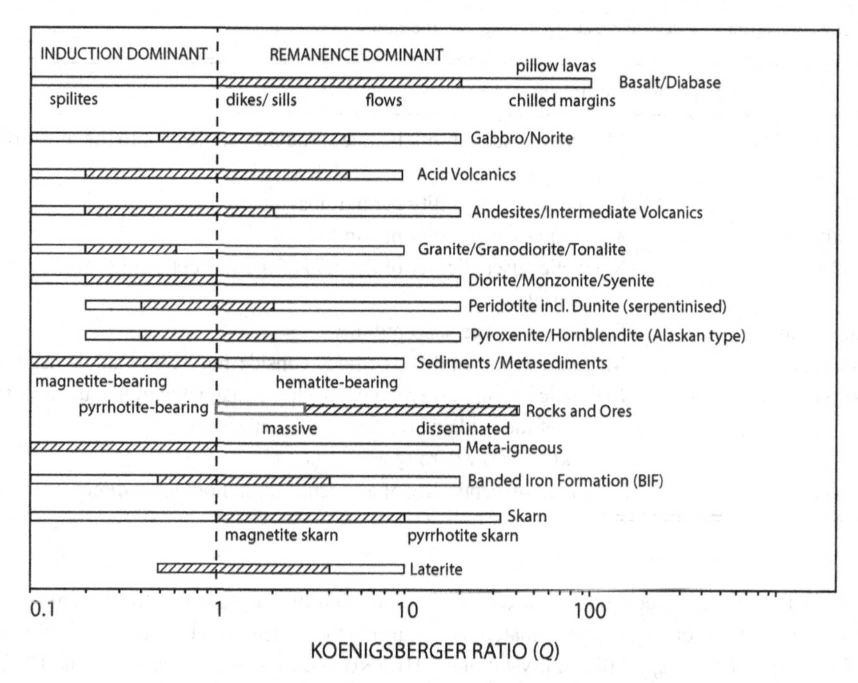

FIGURE 10.5 Summary of the range of the Koenigsberger ratio ($Q = J_{rem}/J_{ind}$) for a variety of rock types. Induced magnetization is dominant below $Q = 1.0$. Diagonally ruled segments denote the typical ranges observed in nature. After CLARK and EMERSON (1991).

order of 10^{-5} SIu, and air and hydrocarbon gases have a susceptibility roughly two orders of magnitude smaller.

10.5 Magnetic susceptibility

Several magnetic properties are reported to a greater or lesser extent for rocks, minerals, and other terrestrial materials in physical property compilations. The most common of these and the one that has the most general significance to geophysical exploration is magnetic susceptibility, the ease with which a substance is magnetized. This property controls the induced magnetization in rocks and other materials. It is the principal property of diamagnetic and paramagnetic materials as well as a critical parameter of ferrimagnetic materials. Induced magnetization (J_{ind}) is generally considered to dominate remanent magnetization (J_{rem}) in rocks, particularly plutonic and sedimentary rocks. This is illustrated in Figure 10.5 which shows the range of the Koenigsberger ratio ($Q = J_{rem}/J_{ind}$) of a variety of types of rocks. Diagonally ruled segments denote the most common values observed in nature. Volcanic rocks and mafic plutonic rocks (e.g. gabbro, norite) generally have remanent values exceeding their induced magnetization. Intense and stable remanence predominates in rocks that contain fine-grained magnetite (generally less than 20 μm). Induced magnetization dominates in granitic and metamorphic rocks containing predominately coarse-grained magnetite that make up the vast majority of the upper basement rocks of the continents.

Table 10.2 lists several types of magnetic susceptibilities that in practice are used in explaining the relatively simple property that relates induced magnetization of a substance to the ambient magnetic field (e.g. HANNA, 1977). Magnetic susceptibility varies with the intensity of the ambient field, so measurements of this property are made at weak fields of the order of the terrestrial field. Accordingly, magnetic susceptibilities reported in geophysics are all weak-field susceptibilities. Susceptibilities can be given in terms of their mass or volume. In classical physics, susceptibility is generally given in terms of mass or the equivalent specific susceptibility. However, in geophysics, susceptibilities generally are given in terms of their volume, because calculations of magnetic fields of bodies are based on considerations of their volume rather than their mass. Accordingly, susceptibilities presented in the geophysical literature should be assumed to be weak-field volume magnetic susceptibilities unless noted otherwise. The mass susceptibilities are converted to volume susceptibility by multiplying by density.

The susceptibility reported in most geophysical literature is the weak field, true or intrinsic, magnetic susceptibility, k. This is sometimes referred to as the bulk magnetic susceptibility, as it is the sum of the susceptibilities of the constituent materials. In most terrestrial materials this

TABLE 10.2 Types of magnetic susceptibilities used in magnetic methods.

Type	Symbol	Definition
Weak-field susceptibility		Magnetic susceptibility in a field roughly equivalent to the terrestrial field ($\approx$ 40 A/m)
Mass susceptibility	k_m	Magnetic susceptibility per unit mass
Volume susceptibility	k	Magnetic susceptibility per unit volume
True susceptibility	k	Magnetic susceptibility of the sum of the susceptibilities of the constituent materials
Intrinsic susceptibility	k	Equivalent to true susceptibility
Apparent susceptibility	k_a	Magnetic susceptibility of a body considering its internal demagnetization
Effective susceptibility	k_e	Magnetic susceptibility which produces magnetization equivalent to the scalar sum of induced and remanent magnetizations [$= k(1 + Q)$ where $Q = J_{rem}/J_{ind}$]
Crystalline susceptibility		Magnetic susceptibility of a specific crystallographic direction in a substance

is dominated by ferrimagnetic components, but it can be influenced by the contributions of paramagnetic materials (Clark and Emerson, 1991) especially in crystalline mafic rocks.

The true susceptibility normally reported in the geophysical literature disregards internal demagnetization effects. Demagnetization arises from an internal field due to the presence of magnetic poles on the surface of a magnetic body. This occurs wherever the lines of the magnetic field cross a magnetic boundary. The internal field decreases the ambient field within the magnetic body, effectively decreasing the magnetization, and thus the magnetic susceptibility. The susceptibility of a body considering its internal demagnetization is the effective magnetic susceptibility, $k_e = k/(1 + Nk)$, where N is the demagnetization factor. The demagnetization factor is a function of the shape of the magnetic volume and direction of the ambient magnetizing field, varying from 0 for magnetization along the long axis of a needle-shaped body to 4π for magnetization across a flat surface, in CGSu. The demagnetization factor in SIu is denoted by $D = N/4\pi$. In practice, demagnetization is negligible where true susceptibility values are less than about $1{,}250 \times 10^{-4}$ SIu or $10{,}000 \times 10^{-6}$ CGSu.

Another useful form of magnetic susceptibility is apparent magnetic susceptibility, k_a, which considers the magnetization induced by the ambient field, as well as the remanent magnetization assuming it parallels the ambient magnetic field. This latter assumption is approximated where the remanent magnetization is primarily VRM. Effective magnetic susceptibility can be calculated from $k_e = k(1 + Q)$. This apparent susceptibility should not be confused with "apparent magnetic susceptibility (permeability)" that is used to describe the susceptibility computed from the in-phase electromagnetic response of the Earth in the absence of electrical conduction currents (e.g. Huang and Fraser, 2000; Won and Huang, 2004).

A complication to magnetic susceptibility which is a potential source of dispersion in the tabulation of measurements of a specific rock type is magnetic susceptibility anisotropy. This anisotropy originates from demagnetization effects related to the shape of the grains and to magnetocrystalline anisotropy. The latter is the result of magnetic domains in crystalline materials being more easily magnetized along some crystallographic directions than others. For example, magnetite which has a cubic crystalline structure is more easily magnetized along one of its diagonals. The effect in magnetite is negligible at terrestrial field intensities. This is not the case for the components of the titanohematite and pyrrhotite series which have a minimum susceptibility normal to their basal plane. Magnetic susceptibility is a second-ranked tensor, and thus consists of nine components which can be reduced to six by invoking the law of conservation of energy.

Shape anisotropy can have a marked effect on the susceptibility of rocks which have a strong orientation of the long axes of the ferrimagnetic minerals, a common occurrence in some sedimentary and metamorphic rocks. For example, the long dimensions of ferrimagnetic minerals tend to lie in the bedding planes of detrital sediments and in the foliation planes of metamorphic rocks. These minerals are magnetized more easily in the long directions because of the minimal effect of internal demagnetization. As a result, the maximum magnetic susceptibility is parallel to the planar structures and the direction of lineation. This leads to the use of anisotropic magnetic susceptibilities in petrofabric studies of flow fabrics in volcanic rocks, paleocurrent investigations in sediments and sedimentary rocks, and foliation of igneous and metamorphic rocks. However, the interpretation of magnetic anomalies from highly laminar formations also needs to consider magnetic susceptibility anisotropy (e.g. BATH, 1962).

The degree of magnetic susceptibility anisotropy is given simply as the ratio of the maximum to the minimum susceptibility. Although in most geophysical applications we assume this value is 1 (i.e. the susceptibility is isotropic), measurements of this property of sedimentary, volcanic, and metamorphic rocks vary from 1.0 to 1.2 or more with increasing parallelism of the ferrimagnetic minerals. Strongly foliated metamorphic rocks may reach ratios of 2.0, and banded iron formations may have anisotropies of 4 or more (JAHREN, 1963). In contrast, plutonic rocks seldom have measurable anisotropies.

Magnetic susceptibility is a dimensionless quantity. Traditionally, however, in CGSu, volume magnetic susceptibility has been incorrectly expressed as EMu/cm^3, rather than in the correct form which is simply CGSu. In SIu, each CGSu of susceptibility is equivalent to 4π (e.g. PAYNE, 1981; SHIVE, 1986).

The volume percentage of magnetite in a rock can be useful for estimating its magnetic susceptibility because magnetite is the principal source of magnetic susceptibility in Earth materials. Numerous studies have been made of this relationship with somewhat different results. The magnetite content of rocks is either estimated by magnetic separation of the magnetite or calculated from chemical analyses of the Fe_2O_3 and FeO content of the material. The differences in the results of these studies reflect variations in magnetic susceptibility of magnetite largely due to the presence of TiO_2 within the magnetite crystalline structure, to grain size and shape, and to the interaction of the magnetic field of the magnetite grains. These factors are a function of rock type and geologic history of the material. Therefore, observed variations in the results of empirical studies of the relationship between magnetite content and magnetic susceptibility are to be expected.

The volume magnetic susceptibility of magnetite as indicated in Figure 10.6 varies from a value in SIu of near unity to 10 or more depending upon the grain size and the chemical composition. Higher values are associated with pure magnetite and coarser grains. However, the apparent magnetic susceptibility of magnetite as measured on specimens or *in situ* is considerably less than indicated in Figure 10.6. SLICHTER (1929) found that the apparent susceptibility k_a of magnetite can be represented in the form

$$k_a = \frac{k_m V}{1 + k_m N(1 - V)}, \qquad (10.2)$$

where k_m is the true magnetic susceptibility of magnetite, and V is the fraction volume of magnetite. For a rock with 1% magnetite by volume where the demagnetization factor $N = 3$, which corresponds to grains of a prolate spheroidal shape, and $k_m = 1.0$, the apparent volume magnetic susceptibility k_a is 0.0314 SIu or 0.0025 in CGSu. STACEY (1963) suggested that typically magnetite grains are in the form of a prolate spheroid with a mean axes ratio of 1.5, which leads to a value of $N = 3.9$ where the grains are distributed randomly. Using Slichter's formula, the apparent susceptibility is 0.0264 SIu (= 0.0021 CGSu). In contrast, the slope between magnetic susceptibility and magnetite content of 0.0377 SIu (or 0.003 CGSu) has been widely used in dealing with Earth materials which have volumes of magnetite of up to several percent distributed throughout them as shown in Figure 10.7 (e.g. NETTLETON and ELKINS, 1944; DOBRIN and SAVIT, 1988). The relationship is roughly linear for up to only several volume percent, and becomes more complicated at larger concentrations, probably owing to the interactions among the magnetite grains.

Figure 10.6 presents a useful compilation of the magnetic susceptibilities of a variety of rock types and pertinent minerals. The susceptibility that is given is the true weak-field volume magnetic susceptibility, which is simply referred to as the magnetic susceptibility. This is the susceptibility used in the calculation of magnetic anomaly fields assuming that the remanent magnetization is negligible and the susceptibility value is weak ($<1{,}250 \times 10^{-4}$ SIu).

The previously described generalizations regarding magnetic susceptibilities and rock types are clearly evident in Figure 10.6. The shaded areas in the susceptibility ranges indicate typically occurring values. As CLARK and EMERSON (1991) point out, there are two shaded areas for numerous rock types, indicating a bimodal distribution of susceptibilities caused by the presence or the absence of ferrimagnetic minerals. These bimodal distributions complicate the identification of a specific susceptibility with a rock type. Further information is required on the geochemistry of the rock, reflecting its origin and subsequent geological and thermal history, to decide which grouping of susceptibility should be assigned to a specific rock unit.

10.6 Magnetization of rocks and soils

The preceding description testifies to the potential complexity of the magnetization of Earth materials. Magnetization is a function not only of the chemical composition and origin of the material, but of its geological and thermal history.

The primary rock-forming minerals are either diamagnetic or paramagnetic, and thus contribute in only minor

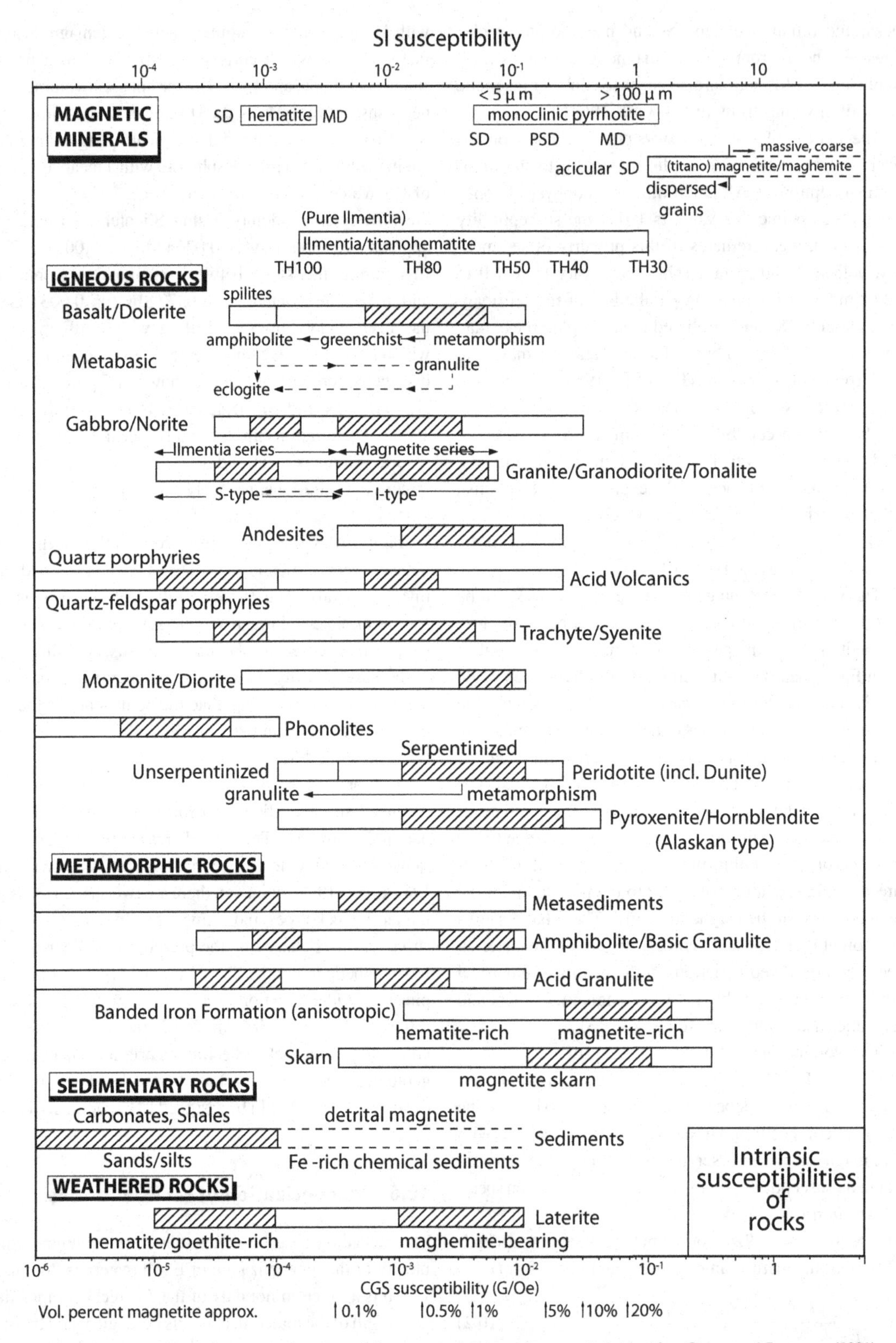

FIGURE 10.6 True or intrinsic volume magnetic susceptibility as measured in weak magnetic fields. After CLARK and EMERSON (1991).

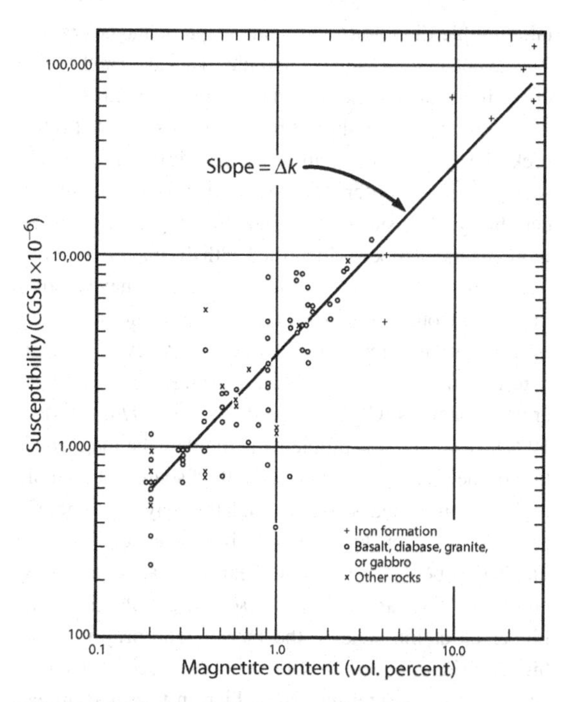

FIGURE 10.7 Data showing the linear relationship of magnetic susceptibility with up to several percent volume concentration of magnetite. The slope of the line indicates that the susceptibility, k, is approximately 0.003 CGSu (0.0377 SIu) units for each percent concentration of magnetite up to several percent in a rock. After MOONEY and BLEIFUSS (1953).

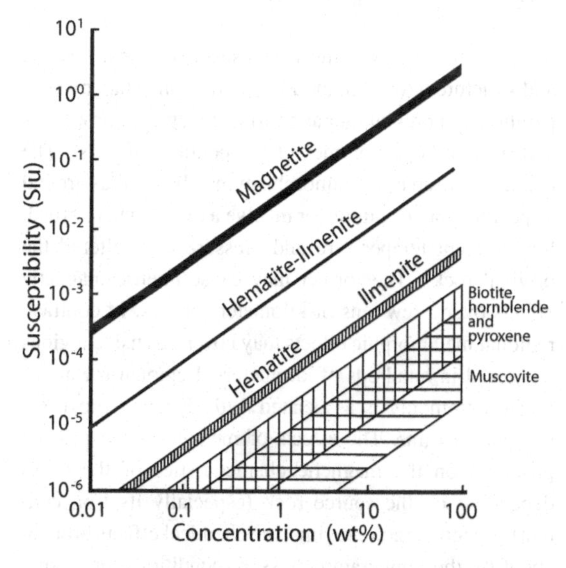

FIGURE 10.8 Mineral contributions to rock susceptibility as a function of their concentration by percent weight. From TARLING and HROUDA (1993), after SCHÖN (1996).

ways to the magnetism of the vast majority of rocks where minerals of the titanomagnetite or titanohematite series or magnetic pyrrhotites are present. This is well illustrated in Figure 10.8, which shows schematically the contribution of minerals to susceptibility of a rock as a function of their weight percent (wt%) concentration within the rock. In rocks with only very small quantities of ferromagnetic iron oxide or sulfide minerals, paramagnetism associated with minerals consisting of iron and magnesium such as biotite, hornblende, pyroxene, olivine, and garnet will significantly contribute to magnetic susceptibility. A number of studies have investigated the general magnetic character of rocks and the factors that cause variations in these properties (e.g. STACEY and BANERJEE, 1974; HAGGERTY, 1979; O'REILLY, 1984; GRANT, 1984a,b; WASILEWSKI, 1987; SHIVE *et al.*, 1988; REYNOLDS *et al.*, 1990; HENKEL, 1991; SCHÖN, 1996).

10.6.1 Igneous rocks

Igneous rocks, which originate from the solidification of magma, make up a prominent part of the continental crust and the vast majority of the oceanic crust as well as underlying mantle. Thus, they are a significant source of magnetization in the Earth where the temperature is below the Curie point of the constituent magnetic minerals – i.e. the temperature above which these minerals revert to a paramagnetic state. Igneous rocks may occur either as plutonic rock that has solidified within the Earth or as volcanic rock that originates in either subaerial or subaqueous conditions. The bulk composition of igneous rocks varies greatly depending on the source of the molten rock and the nature and degree of differentiation of the components during the intrusion, extrusion, and solidification of the magma.

The primary classification of igneous rocks is on the basis of their mineralogy reflecting their bulk chemical composition. Those that are rich in silica are referred to as felsic (or acidic) with granite and rhyolite (the volcanic equivalent of granite) being the primary types. These rocks occur primarily in the upper crust of the continents. At the other end of the composition spectrum are mafic (or basic) rocks rich in iron and magnesium that make up the great majority of the oceanic crust and the lower continental crust. Ultramafic rocks which are relatively higher in iron and magnesium and lower in silica make up the upper mantle and occur only rarely in the crust. The magnetic properties of igneous rocks depend on the bulk rock composition, but also their oxidation state, hydrothermal alteration, and metamorphism. As a result the magnetic properties are related in complex ways to the source of the igneous rocks, geological setting, and history of subsequent geochemical processes (Clark, 1999).

Mafic rocks contain greater amounts of iron and titanium oxides than felsic rocks (roughly 5% versus 1%), and thus commonly have greater magnetizations. However, this generality is subject to error from factors other than chemical composition. For example, titanomagnetite minerals occurring within mafic rocks tend to be higher in titanium, and thus have weaker magnetizations and lower Curie point temperatures. The rates of cooling and the availability of oxygen have a profound effect on the solubilities of the magnetic oxide solid solutions, and thus the resulting magnetic minerals, their grain size, and magnetization. Components of the Fe–Ti–O solid solution series undergo spontaneous separation (exsolution) as the rocks cool. This separation is important because it results in increasing magnetization, Curie temperatures, and magnetic susceptibilities. During the cooling process, rocks undergo alteration with the oxidation of ulvöspinel toward ilmenite and the formation of strongly magnetic magnetite. If high-temperature oxidation continues to an advanced stage within the solid solution series, as it can in subaerial mafic volcanic rocks, magnetization may be decreased by the formation of non-ferrimagnetic components. Additionally, subsequent low-temperature oxidation will convert magnetite to low-magnetization hematite.

The magnetic components of igneous rocks and their magnetic properties are also affected by the rate of cooling of the molten magma. In mafic volcanic rocks (e.g. basalt) the rate of cooling is high; volatiles, including oxygen, readily escape and the iron/titanium oxides tend to remain in solid solution, resulting in lower magnetic susceptibilities. However, the small grain sizes of the iron/titanium magnetic oxides are magnetically stable, and thus have high coercivities and remanent magnetization stabilities. This situation is evident in the integrated magnetization of the roughly 6 km of mafic rocks of the oceanic crust (Harrison, 1987) which has an estimated induced component ranging from 0.15 to 0.34 A/m and a remanence of 0.90 to 3.71 A/m. Typically the Koenigsberger ratio of volcanic rocks is greater than those of more coarse-grained slower-cooled plutonic rocks (Figure 10.5) owing to their smaller grain size resulting from higher cooling rates. However, viscous remanent magnetization may be an important part of the magnetization of coarse-grained plutonic rocks.

In addition to the effect of grain size on magnetization, the shape of magnetic mineral grains in all types of rocks affects the overall magnetization because of the anisotropic nature of ferromagnetic minerals. Anisotropy arises from intrinsic variation in the ease of magnetization with crystallographic directions and from shape demagnetization. This is not a factor in most plutonic rocks, but can have an effect in volcanic rocks which have flow structures.

A common generality of rock magnetism is that felsic rocks, i.e. granitic rocks, are less magnetic than more mafic rocks. However, observations show that the magnetic susceptibility of felsic rocks is bimodal (Figure 10.6). The more magnetic mode is identified with the magnetite series of granites that are of the I-type mode which have a dominantly igneous or meta-igneous source, while the lower-magnetization granites, the S-type, are believed to originate in sedimentary rocks with a strong interaction with crustal materials (Chappel and White, 1974; Clark, 1999). However, it is problematic to relate the magnetization of the granite to its origin, depth of source, or tectonic regime in any global sense although this may be possible in a restricted regime based on localized investigations. The distinction between magnetite-bearing (magnetite series) and magnetite-barren (ilmenite series) granites is in the higher oxidation state of the former. The origin of this higher oxidation state may be primary or secondary processes, or the source material which in turn may reflect differing tectonic regimes. The magnetization of rhyolitic rocks, the volcanic equivalent of granitic rocks, also may be either high or low. In this case the high magnetization is associated with selective intense thermal remanent magnetization.

10.6.2 Metamorphic rocks

Metamorphic rocks – those whose mineralogy, texture, and structure have been modified in the solid state by temperature and pressure – have variable magnetic properties and occupy large volumes of the continental crust. The changes caused by metamorphism may be local as around a specific igneous pluton, or involve a region where broad, low-gradient temperature and pressure have altered the original rocks. The former may cause intense magnetic effects over a few tens of kilometers or less. In contrast, regional metamorphic effects may cover extensive regions measured in hundreds of kilometers. Regional metamorphism commonly is associated with dynamic processes of plate margins. The effects of metamorphism may be profound on the magnetic characteristics of the rocks depending on the source rock (especially its iron content), the temperature regime, the chemical effects brought about by the temperature/pressure conditions, and especially the oxidation state. The latter controls the amount and type of iron oxides that form and the partitioning of the iron between oxides and silicates. Rocks of similar bulk chemistry may have quite different magnetizations as a result of metamorphic processes. PULLAIAH *et al.* (1975) show how magnetization of the magnetic minerals in rocks is gradually lost during heating accompanying burial and metamorphism; the reverse occurs during uplift and cooling with the acquisition of viscous remanent magnetization.

HAGGERTY (1979) notes that metamorphism usually decreases the magnetization of igneous rocks. For example, little remanence survives greenschist facies metamorphism. However, there are notable exceptions to this generality. For example, GRANT (1984a,b) describes the production of magnetite by the breakdown of hydrous iron/magnesium-bearing silicates with increasing temperature. While under intense metamorphism which produces granulite-grade rocks, iron/titanium oxides recombine to form magnetite–ilmenite solid solutions, causing a decrease in overall magnetization. There is a tendency for an increase in magnetic susceptibility and viscous remanent magnetization with increasing metamorphic grade as a result of increasing grain size which permits easier domain-wall movement. Viscous remanent magnetization may increase with temperature up to the vicinity of the Curie temperature, with Koenigsberger values of the order of one and direction similar to the ambient terrestrial magnetic field.

Hydrothermal alteration of ultramafic rocks such as peridotite and dunite leads to serpentinization and is normally accompanied by a marked change in magnetization (SAAD, 1969a,b). The remanent magnetization developed as a result is mainly chemical remanent magnetization (CRM). During serpentization at low temperatures (~300–400 °C) the iron atoms released from the silicate structure of paramagnetic olivine and pyroxene, are oxidized to form ferrimagnetic magnetite. The intensity of both remanent and induced magnetization increases exponentially with serpentinization. At the early stages of serpentinization, the magnetite grains produced are fine-grained and in most cases single domain, producing a very stable CRM. As the degree of serpentinization increases, the magnetic minerals grow from single to multi-domain size. The growth in grain size is accomplished by coagulation of the earlier fine grains rather than by a nucleation process. The remanent magnetization becomes highly unstable as the rock becomes intensely serpentinized due to the growth of magnetite grains to multidomain size and their oxidation, in some cases, to maghemite. The intensity of magnetization depends on mode of occurrence and state of oxidation of the magnetite, as well as on original rock composition. Serpentinized dunite was found by SAAD (1969a) to be less magnetic than equally serpentinized peridotite because the olivine in the dunite has a lower iron content than the olivine and pyroxene in the peridotite. Therefore, it is expected that serpentinized ultramafic bodies will have highly variable associated magnetic anomalies with those of dunitic composition having lower magnetic anomalies. The effect of serpentization of ultramafic rocks on magnetic susceptibility (k), remanent magnetization (J_{rem}), and Koenigsberger ratio (Q) is illustrated in Figure 10.9.

With increasing prograde metamorphism of serpentinized ultramafics, Mg and Al may be substituted into the magnetite causing a decrease or destruction of magnetization (CLARK and EMERSON, 1991). Metamorphism progressively can demagnetize serpentinites changing them to paramagnetic at granulite grade. Subsequent retrograde serpentinization, however, may cause the rock to become magnetic again. SHIVE *et al.* (1988) found that susceptibilities generally decrease with increasing metamorphic grade. They explained that to be due to the production of increasing amounts of chrome-rich spinel which dilutes the magnetite component.

The effect of metamorphism upon sedimentary rocks generally is minimal except where iron-rich minerals are present in the source rocks. For example, in ferric oxide-bearing iron formations, metamorphism may cause reduction of the oxides to magnetite. The result is a metamorphosed iron formation with both intense magnetic susceptibility and remanent magnetization. However, under continuing intense metamorphism, the iron oxides may be further modified to silicates that have minimal magnetic properties. In a similar manner, rocks containing pyrite may be converted to magnetite as a result of high-temperature processes.

10.6.3 Lithospheric rocks

Long-wavelength magnetic anomalies observed in satellite magnetic measurements and low-pass filtering of regional and continental-scale magnetic data sets have focused attention on magnetic sources within the lower crust and in the lithosphere in general (e.g. FROST and SHIVE, 1986; SHIVE *et al.*, 1992). Analysis of these anomalies, considering the best estimates of the thickness of the magnetic lithosphere, indicates overall magnetizations of the magnetic lithosphere ranging from 2 to 10 A/m, with 4 A/m a typical value. These magnetizations are of the order of 10–100 times the average value of upper crustal rocks as well as uplifted lower crustal rocks. Accordingly, additional sources of magnetization are required in the deep magnetic lithosphere.

Studies of deep rock fragments (xenoliths) and uplifted lower crustal rocks as well as thermodynamic considerations have led most investigators to the conclusion that

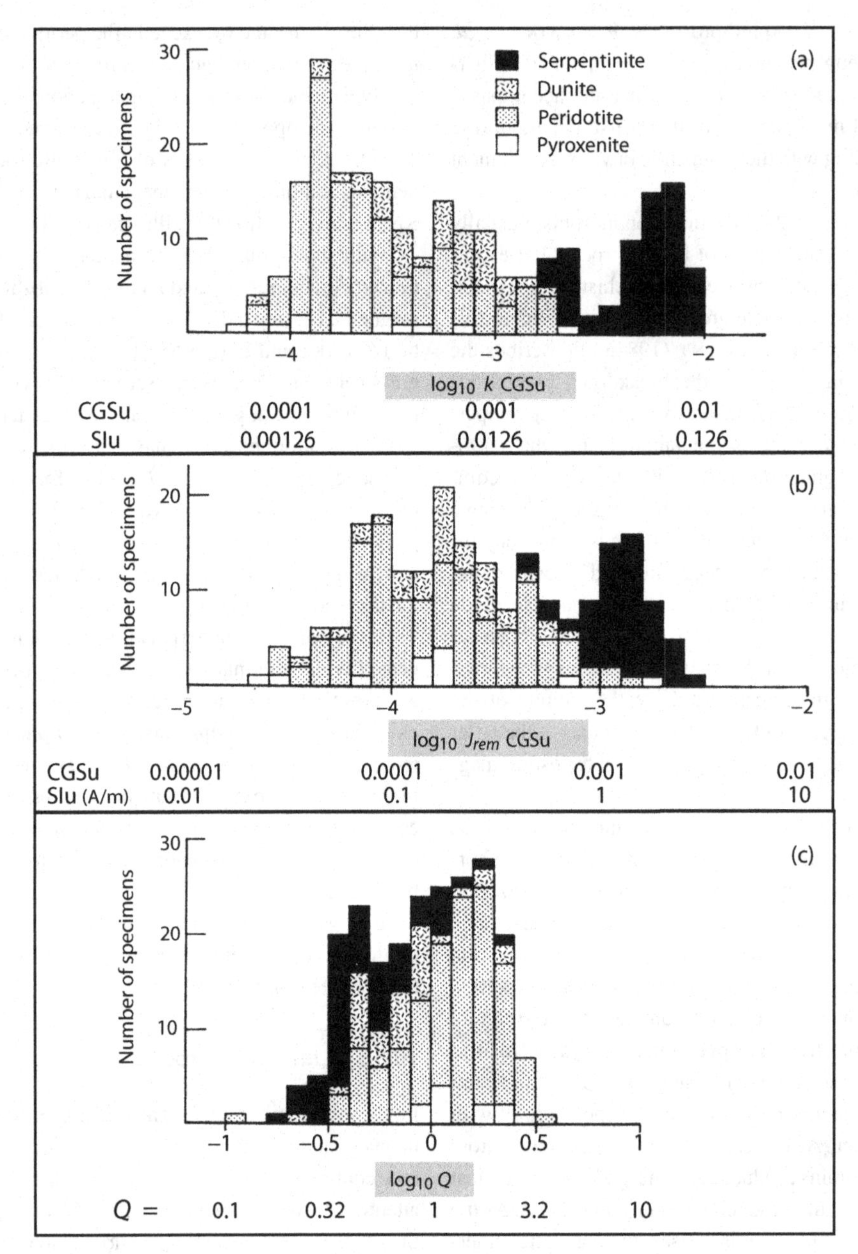

FIGURE 10.9 Histograms of (a) magnetic susceptibility (CGSu), (b) intensity of natural remanent magnetization (CGSu), and (c) Koenigsberger ratio of ultramafic rocks and their metamorphic product, serpentinite (SAAD, 1969a).

magnetite is probably the major source of magnetism in the lower crust (e.g. WASILEWSKI and MAYHEW, 1982; KELSO *et al.*, 1993). The depth extent of this magnetism is controlled by the Curie temperature of magnetite, which is about 600 °C in the deep crust owing to pressure effects. This temperature generally is reached near the base of the crust at the Mohorovičič discontinuity (or Moho) in the continents and slightly below the Moho in the oceans. KELSO *et al.* (1993) have measured the magnetic properties of granulitic rocks originally located in the deep crust and now cropping out in the Proterozoic Arunta Block of central Australia and found a median natural remanent magnetization of 4.1 A/m and a Q of 7.2. Laboratory investigations indicate that nearly 50% of the remanence persists after 400 °C demagnetization, suggesting that thermo-viscous remanent magnetization along the ambient magnetic field may dominate the magnetization in lower crustal rocks. The magnetic susceptibilities of these rocks are only capable of producing magnetizations of less than 1 A/m and the susceptibility of the primary magnetic mineral, magnetite, is constant to within 30 °C of the Curie temperature where it decreases rapidly. WASILEWSKI *et al.* (1979) present evidence that the Moho is a magnetic boundary, because iron is dominant in highly magnetic oxide phases within the crust and in non-magnetic spinels below the crust. HAGGERTY (1978) alternatively has suggested that metallic iron and iron alloys produced during serpentization of the upper mantle are a source of magnetism within the upper mantle. However, FROST and SHIVE (1986) have argued against this as a major source of magnetism on geologic, petrologic, and thermodynamic grounds.

A number of other sources of deep magnetization have been considered as the source of long-wavelength magnetic anomalies. The evidence that monoclinic, ferromagnetic pyrrhotite is the major carrier of magnetism in the deep (9 km) drill hole in Bohemia offers a possible source of deep magnetization. Further, studies by KLETETSCHKA *et al.* (2000a,c) of the intensity of thermal remanent magnetism in multidomain-sized grains of titanohematite suggest that titanohematite could be an important source of remanent magnetization in lower crustal rocks, in contrast to the general assumption of prevailing induced magnetization of these rocks. This is supported by the work of MCENROE and BROWN (2000) showing that metamorphic titanohematite carrying intense remanent magnetization is commonly present in granulites formed in lower crustal metamorphism. KLETETSCHKA *et al.* (2000b) have also suggested that the thermal remanent magnetization of multidomain hematite and titanohematite may be a significant source of crustal magnetic anomalies on Mars. Regardless of the origin of lithospheric magnetization, a major source of magnetization is present in the lower crust, and the continental crust is more magnetic on average than the thinner oceanic crust as a result of its greater thickness.

Over the past few decades studies of the Moon and Mars have shown that these planetary bodies have local magnetic fields due to crustal magnetization, despite the lack of a global magnetic field which causes induced magnetization. On the Moon, about 4.2 billion years ago, the loss of heat caused the liquid core to disappear and along with it the core dynamo action which had produced an early, generally axially oriented global magnetic field. However, rocks formed prior to the termination of the field acquired a weak normal remanent magnetization which is still preserved in local regions of the Moon, particularly on its far side. This magnetization likely decreased with time and may have been altered by shock magnetization associated with impacts and perhaps other local processes. This conclusion is based on analysis of the measured magnetic field of the Moon, which reaches intensities of only a few hundred nanoteslas, and measurements of lunar rocks (MACKE *et al.*, 2010).

Similarly, Mars, which lost its global magnetic field on cooling more than 4 billion years ago, has local magnetic fields, particularly in the southern highlands. However, in contrast to the Moon, the magnetization of Martian, crustal rocks is intense, as determined from inversion of magnetic anomalies and measurements of meteorites originating on Mars. This magnetization is roughly 10 times that of the Earth. Although several processes have been suggested for the magnetization of the Martian crust, it is likely that the source is remanent magnetization acquired during the span of time when Mars had a global magnetic field. A candidate for this intense remanence is lamellar magnetization (MCENROE *et al.*, 2007) associated with micrometer to nanomicrometer intergrowths of ilmenite and hematite, leading to intense remanent magnetizations with high coercivity. This would explain the strong normal remanent magnetization which remains after more than 4 billion years. Similar magnetizations are observed on Earth in anorthositic ilmenite deposits which can carry strong remanent magnetizations that originated in Precambrian time (ROBINSON *et al.*, 2004), and in hematite ore bodies in which very small amounts of magnetite and maghemite intergrowths within the hematite lead to normal remanent magnetization of strong coercivity (SCHMIDT *et al.*, 2007). In the latter case the intergrowths are not exsolution lamellae, but were formed when the rocks were exposed to a high-temperature period. These studies of extraterrestrial bodies are providing new information that is potentially helpful in understanding terrestrial magnetic properties and processes.

10.6.4 Sedimentary rocks

Characteristically sedimentary rocks, both chemical and detrital, are significantly less magnetic than crystalline rocks, owing to the lack of ferrimagnetic minerals. These minerals may be absent as a result of non-deposition or because of post-depositional geochemical alteration which transforms them into non-magnetic minerals. The type and amount of ferrimagnetic minerals deposited in sediments is a function of the type of source rocks. Typically they are more abundant where the source is crystalline rocks, but the ferrimagnetic minerals commonly are oxidized to non-magnetic forms during the weathering and erosion of these rocks and deposition of the detrital particles into sediments. Nonetheless, there are numerous examples of clastic sedimentary rocks that contain measurable amounts of ferrimagnetics giving rise to small, low-amplitude magnetic anomalies (e.g. MUSHAYANDEBVU and DAVIES, 2006; HUDSON *et al.*, 2008; GRAUCH and HUDSON, 2011). These anomalies are difficult to map, except where they are steeply folded or faulted, without high-resolution and -sensitivity magnetic surveying because thin, shallow-dipping magnetic sedimentary units produce only minor short-wavelength anomalies at their edges (e.g. GRAUCH and HUDSON, 2007). GRAUCH and HUDSON (2011) have compiled the volume magnetic susceptibility of sedimentary rocks from numerous regions over North America. They find that, despite a wide variation in susceptibility within rock types, the values are generally less than 2×10^{-3} SIu (160×10^{-6} CGSu) with the higher values for sandstones and shales primarily from detrital magnetic minerals, and lower values for chemically precipitated rocks such as carbonates and salts.

In contrast to the minor quantity of detrital ferrimagnetic minerals in most sedimentary rocks, significant quantities of magnetite (commonly $>10\%$ by weight) occur in Archean and Early Proterozoic iron formations in which ferrous iron has precipitated from sea water before the advent of an oxidizing atmosphere and oceans. These chemical sediments are among the most magnetic rocks (e.g. BATH, 1962; JAHREN, 1963). Another specialized sedimentary rock in terms of potential magnetic properties is coal. Coal, which consists primarily of preserved organic material, also contains minor quantities of pyrite (FeS_2) because of the localized reducing environment and activity of sulfate-reducing bacteria. Although pyrite is paramagnetic, and thus only a minor magnetic mineral, it is transformed into significant magnetic minerals, such as magnetite, maghemite, and hematite, as well as perhaps pure metallic iron (DE BOER *et al.*, 2001), when subject to thermal oxidation during fires and subsequent cooling below the Curie point of the magnetic minerals. The transformed mineral phases depend on a variety of conditions including the bulk composition of the original sedimentary rock and the environment during the thermal oxidation. DE BOER *et al.* (2001) find enhanced magnetic properties of thermally altered coals from northwest China that range from 0.1 to 10 A/m for the TRM and susceptibilities of 10^{-2} to 10^{-1} SIu. Adjacent sedimentary rocks, too, may suffer oxidation of indigenous pyrite due to the high temperatures (800–1000 °C) reached by the nearby burning coal. Additionally, SCHUECK (1990) indicates that pyrite in coal is chemically oxidized into magnetic minerals, and thus magnetic mapping may be used to isolate sources of acid mine drainage within coal mines and dumps.

Highly magnetic minerals also may occur in minor amounts in sediments as a result of post-depositional processes which lead to localized magnetic oxides and sulfides, and preservation of detrital ferrimagnetic particles. The localized regions may be associated with geochemical or biological activity promulgated by mineral and hydrocarbon deposits. The possible generation of magnetic ferrous iron oxides and sulfides in sedimentary rocks overlying hydrocarbon deposits (e.g. DONOVAN *et al.*, 1979, 1984, 1986; HENDERSON *et al.*, 1984; MACHEL, 1996), and thus the direct detection of buried hydrocarbons, has led to sporadic studies of the source of this possible relationship (REYNOLDS *et al.*, 1990; PHILLIPS *et al.*, 1998).

Magnetite is the most common ferrimagnetic mineral present in sedimentary rocks, but titanohematite may be the dominant magnetic oxide in some rocks (REYNOLDS, 1977). Magnetite is deposited in sediments along with other high-density minerals where moving water can no longer sustain the transport of these heavy minerals owing to a decrease in the velocity of the water. This leads to concentrations of magnetite associated with, for example, shoreline depositional features and changes in stream or river character. Ferrimagnetic sulfides may originate in sedimentary rocks as a result of microbial activity. They are uncommon, but may occur as monoclinic pyrrhotite or cubic greigite in either oxidizing or reducing environments. The resulting mineralization may be sufficient to produce minor magnetic anomalies.

10.6.5 Soils

Soils form a special component of the investigation of the magnetic properties of Earth materials because of the unique nature of soil processes and the human impact upon them. Their magnetic properties are important to the use of the magnetic method in archaeological, environmental, and other near-surface studies.

Generally, the magnetic properties of soils bear little relation to their parent rocks (COOK and CARTS, 1962) because during the chemical breakdown of the rocks the iron-rich minerals are converted to non-magnetic phases. However, under certain conditions they will reflect those of the rock types at the source or their provenance. Such is the case where the soils contain numerous lithic fragments and minerals whose magnetic properties have not been altered, owing to insufficient time or prevailing environmental conditions.

In the north-central USA, for example, soils have developed on glacial debris and sediments transported by Pleistocene continental glaciers from the Precambrian Shield of Canada into the United States (GAY, 2004). These soils contain numerous crystalline (igneous and metamorphic) rock materials which maintain their magnetic character. The distribution of these magnetic components varies with the rock types of their provenance, the distance from the source, and the effects of local glacial-related processes. The former two effects alter the magnetic properties on a regional scale depending on the source area of the glacial lobe which deposited the material and the distance from the source, whereas the glacial processes are effective locally within the glacial deposits of any particular lobe (Hinze, 1963; Gravenor and Stupavsky, 1974). An example of the latter is the concentration of magnetic materials within deposits (eskers) associated with streams flowing on or within glaciers during melt periods (Onesti and Hinze, 1970). The irregular distribution of the magnetic components within glacial materials and glacially derived sediments results in randomizing of any remanent magnetization and highly variable magnetic susceptibilities.

An example of magnetic properties derived indirectly from the parent rock is the strong magnetization observed in lateritic soils formed in tropical climates on mafic rocks (e.g. Resende *et al.*, 1986). The magnetic phase of these soils is titanomaghemite, which presumably is derived from primary titanomagnetite originating in the underlying bedrock.

Topsoils generally exhibit substantially higher magnetizations than the underlying subsoils (Le Borgne, 1955, 1960). Magnetic susceptibility is enhanced in the uppermost layers, including the surficial A-horizon and well into the B-horizon, decreasing gradually an order or two of magnitude over roughly the first meter. This enhancement has been ascribed to the conversion of iron oxides from the weakly ferrimagnetic hematite to the strongly ferrimagnetic magnetite and maghemite by soil combustion from surface fires and fermentation from seasonal surface reduction and oxidation processes. Additional, more limited, contributions may come from magnetite created by microbial activity, either internally in magnetotactic microorganisms, which are bacteria that align with the ambient magnetic field, or externally in iron-reducing bacteria (Bazylinski and Moskowitz, 1997), and from atmospheric deposition of anthropogenic particulate pollution (Maher, 1986).

Looking at this in more detail, enhancement of topsoil magnetization is principally associated with successive periods of reduction and oxidation in soils. In this process the non-ferrimagnetic oxides naturally occurring in soils, such as hematite, are reduced to magnetite or a phase close in composition to magnetite and subsequently may be oxidized to maghemite which has approximately twice the magnetic susceptibility of the original iron oxides (Aitken, 1972). This process can be initiated by the reduction associated with natural or human-induced burning of organic material in soils that is followed by the reoxidation of magnetite as air enters the soil pore spaces after the fire. A secondary origin of this process which is much more widespread, but produces lower magnetization enhancement, is the fermentation process (Mullins, 1977) which consists of relatively rapid alternating oxidation and reduction periods associated with wetting and drying climatic cycles.

Soil fermentation involves reducing hematite to magnetite at ordinary temperatures by decay of organic matter during wet, oxygen-starved or anaerobic periods. Alternating anaerobic and dry, oxygen-rich or aerobic conditions must be fairly rapid because a prolonged period of high humidity can cause the removal of oxygen molecules by the soil microbes, thereby allowing the water-soluble iron to leach out where sufficient soil drainage occurs. The characteristics and consequences of the fermentation mechanism are not well known because attempts to produce magnetic enhancements by accelerated fermentation in the laboratory have been relatively unsuccessful (Le Borgne, 1960; Scollar *et al.*, 1990). However, the fill from archaeological pits usually is more magnetic than the surrounding topsoil, which suggests that fermentation enhancements may be significant on archaeological timescales (Aitken, 1972).

Repeated burning of the topsoil by natural and domestic fires is probably the primary way that the hematite–magnetite–maghemite conversion is achieved. Burning organic material produces both a strong reducing environment of carbon monoxide and a temperature increase to accelerate the reduction process. Reoxidation to maghemite occurs during subsequent cooling when air again reaches the soil. In general, a fire magnetically enhances the soil to a depth of only a few centimeters because of the low thermal conductivity of soils. Subsequent worm and other animal burrowing action, cultural activities, and soil sedimentation processes disperse the more magnetized soil layer underlying the fire to greater depths.

The magnetic susceptibility enhancement from soil combustion can be studied in the laboratory by oven-heating the soil to temperatures typical of archaeological fires (e.g. 350–650 °C) in a reducing environment of nitrogen and then cooling it in the oxygenated environment of air (Tite and Mullins, 1971). These studies show a marked linear enhancement of specific magnetic susceptibility with increasing concentrations of iron up to several weight percent. Typical behavior is illustrated in Figure 10.10 where the specific (mass) susceptibilities

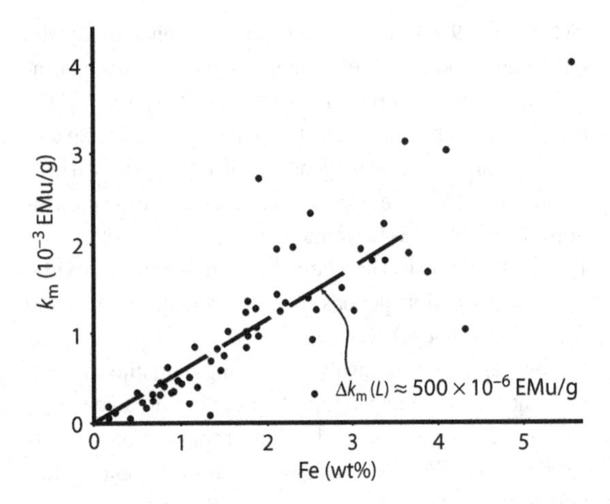

FIGURE 10.10 Typical relationship between specific (mass) magnetic susceptibility (k_m) and the weight percent (wt%) of iron (Fe) in soils oven-heated in a reducing (nitrogen) environment and then cooled in an oxidizing (air) environment to simulate the effects of archaeological fires. The slope $\Delta k_m(L)$ of this relationship for the first several weight percent is also indicated. After GRAHAM (1976).

(k_m) of laboratory heated soils were found to follow a slope of about $\Delta k_m(L) = 0.005$ EMu/g over the first several weight-percent. Beyond several weight percent, the relationship is no longer linear, possibly owing to complex magnetic interactions of the iron oxide grains. Thus, the susceptibility enhancement from soil combustion is relatively predictable because most soils have iron oxide weight concentrations of several percent or less (e.g. TITE and MULLINS, 1971; TITE, 1972; AITKEN, 1972; MULLINS, 1974; ROTH *et al.*, 1974; GRAHAM, 1976; LATZ *et al.*, 1981; SCOLLAR *et al.*, 1990).

For non-agricultural soils with organic concentrations less than 2%, significant magnetic enhancement during combustion also results from increasing the organic content (e.g. MULLINS, 1974; GRAHAM, 1976). Susceptibility increases strongly as the organic matter content in the soil increases up to about 2% where the reduction effect is apparently maximum and the fired soil shows no further susceptibility enhancement with greater organic content. Human activities increase the soil's organic content, which in a low-organic soil containing iron oxides can significantly enhance the magnetism of fired soil and the role of the magnetic method for mapping these activities.

Magnetic susceptibility enhancement of topsoils at archaeological sites, however, is due primarily to the amount of soil fired, which is proportional to the number of fires that the site has experienced. Thus, the soils most conducive to magnetic exploration are organic-rich with significant concentrations of iron oxides and an extensive history of occupation. Glacial soils of the US midcontinent, for example, commonly exhibit high (>2 wt%) iron oxide concentrations (e.g. ROTH *et al.*, 1974; LATZ *et al.*, 1981), and thus tend to be relatively well suited to magnetic exploration for fired soil artifacts.

The amount of iron oxides actually converted to the more magnetic phases in soils can be estimated from the conversion factor $CF = [k_m(F)/k_m(L)]$, where $k_m(F)$ is the susceptibility measured from field samples of the soil, and $k_m(L)$ is the laboratory-measured maximum susceptibility of the soil when presumably the iron oxides have all been converted by oven-baking the soil in the reducing-then-oxidizing environment (TITE and MULLINS, 1971). In English soils, for example, the CF for iron oxides from archaeological soil features was found to vary from 10% to 35% with an average of $<CF> \approx 11\%$, whereas for soils at non-archaeological sites, the CFs ranged typically around 2–3% (e.g. TITE, 1972; MULLINS, 1974). Thus, an empirical CGSu estimate of the magnetic volume susceptibility for UK soils at archaeological sites may be obtained from

$$k \approx \Delta k_m(L) \times <CF> \times \text{wt}\% \times \sigma$$
$$= 5.5 \times \text{wt}\% \times 10^4 \qquad (10.3)$$

for the soil density σ in g/cm^3. However, in practice the strong viscous magnetic remanence of soils may amplify this susceptibility by 2–4 times.

The enhancement of soil magnetization from iron oxide conversions can grow significantly with time owing to magnetic viscosity that produces typically strong viscous remanent magnetizations (VRM) in soil. As the result of magnetic viscosity, the magnetization, J_t, at any given time, t, relative to its magnetization, J_{t_0}, at an earlier time, t_0, is

$$J_t = J_{t_0}\left[1 + (V_c)\log\left(\frac{t}{t_0}\right)\right], \qquad (10.4)$$

where the proportionality constant V_c is the viscosity constant, which can be measured from a sample of the soil (SCOLLAR *et al.*, 1990). Typical values for V_c range between 3–6% so that the increase in the magnetization of a soil feature over typical archaeological timescales can be considerable. For example, a 5% value for V_c results in a 45% net increase in magnetization between $t_0 = 100$ s and $t = 3{,}000$ yr. Thus, the magnetization enhancement from the conversion of iron oxides in topsoils may well more than double when magnetic viscosity is taken into account. SCOLLAR *et al.* (1990) provide additional details on the magnetic properties of soils at archaeological sites and their measurements.

Soils overlying sanitary landfills also have higher magnetic susceptibilities. ELLWOOD and BURKART (1996) ascribed this to the growth of magnetite-producing bacteria prompted by reducing conditions from methane gas rising from decaying organic matter and the periodic infiltration of surface precipitation. Under these conditions iron is transported deeper in the soil where it is reprecipitated and oxidized into the magnetic phase, maghemite, during dry summers in much the same manner as occurs in normal soils. In a similar manner, plumes of rising methane above hydrocarbon deposits produce conditions favorable for the occurrence of magnetic ferrous iron oxides and sulfides, both monoclinic pyrrhotite and greigite (e.g. REYNOLDS *et al.*, 1990; MACHEL, 1996).

In summary, the volume magnetic susceptibility of the uppermost soils can vary widely (e.g. BUTLER, 2003), generally over the range of 0.001–0.01 SIu with most values occurring within the range of 0.002–0.006 SIu. Archaeological disturbances of organic, iron-oxide rich soils can significantly enhance topsoil magnetizations, because of weakly magnetic hematite being reduced to strongly magnetic magnetite which oxidizes to strongly magnetic maghemite. The NRM of magnetic soils is likely to be appreciable because magnetite occurs in ultra-fine grains susceptible to VRM and the effects of heating due to surface fires. The Koenigsberger ratio can be 2 or more in surface soils.

10.6.6 Fired clay

Like soils, fired clay may be significant and useful in archaeological engineering, and other near-surface investigations. Fired clay objects like pottery, tiles, bricks, kilns, hearths, and fire pits are subjected to temperatures that often exceed the Curie point of their constituent iron minerals (e.g. 680 °C for hematite). As a result, these materials acquire a TRM that is stable and intense and directed in the ambient magnetic field. Where temperatures do not reach the Curie points, PTRM will be acquired. These magnetizations often are studied to date the firing of the clay as well as for other archaeological purposes. The magnetizations range from roughly 0.3 A/m for red oxidized clays to 300 A/m for gray reduced or gleyed clays (AITKEN, 1972; SCOLLAR *et al.*, 1990). Fired clay objects in general tend to be characterized by relatively strong magnetizations and magnetic anomalies with intensities second only to those of iron objects.

10.6.7 Iron objects

Iron objects from metallic iron and alloys of Fe–Ni–Co are the most magnetic materials that are likely to be encountered in the near-subsurface. Metallic iron is rare in nature, and thus iron objects are related to human activities where their occurrence in the subsurface may be either deliberate or inadvertent. Their magnetic properties are highly variable depending on their chemical constituents and thermal, mechanical, and chemical history. The volume magnetic susceptibility of iron objects is commonly reported between 10 and 125 SIu, while the magnetic moment per unit of weight, which is useful to consider in many applications, is given by BREINER (1973) as 10^5 to 10^6 CGSu (EMu) per 1,000 kilograms or 10^8 to 10^9 A×m^2 in SIu. RAVAT (1996) found that the effective apparent magnetic susceptibility of unrusted steel drums is of the order of 100 SIu and that the Koenigsberger ratio generally is less than 0.5. In contrast, BREINER (1973) and others report ratios of 10 to 100 for iron and steel objects. These high values show that strong remanent magnetizations are possible, depending on the mechanical history of the object and its exposure to direct current electrical fields. RAVAT (1996) found inconsistent, but only minor modifications of the magnetic properties of steel drums exposed to rusting over a several-year span of time, indicating complex magnetic changes at least in the early stages of natural weathering.

10.6.8 Summary

The above generalizations illustrate that great care must be taken in specifying the magnetic properties of classes of rocks, soils, and anthropogenic features. Magnetic properties commonly depend on factors involving the geological, geochemical, and thermal history of the materials which may not be readily apparent without detailed study of the composition, texture, and magnetic properties of magnetic as well as the non-magnetic minerals, and the structural and stratigraphic relationships between rock units.

A useful summary of the crustal distribution of the aeromagnetically important minerals is shown in Figure 10.11. Magnetite, the most important magnetic mineral, is produced and destroyed under a complex array of environments and conditions that lead to strong magnetic susceptibility and may cause low remanent magnetization, or weak susceptibility and high remanent magnetization. Furthermore, generalizations are difficult because some rocks contain strongly magnetic minerals other than magnetite. The result is that prediction of magnetic properties on the basis of Earth material type can only be made in broad terms.

Intense magnetic properties are associated with certain soils, fired iron-rich clays, and iron objects. As a result localized magnetic anomalies commonly are related to anthropogenic sources which make them useful to

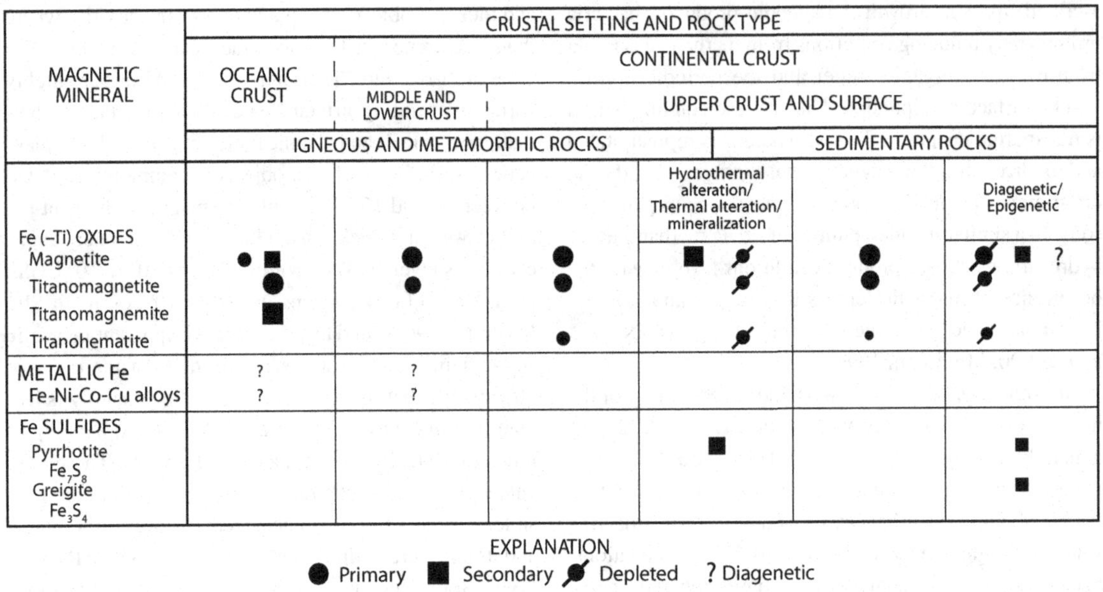

MAGNETIC MINERAL	CRUSTAL SETTING AND ROCKTYPE					
	OCEANIC CRUST	CONTINENTAL CRUST				
		MIDDLE AND LOWER CRUST	UPPER CRUST AND SURFACE			
	IGNEOUS AND METAMORPHIC ROCKS				SEDIMENTARY ROCKS	
				Hydrothermal alteration/ Thermal alteration/ mineralization		Diagenetic/ Epigenetic
Fe (–Ti) OXIDES						
Magnetite	● ■	●	●	■ ⌀	●	⌀ ■ ?
Titanomagnetite	●	●	●	⌀	●	⌀
Titanomagnemite	■					
Titanohematite			●	⌀	●	⌀
METALLIC Fe	?	?				
Fe-Ni-Co-Cu alloys	?	?				
Fe SULFIDES						
Pyrrhotite Fe_7S_8				■		■
Greigite Fe_3S_4						■

EXPLANATION

● Primary ■ Secondary ⌀ Depleted ? Diagenetic

FIGURE 10.11 Aeromagnetically important minerals of the crust in terms of their distribution and origin. The minerals are divided into primary (dot) and secondary (square) origins. Primary minerals are (1) crystallized in magma, (2) deuteric alteration products in igneous rocks or metamorphic products in metamorphic rocks, or (3) detrital minerals in sedimentary rocks (including chemically precipitated magnetite, such as that deposited in banded iron formations). The dot-with-slash symbol denotes settings or conditions where minerals may be depleted. Diagenetic magnetite is denoted by a bold question mark. Secondary minerals include those formed by replacement of earlier magnetic precursors (e.g. titanomaghemite from titanomagnetite in oceanic crust) and those formed by nucleation or from a non-magnetic precursor. The size of the symbols crudely indicates their relative abundance within a single column and cannot be meaningfully compared between columns. After Reynolds *et al.* (1990).

archaeological investigations and other studies where the surface has been disturbed, and in some cases where foreign materials have been placed in the near-subsurface.

Much of the information on near-surface magnetic properties is disseminated in the literature on studies by soil scientists, archaeogeophysicists, and environmental geomagnetists, and thus is not readily available in a form that is immediately useful to geophysical site characterization. Environmental magnetism, the application of rock and mineral magnetic studies to the environmental impacts of the transport, deposition, and alteration of magnetic materials, is being used for a wide range of environmental problems including the effects of global change, climatic change, and human impact on the environment (e.g. Verosub and Roberts, 1995). These studies are an increasing source of near-surface magnetic properties.

10.7 Magnetic property measurements

As in the case of density, the magnetic properties of subsurface materials used in planning and interpretation of surveys are commonly based on values extracted from published compilations of magnetic property measurements. However, the wide range of measured magnetic properties for any Earth material type means this can lead to less than satisfactory results. The use of data tabulations often is a necessity because of the lack of samples for measurements or of appropriate instrumentation, but wherever possible direct measurement of samples is the desirable course of action. Care must be taken to insure that the samples are representative and that the number of measurements is adequate to test the variability of the property within any one subsurface unit. Guidelines provided in Section 4.6.1 on density measurements for the selection and number of samples are useful for magnetic property measurements as well.

Two magnetic properties characteristically are of interest for magnetic exploration purposes: magnetic susceptibility and the remanent magnetization, both its intensity and its direction. As explained previously, the former controls induced magnetization and the latter is the magnetization that is present in a zero external field. The two combine to make up the total magnetization that is the source of magnetic anomalies.

Magnetic susceptibility has been measured over the years by numerous techniques, including magnetic balances and inductance bridges. Alternatively, measurements can be simply made by determining the effect that a rock sample has on the position of a magnetic needle. A similar method is to measure the effect of rotating a specimen near a resonance magnetometer. JAHREN and BATH (1967) and BREINER (1973) describe such a method that provides a rough measure of both induced and remanent magnetization components. The procedure involves rotating a roughly equidimensional rock sample at a specified distance from a total intensity magnetometer in either the Gauss A (i.e. parallel to ambient magnetic field direction) or Gauss B (i.e. perpendicular to the field direction) position while observing how the magnetic field changes from the ambient field without the sample present. The magnetic moment of the sample in the induced case is equal to the product of the magnetic susceptibility, the ambient magnetic field, and the volume of the sample, all in common units. The field for measuring the remanent component is equal to one-half the difference between the maximum and minimum measurements of the magnetometer observed while rotating the sample. For measuring the induced component (magnetic susceptibility), the field is the difference between the ambient field measurement and the average of the observed maximum and minimum observed fields.

More commonly and more accurately, measurements are made on the basis of the effect of a magnetic rock sample on an inductance bridge. The rock sample is placed within a winding of one leg of the bridge, or close to it near proximity, and the imbalance of the bridge caused by the sample is determined and related to the magnetic susceptibility through calibration using materials or fluids of known magnetic susceptibility. The magnetic field to which the specimen is exposed is of the same order of the magnetic field at the surface of the Earth, 30,000–60,000 nT. The frequency of alternating fields used in these balances is kept low (<5 kHz) to avoid conductive responses from the high electrically conductive magnetic minerals. The measurements are affected not only by the susceptibility, but also by the viscous magnetization retained by the sample in the short duration of the alternating field reversal. In some instruments, the viscous magnetization of very fine-grained magnetic minerals is identified by determining the susceptibility at two frequencies a decade apart (CLARK and EMERSON, 1991). A lower susceptibility is measured at the higher frequency if viscous magnetization affects the results.

Inductance bridges are available that measure the susceptibility of a set volume of material in the shape of a cylindrical sample obtained from drill core, or granular samples placed in a cylindrical specimen holder for measurement. In the latter case, the porosity of the sample must be determined to adjust the measurement to the actual volume of the sample. For measurements of susceptibilities on individual samples or outcrops of either consolidated rocks or unconsolidated sediments and soils, susceptibility bridges are modified to include the Earth material within the effective region of the field of the coil. The size of the coil controls the depth range which is tested by the bridge. Typically these depths are less than several centimeters (LECOANET and SEGURA, 1999); nonetheless, *in situ* bridges provide rapid, effective measurements which do not require the collection of samples. Particularly useful are portable hand-held, induction bridge magnetic susceptibility meters that store numerous measurements for later downloading into a computer for analysis.

Other variations on these bridges enable remote measurements to be made by probes inserted into soils and downhole logging devices. For example, DANIELS and KEYS (1990) describe a downhole logging sensor with a sensitivity of 10^{-6} SIu, which consists of a solenoid wound on a high permeability core connected to an inductance bridge. The magnetic susceptibility of the volume of rock surrounding the sensor is determined from the amplitude of the quadrature (out of phase) component of the bridge output signal. For specialized purposes, it may be useful to measure the directional properties of magnetic susceptibility, the magnetic anisotropy (e.g. HANNA, 1977; BORRADAILE and HENRY, 1997).

The remanent magnetic properties of Earth materials are also determined by a number of methods. All require carefully selected samples that are oriented prior to their extraction from the Earth. As mentioned above, remanent magnetization can be determined by measuring the effect of rotation of an appropriately oriented nearby sample on the observations of a magnetic sensor (e.g. BREINER, 1973; RAVAT, 1996). However, most remanent magnetizations of significance to geophysical exploration are measured by the spinner magnetometer, or rock generator. This magnetometer spins an equidimensional, oriented sample next to a pickup coil or magnetic sensor. The amplitude and phase of the recorded signal are used to determine the intensity and direction of the remanent component orthogonal to the spin axis. By spinning the specimen around various axes and measuring the resultant sinusoidal output, the total remanent component and its direction relative to the sample, and thus to the Earth, are determined.

10.8 Magnetic property tabulations

Geophysical texts generally provide brief tabulations of the range of magnetic susceptibilities of common rock types. In addition, several comprehensive compilations of magnetic properties, primarily magnetic susceptibilities, have been published (e.g. SLICHTER, 1942; LINDSLEY *et al.*, 1966; PARASNIS, 1971; DORTMAN, 1976; HENKEL, 1976; STRANGWAY, 1981; CARMICHAEL, 1982; CLARK, 1983; CLARK and EMERSON, 1991; HUNT *et al.*, 1995; SCHÖN, 1996; DUNLOP and ÖZDEMIR, 2007; CHANDLER and LIVELY, 2011; GRAUCH and HUDSON, 2011). Detailed measurements of rocks from specific regions or of particular rock types also are presented in numerous journal publications (e.g. WERNER, 1945; MOONEY and BLEIFUSS, 1953; BATH, 1962; JAHREN, 1963; LIDIAK, 1974; ISHIHARA, 1979; KRUTIKHOVSKAYA *et al.*, 1982; CRISS and CHAMPION, 1984; LAPOINTE *et al.*, 1984; HAHN and BOSUM, 1986).

As we said, using tabulations of magnetic properties requires care because the results are often presented in the form of histograms showing the frequency of occurrence of values within a specified range. The range of values is given in some reports in a linear manner, while others are shown on a logarithmic scale. The latter are common because of the broad range of magnetic properties which often cover several orders of magnitude (LARSSON, 1977). Normal distributions of values will be skewed when presented on a log scale. This distortion can be misleading if the significance of the logarithmic scale is not fully appreciated.

Typical examples of log–normal frequency plots are shown in Figure 10.12 for 4,621 measurements of magnetic susceptibility of drill cuttings of all rock types, gneisses, and metabasites (metamorphosed mafic rocks) from a 9,100 m deep drill hole in Bohemia (KTB-HB). Figure 10.12(a) shows the two maxima of the histogram for all rock types due to the two main lithologies in the (b) and (c) panels. RAUEN *et al.* (2000) ascribe the tail of the histograms toward increased susceptibilities to larger contents of ferromagnetic phases such as pyrrhotite. They find that the increased values of the metabasites over the gneisses are due to the greater abundance of strongly paramagnetic minerals such as hornblende in the metabasites. Although monoclinic ferrimagnetic pyrrhotite is the primary strongly magnetic mineral in the rocks over most of 9,100 m depth of the hole, the magnetite-rich zones are responsible for the higher susceptibility values.

Generalized tabulations of magnetic properties, although informative in a general way, are less useful than tables of rock densities and often of limited assistance in evaluating the magnetic character of rocks in a specific area. Magnetic properties are characteristically heterogeneous within a rock type at all scales, measured from sub-centimeter to kilometer distances. This is true because

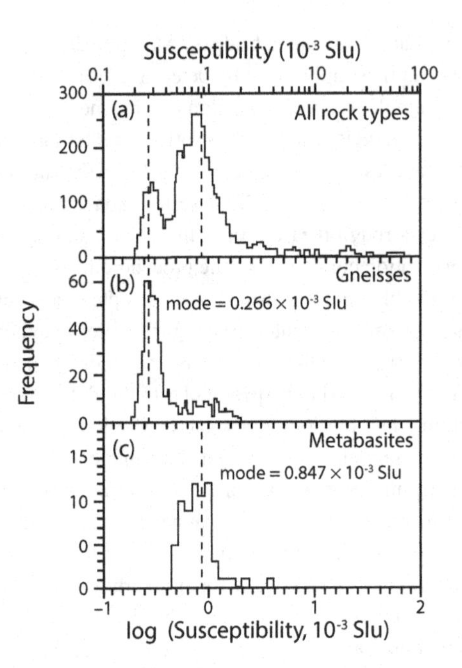

FIGURE 10.12 Log–normal frequency plots of magnetic susceptibility measurements from drill cuttings from the main deep drill hole in Bohemia (KTB-HB). (a) All rock types (4621 measurements); (b) gneisses (503 measurements); (c) metabasites (67 measurements). The modes of the gneisses and metabasites histograms are marked and specified. After RAUEN *et al.* (2000).

of the irregular distribution of ferrimagnetic minerals and the inconsistent distribution of the effects of secondary processes. This heterogeneity, referred to by NETTLETON (1976) as the Principle of Infinite Detail, generally increases with increasing magnetization of the rock, and the variations are greater between rock types from different geological provinces than for a rock type within a specific province because of variable geologic and thermal histories of provinces.

The distribution of magnetic properties within a range for a specific rock type may not be characteristic of the rocks in a specific region, owing to localized effects that produce or destroy ferrimagnetic components. As a result, tabulations that provide only a range can lead to inappropriate conclusions if the simple mean of a range is used in analyses. Clearly, tabulations that provide results from the study area are most desirable; where these are not available, the user of general tabulations needs to be concerned with the statistical attributes of the tables.

10.9 Key concepts

- Magnetization, or in actuality the magnetization contrast, the controlling physical property of the magnetic method, is the vectorial resultant of the induced and remanent magnetizations. Induced magnetization is a function of the magnitude and direction of the ambient magnetic field and of the magnetic susceptibility of the material, the ease with which the material is magnetized. Remanent magnetization is a permanent magnetization that reflects the history and origin of the material.

- The magnitude of remanent magnetization is generally much less than induced magnetization, but the exceptions are numerous and noteworthy, and the direction of the remanent magnetization is commonly not in the orientation of the ambient field. As a result, it is difficult to estimate magnetization except in very general terms that are subject to numerous exceptions. The resulting uncertainty is amplified by the fact that magnetizations commonly vary over a range of 10^3 or more and that the magnetization of most Earth materials is controlled by a few minerals which only occur as accessory constituents.
- Measurements of magnetic properties generally are made on samples extracted from the Earth, although under special conditions magnetic susceptibility can be measured *in situ* on outcrops. Magnetic susceptibility is measured by determining the effect of the sample on an induction bridge, and remanent magnetization, both direction and amplitude, is observed by measuring the effect of rotation of an oriented sample on a magnetic sensor.
- Magnetizations for purposes of geophysical exploration are determined per unit volume. The origin of magnetization in a few accessory minerals often leads to strong variability within rock units. Thus, care must be taken to insure that a representative suite of samples is measured in determining the magnetization of a subsurface unit.
- The magnetization of igneous and metamorphic rocks generally increases as their mafic content increases, as a result of an increasing proportion of magnetite and to a lesser extent other ferromagnetic and paramagnetic minerals. The remanent magnetizations of these rocks are low, commonly less than the induced magnetization. Volcanic rocks, particularly the more mafic volcanic rocks, are the exception: they carry a strong remanent magnetization as a result of their rapid cooling and fine-grained texture.
- Sediments and sedimentary rocks generally have negligible magnetizations because of very low contents of magnetic minerals. However, there are notable exceptions to this generality which reflect both primary detrital and chemically precipitated minerals and secondary ferromagnetic minerals due to post-depositional processes.
- The magnetization of soils may reflect their parent material where surface processes have not altered the constituent magnetic minerals, but surface layers may have an enhanced magnetization due to processes that control oxidation and reduction of their magnetic components. These include surface fires, fermentation associated with successive periods of oxidation and reduction, and microbial activity. Additionally, strong magnetizations can occur in surficial materials from the presence of fired clay that has acquired a thermal remanent magnetization, and ferrous objects with both induced and remanent magnetizations.

Supplementary material and Study Questions are available on the website www.cambridge.org/gravmag.

11 Magnetic data acquisition

11.1 Overview

The acquisition of magnetic data is relatively simple, rapid, and less complex than are the observations of data of most geophysical methods. Significant improvements continue to be made in magnetic instrumentation which facilitate accurate observation of the geomagnetic field on the Earth's surface as well as on a variety of airborne platforms, and from satellites of the Earth, Moon, and planets of the solar system. Most observations are made of the scalar, total intensity of the field, with alkali-vapor (resonance) magnetometers which readily achieve a sensitivity of better than a nanotesla with rates of several observations per second from a moving platform. These measurements are supplemented for special purposes by measurements of gradients, vectors, and tensors. Vector and tensor measurements are made with flux-gate magnetometers and increasingly with the highly sensitive superconducting quantum interference device magnetometers.

Surface or near-surface surveys are conducted on grids or parallel lines to map with high resolution the near-surface, local magnetic anomalies associated with a variety of archaeological, engineering, and environmental problems, but most magnetic surveys are conducted from a wide variety of airborne platforms. Although helicopter surveys may use outboard sensors placed in an aerodynamically stable housing at the end of a cable towed by the helicopter to place the sensor close to the surface to achieve the highest possible resolution, most airborne surveys use inboard sensors which require that extraneous magnetic effects of the aircraft are compensated by passive and active systems.

Surveys are flown along parallel lines separated by a few hundred meters to several kilometers at altitudes generally ranging from 150 m to several hundred meters. For geological mapping purposes the altitude is placed as low as can be flown safely, considering the topography and the flight characteristics of the aircraft, to improve the resolution of individual anomalies. To achieve this end most surveys are flown at a constant mean terrain clearance, although this procedure can lead to distortion of anomalies by magnetic terrain effects and major variations in the altitude of the aircraft above the surface as a result of limitations in the flight characteristics of the aircraft. Satellite observations of the Earth and extraterrestrial bodies are continuing to improve, providing useful information on planetary fields and, under special conditions, characteristics of the lithosphere.

11.2 Introduction

The acquisition of magnetic data is relatively simple and rapid compared with most geophysical observations. Furthermore, magnetic instrumentation is relatively inexpensive and easy to operate, and the planetary effects on the magnetic observations are generally minor compared with the geological magnetic anomalies that are of interest in exploration surveys. The latter, which simplifies the reduction of magnetic data, is true if observations are not taken during periods of intense temporal variations of the geomagnetic field.

Prior to the late 1940s, magnetic surveying was conducted primarily with ground-based instrumentation which measured a vector component of the field to an accuracy approaching a few nanotesla with tripod-mounted mechanical balance magnetometers. This was satisfactory because the intensity of anomalies of interest was largely a few tens of nanotesla (nT) or more. Since then the availability of more precise and accurate electronic instruments which are insensitive to motion and orientation has led to rapid and efficient surveying on mobile platforms, especially from aircraft, with accuracies of the order of 1 nT (or 1γ) which is roughly 2×10^{-5} the surface geomagnetic field. More recently an accuracy of 0.1 nT is sometimes targeted in high-resolution surveys. These levels of accuracy are readily achieved by available instrumentation and can be approached in data reduction depending on the noise envelope of the observations. Minimization of the errors in surveys designed to map at these accuracies requires optimum instrumentation and exacting, but standardized, procedures for observing terrestrial magnetic fields.

Regional magnetic anomaly data covering most continental areas and large segments of the oceans have been assembled in commercial and governmental data repositories and into publicly available maps. However, the resolution and accuracy possible in current mapping of magnetic anomalies are usually not met in these public-domain data, requiring new surveys designed specifically for a particular exploration objective. These surveys involve mapping at the extremes of the magnetic anomaly spectrum: that is, near-surface, high-resolution surveys for anomalies small in size (often meters or a few tens of meters) and amplitude, which may involve magnetic gradient and tensor measurements; and, at the opposite end of the spectrum, satellite magnetic surveys which map scalar and vector components of very long wavelength anomalies of the order of 500 km or more. The potential higher resolution capabilities of magnetic anomalies over gravity anomalies from common sources is a major advantage of the magnetic method, but it also means that magnetic anomalies are smaller in size which requires significantly greater density of observations in the magnetic method. As a result care must be exercised in designing magnetic surveys to capitalize on their potential resolution.

11.3 Instrumentation

As mentioned in Section 8.3, a variety of techniques have been used to measure the angular relationships and magnitudes of the vector components of the Earth's magnetic field both absolutely and relatively. Prior to World War II magnetic surveys involved either pivoted needle instruments or magnetic variometers. Hand-held needle instruments which measure anomalous fields with amplitudes in excess of several hundred nanotesla generally use a measurement of the oscillation or rest position of a counter-weighted magnetic needle pivoted on a horizontal axis either parallel (measuring the combination of the amplitude and dip of the Earth's field) or perpendicular (measuring the vertical component of the field) to the magnetic meridian. These instruments were largely restricted to mapping where the bedrock consisted of highly magnetic rocks, such as iron formations and volcanic rocks. Magnetic variometers measure either the deflection of a counter-balanced magnetic system that oscillates in the vertical plane perpendicular to magnetic north (e.g. the Schmidt-type variometer) or the force needed to bring an unoriented magnet system rotating in the magnetic meridian to a horizontal position (e.g. the torsion magnetometer (Haalck, 1956). Although magnetic variometers could reliably measure field changes approaching several nanotesla, they were slow and cumbersome to use because they required a tripod mount for leveling and, in some types, orientation. As a result they were replaced almost entirely by nuclear resonance magnetometers by the mid-1950s, with the vast majority of current magnetic surveying employing atomic resonance sensors.

Resonance magnetometers are scalar in nature, measuring the absolute total intensity of the magnetic field without consideration of the field's direction, and thus require no accurate orientation or leveling of the instrument. This is an important advantage that greatly increases their utility by making accurate observations possible on mobile platforms and explains their broad use in exploration magnetics. Nuclear or resonance magnetometers include the proton-precession, alkali-vapor, and Overhauser devices. They are used either alone to measure the total field or in pairs, involving a variety of configurations, to measure the total field gradient in various directions.

Another type of magnetometer is the flux-gate that was widely applied to observations on aircraft, ships, satellites, and other moving platforms for several decades. As the first airborne magnetometer, the flux-gate was originally designed and used to detect submarines during World War II. It is essentially a direction-dependent variometer that can measure either the vertical, horizontal, or total field and as such, even though largely replaced in airborne operations, has a role in measuring vector magnetic fields. Similarly, the high-sensitivity superconducting quantum interference device (SQUID) has an increasing role in exploration magnetics.

All modern magnetometers record digitally, and several have built-in provisions for mapping using GPS surveying and displaying the surveyed observations. They have individual advantages and limitations which need to be considered when selecting an instrument for a particular survey.

11.3.1 Resonance magnetometers

Nuclear resonance magnetometers such as proton-precession, Overhauser, and alkali-vapor devices have sensors containing fluids or gases, with atomic properties that are sensitive to changes of the geomagnetic field. The operational principles of these devices are described below.

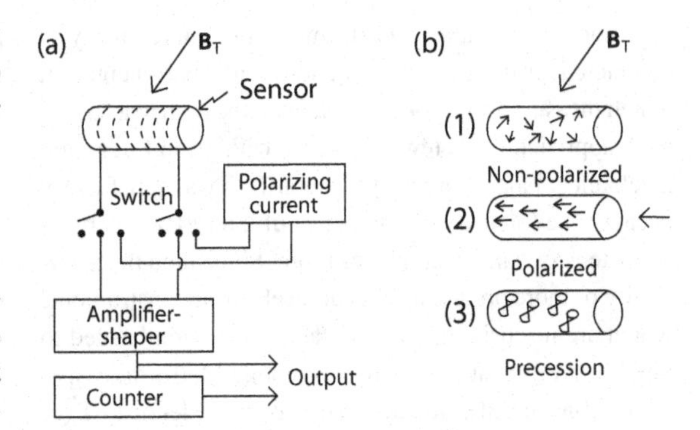

FIGURE 11.1 Simplified proton-precession magnetometer instrumentation with (a) the axis of the hydrogen-rich sensor oriented at a high angle to the magnetic field, $\mathbf{B}_T$. (b) Alignment of protons in the sensor in normal magnetic field (1), upon polarization by applied magnetic field along the axis of the sensor (2), and in precession mode (3) upon termination of applied axial magnetic field.

Proton-precession magnetometers

The proton free-precession magnetometer, which has an accuracy of the order of 0.1 to 1 nT, measures the precession frequency, the Larmor frequency, of protons in a hydrogen-rich fluid that have been oriented at a large angle to the geomagnetic field ($\mathbf{B}_T$) by a strong d.c. magnetic field ($\mathbf{B}_P$) originating from a current passing through a wire coil wound around the fluid container, the sensor (Figure 11.1(a)). Upon termination of the current through the coil and relaxation of the applied magnetic field along which the protons of hydrogen align (Figure 11.1(b.2)), the protons acting as tiny, spinning bar magnets precess around the ambient magnetic field (Figure 11.1(b.3)) at a frequency dependent on the intensity of the ambient magnetic field through a well-known constant, the gyromagnetic ratio of protons. The precessing protons induce a current in the surrounding wire coil which decays exponentially owing to thermal motions with loss of a coherent signal within a few seconds. The frequency of the induced signal due to the coherent precession of the protons is of the order of a few thousand hertz and is readily measured with a frequency counter (Figure 11.1(a)). The amplitude of the induced precession is directly dependent on the orientation of the axis of the sensor coil and the direction of the total field. No signal is received if the sensor axis and the field are parallel, and it is maximum with a right angle relationship between them. As a result the sensor requires crude orientation depending on the inclination of the geomagnetic field.

The proton-precession magnetometer, which was first suggested by the work of PACKARD and VARIAN (1954), has several advantages over previously used mechanical magnetic sensors and is readily produced as a robust, portable, and relatively inexpensive instrument. However, it does have disadvantages. Observations of the magnetic field are not continuous; rather, they require a finite period of time for orienting the protons by the current passing through the coil and for measuring the precession frequency. Decreasing the reading interval will decrease the sensitivity of the instrument. This is not a significant problem for static-mode land surface measurements, but may cause difficulties for measurements from a moving platform where a choice must be made between data interval and sensitivity. Furthermore, the proton-precession magnetometer is subject to erroneous observations caused by a.c. power interference and large magnetic gradients which produce an incoherent precession signal across the fluid container. Gradients of about 300 nT/m or more, which may occur near highly magnetic natural or man-made materials, will cause the instrument to give spurious readings. Errors may also result from rotation of the sensor during the counting period, especially in towed sensor systems. An additional disadvantage is the need for considerable power for the current to orient the protons along the axis of the sensor. As a result of these limitations the proton-precession magnetometer has been largely replaced for airborne measurement applications, as well as in surface surveying, by other resonance magnetometers.

Alkali-vapor magnetometers

Alkali-vapor or optical absorption magnetometers, which also measure absolute total magnetic field intensity, overcome some of the difficulties inherent in proton-precession sensors, and thus are the current instrumentation of choice in airborne magnetometry. Alkali-vapor sensors (Figure 11.2) are miniature atomic absorption instruments with a signal response proportional to the ambient magnetic field. They are capable of an order of magnitude greater sensitivity, have a shorter cycling time and thus much higher sampling rate, and are tolerant of much higher magnetic gradients than proton-precession magnetometers. Accordingly, they have been used in specialized static observations requiring high sensitivity and in mobile platform applications.

These instruments operate by the principle of optical pumping, together with optical monitoring, that is used in radio-frequency spectroscopy. A beam of polarized radio-frequency light corresponding to a specific line in the alkali-vapor spectrum irradiates a cell containing an alkali vapor such as cesium, rubidium, potassium, or helium (Figure 11.2(a)). This causes the electrons in that energy level in the gas to be pumped to a higher energy level (Figure 11.2(b)) which is split into close magnetic (Zeeman) levels in the presence of a magnetic field such as the Earth's field where the light is absorbed. However, the electrons spontaneously decay to a lower energy state that cannot absorb light, and the cell becomes transparent, triggering a radio-frequency magnetic field that causes the electrons to shift to a state where the cell once again absorbs the light (Figure 11.2(b)). The frequency of the signal required to achieve transparency of the cell, the Larmor frequency, is a function of the ambient magnetic field and is much higher than the frequency involved in the proton-precession magnetometer. Thus, the magnetic field is measured by varying the radio-frequency signal and recording the frequency when the cell once again absorbs light.

Alkali-vapor magnetometers have an order of magnitude or more greater sensitivity than proton-precession magnetometers, with sensitivities reported to be in the range from 0.001 to 0.01 nT. Sampling intervals vary depending on the design of the instrument, but rates of up to 20/second (i.e. 20 Hz) are used. Unfortunately, optically pumped instruments have a small inherent heading error of roughly 1 nT and dead zones, in which the signal is lost, which are related to the orientation of the instrument with respect to the magnetic field. These problems have been largely overcome in commercial applications with a variety of methods including use of multi-celled sensor configurations, mechanical orientation of the sensor to maintain an optimum angle with the magnetic field, use of the split-beam cesium sensor (Hardwick, 1984b), and use of a vapor which limits heading errors to less than 0.1 nT. The characteristics of this type of magnetometer can be modified somewhat depending on the alkali vapor used in the sensor cell, but all are much less sensitive than proton-precession instruments to errors due to high gradients. Current commercial magnetometers are available with cesium, potassium, or helium vapor as the sensor. Both potassium and helium have higher Larmor frequencies, and thus have potentially higher sampling rates for a given sensitivity.

Overhauser magnetometers

The third type of resonance magnetometer is the Overhauser or spin-precession magnetometer (Acuña, 2002). This instrument is an enhanced proton-precession magnetometer that uses a wire-wrapped container filled by a special hydrogen-rich fluid with added paramagnetic ions (e.g. free unpaired electrons) as its sensor. It derives its name from the Overhauser effect which describes the transfer of spin energy of orbital electrons to the protons of hydrogen atoms in a fluid that contains a special dissolved salt (Overhauser, 1953).

A radio-frequency electromagnetic field excites the system rather than a discrete polarizing pulse. The very high-frequency electromagnetic resonance of the paramagnetic ions causes the protons to precess continuously for a nearly continuous sampling of the magnetic field's variations at rates up to 5 hz or sometimes more. Relative to the standard proton-precession magnetometer, the Overhauser device requires less power and measures nearly continuous field variations at greater sensitivities (0.1 to 0.01 nT) with 100–1,000 times the strength of the discrete signal, and hence has a very broad operating range. Thus, Overhauser measurements have significantly improved signal-to-noise and reduced measurement uncertainties. In addition, the Overhauser instruments are without heading error or dead zones where they are inoperative.

11.3.2 Flux-gate magnetometers

In specialized applications of the magnetic method, it may be desirable to obtain the directional attributes as well as the total magnitude of the magnetic field (see Section 11.3.5). The scalar intensity of the field is determined by resonance-type magnetometers, but flux-gate magnetometers are used to measure vector components of the field. The flux-gate instrument, which is no longer regularly used for measuring the total field variations in airborne magnetometry, measures only the relative change, not the absolute value, of the geomagnetic field and is directionally dependent.

Construction details vary among flux-gate instruments, but, in the classical design, this magnetometer consists of a pair of identical, but oppositely wound inductive coils with cores of the same high magnetic permeability material along their axes (e.g. cores A and B in Figure 11.3). The cores are magnetized to saturation by the induced fields from an alternating current passed through the windings of the coils in opposite directions. In the

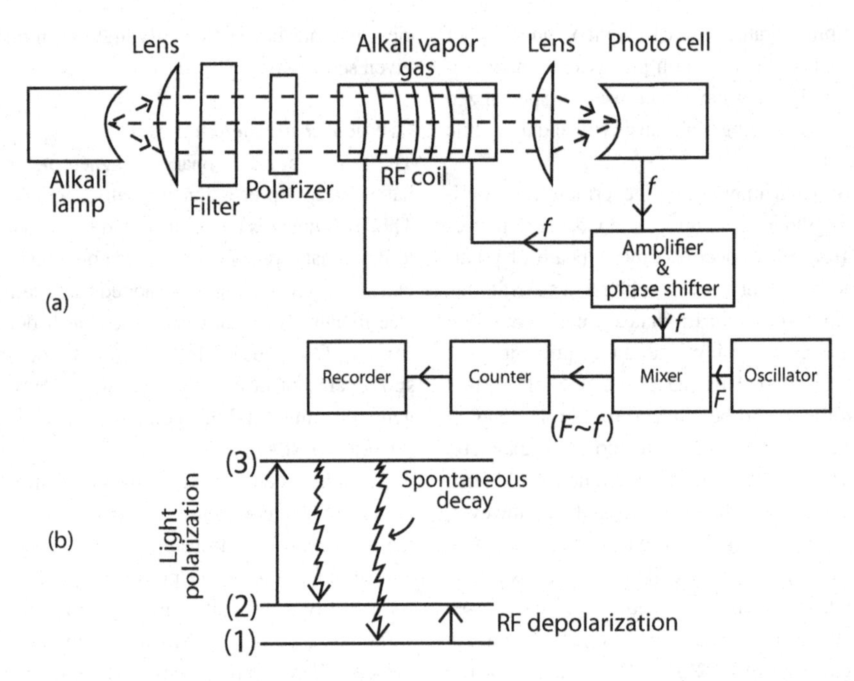

FIGURE 11.2 (a) Schematic diagram of a self-oscillating alkali-vapor magnetometer. (b) Schematic illustration of alkali metal energy levels used in the alkali-vapor magnetometer. Spontaneous decay of unstable electrons of an alkali metal and excitation of electrons from level 2 to 3 only by light polarization of a specific wavelength leads to a fully depopulated level 2. In this state the vapor cell stops absorbing the alkali-vapor light and becomes transparent to the light. Unstable electrons in level 3 decay to levels 1 and 2, eventually causing a full population of level 1 and depopulation of level 2. The unit detects the fluctuation of light intensity as the cell moves from transparent to opaque with the change in population levels and measures the frequency of the corresponding depolarization field. The radio-frequency (RF) depolarization field moves electrons from level 1 back to level 2 at a frequency directly proportional to the ambient magnetic field. Adapted from Hood and Ward (1969).

absence of an ambient magnetic field $\mathbf{B}_T$, there will be no difference in the induced fields of the cores other than polarity in the voltages induced in the secondary coils wound around the induction coils. This is true only where the magnetic properties of the cores are equivalent. However, as illustrated in Figure 11.3(b), a static magnetic field $\mathbf{B}_T$ in the direction of the cores will reinforce the magnetization in one core, causing it to reach saturation first, and oppose it in the oppositely wound induction coil to shift the phases of the two secondary coil induced fields $\mathbf{B}_A$ and $\mathbf{B}_B$ so that the resulting coil voltages V_A and V_B sum as alternating positive and negative voltage spikes. The sum of these voltages is proportional to the intensity of the magnetic field in the direction of the sensors (Figure 11.3(a)) and can be determined by nulling the voltage with a current in the secondary coil calibrated in terms of the ambient magnetic field. A d.c. bias resistor is placed across one of the cores' elements (Figure 11.3(a)) which unbalances the circuit producing voltage pulses of equal intensity in the presence of a zero magnetic flux, preventing drift of the readings with time. More recent versions of these magnetometers use only the prominent second harmonic component of the unbalanced voltage.

Three mutually perpendicular elements appropriately oriented can measure the individual vector components of the field, and thus their angular relationships and by calculation the total field intensity. Sensitivities of flux-gate magnetometers are usually about 1 nT up to the order of 0.1 nT, in specially constructed instruments. A wide variety of alternative designs of the flux-gate magnetometer have been developed including the cylindrical core flux-gate and the ring-core design which has found considerable use in space measurements because of its low mass and simplicity. These and the history of vector magnetic measurements in space are described by R. C. Snare in the website http://www-ssc.igpp.ucla.edu/aersonnel/russel/ESS265/history.html.

Airborne flux-gate magnetometers, which were the first to measure the crustal fields from moving platforms, generally measure the total magnetic field to minimize problems in orienting the sensor to measure the field to a sensitivity required for magnetic exploration. The axis of the flux-gate element is accurately aligned in the direction

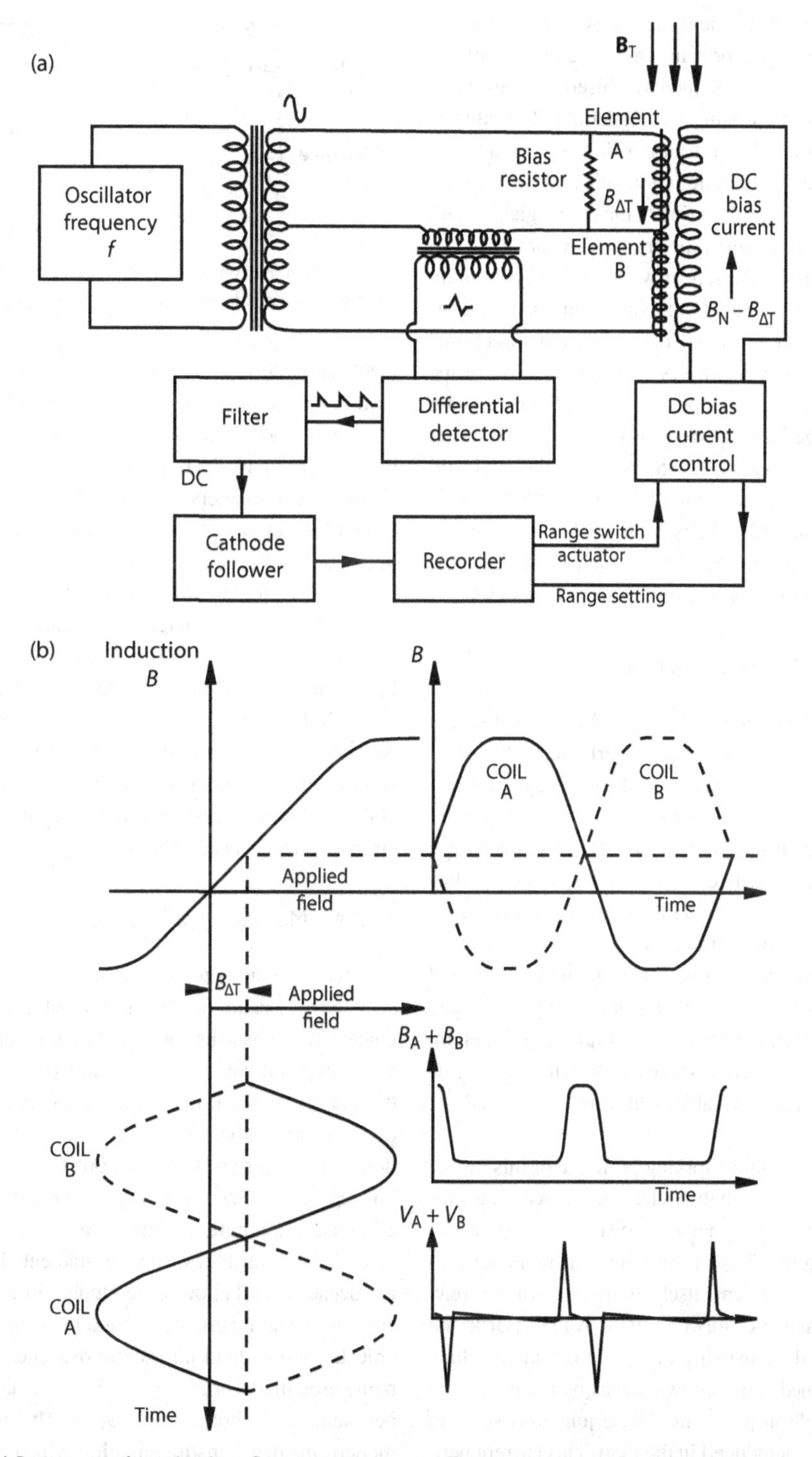

FIGURE 11.3 (a) Basic circuit of a peak voltage flux-gate magnetometer. The sensor consists of two identical high magnetic permeability ferromagnetic cores within identical, but oppositely wound inductive coils (elements A and B). (b) Illustration of the effect of the ambient magnetic total field $\mathbf{B}_T$ on the magnetic saturation of the ferromagnetic cores by the applied field which causes a shift in the phase of the current induced in the secondary coil. The resulting voltages V_A and V_B sum as alternating positive and negative voltage spikes. The sum of these voltages is proportional to the intensity of the ambient magnetic field in the direction of the elements. Adapted from Hood and Ward (1969).

of the Earth's magnetic field for measurement using a three-axis movable platform in which the outputs of the supplemental orthogonal elements are used to position the primary element for sensing the total field. The supplemental elements are driven to a position of zero field by a servo-motor feedback system, thus placing the primary sensor accurately in the direction of the geomagnetic field. Handheld flux-gate magnetometers can measure the vertical field by orienting the sensor with a gimbal pendulum assembly. Three orthogonal component flux-gate magnetometers with sensitivities of 0.1 nT are used in directional magnetic mapping in drill holes where orientation of the sensors is achieved gyroscopically and inclinometers are used to determine the tilt of the sensor probe (e.g. BOSUM *et al.*, 1988; MORRIS *et al.*, 1995). Alkali-vapor total field intensity magnetometers are used also in drill holes, even in shallow drill holes for the purpose of detecting ferrous metals, but they lack the directional sensitivity that is possible with three-component flux-gate magnetometers.

11.3.3 SQUID magnetometers

The most sensitive magnetometer available for exploration is the superconducting quantum interference device or SQUID, which is sometimes called the cryogenic magnetometer because it uses measurements associated with superconducting metals and alloys which require very low temperatures, that is cryogenic temperatures (below $-150\,^{\circ}$C). The need to maintain the sensor at cryogenic temperatures is an obvious disadvantage, but this problem has been relieved to some extent by the use of liquid nitrogen ($-196\,^{\circ}$C) as the coolant in recently developed SQUIDs rather than the lower-temperature liquid helium that was previously used. As a result current instruments are smaller and more portable than those of a decade or more ago.

An explanation of the working principle of this instrument is more complex than nuclear resonance magnetometers because of the need to deal with the wave-mechanical nature of electrons and superconductivity. However, the instrument itself is rather simple (e.g. ZIMMERMAN and CAMPBELL, 1975; WEINSTOCK and OVERTON, 1981) consisting of a superconducting loop which is weakened at one or two places by point contacts known as Josephson junctions. These junctions serve to amplify the current induced in the loop. This current periodically exceeds the critical current at which the loop takes on a finite electrical resistance, in contrast to the zero resistance of the normal superconductivity state. As a result a voltage appears in the loop that is a measure of the external magnetic field perpendicular to the loop. A negative

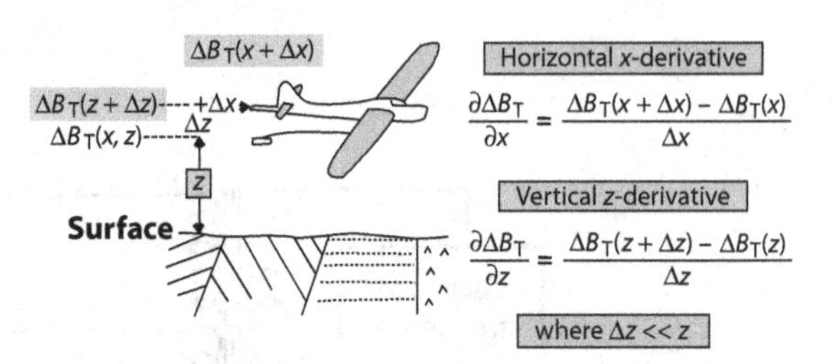

FIGURE 11.4 Measuring the first vertical and horizontal derivatives of the total intensity of the Earth's magnetic field.

feedback circuit is used to detect the voltage and null the field. The output of this circuit is a measure of the field cutting the loop which can be determined to a sensitivity of the order of 10^{-5} nT. Thus, SQUIDs are very sensitive vector magnetometers, which with appropriate combinations of superconducting loops can be used to measure magnetic gradients.

Several different types of SQUIDs have been developed for measuring the geomagnetic field and its gradients. The high sensitivity of SQUIDs makes them particularly useful in measuring gradients. As a result they have been used successfully with multiple sensors to observe the full tensor (gradients of the three mutually perpendicular vector components) of the magnetic field (e.g. SCHMIDT *et al.*, 2004) including measurements of magnetic tensors from aircraft (STOLZ *et al.*, 2006).

11.3.4 Magnetic gradiometers

Magnetic gradiometers are simply two magnetic sensors, either scalar or vector, separated by a fixed distance with the measurements made simultaneously in both and compared for the observation (Figure 11.4). As in the gravity method, measurement of magnetic gradients offers several advantages over amplitude field observations (SCHMIDT and CLARK, 2006) that include minimizing regional effects, increasing anomaly resolution, and eliminating temporal magnetic variations. Taking advantage of these attributes, surface gradients have been used for decades in shallow zone studies in association with archaeological investigations and the detection of ferrous objects such as land mines and ordnance. They are also being used increasingly in airborne and marine studies for petroleum and mineral exploration. This is facilitated by modern magnetic instrumentation which allows efficient observation of fields at the close spacing required to measure the short wavelengths of gradients.

Although horizontal gradients are used occasionally, vertical gradient measurements are particularly useful in improving the resolution. Interest in gradients is increasing because of their use in determining the full tensor of the magnetic field which offers interpretational and acquisition advantages over the simple scalar field usually measured (Schmidt and Clark, 2006). Additionally, Woolridge (2004) notes that gradient measurements can be used to level magnetic field data without tie lines using generalized 3D Hilbert transforms. They also improve gridding using the lateral gradient as a constraint.

Vertical gradient measurements may be made using a single magnetometer with observations at two levels or, more commonly and more efficiently, with the simultaneous use of two magnetic sensors fixed at a constant vertical separation of the order of a meter or two. Vertical gradients of the geomagnetic field are of the order of 0.5 nT/30 m, but crustal vertical magnetic gradients are commonly 10 nT or more per 30 m. As a result where the magnetic gradient is measured as the difference between two vertically separated magnetometers, high sensitivity instrumentation and minimal errors are needed (Hardwick, 1984a). Hood *et al.* (1979) showed that a magnetometer sensitivity of at least 0.01 nT is required for airborne mapping of vertical gradients with a 2 m sensor separation in crystalline bedrock terranes. As a result of these sensitivity requirements, the high-sensitivity SQUID magnetometer is especially suited for gradient measurements, although other resonance magnetometers which measure the total magnetic field have been extensively used for this purpose, especially in crystalline bedrock terranes and in the exploration for ferrous objects in the near surface where the anomaly gradients are relatively high (e.g. Billings and Wright, 2009).

Alternatively, it is possible to compute the vertical gradient of the total magnetic field from observations of the total field at one level providing that the field is fully sampled. This can never be done completely accurately, but with high-density observation coverage, it can be approximated. For example, Roberts *et al.* (1990) and Billings and Wright (2009) show a near equivalence of observed and calculated vertical gradients where the data density at a single elevation is sufficient to approximate the field at that level. However, calculated gradients do not eliminate the effect of temporal magnetic variations as do measured gradients, a significant advantage of measured gradients.

11.3.5 Vector versus scalar magnetic measurements

Before the mid-1940s, essentially all exploration magnetic observations were vector components, primarily the vertical magnetic anomaly field. However, since the advent of the airborne flux-gate and resonance magnetometers, essentially all exploration magnetic observations have been made of the scalar field. It was realized that the total magnetic anomaly field is in error because the amplitude of the total geomagnetic field is subtracted in a scalar sense from the observed field to determine the anomaly regardless of the orientation of these components. Nonetheless, as shown previously in Table 8.3 the error is minor for most crustal magnetic anomalies and only approaches 10% as the anomalies exceed 10,000 nT. As a result the scalar total magnetic intensity has been and continues to be the acceptable magnetic anomaly component for exploration purposes.

Despite the universal acceptance of the scalar component in exploration, interest continues in measuring vector components and their gradients as either a replacement for or supplement to total field measurements. It is of course possible to calculate the vector components from the total field in a variety of ways (e.g. Purucker, 1990), but limited sampling of the total field restricts the accuracy of these calculations. As a result measurements of the vector components of the geomagnetic field continue to be made with either flux-gate or SQUID magnetometers. Vector measurements are made in satellite surveying to identify fields derived from the ionosphere and meridional currents (Maeda *et al.*, 1982; Ravat *et al.*, 1995). Also, slight differences between observed vector data and those computed from scalar data have been used to identify strong remanent magnetization in crustal anomaly sources (Girdler *et al.*, 1992; Taylor and Ravat, 1995; Purucher and Wonik, 1997; Langel and Hinze, 1998). In addition, a joint inversion of vector and scalar components has been used to stabilize and improve the interpretation of magnetic anomaly data (Purucker, 1990; Li and Oldenburg, 2000).

Vector magnetic measurements have also been used by Blakely *et al.* (1973) to distinguish 2- from 3D magnetic anomaly sources and detect off-track sources in mapping marine magnetic stripes. They found that the magnetic component parallel to the magnetic stripes observed only anomalies from 3D sources and not the 2D magnetic stripe anomalies. In addition, gradients of vector components are finding increasing application in both mineral and petroleum exploration. Schmidt and Clark (2006) have reviewed the properties of the magnetic gradient tensor and outlined their merits. The tensor combines the advantages of gradient as well as vector component measurements in improving resolution and interpretation as well as minimizing errors in the data acquisition.

11.4 Survey design and procedures

Magnetic observations are made in drill holes, mines and tunnels, and on the terrestrial surface, as well as from submarine, ship, airborne ($\leq$20 km), balloon ($\leq$40 km), and satellite ($\geq$150 km) platforms. Magnetic measurements close to the surface are the norm in near-surface geophysical studies because of their higher resolution than airborne observations. However, airborne measurements are particularly efficient and useful in studies of extensive areas, especially in regions where surface access is problematic. Drillhole measurements are made with specialized instrumentation using either directional magnetometers or total field instruments. Directional measurements in bore-holes require more sophisticated instrumentation because of the need for orientation of the sensors, but they can be powerful tools to pinpoint the location of buried magnetic sources with a high degree of resolution. Survey design of and procedures for drillhole measurements are similar to those of near-surface measurements. Satellite magnetic observations have taken on an increasing role in studying the main geomagnetic field and its variations and, in specialized cases, mapping regional lithospheric magnetic sources.

The design of a survey and the procedures used in data acquisition are determined by factors including the objectives of the survey, the size, location, and physical characteristics of the study area, the anticipated sources and nature of magnetic anomalies and noise, and the resources available. In several respects the process is similar to that described in Section 5.4 for the gravity method. However, fundamental differences do exist because of the dipolar nature of the magnetic field, the sources of magnetic anomalies and noise, and the measurement principles. For example, the size of the area of a survey or the length of an observation profile depends primarily on the maximum depth of the anticipated sources and the nature of the regional anomalies. The source depth controls the extent of the anomaly, and the coverage must be sufficiently extended and dense to permit isolation of the target anomaly from regional anomalies derived from broader, deeper geological sources. The interference of regional with local anomalies is much greater in gravity than in magnetics because magnetic anomalies have a significantly smaller extent than do gravity anomalies from the same source. As a result, the coverage required in magnetic surveys is less extended, but greater data density is required than in gravity surveying.

A common assumption in magnetic surveying is that the coverage should extend beyond the limits of the area of interest for a minimum distance equal to the maximum depth of the target sources. The spacing between measurements should not exceed the minimum depth of the target sources. This is based on the generality that the anomaly from a concentrated source decreases to 10% of its maximum amplitude at a distance equivalent to the depth to the center of a concentrated magnetic source, and contrasts with the three-fold greater width for the gravity anomaly from a similar source. The coverage suggested by this generality should be considered minimum, especially where regional anomalies are dominant and complex and the sources have a vertical extent of several times the depth to their top surfaces.

FIGURE 11.5 The layout of ground surveys involves making appropriate choices of the magnetometer, survey altitude, and grid spacing, as in this example from an archaeomagnetic survey that uses a proton-precession magnetometer at 0.5 m above the ground surface at a station grid interval of 1.5 m.

11.4.1 Land surface surveys

Surface (or near-surface) surveying involves obtaining magnetic observations at appropriate accuracies, intervals and resolution to map the effective signatures of the subsurface targets or sources of interest (Figure 11.5). Observations commonly are made at specified locations using handheld instruments with the sensor located at the top of a staff of a half a meter or more above the ground surface to minimize the effects from local variations in the magnetization of the soil and magnetic fields derived from currents induced in electrically conductive soils from fields originating within the magnetometer. Stations are either organized on a grid pattern or, in mapping linear patterns of anomalies, on parallel traverses perpendicular to the anticipated anomaly strike. The separation between traverses depends on the continuity of the linear anomalies and the detail required for the purposes of the survey, but traverses generally can be placed at a distance of several times the station interval without encountering problems in mapping anomalies. Grids usually are orthogonal with the directions controlled by local surface conditions or placed parallel and perpendicular to magnetic north.

The grid interval or the station spacing along traverses depends on the objective of the survey. If the survey is intended to only detect the presence of anomalies, the intervals can be large with only one or two observations that occur within the area of the anomaly. The dimension of the anomaly above any prescribed magnetic noise level can be determined by calculating the anomaly from the characteristics of the target anomaly source and the nature of the ambient terrestrial magnetic field. However, the noise characteristics may be difficult to establish beforehand, so it is desirable to conduct the survey in a manner that will minimize the effects of the various sources of magnetic noise. A general rule is that the station interval should not be greater than the minimal depth of the sources, assuming that they are concentrated or linear-concentrated sources. This generality suggests that at least one station will be within the area or length in which the anomaly exceeds 50% of its maximum amplitude.

A station interval equal to the depth is not adequate for mapping anomalies for analysis and modeling. For these purposes the highest horizontal gradients that are anticipated should be mapped. As a generality, assuming a concentrated or a linear-concentrated source, a station interval of 0.2 of the depth to the center of the source will map the highest gradients and the amplitude of the anomaly within 10% of its maximum value. As the depth extent of the source increases, the appropriate station spacing may increase from this generality.

As explained in Chapter 10, there are numerous sources of magnetization, both natural and anthropogenic, in the subsurface. In some surveys, these are the objective of the mapping, but in others they distort the magnetic anomalies derived from deeper, larger sources. Many sources are equidimensional and, although intensely magnetic, produce anomalies that decrease rapidly with distance. Near-surface effects can be minimized in the magnetic observations by increasing the height of the sensor above the ground surface. The presence of localized, near-surface sources can be detected by making several observations over a short time period measured in minutes within a limited area at significantly smaller intervals than the station spacing of the survey. The average value of these measurements can be used as representative of the station, but if the variations are considerably greater than anticipated from the anomalies derived from the target sources, the station should be eliminated from the survey. Of course, care must be used to avoid obvious magnetic effects from ferrous objects and electrical fields either on the instrument operator or in the vicinity of the station. The required separation between observations and potential sources of noise depends on the source and shape of the extraneous source, its orientation in the terrestrial magnetic field, and the survey requirements. Distances of 50 to 150 m are desirable from ferrous objects and a few hundred meters from power lines. Extreme examples of cultural effects which produce surface magnetic anomalies of hundreds of nanoteslas or more are cathodically protected pipelines and in-place drillhole casings.

Specific situations may dictate the need to acquire high-resolution, near-surface survey data over extensive regions involving hundreds or thousands of observations. To minimize the time and costs involved in surveys of this size, magnetometers have been mounted on various types of transport including motor vehicles, bicycles, balloons, helicopters, small remote-controlled unmanned aircraft, and horses. These are most effective if the magnetometer automatically makes observations at a prescribed time or distance interval and is tied to a GPS mapping system that locates the geographic position of each measurement. All of these observations including the time of each measurement are digitally recorded and downloaded on a daily basis to a computer for data storage, processing, and presentation. For accurate observations, care must be used to isolate the sensor from the magnetic fields associated with the transport or compensate for these fields. Similar procedures are used for shipborne surveys of water-covered regions. In these cases the magnetic sensor is normally trailed behind the vessel in a streamlined container to avoid the vessel's magnetic effects.

An important consideration in the design of a magnetic survey is the procedure that will be used to monitor the temporal variations in the magnetic field caused primarily by the interaction of electromagnetic and corpuscular radiation from the Sun with the Earth's ionosphere and by tidal forces acting upon the ionosphere. As discussed in Section 8.3.4, these fields, which are of sufficient intensity to interfere with precise magnetic mapping of geologic sources, are notably unpredictable, preventing their modeling on theoretical grounds. Accordingly, the magnetic field is monitored either by reoccupation of a base station or by a base magnetometer that periodically measures the ambient magnetic field. The latter is the preferable approach, but requires a separate, special instrument and a secure location for the magnetometer, preferably within several tens of kilometers of the survey area. These instruments include precise timing and a reference time base for correlation with the recorded time of the field observations. Repeated observations with these stationary magnetometers are typically measured in minutes for most mapping purposes. Base-station magnetometers need to be placed in locations that are remote from ferrous or highly electrically conducting materials, including rocks, sources of direct electrical current, and instrumentation that involves switching of intense alternating electrical fields. They are a feasible method of monitoring temporal variations in magnetic fields because the inherent drift of modern magnetometers, unlike gravimeters, is negligible considering the sensitivity of the instruments. However, it is prudent to supplement the base magnetometer observations with reoccupations of a base station to insure that the magnetometers are operating correctly. Real or virtual reoccupation of base stations is an alternative method of monitoring the temporal variations in the magnetic field. The siting of magnetic base stations is particularly important. Care must be used to avoid high gradient and anomalous magnetic background levels and local fields from ferrous materials and electrical currents, including variable effects from traffic on nearby roadways. These stations must be well described and easily recognized so that reoccupations will be in precisely the same location.

The time interval between reoccupations is a function of the variability of the magnetic field, both its gradient and amplitude, and the requirements of the survey. During magnetic quiet times, especially at night, the variation in the magnetic field generally is less than 10 nT/hour, and thus can readily be monitored at hour-long intervals with the assumption of linear variation between observations. However, during magnetic storm periods when the Earth is exposed to intense bombardment by corpuscular radiation from the Sun, the fluctuations of the magnetic field may reach several 100 nT over intervals measured in minutes. During these periods, it is impossible to monitor by either a base magnetometer or station reoccupation the variations to a precision that permits accurate spatial mapping of the terrestrial magnetic field. Generally, it is advisable to reoccupy the base station at least at the beginning and end of the day when monitoring the magnetic variations with a base magnetometer and on a semi-hourly basis when using the reoccupations for monitoring of the field. However, this may vary depending on numerous factors including the magnetic latitude, the state of the terrestrial field, and the requirements of the survey.

More intense and variable magnetic fields are anticipated within roughly five degrees of the magnetic equator owing to the equatorial electrojet and at magnetic latitudes of from 60° to 80° in the auroral regions. During extended magnetic surveying periods, it is advisable to monitor forecasts of the occurrence of magnetic storms made by national geophysical agencies, thus avoiding surveying during periods of rapid and intense magnetic field variations. The variation in the magnetic field, either measured or predicted, is often described by magnetic indicies which are discussed in Chapter 12.

11.4.2 Marine surveys

Marine magnetic surveys have produced much of the evidence for plate tectonics and facilitated the exploration of the seafloor for mineral and energy resources, as well as archaeological, engineering, and military applications. These surveys are typically conducted by towing a watertight, aquadynamically stable housing or fish that contains the magnetometer's sensor.

To avoid the ship's magnetic effects, the fish is towed roughly 3 m below the surface at several ship lengths on a cable that includes electrical conductors to power and operate the sensor. Typically, the sensor consists of an alkali-vapor magnetometer, but various types and configurations of magnetometers are employed to measure gradients as well as vector components which are used to facilitate identification of anomalies and removal of extraneous components (e.g. Engels *et al.*, 2008). In addition, surveys may be made close to the ocean bottom to improve the resolution of the observations. The survey tracks are typically perpendicular to the magnetic or structural fabric of the seafloor. The track spacing is generally dictated by the minimum magnetic source depth which is commonly taken as the water depth. Seafloor depths typically range over 2–6 km so that track spacings of 2–10 km are normally sufficient for most oceanic geology applications. However, marine magnetic surveying is often a secondary component of bathymetric, gravity, seismic, or other geophysical/oceanographic surveys that accordingly dictate the spacing and direction of the magnetic data tracks.

Temporal variations in the geomagnetic field may be checked from the records of nearby geomagnetic observatories or base stations anchored to the seafloor within the survey area with time synchronization between the marine sensor and the base magnetometer preferably to the nearest second. However, the need for magnetic base observations is eliminated when surveying with magnetic gradiometers. In general, the major source of error in modern marine magnetic surveys is the quality of the control on the temporal variation in the field. Prior to the availability of GPS data in the mid-1980s, navigation errors contributed the principal uncertainties in marine magnetic data.

11.4.3 Airborne surveys

Airborne magnetic surveying is preferred for most geophysical applications except for those requiring the highest resolution of anomaly sources. Airborne surveying is economical, rapid, and efficient in studying extensive regions and minimizes the effects from cultural features, temporal variations, and near-surface geologic sources. However, these advantages are achieved only with great care in planning and conducting the survey. Numerous journal, contractor, and governmental publications describe the advances in airborne magnetic surveys since the first surveys for geological purposes in the 1940s. Particularly useful is the report of the Geological Survey of Canada on aeromagnetic specifications and contracts (Aeromagnetic Standards Committee, 1991) and the more recent web-based books by REEVES (2005) and REEVES and BULLOCK (2005).

Objectives

The design and conduct of aeromagnetic surveys is very much dependent on the objectives of the project. Generally, objectives can be defined as reconnaissance, regional, or detailed (Aeromagnetic Standards Committee, 1991). Reconnaissance surveys are those with widely spaced flight paths that are conducted to obtain broad tectonic and geologic characteristics of an extensive region at a minimum of cost. They are often used to map selected characteristics of a region to isolate limited areas for more extensive study: for example, faulted regions, areas of deep basement, or volcanic terranes may be identified. Flight line spacings typically are a kilometer or more and the altitude of the surveys less than the flight line spacing depending on the specific survey objectives. Many nationwide surveys belong to this category.

Regional surveys provide a more comprehensive and detail view of the geology and tectonics of a region than reconnaissance surveys. Often they are used in geological mapping at scales of the order of 1:100,000 based on measurements at altitudes of a few to several hundred meters and line spacing/flight altitude ratios of 1.5 to 5.0. They too are used to select regions for more detailed, high-resolution study based on attributes of special interest such as intrusive contacts, alteration zones, and basin structures. These detailed surveys are flown as close to the source of anomalies as possible. Safety concerns for surveys where the sources are close to the surface generally limit the flight altitudes to several tens of meters and use flight altitude/line spacing ratios of 2 to 1.

Aircraft

An important part of designing an aeromagnetic survey is the selection of the appropriate aircraft. Surveys have been flown with every conceivable aircraft from small, low-power aircraft to wide-ranging four-engine planes, as well as helicopters (Figure 11.6) and lighter-than-air platforms. The selection of a survey aircraft is a tradeoff among numerous factors including the flight duration, speed, stability, cost effectiveness, distance of survey from airports, instrumentation, the terrain of the region, and the required power characteristics of the aircraft. The latter is especially important where the flight altitude is established as constant above mean terrain to achieve high resolution and consistency in altitude above magnetic sources that are close to the surface.

In rugged terrain where the topography is steep and surface elevations vary in excess of a few tens of meters, a low-powered aircraft cannot safely maintain a constant elevation above the surface. Typically flights in these terranes are too low above topographic highs and too high above topographic gradients because of safety considerations. Furthermore, these altitude deviations may vary depending on the direction of the aircraft. These altitude variations introduce changes in the magnetic measurements that are unrelated to magnetic sources and are difficult to remove from the observations. In addition, where flight direction is a factor in determining the safe altitude of the aircraft, abnormal changes in the magnetic measurements may vary from line to line. As a result of these problems every effort is made to insure the safety of the aircraft while maintaining as much as possible the design specifications of the survey. This necessitates close coordination between survey specifications and the aircraft flight characteristics in selecting the appropriate aircraft (COWAN and COOPER, 2003).

Flight specifications

Observations are made along parallel flight lines generally directed perpendicular to the strike of the prevailing magnetic anomalies or, where there is no dominant strike to the anomalies, in a direction that will facilitate the surveying procedure. For example, the flight lines may be oriented to take into consideration the terrain in the survey area and the need to maintain a constant flight height above the surface. If not observed in a particular direction for a compelling geologic reason, it is advantageous from an interpretational point of view at high geomagnetic inclinations ($\geq 70°$) to fly north–south tracks so as to map the full dipolar effect of anomalies including both the positive and negative components of the anomaly. Similarly, at low geomagnetic inclinations ($\leq 25°$), east–west flight

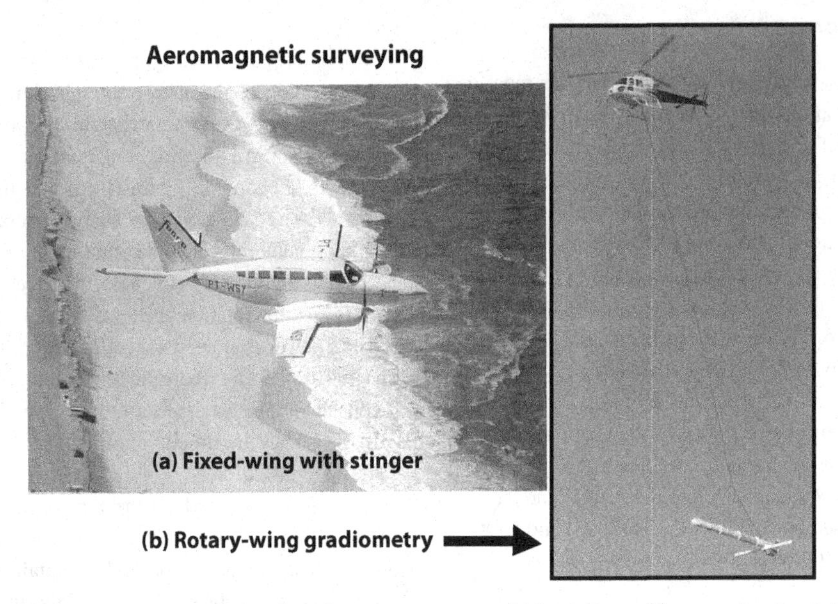

FIGURE 11.6 Airborne magnetic surveys commonly deploy magnetometers on (a) fixed-wing or (b) rotary-wing aircraft. In (a), the magnetometer is housed in the tail stinger to map total field anomalies. In (b), the vertical and horizontal magnetic gradiometer is suspended by cable from the helicopter. Courtesy of Fugro, Inc.

lines have limited significance because of the absence of magnetic anomalies from north–south striking anomalies near the equator (LILLEY, 1968). For most exploration purposes it is desirable to conduct surveys north–south or within about 30° of the perpendicular to the prevailing strike of the anomalies. In mapping large regions that have marked changes in the strike of the anomalies, it may be necessary to break the region into blocks with different flight directions to maintain an appropriate flight direction.

The spacing of the traverses depends on the objective of the survey and the depth to the magnetic sources. REID (1980) showed that to avoid aliasing errors in the data which will deteriorate the identification of anomalies and their interpretation, flight lines should have a maximum spacing of twice the depth to the target sources and that this should be decreased to the depth of the sources for comprehensive analysis and modeling of the data. Figure 11.7 shows four total field magnetic anomaly maps over the Marmora magnetic anomaly in southern Ontario, Canada, based on flight line spacings of 1, $^3/_4$, $^1/_2$, and $^1/_4$ miles along north–south flight lines observed at 500 feet above the ground surface (AGOCS, 1955). The $^1/_4$ mile line spacing provides a much improved resolution of the anomaly over the other maps, but even this map is significantly aliased because the depth to the source of the anomaly is 600–700 feet below the observation level, indicating that at a minimum the flight lines should be of the order of $^1/_8$ mile.

Gradient measurements are generally applied to detailed surveys, and thus require a minimum spacing of the depth to the sources. The spacing of flight lines is constant over a survey or a survey block so as to achieve consistency in magnetic patterns over similar geology. However, in reconnaissance surveying, banded flight tracks may be used as a cost-saving measure. In these surveys groups of two or three flight tracks, a band, are measured at a closer separation than the flight track separation between a much larger number of intervening survey lines. The closer flight line spacing in the bands provides for more detailed interpretation and definition of the strike of the magnetic anomalies, yet a broad region can be covered and studied for selecting limited regions for more detailed analysis in an economical manner. Another flight line pattern variation is to establish an orthogonal grid of flight lines. This pattern has been used in some detailed studies especially where the target magnetic anomalies do not have a consistent strike, but generally, considering the additional cost, the extra magnetic observations of the orthogonal grid are unwarranted.

The sampling interval along the flight lines can be synchronized to a range of distances along the track or time-interval depending on the resolution requirements of the survey and the speed of the aircraft. Generally, several

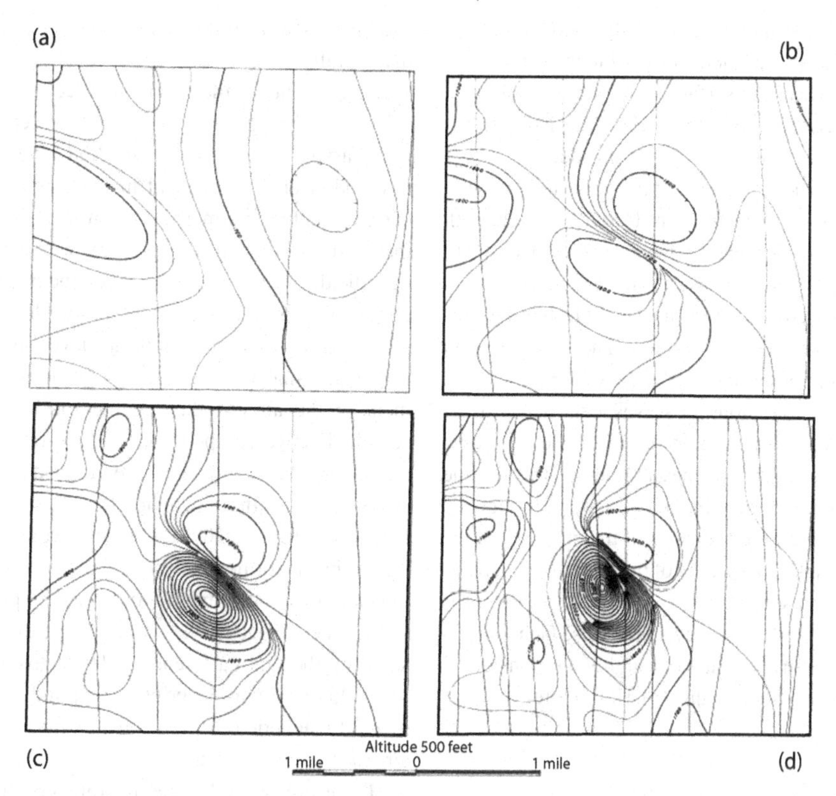

FIGURE 11.7 Comparison of total field magnetic anomaly maps observed over the Marmora magnetic anomaly of southern Ontario, Canada at an elevation of 500 feet with flight line spacings of (a) 1 mile, (b) 3/4 mile, (c) 1/2 mile, and (d) 1/4 mile. Note the much better definition of the anomaly at decreasing flight line separation. Adapted from AGOCS (1955).

observations are made per second, resulting in a separation between observations of the order of a few tens of meters, although smaller intervals are possible depending on the instrumentation and speed of the aircraft and the survey objective.

The altitude of the surveying is dictated by several factors including the objectives of the survey, the geology including the depth and depth extent of the sources, the surface relief, flight safety considerations, and the presence of cultural features. However, the principal concern in most surveys is anomaly resolution, thus the flight altitude is made as low as possible commensurate with safety considerations. Typically, for geological mapping this is an altitude of 150 m with a constant mean terrain clearance to achieve maximum resolution, but by using helicopters for surveying the altitude can be decreased. In some surveys the altitude can be lowered to only a few meters in flat terrain to achieve maximum resolution for engineering and environmental purposes. A useful generality is that the altitude is roughly equivalent to the minimum separation of features which can be resolved. The term "mean" is used to acknowledge that aircraft cannot actually maintain a constant altitude above terrain because of flight safety and limitations in maneuverability. Concern has been raised about interpreting data obtained from a mean terrain clearance flight mode in terranes where the surface rocks are magnetic because of the impact of magnetic terrain effects on the interpretation of the data and the difficulty in accurately determining these effects (GRAUCH and CAMPBELL, 1984; REFORD, 1984; UGALDE and MORRIS, 2008).

An alternative to flights at a constant terrain clearance is a survey which is loosely draped over the surface so as to achieve maximum resolution while minimizing magnetic terrain effects and the variations in the flight surface. In modern surveys, flights can be pre-planned for an appropriate drape over the surface using digital elevation models and the flight characteristics of the aircraft. This plan is loaded into the GPS navigation system together with the planned survey tracks. The navigational system guides the aircraft crew to maintain the planned horizontal and vertical position of the aircraft along the flight line.

Another alternative flight mode is to conduct surveys at a constant elevation, generally at a constant barometric elevation. This procedure is commonly used in surveys to map the depth to basement for petroleum exploration and in regions where the surface rocks are not magnetic and/or the surface relief is very rugged making it difficult to maintain constant terrain clearance. Considering average horizontal and vertical gradients of the geomagnetic field, in order to maintain an accuracy of 0.1 nT the flight path deviations in the vertical and horizontal directions need to be restricted to roughly 7 m and 30 m, respectively.

An important consideration in the flight-line specifications is the distribution of tie-lines. The magnetic data at the intersection of tie-lines and survey flight tracks are used to minimize variations in survey data due primarily to the temporal variations in the magnetic field over the course of the flight line measurements. The tie-line spacing must consider the temporal variation of magnetic field and the accuracy requirements of the objective of the survey. Accordingly, the spacing of tie-lines flown perpendicular to the flight survey pattern will vary with geomagnetic latitude as well as survey objectives. Generally, a tie-line/flight-line spacing ratio of roughly 3 is considered optimum, but it may reach values of 10 or more.

Instrumentation

Airborne measurements typically are made with alkali-vapor magnetometers with the sensor placed in a tail-stinger or in wingtip pods to minimize the need for compensation of the magnetic effects from the aircraft. The use of sensors in an aerodynamically stable housing that are trailed behind the aircraft on a cable at a distance beyond the meaningful magnetic effect of the aircraft has largely been eliminated except in helicopters. Inboard installations typically have a greater signal-to-noise ratio and are more convenient, safer, and less subject to oscillations during flights which degrade the accuracy of the observations. Thus, total field, vector, and gradient measurements are now made with inboard installations. Nonetheless, achieving the desired accuracy with inboard installations is a challenge because of magnetic effects of the aircraft (e.g. HARDWICK, 1984b).

There are several sources of magnetic noise from the aircraft, but the primary effect is from the magnetic components of the aircraft engine(s). As a result the sensor is placed as far from the engine as practicable, either as a stinger, which is most common, or as a wing pod. However, there are numerous other sources of aircraft magnetic noise such as induced magnetic effects in soft iron components in the aircraft, eddy currents arising from the currents in the airframe due to motion of this conductive medium through a magnetic field, and the magnetic effects of electric currents in electric and electronic components onboard the aircraft.

Adjustment for the magnetic effects of the aircraft is achieved either through passive and/or active compensation. Passive compensation, which involves placement by trial-and-error of permanent magnets or coils carrying a current and high permeability iron straps near the sensor that will cancel both the permanent and induced magnetic fields of the aircraft, has been the primary source of eliminating aircraft fields at the sensor. However, compensation can be more efficiently achieved using an analytical model that corrects the observations for the movement of the aircraft through the Earth's field. This is so-called active compensation. The active compensation model and its coefficients are derived from empirical observations of the magnetic effects of the aircraft (e.g. PICOENVIROTEC, 2009) when subject to maneuvers and driven by an orthogonal set of three flux-gate magnetometers that sense the variations in the pitch, yaw, and roll by the change in the amplitude of the field as measured by the flux-gate sensors. The active compensation methodology is now the principal method of compensating inboard magnetic sensors for the extraneous fields of the aircraft.

There are several other instruments that are necessary auxiliary equipment to the magnetic sensors. These include a data system for the preservation of the magnetic data and registering the coordinates and time of the data observations. Essentially all positioning of survey aircraft is achieved through differential GPS, which may be supplemented with Doppler and inertial navigational systems. Differential GPS achieves an accuracy of 1 to 5 m in position. In addition, some combination of laser, radio, radar, and barometric altimeters is used to establish absolute elevation and altitude of the aircraft above the ground surface. The accuracy of these instruments is quite different, with the high-precision laser altimeters achieving an accuracy of a few centimeters.

In addition to the onboard instrumentation, a magnetometer base station is established at a magnetically quiet site in the vicinity of the base of the survey operations to monitor the time variations in the field. If the survey is large or remote (≥ 50 km) from the survey base, an additional base magnetometer may be set up so that the survey operations are not much more than 50 km from a base magnetometer. Digital observations of this usually scalar instrument at a minute or second interval are recorded for later use in removing temporal variations in the magnetic field and to track the onset and dissipation of magnetic storm activity. Surveys are normally aborted if the monotonic variation in the field exceeds 2–5 nT over a 5 minute period or pulsations of the field vary by several nanoteslas over periods of 5 to 10 minutes.

Pre-survey airborne tests
Three airborne tests of the magnetometer system are normally made prior to conducting a survey to insure the accurate measurement and location of anomalies. The first of these is the figure-of-merit (FOM) which determines the variation in the magnetic readings in an aircraft due to movements of the aircraft as it travels in different directions. As such the FOM evaluates the errors remaining in the observations after the sensor system is compensated for extraneous magnetic fields derived from the aircraft. In this test the aircraft repeatedly flies over a specific location where the magnetic anomalies are insignificant, going into $\pm 5^\circ$ pitches and yaws and $\pm 10^\circ$ rolls while flying north, south, east, and west during a period of stable temporal variations and over a time period of approximately 5 seconds. The sum of the difference in the amplitudes, without regard to sign, at the common point for the 12 different measurements is the FOM, which is commonly less than 1 nT.

Another test is the cloverleaf test which checks to be certain that there is no heading error in the observations. In this test, flights are made in a cloverleaf pattern in a region of low magnetic anomalies. The amplitude of the measurement of the center-point of the cloverleaf flown in the four cardinal directions should be the same if there is no heading error.

Finally, a lag test is performed to determine if there is a lag in the observation recovery system such that the data position is displaced from its true position. This is accomplished by making observations in opposite directions over an easily identified, isolated anomalous source such as a surface cultural source. The difference in the location of the anomaly on flights in the opposite direction is twice the lag of the system. It reflects lag in the electronics of the airborne system and different positioning of instrumentation in the aircraft. This lag, which is often of the order of several meters, should be taken into account in the positioning of the data observation.

11.5 Magnetic measurements from space

Satellite magnetic surveys provide unique constraints on the compositional, structural and thermal properties of the lithosphere to enhance the geological utility of near-surface surveys. Satellite data are also obtained essentially for the entire Earth and hardly any region outside about $\pm 87^\circ$ latitude is too remote for observation. Thus, patterns difficult to measure or perceive in near-surface surveys are delineated by the broad data coverage. Additionally, satellite observations form a consistent data set free from non-uniformity caused by secular variations. They also are largely free from the effects of near-surface geologic sources that tend to mask or distort the signatures of deep, broad magnetic variations of the lithosphere. However, in combination with near-surface anomaly data, satellite magnetic observations provide important boundary conditions to enhance lithospheric modeling of anomaly variations from at or near the Earth's surface to satellite altitudes.

The realization of the importance of satellite magnetic observations for extending the geological utility of near-surface magnetic surveys and constraining crustal magnetization variations measured in hundreds to thousands of kilometers has led to increased availability of low Earth-orbiting (LEO) satellite magnetic surveys. The polar-orbiting Ørsted and CHAMP missions, which operated at altitudes of about 600–700 km and 300–450 km, respectively, provide two unique geomagnetic field boundary conditions to complement analyses of near-surface surveys. The data from these satellites are scheduled to be augmented around 2013 by a constellation of three near-polar-orbiting satellites from the Swarm mission. The Swarm measurements at altitudes of 450–550 km can be converted to gradient anomalies that will improve crustal anomaly estimates, especially in the polar regions where the distorting effects of the temporally varying external magnetic fields are strongest.

The instrumentation employed on most satellites for measuring the magnetic field is described by Langel and Hinze (1998). It consists of a boom-mounted triaxial flux-gate magnetometer oriented by a star camera for measuring the relative vector components of the fields. In addition an alkali-vapor magnetometer is used to measure the absolute scalar magnetic field.

Magnetic measurements taken by satellites are dominated by the regional core field ($\sim$98%) and external field ($\sim$2%) components with minor contributions from the lithospheric field ($\leq$0.2%). These field contributions with source regions within and outside the Earth are illustrated in Figure 11.8. Satellite magnetic observations are commonly averaged spatially and rendered into global spherical harmonic models that are effective for core and external field studies. This approach is limited, however, for resolving anomaly detail in lithospheric studies, which tend to be more local in scope and conducted over finite spherical patches of the planetary surface. Furthermore, spherical harmonic model coefficient errors from gaps and uneven coverage in the global data set, and variable data measurement and diurnal field reduction

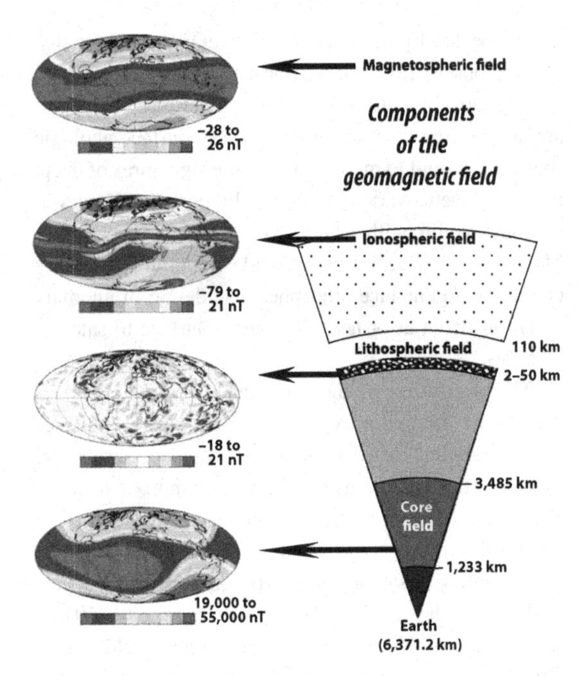

FIGURE 11.8 Satellite magnetic measurements are dominated by the effects of the main field (bottom) and the external fields (top), which include the ionospheric and magnetospheric fields. Lithospheric fields (second from bottom) make only minor contributions to satellite observations. A color version of the figure is available at www.cambridge.org/gravmag. Adapted from MANDEA and PURUCKER (2005).

errors can complicate local anomaly predictions. However, the regional scale of satellite observations greatly favors the use of spherical over linear Cartesian coordinates for efficient processing and analysis of the lithospheric anomalies.

For spherical patch applications, least-squares equivalent point source (EPS) modeling is effective for representing satellite magnetic observations in spherical coordinates. EPS modeling involves relating the satellite anomalies to the magnetic moments of a spherical distribution of equivalent point dipoles (Section 9.7.6) by least-squares matrix inversion (Appendix A.4.2). The spherical magnetic dipole is the simplest numerical source function to invoke, but spherical coordinate prisms and other simple equivalent sources with least-squares estimated magnetizations have also been used. The equivalent source models in turn can be processed for the magnetic potential, anomaly gradients, vector components, continuations and other signal properties because they are all linearly related as explained in Chapter 9.

Lithospheric studies require the extraction of the relatively minor lithospheric components in satellite magnetic observations that are strongly dominated by core and external field effects. However, local petrologic, structural, thermal, and other lithospheric constraints can be effectively invoked for mapping lithospheric components which in turn can enhance local core and external field components, as described below.

11.5.1 Identifying lithospheric magnetic anomalies

The identification of the rather minuscule lithospheric components of the magnetic field at satellite elevations from the core-derived and time-varying ionospheric and magnetospheric components is a significant challenge. The amplitude of lithospheric magnetic anomalies is one thousandth or less of the core-derived field. Numerous methodologies have been described to achieve this goal (e.g. LANGEL and HINZE, 1998). The coherent or static properties of the lithospheric and core components, for example, can be exploited by the procedures in the flow chart of Figure 11.9 to extract lithospheric anomalies with considerable detail from satellite observations. This approach is similar to the efforts described in Section 5.5.2 that used spectral correlation theory (Appendix A.5.1) for extracting marine gravity anomalies from satellite altimetry. In satellite magnetic applications, spectral correlation theory can help to differentiate spatially and temporally static lithospheric and core components from dynamic external field effects. An additional separation of core and lithospheric components is also possible using their correlations with the magnetic effects of the crust's thickness variations modeled from seismic data, or isostatic analyses of the free-air and crustal terrain gravity effects (e.g. VON FRESE *et al.*, 1999a; LEFTWICH *et al.*, 2005) and other constraints.

Because the orbits are near-polar, they may be separated into the ascending and descending sets of subparallel tracks across the spherical patch of interest. Each data track includes overlapping long-wavelength core and lithospheric components that mostly occur in degrees 11–15 as shown in Figure 11.10. Accordingly, the separation of these fields at satellite altitudes is more problematic than at or near the crustal surface where the displacement distances of the sources are significantly smaller (e.g. LANGEL and ESTES, 1982; MEYER *et al.*, 1985; VON FRESE *et al.*, 1999b).

To facilitate the separation of the core and lithospheric components, each track is reduced for two core field models to isolate, for example, the components in the degree 11–15 band. Track residuals obtained by removing the presumed core field through the lower degree (e.g. degree 11) contain crustal magnetic components of degree 12 and greater and the core field components essentially up to degree 15 (Figure 11.10).

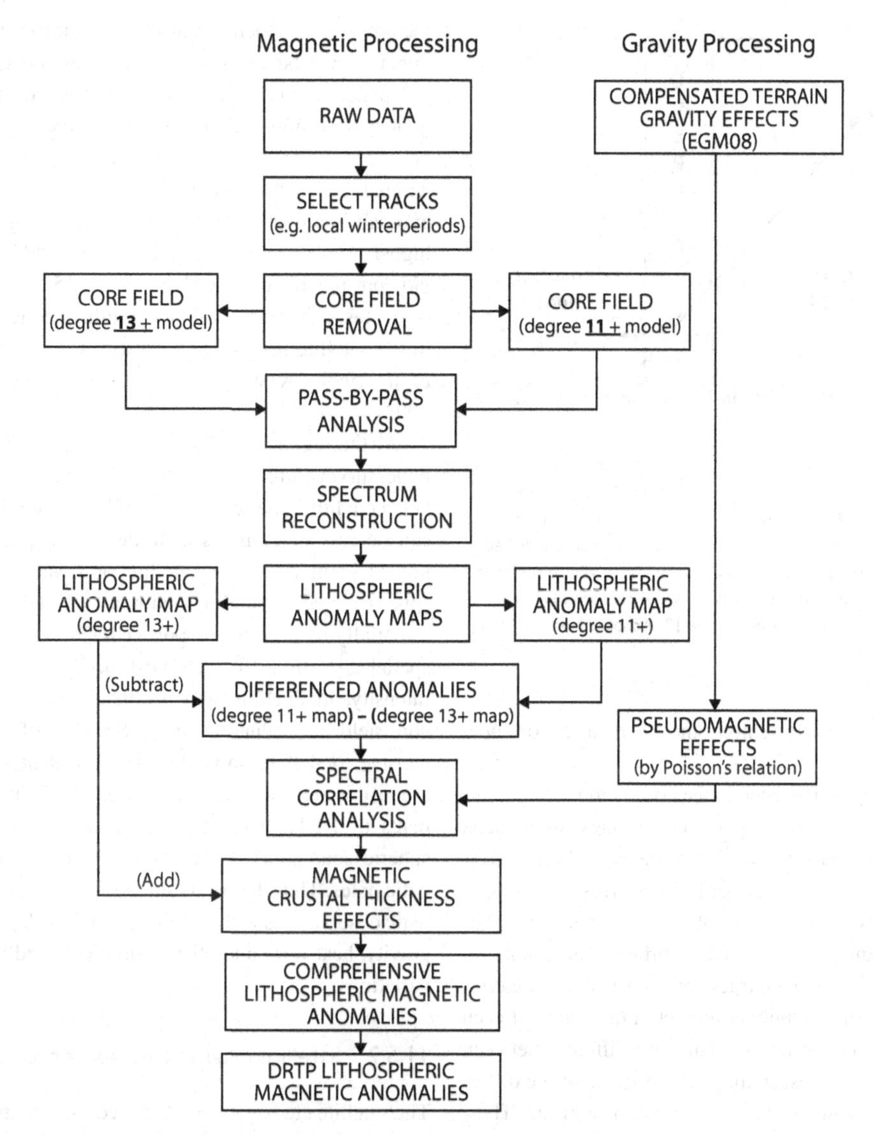

FIGURE 11.9 Data reduction scheme for extracting lithospheric anomalies from satellite magnetometer observations; differentially reduced-to-the-pole (DRTP); see text. Adapted from VON FRESE and KIM (2003).

On the other hand, removing presumed core field estimates through degree 15 yields residuals that contain essentially crustal components with minimal core field effects.

Each set of residuals is processed for separate estimates of the static core and lithospheric field components. This processing involves correlation-filtering the residuals between neighboring tracks for correlative data features that can reflect the static core and lithospheric effects. Common core field and lithospheric anomaly components will be registered on neighboring passes separated by small distances relative to altitude, whereas data components that do not correlate between these passes must involve non-lithospheric and -core effects. The resulting enhancement of the correlation coefficient between the filtered neighboring residual data tracks improves the lithospheric anomaly signal to non-lithospheric noise according to Equation A.72.

The correlation-filtered residual ascending and descending data tracks are then processed into maps at common altitude and spherical coordinates by least-squares collocation (GOYAL *et al.*, 1990), or EPS inversion, or some other procedure that accounts for the large altitude variations in the orbital data. The co-registered ascending and descending maps for each residual data set are also correlation filtered against each other for common

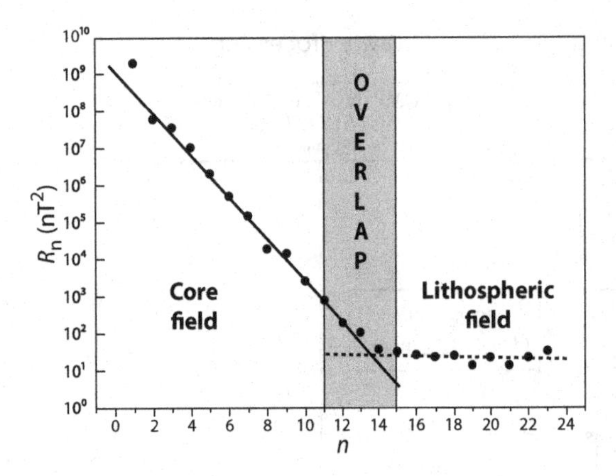

FIGURE 11.10 The Lowes–Mauersberger (R_n) spectra of the geomagnetic field at the Earth's surface, where R_n is the mean square amplitude of the magnetic field over a sphere produced by harmonics of degree n. Magsat magnetic observations established the presence of the prominent break in the slopes where the core and crustal field components overlap in spherical harmonic degrees $n = 11$–15 centered roughly on degree 13. Adapted from LANGEL and ESTES (1982).

features to further reduce non-lithospheric and -core field noise.

The principal difference in the correlation filtered output maps is the corrugation or track-line noise that results from the along-track processing of the orbital data (KIM *et al.*, 1998b). This washboard effect gives each map a corrugated texture that trends along the respective strikes of the ascending and descending orbital data tracks. In the wavenumber domain, track-line noise effects are typically concentrated in only two spectral quadrants of each map, and the corrupted quadrants are different between the ascending and descending maps because of the different strike orientations of the respective data orbits. Thus, the mutually exclusive pairs of cleaner quadrants from the ascending and descending maps can be combined into a reconstructed spectrum that, when inversely transformed, yields a magnetic map with minimal track-line noise of the static components presumed for the core and lithosphere.

To separate the lithospheric and core components, the differences obtained by subtracting the static lower-degree (e.g. degree 11) core field residuals from the higher-degree (e.g. degree 15) core field residuals are considered as shown in the bottom half of the flow chart in Figure 11.9. In terms of the overlap generalized in Figure 11.10, for example, the differences include the mixed core and crustal components in the core field model through degree 15 plus the static degree 11–15 components from the track data that presumably reflect additional core and lithospheric contributions. As indicated in the flow chart (Figure 11.9), the crustal components in the residual degree 11–15 components may be extracted by correlation filtering using independent crustal field models derived from collateral geophysical information. The extracted components may be added to the static degree 15 and higher lithospheric anomaly estimates to obtain a comprehensive lithospheric anomaly map for the spherical patch in degrees 11 and higher. The remaining non-lithospheric differences provide presumably enhanced degree 11–15 core field model estimates consistent with local lithospheric constraints in the satellite magnetic observations (e.g., VON FRESE *et al.*, 1999b; KIM *et al.*, 2002; VON FRESE and KIM, 2003).

At the regional scales and altitudes of satellite magnetic surveys, displacements in the spatial relationships between lithospheric sources and total field anomalies can be severe. Thus, as a further processing step (Figure 11.9) in preparing satellite observations for lithospheric analysis, satellite total field anomalies can be differentially reduced-to-the-pole (DRTP) to minimize these spatial distortions. The DRTP procedure adjusts for the intensity, inclination, and declination variations of the core field by evaluating the EPS model of the comprehensive total field anomalies at vertical inclination and a constant magnetization intensity. The induced components of DRTP anomalies are centered on their lithospheric sources as if the sources were at the geomagnetic poles. Thus, DRTP anomalies facilitate linking magnetic observations with the lithospheric geology, as well as gravity, heat flow, and other source-centered geophysical variations.

11.5.2 Satellite magnetic mapping progress

The satellite era began with the successful launch of Sputnik 3 in 1958 which carried flux-gate magnetometers that mapped portions of the geomagnetic field to an accuracy of about 100 nT over altitudes of roughly 225–1,900 km. Since Sputnik, some 23 additional low Earth-orbiting (LEO) satellites have been operated mostly to map and monitor the behavior of the Earth's core field (e.g. LANGEL and HINZE, 1998).

A series of satellites operated over 1967–1971, called Polar Orbiting Geophysical Observatories (POGO), obtained total field data with measurement accuracies of about 6 nT that identified the intense Bangui anomaly from crustal rocks of west central Africa (REGAN *et al.*, 1975). This finding prompted efforts to develop a satellite mission dedicated to mapping lithospheric magnetic fields. Accordingly, the 7 month Magsat mission was launched in November 1979 to extend POGO's lithospheric

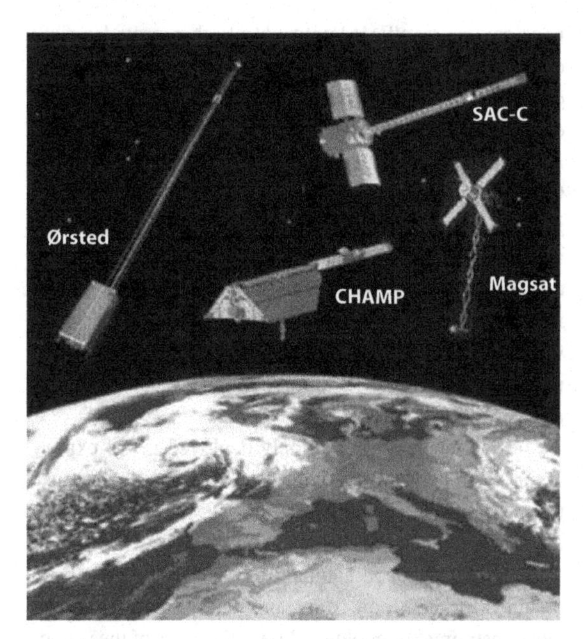

FIGURE 11.11 Low Earth-orbiting (LEO) geomagnetic field mapping satellites. Booms extend the magnetometers away from the electromagnetic environments of the satellites. A color version of the figure is available at www.cambridge.org/gravmag.

findings by mapping not only the field's magnitude but also its direction, and thus the magnetic vector components. The measurement accuracies in the total field and vector components were 3 and 6 nT, respectively. In addition, Magsat operated at more circular and lower altitude (~350–450 km) orbits than the POGO satellites to better resolve magnetic anomalies of the lithosphere. However, the utility of Magsat data for lithospheric studies of the southern hemisphere was limited because the mission flew during austral summer and fall when the corrupting effects of the south polar external fields on the measurements were maximum.

The Magsat data have been essentially replaced for lithospheric studies by the more accurately measured and temporally extensive data sets mapped by the Ørsted (1999–2009) and CHAMP (2002–2010) missions at respective altitude ranges of roughly 600–750 km and 300–450 km (Figure 11.11). The latter two missions carried resonance and flux-gate magnetometers that provided both total and vector component data to measurement accuracies of about 0.5 nT. Deployed on optical benches or booms, the magnetometers were monitored for their in-flight temporal and spatial orientations using onboard GPS receivers and star cameras. The LEO observations from the Magsat, CHAMP, and Ørsted satellites have yielded significant constraints on the regional petrologic variations of the crust and upper mantle (i.e. the lithosphere), and crustal thickness and thermal perturbations. These observations know no political boundaries, have uniform accuracy and spatial distribution, and were acquired in time spans short enough that the time variation of the Earth's main field is not a limiting factor to interpretation.

The above missions were all launched into near-polar orbit at inclinations no greater than about 87.5°, and so the orbits track tangentially to the outer margins of polar holes or gaps in coverage with diameters no smaller than about 5° across the poles. Thus, the track coverage density increases from the equator, where the tracks are essentially parallel and trend N–S, to the polar gaps where the number of track cross-overs is maximum. However, in terms of the internal field components from the core and lithosphere, the minimum resolvable full wavelength is roughly at the orbital altitude scale. Thus, the CHAMP mission, for example, resolves lithospheric anomalies at wavelengths no smaller than about 300–400 km.

Efforts to remove diurnal effects are basically relegated to selecting orbital data tracks measured over magnetically "quiet" periods as indicated by the geomagnetic observatory-measured activity indices. This approach, however, is suspect because the ground-based indices often show little or no correlation to satellite-altitude disturbances of the data variances in the orbital measurements. Each orbit takes about 96 minutes to complete, so that the spatially and temporally variable diurnal fields tend to be dispersed over the orbital measurements as mostly random effects relative to the coherent effects of the core and crust. Thus, correlation filtering of neighboring orbital data tracks and anomaly maps at different local magnetic times can help to suppress the predominantly uncorrelated effects of the dynamic external fields in the satellite measurements as described previously in Section 11.5.1.

Given the vast amounts of magnetic observations collected by the multi-year Ørsted and CHAMP missions, anomaly maps are commonly made by isolating the orbital data obtained during magnetically quiet periods, and adjusting the extracted data using first-order models of the ring current, magnetopause current, tail current, field-aligned currents, and other external field effects (Figure 8.3) that can be calibrated by the mission data (e.g. LANGEL and HINZE, 1998). The reduced anomaly data are next gridded in spherical coordinates using the 3D binning process described previously in Section 5.5.3 for gridding satellite gravity observations. The magnetic anomaly grid is then transformed for analysis and interpretation into a set of spherical harmonic coefficients.

The spherical harmonic series for the magnetic potential function, for example, may be written as

$$V(r,\theta,\phi) = a_e \sum_{n=1}^{\infty}\sum_{m=0}^{n}\left(\frac{a_e}{r}\right)^{n+1} P_{n,m}[\cos(\theta)][g_{n,m}\cos(m\phi) + h_{n,m}\sin(m\phi)] + a_e \sum_{n=1}^{\infty}\sum_{m=0}^{n}\left(\frac{r}{a_e}\right)^{n} P_{n,m}[\cos(\theta)] \times [q_{n,m}\cos(m\phi) + s_{n,m}\sin(m\phi)], \quad (11.1)$$

where the variables are the same as for the spherical harmonic gravity potential in Equation 5.15 except that the $P_{n,m}[\cos(\theta)]$ are the Schmidt quasi-normalized forms of associated Legendre functions of degree n and order m, and the $g_{n,m}$, $h_{n,m}$, $q_{n,m}$ and $s_{n,m}$ are the Gauss coefficients of **B** relative to $P_{n,m}[\cos(\theta)]$ (e.g. LANGEL and HINZE, 1998). The $g_{n,m}$ and $h_{n,m}$ coefficients describe fields originating within the Earth, whereas the $q_{n,m}$ and $s_{n,m}$ coefficients describe fields originating outside the Earth. The internal **B** contains contributions from the Earth's core, lithosphere, and induced currents of the subsurface.

As discussed previously for spherical harmonic representations of satellite gravity data in Section 5.5.3, the effective degree and order of the spherical harmonic expansion is fundamentally constrained by the altitude of the observations, assuming the observations are spaced laterally at intervals smaller than the altitude. Thus, the internal field effects observed over an altitude of 400 km, for example, may be reliably modeled in principle to spherical harmonic degree and order of roughly 100, although in practice the higher-degree terms tend to be increasingly corrupted by measurement errors, external field components, and other noise in the observations. Based on the coefficient powers, the components through about degree and order 13 are commonly ascribed to the core field with the higher-degree and -order terms assigned to the lithospheric field. However, as described previously, this simple interpretation of the lithospheric anomalies is greatly complicated by their spectral overlap with the core field components and contamination by external field effects, because both non-lithospheric sources produce magnetic effects that are typically orders of magnitude larger than the lithospheric anomalies.

Figure 11.12 gives a global total field Earth Magnetic Anomaly Grid (EMAG2) that incorporates satellite, airborne, and ship magnetic survey data, as well as extrapolated marine anomalies based on oceanic crustal age modeling (MAUS *et al.*, 2009). The grid is the basis for the 720 degree and order spherical harmonic NGDC720 geomagnetic model (http://www.ngdc.noaa.gov/geomag/EMM/emm.shtml) that provides field components at a resolution of about 15 arcmin. However, care must be taken to check the reliability of model predictions which can be highly problematic at points located a hundred or more kilometers from measured anomalies. Over the large unsurveyed blank areas of the Antarctic grid in Figure 11.12, for example, the model predictions are most reliable at satellite altitude, but of little or no consequence at the 4 km altitude of the grid because of magnetic anomaly measurement and reduction errors, as well as the fundamental non-uniqueness of the underlying magnetic model and its predictions in unsurveyed areas (e.g. VON FRESE *et al.*, 2005).

Magnetometers are routinely deployed in the exploration of the planetary bodies of the solar system because they acquire with relative ease measurements of considerable geological and planetary significance. As a result, a variety of magnetic anomaly measurements have been obtained for the Moon, Mars, and Venus that offer important constraints for geological investigations of these bodies and their satellite gravity, topography, and other remote sensing observations (Section 7.4).

Magnetometers onboard the lunar Apollo subsatellites, for example, showed that the Moon's external magnetic field is very weak compared with the Earth's field. The Moon lacks a core field, although one may have operated in its early history to produce the remanent magnetization observed in the returned lunar rocks, and the crustal anomalies mapped by magnetometers on the lunar surface and by the electron reflectometers and magnetometers onboard the Apollo subsatellites (e.g. FULLER and CISOWSKI, 1987). The Apollo data mapped numerous crustal magnetic anomalies with wavelengths of several to hundreds of kilometers and amplitudes of less than a nanotesla to greater than 100 nT. These results were confirmed and further detailed and extended in coverage by the magnetometer data from the Lunar Prospector (e.g. HOOD *et al.*, 2001) and SELENE (e.g. TSUNAKAWA *et al.*, 2010) missions. For the most part, lunar crustal anomalies appear to reflect mostly the crustal demagnetizing and magnetizing effects of large meteorite impacts at the impact basins and their antipodes, respectively.

The fact that Mars also lacks a core field was first inferred as a result of the Mariner-4 spacecraft flyby in 1965. However, a core field may have operated in its early history to produce the remanently magnetized meteorites found on Earth that are presumed to have come from Mars, and the crustal magnetic anomalies mapped by the magnetometer and electron reflectometer onboard the Mars Global Surveyor (MGS). Remarkably strong crustal anomalies occur over the ancient cratered southern

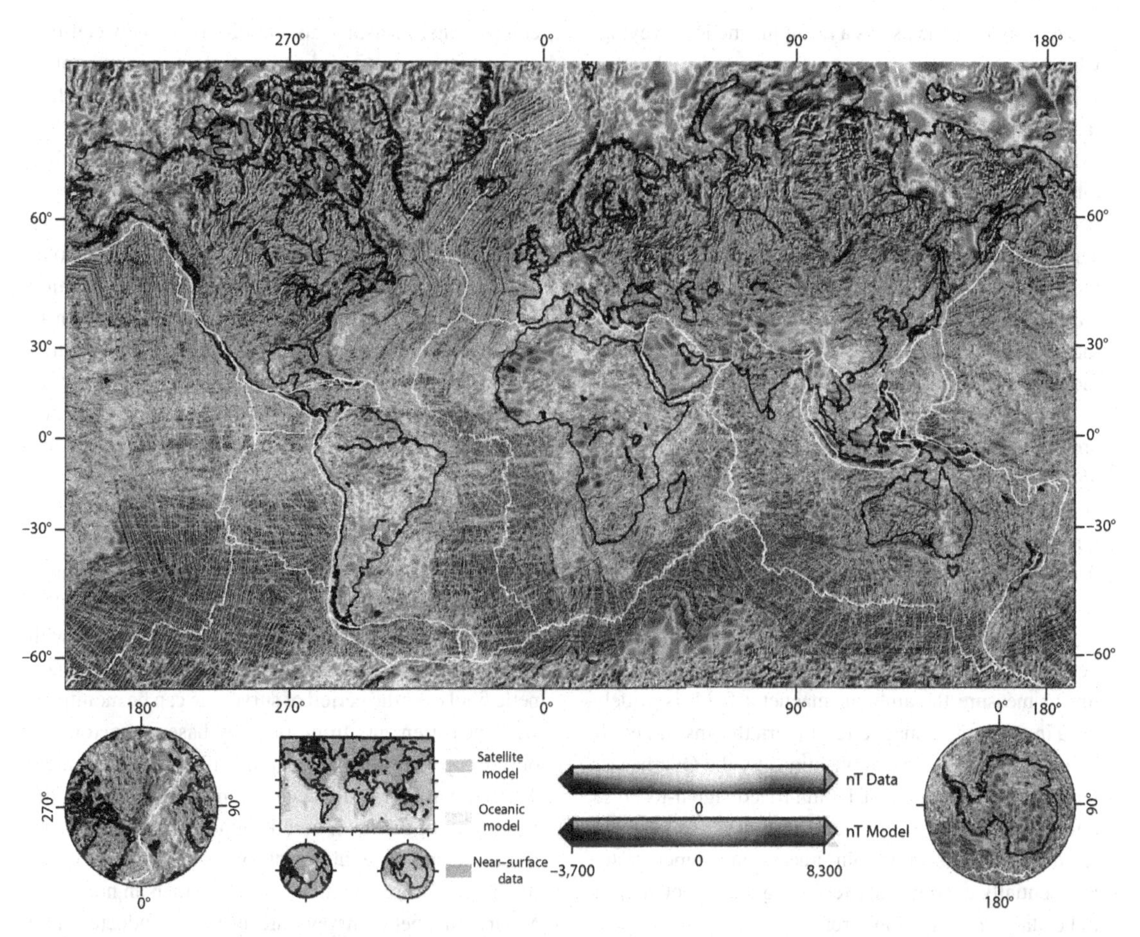

FIGURE 11.12 Mercator and polar stereographic projections (> 40° latitude) of the EMAG2 global grid. The grid has a resolution of 2 arcmin and is referenced to an elevation of 4 km above the geoid. A color version of this figure is available at www.cambridge.org/gravmag. Adapted from MAUS *et al.* (2009).

hemisphere crust with amplitudes up to 1,500 nT at elevations of about 100–200 km, whereas much weaker anomalies with amplitudes < 50 nT characterize younger impact-modified crust such as covers the northern hemisphere (e.g. ACUÑA, 2001). Crustal modeling of the MGS magnetic anomalies is subject to ongoing investigations, but would require intensely magnetized rocks with average natural remanence of about 20 A/m over large 30 km thick crustal slabs with minimum horizontal dimensions no smaller than about 100 km (e.g. ACUÑA *et al.*, 1999).

The most definitive measurements to date of the Venusian magnetic field were obtained by the Pioneer Venus Orbiter mission during its first years of operation (1979–1981). The spacecraft made repeated passes at altitudes of about 150 km which showed that Venus also lacks a significant core field. In addition, crustal surface temperatures of around 450 °C have been measured that suggest temperatures within the crust are probably above the Curie point. Given the apparent absence of a core field and inferred high crustal temperatures, it seems unlikely that the crust can be a source of magnetic fields. Thus, the relatively weak variations (10–100 nT) in the Venusian magnetic observations have been ascribed mostly to interactions of the solar wind with the ionosphere of Venus (e.g. LUHMANN, 1986).

11.6 Key concepts

- Magnetic instrumentation and procedures have greatly improved since the late 1940s when most exploration observations were made with tripod-mounted mechanical instruments which measured a vector component of the magnetic field. These improvements have made it possible to routinely make measurements to an accuracy of 0.1 nT from airborne, marine, and satellite platforms. In addition, measurements are not only made of the total field, but now regularly include vector, gradient, and tensor components. As a result magnetic surveying, which is inexpensive and rapid compared with other geophysical measurements, is extensively used over a range of scales of applications. These investigations extend from studies of near-surface anthropogenic features to regional geological and tectonic features of the continents, oceans, and extraterrestrial bodies.

- The majority of magnetic measurements are made of the total field with nuclear magnetometers which do not have to be accurately oriented, and thus can be used on moving platforms. The earliest nuclear magnetometer, the proton-precession magnetometer, measures the frequency of the precession of the protons in a fluid sensor around the geomagnetic field after being displaced by an applied magnetic field. The frequency of this precession is related to the total (scalar) magnetic field through a well-known constant. However, limitations of the proton-precession magnetometer have caused it to be largely replaced by the more sensitive alkali-vapor magnetometer. The cesium-vapor version of this magnetometer, which uses the principle of optical pumping to measure the ambient magnetic field, is widely used in exploration magnetics. In turn this instrument is being replaced in some applications by the Overhauser magnetometer because of its improved signal-to-noise ratio and reduced measurement uncertainties and power requirements. Also, this spin-precession magnetometer is essentially continuously recording and is not subject to heading or orientation errors.
- The first magnetic instrument usable in aircraft for geological exploration was the flux-gate magnetometer. Although it is a vector magnetometer, it was used in a configuration on moving platforms that constantly oriented it in the direction of the Earth's field. Thus, it measured the total field but in a relative rather than an absolute sense as nuclear magnetometers do. It has been superseded in most exploration applications by nuclear magnetometers except for measuring vector components of the field and their gradients. A significantly more sensitive vector magnetometer is the SQUID which operates in a cryogenic environment. It is receiving increasing attention as an exploration instrument because of its high sensitivity which makes it particularly useful in tensor measurements.
- The vast majority of exploration magnetic measurements are the total field component. However, vector measurements are still of interest because they have advantages over total field values in isolating Earth-derived fields in special cases and they can be used to improve the quality and stability of magnetic interpretation. Also, gradients of both the total field and its vector components, measured as the difference between two magnetometer sensors, have important uses in exploration magnetics. They have the advantage of being free of extraneous effects of temporal variations and they increase the resolution of the measurements, and thus their ability to isolate anomaly sources.
- Ground magnetic surveys are used to map near-surface natural or cultural features with the highest resolution and an appropriately high data density. Station separation should not exceed the anticipated depth to the anomaly sources and closer intervals are required for satisfactory quantitative analysis. The proximity of extraneous ferrous metals and electric currents which generate magnetic fields in ground surveys necessitates care in the siting of observations to minimize effects from these sources. Near-surface surveys of extensive regions are expedited with measurements using various forms of transport including helicopters, unmanned remotely controlled aircraft, and surface vehicles. The positioning of observations is typically established with GPS mapping systems. Temporal variations in the magnetic field over the period of surveying can be monitored with measurements from a nearby base magnetometer or reoccupation of a base station within the survey area. The interval between reoccupations is a function of the variability of the field, both its gradient and amplitude, and the accuracy requirements of the survey. Reoccupation intervals are commonly no more than an hour.
- Marine magnetic surveys are usually conducted as a supplement to other geophysical/oceanographic surveys with a sensor towed beneath the sea surface at a distance which excludes the effects of the towing vessel on the sensor. Difficulties arise with temporal magnetic variations because base magnetometers and tie lines are often limited or of restricted quality. As a result of this problem, gradiometer configurations are sometimes employed at sea. In areas of deep water, it may be necessary to track the sensor close to the water bottom to increase resolution.
- Airborne magnetic survey is the principal source of magnetic anomaly data because it is economical, rapid, and efficient in studying extensive regions, and minimizes the effects of temporal variations and cultural features. Surveys are conducted with a wide variety of operational and survey procedures taking into account the objectives of the survey and the terrain. Data are acquired at high density along parallel flight lines generally separated by a distance roughly equivalent to the depth to magnetic sources, although this will vary with the objective of the survey. Flights for geological mapping of sources close to the surface are normally flown at roughly 150 m

above mean terrain to achieve high resolution without compromising the safety of the aircraft. Other flight modes include surveys draped over mean terrain and constant flight altitudes. The latter are used especially in petroleum surveys where the objective of the magnetic survey is mapping of buried basement. Temporal variations during flight operations are determined in airborne surveys with base magnetometers and tie-lines flown at distances of a few to 10 times the survey flight lines. Most surveys are conducted with alkali-vapor magnetometers with inboard installations that are compensated for a variety of extraneous magnetic fields from the aircraft. An accuracy of 0.1 nT is attained in many surveys making them useful in high-resolution studies. Total field measurement may be supplemented with gradient, vector, and tensor observations in airborne surveys to improve the quality of the interpretation.

- Satellite magnetic missions have been flown for several decades to study the nature and configuration of the main geomagnetic field, but increasingly these measurements have provided long-wavelength magnetic anomaly (> 500 km) information useful in the study of continental-scale geologic sources of the lithosphere. Recent satellite missions at altitudes of roughly 500 km have achieved an accuracy of 0.5 nT in the measurement of both scalar and vector anomalies. Satellite measurements are especially useful because they provide a synoptic view of the magnetic field, and the repetitive measurements of overlapping orbits can be used to minimize extraneous magnetic fields of the ionosphere from lithospheric anomalies. These measurements are being used to provide important information about the nature and history of other terrestrial bodies.

Supplementary material and Study Questions are available on the website www.cambridge.org/gravmag.

12 Magnetic data processing

12.1 Overview

As in the gravity method, magnetic measurements include effects from a wide variety of sources – from terrain, natural and man-made surface features, as well as instrumental, geological, and planetary sources. These sources have a range of amplitudes and periods in the case of temporal variations and wavelengths in spatial contributors which can mask or at least distort magnetic effects from subsurface sources of interest in a survey. Accordingly, observed data are processed to eliminate or at least minimize these effects. The product of this processing is the magnetic anomaly. Fortunately, in most magnetic survey campaigns these extraneous variations are considerably less effective in distorting effects of subsurface sources than those of the gravity method. Thus, the requirements for auxiliary data and reduction procedures are much less stringent for the magnetic method than in the gravity method. The corrections for these effects are largely obtained empirically from measurements of changes in temporal and spatial characteristics of magnetic fields.

Spatial extraneous variations directly perturb the distribution of subsurface effects over the Earth's surface, while extraneous temporal variations cause errors in the measurements. Processing in magnetic exploration removes the extraneous variations directly from the observations, rather than modeling the field at the observation site as in gravity data processing. However, the result is the same. It is equivalent to the anomaly obtained by subtracting the theoretical field from the observations. Magnetic anomalies for interpretational purposes are commonly transformed into the equivalent anomaly that would be observed at the geomagnetic pole. This reduction-to-the pole shifts the anomaly to a position directly over the source and increases the anomaly's symmetry, simplifying the magnetic interpretation.

The objective of the calculation of the magnetic anomaly is to isolate the magnetic expression of geological sources. However, because of the cumulative nature of the measurements there is overlap from the numerous and variable sources within the Earth. An important part of processing is thus the identification and isolation of the anomalies derived from the sources of particular interest in the survey. This is less of a problem in magnetic exploration than in gravity because of the nature of magnetic anomalies from deep and broad subsurface sources. The spectral overlap of anomalies is considerably less from magnetic than from gravity sources. Specifically, magnetic anomalies are more sensitive to the geometric factor manifested in the inverse distance function than are gravity anomalies. Thus, there is much greater spectral separation in magnetic anomalies as a function of their depth. Furthermore, the amplitude of anomalies of long, wide sources in gravity is independent of depth, which is not the case for similar sources in magnetics. The amplitude of the magnetic anomalies decreases with depth and for long, wide sources it approaches zero. As a result the deep, broad geological features of the Earth which produce distorting regional anomalies are much less of a problem in magnetics. Isolation and enhancement techniques primarily in the frequency domain are used to identify and separate out the magnetic anomalies of interest for further interpretation.

12.2 Introduction

Raw or field magnetic observations, as in the case of gravity observations, are processed to put the data into a form that represents the subsurface features of primary interest in the survey. This is a goal that is never completely attained because the measured magnetic field is the superposition of numerous extraneous sources whose spectra overlap with the wavelength composition of the magnetic fields of the subsurface sources of interest. However, with suitable care in data acquisition and processing, the resulting processed data can be made useful for interpretation of subsurface features and processes. In surface or near-surface surveys such as most aeromagnetic surveys this is generally a simpler problem than is the comparable challenge in gravity exploration, because the ratio of subsurface fields to planetary fields is commonly roughly a thousand fold greater for magnetic sources than for gravity sources. Furthermore, extraneous magnetic fields are usually less than the anomalous field in magnetic exploration in contrast to the situation in gravity exploration. However, this is not the case for magnetic measurements at satellite elevations because of the small amplitudes of lithospheric anomalies in comparison to the Earth's field due to the geometric effect that dominates potential field responses and the potential adulterating effect of large-amplitude, extraneous temporal variations. Historically, processing of observed data is much more recent in magnetic exploration than in gravity exploration because the primary use of magnetics in exploration a century ago was in mineral exploration where the magnetic fields of the sources are commonly many times greater than non-geological sources, and thus required limited processing to be useful.

Processing of magnetic data involves removal of extraneous spatial and temporal variations in the magnetic field measurements. These components of the geomagnetic field are summarized in Section 8.3 in the context of the origin and characteristics of the Earth's field. Spatial extraneous variations directly perturb the distribution of subsurface effects over the Earth's surface, while temporal extraneous variations cause errors in the measurements because observations are made during a finite period of time by one of a variety of magnetometer systems. General processing practices in magnetic exploration involve the removal of the extraneous variations directly from the observations rather than the calculation of a model of the field at the observation site as in gravity data processing. However, the result is the same. It is equivalent to the anomaly obtained by subtracting the theoretical field from the observations.

In gravity exploration several types of anomalies have been established considering their potential use and the variations in the components considered in the modeled theoretical field. This is not the case in magnetic exploration. The term magnetic anomaly is used when corrections include two primary components, the main magnetic field and its spatial and temporal variations as well as temporal perturbations caused primarily by corpuscular and electromagnetic radiation from the Sun interacting with the ionosphere and magnetosphere of the Earth's main field. Several extraneous magnetic components, such as terrain and elevation effects which are prominent in gravity data processing, are not involved in the processing of the vast majority of magnetic surveys. However, consideration is given to changes in the core-derived main magnetic field of the Earth, the planetary field, because this field varies continually at a rate which imposes variable spatial values over the Earth which may distort magnetic signals of significance to geological interpretation. This is quite in contrast to the Earth's planetary gravity field which for gravity purposes is assumed to be invariant for timescales associated with observations during a survey.

The objective of the calculation of the magnetic anomaly is to isolate the magnetic expression of geological sources. However, because of the cumulative nature of the measurements there is overlap from many different sources, so an important part of the processing of magnetic data is the identification and isolation of the anomalies derived from the sources of interest. This is less of a problem in magnetic exploration than in gravity, as the spectral overlap of anomalies is considerably less. Specifically, magnetic anomalies are more sensitive to the geometric factor manifested in the inverse distance function than are gravity anomalies. Thus, there is greater spectral separation in magnetic anomalies as a function of their depth than in the gravity case. Furthermore, the amplitude of anomalies of long, wide sources in gravity is essentially independent of depth, whereas for similar sources in magnetics it decreases with depth to roughly zero. Thus the deep, broad geological features of the Earth which produce distorting regional gravity anomalies are a problem in magnetics. Nonetheless, because of spectral and spatial overlap in magnetic fields, magnetic anomalies are processed by isolation and enhancement techniques to identify and separate out the magnetic anomalies of interest to the survey. This is achieved by a variety of filtering processes in the spatial domain or, more frequently, in the wavenumber domain. The background and analytical formulation of these spectral filtering methods are explained in Appendix A.

In this chapter the sources of spatial and temporal magnetic variations that may be considered in the calculation of anomalies useful in exploring the Earth are described, and strategies for their removal are discussed and evaluated. Additionally, the methods and relative merits of the filtering procedures useful in isolating and enhancing magnetic anomalies are explained.

12.3 Extraneous magnetic variations

Magnetic measurements include effects from terrain, natural and man-made surface, geological, instrumental, and planetary sources. These sources have a range of amplitudes and periods in the case of temporal variations and wavelengths in spatial contributors which can mask or distort magnetic effects from subsurface sources of interest. Raw or observed data are therefore processed to eliminate or at least minimize these effects. The result of this processing stage is the magnetic anomaly. Fortunately, in most magnetic survey campaigns these extraneous variations distort the effects of subsurface sources considerably less than in the gravity method. Thus, the requirements for auxiliary data and removal procedures are much less stringent for the magnetic method. Corrections are largely obtained empirically from measurements of changes in magnetic fields, either temporal or spatial. The process of calculating the magnetic anomaly is commonly referred to as correcting or reducing of the magnetic data. As before, correction does not suggest that the observed data are in error and reduction does not imply the data of multiple observations are reduced or continued to a common vertical datum. Rather both terms refer to the conversion of raw magnetic observations to a useable form – an anomaly – which is the difference between the observed magnetic field and the predicted value of the field at the observation site.

The significance of unwanted magnetic field variations depends on their attributes, the nature of the survey, and the characteristics of the anomalies of interest which in turn depend on the objective of the survey. As a result the significance of the individual extraneous variations differs considerably from surface, aeromagnetic, and satellite surveys. For example, magnetic terrain corrections are not required in satellite surveys, are seldom used in surface surveys covering limited areas, and are usually neglected in aeromagnetic surveys except where the topography is rugged and crystalline rocks make up the terrain. As magnetic methods are more directed at the extremes of the wavelength of magnetic anomalies, either long wavelengths associated with broad-scale crustal sources or short wavelengths derived from near-surface anthropogenic sources, extra care is necessary to remove the extraneous variations because of their overlap with the anomalies of interest. For example, separation of the main field from crustal-scale anomalies is particularly important in regional geological studies, and in mapping near-surface low-amplitude sources care is needed to monitor and remove micropulsations in the ambient magnetic field.

The following description of extraneous magnetic variations and their elimination from magnetic surveys is not comprehensive, but is sufficiently complete for most survey objectives. Extensive references are provided to direct the reader to more complete descriptions of variations and their removal from observed data.

The precision and accuracy requirements of the survey are important factors in selecting the corrections and the methods used in their determination. In the following sections, the predictable or measurable sources of variations of the magnetic field are explained, as are the methods of determining corrections for them to prepare anomaly data sets for interpretation. Complementary, detailed, modern descriptions of the processing of magnetic data are presented, for example, by Reeves (2005).

12.3.1 Spatial magnetic variations

Depending on survey objectives, a variety of spatial magnetic variations may be considered in deriving the anomaly of interest. These can include the regional variations of the main magnetic field, differences in the observation elevations, magnetic terrain effects, and the surface and near-surface effects of natural and cultural sources that distort anomaly mapping.

Global magnetic variations

As explained in Section 8.3, the main magnetic field of the Earth is derived from electrical currents developing a dynamo action in the conducting, liquid outer core of the Earth. This core-derived field, which accounts for roughly 98% of the geomagnetic field, closely duplicates the field from a small, Earth-centered, intensely magnetized dipole tilted roughly 11° from the Earth's rotation axis. Thus, the field is not axially symmetric like the gravity field, but varies with both latitude and longitude over the Earth's surface (see Figure 8.4). The change in the magnetic field from the geomagnetic equator to the poles is roughly 100%, varying in a non-linear manner from approximately 30,000 nT at the equator to 60,000 nT at the poles (Figure 8.6). The strength of the field decays as a function of the inverse cube of the distance from the center of the dipole or roughly the center of the Earth. At distances of 8 to 10 Earth radii the field becomes sufficiently weak that solar plasmas, the interplanetary magnetic field, can significantly distort its character. The intensity of the main field also varies with time, changing in both intensity and morphology. For example, the current geomagnetic field has decreased 10% over the past 150 years (OLSON and AMIT, 2006), with the rate of decrease apparently increasing. Additionally, the main geomagnetic field drifts westerly at a rate that averages about 0.18°/year. Superimposed on the main field are secular changes in the field with global root-mean-square values of roughly 80 nT/year. These changes also drift westward at approximately twice the rate of the main field drift and have a lifetime measured in several hundred to several thousand years. These secular components are believed to be associated with chaotic motions in the Earth's electrically conducting outer core and produce both latitudinal and longitudinal changes in the magnetic field measured in several to tens of nanotesla per kilometer over much of the Earth.

In addition, the magnetic field reverses polarity over a period measured in thousands of years or more at irregular intervals, the latest being about 780,000 years ago. These reversals are significant to geomagnetic exploration because reversals are recorded in the remanent magnetization of rocks, especially volcanic rocks which have cooled through the Curie point of the constituent magnetic minerals during these periods, and reverse their anomaly amplitudes. On a much shorter timescale, that is periods of the order of a few months or years, the magnetic field can undergo abrupt changes or jerks in the secular variation. These jerks are of sufficient amplitude (~ 10 nT) to be of interest in the processing of data in magnetic exploration.

As a result of the displacement of the main field from the rotation axis of the Earth, the field's westward drift, and the secular variation of the magnetic field, the geomagnetic field is neither axially symmetric nor static, and this significantly complicates the specification of the spatial variations in the field. Classically the spatial changes in the magnetic field were determined for exploration magnetic studies from regional magnetic charts of the main field and predictions of the secular variation which were prepared every 5 years largely from measurements of the magnetic field at magnetic observatories plus supplementary data from repeat stations. Data from these charts were used to remove the Earth's field for calculation of the magnetic anomaly. However, the need for a digital model of the normal field rather than the analog charts, as well as global satellite magnetic coverage, led to an international effort to define an acceptable worldwide model. This effort was performed by the International Association of Geomagnetism and Aeronomy (IAGA) (LANGEL, 1987) using a spherical harmonic representation of the global field in four dimensions – latitude, longitude, geocentric distance (altitude), and time.

The parameters of the global model first released in 1969, the International Geomagnetic Reference Field (IGRF-65), were based on weighted averages of input from a group of international geomagnetists. Subsequently, the IAGA has derived an updated IGRF model for each 5 year epoch which has attained wide acceptance for magnetic exploration processing. These models use refined data of the main field from ongoing satellite measurements and regional coverage from extensive aeromagnetic surveys plus secular variation data from roughly 150+ geomagnetic observatories. The model of each epoch makes a prediction of the changes in the geomagnetic field for the next 5 years from current trends in the secular variation. These predictions are subject to significant error. Nonetheless, the IGRF is a useful interim model for removing the geomagnetic field from magnetic observations made for exploration purposes. However, to improve the accuracy of the interim anomaly values, the IAGA also prepares every 5 years a Definitive Geomagnetic Reference Field (DGRF) which incorporates the actual temporal (secular) change in the geomagnetic field over the previous 5 years. Removal of the DGRF from the observed magnetic field leads to the best estimate of the magnetic anomaly. Thus, survey data observed over a range of years can be recalculated using the appropriate DGRF. These data can then be more accurately combined into a single data set which will not reflect the boundaries of surveys made at different times because of the use of an incorrect model for the global field in calculating the anomalies. Prior to this advance, inconsistencies in the predicted geomagnetic field model removed from observed magnetic data could cause significant errors in anomaly data sets, particularly in the longer-wavelength components of the anomalies (REGAN and CAIN, 1975).

Spherical harmonics which are used in the IGRF and DGRF models are a convenient and widely accepted manner of representing the variation of a quantity over a spherical body such as the Earth. The advent of magnetic observations from satellites and the computational ease of computers have made it possible to implement this method of representing the geomagnetic field. The IGRF and DGRF models are based on the internal part of the comprehensive geomagnetic scalar potential V of Equation 11.1 given by

$$V_{\text{int}}(r, \theta, \phi) = a_e \sum_{n=1}^{\infty} \sum_{m=0}^{n} \left(\frac{a_e}{r}\right)^{n+1} P_{n,m}[\cos(\theta)] \times [g_{n,m} \cos(m\phi) + h_{n,m} \sin(m\phi)], \quad (12.1)$$

where the variables are the same as in Equation 11.1, and the respective northward, eastward, and vertical components of the normal field (Figure 8.4) are

$$B_N(X) = \frac{1}{r}\frac{\partial V_{int}}{\partial \theta}, \quad B_N(Y) = \frac{-1}{r\sin\theta}\frac{\partial V_{int}}{\partial \phi}, \quad \text{and}$$

$$B_N(Z) = \frac{\partial V_{int}}{\partial r}. \tag{12.2}$$

The IGRF models also incorporate the effect of the oblate shape of the Earth. The Gauss coefficients $g_{n,m}$ and $h_{n,m}$ are determined by fitting the harmonics in a least-squares sense to the global magnetic data. The first-order change of the field with time is introduced into the model by expanding the Gauss coefficients in a truncated finite Taylor series about some mean time of the data set (e.g. LANGEL and HINZE, 1998). Great care is exercised in selecting the data which are used in specifying the Gauss coefficients to obtain uniform coverage over the Earth and to employ only magnetically quiet time data exclusive of magnetic storm events (e.g. SABAKA *et al.*, 2004; MANDEA and PURUCKER, 2005; THOMSON and LESUR, 2007).

The spherical harmonic expansion of the geomagnetic field has been calculated up to degree 100 for preparation of a candidate model for the World Digital Magnetic Anomaly Map (WDMAM) to illustrate the lithospheric anomaly field (HAMOUDI *et al.*, 2007) and up to degree 720 in the US National Geophysical Data Center's model NGDC720 (http://www.ngdc.noaa.gov/geomag/EMM/), which excludes all components below degree 16, and thus provides a waveband of approximately 2,500–56 km. However, only the lower-degree spherical harmonics map the core-derived field, but the actual upper limit of the core-derived components is not a sharp cutoff because of the overlap with components derived from the lithosphere (Figure 11.10).

Since the year 2000, the IGRF has been truncated at degree 13 and the secular variation at degree 8. This is the normal field adjusted to the time and altitude of the survey that is subtracted from the observed magnetic field to obtain the magnetic anomaly. A computer program for calculating magnetic field components from IGRF coefficients and further information on the IGRF is available from the US National Geophysical Data Center (http://www.ngdc.noaa.gov/IAGA/vmod/igrf.html). An additional and similar global magnetic model, the World Magnetic Model (WMM) prepared by the US National Geophysical Data Center and the British Geological Survey, is the standard navigational model for the US and UK Departments of Defense and NATO. In addition, the model (http://www.ngdc.noaa.gov/geomag/models.shtml) is used in a variety of civilian applications including magnetic exploration. The main field spherical harmonics in this model are truncated at degree 12 as are the secular variation components. Still another global magnetic model which is briefly described in Section 11.5.2 incorporates both internal and external components of the terrestrial field. This Comprehensive Magnetic Field (CMF) (LANGEL *et al.*, 1996; SABAKA *et al.*, 2004) has been used successfully in isolating the lithospheric component of the geomagnetic field (e.g. RAVAT *et al.*, 2003, 2009).

In local surveys for near-surface investigations, station locations may be only relative to each other without specification of latitude and longitude. In these surveys, the north–south and east–west gradients of the IGRF are determined at a central location of the survey and assumed to be linear over the survey area. Typically, gradients are less than 10 nT/km except in the vicinity of the geomagnetic poles. World and regional maps of the various components of the magnetic field and its secular variation are available on a number of national web sites. An example is the US Geological Survey geomagnetic website (http://geomag.usgs.gov), which also provides for the calculation of the magnetic field anywhere on the Earth at locations below, on, and above sea level.

Elevation effects

Differences in observation elevations normally are not considered in processing near-surface survey data for anomalies because their effects are mostly negligible relative to the fields from subsurface sources. To understand this, consider the approximate vertical gradient of the vertical dipolar magnetic field at the Earth's surface derived from the vertical (radial) magnetic field $B_N(r_a)$ at sea level a and the field $B_N(r_h)$ at the elevation $a + h$ according to Equation 3.102. Thus, the ratio of these two fields separated by the elevation h is

$$\frac{B_N(r_h)}{B_N(r_a)} = \frac{r_h^3}{r_a^3} = \frac{(r_a + h)^3}{r_a^3}, \tag{12.3}$$

so that

$$B_N(r_h) = B_N(r_a)\left(1 + \frac{3h}{r_a} + \frac{3h^2}{r_a^2} + \frac{h^3}{r_a^3}\right). \tag{12.4}$$

Neglecting second-order and higher terms yields

$$B_N(r_h) \simeq B_N(r_a)\left(1 + \frac{3h}{r_a}\right)$$
$$= 60{,}000\,(\text{nT}) + 0.0282h\,(\text{nT/m}), \tag{12.5}$$

assuming h is in meters, and $B_N = 60{,}000$ nT at the sea level radius $r_a = 6{,}371 \times 10^3$ m, so that

$$\frac{\partial B_N(r_h)}{\partial h} = 0.0282\,(\text{nT/m}) = 0.0086\,(\text{nT/ft}). \qquad (12.6)$$

Accordingly, the normal vertical gradient of the main magnetic field is only of the order of 0.03 nT/m, and variations in the magnetic field from differences in observation elevations are generally negligible considering the magnitude of the field from sources of interest.

Magnetic terrain effects

Terrain effects are a prominent concern in the strategies for acquiring gravity data for geological purposes and in the computation of most gravity anomalies. However, this is not the case in most magnetic surveys. Magnetic terrain effects are usually neglected because the effects are negligible in most geological terrains owing to surface materials of low to negligible magnetization, such as sediments and sedimentary rocks. However, in geologic provinces where crystalline rocks are at the surface and the terrain is rugged, magnetic terrain effects often cannot be disregarded because they attain sufficient amplitude to interfere with the anomalies derived from subsurface sources. Even in these latter terranes, the calculation of the magnetic effect of terrain is problematic because of the common heterogeneity of the magnetization within terrain materials and the difficulty of defining their magnetization properties (Gupta and Fitzpatrick, 1971). This difficulty is especially troublesome where the surface rocks have significant remanent magnetization, such as modern volcanic rocks, because of the need not only to determine the magnetization, but also the direction of magnetization of the terrain materials.

In surface surveys in crystalline rock terranes the magnetic effect of positively magnetized rocks may be either positive or negative depending on whether the terrain is above or below the observation location (Heiland, 1940; Ugalde and Morris, 2008). Examples of the calculated total magnetic intensity at near ground level over a synthetic model of a 250 m deep valley cut into magnetized rock with both vertical and inclined magnetization as illustrated in Figure 12.1 testify to the potential complexity of the magnetic fields from terrain effects. Positively polarized rocks above the observation site will decrease the magnetic field at the site, and *vice versa*, and the situation will be reversed for negatively polarized rocks. Similarly, aeromagnetic surveys which fly below the level of adjacent highlands will observe the effects of both the highlands above the site and the missing material below the observation. In aeromagnetic surveys flown at a constant elevation in rugged topography regions with positively magnetized crystalline surface materials, the terrain effect will roughly mimic the topography at high magnetic latitudes ($\geq 70°$) (e.g. Allingham, 1964), but will vary from this condition at lower magnetic latitudes where the horizontal component of the magnetic field becomes prominent unless the magnetic field is reduced to the pole. The amplitude of the effect will be most prominent where the topographic relief changes rapidly with distance. In the case of satellite surveys, topographic relief is not of concern because of the low amplitude of the effect where there is a small ratio of topographic relief to observation altitude.

Numerous procedures have been devised for calculating the anomalous effect of magnetic terrain with varying degrees of success depending on the degree of accuracy of the simplifying assumptions that are made to implement them. Many of these have been summarized and evaluated by Grauch (1987). Generally these methods are employed as an initial step in the interpretation stage rather than when the observed data are processed to anomaly form. The most direct method is to calculate the 3D magnetic anomaly from the mapped topography, assuming homogeneous magnetic properties for the terrain. Variable magnetizations over the survey region can be accounted for by increasing the detail of the input to the computational procedure. A variety of procedures have been used in these calculations, from equations of the magnetic response of simplified geometric shapes to those which approximate the actual topographic relief (e.g. Marsh, 1971; Plouff, 1976; Wu *et al.*, 2009). Demagnetization effects may have to be considered in these calculations if the rock magnetizations are high and the relief is rugged.

A more indirect approach to determining terrain effects involves statistical correlation of topography with the observed magnetic field of a survey, assuming that the subsurface anomaly sources are spatially unrelated to the topography (e.g. Blakely and Grauch, 1983). A useful form of this method has been developed by Grauch (1987) for minimizing terrain effects in aeromagnetic surveys of variable magnetization, high-relief topography. In this method, the correlation of the observed magnetic anomaly and the magnetic response of the terrain is calculated on a grid point by point using the data within a moving window. The magnetic response of the terrain is calculated using appropriate magnetic parameters for the geologic terrane and the geomagnetic location of the survey. An iterative process based on the computation of the correlation coefficient is used to determine the optimum magnetization of the surface material that will minimize the residuals between the observed and calculated magnetic fields of the terrain. By moving the correlation

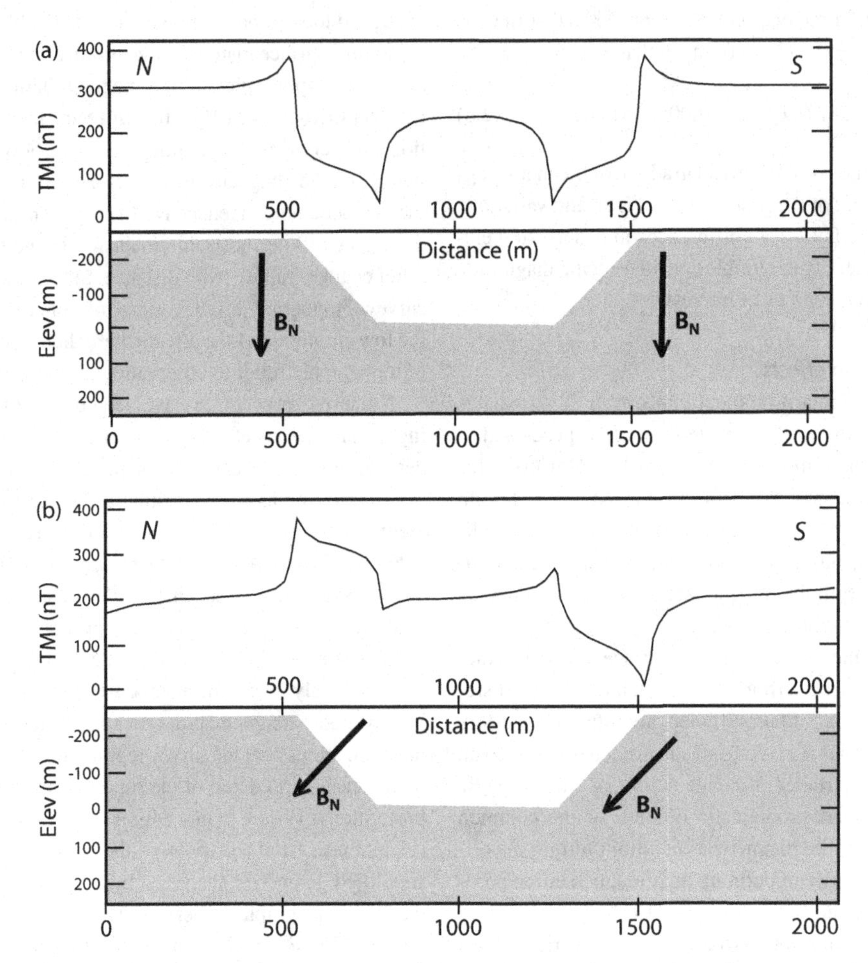

FIGURE 12.1 Total magnetic intensity profiles at 2 m above ground level over a model of a 250 m deep valley consisting of surface material with a magnetic susceptibility of 0.01 SIu with both vertical and 45° inclination magnetization. Adapted from UGALDE and MORRIS (2008).

window over the entire survey grid, the terrain effect over the survey region can be determined despite variable surface magnetization. The method requires the basic assumption of a lack of the spatial coincidence of topography and subsurface sources and a number of other subjective decisions. Nonetheless, Grauch has found the method more suitable than the more direct approach to determining the terrain effect and the method based on linear filtering which removes the component of magnetic data correlative to the terrain of the region (CLARKE, 1971).

Alternatively, the effect of terrain in magnetic survey data can be minimized by low-pass filtering of the observed anomaly data where the spectra of sources of interest do not overlap with the higher-wavenumber terrain effects. The upward-continuation filter is particularly useful in this regard while maintaining the geophysical character of the magnetic anomalies for the elevation of the projection.

Surface and near-surface effects

Surface and near-surface magnetic sources, both natural and cultural features, may cause anomalies which interfere with mapping geological magnetic sources. These anomalies, which are local in nature but may have a wide range of intensities, are either minimized or eliminated by appropriate observational procedures or post-acquisition processing by filtering or specific localized removal. The removal of surface and near-surface effects prior to processing and interpretation is fundamental to successful anomaly interpretation. One of the advantages to aeromagnetic surveying is the attenuation of these effects with distance. This is particularly important in avoiding magnetic noise effects on gradient surveys which are especially sensitive to local, high-frequency anomalies of surface and near-surface sources. Many of the noise sources are essentially point sources, such as localized concentrations of highly magnetic minerals in the surface soils or ferrous objects, so that their associated anomalies decay rapidly with increasing observation distance as the inverse cube factor. However, aeromagnetic surveys typically use the lowest possible survey elevation to increase resolution, so surface and near-surface sources are mapped in these surveys and may distort the anomalies of interest. This is especially the case in populated and industrial regions which are prone to cultural noise effects, but natural magnetic effects may be a problem even in remote, sparsely populated regions (GAY, 2004).

Natural surface and near-surface effects are derived from a variety of sources (see Chapter 10). It has been long recognized (e.g. HEILAND, 1940) that magnetic minerals, primarily magnetite, and magnetite-rich rock fragments, are heterogeneously distributed in residual soils and, particularly, in glacial till deposited at the surface during Pleistocene glaciation in the northern portions of North America and Eurasia. The magnetic-rich components are derived from movement of glaciers over crystalline bedrock and their residual soils and are deposited in glacial tills and related fluvial sediments (Figure 12.2). Their magnetic effects are especially noticeable when observations are made at the surface where the bedrock underlying the glacial deposits is essentially non-magnetic sedimentary rocks. These extraneous effects can be minimized in surface surveys by avoiding local concentrations of magnetic materials, which can be detected by multiple observations made over a limited period of time at sites separated by distances short compared with the gradients of interest in the surveying. Magnetic gradients exceeding those anticipated from subsurface sources should be avoided, or magnetic measurements made at several locations within a limited area can be averaged to obtain a representative measurement of the magnetic field derived from the subsurface (HINZE, 1963).

Surface and near-surface natural magnetic effects are muted at the elevation of even the lowest-altitude aeromagnetic surveys, but there is increasing evidence that they can be observed in aeromagnetic data. In some cases these effects are the subject of interest in the surveying, as for example in the detection of magnetic minerals originating from the geochemical effects associated with petroleum deposits (e.g. STONE *et al.*, 2004) and local anomalies over small igneous intrusives into sedimentary rocks (e.g. RAMO and BRADLEY, 1982; SPARLIN and LEWIS, 1994). However, in other cases these anomalies are troublesome. Surface and near-surface magnetic anomalies from a variety of glacial sediments in the glaciated portions of North America are cited by GAY (2004) that interfere with subsurface magnetic anomalies. Removal of these effects can be done manually by smoothing of the anomaly field through the local magnetic anomaly or by automated techniques that remove the high-frequency components of the anomaly field.

Cultural sources of magnetic anomalies which interfere with subsurface magnetic anomalies include ferrous objects, such as drill holes lined with steel casing, steel buildings, pipelines, bridges, tank farms, and ships, as well as electrical direct currents. An example of the magnetic effects of drill platforms and related ferrous infrastructure on the North Slope of Alaska is shown in Figure 12.3. These are particularly troublesome sources in surface surveys, and thus are avoided by distancing the magnetic measurements from potential sources. This is not usually an option in aeromagnetic surveying, but numerous examples confirm that these sources are visible in the results of aeromagnetic data, particularly data observed in high-resolution surveys flown at elevations of 150 m or less, and are potentially deleterious to magnetic anomaly interpretation. Aluminum smelting facilities which use high-amperage direct currents in the smelting process will produce intense magnetic anomalies at aeromagnetic surveying elevations, and direct currents used for cathodic protection of underground pipelines produce observable anomalies which are one or two orders of magnitude greater than the effect of the steel in the pipelines (GAY, 1986). These are two examples of anomalous effects from surface or near-surface direct currents. However, the majority of cultural magnetic effects arise from ferrous objects. These effects are noted for their high-frequency components and high intensity which commonly produce spike-like anomalies. Their removal from the magnetic anomaly data is problematic because simple linear or non-linear filtering of high-frequency components (spike filters) from the anomaly field also removes these components from the anomalies of interest, which can lead to erroneous interpretations. Alternatively, the cultural noise can be eliminated manually by removing obvious anomalies that are directly related to features observed on the Earth's surface. However, this procedure is time-consuming and expensive. As a result, a number of semi-automatic techniques have been developed to remove cultural magnetic noise (e.g. MUSZALA *et al.*, 2001; HASSAN and PIERCE, 2005; SALEM *et al.*, 2010). These techniques generally use a combination of identifying potential cultural magnetic anomaly sources and their anomaly characteristics together with automated or semi-automated procedures for identifying these characteristics in the anomaly field which can then be automatically removed from the anomaly data.

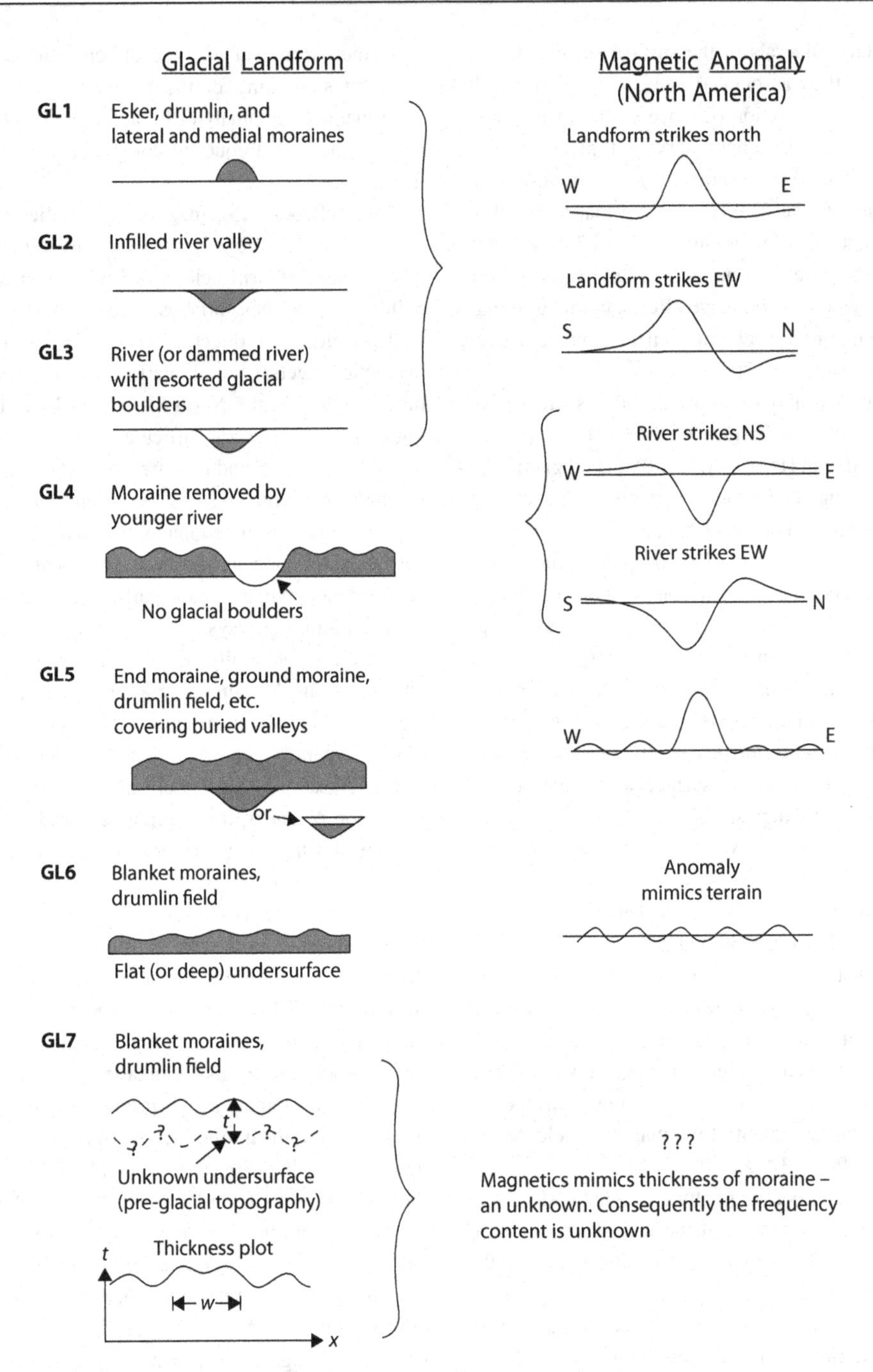

FIGURE 12.2 Types of total field magnetic anomalies observed over various Pleistocene glacial landforms in northern North America due to deposition or removal of crystalline rock-rich sediments. The anomalies will vary with strike as well as landform and contrasting magnetic susceptibilities. Adapted from GAY (2004).

12.3.2 Temporal magnetic variations

In addition to the long-term changes in the main magnetic field (that is, the secular changes in the field that are described above), the magnetic field of the Earth includes a host of temporal changes with periods ranging from a second to roughly a day (LOVE, 2008) that must be considered in isolating the magnetic effects of the subsurface in magnetic measurements (REFORD, 1979). Although the periods are diverse, they are commonly identified in

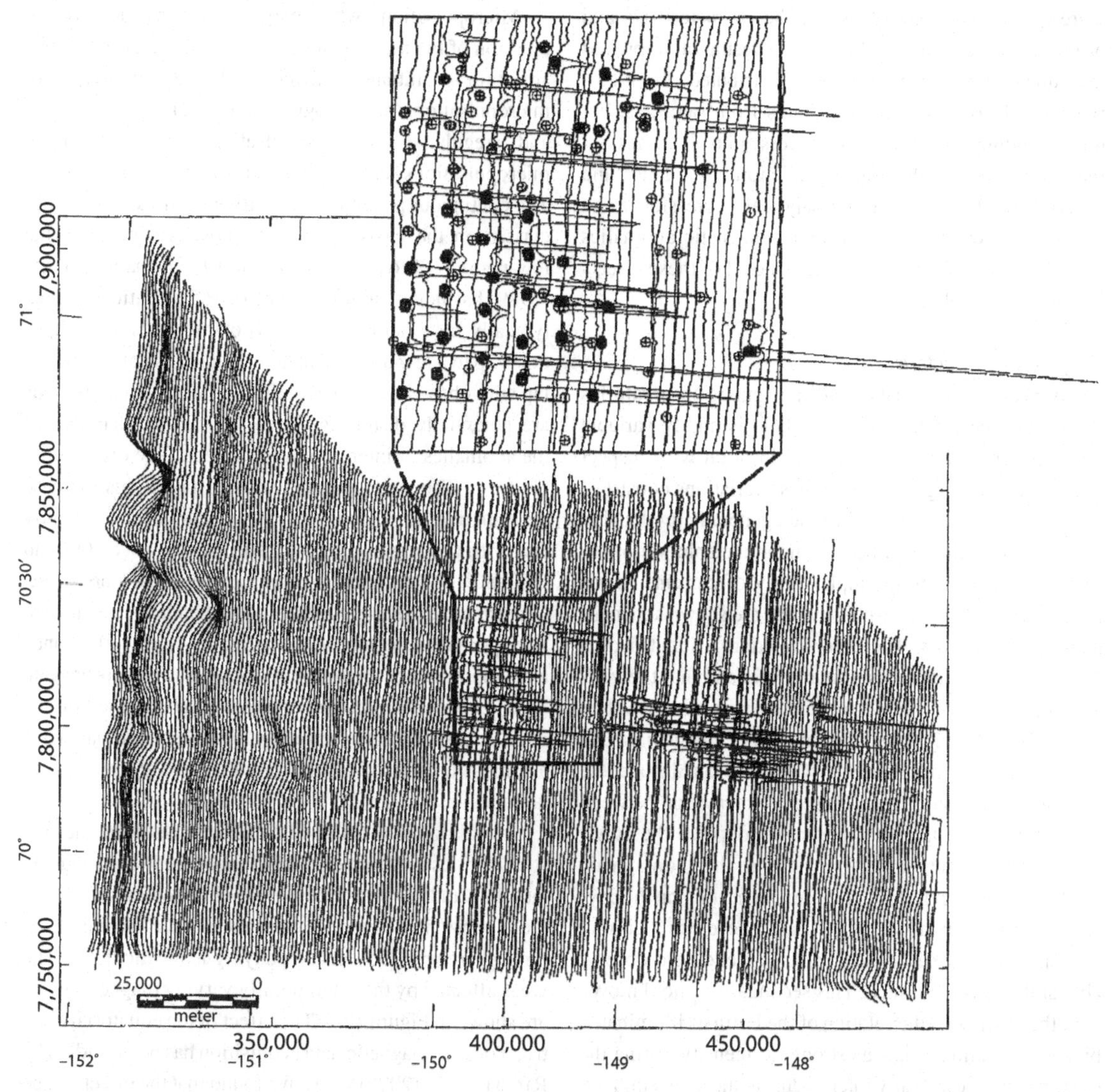

FIGURE 12.3 Detrended aeromagnetic total magnetic anomaly profiles observed over drill platfroms and related ferrous infrastructure on the North Slope of Alaska. Insert map (not to scale) shows the location of wells, drill platfroms, and other cultural features which closely correlate with spike-like magnetic anomalies. Adapted from MUSZALA *et al.* (2001).

exploration analyses as "diurnal variations", with the full realization that they include more than daily periods. Typically these temporal changes are of sufficient intensity to interfere with the high-sensitivity magnetic surveying which is the norm today, and thus they must be removed from field observations before these are useful in subsurface interpretation. The diurnal fields are derived externally from electrical currents and perturbations of the ionosphere and magnetosphere, but they also include secondary fields derived from induced currents in the subsurface. The ionosphere, extending from 50 to 600 km above the Earth, is the locus of strong, eastward-flowing electrical currents that circulate in the sunlit hemisphere. The magnetosphere is the realm of the Earth's magnetic field that interacts with the interplanetary magnetic field, which is the solar magnetic field carried by the solar wind, a plasma, within our planetary system. The magnetosphere is highly asymmetric around the Earth as shown in Figure 8.4, extending outward to several tens of Earth radii in the opposite direction of the Sun. A useful review of the current status of knowledge of these fields is given by CAMPBELL (2003).

Temporal variations in the magnetic field have been observed for over a century at fixed magnetic observatories, providing a wealth of historical data. Additionally, there is increasing knowledge of the characteristics and origin of these temporal fields as a result of improved instrumentation and more widely distributed observatories as well as global measurements from satellites, but understanding of the physical processes involved has not reached a level which permits prediction of the fields and description of them with an analytical expression at an accuracy necessary for determining magnetic anomalies from magnetic observations. The removal of these effects is therefore largely empirical.

Origin and characteristics

Analysis of the historical record at magnetic observatories shows that the external magnetic field is highly variable not only temporally, but also with location with respect to the main geomagnetic field. These variations consist of two primary components – the quiet, generally predictable variation, and the disturbed, or unpredictable variation. They originate from a complex array of currents in the ionosphere and magnetosphere which are shown schematically in Figure 8.4. The quiet variation is identified as S_q, recognizing that it has a solar origin and that the variations are relatively minor compared with the disturbed field. This field originates from motion of charge particles in the ionosphere from roughly 100 to 130 km altitude, the E-region of the ionosphere, caused by interaction between solar X-rays and far-ultraviolet radiation with the atmosphere. The motion of these particles in the geomagnetic field, which is largely due to differential heating on the dayside versus the nightside of the Earth, develops a current system that produces a secondary magnetic field observable at the Earth's surface. This secondary S_q field moves over the Earth with the rotation of the Earth as illuminated by solar radiation. It has its strongest intensity during the daytime and is especially intense during the local summer, reaching amplitudes ranging from 10 to 50 nT or more. It is highly variable with geomagnetic latitude and generally changes phase from the northern to the southern geomagnetic hemispheres.

The ionosphere is subject to tidal effects primarily from the Moon. As a result the currents move with the tides producing an effect with a period of roughly 25 hours, the L_q component, which has an intensity of less than 10% of S_q. These magnetic field components are perturbed by micropulsations that have much shorter periods (1 second to 10 minutes) and generally lower amplitudes (a fraction of a nanotesla to several hundred nanotesla in the auroral zone) than the normal quiet-time variations (LAM, 1989). As a result they will interfere with the mapping of magnetic anomalies derived from the subsurface (e.g. MENK, 1993) that are observed on a platform moving over the surface.

Micropulsations, which may be caused by sudden compression of the magnetosphere as well as other processes, are abruptly initiated and felt worldwide. Two types of micropulsations are recognized: regular or continuous, and irregular. Continuous pulsations, P_c, are trains of micropulsations lasting for several hours to a few days. The continuous pulsations are further subdivided into subclasses on the basis of their periods, ranging from the low-period P_c1 to the longest-period P_c5 pulsations. LAM (1989) has noted the importance of P_c5 pulsations, which typically have periods of 150 to 600 seconds and amplitudes up to a few hundred nanotesla, because of their overlap with observed aeromagnetic anomalies, but all periods are potentially distorting depending on the character of the anomalies of interest and the speed of the observation platform. The average amplitudes of pulsations increase with increasing period, and they occur most frequently and with the largest amplitudes in auroral zones (around $\pm$60–80° geomagnetic latitude) and at local noon time. The irregular micropulsations, P_i, which are nocturnal, have periods generally in the range of 40 to 100 seconds and amplitudes of up to several nanotesla. They propagate over hemispherical distances and can be observed on the dayside of the Earth even though they are nocturnal. They are most intense at auroral latitudes and decrease with latitude.

In addition to the normal S_q variations, during normal solar periods there are equatorial (about $\pm$5° geomagnetic latitude) and auroral (about $\pm$60–80° geomagnetic latitude) horizontal electrojet currents in the E-region of the ionosphere with their accompanying magnetic fields. The areas affected by these temporally varying magnetic fields are shown in Figure 12.4. The effect of the equatorial electrojet on aeromagnetic data acquisition has been studied by RIGOTI *et al.* (2000) who have found that the effect can be treated by normal leveling procedures used in processing aeromagnetic observations. The auroral electrojet effect is particularly intense and persistent because of the high conductivity of the auroral ionosphere and the presence of large horizontal currents. These effects are normally limited to the auroral belt, but during times of disturbance the effects extend beyond the belt.

The second component of temporal variation, the disturbed or unpredictable variation of the geomagnetic field, is highly variable and much more intense than quiet-time variations, and as a result commonly hides the normal solar variations. Disturbances are derived from interaction of a relatively dense cloud of plasma made up of charged particles ejected from the Sun, the solar wind, with the Earth's magnetosphere. The intense plasma cloud, which originates from solar flares observed as sunspots, travels

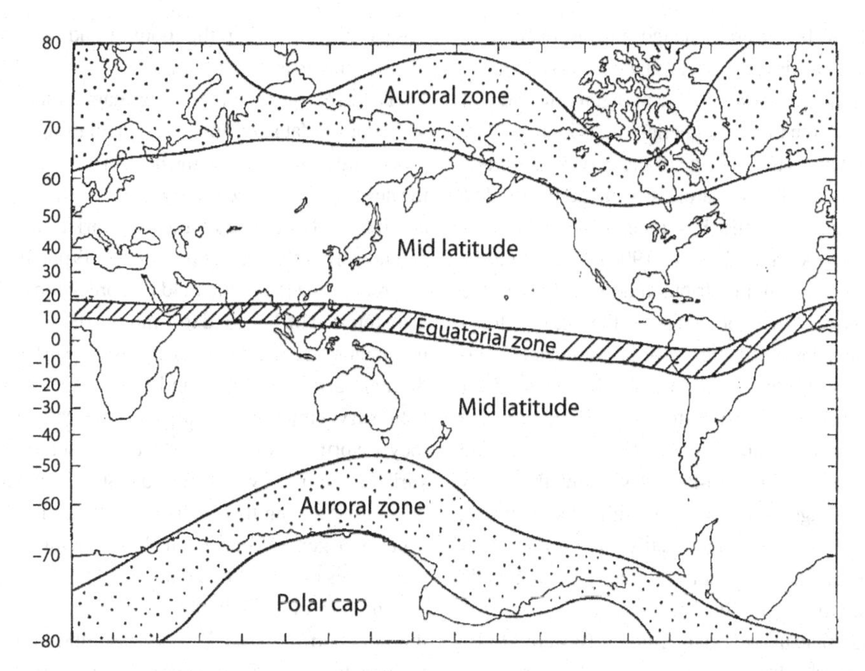

FIGURE 12.4 Approximate locations of the auroral and equatorial electrojets where external temporal variations in the Earth's magnetic field reach maximum values. Adapted from REGAN and RODRIGUEZ (1981).

through space at speeds of 300 to 800 km/s. On reaching the environs of the Earth, the plasma causes a sudden compression of the magnetosphere and generates complex electrical currents in the lower ionosphere, initiating highly irregular magnetic fields which have amplitudes of the order of hundreds of nanotesla or more. The most intense disturbances occur in the auroral zones (Figure 12.4) where the charged particles spiral into the Earth's atmosphere along magnetic lines of force. Their interaction with ions causes auroral light displays in the polar regions of both hemispheres. The largest of these disturbed periods may last for several days and occur simultaneously over the Earth. Disturbed periods of activity reach a climax during the maximum of the sunspot cycle which occurs on an 11-year cycle. The largest of the disturbances have a 27-day period corresponding to the rotation period of the Sun, causing the Earth to be illuminated by the plasma being ejected from long-lived sunspots. During these periods the complex, intense, and unpredictable nature of the temporal field prevents acquisition of magnetic data for most geological purposes.

An additional source of temporal variations in the geomagnetic field is magnetic fields generated by secondary currents in the Earth caused by the varying external field. These fields may have amplitudes of the order of 20% of the external originating field, and thus have a significant effect on measured fields, especially during periods of disturbance. However, even during quiet external field periods, these secondary magnetic fields including micropulsations may perturb the measurements if the survey involves a region of varying Earth conductivities. The varying conductivity of the subsurface channels the location of induced currents, causing changes in the intensity of the secondary field over distances measured in a few tens of kilometers or less (e.g. LILLEY, 1982; LILLEY *et al.*, 1999). Especially important are the changes in conductivity at oceanic coastlines (PARKINSON and JONES, 1979). Additionally, temporal variations in the magnetic field may be encountered over the ocean owing to wave action within the geomagnetic field. LILLEY *et al.* (2004) and LILLEY and WEITEMEYER (2004) have shown that peak-to-trough magnetic fields of 5 nT are related to ocean swells and that these signals are observed at altitudes used in aeromagnetic surveying.

Geomagnetists find it useful and efficient to use a shorthand for the intensity of temporally varying magnetic fields in the form of magnetic indices. Several indices have been defined. Some are used to characterize specific current systems within the space environment and others relate to general variations in the field. The most widely used is the K-index which is a measure of the variability of the field over a 3-hour Universal Time period. It has 10 levels varying from 0 for the lowest activity to 9 for the highest with subscripts of 0, +, and − for intermediate units in thirds. K-levels are selected so that each level represents an approximate 10 nT change. K-values are given for a specific observatory, but a planetary measure of the temporal variation in the magnetic field is identified by the K_p-index which is the average of the K-values for any 3-hour period from 13 observatories over the Earth. The procedure for calculating the K_p-values is described, for example, by SIEBERT and MEYER (1996). Unfortunately these observatories are not distributed uniformly over the Earth, and thus K_p may not represent the actual global variations. In general K_p-values have amplitudes equivalent to 2_0, values of 5 and above are usually an indicator of magnetic storms, and major storms have indices of 8 and above. K_p-values can be converted to typical field strengths in nanotesla at mid-latitudes using the index a_p. The daily average of a_p-values at eight observatories, A_p, represents the equivalent daily amplitude over the Earth. Information on current and past magnetic activity is available from the World Data Centers (see http://www.wdc.rl.ac.uk/wdcmain/guide/gdsystem.html) such as the World Data Center for Solar-Terrestrial Physics, Boulder, Colorado (http://www. nngdc.noaa.gov/stp/WDC/wdcstp.html).

Removal of temporal variations

Temporal variations are routinely removed from magnetic observations to isolate magnetic anomalies derived from the subsurface Earth. Numerous methods have been and are being used to accomplish this objective depending on the type and purpose of the magnetic survey. They vary from subtraction of variations in the field as measured by reoccupation of observation sites to highly analytical techniques for high-sensitivity aeromagnetic surveys.

The simplest of procedures involves monitoring of magnetic variations by reoccupation of observation sites in land surface surveys, at intervals dictated by the likely variation of the field, dependent on magnetic latitude and magnetic disturbance activity and the accuracy requirements of the survey objectives. Typically, the time interval between reoccupations is of the order of an hour or less in mid-latitudes for surveys requiring an accuracy of only a few nanotesla. The measured variations in the field from the reoccupations are subtracted from the field observed at the distribution of stations. Reoccupation of the survey base on a daily basis will provide the information for adjusting values which are likely to change from day to day. A much more precise procedure in surface surveys on both land and water bodies is to monitor the continuing variation in the field with a recording base magnetometer. The measurements of the monitor are removed by subtraction from the station observations as a function of time synchronized between the monitor and the magnetometer measurements at the stations.

The use of a base recording magnetometer to monitor and acquire data for anomaly calculations is an effective procedure providing that the base monitor is located within the near vicinity of the survey and removed from externally and internally derived extraneous variations which do not extend across the area of the survey. Typically these monitors record the magnetic field to a precision of 1 nT or less, at an interval of 1 second to 1 minute. Base recording magnetometers are used in land and water surface surveys for diurnal variation control and less frequently in aeromagnetic surveying for this purpose. In aeromagnetic surveys they are primarily used to detect the onset of intense highly variable temporal variations so that survey operations can be aborted during these periods of time. They can also be used as a general check on the variations calculated by procedures using tie-line data.

The use of a stationary recording magnetometer for diurnal control is based on the assumption that the observations from the base magnetometer are coherent over the entire survey region. This is an acceptable assumption for local ground or lake and river surveys, and for aeromagnetic surveys conducted within a few tens of kilometers of the recording base. However, there is no definitive maximum distance over which a recording magnetometer is appropriate (e.g. REGAN and RODRIGUEZ, 1981). REEVES (1993) suggests a maximum separation of 50 km between the base magnetometer and the most distant of the survey area, but notes that this distance cannot assure coherency in the field over this distance. Strongly influencing this distance is the state of the external field, with variations less predictable during intense activity periods and also in regions where there are strong electrical conductivity contrasts leading to spatially variable secondary magnetic fields from electrical currents induced in the Earth. The latter are especially evident owing to the coastal effect from the difference in the conductivity of the ocean waters and the adjacent coastline. This is a source of difficulty in the use of land-based stationary magnetometers to correct marine magnetic surveys. Base magnetometers must also be placed to avoid local extraneous fields, at least 100 m from large ferrous objects and travelled roads and roughly 500 m from power lines. They should also be located in areas free from strong electrical conductivity contrasts in the subsurface to avoid secondary magnetic fields.

Consideration of the temporal field variations on satellite magnetic data has taken the approach of selecting quiet-time data based on the monitoring of the magnetic field at stationary observatories (LANGEL and HINZE, 1998; MANDEA and PURUCKER, 2005) using a variety of geomagnetic indexes. Generally, data have only been considered when the K_p-index was less than or equal to 1+ or 2 and the prior 3-hour K_p-value was 2+ or less. The latter restriction is to prevent leakage from previous highly variable fields into the succeeding 3-hour period with an acceptable K_p. However, a number of other approaches have been used to select data that are only minimally adulterated with temporal variations or to minimize the temporal variations. For example, ALSDORF *et al.* (1994) found the variance of the anomalous field along orbital passes in the south polar region was a better indicator of the quietness of the data, and RAVAT *et al.* (1995) rejected any orbital pass data for which the variance exceeded 80 nT^2. Further description of the removal of temporal variations from satellite data is given in Section 11.5.1. Clearly, these variations are a primary controlling factor on the applicability of satellite observations to the study of the Earth because of the lack of control on the temporal magnetic variations in these observations.

Minimization of the temporal effect is a primary concern in modern aeromagnetic surveying, especially in the processing of high-resolution data. Periods of disturbed magnetic fields initiated by sunspots and their related magnetic storms are unsuitable for observing data. This is true over all geomagnetic latitudes, but particularly in auroral regions where the disturbances are most intense. Base station monitors in the local area of the surveying or reports from magnetic observatories identify periods of magnetic disturbances. Micropulsations can be locally monitored on base magnetometers and directly removed from aeromagnetic observations, but there is ample evidence that in many regions this is not possible because of their change in amplitude and phase over a few tens of kilometers (e.g. MILLIGAN, 1995; WANLISS and ANTOINE, 1995). This lack of correlation is generally attributed to the effect of variable secondary magnetic fields caused by currents induced in the Earth. O'CONNELL (2001) proposed a method for eliminating micropulsations from airborne magnetic data despite the differences in base and airborne observations, using the correlation of the data sets. The method includes several steps including identical high-pass filtering of the base and airborne data followed by time-shift cross-correlation of the filtered data to achieve maximum correlation within a moving window, and then calculating a gain factor which is used to match the airborne data to the base data. The base data, appropriately time shifted and multiplied by both their correlation coefficient and gain factor, are subtracted from the airborne magnetic data to remove the micropulsations.

During normal magnetic variation periods, aeromagnetic surveying largely depends on using mis-ties in the observed data at intersections (cross-overs) between tie-lines and the survey traverses for determining the diurnal variations that have occurred during the period of the survey. A number of analytical leveling procedures have been used. A comprehensive description of the leveling process has been provided by, for example, MAURING *et al.* (2002). There are several sources of the mis-ties, including positional errors and transient cultural and instrumental effects. In some processing the locations of the ties are adjusted to correct for navigational errors decreasing the amplitude of the mis-ties, but this procedure necessarily must be closely monitored to avoid unreasonable adjustments to the location of the crossover points.

Leveling of the magnetic data using mis-tie information assumes that the source of the difference in the observations is diurnal variations in the measured magnetic field which occur during the period between observations on the survey lines at successive tie points. Prior to implementation of the leveling process, spatial and secular variations in the magnetic field are removed by subtracting the appropriate IGRF from the observations. It is noted that secular variations described by the IGRF during the measurement epoch may be in error because they are based on extrapolation. These errors will cause shifts in the reduced data of a survey observed over an extended period of time which commonly reaches several weeks in length. Errors also may be related to lag in the adjacent survey lines flown in opposite directions, as is common in aeromagnetic surveying. Furthermore, the mis-ties commonly require editing to eliminate spikes in the data and to identify mis-ties which are unreasonable considering the anticipated diurnal variations. These unreasonable values can be caused by several factors other than the temporal variations. One of the more significant factors is the lack of spatial coincidence of the observations because of positional errors. This is particularly important in regions of high horizontal magnetic gradients. As a result it may be necessary to remove mis-ties in these areas or to weight the mis-ties based on the amplitude of the horizontal gradient. Editing may be performed manually, but this is a time-consuming process, so automatic schemes are used wherein the mis-ties can be weighted based on horizontal gradients of the measured magnetic field, and those that exceed a few standard deviations of the preliminary determination of the mis-ties are eliminated.

There are two primary methods of leveling the magnetic data once the mis-ties to be used in the process have been selected or weighted (GREEN, 1983; LUYENDYK, 1997). In one class of methods, the mis-ties along a line or over a region are used to calculate a low-order polynomial that best fits in a least-squares sense the mis-ties (e.g. YARGER *et al.*, 1978; BANDY *et al.*, 1990; and

Appendix A.5.1). This polynomial can then be used to adjust the observations assuming that the diurnal is a slowly varying field. BEIKI *et al.* (2010) have successfully used a differential polynomial filtering procedure which is applicable to surveys with or without tie-lines. NELSON (1994) has also used measured horizontal magnetic gradients to level aeromagnetic data without using tie-lines for minimizing the diurnal effect, and MAURING and KIHLE (2006) have used a moving differential median filter for leveling aeromagnetic data without tie-lines. HUANG (2008) has also proposed a method for leveling without tie-line measurements based on line-to-line correlations by adjusting all lines in a least-squares sense to a reference survey line.

The second main class of methods is to minimize the closure errors in a network of intersecting loops using a least-squares procedure. This is equivalent to the network adjustments made in minimizing the error in surface elevations observed in loops.

Unfortunately, tie-line leveling will not remove all the problems in the airborne magnetic data set. Often a final process is needed to remove line-to-line differences which cause the data to have a herringbone or corrugated appearance (URQUHART, 1988). This process, which is conducted only after leveling of the data from the mis-ties at the crossover positions, is referred to as microleveling or decorrugation. In this procedure (e.g. MINTY, 1991) the short-wavelength components are removed from data for each survey line, then a directional filter in the survey line direction with a bandpass designed to eliminate wavelengths less than twice the line spacing is applied to a grid obtained from the resulting residual data. The long-wavelength component of each line is then interpolated from the filtered grid and added to the short-wavelength components of each survey line. Accordingly, the short wavelengths of the survey lines are retained but the line-to-line short wavelengths causing the corrugation are removed. Care is necessary in applying this filtering procedure because long-wavelength anomalies in the direction of the survey lines can be modified.

The procedures used to remove temporal variations from marine magnetic observations can be treated in the same manner as airborne observations provided that tie-line data and or nearby base monitoring magnetometers are available. Unfortunately this is not a common situation. Furthermore, leveling of survey data using tie-line data is not as efficient for marine data because of the longer period of time between measurements at crossover positions due to the slower speed of marine vessels, and the coastal effect temporal variations that operate on base magnetometers located at marine coastlines as described above.

12.4 Anomaly isolation and enhancement

Magnetic anomalies combine the effects of all horizontal variations in magnetic polarization within the subsurface. Thus, the residual anomalies of interest in an application are commonly distorted by interfering regional effects of the deeper and broader sources, as well as shorter-wavelength noise of the shallower sources. As for the gravity method, the extraction of the residual anomaly is a critical problem that controls the accuracy of the interpretation process, but the problem is not as severe for the magnetic method. The subjectivity of the residual determination process makes it potentially one of the major limitations in applying the magnetic method.

12.4.1 Fundamental principles

As in the gravity method described in Section 6.5, the residual–regional separation problem in magnetics involves methods that either focus on isolating or on enhancing the residual anomaly. Isolation techniques eliminate from the anomaly field all anomalies that do not have the prescribed characteristics of the residual anomaly. Thus, the residual is presumed to be minimally modified and useful for quantitative analysis and modeling. Enhancement techniques, by contrast, accentuate the characteristics of prescribed anomalies to increase their perceptibility. The enhancements distort the anomalies, making them of limited general use in quantitative modeling. Instead, they are primarily used in qualitative visual inspection analysis and interpretation.

The basis for separating anomalies involves consideration of characteristics of the anomaly field such as anomaly amplitudes, horizontal dimensions, gradients, directional attributes, and correlations with geological or other geophysical variations. Numerous techniques devised to address the residual separation problem range from simple graphical smoothing procedures to analytical grid calculation systems (e.g. PETERS, 1949; NETTLETON, 1954), and frequency-domain processing schemes that take advantage of the computational speed and storage capacity of modern computers (e.g. DEAN, 1958; CLEMENT, 1973). Figure 12.5 illustrates the spatial distortion of overlapping magnetic anomalies from five shallow sources due to the magnetic effect of a deeper source. Note that the amplitudes of the shallow-sourced, residual anomalies are considerably affected by the deeper, regional anomaly. However, their gradients are affected to only a limited degree because of the considerable separation in the depth of the sources and the relatively much lower gradients of the deeper source.

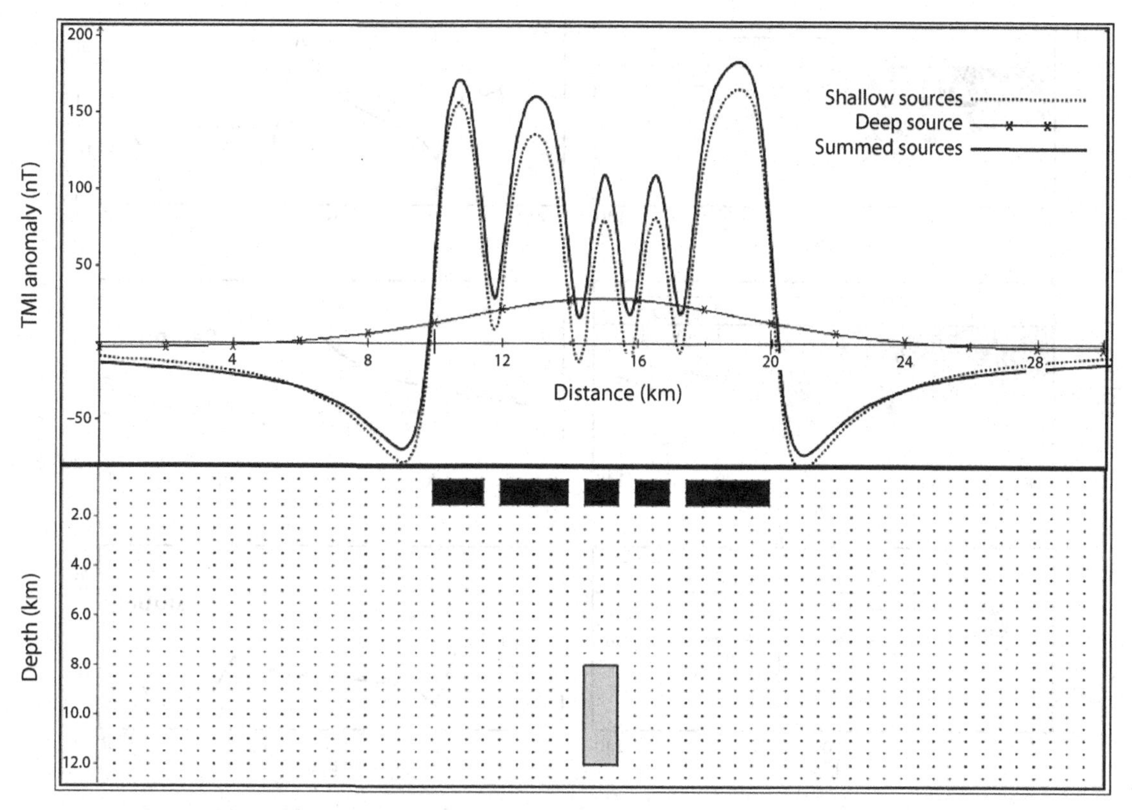

FIGURE 12.5 Total magnetic intensity (TMI) anomalies over five shallow 2D sources, and the interference effect from a deep magnetic anomaly source. Note the minor effect of the deep source.

The following description of methods of isolation and enhancement of magnetic anomalies builds upon the analytical treatment of the methods in Appendix A.5.1. In addition, because the application of the regional-residual separation techniques to magnetic anomaly data closely parallels their use in the gravity method, the reader is directed to Section 6.5 for the foundations of the methods and their use. Although the application of magnetic methods of isolating and enhancing anomalies has much in common with the gravity method, there are differences between them that need to be considered.

Magnetic anomalies have a larger dynamic range in their amplitudes because the breadth of the physical property contrast of the magnetic method, magnetization, is much greater than that of the density contrast in terrestrial conditions. Furthermore, the magnetic anomaly amplitude varies one power faster in the inverse distance power function than in the gravity method for sources of the same geometry. As a result, the character of magnetic anomalies changes much more rapidly with depth than gravity anomalies from equivalent sources, making the overlap in the spectra of magnetic anomalies less than for gravity anomalies from the same sources. In addition, because of the dipolar effect of magnetic sources, magnetic anomalies derived from long, wide sources only produce anomalies along their edges, quite in contrast to gravity anomaly amplitudes of the same sources. The gravity anomaly amplitudes derived from these large regional sources are independent of depth and thus are a major source of regional gravity anomalies, which must be separated from the desired residual anomalies. The significant simplification of the regional-residual problem for magnetic anomalies is illustrated in Figure 12.6 which shows the total anomaly (top right panel) from the superposition of a regional anomaly (middle right) due to two large basement sources (middle left) and a residual anomaly (lower right) from two near-surface sources (lower left) with vertical magnetic polarization. The upper left panel shows the geologic cross-section involving the four sources that give the combined total anomaly. The anomaly profiles show that the basement sources are only evident in the anomaly at the contact of the two sources. Although this contact anomaly has a large amplitude because the contact reaches the surface, it does not highly distort the residual

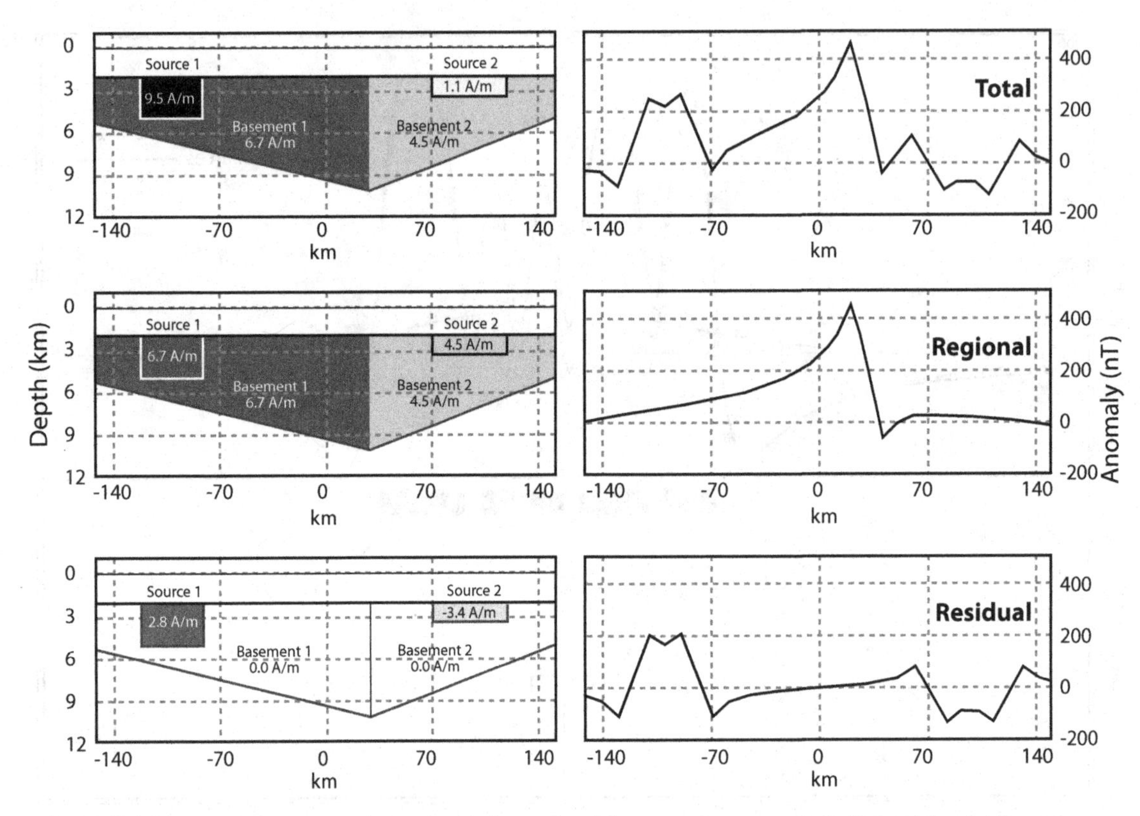

FIGURE 12.6 Magnetization contrasts and related residual anomalies of the near-surface sources in the bottom left and right panels, respectively, are defined by subtracting the regional basement effects (middle panels) from the total effects (top panels) – i.e. the total magnetizations and anomalies are the superposition of regional and residual effects in both source and signal space, respectively.

anomalies due to the surface sources which are shown in the lower panel of the figure.

12.4.2 Isolation and enhancement methods

This section explains and demonstrates methods of isolation and enhancement in application to magnetic anomalies. Several of the methods are illustrated by their application to a geologic model which has also been used to show the application of the methods to gravity anomalies in Section 6.5.3. This permits a comparison between the application of the methods to gravity and magnetic anomalies. In addition, a description is provided for enhancing magnetic anomalies by eliminating the directional effect of magnetization on them. This reduction-to-the-pole methodology calculates anomalies by shifting the ambient magnetization to vertical, as it is at the magnetic poles.

Geological, graphical, and visual inspection methods

In principle, the geological method for stripping out regional and noise anomalies which has important uses in dealing with gravity anomalies is applicable to magnetic anomalies. However, it is seldom used because of the lack of information on magnetic polarization contrasts and the heterogeneous distribution of magnetic minerals in many rock types. Similarly, graphical methods as well as the related trend surface analysis techniques have limited use in isolating anomalies of interest in an anomaly field because they are best at removing broad, strong gradients which are not commonly observed in magnetic anomaly fields. In those situations where residual anomalies occur on gradients, and hand or computed trends are used to isolate the residuals, care is necessary to incorporate the dipolar effect of anomalies and the requirement that according to Gauss' law the isolated anomaly should have equivalent positive and negative power. The equivalence of the negative and positive portions of the anomaly is a useful check on the validity of the candidate regional anomaly.

The ability to visually identify and isolate residual anomalies in a magnetic anomaly field is often improved by image processing of the data which may take a variety of forms. Visual inspection is generally more useful in the magnetic method than in the gravity method because

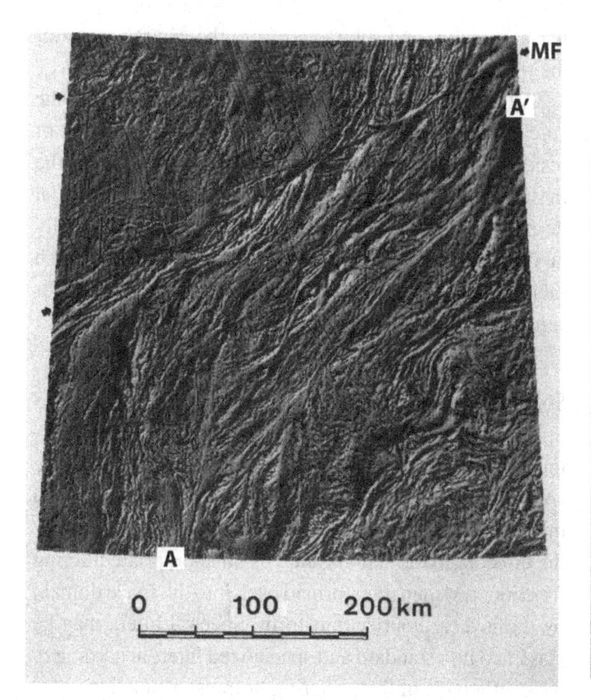

FIGURE 12.7 Shaded relief magnetic anomaly map of a portion of the Northwest Territories of Canada with illumination at an inclination of 30° and a declination of 135°. MF is the MacDonald fault, AA′ is the position of a major geologic contact, and the arrow in the upper left indicates the location of a subtle east-striking linear magnetic anomaly. Adapted from Dods *et al.* (1985).

FIGURE 12.8 Shaded relief magnetic anomaly map of the area shown in Figure 12.7 with illumination at an inclination of 30° and a declination of 45°. Adapted from Dods *et al.* (1985).

of the dynamic range of magnetic anomaly amplitudes and their strong gradients compared with gravity anomalies. The fine detail that is commonly measured in the magnetic method may be lost or at least minimized by contouring the data with equal magnetic anomaly intensity lines. The gridding of the anomaly data and mapping process effectively acts as a high-wavenumber cut filter (e.g. McIntyre, 1981). Accordingly, it is appropriate, where small, high-wavenumber components are present in an anomaly field and of interest in interpretation, to display the anomaly profiles as a series of stacked profiles with the location of the survey line as the baseline of the anomalies. Skilbrei and Kihle (1999) illustrate the usefulness of this approach to studying high-pass filtered aeromagnetic profiles, together with shaded relief presentations showing sedimentary layering and structure on the Norwegian continental shelf.

Shaded relief maps as discussed in Section 6.5.3 for the gravity method have an increasingly important role in the visual analysis of potential field maps. This image processing technique is particularly useful in the analysis of magnetic anomaly maps because the high gradients typical of magnetic anomalies are difficult to portray in contour maps, even colored contour maps, but are emphasized in shaded relief maps. Magnetic anomalies are clearly presented on shaded relief maps if the observations are sufficiently dense to map the anomalies in two dimensions and the illumination angle is roughly orthogonal to the strike direction of the anomaly pattern. Under these conditions, the presentation strongly emphasizes the anomaly patterns directed orthogonal to the illumination angle. Thus, these maps can be used to enhance specific components of magnetic anomaly maps depending on their strike direction, as illustrated in Figures 12.7 and 12.8.

These figures show the magnetic anomalies of a regional area in the Precambrian Shield west of Slave Lake in the Northwest Territories of Canada. Striking obliquely across the area is a major geological structure, the MacDonald fault (MF), which is clearly visible on Figure 12.7, a shaded relief magnetic anomaly map illuminated from an inclination of 30° and a declination of 135°. Numerous other anomalies are observed on this map including the subtle east–west striking linear features in the northeastern corner of the map.

In contrast, Figure 12.8, which is a map of the same magnetic anomaly data illuminated from an inclination of 30° and a declination of 45°, enhances the northwest-striking linear anomalies associated with a swarm of magnetic dikes that cut across the regional geology. Circular or arcuate features are emphasized by removing the directional enhancement, that is by positioning the illumination directly overhead. The use of these shaded relief maps is scale-independent, but data for their construction are required at a high density over the mapped area.

Shaded relief magnetic anomaly maps are useful in identifying lithologic-geologic domains of basement crystalline rocks that have a common geologic and metamorphic history. The magnetic anomalies of the domains have characteristic anomaly attributes which distinguish the region from adjacent areas (e.g. PILKINGTON *et al.*, 2000) when observed at a constant altitude above the basement rocks or when the data are projected computationally to the same altitude above the basement. A constant altitude is needed to minimize the profound effect that source to observation distance can have on the characteristics of the anomalies. Numerous attributes of magnetic anomalies within a region such as their amplitude, direction, and gradients which specify the texture of anomalies, that is the spatial/statistical distribution of the amplitudes of the anomalies, can be used to delimit the domains using visual inspection. This process can be enhanced by introducing derived quantities such as fractal measures and entropy that are obtained by textural filtering (e.g. DENTITH, 1995) of the magnetic anomaly data. Mapping of these attributes can be achieved by filtering the data within a window that is moved across the anomaly map. Statistical analysis of the combined direct and derived components can be used to define regions of similar anomaly amplitudes and texture in a more objective manner than by visual inspection. Numerous assumptions and decisions must be made to automate the domain mapping procedure which can strongly affect the results. To insure high-quality mapping, great care must be exercised in making these decisions and assumptions and verifying the results with known geologic information.

Analytical grid methods

As described for gravity applications in Section 6.5.3, analytical grid methods were used extensively before the dawn of modern electronic computing in the 1950s. A simple profile application is to compute the running mean where data for an equal distance on either side of the central value are averaged with the central value for the regional at the center point. The regional minus the central value provides the residual value. In two dimensions, an imaginary grid or mask operator is moved or lagged across the map. At each data point or lag position, the regional value at the center of the mask is estimated from the average or some weighted average of the observed values covered by the mask operator, and subtracted from the central value for the residual anomaly.

Although more objective than graphical methods, the so-called grid approach can lead to serious distortion of residual anomalies depending on the coefficients of the mask operator and the size of the grid. These spatial domain convolution operations are also computationally laborious, and thus have been largely supplanted by much more efficient and accurate equivalent spectral filters for separating regional and residual magnetic anomalies.

Spectral filtering methods

Spectral filtering is the predominant method of isolating and enhancing magnetic anomalies. In this procedure (Appendix A.5.1), the input data set is Fourier-transformed into the gridded frequency or wavenumber domain by the direct transform, multiplied by an appropriate filtering function, and then transformed back into the space domain by the inverse Fourier transform. Spectral filters may be classified into standard and specialized filters according to the coefficients of their filtering or transfer functions.

Standard filters have transfer functions with coefficients of 1s and 0s that can be applied to any gridded data set to study its wavelength, directional, and correlation properties. Specialized filters, by contrast, have non-integer coefficients for evaluating the horizontal derivative and integral attributes of the gridded data set, as well as the vertical derivatives and integrals implied by Laplace's equation (Equation 3.6). Of course, other solutions of Laplace's equation can also be taken by specialized filters, including the anomalous potential, vector components, tensor gradients, and spatially continued and reduced-to-pole anomalies. In general, specialized filters extend the utility of the mapped magnetic data for subsurface exploration because they facilitate the efficient and accurate implementation of Laplace's equation, Poisson's theorem, and other characterizations of the potential's spatial attributes.

(A) Wavelength filters Wavelength filters can be very valuable in delineating residual anomalies in the typically complicated anomaly patterns of magnetic field observations. As in gravity anomaly fields (Section 6.5.3), the standard transfer functions can be designed for high-pass/cut filters to enhance/suppress higher-frequency (shorter-wavelength) noise of smaller, near-surface sources; low-pass/cut filters to enhance/suppress lower-frequency (longer-wavelength) regional anomalies; and band-pass/reject filters to enhance/suppress anomalies over a band of frequencies or wavelengths.

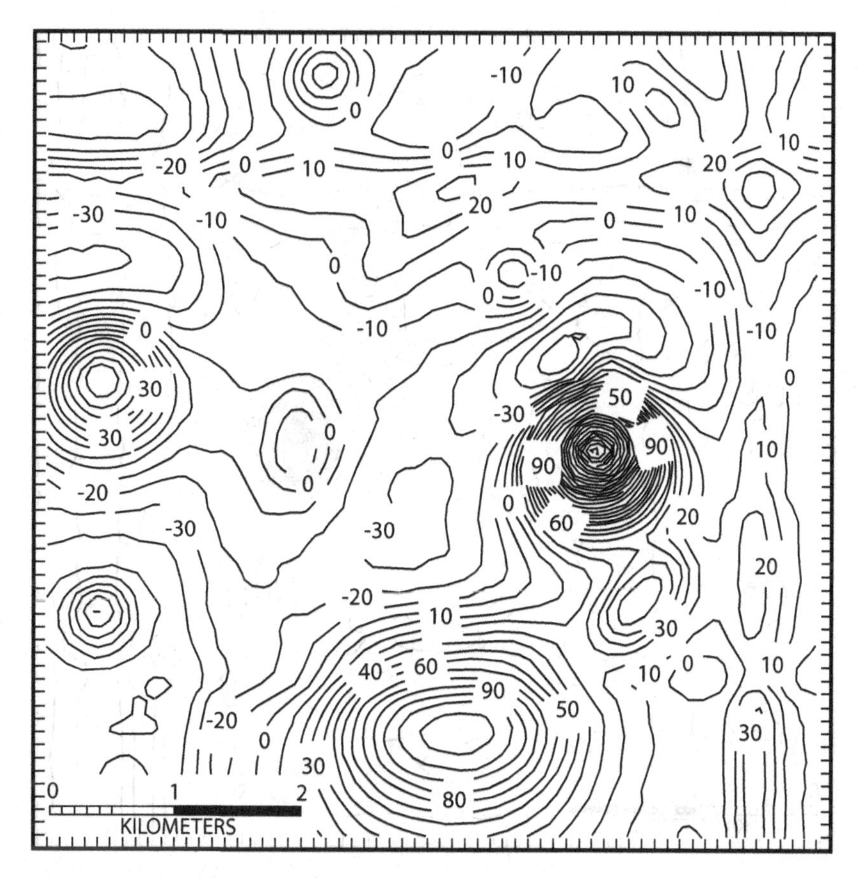

FIGURE 12.9 Total magnetic intensity anomaly map, which is a composite of several residual and regional anomalies plus noise. This is the magnetic equivalent of the gravity anomaly map shown in Figure 6.23. Contour interval is 10 nT. Adapted from Xinzhu and Hinze (1983).

The application of wavelength filtering is illustrated using the magnetic anomaly map of Figure 12.9, which is derived from the same sources that produced the gravity anomaly map in Figure 6.23. Comparisons of these two maps are instructive in demonstrating the relative attributes of gravity and magnetic anomaly fields from common sources. Of particular note are the distorting effects of low-order regional fields on the gravity anomalies relative to the minimal effect of these fields on the magnetic residual fields. In addition, the residual magnetic anomalies are considerably more perceptible than the residual gravity anomalies. The elongated, linear anomalies due to intersecting vertical, tabular bodies paralleling the right and top margins of the map, for example, are much more discernible in the magnetic anomalies than in the gravity data. These and similar differences in the maps have a significant effect on the use of wavelength filtering in the respective fields.

The effect of wavelength filtering on the magnetic anomaly array of Figure 12.9 is shown in Figures 12.10, 12.11, and 12.12. The higher wavenumber components of the magnetic data are emphasized in Figure 12.10 where wavelengths of less than 1,000 m are passed. In this figure the anomalies of higher gradients derived from shallower sources are clearly discerned at the expense of the longer-wavelength anomalies. The effect of the band-pass filter which passes wavelengths between 2,000 and 4,000 m is shown in Figure 12.11. In this case, only the major equidimensional anomalies are emphasized. Finally, the effect of a low-pass/high-cut filter is shown in Figure 12.12 which passes only wavelengths greater than 4,000 m. This map also focuses on three major anomalies derived from separate sources occurring in the lower center, and the right and left centers of the map. The difference in their response reflects variations in the magnetic polarization contrast and size of the sources. It is important to emphasize that wavelength filtering, although useful for increasing the perceptibility of residual anomalies, also generally distorts the anomaly and limits its use in quantitative analysis.

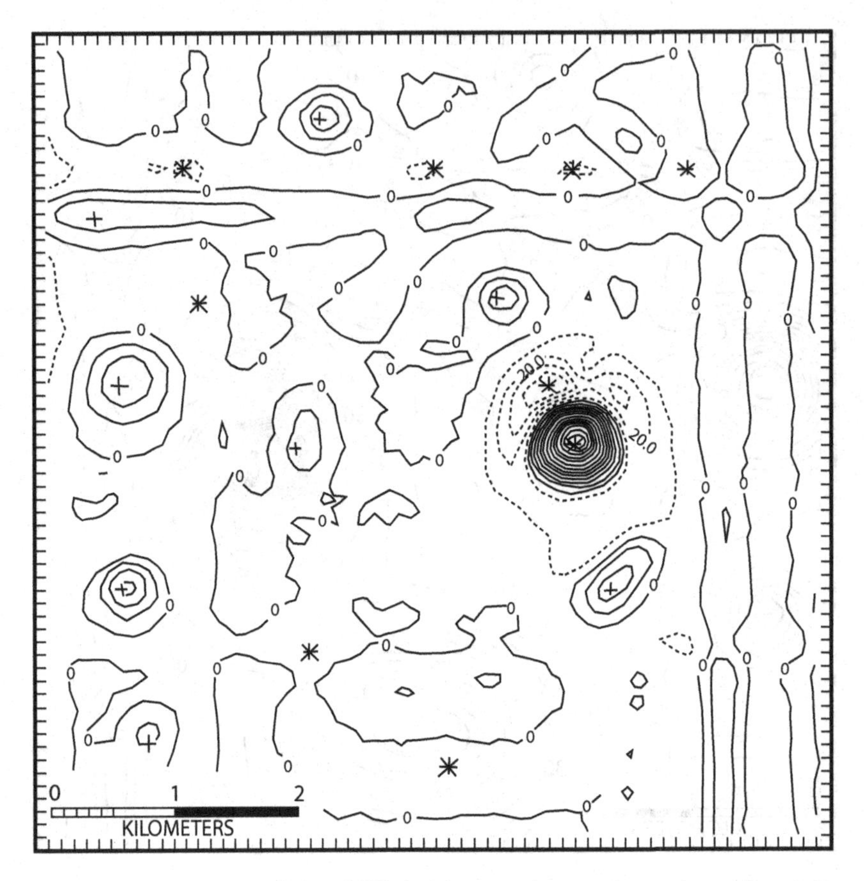

FIGURE 12.10 High-pass filtered magnetic anomaly ($\lambda < 1{,}000$ m) of the composite magnetic anomaly map shown in Figure 12.9. Contour interval is 10 nT. Dashed contours are negative. Adapted from XINZHU and HINZE (1983).

Source depth is a major factor affecting the spectrum of magnetic anomalies, and thus one of the important objectives of wavelength filtering is to focus on anomalies derived from a specific depth range. But, just as it is for gravity anomalies (Section 6.4), this depth filtering can only be achieved very broadly because of the breadth of the anomaly spectrum, and the effect of geometry and size of sources on the spectrum. In particular, as in gravity filtering, a useful minimum-wavelength approximation may be established by assuming that the sources are concentrated volumes. This rule of thumb states that the anomaly from a concentrated (e.g. spherical) source at the depth z_c to its center has the wavelength $\lambda \geq 4 \times z_c$. This is derived from the half-width relationships of the anomaly as in the case of gravity anomalies (Section 7.4.1). Using this relationship, for example, to perform wavelength filtering to obtain the residual anomalies from concentrated sources such as steel drums buried at less than 5 m depth, we would high-pass/low-cut filter the anomaly data for wavelengths of about 20 m and less.

(B) Directional filters Strike-sensitive or directional filters find occasional application in magnetic anomaly map analysis to selectively filter wavenumber components of anomalies which trend over a specific range of directions. With this standard strike-sensitive filter, anomalies originating from geologic sources that have a particular strike can either be enhanced or attenuated. However, all anomalies, even circular anomalies, are stretched out in the pass direction and highly distorted. Thus, this filter must be used with caution. Anomalies that occur within a band of azimuths can either be rejected or passed depending on the objective of the analysis. Generally it is advisable to use the filter in a strike-reject mode because the results are less distorted, and to use a pass band that is no narrower than roughly 30° on either side of the central azimuth. This filter attenuates anomalies striking within a range of directions, and thus permits the interpreter to observe continuity of trends free from a dominant conflicting anomaly strike direction. The directional properties of the filtered anomalies can be highlighted further by shaded relief imaging

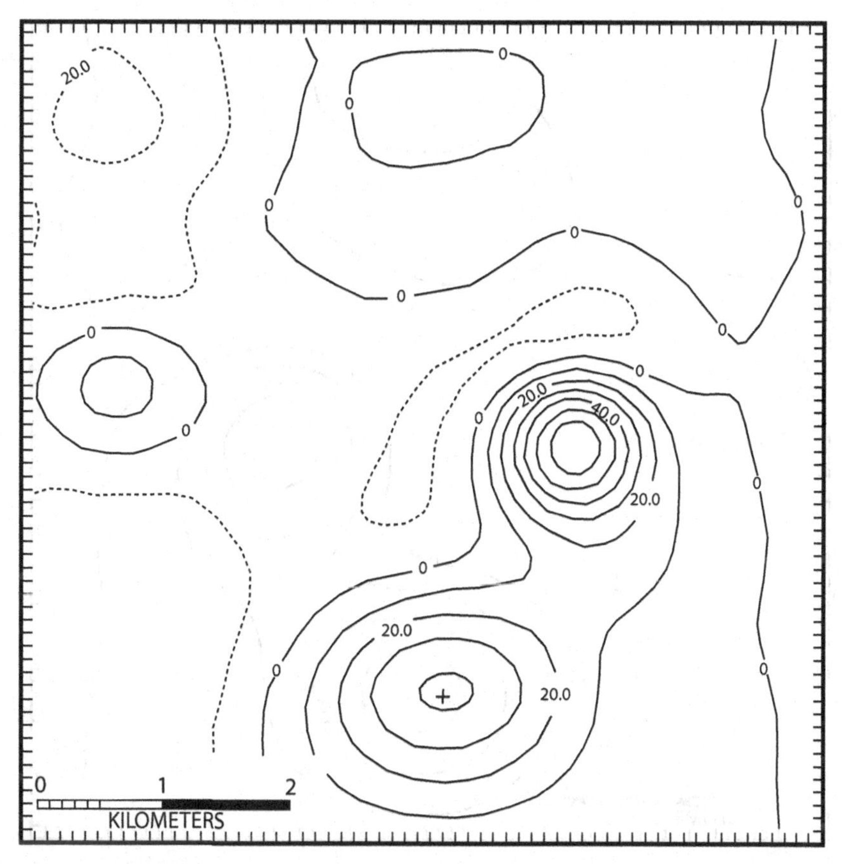

FIGURE 12.11 Band-pass filtered magnetic anomaly (2,000 m $< \lambda <$ 4,000 m) of the composite magnetic anomaly map shown in Figure 12.9. Contour interval is 10 nT. Dashed contours are negative. Adapted from XINZHU and HINZE (1983).

with the illumination angle taken perpendicular to the predominant strike direction.

An example of strike-pass filtering over azimuths 60° to 120° E is shown in Figure 12.13 for the modeled anomalies given in Figure 12.9. The linear anomaly across the map near the upper (northern) edge is emphasized and the similar anomaly striking north along the right edge is eliminated, but the other anomalies that are roughly equidimensional in the observed anomaly map have been distorted by being stretched out in the east–west direction.

(C) Correlation filters Correlation filters invoke standard filter coefficients that isolate anomalies in the context of the correlations that their kth-wavenumber components exhibit with the kth-wavenumber components of other co-registered data sets (VON FRESE *et al.*, 1997a). As described in Appendix A.5.1, correlation filters typically are designed to emphasize either positive ($0 < r(k) \leq 1$), negative ($-1 \leq r(k) < 0$), or null ($r(k) \approx 0$) correlations between data sets. Given how source ambiguities and anomaly superposition effects hinder magnetic anomaly interpretation, the correlation of magnetic anomalies with outcrop, topographic, and other geophysical features can be useful in isolating residuals. Magnetic and gravity anomaly correlations, in particular, are commonly considered for minimizing these interpretational limitations.

Figure 12.14 gives a spectral correlation filtering example for two ascending (i.e. dusk) orbits of Magsat magnetic observations at roughly 330 km altitude across the Arctic from northeastern Greenland to southwestern Finland. Panel (a) shows the magnetic anomalies after removal of the core field model (ALSDORF *et al.*, 1994). The orbits are only about 7 km apart, and thus record essentially the same magnetic effects of the lithosphere because they are so close together compared with the depth of the lithospheric sources. However, the anomalies are relatively poorly correlated ($r = 0.479$) because they also include the spatially and temporally dynamic effects of the auroral external fields, measurement errors, and other nonlithospheric sources. Thus, spectral correlation filtering

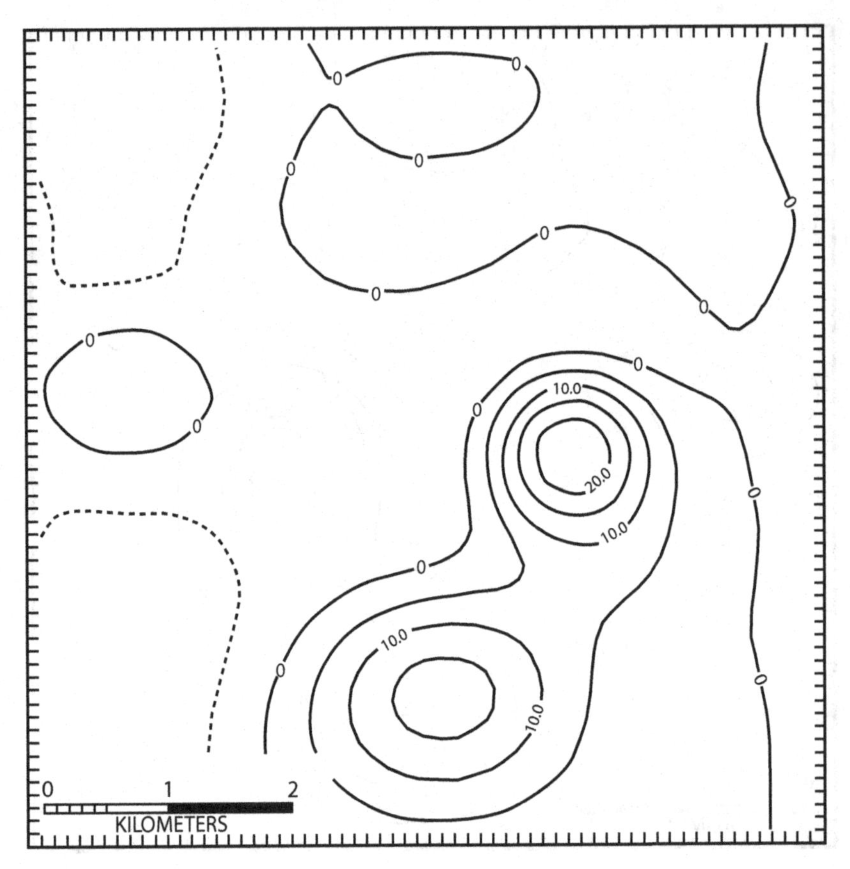

FIGURE 12.12 Low-pass filtered magnetic anomaly ($\lambda > 4{,}000$ m) of the composite magnetic anomaly map shown in Figure 12.9. Contour interval is 10 nT. Dashed contours are negative. Adapted from XINZHU and HINZE (1983).

was applied to help suppress these incoherent features and enhance the presence of the coherent lithospheric signals in the two orbital data sets.

Using the correlation spectrum for the signals in panel (a), all wavenumber components with $r(k) \geq 0.5$ were inversely transformed for the estimates in panel (b) of their lithospheric components. The enhanced correlation ($r = 0.934$) obtained by these results can be broadly interpreted for improvements in the noise-to-signal (N-to-S) ratio from Equation A.72. Accordingly, relative to the raw residuals in panel (a), the correlation-filtered signals reflect roughly a 75% reduction in the non-lithospheric noise effects.

Panel (c) shows the components of the raw residuals that were rejected by the correlation filters. Although partly correlated, they reveal little apparent sensitivity for the regional magnetic effects of the crustal geology (e.g. ocean-continent margins, ocean ridges, basins). They appear more readily caused by non-lithospheric effects (external field effects, core field reduction errors, etc.) and hence were discarded. Accordingly, panel (d) gives the least-squares estimate of lithospheric anomalies from the point-by-point averages of the coherent signals in the second panel. Here, the differences between the coherent signals are also given as root-mean-square errors (RMSE) to constrain interpretations of the lithospheric anomaly estimates.

Poisson's theorem gives the quantitative basis for the correlation between the magnetic and gravity effects of a common source. Figure 12.15 shows Poisson's relation as implemented on co-registered profiles of total field magnetic anomaly (top left) and gravity anomaly (bottom left) data due to a common spherical source. For correlation analysis, the phases of the anomalies must be the same according to Poisson's relation, which in geophysical survey practice can be achieved most expediently by reducing the total field magnetic anomalies to the pole (top right) and taking the first vertical derivative gravity anomaly (bottom right). Both anomaly transformations are readily taken using specialized spectral filters as described in

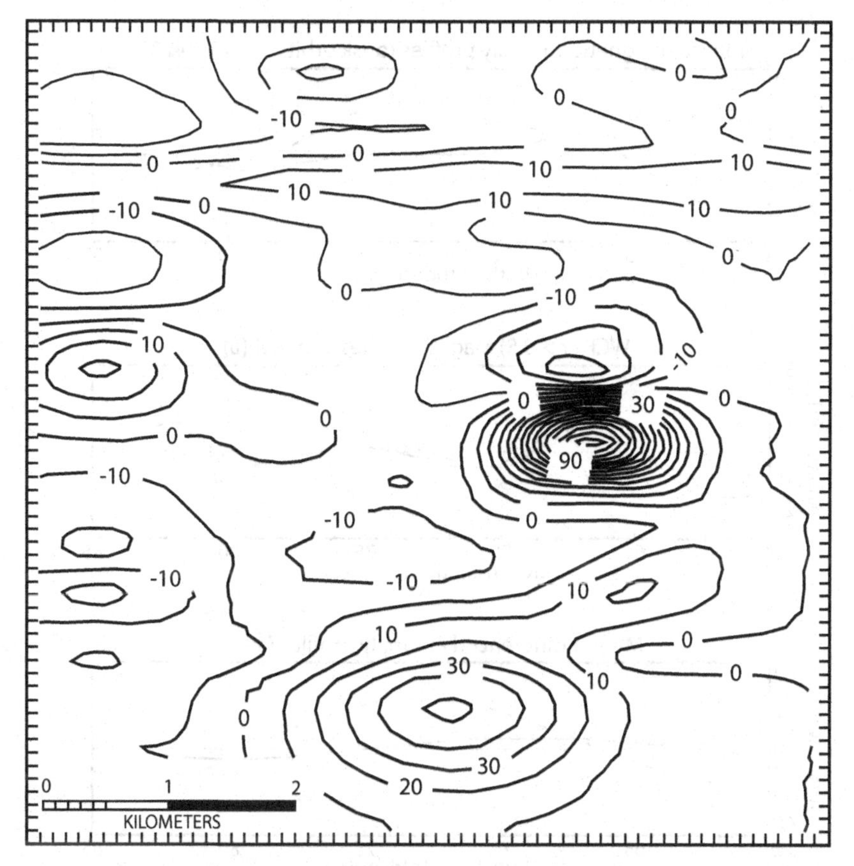

FIGURE 12.13 Strike-pass filtered magnetic anomaly components with azimuths in the range 60–120° E of the composite magnetic anomaly map shown in Figure 12.9. Contour interval is 10 nT. Adapted from XINZHU and HINZE (1983).

the subsubsections below. Examples of the use of these anomaly transformations for correlation filtering are given by VON FRESE *et al.* (1997c) and DE RITIS *et al.* (2010).

(D) Derivative and integral filters Derivative and integral filters use specialized transfer functions that are inverses of each other to evaluate the pth-order derivative and integral components of gridded data as described in Appendix A.5.1. The derivative methods in particular have had considerable application in enhancing local magnetic anomalies, but require high-quality data. Data errors are enhanced with increasing order of differentiation. As a result, the vertical gradient method (i.e. the first vertical derivative) is less subject to problems from errors in the data than is the second vertical derivative method, and thus has been widely applied.

Derivative methods, both horizontal gradient and second vertical derivative, are used to approximate the margins of magnetic sources (e.g. VACQUIER *et al.*, 1951). As illustrated by VACQUIER *et al.* (1951), the zero second vertical derivative contour approximates the plan view outline of the margin of a magnetic anomaly source that is broad in comparison to the height of the observations above the source (Figure 12.16). However, as pointed out for gravity anomalies, the second derivatives of magnetic anomalies are typically useful only in increasing the perceptibility of residual anomalies, and thus are limited in quantifying the source of the anomaly.

An example of a first vertical derivative magnetic anomaly map is shown in Figure 12.17, after XINZHU and HINZE (1983). This map, which was prepared from the total magnetic intensity anomaly map shown in Figure 12.9, does an excellent job of isolating the location of the residual anomalies. Figure 12.18 shows the zero second vertical derivative contour of Figure 12.9 after low-pass filtering the data by upward continuing to an elevation of 100 m to eliminate the high-wavenumber noise components. The zero contour closely approximates the location of the margins of the residual anomaly sources.

The horizontal curvature or second vertical derivative of magnetic anomalies is also used to prepare terrace maps that delineate large, sharply defined regions of similar

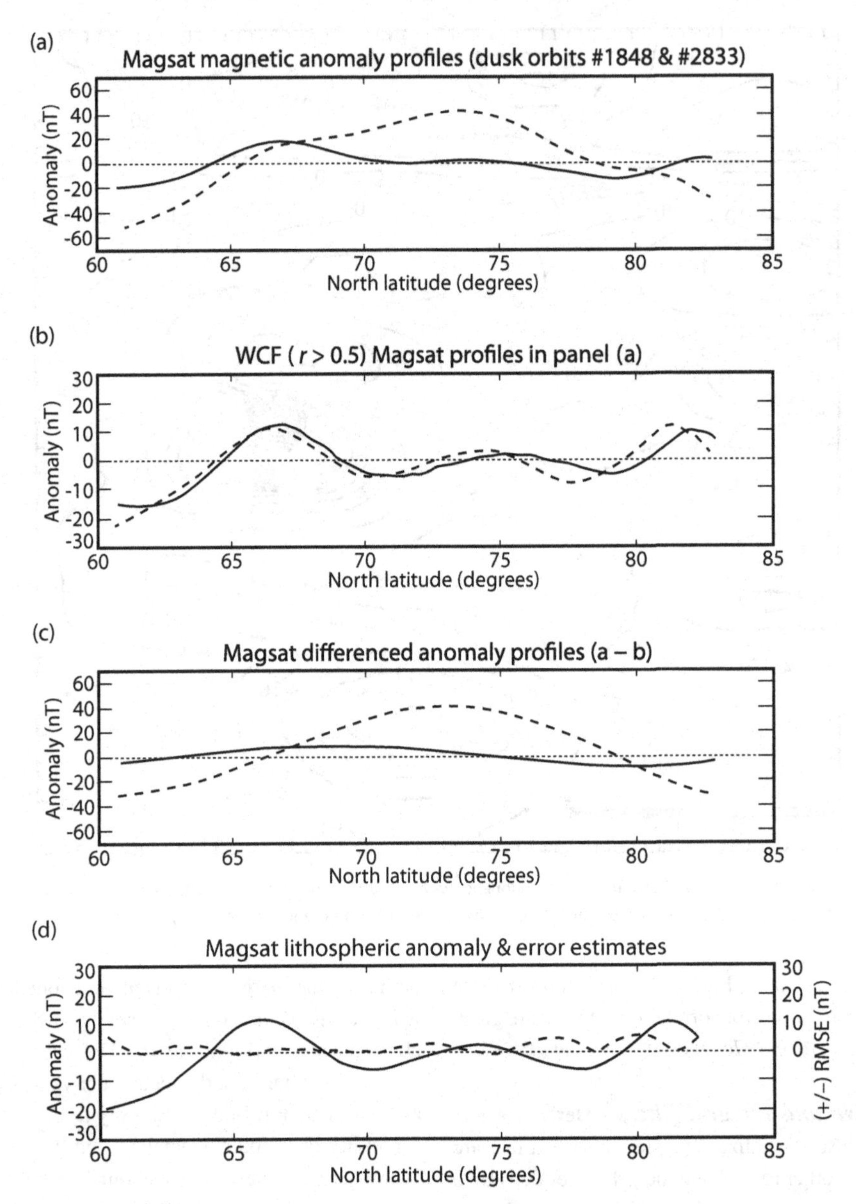

FIGURE 12.14 Wavenumber correlation filtering (WCF) of two nearest-neighbor orbits of magnetic observations from the Magsat mission. In panels (a), (b) and (c), the solid and dashed lines respectively give the magnetic anomalies for orbits #1,848 and #2,833, which are separated by an average distance of about 7 km at altitudes of roughly 400 km. The least-squares anomaly estimates from averaging the WCF profiles in (b), as well as the related root-mean-squared errors (RMSE) are plotted as the respective solid and dashed lines in panel (d). Adapted from VON FRESE *et al.* (1997a).

magnetization properties (CORDELL and MCCAFFERTY, 1989). The terrace map roughly approximates a geological map of the area with the identified boundaries correlating with the position of steeply-dipping lithologic boundaries and faults. This method is based on processing the data of a profile or map with a curvature (second derivative) operator that classifies each data point within a specific domain depending upon the algebraic sign of the local curvature. The process is repeated iteratively on the data resulting from the previous processing until the output consists of a series of horizontal segments separated by vertical boundaries. The horizontal segments, or domains, can be scaled to physical property units by making assumptions regarding the vertical extent of the physical property

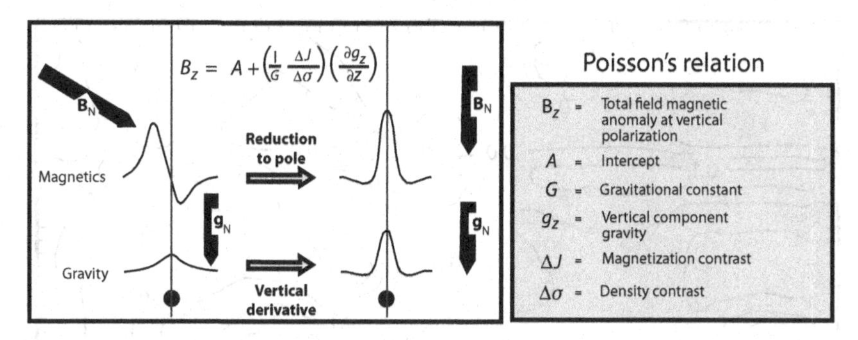

FIGURE 12.15 Poisson's relation between magnetic and gravity anomalies for a point source (dot) with positive density and magnetization contrasts.

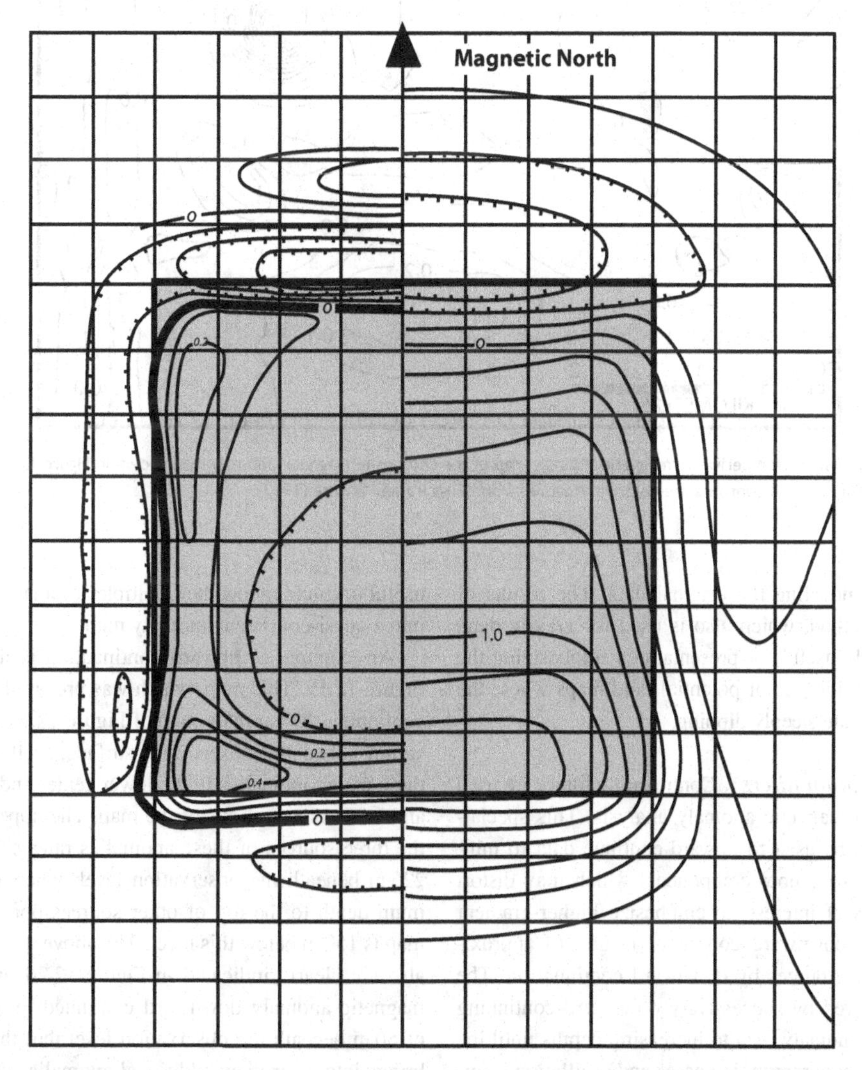

FIGURE 12.16 Normalized total intensity magnetic anomaly (BT/J) of a prismatic source contoured in the right half of the grid with the second vertical derivative (nT/[grid interval]^2) of the total intensity magnetic anomaly contoured in the left half of the grid. The bold contour marks the zero second vertical derivative that helps to delineate the boundary of the source. The grid interval is in units of depth to the top of the prism, and the inclination of the geomagnetic field and the magnetization is 60°, and the declination is 0° so that $I = I' = 60°$ and $D = D' = 0°$. Adapted from VACQUIER *et al.* (1951).

FIGURE 12.17 First vertical derivative magnetic anomaly map of the composite magnetic anomaly map shown in Figure 12.9. Contour interval is 0.1 nT/m. Dashed contours are negative. Adapted from XINZHU and HINZE (1983).

contrast and modeling the original data. The results of the terrace method, which also is used for gravity data, are particularly useful for presentations emphasizing the geological significance of potential field maps where the geologic units are steeply dipping.

(E) Continuation filters Continuation filters are used extensively in magnetic anomaly analysis. This specialized filter can be used to upward-continue data to minimize high-wavenumber components which may distort the anomalies of interest, to emphasize higher-gradient anomalies by downward continuation, and to approximate depths to sources by downward continuation. The latter is achieved by successively downward-continuing the observed anomaly data to increasing depths until the projected anomaly rapidly increases and oscillation is initiated between negative and positive values around the central value. This is the approximate maximum depth of the source of the anomaly. Also, continuation of anomaly data observed at different altitudes to a common altitude is useful in stitching together multiple aeromagnetic surveys into a quasi-consistent anomaly map.

An example of upward continuation is illustrated in Figure 12.19. This map, which was prepared by upward-continuing the anomaly map of Figure 12.9 to 300 m elevation above the observation surface, well isolates the three major anomalies in the lower center and in the right and left center portions of the map. The upper surface of the three sources of these anomalies ranges from 500 to 225 m beneath the observation level, whereas the minimum depth to the top of other sources portrayed in the map is 150 m below this level. The above three anomalies also are clearly indicated in Figure 12.20, which is the magnetic anomaly downward continued to an elevation of 50 m beneath the observation level, but this map also brings into clear view additional anomalies derived from other near-surface sources.

Although continuation filters are attractive for isolating anomalies of predominately long or short wavelengths because their response varies slowly with wavelength, care

FIGURE 12.18 Second vertical derivative magnetic anomaly map at an elevation of 100 m of the composite magnetic anomaly map shown in Figure 12.9. Only the zero second vertical derivative contour is plotted. Adapted from XINZHU and HINZE (1983).

is needed in their application because there are limits to the level of continuation, either upward or downward, that can be done accurately. Downward continuation is only theoretically possible where the source of the anomaly lies below the downward-continuation level. If the level is within the source, the resulting downward-continued anomaly rapidly increases in amplitude and oscillates with alternating positive and negative values around the central value. These oscillating signatures serve no interpretational purpose, but they are the basis for using downward continuation as a crude depth determination procedure. Also downward continuation, which passes only the high-wavenumber components of the anomaly field, is subject to extreme distortions when the anomaly field contains high-wavenumber noise. If the noise can be identified and isolated from the anomaly components it is possible to subject the anomaly data to a high-cut noise filter prior to downward continuation, but overlap between the noise spectra and the anomaly spectra will lead to distortion of the downward-continued anomalies. Downward continuation to great depths will only accentuate the longer-wavelength components because the shorter wavelengths are lost in the noise.

Care must also be used in upward continuation of anomaly fields. Theoretically it is possible to project an anomaly accurately to any higher altitude. However, this is possible only when the observed data map the entire spectra of an anomaly with a high degree of accuracy. Errors in the anomalies may lead to adulterating anomalies or the inability to map anomaly spectra. Furthermore, the mapped areas depending on the depth to the sources must be large enough to accurately map the long wavelengths of the anomaly spectra. Insufficiently large areas will prevent accurate mapping of the spectra for filtering to higher elevations above the level of the observations. Accordingly care must be used in selecting the elevation of upward continuation to a level which will provide valid results. Model studies based on theoretical anomalies, the depth to their source, the anticipated noise, the size of the survey region, and the level of the desired upward

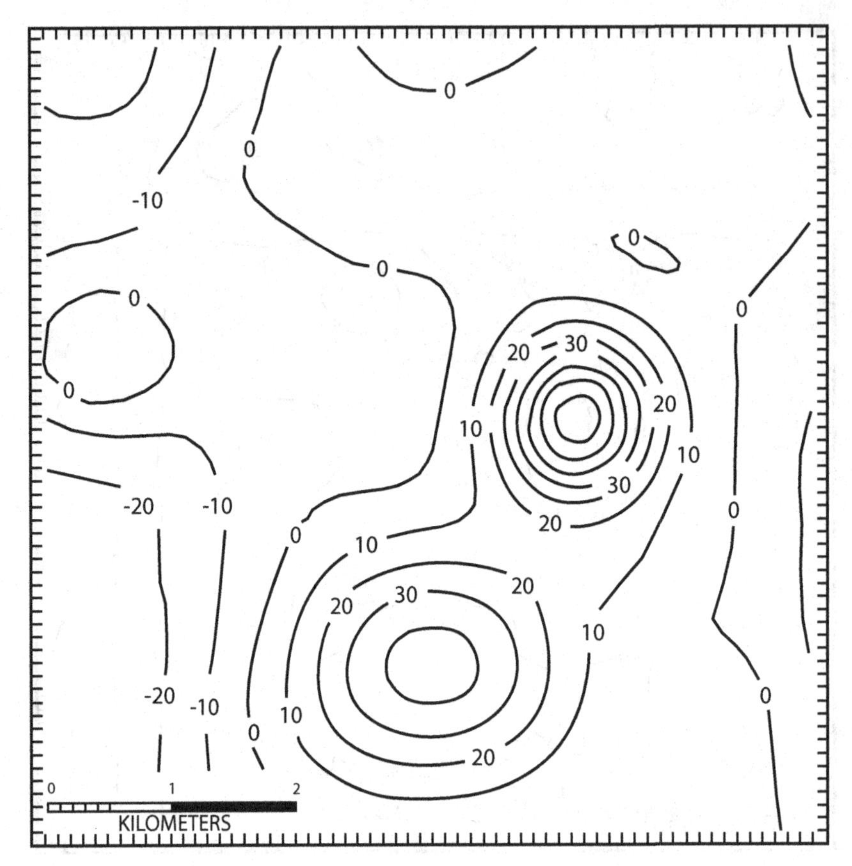

FIGURE 12.19 Upward continued magnetic anomaly to an elevation of 300 m of the composite magnetic anomaly map shown in Figure 12.9. Contour interval is 10 nT. Adapted from Xinzhu and Hinze (1983).

continuation provide useful guidelines for the limits of the upward-continuation filtering. Generally upward continuation is possible with a reasonable degree of accuracy to levels several times the depth to the sources below the observation level.

Combinations of upward- and downward-projected anomalies and the observed anomalies may be used to process the data into a more interpretable form. For instance, Jacobsen (1987) has shown that upward continuation makes a useful filter for roughly isolating magnetic anomalies derived from a specific level within the Earth. This is achieved by subtracting from the original-level anomaly data, the data obtained by upward-continuing the original data to twice the specified level within the Earth. Also, to minimize the noise in an anomaly field, it is possible to upward-continue the observed data to a level where high-wavenumber components are eliminated by rounding off the anomaly values at the projected altitude and then downward-continuing the upward-continued anomaly field to the elevation of the observed anomalies.

(F) Reduction-to-pole filters Owing to the vectorial nature of magnetization and the variation in inclination and declination of the geomagnetic field from the geomagnetic equator to the pole, the maxima of magnetic anomalies due to subsurface sources, unlike those of gravity fields which are due to scalar density variations, are shifted from directly over the source and the anomaly becomes highly asymmetric. This complicates the interpretation of magnetic anomalies, especially at lower geomagnetic latitudes. To overcome this effect, Baranov (1957) suggested a method of transforming magnetic anomalies observed at any geomagnetic latitude to the anomalies that would be observed at vertical magnetization and vertical magnetic field based on Poisson's relation. This so-called reduction-to-pole has become an important step in many magnetic interpretational procedures. The assumption is made that the magnetization in the source is constant and directed in the magnetic field either as a result of induced magnetization by the Earth's field or by viscous magnetization. The process involves a linear transformation of the observed anomaly data either in the wavenumber or space domain to

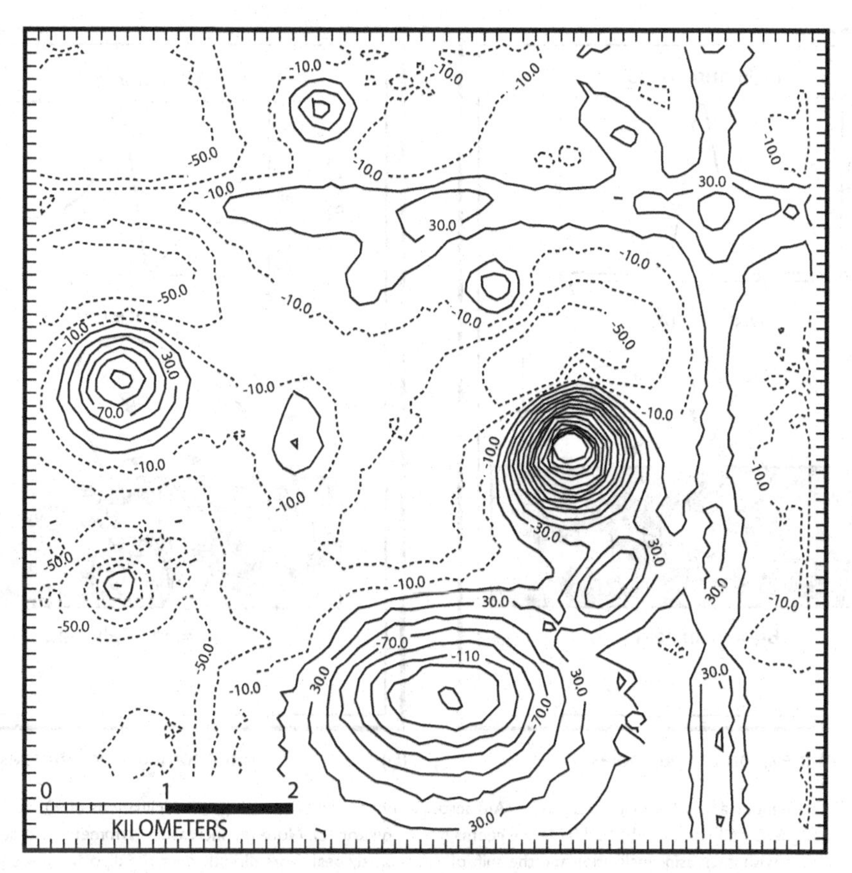

FIGURE 12.20 Downward continued magnetic anomaly to an elevation of −50 m of the composite magnetic anomaly map shown in Figure 12.9. Contour interval is 20 nT. Dashed contours are negative. Adapted from XINZHU and HINZE (1983).

the data as would be observed at either of the geomagnetic poles (GUNN, 1995), and thus the procedure is a filtering operation.

Reduction-to-pole (RTP) filters adjust magnetic anomalies for the complicated effects of the inclined magnetization of the source and the inclination of the ambient magnetic field. The shape of a magnetic anomaly relative to its source is distorted by these parameters causing a decrease in the amplitude of the anomaly, asymmetry in the anomaly, and shift of the peak of the anomaly to the south in the northern geomagnetic hemisphere and the converse in the southern geomagnetic hemisphere. This complicates the interpretation of the shape of the source, the amplitude of its magnetic polarization, and the location of the anomaly source. These effects are illustrated in the total magnetic anomaly profiles of Figures 12.21 and 12.22 across a concentrated source magnetized in an ambient field of varying inclination from the geomagnetic poles to the equator.

In an attempt to minimize this complication, magnetic anomaly fields commonly are transformed to vertical polarization of the magnetic pole. This assumes that all the magnetic anomaly fields are due to magnetization contrasts in the Earth's magnetic field. The RTP-calculation is based on Poisson's relation (Equation 3.106) that connects the gravity U and magnetic V potentials at the observation point (x, y, z), for a common source of density σ, and magnetization with intensity J, in the direction, $i = \sqrt{(\hat{u}'_x)^2 + (\hat{u}'_y)^2 + (\hat{u}'_z)^2}$, by

$$V(x, y, z) = \frac{-C_m J}{G\sigma}\left(\frac{\partial U(x, y, z)}{\partial i}\right), \tag{12.7}$$

where G is the gravitational constant and $\hat{\mathbf{u}}'$ is the unit magnetization vector (Equation 9.11).

Thus, Poisson's relation relates the magnetic potential of an arbitrary volume to the derivative of the gravitational potential of the same volume with respect to the direction of magnetization of the body. If vertical magnetic polarization is assumed where $i = z$ like at the geomagnetic poles, for example, and remembering that the derivative of the gravitational potential in the vertical z-direction is

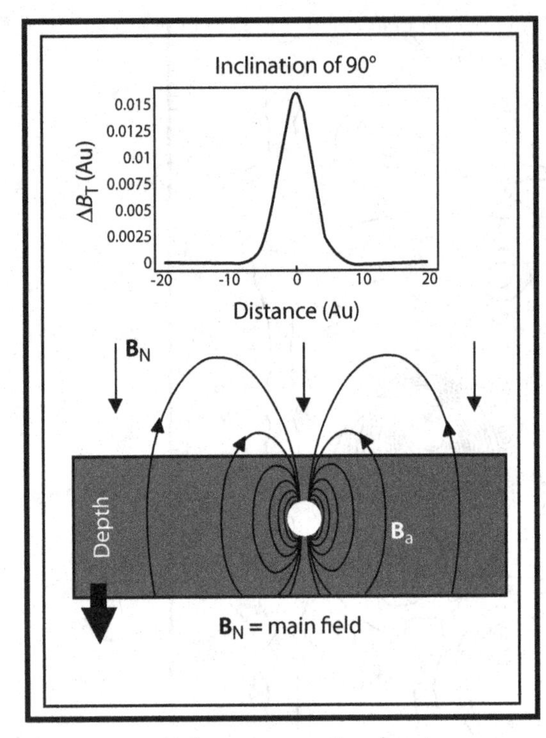

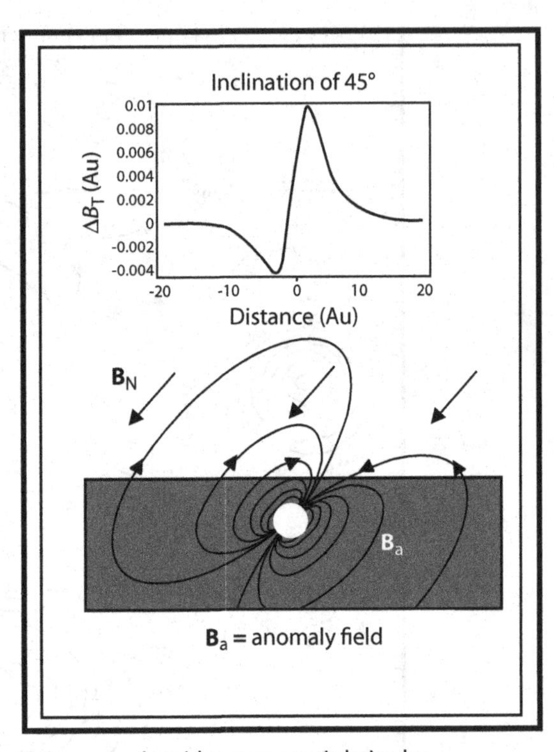

FIGURE 12.21 Total magnetic anomalies in arbitrary units (Au) across a spherical source at magnetic inclinations of 90° at the geomagnetic poles (a), 45° at geomagnetic mid-latitudes (b), and 0° at the geomagnetic equator (c). Note the increasing asymmetry and decreasing amplitude of the anomaly with decreasing inclination and the shift of the anomaly peak from directly over the dipolar source at the geomagnetic poles towards the geomagnetic equator where the peak is split on either side of the source along geomagnetic declination.

the gravitational acceleration, g, then the vertical magnetic intensity is

$$B_z = -\frac{\partial V}{\partial z} = \frac{C_{\mathrm{m}} J}{G\sigma}\left(\frac{\partial^2 U}{\partial z \partial i}\right) = \frac{C_{\mathrm{m}} J}{G\sigma}\left(\frac{\partial g}{\partial z}\right). \quad (12.8)$$

Where the magnetization is in the direction of the ambient geomagnetic field, a rewrite of Equation 12.8 gives the more commonly measured total field anomaly as

$$B_{\mathrm{T}} = \frac{C_{\mathrm{m}} J}{G\sigma}\left(\frac{\partial^2 U}{\partial i^2}\right). \quad (12.9)$$

Thus, the pseudogravity potential U due to the magnetic field B_{T} in the direction i is

$$U = \frac{G\sigma}{C_{\mathrm{m}} J}\int_{-\infty}^{\infty}\int_{-\infty}^{\infty}(B_{\mathrm{T}})\partial i \partial i. \quad (12.10)$$

Using Equation 12.10 in Equation 12.9 and differentiating with respect to the vertical z-direction obtains the vertical intensity from the total intensity by

$$B_z = \frac{\partial^2}{\partial z^2}\left(\int_{-\infty}^{\infty}\int_{-\infty}^{\infty}(B_{\mathrm{T}})\partial i \partial i\right). \quad (12.11)$$

Equation 12.11 gives the total intensity reduced-to-the-pole anomaly (RTP-anomaly) in the spatial domain which requires a numerical evaluation of the double integral (e.g. BARANOV and NAUDY, 1964). More commonly, the RTP-anomaly, B_z, is evaluated in the spectral domain using the fast Fourier transform, $\mathcal{B}_T$, of the total field anomaly, B_{T}. In particular, assuming the magnetization is all induced in the geomagnetic field with direction i obtains the Fourier transform pair

$$B_{\mathrm{T}} = -\partial V/\partial i \Longleftrightarrow -\mathcal{FPD}_{\mathrm{i}} \times \mathcal{V} = \mathcal{B}_T, \quad (12.12)$$

where $\mathcal{FPD}_{\mathrm{i}}$ is the Fourier transform operator for taking the first partial derivative with respect to i. To evaluate the operator, consider the chain rule expansion of the partial derivative given by

$$\frac{\partial}{\partial i} = \frac{\partial}{\partial x}\left(\frac{\partial x}{\partial i}\right) + \frac{\partial}{\partial y}\left(\frac{\partial y}{\partial i}\right) + \frac{\partial}{\partial z}\left(\frac{\partial z}{\partial i}\right), \quad (12.13)$$

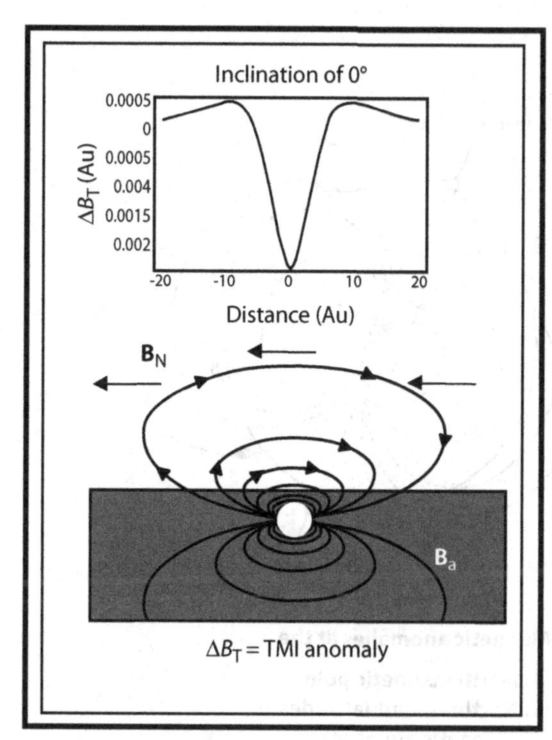

(c) At the geomagnetic equator

FIGURE 12.21 *(cont.)*

where

$$x = i \times \cos I \cos D \ni \partial x/\partial i$$
$$= \cos I \cos D \text{ and } \partial/\partial x \Longleftrightarrow 2\pi j(k_x)$$
$$y = i \times \cos I \sin D \ni \partial y/\partial i$$
$$= \cos I \sin D \text{ and } \partial/\partial y \Longleftrightarrow 2\pi j(k_y)$$
$$z = i \times \sin D \ni \partial \Delta z/\partial i$$
$$= \sin D \text{ and } \partial/\partial z \Longleftrightarrow 2\pi\sqrt{k_x^2 + k_y^2}. \quad (12.14)$$

Using the results of Equation 12.14 in Equation 12.13 yields the spectral operator for the first partial derivative given by

$$\mathcal{FPD}_i = 2\pi \left[jk_x \cos(I)\cos(D) + jk_y \cos(I)\sin(D) + \sqrt{k_x^2 + k_y^2}\sin(I) \right]. \quad (12.15)$$

Now, the potential V in Equation 12.12 is obtained by integrating the total field B_T in i. However, this spatial integration is equivalent in the spectral domain to multiplying the transform $\mathcal{B}_T$ by the inverse derivative operator $(-\mathcal{FPD}_i)^{-1}$ (Appendix A.5.1) so that the transform pair

$$B_z = -\partial V/\partial z \Longleftrightarrow -\mathcal{FPD}_z \times \mathcal{V}$$
$$= (\mathcal{FPD}_z \times \mathcal{B}_T)/\mathcal{FPD}_i = \mathcal{B}_z \quad (12.16)$$

holds, where the spectral first partial derivative operator with respect to z is $\mathcal{FPD}_z = 2\pi\sqrt{k_x^2 + k_y^2}$ as given in Equation 12.14.

Thus, inversely transforming $(\mathcal{FPD}_z \times \mathcal{B}_T)/\mathcal{FPD}_i$ gives the total field anomaly B_T reduced to its equivalent vertical effect B_z at the geomagnetic pole. Although the RTP anomaly B_z was derived assuming induced and/or viscous magnetization, known remanent components can also be accommodated because the total magnetization is the superposition of induced and remanent magnetizations. In addition, this procedure clearly can be adapted to obtain any integral or derivative component of the magnetic potential from any one of the other components. Additional applications of RTP anomalies are considered in Section 13.4.

Unfortunately, RTP methods are unstable at low magnetic latitudes, especially within ±15° of the geomagnetic equator. This results in the amplification of the noise and anomaly content of the data in the north–south direction, rendering the results unsuitable for interpretation. Numerous techniques, both in the spatial and wavenumber domain, have been developed to minimize the instability problem (e.g. Silva, 1986; Gunn, 1995; Swain, 2000; Lu *et al.*, 2003; Arkani-Hamed, 2007). An alternative approach to eliminating the low-magnetic latitude instability is to use the equivalent source procedure described below. This procedure is computationally intensive, but it is readily adapted to differential reduction-to-the-pole where the direction of magnetization and the field are changed across an anomaly field commensurate with the geomagnetic field variation. This procedure readily accounts for the change in the intensity of the field with geomagnetic latitude, with its related change in the induced magnetization and the amplitudes of the anomalies. Still another alternative to reducing magnetic data to the pole at low magnetic latitudes is to reduce the anomalies to the equator (e.g. Aina, 1986). However, this alternative has received limited attention because of the difficulties inherent in interpreting magnetic anomalies in the horizontal geomagnetic field which is present at the magnetic equator. A constant geomagnetic field is generally used for RTP processing of limited sized regions measured in no more than a few hundred kilometers. The potential error in using a constant field over an area can be determined by considering the change in the anomalies over the range of inclination and declination of the survey.

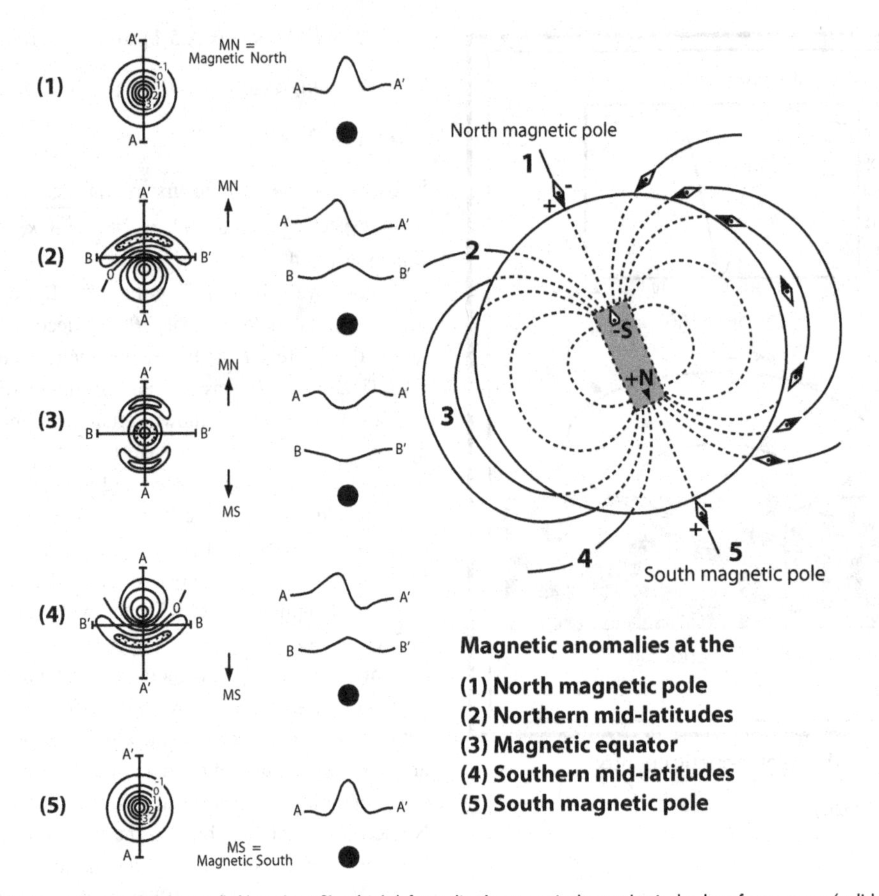

FIGURE 12.22 Magnetic anomaly maps (left) and profiles (right) for a dipole or equivalent spherical subsurface source (solid dot) with constant magnetization intensity along zero declination and varying inclination between the magnetic north (MN) and south (MS) poles, and magnetic equator (ME). The magnetic anomaly can be reduced to the pole (RTP) or the equator (RTE) or any polarization in between using pseudoanomaly transformations. However, the RTP anomaly is the most widely used in practice for comparisons with gravity, heat flow, seismic and other source-centered imagery of the subsurface. Adapted from BREINER (1973).

Figure 12.23 shows the RTP magnetic anomaly map of the observed magnetic anomaly data of Figure 12.9 which was calculated for an inclination of 60° and a declination of 10° W. The RTP anomalies are shifted to the north with a larger shift for the deeper-sourced (longer-wavelength) anomalies. The minima to the north (top) of the anomalies are also highly attenuated. Overall the RTP magnetic anomalies are much more correlated with the gravity anomalies of the model which are shown in Figure 6.23. Figure 12.24 shows the observed total field magnetic anomalies of a portion of east China (a) observed at a mean altitude of roughly 350 m which has been differentially reduced-to-the-pole (DRTP) in the spatial domain (b) using the finite-impulse-response filter (LU, 1998). The mean declination of the geomagnetic field in the area of the map is roughly 5° E and the inclination has a mean value of 50° and varies approximately 20° over the region from south to north. The anomalies of 25 km dominant wavelength in the northern portion of the region are shifted 6 km north on the DTRP map and even greater in the southern portion, and all anomalies of greater than 25 km wavelength are shifted even more. Furthermore, the skewness of the magnetic anomalies is decreased in the DTRP map, making the geological interpretation of the anomalies easier and better.

(G) Pseudogravity filters Psuedogravity maps may be similarly calculated from the magnetic anomaly map. Consideration of Equation 12.10, for example, shows that the gravity anomaly g of a mass variation that also produces a total field anomaly B_{T} can be obtained from

$$g = \frac{\partial U}{\partial z} = \left(\frac{G\sigma}{C_{\mathrm{m}}J}\right)\frac{\partial}{\partial z}\int_{-\infty}^{\infty}\int_{-\infty}^{\infty}(B_{\mathrm{T}})\partial i \partial i. \qquad (12.17)$$

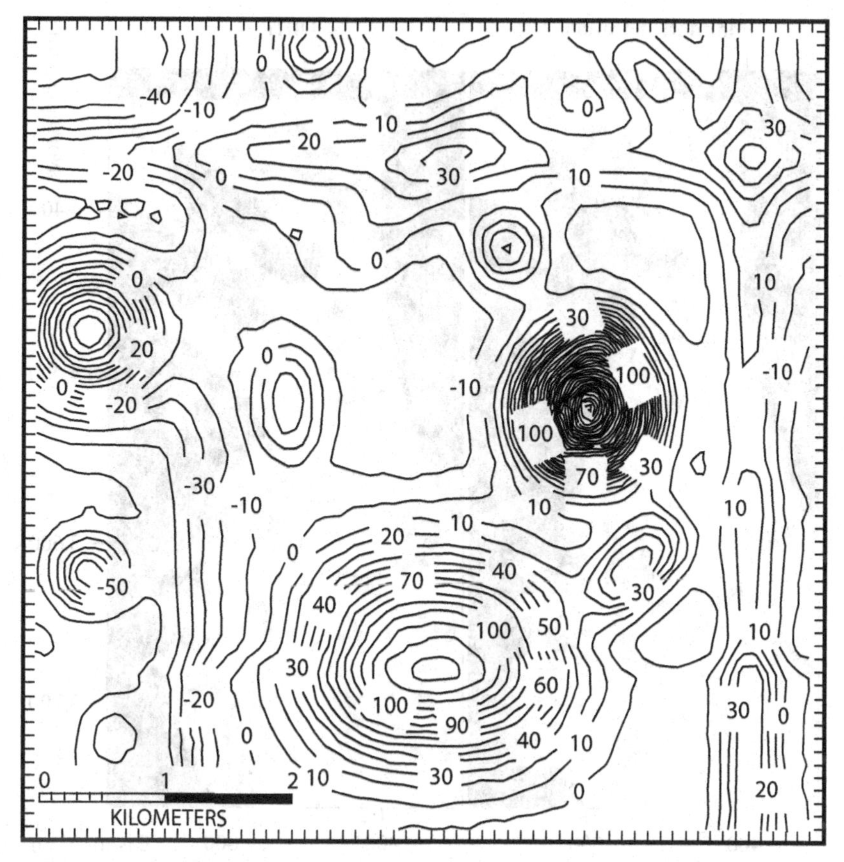

FIGURE 12.23 Reduced-to-pole magnetic anomaly map of the composite magnetic anomaly map shown in Figure 12.9. Contour interval is 10 nT. Adapted from XINZHU and HINZE (1983).

by taking the double integral numerically in the spatial domain (BARANOV, 1957).

In practice, however, the more common approach is to spectrally filter the pseudogravity effects from magnetic effects (e.g. CORDELL and TAYLOR, 1971). Taking the Fourier transform of Equation 12.9, for example, obtains the Fourier transform pair

$$B_T = \frac{C_{\mathrm{m}} J}{G\sigma}\left(\frac{\partial^2 U}{\partial i^2}\right) \Longleftrightarrow \frac{C_{\mathrm{m}} J}{G\sigma}(\mathcal{FPD}_{\mathrm{i}}^2 \times \mathcal{U}) = \mathcal{B}_T, \qquad (12.18)$$

where the square of $\mathcal{FPD}_{\mathrm{i}}$ (Equation 12.15) is the spectral operator for taking the second partial derivative with respect to i^2. Thus, the pseudogravity anomaly forms the transform pair

$$g = \frac{\partial U}{\partial z} = \frac{G\sigma}{C_{\mathrm{m}} J}\left(\frac{\partial}{\partial z}\int_{-\infty}^{\infty}\int_{-\infty}^{\infty}(B_{\mathrm{T}})\partial i \partial i\right)$$
$$\Longleftrightarrow \frac{G\sigma}{C_{\mathrm{m}} J}(\mathcal{FPD}_{\mathrm{z}} \times \mathcal{B}_T)/\mathcal{FPD}_{\mathrm{i}}^2 = \mathcal{G}. \qquad (12.19)$$

Inversely transforming $\mathcal{G}$ obtains the pseudogravity anomaly g in the spatial domain from the wavenumber spectrum $\mathcal{B}_T$ of the total field magnetic anomaly B_{T} due to a source of uniform density σ and magnetization with intensity J and direction i. This approach, of course, can be used to estimate pseudogravity effects from any derivative and integral component of the magnetic potential. Additional pseudogravity anomaly applications are considered in Section 13.4.

Equivalent source methods

Equivalent source modeling also has a role in magnetic anomaly analysis similar to its role in gravity analysis (Section 6.5.3). As opposed to the use of fictitious gravity point poles in gravity anomaly analysis, estimates of the magnetic anomaly components are developed based on modeling the anomaly as the summation of effects from fictitious point dipoles. These derived fields include interpolated and continued fields, derivatives, integrals, pseudogravity anomalies, differentially reduced-to-pole anomalies, and

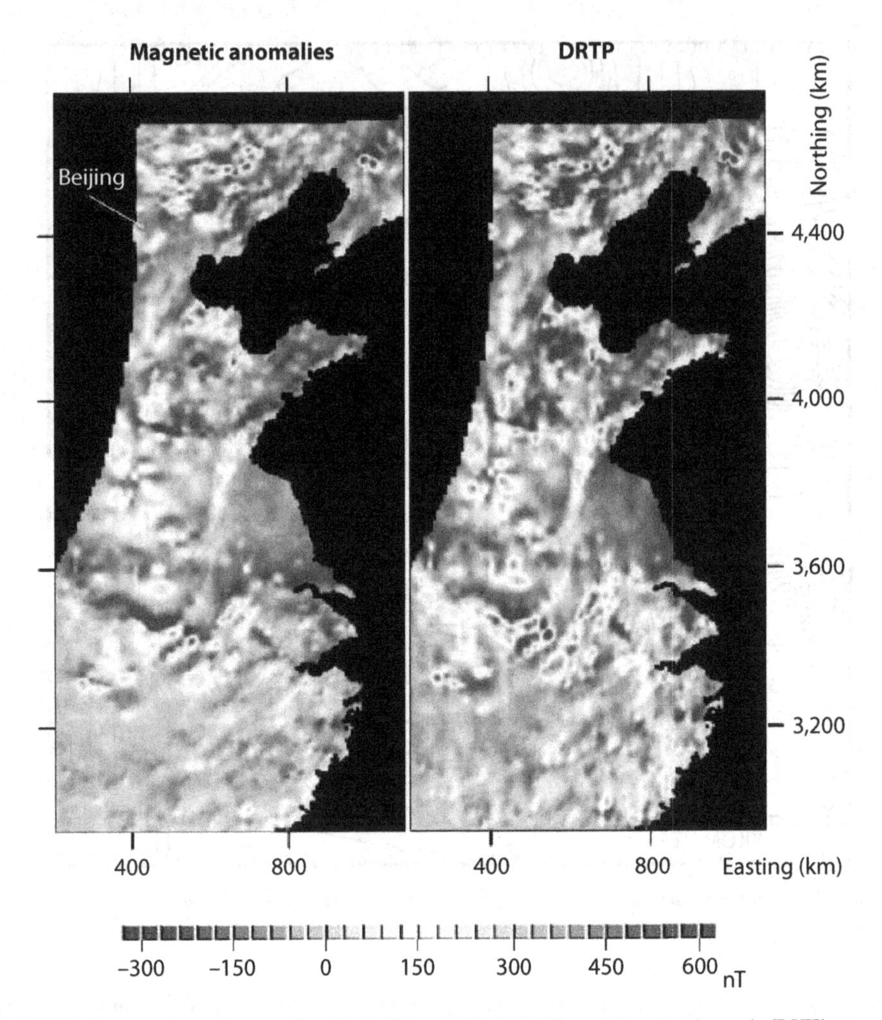

FIGURE 12.24 (a) Total intensity magnetic anomalies of east China and (b) their differentially reduced-to-pole (DRTP) anomalies. Adapted from LU *et al.* (2003).

other attributes of the equivalent sources to isolate and enhance residual-regional components.

As in regional gravity studies, equivalent point source (EPS) inversion is a very effective alternative to spherical harmonic modeling for analyzing magnetic anomaly data registered in spherical coordinates over finite spherical patches of the Earth and other planetary bodies. EPS modeling in general offers greater sensitivity for the local data coverage details, and the locally available geological, geophysical and other subsurface constraints for anomaly analysis. As for the gravity analysis described in Section 6.5.3, EPS inversion relates the magnetic data (i.e. $b_{i,1} \in \mathbf{B}$) to the magnetic moments (i.e. $x_{1,j} \in \mathbf{X}$) of a specified spherical distribution (i.e. $a_{i,j} \in \mathbf{A}$) of magnetic point dipoles by least-squares matrix inversion (i.e. $\mathbf{X} = (\mathbf{A}^t\mathbf{A})^{-1}\mathbf{A}^t\mathbf{B}$) so that $\mathbf{AX} \simeq \mathbf{B}$ (Appendix A.4.2). Thus, a desired data processing objective is achieved by invoking another forward model $\mathbf{A'X}$ where the design matrix $\mathbf{A'}$ replaces the initial design matrix $\mathbf{A}$ in the original forward model $\mathbf{AX}$ that was presumed to account for $\mathbf{B}$.

Spherical coordinate magnetic point dipole potential, vector, and total field components were worked out in Equations 9.62–9.64 for regional magnetic anomaly analysis. They are easily evaluated because they depend only on the spherical (r', θ', ϕ') coordinates and the magnetic moment inclination (I'), declination (D'), and intensity (j) values at the source point, and the (r, θ, ϕ)-coordinates of the observation point. For total magnetic field computations (Equation 9.65), the inclination (I) and declination (D) of the main field at the observation point are also involved. ASGHARZADEH *et al.* (2008) gives additional details on these spherical magnetic field components, associated gradient tensors, and their implementation for modeling the magnetic effects of extended bodies.

EPS modeling is commonly used to process anomaly data observed at differential spatial coordinates into constant elevation data grids and desired derivative and integral anomaly components for interpretation. For magnetic data over finite spherical patches, it is readily employed for isolating and enhancing anomalies for subsurface analysis. In Cartesian space, EPS inversions provide alternate approaches to achieving the anomaly isolation and enhancement objectives that are obtained from the previously described analytical and spectral methods.

For example, the above described reduction-to-pole filter in the spatial and wavenumber domains is implemented assuming that the total field observations are induced by a normal field with a single, usually averaged, value of inclination I', declination D', and intensity B_N where also $I = I'$ and $D = D'$, respectively. This assumption is reasonably appropriate for near-surface magnetic surveys ranging up to several tens of kilometers in scale, but it breaks down for larger-scale coverage and satellite surveys where hundreds of kilometers separate the crustal sources and magnetometer observations. However, modeling the more regional magnetic anomalies by Equation 9.65 can effectively reduce them for the differential effects of the normal field at the source and observation points.

In particular, the magnetic anomalies are fitted to a spherical array of crustal dipoles by solving Equation 9.65 for least-squares values of susceptibility k (Equation 9.60) where an IGRF or some other appropriate field model has been evaluated at each source point for relevant (I', D', B_N)-values and each observation point for the (I, D)-values. Re-evaluating Equation 9.65 using the derived susceptibilities and setting $I' = I = 90^\circ$ and the normal field intensity to a constant value (i.e. $B_N \equiv B_K$) estimates the anomaly data reduced to the pole for the normal field's variable attitudes and intensities over the source points and variable attitudes at the observation points. This simplifying transformation of Equation 9.65 is

$$B_T(\text{DRTP}) = \left(\frac{-C_m k \times B_K}{R^3}\right)\left[\cos(\delta) - \frac{3CC'}{R^2}\right], \tag{12.20}$$

where $\cos\delta = \cos\theta\cos\theta' + \sin\theta\sin\theta'\cos\phi - \phi'$ (Figure 3.15). These anomalies are called differentially reduced-to-pole (DRTP) magnetic anomalies to signify that their variable inclination and intensity components have been reduced to a uniform vertical inclination and constant intensity of magnetization (von Frese *et al.*, 1981a).

This approach also accommodates remanent magnetizations where known, so that DRTP magnetic anomalies in principle are centered on the subsurface magnetization variations in the same way that their vertical derivative gravity anomalies would be if the magnetizations also defined mass contrasts. Indeed, for the dipole with point mass m and radial gravity effect g, the EPS version of Poisson's relation in Equation 6.57 can be reworked for the pseudogravity effect of the magnetized point mass given by

$$\begin{aligned} g &= B_T(\text{DRTP})\left(\frac{G \times m}{-C_m k \times B_K}\right)\left[\frac{C \times R^2}{R^2 - 3C^2}\right] \\ &= \left(\frac{-G \times m}{3C'R}\right)\left[\cos(\delta) - \frac{R^3 B_T(DRTP)}{C_m k \times B_K}\right], \end{aligned} \tag{12.21}$$

where the second expression above was obtained from Equation 12.20 by noting that the gravity parameters and the coefficient C in Equation 3.91 are related by

$$C = \frac{g \times R^3}{G \times m}. \tag{12.22}$$

In general, EPS analysis facilitates anomaly isolation and enhancement in any coordinate system. Whatever can be done analytically on the point dipole can be done with an arbitrary magnetic anomaly to the degree it can be modeled by a distribution of dipoles. Usually the dipoles are distributed in an array or some other configuration that facilitates the analysis. Additional details on using EPS analysis for interpreting magnetic anomalies are given in Section 13.4.

12.4.3 Summary

The summary of the discussion of the isolation and enhancement of gravity anomalies (Section 6.5.4) is generally applicable to magnetic methods of data processing. Numerous methods are available, but the choice of the optimum method is somewhat arbitrary depending on the objective and spectra of magnetic anomalies present. Thus, it is advisable to use more than one technique to determine the range of possible solutions. In the analysis of magnetic methods, the emphasis commonly is on enhancement rather than isolation of anomalies because of the minimal effect upon residual magnetic anomalies from regional sources. This is a result of the dipolar source of the magnetic field which leads to higher-wavenumber components in magnetic anomalies as compared with gravity anomalies from equivalent sources. In all cases, it is necessary to compare the output with the input and residual effects to assess the veracity of these procedures in anomaly analysis.

12.5 Key concepts

- Extraneous magnetic sources have a range of amplitudes and periods in the case of temporal variations and wavelengths in spatial contributors which can mask or at a minimum distort magnetic effects from subsurface sources of interest in a survey. Accordingly, raw or observed magnetic data are processed to eliminate or minimize these effects. The result of this processing is the magnetic anomaly. Fortunately, in most magnetic survey campaigns these extraneous variations are considerably less effective in distorting effects of subsurface sources than in the gravity method. Thus, the requirements for auxiliary data and removal procedures are much less stringent for the magnetic method.
- The spatial variation in the geomagnetic field is largely controlled by the dipolar field originating in the Earth's core which undergoes a 100% variation in field strength from the geomagnetic equator to the pole. This variation involves both a latitude and longitude change due to the displacement of the axis of the field from the axis of rotation of the Earth. Temporal variations in the Earth's main field that are internally derived further complicate spatial variations of the field. These secular variations, which are of the order of a few to tens of nanotesla per kilometer per year, are superimposed on a general westward drift of the main field and its long-term waxing and waning.
- Measurements of the magnetic field at stationary observatories and near-surface and satellite surveys have supplied the basic data for establishing the International Geomagnetic Reference Field (IGRF), which predicts the components of the field in space and time over the Earth for a 5-year epoch. Subtraction of this field from the observed data removes the principal main and secular spatial variations in the geomagnetic field. Each 5-year epoch of this field is updated for the measured change over the preceding 5 years. The resulting definitive international geomagnetic reference field (DGRF) estimates the best-fitting field over the previous epoch which can be used for improving the calculation of the crustal field from the main and secular fields.
- There are several sources of local spatial variations in the magnetic field that may cause distortion in the magnetic anomalies of interest in a survey. For example, surface terrain consisting of crystalline igneous and metamorphic rocks may cause sufficient magnetic effects that they need to be evaluated and removed from the data of high-sensitivity surveys. Their effects can be calculated by direct consideration of their magnetic effects knowing the configuration of the terrain and the magnetization of the surface rocks. However, limited knowledge of the surface magnetic properties has led to more indirect approaches using statistical correlation of topography with the observed magnetic field of a survey to determine the surface magnetization properties assuming that the subsurface anomaly sources are spatially unrelated to the topography.
- Surface and near-surface magnetic sources of both natural and cultural features also may cause anomalies which interfere with mapping subsurface magnetic sources. As a result these effects, which are local in nature but may have a wide range of intensities, are either minimized or eliminated by appropriate observational procedures or post-acquisition processing by filtering or specific localized removal of anomalies. The removal of surface and near-surface effects prior to processing and interpretation is fundamental to successful anomaly interpretation. One of the advantages to aeromagnetic surveying is the attenuation of these effects by virtue of the geometric factor describing the relationship between source and observation site, but in high-sensitivity aeromagnetic surveys it is not uncommon to map both surface natural and cultural features.
- Temporal variations in the magnetic field, with periods varying from seconds to days and intensities reaching more than hundreds of nanotesla interfere with the measurement of magnetic field derived from subsurface sources, and thus must be removed from field observations before they are useful for interpretation. These fields are largely externally derived from electrical currents and perturbations of the ionosphere and magnetosphere which are caused by flux of radiation and charged particles from the Sun, but they also include secondary fields derived from currents induced in the subsurface by the varying magnetic field and tidal effects of the Moon acting upon the ionosphere.
- Based on a long history of measurements as a function of time at stationary magnetic observatories across the Earth, two types of external variations have been identified. The normal variation is the solar quiet-time variation which moves across the Earth as it rotates in the radiation from the Sun and is caused largely by ionospheric currents. These variations are generally measured in tens of nanotesla, but they vary with time, season, and location on the Earth's surface such that they are difficult to predict for eliminating from magnetic measurements. The second type of temporal

variation is the disturbed variation due to the effect of intense corpuscular radiation from sunspot activity. These disturbances are highly unpredictable and may reach intensities of several hundred nanotesla or more especially in the equatorial and auroral belts of the Earth. Magnetic surveying is so profoundly affected during these disturbed periods, which reach a climax during the 11-year cycle of sunspot activity, that surveying is normally aborted during these periods.

- Removal of temporal magnetic variations takes several approaches depending on the type of survey and its accuracy requirements. Land surveys may use data acquired by reoccupation of survey locations, but in most applications, the temporal field monitored at nearby stationary recording magnetometers are used to determine the variations in the field which are subtracted from the survey observations. Base station observations may also be used to remove temporal variation from aeromagnetic surveys, but generally they are only used to detect abnormally intense magnetic variations which suggest that surveying be aborted. Rather, temporal variations during aeromagnetic surveys are determined by comparison of observations at locations where survey lines are crossed by tie-lines. These mis-ties are used to level the survey line data.
- Numerous techniques are available to perform residual–regional separation in the analysis of magnetic anomalies. However, in contrast to the analysis of gravity anomalies, residual magnetic anomalies are generally easier to isolate for interpretation because of the larger dynamic range of magnetic anomalies and the lack of major regional anomalies. Nonetheless, a variety of enhancement techniques have been extensively used to increase the perceptibility of anomalies. These range from simple image enhancement methods such as shaded-relief presentations to derivatives of the observed anomaly field, to continuation of anomaly fields to different levels, to reduction-to-the-pole of magnetic anomalies. The latter have been shown to be particularly valuable for interpretational purposes of fields mapped at mid to low latitudes where anomalies are significantly displaced from their sources and distorted.
- There is no rigorous rule for selecting the optimum procedure for localizing the residual magnetic anomaly. The choice is based largely upon the nature of the anomaly field, the objective of the survey, the quality and coverage of the data, and the experience of the analyst. Analytical methods are generally preferred because they minimize personal bias and are readily applied to mass calculations on complex anomaly fields. The application of more than one procedure is often desirable, permitting comparison of results that aids in the interpretation process.

Supplementary material and Study Questions are available on the website www.cambridge.org/gravmag.

13 Magnetic anomaly interpretation

13.1 Overview

Magnetic interpretation, as in gravity interpretation, operates at several levels of complexity. It can range from simple identification and location of anomalous magnetic bodies in the subsurface to three-dimensional modeling leading to complete characterization of anomaly sources. However, there are numerous differences in the interpretation of these two potential fields. For example, magnetic anomalies in most investigations are primarily caused by contrasting crystalline rocks containing variable amounts of the trace mineral magnetite. This limits the range of possible anomaly sources that need to be considered, although as in all potential field interpretations no interpretation is unique.

Magnetic interpretation is complicated by the dipolar nature of the magnetic field, resulting in both attractive and repulsive effects from an anomaly source, and by the large range of variables that enter into determining the character of a magnetic anomaly. Anomaly characteristics vary significantly with the location and orientation of the source in the geomagnetic field and may be further complicated by the effects of remanent magnetization and internal demagnetization. Nonetheless, various interpretational techniques have been developed for dealing with these complications and shown to be successful in identifying and characterizing magnetic sources. Magnetic anomalies are particularly sensitive to their depth of origin, so special emphasis is placed on depth determination methods. Although these methods generally involve simplifying assumptions of theoretical formulations, with care errors commonly can be limited to roughly 10%.

Interpretation usually begins with isolation and identification of anomalies, generally a simpler process than in the gravity method because of the limited range of anomaly sources in the magnetic method and the higher wavenumber characteristics of magnetic anomalies. The magnetic anomaly pattern and correlation of anomalies with geological information may be sufficiently diagnostic to permit qualitative identification of anomaly sources and extrapolation of geological formations in three dimensions. Although interpretation may be terminated at this stage, more quantitative interpretation is often warranted. Quantitative interpretation may be limited to simple depth determination, especially on profile data assuming two-dimensional sources using methods primarily based on evaluation of the gradients of the anomalies. Or it may involve a more comprehensive investigation using conceptual models of sources based on the character of the anomalies and collateral geological and geophysical information. The more comprehensive interpretation may be based on iterative forward modeling using idealized or arbitrarily shaped sources, or on inverse modeling based on the nature of the anomalies. Both approaches compare the predicted magnetic anomaly with the observed anomaly to obtain a solution.

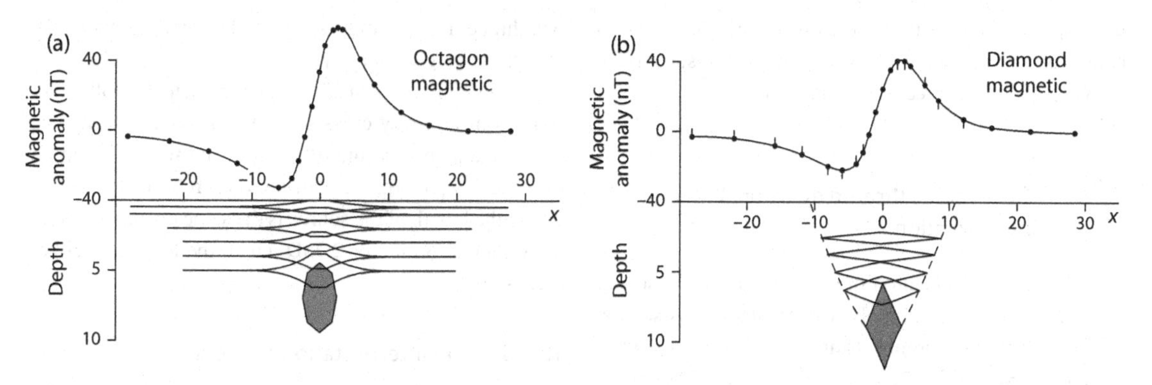

FIGURE 13.1 Equivalent total magnetic anomaly profiles at 60° geomagnetic field inclination derived from multiple sources shallower than an octagonal horizontal prism (a) and a diamond shaped horizontal prism (b). Adapted from JOHNSON and VAN KLINKEN (1979).

13.2 Introduction

Interpretation of magnetic measurements has many similarities with the interpretation procedures of gravity measurements that are described in Chapter 7. However, there are important differences. The magnetic method, for example, is based on a dipolar field rather than the monopolar gravity field. The result is that magnetic signatures are the summation of both attractive and repulsive forces, which complicates the interpretation process. However, the dipole nature of the magnetic field also causes magnetic anomalies to be more sensitive to depth than are gravity anomalies. In addition, gravity measurements are largely focused on changes of the vertical component, whereas the magnetic method offers a variety of measurements for interpretation ranging from the simple total field anomaly (which is the normal situation) to vector, gradient, and tensor components obtained with appropriately oriented sensors operated on the ground, and from ships, aircraft, and satellites. Thus, magnetic measurements offer greater versatility in interpretation than most gravity studies.

Magnetic measurements are commonly acquired along a series of parallel tracks. Thus, the data can be viewed and interpreted either as profiles, or in 2D as a map of isoanomaly contours. Both profile and map interpretations have their peculiar attributes and advantages. Map interpretation presents the data in a form that focuses on the 2D spatial distribution of anomalies. However, contouring of data is in effect a high-cut filtering process, and hence the low-amplitude, higher-frequency anomalies are lost or at least diminished in anomaly maps. To eliminate this problem and to focus on gradients, which are very important in interpretation, data are often presented and analyzed in profile form. This is particularly appropriate for magnetic data which are generally observed at a closer spacing along profiles than the distance between tracks.

Numerous procedures, both empirical and theoretical, have been devised for interpreting magnetic anomaly data using various attributes of the anomaly. Many of these have restricted use because of limiting assumptions and lack of wide applicability. In contrast, the methods described in this chapter are more generally used and lend themselves to rapid digital computations.

13.2.1 Ambiguity in magnetic interpretation

A key element in successful magnetic interpretation is the quality and use of available ancillary data and information regarding the subsurface. Interpretation is ambiguous even in the case of high-quality data because the observed anomaly can be reproduced by an infinite number of source distributions shallower or possibly deeper than the actual source of the anomaly. This is well illustrated in Figure 13.1 where the total intensity magnetic anomaly at 60° inclination is shown due to a 2D octagonal cross-section body striking east–west, and then due to a series of shallower bodies of greater width that produce the equivalent anomaly assuming a constant magnetization contrast with the adjacent rocks. The ambiguity is not relieved by additional measurements of the magnetic field and its various components or by observations at different levels, because these are not independent measurements. However, if the depth or shape of the anomaly source can be defined, its magnetic effects can be uniquely obtained (e.g. ROY, 1962). Thus, ancillary data, especially direct geologic data, provide important, often critical, constraints to hypothesizing conceptual source models used in magnetic interpretation. One of the most successful methods of minimizing ambiguity in interpretation of anomaly sources in the subsurface is to extrapolate the source by way of continuity in the anomaly pattern from direct geologic information obtained from outcrops or drill holes. Even better is to interpolate into the unknown subsurface between obvious sources of the anomaly.

13.2.2 Two- versus three-dimensional interpretation

Profile or 2D interpretation in which the anomaly source is assumed to extend infinitely with constant cross-section perpendicular to the profile, as discussed for gravity interpretation in Section 7.2.2, is an efficient process in comparison to map or 3D interpretation. Thus, interpretation based on these so-called principal profiles which pass through the maximum amplitude of the anomaly are often used initially in modeling with the result obtained being a starting point for 3D interpretation. However, many interpretation techniques, especially those used for determination of source depths from observed profiles, assume that the sources are 2D, which greatly simplifies computational procedures. In these methods, the depth determinations are generally greater than the correct value by roughly 10% (Åm, 1972).

The differences between strike-finite and infinite source magnetic anomalies are difficult to generalize because of the complicating effects of the dip and the length-to-width ratio of the source, and the azimuth and inclination of the magnetic polarization. Nonetheless, experience and theoretical modeling indicate that two-dimensionality is a reasonable assumption in most situations because observations are typically made along profiles directed perpendicular to the strike of the prevailing anomalies. However, if the anomaly strike is more than roughly 35° from the perpendicular to the strike of the anomaly, analyses by 2D techniques are questionable (Bean, 1965). Results of interpretation (depth and width of source) need to be multiplied by the cosine of the angle between the profile and the perpendicular to the anomaly. Anomalies that lie at an acute angle to the profile will be broader with lower gradients than those of principal profiles. Anomaly sources and depths will therefore be interpreted as greater than the actual situation. Furthermore, for purposes of enhanced anomaly resolution, observations are generally made as close to the level of the sources as possible. As a result, the perpendicular strike lengths of the anomaly sources commonly are considerably greater than the depths to the sources, and thus the sources approach two-dimensionality. A rough empirical observation is that the anomalies of sources with lengths perpendicular to the profile are about 90% of the infinite anomaly if their half length is greater than four times the depth to the source. The anomaly of a finite length source tends to be sharper with higher maximum gradients and lower amplitude than an infinite length anomaly.

In modeling anomalies along principal profiles of sources of arbitrary cross-section, it is common practice to assume a specific finite length of the source based on the anomaly pattern. The calculation is referred to as 2.5D where the length of the source is the same on either side of the principal profile or 2.75D where the length on either side is unequal.

13.2.3 The interpretation process

Interpretation is commonly performed iteratively, utilizing previous interpretations as well as other geologic and geophysical data to develop improved models for testing until an interpretation is achieved that is compatible with all data sets. The general approach to magnetic interpretation considers the same basic steps that were described in Section 7.2.3 for gravity interpretation.

For example, once the data are assembled and conceptual models of the anomaly sources developed, residual anomalies are located using enhancement techniques applied to anomaly data. Qualitative interpretation such as may satisfy the objectives of reconnaissance and regional geological mapping surveys will terminate here. The qualitative approach typically involves classifying the residual anomalies in contour or image format in terms of variations in: (1) amplitudes that reflect the distribution of magnetization changes associated with lithological and structural variations; (2) wavelengths, mostly to discriminate the shallower-body, high-frequency effects from the deeper-sourced, low-frequency effects, but which may also include textural changes in the distribution of rocks of varying magnetic polarization; (3) steep gradients that map geologic contacts, lineaments, faults and other geologic boundaries; and (4) directional trends that distinguish the strike of geologic units and structures. However, in this simplified approach care is needed to separate out negative anomalies caused by the bottom effect of a positive magnetization source from those minima associated with negative magnetic polarization contrasts with the surrounding rocks.

The anomaly classifications may be correlated directly with known geology to extrapolate the geology into areas where it is unknown. Figure 13.2 gives an example from a Canadian Shield aeromagnetic survey where the magnetic anomalies suggest the NW–SE anomalies associated with volcanic rocks are offset by a fault. The truncation of the magnetic anomalies is indicative of the NE–SW striking, right lateral slip fault with an offset of roughly one mile. Another example in Figure 13.3 shows the qualitative

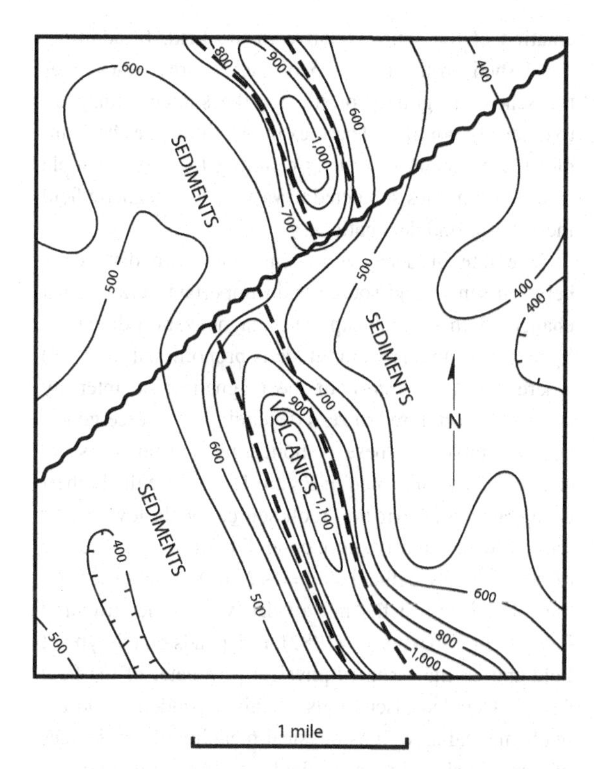

FIGURE 13.2 Qualitative geological interpretation of total field aeromagnetic anomalies (nT) in Newfoundland, Canada, showing the right-lateral displacement by a fault (wavy line) of a volcanic formation (dashed lines) within sediments. Adapted from Scott (1956).

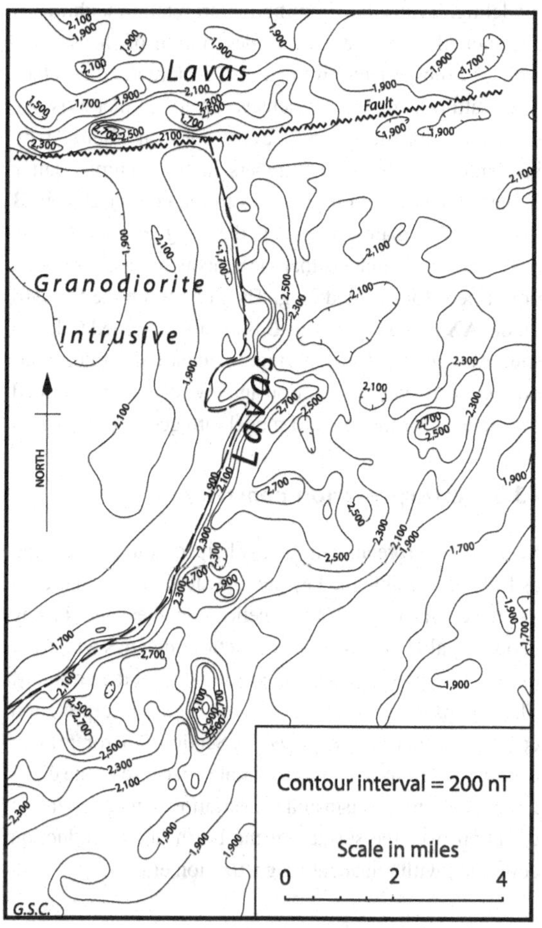

FIGURE 13.3 Qualitative geological interpretation of total field aeromagnetic anomalies (nT) in Newfoundland, Canada, of a granodiorite intrusion into a sequence of lavas. The contact margin is outlined by the dashed line that is terminated to the north by a slightly NE–SW trending fault marked by the dashed wavy line. Aeromagnetic contours show how a large granodiorite intrusive can be distinguised from the surrounding lavas. The linear magnetic feature across the top of the map is indicative of the fault which occurs there. Adapted from Scott (1956).

interpretation of regional long-wavelength anomalies for a relatively uniformly magnetized granodiorite intrusive in Newfoundland ringed by the shorter-wavelength effects of the enhanced magnetizations resulting from thermal effects on the rocks at the margins of the intrusion. The sharp change in anomaly trends to the north is indicative of a fault known from outcrop geology.

Qualitative analyses of magnetic isoanomaly maps are greatly aided by coloring areas of the map that have a range of anomaly values with the blue end of the spectrum at the lowermost values and the red end at the highest values. Generally, it is desirable to have an equal difference in range represented by color shade, but because of the large range of magnetic anomaly values commonly encountered and the need to focus on a particular range of values, it may be necessary to use a non-linear color shading scheme. Interpretation is often enhanced by the use of stacked profile, isometric, and shaded-relief maps, described in Section 12.4.2. The shaded-relief maps, particularly when they are combined with colored contour intervals, are especially useful in identifying shorter-wavelength features that are positioned orthogonal to the direction of the illumination source. Accordingly, an iterative process is required using a range of inclination and direction angles to capture all the significant anomalies of the map.

Quantitative interpretation may take a variety of forms depending on the objectives of the process. It may simply consist of calculating depths to sources, or it may involve interpretation of the subsurface magnetic units from a magnetic profile using inverse or forward modeling techniques or a combination of both, or it may consist of isolation of a residual magnetic anomaly which is interpreted by modeling. Typically, simplified interpretation techniques are applied for a preliminary description of the source, which is subsequently refined into a more detailed characterization using comprehensive inversion methods. As shown in Appendix A.5.2, every solution obtained by inversion is a set of m numbers in the column matrix **X** that relates the n observations in the column matrix **B** to the product matrix **AX** where the design matrix **A** is the $n \times m$ array of numbers that the investigator assumed (i.e. prescribed) for the modeling. The fit between the forward model **AX** and observations **B** is usually obtained to least-squares accuracy. The final stage in quantitative interpretation is to translate the solution and investigator-prescribed coefficients for the forward model into geological terms.

13.3 Interpretation constraints

Magnetic interpretation is greatly helped by understanding as fully as possible the key geological variables that control the characteristics of magnetic anomalies, and the key geophysical variables of the magnetic anomalies which are used to interpret the source characteristics of the anomaly. These considerations are analogous to the interpretation constraints that are described in Section 7.3 for gravity anomalies, but they are more complex because of the dipolar nature of magnetism and the variation in magnetization direction over the surface of the Earth due to induction combined with remanent magnetization effects.

13.3.1 Key geological variables

The amplitude and character of anomaly signatures from magnetic sources in the subsurface vary with source geometry including volume and shape, and source depth. Magnetic anomaly amplitude is a function of the magnetic property contrast of the source relative to the enclosing rock formations which may vary in a complex manner because of the combination of magnetization induced by the geomagnetic field and the intensity and direction of inherent remanent magnetization. Thus, in assessing typical magnetic anomaly signatures, consideration must be given to the relative orientation of the source in the geomagnetic field, the directional and intensity attributes of the geomagnetic field, and the source's and surrounding rock's total magnetization given by the sum of the induced and remanent magnetizations.

Examples of typical total magnetic intensity anomaly profiles are schematically presented by Breiner (1973) in Figure 13.4 for a variety of geometric shapes with positive susceptibility contrasts under five geomagnetic conditions, considering both inclination of the field and orientation of the source relative to the field. In addition, the dashed profile in row 5 of each source panel gives the schematic gravity profile for the source with positive density contrast. These examples illustrate the complicated magnetic signatures resulting from even simple source geometries subjected to varying geomagnetic field inclinations and declinations.

The rate of decay n in amplitude with distance r between sensor and source is an important variable that changes with source shape and magnetization direction. Specifically, anomaly amplitude is proportional to $(1/r^n)$ where the *decay rate* n of the magnetic field intensity decreases from three to zero depending on source geometry. Examples of these complicating dependencies are schematically illustrated in Figure 13.5 where the highest decay rate is related to the dipole source and the lowest over the slab with extensive horizontal dimensions. The degree of homogeneity defined as $n = -N$ of the field specifies the relationship of the rate of decay n to the structural index N of Thompson (1982). It depends on the type of field and on the assumed physical parameter of the field source. Detailed discussions on this dependence and the mathematical as well as physical meaning of the degrees of homogeneity of potential fields and structural indices of Euler deconvolution can be found in Stavrev and Reid (2007) and are further described for depth estimation by Euler deconvolution in Section 13.4.1.

Examples of the parameters that cause variations in magnetic anomalies are illustrated in Figures 13.6 through 13.16 for a variety of vertical slabs. Figure 13.6 shows the change in a north–south profile for a vertical slab 100 m thick whose top increases in depth from 10 to 100 m. Here, the slab has a magnetic susceptibility contrast of 0.0025 CGSu ($\times 4\pi = 0.0157$ SIu) and is magnetized by a geomagnetic field of 58,000 nT oriented along geographic north at an inclination of 75° in the northern geomagnetic hemisphere. Relative to the centerline of the source, the anomaly maxima are displaced towards the geomagnetic equator and the minima shifted towards the geomagnetic pole. The amount of shift is dependent on not only the inclination of the magnetization and the depth of the source, but also the shape of the source. For the case of a dipole source, the displacement is approximately linear with changing inclination from 90° to 50° inclination at a rate of roughly 9% of the depth to the center of the dipole per 10° of inclination. Thus, for example, the peak of the magnetic anomaly of a concentrated source at a depth of 1 km located at 70° inclination will be shifted approximately 180 m to the south in the northern geomagnetic hemisphere. In the case of a point source anomaly the displacement is slightly less than this,

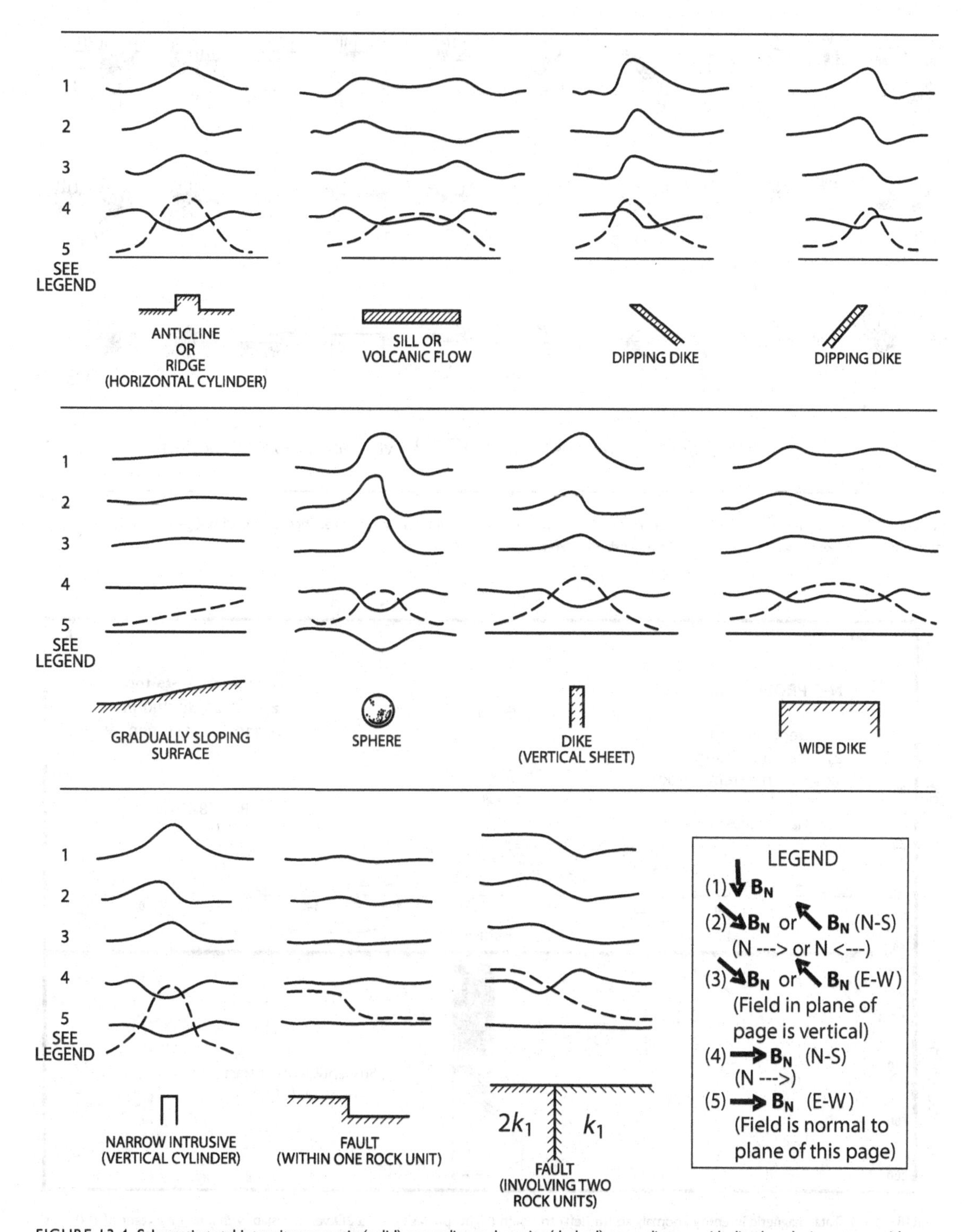

FIGURE 13.4 Schematic total intensity magnetic (solid) anomalies and gravity (dashed) anomalies over idealized geologic sources with positive physical property contrasts (symbolized by k_1 for the fault between two rock units). The legend gives the geomagnetic field ($\mathbf{B}_N$) and anomaly profile orientations. Adapted from Breiner (1973).

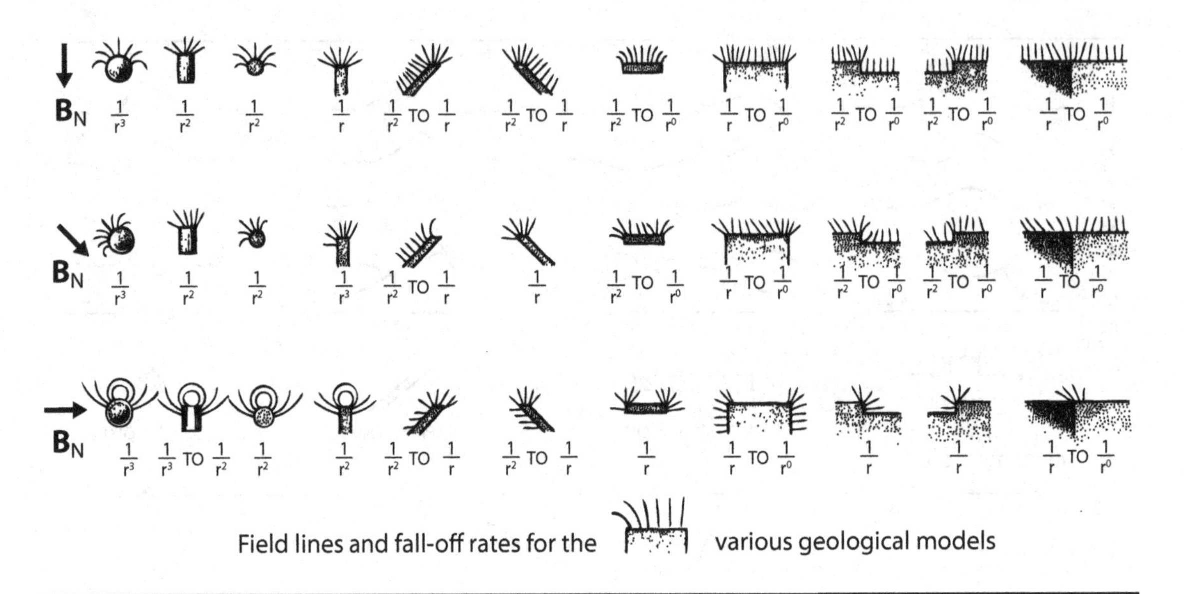

FIGURE 13.5 Field lines and amplitude decay rates ($\frac{1}{r^n}$) for various geologic models at the three inclinations of the geomagnetic field, shown by the $\mathbf{B}_N$-arrows at the left of the figure. Adapted from Breiner (1973).

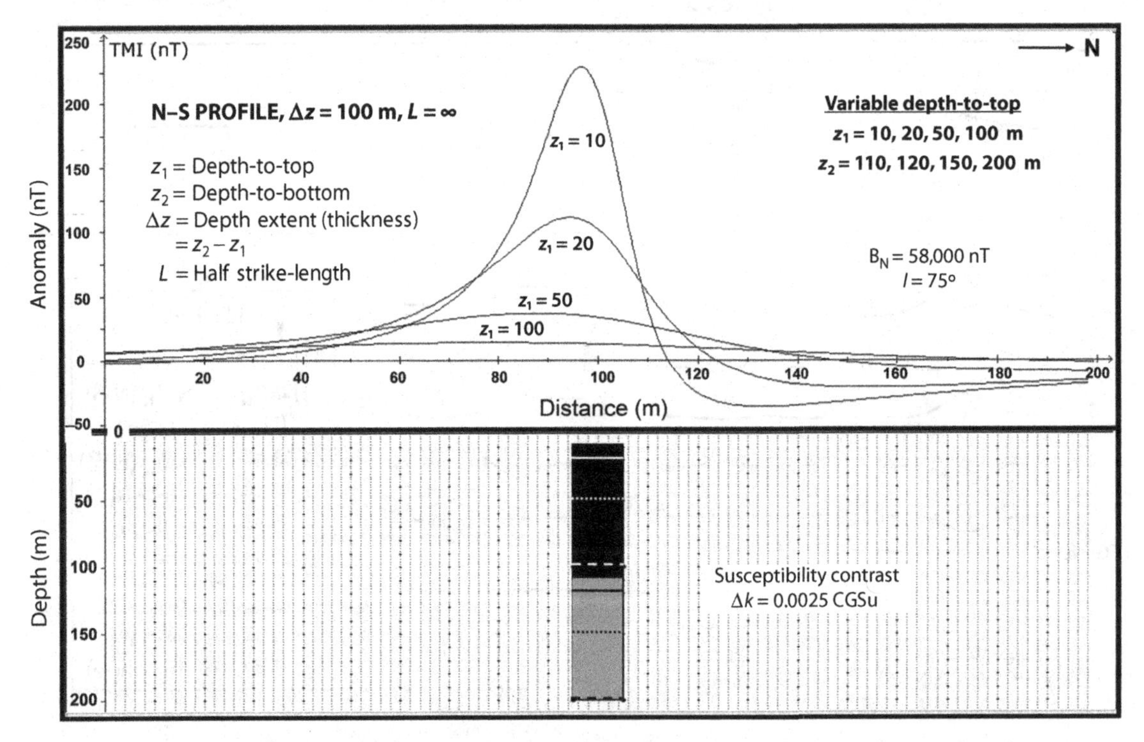

FIGURE 13.6 Total magnetic intensity anomaly south (left) to north (right) profiles over a 2D vertical slab with a depth extent of 100 m as a function of the depth-to-top of 10, 20, 50, and 100 m. The susceptibility contrast is 0.0025 CGSu in an inducing field of 58,000 nT inclined 75° below the horizontal. The source extending from 10 to 110 m is shown by black fill with the top and bottom of the other sources shown by solid, dotted or dashed lines. Note the profound effect on the amplitude and gradients of the anomaly with increasing depth to the source of the anomaly.

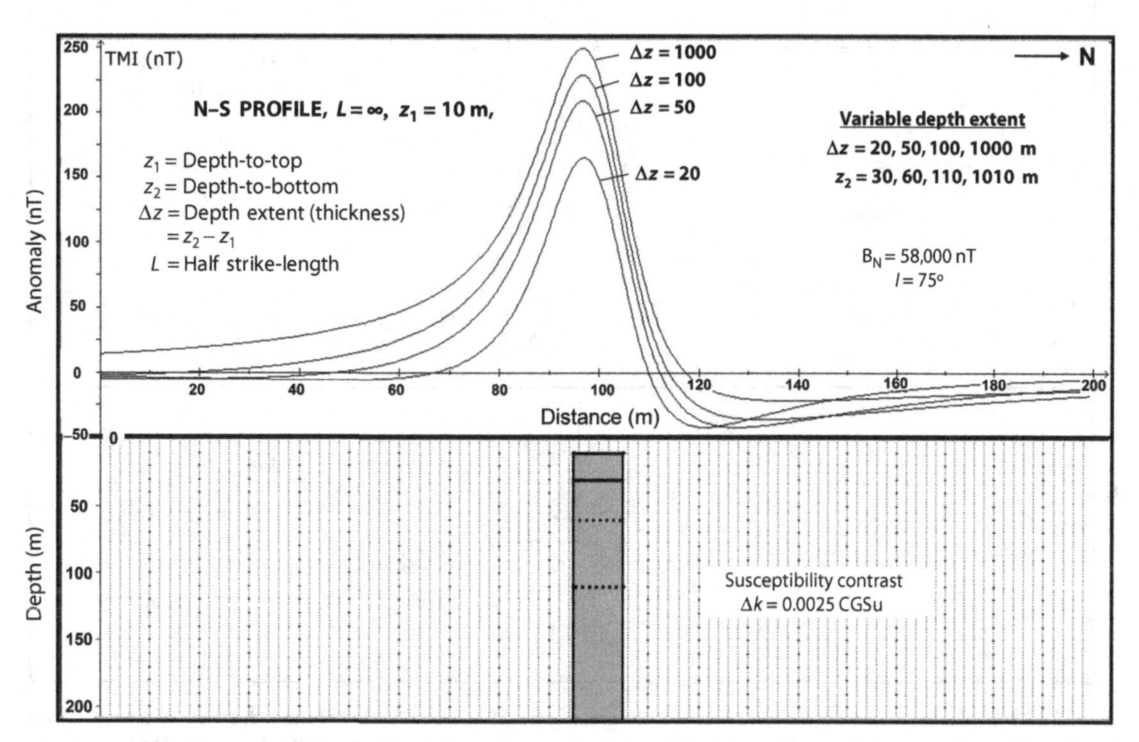

FIGURE 13.7 Total magnetic intensity anomaly south (left) to north (right) profiles over a 2D vertical slab with a depth-to-top of 10 m and variable depth extents Δz of 20, 50, 100, and 1,000 m. The magnetization parameters are the same as in Figure 13.6. Note the similarity in the gradients of the anomalies regardless of depth extent and the insensitivity of the amplitude to increasing depth extent.

approximately 6% of the depth per 10° of inclination (SMELLIE, 1956). The rates of shift increase non-linearly from 50° of inclination to the geomagnetic equator. In the southern geomagnetic hemisphere the positions of the peak and minimum are reversed: that is, the minimum is south of the peak. Note also the profound decay in amplitude and broadening of the anomaly with increasing depth to the source.

The effect of increasing the depth extent or slab thickness Δz is illustrated in Figure 13.7 considering a constant depth to the top of the slab. Here, the amplitude increases very slowly with increasing depth extent to show the dominance of the magnetic effects of the shallower source components in the anomaly. The change in the maximum amplitude of the anomaly is essentially negligible where the depth extent exceeds approximately 10 times the depth to the top of the source (HINZE, 1966) and the maximum gradients vary in only a minor way.

The effect of strike-length on a north–south profile over a slab of constant cross-section is shown in Figure 13.8. The anomaly for a half strike-length of four times the depth to the top of the slab is essentially equivalent to the case for the infinite strike-length of the 2D slab.

Figure 13.9 shows the effect of the orientation of the slab with respect to the magnetic meridian. The anomaly striking N–S as shown by the W–E profile in Figure 13.9 is symmetrical about the centerline of the source. The asymmetry of the anomaly increases as the strike of the source moves from N–S to E–W. Variations in geomagnetic inclination also invoke anomaly asymmetries that increase from inclinations of 90° to 45° as shown in Figure 13.10.

Similarly, the dip of the source will have a profound effect on the magnetic anomaly. This is illustrated in Figure 13.11 which shows the total magnetic field intensity anomaly for a dike/sill source located at a magnetic field inclination of 60° in the northern geomagnetic hemisphere. The magnetic response for a vertical and 45° dip dike are quite different from the same source oriented as a horizontal sill. An even more complex total magnetic field anomaly profile shown in Figure 13.12 is derived from the combined effect of an igneous sill and a steeply dipping feeder dike occurring within a sedimentary basin where the Earth's magnetic field is directed at 60° from the horizontal in the northern magnetic hemisphere.

The effect of observation elevation on magnetic anomalies from a series of sources oriented with and

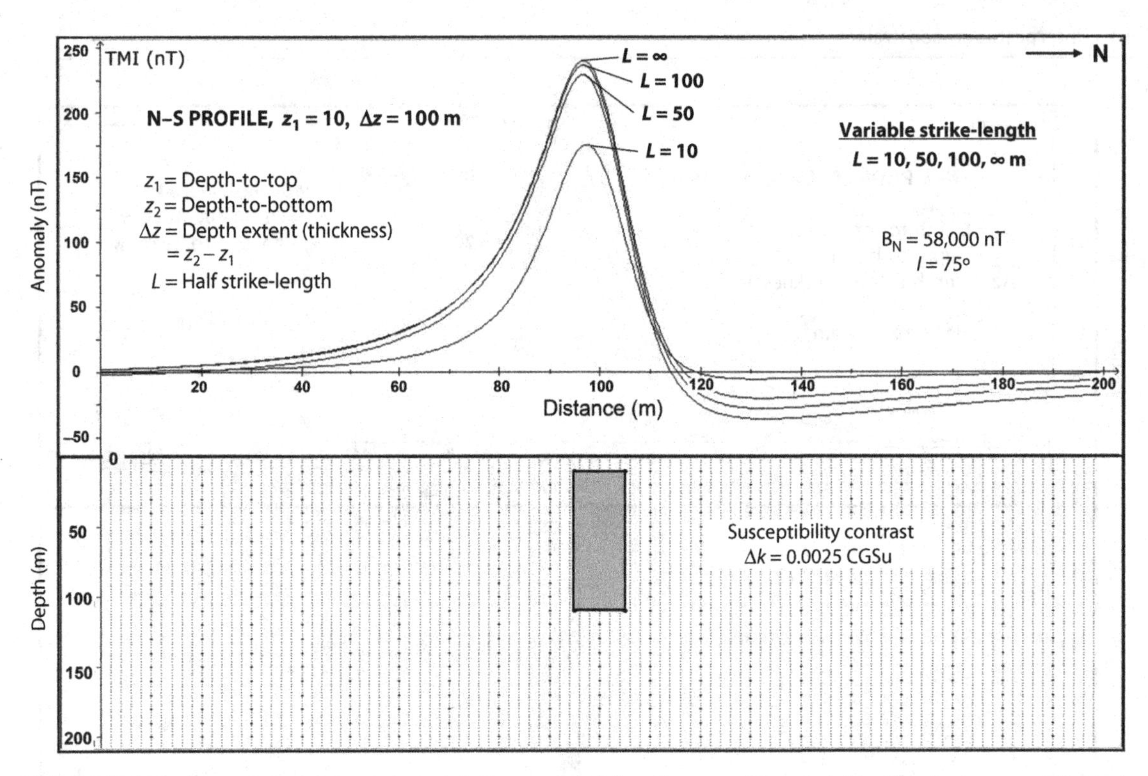

FIGURE 13.8 Total magnetic intensity anomaly south (left) to north (right) profiles over a 2D vertical slab with a depth-to-top of 10 m and depth extent of 100 m, as a function of the slab's strike length perpendicular to the profiles of 20, 100, 200 m, and infinity. The magnetization parameters are the same as in Figure 13.6. Note the minor change in the anomaly when the half-length is five or more times the depth to the top of the source of the anomaly.

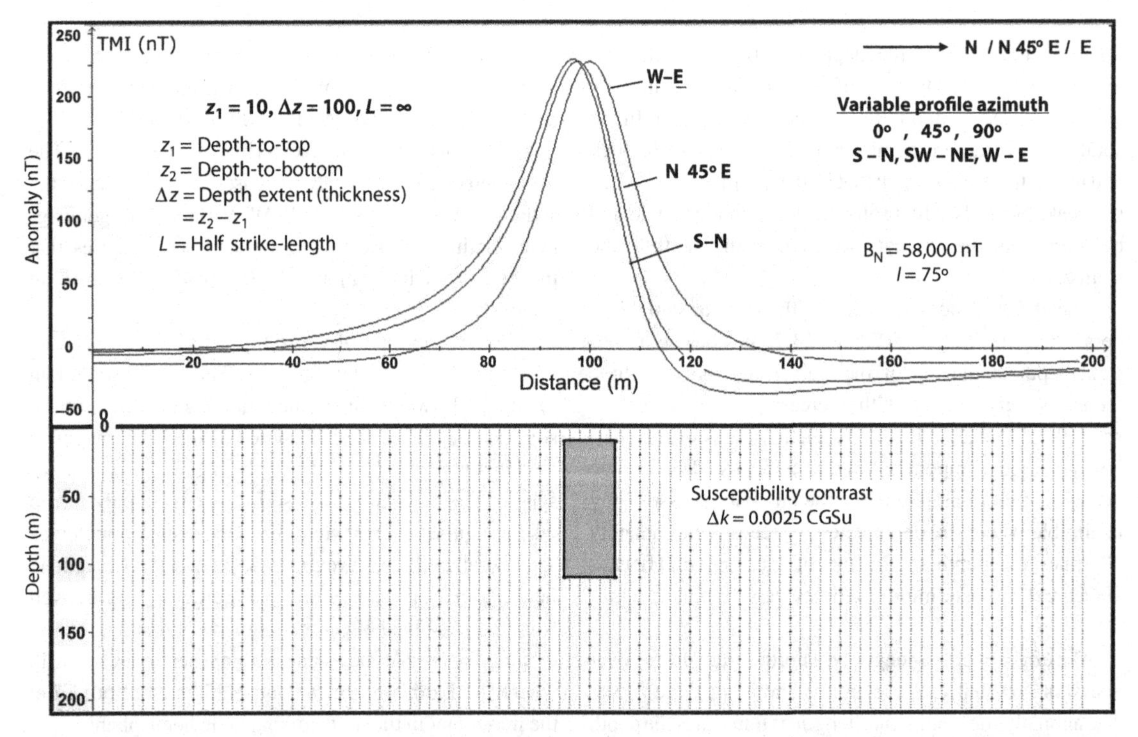

FIGURE 13.9 Total magnetic intensity anomaly profiles (S–N, N 45° E, W–E) over a 2D vertical slab with a depth-to-top of 10 m and depth extent of 100 m. The magnetization parameters are the same as in Figure 13.6. Note the increase in asymmetry of the anomaly with change in strike of the source of the anomaly from north–south to east–west.

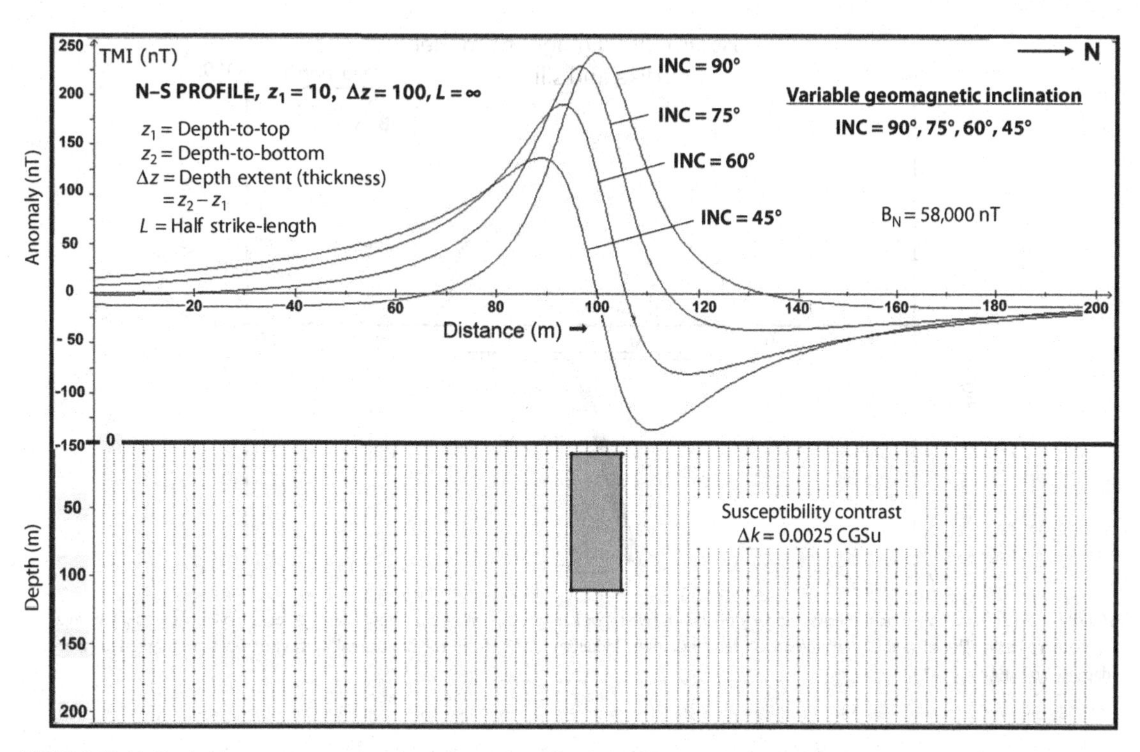

FIGURE 13.10 Total magnetic intensity anomaly south–north profiles over a 2D vertical slab with a depth-to-top of 10 m and depth extent of 100 m. The magnetization parameters are the same as in Figure 13.6, except that the Earth's magnetic field is inclined at 90, 75, 60, and 45 degrees below the horizontal. Note the profound effect of magnetic inclination on the amplitude, asymmetry, and location of the peak of the anomaly referenced to the source position.

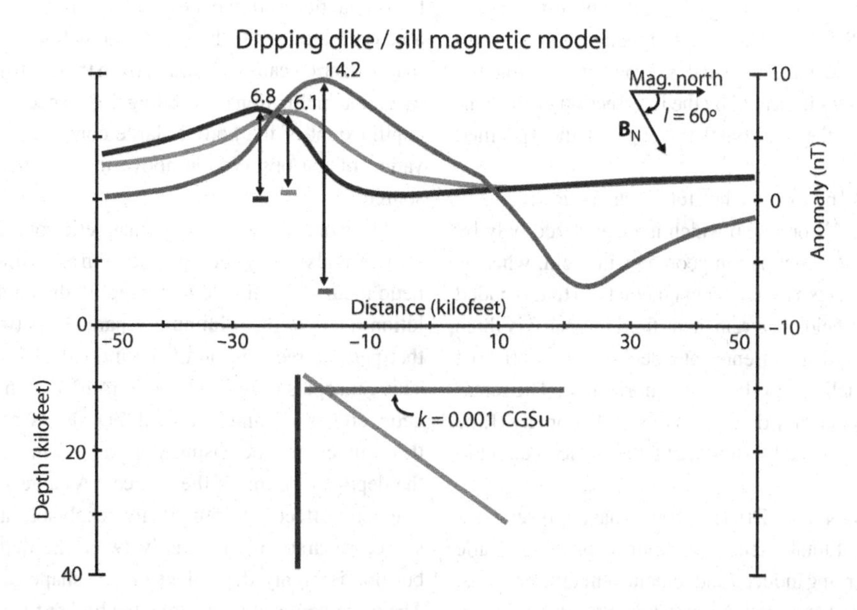

FIGURE 13.11 Total magnetic intensity anomaly profiles of a 2D positively magnetized slab dipping at 0°, 45°, and 90° in a 60° inclination inducing magnetic field. Adapted from Prieto (1998). Note the profound change in the anomaly with the dip of the anomaly source.

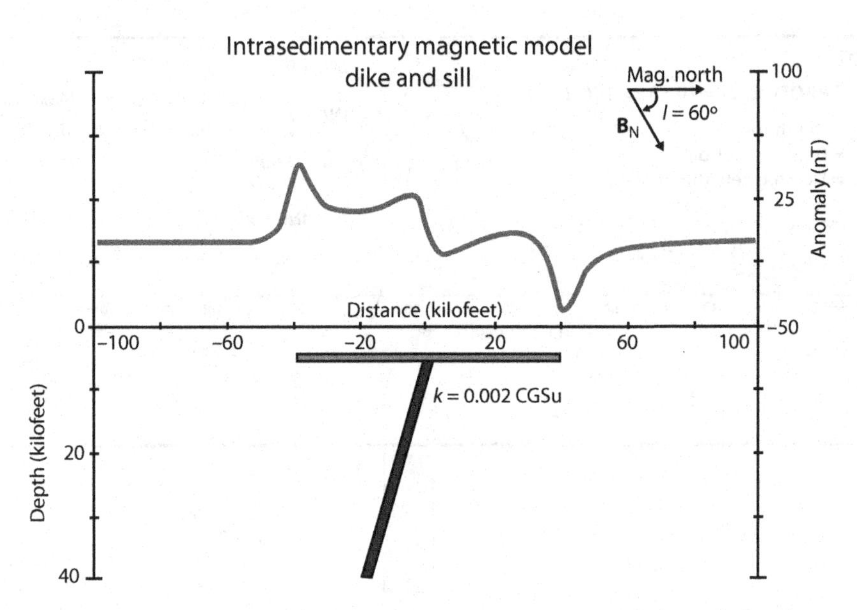

FIGURE 13.12 Total magnetic intensity anomaly profile derived from an igneous sill and a steeply dipping feeder dike occurring within a sedimentary basin. The sill and dike are positively magnetized by induction at 60° from the horizontal in the northern magnetic hemisphere. Adapted from PRIETO (1998).

perpendicular to the magnetic meridian at magnetic inclinations of 45° and 75° in the northern geomagnetic hemisphere is well illustrated in Figures 13.13, 13.14 and 13.15. These figures show vertical contour maps of the total intensity magnetic anomaly in nanoteslas for the specified source with the magnetic susceptibility contrast of 0.0025 CGSu. The change in magnetic anomaly with elevation can be readily visualized by considering the anomaly traverses indicated by the intersection of the contour lines with the horizontal traverses at the specified elevations.

Figure 13.16 shows the total magnetic intensity anomaly (Line 1) for a slab which is magnetized only by induction in a 75° inclination geomagnetic field, whereas Line 2 incorporates a remanent magnetic field that is added to the induced field. The remanent field intensity is three times the induced and oriented at a declination of 60° east of north and inclined 45° below the horizontal. The remanent field effect clearly dominates the total anomaly (Line 2) and can be isolated by subtracting the induced effect in Line 1.

ANDREASEN and ZIETZ (1969) have presented a large number of total intensity contour maps of prismatic bodies with varying induced and remanent magnetizations. They have noted from these maps, knowing the geomagnetic location of the body, that the orientation of the center line connecting the maximum and minimum of the anomaly and the ratio of their amplitudes can be used in some situations to estimate the directional attributes of the total magnetic polarization. This has been further studied by SCHNETZLER and TAYLOR (1984), who found that these two attributes are a complicated function of the geomagnetic field and body magnetization inclination and of the orientation and thickness of the prism, but that reasonable estimates of the direction and inclination of the total magnetization can be obtained from these attributes at geomagnetic inclinations exceeding 30°, especially where the depth extent of the body is large compared with the elevation of the observation above the top of the anomaly source.

The isolation of a specific magnetic source from other sources is also a key geologic factor in assessing if the magnetic method is suitable for mapping the anomaly. Resolution refers to the minimum separation between sources that permits recognition of the individual source effects. This concept is illustrated in Figure 13.17, where the magnetic effects of finite vertical 2D sheets are shown as the sources are increasingly separated in increments of the depth to the top of the sources. As a general rule, the magnetic effects of sources are resolvable at horizontal source separations of roughly twice the depth or more, but this is highly dependent on the shape of the source. The more compact the source, the higher the resolution or the smaller the distance between individual sources that can be recognized in the anomaly pattern. For example, thin vertical sheets, such as dikes, can be resolved at a

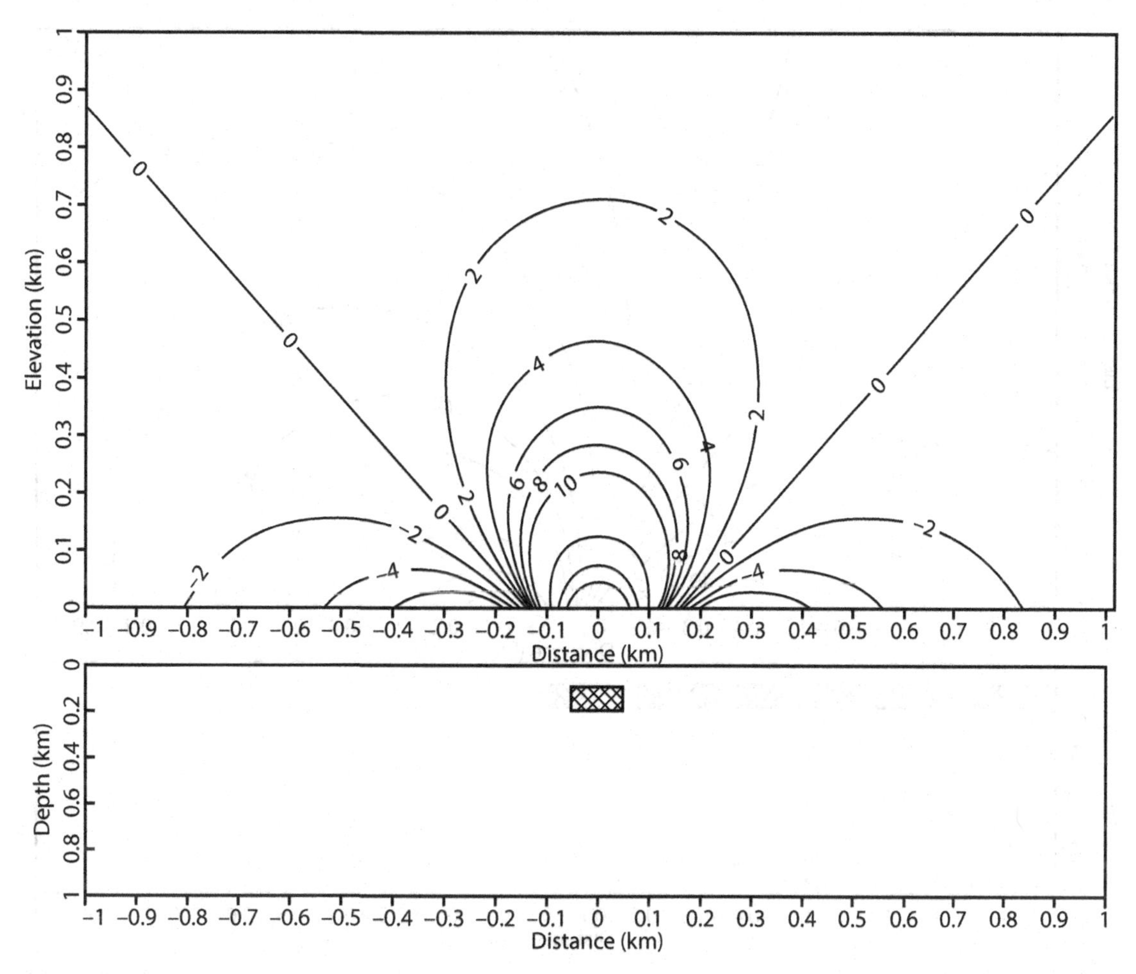

FIGURE 13.13 Vertical contour map of total magnetic intensity anomaly in nT for the source cross-section shown with infinite strike length perpendicular to the west-to-east (W–E) section in the northern geomagnetic hemisphere. The magnetization is induced for a 0.0025 CGSu magnetic susceptibility contrast in a geomagnetic field with intensity of 60,000 nT and inclination of 45° below the horizontal. Note the variable contour interval.

minimum separation of approximately 1.2 times the depth to the top of the dikes. Resolution of two bodies is difficult unless the upper source is small in size with respect to the lower source so that its spectrum is dominated by relatively short-wavelength components which will make the anomaly observable in the background longer wavelengths of the lower source, and it is relatively highly magnetized so that the anomaly is readily discernable.

13.3.2 Key geophysical variables

The key geophysical variables in magnetic interpretation relate to the anomaly attributes that are principally used for mapping subsurface magnetic property variations. These variables include the amplitude, shape, sharpness, and susceptibility of the magnetic anomaly, as well as its correlation with other geophysical and geological features.

The amplitude of a magnetic anomaly at an observation point is the product of the magnetization contrast of the source and the geometric parameters describing the volume of the source and its distance from the observation point. Its polarity is controlled by the sign of the related magnetization contrast that results from subtracting the magnetization of the surrounding material from the source's magnetization. However, it also reflects the magnetization's orientation, as indicated by the examples in Figure 13.4. Thus, for magnetization induced by the geomagnetic field, for example, decreasing magnetic inclination from vertical in either geomagnetic hemisphere increasingly shifts the anomaly peak (i.e. maximum or minimum) from directly over the center of the source

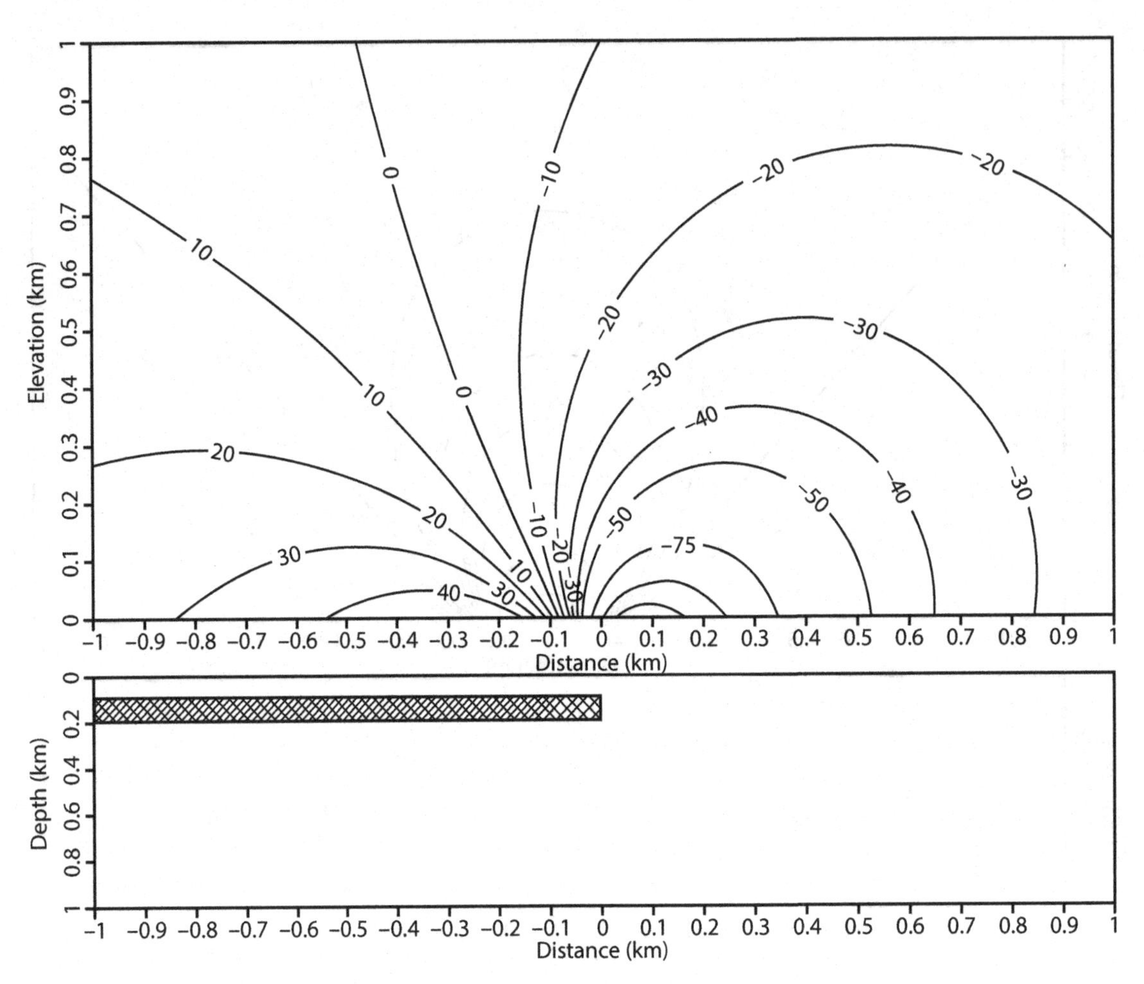

FIGURE 13.14 Vertical contour map of total magnetic intensity anomaly in nT for the source cross-section shown with infinite strike length perpendicular to the south-to-north (S–N) section in the northern geomagnetic hemisphere. The magnetization is induced for a 0.0025 CGSu magnetic susceptibilty contrast in a geomagnetic field with intensity of 60,000 nT and inclination of 75° below the horizontal. Note the variable contour interval.

along the direction of the declination of the magnetization towards the geomagnetic equator. Furthermore, at zero inclination, the anomaly is again centered on the source, but with opposite polarity to that of the anomaly at vertical inclination.

Relative to density which is a ubiquitous bulk property of geological materials and structures, magnetization is a property derived primarily from the distribution of the trace mineral, magnetite. As discussed in Section 10.6, this relatively minor mineral constituent in mostly igneous and metamorphic rocks can cause the magnetization of the rock to vary over orders of magnitude, thus greatly complicating efforts to interpret amplitude variations into subsurface rock types and structures.

These effects are further complicated by the fact that the anomalous magnetization is the superposition of the induced component produced in the source by the applied core field plus the remanent components imprinted in the source since its formation (e.g. Figure 10.1). However, with sufficient drillhole, gravity, seismic, or other ancillary geological and geophysical constraints on the subsurface geometric and magnetic susceptibility properties of the source, an effective estimate of the induced anomaly may be made and subtracted from the total anomaly to resolve the remanent component for additional insights on the possible geological history and significance of the source.

By Poisson's theorem for the magnetic and gravity potentials of a common anomaly source, the depth to the source affects the magnetic anomaly at a decay rate that is one power higher than for the gravity anomaly of the same source. As was noted in Section 7.3.2, the decreased sensitivity of the gravity anomaly to depth enhances its

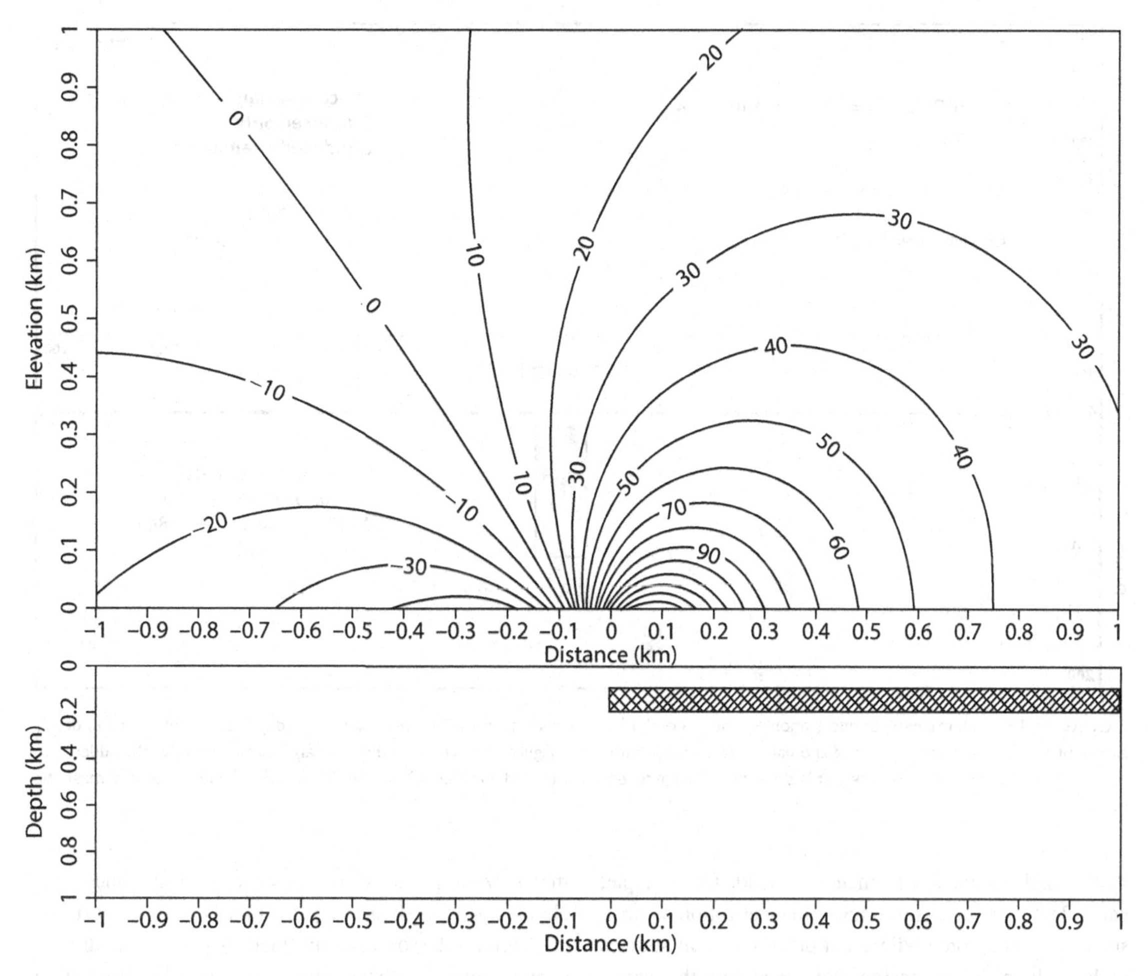

FIGURE 13.15 Vertical contour map of total magnetic intensity anomaly in nT for the source cross-section shown with infinite strike length perpendicular to the south-to-north (S–N) section in the northern geomagnetic hemisphere. The magnetization is induced for a 0.0025 CGSu magnetic susceptibilty contrast in a geomagnetic field with intensity of 60,000 nT and inclination of 75° below the horizontal. This is the magnetic effect of the south edge of the source shown in Figure 13.14. Note the variable contour interval.

sensitivity to the depth extent of the source at the expense of the shorter-wavelength effects of its shallower features. For the magnetic anomaly, however, these depth sensitivities are reversed, as suggested by comparing the examples in Figures 13.6 and 13.7.

A further example of the profound influence of source depth on magnetic anomaly mapping is illustrated in Figure 13.18. The solid profile corresponds to the airborne magnetic observations at a constant altitude above the surface. The anomalies at either end of the profile are dominated by short-wavelength components due to the proximity of the basement crystalline rocks. In contrast, the center of the profile is limited to longer-wavelength components originating from the deep basement beneath the sedimentary basin that consists of essentially non-magnetic sedimentary rocks. However, the entire dashed profile, which would be obtained if the magnetic anomalies were observed at a constant high altitude above the surface of the basement crystalline rocks has consistently longer-wavelength anomalies along its length. The dotted anomaly profile is the same as would be observed at low altitude along the profile if the sedimentary basin was not present and the crystalline rocks cropped out along its entire length.

Another obvious feature of a magnetic anomaly is its shape, which broadly indicates how magnetization may be concentrated in the subsurface in the context of the magnetization orientation. Although source shape may not be revealed very well by the anomaly shape, magnetic anomalies are better at reflecting source shape than gravity anomalies because of their inherent higher resolving power. The circular, radially symmetric anomaly induced

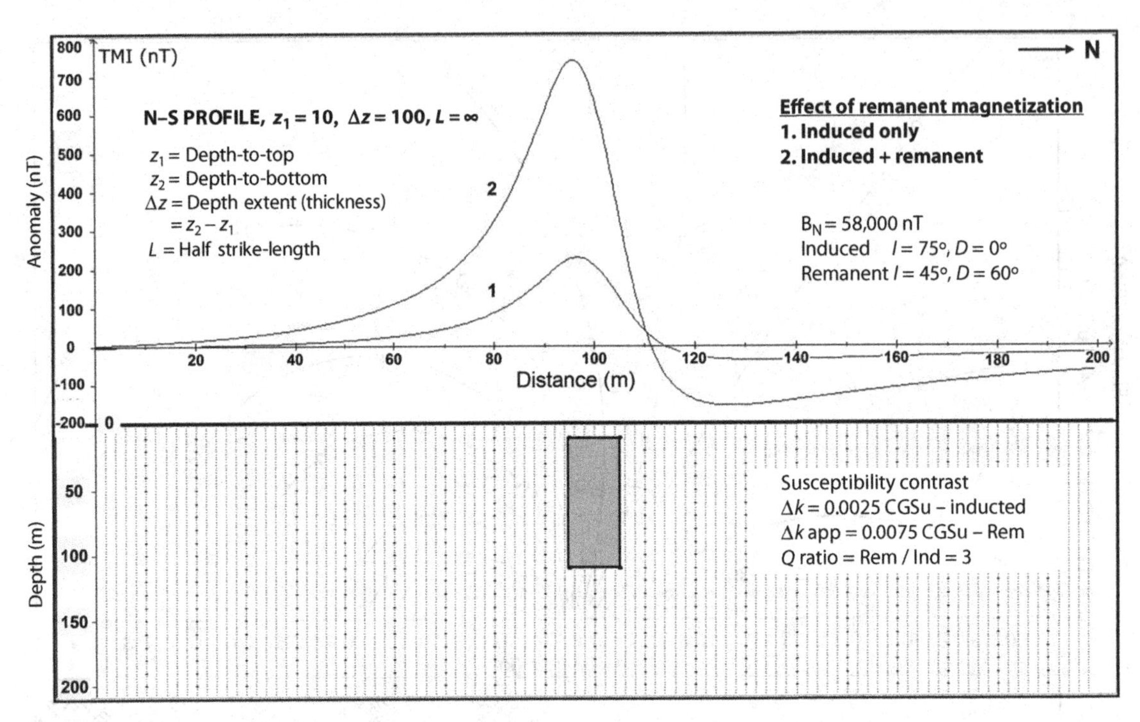

FIGURE 13.16 Total magnetic intensity anomaly south–north (S–N) profiles over a 2D vertical slab with a depth-to-top of 10 m and depth extent of 100 m. The magnetization of anomaly 1 profile is the same as in Figure 13.6, whereas the anomaly 2 profile includes the added remanent component with intensity that is three times the induced intensity and directed 45° below the horizontal and 60° east of magnetic north.

by the high-inclination geomagnetic field, for example, can reflect small concentrations of magnetization contrast such as caverns, ore bodies, and other spherical sources with small spatial dimensions compared with the depths of their centers. An example of this type of anomaly is profiled in traverse 1 for the sphere in Figure 13.4, with the other traverses showing how the circular anomaly changes shape as geomagnetic inclination decreases. However, Figure 13.4 also shows that the circular anomaly can mark the presence of long vertical pipe-like magnetic sources such as kimberlite pipes, thin intrusive plugs, and other vertical cylinder-like bodies with small horizontal dimensions relative to the depths to their tops and their depth extent. A positive magnetic anomaly from a vertically limited body should include a moat-like negative anomaly encircling the high, but commonly this feature is not discerned because of superimposed positive anomalies and lack of adequate definition of the base or zero level of the geomagnetic field in the region. Clearly, the correct interpretation of these anomaly shapes requires an appreciation for the relevant geology and its capacity to host the magnetic sources, and careful isolation of the components of the anomaly.

The elliptical anomaly is another prominent and important signature. It can mark the presence of linear magnetization contrasts related not only to anticlines, synclines, horsts, grabens, volcanic tubes and other horizontal cylinder-like sources with small cross-sectional dimensions compared with the central axis depth, but also to thin dikes and other vertical thin sheet bodies with thicknesses that are small relative to the depth to the top and lengths that are several times their widths (Figure 13.4). Similar-appearing anomalies with which they may be confused occur at the margins of wide horizontal thin and thick tabular sources, such as shown in Figure 13.4 for the sill and thick dike.

Anomaly flexures and gradient changes indicate disruption in the continuity of a source, for example by faults (Figure 13.4), as well as contacts between folded rock units of contrasting magnetizations (Figure 13.3), or the curvature of a source due to folding. Magnetic lineaments are commonly recognized as linear zones where anomalies terminate (as for the fault in Figure 13.2) or the anomaly texture abruptly changes orientation (as at the fault in Figure 13.3).

Another key parameter in magnetic interpretation is anomaly sharpness, which is a measure of the largest horizontal gradients of the anomaly. It is a function of the depth and geometric configuration of the source at a

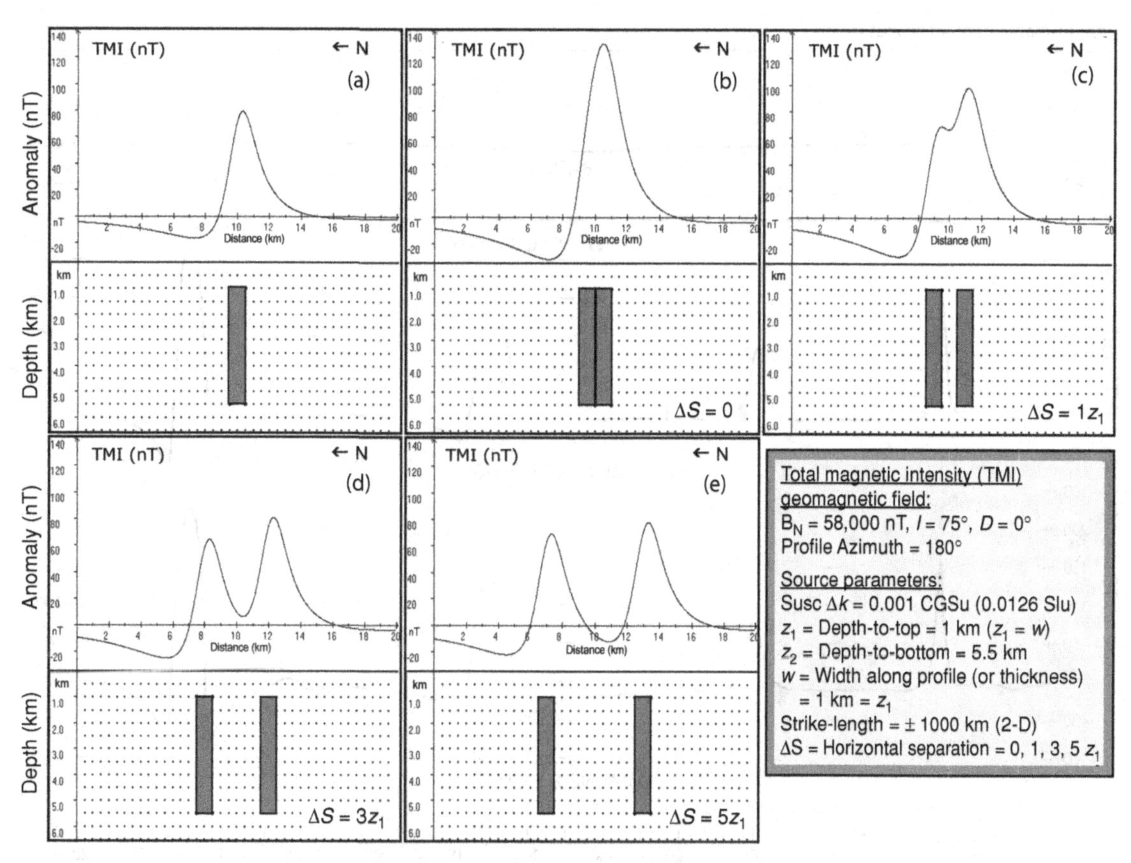

FIGURE 13.17 Magnetic anomaly resolution in terms of the horizontal separation of 2D vertical sheet sources with susceptibility of 0.001 CGSu and horizontal thickness equal to the depth to the tops. The depth to the top of the sources is 1 km. (a) Shows the total intensity magnetic anomaly of a single sheet and (b), (c), (d), and (e) show the combined anomalies of two of these sheets at increasing horizontal separation. The magnetization of the sources is induced by a 58,000 nT field that has an inclination of 75° and a declination of 0°.

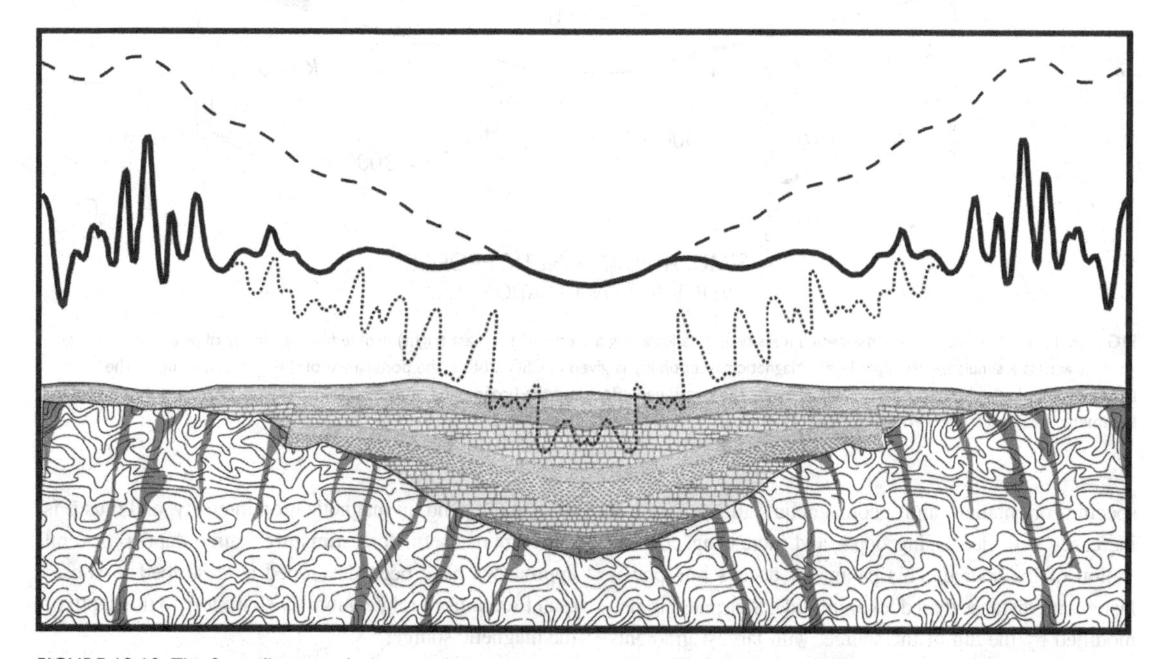

FIGURE 13.18 This figure illustrates the increasing aeromagnetic anomaly wavelengths with increasing magnetic basement depth. The solid profile shows the basement magnetic anomalies at a constant surface terrain clearance, whereas the dotted and dashed profiles display the anomalies at constant basement terrain clearances at low and high altitudes, respectively, above the top of the basement surface.

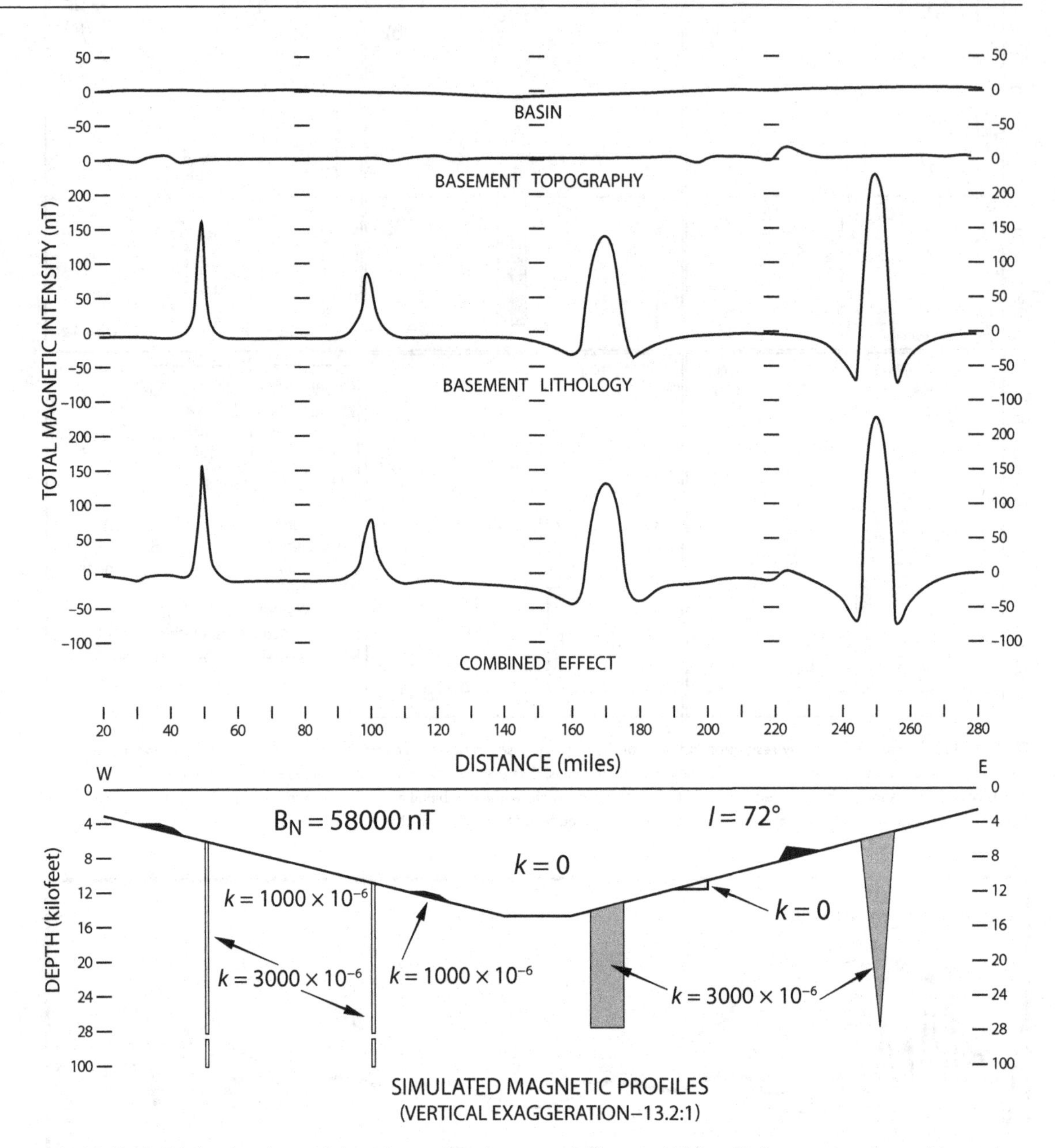

FIGURE 13.19 Calculated total magnetic intensity anomalies along a west (left) to east (right) profile from a variety of potential anomaly sources across a simulated Michigan basin. Magnetic susceptibility is given in CSGu. Note the dominance of the anomalies due to the basement lithologic changes in the profile showing the combined effect and the broadening of these anomalies with increasing depth to the basement. Adapted from Hinze and Merritt (1969).

specific orientation and geomagnetic latitude, and it increases with decreasing depth and increasing concentration of the source. This parameter decays one power faster with distance than does amplitude. Sharpness is also modified by the dip of the source, with largest gradients generally on the margin of the anomaly opposite the direction of dip (e.g. Figure 13.3). Unlike amplitude, sharpness does not depend on magnetization intensity, and thus it is much more useful for estimating source depths. Several approaches have been developed as described in subsequent sections to relate anomaly sharpness to the depth of the magnetic source.

Anomaly perceptibility refers to the ability to discern an anomaly from a specific source in a pattern of magnetic anomalies. It depends on the anomaly's amplitude, sharpness, shape, and isolation from other anomalies of similar wavelength characteristics. Isolation is a measure of the resolution of sources which is the minimum distance between similar sources that still permits recognition of the individual anomalies and their sources. The horizontal resolution of individual sources is complex, depending on several properties of the source and its magnetization, but generally a source separation of the order of one to two times the depth of the source permits identification of individual anomalies. Generalizations regarding vertical resolution of sources are difficult to make because of the broad range of possible anomaly sources.

Patterns of anomalies are subject to isolation and enhancement procedures to improve the perceptibility of anomalies of specific characteristics such as amplitude, sharpness, direction, and correlation with geological, geophysical, and other data. An example of the variation in the magnetic anomalies associated with different sources and their superposition is illustrated in Figure 13.19. This figure includes the computed total magnetic intensity anomaly west–east profiles of the major sources of anomalies in the Paleozoic Michigan basin, which has a maximum depth of approximately 5 km. The source of the anomalies is indicated at the bottom of the figure with magnetic polarization contrasts in CGSu. The anomalies from the basin, lower crustal geologic sources, and downwarp of the crust by a few kilometers are less than 50 nT in amplitude and are so broad that they are difficult or impossible to discern in the combined anomaly.

The principal anomalies of the combined anomaly profile are the intrabasement sources of the upper crust and, to a lesser extent, the suprabasement topography. The intrabasement anomalies are sufficiently isolated and defined that they are useful for depth determination analysis. The suprabasement anomalies are difficult to isolate and distinguish from minor variations in the properties of the intrabasement sources. Other sources of anomalies of the order of a few nanoteslas or less that might be detected by high-precision surveys are intrabasin sources associated with variations in the magnetic properties of non-horizontal sedimentary rocks and overlying sediments not shown in this magnetic simulation. Additionally, igneous intrusions which are unknown in the Michigan basin but are known from the adjacent Illinois basin could be a source of significant detectable anomalies with amplitudes reaching a few hundred nanotesla.

Figure 13.20 illustrates total intensity anomalies at two different geomagnetic inclinations, involving the combined effect of several intrasedimentary (C and D), suprabasement (B), and intrabasement (A) sources. Changes in the depth of the Curie point isotherm could also be the source of anomalies, but generally the minor variations anticipated in this isotherm in cratonic regions are extremely difficult to isolate in the complex pattern of anomalies.

13.4 Interpretation techniques

After residual anomalies are identified and located, the interpretation process commonly involves the use of simplified inversion techniques to ascertain critical characteristics of the source of the anomalies in general accordance with the flowchart in Figure 7.4 for gravity anomaly interpretation. These techniques involve relatively simple first-order depth determination methods applied to magnetic anomaly profile and map data. The successful application of these techniques requires a thorough understanding of their basis, theoretical or empirical, and the assumptions used in their development.

13.4.1 Depth determination methods

In most magnetic interpretations it is desirable to determine the characteristics of an anomaly source, its magnetization contrast, geometry and size, and its depth. However, owing to inherent ambiguity, simultaneous determination of all these characteristics is impossible without supplemental geological or geophysical information. As a result magnetic interpretation is often centered on limited objectives, for example to determine the depth to the sources of anomalies. Magnetic anomalies as described in the previous sections are particularly sensitive to the distance from the source. Furthermore, magnetic measurements can be made relatively rapidly and inexpensively. Thus the magnetic method is extensively used for determining the depth to magnetic crystalline rocks and especially the basement underlying surface sediments and sedimentary rocks, which are essentially transparent to magnetic fields.

Numerous techniques have been developed for magnetic depth determination. Nabighian *et al.* (2005b) provide a comprehensive history and review of methods of magnetic interpretation including those of depth determination. The earliest methods were relatively simple and easily applied manually to single magnetic anomalies, but with the need to evaluate magnetic data with large numbers of complex anomalies and improvements in computation efficiency, the methods have become more automated. Many of these digital methodologies incorporate identification of the location and configuration of anomaly sources as well as their depth and in some cases their magnetic polarization contrasts.

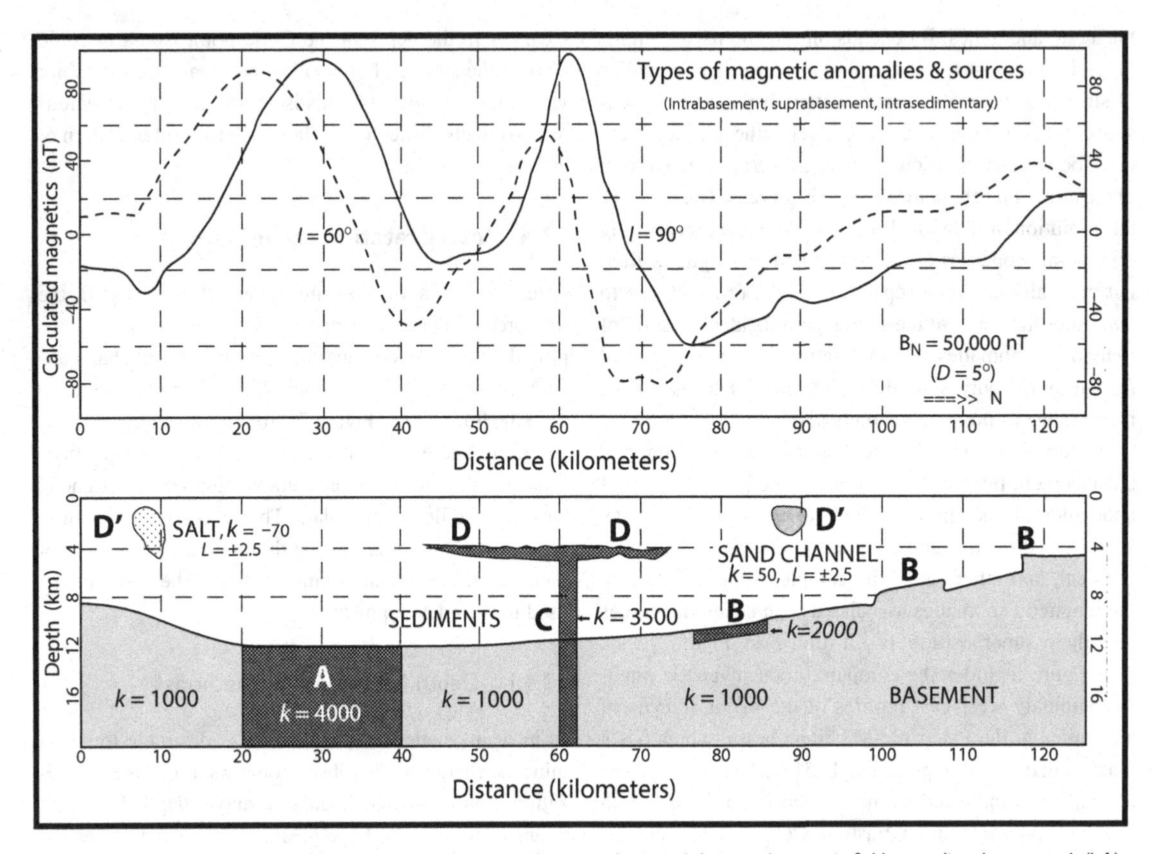

FIGURE 13.20 Potential sources of magnetic anomalies in a sedimentary basin and their total magnetic field anomalies along a south (left) to north (right) profile due to a magnetic field of 50,000 nT with a declination of 5° and inclinations of 90° and 60°. The magnetic susceptibility contrasts of the sources (k) are given in CGSu × 10^6. The sources are strike-infinite unless the strike half-length (L) in kilometers on either side of the profile is given for the source.

Depths in one way or the other are based on the maximum gradient of the anomaly, which is the horizontal change in the anomaly amplitude. The methods range from simple visual measurements of these gradients to complex computer algorithms. Some are designed to operate on single, isolated anomalies while others are intended to interrogate many complex anomalies in either profile or map format. Some of the methods are empirical, others are derived from theoretical relationships. Regardless of the method, their successful use is highly dependent on the experience of the interpreter, which is obtained by applying the method to computed or observed anomalies, and on an understanding of the assumptions of the method and how well they are met by the sources and anomalies of the survey.

Choice of the appropriate method to be used in a particular study appears to be a highly personalized decision by the interpreter. However, important criteria in this decision are the experience of the interpreter, the objective of the survey, the quality of the measurements, the nature of the noise and regional anomalies that disturb the anomalies of interest, and the available financial and computational resources. Although the simple graphical methods continue to have a role in some analyses, the interpretation of most magnetic surveys today incorporates theoretically based techniques which not only operate on the horizontal derivative (gradient), but may also involve the vertical derivative and higher-order derivatives of the field. These are either measured directly or computed from the observed total magnetic anomaly.

Although in theory the magnetic method can be used to determine the depth to any isolated magnetic source that produces an observable anomaly, it is most commonly applied to determining the depth to the top of intrabasement sources which are assumed to subcrop at the basement surface. These depths are then used to map the configuration of the basement surface. This approach is used because the relief on the basement surface generally does not produce anomalies that are readily interpreted in terms of their source depth.

Magnetic depth determinations, especially when using simplified techniques, are corroborated by using more than one technique. Consistent results do not assure accuracy, but they increase confidence in the calculated depths. A common practice when analyzing single or multiple anomalies is to check the results of the depth determination by comparing the observed anomalies with a forward calculation based on the results of the interpretation of the anomaly. Special care is given to comparing the gradients of the anomalies which are most sensitive to depth. Lack of correlation of the observed and calculated anomalies suggests that the analysis be redone.

Selecting suitable anomalies and profiles

In making an effective depth estimate, a critical decision is the selection of an anomaly for analysis. A suitable anomaly is isolated from the effects of other anomalies. Where no isolated anomaly exists but an estimate is desired, care must be exercised to choose the flank of an anomaly that is not significantly influenced by another anomaly, or the anomaly must be isolated before analysis from interfering anomalies. The effect of interfering anomalies may cause the depth determinations to be either too deep or shallow depending on the interference pattern. Interference is a problem in the magnetic method, but is not as severe as in the gravity method because of the longer wavelengths associated with the gravity anomaly from the same source. It is also advisable to avoid magnetic anomalies derived from sources of limited depth extent and finite strike length because these characteristics violate the assumptions of many depth determination methods.

In depth determination of the basement in sedimentary basins, it is generally advisable to select the anomalies with the highest gradients, provided intrabasin anomalies are not present based on knowledge of the geological history of the region, because anomalies with lower gradients are likely to originate from dipping sources, sources with sloping upper surfaces, or sources that do reach the basement surface. Analysis of these lower gradient anomalies will result in erroneously deep sources.

The profiles used in analysis should be perpendicular to the contours where they are parallel to each other and as nearly through the center of the anomaly as interference effects will permit. In the contour anomaly map of Figure 13.21, profiles (A) and (D) are most satisfactory for estimating depths. Profile (B) includes interference effects that seriously degrade its usefulness for estimating depths. In addition, this profile is near the end of the anomaly where the 2D assumption is violated. In general, the least desirable place for a profile is near the end of an anomaly where contours are nearly perpendicular to strike, because the depth estimates from these profiles include large, uncorrectable errors. This example illustrates the advantage of using a spatial view of the magnetic anomalies as provided by a map to select profiles for analysis.

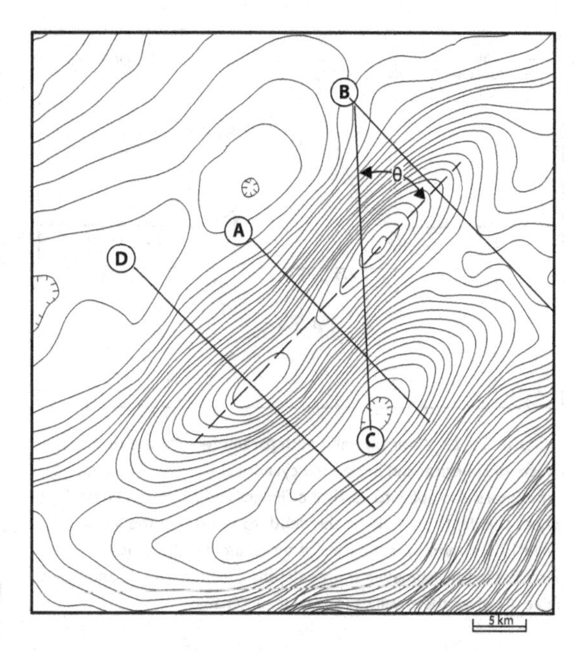

FIGURE 13.21 Selection of suitable anomaly and profiles for magnetic depth estimation is an important step in the interpretation process. Profiles A and D are optimum for investigating the positive anomaly whose axis is marked by the dashed line.

Graphical techniques

A large number of graphical techniques have been developed for estimation of the source depth of magnetic anomalies (e.g. Åm, 1972). Although some are empirical, for the most part they are derived from using simplifying assumptions applied to the theoretical magnetic responses of idealized sources. Nonetheless, these methods are useful for rapid or preliminary analysis of data and are often all the interpretation that is justified by the objective of the survey or the data. They are generally applied to single anomalies manually, but they can be based on automated analysis of the anomalies along profiles. The most rapid and simplistic of these methods of estimating depths involve the width of the anomaly at a specified proportion of its amplitude, such as the half-width methods, or measurements of distance related to the horizontal maximum gradient of the anomaly, such as straight slope rules.

(A) Half-width depth rules Half-width depth rules are generally applied only to anomalies that can be

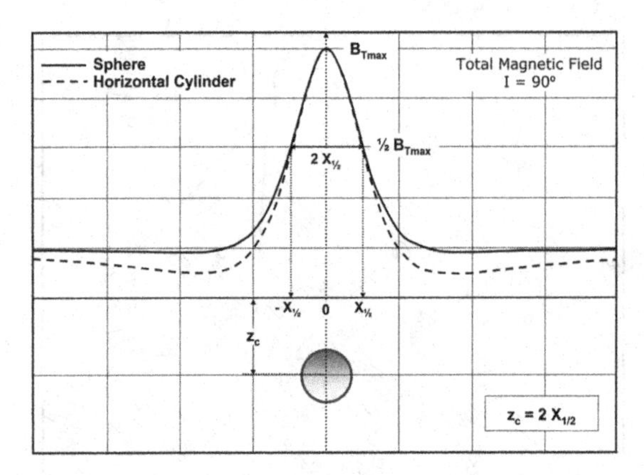

FIGURE 13.22 Anomaly half-width parameters for the magnetic effects of spherical and horizontal cylindrical sources. The anomaly parameters also apply to vertical cylinders and narrow vertical dikes where the depth z would be the depth to the top of the source. Table 13.1 summarizes the uses of these anomaly half-width parameters for estimating source depths z from magnetic anomalies. Table 7.1 gives these values for gravity anomalies.

TABLE 13.1 Vertical magnetic half-width depths either to the tops (z_t) or the centers (z_c) of various idealized sources. The meanings of the half-width parameters $X_{1/2}$ for the sphere and cylinders are illustrated in Figure 13.22, whereas the parameter $X^*_{1/2}$ for the semi-infinite horizontal sheet is given in Figure 13.23. The complementary gravity half-width depths for these sources are given in Table 7.1.

Source	Magnetic depth
Sphere	$z_c \leq 2.0 \times X_{1/2}$
Thin horizontal cylinder	$z_c \leq 2.0 \times X_{1/2}$
Deeply extending vertical cylinder	$z_t \leq 1.3 \times X_{1/2}$
Narrow vertical dike	$z_t \leq 1.0 \times X_{1/2}$
Vertical fault	$z_c \leq 1.0 \times X^*_{1/2}$

identified as approximating a simple, single idealized source at high magnetic latitudes (>70°) where vertical magnetization is approximated. The depth is related to one-half the horizontal dimension of the anomaly at one-half the amplitude (Figure 13.22) for sources approximated by spheres, vertical and horizontal cylinders, and vertical sheets. Commonly the half-width is determined by taking one-half of the total width of the anomaly at half maximum. This minimizes errors from anomaly interference and from the asymmetry of the anomaly owing to non-vertical magnetic polarization. For a semi-infinite horizontal thin sheet or fault magnetic anomaly, the half-width is half the distance between the maximum and minimum values of the anomaly (Figure 13.23). The results from these horizontal measurements are the maximum depths possible assuming the idealized source geometry. The half-width rules for a variety of magnetic sources are summarized in Table 13.1. The half-width parameters and these relationships are shown in Figures 13.22 and 13.23. Additional half-width depth relationships are given for other source configurations by PARASNIS (1986).

The accuracy of the half-width depth rules is limited by non-vertical magnetization, estimation of the base level

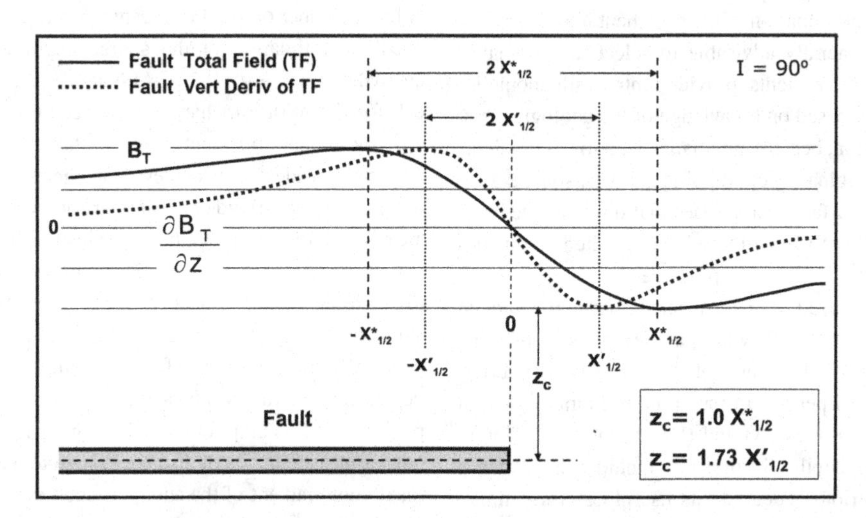

FIGURE 13.23 Anomaly half-width parameters for the magnetic effects (total field and its vertical derivative) of semi-infinite horizontal thin sheet sources (faults). Table 13.1 summarizes the use of these half-width parameters for estimating source depths z from magnetic anomalies. Table 7.1 gives these values for gravity anomalies.

of the anomaly, interference from other anomalies, and misidentification of the source geometry. The latter is particularly problematic. A circular anomaly, for example, can be interpreted as the effect of both a concentrated three-dimensional spherical source (magnetic dipole) and the vertical cylinder or plug (magnetic monopole) where the horizontal dimensions of both sources are small compared to their depths. The elliptical anomaly, on the other hand, can reflect the effects of a horizontal cylinder source (a line of magnetic dipoles) and a vertical thin sheet or dike where the long horizontal dimension of both sources is several or more times greater than the orthogonal short dimension. Clearly, for reliable depth determination, the anomaly must be ascribed to the appropriate subsurface source.

(B) Straight slope depth rules The straight slope rules for depth determination are based on empirical studies of measurable horizontal dimensions of the magnetic effects of a 2D tabular body (dike) with a flat top (Figure 13.24) and prism-shaped sources in varying magnetic fields. These rules enjoy widespread application, providing effective results in the hands of the experienced interpreter. They rely on measurements that are generally insensitive to the relative size and attitude of the source and that avoid significant interference from overlapping anomalies. They are particularly advantageous because they do not require estimation of a base level, as this may be difficult to evaluate where anomalies are subject to significant interference from other anomalies or are affected by gradients from regional anomalies derived from deep, broad sources with high magnetization contrast with the adjoining rock formations.

(1) Straight slope (maximum slope) method. This purely empirical technique (VACQUIER *et al.*, 1951; STEENLAND, 1963) is based on the optical illusion that the anomaly approximates a straight line near its inflection point. Mathematically, the anomaly has no straight slopes, but the steepest part of the anomaly curve is very nearly a straight line. In practice, the interpreter draws the tangent to the steepest gradient on the anomaly curve, and measures the horizontal distance SSL between the points at which the tangent departs from the curve. Thus, the length of the horizontal projection of the steepest slope can be measured as the straight slope line (SSL) and related to depth z of the source (Figure 13.24) by

$$SSL = f_1 \times z \text{ or } z = SSL/f_1, \tag{13.1}$$

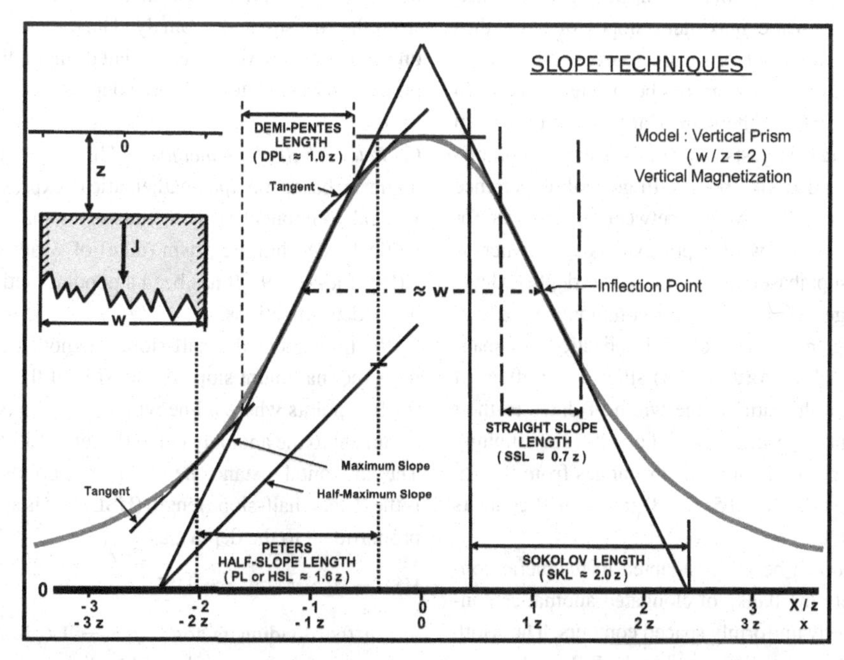

FIGURE 13.24 Popular magnetic depth (z) estimators using straight slope, half-slope, Sokolov, and demi-pentes length parameters of magnetic anomalies due to infinite vertical prism sources assuming vertical magnetization. Numbers shown in the figure are commonly used values of the dividing factors to obtain depth.

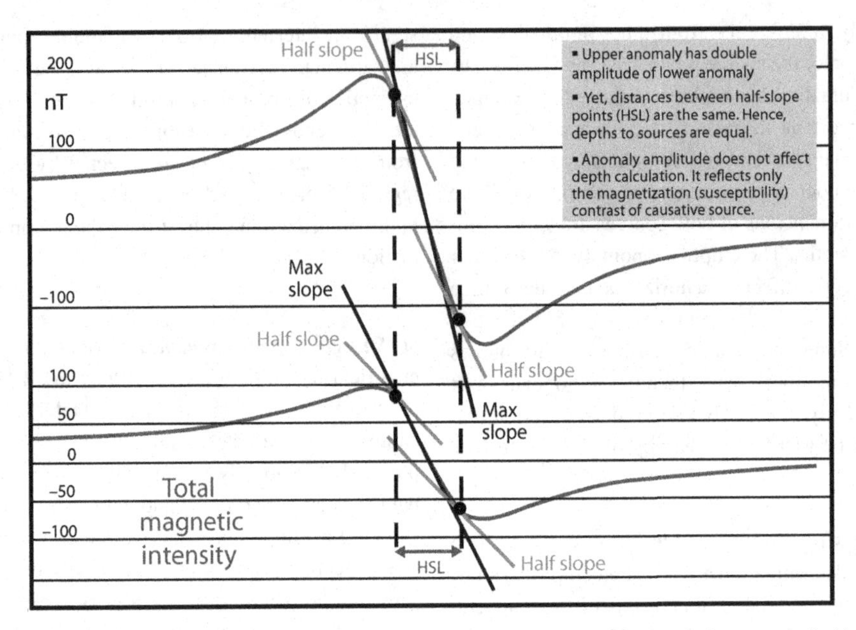

FIGURE 13.25 Illustrations of Peters' half-slope parameters for a semi-infinite horizontal thin magnetic sheet source (fault). HSL parameter depends on slope of anomaly flanks, but does not depend on anomaly amplitude. This, in general, applies to other magnetic sources.

where the proportionality factor f_1 is an approximation depending on the characteristics of the source geometry and magnetization. VACQUIER *et al.* (1951) arrived at an average proportionality factor of 1.1 from measurements made on the maximum slopes of anomalies derived from a variety of prismatic sources. However, a range of proportionality factors has been determined from other measurements of theoretical anomalies of known depth. ÅM (1972) suggests that 0.7 is a more appropriate factor for 2D dike-like bodies. In general, experience shows that f_1 typically varies between 1.0 to 0.8 for intrabasement sources with depth extents $\gg z$, whereas $f_1 \approx 0.7$ for suprabasement sources with depth extents $\leq z$. The average $<f_1> = 0.8$ is a commonly used factor for studying the general depth implications of magnetic anomalies. SKILBREI (1993) studied the effect of depth extent and the ratio of the width of dikes to their depth on straight slope lengths and found that for infinitely deep dikes the proportionality factor ranges from 0.5 for thin dikes (width/depth <0.5) to 1.0 for vertical contacts (width/depth $\approx$ 8).

This method can be easily adapted to magnetic contour maps where the flanks of elongated anomalies contain bands of fairly uniformly spaced contours. The width of the uniform equi-gradient band is the SSL and a measure of z. Specifically, the SSL is the distance perpendicular to the contour lines over which they are uniformly spaced.

In general, experience is necessary to use the SSL method with a precision approaching 10%. The results are subject to the biases of individual interpreters and greatly influenced by interfering anomalies if they are not removed from the investigated anomaly. Thus, the method should only be used on very well isolated anomalies to obtain preliminary estimates of source depths.

(2) Peters' half-slope method. This non-empirical technique is based on the mathematical expression for the vertical component of the magnetic anomaly due to a 2D vertical-sided infinite prism (dike) of width w and depth z (PETERS, 1949). It has been a popular method for rapid depth determinations.

To implement the half-slope method, the interpreter finds the maximum slope on the side of the anomaly and the two points where a line with half the maximum slope is tangent to the anomaly curve (Figures 13.24 and 13.25). The horizontal distance between these points of tangency is the Peters' half-slope length (HSL) that is approximately proportional to the depth z as

$$\text{HSL} = f_2 \times z \ni z = \text{HSL}/f_2, \tag{13.2}$$

where the dividing factor $< f_2 > = 1.6$ is a commonly used value for many geological bodies.

Although the average dividing factor is often applied in practice, it is applicable only for dikes with width-to-depth ratio $(w/z) = 2.2$. In general, theoretical

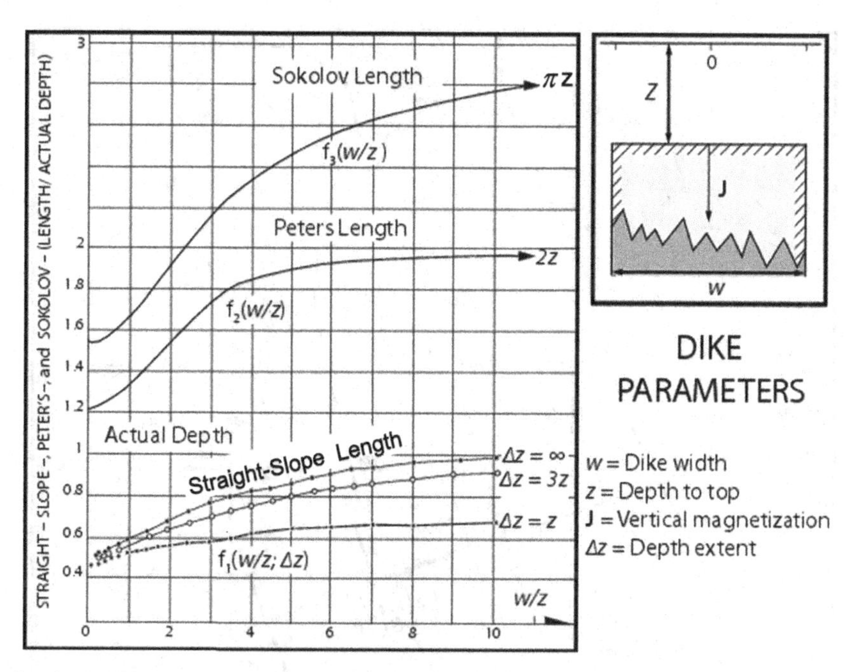

FIGURE 13.26 Straight-slope, Peters' half-slope, and Sokolov length variations with varying dike widths (w) and depths (z). For the straight-slope lengths, the curves for varying depth extents (Δz) of the dike are also shown. Adapted from Skilbrei (1993). The curves for the Peters' and Sokolov lengths are from Åm (1972).

considerations indicate that f_2 varies with the dike's ratio (w/z) as shown in Figure 13.26. Thus, the dividing factor ranges from $f_2 = 1.2$ for thin sheets with $(w/z) \longrightarrow 0$, to $f_2 = 1.6$ for dikes with $(w/z) \approx 2$, and $f_2 = 2.0$ for very wide dikes with $(w/z) \longrightarrow \infty$. Åm (1972) found that more precise determinations are obtained on the gradual rather than the steep side of an asymmetric anomaly providing that the slope of the gradual side is not adulterated by interference from adjacent magnetic sources.

An improved dividing factor can sometimes be derived for isolated anomalies to refine the depth estimates. The procedure involves obtaining a preliminary depth z_1 using $f_2 = 1.6$ and estimating the half-width of the dike $(w/2)$ by measuring the distance from the maximum (crest) of the anomaly to the point of maximum slope (inflection point). The preliminary ratio (w/z_1) is next used to look up a better dividing factor from Figure 13.26 which is then multiplied against the HSL for an improved depth estimate z. This process usually can be iterated once or twice to refine the depth estimate further.

Peters' half-slope method can be applied easily to anomaly contour maps without extracting profiles from the map. As an example, consider the suitable anomaly in the total intensity aeromagnetic anomaly map of Figure 13.27. By using the band of contours between the magnetic high and low of the anomaly, the point is located at which the distance between adjacent contours is the shortest. This point of maximum slope (i.e. the inflection point) is found usually near the center of the contour band. The distance between the two adjacent contours at maximum gradient where the contour spacing is minimum is then measured and the points on either side of the maximum gradient point are located where the contour spacing is twice that of the smallest spacing of the maximum gradient. The horizontal distance between these points is the HSL length used to determine the depth to the source.

The critical points used to implement Peters' method can be easily identified on curves of the horizontal derivative of the total magnetic field anomaly. These simple correspondences are illustrated in Figure 13.28 and summarized in Table 13.2. The half-slope measurements can also be made on horizontal gradient contour maps. The HSL is the distance across the crest of the horizontal gradient anomaly between the contours having half the value of the peak, which is the half-amplitude width of the horizontal gradient anomaly.

The classical Peters' method assumes a vertical magnetic field and vertical dike. However, it can also be modified for other geomagnetic inclinations and inclined dikes

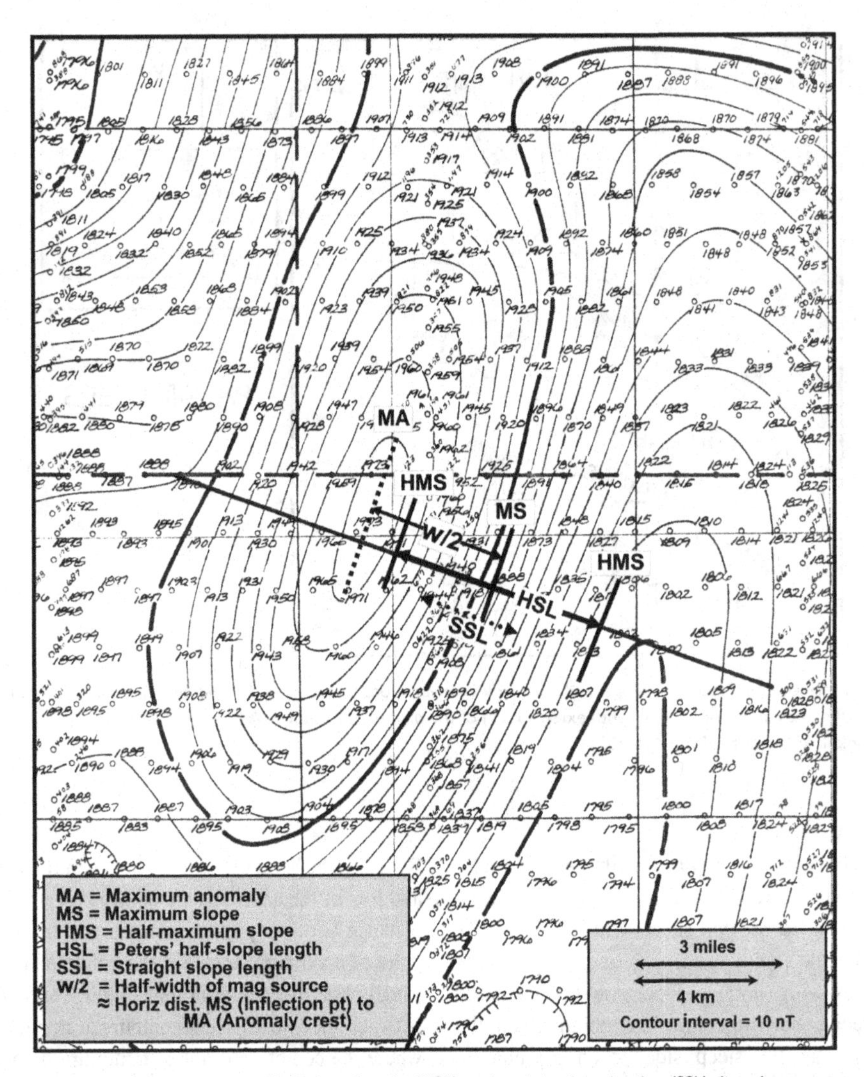

FIGURE 13.27 Illustrations of the estimation of half-slope lengths (HSL) and straight-slope lenghts (SSL) directly on aeromagnetic total intensity anomaly contour maps for depth determinations.

at any strike azimuth. Under these conditions, the dividing factor f_2 will show dependence on the effective inclination that combines the effects of inclination and the dip and strike of the dike. An appropriate dividing factor can be established by calibrating the depth estimates against depths known from drilling, seismic, or other geological and geophysical constraints.

A similar method, the demi-pentes method, which also uses the horizontal distance between slope tangents, has been suggested by Giret and Naudy (1963). In this method the horizontal distance between the intersections of half maximum slope tangents of anomaly profiles and the tangent to the slope of the inflection point is roughly equal to the depth to the top of the anomaly source.

(3) Sokolov's method. Sokolov's method is a rapid, easily applied technique for estimating depth to the source from the slope of the magnetic anomaly that is based on the same assumptions as the Peters' method (Åm, 1972). The Sokolov length (SKL) is the horizontal distance between where the maximum slope intersects the anomaly's zero axis and the tangent line of its central peak (Figure 13.24).

As in the Peters' method, the SKL can be made proportional to depth z by

$$\mathrm{SKL} = f_3 \times z \ni z = \mathrm{SKL}/f_3. \tag{13.3}$$

However, the SKL is more sensitive than the Peters' length HSL to the effective inclination and the width-to-

TABLE 13.2 Comparison of critical points on the total field anomaly used for the Peters' half-slope depth estimation and their horizontal derivatives. These critical points and their horizontal derivatives are illustrated in Figure 13.28.

Total magnetic field anomaly	Horizontal derivative anomaly
Maximum or minimum (peak or trough)	Zero (zero slope)
Inflection point (max. +ve slope for anomaly increasing with distance x)	Maximum (anomaly peak)
Inflection point (max. −ve slope for anomaly decreasing with distance x)	Minimum (anomaly trough)
Horizontal distance between half-slope points (HSL)	Anomaly half-width at half-maximum amplitude (or at 1/4 the total peak-to-trough amplitude)

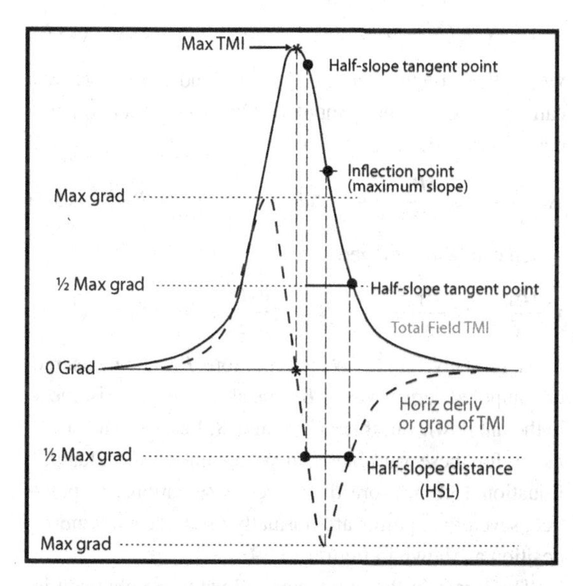

FIGURE 13.28 Relationships between the half-slope length (HSL) and characteristic points on the total field anomaly (solid curve) and its horizontal derivative (dashed curve) at an inclination of 90°. These relationships are summarized in Table 13.2.

depth ratio (w/z) of the dike (Figure 13.26). In general, $f_3 = 1.54$ for thin dikes with $w/z \longrightarrow 0$, $< f_3 > = 2.0$ for dikes with $w/z \approx 2.2$, and $f_3 = 3.14$ for very wide dikes with $w/z \longrightarrow \infty$. Solokov's method is easier to apply, however, and also generally less sensitive to interfering anomalies, although the latter can affect the determination of the zero level of the anomaly. It also has the limitation that the zero level of the anomaly must be determined which may be difficult in many anomaly patterns.

Semi-automatic approaches

With the advent of digital computing, automated procedures were developed for calculating subsurface source parameters by the deconvolution or inversion of windowed anomaly data. The source parameters are deconvolved from the anomaly data covered by the window operator, which is moved over the data set at intervals that may range from the data grid interval where the windows have maximum overlap to intervals where no overlap occurs. Varying interpretational strategies and window sizes lead to a large number of often diverse results in this automated procedure. Selection of the optimum source characteristics from these multiple results requires considerable judgment by the interpreter based on experience with the method and knowledge of the geologic framework of the region. The multiple results can be presented in tabular or more commonly graphical form for analysis. The pattern of the results in graphical form is particularly useful to the interpreter in reaching decisions. Thus, interpretation is truly semi-automatic with the interpretation reached by a human from electronically computed results. Commonly these methods are used to determine the location of the anomaly sources and depth and, in some methods, the configuration of the sources and their physical property contrast.

Numerous techniques using either manual or semi-automatic approaches have been described for specifying the source of an isolated anomaly including its depth (e.g. GAY, 1963; BEAN, 1965; GRANT and MARTIN, 1966; NAUDY, 1971; PARASNIS, 1986) based on the characteristics of anomalies from a variety of idealized sources such as spheres, horizontal and vertical cylinders, prisms, faults, and thin and thick dikes. However, these methods have largely been replaced by automated techniques that rapidly interpret the anomalies of large data sets using either the anomaly values or their derivatives. In general, these methods are usually implemented on magnetic survey profiles assuming that the source of the anomalies can be considered to be two-dimensional or, less frequently, on gridded map data sets. Representative automated techniques that are widely used are briefly explained in the following descriptions, but the reader seeking to implement these methods will want to refer to the cited original references for further, specific details.

(A) Euler deconvolution A function such as the magnetic potential $V(x, y, z)$ or any of its spatial derivatives is homogeneous of degree n if

$$V(tx, ty, tz) = t^n \times V(x, y, z). \qquad (13.4)$$

Thus, it satisfies the differential equation known as Euler's homogeneity equation or simply Euler's equation given by

$$x\frac{\partial V}{\partial x} + y\frac{\partial V}{\partial y} + z\frac{\partial V}{\partial z} = n \times V. \qquad (13.5)$$

A point dipole located at the source point (x', y', z'), for example, has effects that are homogeneous of degree $n = 3$ in the variables $\Delta x = x - x'$, $\Delta y = y - y'$, and $\Delta z = z - z'$, where (x, y, z) is the observation point coordinates.

Euler's equation has been used in various ways to interpret magnetic anomaly data, but it was the use of this method to determine the depth to and location of anomaly sources by THOMPSON (1982) and subsequent developments that have led to its widespread use. The original technique was referred to as the EULDPH method, but with further development, it has become known as Euler deconvolution and more recently as the extended Euler deconvolution technique (MUSHAYANDEBVU *et al.*, 2001) when the method is used to calculate properties of the source as well as its location and depth.

The significant advantage of this technique is that the results are independent of the direction of magnetic polarization and geologic knowledge of the source. Thus, no *a priori* assumption must be made of the geometry of the source of the anomaly and the data do not have to be reduced-to-the-pole for analysis. Furthermore, it is unnecessary to define a base level for an anomaly. Theoretically, the technique is restricted to the few bodies with integer values of the decay rate factor n, which in this application is also the negative of the structural index (N), but in practice it is used for all source configurations.

The method has generally been applied to regularly spaced profile data, but has been expanded for use on gridded map data (REID *et al.*, 1990). Since its inception, the application of the technique has been improved with better understanding of methods for selecting the optimum solution of Euler's equation from a diverse range of results, with development of procedures for minimizing the effects of interference among anomalies, and with expansion of the technique to determination of the dip and magnetic susceptibility contrast of the source as well as to gravity data (e.g. RAVAT, 1996; NABIGHIAN and HANSEN, 2001; HSU, 2002; SILVA and BARBOSA, 2003; MUSHAYANDEBVU *et al.*, 2001, 2004; COOPER, 2006; STAVREV and REID, 2007).

For the total field magnetic observation $B_T(x, y, z) = \Delta B_T + B$ with anomaly ΔB_T and a base level or constant background field B, Euler's equation becomes

$$\Delta x\frac{\partial B_T}{\partial x} + \Delta y\frac{\partial B_T}{\partial y} + \Delta z\frac{\partial B_T}{\partial z} = n \times \Delta B_T = N \times (B - B_T), \qquad (13.6)$$

where the structural index $N = -n$ and the derivatives can be measured or computed. On profile data, Euler's equation reduces at $z = 0$ to

$$\Delta x\frac{\partial B_T}{\partial x} - z'\frac{\partial B_T}{\partial z} = n \times \Delta B_T = N \times (B - B_T), \qquad (13.7)$$

which can be rearranged to give

$$x'\frac{\partial B_T}{\partial x} + z'\frac{\partial B_T}{\partial z} + N \times B = x\frac{\partial B_T}{\partial x} + N \times B_T. \qquad (13.8)$$

Thus, for a given index N, observations B_T, and measured or computed derivatives of B_T, the above equation is linear in the unknown variables x', z', and B. Least-squares estimates for these variables can be obtained by evaluating Equation 13.8 at more than three observations. In practice, seven data points are normally used at each window position as shown in Figure 13.29.

The errors in the parameter estimates are obtained in terms of their standard deviations which commonly are also normalized by the estimated depth and expressed as a percentage. Deconvolving the profile data with a series of operators or window sizes W yields a sequence of (x', z') locations and depths that are plotted for a given structural index N or a series of structural indices. Effective solutions cluster or group together well and have a small relative error compared with other indices.

Empirical criteria have been developed that are effective in accepting or rejecting depth estimates for analysis. For example, given a structural index N, window size W, and standard deviation $\sigma_{z'}$, THOMPSON (1982) suggested tolerances (TOL) be set such that z' is accepted for

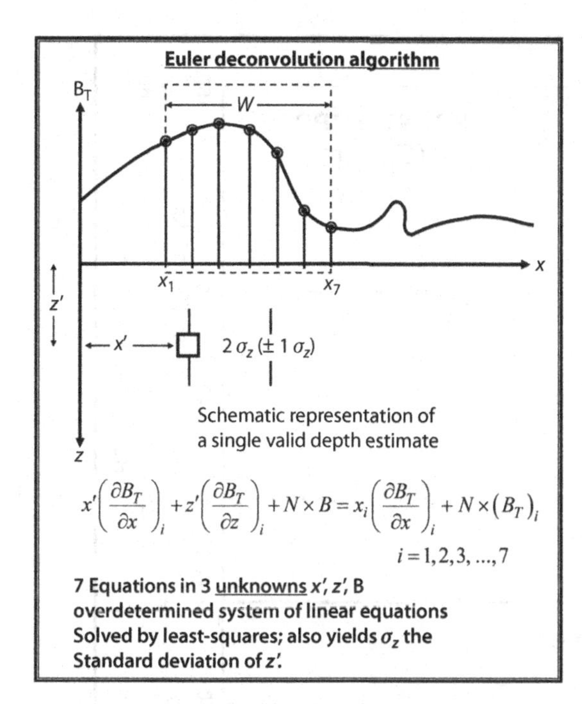

FIGURE 13.29 Euler deconvolution applied to a seven point moving window operator W for locating horizontal x' and depth z' coordinates of a source with structural index N from seven magnetic anomaly observations B_T. Adapted from THOMPSON (1982).

plotting if TOL $\leq (z'/[N \times \sigma_{z'}])$ and rejected otherwise. The pass criterion favors depths with small $\sigma_{z'}$ and/or small N, whereas the reject criterion screens out unreliable depths with large $\sigma_{z'}$ and/or large N.

The tolerance clearly depends on the estimated depth z' where the deeper estimates have greater uncertainty. Most magnetic anomalies are dipolar, and thus accommodate geologic sources with higher structural indices. However, lower structural indices are usually more directly related to the depth to the top of intrabasement sources, and thus are better depth indicators. The tolerance is also inversely weighted by the estimate $\sigma_{z'}$ of the standard deviation of z' and the structural index N.

The tolerance value implemented in actual applications reflects the quality of the data and the interpreter's experience. Noisy data commonly call for values as low as TOL $\approx 5 - 10$, whereas values as large as TOL $\approx 50 - 80$ are used for noise-free model data and high-resolution aeromagnetic observations. Experience suggests that TOL ≈ 20 is a good starting default value that must be increased if too many depth estimates are plotted and *vice versa*.

In this approach for the Euler estimates to be most effective, the correct structural index must be selected. The deconvolution assumes that geological formations can be approximated by magnetic sources of simple geometries such as spheres ($N = 3$), vertical and horizontal cylinders ($N = 2$) and sheets ($N = 1$), and the vertical edges ($N = 0$) of thick slabs (Figure 13.30). In practice, of course, the structural index is usually not known and the application may also include multiple source effects with a variety of structural indices. Thus, a range of N are typically implemented in practice and the relative clusters of Euler estimates studied for possible source associations.

EULER Structural index N for simple sources	POINT DIPOLE PD	POINT POLE PP	LINE DIPOLES LD	LINE POLES LP	SHEET DIPOLES SD	SHEET POLES SP
MAGNETIC & GRAVITY MODELS	SPHERE	VERT CYLINDER	HORIZ CYLINDER	VERT SHEET	HORIZ SHEET SEMI-∞ SLAB	CONTACT (INTERFACE)
	PT. MASS	VERT ROD	HORIZ ROD LINE MASS	THIN DIKE	FAULT (SMALL THROW)	
MAG FIELD	3	2	2	1	1	0
MAG 1st DERV	4	3	3	2	2	1
MAG 2nd DERV	5	4	4	3	3	2
GRAV FIELD	2	1	1	0	0	−1 ?
GRAV 1st DERV	3	2	2	1	1	0
GRAV 2nd DERV	4	3	3	2	2	1

FIGURE 13.30 Structural indices N for simple sources of magnetic and gravity fields and their first and second vertical or horizontal derivatives.

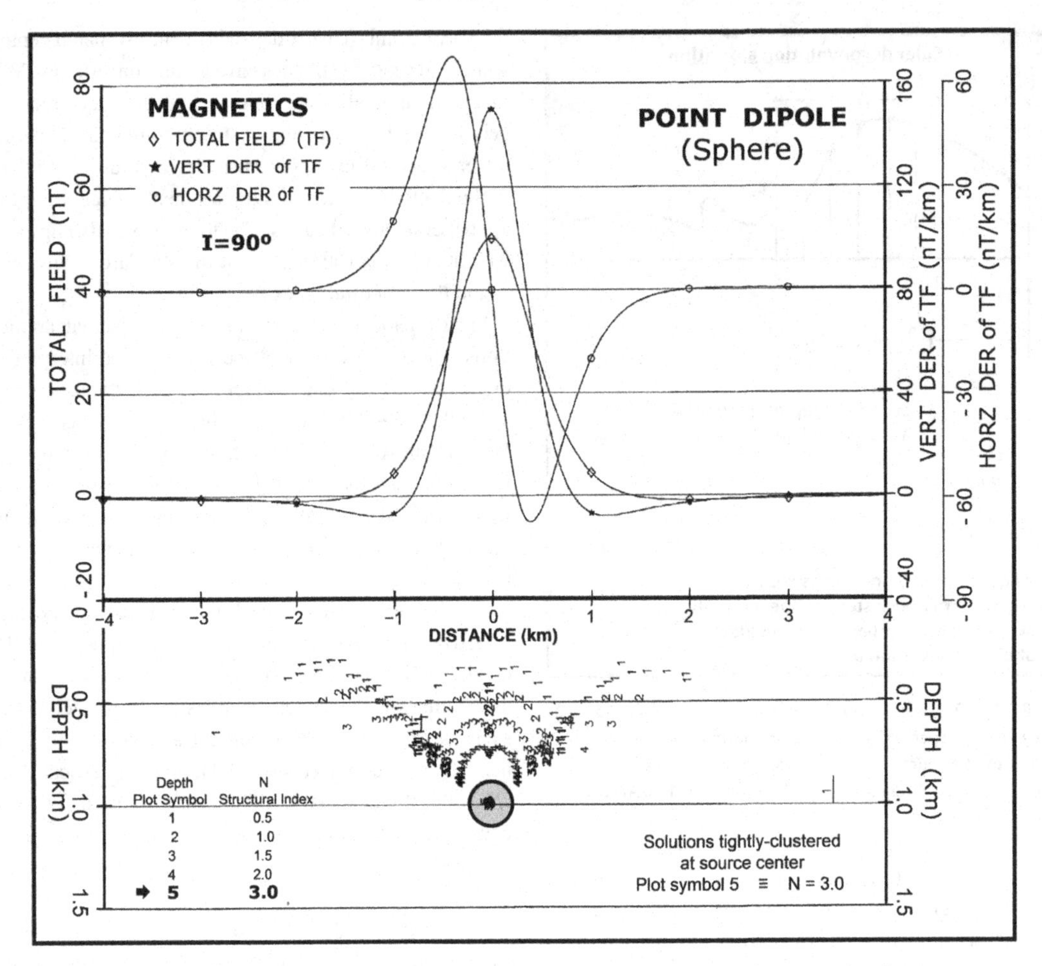

FIGURE 13.31 Euler deconvolutions of the magnetic effects of a buried dipole or a sphere. The solutions are based on two passes employing five structural indices with TOL = 30.

Horizontal source locations x' are usually well determined by Euler deconvolution, so that an incorrect choice of structural index N primarily leads to errors in the estimated source depths z'. The structural index applies to anomalies from sources with a single characteristic location and depth. It is not valid for complex bodies with results that include theoretically impossible negative indices.

Examples of the Euler estimates for a point dipole or magnetized sphere (Figure 13.31), line of dipoles or horizontal cylinder (Figure 13.32), sheet of dipoles or semi-infinite thin slab or vertical fault (Figure 13.33), and sheet of monopoles or infinite contact or interface (Figure 13.34), illustrate the relative advantages and limitations of the deconvolutions. The estimates were obtained for the five structural indices $N = 0.5$, 1.0, 1.5, 2.0, and 3.0 with the respective solutions plotted using the numeric symbols 1, 2, 3, 4, and 5. Each figure indicates the TOL used to minimize the scatter in the solutions, as well as the number of passes and window sizes implemented in the analysis.

The results in Figures 13.31–13.34 show that the solutions are accurate and well-clustered when the correct structural index is applied. For example, for the infinite vertical or dipping contact or interface (Figure 13.34), all the non-zero values of N yield deeper scattered solutions, but the solutions for $N = 0$ are tightly clustered and accurate. Smaller values of N typically produce shallower scattered depths, whereas larger values yield deeper scattered solutions. However, the shallower and deeper solutions tend to converge or trend toward the correct location of the source.

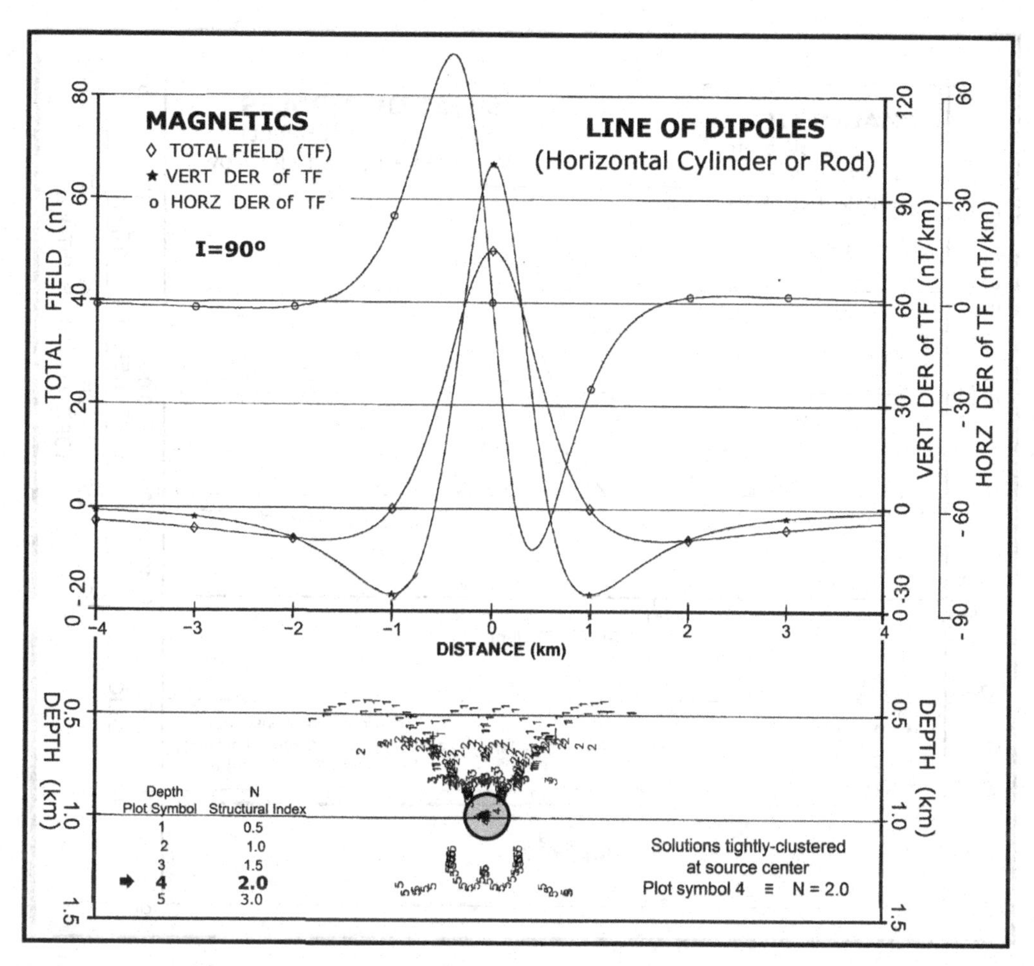

FIGURE 13.32 Euler deconvolutions of the magnetic effects of a buried line of dipoles or a 2D horizontal cylinder. The solutions are based on two passes using five structural indices with TOL = 30.

Alternatively, it is possible to solve the Euler equation in a least-squares sense for not only the location of the source and its depth, but for its structural index. As shown, for example, by HSU (2002) and illustrated in Figure 13.35 the structural index of a thin horizontal source (a step-like structure with a large depth-to-throw ratio) and its depth is located by a concentration of results and increased accuracy with a low standard deviation of the depth solutions from the Euler deconvolution plotted on the figure.

MUSHAYANDEBVU *et al.* (2001) and COOPER (2006) have extended Euler deconvolution to the determination of the dip and the susceptibility–thickness product of the thin dike model and the dip and susceptibility of the contact source. COOPER (2006) implements the Hough transform, which is used to locate features such as circles and lines in digital data sets, to execute this calculation after first locating the position and depth of the source. The optimum dip and susceptibility (-thickness) is selected on the basis of results from a range of data obtained from the Hough transform values summed over a moving window. Successful application of this methodology requires experience with the technique based on analysis of theoretical models which duplicate the general geology of the region.

In general, Euler's equation can be solved in either data profile or map formats for the position of the source and the structural index, provided that the magnetic anomaly and its derivatives can be measured or calculated accurately from the magnetic observations (THOMPSON, 1982; REID *et al.*, 1990). Solution is readily achieved through matrix inversion techniques applied to a

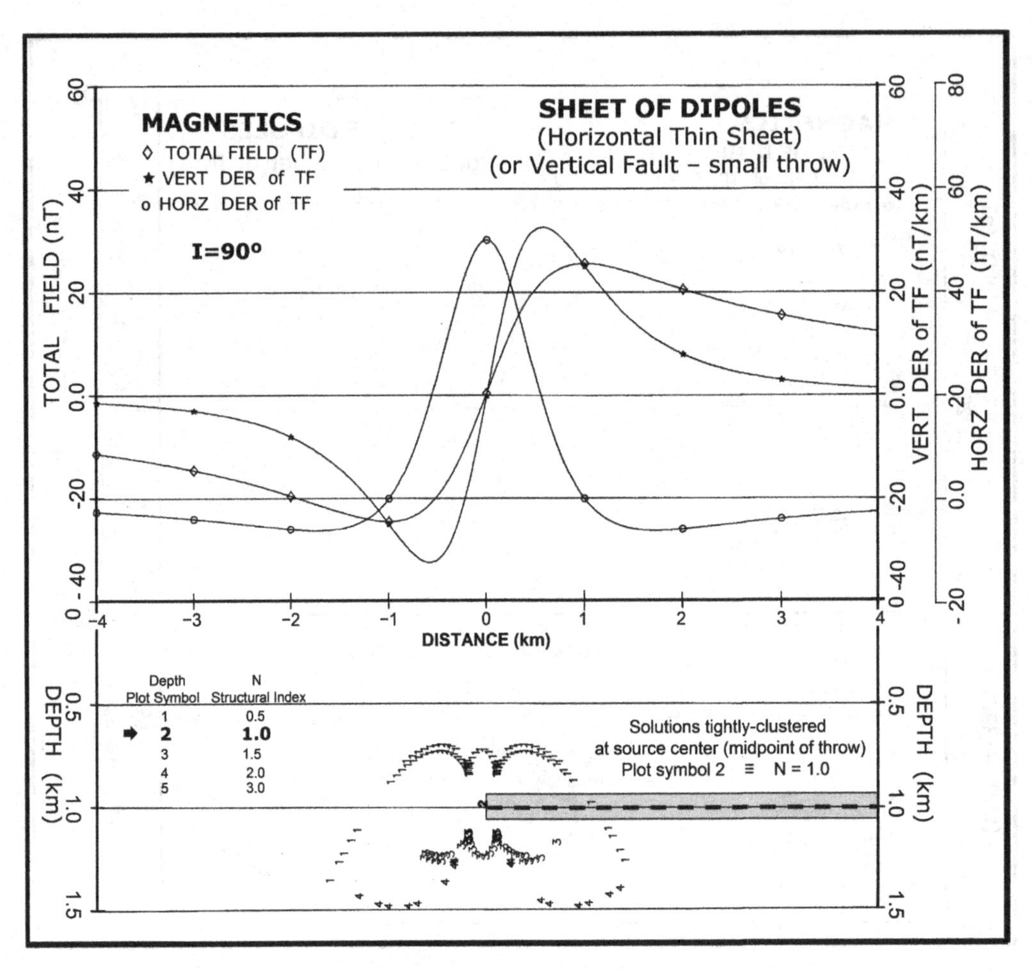

FIGURE 13.33 Euler deconvolution of the magnetic effects of a buried sheet of dipoles, a 2D horizontal thin sheet, or a vertical fault with small throw. The solutions are based on two passes using five structural indices with TOL=100.

windowed series of anomaly values. The method is particularly effective where the anomalies are well isolated and derived from simple concentrated sources. It is inappropriate for more complex shapes where the structural index varies with distance from the source (e.g. Ravat, 1996).

Hansen and Suciu (2002) extended the single-source Euler deconvolution method to multiple sources both for line-oriented data (2D case) and gridded data (3D case). This approach better accounts for the overlapping effects of nearby anomalies and hence can improve the accuracy of depth estimation.

(B) Statistical spectral techniques Spectral analysis is used in a variety of ways in the analysis of magnetic anomaly data. It can be used in the design and application of filters of various types to isolate or enhance particular attributes of anomalies, as described in Appendix A.5.1, and it also can be used in the inversion of individual or groups of anomalies. The method has been especially useful in determining the average depth to an ensemble of magnetic sources observed on either profiles or maps (e.g. Spector and Grant, 1970; Treitel *et al.*, 1971). The sources of magnetic anomalies within a region are assumed to average out so that spectral properties of an ensemble of sources are equal to those of the average of the ensemble.

This methodology is advantageous because it is statistically oriented, averaging source depths over a region containing complex patterns of anomalies. It also is less affected by interference effects due to overlapping anomalies and high-wavenumber noise than other methods

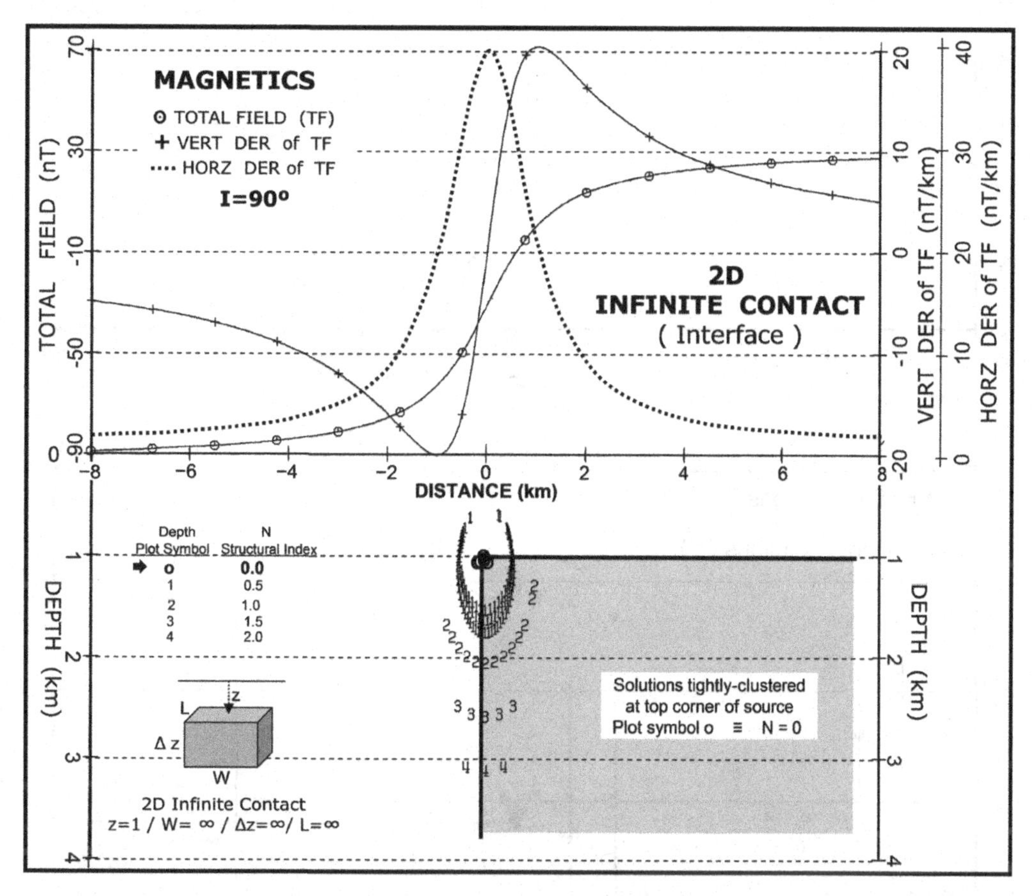

FIGURE 13.34 Euler deconvolutions of the magnetic effects of a buried sheet of monopoles or a 2D vertical contact. The solutions are based on three passes using five structural indices with TOL=60.

because it is based entirely on analyzing the wavelengths of the anomalies. Furthermore, it is independent of the directional attributes of the magnetization of the sources and the geomagnetic field, and it can be used to study a wide range of depths by varying the window involved in the data analysis.

Numerous methods have been used to calculate the spectra of anomalies (Appendix A.4.3) including the efficient and widely used fast Fourier transform (FFT). This method obtains the power spectral density by taking the squared amplitude of the Fourier transform of the infinitely extended data sequence, assuming periodic extension of the data within the window, with appropriate statistical averaging. A critical step in the procedure is selection of the size of the data window. The maximum depth of penetration is generally assumed to be roughly 20 to 25% of the window length or diameter in the case of radial averaging of the spectra (Saad, 1977b). A window that is too small will not capture sufficient anomalies for averaging and one that is too large will average the depths over too great a length or region to have geological meaning.

To circumvent this problem, a maximum entropy method can be used in spectral estimation (e.g., Saad, 1977a,b; Blakely and Hassanzadeh, 1981). The maximum entropy method is attractive because it allows smooth spectra of high resolution to be determined directly from the data, thus avoiding undesirable effects from data truncation. It is superior to other methods where small windows are involved. This permits smaller windows with resulting higher horizontal resolution in the depth determinations. Further description of the maximum entropy method is contained in Appendix A.4.3. The accuracy of the magnetic interpretation by the spectral technique is highly dependent on the accuracy of the calculated spectrum, thus considerable care is warranted in the spectral calculation.

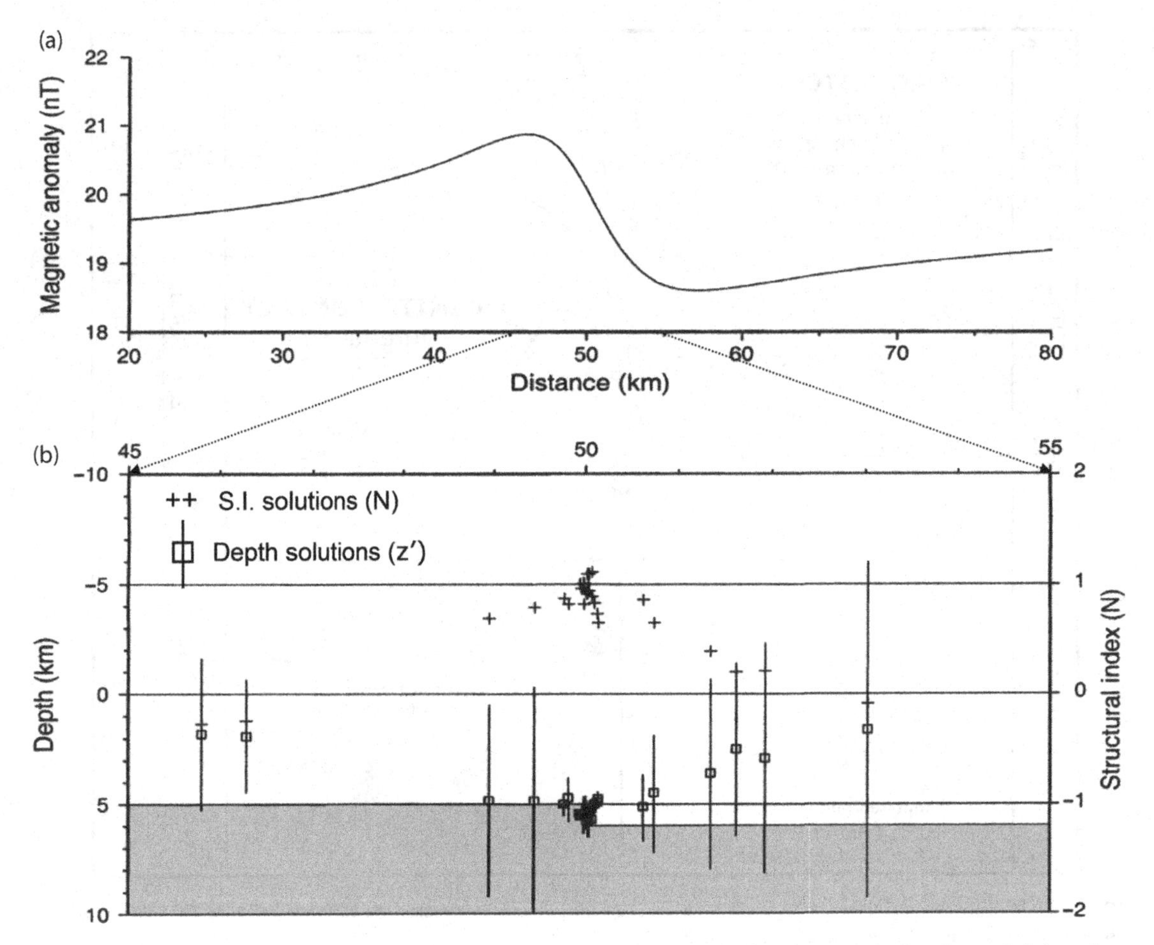

FIGURE 13.35 Simultaneous solutions of Euler's equation for the horizontal location, depth, and structural index of a step fault from the vertical derivatives of its magnetic anomaly. The solutions are based on two passes over first and second vertical derivative profiles using 4-point windows. Adapted from Hsu (2002).

Estimation of depth to the top of magnetic sources follows directly from considering the power spectrum (PS) due to an ensemble of magnetic sources that are at an average depth (z) from the observation elevation given by

$$\mathrm{PS}(k) = Ce^{-4\pi zf} = Ce^{-2zk}, \tag{13.9}$$

where f is the frequency ($= 1/\lambda$ where λ is the wavelength) in cycles/unit distance that corresponds to wavenumber $k = 2\pi f = 2\pi/\lambda$ in radians/unit distance, and C is a constant which includes field parameters and magnetic properties. This equation assumes random magnetization (i.e. its PS is constant) and that the size and thickness factors of the source are negligible, or that the power spectrum has been corrected for these factors by removing the respective power spectral components.

In general, the exponential depth factor is the dominant factor in the PS expression for either a single source or an ensemble of sources, permitting the assumption that the size and thickness factors are negligible. Taking the natural logarithm of both sides of Equation 13.9 expresses the power spectrum in the linear equation

$$\ln[\mathrm{PS}(k)] = \ln[C] - 4\pi zf = \ln[C] - 2zk, \tag{13.10}$$

where $\ln[C]$ is the intercept and $4\pi z$ or $2z$ is the slope of the straight line plot of the $\ln[PS]$ against f or k, respectively. Thus the average depth to magnetic sources is computed from the slope as

$$\begin{aligned} z(f) &= |\mathrm{slope}|/4\pi \\ &= |\mathrm{slope}| \times 0.08 \text{ cycles/unit distance, or} \\ z(k) &= |\mathrm{slope}|/2 \\ &= |\mathrm{slope}| \times 0.5 \text{ radians/unit distance.} \end{aligned} \tag{13.11}$$

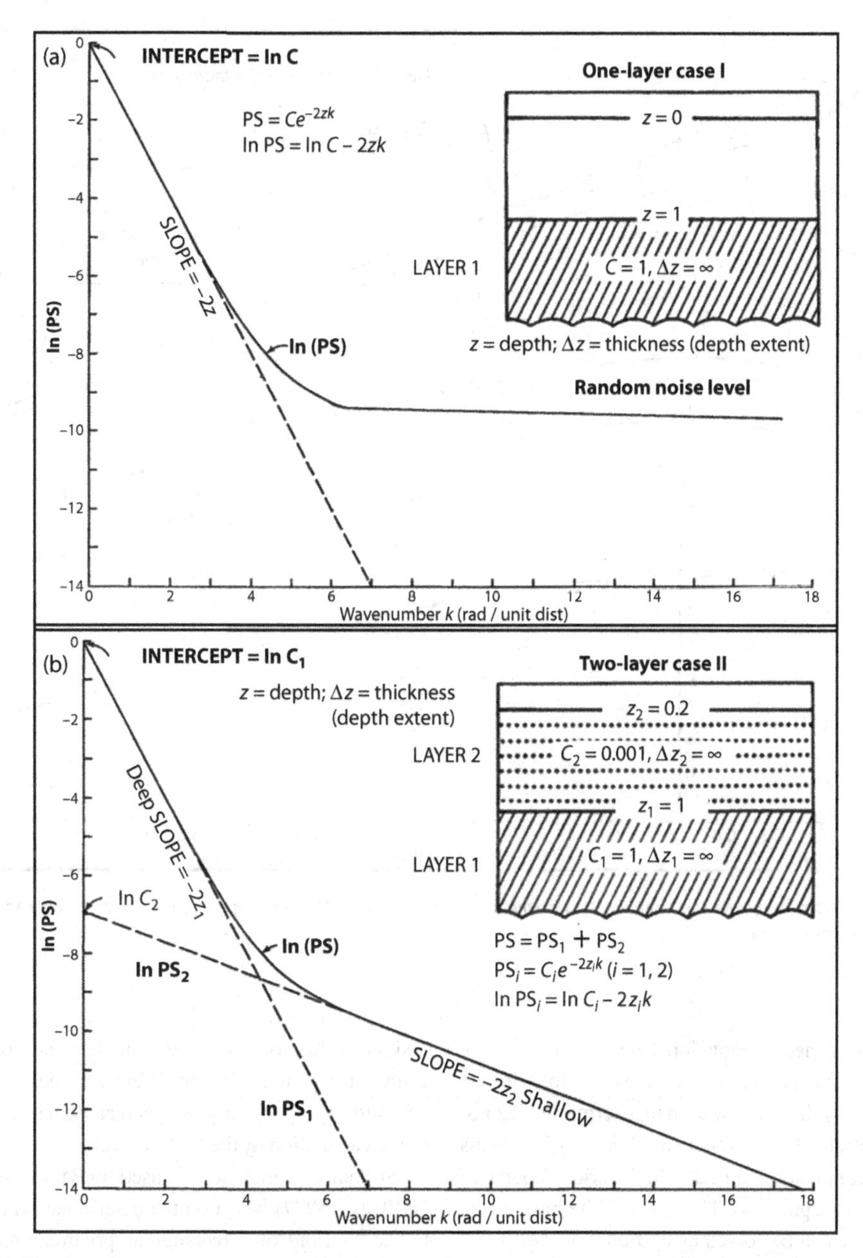

FIGURE 13.36 Theoretical power spectrum due to (a) a single ensemble of sources at an average depth z and (b) a double ensemble at depths z_1 and z_2.

Figure 13.36(a) shows the theoretical power spectrum due to a single ensemble of sources at an average depth z and, because the concept can be extended to multiple ensembles, a double ensemble at depths z_1 and z_2 is shown in Figure 13.36(b).

The PS also contains information on the thickness of the magnetic sources which can be used to determine the base of the source. In this regard, the method has been widely used to map the depth of the Curie point isotherm (~580 °C) where magnetite loses its ferrimagnetic properties. The theoretical power spectrum of the magnetic field due to a magnetic source between depths to the top z_t and the base z_b of the source can be expressed as

$$PS(k) = C(e^{-z_t k} - e^{-z_b k})^2, \tag{13.12}$$

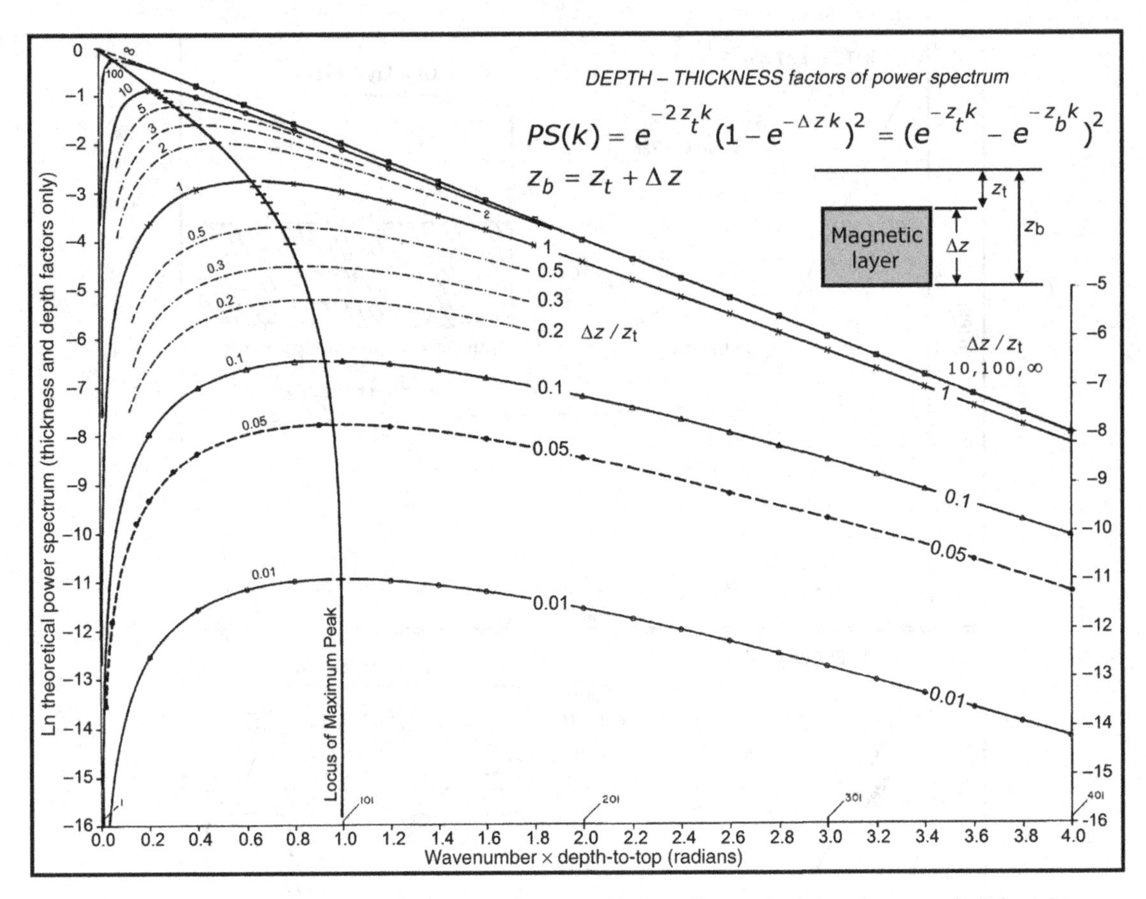

FIGURE 13.37 Theoretical power spectrum due to a magnetic source with top and bottom depths z_t and z_b, respectively, for various values of thickness-to-depth ratios ($\Delta z/z_t$).

where C is as defined in Equation 13.9 (SAAD, 1977b). Assuming the crustal magnetization is random in the region sampled by the window used in determining the PS, a plot of Equation 13.12 as shown in Figure 13.37 reveals that the spectrum reaches a peak which varies depending on the thickness/depth ratio. The depth to the base of the magnetic source can be solved using Equation 13.12 and the value of k at the maximum and an estimate of the depth to the top of the source from the straight-line slope of the power spectrum. For a bottomless source the $k(\max) = 0$ and is equal to $1/z_t$ for thin sheet-like sources.

Alternatively, the bottom depth can be determined by forward modeling of the spectral peak using Equation 13.12 (e.g. ROSS *et al.*, 2006; RAVAT *et al.*, 2007). In this approach the observed spectrum is iteratively matched by the theoretical spectrum. The constant C can be modified to move the model curve up or down and the depth to the top adjusted to fit the observed spectra. The results are inherently more robust than selecting a single value for $k(\max)$ to calculate the bottom of the source. It is noted that a small error in the estimate of the depth to the top of the source can cause a significant error in the calculation of the bottom depth.

Still another method was used by BHATTACHARYYA and LEU (1977) based on the determination of the depth to the centroid of a rectangular prism source using the radially averaged power spectrum determined over a data window and an estimate of the depth to the top of the prism.

It is also theoretically possible to determine the width of a magnetic source from its spectrum because for a single source the width effect causes a series of damped oscillations with equally spaced troughs and peaks. The first trough occurs at $k(0) = \pi/a$ for the source of width $= 2a$, whereas other troughs occur at the harmonics of this fundamental wavenumber. Thus, the width of the source is equal to $2\pi/\Delta k$ where Δk is the wavenumber between successive troughs or peaks in the spectrum.

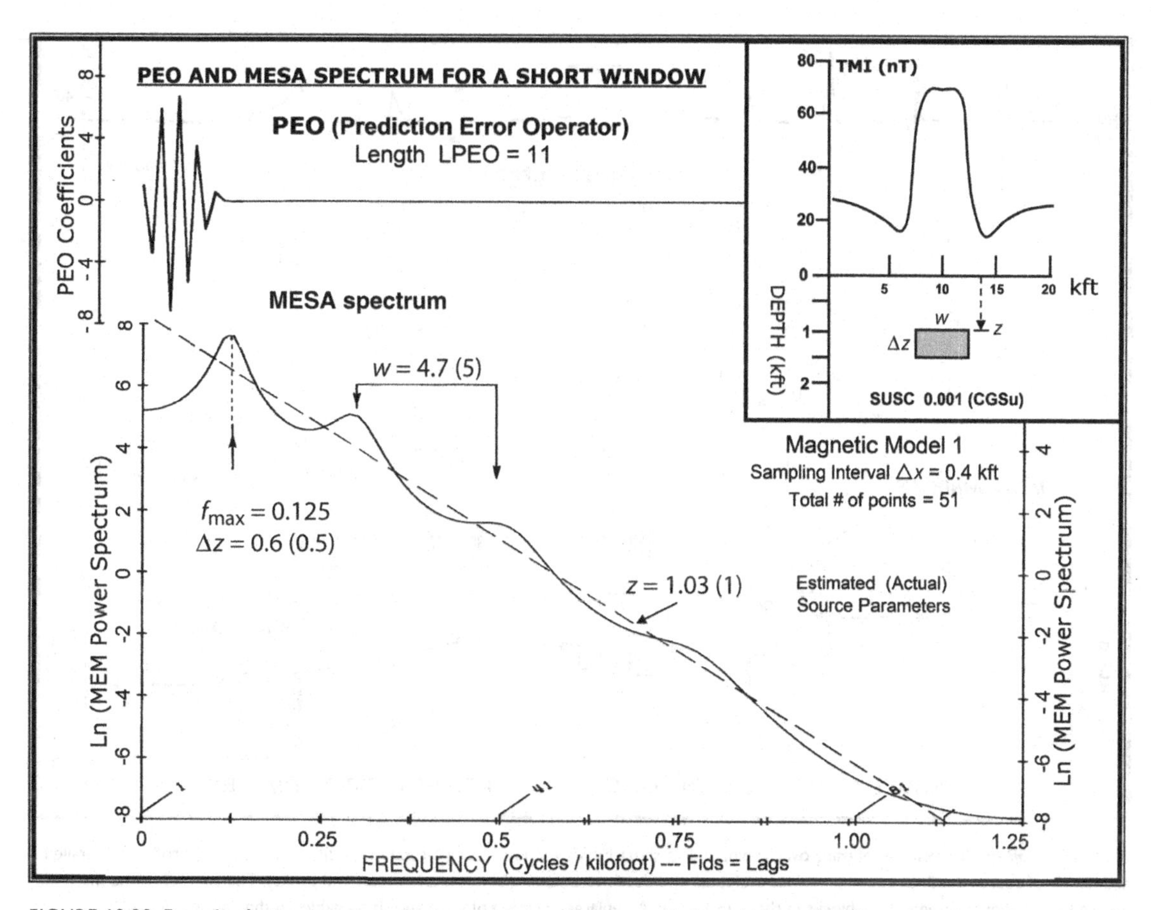

FIGURE 13.38 Example of estimating source depth-to-top ($z = z_t$) from spectral slope, thickness (Δz) from spectral maximum peak, and width (w) from frequency separation of successive peaks in the power spectrum. Numbers in parentheses are actual values.

An example of estimating the width from the separation of successive peaks in the MESA power spectrum of a single prismatic source is shown in Figure 13.38. The ME (maximum entropy) spectrum is computed from the short window of the total magnetic intensity (TMI) profile (insert) using a prediction error operator (PEO) of length LPEO = 11 coefficients. Also shown are the estimation of depth to the top-of-source from the slope of the linear logarithmic spectrum, and thickness z from the location of maximum peak f_{max}. The estimates of the source parameters show good agreement with the actual values listed in parentheses.

Figure 13.39 shows a model example of a TMI profile over an ensemble of prismatic sources with variable depths, thicknesses, and widths used for detailed estimation of source parameters by the maximum entropy spectral method. The MESA power spectra were computed using overlapping sliding short windows to obtain detailed information regarding variations in source parameters from one end of the profile to the other. In this figure, the spectral windows are plotted at their respective locations along the model. Depths-to-top obtained automatically from the spectral slopes are plotted as circles over the model at the center of the corresponding window. Additional depth estimates plotted as squares are from other MESA windows not shown in this figure. The depth-to-bottom of sources, estimated from position of maximum spectral peak when available, is shown by the short lines connected by dots; the results are sometimes questionable for wide sources.

Figure 13.40 illustrates using the high-resolution MESA technique to analyze TMI profile data from the NURE Survey over the National Petroleum Reserve in Alaska's North Slope (SAAD, 2009a,b). The TMI profile data in the insert display short-wavelength sedimentary micro-magnetic anomalies embedded in regional basement anomalies and high-frequency noise. The computed MESA power spectrum is interpreted to show the three

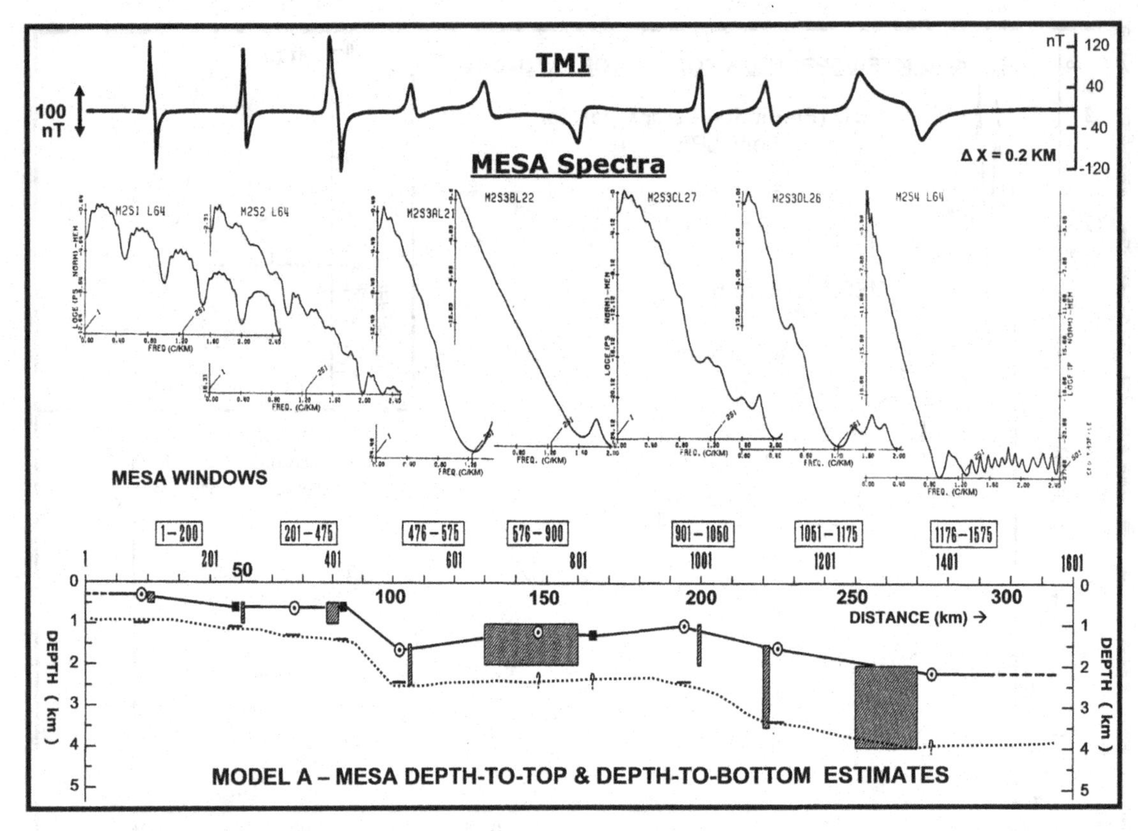

FIGURE 13.39 Model example of using overlapping sliding short MESA windows along a total magnetic intensity (TMI) profile for detailed estimation of source depth-to-top (from spectral slopes) and depth-to-bottom (from maximum spectral peaks). The sources of the magnetic anomalies are shown in the shaded blocks in the lower panel. An enlarged version of this figure is available on the website, www.cambridge.org/gravmag, for better viewing.

main frequency components in the data. The desired signal associated with the short-wavelength sedimentary magnetic anomalies appears to be predominant within the wavelength interval of about 0.5 to 3 or 4 km. Based on this spectrum, a band-pass filter passing 0.7–9 km wavelengths was applied to the TMI data to isolate and enhance the micro-magnetic anomalies of interest shown in Figure 13.41. Using the spectral slope method, the average depth to the sources of these anomalies is about 200 meters below surface. The spatial band-pass data in the insert of Figure 13.41 clearly show several high-frequency sedimentary magnetic anomalies which appear to correlate with oil and gas production or oil seeps in the area.

(C) Werner deconvolution From the perspective of magnetic anomalies measured along profiles, lateral margins or contacts between subsurface magnetic sources of anomalies can be interpreted by the collected effects of thin sheet-like bodies. In particular, WERNER (1953) showed that the total magnetic field anomaly profile (ΔB_T) across a 2D thin dike (i.e. thin sheet) extending infinitely deep (i.e. with width less than the depth z' to the top of the sheet) can be expressed as

$$\Delta B_T(x) = \frac{A(x - x') + (C \times z')}{(x - x')^2 + z'^2}, \tag{13.13}$$

where x is the distance along the profile, and x' is the lateral position of the top of the dike (Figure 13.42). The constants A and C are functions of the susceptibility, magnetic inclination and declination, and the dip of the dike as defined by WERNER (1953).

As identified by WERNER (1953) for dikes perpendicular to the profile, the above equation can be rewritten in a linear form

$$a_o + a_1 x + c_o \Delta B_T + c_1(x \Delta B_T) = x^2 \Delta B_T, \tag{13.14}$$

with

$$a_o = -(A)x' + (C)z'; \; a_1 = A;$$

$$c_o = -(x')^2 - (z')^2; \text{ and } c_1 = 2x', \tag{13.15}$$

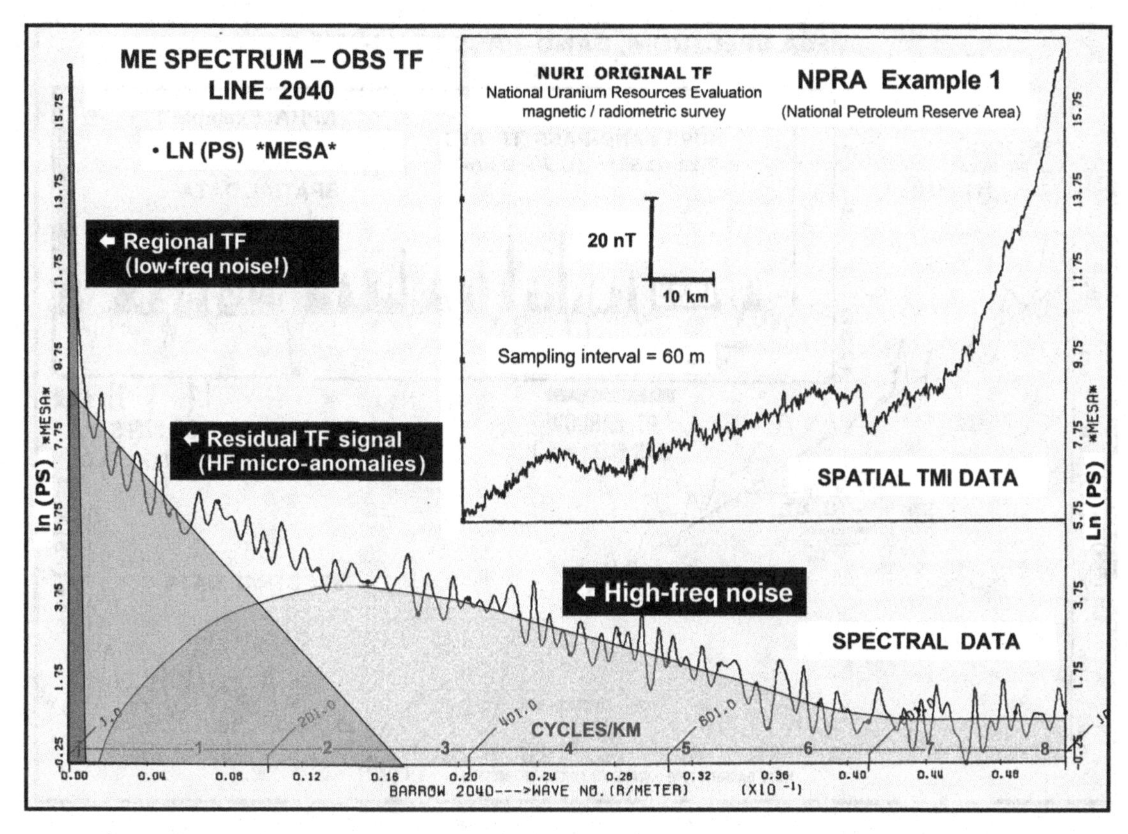

FIGURE 13.40 Maximum entropy spectral analysis (MESA) of the components of a NURE total magnetic intensity (TMI) profile over the National Petroleum Reserve (NPRA) in Alaska's North Slope. Adapted from SAAD (2009a,b).

where

$$x' = \frac{c_1}{2};\ z' = 0.5\sqrt{-4c_o - c_1^2};\ \text{and}\ C = \frac{(2a_o + a_1c_1)}{\sqrt{-4c_o - c_1^2}}. \tag{13.16}$$

Equation 13.14 is linear in the four variables a_o, a_1, c_o, and c_1 so that a mathematically unique solution can be found for them from evaluating the equation at four $(x, \Delta B_T)$ points (Figure 13.42). In turn, the dip of the thin dike and its magnetic susceptibility contrast are calculated from A and C, the assumed dike thickness, and the inclination and declination of the inducing magnetic field. HARTMAN *et al.* (1971) have described the use of this method not only for dikes but also for intrabasement contacts.

The procedure is most effective for well-isolated anomalies. However, in an actual survey, the anomalous field ΔB_T is rarely measured without interfering anomaly fields. For example, the measured field B_T may include the anomalous field ΔB_T plus linear interference with intercept and slope coefficients D_o and D_1, respectively, so that

$$B_T = \Delta B_T + D_o + D_1 x. \tag{13.17}$$

Thus, solving for ΔB_T and substituting into Equation 13.14 yields

$$e_o + e_1 x + c_o B_T + c_1 x B_T + e_2 x^2 + D_1 x^3 = x^2 B_T, \tag{13.18}$$

where

$$e_o = -Ax' + Cz' + C_o[(x')^2 + (z')^2];$$
$$e_1 = A - 2D_o x' + D_1[(x')^2 + (z')^2];$$
$$\text{and}\ e_2 = D_o - 2D_1 x'. \tag{13.19}$$

Accordingly, from six observations (x, B_T), Equation 13.18 may be solved by standard matrix inversion (i.e. deconvolution) methods for the six unknown coefficients e_o, e_1, e_2, c_o, c_1, and D_1. These results, in turn, can be used

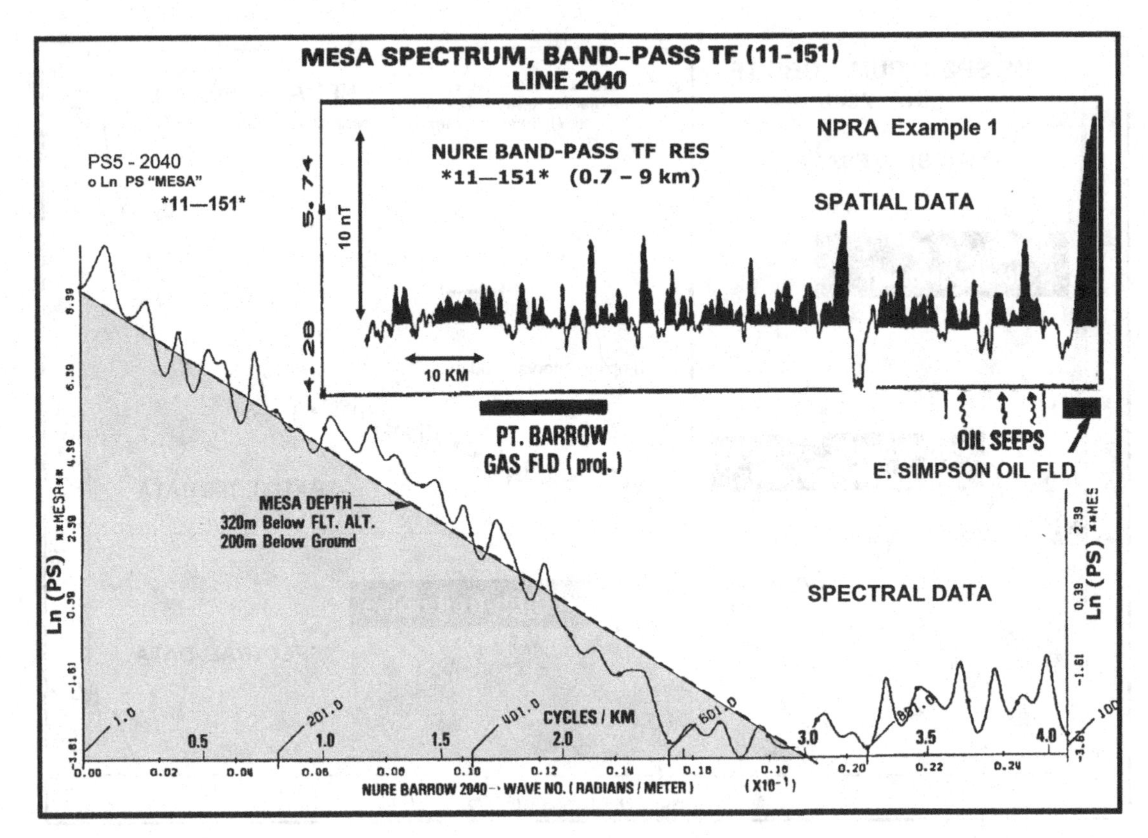

FIGURE 13.41 High frequency sedimentary magnetic anomalies extracted by MESA from the total magnetic intensity (TMI) profile in Figure 13.40 over the National Petroleum Reserve (NPRA) in Alaska's North Slope. Adapted from SAAD (2009a,b).

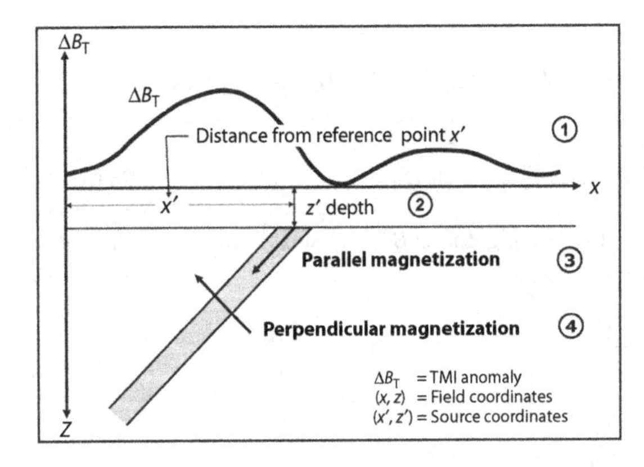

FIGURE 13.42 Werner deconvolution requires a minimum of four anomaly values to solve for the four basic physical variables of the 2D thin dike as specified in the circled (1–4) elements of this illustration.

to obtain various dike parameters such as

$$x' = \frac{c_1}{2}; \quad A = a_1; \quad C = \frac{(2a_o + a_1 c_1)}{\sqrt{-4c_o - c_1^2}}; \text{ and}$$

$$z' = 0.5\sqrt{-4c_o - c_1^2}. \tag{13.20}$$

Typically, the dike estimates are established over six consecutive data points with the center of the window or six-point operator shifted or lagged to the next data point until all anomaly data have been covered and processed. In practice, additional passes are made with increasing point spacing or lags on low-pass filtered versions of the total field anomalies to focus on anomalies at increasing depths. Calculated depths are examined for consistency and inconsistent depth estimates are rejected.

Werner deconvolution is not only effective for determining the position, depth, dip, and magnetic susceptibility contrast times thickness product of thin vertical or dipping sheets, but can also be employed to determine the characteristics of 2D contacts or interfaces of prismatic bodies. The latter is based on the equivalence of the horizontal gradient of an interface anomaly and the anomaly of a thin dike, as shown in Figures 13.43 and 13.44. Note the similarity between Figures 13.43(a) and 13.44(b) for the dipping thin sheets and dipping contacts. Horizontal derivatives may be observed in the field, but more commonly they are calculated from the difference in total intensity values at a prescribed horizontal distance.

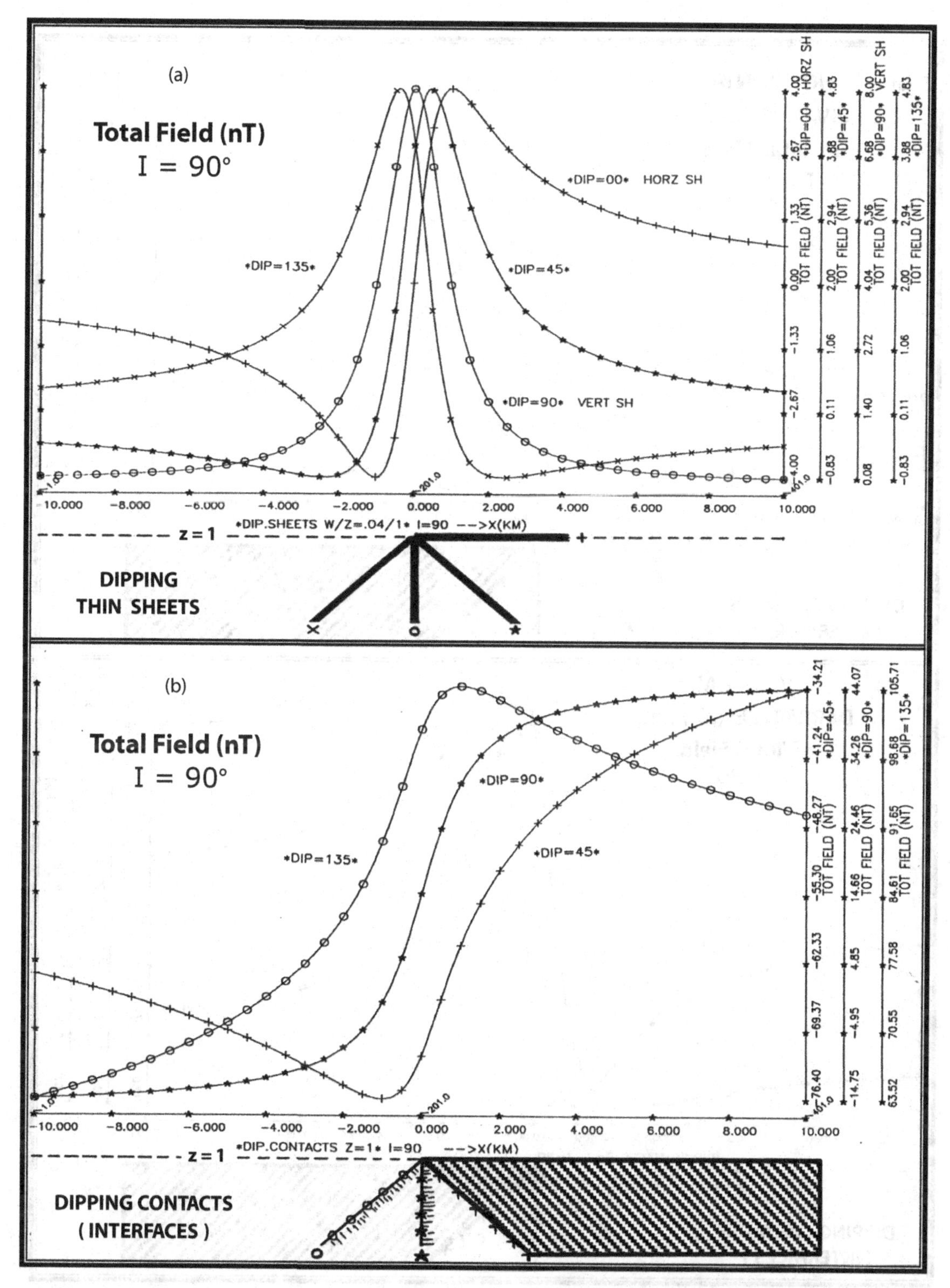

FIGURE 13.43 Total magnetic intensity (TMI) anomalies of horizontal and dipping thin sheets (a) and dipping contacts or interfaces (b).

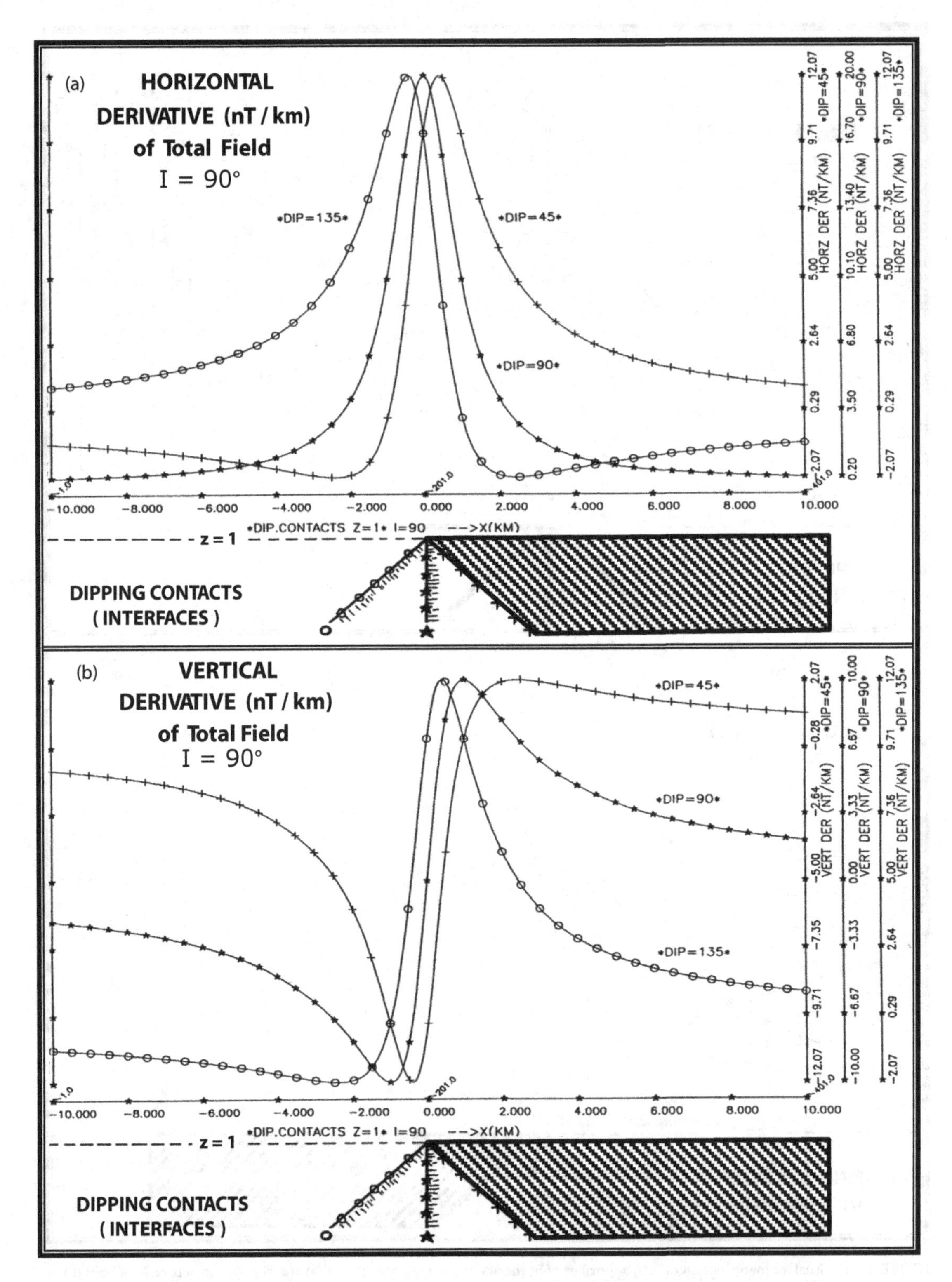

FIGURE 13.44 Horizontal (a) and vertical (b) derivatives of the total magnetic intensity (TMI) anomalies due to dipping contacts or interfaces. Mathematically, horizontal derivatives of the TMI of dipping contacts are equivalent to the TMI of dipping sheets.

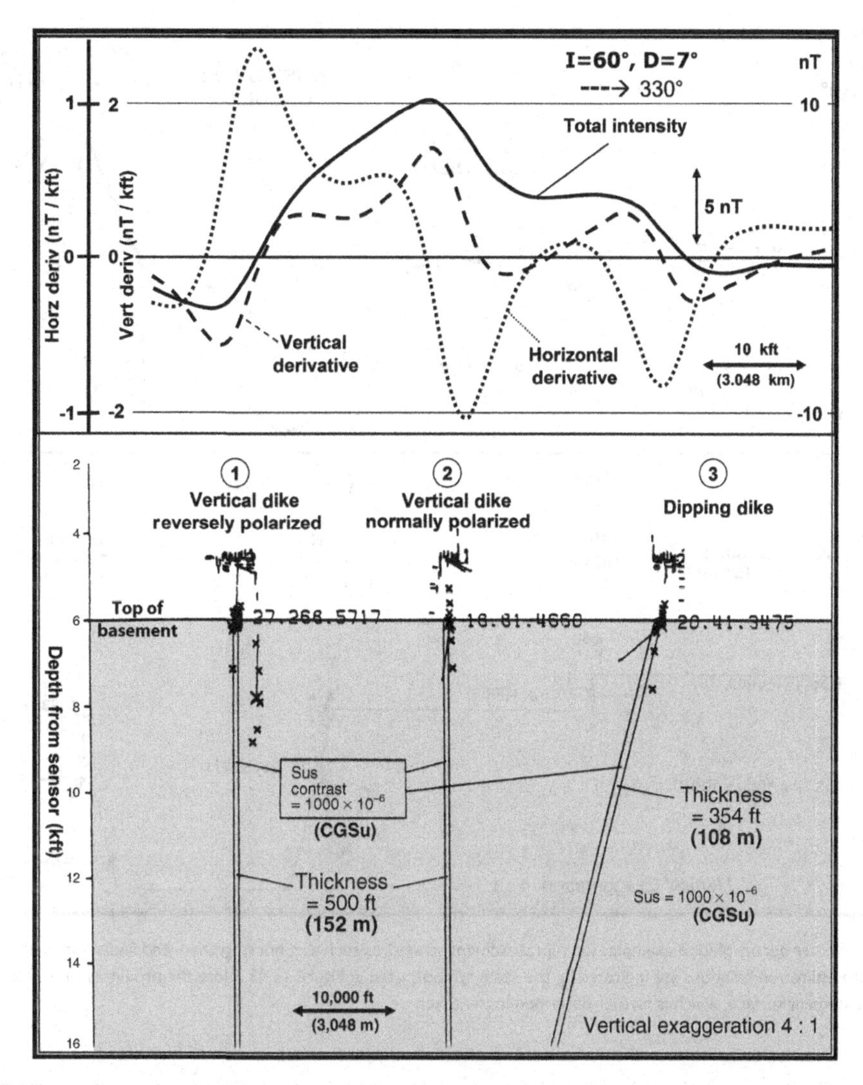

FIGURE 13.45 Werner deconvolution interpretation examples of vertical and dipping thin sheets with their upper surface at the depth indicated by the horizontal line. Deeper clustered solutions are thin sheet total field solutions marked by (×)-symbols. Shallower scattered estimates are interface horizontal gradient solutions marked by (–), and vertical gradient solutions are mapped by (|)-symbols. Also shown are dip and susceptibility or (susc × thickness) data. The optimum solution marked by the cluster of total field solution is positioned at the actual depth of the model sources.

The depth and position of the contact are independent of the direction of magnetization, and thus are unaffected by remanent magnetization. Experience shows that the depths from the gradient calculations are typically of the order of 20% shallower than those obtained from the total field (e.g. see the model examples in Figures 13.45 and 13.46). The interpreter must decide from the appearance of the anomalies, the calculated dip and susceptibility contrast, and collateral information whether a thin dike or edge effect is the source of the anomaly, and thus which calculated depths are appropriate. Once the depth picks are made, the interpreter must establish the geological significance of the results. Clearly, considerable experience and judgment enter into successful interpretation by this procedure despite its automated nature.

Because of the linearity of the gradients with the inverse distance, the deconvolution can also easily be applied to measured or calculated vertical gradient data. For example, Figures 13.44(b) and 13.43(a) show the similarity between the vertical derivative of the TMI anomaly of a vertical contact and the TMI anomaly of a horizontal thin sheet. The vertical derivative is advantageous because it is sensitive to the edges of wide dikes. Applications of Werner

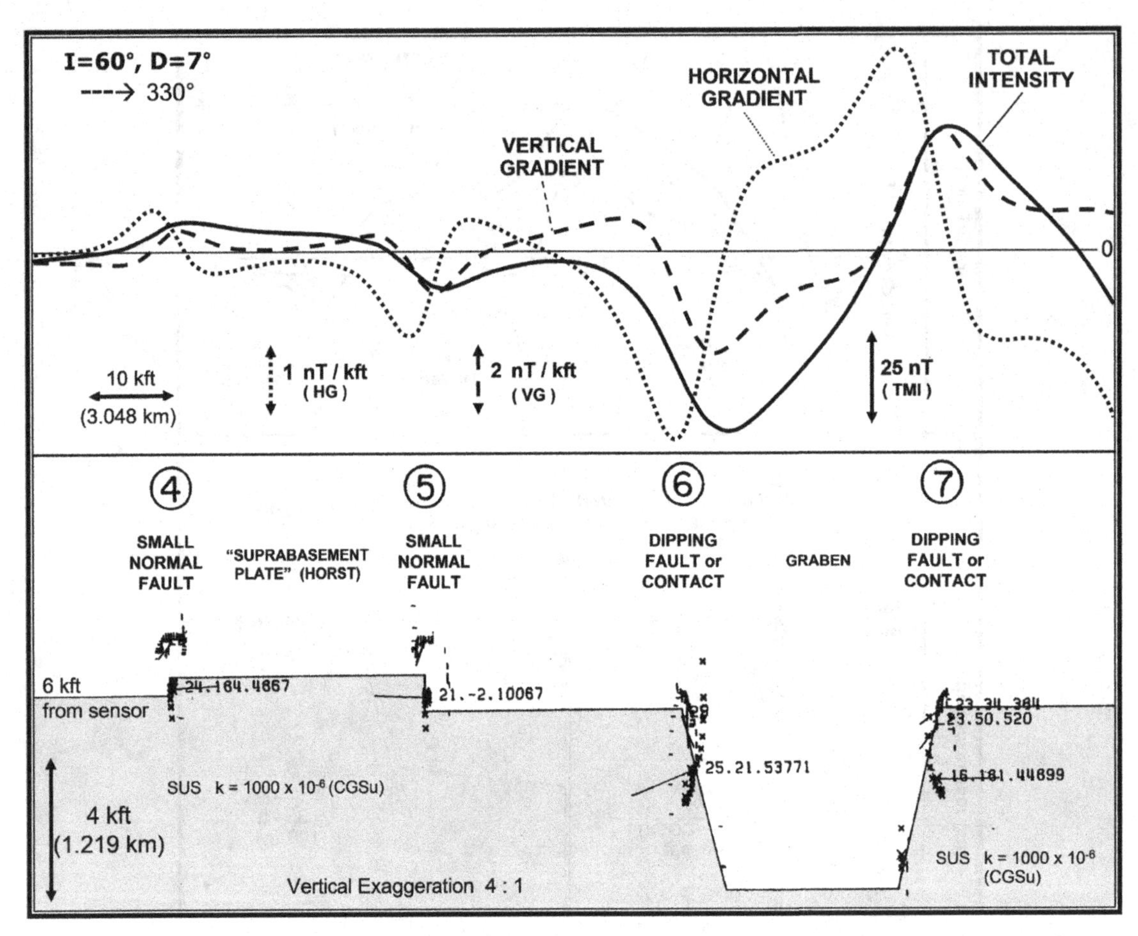

FIGURE 13.46 Werner deconvolution examples for suprabasement structures such as a horst, graben, and faults with small and large throws. The depth estimation solutions are indicated by the same symbols used in Figure 13.45. Note the proximity of the depth solutions to the modeled basement surface, which is further described in the text.

deconvolution to the calculated vertical gradient may result in superior separation of neighboring anomalies, but loss of depth resolution for deep sources may also occur.

The Werner deconvolution depth calculations in Figures 13.45 and 13.46 show that the derivative solutions are generally about 20% shallower than total field solutions. The total field solutions with the (×)-symbols are well clustered and accurate for thin dikes or sheets, small-throw faults, horst or low-relief suprabasement structures as compared with the shallower scattered derivative solutions marked by the (−, |)-symbols. For contacts and large-throw faults, however, the opposite is true with the shallower derivative solutions being more clustered and accurate than the total field solutions. Also shown in these figures are dip and susceptibility data in the case of interface solutions or (susceptibility × thickness) product data for thin sheet solutions.

An example of the applications of Werner deconvolution and MESA to a profile of TMI data over the Beaufort Sea area is shown in Figure 13.47. The TMI profile indicates the presence of at least two distinct magnetic horizons at different depth levels. The southern one-third of the profile is on land characterized by high-frequency magnetic anomalies with 10–30 nT amplitudes that obviously have a near-surface origin, averaging about 400 m sub-sensor (i.e. 100 m below surface) as suggested by the Werner depths and MESA. The broad large-amplitude (>100 nT) anomaly in the center of the profile is due to an offshore, basement feature or a basic intrusive body at a depth of about 6.4 km from the Werner solutions and MESA calculations. In the Werner solutions and MESA, the high-frequency anomalies seem to persist northward offshore becoming broader or deeper and weaker away from land. These anomalies are believed to originate from

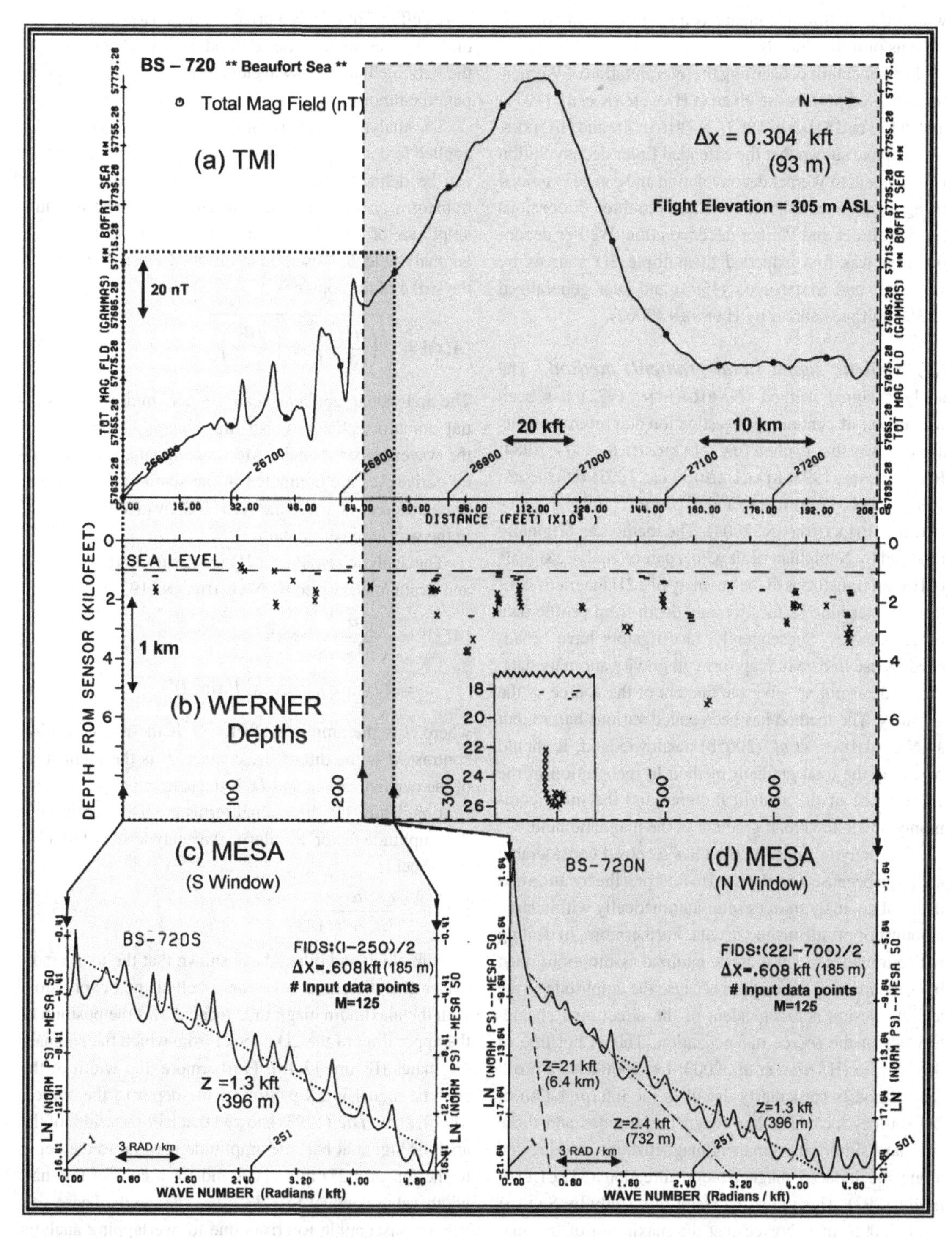

FIGURE 13.47 Total magnetic intensity profile over the Beaufort Sea area (a) analyzed for Werner deconvolution depths (b), and MESA depths for the higher frequency sources in window (c), and the regional source in window (d).

within the shallow sediments as a result of high concentrations of iron minerals.

Further details concerning the interpretation of Werner-deconvolved profiles are given in HARTMAN *et al.* (1971) and in KU and SHARP (1983). NABIGHIAN and HANSEN (2001) have shown that the extended Euler deconvolution is equivalent to Werner deconvolution and can be extended using generalized Hilbert transforms to three dimensions for both Euler and Werner deconvolution. Werner deconvolution was first extended to multiple 2D sources by HANSEN and SIMMONDS (1993) and later generalized to 3D multiple sources by HANSEN (2002).

(D) Analytic signal (total gradient) method The analytic signal method (NABIGHIAN, 1972) has been the subject of continuing investigation and improvements since it was first applied (e.g. NABIGHIAN, 1974, 1984; ROEST *et al.*, 1992; MACLEOD *et al.*, 1993; HSU *et al.*, 1996, 1998; DEBEGLIA and CORPEL, 1997; KEATING and PILKINGTON, 2004). The method as originally reported by Nabighian dealt with a pair of analytic signals or Hilbert transforms of the anomaly of a 2D magnetic contact to determine its location and depth from profile data of 2D sources. Subsequently, investigators have broadened its use to data in map form, to gravity anomaly data, and to determining other parameters of the source of the anomaly. The method has been called various names, but as NABIGHIAN *et al.* (2005b) acknowledged, it should be called the total gradient method in recognition of the equivalence of the analytical signal and the more commonly understood total gradient of the magnetic field.

The analytic signal method has received considerable attention because it can be used to interpret the location and depth of anomaly sources semi-automatically within large amounts of profile magnetic data. Furthermore, in dealing with anomalies of 2D sources, minimal assumptions must be made in using the method because the amplitude of the analytical sign is independent of the directional characteristics of the source magnetization. This is not true of 3D sources (HANEY *et al.*, 2003; LI, 2006). As a result the method is particularly useful in the interpretation of 2D sources located at low magnetic latitudes and those that have significant remanent magnetization not directed along the induced magnetization direction (MACLEOD *et al.*, 1993). This too is the result of a study by SALEM *et al.* (2002) that showed that the maximum of the analytic signal amplitude of a dipolar source is not necessarily located directly over this source. The analytic signal method is subject to problems from interference among anomalies, but its use of derivatives of the measured field enhances the resolution of independent sources and minimizes effects of deeply buried sources. However, the use of derivatives subjects the method to considerable error if the data include short-wavelength observational or computational noise.

The analytic signal of magnetic anomalies as originally applied to potential field data is a complex quantity which can be defined either by the total field and its Hilbert transform or by its orthogonal derivatives. The absolute amplitude of the 2D analytic signal of the total magnetic anomaly field B_T observed at distance x perpendicular to the strike of the source is

$$|A(x)| = \sqrt{\left(\frac{\partial B_T}{\partial x}\right)^2 + \left(\frac{\partial B_T}{\partial z}\right)^2}. \qquad (13.21)$$

The individual gradients can be determined with spatial domain techniques, but are generally calculated in the wavenumber domain. More commonly, the horizontal derivatives are computed in the spatial domain using finite differences, while the vertical derivative is computed in the wavenumber domain.

The analytic signal of a 2D magnetic contact at $x = 0$ and depth h as derived by NABIGHIAN (1972) is

$$|A(x)| = \frac{\alpha}{\sqrt{h^2 + x^2}} \quad \text{with}$$
$$\alpha = 2J(\sin d)(1 - \cos^2 I' \sin^2 D'), \qquad (13.22)$$

where α is the amplitude factor, J is the magnetization contrast, d is the dip of the contact, I' is the inclination of the magnetization, and D' is the azimuth of the magnetization. Thus, the directional terms are included in only the amplitude factor. Similarly, the analytic signal of a 2D thin sheet is

$$|A(x)| = \frac{\alpha}{(h^2 + x^2)}. \qquad (13.23)$$

Nabighian and others have shown that the total gradient or analytic signal takes on a bell-shaped expression, with the maximum magnitude located over the position of the upper limit of the 2D contact from which the anomaly originates (Figure 13.48). Furthermore, the width of the analytic signal is a function of the depth to the source. MACLEOD *et al.* (1993) showed that half the width of the analytic signal at half the amplitude is equal to the depth to the top of a 2D thin sheet and for a contact this half width value is equal to 1.73 times the depth. These values are susceptible to errors due to overlapping analytic signals from interfering anomalies and require a correct determination of the source of the anomaly, and thus its structural index. HSU *et al.* (1996, 1998) have used an alternative measure of depth for 2D sources based on the ratio of the analytic signal to the higher-order analytic

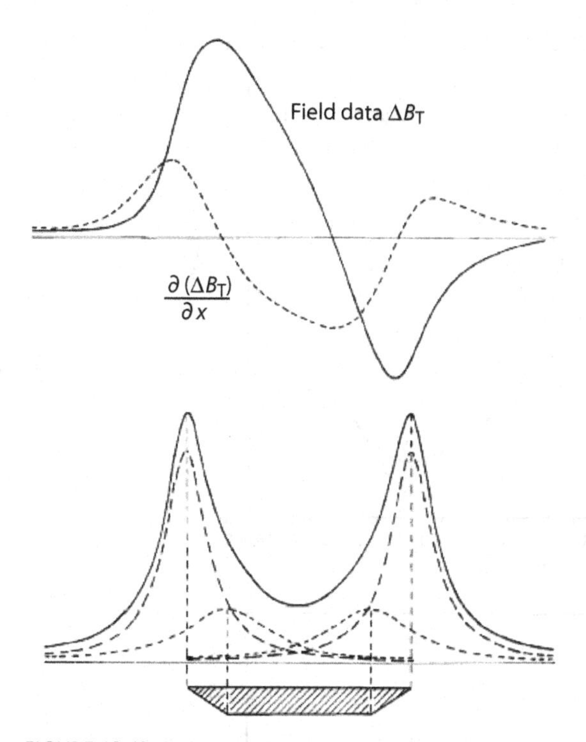

FIGURE 13.48 Analytic signal example showing the total intensity and horizontal derivative of a 2D trapezoidal source in the upper figure, and the amplitude signal curve (solid line) and its symmetric components (dashed lines) in the lower figure. Note the centers (maxima) of the bell-shaped components are located directly over the corners of the trapezoidal source, and half-width values are proportional to the depth to the corners of the source. Adapted from NABIGHIAN (1972).

signal, and SALEM *et al.* (2002) have used a similar approach for estimating the depth to dipolar sources. Additionally, KEATING and PILKINGTON (2004) have used the Euler deconvolution method on the analytic signal to determine source location and an estimate of the structural index and in turn the depth; finally, in their method, α is calculated, and thus the dip and magnetization contrast determined from relationships involving the analytic signal and the vertical gradient of the magnetic field.

The total gradient has advantages over the use of horizontal gradients (BLAKELY and SIMPSON, 1986) for locating the position of a two-dimensional source because the results of the former are independent of dip and magnetization direction. However, the difference in the location of the peak of the two signals can be used to identify dip direction (PHILLIPS, 2000) or remanent magnetization (PILKINGTON and KEATING, 2004). Figure 13.49 illustrates the results of Euler deconvolution of the analytic signal along a total magnetic intensity anomaly profile observed over the Canadian Precambrian shield in northeast Ontario, Canada. The figure shows the total magnetic intensity anomaly profile of a thin magnetic sheet, a dike, and its analytic signal. The results of deconvolution of the analytic signal show a tight grouping and indicate a source depth of roughly 130 m with a structural index of 2.15, which closely approximates the theoretical index of 2 for the analytical signal of a thin magnetic sheet. The dip of the dike is calculated to be 104°. The depth and dip approximate the parameters of the dike as obtained from geological information.

(E) Source parameter imaging™ (local wavenumber technique) The application of the complex analytic signal to potential fields by NABIGHIAN (1972) has spawned a number of variations to broaden its application and to increase the efficiency and accuracy of the method. One of these variations is the local wavenumber or Source Parameter Imaging™ method initially described by THURSTON and SMITH (1997) and subsequently further developed by others (e.g. THURSTON *et al.*, 1999, 2002; SMITH *et al.*, 1998; SALEM *et al.*, 2005). This method, which requires second derivatives of the anomaly field, employs a local wavenumber term, defined as the spatial derivative of the local phase, to determine the depth to 2D magnetic sources. As a result the method is also referred to as the local wavenumber technique. Wavenumber as used in this method is not to be confused with the term wavenumber used in describing a gridded signal in the spectral domain.

NABIGHIAN (1972) showed that the analytic signal of the magnetic anomaly due to a 2D magnetic source can also be expressed by the complex number

$$A_1(x,z) = \frac{\partial B_T(x,z)}{\partial x} - j\frac{\partial B_T(x,z)}{\partial z}, \tag{13.24}$$

where $B_T(x,z)$ is the total magnetic anomaly observation point at a distance x along the principal profile and a vertical distance z above the source, and j is the imaginary number ($= \sqrt{-1}$). Further, he showed that the horizontal and vertical derivatives which make up the real and imaginary parts of the analytic signal are related through the Hilbert transform – i.e. $\partial B_T(x,z)/\partial x \Longleftrightarrow \partial B_T(x,z)/\partial z$.

THURSTON and SMITH (1997) define the local wavenumber k as the derivative of the change in phase in the x-direction, that is

$$\begin{aligned} k_x &= \frac{\partial}{\partial x}\tan^{-1}\left[\left(\frac{\partial B_T}{\partial z}\right)\Big/\left(\frac{\partial B_T}{\partial x}\right)\right] \\ &= |A|^{-2}\left[\frac{\partial^2 B_T}{\partial x \partial z}\left(\frac{\partial B_T}{\partial x}\right) - \frac{\partial^2 B_T}{\partial x^2}\left(\frac{\partial B_T}{\partial z}\right)\right], \end{aligned} \tag{13.25}$$

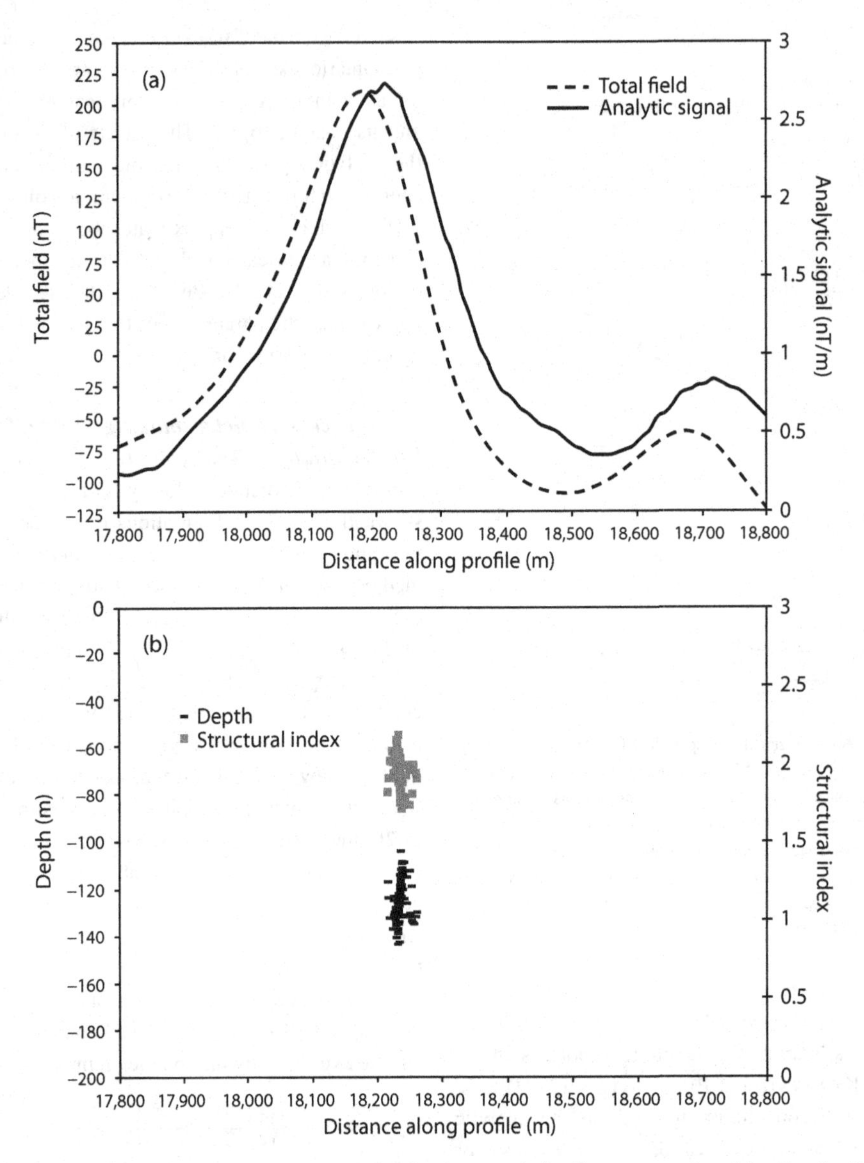

FIGURE 13.49 (a) Total magnetic intensity anomaly and calculated analytic signal of a dike over a portion of the Canadian Precambrian shield in northeast Ontario, Canada. (b) Results obtained from Euler deconvolution of the analytic signal, showing a tight cluster of depth and structural index solutions indicating a thin magnetic dike (structural index of 2.15) at a depth of roughly 130 m. Adapted from KEATING and PILKINGTON (2004).

where A is the amplitude of the analytic signal from Equation 13.21. The local wavenumber of the magnetic anomaly of simple sources with horizontal location x' and depth z' is

$$k_x = \frac{(N+1)(z'-z)}{(x-x')^2 + (z-z')^2}, \qquad (13.26)$$

where N is the structural index which is an integer characterizing the fall-off of the anomaly amplitude with depth (i.e. $N = 0$ for contacts, 1 for a thin sheet, and 2 for a horizontal cylinder) (SMITH *et al.*, 1998; SALEM *et al.*, 2005). Equation 13.26 is obtained by substituting the expressions for the various derivatives of the total fields due to the simple models into Equation 13.25 and simplifying. Assuming a value for N based on the character of the anomaly pattern and geologic information, the depth to the source can be calculated from the above equation and the calculation of the local wavenumber from the derivatives of the anomaly field in Equation 13.25 approximated in the wavenumber domain. Furthermore, the dip and magnetization contrast of the contact and thin-sheet sources can be determined from relationships given by THURSTON and SMITH (1997) assuming the sources do not have significant remanent magnetization. As is the case of the analytic signal method, for 2D sources the results are independent of source magnetization directional characteristics.

An improvement and expansion of the method to include horizontal cylinder sources in addition to sloping contacts and thin sheets (THURSTON and SMITH, 1997) has been described by SMITH *et al.* (1998) and THURSTON *et al.* (2002). These revisions consider the properties of the basic analytic signal and the derivatives of the analytic signal responses from which the appropriate model can be determined, permitting the correct N-value to be used in Equation 13.26 to calculate depth.

Additionally, SALEM *et al.* (2005) have suggested that the local wavenumber method can be enhanced (referred to as the ELW method) using a combination of the local wavenumber field (also known as the horizontal local wavenumber) and its phase-rotated version, the vertical local wavenumber field. The latter can be calculated readily from the amplitude of the analytic signal and derivatives of the magnetic anomaly (SALEM *et al.*, 2005) as given by

$$k_z = \frac{\partial}{\partial z}\tan^{-1}\left[\left(\frac{\partial B_{\mathrm{T}}}{\partial z}\right)\Big/\left(\frac{\partial B_T}{\partial x}\right)\right]$$
$$= -|A|^{-2}\left[\frac{\partial^2 B_{\mathrm{T}}}{\partial x \partial z}\left(\frac{\partial B_{\mathrm{T}}}{\partial z}\right) - \frac{\partial^2 B_{\mathrm{T}}}{\partial z^2}\left(\frac{\partial B_T}{\partial x}\right)\right]. \tag{13.27}$$

Similar to Equation 13.26, the vertical local wavenumber for simple sources is related to the structural index N by

$$k_z = \frac{(N+1)(x-x')}{(x-x')^2 + (z-z')^2}, \tag{13.28}$$

where k_z and k_x form a Hilbert transform pair. Division of the horizontal local wavenumber equation by the vertical local wavenumber equation results in the linear equation

$$k_x x + k_z z = k_x x' + k_z z'. \tag{13.29}$$

This equation does not require estimation of the source parameter N for the calculation of depth as is necessary in Equations 13.26 and 13.28. It can be solved for source location (x', z') using conventional methods of matrix inversion. With the source location known, the structural index N can be found from either Equation 13.26 or 13.28.

The local wavenumber method is particularly subject to noise because of the use of higher-order derivatives in the calculations, and PHILLIPS (2000) reports that interference effects from the margins of thick sheet-like bodies leads to erroneous results as they do in the analytic signal method.

(F) Tilt angle (tilt derivative) method A refinement of the analytic signal method to determine the location and depth of vertical magnetic contacts without prior information on the source configuration has been suggested by MILLER and SINGH (1994) and by VERDUZCO *et al.* (2004) using the horizontal gradient amplitude (first horizontal derivative) of the tilt angle. The method, which has been further developed by others including SALEM *et al.* (2007, 2008) and FAIRHEAD *et al.* (2008), has received considerable interest because of its fundamental and practical simplicity.

The local phase or tilt angle, or the tilt derivative, is defined as

$$TA \equiv \tan^{-1}\left[\left(\frac{\partial B_T}{\partial z}\right)\Big/\left(\frac{\partial B_{\mathrm{T}}}{\partial h}\right)\right], \text{ where}$$
$$\frac{\partial B_{\mathrm{T}}}{\partial h} = \sqrt{\left(\frac{\partial B_{\mathrm{T}}}{\partial x}\right)^2 + \left(\frac{\partial B_{\mathrm{T}}}{\partial y}\right)^2}, \tag{13.30}$$

and $(\partial B_{\mathrm{T}}/\partial h)$, $(\partial B_{\mathrm{T}}/\partial x)$, $(\partial B_{\mathrm{T}}/\partial y)$, and $(\partial B_{\mathrm{T}}/\partial z)$ are the derivatives of the total magnetic field B_{T} in the $(h; x, y, z)$-directions with h being any horizontal direction in the (x, y)-plane. For a 2D vertical contact that is vertically polarized, SALEM *et al.* (2007) have shown that the tilt angle, which ranges between $\pm 90°$, is zero directly over the contact, thus locating the position of the contact. Furthermore, the horizontal distance from the 45° to the 0° position of the tilt angle is equal to the depth to the top of the contact. Assumptions require that the magnetic field be reduced-to-the-pole and that the remanent magnetization of the source be negligible. Deviations from the vertical dip of the contact assumed in the calculations are generally not a significant limitation of the method (e.g. GRAUCH and CORDELL, 1987; FAIRHEAD *et al.*, 2008).

Although, as is the case with other methods, results are subject to error where the anomaly patterns of contacts interfere with each other, the method is advantageous because it is not dependent on higher-order derivatives, and thus is less subject to noise than other methods requiring higher-order derivatives. Additionally, the method uses normalized values based on the ratio of derivatives, so it is particularly suitable to situations where both shallow and deep sources are being investigated or where the magnetic anomaly amplitude varies over a wide range owing to extreme variations in the magnitude of the magnetic moment of the sources. The method is not based on the moving window study approach as in most other

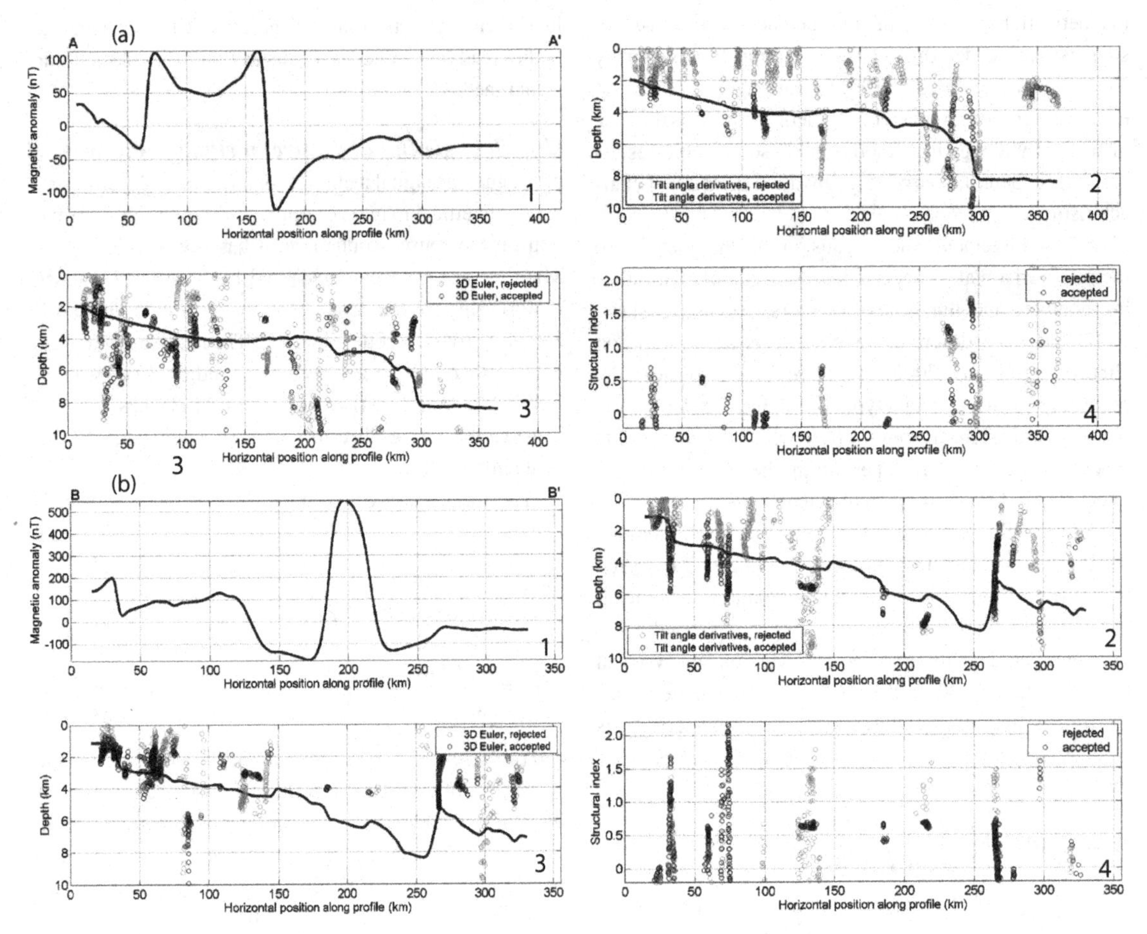

FIGURE 13.50 The A–A′ and B–B′ sections in (a) and (b) respectively show a magnetic data profile (panel 1) with estimated source locations derived from tilt angle derivatives (panel 2) and 3D Euler deconvolution (panel 3), with the true basement in black line. Panel 4 of each section shows the estimated structural indices. Adapted from SALEM *et al.* (2008).

semi-automated methods, and thus is not subject to errors due to improperly selected window size.

SALEM *et al.* (2008), in an expanded use of the method, have shown that derivatives of the tilt angle can be used on gridded magnetic anomaly data to estimate the location and depth of sources without prior knowledge of the geometry of the source. The structural index is estimated using the second-order derivatives of the field. The procedure involves developing a linear equation in terms of the wavenumbers similar to that used in the 3D Euler equation which does not require specification of the structural index. A moving window approach is used to describe simultaneous equations that are solved in a least-squares sense using matrix theory for the location of the source of the anomaly, and the structural index is estimated from an equation obtained by taking the derivative of the 3D Euler equation. The data window size is selected on the basis of the noise imposed on the data and the interference among anomalies. Larger windows minimize noise, while increasing the risk of errors due to anomaly interference.

Figure 13.50 presents the results of a test of the interpretation method described by SALEM *et al.* (2008) in an area known as the Volcanic Tablelands north of Bishop, CA, an area that has been previously used in testing magnetic interpretation methods. The surface magnetic rocks are assumed to be the magnetic basement configuration. Figure 13.50(a) is a NW–SE total magnetic intensity profile in the area that is perpendicular to the topographic trends of the region. The results shown in profile form in Figure 13.50(b) are based on analysis by 3D Euler deconvolution (REID *et al.*, 1990) and the above-described methodology of SALEM *et al.* (2008) using a 15×15 grid with a grid interval of 0.5 km. All results obtained within

4 km on either side of the profile are projected onto the profile. Three-dimensional Euler deconvolution was performed assuming a structural index of zero, equivalent to that of a vertical contact. Acceptance criteria for solutions considered only results within 2.5 km of peaks in the wavenumber and depths less than 15 km and with the standard deviation of the estimated depths less than 10% of the estimated depth. Rejected solutions are shown in gray and accepted results are in black on the profiles. The actual topography or in this case the magnetic basement is shown by the heavy bold line. Clusters of depth estimates occur on either side of the true basement surface, with tight clusters at $x = 70$ and 170 km slightly deeper than the actual surface. The structural indices of these clusters are approximately 0.5, slightly larger than expected for the actual sources. In general, the results of 3D Euler deconvolution show a greater degree of scatter around the true depths than the tilt derivative solutions.

FAIRHEAD *et al.* (2008) have further modified the tilt-angle method of interpreting magnetic anomalies to include determination of dip and the magnetic susceptibility contrast across a 2D vertical contact source. The dip, or direction of change from high to low magnetic susceptibility, of the contact is estimated from the slope of the tilt-angle derivative: that is, the dip direction from positive to negative tilt across the contact. The magnetic susceptibility is estimated from the maximum of the analytic signal which occurs over the edge of the vertical contact. The procedures are all based on reduced-to-pole magnetic anomalies reduced-to-the-pole. LAHTI and KARINEN (2010) investigated the zero contours of the tilt derivative of a reduced-to-the-pole magnetic anomaly, and showed that the changing symmetry of these contours, upward-continued to a series of levels, can be used to estimate the direction of dip of a prismatic body. The spreading of these zero contours at increasing continuation level is asymmetric, with the dip being in the direction of the expanded distance between the zero contours.

(G) Other interpretation techniques In addition to the above techniques, a number of other methodologies have been described which have found some application to magnetic interpretation including depth determination. For example, O'BRIEN (1972) developed a frequency-domain method involving spatial autoregression on the horizontal derivative of the total field for mapping the margins of and depths to 2D magnetized prisms (CompuDepth method). The method is based on successive frequency shifting of the Fourier spectrum, linear phase shifting, and analysis of a system of equations to solve for the location of the edges of the sources and their depths. The method is designed to separate the components of overlapping sources.

A more recent method, also based on evaluation of the spectra of the magnetic anomalies, is the one-dimensional wavelet transform method which evaluates the derivatives of the upward-continued magnetic field. This approach has led to useful interpretation of both the shape and the depth and location of homogeneous sources (e.g. MOREAU *et al.*, 1999; SAILHAC *et al.*, 2000; MARTELET *et al.*, 2001). It is based on geometrical analysis of the real and imaginary parts of the complex wavelet coefficients which converge to the location of the source. Furthermore, the convergence is at the center or top of the source depending on the source configuration. The phase of the wavelet coefficients indicates the apparent inclination of magnetization of the source.

Depth estimation corrections

Several simple corrections can be applied to the magnetic depth estimates to account for various survey conditions in the data and anomaly departures from the 2D assumption. For example, the scale correction adjusts magnetic data sampled in time to their spatial coordinates to account for the ground speed of the shipborne, airborne, and satellite magnetometers. Scale correction errors can strongly affect anomaly gradients and the related depth estimates. The flight elevation correction subtracts the flight elevation from the computed depth to adjust the depth estimates to the ground elevation or sea level or some other desired flight-independent elevation datum.

Where depth determinations are made directly from observed magnetic profiles that are not directed perpendicular to the strike of the anomalies, a correction factor may be needed to account for the erroneously deep depth obtained in the analysis. That is, the calculated depth is multiplied by the cosine of the angle between a principal profile (perpendicular to the strike of the anomaly) and the actual direction of the profile. Thus, a depth calculated from a profile striking 30° from the perpendicular to the profile is multiplied by the cos 30° or 0.87. This correction must be considered only approximate and should not be used for angles exceeding 45°.

Depth estimation techniques, whether graphical or semi-automatic, usually make the assumption that the source of the anomaly is 2D or sufficiently elongated that it can be considered to meet this assumption. A general rule is that a prismatic source can be considered 2D provided that the length exceeds roughly four times its width and the depth is of the order of the width or less. For depth determinations made on circular anomalies using 2D techniques, the depths will be too shallow. A general rule is to

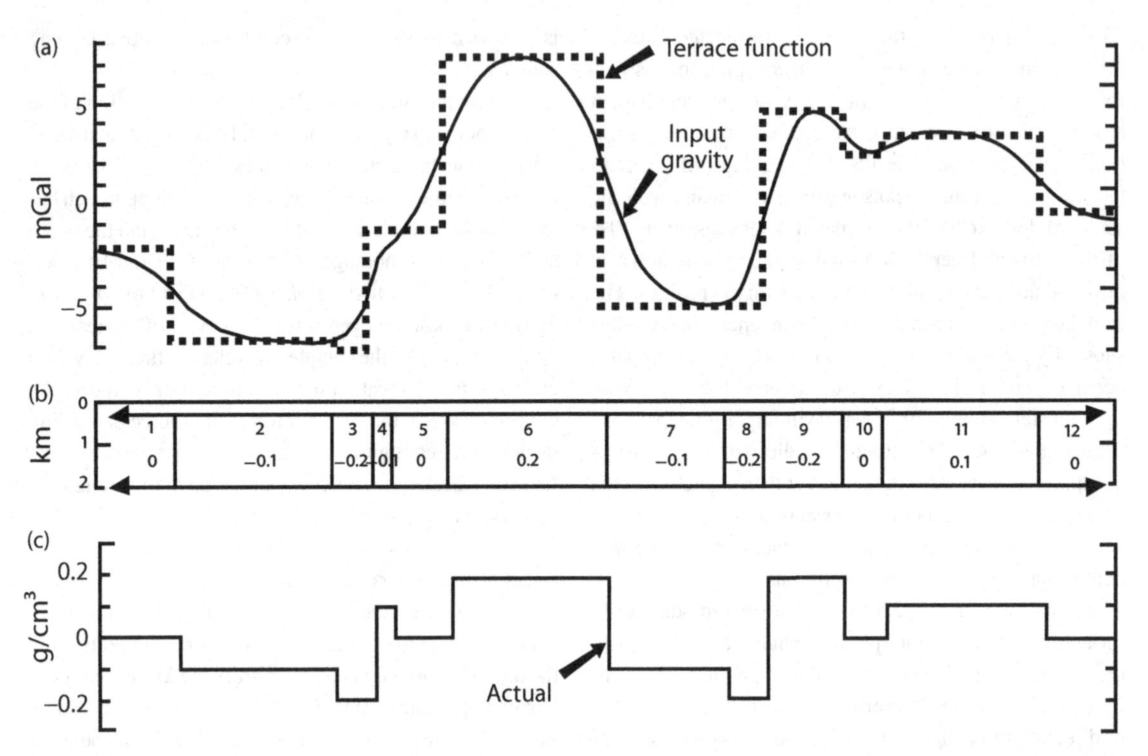

FIGURE 13.51 (a) Input gravity anomaly profile and the final terrace function showing the location of the identified sources of the anomalies. (b) Two-dimensional blocks used to calculate the input showing the number of the block (above) and the density contrast in CGSu (below). Horizontal scale equals vertical scale. (c) Model density contrast function. Adapted from CORDELL and MCCAFFERTY (1989). This methodology is applicable to magnetic anomalies after they have been converted to psuedogravity anomalies to remove the dipolar effects of magnetic fields.

increase the depth estimate by 10 to 20%, but clearly the result is likely to be in error.

13.4.2 Interpretation of magnetic maps

Several of the semi-automatic magnetic anomaly interpretation methods are useful for not only determining depths to sources, but other attributes of the anomaly sources. These methods can be applied not only to observed magnetic profiles, but to magnetic maps as well. However, there are several additional methods that are potentially important in the interpretation of magnetic maps. This section briefly describes these methods, but the interested reader is directed to more comprehensive treatments in the relevant original descriptions.

Terracing

A conceptually simple method of magnetic map interpretation is the stepped or terracing method of physical property mapping (CORDELL and MCCAFFERTY, 1989). The method uses a so-called terracing operator to recast either gravity or magnetic anomaly maps into a simulated-geologic map format. The terracing operation defines abrupt boundaries between geologic domains of contrasting magnetic properties from the magnetic anomalies using an iterative process.

The terracing method is based on the sense of the curvature of the local anomaly as determined within a window which is moved across a grid of anomaly values. The curvatures are evaluated to determine uniform domains of the field which are separated by abrupt boundaries. The anomaly at the center of the window is modified depending on the sign of the local curvature. The moving window process is repeated on the modified anomaly values until the output is a series of constant values or horizontal segments representing psuedo-geologic units mapped by the anomaly field.

The results of the methodology on a computed gravity profile are shown in Figure 13.51 where (a) illustrates the input gravity anomaly profile and the final terrace function showing the location of the identified sources and (b) gives the 2D blocks used to calculate the input, showing the number of the block (above) and the density contrast (below). The illustration shows the effectiveness of the method in identifying the major sources of the gravity anomaly profile. The terracing method is equally useful in identifying magnetic sources and in application to magnetic maps, but only after the magnetic data are transformed to pseudogravity to remove the dipolar effects and directional attributes of the magnetization. In applying the method to a map based on a 2D square grid of values, the local curvature is calculated from the four grid values surrounding the central point.

Apparent magnetic susceptibility mapping

The terracing method described above is akin to "susceptibility mapping" as defined by GRANT (1973) and refined by several others. The method as reviewed in YUNSHENG *et al.* (1985) prepares an "apparent susceptibility" map that is particularly useful for geologic interpretation in regions such as Precambrian shields, where strong magnetic contrasts occur in exposed or near-surface crystalline rocks or buried beneath non-magnetic sedimentary rocks. The calculation is based on application of Fourier transform methods of analysis of data in the wavenumber domain from vertical-sided prisms. The method assumes that the magnetic field is derived from an assemblage of infinitely extending, vertically sided prisms, one grid cell in dimension, with a prism centered on each grid point of the magnetic map. Further, the magnetization of the prisms which is allowed to vary is only due to induction in the geomagnetic field, and the depth to the top of the prisms is constant. The magnetic anomaly data are reduced-to-the-pole and downward-continued to the top of the prisms prior to analysis in the wavenumber domain.

The superposition of the magnetic anomalies of the prisms in the wavenumber domain, assuming the magnetization, depth, and size factor are constant over the map grid and dividing by these factors, is

$$\mathcal{B}_{Tn}(u, v) = 2\pi B_{\mathrm{N}} \sum_{j=1}^{n} k_j \mathcal{D}_j(x, y; u, v), \qquad (13.31)$$

where the complex $\mathcal{B}_{Tn}$ is given as an array of discrete values corresponding to the fast Fourier transform of the total field magnetic data grid, (u, v) are wavenumbers in the (x, y) directions, n is the number of prisms, j is the index of each prism, B_{N} is the intensity of the normal geomagnetic field, k_j is the magnetic susceptibility of the jth prism, and the displacement factor is $\mathcal{D}(x, y; u, v) = \exp[-i(ux + vy)]$. Taking the inverse Fourier transform of the above equation gives

$$B_{Tn}(x, y) = 2\pi B_{\mathrm{N}} \sum_{j=1}^{n} k_j D_j(u, v; x, y), \qquad (13.32)$$

where $D_j(u, v; x, y)$ is the inverse Fourier transform of $\mathcal{D}_j(x, y; u, v)$. Each data point is close to the top surface of the corresponding prism and at its center, and thus it is assumed that surrounding prisms have negligible effect on the value at each data point. Accordingly, Equation 13.32 becomes

$$B_{Tn}(x, y) = 2\pi B_{\mathrm{N}} k(x, y). \qquad (13.33)$$

Dividing the inverse Fourier transform by $2\pi B_{\mathrm{N}}$ results in the apparent susceptibility of each data point of the grid. YUNSHENG *et al.* (1985) show that the effect of dip of the prisms is to reduce the calculated susceptibility by a function of the cosine of the angle of deviation from the vertical. Thus, unless the dip of the geological elements is less than roughly 70°, the effect on the calculated susceptibility is less than 10%. The increased resolution of the apparent susceptibility map over the total magnetic anomaly field map is shown in Figure 13.52 for an area in Quebec, Canada.

Electromagnetic apparent susceptibility

The apparent magnetic susceptibility in the previous section should not be confused with the "apparent susceptibility" of the near-surface Earth that is determined electromagnetically (e.g. HUANG and WON, 2001). It is well known that a magnetic field is generated within the Earth by an applied magnetic field derived from a current flowing in a coil, and that this field is proportional to the magnetic susceptibility of the Earth (e.g. HUANG and FRASER, 2000). In taking advantage of this process a procedure is used that generally assumes that the inducing, primary field interacts with a uniform half space, and thus the qualifying term "apparent" is used (WON and HUANG, 2004).

The apparent susceptibility is determined electromagnetically by measuring the Earth's response to a magnetic field from a vertical dipole source generated by a horizontal coil within which a variable multi-frequency current causes induction of an electric current. This coil system traverses across the Earth, generally from a near-surface ($\leq$30 m) airborne platform. The secondary electrical magnetic fields caused by both the varying eddy currents in the Earth conductors and induction in the magnetic components are observed in a common receiving coil located near the source coil.

The effect of strong magnetic susceptibility is to reduce the in-phase component measured in the receiving coil which is opposite to the effect of the response due to the electrical conductivity effect. Multi-frequency measurements are used to solve for common unknowns. Low-frequency currents are used to enhance the depth of penetration of the primary field, and because at low frequencies

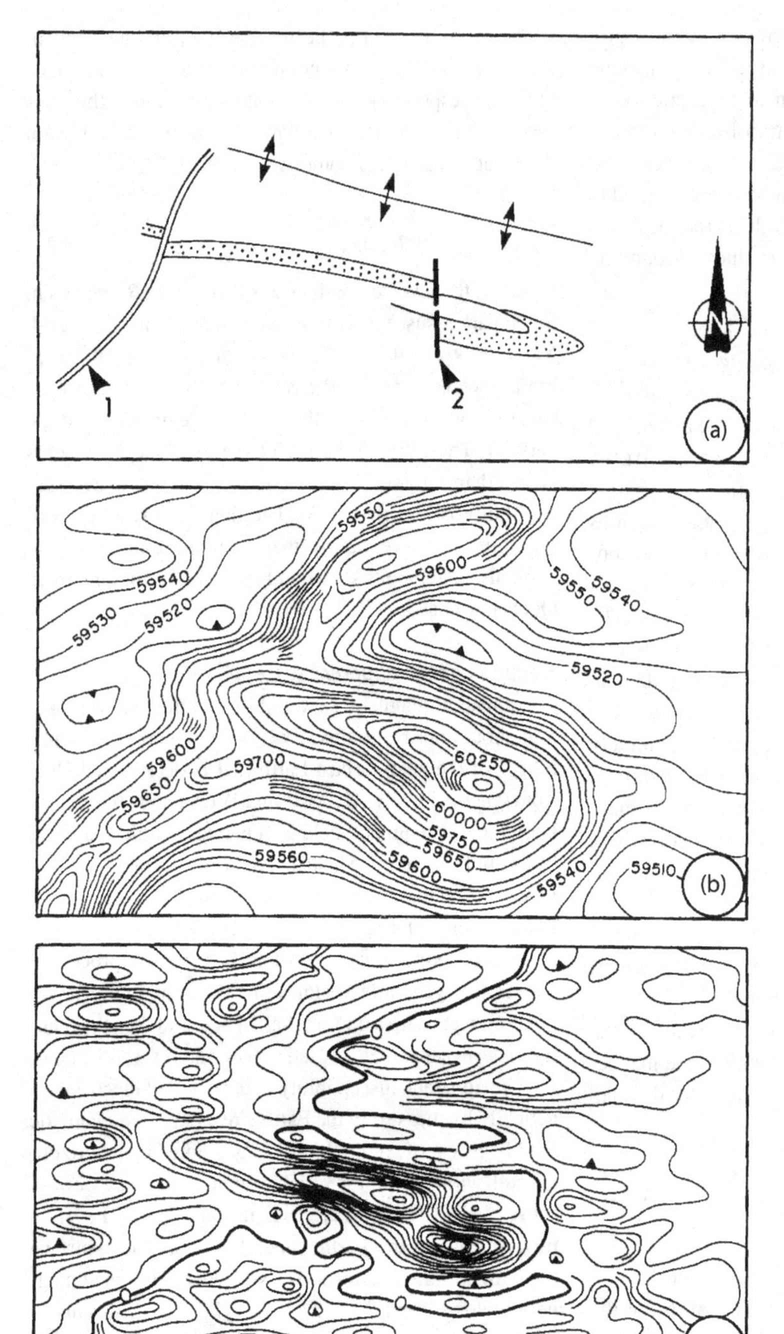

FIGURE 13.52 (a) Geology of a portion of the Precambrian crystalline shield in Quebec, Canada (1 is a dike and 2 is a fault). (b) Observed total magnetic field processed by the apparent magnetic susceptibility method. (c) Apparent magnetic susceptibility map showing the dike cutting northeasterly across the western portion of the region and the fault displacing the dotted geologic unit near the center of the region. Adapted from YUNSHENG *et al.* (1985).

in resistive rock regions, the eddy current secondary effects are small, the magnetic induction effects are more noticeable. Notably, the electromagnetic "apparent susceptibility" is only a function of the susceptibility and not the combined effect of the induced and remanent magnetization as in measurements of the magnetic field with magnetometers. Thus, it is theoretically possible to distinguish magnetic anomalies that are derived largely from induction (susceptibility) effects by comparing apparent susceptibility maps determined from inversion of magnetic

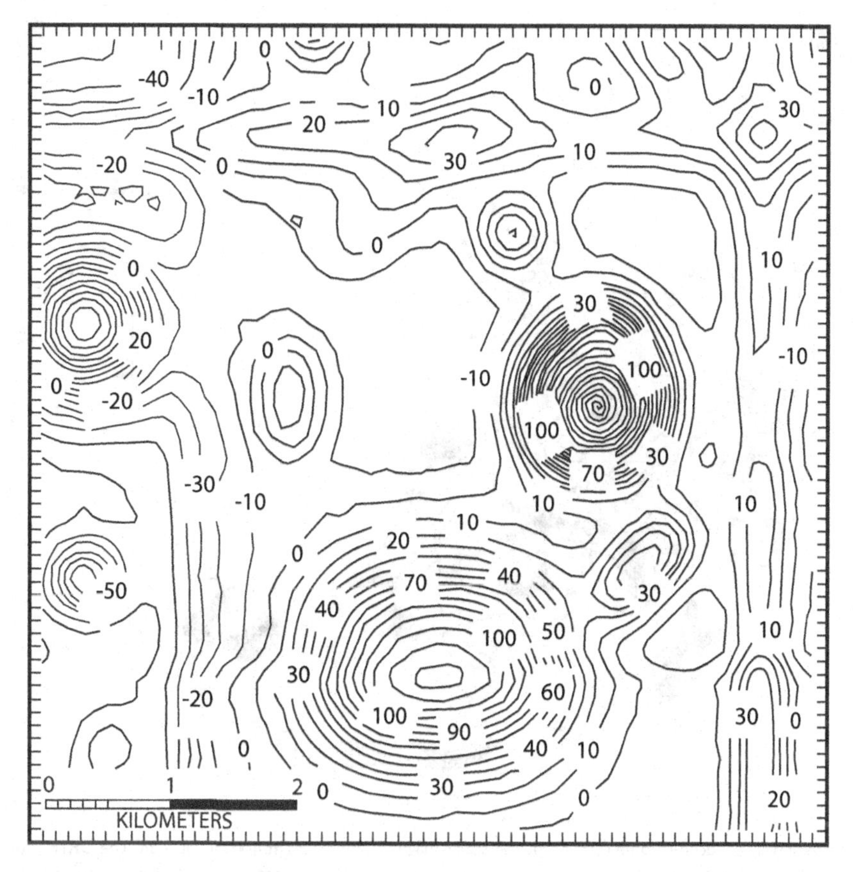

FIGURE 13.53 Reduced-to-pole magnetic anomaly map of the composite magnetic anomaly map shown in Figure 12.9. Contour interval is 10 nT. Adapted from XINZHU and HINZE (1983).

anomalies and those measured electromagnetically (WON and HUANG, 2004).

Reduced-to-pole magnetic anomalies

As described in Section 12.4.2(F), a magnetic anomaly with known direction of magnetization can be readily transformed into its equivalent at another specified polarization direction. Assuming all polarization is in the direction of the Earth's magnetic field, the total field anomaly at the geomagnetic poles is the vertical magnetic anomaly centered on its subsurface source, whereas at lower latitudes the polarization inclination and declination variations complicate the spatial relationship of the total field anomaly to its source (Figure 12.22). Transforming the total field magnetic anomaly from its actual inclination to the vertical (e.g. Equation 12.16) results in the reduced-to-pole (RTP) magnetic anomaly, which by Poisson's theorem is centered on its source in the same way that the source is centered within its first vertical derivative gravity anomaly (Section 9.10).

The example in Figure 13.53 gives the reduced-to-pole transformation of the magnetic anomaly map in Figure 12.9 that was modeled from sources magnetized by induction only in an ambient geomagnetic field of the northern hemisphere with intensity 50,000 nT, inclination $I = 60°$, and declination $D = 10°$ W. Careful examination of these two maps shows the modification of the anomaly pattern and slight northward shift of the magnetic anomalies in the RTP map. The asymmetry of the anomalies on the RTP map is due to the sources and the interference of anomalies, and not greatly affected by the inclined magnetization in the ambient magnetic field. Although the difference in the maps is not great, greater confidence in analyses can result when the RTP map is also used for interpretation purposes where the magnetization of the sources can be assumed to be only by induction in the geomagnetic field.

In general, as the inclination of the magnetic field becomes more oblique to the vertical, the spatial distortion of the anomaly relative to its sources greatly increases and the transformation to the vertical becomes more important.

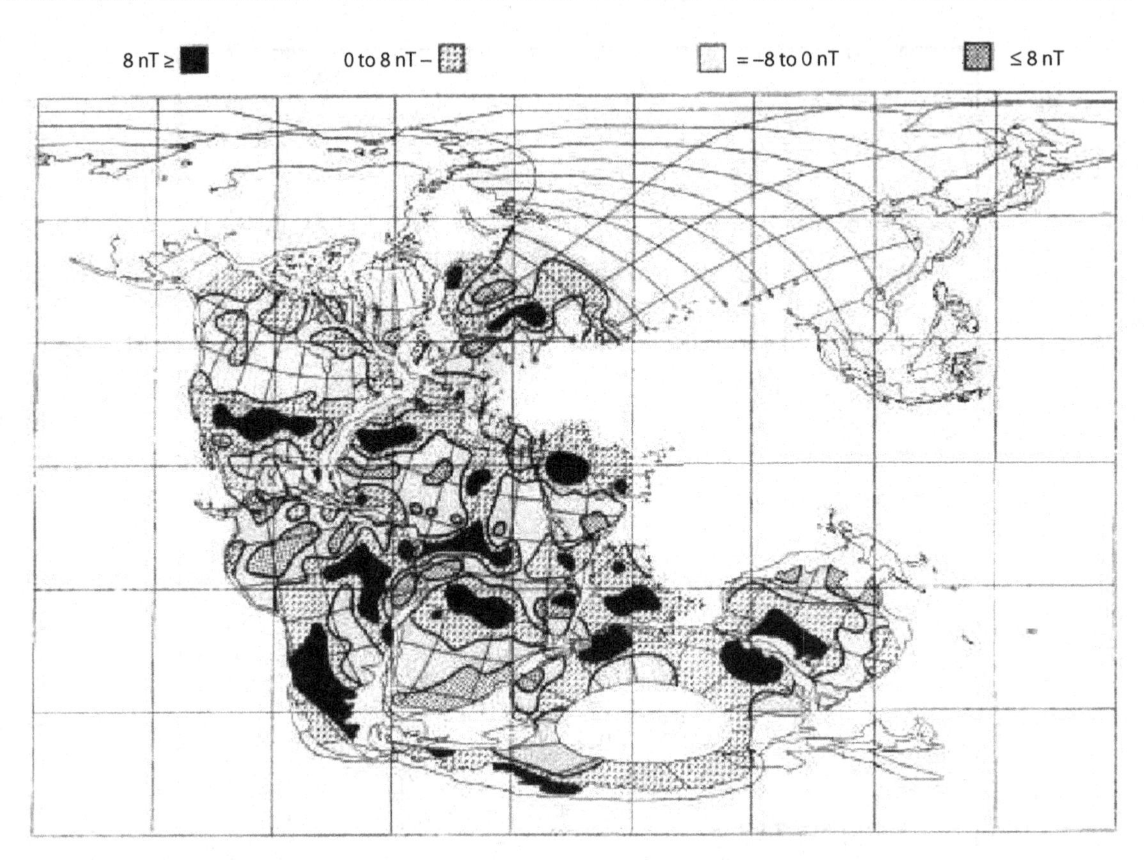

FIGURE 13.54 Differentially reduced-to-pole (DRTP) Magsat scalar magnetic anomalies at 400 km altitude. The DRTP anomalies were obtained for a constant normalization intensity $B_K = 60{,}000$ nT and plotted on a Late Triassic continental reconstruction from SMITH *et al.* (1981) (cylindrical equidistant projection). Adapted from VON FRESE *et al.* (1986).

Unfortunately, the transformation becomes unstable and generally unacceptable at low magnetic latitudes where the inclination is less than a few tens of degrees owing to noise in the magnetic anomalies. This problem can be eliminated at these magnetic latitudes by reducing the magnetic anomaly map to the magnetic equator, rather than the pole as described in Section 12.4.2(F). However, reduced-to-equator (RTE) anomalies have seen little practical application because of the difficulties in interpreting magnetic anomalies at low magnetic latitudes.

The above development of RTP/RTE anomalies is appropriate at spatial scales where the ambient geomagnetic field properties are essentially constant (generally less than $\pm 5°$) at all source and observation points and an average direction of the field can be assumed. However, this assumption seriously breaks down for satellite-measured lithospheric anomalies and regional airborne and marine magnetic survey compilations. To correct for the directional and intensity changes of the normal field, the regional magnetic anomalies can be differentially reduced-to-pole (DRTP) by EPS inversion (VON FRESE *et al.*, 1981b). The procedure involves least-squares matrix inversion of the magnetic anomalies using an appropriate geomagnetic reference field model to determine magnetic susceptibilities for a spherical array of point dipoles at crustal altitude. The DRTP anomalies are obtained by simply recomputing the EPS anomalies for radial inclination at both source and observation points, and polarizing the dipoles by a uniform field strength B_K (e.g. Equation 12.20).

If remanent magnetization can be ignored as a primary source of anomalies in regional magnetic data, then the DRTP anomalies are in principle centered over their lithospheric sources and directly map out their magnetic susceptibility variations. This result clearly helps in relating regional magnetic observations to the crustal geology, and remote sensing, heat flow, gravity, seismic, and other source-centered geophysical data. It also has important implications for continental- or global-scale tectonic analyses, as geologic source regions can be compared directly in terms of their DRTP anomaly signatures.

In Figure 13.54, for example, the DRTP magnetic anomalies from the Magsat mission show remarkable correlation of regional lithospheric magnetic sources across rifted continental margins when plotted on a Pangea reconstruction. First observed by NASA's satellite magnetic programs, these anomalies provide another fundamental perspective on the geological evolution and dynamics of the continents and oceans.

A wavenumber domain implementation of variable reduction-to-pole has also been developed (e.g. ARKANI-HAMED, 1988; SWAIN, 2000), but these DRTP anomalies are differentially adjusted only to vertical inclination and not for intensity variations in the polarizing field. The spectral DRTP approach allows the investigator to apply RTP without the requirement of breaking up a large block into smaller areas of relatively uniform inclinations. However, the spectral method has found limited use in practice because of its extensive data handling requirements, which are greatly exacerbated by the iterative nature of the algorithm. In addition, the process, like other RTP procedures, is unstable at low geomagnetic latitudes, but a modification of the method by ARKANI-HAMED (2007) improves the results at all but very low geomagnetic latitudes.

Pseudogravity anomalies

As described in Sections 3.9, 7.4.6, and 12.4.2(G), Poisson's theorem can be used to investigate a source's magnetic effects for possible gravity effects, assuming that the source has correlative magnetization and density variations. Pseudogravity maps appear as gravity maps, and thus are free from the distortion in magnetic anomaly maps caused by inclined magnetic polarization and fields. They are particularly useful in enhancing regional anomalies of magnetic maps, which are often difficult to discern in the magnetic field, and also in making it possible to use gravity methods for interpretation of magnetic maps.

Although pseudogravity maps appear as gravity maps, they only reflect the magnetic properties of the rocks of the area, not density. The pseudogravity transformation requires assumptions concerning the direction of magnetization within the rocks of the mapped area, as well as a scaling factor given by the ratio of the density to magnetization of the rocks. The latter assumption is arbitrary, but the former is usually based on visual analysis of the character of the magnetic anomalies and the direction of the ambient field. Generally, the magnetization is assumed to be solely by induction, and thus in the direction of the ambient field.

Poisson's theorem yields usable linear equations (e.g. Equations 12.8 and 12.9) relating measured values of the magnetic field to derivatives of the gravitational field. Assuming that a sufficient number of observations are available over a finite homogeneous source body, an overdetermined system of linear equations can be set up to solve for the three independent unknowns which include the ratio (J/σ) and the inclination I' and declination D' of the total magnetization vector $\mathbf{J}$. The inducing field strength, B_N, at the source and its direction at the observation point (I, D) must also be known to solve this system of equations.

The ratio (J/σ) and the direction of $\mathbf{J}$ in theory can be uniquely resolved given the assumptions implicit in applying Poisson's theorem. In particular, where remanent magnetization is negligible, the magnetization is induced so that $J = J_{ind} = k \times B_N$. Thus, given appropriate bounds on σ, the effective range of the susceptibility k can be calculated from

$$\frac{\sigma}{J} = \left(\frac{1}{G \times B_T}\right)\frac{\partial^2 U}{\partial i^2} = \left(\frac{1}{G \times B_z}\right)\frac{\partial g}{\partial z}. \qquad (13.34)$$

The second expression above is for total magnetization induced in the direction i, whereas the last expression is for vertical magnetic field and vertical magnetization.

More generally, $\mathbf{J}$ cannot be determined uniquely because it is the vector sum of the induced and remanent magnetizations. However, the magnetizations are related by the Koenigsberger ratio $Q = J_{rem}/kB_N$. Although various values of J_{rem} and kB_N and their associated directions may satisfy Poisson's theorem, the minimum value of Q is where the remanent and total magnetization vectors are perpendicular (e.g. GROSSLING, 1967; CORDELL and TAYLOR, 1971). Therefore, solving Equations 12.8 and 12.9 theoretically can yield unique determinations of (J/σ), the inclination and declination of $\mathbf{J}$, and the minimum Koenigsberger ratio.

CORDELL and TAYLOR (1971), for example, determined (J/σ) and the direction of the magnetization vector of a North Atlantic seamount by least-squares inversion of gravity and magnetic data. Using reasonable estimates of density, the range of susceptibility contrasts was estimated as well as the ranges of possible remanent magnetization directions and intensities and the minimum Q value. Paleomagnetic pole positions were then calculated based on the possible Q values.

Formation boundaries

An important step in the interpretation of magnetic anomalies and in developing conceptual models for analysis is the determination of the approximate formational boundaries of anomalous subsurface masses from the magnetic anomalies. As in the gravity method (Section 7.4.4), this is primarily achieved by relating the boundaries to the inflection point of the marginal gradients of the anomalies. Inflection points, or their equivalent zero second vertical derivative or maximum horizontal gradient, have been used in this regard for both gravity and magnetic interpretation by numerous investigators (e.g. VACQUIER *et al.*,

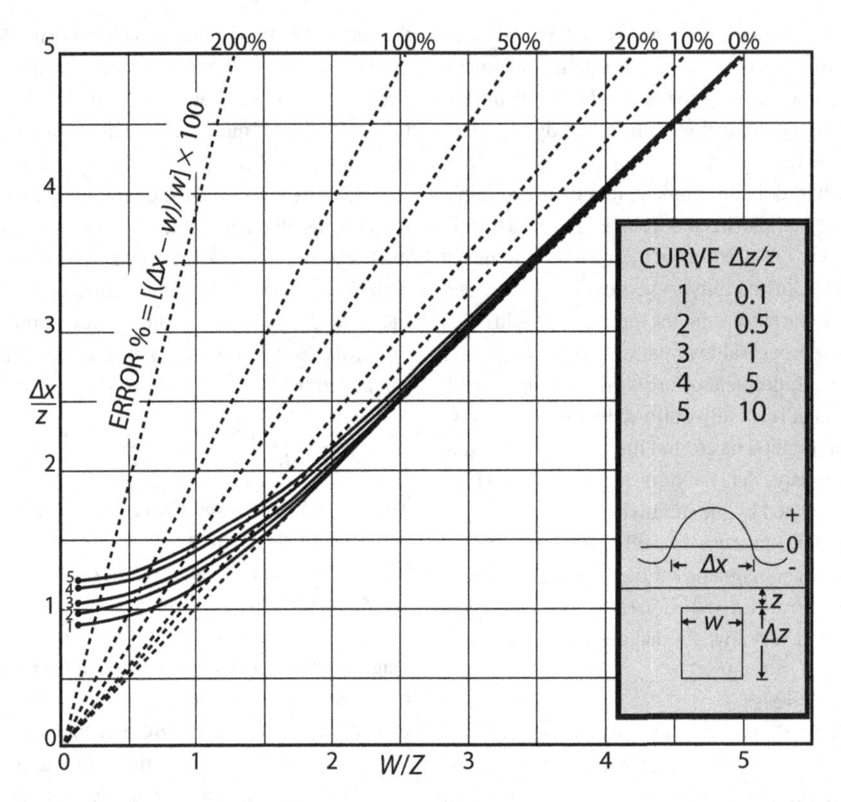

FIGURE 13.55 Normalized distance between zero second vertical derivative total magnetic anomaly values versus normalized width of anomalous 2D, vertical-sided, vertically polarized prisms for various depth extents of the prisms. Dotted lines are percentages of errors for source widths estimated from the distances between zero second vertical derivative values. Modified from data in SPURGAT (1971).

1951; RUDMAN and BLAKELY, 1965; SPURGAT, 1971; CORDELL, 1979; BLAKELY and SIMPSON, 1986). The technique is readily applied to both profile and map magnetic anomaly data.

The correlation between the inflection point of the marginal gradients and the formation boundaries generally makes the assumption that the boundary is vertical and the width of the anomaly source is large with respect to depth. GRAUCH and CORDELL (1987) have studied limitations to the correlation considering various attributes of the anomaly sources. The effect of width and depth extent of the source as a function of the depth to the top of the source on the zero second vertical derivative is illustrated in Figure 13.55 assuming vertical polarization. The vertical polarization requires that the anomaly be located at high magnetic latitudes or that the data be reduced-to-pole for this figure to be applicable.

The results presented in Figure 13.55 show that the error in the position of the zero second vertical derivative increases slowly with increasing depth extent and the error is limited ($<10\%$) for width-to-depth ratios of 2 or more. Comparison of this figure with the comparable figure for gravity anomalies (Figure 7.24) indicates that the error in mapping the formational boundaries is much less for magnetic anomalies than gravity anomalies. Thus, where only gravity anomalies are available, converting them to pseudomagnetic anomalies should improve the mapping of formational boundaries from the inflection points of the marginal anomalies.

A useful alternative technique for determining source formation boundaries is the tilt-angle (tilt-derivative) method described above. Other techniques for mapping contacts, magnetic boundaries and lineaments include the application of the enhanced horizontal derivative (FEDI and FLORIO, 2001), analytic signal (HSU *et al.*, 1996), local wavenumbers, and curvature (ROBERTS, 2001; HANSEN and DERIDDER, 2006; PHILLIPS *et al.*, 2007). PHILLIPS (2000) and PILKINGTON and KEATING (2004) describe and compare some of these techniques in further detail.

13.5 Modeling anomaly sources

As described in Section 7.5 on gravity anomaly modeling and in Appendix A.5.2 on the general anomaly modeling process, anomaly inversion is the fundamental procedure by which quantitative parameters of the subsurface are estimated from anomaly data. For example, the inversion of residual anomalies is used to determine shape attributes, volume, depth, depth extent, magnetic properties, and other features of the subsurface sources. Solving the inverse problem requires postulating a mathematical forward model of the source with parameters that can be related to variations in the anomaly data.

The solution of the inverse problem is based on how well the anomaly predictions derived from the forward model compare with the residual anomaly. However, a close match between predicted and residual anomalies does not guarantee a unique solution; rather, the solution is only one of a family of possible interpretations. This is because of the source ambiguity of potential fields, solution uncertainties related to errors in the residual anomaly values, and errors in postulating the mathematical model which represents a simplified approximation of the actual source.

Magnetic anomaly modeling approaches are analogous to the gravity anomaly modeling methods described in Section 7.5, where forward modeling, for example, involves establishing a solution by trial-and-error adjustments of the unknown parameters of a conceptual forward model until the model's predictions reproduce the observed data within a specified margin of error. Inverse modeling, on the other hand, determines the unknown parameters directly from the observed data using an optimization process to match the predicted data with the anomaly observations. Several of the methods of magnetic interpretation described above, such as apparent magnetic susceptibility mapping, Werner deconvolution, and Euler deconvolution, are inversion methods which solve for one or more attributes of the anomaly source directly from the observed residual.

13.5.1 Forward modeling

This time-honored approach to anomaly inversion iteratively alters the parameters (i.e. the unknowns) of the forward model by trial-and-error until its predictions match the observed anomaly values. The classical forward modeling approach is perhaps the most used and successful method of interpreting magnetic anomaly data. An analytical representation of a source is dictated by a conceptual model of the subsurface that commonly includes preliminary assessments of the anomaly for source features using simplified rule-of-thumb methods. The forward modeling refines the preliminary estimates and provides insights on the sensitivity of the anomaly data for the details of the source in the subsurface.

In forward modeling, the anomaly from an assumed model is iteratively re-computed based on modifications of the model characteristics until a "close" correlation with the residual anomaly is achieved. "Close" correlation between the residual and calculated anomalies is highly subjective, with criteria for judgment likely to vary with interpreter, amount of geologic and geophysical control, purpose, and resources. However, the most critical aspect of modeling is the similarity of horizontal gradients which can be checked by calculating and comparing the observed and modeled horizontal derivatives. The amplitude of the modeled anomaly can be matched at the final stage by adjustment of the magnetic property contrast.

The forward modeling approach is feasible as long as the number of source parameters which must be considered in the inversion is small. However, for problems with many unknowns, it becomes difficult to know which of the variables to modify in each iteration, and convergence to a solution can become impractically slow. The classical inverse modeling method is commonly adopted to solve these more complex problems involving many unknowns and anomaly observations. Inverse modeling most commonly invokes matrix algebra (Appendix A.4) to determine the unknown model parameters using least squares or some other objective measure or norm for the fit between the model's predictions and the residual anomaly.

13.5.2 Inverse modeling

As discussed in Section 7.5 and Appendix A.5.2, both the forward and inverse modeling methods involve the same inverse problem expressed by the generic matrix system $\mathbf{AX} = \mathbf{B}$. Here, $\mathbf{B}$ is the column matrix of the anomaly observations, and $\mathbf{AX}$ is the forward model that the interpreter has assumed to account for the anomaly observations. The forward model consists of the known or assumed parameter values in the design matrix $\mathbf{A}$ and the unknown parameters in $\mathbf{X}$ that must be determined by the inversion. Forward modeling evaluates the unknown parameters in $\mathbf{X}$ by trial-and-error, whereas inverse modeling obtains more objective estimates of the unknowns such as from the least-squares solution $\mathbf{X} = (\mathbf{A}^{\mathrm{T}}\mathbf{A})^{-1}\mathbf{A}^{\mathrm{T}}\mathbf{B}$.

Inversion estimates the unknown parameters (i.e. $\mathbf{X}$) of the presumed forward model from its assumed or known parameters (i.e. $\mathbf{A}$) and the measured anomaly values (i.e. $\mathbf{B}$). Thus, to implement inversion for quantitative interpretation, the interpreter must specify the mathematical details of the forward model. The details for modeling the scalar potential, vector field, total field, and tensors of any uniformly magnetized 3D and 2D source are described in Section 9.5.

In general, the quantitative complexity of magnetic modeling is dictated by the configuration of the body just as it is in gravity modeling (Section 3.6). The idealized source, for example, has the simple symmetric form (e.g. spherical, cylindrical, prismatic) that can be volume integrated in closed analytical form for its magnetic effect at an observation point. The general source, on the other hand, has an arbitrarily shaped volume that must be filled in by idealized bodies to numerically estimate the magnetic effect from the summation of the idealized body effects at the observation point. The next two subsections respectively detail further analytical properties and uses of idealized dike and other body equations in magnetic interpretation, and outline the role of the commonly employed prism for gridded magnetic anomaly modeling.

A variety of methods are used to minimize the problem of the inherent ambiguity of the interpretation of potential fields. A common approach to this problem is to correlate magnetic and gravity anomaly fields under the assumptions that the anomalies have similar geometric sources and consistent variations in their physical property contrasts, magnetic polarization and density. Correlation can take different forms from simple visual correlation of anomalies on profiles or maps to forward modeling of the source of coincident magnetic and gravity anomalies to joint inversion of the anomalies. The latter is particularly effective if the assumptions are generally correct. Several approaches to joint interpretation magnetic and gravity data are described in the literature and improving the methodologies is an active research area. The methods of joint inversion (e.g. Bott and Ingles, 1972; Menichetti and Guillen, 1983; Zeyen and Poous, 1993) have been reviewed and summarized by Pilkington (2006). They are based on a combined solution of the appropriate set of equations that describe the common source of both anomalies. This often takes the form of joint inversion of the two data sets by the damped least-squares approach described above and detailed in Appendix A.4.2.

13.5.3 Idealized body interpretation

The magnetic effects of idealized 3D and 2D bodies are modeled by setting the appropriate limits and executing the desired integrals in Equations 9.25–9.30 and 9.40–9.45, respectively. The 2D dike equations are perhaps the most widely used idealized body effects in magnetic anomaly interpretation. A set of idealized body magnetic equations due to Nettleton (1942) and others is also widely applied in magnetic exploration.

Two-dimensional dike

The most comprehensively studied source of magnetic anomalies is the dike with a flat top and parallel sides extending to infinity down dip and along strike (e.g. Gulatee, 1938; Cook, 1950; Werner, 1953; Hall, 1959; Gay, 1963; Reford, 1964, 1986; Parasnis, 1986; Reeves, 2005). In addition, equations for magnetic anomalies have been derived for modified forms of the dike including those with ranges of dip as well as finite depth and strike length. Two basic types of dike equations have been developed: one for thin dikes where the depth to the top is more than twice the width of the dike, and a general, more complex form which is for a thick dike with no depth to width restriction. The derivation of these equations assumes a uniform magnetization contrast with the enclosing medium without self-demagnetization and without remanent magnetization unless specified in the total magnetization vector. The total intensity anomaly is calculated assuming the anomaly is small with respect to the inducing field and is the sum of the horizontal and vertical magnetic fields projected onto the direction of the inducing magnetic field.

The dike equation is widely studied because it directly approximates the shape of numerous geological features such as layered rocks, veins, and dikes and, where modified forms of the dike are considered, geometric shapes such as horizontal slabs, prisms, and faults can be modeled. Furthermore, more complex geometric forms can be approximated by the summation of the effects from numerous dike-like forms. Consideration of the dike equation can also be useful in developing methods for and constraints on magnetic interpretation (Reeves, 2005).

The equation for the magnetic anomaly due to a 2D dike has been derived by numerous authors including those listed above. Reford (1986) has presented a comprehensive derivation based on determining the magnetic effect of a flat-topped surface and a bounding edge of a magnetic volume using Green's equivalent layer theorem (e.g. Grant and West, 1965; Blakely, 1995). Considering the magnetic components derived from the magnetic potential of an edge dipping at an angle (d) from the horizontal and from the flat top of the edge, the vertical magnetic intensity ($\Delta\mathcal{B}_z$) and the horizontal intensity ($\Delta\mathcal{B}_x$) perpendicular to the strike of the edge extending to an infinite depth (Figure 13.56(a)) are respectively given by

$$\Delta\mathcal{B}_z = (2JF\sin d) \times [\sin(i'-d) \times \ln R + \cos(i'-d)\tan^{-1}(x/z)] \qquad (13.35)$$

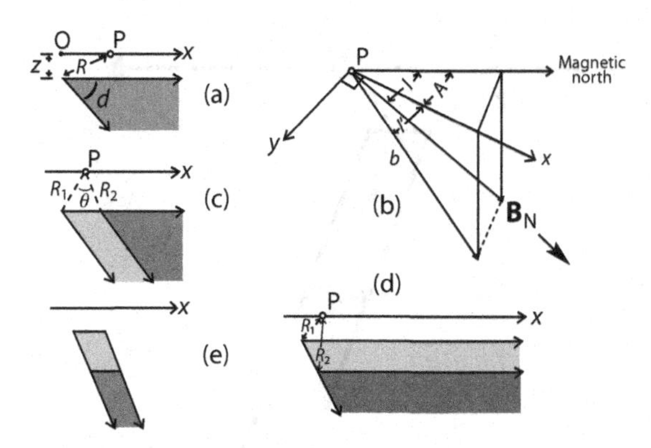

FIGURE 13.56 Illustrations for formulation of equations for 2D dikes. (a) Edge effect cross-section which forms the basis of the dike equations. (b) Angular relationships between the geomagnetic field and the magnetization of the 2D dike along x, which is perpendicular to the strike of the dike. (c) Cross-section of a 2D dike obtained from the difference between two equivalent but horizontally offset edges. (d) Cross-section of 2D horizontal slab obtained from the difference between an edge at two different depths. (e) Cross-section of finite depth extent, 2D dike obtained from the difference of a dike at two different depths. Adapted from REFORD (1986).

and

$$\Delta\mathcal{B}_x = (2JF\sin d) \times [-\cos(i' - d) \times \ln R + \sin(i' - d)\tan^{-1}(x/z)], \quad (13.36)$$

where J is the magnetization intensity (J = magnetic susceptibility (k) × the magnetizing field intensity (B_N) for induced magnetization only); x is the horizontal distance from the observation point to the apex of A of the edge; A is the angle between the positive x-axis perpendicular to the strike of the dike and magnetic north (Figure 13.56(b)); a is the angle between the positive x-axis and the horizontal projection of the direction of magnetization ($a = A$ for induced magnetization only); I is the inclination of the Earth's magnetic field from the horizontal (which is negative in the southern hemisphere); i is the inclination of the direction of magnetization from the horizontal ($I = i$ for induced magnetization only); I' is the projection of I onto the xz-plane perpendicular to the strike of the edge; i' is the projection of i onto the xz-plane perpendicular to the strike of the edge; b is the component in the xz-plane of a unit vector in the total magnetic field direction $= \sin I/\sin I' = \cos I\cos A/\cos I' = (1 - \cos^2 I \sin A)^{\frac{1}{2}}$; F is the component in the xz-plane of a unit vector in the magnetization direction $= \sin i/\sin i' = \cos i \cos a/\cos i' = (1 - \cos^2 i \sin a)^{\frac{1}{2}}$ (with $b = F$ for induced magnetization only); and z is the vertical distance between the observation point and the top surface of the dike.

Assuming these magnetic effects are small in comparison to the normal magnetic field (B_N), the total magnetic anomaly due to the edge is

$$\begin{aligned}\Delta\mathcal{B}_T &= \Delta\mathcal{B}_x \cos I \cos A + \Delta\mathcal{B}_z \sin I \\ &= (2JFb\sin d) \times [-\cos(I' + i' - d)\ln(R) \\ &\quad + \sin(I' + i' - d)\tan^{-1}(x/z)]. \quad (13.37)\end{aligned}$$

Thus, if the magnetization is induced only

$$\Delta\mathcal{B}_T = (2kB_N b^2 \sin d) \times [-\cos(2I' - d) \times \ln R + \sin(2I' - d)\tan^{-1}(x/z)]. \quad (13.38)$$

The total magnetic intensity anomaly for a strike and depth infinite dike can then be obtained by determining the difference between the effects of two parallel edges where the angular distances from the observation point to the apex of the bounding edges of the dike are R_1 and R_2 which intersect at angle θ (Figure 13.56(c)) so that

$$\Delta\mathcal{B}_T = (2kB_N b^2 \sin d) \times [\cos(2I' - d) \times \ln(R_2/R_1) + \sin\{\theta(2I' - d)\}]. \quad (13.39)$$

The equation for the total intensity anomaly of a horizontal slab is the same as the dike equation (Figure 13.56(d)), and the equation of a finite depth dike is the difference between the dike effect of a dike extending infinitely downward from the top surface of the dike and the effect of a dike extending to infinity from the flat bottom of the dike (Figure 13.56(e)). The magnetic anomalies of dikes with remanent magnetization can be calculated by determining the vectorial resultant of the magnetization due to both induced and remanent magnetization and using these components to determine the value of i, i' and F, where F is the magnetization projected into the xz-plane.

These equations consist of three terms. The first term is the amplitude term which is dependent on the magnitude, azimuth, and inclination of the magnetization (or the magnetization contrast with the surrounding formations), the azimuth and dip of the dike, and the inclination of the ambient magnetic field. Within the brackets there are two terms, the logarithmic and angle terms which are a function of the distance from the observation site to the apex of the edges which bound the magnetic volume and the angular relationships of the dike with respect to the ambient magnetic field and the magnetization of the volume. The angle term is symmetric about the center of the top of an infinitely deep dike, while the logarithmic term is antisymmetric about this location.

The combination of the two terms within the first term of the dike equation which varies with the directional attributes of the magnetization determines the shape of the magnetic anomalies. For example, considering induced positive magnetization of a steeply dipping dike striking perpendicular to magnetic north, the shape of the total intensity magnetic anomaly profile is positive and symmetric at the magnetic poles, antisymmetric at 45° inclination, and symmetric and negative at the magnetic equator. All these relationships are determined by the cosine and sine of the angular relation $(2I' - d)$ involved in the logarithmic and angle terms of Equation 13.39. In addition, the combination of these terms explains the negative anomaly to the north in the northern geomagnetic hemisphere of a steeply dipping, east/west striking dike and to the south in the southern geomagnetic hemisphere.

For steeply dipping dikes in high geomagnetic latitudes, the dike can be considered to be infinitely deep provided that the depth to the bottom is roughly 10 times the depth to the top. At lesser depth extents, the effect of the lower boundary of the dike enhances the negative portion of the anomaly and decreases its overall amplitude. In addition, the effect of the lower boundary may lead to negative anomalies on both sides of a positive anomaly associated with a steeply dipping, east–west striking source. Where the edges of a dike are widely separated compared to the depth to the upper surface, the anomalies of each edge will be sufficiently separated so that individual anomalies can be identified. The anomalies of the northern and southern margins of the anomalies will be quite dissimilar depending on whether the magnetic source lies north or south of the edge. These few examples illustrate the use of the dike equation in developing guidelines for magnetic interpretation.

For the case of a thin dike (Figure 13.57) where the width (w) of the dike is small compared with the depth to the top (z), the total intensity magnetic anomaly is

$$\Delta B_T = (2kB_N b^2 w \sin d) \times [(z/R^2) \times \sin(2I' - d)(x/R^2)\cos(2I' - d)], \quad (13.40)$$

where R is the angular distance between the observation point and the center of the flat top of the thin dike, and w is the horizontal width of the dike, not its true width perpendicular to the edges of the dike. The shape of the thin dike anomaly is relatively insensitive to the width of the source, so it is impossible to determine the width of the source from the anomaly. The amplitude of the anomaly is largely a function of the magnetization of the dike.

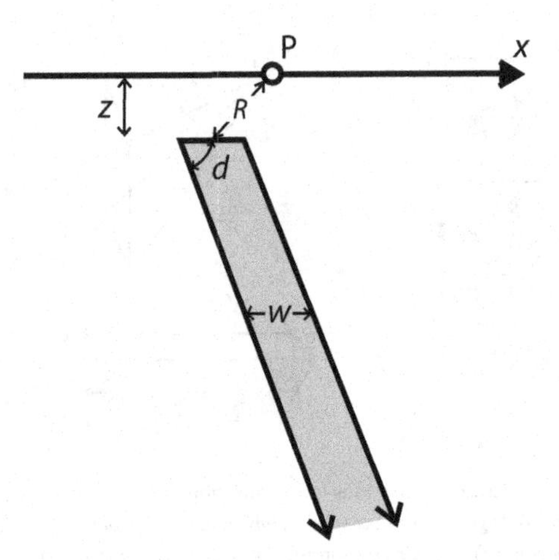

FIGURE 13.57 Cross-section of a thin 2D dike used in the formulation of magnetic field equations.

REFORD (1964) has prepared a series of illustrations (Figure 13.58) showing the shape of anomalies over thin dikes dipping at a range of angles for variable orientations with respect to magnetic north at inclinations from the equator to the pole. Anomalies from dikes at angular values intermediate to those shown in this figure can be approximated from those illustrated. These profiles are useful in showing the general shape of total intensity magnetic anomalies over a range of dike dips in both the southern and northern geomagnetic hemispheres, striking at varying angles to the magnetic meridian at ambient Earth's field inclinations of 0°, 30°, 60°, and 90°. For example, no magnetic anomaly is associated with dikes, thin or thick, that strike north–south at the magnetic equator (see Figure 13.58, third profile from the top) except for anomalies at the ends of the dikes due to magnetization of the dike's end faces.

Other idealized sources

In general, the idealized body magnetic effects in Tables 9.1 and 9.2 and Figure 9.5 are useful for interpreting simple anomalies and developing preliminary conceptual models. These effects form the basis for estimating source depths from anomaly half-widths (Table 13.1) and slopes. A simple form of Euler's Equation 13.5 shows that the source depth can also be estimated from the ratio of the magnetic central amplitude (MCA) and its vertical derivative ($\partial(\text{MCA})/\partial z = \text{MCA}'$).

Specifically, the central amplitude (MCA) and depth z are related by evaluating the basic expression $B_T = B_z = (J \times \text{Vol})/r^n$ of the idealized source equations (Table 9.1)

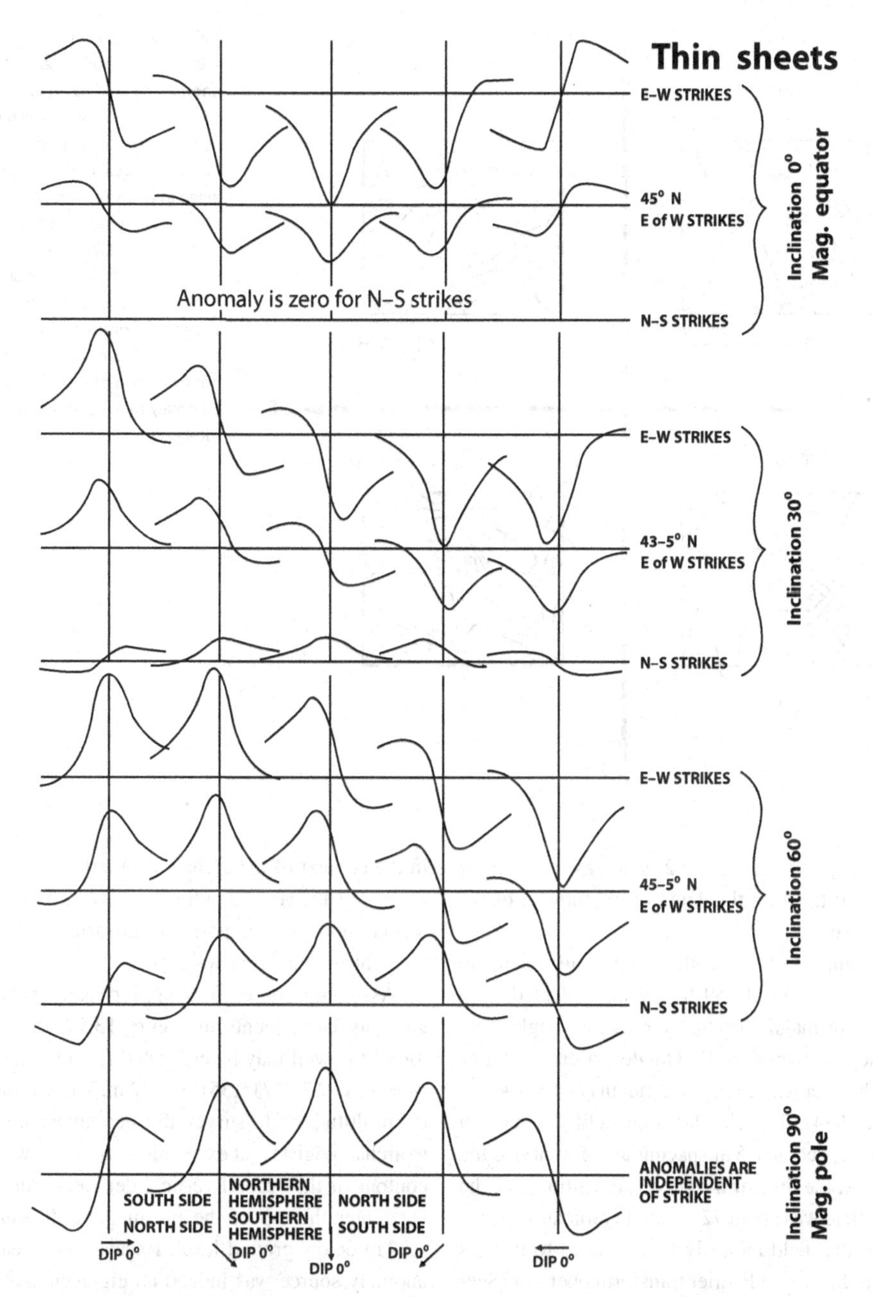

FIGURE 13.58 Total field magnetic anomalies caused by a thin sheet dipping at various angles and orientation with respect to magnetic north. The north side of the anomaly profile is to the right in the northern geomagnetic hemisphere and to the left in the southern geomagnetic hemisphere. Adapted from REFORD (1964).

at $x = 0$ so that

$$\mathrm{MCA} = \frac{\mathrm{MM}}{z^n} \ni \frac{\partial(\mathrm{MCA})}{\partial z} = -n\frac{\mathrm{MM}}{z^{n+1}}$$

$$= -n\frac{\mathrm{MCA}}{z} \rightarrow z = \frac{-n \times (\mathrm{MCA})}{\partial(\mathrm{MCA})/\partial z}, \qquad (13.41)$$

where MM $= (J \times \mathrm{Vol})$ is the magnetic moment of the source. Thus, to make an effective estimate of the depth to the center z_c of the spherical source, for example, the interpreter inserts into Equation 13.41 reliable values of the MCA and $\partial(\mathrm{MCA})/\partial z$ and uses $n = 3$. However, depth estimates both to the top (z_t) and the central axis (z_c) of vertical and horizontal cylinder sources, respectively,

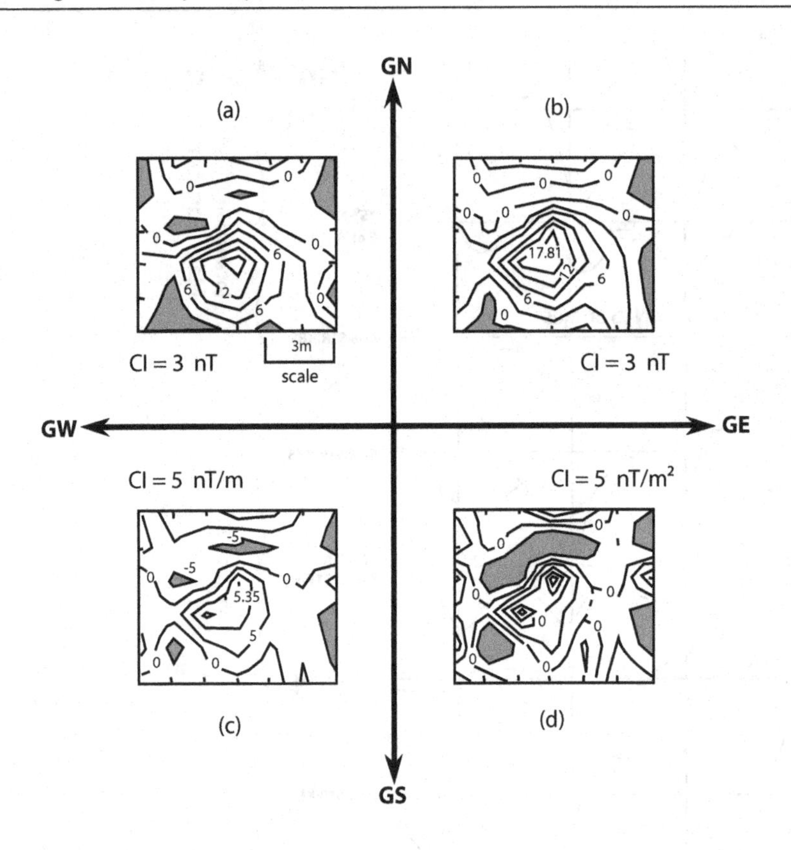

FIGURE 13.59 (a) Ft. Ouiatenon, Indiana monopolar (vertical cylinder) anomaly related to an eighteenth century well in the total field which has been reduced-to-pole in map (b). The first and second vertical derivatives of map (b) are given in maps (c) and (d), respectively. Peak amplitudes for maps (b) and (c) are annotated to facilitate the depth-to-source calculation. Because of local noise, the zero second vertical derivative contour in map (d) was closed by the dashes to make the subsurface configuration of the well more explicit. Anomaly minima are shaded. CI, contour interval.

require the decay rate factor $n = 2$, whereas $n = 1$ must be invoked for estimating the depth of the top (z_t) of the thin vertical sheet.

As an example, consider the total field magnetic anomaly data in Figure 13.59 that were collected by a proton-precession magnetometer over the early eighteenth century French settlement of Ft. Ouiatenon on the banks of the Wabash River near Lafayette, Indiana (von Frese and Noble, 1984). The circular total field anomaly in map (a) was mapped at 1.5 m spacing at 0.5 m above the ground level. At the time of the survey, the attitude of the geomagnetic field was about 72 °N inclination and 0° declination. The total field anomaly was reduced-to-pole as shown in map (b) using a Fourier transform operator (Section 12.3.2). Spectral filters were also applied to the RTP anomaly to take the first and second vertical derivatives in maps (c) and (d), respectively.

The derived anomaly components in maps (b), (c), and (d) were obtained without reference to the anomaly source. However, a source assessment is necessary if these results are to be used for depth estimation by methods such as the half-width rules in Table 13.1 or Equation 13.41. In general, the anomaly's circular configuration is consistent with the effects of both a relatively deep spherical source (i.e. source 2 with $n = 3$) and a vertical cylinder-like magnetization concentration (i.e. source 5 with $n = 2$). However, in the context of the archaeomagnetic exploration targets of the site, the vertical cylinder option is preferred because it accommodates the possible subsurface configuration of an eighteenth century water well.

Assuming source 5 is appropriate for modeling the anomaly components in Figure 13.59, the depth to the top of the well may be estimated from Equation 13.41 as $z_t = -2(-17.81/15.35) = 2.32$ m. Thus, adjusting for the 0.5 m altitude of the survey, the recommendation was made to archaeologists that excavations centered within the zero contour of the second vertical derivative map (d) should encounter the top of the presumed well within roughly 1.82 m below ground level. Excavation revealed that the anomaly source was indeed an eighteenth century water well that had been intentionally filled with magnetically enhanced topsoil and capped with a clay layer.

Another idealized source shape which is useful in determining the amplitude of anomalies from structures such as suprabasement features (relief, horsts, grabens etc.) as well as intrabasement contacts is the vertical fault. In this case, the displacement (Δz) on a vertical fault can be determined provided that the depth to the top (z) of the slab and the magnetic susceptibility contrast (Δk) can be estimated (Figure 13.60). The vertical fault can be approximated by a semi-infinite horizontal slab of finite thickness (t). For ease in calculation, the assumptions are

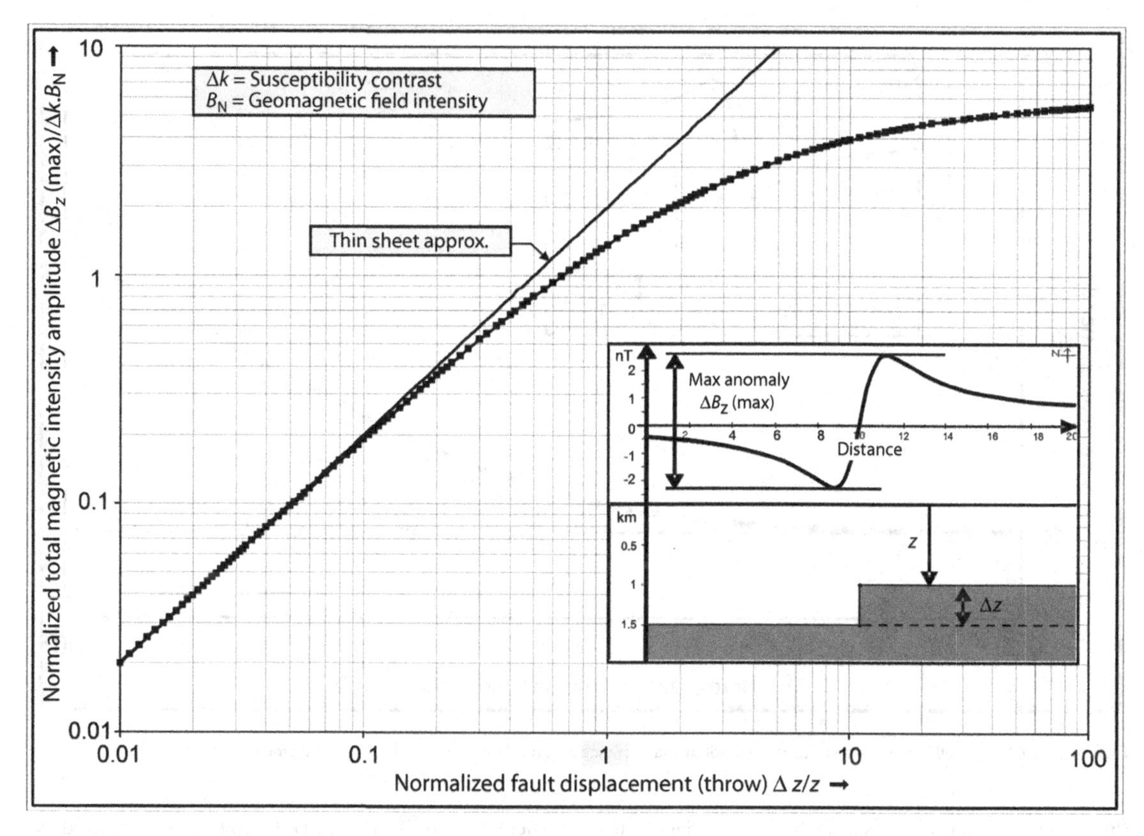

FIGURE 13.60 Chart for estimating maximum magnetic anomaly amplitude due to vertical fault displacement, assuming vertical or reduced-to-pole fields and vertical magnetization.

made that the magnetization is vertical as in reduced-to-pole magnetic fields or as encountered in induced magnetization at high magnetic latitudes in either hemisphere, and that the magnetization is only induced by the geomagnetic field, $\mathbf{B}_N$. Using the vertical magnetic effect of the vertically faulted horizontal slab (Reford and Sumner, 1964), the normalized amplitude difference $\Delta B_z(\max) = B_z(\max) - B_z(\min)$ between the peak ($B_z(\max)$) and trough ($B_z(\min)$) is

$$\frac{\Delta B_z(\max)}{\Delta k \times B_N} = 4\left[\tan^{-1}\left(\sqrt{1+[\Delta z/z]}\right) - \tan^{-1}\left(\frac{1}{\sqrt{1+[\Delta z/z]}}\right)\right]. \tag{13.42}$$

For small displacements where $\Delta_z \ll z$, the vertical fault can be approximated by a thin horizontal sheet so that Equation 13.42 reduces to

$$\frac{\Delta B_z(\max)}{\Delta k \times B_N} = 2\left(\frac{\Delta z}{z}\right). \tag{13.43}$$

In this case, z is the depth to the thin sheet or the mean depth of the semi-infinite slab. For large displacements where $\Delta z \gg z$, the fault can be regarded as a contact or an interface between intrabasement blocks having different susceptibilities with a contrast of Δk.

Based on the above equations, Figure 13.60 charts the normalized maximum amplitude of the magnetic anomaly as a function of the throw-to-depth ratio. To obtain the maximum magnetic anomaly in nT, the ordinate value is multiplied by the magnetization intensity $J = \Delta k \times B_N$ with the susceptibility contrast Δk in CGSu and the inducing field intensity B_N in nanoteslas. The chart can be used to obtain an estimate of the maximum amplitude due to a vertical fault or a large vertical contact or inversely to determine the throw of the fault knowing the depth top. In areas of the U.S. Gulf Coast, for example, where $B_N \approx 50{,}000$ nT, typical J-values to be used with the chart range (a) for felsic igneous rocks (e.g. granites) over 5 to 10 (corresponding to Δk of 100 to 200 micro-CGSu), (b) for basic igneous rocks (e.g. gabbros) over 50 to 150 (corresponding to Δk of 1,000 to 3,000 micro-CGSu), (c) for shale structures and sand channels over 0.5 to 2.5 (corresponding to Δk of 10 to 50 micro-CGSu), and (d) for salt structures over -2 to -3 (corresponding to Δk of -40 to -60 micro-CGSu). Figure 13.60 also shows that the

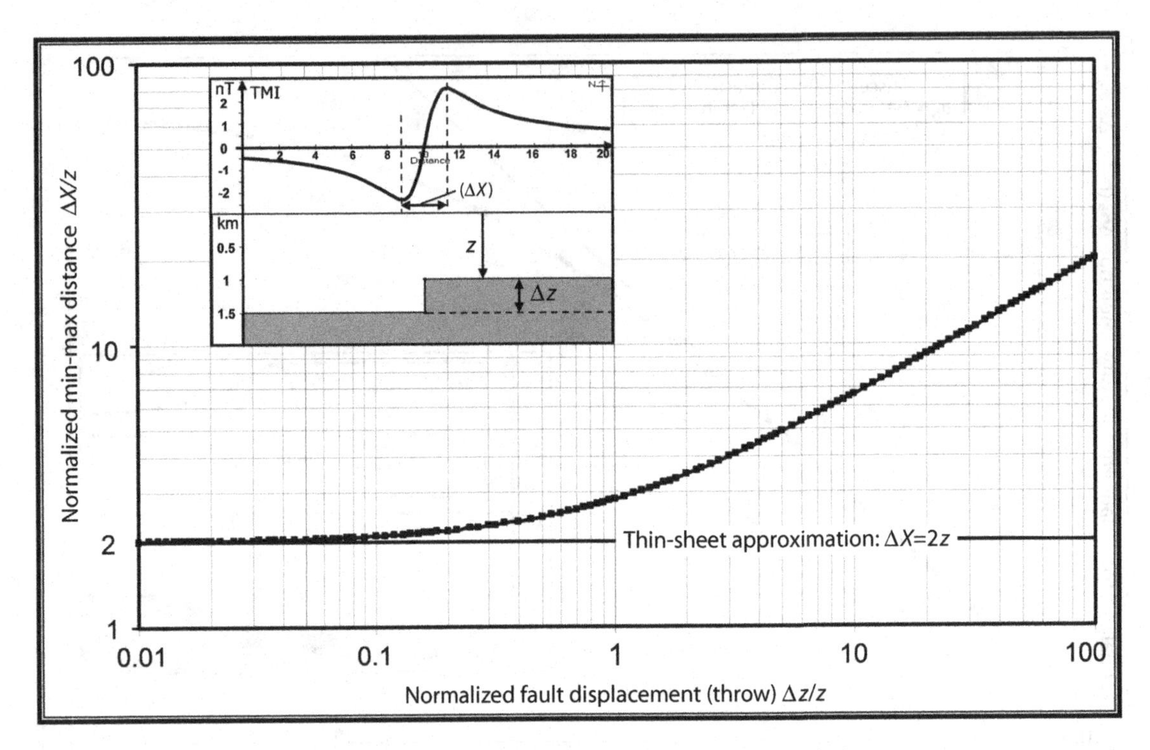

FIGURE 13.61 Magnetic anomaly peak-to-trough horizontal distance as a function of vertical fault displacement.

thin-sheet approximation is satisfactory for displacements of 20% or less of the depth to the slab.

The horizontal distance from peak to trough or maximum to minimum of the magnetic anomaly, $\Delta x = x_{\max} - x_{\min}$, depends on both the depth-to-top and depth-to-bottom or the throw of the fault given by

$$\frac{\Delta x}{z} = 2\left[\sqrt{1+(\Delta z/z)}\right]. \tag{13.44}$$

The thin-sheet approximation $\Delta x/z = 2$ holds for fault displacements of 20% or less as shown by the chart in Figure 13.61.

Similarly, the maximum amplitude of the vertical magnetic anomaly anticipated from a broad variety of geological sources located at a specified depth can be approximated using two simple end-member geometric models. The 2D vertical prism of infinite depth extent and width w approximates an infinitely long source, whereas the 3D vertical cylinder of infinite depth extent and radius R represents a source of limited strike length (Figure 13.62). The normalized maximum vertical magnetic intensity of the 2D source is proportional to twice the plane angle Θ subtended by the top of the prism at the position directly over the center of the body, and for the 3D source the maximum vertical magnetic field is given by the solid angle Ω subtended by the flat top at the position over the axis of the vertical cylinder. Thus, for 2D structures, approximated by vertical prisms, the normalized maximum (vertical) magnetic intensity is

$$\frac{\Delta B_z(\max)}{\Delta k \times B_{\mathrm{N}}} = 2\Theta_{\max} = 4\tan^{-1}\left(\frac{w/2}{z}\right), \tag{13.45}$$

and for 3D structures approximated by vertical cylinders, it is

$$\frac{\Delta B_z(\max)}{\Delta k \times B_{\mathrm{N}}} = \Omega_{\max} = 2\pi\left[1 - \frac{z/R}{\sqrt{1+(z/R)^2}}\right]. \tag{13.46}$$

Based on the above equations, Figure 13.62 charts the normalized maximum amplitude of the magnetic anomaly as a function of the depth-to-width ratio for 2D structures or the depth-to-diameter ratio for 3D structures. The maximum magnetic effect in nanotesla is obtained from multiplying the ordinate values by the induced magnetization intensity $J = \Delta k \times B_{\mathrm{N}}$. The chart can be used to obtain an approximation of the maximum magnetic effect due to any 2D or 3D diapiric salt or shale structure, and sand channels with large or limited depth extent involving depths-to-top z_{t} and -bottom z_{b}. Anomalies derived

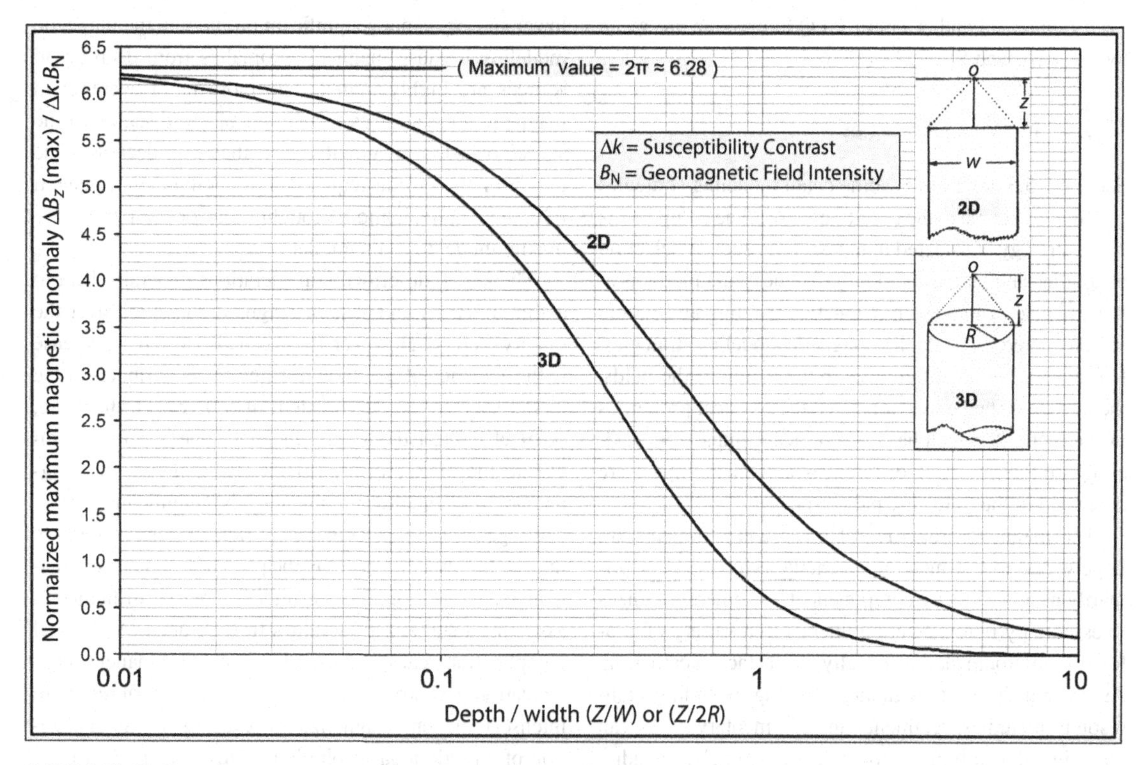

FIGURE 13.62 Chart for estimating maximum magnetic anomaly amplitude due to 2D and 3D structures assuming vertical magnetizations and fields.

from features of limited depth extent such as sand channels can be obtained from the chart by simply subtracting the maximum effect for the depth z_b from the effect for the depth z_t. Furthermore, for 2.5D structures (with limited strike length), the maximum effect should fall somewhere between the two curves for the 2D and 3D cases in the chart, depending on the strike length relative to the width (or diameter) of the source. This chart can also be used to estimate the maximum magnetic effect due to 2D or 3D suprabasement structures, igneous intrusions, or other magnetic sources that can be approximated by either of the two geometric models.

13.5.4 Gridded anomaly modeling

Magnetic anomaly data and related complementary geological and geophysical data are increasingly being formatted and made available as grids in profile, map, and volume dimensions. Thus, the magnetic effects of the prism are finding extensive uses for magnetic anomaly modeling in Cartesian coordinates (e.g. Sharma, 1986; Blakely, 1995; Pignatelli *et al.*, 2011) and spherical coordinates (e.g. von Frese *et al.*, 1981b; Asgharzadeh *et al.*, 2008). In addition, because all integration limits are fixed by the specified dimensions of the prism, numerical modeling of the prism's magnetic effects is readily implemented by Gauss–Legendre quadrature (GLQ) integration (Section 9.7.3).

The accuracy of GLQ modeling is controlled by adjusting the number, and hence spacing, of the Gauss–Legendre nodes relative to the elevation of the calculation point. In practice, the adjustments can result from subdividing the body into smaller prisms, or increasing the number of equivalent point dipoles, or increasing the distance between the calculation point and body. The latter consideration suggests that the GLQ formulation is well suited for satellite altitude magnetic modeling, owing to the orbital elevations of the measurements.

Magnetic GLQ modeling on the surface of a body also lacks singularities because the nodes are displaced away from the surface into the body's interior (e.g. Ku, 1977; von Frese *et al.*, 1981b). Furthermore, it always yields least-squares estimates for any selected number (≥ 2) of point dipoles (Stroud and Secrest, 1966).

As described in Section 9.7.3, point dipole magnetic effects can be readily incorporated into GLQ formulae for modeling the magnetic effects of any conceivable distribution of magnetization. Additional details concerning the theory, practice, and errors of GLQ integration are given by Stroud and Secrest (1966).

13.5.5 Modeling strategies

Section 7.5.5 describes in some detail modeling strategies that are applicable to the gravity method. The components of the strategy considered in that section, such as selection of profiles and length and depth of conceptual model profile, are also of interest in magnetic modeling and should be considered in developing magnetic modeling strategies. However, there are differences in the nature of the fields, the inverse distance functions, and the physical properties of the methods which have a significant impact on modeling strategies. These differences and their impact are described in the following paragraphs.

The magnetic properties of rocks and sediments are largely due to the mineral magnetite which occurs primarily as an accessory component. This mineral oxidizes to essentially non-magnetic minerals in a variety of geological environments, especially in surface weathering. As a result, in most magnetic modeling excluding high-resolution studies, sediments and sedimentary rocks can be assigned a zero magnetization. Accordingly, the adjacent formations can be given a magnetization equal to their assumed value rather than the physical property contrast.

A major advantage of the magnetic method over the gravity method in interpretation is the increased sensitivity to the depth of anomaly sources, because magnetic anomaly amplitude decreases with anomaly source depth more rapidly than gravity anomalies. As a result, greater care must be used to assigning depths to model sources, but this also means that magnetic modeling potentially can be more definitive concerning anomaly source depth than the gravity method. Furthermore, magnetic modeling is much less sensitive to the depth extent of anomaly sources. This can significantly simplify the modeling procedure because the observed anomalies can be more readily duplicated with a range of depth extents and because it limits the depth of the conceptual magnetic model that needs to be considered in the modeling.

An additional impact of the inverse distance function of the magnetic method is that the magnetic anomaly is not as broad as the gravity anomaly from the same source. The result is the higher resolution of the magnetic method, leading to less overlap of anomalies from adjacent sources. This leads directly to improved definition of individual anomalies which enhances the modeling procedure. The resolution of the magnetic method and isolation of magnetic anomalies is also improved because the effect of deep broad anomaly sources which cause gravity anomalies that commonly mask or distort anomalies from shallower sources is not a significant problem in the magnetic modeling because magnetic anomalies from deep broad sources are highly muted. The more rapid decrease in magnetic anomalies as a function of the distance to the source over that of gravity anomalies also decreases the length of the profile required in modeling. As a general rule, the length should be roughly four or more times the maximum depth of the model.

The above described considerations generally lead to a simplification and ease of magnetic modeling over the gravity method. However, these can also lead to a decrease in the accuracy of mapping of some of the attributes of the model such as the depth extent of sources. Furthermore, other characteristics of the magnetic method may complicate the modeling procedure. For example, the magnetization of geologic units may be significantly more heterogeneous than their density and the magnetization may be highly influenced by the presence of intense remanent magnetization. The result is increased difficulty in assigning magnetization values to rock units. Additional complications can be caused by reversed remanent magnetization and structural disturbance of formations with intense remanent magnetization causing the magnetization of an individual geologic unit to have quite variable magnetization values and directions.

Still another complication in magnetic modeling is the dipolar nature of the magnetic field. The dipolar nature of these fields causes negative anomalies to be associated with positive magnetizations and vice versa. Therefore, unless care is used in modeling, the negative anomalies can be misinterpreted as related to sources that have a negative or reversed magnetization.

Magnetic modeling is also different from gravity modeling in the treatment of units of length that are used consistently in the model. Gravity modeling leading to calculation of the anomaly in milligals requires consideration of the length units used in constructing the model. However, this is unnecessary in magnetic modeling. The anomaly calculations result in magnetic units, nanoteslas, which are independent of the length units.

Matching observed and modeled magnetic anomalies

As in the case of gravity modeling (Section 7.5.5), to obtain a better fit between observed and calculated magnetic anomalies, modeling parameters can be changed according to the characteristics of the mismatch that must be rectified. For example, to increase or decrease anomaly amplitude without changing the shape of the anomaly requires, respectively, increasing or decreasing the physical property (i.e. magnetization), which is a linear function of amplitude. On the other hand, to change the shapes or flanks (slopes) of anomalies, the geometry and/or depth of the model must be changed, and these are non-linear functions of amplitude. For example, the deeper the source, the broader the anomaly it produces with longer wavelength and lower slopes and frequency. Thus, to make an anomaly's flank steeper or flatter, the source must be placed shallower or deeper, respectively. Additional parameters that affect the shape and amplitude of magnetic anomalies include the presence of remanence, dip of magnetization, dip of the source, and inclination of the normal field. Gradient and tensor modeling requires greater attention to and refinement of depth to sources, while taking advantage of the higher resolution of these components in defining individual anomaly sources.

13.5.6 Demagnetization effects

In most magnetic interpretation methods the assumption is made that magnetization is uniform throughout the source volume (e.g. VACQUIER *et al.*, 1951) and a function of its true susceptibility. This is a reasonable assumption in most geological situations because, considering the source-to-observation distance in most magnetic surveys, the distribution of magnetic minerals can be considered homogeneous and the effects of magnetic crystalline anisotropy are negligible. However, this assumption neglects the possible effects of the internal demagnetization field of magnetic sources which causes a decrease in and deflection of the magnetizing field. As a result, the effective magnetic susceptibility of a magnetic anomaly source is decreased from the true susceptibility as described in Section 10.5.

Demagnetization, sometimes referred to as self or shape demagnetization, occurs wherever free magnetic poles occur on an interface or a discontinuity such as the boundary of a magnetic source. It is a result of the dipole interaction between the magnetic moments originating at a boundary. Thus, both induced and remanent magnetizations are affected by demagnetization. The effect of demagnetization is dependent on the shape of the source and the direction of the ambient inducing magnetic field. It is negligible at true susceptibilities of roughly less than 1,250 SIu (= 10,000 CGSu). Demagnetization is therefore neglected in all geological situations except highly magnetic anomaly sources such as banded iron formations, ultramafic rocks, and certain types of ore bodies.

The basic theory of demagnetization is that the internal magnetic field (B_{int}) of a homogeneous material without remanent magnetization is

$$B_{int} = B_N - B_d, \tag{13.47}$$

where B_N is the ambient magnetic field and B_d is the demagnetization field which is proportional to the internal magnetization and directed opposite to the ambient Earth's magnetic field. The internal magnetization of the source volume which takes into account the internal demagnetization is called the effective magnetization (J_e). It is the source of the magnetic anomaly of the volume so that

$$B_d = N \times J_e, \tag{13.48}$$

where N is the demagnetization factor in the direction of the magnetization for the particular source body shape. Other demagnetization factors apply to magnetizations directed in the two principal orthogonal axes. Using Equations 13.47 and 13.48, the effective magnetization (GUO *et al.*, 1998) is

$$J_e = \frac{k \times B_N}{(1 + Nk)} = k_e \times B_e, \tag{13.49}$$

where k is the true susceptibility and k_e is the effective susceptibility. Thus, the effective susceptibility is

$$k_e = \frac{k}{(1 + Nk)}. \tag{13.50}$$

Demagnetization factors (N) for lines of poles, dipoles, and other simple sources with shapes bounded by second-order surfaces can be approximated by the demagnetization factors of ellipsoids (e.g. JOSEPH, 1976; SHARMA, 1968; EMERSON *et al.*, 1985). This approximation is valid because ellipsoids are bounded by second-order surfaces, and thus have a uniform magnetization. The uniform magnetization assumption fails in the near-body region close to edges and corners where the second-order bounding surface breaks down.

Numerous schemes have been used to determine demagnetization factors of more complex volume sources. ZIETZ and HENDERSON (1956) have used physical modeling involving laboratory measurements, but analytical methods (e.g. VOGEL, 1963; SHARMA, 1966; ESKOLA and TERVO, 1980; VEITCH, 1980; KOSTROV, 2007) are much more efficient. The basic procedure of analytical methods is to perform numerical integration of the effective magnetizations and related composite magnetic field of multiple volume elements arranged to approximate the source volume assuming an average uniform magnetization for each element of the volume or for the entire source body (e.g. VOGEL, 1963; JOSEPH, 1976). The resulting anomaly field is the summation of the effects of the volume elements. This approach has been modified by ESKOLA and TERVO (1980) and LEE (1980) to employ a surface element integration method which is more computationally efficient than the integration of the volume elements.

SHARMA (1968) has used small prismatic cells to approximate the source volume, and BLOKH (1980) has used an array of horizontal circular cylinders to approximate 2D sources. Figure 13.63 shows that the measured north–south anomaly over a horizontal rectangular plate

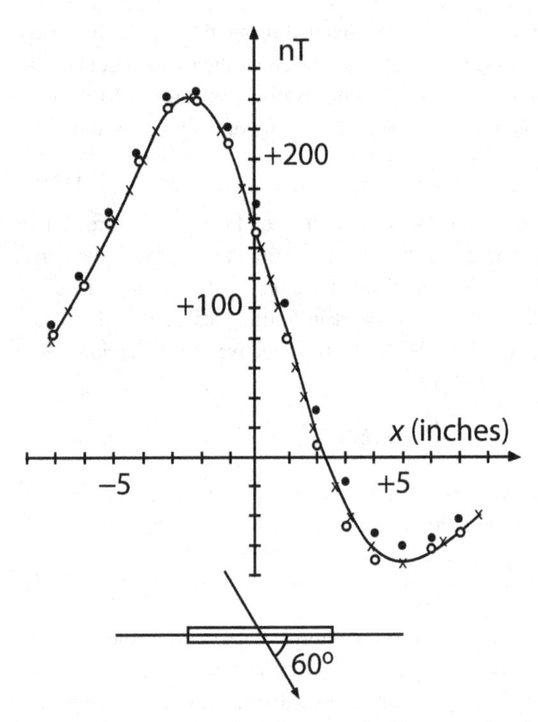

FIGURE 13.63 Comparison of the measured and computed total intensity values for the mid north–south profile over a flat, horizontal source. The crosses are the measured values from Figure 4 of ZIETZ and HENDERSON (1956). The solid circles are computed values from Figure 3 of VOGEL (1963). The open circles are computed values based on an average demagnetization factor from SHARMA (1968). Adapted from SHARMA (1968).

magnetized in a field of 53,400 nT at a 60° angle from the horizontal is closely duplicated by the field determined from applying an averaged demagnetization factor ($N = 0.816$) to the calculated anomaly without demagnetization (SHARMA, 1968) and also by using the analytical determination of demagnetization based on the method of VOGEL (1963). More recently, KOSTROV (2007) has used triangular cells to accurately delineate the shape of complex shapes with relatively few volume elements. GUO *et al.* (2001) and KOSTROV (2007) present comprehensive descriptions of the methodologies of determining the effects of demagnetizations and their potential errors.

The effect of demagnetization on the nature of anomalies derived from strongly magnetic sources is to decrease the anomaly amplitude and rotate the direction of magnetization toward the major axis of flattened sources (e.g. GAY, 1963; PARASNIS, 1986). The result is that the interpretation of the source of anomalies derived from highly magnetic bodies will be in error if demagnetization is not taken into account. Magnetization, dip, and depth of sources are likely to be erroneous. However, PLENGE (1978) shows that the impact of demagnetization will decrease with elevation of the observations above the source.

An example of the effect of demagnetization on an iron ore body located in western Gansu Province, China (GUO *et al.*, 1998, 2001), is presented in profile shown in Figure 13.64. The anomaly curve labeled (3), which is calculated from the interpreted source considering self demagnetization, closely matches the observed vertical intensity anomaly (1). The anomaly (2) which does not match the observed anomaly is calculated without considering demagnetization. The directions of magnetizations shown in the figure illustrate the change in direction of the total magnetization towards the plane of the iron body, owing to demagnetization.

13.5.7 Remanent magnetization effects

As described in Section 10.3.5, essentially all rocks have remanent magnetization to some degree. However, in most rocks this component of magnetization is minor compared with the induced magnetization so that the Koenigsberger ratio (Q) is less than 1 (Figure 10.5). As a result, a common assumption in magnetic interpretation is that the remanent magnetization is negligible or is a result of viscous remanent magnetization, and thus is directed in the Earth's magnetic field parallel to the induced magnetization. Nonetheless, numerous rocks also have Koenigsberger ratios of one or more (Section 5.4.2), which invalidates this assumption. This is particularly true of mafic volcanic rocks, many mafic and ultramafic plutonic rocks, and numerous specialized rocks such as ilmenite-bearing anorthositic rocks.

Although the impact of remanent magnetization on observed magnetic anomalies has long been known (e.g. VACQUIER *et al.*, 1951; GREEN, 1960; BOOKS, 1962; ZIETZ and ANDREASEN, 1967), it has been only recently that major efforts have been made to interpret these anomalies. These efforts have been driven by the desire to interpret the source of magnetic anomalies in ocean basins which are largely produced by remanent magnetization, by interest in deep crustal sources which may originate at least in part by permanent magnetization, and by the need to quantitatively characterize ore bodies and buried localized iron and steel objects that often have intense magnetic remanence.

The total magnetization of sources which contrasts with the adjacent rock formations to produce magnetic anomalies is the vectorial summation of the induced and the remanent magnetization components. Accordingly, the impact of remanent magnetization is a function of both its magnitude and direction (azimuth and inclination) relative to

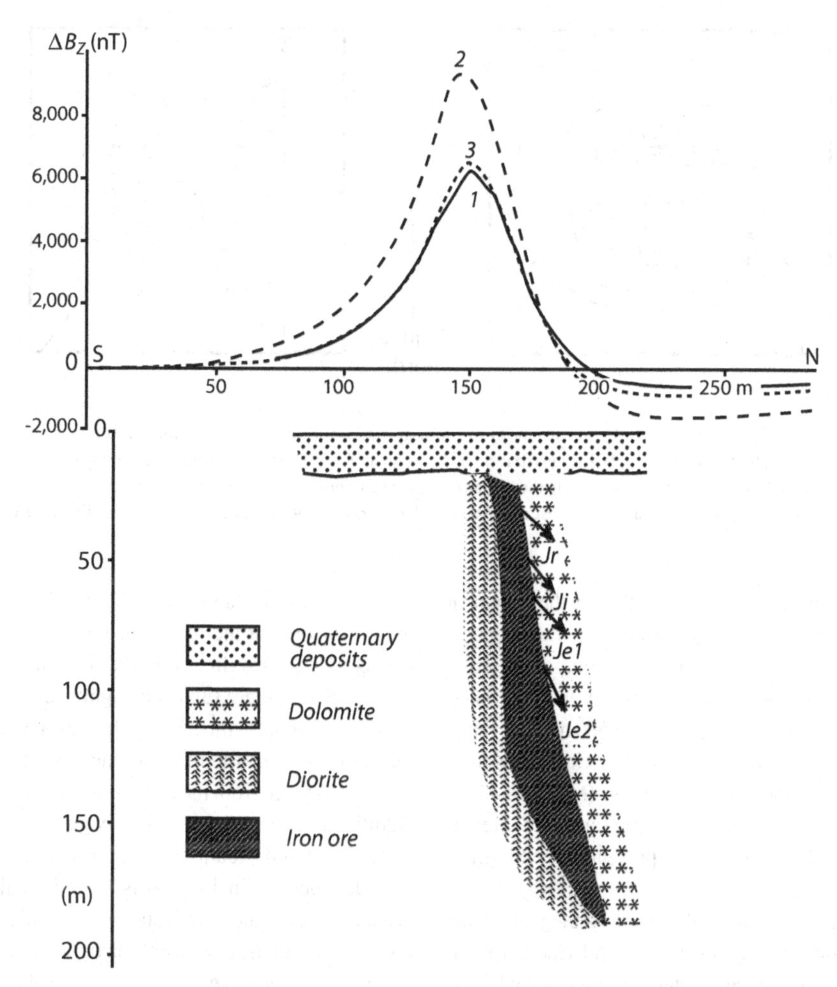

FIGURE 13.64 Results of quantitative modeling of a vertical magnetic intensity profile over an iron ore body in western Gansu Province, China. Anomaly profile (1) is the observed anomaly, profile (2) is the calculated anomaly without considering the demagnetization correction of the ore body, and profile (3) is the calculated anomaly after applying the demagnetization factor. The value Ji is the intrinsic (true) induced magnetization, Jr is the intrinsic remanent magnetization, $Je1$ is the effective magnetization without applying demagnetization, and $Je2$ is the effective magnetization after applying the demagnetization correction to the ore body. The inclination of the effective magnetization changes from 56° to 62° on applying the demagnetization correction. The inclination of the Earth's magnetic field in the region is 58°. Adapted from Guo *et al.* (2001).

the induced component. Most remanence is a function of the direction of the Earth's magnetic field at the time the magnetization was acquired. Thus, because of reversals in the Earth's magnetic field, continental drift, and magnetic polar wandering, the direction of the remanent magnetization in magnetic sources will vary significantly as a function of the age at which the magnetization was acquired. The end members of the possible directions of magnetizations are the direction of the induced and the remanent magnetizations. As shown by Morris *et al.* (2007), these end members are effectively reached where $Q < 0.1$ and > 10, respectively. At intermediate Q values, the total magnetization is related to both magnitude and direction of the two components. A special case is where the induced and remanent components are oppositely directed and near equal magnitude, resulting in essentially a null total magnetization and no observable magnetic anomaly.

Identifying and characterizing remanence

An initial step in interpretation of magnetic anomalies is the identification and characterization of the total magnetization contrast. Where geological samples of the source are available, the presence and nature of the remanent magnetization can be determined by direct measurement. Alternatively, where samples are unavailable but the age of the remanent magnetization can be approximated, it

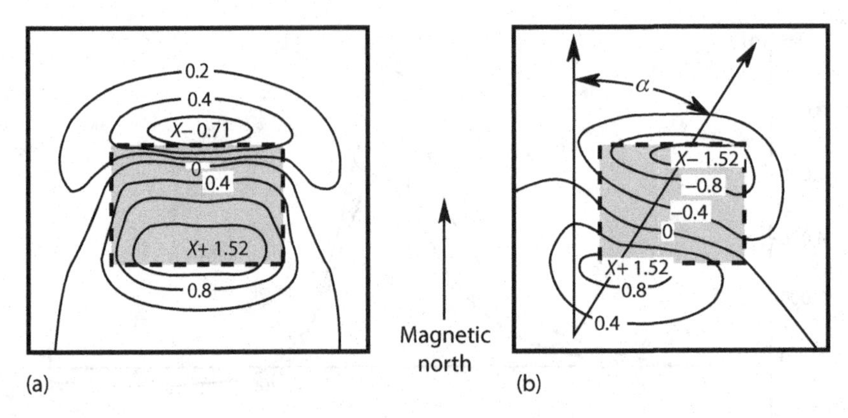

FIGURE 13.65 Modeled fields from $4 \times 6 \times 1$ unit size body at geomagnetic field inclination of $60°$. (a) Induced magnetization with azimuth $D' = 0°$ and inclination $I' = 60°$. (b) The magnetization includes an additional remanent component oriented at an azimuth $D'_{rem} = 30°$ and an inclination $I'_{rem} = 0°$. The angle α is the azimuth of the line connecting the maximum and minimum of the anomaly, which are located by the symbols $x+$ and $x-$. Adapted from SCHNETZLER and TAYLOR (1984) and ANDREASEN and ZIETZ (1969).

may be possible to estimate the direction of the remanent magnetization knowing the apparent polar wander paths (APWP) for the anomaly region since it acquired the remanence. These paths are known sufficiently well for magnetic interpretation purposes for the past 500 million years, and even into Precambrian time for more intensely studied regions such as the Canadian Shield (MORRIS *et al.*, 2007). Further details on using geologic samples for remanent magnetization estimates are presented by MORRIS *et al.* (2007).

In the absence of geological control on the source, the nature of the magnetic anomaly may provide information on the lack or presence of intense remanence which is directed so that the total magnetization is not parallel to the ambient magnetic field, particularly at intermediate geomagnetic latitudes. ANDREASEN and ZIETZ (1969) have addressed this topic by calculating the magnetic anomaly of vertical-sided prisms with a range of remanence directions at geomagnetic inclinations ranging from 0° to 90°. Figure 13.65 shows the change in the total magnetic intensity anomaly from a prism outlined in the thick line due solely to induced magnetization (map (a)) and the anomaly due to the combined effect of induced and equivalent remanence that is directed horizontally 30° E of the induced magnetization (map (b)). The position and relative intensity of the maximum and minimum of the anomaly have been significantly modified by the remanence. The line connecting the maximum and minimum is skewed by an angle (α) equivalent to the azimuth of the remanence, and the ratio of the maximum to minimum has changed from approximately 2 to 1 suggesting a flattened inclination total magnetization vector from the induced magnetization anomaly.

SCHNETZLER and TAYLOR (1984) have shown that these measures of the remanence, the azimuth of the line connecting the maximum and minimum and the ratio of their amplitudes, are subject to significant complexity, and thus are potentially misleading. However, as a general rule the ratio of the maximum to minimum increases in sensitivity to remanence from the geomagnetic equator to the pole. Additionally, the azimuth of the magnetization can be estimated in limited situations, especially at mid-geomagnetic latitudes, but the inclination is very difficult to estimate because of difficulties in isolating anomalies which cause problems in relating adjacent maxima and minima and in specifying the zero-level of the anomaly. Furthermore, the ratio of maximum to minimum will change with the dip of the source. For example, sources with infinite extent in more than one direction, such as 2D dikes, with different dips can produce identical anomalies depending on the direction of the total magnetization (BRUCKSHAW and KUNARATNAM, 1963). If the dip of the dike is rotated by the same amount as the total magnetization is rotated in the same direction, the anomalies will be the same. Accordingly, viewing the nature of the magnetic anomaly may identify the presence of end members of the possible directions of the total magnetization where anomalies are well defined and at mid- to high-magnetic latitudes, but care must be used in attempting to characterize the directional attributes of the magnetization from the observed anomaly.

Several investigators have used gravity anomalies in conjunction with magnetic anomalies to determine the remanent magnetization of the anomaly source. SHURBERT *et al.* (1976) have successfully analyzed the remanence of basement sources in northwest Texas, including the much-studied Crosbyton anomaly, by calculating the source from the gravity anomaly and then using iterative forward modeling of the source to find the remanence that provides the closest fit to the observed magnetic anomaly. This is similar to the method used to study the remanence of sea mounts where the bathymetry defines the geometry of the source of the magnetic anomaly that is compared to the observed anomaly over the sea mount (e.g. VACQUIER, 1962; UYEDA and RICHARDS, 1966; MERRILL and BURNS, 1972].

CORDELL and TAYLOR (1971) use a method based on Poisson's theorem relating gravity and magnetic fields due to a common source. A system of linear equations based on the observed anomalies involving magnetization and density of the source is solved for the components of the total magnetization vector and the minimum value of the Koenigsberger ratio. If the density of the source can be assumed from other information, it is possible to determine the remanent magnetization direction uniquely. A similar method based on Poisson's theorem was used by BOTT *et al.* (1966) to determine the magnetization direction of a source without making assumptions about the shape of the source. In this method the pseudogravity anomaly is calculated from the observed magnetic anomaly using a range of possible remanent magnetization directions. The optimum values of magnetization yield positive values of the pseudogravity anomaly over the entire anomaly. This is based on the assumption that the source density and computed pseudogravity anomaly above the source have the same sign. All these methods assume that the source is finite and the physical property contrast is uniform over the source of the anomalies.

Another method of identifying remanent magnetization in magnetic anomalies has been used by ROEST and PILKINGTON (1993) using the reduction-to-pole transformation and assuming the 2D magnetic sources have near-vertical sides that are separated by at least the depth of the top of the source from the observations. In their method, sometimes referred to as the cross-correlation method, they compare the analytic signal and the horizontal gradient of the pseudogravity calculated from the magnetic anomaly reduced-to-the-pole using ambient Earth field parameters. Both of the calculated components exhibit maxima over the margins of the source and they are coincident if there is no significant remanence in the source. If the maxima do not coincide, reduction-to-the-pole is recomputed with revised parameters until they do correlate. These revised parameters are used to characterize the remanence of the source. DANNEMILLER and LI (2006) have extended this method to find the magnetization of 3D sources which leads to the maximum symmetry of the reduced-to-pole transformations. This method requires stable reduced-to-pole transformations, and this is challenging at low-magnetic latitudes. Further it is more sensitive to inclination as the poles are approached, but the azimuth becomes less important in these regions.

Two additional methods of analytically determining remanence in the source of an anomaly have been described and compared with the cross-correlation method by LI *et al.* (2010). LOURENCO and MORRISON (1973), building on integral relationships of magnetic moments developed by HELBIG (1962) that employ the three orthogonal components of the magnetic field, have suggested a method of determining remanence directions of the anomaly source. The three orthogonal components can either be measured or approximated from the observed total field anomaly from the appropriate wavenumber domain operators. Additionally, a method of determining the remanence of 2D anomaly sources has been described by HANEY and LI (2002) using multiscale edges of the anomaly source obtained by a continuous wavelet transform. The location of the edges is a function of the magnetization direction, and thus tracking of the multiscale edges can be used to estimate the inclination of the magnetization.

Interpretation with remanence

As described in Chapter 9 and more specifically treated in terms of magnetic anomalies in this chapter, the interpretation of anomalies is largely based on either forward or inverse modeling. In forward modeling, which involves comparison of the anomalies of conceptual subsurface sources with the observed anomalies, remanent magnetization is introduced by vectorially combining the induced and remanent magnetization to obtain the total magnetization, both its directional attributes and magnitude, and using these parameters in the calculation of the anomaly components. This assumes that the magnetization is uniform over the source. However, this assumption is not always valid, as for example in the case of volcanic rocks with intense remanent magnetization that have been structurally disturbed, causing the remanent magnetization vector to be variable in its direction over the extent of the source. MARIANO and HINZE (1993) and MARIANO and HINZE (1994) describe a method based on the equivalent source procedure of KU (1977), where the source is subdivided into an aggregate of equivalent sources consisting of lines of dipoles whose magnetizations are rotated taking into account the structural disturbance of the remanence vector of the source rock combined with the induced magnetization in the direction of the ambient geomagnetic field. The magnetic fields of the individual lines of dipoles,

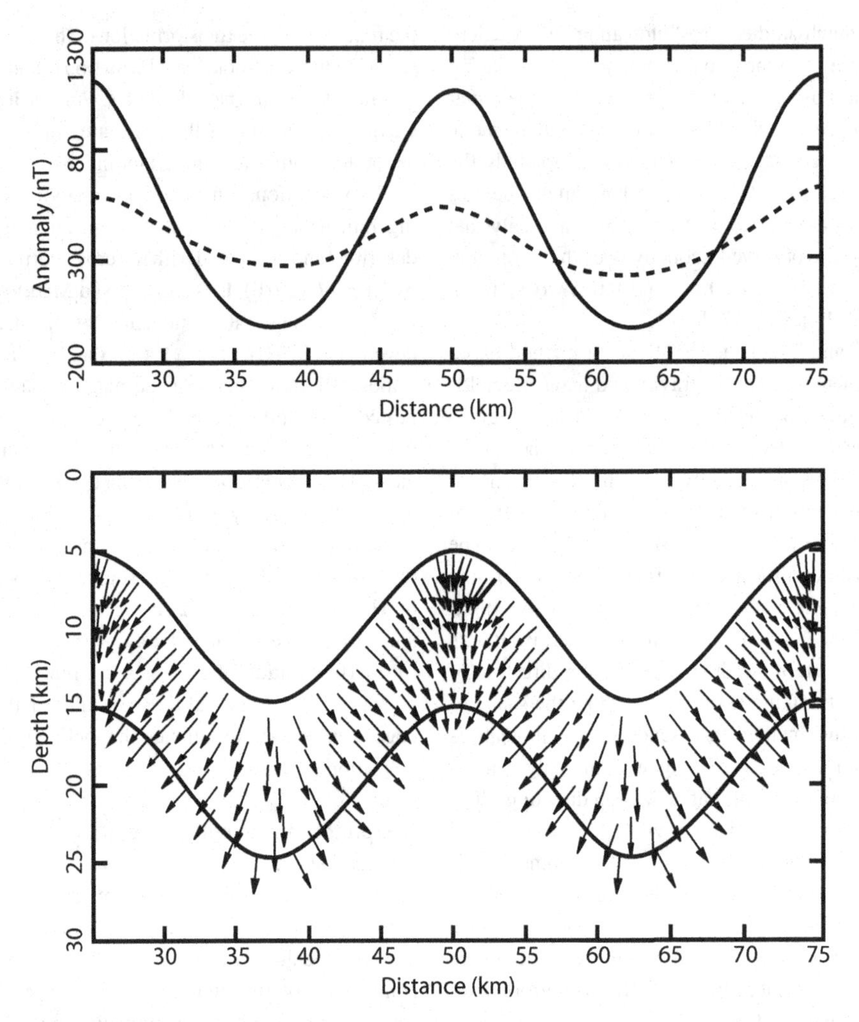

FIGURE 13.66 Total field magnetic anomalies calculated assuming homogeneous (solid) and heterogeneous magnetization (dashed) due to gradational rotation of the remanent vector of a folded formation with homogeneous remanent magnetization prior to folding. The inclination of the induced and unrotated remanent magnetization vectors is 90°. The Koenigsberger ratio is 8. Adapted from MARIANO and HINZE (1993).

each with different magnetization, are summed over the entire body to determine the magnetic anomaly of the source. The error resulting from the use of an aggregate of lines of dipoles is kept to a few percent by maintaining the distance between the line sources to less than one-tenth their depth below the observation level. An example of this method is given in Figure 13.66, which shows the impact on the total magnetic intensity anomaly associated with a folded remanently magnetized formation assuming homogeneous magnetization (solid line) as well as remanent magnetization rotated by the folding of the formation (dashed line).

Current procedures of inversion of magnetic anomalies for distributions of magnetic susceptibility require information on the total magnetization direction, including the effect of remanent magnetization. LI *et al.* (2010) have reviewed this requirement and approaches to meet it, and they have developed two general procedures for solving the problem. The first is to estimate the remanent magnetization from either geological information or analytical analysis of the observed magnetic anomaly as described above. This magnetization combined with the induced magnetization to obtain the direction of the total magnetization vector is supplied to the inversion procedure, such as that of LI and OLDENBURG (1996), to interpret the anomaly. The second method acknowledges that the direction of magnetization is unknown, and thus the inversion operates upon the amplitude of the magnetic anomaly vector which is only weakly dependent on the magnetization direction. Direct inversion of amplitude data is described by Shearer (2005), and examples of its application to theoretical and field anomalies are presented by Li *et al.* (2010).

13.6 Key concepts

- Magnetic interpretation is never unique, owing to the presence of data errors, effects from geologic inhomogeneities, and the theoretical source ambiguity of potential fields. Subsurface constraints from drillholes, geology, geophysical, and other observations can limit interpretational ambiguities of magnetic analysis, but not eliminate them completely.
- Magnetic interpretations range from qualitative analysis, where objectives are satisfied by identification of anomalies and anomaly patterns, to comprehensive quantitative modeling of the anomaly data. Quantitative interpretation involves relatively complex 3D modeling efforts of anomaly maps or simpler 2D modeling applied to profiles across elongated magnetic anomalies.
- Effective interpretation is dependent on knowledge of the key geological variables that control the amplitude and spatial characteristics or geometry of the magnetic anomaly. Anomaly characteristics vary with source geometry including volume and shape, source depth, isolation of the source from other sources, magnetization contrast and direction of the source relative to the adjacent formations, and orientation of the source with respect to the ambient magnetic field. Knowledge of the key geophysical variables also is necessary to effectively address the inverse problem of estimating geological parameters from the anomaly data. Key geophysical variables used in interpretation include the amplitude, sharpness, shape, and perceptibility or isolation of the magnetic anomaly.
- Magnetic interpretation is commonly initiated using simplified techniques to ascertain critical characteristics of the anomaly source that are further detailed by more comprehensive follow-up anomaly modeling efforts. Anomaly half-widths, straight slopes along anomaly flanks, and the horizontal anomaly derivatives are widely used graphical methods to obtain first order estimates of source depths. Semi-automated approaches consider possible source depths within moving windows of data by methods such as Euler and Werner deconvolution, as well as depths from analytic signal methods and the slopes identified in the power spectrum of the anomaly data. Magnetic dike approximations are useful for interpreting depths and dips of 2D contacts, whereas the inflection points on the flanking gradients of reduced-to-pole and pseudogravity anomalies aid in locating the lateral margins of magnetic sources.
- Magnetic modeling proceeds either through forward modeling using trial-and-error methods where only a few unknown parameters of a model must be determined, or by inverse modeling using usually least-squares inversion methods where a greater number of unknowns must be estimated. Both approaches require specifying a forward model that produces predictions to compare with the gravity anomaly observations.
- Forward models with closed form analytical magnetic effects involve idealized sources that have simple symmetric shapes (spherical, cylindrical, prismatic, etc.) which can be readily volume integrated. The effects of a more complex general source with irregularly shaped volume must be numerically estimated at the observation point by summing the effects of idealized sources that fill the volume.
- Arbitrarily shaped 2D and 3D sources are widely modeled using, respectively, vertical laminas and vertical stacks of horizontal laminas to approximate the irregularly shaped body. In both cases, the lamina is approximated by a polygon with the magnetic effect which is completely described by the polygon's magnetization and the end-point coordinates or vertices of the bounding line elements relative to the observation point.
- An alternate approach, which offers the additional advantage of obtaining least-squares estimates of the magnetic effects of the general source, is to distribute the idealized bodies throughout the irregularly shaped body according to the Gauss–Legendre quadrature decomposition of its volume. The approach using point dipoles is well suited for machine computations of the least-squares magnetic potential, related vector components, and the spatial gradients to any order for any conceivable distribution of magnetization.
- In most magnetic interpretation methods the assumption is made that magnetization is uniform throughout the source volume and a function of its true susceptibility. This is a reasonable assumption in most geological situations. However, it neglects the possible effects of the internal demagnetization field of magnetic sources. Demagnetization occurs wherever free magnetic poles occur on the boundary of a magnetic source and is a result of the dipole interaction between the magnetic moments originating at a boundary. Thus, both induced and remanent magnetizations are affected by demagnetization. The effect of demagnetization is dependent on the shape of the source and the direction of the ambient

inducing magnetic field. It is negligible at true susceptibilities of less than about 1,250 SIu (= 10,000 CGSu). As a result, demagnetization is neglected in all geological situations except highly magnetic anomaly sources such as banded iron formations, ultramafic rocks, volcanic rocks with intense remanent magnetization, and certain types of ore bodies.

- Essentially all rocks have remanent magnetization to some degree, but in most rocks this component of magnetization is minor compared with the induced magnetization. A common assumption in magnetic interpretation, therefore, is that the remanent magnetization is negligible or is directed in the Earth's magnetic field parallel to the induced magnetization. Nonetheless, numerous rocks such as young mafic volcanic rocks, many mafic and ultramafic plutonic rocks, and numerous specialized rocks such as ilmenite-bearing anorthositic rocks have intense remanent magnetization directed at an angle from the induced magnetization which modifies the induced component of the magnetic anomaly. In the absence of direct paleomagnetic and geologic data, a variety of methods have been developed for recognizing the presence of intense remanent magnetization in magnetic anomalies and interpreting the attributes of these combined induced and remanent magnetization anomaly sources.

Supplementary material and Study Questions are available on the website www.cambridge.org/gravmag.

Part III
Applications

14 Applications of the gravity and magnetic methods

14.1 Introduction

The nature of terrestrial gravity and magnetic fields, the underlying principles of applying these fields to exploring the Earth, and an overview of the current practices for using these methods form the subject of the previous chapters. In addition to these principles and practices, a review of the applications of these methods of exploration is important to understanding them and their role in determining the nature, composition, and structure of the Earth. In this chapter, a brief introduction is given of the application of gravity and magnetic methods to subsurface exploration. It serves as an introduction to a website that accompanies this book, which can be accessed at http://www.cambridge.org/gravmag. The website presents an expanded explanation of the applications of the methods and examples of their application largely from published case histories. Providing the case histories on a website decreases the length of the book, allows the inexpensive use of color in illustrations, and makes it easier to update the material with new applications. The reader is urged to visit the website and study the representative examples of the application of gravity and magnetics as a useful adjunct to understanding the methods.

14.2 General view of applications

Geophysical methods have had a seminal role in determining the nature of the Earth, leading to our current state of knowledge about the origin and evolution of our planet and the processes involved in the Earth system. Additionally, they have helped to improve the quality of life for the Earth's inhabitants by investigating natural potential hazards and locating exploitable resources. Gravity and magnetic methods are among the important geophysical techniques for studying the unseen Earth.

Initially, gravity and magnetic methods were directed almost entirely to global-scale problems. Subsequently, for nearly the past century, they have also been used to search out and investigate natural resources and to study potential natural hazards. This move was largely driven by human needs and the development of improved technology, beginning with the search for hydrocarbons because of the marked increase in demand for oil and gas since the 1920s. Advances in technology, which continue today, now contribute to academic studies of the Earth as well as applied investigations.

Gravity and magnetic methods have numerous attractive attributes. They are generally less expensive than other geophysical methods and can be observed at a density and precision satisfactory for applications over a range of scales from drill holes to marine, airborne, and satellite platforms. However, their measurements infrequently give a complete answer to an exploration or geological problem. The spatial resolution of both methods, especially the gravity method, is limited and they seldom provide an unambiguous solution. As a result a hallmark of successful application of gravity and magnetic methods is their synergistic use in combination with auxiliary geophysical data and collateral geologic information to reduce their ambiguity (Saltus and Blakely, 2011). Another important attribute that broadens their application is the availability of commercially and publicly available, relatively precise data sets covering essentially the entire Earth. Although the data density and, in many cases, the quality of these data sets are insufficient for localized and special studies, they are useful in many regional investigations or serve as background for the more localized studies.

For purposes of presentation and discussion in the website accompanying this book, applications of gravity and magnetic methods are divided into four major categories: near-surface studies, energy resource applications, mineral resource exploration, and lithospheric investigations. Each of these categories is described in a chapter of the website. Near-surface studies are concerned with the application of gravity and magnetic methods to Earth-related investigations for engineering, environmental, and archaeology problems which largely are focused on the near-surface Earth. Energy resource applications are concentrated on hydrocarbon exploration and exploitation, but also include a description of examples of the use of gravity and magnetics in coal and geothermal investigations. Mineral resource exploration encompasses both indirect and direct studies largely for metallic ore bodies, although examples also are presented of the use of the methods in diamond and uranium exploration. Finally, applications to lithospheric investigations cover the broad use of gravity and magnetic studies in the continental and oceanic crust and the uppermost mantle which make up the rigid lithosphere of the Earth. Generally these studies are of an academic nature, but the results help to provide an understanding of the nature of the Earth and how Earth processes operate, both of which are important to exploration for more utilitarian purposes.

The examples discussed in the accompanying website are representative rather than comprehensive. They illustrate the range of scales and amplitudes of anomalies derived from various geologic sources associated with a specific application. The description of the examples generally incorporates both the geology, with either an explicit or implicit consideration of the physical property contrast, and the nature of the geophysical anomalies. Where pertinent, methodologies for processing and interpretation of the data are described to illustrate application of procedures to a particular problem. The examples range from studies in the early use of the methods to modern examples that are generally publicly available. Some of the early applications are particularly useful in demonstrating the principles of the methods, whereas recent case histories show the vitality of the methods in studying today's complex view of the Earth.

We can anticipate that, spurred by improving technology and more readily accessible, high-quality data bases, new applications will be discovered for the use of the methods. For example, increasing precision in gravity observations has led to new applications related to the temporal variations in fluid and gas movements in the subsurface, and improved magnetic measurements and processing are giving new life to investigating low-amplitude anomalies associated with sedimentary basin fill as well as long-wavelength anomalies related to the base of the magnetic crust. One of the most notable characteristics of the case histories is that they illustrate the breadth of the scale of the applications of the gravity and magnetic methods. Wavelengths of use in the study of the Earth range from under a meter to anomalies measured in thousands of kilometers associated with extensive, deep sources. In the same way, amplitudes of anomalies of interest today cover several orders of magnitude.

14.3 Near-surface studies

Investigation of the nature of the subsurface is an important part of the siting of man-made structures because of the need to insure the long-term stability of the construction. The breadth and intensity of these investigations has greatly increased in recent decades because of the increased potential risk to humans and the environment from critical structures such as dams, bridges, waste repositories, and nuclear facilities. The risks are exacerbated by the expanding global population and development of urban centers, forcing building of structures in proximity to humans where subsurface conditions exist that may not be favorable for construction purposes.

Near-surface studies have also taken on an increasing role in determining the characteristics of the subsurface that control the occurrence, movement, and quality of subsurface water for human, agricultural, and industrial uses. In addition, such studies have expanded greatly in response to environmental concerns over identification of subsurface regions that have been adulterated by humans and require remediation and reclamation. To a lesser extent the need to identify and map archaeological sites of prehistoric human activity has fostered increasing near-surface investigations.

Gravity and magnetic methods have an important role in such near-surface investigations. Recent technological improvements in observing, processing, and analyzing these data have made these methods more efficient and more sensitive, and thus applicable to an increasing range of problems. The methods are especially useful in mapping the location of steep contacts between formations of different germane properties that may be a significant hazard to the integrity of critical structures. Accordingly, they are commonly used in the reconnaissance stage of investigations to localize regions that include steeply dipping faults and lithologic contacts for intensive studies with higher-resolution methods that require more expensive, time-consuming field surveying procedures and processing. This use is particularly appropriate for the magnetic method which is especially efficient in both acquisition and processing of data.

Gravity and magnetic methods are especially useful in the near-subsurface where the search objects are small compared with the volume of the study region, but have high property contrast with the surrounding soils or rocks. Examples are the search for buried unexploded ordnance, metallic archaeological objects, and subsurface solution voids in carbonate or volcanic rocks. These methods are also useful where it is desirable to investigate a volume of the near surface non-invasively, such as in landfills and repositories. Also, they are particularly sensitive to near-vertical subsurface interfaces which are difficult to map with the higher-resolution ground-penetrating radar and seismic reflection methods which are used in near-surface investigations.

Additional information on the applications of geophysics to near-surface studies is available in geophysical journals and in specialized books and monographs, including WARD (1990), SHARMA (1997), BUTLER (2005), BURGER *et al.* (2006), WITTEN (2008), OSWIN (2009), MILSON and ERIKSEN (2011), and REYNOLDS (2011).

14.4 Energy resource applications

Geophysical methods have a significant role in increasing the efficiency of the exploration and exploitation of solid-Earth energy resources including petroleum and natural gas, coal, and geothermal power sources. These indirect methods are particularly important in the search for and production of hydrocarbon resources because of the continuing need to discover new reservoirs which largely occur at increasing depths and with limited or no observable surface expression.

Since the earliest days of modern exploration, gravity and magnetic methods have had an important role in the search for oil and gas. In the first half of the twentieth century, the gravity method in particular played a central role in developing Gulf Coast, US oil fields associated with salt diapirs because of intense gravity anomalies resulting from the strong density contrast between the salt and the intruded sedimentary rocks. The first successful use of geophysics to hydrocarbon exploration was the mapping of the Nash Dome in Texas by the gravity method in 1924.

The early development of portable, high-sensitivity gravimeters and magnetometers was largely in response to their role in locating new petroleum reservoirs. These instruments were used in regional investigations and localized studies in special geological situations. However, improvements in seismic instrumentation, recording, and processing in the second half of the past century brought the seismic reflection method into prominence over gravity and magnetic methods in hydrocarbon exploration. Nonetheless, the gravity and magnetic methods retain a role in providing regional tectonic and geological information for exploration planning, complementing seismic reflection interpretation, and identifying detailed information on specialized geological features of interest to oil and gas exploration. The gravity method in particular has an important role in supporting and refining interpretation of seismic reflection studies.

Gravity and magnetic methods have proven especially useful in hydrocarbon exploration since the development of precision methods for measuring gravity and magnetic fields from ship and airborne platforms and for locating the precise position of observations. Airborne gravity and magnetic gradiometry, GPS positioning, and satellite-derived ocean gravity measurements are proving to be particularly effective in extending the range and effectiveness of gravity and magnetic methods, and thus are widely used in hydrocarbon exploration.

A number of conditions must be present to permit the accumulation of hydrocarbon deposits – sufficient volume of appropriate source rocks, thermal conditions for the generation of oil and gas from organic-rich rocks, pathways for their upward migration to reservoirs, void space in reservoir rocks, and geological features that will trap the lighter-than-water oil and gas ascending through the Earth. As a result, essentially all hydrocarbon deposits occur in thick sequences of sedimentary rocks or much less frequently in igneous rocks within the sedimentary rock column.

The gravity and magnetic methods can provide useful information bearing on all these conditions for the accumulation of hydrocarbons, although it is only in specific situations that this information is definitive from either of the methods alone. These methods are also useful in locating and evaluating reservoir rocks that will entrap natural gas and carbon dioxide that are pumped into them, providing natural underground storage facilities. Geologic storage of carbon dioxide could be an important element in mitigating greenhouse gases, and underground storage provides flexibility in natural gas distribution to consumers.

Regional gravity and magnetic surveys are relatively inexpensive ways for mapping the depth and configuration of the basement crystalline rocks. Magnetic methods, with their higher resolving power, are especially effective in this regard because they can be used to map the top of intrabasement magnetic sources. Mapping of the surface of the basement gives information on the overlying volume of sedimentary rocks, and thus the petroleum-bearing potential of sedimentary basins. Also, mapping of basement geology is important because intrabasin structures and stratigraphic features of interest to petroleum exploration commonly are related to reactivation of basement features or as a result of differential compaction within the sedimentary sequence with varying depths. Furthermore, deep crustal and upper mantle studies with gravity and magnetic methods may provide clues to major tectonic controls and the distribution of temperatures which relate to the maturation of hydrocarbon deposits.

High-resolution, precision gravity and magnetic studies focused on higher-wavenumber-components of the fields are useful in studying gravity and magnetic anomalies derived from intrabasin faults, diapirs, stratigraphic variations, and alteration zones. These can provide useful information related to the migration path of hydrocarbons, reservoir rock parameters, and especially structures in the sedimentary sequence that provide traps for ascending oil and gas. Furthermore, over the past few decades the gravity method has been investigated as a tool for monitoring fluid and gas movements in reservoirs with time lapse high-sensitivity measurements. The viability of both methods is in large part due to their cost effectiveness in enhancing solid-Earth exploration, but also to their unique role in mapping the subsurface.

Although the primary use of gravity and magnetic methods in solid-Earth energy resource applications is in the oil and gas industry, they are also used in other energy resource studies, for example in the exploitation of coal resources primarily by mapping the structure of formations which may be important in mining operations. Furthermore, the magnetic method can be useful in detecting areas of previous subsurface coal burns, because the magnetic properties of the formations are enhanced by the burning process. Also, the magnetic method has been used successfully to localize and study regions of high terrestrial heat flow that may serve as potential geothermal power sources. The high temperatures associated with these regions produce mappable modifications to rock magnetic properties.

Additional information on the application of gravity and magnetic methods to energy resources is presented in numerous case histories in exploration geophysics journals, in the compilations by SAAD (1993), GIBSON and MILLEGAN (1998), and BIEGERT and MILLEGAN (1998), and in modern geophysical exploration texts.

14.5 Mineral resource exploration

The rising population of the Earth and the desire to improve the quality of life leads to increasing need for mineral resources. The demand has been and is likely to be increased by the continuing improvement in living standards in developing nations. This demand must be met at least in part by new mineral discoveries which will supplement current mineral reserves. These new deposits are likely to show little if any direct surface evidence of their existence, and thus will need to be discovered by geophysical techniques that remotely sense the deposits at depth.

Over most of history, mineral and metal deposits were found by "prospectors" searching for evidence of mineralization in surface materials. In more recent times, ore deposits have been found through geological mapping backed up by geochemical and geophysical exploration. The ores of mineral resources in the broadest sense include all commodities that can be economically extracted from the Earth, but the description of applications in the chapter on mineral resource exploration in the accompanying website is restricted to ores, both metallic and non-metallic, that can be extracted for a profit and excludes such valuable Earth commodities as ground water, sand and gravel, oil and gas, and coal, which are treated elsewhere.

The breadth of the subject matter of geophysics as applied to mineral resources is great, ranging from the search for ores of aluminum to zinc through to the investigation of the nature of these deposits to enhance their exploitation. Ores have a wide diversity of physical properties, and they can occur in a variety of geologic settings that include essentially all types of rocks, sediments, soils, and alluvium. As a result a broad range of geophysical methods are used in mineral resource applications.

Gravity and magnetic methods, particularly the magnetic method, have a long and distinguished history of application to mineral resource exploration and exploitation. These methods have taken on increased importance because they lend themselves to observation from airborne platforms, and this has stimulated their use in investigating broad areas of the Earth over a range of scales. However, their role in mineral resource applications is somewhat different than that for oil and gas exploration. As described previously, these methods are primarily used in reconnaissance studies and specialized detailed investigations in hydrocarbon exploration and exploitation. In contrast, gravity and magnetic methods are extensively used in a range of scales in mineral resource applications from continental-sized regions to individual prospects. Most ore bodies occur within or in the proximity of crystalline rocks, including igneous, both volcanic and plutonic, and metamorphic rocks. Even if the ore bodies are not directly detectable, gravity and magnetic methods are extensively used in these terranes as a guide to geologic mapping. They are particularly useful in structure mapping and, under appropriate conditions, can be used to isolate specific geologic features and identify rock types because of the strong magnetization and density contrasts in crystalline rock terranes. However, care is used in identifying rock types from magnetic and gravity anomalies solely on the basis of anomaly amplitudes and patterns, because of overlap in their characteristics.

Ore bodies may be directly related to both gravity and magnetic anomalies which, depending on the physical property contrasts with the enclosing rocks, may be either positive or negative. However, these anomalies are seldom diagnostic by virtue of their amplitude, shape, or dimensions. For example, circular-form magnetic minima may originate from felsic plutons associated with significant mineralization because of the negative magnetic susceptibility contrast with the intruded country rock, or originate by alteration and destruction of magnetite within the plutons, or by reversed remanent magnetization of the source. Ample field investigations show that the nature of ore deposit anomalies can be duplicated by a wide variety of extraneous geological sources and can be hidden in errors in the observation, processing, and interpretation of both gravity and magnetic data. This is particularly the case because of the low amplitudes of ore body anomalies, which are rarely over a few milligals and a few hundred nanoteslas, and their limited areal dimensions.

Several additional factors complicate the identification of the specific ore bodies by their gravity and magnetic anomalies. Anomaly amplitudes decrease with the depth of the source so that deep ore bodies (those of more than a few hundred meters depth) are seldom large enough or have intense enough physical property contrasts to produce identifiable anomalies at the observation level. The relationship of amplitude with depth is complex, as described elsewhere in the treatment of the individual methods, but the magnetic anomaly amplitude decreases one power faster than the gravity anomaly for the same source configuration and the anomalies of concentrated, compact sources such as ore bodies decrease more rapidly with depth than sources that have long horizontal dimensions. Also, anomalies from deeper sources have greater energy in the longer wavelengths. These factors complicate the identification of ore body anomalies in a complex anomaly field.

Although ore minerals generally have a density around twice that of common rock-forming minerals, and the magnetic properties of some ores and associated minerals may be several orders of magnitude greater than that of the enclosing rocks, these ore minerals are usually disseminated through the ore body to an extent that the overall physical property contrast will be limited and heterogeneous, minimizing the anomaly amplitudes.

The broad diversity in the signature of gravity and magnetic anomalies associated with ore bodies severely complicates their identification and interpretation. Generally, exploration is focused on a single or a few types of mineral resources, and implementation of the survey and interpretation of the results are developed around conceptual geologic models of the ore bodies and their context within the geologic terrane. However, the ambiguity of interpretation of both magnetic and gravity methods leads to considerable uncertainty in the validity of interpretations. Accordingly, gravity and magnetic methods are seldom used without sensitivity analyses in the interpretation to evaluate the credibility of the results and without integrating the interpretation with geologic and physical property data and the results of other geophysical measurements.

The primary application of gravity and magnetics in mineral resource applications is in the mapping of favorable geology for specific ore bodies and under suitable, usually restricted, conditions in the direct detection of ore bodies. However, with the increasing resolution, precision, and accuracy of the observation, processing, and interpretation of gravity and magnetic data, these methods are taking on an increasingly important role in the exploitation of identified ore bodies. Inverse modeling of isolated anomalies is being used to delineate the extent of ore bodies, thus guiding development drilling and maximizing ore recovery. Gravity anomalies are generally more useful in quantitative analysis than magnetic anomalies because of their lesser sensitivity to depth. Also, gravity anomalies can be used to estimate the tonnage of an ore body, but seldom is it possible to make inferences of the grade of an ore body from the geophysical anomalies.

In addition, gravity and magnetic data can be used to evaluate potential safety and environmental concerns in exploiting mineral resources. For example, both methods can be used to identify rock types and the location of faults and alteration zones which affect rock mass characterization. This may provide estimates of the strength of the rocks of the ore bodies, as well as the nearby rocks, which are important to optimizing the mining design and improve safety in the mine operation. Gravity and magnetic data may also be used to map geological formations that are susceptible to collapse over underground mining operations. Finally, gravity surveying can be used to map bedrock topography that may control the movement of waters from the mine workings to the surrounding environment.

Useful collections of case histories that provide further examples are available in SEG (1966), MORLEY (1970), HOOD (1979), HINZE (1985), GARLAND (1989), GUBINS (1997), GIBSON and MILLEGAN (1998), and MOON *et al.* (2006). Particularly useful reviews of geophysical signatures of mineral resource deposits are given by DENTITH *et al.* (1994) for Western Australia and HILDENBRAND *et al.* (2000) for the western United States. Additional case histories are published in relevant journals and textbooks.

14.6 Lithospheric investigations

Geophysical investigations have largely been responsible for defining the lithosphere of the Earth and its attributes. The lithosphere, which generally consists of the crust of the Earth and the underlying uppermost portion of the mantle, is the rigid outer layer of our planet which overlies the mobile asthenosphere. It is not a petrologic boundary, but rather is the lower limit of the relatively strong outer shell of the Earth which differs in thickness between oceanic and continental plates. Along the axis of the mid-ocean ridges it may be as thin as the crust, but generally in the oceans it is from 50 to 100 km thick. In continental regions it typically is thicker, ranging from 100 to 200 km. This zone preserves the history of the Earth processes, both the products of and structure resulting from asthenospheric movements and thermal perturbations that interact with the overlying plates, the erosion and modification of the plates, and the sedimentation of detritus resulting from erosion. Accordingly, geologists study the nature of the lithosphere, its petrology, structure, and processes to decipher the Earth's history and potential hazards to humans.

The limited deep drilling and widespread surficial cover of sediments derived from erosion as well as terrestrial waters make it necessary to sense the nature of the lithosphere remotely by geophysical methods. Essentially all geophysical methods have contributed to our knowledge of the lithosphere. Although gravity and magnetic methods were among the first used for this purpose, they have taken a secondary role to seismic methods over the past several decades because of their limited resolving power. Nonetheless, gravity and magnetic methods continue to have an important role, particularly in structural studies of the lithosphere.

There are notable variations in the density and magnetization properties of the crust and upper mantle that make up the lithosphere. Accordingly, the gravity and magnetic methods have a rich history and important uses today in studying the nature, composition, and structure of the Earth. This is particularly true in the crystalline portion of the lithosphere where structural deformation and intrusive events are more likely to result in near-vertical contacts between rocks of varying physical property contrast than exist in the largely near-horizontal layering of the overlying sediments and sedimentary rocks. These near-vertical contacts lend themselves to producing much more identifiable gravity and magnetic anomalies than do horizontal interfaces.

The history of the development and use of the gravity and magnetic methods is quite different in terms of lithospheric investigations. Large regional geologic features produce relatively intense and readily mapped and identified gravity anomalies. Thus, historically, gravity anomalies have been used to map continental and subcontinental-scale geologic structures, such as orogens, sedimentary basins, and intrusive complexes, despite less sensitive measurements made at wide spacing that were dictated by the limits of historic instrumentation. Advances in gravity data acquisition in terms of sensitivity and the use of marine and airborne platforms, as well as in anomaly processing, have made it possible to investigate ever smaller and more detailed geological features and achieve dense coverage over regions not favorable for ground-based, dense measurement networks. Furthermore, these investigations are being made efficiently and effectively using gradient and tensor observations as well as vertical acceleration measurements. As a result, new applications have opened up for gravity investigations which include more detailed studies of individual geologic features.

In contrast, the early magnetic investigations were largely limited to detailed studies of localized geologic features, but with the increasing sensitivity of magnetic instrumentation particularly in shipborne, airborne, and satellite platforms, and major improvements in processing of the data, there has been wider application of the method to regional studies. These regional studies have made it possible in marine and airborne measurements to map detailed features over broad regions of the Earth. This is particularly advantageous because of the high resolution obtainable from magnetic measurements, especially when using measurements of gradients and tensors. Thus, for example, magnetic mapping of the oceanic crust from both ships and aircraft has observed the magnetic lineations and their disruptions that were the prime driver in developing the plate tectonics paradigm and in deciphering the history of the oceans and their interaction with the continents. Furthermore, these regional magnetic studies have found important applications in mapping ancient crustal sutures, individual igneous plutons, faults, volcanoes, ancient impact structures, and a broad variety of other structures and geologic terrains that have had a marked impact on our knowledge of the make-up of the Earth, its history, and geologic processes.

Examples of the use of gravity and magnetic methods to lithospheric studies are extensively treated in geophysical and geological journals, in the reports of governmental agencies, and in numerous books and monographs. A few of the prominent references citing this application of gravity and/or magnetic methods include: Woollard (1943), Simpson and Jachens (1989), Blakely and Connard (1989), Simpson *et al.* (1986), Hinze (1985), Mishra (2011), and the references on specific regions included in these publications.

Appendix A Data systems processing

A.1 Overview

Gravity and magnetic anomaly data are commonly expressed in standard formats for electronic analysis and archiving in digital data bases. A voluminous literature full of application-specialized jargon describes numerous analytical procedures for processing and interpreting anomaly data. However, when considered from the electronic computing perspective, these procedures simplify into the core problem of manipulating a digital forward model of the data to achieve the data analysis objectives. The forward model consists of a set of coefficients specified by the investigator and a set of unknown coefficients that must be determined by inversion from the input data set and the specified forward model coefficients. The inversion typically establishes a least-squares solution, as well as errors on the estimated coefficients and predictions of the solution in terms of the data and specified model coefficients. The inversion solution is never unique because of the errors in the data and specified model coefficients, the truncated calculation errors, and the source ambiguity of potential fields. Thus, a sensitivity analysis is commonly required to establish an "optimal" set or range of solutions that conforms to the error constraints. Sensitivity analysis assesses solution performance in achieving data analysis objectives including the determination of the range of geologically reasonable parameters that satisfy the observed data.

For electronic analysis, anomaly data are typically rendered into a grid that is most efficiently modeled by the fast Fourier transform (FFT) which has largely superseded spatial domain calculations. This spectral model accurately represents the data as the summation of cosine and sine series over a finite sequence of discrete uniformly spaced wavelengths. These wavelengths correspond to an equivalent discrete sequence of frequencies called wavenumbers which are fixed by the number of gridded data points. The FFT dominates modern gravity and magnetic data analysis because of the associated computational efficiency and accuracy, and minimal number of assumptions.

The purpose of the inversion process is to obtain a model that can be interrogated for the data's derivative and integral components to satisfy data analysis objectives. These objectives tend to focus mainly on isolating and enhancing gravity and magnetic anomalies using either spatial domain convolution methods or frequency domain (i.e. spectral) filtering operations, as well as on characterizing anomaly sources by forward modeling using trial-and-error inversion and inverse modeling by linear matrix inversion methods. Effective presentation graphics to display the data and data processing results also are essential elements of modern data analysis.

A.2 Introduction

Gravity and magnetic anomaly data consist of discrete samples of continuous fields and related auxiliary information such as time of observation, location, and elevation. These data commonly are electronically recorded in digital form for later computer analysis. The observations are subjected to intensive processing to prepare them for interpreting the location and nature of subsurface sources of the respective relevant physical properties. This processing involves a variety of steps including data transformation, reduction to anomaly form, isolation and enhancement of specific anomalies, inversion of the anomalies by modeling, and presentation of the data and its interpretation. Except for reduction of the observations to anomaly form, the processes are broadly generic and thus applicable to both gravity and magnetic data as well as to other geophysical data. Reductions of gravity and magnetic observations to anomaly form are specific to the measured field and thus are treated in separate chapters, whereas generic data analyses for these fields are described in this appendix.

With advances in data acquisition, improvements in analysis and interpretation techniques, and continuing increases in the speed and storage properties of digital computers, data analysis has become increasingly sophisticated and comprehensive. Software for gravity and magnetic data analysis has become generally available to the community and can be used with only a rudimentary knowledge of the computational techniques and their foundations which are the basis of the software. However, proper application to and useful results from data analysis require an understanding of the fundamentals of the procedures and their relative advantages and limitations. In this appendix, the focus is on the basic theory and use of the more generally used data analysis methodologies in gravity and magnetic exploration of the subsurface Earth.

Several principal topics are described in this appendix including (1) data bases and standards, (2) mathematical methods, and their applications to (3) regional/residual anomaly separation, (4) anomaly source modeling, and (5) data graphics and presentation. Numerous mathematical methods are employed in the processing of potential-field data, but the two that are most prominent and described herein are matrix methods and spectral analysis. The regional/residual anomaly separation section describes methods of isolating and enhancing specific anomaly components. These are further detailed in the data processing and interpretation chapters of the gravity and magnetic exploration sections. The segment dealing with anomaly source modeling considers the mathematical fundamentals of both forward and inverse modeling which are applied in the potential theory and data interpretation chapters of the gravity and magnetic exploration sections. Finally, the segment on data graphics and presentation presents data gridding, interpolation, and mapping procedures and presentation of anomaly data in a variety of formats. All topics include numerous references that supplement and complement the text, which will assist the reader interested in greater detail.

A.3 Data bases and standards

Vast quantities of gravity and magnetic data are available in varying degrees of detail and accuracy covering much of the Earth. These data are derived from extensive land, marine, airborne, and satellite surveys primarily over the past half century that have been conducted for resource as well as regional and global geological investigations. Most of these data, either original observations and auxiliary information or gridded data sets, have been compiled into electronic data bases which reside in a variety of data centers. These data centers are maintained by geoscience governmental organizations, professional societies, universities, research institutes, and commercial enterprises.

Data sets residing in non-commercial centers generally are of a regional nature with data intervals of a kilometer or two or more and are available at no charge or a modest cost to cover preparation and distribution of the data. Commercial centers focus on proprietary data sets of a generally more detailed nature covering limited regions, but in recent years significant strides have been made to acquire data covering continental regions and to compile them into a single data set (e.g. FAIRHEAD *et al.*, 1997). Additionally, airborne and satellite observations of gravity and magnetic fields have been useful in compiling data sets, particularly over marine regions, and in preparing global data sets of increased resolution and accuracy. These data sets are readily available and downloadable from World Wide Web sites for analysis and interpretation purposes.

Gravity and magnetic data sets are commonly used to prepare maps of these data for visual inspection and evaluation. Whether in observed or anomaly form, these data are made available in either point, line, or grid form for distribution as well as mathematical processing and analysis that are necessary in the reduction and interpretation of the data. Gravity data have traditionally been presented in a format that consists of the principal facts of the point observation which includes the observed gravity, elevation relative to sea level, latitude and longitude of the station, terrain correction, and typically the free-air and Bouguer gravity anomalies. The data file may also include additional anomaly types and relevant information on data acquisition. Magnetic data files are generally simpler than those of gravity because of the fewer parameters involved in magnetic data reduction. Magnetic anomaly data sets based on data acquired prior to the advent of digital recording in the 1970s are field values that have been digitized from analog records or anomaly values obtained by digitization of anomaly maps. Data bases derived from digitally recorded magnetic field values typically specify the observation value, the location of the observation, the time of the measurement, the geomagnetic reference field value, and the computed anomaly value after leveling of the observations.

Although data files available from data centers may vary, efforts have been made by professional organizations and others to standardize the data sets to improve the efficiency and effectiveness of exchanging and archiving gravity and magnetic anomaly data. REID (2001) has considered the worthwhile attributes of a standard. These have generally been met for gridded data sets in the Grid eXchange Format GXF-Rev 3 of 2001 that has been adopted by the Gravity and Magnetics Committee of the Society of Exploration Geophysicists (SEG). Revision 3 of the exchange format is a standard ASCII file format that acknowledges the variability of map coordinate systems and implements the exchange of 2D coordinate system information. Supporting software of the standard allows properly formatted GXF files of grids to be imported and exported.

A related standard has been developed by the Australian Society of Geophysicists (ASEG) Standards Committee, the ASEG-GDF2 (Draft 4) published in 2003. Detailed information on this standard as well as the GXF standard is available from reports on the web. The ASEG-GDF2 standard is an ASCII data exchange and archive standard for both point and line data. The latter is particularly useful for airborne data collected along flight lines. It is based on a self-defining format that permits data to be identified and loaded into computer applications. The standard consists of a minimum of four files. Three describe the data, with one defining the format of the data, another containing a text description of the data and survey contents, and the third holding associated metadata regarding the map datum and the projection of the geophysical data as specified in the GXF standard. The fourth file contains the data values.

The metadata that accompanies the data is a critical part of the standard assuring that the data and information related to the data are not separated from each other and insure the proper use of the data. "Metadata" has a somewhat different meaning depending on what data it is applied to, but in general it is "information" about the "data". It provides the critical information that includes a description of the data, its format, technical information about the data acquisition and processing, the origin and distribution of the data, and the point of contact for the data. Metadata files provide a template to assure that all the necessary information pertaining to the data is permanently archived with the data, although the details of these templates vary among data providers.

Two significant concerns about gravity and magnetic anomaly surveys are the resolution and accuracy of the data. Conventional metadata provides information on the spacing of the observations that relate to resolution, but few metadata files provide information that is useful in defining the accuracy of the data. However, some guidance is provided with information on the precision of the instrumentation and, in the case of gravity data, information is commonly provided on the methods of determining the station elevation that can be a guide to the accuracy of the elevation which is a major factor in determining the accuracy of gravity anomalies.

A.4 Mathematical methods

Electronic computations are digital and carried out by linear series manipulations on linear digital models of the input data. The process of modeling gravity and magnetic data is commonly called inversion where the ultimate objective is not simply to replicate the data, but rather to establish a model that can be analytically interrogated for new insights on the processes causing the sources of data variations. The digital data model is the basis for achieving any data processing objective by electronic computing. It can be demonstrated with the problem of relating n-data values ($b_i \in \mathbf{B}$, $\forall i = 1, 2, \ldots, n$) to the simultaneous system of n-equations

$$\begin{aligned} a_{11}x_1 + a_{12}x_2 + \cdots + a_{1m}x_m &= b_1 \\ a_{21}x_1 + a_{22}x_2 + \cdots + a_{2m}x_m &= b_2 \\ \ddots \qquad\qquad \ddots \qquad &= \vdots \\ a_{n1}x_1 + a_{n2}x_2 + \cdots + a_{nm}x_m &= b_n, \end{aligned} \tag{A.1}$$

which is linear in the m-unknown variables ($x_j \in \mathbf{X}$, $\forall j = 1, 2, \ldots, m$) and the $(n \times m)$-known variables ($a_{ij} \in \mathbf{A}$) of the forward model ($\mathbf{AX}$) that the investigator assumes will account for the observations. In matrix notation, this linear equation system becomes

$$\mathbf{AX} = \mathbf{B}, \tag{A.2}$$

where $\mathbf{B}$ is the n-row by 1-column matrix or vector containing the n-observations, $\mathbf{X}$ is the m-row by 1-column vector of the m-unknowns, and the design matrix $\mathbf{A}$ is the n-row by m-column matrix containing the known coefficients of the assumed model $\mathbf{AX}$.

The forward model is the heart of any analysis and recognizing its attributes is essential to productive gravity and magnetic anomaly analysis. Linear models dominate modern analysis, but non-linear forward models also can be invoked and were popular in inversion before the advent of electronic computing because of their relative ease in manual application. For computer processing, the non-linear model is linearized either by the application of the Taylor series or divided up numerically into array computations.

In general, any linear or non-linear inversion involves only three sets of coefficients with one set taken from the measured gravity or magnetic data (**B**), another set imposed by the interpreter's assumptions (**A**), and the final set of unknowns (**X**) determined by solving Equation A.2. Electronic computers process these coefficients as data arrays with basic matrix properties so that matrix inversion is the predominant mode for obtaining the solution (**X**). However, the most widely used implementation of matrix inversion in anomaly analysis invokes the spectral forward model because it offers maximum computational efficiency with minimum assumptions.

A.4.1 Basic matrix properties

A matrix is a rectangular array of numbers such as given by

$$\mathbf{A} = \begin{pmatrix} a_{11} & a_{12} & a_{13} \\ a_{21} & a_{22} & a_{23} \\ a_{31} & a_{32} & a_{33} \\ a_{41} & a_{42} & a_{43} \end{pmatrix}. \tag{A.3}$$

The elements of **A** are a_{ij}, where the i-subscript denotes its row number ($\rightarrow$) and the j-subscript is its column number ($\downarrow$). Thus, the matrix **A**-above has dimensions of 4 rows by 3 columns, or is of order 4×3 which also may be indicated by the notation $\mathbf{A}_{(4\times3)}$. The matrix is even-ordered or square if the number of rows equals the number of columns, otherwise it is odd-ordered or non-square.

The matrix **A** is symmetric if it is even-ordered and $a_{ij} = a_{ji}$, like in

$$\mathbf{A} = \begin{pmatrix} 1 & 3 & 6 \\ 3 & 4 & 2 \\ 6 & 2 & 5 \end{pmatrix}. \tag{A.4}$$

The symmetric matrix of order n can be packed into a singly dimensioned array of $p = n(n+1)/2$ elements. Thus, symmetric matrices are convenient to work with because they minimize storage requirements in computing.

The transpose of a $n \times m$ matrix **A** is the $m \times n$ matrix $\mathbf{A}^t$, where $a_{ij} = a^t{}_{ij}$. Thus, in taking the transpose of a matrix, the rows of the matrix are taken for the columns of its transpose, or equivalently the columns of the matrix for the rows of its transpose. For example,

$$\mathbf{A} = \begin{pmatrix} 1 & 2 & 3 \\ 4 & 5 & 6 \end{pmatrix} \quad \text{has the transpose} \quad \mathbf{A}^t = \begin{pmatrix} 1 & 4 \\ 2 & 5 \\ 3 & 6 \end{pmatrix}. \tag{A.5}$$

Matrix addition and subtraction, $\mathbf{C} = \mathbf{A} \pm \mathbf{B}$, involves the element-by-element operations defined by $c_{ij} = a_{ij} \pm b_{ij}$. For the operations to be defined, the matrices **A**, **B**, and **C** must all have the same dimensions. These operations are associative because $(\mathbf{A} \pm \mathbf{B}) \pm \mathbf{C} = \mathbf{A} \pm (\mathbf{B} \pm \mathbf{C})$, and commutative because $\mathbf{A} \pm \mathbf{B} = \mathbf{B} \pm \mathbf{A}$.

Matrix multiplication, $\mathbf{C} = \mathbf{AB}$, involves the row-by-column multiplication of matrix elements given by $c_{ij} = \sum_{k=1}^{n} a_{ik}b_{kj}$. Matrix multiplication is possible only between conformal matrices where the dimensions of the product matrices satisfy $\mathbf{C}_{(m\times n)} = \mathbf{A}_{(m\times k)}\mathbf{B}_{(k\times n)}$. Clearly, $\mathbf{C} = \mathbf{AB}$ does not imply that $\mathbf{C} = \mathbf{BA}$, and hence matrix multiplication is not commutative. As an example, the products of the matrices in Equation A.5 are

$$\mathbf{AA}^t = \begin{pmatrix} 14 & 32 \\ 32 & 77 \end{pmatrix} \quad \text{and} \quad \mathbf{A}^t\mathbf{A} = \begin{pmatrix} 17 & 22 & 27 \\ 22 & 29 & 36 \\ 27 & 36 & 45 \end{pmatrix}. \tag{A.6}$$

Note that $\mathbf{A}^t\mathbf{A}$ is a symmetric $(m \times m)$ matrix, whereas $\mathbf{AA}^t$ is a symmetric $(n \times n)$ matrix. These product matrices are important because matrix inverses are defined only for square matrices.

For matrices **A** and **B** that are square in the same order, **B** is the inverse of **A** or $\mathbf{B} = \mathbf{A}^{-1}$ if $\mathbf{BA} = \mathbf{I}$, where

$$\mathbf{I} = \begin{pmatrix} 1 & 0 & \cdots & 0 \\ 0 & 1 & \cdots & 0 \\ \vdots & \vdots & \ddots & \vdots \\ 0 & 0 & \cdots & 1 \end{pmatrix} \tag{A.7}$$

is the identity matrix that has the special property $\mathbf{BI} = \mathbf{IB} = \mathbf{B}$. The inverse $\mathbf{A}^{-1}$ can be found by applying elementary row operations to **A** until it is transformed into the identity matrix **I**. The row operations consist of adding and subtracting linear multiples of the rows from each other. The effects of these row operations on the corresponding elements of the identity matrix give the coefficients of the inverse $\mathbf{A}^{-1}$. However not every square matrix **A** has an inverse. Indeed, the inverse exists only if the determinant of **A** is not zero or $|\mathbf{A}| \neq 0$. If the determinant is zero then the matrix **A** is said to be singular. Matrix singularity occurs if a row or column is filled with zeroes or two or more rows or columns are linear multiples of each other.

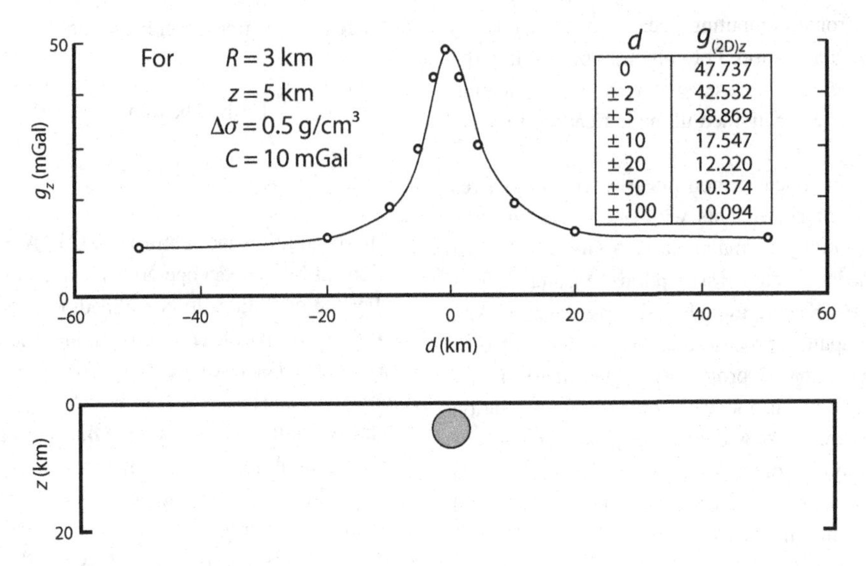

FIGURE A.1 Profile of the gravity effect ($g_{(2D)z}$) across a buried horizontal 2D cylinder striking perpendicular to the profile superimposed on a constant regional field C. This profile is used to illustrate the anomaly inversion process. See the text for additional modeling details.

A.4.2 Matrix inversion

Using the elementary matrix properties described above, the common array methods for solving linear systems of simultaneous equations are readily developed. To illustrate these methods, the gravity profile in Figure A.1 is used that crosses perpendicularly over a 2D horizontal cylinder anomaly source in the presence of a constant regional gravity field. Each observation along the profile is the gravity effect of the cylinder plus the regional field computed from

$$g_{(2D)z} = 41.93\Delta\sigma\left(\frac{R^2}{z}\right)\left(\frac{1}{(d^2/z^2)+1}\right) + C, \quad \text{(A.8)}$$

where $g_{(2D)z}$ is the vertical component of gravity in mGal and d is the distance along the profile in km for the horizontal cylinder with radius $R = 3$ km, depth to the central axis $z = 5$ km, and density (or density contrast) $\Delta\sigma = 0.50$ g/cm^3, in the regional gravity field of amplitude $C = 10.0$ mGal. Note that in Chapter 3, the symbols $\mathcal{F}_z$ and $\mathcal{G}$ are also used to describe the 2D gravity effect (i.e. $g_{(2D)z} \equiv \mathcal{F}_z \equiv \mathcal{G}$).

Equation A.8 involves two linear terms so that an interpreter can estimate a variable in each of the terms from the remaining variables and the observations. Consider, for example, determining $\Delta\sigma$ and C assuming that R, z, and d are known and the model (Equation A.8) is appropriate for explaining the variations in the $\mathcal{G}$ observations. To evaluate the two unknowns, at least two observations are needed which may be modeled by the linear system of equations

$$b_1 = a_{11}x_1 + a_{12}x_2$$
$$b_2 = a_{21}x_1 + a_{22}x_2, \quad \text{(A.9)}$$

where the design matrix coefficients are

$$a_{11} = [41.93(R^2/z)]/\left[(d_1^2/z^2)+1\right],\ a_{12} = 1.0,$$
$$a_{21} = [41.93(R^2/z)]/\left[(d_2^2/z^2)+1\right],\ a_{22} = 1.0. \quad \text{(A.10)}$$

The design matrix coefficients represent the forward model in Equation A.8 evaluated with the known parameters (R, z, d) set to their respective values and the unknown parameters ($\Delta\sigma$, C) numerically set to equal unity.

In mathematical theory, the two equations in two unknowns can be solved "uniquely" from any two observations. However, if the two observations are located symmetrically about the anomaly peak, then the two equations in the above system are co-linear and $|\mathbf{A}| = 0$ so that no solution is possible. Thus, in the context of the assumed model, the solution can be completely obtained from the peak value and the observations either to the right or left of the peak. In other words, the interpreter's choice of the model (Equation A.8) established the interdependency that made roughly 50% of data in Figure A.1 redundant for subsurface analysis.

Consider now the anomaly values on the peak at $d_1 = 0$ km and the flank at $d_2 = 20$ km so that the system becomes

$$47.737 = 75.474x_1 + x_2$$
$$12.220 = 4.440x_1 + x_2. \quad \text{(A.11)}$$

The design matrix and its determinant and inverse for this system are, respectively,

$$\mathbf{A} = \begin{pmatrix} 75.474 & 1.000 \\ 4.440 & 1.000 \end{pmatrix}, \quad |\mathbf{A}| = 71.034,$$

$$\text{and} \quad \mathbf{A}^{-1} = \begin{pmatrix} 0.0141 & -0.0141 \\ -0.0625 & 1.0625 \end{pmatrix}. \tag{A.12}$$

Cramer's Rule is a relatively simple method for obtaining the solution $\mathbf{X}$ whereby $x_j = |\mathbf{D_j}|/|\mathbf{A}|$. Here, the augmented matrix $\mathbf{D_j}$ is obtained from $\mathbf{A}$ by replacing the j-column of $\mathbf{A}$ with the column vector $\mathbf{B}$. Thus, by Cramer's Rule,

$$x_1 = \frac{35.517}{71.034} = 0.50 \quad \text{and} \quad x_2 = \frac{710.34}{71.034} = 10.0, \tag{A.13}$$

which are the correct values. Cramer's Rule, however, is practical only for simple systems with relatively few unknowns because of the labor of computing the determinants. For larger systems, a more efficient approach uses elementary row operations to process $\mathbf{A}$ for its inverse so that the solution to Equation A.11 can be obtained from

$$\mathbf{X} = \mathbf{A}^{-1}\mathbf{B} = \begin{pmatrix} 0.50 \\ 10.0 \end{pmatrix}. \tag{A.14}$$

Even more efficient approaches directly diagonalize the linear equations in place for the solution coefficients. Gaussian elimination uses elementary row operations to sweep out the lower triangular part of the equations. As an example, the system A.11 reduces to

$$\begin{bmatrix} 47.737 = 75.474x_1 + x_2 \\ 12.220 = \ \ 4.440x_1 + x_2 \end{bmatrix}$$

$$\sim \begin{bmatrix} 47.737 = \ 75.474x_1 + x_2 \\ 9.412 \ = 0.0 + 0.9412x_2 \end{bmatrix},$$

from which clearly the solution coefficients are again

$$x_2 = \frac{9.412}{0.9412} = 10.0 \quad \text{and} \quad x_1 = \frac{(47.737 - 10.000)}{75.474} = 0.50, \tag{A.15}$$

where x_1 was obtained by back-substituting the solution for x_2. Instead of back substitution, the row operations can be continued to also sweep out the upper triangular part of the system. This procedure, called Gauss–Jordan elimination, obtains the solution coefficients explicitly along the main diagonal.

These matrix inversion methods are applicable for systems where the number of unknowns m equals the number of input data values n. However, for most gravity and magnetic inversions, $m < n$ and the system of equations is always related to an incomplete set of input data containing errors, as well as uncertainties in estimating the coefficients of the design matrix $\mathbf{A}$ and other errors. Thus, the least-squares solution is commonly adopted that minimizes the sum of squared differences between the predicted and observed data.

This result involves solving the system of normal equations derived from differentiating the sum of squared differences with respect to the model parameters being estimated in the inversion. Setting the normal equations to zero shows that the system $\mathbf{AX} = \mathbf{B}$ is equivalent to the $\mathbf{A}^t$-weighted system $\mathbf{A}^t\mathbf{AX} = \mathbf{A}^t\mathbf{B}$, which has the least-squares solution

$$\mathbf{X} = (\mathbf{A}^t\mathbf{A})^{-1}\mathbf{A}^t\mathbf{B} \tag{A.16}$$

if $|\mathbf{A}^t\mathbf{A}| \neq 0$.

Of course, the solution is found more efficiently by processing the weighted system in place with Gaussian elimination than by taking the inverse $(\mathbf{A}^t\mathbf{A})^{-1}$ directly and multiplying it against the weighted observation vector $(\mathbf{A}^t\mathbf{B})$. This method works for any set of equations which can be put into linear form. No derivatives need be explicitly taken for the least-squares solution, which contains only the known dependent and independent variable coefficients from $\mathbf{B}$ and $\mathbf{A}$, respectively.

The matrix $\mathbf{A}^t\mathbf{A}$ is symmetric and positive definite with $|\mathbf{A}^t\mathbf{A}| > 0$, and thus can be factored into an explicit series solution that is much faster and more efficient to evaluate than the conventional Gaussian elimination solution. In particular, the matrix $\mathbf{A}^t\mathbf{A}$ can be decomposed into $(m \times m)$ lower and upper triangular Cholesky factors $\mathbf{L}$ and $\mathbf{L}^t$, respectively, so that $\mathbf{LL}^t = \mathbf{A}^t\mathbf{A}$. The coefficients of $\mathbf{L}$ are obtained from

$$\begin{aligned}
l_{11} &= \sqrt{a_{11}}, && \forall i = j = 1 \\
l_{j1} &= a_{j1}/l_{11}, && \forall j = 2, 3, \ldots, m \\
l_{ii} &= \sqrt{a_{ii} - \sum_{k=1}^{i-1} l_{ik}^2}, && \forall i = 2, 3, \ldots, (m-1) \\
l_{ji} &= 0 && \forall i > j \\
&= \left(\frac{1}{l_{ii}}\right)\left(a_{ji} - \sum_{k=1}^{i-1} l_{ik}l_{jk}\right), && \forall i = 2, 3, \ldots, (m-1) \\
& && \textit{and} \\
& && j = (i+1), (i+2), \ldots, (m), \textit{and} \\
l_{mm} &= \sqrt{a_{mm} - \sum_{k=1}^{m-1} l_{mk}^2}.
\end{aligned} \tag{A.17}$$

Note that in the above Cholesky elements, the summations were effectively taken to be zero for $i = 1$.

In general, the Cholesky factorization allows x_m to be estimated from a series expressed completely in the known coefficients of **L**, **A**, and **B**, which is back-substituted into the series expression for x_{m-1}, which in turn is back-substituted into the series for x_{m-2}, etc. The Cholesky solution is much faster than Gaussian elimination approaches that require twice as many numerical operations to complete. The Cholesky solution also has minimal in-core memory requirements because the symmetric $\mathbf{A}^t\mathbf{A}$ matrix can be packed into a single-dimensioned array of length $m(m+1)/2$. For systems that are too large to hold in active memory, updating Cholesky algorithms are available that process the system on external storage devices (e.g. LAWSON and HANSON, 1974).

The Cholesky factorization is problematic, however, in applications where the coefficients $l_{i,j} \longrightarrow 0$ within working precision. In these instances, the products and powers of the coefficients are even smaller so that solution coefficient estimates become either indeterminant or blow up wildly and unrealistically. Indeterminant or unstable solutions respectively signal singular or near-singular $\mathbf{A}^t\mathbf{A}$ matrices. However, these ill-conditioned matrices can still be processed for effective solutions using the error statistics of the inversions to help suppress their singular or near-singular properties.

Error statistics

In general, reporting of gravity and magnetic inversion results routinely requires quantifying the statistical uncertainties of the analysis. These assessments commonly focus on the variances and confidence intervals on the solution coefficients and predictions of the model, and the coherency and statistical significance of the fit of the model's predictions to the data observations.

For example, applying variance propagation (e.g. BEVINGTON, 1969) to the linear function $(\mathbf{A}^t\mathbf{A})^{-1}\mathbf{A}^t\mathbf{B}$ shows that the solution variance can be obtained from

$$\sigma_{\mathbf{X}}^2 = [(\mathbf{A}^t\mathbf{A})^{-1}\mathbf{A}^t]^2\sigma_{\mathbf{B}}^2 = (\mathbf{A}^t\mathbf{A})^{-1}\sigma_{\mathbf{B}}^2, \tag{A.18}$$

where the statistical variance in the observations $\sigma_{\mathbf{B}}^2$ is specified either *a priori* or from

$$\sigma_{\mathbf{B}}^2 \simeq [\mathbf{B}^t\mathbf{B} - \mathbf{X}^t(\mathbf{A}^t\mathbf{A})\mathbf{X}]/(n-m) \ \forall\ n > m, \tag{A.19}$$

which is an unbiased estimate if the model **AX** is correct. Equation A.18 is the symmetric variance–covariance matrix of **X** with elements that are the products $\sigma_{x_i}\sigma_{x_j}$ of the standard deviations of x_i and x_j. Thus, the elements along the diagonal where $i = j$ give the variances $\sigma_{x_i}^2 = \sigma_{x_i}\sigma_{x_j}$ and the off-diagonal elements give the covariances that approach zero if x_i and x_j are not correlated.

The $100(1-\alpha)\%$ confidence interval (or equivalently the $\alpha\%$ significance interval) on each $x_i \in \mathbf{X}$ is given by

$$x_i \pm (\sigma_{x_i})t(1-\alpha/2)_{n-m}, \tag{A.20}$$

where $t_{n-m}(1-\alpha/2)$ is the value of Student's t-distribution for the confidence level $(1-\alpha/2)$ and degrees of freedom $\nu = (n-m)$. The individual confidence intervals for x_i can also be readily combined into joint confidence regions (e.g. JENKINS and WATTS, 1968; DRAPER and SMITH, 1966), but they are difficult to visualize for $m > 3$.

Application of the variance propagation rule to **AX** shows that we can estimate the variance on each prediction $\hat{b}_i \in \hat{\mathbf{B}}(= \mathbf{AX} \simeq \mathbf{B})$ from

$$\sigma_{\hat{b}_i}^2 = \mathbf{A}_i(\mathbf{A}^t\mathbf{A})^{-1}\mathbf{A}_i^t\sigma_{\mathbf{B}}^2, \tag{A.21}$$

where $\mathbf{A}_i$ is the row vector defined by $\mathbf{A}_i = (a_{i1}\ a_{i2}\ \cdots\ a_{im})$. Thus, the $100(1-\alpha)\%$ confidence limits or error bars for the $\hat{b}_i$ at $\mathbf{A}_i$ can be obtained from

$$\hat{b}_i \pm \sigma_{\mathbf{B}}\left[\sqrt{\mathbf{A}_i(\mathbf{A}^t\mathbf{A})^{-1}\mathbf{A}_i^t}\right]t(1-\alpha/2)_{n-m}. \tag{A.22}$$

Another measure of the fit is the correlation coefficient r which when squared gives the coherency that indicates the percent of the observations b_i fitted by the predictions $\hat{b}_i$. The coherency in matrix form is given by

$$r^2 = (\mathbf{X}^t\mathbf{A}^t\mathbf{B} - n\bar{b}^2)/(\mathbf{B}^t\mathbf{B} - n\bar{b}^2), \tag{A.23}$$

where $\bar{b}$ is the mean value of the $b_i \in \mathbf{B}$. In general, our confidence in the model **AX** tends to increase as $r^2 \longrightarrow 1$.

The coherency (Equation A.23) reduces algebraically to

$$r^2 = \left[\sum_{i=1}^{n}(\hat{b}_i - \bar{b})^2\right] \Big/ \left[\sum_{i=1}^{n}(b_i - \bar{b})^2\right], \tag{A.24}$$

where $\sum_{i=1}^{n}(\hat{b}_i - \bar{b})^2$ is the sum of squares due to regression and $\sum_{i=1}^{n}(b_i - \bar{b})^2$ is the sum of squares about the mean. In general, the sum of squares about the regression is $\sum_{i=1}^{n}(b_i - \bar{b}_i)^2 = \sum_{i=1}^{n}(\hat{b}_i - \bar{b})^2 + \sum_{i=1}^{n}(b_i - \hat{b}_i)^2$ so that if **AX** is correct, then $[\sum_{i=1}^{n}(\hat{b}_i - \bar{b})^2] >> [\sum_{i=1}^{n}(b_i - \hat{b}_i)^2] \ni r^2 \longrightarrow 1$. Thus, an analysis of variance or ANOVA table can be constructed to test the statistical significance of the fit between $\hat{b}_i$ and b_i as shown in Table A.1. The null hypothesis that the model's predictions ($\hat{\mathbf{B}}$) do not significantly fit the observations (**B**) is tested by comparing the estimate $F = \mathrm{MS_R}/\mathrm{MS_D}$ from the ANOVA table with the critical value (F_c) from the Fisher distribution with degrees of freedom $\nu = (m-1), (n-m)$ at the desired level α of significance (or alternatively $(1-\alpha)$ of confidence). The hypothesis is rejected if $F > F_c$; otherwise there is no reason to reject the hypothesis based on the F-test.

TABLE A.1 The analysis of variance (ANOVA) table for testing the significance of a model's fit to data.

Error source	ν	Corrected sum of squares (CSS)	Mean squares	F-test
Regression error	$m-1$	$\mathbf{X}^t\mathbf{A}^t\mathbf{B} - \left(\sum_{i=1}^{n} b_i\right)^2/n$	$MS_R = CSS/\nu$	$F =$
Residual deviation	$n-m$	$\mathbf{B}^t\mathbf{B} - \mathbf{X}^t\mathbf{A}^t\mathbf{B}$	$MS_D = CSS/\nu$	MS_R/MS_D
Total error	$n-1$	$\mathbf{B}^t\mathbf{B} - \left(\sum_{i=1}^{n} b_i\right)^2/n$		

Sensitivity analysis

The above sections have illustrated how inversion determines an unknown set of discrete numbers (**X**) that relates a set of discrete numbers (**B**) observed from nature to a given set of discrete numbers (**A**) from the model (**AX**) that the interpreter presumes can account for the observations (**B**). Of course, once the forward model has been postulated, the solution (**X**) and its error statistics are routinely established in the least-squares sense from Equations A.16 and A.18, respectively.

The objective of inversion, however, is not simply to model the observations, but to develop a model that is effective in estimating additional key properties of the observations such as the related derivatives, integrals, values at unsurveyed coordinates, physical property variations, and other attributes. The solution in practice is never unique, but one of a range of solutions that is compatible with the ever-present error bars on the **A**- and **B**-coefficients and the round-off or truncation errors of the computations. Thus, a sensitivity analysis is commonly implemented to sort out the more effective solutions for a particular processing objective.

The assumed **A**-coefficient and computing errors can be particularly troublesome because an important requirement for obtaining the least-squares solution is that ($\mathbf{A}^t\mathbf{A}$) is non-singular or equivalently that $|\mathbf{A}^t\mathbf{A}| \neq 0$. Singularity of the system occurs if a row (or column) contains only zeros, or if two or more rows (or columns) are linear multiples of each other and thus co-linear. The more common situation in practice, however, is for the system to be near-singular or ill-conditioned, where in the computer's working precision the elements of a row (or column) are nearly zero, or two or more rows (or columns) are nearly linear multiples of each other and thus have a high degree of co-linearity.

For poorly conditioned systems, $|\mathbf{A}^t\mathbf{A}| \longrightarrow 0$ so that by Cramer's rule $\mathbf{X} \longrightarrow \infty$. Thus, near-singularity of the linear system is characterized by a solution with large and erratic values or large variance. In this instance, the solution is said to be unstable and its linear transformations beyond predicting the observations will be largely meaningless, even though it models the observed data with great accuracy.

An effective numerical approach for stabilizing solutions is to add a small positive constant (EV) to the diagonal elements of ($\mathbf{A}^t\mathbf{A}$) to dampen the effects of nearly co-linear or zero rows in the system. This approach is equivalent to obtaining the damped mean-squared error solution (Hoerl and Kennard, 1970a,b) given by

$$\mathbf{X_d} = (\mathbf{A}^t\mathbf{A} + (\mathrm{EV}) \times \mathbf{I})^{-1}\mathbf{A}^t\mathbf{B}, \tag{A.25}$$

with the variance (Marquardt, 1970)

$$\sigma^2_{\mathbf{X_d}} = (\mathbf{A}^t\mathbf{A} + (\mathrm{EV})\mathbf{I})^{-1}(\mathbf{A}^t\mathbf{A})(\mathbf{A}^t\mathbf{A} + (\mathrm{EV})\mathbf{I})^{-1}. \tag{A.26}$$

This approach is frequently called ridge regression and the constant (EV) is known by several names including damping factor, Marquardt parameter, and error variance. The last term reflects the fact that Equation A.25 is equivalent to adding into each **A**-coefficient random noise from the normal distribution with zero mean and standard deviation $\sqrt{\mathrm{EV}}$. Thus, the noise products in the off-diagonal elements of ($\mathbf{A}^t\mathbf{A}$) average out to zero, whereas along the diagonal they equal the noise or random error variance EV.

To reduce solution variance and improve the mean-squared error of prediction of ill-conditioned systems, a value for the error variance must be found that is just large enough to stabilize the solution, but small enough to maximize the predictive properties of the solution. A straightforward approach to quantifying the "optimal" value $\mathrm{EV_{opt}}$ is to develop trade-off diagrams like the example in Figure A.2. Here, Curve 1 compares the sum of squared residuals ($\mathrm{SSR}_j[\mathbf{B}, \hat{\mathbf{B}}(\mathrm{EV}_j)] = \sum_{i=1}^{n}[b_i - \hat{b}_i(\mathrm{EV}_j)]^2 = C$) between the observed and modeled data for a range of error variances (EV_j) that usually is established by trial-and-error. For any application of the model, solutions with predictions that do not deviate significantly from the observations are clearly desired such as those $\leq \mathrm{EV_{opt}}$.

Curve 2, on the other hand, is developed to test the relative behavior of the solution in estimating derivatives, integrals, interpolations, continuations, or other attributes $\hat{b}'(\mathrm{EV}_j)$ of the observations as a function of EV_j. This

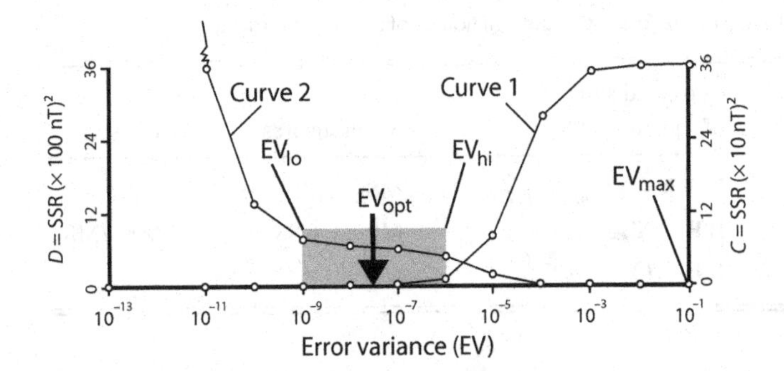

FIGURE A.2 Using performance curves to estimate the "optimal" value of error variance (EV_{opt}) for effective anomaly inversion. In this example, curves 1 and 2 with respective C- and D-ordinates were established to determine the EV-range (shaded) over which regional magnetic anomalies of southern China and India could be effectively reduced to the pole. Adapted from VON FRESE *et al.* (1988).

performance curve may be evaluated from norms such as $SSR_j[\hat{\mathbf{B}}'(EV_j), \hat{\mathbf{B}}'(EV_{max})] = D$, where $\hat{\mathbf{B}}'(EV_{max})$ are the predictions at an excessively damping and dominant error variance EV_{max}. Another useful form of this curve tests the behavior of the differences between successive predictions by $SSR_j[\hat{\mathbf{B}}'(EV_j), \hat{\mathbf{B}}'(EV_{j-1})]$. Curve 2 targets the range $EV_j \leq EV_{lo}$ where small changes in the **A**-coefficients yield radically changing, and thus highly unstable and suspect, predictions $\hat{\mathbf{B}}'(EV_j)$. The more stable and robust predictions over the range $EV_j \geq EV_{hi}$, on the other hand, are increasingly limited because they are increasingly less constrained by the observations as shown in Curve 1.

Thus, the trade-off diagram in Figure A.2 marks the range of error variances $EV_{lo} \leq EV_j \leq EV_{hi}$ for obtaining reliable stable predictions, where the "optimal" predictions $\hat{\mathbf{B}}'(EV_{opt})$ are taken for the error variance at the range's midpoint. In addition, constraints on the errors of $\hat{\mathbf{B}}'(EV_{opt})$ may be obtained by considering the difference map $[\hat{\mathbf{B}}'(EV_{lo}) - \hat{\mathbf{B}}'(EV_{hi})]$ (e.g. VON FRESE *et al.*, 1988). Note, however, that the optimal error variance for one application does not necessarily hold for other applications. Thus, the most successful inversion requires constructing the performance curve that is specific to the application (e.g. Curve 2) for comparison with Curve 1.

Note, however, that the range of acceptable error variances for an application establishes error bars on the solution coefficients via Equation A.26. Thus, the sensitivity analysis can be extended to investigate the range of geologically reasonable solution coefficients that satisfy the observed data.

In summary, the ridge regression approach enhances the conditioning of the complete $\mathbf{A}^t\mathbf{A}$ matrix by adding EV to the diagonal elements of the nearly co-linear or zero rows that is just large enough to stabilize the solution, yet small enough to maximize its predictive power. This approach substantially improves the accuracy of model predictions and optimizes available computer time and storage by implementing fast, efficient matrix decomposition algorithms like Cholesky factorization. Roughly 90% of the computational labor of inversion is taken up in evaluating the forward problem for the **A**-coefficients, whereas the Cholesky factorization of ($\mathbf{A}^t\mathbf{A}$) and subsequent determination of **X** account for the remaining 10%. Thus, for the cost of computing and storing the **A**-coefficients, roughly nine sets of solutions can be prepared for sensitivity analysis.

A contrasting approach for stabilizing the solution for improved prediction accuracy is to trim the $\mathbf{A}^t\mathbf{A}$ matrix of its offending rows by singular value decomposition. This approach involves the eigenvalue-vector decomposition of the $\mathbf{A}^t\mathbf{A}$ matrix to identify its non-zero rows that are effectively linearly independent within working precision. The improved solution is obtained from the generalized inverse that is constructed from the maximum subset of linearly independent rows in the system (e.g. LANCZOS, 1961; JACKSON, 1972; WIGGINS, 1972; GUBBINS, 2004). However, the computational labor in implementing this approach is several times greater than that required to process the full $\mathbf{A}^t\mathbf{A}$ matrix with fast linear equation solvers. Hence, this approach is often impractical to apply at the large scales of most modern gravity and magnetic inversions and is not considered further.

A.4.3 Spectral analysis

The above matrix procedures can be extended to the inversion of gridded gravity and magnetic anomaly data for their representative sine and cosine or spectral models that, in turn, can be evaluated for the differential and integral properties of the geopotential data. Spectral analysis requires that the data satisfy the Dirichlet conditions, whereby the signal $\mathbf{B}(t)$ in the interval (t_1, t_2) must be single-valued, and have only a finite number of maxima, minima, and discontinuities, and finite energy (i.e. $\int_{t_1}^{t_2} |\mathbf{B}(t)|\, dt < \infty$). Note that t is a dummy parameter representing space or time or any other independent variable against which the dependent variable **B** is mapped. Most mappable data sets including gravity and magnetic fields satisfy the Dirichlet conditions, and thus can be converted by inversion into representative sine and cosine series for analysis.

Spectral analysis includes the inverse processes of Fourier analysis and Fourier synthesis. Fourier analysis decomposes the t-data domain signal $\mathbf{B}(t)$ into representative wave functions in the f-frequency domain by taking the Fourier transform (FT) of $\mathbf{B}(t)$ given by

$$\mathcal{B}(f) = \int_{-\infty}^{\infty} \mathbf{B}(t)e^{-j(2\pi f)t}\,dt = \int_{-\infty}^{\infty} \mathbf{B}(t)\cos(\omega t)\,dt - j \times \int_{-\infty}^{\infty} \mathbf{B}(t)\sin(\omega t)\,dt, \quad (A.27)$$

where the imaginary number $j = \sqrt{-1}$. The wave expression of the exponential follows Euler's formula

$$e^{\pm j\omega} = \cos(\omega) \pm j\sin(\omega), \quad (A.28)$$

with the angular frequency $\omega = 2\pi f = 2\pi/\lambda$ for linear frequency f and wavelength $\lambda = 1/f$.

The spectral analysis literature is complicated by the variety of notations used to describe the properties of the Fourier transform. For example, the cosine and sine integrals in Equation A.27 are also known as the cosine transform (CT) and sine transform (ST), respectively, so that the Fourier transform may be expressed as the complex number

$$\mathcal{B}(f) = \mathrm{CT}(f) - j\mathrm{ST}(f). \quad (A.29)$$

As shown in Figure A.3, $\mathcal{B}(f)$ can be uniquely mapped into the (CT, ST)-plane using polar coordinates (r_f, θ_f), where

$$\mathrm{CT}(f) = r_f\cos(\theta_f)$$

$\Longrightarrow$ **Cosine transform** or the real part ($\equiv R(f)$) of $\mathcal{B}(f)$,

$$\mathrm{ST}(f) = r_f\sin(\theta_f)$$

$\Longrightarrow$ **Sine transform** or the imaginary part ($\equiv I(f)$) of $\mathcal{B}(f)$,

$$r_f^2 = |\mathcal{B}(f)|^2 = \mathrm{CT}^2(f) + \mathrm{ST}^2(f)$$

$\Longrightarrow$ **Power spectrum** of $\mathcal{B}(f)$,

$$r_f = |\mathcal{B}(f)| = \sqrt{\mathrm{CT}^2(f) + \mathrm{ST}^2(f)}$$

$\Longrightarrow$ **Amplitude spectrum** of $\mathcal{B}(f)$, and

$$\theta_f = \tan^{-1}[-\mathrm{ST}(f)/\mathrm{CT}(f)]$$

$\Longrightarrow$ **Phase spectrum** of $\mathcal{B}(f)$. (A.30)

From these relationships, the Fourier transform of Equation A.27 also has the complex polar coordinate expression

$$\mathcal{B}(f) = r_f[\cos(\theta_f)] - j[\sin(\theta_f)] = r_f e^{-j\theta_f} = |\mathcal{B}(f)|e^{-j\theta_f}. \quad (A.31)$$

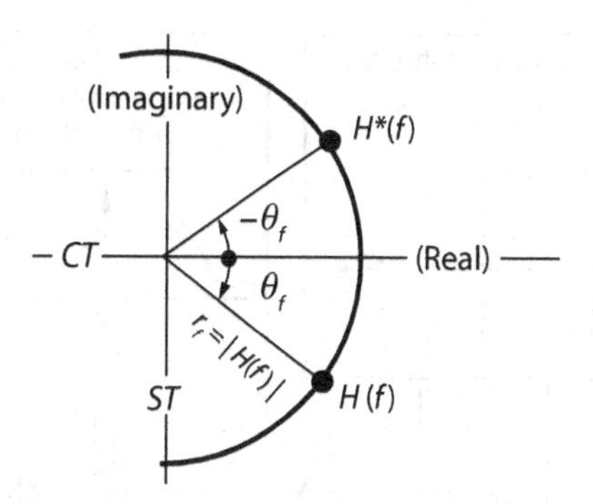

FIGURE A.3 The projection of the Fourier transform $H(f)$ and its complex conjugate $H^*(f)$ in the polar complex coordinate (CT, ST)-plane.

The phase angle θ_f satisfies $-\pi < \theta_f \leq \pi$ so that any integral multiple of 2π can be added to θ_f (in radians) without changing the value of $\mathcal{B}(f)$.

The complex conjugate $\mathcal{B}^*(f)$ is just the transform $\mathcal{B}(f)$ with the polarity of its imaginary component reversed by its reflection about the real axis, as shown in Figure A.3. Thus, in polar coordinate form,

$$\mathcal{B}^*(f) = \mathrm{CT}(f) + j\mathrm{ST}(f) = r_f\cos(\theta_f) + j\sin(\theta_f) = r_f e^{j\theta_f} = |\mathcal{B}(f)|e^{j\theta_f}, \quad (A.32)$$

where $|\mathcal{B}^*(f)| = |\mathcal{B}(f)| = r_f$ and $\theta^* = \tan^{-1}[\mathrm{ST}(f)/\mathrm{CT}(f)] = -\theta_f$. In addition, $\mathcal{B}^*(f)\mathcal{B}(f)$ and $\sqrt{\mathcal{B}^*(f)\mathcal{B}(f)}$ are the respective power and amplitude spectra.

Fourier synthesis, on the other hand, is the inverse operation that reconstitutes the t-data domain signal $\mathbf{B}(t)$ from its wave model $\mathcal{B}(f)$ in the f-frequency domain. Fourier synthesis is achieved by taking the inverse Fourier transform (IFT)

$$\mathbf{B}(t) = \int_{-\infty}^{\infty} \mathcal{B}(f)e^{+j(2\pi t)f}\,df. \quad (A.33)$$

Figure A.4 schematically illustrates the analysis of the simple sinusoid in the left panel into the two sine waves in the middle panel that when summed synthesize a good facsimile of the original signal. The spectral properties of the analyzed sine wave components are commonly presented in a stick diagram or periodogram like the one illustrated in the right panel, which gives the amplitudes of the positive and negative half-periods of the component sine waves. Note that the symmetry in these amplitudes translates into considerable computational efficiency because only half of

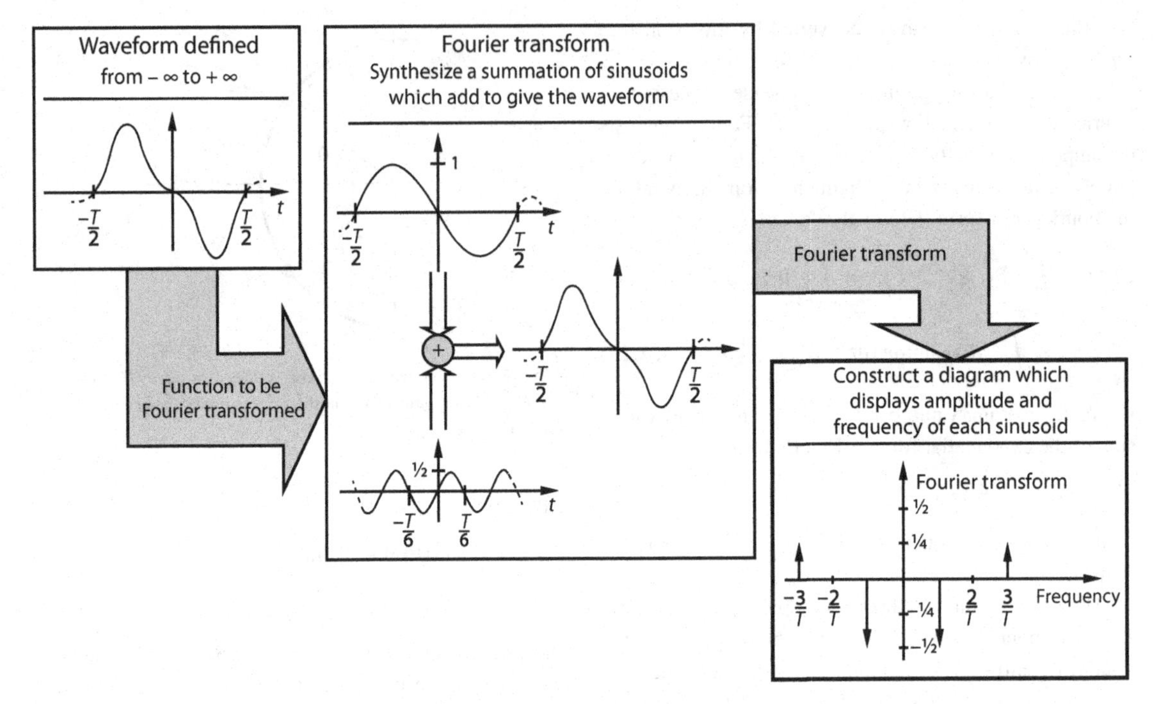

FIGURE A.4 Interpretation of the Fourier transform for a simple sinusoidal wave. After BRIGHAM (1974), by permission of Pearson Education, Inc., Upper Saddle River, New Jersey.

them are necessary for establishing the full spectral model of the data.

Notational complexities in the spectral analysis literature can result, however, from the different forms of normalization that may be applied to the forward and inverse Fourier transforms. For example, the generic transforms $\mathcal{B}(\omega) = a_1 \int_{-\infty}^{\infty} \mathbf{B}(t)e^{-j\omega t}dt$ and $\mathbf{B}(t) = a_2 \int_{-\infty}^{\infty} \mathcal{B}(\omega)e^{j\omega t}d\omega$ require that $a_1 \times a_2 = 1/2\pi$. This normalization in the literature is variously achieved by setting $a_1 = 1$ and $a_2 = 1/2\pi$, or $a_1 = a_2 = 1/\sqrt{2\pi}$, or $a_1 = 1/2\pi$ and $a_2 = 1$.

Analytical transforms

In the previous section, it was shown that the analytical integral transform, and its cosine and sine transforms (CT, ST), and amplitude and phase spectra ($|\mathcal{B}(f)|, \theta_f$) provide three equivalent representations of data. To illustrate these and other essential properties of the analytical transform that carry over to gravity and magnetic data analysis, consider now the simple example of the t-data domain function in Figure A.5(a) given by

$$\mathbf{B}(t) = \beta e^{-\alpha t} \ \forall \quad t \geq 0$$
$$= 0 \quad \forall \quad t < 0. \tag{A.34}$$

Taking the analytical Fourier transform of $\mathbf{B}(t)$ gives

$$\mathcal{B}(f) = \int_0^\infty \beta e^{-\alpha t} e^{-j2\pi f t} dt = \beta \int_0^\infty e^{-(\alpha + j2\pi f)t} dt.$$
$$= -[(\beta e^{-(\alpha + j2\pi f)t})/(\alpha + j2\pi f)]|_0^\infty = \frac{\beta}{(\alpha + j2\pi f)}. \tag{A.35}$$

The last expression above is the most compact form of the transform and thus the most likely one to be reported in the literature. To expand it into its equivalent (CT, ST) and ($|\mathcal{B}(f)|, \theta_f$) representations, the integral transform in Equation A.35 must be separated into its real and imaginary components ($R(f), I(f)$). This can be done by multiplying the integral transform with the special form of unity given by the ratio of the complex conjugate of the denominator to itself (i.e. $[\alpha - j2\pi f]/[\alpha - j2\pi f]$) so that

$$R(f) = \beta\alpha/(\alpha^2 + (2\pi f)^2) \quad = \mathrm{CT}(f),$$
$$I(f) = -2\pi f\beta/(\alpha^2 + (2\pi f)^2) = \mathrm{ST}(f),$$
$$|\mathcal{B}(f)| = \beta/\sqrt{\alpha^2 + (2\pi f)^2} \quad = \textbf{amplitude spectrum}, \text{ and}$$
$$\theta_f = \tan^{-1}(-2\pi f/\alpha) \quad = \textbf{phase spectrum}. \tag{A.36}$$

Figures A.5(b) and A.5(c) respectively illustrate the equivalent (CT(f), ST(f))- and ($|\mathcal{B}(f)|, \theta_f$)-representations of the integral transform. Note also that the cosine transform

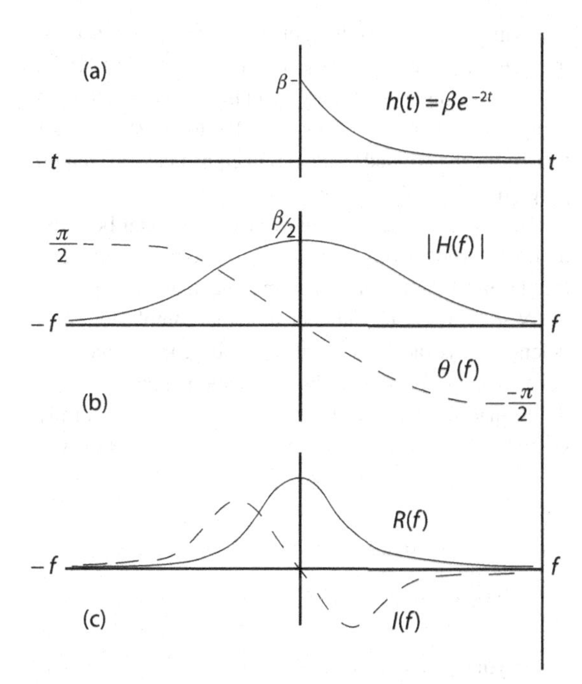

FIGURE A.5 Data and frequency domain representations of Equation A.36.

(CT(f)) and amplitude spectrum ($|\mathcal{B}(f)|$) are both a symmetric or even function with $R(f) = R(-f)$, whereas the sine transform (ST(f)) and phase spectrum (θ_f) are an antidiametric or odd function with $I(f) = -I(-f)$ that integrates to zero over symmetrical limits $(-f, f)$.

Consider now the Fourier synthesis of $\mathbf{B}(t)$ from its transform $\mathcal{B}(f)$ in Equation A.35, which can be expanded as

$$\mathcal{B}(f) = \frac{\beta}{(\alpha + j2\pi f)} = \frac{\beta\alpha}{\alpha^2 + (2\pi f)^2} - j\frac{2\pi f\beta}{\alpha^2 + (2\pi f)^2}. \tag{A.37}$$

Taking the inverse Fourier transform (IFT) of $\mathcal{B}(f)$ and expanding by Euler's formula gives

$$\hat{\mathbf{B}}(t) = \int_{-\infty}^{\infty}\left[\frac{\beta\alpha\cos(2\pi ft)}{\alpha^2 + (2\pi f)^2} + \frac{2\pi f\beta\sin(2\pi ft)}{\alpha^2 + (2\pi f)^2}\right]\mathrm{d}f$$
$$+ j\int_{-\infty}^{\infty}\left[\frac{\beta\alpha\sin(2\pi ft)}{\alpha^2 + (2\pi f)^2} - \frac{2\pi f\beta\cos(2\pi ft)}{\alpha^2 + (2\pi f)^2}\right]\mathrm{d}f. \tag{A.38}$$

Now, the imaginary integral above goes to zero because the product of an even and odd function or vice versa is an odd function. However, the real integral contains two even functions because the product of two even functions or two odd functions is even. Thus, the IFT reduces to the real term

$$\hat{\mathbf{B}}(t) = \int_{-\infty}^{\infty}\left[\frac{\beta\alpha\cos(2\pi ft)}{\alpha^2 + (2\pi f)^2} + \frac{2\pi f\beta\sin(2\pi ft)}{\alpha^2 + (2\pi f)^2}\right]\mathrm{d}f. \tag{A.39}$$

Standard tables of integrals show that

$$\int_{-\infty}^{\infty}\frac{\cos(ax)}{b^2 + x^2}\mathrm{d}x = \frac{\pi}{b}\mathrm{e}^{-ab}\ \forall\ a > 0, \text{ and} \tag{A.40}$$

$$\int_{-\infty}^{\infty}\frac{x\sin(ax)}{b^2 + x^2}\mathrm{d}x = \pi\mathrm{e}^{-ab}\ \forall\ a > 0,$$

so that the IFT becomes

$$\hat{\mathbf{B}}(t) = \frac{\beta\alpha}{(2\pi)^2}\left[\frac{\pi}{(\alpha/2\pi)}\mathrm{e}^{-(2\pi t)(\alpha/2\pi)}\right] + \frac{2\pi\beta}{(2\pi)^2}\left[\pi\mathrm{e}^{-(2\pi t)(\alpha/2\pi)}\right] \text{ or}$$
$$= \frac{\beta}{2}\mathrm{e}^{-\alpha t} + \frac{\beta}{2}\mathrm{e}^{-\alpha t} = \beta\mathrm{e}^{-\alpha t} \quad \forall\ t > 0. \tag{A.41}$$

Accordingly, $\mathbf{B}(t)[\simeq \hat{\mathbf{B}}(t)]$ and $\mathcal{B}(f)$ form a Fourier transform pair that is commonly indicated by the double-arrow notation

$$\mathbf{B}(t) \Longleftrightarrow \mathcal{B}(f) \quad \text{or} \tag{A.42}$$

$$\beta\mathrm{e}^{-\alpha t} \Longleftrightarrow \frac{\beta}{\alpha + j2\pi f} = \frac{\beta}{\sqrt{\alpha^2 + (2\pi f)^2}}\mathrm{e}^{j[\tan^{-1}(-2\pi f/\alpha)]}.$$

Table A.2 summarizes additional analytical properties of the Fourier transform (FT). The mathematical operations are obtained by taking the FT of the t-data domain operation or the IFT of the f-frequency domain operation. They are useful in considering the design of transforms and the related filtering operations. For example, any set of gravity or magnetic anomalies describes a real function, which accordingly has the transform consisting of both real (i.e. cosine transform) and imaginary (i.e. sine transform) coefficients. The averaging operator for smoothing anomaly data, on the other hand, is an even function of real numbers where only the real coefficients of the transform are required in the design and application of this filter.

Numerical transforms

The bulk of the extensive literature on Fourier transforms deals with their applications to continuously differentiable analytical functions. Most gravity and magnetic applications, however, involve inaccurately measured, discrete, gridded samples of unknown analytical functions that require the application of the numerical discrete Fourier transform (DFT). Developed mostly in the latter half of the twentieth century with the advent of digital electronic computing, the DFT can model any gridded data set. This

TABLE A.2 Analytical properties of the Fourier transform.

Data domain	$\Longleftrightarrow$	Frequency domain
Addition	$\Longleftrightarrow$	Addition
$\mathbf{B}(t) \pm \mathbf{C}(t)$	$\Longleftrightarrow$	$\mathcal{B}(f) \pm \mathcal{C}(f)$
Multiplication	$\Longleftrightarrow$	Convolution
$\mathbf{B}(t) \times \mathbf{C}(t)$	$\Longleftrightarrow$	$\mathcal{B}(f) \otimes \mathcal{C}(f)$
Convolution	$\Longleftrightarrow$	Multiplication
$\mathbf{B}(t) \otimes \mathbf{C}(t)$	$\Longleftrightarrow$	$\mathcal{B}(f) \times \mathcal{C}(f)$
Differentiation	$\Longleftrightarrow$	Differentiation
$\partial^p \mathbf{B}(t)/\partial t^p$	$\Longleftrightarrow$	$[j2\pi f]^p \mathcal{B}(f)$
Integration	$\Longleftrightarrow$	Integration
$\underbrace{\int\int\cdots\int}_{p} \mathbf{B}(t)\mathrm{d}t^p$	$\Longleftrightarrow$	$[j2\pi f]^{-p} \mathcal{B}(f)$
Symmetry	$\Longleftrightarrow$	Symmetry
$\mathcal{B}(t)$	$\Longleftrightarrow$	$\mathbf{B}(-f)$
Data scaling	$\Longleftrightarrow$	Inverse scale change
$\mathbf{B}(ct)$	$\Longleftrightarrow$	$\frac{1}{\lvert c\rvert}\mathcal{B}\left(\frac{f}{c}\right)$
Inverse scale change	$\Longleftrightarrow$	Frequency scaling
$\frac{1}{\lvert c\rvert}\mathbf{B}\left(\frac{t}{c}\right)$	$\Longleftrightarrow$	$\mathcal{B}(cf)$
Data shifting	$\Longleftrightarrow$	Phase shift
$\mathbf{B}(t \pm t_o)$	$\Longleftrightarrow$	$\mathcal{B}(f)\mathrm{e}^{\pm j2\pi f t_o}$
Data modulation	$\Longleftrightarrow$	Frequency shifting
$\mathbf{B}(t)\mathrm{e}^{\pm j2\pi t f_o}$	$\Longleftrightarrow$	$\mathcal{B}(f \mp f_o)$
Even function	$\Longleftrightarrow$	Real (R)
$\mathbf{B}_e(t)$	$\Longleftrightarrow$	$\mathcal{B}_e(f) = R_e(f)$
Odd function	$\Longleftrightarrow$	Imaginary (I)
$\mathbf{B}_o(t)$	$\Longleftrightarrow$	$\mathcal{B}_o(f) = jI_o(f)$
Real function	$\Longleftrightarrow$	Real even, Imaginary odd
$\mathbf{B}(t) = \mathbf{B}_r(t)$	$\Longleftrightarrow$	$\mathcal{B}(f) = R_e(f) + jI_o(f)$
Imaginary function	$\Longleftrightarrow$	Real odd, Imaginary even
$\mathbf{B}(t) = j\mathbf{B}_i(t)$	$\Longleftrightarrow$	$\mathcal{B}(f) = R_e(f) + jI_o(f)$

numerical modeling approach is also more straightforward to implement than analytical Fourier transforms and thus has wide application in modern data analysis.

The DFT is a streamlined adaptation of the general finite Fourier series (FS) used to represent non-analytical, discretely sampled, finite-length signals. In the general case of non-uniformly sampled data where the FS is applied, the full $\mathbf{A}^t\mathbf{A}$ matrix must be computed. However, in electronic processing, the data are represented as an orthogonal data grid for which the $\mathbf{A}^t\mathbf{A}$ matrix simply becomes the diagonal matrix because the off-diagonal elements involve cross-products of orthogonal wave functions that are zero (e.g. Jenkins and Watts, 1968; Davis, 1986). The language of these computations also becomes comparably abbreviated, invoking the term wavenumber to describe the gridded frequency or equivalent wavelength properties of the data array.

The simplicity of wavenumber analysis can be illustrated by considering the DFT of a finite-length data profile $\mathbf{B}(t_n)$ uniformly spanning the interval $0 \le t_n \le \Delta t(N-1)$. Here, the number of signal samples or coefficients, N, is the fundamental period of the signal, and Δt is the uniform sampling or station interval. Thus, the signal consists of the sequence of N uniformly spaced coefficients $b_n \;\forall\; n = 0, 1, 2, \ldots, (N-1)$ given by

$$\mathbf{B}(t_n) \equiv (b_0 = b[t_0], b_1 = b[t_1], b_2 = b[t_2], \ldots, b_{N-1} = b[t_{N-1}]). \qquad \text{(A.43)}$$

In applying the DFT, it is assumed that the frequency interval of N samples is the fundamental period over which the signal is uniformly defined and that outside this interval the signal repeats itself periodically. The example in Figure A.6 shows the relevant book keeping details for a uniformly sampled signal of $N = 21$ coefficients.

To Fourier analyze the signal $\mathbf{B}(t_n)$, its one-dimensional discrete Fourier transform $\mathcal{B}(\omega_k)$ is estimated using the discrete frequency variable $\omega_k = k\Delta\omega = k(2\pi/N) \;\forall\; k = 0, 1, 2, \ldots, (N-1)$. Here, $k \equiv$ wavenumber at which the coefficients $\mathsf{b}_k \in \mathcal{B}(\omega_k)$ are estimated by

$$\hat{\mathsf{b}}_k = \sum_{n=0}^{N-1} b_n \mathrm{e}^{-j(n)\frac{2\pi}{N}(k)} \qquad \forall\; k = 0, 1, 2, \ldots, (N-1), \qquad \text{(A.44)}$$

where $\hat{\mathsf{b}}_k \simeq \mathsf{b}_k$. This equation effectively samples the continuous Fourier transform $\mathcal{B}(\omega)$ at the uniform frequency interval of $(2\pi/N)$ so that the wavenumber coefficients $\hat{\mathsf{b}}_k$ are actually the Fourier series coefficients to within the scale factor of $(1/N)$.

The wavenumber k relates the various uses of frequency and wavelength in different applications. For example, in gravity and magnetic anomaly analyses, the focus is principally on linear frequency f in the units of cycles per data interval Δt and wavelength (or period) λ $(= 1/f)$ in the units of the data interval Δt, whereas time series analysis commonly invokes the circular or angular frequency ω $(= 2\pi f)$ in the units of radians per data interval Δt. These various concepts are connected in terms of

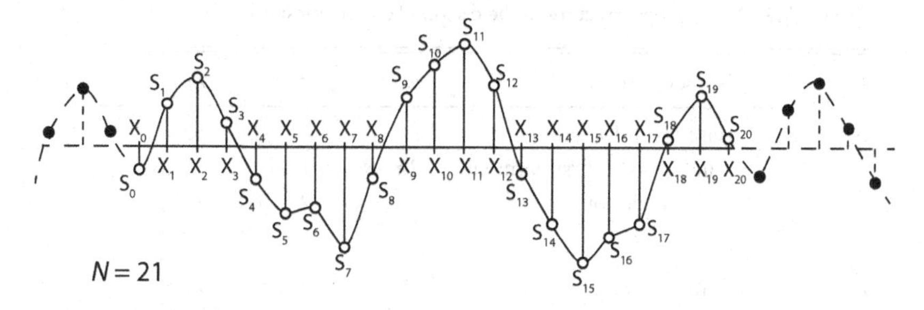

FIGURE A.6 The bookkeeping details in taking the Fourier transform of a 21-point signal. The transform assumes that the data (open circles) are infinitely repeated (black dots) on all sides of the signal.

the wavenumber k by

$$\omega_k = \left(\frac{2\pi}{N}\right)k \longrightarrow f_k = \left(\frac{k}{N}\right) \ni \lambda_k = \left(\frac{N}{k}\right)\Delta t = \frac{2\pi}{\omega_k}. \tag{A.45}$$

The harmonic frequencies defined by k fix the range of frequencies or wavelengths over which the DFT can analyze the signal $\mathbf{B}(t_n)$. The $k = 0$ component, for example, corresponds to the infinite wavelength or zero frequency base level of the signal, which is also sometimes referred to in electrical terms as the direct current or DC component. The mean or a low-order trend surface is commonly removed from the data to make the DC component zero so that the significant figures of the other coefficients are more sensitive to the remaining variations in the data.

The maximum finite wavelength component of the signal is the fundamental wavelength for $k = 1$ given by $\lambda_1 = N\Delta t$. This result seems counter-intuitive because the length of the signal with N coefficients spaced uniformly at the interval Δt should be $(N-1)\Delta t$. However, in applying the DFT, it is assumed that the signal repeats itself one data interval Δt following the position of the last term or coefficient, $b_{(N-1)}$, so that the actual length of the signal is indeed $N\Delta t$.

The highest-frequency (or minimum-wavelength) signal component, on the other hand, is for $k = (N/2)$ where $\omega_{N/2} = \pi$. This limit of frequency (or wavelength) resolution is called the Nyquist frequency (or wavelength). At higher wavenumbers, the coefficients $\mathfrak{b}_{k>(\frac{N}{2})}$ can be found from the lower wavenumber coefficients $\mathfrak{b}_{k\leq(\frac{N}{2})}$ because $\mathfrak{b}_{(\frac{N}{2}+1)} = \mathfrak{b}^{\star}_{(\frac{N}{2}-1)}, \mathfrak{b}_{(\frac{N}{2}+2)} = \mathfrak{b}^{\star}_{(\frac{N}{2}-2)}, \ldots, \mathfrak{b}_{(\frac{N}{2}+\frac{N}{2}-1)} = \mathfrak{b}^{\star}_{(1)}$. This folding or symmetry of coefficients makes for extremely efficient computation because only the relatively small number of coefficients up to and including the Nyquist wavenumber $k = N/2$ needs to be explicitly computed and stored to define the full DFT. Table A.3 summarizes the nomenclature, frequency, and wavelength attributes for each kth wavenumber coordinate of the DFT(1D).

In summary, the wavenumber coefficient estimates $\mathfrak{b}_k$ (Equation A.44) represent the discrete uniform sampling of the continuous transform $\mathcal{B}(f)$ at the interval $(1/N)$. The application of Euler's formula expands these coefficient estimates for their sine and cosine transforms, amplitude and phase spectra, and other components described in Equation A.30.

The discrete signal coefficients $b_n \in \mathbf{B}(t_n)$, on the other hand, are synthesized or modeled from

$$\hat{b}_n = \frac{1}{N}\sum_{k=0}^{N-1} \mathfrak{b}_k e^{+j(n)\frac{2\pi}{N}(k)} \qquad \forall\, n = 0, 1, 2, \ldots, (N-1), \tag{A.46}$$

where $\hat{b}_n \simeq b_n$. To within the scale factor $(1/N)$, $\hat{\mathbf{B}}(t_n)$ represents the Fourier series estimate for the sequence $\mathbf{B}(t_n)$ from the Fourier coefficient estimates $\mathfrak{b}_k$. Hence, $\hat{\mathbf{B}}(t_n)$ is a periodic sequence which estimates $\mathbf{B}(t_n)$ only for $0 \leq n \leq (N-1)$ and $[\mathbf{B}(t_n) \simeq]\ \hat{\mathbf{B}}(t_n) \Longleftrightarrow \hat{\mathcal{B}}(f_k)\ [\simeq \mathcal{B}(f_k)]$.

The discussion of the DFT so far has focused on its application to single-dimensioned profile data. Many studies, however, involve analyzing gravity and magnetic anomalies in 2D arrays or maps. Fortunately, the 1D DFT can be readily extended to the representation and analysis of arrays in two or more dimensions.

Consider, for example, the $(N \times M)$ coefficients $b_{n,m}$ that sample the 2D signal $\mathbf{B}(t_{n,m})$ at uniform intervals Δt_n and Δt_m, where $0 \leq n \leq (N-1)$, $0 \leq m \leq (M-1)$, and N, M, Δt_n, and Δt_m take on arbitrary values. To Fourier analyze the signal $\mathbf{B}(t_{n,m})$, the coefficients $\mathfrak{b}_{k,l} \in \mathcal{B}(f_{k,l})$ are estimated at wavenumber coordinates (k, l) from

$$\mathfrak{b}_{k,l} = \sum_{n=0}^{(N-1)}\sum_{m=0}^{(M-1)} b_{n,m} e^{-j(m)\frac{2\pi}{M}(l)} e^{-j(n)\frac{2\pi}{N}(k)}, \tag{A.47}$$

TABLE A.3 Wavenumber structure of the discrete Fourier transform.

k	Nomenclature	ω_k	f_k	λ_k
0	Base level; DC component	0	0	∞
1	Fundamental; first harmonic	$2\pi/N$	$1/N$	$N\Delta t$
2	Second harmonic	$2\times\omega_1$	$2\times f_1$	$\lambda_1/2$
$\vdots$	$\vdots$	$\vdots$	$\vdots$	$\vdots$
m	m-th harmonic	$m\times\omega_1$	$m\times f_1$	λ_1/m
$(N/2)-1$	$[(N/2)-1]$-th harmonic	$\pi-\omega_1$	$\pi-f_1$	$2\pi/(\pi-\omega_1)$
$N/2$	Nyquist	π	$1/2$	$2\times\Delta t$
$\vdots$	$\vdots$	$\vdots$	$\vdots$	$\vdots$
$N-1$	$[N-1]$-th harmonic	$-\omega_1$	$-f_1$	$-\lambda_1$

for all $0\le k\le(N-1)$ and $0\le l\le(M-1)$ so that $\hat{\mathsf{b}}_{k,l}\simeq\mathsf{b}_{k,l}$. On the other hand, the discrete signal coefficients $b_{n,m}\in\mathbf{B}(t_{n,m})$ are synthesized from

$$\hat{b}_{n,m}=\left(\frac{1}{N\times M}\right)\sum_{k=0}^{(N-1)}\sum_{l=0}^{(M-1)}\hat{\mathsf{b}}_{k,l}\mathrm{e}^{+j(m)\frac{2\pi}{M}(l)}\mathrm{e}^{+j(n)\frac{2\pi}{N}(k)},\tag{A.48}$$

for all $0\le n\le(N-1)$ and $0\le m\le(M-1)$ so that $\hat{b}_{n,m}\simeq b_{n,m}$ and $\mathbf{B}(t_{n,m})\simeq\hat{\mathbf{B}}(t_{n,m})\Longleftrightarrow\hat{\mathcal{B}}(f_{k,l})\simeq\mathcal{B}(f_{k,l})$.

Figures A.7 and A.8 illustrate the basic organizational details of the wavenumber spectra for an array of $(N=16)\times(M=32)$ anomaly values. For these Nyquist-centered spectra, the quadrants are arranged clockwise with the outside corners defining the base level or DC component of each quadrant. Gravity and magnetic data are real numbers so that quadrants 3 and 4 are the complex conjugates of quadrants 1 and 2, respectively (Figure A.8(a)). This symmetry of quadrants means that only one quadrant out of each of the two conjugate quadrant pairs needs to be computed to define the full spectrum. On the other hand, if the real numbers define an even function, then only one quadrant needs to be computed because the other three quadrants are symmetric to it (Figure A.8(b)). Thus, for even, real functions like a radially symmetric averaging operator for data smoothing, for example, the filter's design is considered in only a single spectral quadrant.

The Nyquist-centered arrangement of spectral quadrants facilitates computer processing objectives, but it is not ideal for displaying the spectrum's symmetry properties because the amplitudes near the Nyquist are marginal and difficult to follow visually. Thus, the literature commonly presents the spectrum in the higher-energy, DC-centered format with quadrants 1 and 2 transposed with quadrants 3 and 4, respectively. This type of spectral plot concentrates the higher-amplitude wavenumber components about the center of the spectrum to emphasize better the dominant energy patterns of the data.

Using the linear addition property of transforms (Table A.2), the dominant wavenumber components can be inversely transformed to estimate the corresponding data domain variations and the possible sources of this behavior. The amount of total signal power or energy that these wavenumber components account for is also quantified with the extension of this approach called Parseval's theorem. This theorem expresses the signal's power in terms of its wavenumber components so that for the DFT(1D), for example,

$$\sum_{n=0}^{N-1}b_n^2=(1/N)\sum_{k=0}^{N-1}\mathsf{b}_k^2.\tag{A.49}$$

Thus, if a solution for a subset of wavenumber components is offered that accounts for say 49% of the signal's power, then the solution can also be said to explain this amount of the signal's power or 70% of its energy.

The physical concepts of signal power and energy can also be expressed in terms of statistical variance and standard deviation, respectively, using the $k\ge1$ terms from Parseval's theorem. Specifically, the unbiased estimate of the signal's variance is

$$\sigma_{\mathbf{B}}^2=\frac{1}{N-1}\sum_{n=0}^{N-1}(b_n-\bar{b})^2=\frac{1}{N(N-1)}\sum_{k=1}^{N-1}\mathsf{b}_k^2,\tag{A.50}$$

where $\bar{b}$ is the mean value of the coefficients $b_n\in\mathbf{B}(t_n)$. If the mean has been removed from the signal, then Equations A.50 and A.49 are equivalent to within the scale factor of $[1/(N-1)]$.

The power and amplitude spectra of gravity and magnetic data typically exhibit considerable dynamic range

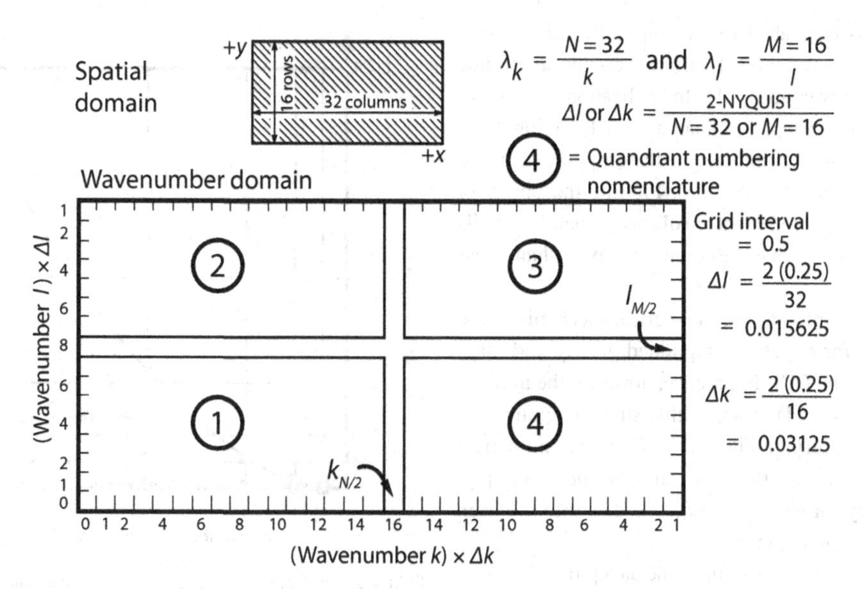

FIGURE A.7 The machine storage format for representing the discrete Nyquist-centered wavenumber transform of a 16×32 data array. Adapted from Reed (1980).

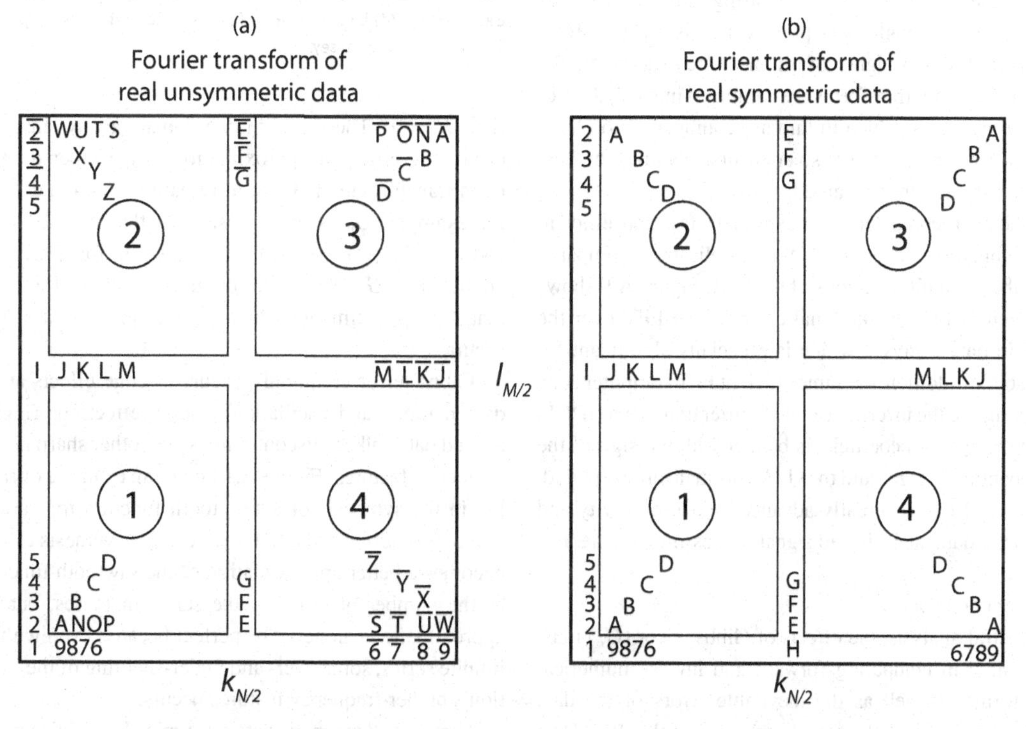

FIGURE A.8 Symmetry properties of the Nyquist-centered wavenumber spectrum for (a) non-symmetric and (b) symmetric real gridded data where nx and ny are the number of data in the x- and y-dimensions of the grid. Adapted from Reed (1980).

that is difficult to map out with linear contour intervals. Thus, to visualize the spectral variations most completely, it is common practice to contour the spectra logarithmically in decibels. Technically, the decibel (dB) is the base-10 logarithmic measure of the power of two signal amplitudes b_1 and b_2 defined by

$$(b_1^2/b_2^2)\,|_{\text{dB}} \equiv 10\log\,(b_1^2/b_2^2) = 20\log(b_1/b_2). \qquad \text{(A.51)}$$

Hence, 6 dB is equivalent to the amplitude ratio of 2 or power ratio of 4, whereas 12 dB reflects the amplitude ratio of 4 and power ratio of 16. In application, each spectral coefficient is simply evaluated with Equation A.51 assuming that the denominator of the amplitude ratio is normalized to unity (i.e. $b_2 \equiv 1$). Plotting the power or amplitude spectral coefficients in dB brings out the details of their variations much more comprehensively than linear contouring.

Up to this point, the basic numerical mechanics have been outlined for representing gridded gravity and magnetic data with the DFT. In practice, however, the numerical operations of the DFT are further streamlined into the fast Fourier transform (FFT). The FFT involves numerical engineering to evaluate the DFT using the computer operation of binary bit reversal to sort the data into the even numbered and odd numbered points.

For the FFT(1D), for example, the data profile is sorted into ordered even (i.e. $b_0, b_2, b_4, \ldots, b_{2n}$) and odd (i.e. $b_1, b_3, b_5, \ldots, b_{2n+1}$) point strings. The sorting is continued on the new data strings until strings are obtained containing only a single data point which is the transform of itself. Using weights of simple functions of N, the (N) single-point transforms are combined into $(N/2)$ two-point transforms, which in turn are combined into $(N/4)$ four-point transforms, and so on recursively until the final N-point transform is obtained.

The FFT computation is extremely fast and efficient involving, for example, $N \ln(N)$ multiplications compared with the N^2 multiplications of the DFT. Figure A.9 shows the savings of computational effort of the FFT over the DFT in data analysis, which is especially significant for $N \geq 64$. In addition, the same efficient FFT computer code can compute the inverse discrete Fourier transform (IDFT) from the spectral coefficients $\mathfrak{b}_k$, but with the sign of the exponential changed and the $(1/N)$ normalization applied. Thus, the FFT has greatly advanced modern gravity and magnetic data analysis and signal processing in general.

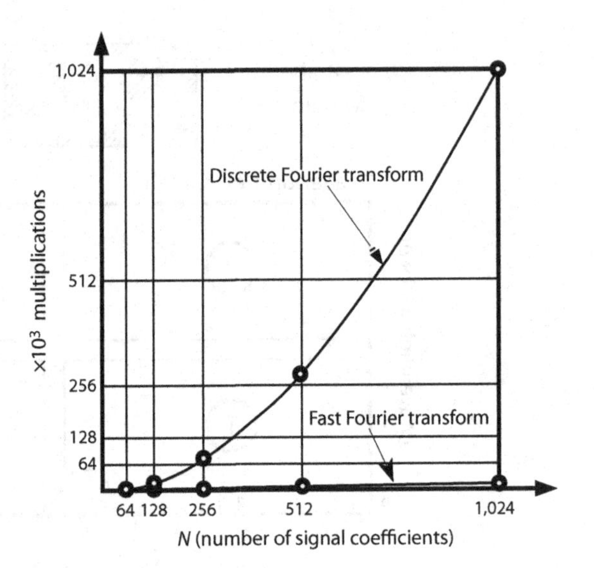

FIGURE A.9 Comparison of the number of multiplications required for calculating the DFT and the FFT. The computational efficiency of the FFT increases dramatically at about 64 signal coefficients and higher. Printed and electronically reproduced after Brigham (1974) by permission of Pearson Education, Inc., Upper Saddle River, New Jersey.

Numerical errors

In spectral analysis, the effects of Gibbs' error (G_e) must be limited in computing forward and inverse numerical transforms, as well as the inevitable errors of the data grid in representing the actual variations of the data. Data gridding errors can result in wavenumber leakage, aliasing, and resolution problems in estimating effective spectral properties.

(A) Gibbs' error Gibbs (1896) showed that even in the limit where $k \longrightarrow \infty$, the errors (G_e) in estimating the forward and inverse numerical transforms are never zero (i.e. $G_e^2 \neq 0$). The signal in one domain always involves oscillating errors of approximation in the other domain upon transformation. Thus, in Equations A.44 and A.46, for example, G_e is ever present so that $\hat{\mathfrak{b}}_k = \mathfrak{b}_k + G_e$ and $\hat{b}_n = b_n + G_e$, respectively. However, for nearly any application, G_e is or can be made sufficiently small that the approximations $\hat{\mathfrak{b}}_k \simeq \mathfrak{b}_k$ and $\hat{b}_n \simeq b_n$ are very accurate.

Gibbs' error commonly occurs as edge effects at the data margins and oscillating shadow effects or ringing around data spikes, discontinuities, and other sharp higher frequency features. Figure A.10 gives an example of ringing in the synthesis of a saw tooth function for various numbers of terms in the IDFT series. The synthesis clearly becomes a better approximation of the sawtooth function as the number of terms in the series increases, but the approximation can never be perfect because even with an infinite series, some over- and under-shooting of the function's higher-frequency features occurs.

Gravity and magnetic data sets typically exhibit smooth enough behavior that Gibbs' error is manifest mostly in edge effects involving a few data rows and columns along the data margins. The severity of Gibbs' error can be checked in practice by examining the residuals $[\mathbf{B}(t_n) - \hat{\mathbf{B}}(t_n)]$. Adding a rind (or border) of several data rows and columns to the data set will help to migrate the edge effects out of the study area. Where no real data are

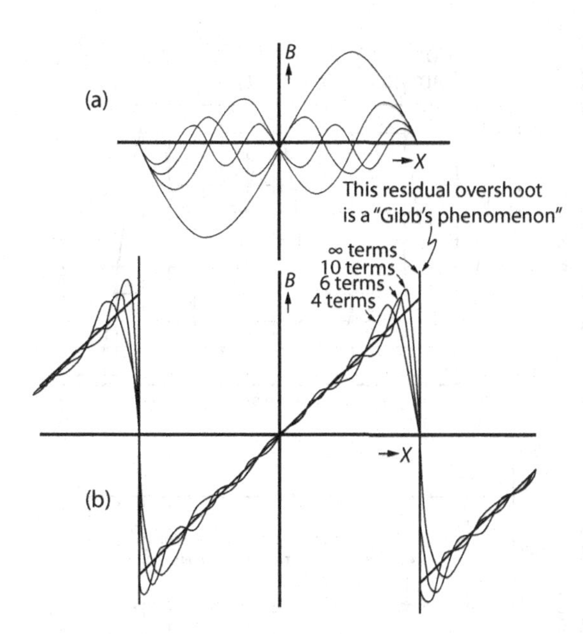

FIGURE A.10 Fourier synthesis of a sawtooth function. (a) The first four Fourier components of a sawtooth are $\sin(t)$, $-(1/2)\sin(2t)$, $(1/3)\sin(3t)$, and $-(1/4)\sin(4t)$. (b) Superposition of the Fourier components reconstitutes the sawtooth function. After SHERIFF (2002).

available, a synthetic data rind constructed by folding out the rows and columns along the margins of the data set can be tried. The most common approach, however, is to use window carpentry to smoothly taper the margins of the data set to zero or some other base level. JENKINS and WATTS (1968) summarize the statistical properties of the Bartlett, Hammming, Hanning, Parzen, Tukey, and other functions for window carpentry applications in spectral analysis.

(B) Wavenumber leakage Window carpentry to taper the edges of the data set also reduces wavenumber leakage or errors in the estimated spectrum that occur when the length of the gridded data set is not an integer multiple of the fundamental period of the data. This condition can result in sharp edge effects in the data domain that introduce significant Gibbs' error into the wavenumber components.

The signals in Figure A.11 illustrate the wavenumber leakage problem. The top left panel shows the 32 points of a single-frequency cosine wave digitized uniformly over the length that is 8 times its fundamental period, and the bottom left panel gives the correct frequency response of the digitized signal in the Nyquist-centered amplitude spectrum from its DFT. In the top right panel, on the other hand, the signal is digitized over the length that is 9.143 times the cosine's fundamental period. The requirement for the digitized signal to repeat itself at the fundamental period of the digitization scheme introduces a sharp edge effect with large Gibbs' error. The bottom right panel shows that this error significantly corrupts all wavenumber components of the amplitude spectrum. These erroneous amplitudes reflect the DFT's response to modeling the spectral properties of the artificial edge effect that the digitization scheme introduced.

Figure A.12 shows the effect of window carpentry for significantly reducing the Gibbs' error from the artificial data edge and improving the estimate of the cosine wave's spectral properties. Here, the upper right panel gives the tapered cosine wave $\mathbf{TB}(t) = \mathbf{B}(t) \times \mathbf{H}(t)$, where the Hanning window function $\mathbf{H}(t)$ is in the upper left panel and $\mathbf{B}(t) = \cos(\frac{2\pi}{9.143}t)$ is the original signal in the upper right panel of A.11. In other words, the tapered cosine wave $\mathbf{TB}(t)$ was obtained by multiplying each coefficient of the signal $\mathbf{B}(t)$ against the corresponding value of the Hanning window $\mathbf{H}(t)$.

The tapering has greatly muted the artificial edge effect and related Gibbs' error so that the DFT of the tapered signal gives the much improved amplitude spectrum in the lower right panel. In general, spectral analyses of gravity and magnetic data commonly employ window carpentry to limit wavenumber leakage, as well as edge effects in the synthesized data, and other Gibbs' errors generated by the margins of the data sets.

(C) Wavenumber aliasing Gridding at an interval that is too large to capture significant shorter wavelength features of the data can also corrupt the spectrum. Shannon's sampling theorem (SHANNON, 1949) points out that the shortest wavelength that can be resolved in a gridded data set is given by the Nyquist wavelength $\lambda_{N/2} = 2 \times \Delta t$, where Δt is the grid interval. In other words, proper gridding of a data set requires that the Nyquist frequency $f_{N/2} = 1/\lambda_{N/2}$ of gridding be at least twice the highest frequency in the data.

Any data feature with frequency greater than $f_{N/2}$ by the amount Δf will be represented in the data grid by an artificial longer wavelength feature with frequency $(f_{N/2} - \Delta f)$. This artificial feature is said to be the alias of the actual shorter-wavelength data feature with frequency $(f_{N/2} + \Delta f)$. Thus, aliasing is purely an artifact of a sampling rate that is too low to represent the higher-frequency data features. Inclusion of aliased features yields an incorrect data grid and a spectrum that result in data analysis errors.

Figure A.13 gives a simple example of signal aliasing from ROBINSON and TREITEL (1980). The solid wave is part of a 417 Hz signal that is to be gridded. The open dots

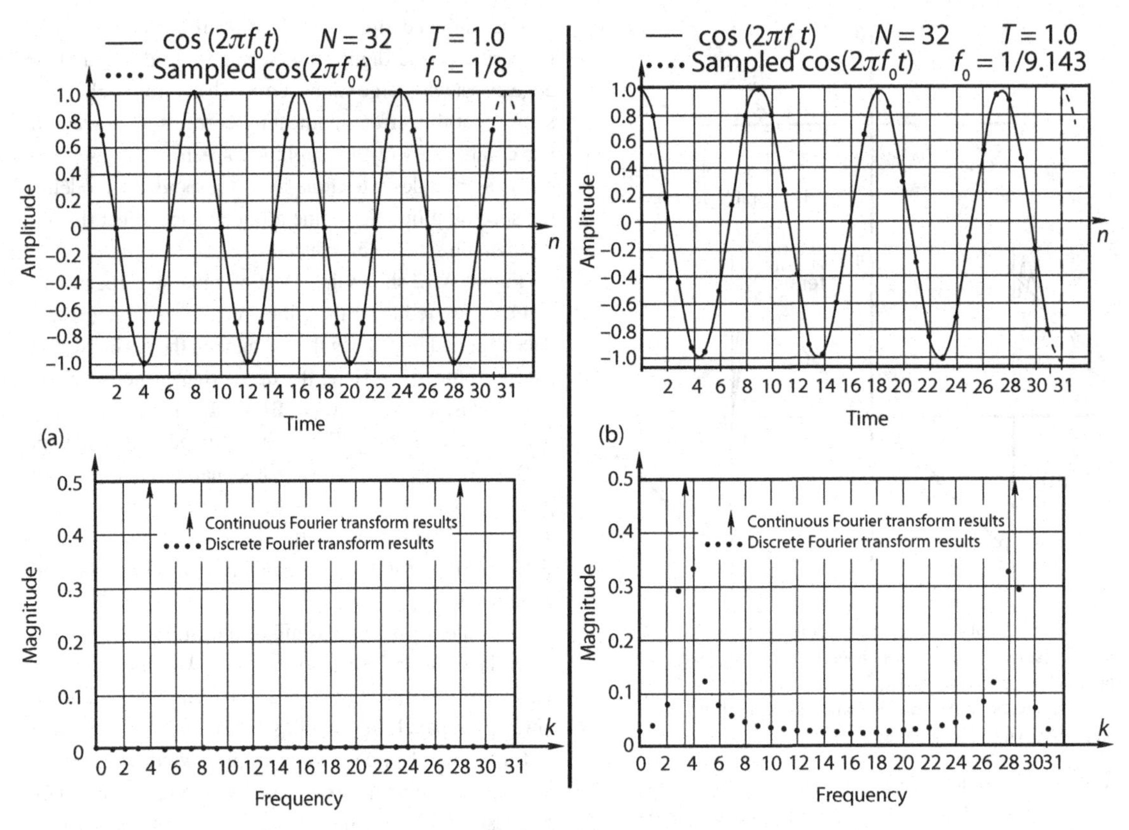

FIGURE A.11 Comparison of the Fourier transform for a cosine waveform truncated over an interval that is an integer multiple of the period (a) and where the truncation interval is not a multiple of the period (b). Printed and electronically reproduced after BRIGHAM (1974) by permission of Pearson Education, Inc., Upper Saddle River, New Jersey.

indicate the grid samples collected at the rate of 500 Hz. However, the grid samples define the completely erroneous dashed wave of frequency 83 Hz because the sampling rate is much too low. The solid line folded in accordion fashion in the lower diagram gives the aliased frequencies for data frequencies above the sampling Nyquist rate $f_{N/2} = 250$ Hz. Thus, to avoid aliasing in gridding the 417 Hz signal, gridding rates of at least 834 Hz and higher are necessary. Doubling the 500 Hz sampling rate, for example, would sample at the open dots and midway between them for an effective grid of the 417 Hz signal.

Establishing an appropriate sampling rate that avoids aliasing for any set of gravity or magnetic observations is difficult in practice because the frequency components in the observations are usually poorly known. Excessive sampling of the data set is always possible, but this can lead to unacceptable data processing requirements. Another, perhaps more effective, approach is to grid and plot the data at increasingly smaller grid intervals. An appropriate grid interval yields a data plot that does not change appreciably at smaller grid intervals.

(D) Wavenumber resolution The wavelength or equivalent frequency components between the harmonics are impossible to resolve with the DFT. Thus, the DFT reduces the degrees of freedom available to the problem by half because it transforms the N data coefficients into $(N/2)$ unique wavenumber coefficients. Most gravity and magnetic data sets are large enough that the degrees of freedom lost in using the DFT are not a problem for achieving the objectives of the studies. However, the loss can seriously limit studies involving a small number ($N \leq 10$) of observations. Here, however, the statistical maximum entropy estimate of the power spectrum can be invoked, which can provide N unique wavenumber coefficients at computational efficiencies that approach those of the FFT (e.g. CLAERBOUT, 1976; PRESS *et al.*, 2007).

Maximum entropy spectral analysis

The theory of maximum entropy estimation of power spectra for geophysical profiles is detailed in BURG (1975), ULRYCH and BISHOP (1975), SAAD (1978) and several other more recent publications. Compared with the

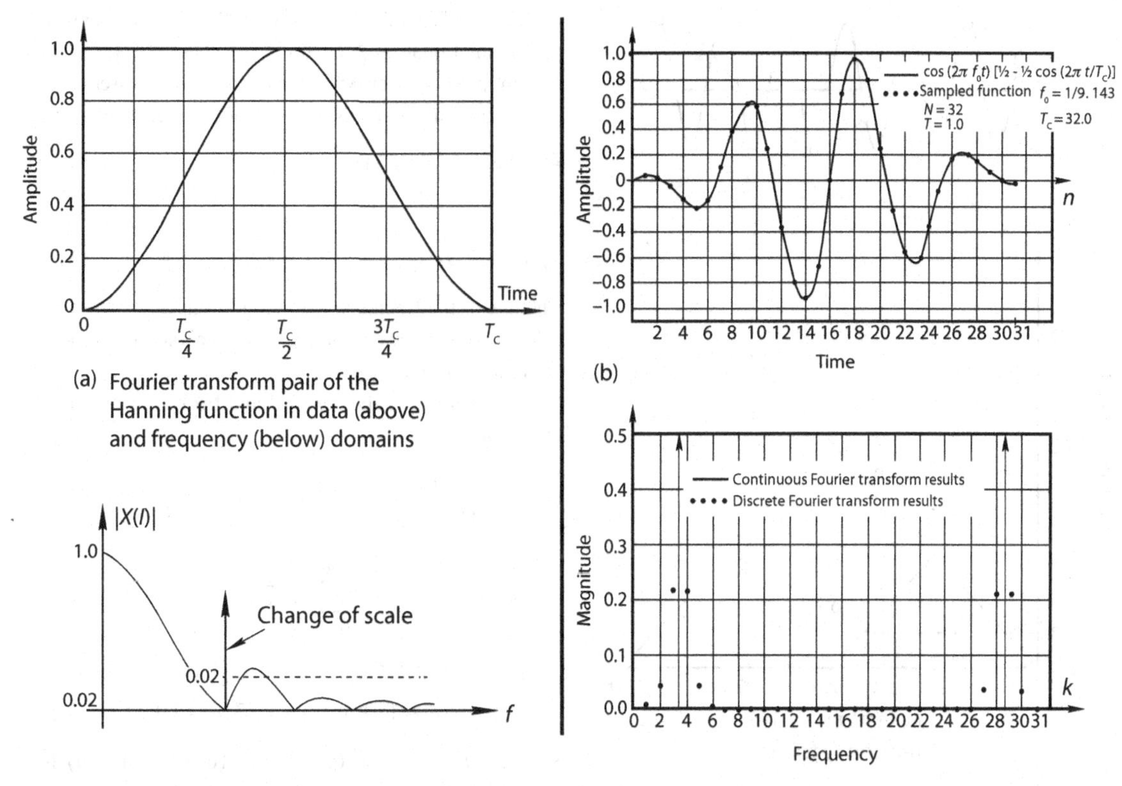

FIGURE A.12 Application of the Hanning function (a) to taper the poorly sampled cosine wave in Figure A.11(b) greatly improves the accuracy of the cosine's frequency response (b). Printed and electronically reproduced after BRIGHAM (1974) by permission of Pearson Education, Inc., Upper Saddle River, New Jersey.

standard fast Fourier transform, maximum entropy spectral analysis (MESA) is as numerically efficient to implement, but can increase the resolving power of data spectral models with no need for window carpentry. In addition, MESA appears to be insensitive to random noise and hence capable of detecting weak signals in the data.

MESA is based on modeling or predicting an observation in a uniformly sampled signal from other observations weighted by the α-coefficients of the prediction filter that maximizes the integrated natural logarithmic power or entropy of the predictions. In particular, the MESA synthesis of the mth observation is

$$b_m = \sum_{i=1}^{N-1} \alpha_i b_{m-i} + \varepsilon_m, \qquad (A.52)$$

where $(\alpha_1, \alpha_2, \ldots, \alpha_{N-1})$ is the prediction error filter (PEF) of length $(N-1)$. Rearranging this equation yields the corresponding prediction error given by

$$\varepsilon_m = b_m - \sum_{i=1}^{N-1} \alpha_i b_{m-i} = \sum_{i=0}^{N-1} \gamma_{i+1} b_{m-i}, \qquad (A.53)$$

where $(\gamma_1, \gamma_2, \ldots, \gamma_N)$ is the PEF of length (N) with $\gamma_1 = 1$ and $\gamma_{i+1} = -\alpha_i$ for all $i = 1, 2, \ldots, (N-1)$.

The maximum entropy (ME) spectrum of the data profile is simply proportional to the reciprocal of the power response of the PEF and is given by

$$PS(k) = \frac{2\Delta t P_N}{|\sum_{i=1}^{N} [\gamma_i(N)] e^{-jk(i-1)\Delta t}|^2}, \qquad (A.54)$$

where Δt is the uniform sampling interval, k is the wavenumber, P_N is the output power of the N-point PEF, and $\gamma_i(N)$ is the ith coefficient of the N-point PEF (SAAD, 1978). This estimate is based wholly on the data observations, and thus neither requires extending the edges of the profile nor applying any particular window function to the data prior to analysis. Furthermore, the degree of spectral smoothness or desired resolution is achieved by simply varying the length of the prediction error filter from which the power spectrum is computed.

According to BURG (1975), the optimum spectrum is determined by maximizing the entropy, expressed as the integral of the natural logarithmic power spectrum, subject to the constraints that the spectrum agrees with the known

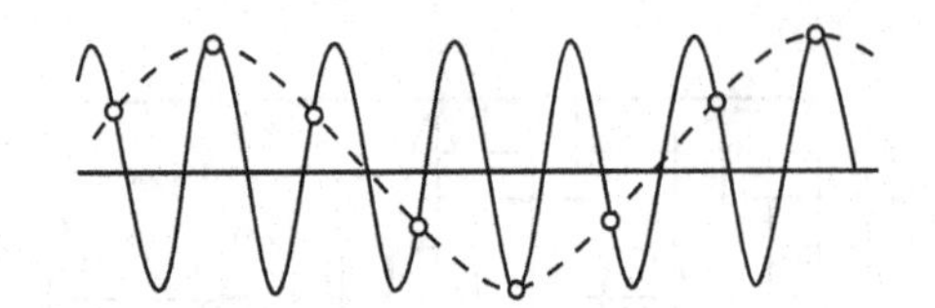

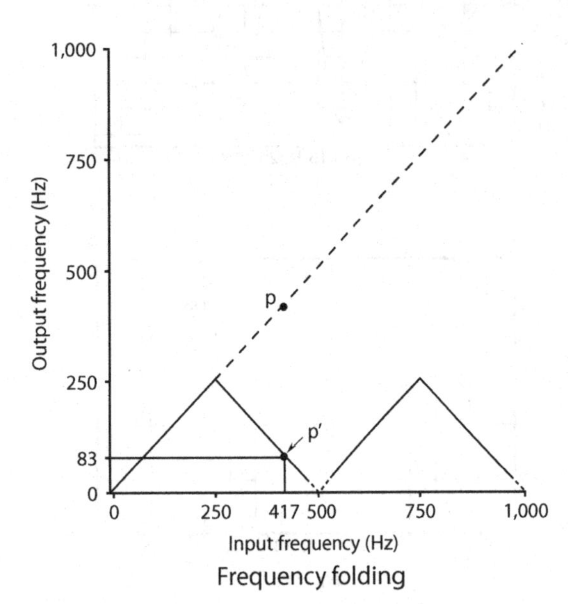

FIGURE A.13 Aliasing of a 417 Hz signal sampled at 500 Hz occurs at the folding frequency of 250 Hz, which is the Nyquist frequency of sampling.

values of the autocorrelation function. The solution of the variational problem using Lagrange's multipliers (which become the PEF coefficients) yields the the MESA spectrum (Equation A.54), as well as the normal equations or the autocorrelation matrix equation, whose solutions using a Levinson recursive approach give the PEF coefficients and the prediction error power required to compute the spectrum.

To estimate the nth PEF coefficient $\gamma_n(n)$ for a filter of any length n, BURG (1975) considered minimization of the average power of the forward and backward prediction errors, $\overrightarrow{\varepsilon_m}$ and $\overleftarrow{\varepsilon_m}$, given respectively by

$$\overrightarrow{\varepsilon_m}(n) = \sum_{i=1}^{n} \gamma_i(n) \times b_{m+n-i} \equiv \sum_{i=1}^{n} \gamma_{n-i+1}(n) \times b_{m+i-1} \tag{A.55}$$

and

$$\overleftarrow{\varepsilon_m}(n) = \sum_{i=1}^{n} \gamma_i(n) \times b_{m+i-1} \equiv \sum_{i=1}^{n} \gamma_{n-i+1}(n) \times b_{m+n-i} \tag{A.56}$$

for all $m = 1, 2, \ldots, (M - n + 1)$. By virtue of Levinson recursion (Equation A.61 for $\gamma_i(n)$, these outputs of the n-point prediction error filter can also be expressed as the linear combination of the $(n-1)$-point filter outputs

$$\overrightarrow{\varepsilon_m}(n) = \overrightarrow{\varepsilon_{m+1}}(n-1) + \gamma_n(n) \times \overleftarrow{\varepsilon_m}(n-1) \tag{A.57}$$

and

$$\overleftarrow{\varepsilon_m}(n) = \overleftarrow{\varepsilon_m}(n-1) + \gamma_n(n) \times \overrightarrow{\varepsilon_{m+1}}(n-1). \tag{A.58}$$

The use of the Levinson recursion in this and subsequent calculations greatly simplifies the analysis and reduces computation times.

The average of the forward and backward prediction error powers is

$$\begin{aligned}\bar{P}_n &= \frac{1}{2(M-n+1)} \sum_{m=1}^{M-n+1} \{[\overrightarrow{\varepsilon_m}(n)]^2 + [\overleftarrow{\varepsilon_m}(n)]^2\} \\ &= \frac{1}{2(M-n+1)} \sum_{m=1}^{M-n+1} \{[\overrightarrow{\varepsilon_{m+1}}(n-1) + \gamma_n(n) \\ &\quad \times \overleftarrow{\varepsilon_m}(n-1)]^2 \\ &\quad + [\overleftarrow{\varepsilon_m}(n-1) + \gamma_n(n) \times \overrightarrow{\varepsilon_{m+1}}(n-1)]^2\}.\end{aligned} \tag{A.59}$$

Solving $\partial \bar{P}_n / \partial \gamma_n(n) = 0$, the optimal coefficient $\gamma_n(n)$ that minimizes $\bar{P}_n$ is given by

$$\gamma_n(n) = \frac{-2 \sum_{m=1}^{M-n+1} \{\overrightarrow{\varepsilon_{m+1}}(n-1) \times \overleftarrow{\varepsilon_m}(n-1)\}}{\sum_{m=1}^{M-n+1} \{[\overrightarrow{\varepsilon_{m+1}}(n-1)]^2 + [\overleftarrow{\varepsilon_m}(n-1)]^2\}} \tag{A.60}$$

for all $n = 1, 2, \ldots, N$.

By the Levinson recursion, the remaining $[\gamma_2(n), \gamma_3(n), \ldots, \gamma_{n-1}(n)]$-coefficients of the n-point prediction error filter are expressed in terms of $\gamma_n(n)$ by

$$\gamma_i(n) = \gamma_i(n-1) + \gamma_n(n) \times \gamma_{n-i+1}(n-1) \tag{A.61}$$

for all $i = 2, 3, \ldots, (n-1)$.

The prediction error power of the n-point filter also is recursively determined from the Levinson relationship

$$P_n = P_{n-1} \times \left[1 - \gamma_n^2(n)\right] = P_1 \Pi_{i=2}^{n} \left[1 - \gamma_i^2(i)\right]. \tag{A.62}$$

The product here is recursive in terms of the zero-lag autocorrelation or power of the input data given by

$$P_1 = \frac{1}{M} \sum_{i=0}^{M-1} b_i^2 \tag{A.63}$$

and the $\gamma_n(n)$ coefficient of the optimal n-point long prediction error filter. Additional details along with computer codes for implementing the above equations are given in ULRYCH and BISHOP (1975), CLAERBOUT (1976), SAAD (1978), and PRESS *et al.* (2007).

All measured data contain errors and thus there exists an optimal length L for the prediction error filter that is not too short to suppress signal resolution or too long to incorporate noise components in the power spectrum estimate. AKAIKE (1969, 1974) found that the optimal length can be determined by monitoring the final prediction error (FPE) and the Akaike information criterion (AIC) given respectively by

$$\mathrm{FPE}(n) = \left(\frac{M+n}{M-n}\right) \times P_n \quad \text{and}$$
$$\mathrm{AIC}(n) = M \times \ln(P_n) + 2n, \tag{A.64}$$

where M is the number of points (or length) of the input data, n is the number of coefficients (or length) of the prediction error filter with $n = 1, 2, \ldots, L$ (= the optimal length of the filter), and P_n is the residual power of the n-point long prediction error filter. Note that $M \gg n$ results in $M \times \ln[\mathrm{FPE}(n)] \longrightarrow \mathrm{AIC}(n)$. In addition, the above expressions assume that the input data have been detrended, otherwise n must be replaced by $n-1$.

The two performance parameters are readily monitored during the recursive computation of the prediction error filter coefficients as shown in Figure A.14. The value of n for which FPE or AIC is minimum is taken as the optimum length L of the prediction error filter. In Figure A.14, for example, the global minimum occurs at $n = 27 \equiv L$, whereas the first local minimum at $n = 14$ would underestimate the length, which in turn would produce an over smoothed, less detailed spectrum. In common practice, $L \ll M$ so that the variations of FPE and AIC are highly correlated and usually point to the same L-value. Thus, it is often sufficient to use just one of the two performance parameters to establish the optimal length of the prediction error filter.

The resolving power of MESA can be exploited over overlapping or non-overlapping windows along a given profile, or over the entire profile to estimate depths and other parameters of multiple anomaly sources at varying depths. Examples of MESA for gravity and magnetic anomaly interpretation are given in Chapters 7 and 13, respectively.

A.4.4 Convolution

The objective of inversion is to obtain a model (**AX**) for the data (**B**) that can be used to predict additional data attributes such as their values at unsurveyed coordinates, gradients, integrals, and other properties. Of course, to realize these additional predictions, new design matrix coefficients are needed. For example, to take the partial derivative of **B** with respect to t, the original coefficients

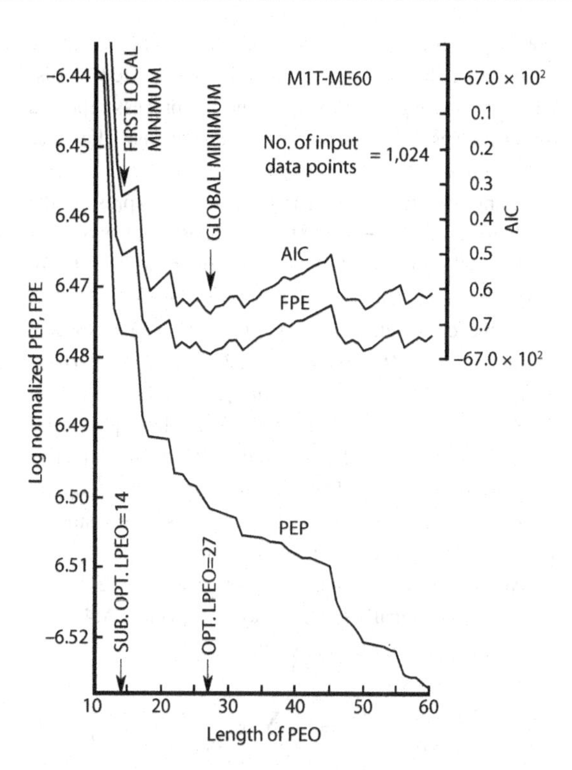

FIGURE A.14 Final prediction error (FPE), Akaike information criterion (AIC), and prediction error power (PEP = P_n) as a function of the length m of the prediction error filter. Adapted from SAAD (1978).

in **A** must be exchanged for another set of coefficients in **D** defined by $\partial\mathbf{A}/\partial t$ because

$$\frac{\partial\mathbf{B}}{\partial t} = \frac{\partial(\mathbf{AX})}{\partial t} = \left(\frac{\partial\mathbf{A}}{\partial t}\right)\mathbf{X} = \left[\left(\frac{\partial\mathbf{A}}{\partial t}\right)(\mathbf{A}^t\mathbf{A})^{-1}\mathbf{A}^t\right]\mathbf{B}$$
$$= [\mathbf{D}]\mathbf{B}. \tag{A.65}$$

Obviously, the solution **X** does not have to be taken explicitly to estimate the derivative and integral properties of **B**.

In electronic processing, **B** is invariably gridded at the uniform interval Δt as noted in Equation A.43. Thus, numerical differentiation and integration is possible using efficient electronic convolution of the signal with the appropriate coefficients of window or mask operators in the t-data domain. For example, effective numerical estimates of $\partial\mathbf{B}/\partial t$ with increasing t are obtained by the convolution of $\mathbf{B}(t_N)$ with the three-point first derivative operator $\mathbf{D}_1(3_t) = (\frac{0.5}{\Delta t}\ 0\ \frac{-0.5}{\Delta t})$.

Convolution involves the mechanical operations of folding, shifting, multiplying, and integrating. The M-point operator, $\mathbf{D}_1(M_t)$, is first folded or reversed about the ordinate axis and then centered on the nth data point b_n where the products of the M overlapping or paired-up coefficients are taken and summed (i.e. integrated) for the estimate $\partial \mathbf{B}(t_n)/\partial t$. The operator is then shifted or displaced by the lag or amount Δt and the process repeated for the next estimate $\partial \mathbf{B}(t_{n+1})/\partial t$, and so on until all values of the signal $\mathbf{B}$ have been processed.

Using the symbol $\otimes$ for convolution expresses the problem as $\partial \mathbf{B}/\partial t = \mathbf{B} \otimes \mathbf{D}_1$ with coefficient estimates $\partial \mathbf{B}(t_n)/\partial t = \partial b_n/\partial t$. The 3-point operator $\mathbf{D}_1(3_t)$ at t_1, for example, gives $\partial \mathbf{B}(t_1)/\partial t = \mathbf{B} \otimes \mathbf{D}_1 = \frac{b_2 - b_0}{2\Delta t} = \partial b_1/\partial t$. Thus, the convolution $\mathbf{B} \otimes \mathbf{D}_1$ gives $(N-2)$ numerical estimates of $(\partial \mathbf{B}(t_n)/\partial t)\ \forall\ n = 1, 2, \ldots, (N-2)$. However, at the end points, b_0 and b_{N-1}, the interpreter must invoke assumptions on the signal's extrapolated properties or other special boundary conditions to obtain effective convolution estimates. Thus, for the M-point mask operator, potentially corrupting edge effects must be considered for $(M-1)$ estimates of the convolution.

Another grid-based approach is to evaluate the derivative from the signal's spectral model Equation A.47. For the gridded signal, t changes with n so that

$$\frac{\partial b_n}{\partial t} \simeq \frac{\partial \hat{b}_n}{\partial t}\frac{\partial t}{\partial n} = \frac{\partial \hat{b}_n}{\partial n} = \frac{1}{N}\sum_{k=0}^{N-1} \mathfrak{b}_k \frac{\partial}{\partial n}\left(e^{j(n)\frac{2\pi}{N}(k)}\right)$$

$$= \frac{1}{N}\sum_{k=0}^{N-1} \mathfrak{b}_k (j\frac{2\pi}{N}k)\left(e^{j(n)\frac{2\pi}{N}(k)}\right) \tag{A.66}$$

with scalar transform coefficients $\mathfrak{b}_k \in \hat{\mathcal{B}}(\omega_k)$ and $j(\frac{2\pi}{N})k = \mathfrak{d}_k \in \hat{\mathcal{D}}_1(\omega_k)$. In other words, Equation A.66 is the IDFT of the product $\hat{\mathcal{B}}(\omega_k) \times \hat{\mathcal{D}}_1(\omega_k)$.

Thus, the IDFT of the products of the signal's transform coefficients $\mathfrak{b}_k$ with the coefficients $\mathfrak{d}_k$ of the transfer function or filter $\hat{\mathcal{D}}_1(\omega_k)$ estimate the first derivatives of the signal along the n-axis. However, by the convolution theorem (Table A.2), which is the basis of modern signal processing, multiplication in one domain is equivalent to convolution in the other domain. Thus, the derivative operations in the t-data and f-frequency domains form the Fourier transform pair $\mathbf{B} \otimes \mathbf{D}_1 \Longleftrightarrow \hat{\mathcal{B}}(\omega_k) \times \hat{\mathcal{D}}_1(\omega_k)$, which involves the coefficient-by-coefficient multiplication of the amplitude spectra and the addition of the phase spectra.

For signal processing, the domain of multiplication is generally preferred over that of convolution because it greatly minimizes the errors and number of mechanical operations involved in achieving the desired result. For example, to taper the edges of a frequency domain filter to reduce Gibbs' error, the ideal filter's response is typically Fourier-synthesized into the data domain where window carpentry is applied to smooth the edges by coefficient-to-coefficient multiplication. The smoothed response is then Fourier-analyzed back into the frequency domain for the desired filtering applications.

In summary, several approaches are available for interrogating gravity and magnetic observations for insight on the subsurface. An explicit linear model $\mathbf{AX}$ of the subsurface can be directly related to the observations, for example, using matrix inversion methods to solve for $\mathbf{X}$. Here, the integral and derivative properties of the data can be explored by exchanging the original design matrix coefficients for the coefficients of the integrated and differentiated model, respectively. However, because of the fundamentally graphical nature of integration and differentiation, these properties are also accessible numerically from convolution operations that spectral filters, in turn, can perform with utmost accuracy and efficiency. The next section considers common gravity and magnetic anomaly applications of these mathematical methods.

A.5 Anomaly analysis

A variety of mathematical techniques, including those explained above, find extensive use in data processing that is an important part of anomaly analysis. In this section, methods are described to help identify specific anomaly components, and perform source modeling. The approaches to these processes are fundamentally the same for gravity and magnetic methods, and thus can be considered together in this chapter, but differences do exist; these are treated in individual chapters dealing with the specific method.

A.5.1 Isolation and enhancement

The measured gravity and magnetic fields consist of the superposition of anomalies from a variety of sources. They include: (1) the so-called residual anomalies, which are the anomalies of particular interest in a study, (2) the longer-wavelength regional components derived largely from deep, large-volume geologic sources, and (3) the shorter-wavelength noise due to observational and data reduction errors and small, shallow sources. The process of removal of the interfering regional and noise components in the anomaly field is the residual–regional problem, which is one of the critical steps in anomaly interpretation. It is solved by either isolating the residual anomaly through elimination or attenuation of the regional and noise anomalies or enhancing the residual relative to the interfering effects.

Isolation techniques attempt to eliminate from the observed anomaly field all components that do not have a certain specified range of attributes that are defined by the anticipated anomalies from the geologic sources of interest, the residual anomalies. Depending on the effectiveness of this procedure the isolated anomaly is amenable to quantitative analysis. In contrast, enhancement techniques accentuate a particular characteristic or group of attributes which are definitive of the desired residual anomalies. The enhancement process increases the perceptibility of the residual anomaly in the anomaly field. As a result the anomalies are distorted from those of the observed anomaly field, limiting their usefulness for quantitative analysis, but improving their use in qualitative visual inspection analysis.

The isolation and enhancement of gravity and magnetic anomalies is a filtering process in which only the spatial wavelengths of interest are passed and the remainder are eliminated or at least highly attenuated. Unfortunately the wavelengths of interfering anomalies, especially gravity anomalies, overlap considerably, complicating the effectiveness of the filtering process. Furthermore, the cut-offs in the characteristics of filters are usually not sharp, leading to possible amplitude and phase distortion in the results. Thus the residual determination process is subjective and is a major potential limitation in anomaly analysis.

Residual anomaly separation may be based on a variety of anomaly attributes such as magnitude, shape, sharpness or gradients, orientation, and correlation with other data. Both spatial and spectral analysis techniques have been used for this purpose. They range from simple graphical techniques employing manual procedures to a host of filtering methods based on the computationally intensive spectral analysis procedures described above. Current usage is focused on spectral analysis of gravity and magnetic fields, but graphical methods and mathematical procedures which emulate them still find use, particularly in the analysis of limited-extent studies in which the residual anomalies are distinctive from the regional and noise components.

Spatial filtering

A number of methods can be applied directly to anomaly maps to isolate and enhance various spatial attributes of anomalies. In the hands of the experienced interpreter, these methods can be very effective in bringing out a wide range of anomaly details for analysis.

(A) Geological methods The most direct and satisfying method of isolating local anomalous signals is to remove the effect of regional geological sources that obscure the local anomaly, by calculating the regional anomaly from sources defined by auxiliary geological and geophysical information. For example, the regional anomaly derived from seismically mapped depths to the base of the crust and bedrock depths constrained by borehole information can be calculated and removed from the observed signals to isolate the local anomaly. Alternatively, regional signals surrounding an observed local anomaly can be modeled to obtain the physical property distribution which best reproduces the observed field, and the regional field derived from this distribution can be subtracted from the observed field to isolate the local anomalous signal. Li and Oldenburg (1998b) have described such an approach and suggest that the advantages of the method are that there is limited distortion of the local anomaly and minimal effects from topography and overlap in the spectra of regional and residual anomalies. Variations on this so-called inversion-based anomaly separation have been suggested by Gupta and Ramani (1980), Pawlowski and Hansen (1990), Pawlowski (1994), and others using least-squares linear Wiener filters designed to remove the spectra of the regional geology as defined by the measured power spectrum of the anomaly pattern, modeling of known geology, or inversion of the regional anomalous field.

(B) Graphical methods Graphical methods are appropriately considered non-linear in the sense that it is impossible to duplicate the results exactly, but they are flexible, permitting the use of the interpreter's experience and auxiliary data in their application. The success of the methods is dependent on the experience of the interpreter (especially in the specific geological terrane), the simplicity of the regional anomaly, and the perceptibility of the residual anomaly. The methods were used extensively before computers were commonly available to perform analytical linear residual–regional separations rapidly and inexpensively.

Graphical methods are labor-intensive and thus are not used except where the survey area is small, the regional is simple and noise subdued, and the residual anomalies relatively easily discerned. In this case, the separation process can be carried out with a minimum distortion of the residual anomaly, and quantitative calculations are readily performed on the isolated residual anomalies. When graphical methods are used, it is advisable to ascertain the geological reasonability of the regional as a means of validating the viability of the separation process.

Profile methods are the simplest of the graphical methods. They visually establish the regional anomaly as a smooth curve through the values of an observed anomaly profile that excludes the anomalous portion of the profile associated with the residual anomaly. The selection of the exclusion segment, of course, is a subjective decision depending greatly upon the analyst's experience and

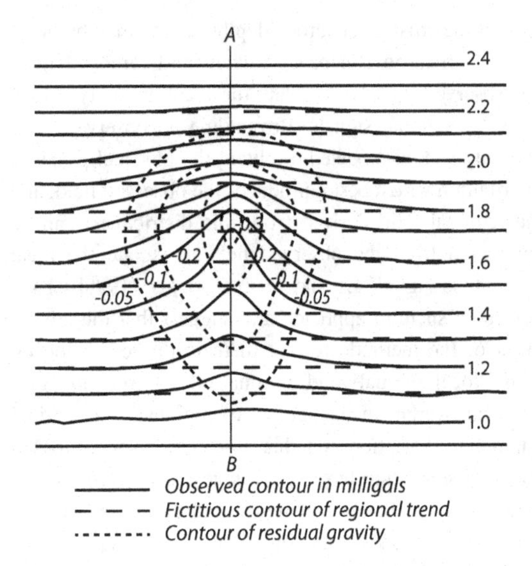

FIGURE A.15 Schematic gravity anomaly contour map of the residual anomaly superimposed on a regional anomaly gradient illustrating the method of isolating residual anomalies by subtracting fictitious smooth contours of the regional. Adapted from DOBRIN and SAVIT (1988).

ability. Successful application of the method requires anomaly data well beyond the limits of the target area. The regional anomaly is subtracted from the observed anomaly to determine the residual anomaly. However, where the regional is complex and the residual less identifiable, the separation of regional from residual is difficult and likely to be non-reproducible.

Where the anomaly fields are in map form, the graphical method can be used to isolate the residual anomalies by establishing the regional anomaly field on the individual profiles which form a grid of intersecting profiles across the map area. Control on the regional is determined by adjusting the regionals so that they are the same at the intersection points. The resulting regional map is subtracted from the observed anomaly to obtain the residual map. Another method is to draw smooth, consistent contours through the anomaly map that represent the regional effects of deep, large sources. These regional contour values are subtracted from the observed anomaly values to obtain the residual anomaly pattern (Figure A.15). The regional anomaly obtained in this manner, as with all graphical methods, should be checked to determine that it is derived from a geologically reasonable source.

(C) Trend surface analysis An analytical approach to residual–regional separation by graphical methods is to represent the residual–regional component anomalies as least-squares polynomial approximations or trend surfaces (e.g. COONS *et al.*, 1967; STEENLAND, 1968). The method is used where the regional anomalies are complex and the differences between the regional and residual anomalies are subtle. It produces objective, automatic computations that facilitate the rapid exploration of observed anomalies for meaningful regional–residual components. However, the application and interpretation of the method is still subject to the concerns regarding subjectivity and non-uniqueness of the graphical methods. As in graphical methods, trend surface analysis can be applied to both profiles and maps.

The more common implementation of the method involves modeling the regional component as a low-order polynomial that fits the observed anomalies so as to minimize the sum of squared residuals. The residual anomaly component then is estimated by subtracting the polynomial model of the regional from the observed anomalies. At the other extreme, possible noise components may be estimated by extending the method to higher-degree polynomials, where the maximum degree possible is $(n-1)$ for n observed anomaly values.

By Equation A.16, the least-squares pth degree polynomial model for the regional–residual component can be established using only the $(n \times 1)$ **B**-matrix of observed anomaly values and the forward polynomial model of unknown coefficients that is presumed to account for the variations in observations. Suppose, for example, that the regional ($\mathbf{B}_R$) may be approximated by the first degree equation of the plane through the observed anomalies mapped in (x, y)-coordinates. Thus, the linear system for establishing the regional is

$$\mathbf{B} = \alpha + \beta x + \gamma y \equiv \mathbf{B}_R \ni \mathbf{B}_r = \mathbf{B} - \mathbf{B}_R, \qquad (A.67)$$

where $\mathbf{B}_r$ estimates the residual anomalies, and the $m = 3$ unknown coefficients, α, β and γ, are respectively the plane's mean value, and the x- and y-directional slopes. Simply evaluating the forward model with the unknown coefficients set to unity gives the $(n \times m)$ coefficients of the design matrix **A** from which the least-squares solution is

$$(\alpha\ \beta\ \gamma)^t = (\mathbf{A}^t\mathbf{A})^{-1}\mathbf{A}^t\mathbf{B}. \qquad (A.68)$$

Of course, the characterizations of the regional $\mathbf{B}_R$ and related residual $\mathbf{B}_r$ are never unique. In Figure A.16, for example, the residual estimate at O may be based on subtracting the average or some estimate from a polynomial or another function fit to the eight values that the interpreter selected to characterize the regional for this application. Clearly, to make the analysis most effective, the interpreter must use available ancillary geological and geophysical data to justify the viability

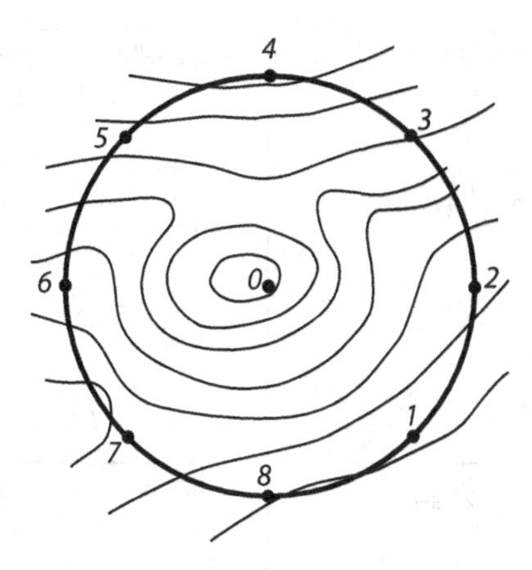

FIGURE A.16 Computation of the residual gravity anomaly at O based on eight regional anomaly values selected by the interpreter. The non-unique regional anomaly at O may be taken as an average of the selected values or some estimate from a polynomial or another function fit to them. Adapted from GRIFFIN (1949).

of the regional, noise, and residual anomalies as fully as possible.

In general, the trend surface analysis technique for regional–residual separation effectively operates like a low-pass filter. However, the wavelength characteristics of the method are poorly understood and only can be controlled in a limited way. The prominent wavelength of the polynomial approximation can be crudely determined by noting that there are $(n-1)$ maxima and minima in a least-squares approximation where n is the maximum degree of the equation. Thus, for example, considering a profile, a regional anomaly with a prominent wavelength of the order or degree 2 can be anticipated that corresponds to the length of the profile divided by $(n-1)$. As a result the degree of appropriate polynomial equation must take into consideration the size of the profile or area as well as the character of the regional anomaly.

The problem, however, is much more complex than suggested by this approximation. For example, the application of trend surface analysis to maps also involves a directional bias depending on the shape of the area involved (e.g. THURSTON and BROWN, 1992). The result is incorrect orientation of the regional field and distortion of the residual anomalies. The distortion of residual anomalies is particularly troublesome where the regional anomalies have high gradients (e.g. STEENLAND, 1968) and near the edges of the survey. However, supplementing the data with rinds of real or simulated data can help mitigate some of these effects.

An alternative but equally subjective approach to least-squares trend surface computations is the use of finite element analysis (MALLICK and SHARMA, 1999). In this method the regional field at map positions is determined to a first- through third-order approximation using a weighted sum of discrete values surrounding the position. The nature of the regional is established by a weighting factor determined by the element shape function based on finite-element analysis.

As a result of the hazards of the trend surface analysis technique, the method must be used with caution in residual–regional separation, especially where anomalies will be subjected to quantitative interpretational procedures. However, the method does serve well in preliminary separation of major regional effects by low-order polynomials.

(D) Analytical grid methods Analytical spatial methods include a variety of techniques for residual–regional separation based on determining the regional value at a location on an equidimensional grid from the surrounding anomaly values. These methods are free of personal bias that limits the graphical procedures, but they are still subjective because the results are highly dependent on the method of determining the surrounding values and the weighting applied to these values. The simplest of procedures is to average the data on a circle of fixed radius around the point at which the regional valued is being approximated and subtract the average from the central value to determine the residual anomaly at the location. This procedure is repeated at successive grid points until the entire map grid is evaluated. Other methods use values obtained on multiple rings surrounding the central location, and the method can be used on profile data as well by using adjacent data to determine the regional value.

The method can be adapted to a variety of filtering approaches including derivatives, continuations to a different elevation, and wavelength filtering. This is accomplished by convolving the input data with a filter designed to pass specified wavelengths (e.g. DEAN, 1958; FULLER, 1967). The filtering function, that is the convolution operator or weights, is designed by specifying the desired wavenumber response and taking its inverse Fourier transform. These transforms are modified from the continuous data case to the discrete data set condition and are made practical by applying a smoothing function causing a gradual cutoff to where the amplitude of the function is negligible. The direct transform can be used to calculate the response of the filter in the wavenumber domain. Although this spatial method is readily applied to rapid analysis of gridded data by digital computers, it has been largely superseded by more efficient filtering using spectral analysis techniques that are described in the next section.

Spectral filters

Spectral filters efficiently process gridded data for their basic geometric properties and the theoretical extension of these properties into other anomaly forms. The routine act of electronically plotting data involves the preparation of a data grid that provides access to these data properties directly via the FFT. Thus, spectral filtering has become widely used for anomaly analysis.

Filtering in the wavenumber domain involves multiplying the Fourier transform of the data by the coefficients of a filter or transfer function that achieves one or more data processing objectives. Standard filters with transfer function coefficients of 1s and 0s can be applied to any gridded data set to study the wavelength, directional, and correlation properties of the anomalies. Wavelength and directional filters suppress or enhance data features based on their spatial dimensions and attitudes, respectively. Spectral correlation filters, by contrast, extract positively, negatively, or null correlated features between two or more data sets.

Specialized filters, on the other hand, have non-integer coefficients that account for the horizontal derivatives and integrals of anomalies and their extensions via potential field theory. For example, horizontal derivative and integral filters estimate signal differentials and integrals with respect to the independent variable t that can be extended to predict data variations along the vertical axis by Laplace's Equation 3.6. Thus, Laplace's equation is used to design vertical derivative and integral filters, as well as upward- and downward-continuation filters to assess the anomaly fields respectively above and below the mapped observations.

(A) Wavelength filters These standard filters consist of patterns of 1s and 0s that respectively pass and reject the signal's transform components $\hat{b}_k$ upon multiplication. The equivalent data domain convolution operator is a real symmetric or even function so that only a single quadrant of the real transform is needed for designing the pass/reject pattern that by symmetry extends to the other real quadrants. The imaginary transform is zero, and thus these filters are called zero phase or minimum phase filters.

Figure A.17 illustrates the low-pass/high-cut filter designed to pass frequencies in a data profile up to the 40th harmonic and reject the higher harmonics. The top right panel is a linear, DC-centered plot of both real quadrants of the ideal filter. The solid dots on the filter edges locate the 50% amplitude levels. The inverse transform of this boxcar filter is the equivalent data domain convolution operator given by the sinc function ($\equiv \sin(t)/t$) in the top left panel. This convolution operator is defined over all data points and clearly is not a simple averaging function.

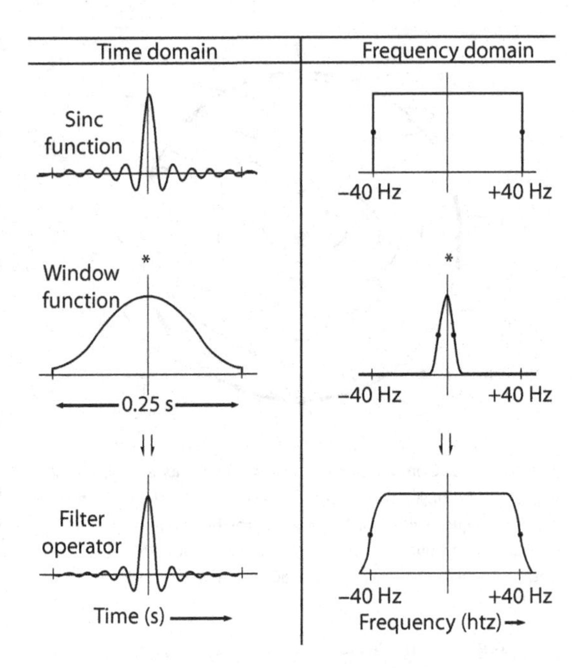

FIGURE A.17 Design of the digital low-pass filter operator.

To minimize Gibbs' error, window carpentry is used to taper the sharp edges of the boxcar function in the data domain (left panels) rather than in the frequency (wavenumber) domain (right panels) where the frequency responses of the window and ideal filter functions must be convolved. The left middle panel shows the Hamming window function that was applied. It consists of a single cycle cosine wave raised on a slight pedestal over the fundamental period of the data profile and has the frequency response shown in the right central panel. The Hamming window was multiplied against the ideal filter operator (left, top panel) to obtain the tapered filter operator (bottom, left panel) with the tapered frequency response given in the lower right panel. The tapered filter has nearly the same performance characteristics as the ideal filter, but with greatly reduced Gibbs' error. Typically, the performance specifications of the tapered filter are reported in terms of its 50% cut-off, which for the low-pass filter in the lower right panel is at the 40th harmonic of the data profile.

Figure A.18 shows the design of digital high-pass/low-cut and band-pass filters for a data profile. Because the wavenumber components in the Fourier transform follow the superposition principle, the 1s and 0s of the low-pass filter are interchanged to obtain the complementary

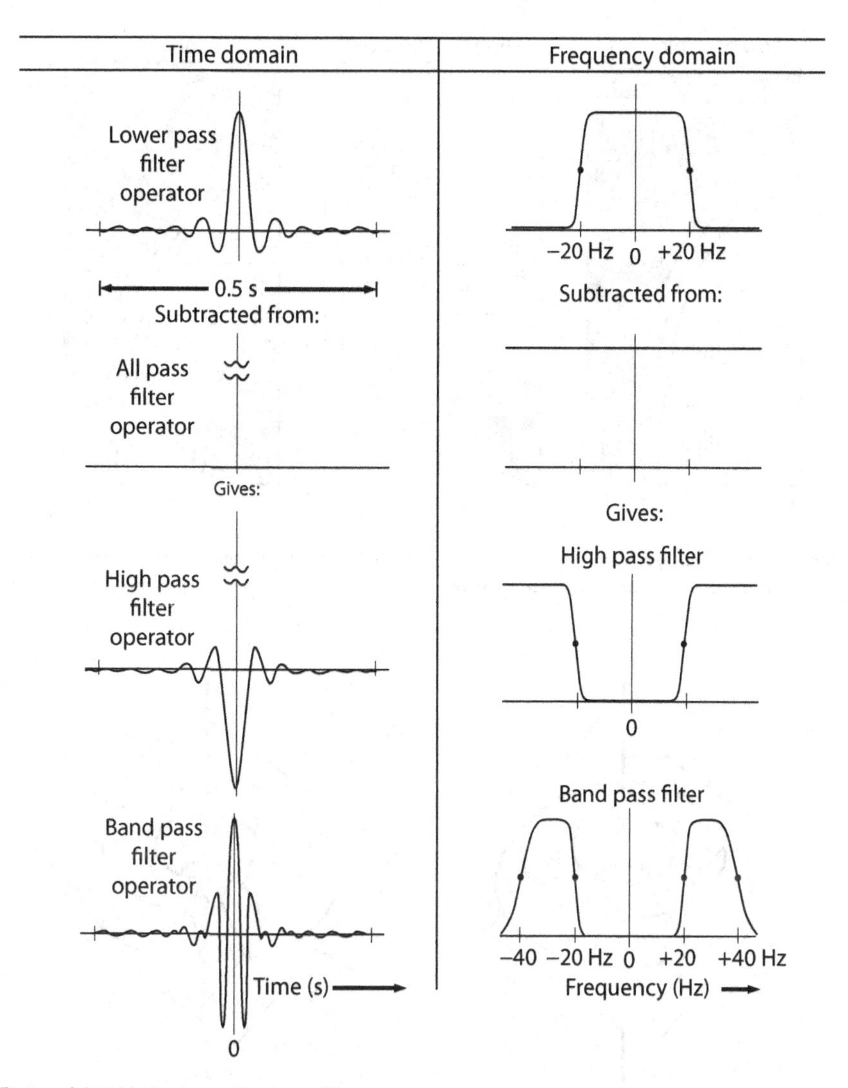

FIGURE A.18 Design of digital high-pass and band-pass filter operators.

transfer function of the high-pass filter. Suppose, for example, that it is desired to pass frequencies above the 20th harmonic of the data profile and sharply attenuate the lower frequencies. The tapered low-pass filter operator is first designed with the 20th harmonic cut-off as shown in the top panels. Then the all-pass filter is designed that consists of the data spike in the data domain (second left panel) with the uniform amplitude components in wavenumber domain (second central panel). Subtracting the low-pass filter from the all-pass filter yields the high-pass filter in the third panels with the 50% cut-off at the 20th harmonic of the data profile. Furthermore, subtracting the high-pass filter from the low-pass filter in the bottom panel of Figure A.17 gives the band-pass filter in the bottom panels of Figure A.18 that passes the wavenumber components essentially between the 50% cut-offs at the 20th and 40th harmonics.

Extending wavelength filtering to the 2D transform with wavenumbers k and l in the respective n- and m-directions is also straightforward. Specifically, the transfer function for the ideal wavelength filter is

$$\begin{aligned} \mathrm{WLF}(k,l) &= 1 \quad \forall \; k_{\mathrm{L}}^2 + l_{\mathrm{L}}^2 < k^2 + l^2 < k_{\mathrm{H}}^2 + l_{\mathrm{H}}^2, \\ &= 0, \text{ otherwise}, \end{aligned} \tag{A.69}$$

where the lowest and the highest wavenumbers to be passed are $(k_{\mathrm{L}}, l_{\mathrm{L}})$ and $(k_{\mathrm{H}}, l_{\mathrm{H}})$, respectively. To taper the filter's response and reduce its Gibbs' error, window carpentry is applied in the data domain to the inversely transformed ideal filter. Figure A.19 gives examples of

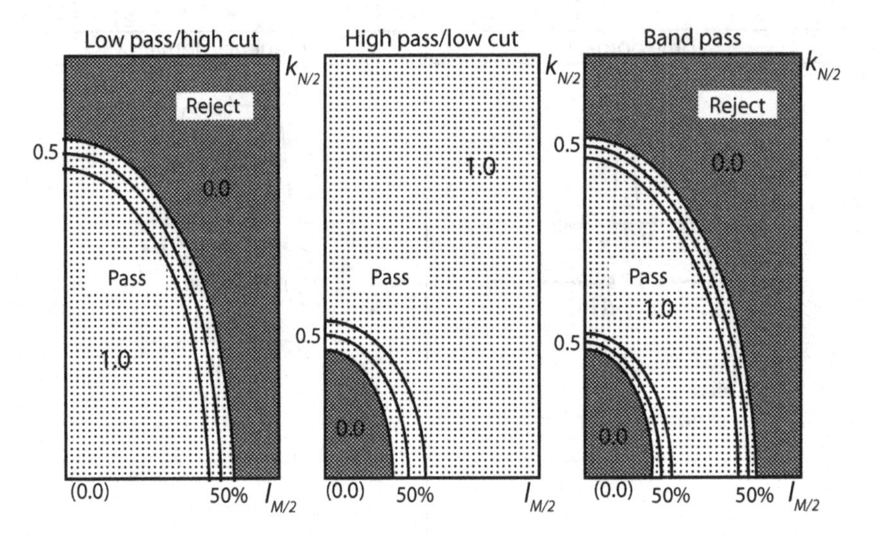

FIGURE A.19 Three idealized wavelength filter operators with edges tapered for enhanced suppression of Gibbs' error. The middle contour is the 50% value of the taper. The design of these filters requires only a single quadrant. Adapted from REED (1980).

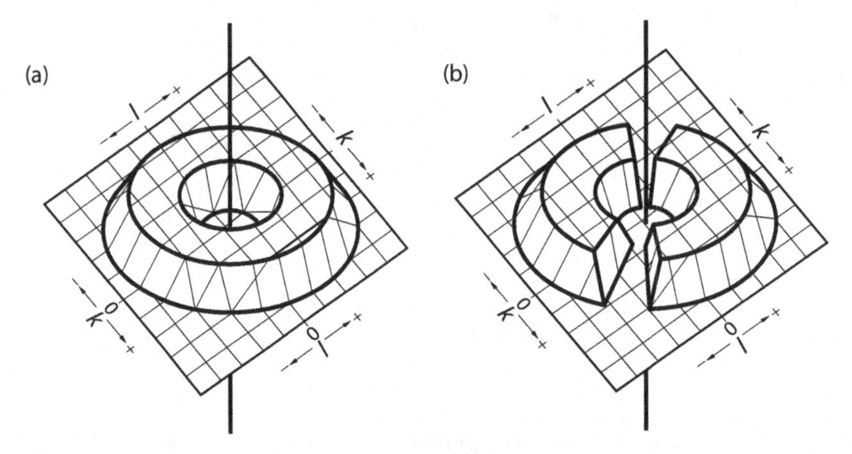

FIGURE A.20 (a) The smoothed or tapered doughnut-shaped response function of a band-pass filter for extracting gridded data features with maximum and minimum wavelengths defined by the wavenumbers at the respective outside edges of the doughnut hole and the doughnut. (b) Response function for the composite band-and-strike-pass filter that passes the maximum and minimum wavelength features which lack components with wavenumbers in the wedge-shaped breaks of the doughnut. Adapted from LINDSETH (1974).

low-pass/high-cut (left panel), high-pass/low-cut (middle panel), and band-pass (right panel) filters designed for the 16×32 data array considered in Figure A.7.

Only a single real quadrant is necessary for designing the essential elements of these wavelength filters, which consist predominantly of 1s and 0s and some in-between values that reflect the smoothing of their edges to minimize Gibbs' error. Figure A.20(a) gives an isometric view of a complete tapered, DC-centered band-pass filter designed to reject data features with wavenumber components corresponding to the 0s of the filter. In the next section, a further adaptation of 1s and 0s is described that brings out the filtered features with specific directional or strike orientations.

(B) Directional filters For gridded 2D data, the ratios of the orthogonal wavenumbers reflect the azimuthal or strike orientations of the data variations. For example, the N/S-striking elliptical anomaly in Figure A.21(a) has an effectively infinite wavelength component in the N/S direction with a significantly shorter wavelength component in the E/W direction. Thus, the spectral energy for this feature

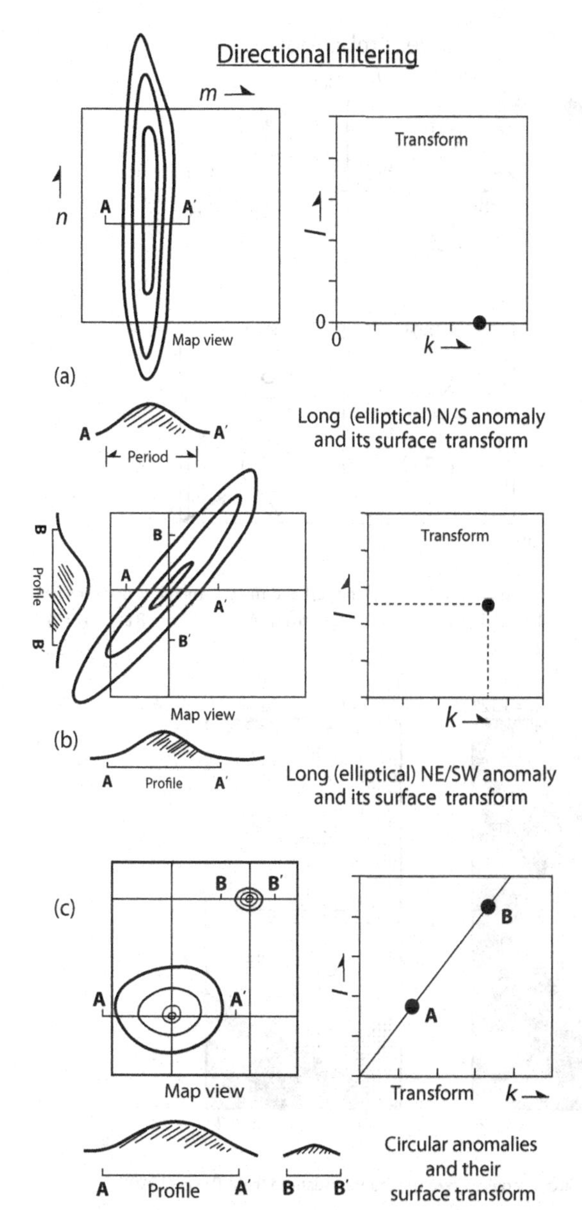

FIGURE A.21 Map and frequency responses for (a) elliptical N/S-striking, (b) elliptical NE/SW-striking, and (c) circular longer and shorter wavelength anomalies. Adapted from LINDSETH (1974).

plots towards the higher-wavenumber end of the k-axis as shown in the top right panel.

The data can be tested for this feature using a standard wedge-pass filter centered on the k-axis, which is also sometimes called a fan or butterfly or pie-slice filter. To pass N/S-striking regional features at all E/W wavelengths, the edges of the wedge filter may be defined for example at 80° and 100° clockwise from the l-axis of the transform. Then 1s and 0s are assigned in-between and outside these azimuths, respectively, to define the ideal strike-sensitive filter, which is next inversely transformed to the data domain for window carpentry to smooth the filter's edges slightly to reduce Gibbs' errors. Transforming the tapered E/W-striking elliptical convolution operator into the wavenumber domain, multiplying it against the data's amplitude spectrum coefficient-by-coefficient and inversely transforming the filtered output into the data domain leads to an estimate of the N/S-trending anomalies in the data.

To pass E/W-striking data features, on the other hand, the wedge filter is simply rotated 90° to center it on the l-axis. At orientations between these two extremes, the wavenumber energies migrate predictably into other parts of the spectrum. For example, the elliptical anomaly in Figure A.21(b) defines wavelength components that are longer in the N/S-direction than in the E/W-direction. Thus, this orientation contributes energy to the portion of the spectrum involving the lower-frequency l and higher-frequency k wavenumbers. Of course, as the anomaly is rotated to strike E/W, the spectral energy migrates towards the Nyquist end of the l-axis.

As shown in Figure A.21(c), the spectral energies of circular features are concentrated along the 45°-line in the transform where $l \simeq k$. Elliptical data features with azimuths at 45° to the map axes also contribute energies in this part of the spectrum. However, these features die out as the wedge is rotated away from the 45°-line in the transform, whereas circular features contribute energies at all orientations of the wedge filter.

To find elliptical features with specific across-strike wavelengths, the wedge filter is multiplied against the appropriate band-pass filter to construct a composite filter that further restricts the range of 1s such as those shown in Figure A.19(b). This approach can also help to discriminate circular features of different wavelengths as shown in Figure A.21(c).

The azimuthal orientations in the spatial and wavenumber domains involve the 90° rotation shown in Figures A.22(a) and (b), where Figures A.22(c) and (d) give the typical computer storage schemes for designing and applying the directional filtering. In these schemes, passing or rejecting feature orientations is considered in the data domain between azimuths $A1$ and $A2$ in the range $0° \leq A1 < A2 \leq 180°$ measured clockwise positive from the left axis of the data set (e.g. Figure A.22(c)). The minimum phase transfer functions are commonly designed with 1s and 0s in the first and second spectral quadrants that extend into the two conjugate quadrants by symmetry. Thus, the transfer function for the ideal strike-pass filter,

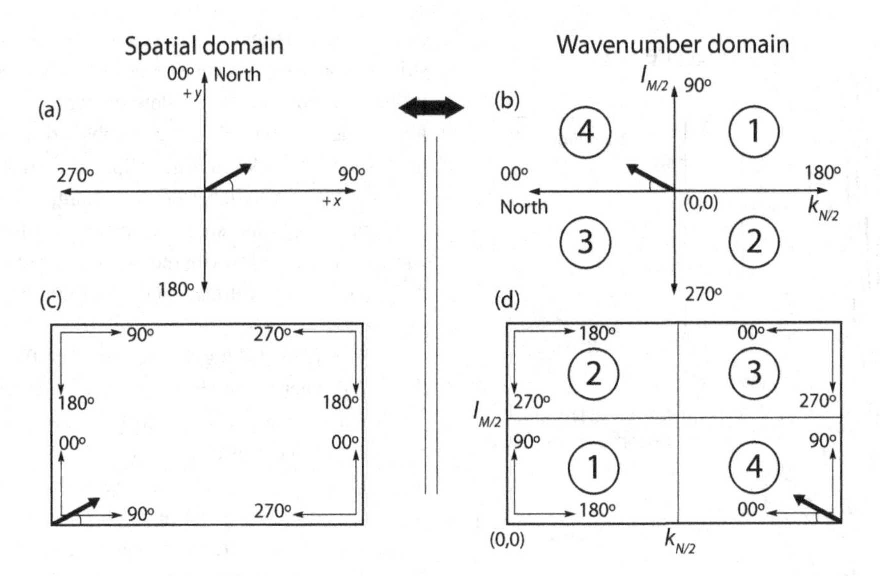

FIGURE A.22 Contrasting representations for the azimuthal orientations in the data domain [(a) and (c)] and the wavenumber domains [(b) and (d)]. Panels (c) and (d) also show the machine storage formats for the data and wavenumbers, respectively. Adapted from REED (1980).

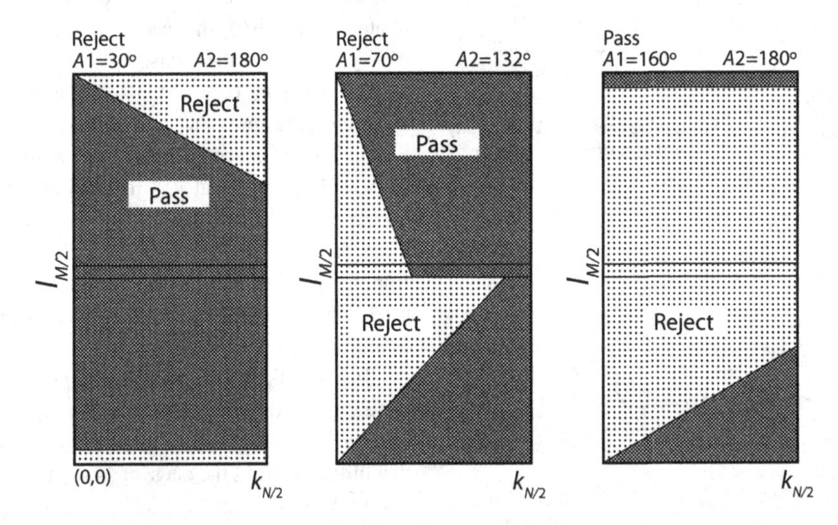

FIGURE A.23 Three idealized directional filters. The design of these filters requires two quadrants. Adapted from REED (1980).

for example, is

$$\mathrm{SPF}(k, l) = 1 \quad \forall \quad -\tan^{-1}(A2) \leq l/k \leq -\tan^{-1}(A1),$$
$$= 0, \text{ otherwise.} \qquad \text{(A.70)}$$

Simply interchanging the 0s and 1s in Equation A.70 gives the transfer function of the ideal strike-reject filter.

In practice, azimuthal differences $(A2 - A1) < 20°$ typically affect relatively marginal levels of the signal energy, and thus wedges of about 20° and wider are usually required to produce useful filtered outputs. As examples, Figure A.23 shows ideal directional filters designed for the 16 × 32 data array in Figure A.7. In the left and middle panels, the filters reject azimuths in the data from 30° to 180° and 70° to 132°, respectively, whereas in the right panel the directional filter passes data azimuths from 160° to 180°.

The complete filters consist of the coefficients designed in quadrants 1 and 2 of the computer-formatted, Nyquist-centered spectrum shown in Figure A.22(d) and mapped by symmetry into the conjugate quadrants 3 and 4, respectively. However, in this machine format, the filter's response is less than intuitive, and thus for presentation purposes, the quadrants are typically

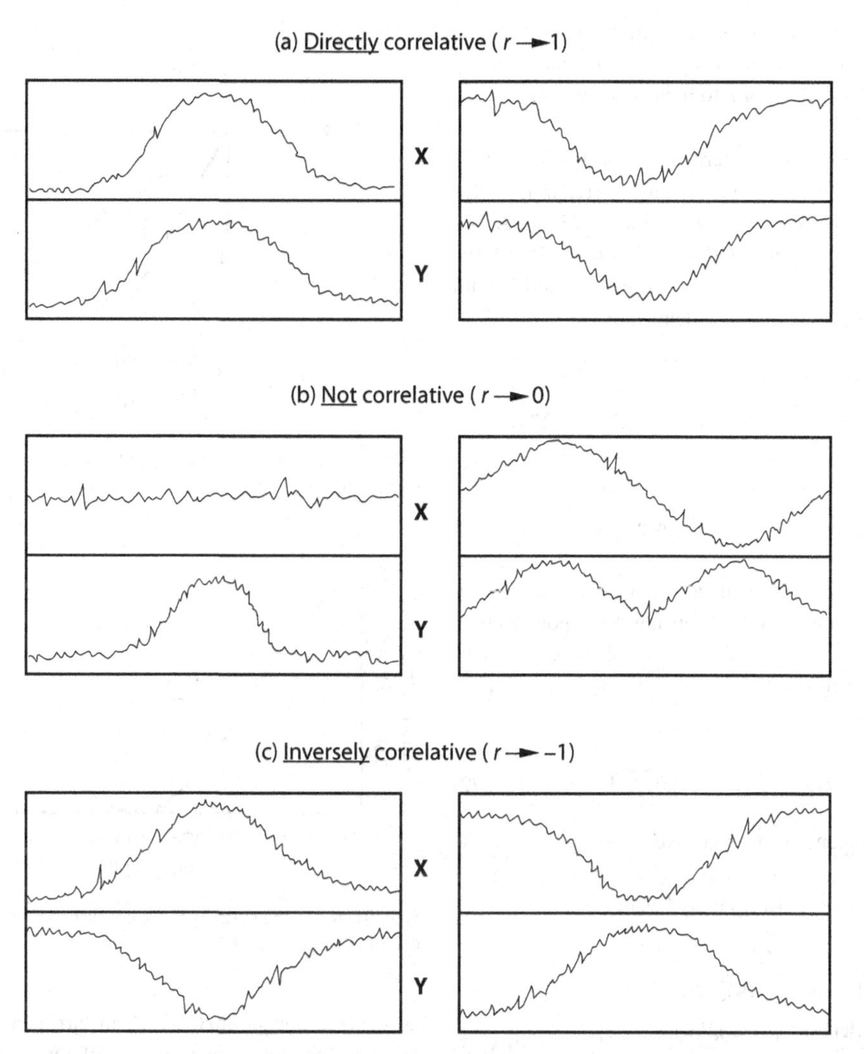

FIGURE A.24 Interpretations of the correlation coefficient r for (a) positive, (b) null, and (c) negative feature associations. Adapted from VON FRESE *et al.* (1997a).

transposed into the DC-centered spectrum like the one in Figure A.20(b).

In general, a good practical test of the performance of a wedge or, for that matter, any filter is to sum the data passed and rejected by the filter for comparison with original input data. Well-performing filters yield sums that closely match the input data with minimal residuals.

(C) Correlation filters Standard correlation filters of 1s and 0s can also be developed to extract correlative features between two or more data sets (VON FRESE *et al.*, 1997a). A specialized form of this filter is the least-squares Wiener filter (WIENER, 1949) designed to extract random signals. Figure A.24 summarizes the feature associations that the correlation coefficient (r) quantifies. In panel (a), for example, peak-to-peak and trough-to-trough feature associations reflect direct or positive correlations where $r \longrightarrow +1$. Panel (c), on the other hand, illustrates peak-to-trough and trough-to-peak feature associations that reflect inverse or negative correlations where $r \longrightarrow -1$.

In general, the above interpretations are unique only if $|r| = 1$, whereas interpretational uniqueness does not hold for other values where $-1 < r < +1$. Panel (b), for example, illustrates two interpretations that quantitatively account for the null correlation where $r \longrightarrow 0$. The left profiles reflect the classical explanation assumed for the lack of correlation, namely that variations in one data set do not match the variations in the other data set. However, the null correlation also results when 50% of the signals directly correlate and the remaining 50% of the signals inversely correlate as in the right set of profiles. This example illustrates the folly of using some arbitrary threshold value ($\neq \pm 1$) of r to indicate how signals actually correlate.

Initial efforts to characterize the spectral properties of correlation focused on the squared correlation coefficient or the coherency r^2 (see Equations A.23 and A.24). Following the pioneering studies of WIENER (1930), the coherency coefficient between two signals **X** and **Y** with Fourier transforms $\mathcal{X}$ and $\mathcal{Y}$ is defined by

$$\begin{aligned} r^2 &\equiv |\mathcal{X}\mathcal{Y}^*|^2/[(\mathcal{X}\mathcal{X}^*)(\mathcal{Y}\mathcal{Y}^*)] \ \forall \quad [(\mathcal{X}\mathcal{X}^*)(\mathcal{Y}\mathcal{Y}^*)] > 0 \\ &\equiv 0 \qquad \forall \quad [(\mathcal{X}\mathcal{X}^*)(\mathcal{Y}\mathcal{Y}^*)] = 0, \end{aligned} \tag{A.71}$$

where $(\mathcal{X}\mathcal{X}^*)$ and $(\mathcal{Y}\mathcal{Y}^*)$ are the power spectra and $(\mathcal{X}\mathcal{Y}^*)$ is the cross-power spectrum.

The coherency coefficient essentially tests signals that differ only in terms of their random noise components (e.g. FOSTER and GUINZY, 1967). Thus, coherency provides very useful estimates of the noise (N)-to-(S) signal ratio given by

$$(N/S) \simeq \sqrt{(1/\sqrt{r^2}) - 1} = \sqrt{(1/|r|) - 1}. \tag{A.72}$$

However, coherency is not sensitive to the sign of the correlation coefficient, or the individual wavenumber components for which it is always unity because

$$\begin{aligned} r^2(k) &= \frac{|\mathcal{X}(\mathbf{k})\mathcal{Y}^*(\mathbf{k})|^2}{[\mathcal{X}(\mathbf{k})\mathcal{X}(\mathbf{k})^*][\mathcal{Y}(\mathbf{k})\mathcal{Y}(\mathbf{k})^*]} \\ &= \frac{([\mathcal{X}(\mathbf{k})\mathcal{Y}(\mathbf{k})^*][\mathcal{X}(\mathbf{k})\mathcal{Y}(\mathbf{k})^*])^*}{[\mathcal{X}(\mathbf{k})\mathcal{X}(\mathbf{k})^*][\mathcal{Y}(\mathbf{k})\mathcal{Y}(\mathbf{k})^*]} = 1. \end{aligned} \tag{A.73}$$

Thus, coherency is implemented predominantly over band-limited averages of the spectra where it resolves only the relatively gross, positively correlated relationships between the signals.

The full spectrum of correlations between digital data sets becomes evident, however, when the transforms are considered as vectors in the complex plane at any given wavenumber, k. Analysis of the polar coordinate expressions for the wavevectors $\mathcal{X}(\mathbf{k})$ and $\mathcal{Y}(\mathbf{k})$ in Figure A.25, for example, shows that their correlation is

$$r(k) = \cos(\Delta\theta_k) = \frac{\mathcal{X}(\mathbf{k}) \cdot \mathcal{Y}(\mathbf{k})}{|\mathcal{X}(\mathbf{k})||\mathcal{Y}(\mathbf{k})|}, \tag{A.74}$$

where the numerator is the dot product between the wavevectors with the phase difference $\Delta\theta_k = [\theta_{\mathcal{Y}(\mathbf{k})} - \theta_{\mathcal{X}(\mathbf{k})}]$ (VON FRESE *et al.*, 1997a). In other words, the correlation coefficient between two wavevectors is their normalized dot product, which in turn is a simple cosinusoidal function of their phase difference. The meaning of this result is evident in Figure A.26, which illustrates the variations of the correlation coefficient computed as a function of $\Delta\theta_k$ as any sine or cosine of period k is lagged past itself.

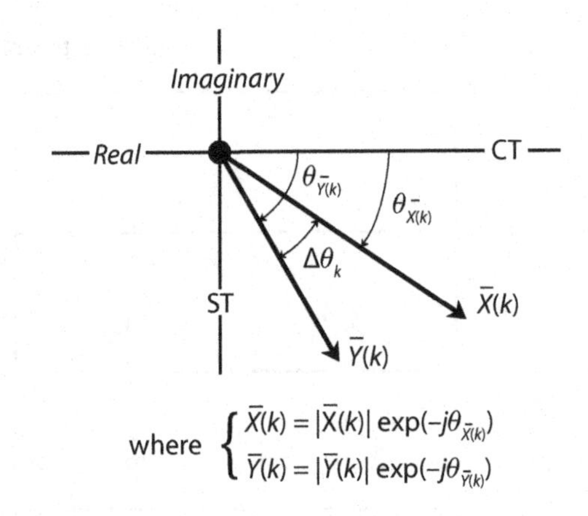

FIGURE A.25 The kth wavevectors for maps **X** and **Y** represented as complex polar coordinates.

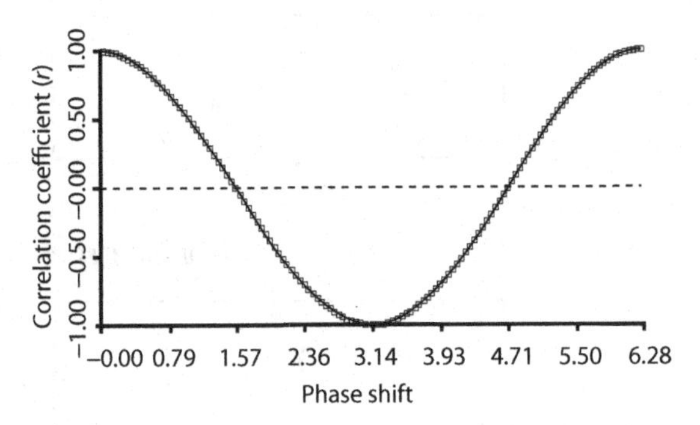

FIGURE A.26 The meaning of the correlation coefficient described by Equation A.74.

This fundamental result relates several important spectral concepts in the literature, including

$$\begin{aligned} \cos(\Delta\theta_k) &= \frac{\sigma^2_{\mathcal{X}(\mathbf{k}),\mathcal{Y}(\mathbf{k})}}{\sqrt{\sigma^2_{\mathcal{X}(\mathbf{k})}\sigma^2_{\mathcal{Y}(\mathbf{k})}}} \\ &\text{(e.g. ARKANI-HAMED and STRANGWAY, 1986)} \\ &= \frac{Re[\mathcal{X}(\mathbf{k})\mathcal{Y}(\mathbf{k})^*]}{\sqrt{[\mathcal{X}(\mathbf{k})\mathcal{X}(\mathbf{k})^*][\mathcal{Y}(\mathbf{k})\mathcal{Y}(\mathbf{k})^*]}} \text{ (KAULA, 1967)} \\ &= Re\left[\frac{\mathcal{X}(\mathbf{k})}{\mathcal{Y}(\mathbf{k})}\right]\left[\frac{|\mathcal{Y}(\mathbf{k})|}{|\mathcal{X}(\mathbf{k})|}\right] \text{ (DORMAN and LEWIS, 1970)}, \end{aligned} \tag{A.75}$$

where the raw, unnormalized ratio $[\mathcal{X}(k)/\mathcal{Y}(k)]$ is commonly called the in-phase response or admittance function between signals **X** and **Y**. DORMAN and LEWIS (1970) and subsequently others (e.g. WATTS, 2001) use the admittance function between elevation or bathymetry and gravity to investigate the lithospheric response to varying topographic and geologic loads.

To investigate **X** and **Y** for correlative features, Equation A.74 is evaluated for all k-wavenumbers to establish the correlation spectrum of the signals. From the correlation spectrum, the wavenumber components are chosen with correlation coefficient values that conform to the desired feature associations. Inversely transforming the selected wavenumber components in both signals maps out the correlative features for analysis.

For example, inversely transforming the wavenumber components for which $r(k) \geq 0.70$ favors the stronger positively correlated data variations between the signals, whereas the components with $r(k) \leq -0.70$ reflect the stronger negatively correlated data features. Inversely transforming wavenumber components for which $-0.30 \leq r(k) \leq +0.30$, on the other hand, enhances the presence of null correlated data features.

Modifying the spectral properties of the data by window carpentry or padding the data margins with zeroes greatly affects the correlation spectrum (JONES, 1988). Thus, these modifications are avoided in correlation filtering, which is very beneficial for other spectral applications. When edge effects are unacceptably severe in the correlation filtered output, every effort is made to expand the data margins with real observations. Where this is not possible, folding out the data rows and columns across the data margins always results in more meaningful correlation filtered output than adding rinds of zeroes to the data sets (JONES, 1988).

In general, meaningful filter outputs account for significant levels of the input energy. Thus, the standard deviations of the input and output data are typically assessed to monitor the energy levels. These efforts also help establish effective r-threshold levels of the correlation filter. For example, filtering for wavenumber components satisfying $r \geq 0.95$ may significantly improve the signal-to-noise ratio by Equation A.72, but the result will be meaningless if the output energy is in the noise level of the data.

As with any standard filter, spectral correlation filtering cannot produce results that are not already evident in the original data sets. Indeed, an important test of the veracity of the correlation filtered anomalies is to verify their existence in the original data sets. Spectral correlation filters can be effective in isolating anomaly correlations in large and complex data sets. However, the occurrence of correlated anomalies is not necessarily indicative of a common subsurface source, owing to the source ambiguity of potential fields. Gravity and magnetic anomaly applications of correlation filtering are considered further in Sections 6.5.3 and 12.4.2, respectively.

(D) Derivative and integral filters To this point, only standard filters constructed from simple patterns of 1s and 0s have been considered. For derivative and integral filters, however, specialized transfer functions involving non-integer coefficients are required that are simple extensions of the signal's wavenumbers. For example, taking the first order partial derivative of the signal **B** along its n-axis in Equation A.66 showed that the transfer function coefficients are $j\frac{2\pi}{N}k = j\omega_k = j2\pi f_k$ so that $\partial\mathbf{B}/\partial n \Longleftrightarrow [j\frac{2\pi}{N}k]\mathcal{B}$ and $\partial/\partial n \Longleftrightarrow [j\frac{2\pi}{N}k]$ form transform pairs. Furthermore, it can be shown that $\partial^p/\partial n^p \Longleftrightarrow [j(\frac{2\pi}{N})k]^p$ by taking the pth order derivative of Equation A.66.

These results also hold for derivatives taken along the m-axis of the 2D gridded data set modeled by the transform Equation A.48 for which $\partial^p/\partial m^p \Longleftrightarrow [j(\frac{2\pi}{M})l]^p$. Now, gravity and magnetic fields follow Laplace's equation

$$\nabla^2\mathbf{B}_{M,N} = \left(\frac{\partial^2}{\partial m^2} + \frac{\partial^2}{\partial n^2} + \frac{\partial^2}{\partial z^2}\right)\mathbf{B}_{M,N} = 0, \qquad \text{(A.76)}$$

which relates the measured field variations along the orthogonal map dimensions m and n to the field's variations in the vertical z direction that is perpendicular to the map. Thus, in theory, the simple transfer function coefficients for the measured horizontal derivatives of the field can be used to predict the field's vertical derivatives.

To see this, take the Fourier transform of Equation A.76 to obtain the linear second-order differential equation

$$\frac{\partial^2\mathcal{B}_{l,k,z}}{\partial z^2} = (2\pi)^2[f_l^2 + f_k^2]\mathcal{B}_{l,k,z} = a\mathcal{B}_{l,k,z} \qquad \text{(A.77)}$$

for the data array at altitude z. This equation has the general solution

$$\mathcal{B}_{l,k,z} = Ce^{+z\sqrt{a}} + De^{-z\sqrt{a}} \qquad \text{(A.78)}$$

with constant coefficients C and D. However, the physically realistic form of the solution requires that $(\partial\mathcal{B}_{l,k,z}/\partial z) \longrightarrow 0$ as $z \longrightarrow \infty$. Invoking these boundary conditions gives

$$\lim_{z\to\infty} \mathcal{B}_{l,k,z} = Ce^{+\infty\sqrt{a}} + De^{-\infty\sqrt{a}} = 0, \qquad \text{(A.79)}$$

which implies that $C \equiv 0$ for Equation A.79 to hold. Thus, the realistic form of the solution is

$$\mathcal{B}_{l,k,z} = De^{-z\sqrt{a}} \qquad \text{(A.80)}$$

so that

$$\frac{\partial\mathcal{B}_{l,k,z}}{\partial z} = \left[-2\pi\sqrt{f_l^2 + f_k^2}\right]\mathcal{B}_{l,k,z}. \qquad \text{(A.81)}$$

TABLE A.4 Spectral filter coefficients for estimating p-order derivatives and integrals in the horizontal (H) and vertical (V) directions.

pth order	Differentiation	Integration	Filter coefficients
mth H-axis	$p > 0$	$p < 0$	$[j2\pi(l/M)]^p$
nth H-axis	$p > 0$	$p < 0$	$[j2\pi(k/N)]^p$
zth V-axis	$p > 0$	$p < 0$	$[-2\pi\sqrt{(l/M)^2 + (k/N)^2}]^p$

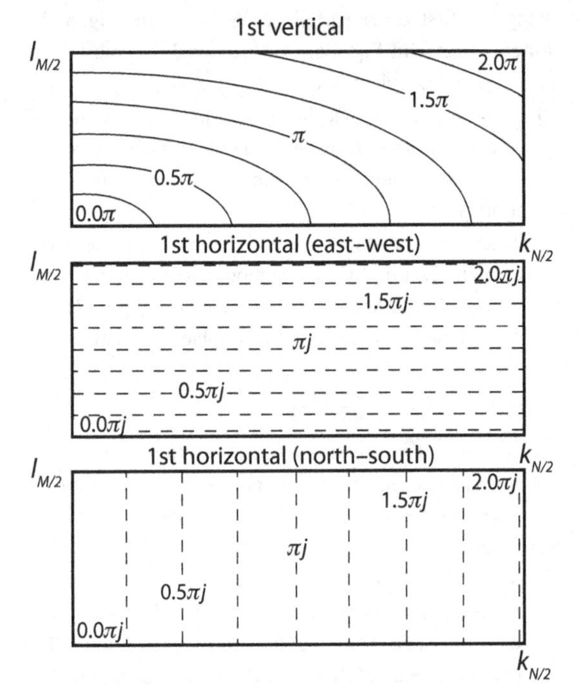

FIGURE A.27 Ideal filter operators for the vertical and horizontal derivatives to order $p = 1$. The design of these filters requires only a single quadrant. Adapted from REED (1980).

In other words, these results show that the vertical derivative of the signal forms the Fourier transform pair $(\partial \mathbf{B}_{M,N}/\partial z) \Longleftrightarrow [-2\pi\sqrt{f_l^2 + f_k^2}]\mathcal{B}_{l,k}$, or more generally that $(\partial^p \mathbf{B}_{M,N}/\partial z^p) \Longleftrightarrow [-2\pi\sqrt{f_l^2 + f_k^2}]^p \mathcal{B}_{l,k}$ holds for any order p of the vertical derivative. Table A.4 summarizes the transfer function coefficients for horizontal and vertical differentiation and the corresponding inverse operations of integration.

As examples, Figure A.27 shows ideal first-derivative filters designed for the 16 × 32 data array in Figure A.7. Figure A.28 gives the ideal vertical second and fourth derivative filters in the left and right panels, respectively. Derivative filters clearly amplify the higher-frequency over the lower-frequency components and thus work like high-pass/low-cut filters. However, the noise amplification in the higher-frequency components due to measurement, sampling, and other errors can limit substantially the veracity of gravity and magnetic derivative estimates, especially those with orders $p > 2$.

For the inverse operation of integration, the derivative transfer function coefficients are simply inverted to obtain the integration filter coefficients as shown in Table A.4 (e.g. JENKINS and WATTS, 1968). Thus, $\underbrace{\int\int\cdots\int}_{p} \mathbf{B}_{M,N}(dn^p \text{ or } dm^p) \Longleftrightarrow$ $([j(\frac{2\pi}{N})k]^{-p}$ or $[j(\frac{2\pi}{M})l]^{-p})\mathcal{B}_{l,k}$ are the Fourier transform pairs for the p-integrals of the signal in the n- or m-directions, respectively. Furthermore, by virtue of Laplace's equation,

$$\underbrace{\int\int\cdots\int}_{p} \mathbf{B}_{M,N}(dz^p) \Longleftrightarrow \left[-2\pi\sqrt{f_l^2 + f_k^2}\right]^{-p} \mathcal{B}_{l,k}$$

for the p-integrals of the signal in the vertical z-direction.

Integration filters amplify lower- over higher-frequency components and thus are like low-pass/high-cut filters. Their outputs are much more stable and less noisy than derivative filter outputs. However, in the application of integral filters, care must be taken to set to zero the zeroth harmonic component (i.e. $\mathsf{b}_0 \equiv 0$), as well as any other wavenumber component that is effectively zero within working precision. In addition, effective numerical performance may require extensive padding of the data set, perhaps out to one or two times the original dimensions of the data set.

As an example, consider the gravity effects of the five horizontal cylinders in Figure A.29 as modeled from Equation A.8 with the regional field set to zero mGal (i.e. $C = 0$). The top panel shows the parameters of the five cylinders, and the middle panel gives their gravity effects over a 90 km simulated survey transect at 1 km station intervals. To help optimize the performance of the derivative and integral filters, the value at each end of the simulated signal was extended another 100 km as shown in the bottom panel.

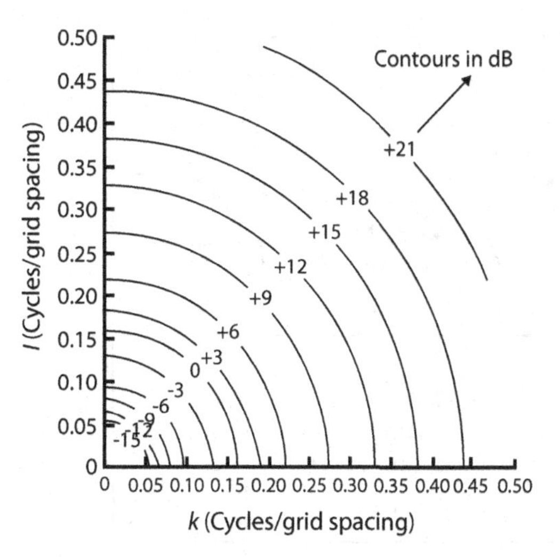

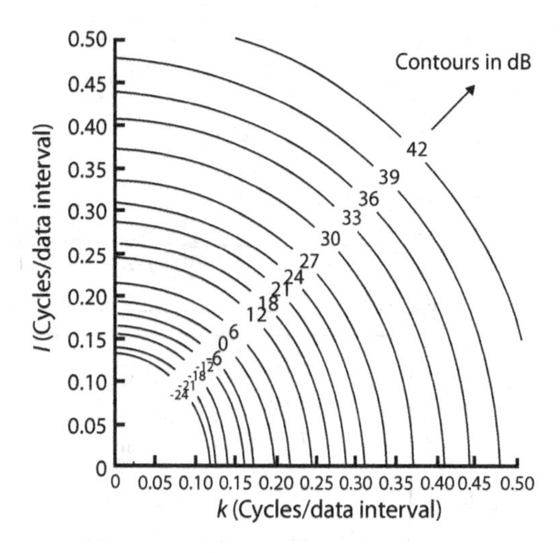

FIGURE A.28 Ideal filter operators for the vertical derivatives to orders $p = 2$ (left) and $p = 4$ (right). The design of these filters requires only a single quadrant. Adapted from FULLER (1967).

Figure A.30 compares spectrally determined vertical derivatives and integrals of the simulated signal against those derived analytically from Equation A.8. In the top panel, the first vertical derivative spectral estimate for $p = 1$ compares quite well with the analytical derivative except for some slight mismatches at the higher gradient peaks and at the edges of the signal. For the second vertical derivative spectral estimate with $p = 2$ in the middle panel, these mismatches are somewhat more severe, but the overall comparison with the analytical second derivative is still very good. In the bottom panel, the analytical second vertical derivative signal was spectrally integrated with $p = -2$ to obtain the original simulated anomaly effects with virtually no error.

Suppose now that the buried horizontal cylinders are also presumed to be all filled by material with a magnetization intensity J induced by a vertically or z-inclined polar geomagnetic field. This assumption can be tested by multiplying the first vertical derivative gravity anomalies in the top panel by $(-J/G\sigma)$ to transform them into the pseudomagnetic vertical component anomalies B_z (Equation 3.110) for comparison against actual magnetic measurements. Similarly, the pseudogravity effects may be obtained from the vertical magnetic anomalies by spectrally integrating B_z with $p = -1$ and multiplying by $(-G\sigma/J)$.

(E) Interpolation and continuation filters Specialized filters can also be developed to estimate data values at unsurveyed locations. For example, to densify the data grid for additional data estimates at one-half the original data interval, a zero can be inserted at the Nyquist end of the spectrum, or if additional estimates at one-quarter the data interval are desired, the transform can be padded with two zeroes, etc. This approach does not modify the spectral properties of the data set and is purely an interpolation scheme. Sometimes slight smoothing by low-pass filtering helps to make the interpolated estimates appear more meaningful.

Because gravity and magnetic observations satisfy Laplace's equation, upward- and downward-continuation filters can be developed to estimate data variations in the vertical z-direction above and below the data grid, respectively. To obtain the transfer function coefficients for upward continuation, for example, Equation A.81 is rearranged into the differential equation $(\partial\mathcal{B}_{l,k}/\mathcal{B}_{l,k}) = [-2\pi\sqrt{f_l^2 + f_k^2}]\partial z$ and integrated to obtain

$$\ln(\mathcal{B}_{l,k})|_{\mathcal{B}_{l,k,z=0}}^{\mathcal{B}_{l,k,z}} = \left[-2\pi\sqrt{f_l^2 + f_k^2}\right] z|_{z=0}^{z=z} \quad \ni$$

$$\ln(\mathcal{B}_{l,k,z}/\mathcal{B}_{l,k,z=0}) = \left[-2\pi\sqrt{f_l^2 + f_k^2}\right] z. \tag{A.82}$$

Taking the exponential of both sides of the above equation gives

$$\mathcal{B}_{l,k,z}/\mathcal{B}_{l,k,z=0} = e^{\left[-2\pi\sqrt{f_l^2+f_k^2}\right]z} \quad \ni$$

$$\mathcal{B}_{l,k,z} = (\mathcal{B}_{l,k,0})e^{\left[-2\pi\sqrt{f_l^2+f_k^2}\right]z} \tag{A.83}$$

so that $\mathbf{B}_{M,N,z>0} \Longleftrightarrow (\mathcal{B}_{l,k,0})e^{[-2\pi\sqrt{f_l^2+f_k^2}]z}$ is the Fourier transform pair for upward continuation.

For downward continuation, the procedure is the same except that the limits of integration are interchanged so

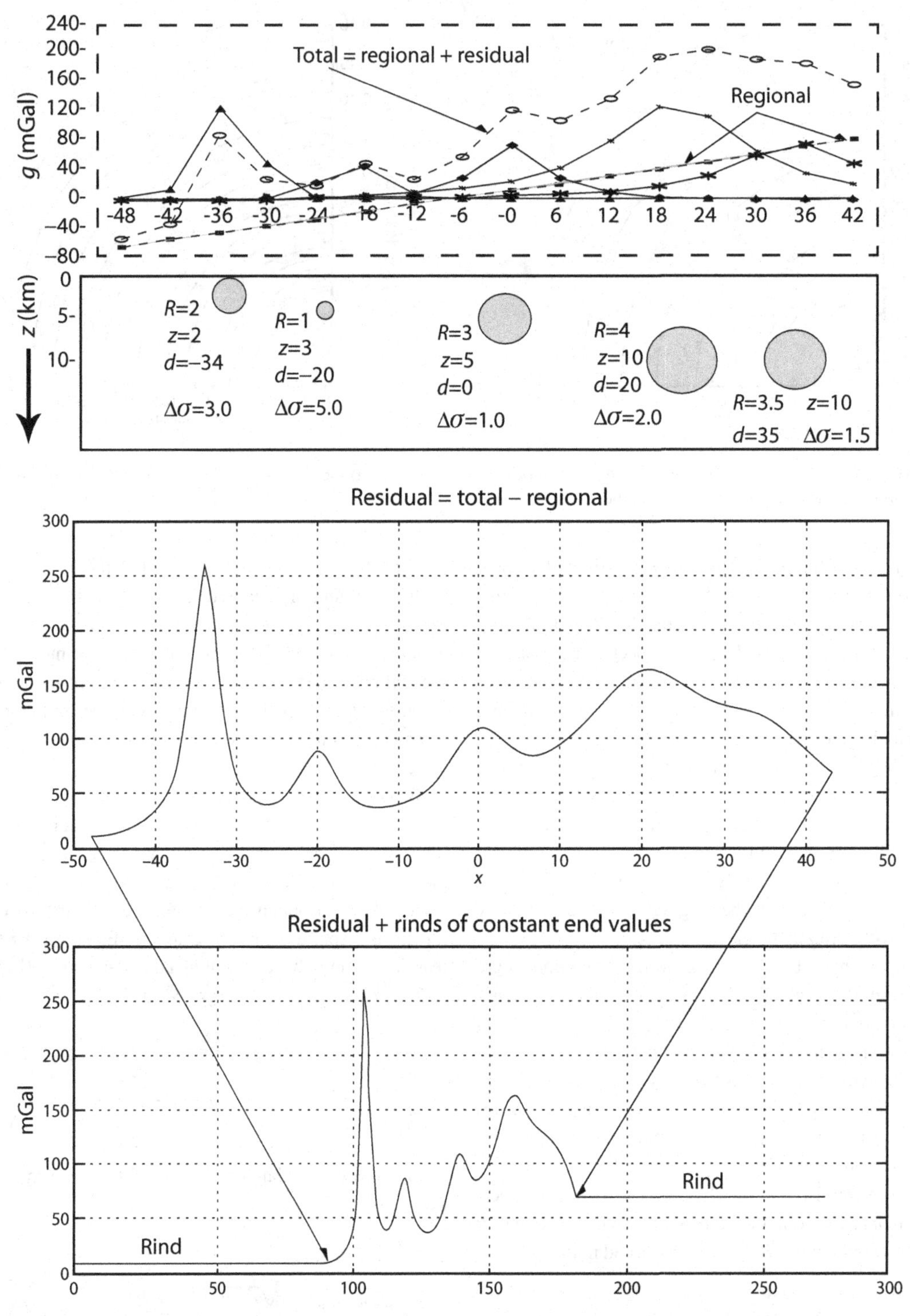

FIGURE A.29 Example of the gravity effects of five horizontal cylinders. The top panel gives the parameters of the cylinders, whereas the middle panel shows the gravity effects over a 90 km survey at 1 km intervals which are extended as a rind of constant values in the bottom panel to enhance the accuracy of the derivative and integral estimates given in Figure A.30.

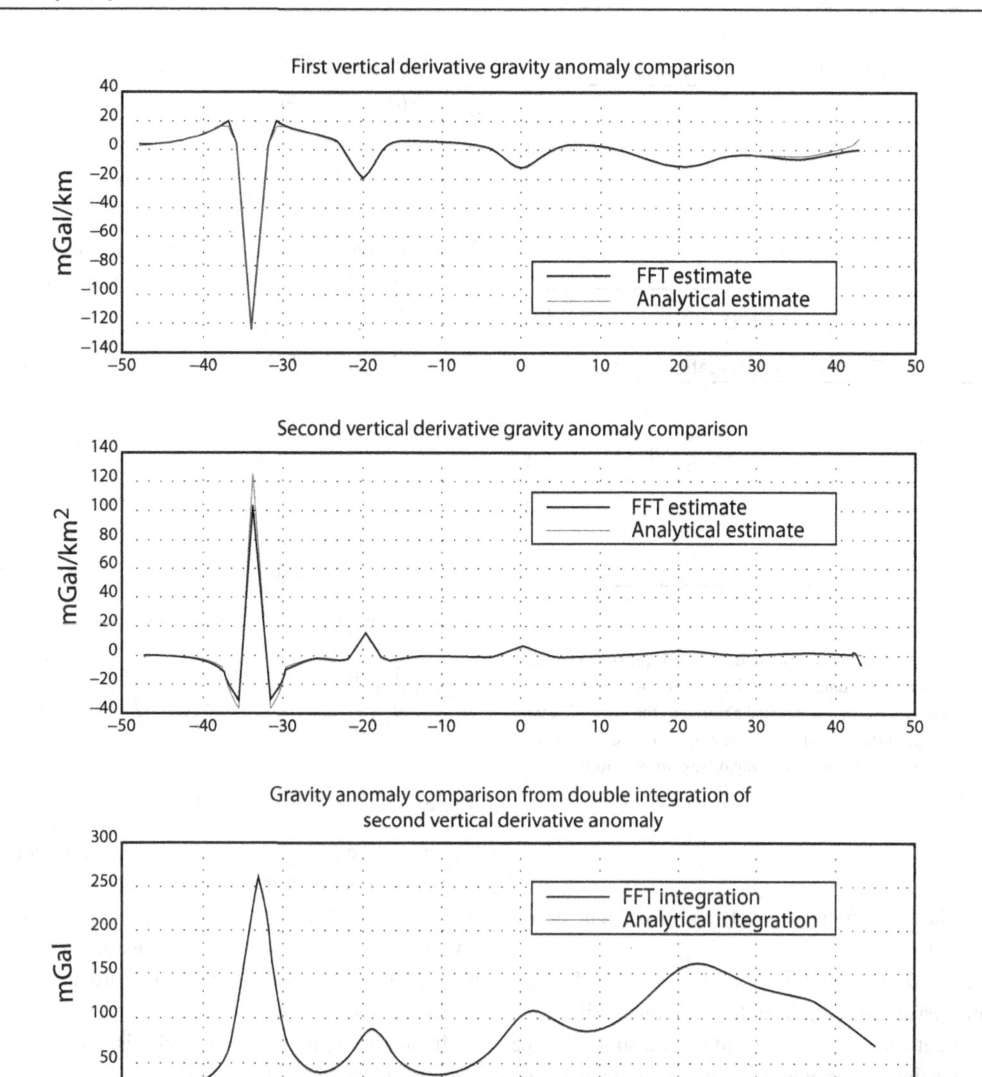

FIGURE A.30 Vertical first (top) and second (middle) derivatives estimated spectrally for the gravity effects in Figure A.29. The comparison (bottom) of the double spectral integration of the derivative in the middle panel to the original gravity effects reveals negligible differences.

that

$$\ln(\mathcal{B}_{l,k})|_{\mathcal{B}_{l,k,z=0}}^{\mathcal{B}_{l,k,-z}} = \left[-2\pi\sqrt{f_l^2 + f_k^2}\right] z|_0^{-z} \ni$$

$$\ln(\mathcal{B}_{l,k,-z}/\mathcal{B}_{l,k,z=0}) = \left[2\pi\sqrt{f_l^2 + f_k^2}\right] z. \quad \text{(A.84)}$$

Taking the exponential again of both sides gives

$$\mathcal{B}_{l,k,-z}/\mathcal{B}_{l,k,z=0} = e^{\left[2\pi\sqrt{f_l^2+f_k^2}\right]z} \ni$$

$$\mathcal{B}_{l,k,-z} = (\mathcal{B}_{l,k,0})e^{\left[2\pi\sqrt{f_l^2+f_k^2}\right]z} \quad \text{(A.85)}$$

so that $\mathbf{B}_{M,N,z<0} \iff (\mathcal{B}_{l,k,0})e^{\left[2\pi\sqrt{f_l^2+f_k^2}\right]z}$ is the Fourier transform pair for downward continuation.

Upward and downward continuations are obviously inverse operations with transfer function coefficients given by

$$\exp\left(\left[-2\pi\sqrt{f_l^2 + f_k^2}\right] z\right), \quad \text{(A.86)}$$

where the elevation of continuation $z > 0$ for upward continuation and $z < 0$ for downward continuation. Figure A.31 shows ideal upward- (upper panel) and downward- (lower) continuation filters to continue the 16×32 data array in Figure A.7 through elevation differences of three times the 1 km grid interval. Figure A.32 gives the ideal frequency responses for upward-continuation filters that

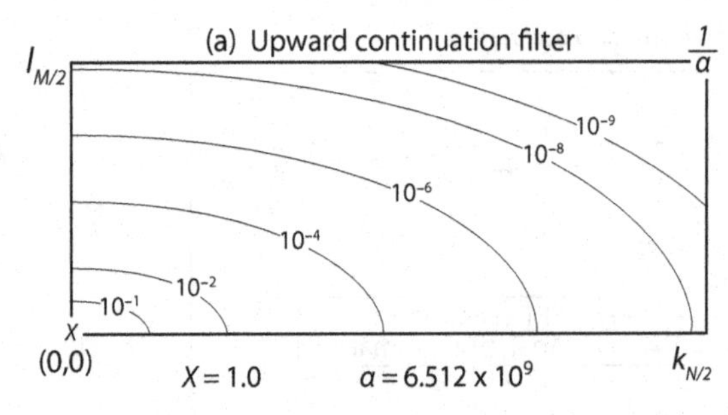

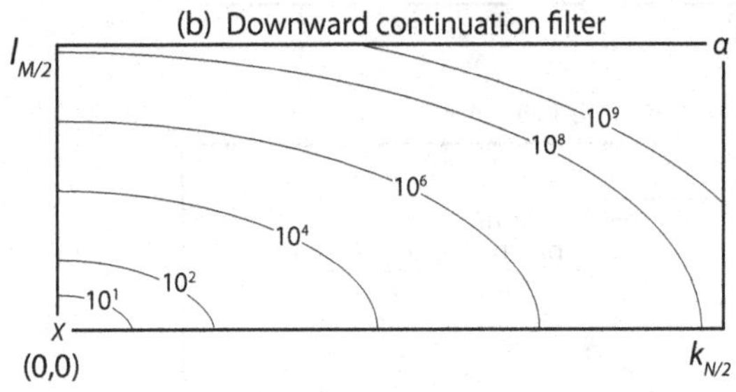

FIGURE A.31 Ideal upward- (a) and downward- (b) continuation filter coefficients for continuations of the data to three station intervals above and below the original altitude of the 16×32 data array. The design of these filters requires only a single quadrant with coefficients ranging from the x to α values shown. Adapted from REED (1980).

continue the data one (left panel) and two (right panel) grid intervals.

Clearly, upward-continuation filters smooth data and emphasize their regional features much like low-pass/high-cut filters do. Downward continuations, on the other hand, tend to promote the higher-frequency features in the data like high-pass/low-cut filters. Continuation estimates are most reliable over elevations within a few grid intervals of the observations because of measurement errors and the non-uniqueness of continuation (e.g. VON FRESE *et al.*, 2005). To be certain of gravity and magnetic field behavior at more distant elevations, there is no recourse but to survey. However, multiple altitude anomaly grids are becoming increasingly available from surface, airborne, and satellite surveys where both the C and D coefficients in Equation A.78 can be applied to evaluate anomaly values at the intermediate altitudes beween the grids.

In particular, let the lower-altitude anomaly grid be referenced to relative zero and the higher-altitude grid referenced to altitude H where the Fourier transforms of $\mathbf{B}(0)$ $(\equiv \mathbf{B}_{M,N,0})$ and $\mathbf{B}(H)$ $(\equiv \mathbf{B}_{M,N,H})$ are $\mathcal{B}(0)$ $(\equiv \mathcal{B}_{l,k,0})$ and $\mathcal{B}(H)$ $(\equiv \mathcal{B}_{l,k,H})$, respectively. Both transforms may be related through Equation A.78 to common C and D coefficients given by

$$C = \frac{\mathcal{B}(0)e^{-z\sqrt{a}} - \mathcal{B}(H)}{e^{-z\sqrt{a}}} \quad \text{and}$$

$$D = \frac{\mathcal{B}(H) - \mathcal{B}(0)e^{+H\sqrt{a}}}{e^{-z\sqrt{a}}}. \tag{A.87}$$

Thus, the interval continued field at an intermediate z-altitude (i.e. $0 \leq z \leq H$) can be obtained by inversely transforming

$$\begin{aligned}\mathcal{B}(z) &= \frac{1}{q}\Big[\mathcal{B}(0)e^{-H\sqrt{a}}e^{+z\sqrt{a}} - \mathcal{B}(H)e^{+z\sqrt{a}} + \mathcal{B}(H)e^{-z\sqrt{a}} \\ &\quad - \mathcal{B}(0)e^{+H\sqrt{a}}e^{-z\sqrt{a}}\Big] \\ &= \frac{1}{q}\Big[\mathcal{B}(0)(e^{+[z-H]\sqrt{a}} - e^{-[z-H]\sqrt{a}}) \\ &\quad + \mathcal{B}(H)(e^{-z\sqrt{a}} - e^{+z\sqrt{a}})\Big]\end{aligned} \tag{A.88}$$

which in compact matrix notation is given by

$$\mathcal{B}(z) = \frac{1}{q}\begin{pmatrix}[e^{+(z-H)\sqrt{a}} - e^{-(z-H)\sqrt{a}}] \\ [e^{-z\sqrt{a}} - e^{+z\sqrt{a}}]\end{pmatrix}^t \begin{pmatrix}\mathcal{B}(0) \\ \mathcal{B}(H)\end{pmatrix} \tag{A.89}$$

with

$$q = e^{-H\sqrt{a}} - e^{+H\sqrt{a}}. \tag{A.90}$$

The interval continuation operator $\mathcal{B}(z)$ fully honors the boundary conditions imposed by multiple altitude slices of the anomaly field, and thus its predictions yield insights on the altitude behavior of anomalies that are not available from standard single-surface upward and downward continuations.

In summary, the spectral model is very powerful in gravity and magnetic data analysis because it has tremendous numerical efficiency and accuracy and does not invoke assumptions about the source of the data variations. For hardly more than the effort of gridding, the spectral model provides access to the basic wavelength, directional, correlative, differential, integral, interpolation, and continuation properties of the data. The chapters in this book describe numerous examples of the above general spectral analysis techniques, as well as more specialized spectral methods for estimating specific attributes of gravity and magnetic anomalies. The next section investigates the critical role of source modeling in analyzing these data.

A.5.2 Anomaly source modeling

A principal objective in mapping the Earth's gravity or magnetic field variations is to develop plausible models of the subsurface from the inversion of the residual anomalies. As illustrated in Figure A.33, this process is a quasi-art

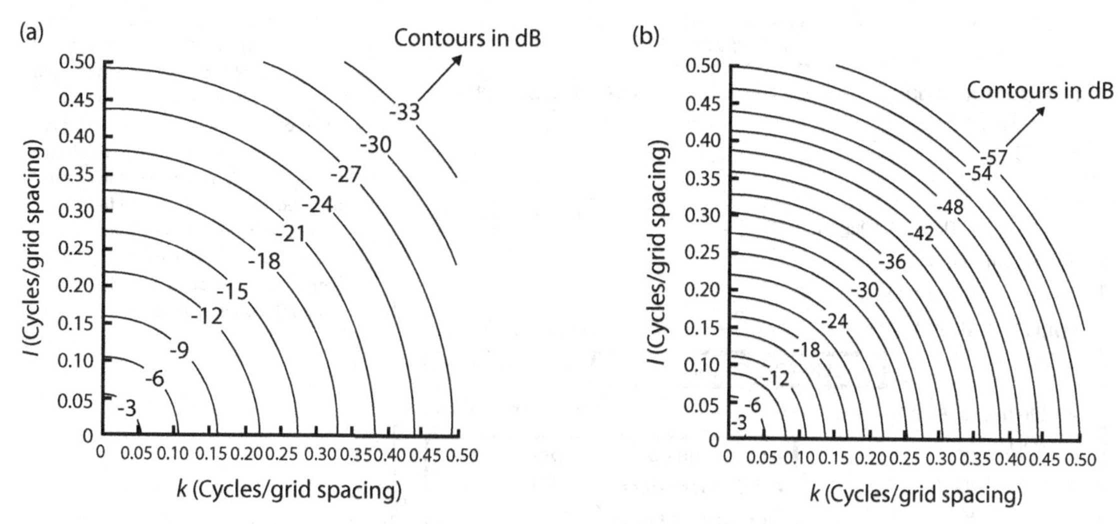

FIGURE A.32 Ideal upward-continuation filters to continue data (a) one- and (b) two-station intervals. Changing the sign of the transfer function coefficients continues the data downward one- and two-station intervals, respectively. Adapted from Fuller (1967).

form that requires the interpreter to conceive a mathematical model ($f(\mathbf{A}, \mathbf{X})$) with known and unknown parameters **A** and **X**, respectively, that can be related to variations in the observed data (**B**). The mathematical model is a simple, usually linear, approximation of the conceptual model that the interpreter proposes to account for the data variations. Assumptions arise both implicitly and explicitly to convert the conceptual model into the simple mathematical model. They critically qualify the applicability of the inversion solution, and thus should be reported with the conclusions and interpretations of the analysis. Figure A.34 illustrates two examples of translating conceptual geological models into simplified mathematical models for inversion.

The forward modeling component of inversion involves evaluating the mathematical model for synthetic data (i.e., $\hat{b} \in \hat{\mathbf{B}}$ in Figure A.33). Gravity forward models of the 3D body were developed in Chapter 3 that provide synthetic estimates of the potential (Equation 3.23), the three vector components (Equations 3.24–3.26), and the nine gradient tensors (Equations 3.28–3.33). Chapter 9 also developed the corresponding 3D forward magnetic models that offer synthetic estimates of the magnetic potential (Equation 9.25), the three vector components (Equations 9.27–9.29) and total field (Equation 9.30), and the nine magnetic gradient tensors (Equations 9.40–9.45).

The veracity of an inversion's solution is judged on how well the predicted data ($\hat{\mathbf{B}}$) compare with the observed data (**B**), based on quantitative criteria such as the raw differences ($\hat{\mathbf{B}} - \mathbf{B}$) or squared differences $(\hat{\mathbf{B}} - \mathbf{B})^2$ or some other norm. In general, a perfect match between the predictions and observations is not necessarily expected or desired because the observational data (**B**) and the coefficients (**A**) assumed for the forward model are subject to inaccuracy, insufficiency, interdependency, and other distorting effects (Jackson, 1972).

Data inaccuracies result from mapping errors, random and systematic, related to imprecision in the measurement equipment or techniques, inexact data reduction procedures, truncation (round-off) errors, and other factors. They scatter the data about a general trend or line (e.g. Figure A.34) so that a range of model solutions yields predictions within the scattered observations. However, as described in Appendix A.4.2, the statistical properties of the observational data (**B**) can be propagated to characterize uncertainties in the solution variables (**X**).

Insufficient data do not contain enough information to describe the model parameters completely. Examples include a lack of data observations on the signal's peak, trough, flank, or other critical regions. The data coverage also may not be sufficiently comprehensive to resolve critical longer-wavelength properties of data, or the station interval may be too large to resolve important shorter wavelength features in the data. Another source of data insufficiency is the imprecision in the independent variables (**A**) that promotes solution biases. In general, data insufficiency issues are difficult to resolve without new data observations or supplemental geologic and geophysical information.

Interdependency in data sets is where the independent variables for two or more data points cannot be distinguished from each other within the working precision of the mathematical model. Examples include data points

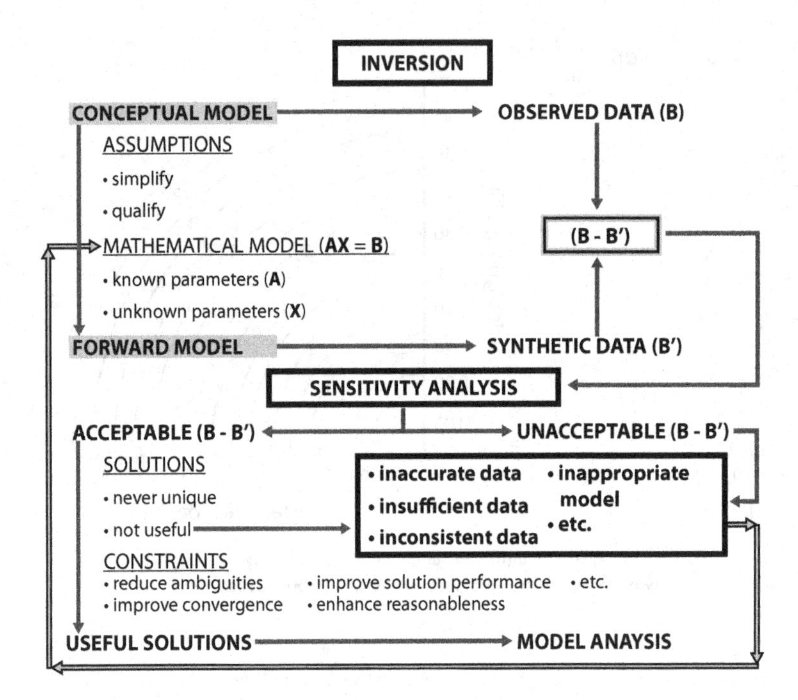

FIGURE A.33 The process of inversion (solid arrows) establishes a forward model (AX) of the observations (B) and performs a sensitivity analysis to find a subset of possible solutions that may give new insights on the processes producing the data variations. The feedback loop (black-bordered gray arrows) from unacceptable residuals and useless solutions to the mathematical model reformulates the forward model until a useful solution is produced.

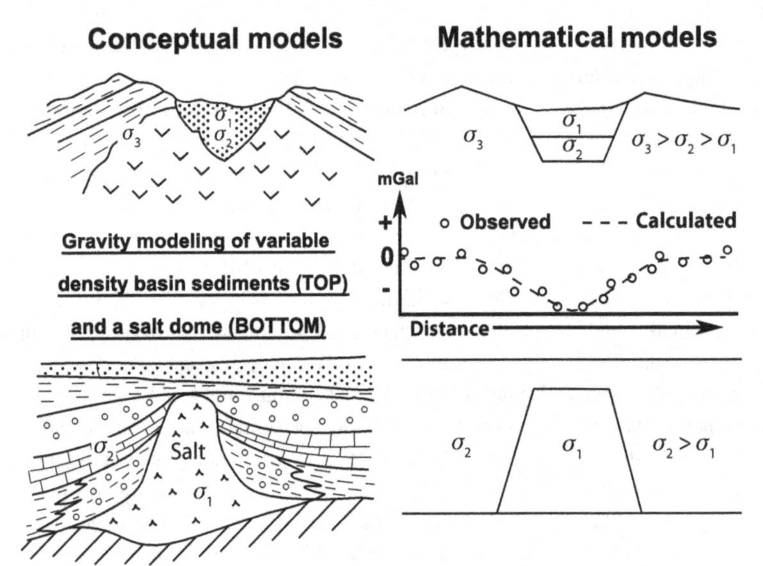

FIGURE A.34 Schematic examples of the inversion of gravity anomaly data for the subsurface distributions of lower-density sediments with densities that increase with depth within a basin surrounded by higher-density rocks (upper row) and a lower-density salt dome displacing higher-density sedimentary rocks (bottom row). Both conceptual models can be mathematically represented by the simplified density models on the right that predict a relative gravity minimum over the center of the sources, as illustrated by the generic gravity anomaly calculated in the middle panel of the right column.

that are too close to each other for the mathematical model to discriminate, or where perfect symmetry exists so that only half of the data observations need to be considered in the inversion (e.g. Figure A.1). Incorporating dependent data into the computations provides no new information and results in considerable computational inefficiency and solutions with erratic and often unreasonable performance characteristics. However, as described in Appendix A.4.2, this problem can be mitigated somewhat using random noise in the **A**-coefficients to optimize the solution for acceptable performance attributes with only marginal growth in the deviation between the observed and predicted data.

The use of an inappropriate model for inversion results in significant negative consequences ranging from invalid solutions to valid solutions produced from excessive or unnecessary effort. Compromised solutions occur, for example, from attempts to fit a 2D model to data that exhibit 3D effects, or a homogeneous model where the effects of anisotropy are evident. In addition, the use of a complex model where a simpler one can achieve the objectives of the inversion leads to unnecessary and often unacceptable expenditures of effort. As we emphasize throughout the book, however, gravity and magnetic fields vary inversely with distance to their sources so that, in theory at least, an infinite number of equivalent source distributions can be fit to these fields. Thus, the inherent non-uniqueness of these solutions allows considerable flexibility in developing appropriate models for effective gravity and magnetic data inversions.

In general, the solutions of gravity and magnetic inverse problems are not unique and require the application of constraints to identify the subset of the more effective solutions for a given objective. Constraints are additional information about the model parameters that when imposed on the inverse problem help reduce interpretational ambiguities, improve convergence to a solution, enhance the reasonableness of the solution and its predictions, and otherwise improve the effectiveness of the solution.

The identification of a useful solution or set of solutions has only established a set of numbers for **X** that map the assumed numbers in **A** into the observed numbers of **B**. These three sets of numbers constitute the core of the inversion and the basis for other investigators to test the modeling efforts. The final stage of the process interprets these numbers for a story and graphics in a report that addresses the veracity of the conceptual hypothesis, the consequences of accepting or rejecting the hypothesis, alternate interpretations, and other issues for applying the inversion results effectively.

In summary, the creative, artful elements of anomaly modeling or inversion involve developing the conceptual and forward models and the final interpretation. The analytical efforts to obtain solutions (**X**), by contrast, are routine and well known since Gauss and Fourier established them at the beginning of the nineteenth century. In practice, the largest analytical challenge is the sensitivity analysis required to identify an optimal set of solution coefficients for the given modeling objective (Appendix A.4.2).

In the respective gravity and magnetic anomaly interpretation Sections 7.5 and 13.5, forward modeling involved establishing the solution **X** by trial-and-error adjustments of the unknown parameters until the predictions of the forward model reproduce the observed data within a specified margin of error. Inverse modeling, on the other hand, referred to determining the unknown parameters directly from the observed data using an optimization process comparing the observed and predicted data. Anomaly inversions involving a relatively small number (usually 5–10) of unknowns are commonly implemented by forward modeling, whereas inverse modeling is typically invoked for greater numbers of unknowns, although it is applicable to any number of unknowns.

A.6 Data graphics

Modern reporting of gravity and magnetic anomaly analyses requires the use of effective graphics to represent the anomalies and their inversions. Graphics may range from simple line plots and cartoons to elaborate 3D color plots and animations of the data and analytical results that are made possible by a variety of efficient computer processing and graphics software and hardware. Simple listing of data is often available, but seldom used. Rather, the data are presented in graphical form in a spatial context or less frequently tied to a temporal base for ease of viewing and analysis. These presentations greatly enhance the use and understanding of the gravity and magnetic anomaly data, but care is required to use graphical formats and parameters that present the data and results with minimal bias and distortion.

In this section, rudimentary principles are outlined for developing useful presentations of the data and their interpretations with modern digital graphics. An important graphical consideration in any application, for example, is the choice of the coordinate reference frame for projecting gravity or magnetic data onto a map. Presenting data in various forms such as profiles or contour maps requires significant application-dependent choices. This section provides guidelines for the most effective imaging of anomaly data for analysis and interpretation.

Transforming gravity and magnetic data into data grids greatly aids electronic data processing and graphics efforts. However, gridding presents critical data issues because modern algorithms operate at a wide variety of performance levels that can make even unacceptable data look quite suitable for analysis. Thus, the interpreter must be proactive in understanding how the data are gridded for analysis and take suitable precautions when dealing with grids generated by inappropriate or unknown algorithms.

The statistical properties of the data facilitate choosing effective graphical parameters for revealing significant gravity and magnetic anomaly variations. These statistical attributes also allow the interpreter to standardize and normalize data sets into common graphical parameters where feature associations and other interpretational insights are more apparent.

A.6.1 Map projections and transformations

Gridded anomaly data of areal gravity and magnetic studies are presented in map form for analysis and interpretation. This requires the projection of data observed on the Earth's spherical surface, or more appropriately an oblate spheroid which represents the true Earth shape, onto a flat surface, that is a map. Literally hundreds of projection methodologies have been developed (e.g. SNYDER, 1987). All of these produce distortions from the actual spherical surface, and thus the choice of projection is based on minimizing a particular undesirable attribute of the map, e.g. area, shape, or scale, and its intended use. However, the map projection may be chosen to duplicate the projection of existing maps which are used in conjunction with the anomaly map.

The geographic position of observations is generally given in degrees of longitude and latitude to seven decimal places except in local surveys in which the Earth is assumed to be flat and observations are measured in Cartesian coordinates. Currently, to achieve consistency in locations, the International Terrestrial Reference Frame (ITRF) in conjunction with the Geodetic Reference System (GRS80) ellipsoid is used for the horizontal datum. The ITRF differs from the 1984 World Geodetic System (WGS84) which is used in specifying horizontal location in the Global Positioning System (GPS) by less than 10 cm.

A widely used projection for gravity and magnetic anomaly maps of limited regions is the transverse Mercator projection which projects from the Earth's surface onto a cylinder whose axis is in the equatorial plane (i.e. transverse to the normal Mercator projection which projects the Earth's surface onto a cylinder whose axis is coincident with the Earth's axis of rotation) and is tangential to the Earth's surface along a meridian of longitude which is termed the central meridian of the map. This projection is conformable: that is, relative local angles are shown correctly. As a result the scale is constant at the equator and at the central meridian, and the parallels of longitude and meridians of longitude intersect at right angles. The distortion of the map and scale increases with the east–west x distance, and thus the projection is not suitable for world maps.

To achieve high scale accuracy in the transverse Mercator projection the central meridian is changed to coincide with the map area. For world-wide consistency in the universal transverse Mercator projection, the Earth between 84° N and 80° S is divided into 60 zones each generally 6° wide in longitude. The central meridian is placed at the center of each zone. True scale is achieved approximately 180 km on either side of the central meridian. Beyond this distance the scale is too great. Locations in a zone are referenced to the intersection of the central meridian of the zone and the equator. This position in the northern hemisphere is given a coordinate of 500,000 m in the x-direction and zero in the y-direction with coordinates increasing in the east and north directions. In the southern hemisphere, x remains 500,000 m and y is 10,000,000 m.

Whatever the projection, it is desirable to place the origin of the coordinates at a specified position that is held constant in producing maps of a variety of anomalies and other gridded data. This so-called registering of the coordinates aids in joint computational analysis of the various gridded data. Grids generally are made equidimensional for simplicity and convenience and to avoid anisotropic effects due to the different spacing in the orthogonal directions. Commonly grids are oriented in the cardinal directions, i.e. N–S, E–W, in dealing with regions measured in distances measured in hundreds of meters. However, with grids measured in kilometers or greater, the spherical shape of the Earth necessitates modification of the grid from the cardinal directions to maintain the orthogonal, equispaced grid. Contour maps from equidimensional grids can be displayed on non-square grids such as latitude, longitude by transforming the grid to the map projection, but consideration should be given to the bias resulting from the non-square grid values. In local surveys where data are acquired along transects, the grid may be oriented so that one coordinate of the grid coincides with the orientation of the transects to capture the detail available in the higher sampling interval along the transects.

A.6.2 Gridding

Uniformly sampled or gridded anomaly data are preferred for electronic data analysis because the data input and storage requirements are minimal. All that is necessary are the signal amplitudes in their proper spatial order, and a reference point and grid interval from which the independent variable coordinates of the anomaly amplitudes can be computed internally. Converting a set of irregularly spaced, finite, discrete anomaly values into gridded format involves critical choices of the grid interval and the interpolation procedure (McDONALD, 1991).

As discussed in Appendix 4.3 above, the grid interval Δt cannot be larger than one-half the shortest wavelength that is present in the anomaly data. That is, the sampling or grid interval must be at least twice the highest frequency anomaly in the data set. Otherwise, the grid will include erroneous anomaly components that corrupt the anomaly analysis. Gridding at an excessively short interval can of course limit sampling errors, but results in large and possibly unacceptable levels of computational effort. A more computationally efficient grid interval can be ascertained by gridding a representative segment of the anomaly data at successively smaller intervals and noting the interval giving anomalies that do not change appreciably at smaller sampling intervals.

Generally, the grid spacing is chosen so that each grid square will contain at least a single point value or it is set at the approximate average spacing between point values. Where point values are closely spaced along widely spaced transects, it may be advantageous to use an anisotropic gridding scheme, or to smooth the data along the transects before gridding to reflect the across-track Nyquist wavelength defined by the track spacing. Further, it may be useful to insert dummy point values based on manual interpolation in regions of the data field with sparse or no data, but this should be done as a last resort in the gridding process with the dummy point values interpolated conservatively and the number of points held to a minimum.

Gridding schemes commonly are classified according to the prediction strategy employed. A local strategy, for example, predicts a grid value from samples restricted to a window surrounding the grid point, whereas a global strategy utilizes all the anomaly data for estimating the grid value. The choice of a gridding method in practice is controlled by such factors as the characteristics of the data, the desired application, and the interpreter's experiences. However, all gridding schemes are beset by problems of avoiding spurious features in regions of sparse data while at the same time representing short-wavelength data features accurately. They also invariably suffer from edge effects or prediction errors due to the diminished data coverage at the data set margins. It is advisable to screen the gridding for errors by visually checking contour maps prepared from the grid for closed contours that do not contain observations, series of elliptical contours centered on a succession of observations, and abnormally high gradients.

A broad variety of interpolation procedures are available that can be invoked for anomaly gridding. Interpolation involves finding the dependent variable value (i.e. prediction) at a specific value of the independent variable (i.e. prediction point) from a local or global set of surrounding data. As an example, Figure A.35(a) illustrates predicting $b_p = b(t_p)$ along a profile using linear (dashed line) and cubic spline (solid line) interpolation functions that pass exactly through the data observations, and a least-squares polynomial function (dotted line) that approximately fits the observations.

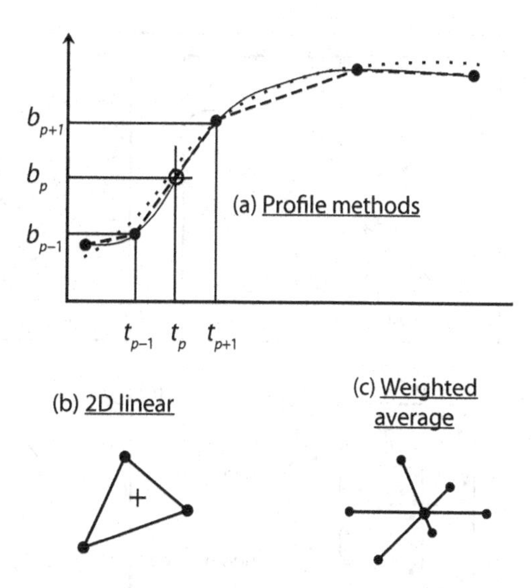

FIGURE A.35 Schematic illustration of predicting the value $b(t_p)$ for (a) a profile between known data points (dots) at the location t_p by linear (dashed line), cubic spline (solid line), and least-squares (dotted line) methods, and for maps using (b) linear interpolation by surrounding triangles and (c) statistical interpolation by inverse-distance weighted averages. Adapted from BRAILE (1978).

Linear interpolation

For profile anomaly data, this local method predicts a grid value based on the two surrounding observations. The prediction is obtained by

$$b_p \equiv b(t_p) = b_{p-1} + \frac{(b_{p+1} - b_{p-1})(t_p - t_{p-1})}{(t_{p+1} - t_{p-1})}, \quad \text{(A.91)}$$

where b_{p+1} and b_{p-1} are the two observations on either side of the prediction b_p as shown in Figure A.35(a).

For anomaly data mapped in 2D spatial coordinates, the linear interpolation can be implemented by fitting the first degree plane to the three nearest observations surrounding the prediction point as shown in Figure A.35(b). Thus, each grid estimate is evaluated via **B** in Equation A.67 from the α, β, and γ coefficients determined in Equation A.68 for the plane through the nearest three surrounding anomaly values. The surrounding triangle method is effective for well-distributed observations, as shown by the example in Figure A.36(a).

The linear interpolation function exactly matches the observations, but invokes no other assumptions concerning the analytical properties of the data and predictions. Thus, the grid predictions provide a relatively honest and pristine view of the data set. However, the method does not smooth the data, and this lack of smoothing may lead to interpolated values which are unsatisfactory for use in analyzing

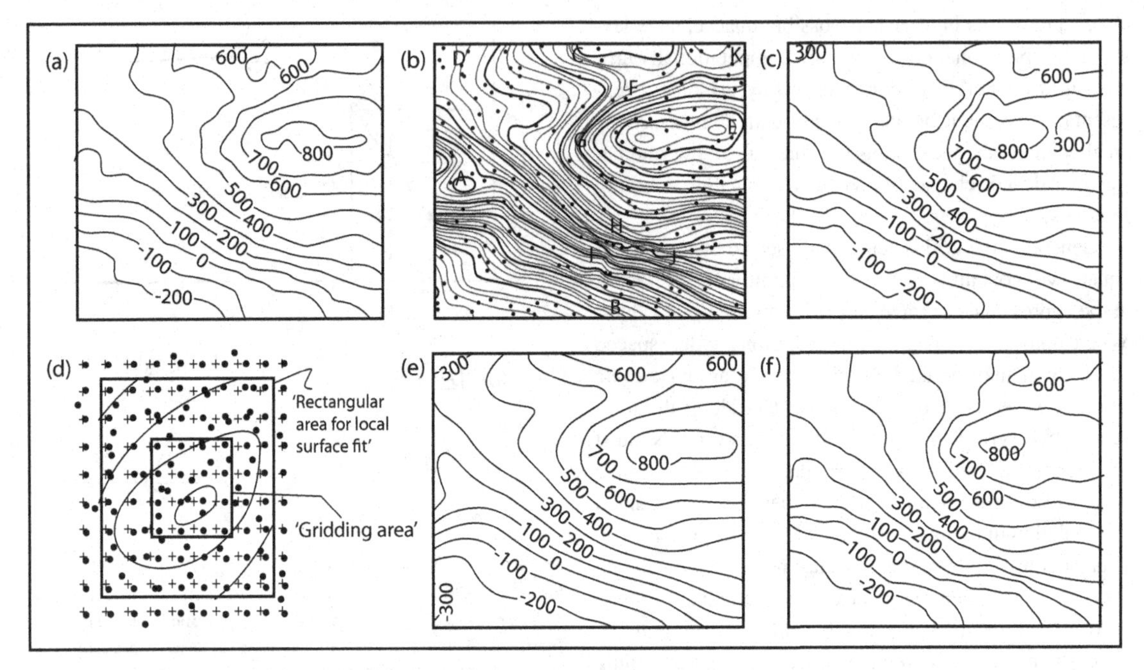

FIGURE A.36 Contour map comparisons of data gridded from 200 randomly located point samples (dots) of the aeromagnetic anomalies in map (b). Map (a) is based on the surrounding triangle method in Figure A.35(b), map (c) is from the intersecting piecewise cubic spline method in Figure A.37(b), and map (f) from the weighted statistical averaging method in Figure A.35(c). Map (e) is from the local polynomial surface fitting method illustrated in map (d). All maps are based on grids of 20×20 values at the interval of 2 km. Contour values are in nT. Adapted from BRAILE (1978).

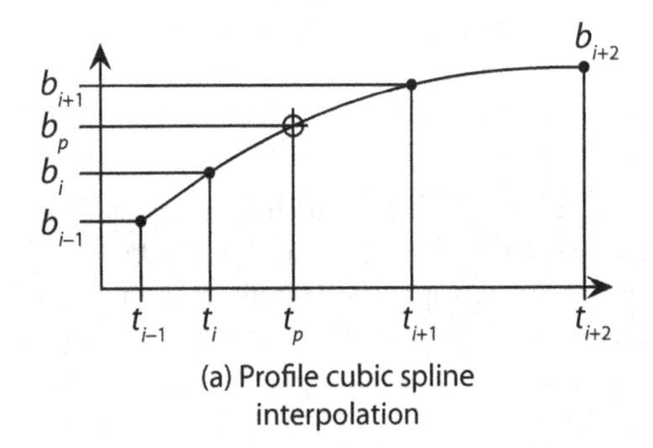

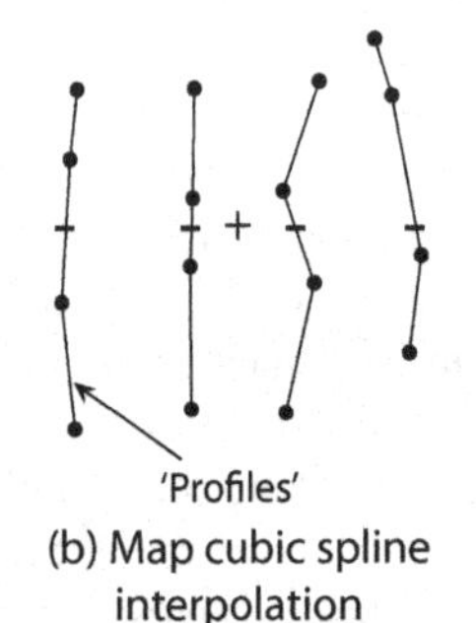

FIGURE A.37 Schematic illustration of predicting (a) a profile value b_p between known data points (dots) at the location t_p by the cubic spline method, and (b) a map value (+) by intersecting piecewise cubic splines.

the higher-wavenumber components of the anomaly data in, for example, high-pass filtering and derivative analyses.

Linear interpolation is a simple, very cost-effective gridding method that imposes no constraints on the data errors. Thus, its utility for rendering the derivative properties of the anomaly data is relatively limited. However, cubic spline interpolation smoothly honors the anomaly observations exactly while also maintaining the continuity of first and second horizontal derivatives at the observations.

Cubic spline interpolation

Cubic splines are piecewise third-degree polynomials that are constrained to maintain continuous first and second horizontal derivatives at each observation point (e.g. DAVIS, 1986; GERALD and WHEATLEY, 1989; MATHEWS and FINK, 2004). For profile data, they are commonly used to predict an anomaly value based on the two observations on either side of the prediction point as shown in Figure A.37(a). Over each ith interval or span between two successive observations b_i and b_{i+1}, a cubic spline function can be determined for the interpolations

$$b_p = c_0 + c_1(t_p - t_i) + c_2(t_p - t_i)^2 + c_3(t_p - t_i)^3$$

$$\forall \quad t_i \leq t_p \leq t_{i+1}. \quad \text{(A.92)}$$

The four coefficients defining the ith spline are

$$c_0 = b_i; \quad c_1 = F_i - \frac{h_i(2S_i + S_{i+1})}{6}; \quad c_2 = \frac{S_i}{2};$$

$$\text{and} \quad c_3 = \frac{S_{i+1} - S_i}{6h_i}, \quad \text{(A.93)}$$

where $h_i = t_{i+1} - t_i$ is the data or span interval, F_i and S_i are the respective first and second horizontal derivatives at the observed anomaly value b_i, and S_{i+1} is the second horizontal derivative at the successive anomaly value b_{i+1}. The horizontal derivatives can be taken numerically so that at b_i, for example,

$$F_i = \frac{\partial b_i}{\partial t} = \frac{b_{i+1} - b_{i-1}}{t_{i+1} - t_{i-1}} = \frac{b_{i+1} - b_{i-1}}{h_i + h_{i-1}}$$

$$\text{and} \quad S_i = \frac{\partial^2 b_i}{\partial t^2} = \frac{F_{i+1} - F_{i-1}}{h_i + h_{i-1}}. \quad \text{(A.94)}$$

To interpolate the N anomaly values to a grid, Equation A.92 must be fitted to each of the $(N - 1)$ spans. However, only $(N - 2)$ derivatives are defined so that the processor must specify the derivatives at the beginning and ending data points of the profile. A common default is to take the natural spline where these derivatives are set to zero. However, the end conditions are arbitrary, and thus are usually taken to minimize spurious grid estimates with wavelengths smaller than the span lengths in the beginning and ending spans. In addition, making the derivatives equal at the first and last points of the profile can force the grid to cover a full period (e.g. GERALD and WHEATLEY, 1989) to help minimize spectral leakage. Simply adding artificial anomaly values at each end of the profile is also effective where the processor can infer the extrapolated properties of the anomaly profile.

As an example, Figure A.38 shows the magnetic field behavior measured at a base station for a ground-based magnetic survey of a prehistoric archaeological site on the north bank of the Ohio River in Jefferson County, Indiana. The natural cubic spline was used to interpolate the drift curve (solid line) from 33 base station observations (dots). The drift corrections were applied relative to the first base station observation of the day as shown by the arrows from the successive base station readings.

For the first half of the day, the cubic spline corrections appear to be well constrained by the observations and thus would be closely matched by simpler linear interpolations. However, over the lunch break, spurious cubic spline corrections were obtained because of the lack of constraining observations. This overshoot was exacerbated by additional spurious geomagnetic field behavior apparently related to electromagnetic emissions from a steamboat that had pulled up onto shore about 100 m from the base station. With the steamboat's departure some 24 minutes later, the cubic spline corrections again became consistent with the observations.

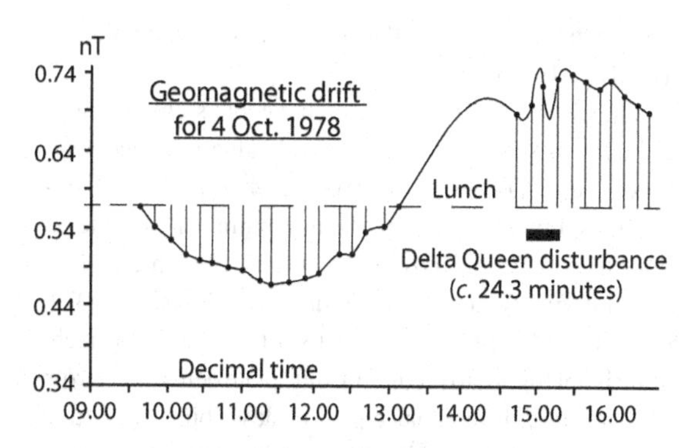

FIGURE A.38 Geomagnetic field drift for 4 October 1978 at a prehistoric archaeological site in Jefferson County, Indiana, on the north bank of the Ohio River across from Louisville, Kentucky. The diurnal field variation curve (solid line) was interpolated from the base station observations (dots) by the natural cubic spline with the mean value of 56, 566 nT removed. The horizontal axis is in decimal time where for example 15.50 = 1530 hours = 3 : 30 p.m.

This application illustrates the relative advantages of using gridding schemes with advanced analytical properties over simpler schemes. For example, to maintain continuity of derivatives at the observation points, the cubic spline produced excessive over- and undershoots that flagged inadequately sampled regions of the drift curve. Simple linear interpolation, on the other hand, is indifferent to poor sampling and thus must be used with due caution in regions of sparse data coverage.

For gridding anomaly data mapped in 2D spatial coordinates, the cubic spline can be implemented by arranging the data into profiles and fitting intersecting splines to the profiles as shown in Figure A.37(b). At each grid node, the nearest two profiles on both sides are identified and four cubic spline estimates made at the profile coordinates marking the intersections of the crossing line from the grid node. The cubic spline estimate at the grid node is then made from the four intersecting line estimates. This approach is simple to implement and is particularly well suited for gridding track-line data from airborne, marine, and satellite surveys. The implementation also does not require that the profiles be straight and uniformly spaced and sampled. Thus, it is effective on randomly distributed data as shown in Figure A.36(c) where the implementation was much faster than for the surrounding triangle method (Braile, 1978).

The piecewise 2D or biharmonic cubic spline can also be implemented to estimate the grid value from the nearest surrounding four (or more) observations. Indeed, this bicubic spline is applied by the minimum curvature technique as perhaps the most widely used current method for gridding gravity and magnetic data (e.g. Briggs, 1974; Swain, 1976; Smith and Wessel, 1990). The technique is analogous to predicting thin-plate deflections from a biharmonic differential equation describing curvature. This equation is solved iteratively using finite differences modeled by bicubic splines with continuous derivatives at the observation points. Convergence is rapid, and grid estimates tend toward the values of the observations as the observations approach the grid node.

The minimization of the curvature eliminates short-wavelength irregularities, providing a smooth result retaining the major features of the field. Generally, this gridding procedure produces acceptable results as long as the arbitrarily located data are well distributed over the field. Prior to gridding, noisy data commonly must be filtered to remove the short-wavelength components because the gridding requires the surface to pass through the observations.

In addition, undesirable spurious oscillations may develop in areas of the field where the data are poorly distributed. However, the results can be greatly improved by adding tension to the elastic sheet flexure equation (Smith and Wessel, 1990), which is equivalent to using error variance (Section 4.2(B)) in determining the spline coefficients. The advantage is illustrated in Figure A.39 where two contour maps of marine gravity data are compared. Map (a), which was produced using the normal untensioned minimum curvature method, shows an unwarranted and unconstrained closed minimum south of the central positive anomaly. Map (b), produced with tensioned splines, on the other hand, eliminates the unconstrained minimum with little effect on the other regions of the mapped field.

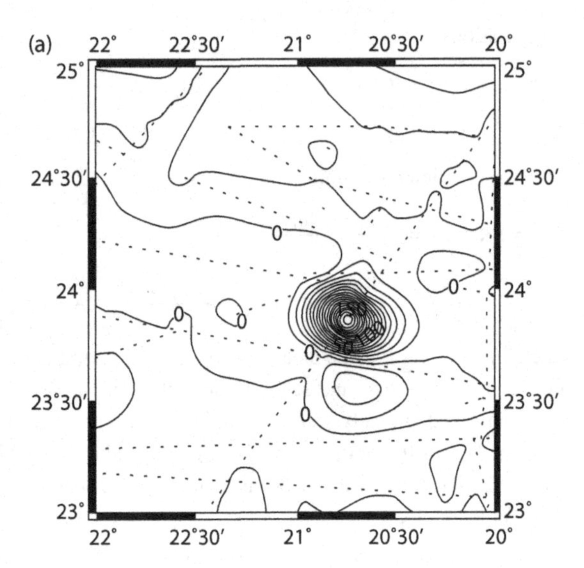

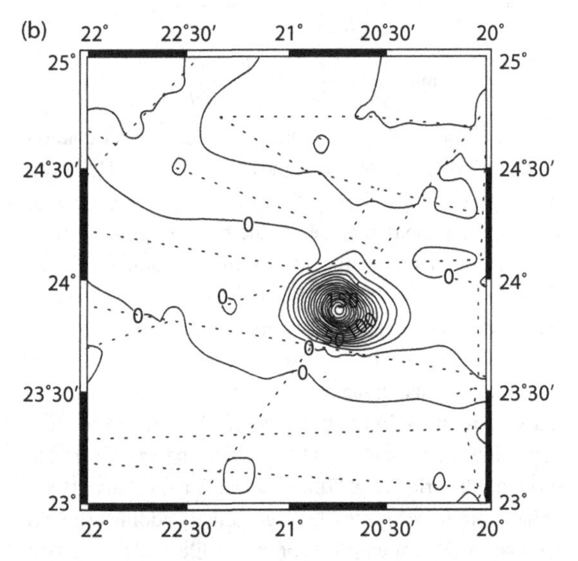

FIGURE A.39 Contour maps of marine gravity anomalies prepared from grids interpolated by the minimum curvature method. Map (a) was prepared from grid values obtained by the normal untensioned minimum curvature method. Map (b) was prepared from grid values estimated by tensioned minimum curvature. Adapted from Smith and Wessel (1990).

Minimum curvature gridding also is problematic when applied to anisotropic data with strong directional biases. Closed contours centered on maxima and minima can be produced along transects that sample field components having much larger length scales oblique to the transects than along them as is the case in most airborne surveying. These undesirable mapping artifacts are common for transects striking across elliptical anomalies. In addition, elevation deviations in flight lines from nominal specifications can produce undesirable line anomalies as shown in Figure A.40(a). Problems like this may be mitigated by low-pass filtering the transect data, or using tensioned splines or another interpolation technique like equivalent source gridding that produced the results of Figure A.40(b).

Equivalent source interpolation

As noted in the inverse modeling section A.5.2, any gravity or magnetic anomaly can be modeled by an indefinite number of equivalent sources. Equivalent source solutions

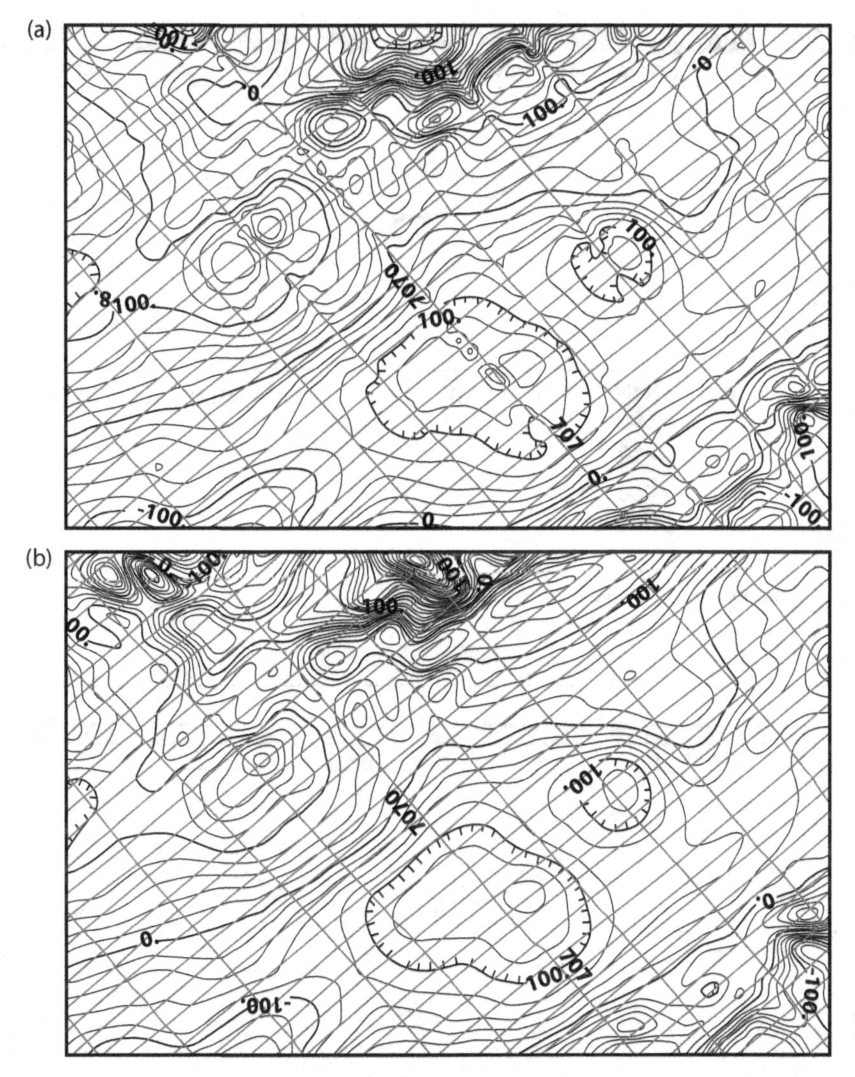

FIGURE A.40 Aeromagnetic survey data gridded (a) using a minimum curvature algorithm and (b) equivalent source interpolation. Both maps were produced at a contour interval of 20 nT. Courtsey of EDCON-PRJ, Inc.

readily accommodate analysis of data that are variably distributed horizontally and in altitude. The solutions of course do not possess a physical reality, but can be used to recalculate the anomaly and its derivative and integral components at any desired spatial coordinates including grids at constant or variable altitudes.

This gridding method typically invokes elementary analytical sources such as the gravity point mass (Chapter 3) or magnetic point dipole (Chapter 9) or their simple extensions as spheres, cylinders, or prisms. A common implementation relates the anomaly data by least-squares matrix inversion to the physical property variations of an equivalent layer comprising uniform constant thickness blocks (e.g. MENDONCA and SILVA, 1995; COOPER, 2000). The gravity profile of the two prismatic 2D sources in Figure A.41(a), for example, is reproduced almost exactly by the variable density effects of the equivalent layer in Figure A.41(b). Assuming negligible conditioning errors for the inversion, the equivalent layer model also produces the anomaly and its derivative and integral components with comparable accuracy at all spatial coordinates on or above the Earth's surface.

In contrast to minimum curvature, equivalent source gridding is typically implemented using all the data, but with more computational labor. However, it accommodates the anomaly gradients more comprehensively because it is based on globally applicable force field

(a)

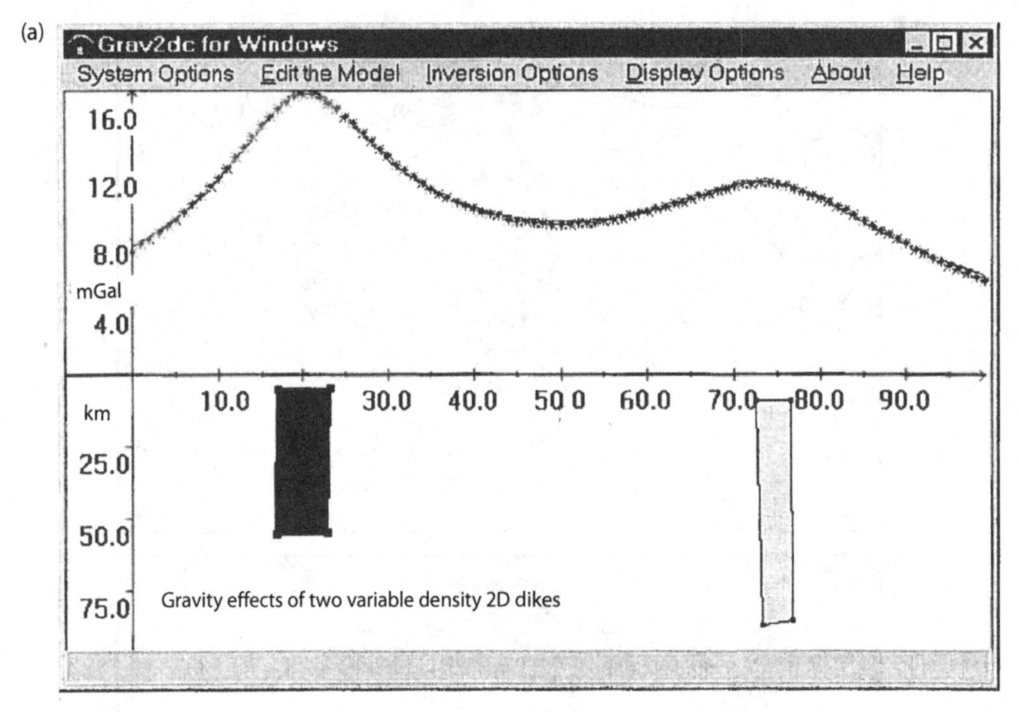

(b)

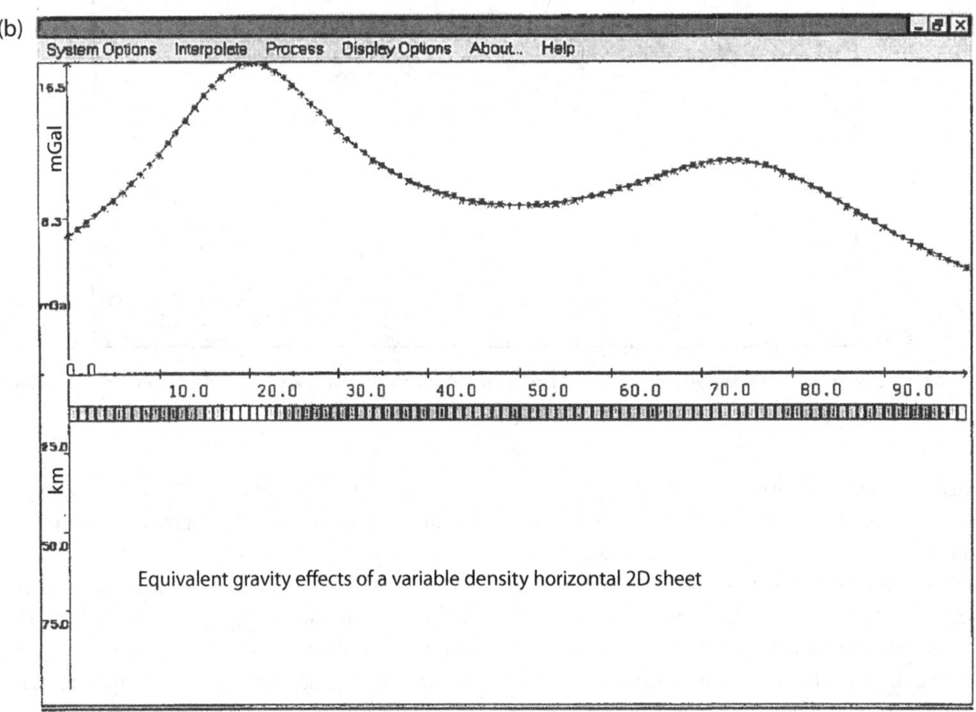

FIGURE A.41 Screen shots of (a) the 2D gravity effects of two prismatic bodies with different uniform densities equivalently modeled by (b) the effects of numerous density contrasts within a uniformly thick layer. Adapted from Cooper (2000).

equations. The artificial line anomalies in Figure A.40(a), for example, result from the ineffectiveness of the minimum curvature gridding to account for the altitude variations in the aeromagnetic data. However, the equivalent magnetic layer gridding readily incorporates these variable altitude effects to virtually eliminate the line anomalies as shown in Figure A.40(b).

In addition, the equivalent source model can always be updated to improve gridding accuracy and resolution. The updating involves fitting a subsequent model to the residual differences between the observations and initial model predictions so that the integrated effects of the models improve the match to the observations (e.g. VON FRESE *et al.*, 1988; GUNN and ALMOND, 1997).

Polynomial interpolation

Profile and surface polynomials of order $n > 3$ can be implemented in both local and global gridding applications. However, the local deployment is commonly preferred because it offers fewer numerical challenges. The global implementation, for example, typically requires a higher-order polynomial where the effective order often is unclear and difficult to justify. Furthermore, the computational labor tends to be significantly greater because of the increased number of polynomial coefficients involved. The 2D polynomial of order n, for example, requires $([(n+1)(n+2)/2]-1)$ coefficients that must be solved for and manipulated to obtain the grid estimates.

The local deployment is based on defining the gridding area within a surface fitting area as shown in Figure A.36(d). All observations (dots) within the surface fitting area are identified and fitted to a polynomial surface of order n. The polynomial is then evaluated at the grid nodes (pluses) within the gridding area where the edge effects are limited by the choice of the surface fitting area. The gridding and surface fitting areas are moved successively until the entire grid has been processed.

The processor controls the gridding procedure through the choices of the sizes of the gridding rectangle and surface fitting area, and the order of the polynomial. As an example of local surface fitting, Figure A.36(e) illustrates the contour map of the 2 km grid obtained from the 200 data points in Figure A.36(b) using the 2D polynomial of order $n = 4$ over the surface fitting area of 6 km × 6 km (BRAILE, 1978).

Statistical interpolation

Statistical procedures have been extensively used for gridding Earth science data (e.g. DAVIS, 1986). Indeed, the weighted average procedure in Figure A.35(c) often is included as a default gridding algorithm in commercial computer graphics packages. The inverse distance averaging procedure in the two (x, y)-spatial dimensions, for example, involves locating a selected number k of data points nearest the prediction point. The distance D_{ip} from the observation point (x_i, y_i) to the prediction point (x_p, y_p) is

$$D_{ip} = \sqrt{(x_p - x_i)^2 + (y_p - y_i)^2} \tag{A.95}$$

so that the grid estimate can be obtained from

$$b_p = \frac{\sum_{i=1}^{k}\left(b_i/D_{ip}^n\right)}{\sum_{i=1}^{k}\left(1/D_{ip}^n\right)}, \tag{A.96}$$

where n is the order of the distance function that the processor selects for the inverse distance weighting. This simple statistical method is effective as shown in Figure A.36(f) where the contour map was prepared from the gridded averages of the nearest $k = 9$ neighbors weighted by the first-order (i.e. $n = 1$) inverse distance function (BRAILE, 1978).

Another widely used statistical approach in the Earth sciences exploits theoretical or observed structure in the covariance function of the observations which reflects how the products of the observations behave in terms of their separation distances. This covariance modeling is known as kriging in mining and geological applications (e.g. DAVIS, 1986) and least-squares collocation in gravity and geodetic applications (e.g. MORITZ, 1980a).

Figure A.42(a) illustrates the covariance modeling of k detrended observations where each observation b_i is taken as the true signal $\hat{b}_i$ plus noise $\mathcal{N}$ or

$$b_i = \hat{b}_i + \mathcal{N}. \tag{A.97}$$

The objective is to estimate the true signal at the grid point given by

$$\hat{b}_p = \mathbf{C}(\hat{b}_p, b_i)^t\mathbf{C}(b_i, b_i)^{-1}\mathbf{B}, \tag{A.98}$$

where $\mathbf{C}(\hat{b}_p, b_i)$ is the $(k \times 1)$ cross-covariance vector between the signal and observations, $\mathbf{C}(b_i, b_i)$ is the $(k \times k)$ covariance matrix of observations, and $\mathbf{B}$ is the $(k \times 1)$ observation vector.

The symmetric covariance matrix contains elements that are the products of all pairs of observations normalized by the total variance of the observations or

$$\mathbf{C}(b_i, b_i) = \mathbf{B}\mathbf{B}^t/\sigma_{\mathbf{B}}^2. \tag{A.99}$$

The elements of the cross-covariance matrix, on the other hand, cannot be determined in this way because the value of $\hat{b}_p$ is not known. However, in many applications, the products show a dependence on offset (i.e. separation

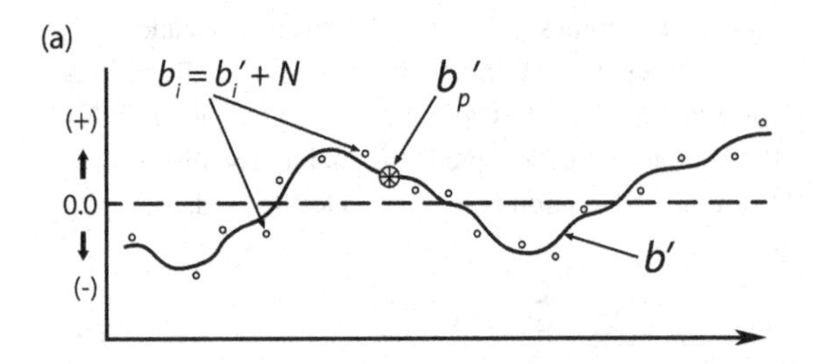

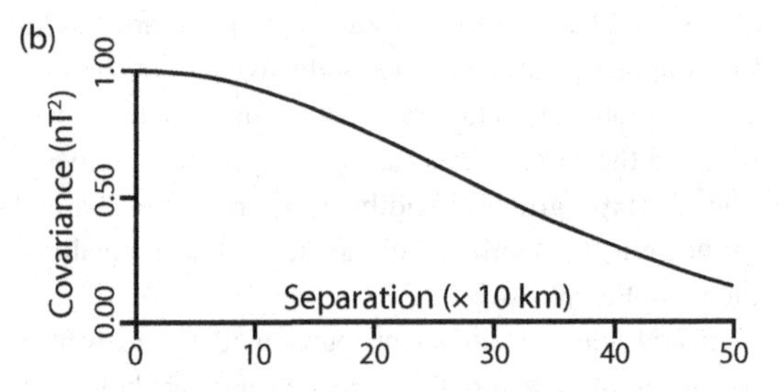

FIGURE A.42 (a) Gridding by least-squares collocation. (b) A typical covariance function adapted to grid satellite-altitude magnetic anomalies. Adapted from GOYAL *et al.* (1990).

distance) between observations as in the theoretical gravity covariance function of Figure A.42(b) from MORITZ (1980a). Thus, when available, the covariance function can be evaluated for the cross-covariance matrix elements in terms of the offsets defined by the observations relative to the prediction $\hat{b}_p$. Indeed, numerically efficient implementations of Equation A.98 commonly invoke the use of the covariance function to establish the elements of both the covariance and cross-covariance matrices.

GOYAL *et al.* (1990) used 3D collocation, for example, to process multi-altitude Magsat crustal magnetic anomalies over South America into a 2°-grid at the uniform altitude of 405 km. The Magsat crustal anomalies were derived from minimally disturbed dawn-side observations collected over the unevenly distributed orbital tracks at altitudes varying from about 345 to 465 km. The procedure involved sorting the Magsat data into $2° \times 2° \times 120$ km bins and processing each bin for the least-squares grid estimate using Equation A.98. However, effective gridding required the use of a trade-off diagram as described in Figure A.42 above to test the conditioning of the covariance matrix inverse. More generally, this practice is recommended whenever a matrix inverse must be taken to protect against using unstable and poorly performing solutions in analysis.

The foregoing descriptions illustrate the range of statistical and other gridding methods available for gravity and magnetic anomaly analysis. However, they cover only a small fraction of the great number of gridding methods that have been developed through the history of numerical data analysis. In general, selecting a gridding method is subjective and depends largely on the experience and capabilities of the processor, as well as the presumed nature of the surface, characteristics of the data set, and use of the gridded values. To assess its effectiveness, however, the grid must ultimately be rendered into a contour map or other appropriate graphic to compare with the input data. Thus, effective graphical parameters must be invoked to reveal the significant anomaly variations.

A.6.3 Graphical parameters and practice

Every graphical presentation of the data set should include statistical attributes that summarize its key variational properties such as the amplitude range (AR = [min; max]) of the minimum and maximum values, amplitude mean (AM) or average value, amplitude standard deviation (ASD) or energy, and amplitude unit (AU) together with the contour interval (CI) and grid interval (GI). The amplitude range, of course, indicates the total amplitude variation, whereas the grid interval defines the Nyquist wavelength of the data set.

However, the most critical map attributes are the amplitude mean, $\mu_{\mathbf{B}}$, and standard deviation, $\sigma_{\mathbf{B}}$, because they set the appropriate contour (or color) range and interval for representing the data. For example, where the data have been effectively detrended, and thus are statistically stationary, roughly 95% of the variability of the data set can be expected to occur within $\mu_{\mathbf{B}} \pm 2\sigma_{\mathbf{B}}$. Thus, choosing a contour interval that fits roughly 20 contour levels into the 95% variability range should reveal the key anomaly features and behavior of the data set.

A simple non-parametric test for stationarity is to check if roughly 50% of the data amplitudes are in the range $\mu_{\mathbf{B}} + 2\sigma_{\mathbf{B}}$. If not, then the data should be explored for fitting and removing a low-order trend surface, as is done to suppress the DC-component in Fourier transform analysis. This approach is especially warranted where the trend surface may be rationalized for unwanted regional effects of the survey parameters and/or subsurface sources that are not the focus of the analysis.

Standardized and normalized data

Another important application of the mean and standard deviation is to standardize data into dimensionless coefficients

$$s_i(\mathbf{B}) = \frac{b_i - \mu_{\mathbf{B}}}{\sigma_{\mathbf{B}}}, \tag{A.100}$$

which have the mean $\mu_S = 0$ and standard deviation $\sigma_S = 1$. However, the standardized coefficients can also be transformed into normalized coefficients

$$n_i(\mathbf{B}) = s_i \times \sigma_N + \mu_N, \quad \text{(A.101)}$$

where the dimensions and mean μ_N and standard deviation σ_N of the normalized coefficients can be set to any desired units and values, respectively.

In practice, however, non-zero means are commonly removed from the coefficients to enhance scaling the residual variations to the working precision of the calculations. Suppressing the non-zero means also simplifies the coefficients where, for example, Equation A.101 becomes $n_i(\mathbf{B}) = (\sigma_N/\sigma_B)b_i$. In presenting the normalized coefficients, the normalization factor

$$\text{NF} = \left(\frac{\sigma_B}{\sigma_N}\right) \quad \text{(A.102)}$$

is typically included with the other statistical attributes so that the corresponding coefficients of the original signal b_i may be readily estimated as the normalized data are considered for meaningful patterns. The normalization modifies the data gradients only by the multiplicative constant NF so that the correlation coefficient between the data set **B** and its transformation by Equation A.101 is unity.

Obviously, standardized or normalized coefficients can be used to compare disparate data sets in common graphical parameters. For example, one might want to study the correlations between Bouguer gravity anomaly estimates (Section 6.4) and the terrain elevations, to test the veracity of the terrain density used in making the Bouguer reduction. As seen in Section 4.6.2, the terrain and gravity anomaly features would be positively correlated where the reduction density is too low, and negatively correlated where it is too high. Thus, to help visualize these feature correlations, the data sets can be normalized to the same statistical properties to map them in a common set of colors or contours. However, the feature correlations can be located even more explicitly with further simple manipulations of the normalized coefficients.

Local favorability indices

To identify correlated features in signals **B** and **D**, the coefficients $b_i \in \mathbf{B}$ (e.g. the Bouguer gravity anomalies) and $d_i \in \mathbf{D}$ (e.g. the digital elevation model) are assumed to have common independent variable coordinates and thus to be co-registered. Merriam and Sneath (1966) introduced the concept of local favorability indices (LFI) for mapping out occurrences of positively correlated features in co-registered data sets from their standardized coefficients. von Frese *et al.* (1997a) generalized the concept to normalized coefficients and extended it to map out the negatively correlated features as well.

Specifically, to map out directly correlated features, the summed local favorability indices (SLFI) are estimated from the point-by-point addition or stacking of the normalized coefficients

$$\text{SLFI}_i = n_i(\mathbf{B}) + n_i(\mathbf{D}). \quad \text{(A.103)}$$

The coefficients satisfying $\text{SLFI}_i > 0$ emphasize the positive features in the two data sets that are correlative (i.e. peak-to-peak correlations), whereas the coefficients satisfying $\text{SLFI}_i < 0$ map out the correlative negative features (i.e. valley-to-valley correlations). In general, the SLFI coefficients tend to emphasize occurrences of directly correlated features and suppress the occurrences of inversely and null correlated features in the data sets.

To map out inversely correlated features, on the other hand, the differenced local favorability indices (DLFI) are evaluated from the point-by-point differences in the normalized coefficients

$$\text{DLFI}_i = n_i(\mathbf{B}) - n_i(\mathbf{D}). \quad \text{(A.104)}$$

The coefficients satisfying $\text{DLFI}_i > 0$ emphasize positive features in **B** that correlate with negative features in **D**, whereas the coefficients $\text{DLFI}_i < 0$ map out the negative features in **B** that correlate with positive features in **D**. In general, the DLFI coefficients tend to emphasize occurrences of inversely correlated features and suppress the occurrences of directly and null correlated features in the data sets.

Figure A.43 gives a simple example of the utility of local favorability indices for mapping out feature correlations. Signals $A(x)$ and $B(x)$, normalized to $\sigma_N = 1.394$ and $\mu_N = 0$ as shown in respective panels (b) and (c), clearly have many correlated features. However, the overall correlation coefficient between them is zero because the respective left and right signal halves correlate directly and inversely.

The SLFI and DLFI coefficients in the respective panels (a) and (d) reveal the feature associations in considerable detail. The shaded pattern delineates the region between $\pm 1\sigma_N$, whereas the SLFI coefficients that are presented in solid black and white, respectively, map out nearly all of the peak-to-peak and valley-to-valley feature correlations. The DLFI coefficients were obtained by subtracting signal $B(x)$ from signal $A(x)$. Thus, the cross-hatched coefficients map out positive features in signal $A(x)$ that correlate with negative features in signal $B(x)$, whereas the diagonally ruled coefficients mark the negative features

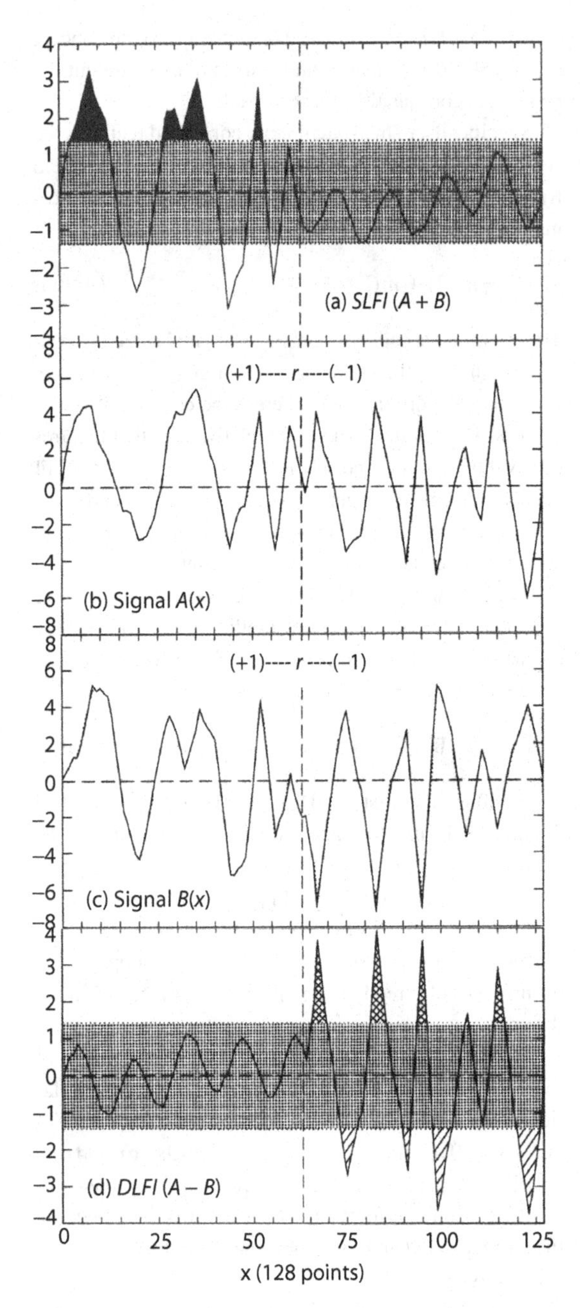

FIGURE A.43 Signals A and B of 128 coefficients each in respective panels (b) and (c) with correlation coefficient $r = 0.0$ have feature associations that are mapped out using the SLFI and DLFI coefficients in the respective panels (a) and (d). Adapted from von Frese *et al.* (1997a).

in the signal $A(x)$ that correlate with positive features in signal $B(x)$.

In general, considering the LFI coefficients from larger to smaller magnitudes allows the correlation analysis to proceed from the relatively simpler representations that involve the strongest feature correlations to increasingly complex representations which include the weaker, more subtle feature correlations. The coefficients also aid in isolating feature correlations in the correlation filtered outputs that are synthesized at all coordinates of two or more data grids.

A.6.4 Presentation of data

Gridded data can be presented in numerous forms that assist the analyst in evaluating and interpreting the data. Traditionally, data are shown in profile or contour form, but the availability of high-speed schemes for computer processing and presentation has greatly increased the alternatives to these graphical formats. Data observed along transects are particularly amenable to presentation in profile form, although profiles can be extracted from gridded data as well and anomaly values connected by line segments positioned by intermediate values using an appropriate interpolation scheme as described previously. Profiles are generally oriented orthogonal to the direction of the strike of elongate features of the data set for interpretation. They can be stacked in their relative spatial position based on a level of the variable representing the position of the profile. The resulting stacked profile map gives a 3D view of the data set focusing on the higher-wavenumber components that may be lost in the normal contouring process. This is particularly true of data observed along the profiles.

Contour maps of anomaly data sets and their various derivatives are the most common way of presenting 2D data. Machine contouring of equispaced, orthogonally gridded data have reached a high degree of accuracy and essentially total acceptance. Numerous schemes have been developed to locate the position of a contour value on a 2D representation based on interpolation along the sides of the grid squares using one of the methods described above. To improve the map presentation, the basic grid squares are divided into smaller subgrids and the interpolation process repeated until an esthetically pleasing presentation is achieved.

The contour interval of the anomaly map is one of the critical decisions in the map making process, depending on the purpose of the map, the map scale, and the range of variable over the mapped area. Generally, the contour interval lies between one-tenth and one-twentieth of the total range and is constant, but exceptions occur. For example, a logarithmic contour interval can be used if the range is extremely large or contour intervals may be small over the range of special interest for the map's purpose with higher or lower values given saturation levels – that is, no contours are presented above or below specified levels. Care is required in visual analysis of maps with non-uniform contour intervals because of distortion in the observed gradients. Color-filled contouring, which is becoming increasingly dominant, wherein the space between adjacent contour levels is colored, provides for rapid evaluation of amplitudes, patterns, and trends. Generally, the colors are shaded as in the electromagnetic spectrum, but with blue colors representing the lows and red colors the high values. Finally, it should be noted that contour maps should be annotated with information on the grid used in the presentation and, if possible, the location of the original point values that were used to develop the grid.

A useful alternative to the contour map which improves visualization of the anomalies is a 3D perspective view. It pictures the map from a specified azimuth and angle above the horizon with variable amplitudes plotted as heights, as in a 3D view of topography. Alternatively, it may view depths for subsurface features in a manner comparable to heights. Complementary to the contour map is the shaded relief map which is useful for enhancing residual anomalies for visual inspection interpretation. This technique is analogous to viewing a topographic surface from an elevated position. This form of presentation is based on viewing the calculated reflection from the anomaly surface due to a fictitious light source at a specified position (HORN and BACHMAN, 1978).

In the shaded relief presentation, the assumption is made that the viewing position is at a great distance from the surface and that the intensity is independent of the angular relation between the normal to the surface element and the viewing site (DODS *et al.*, 1985). As a result, the brightness or intensity is proportional to the cosine of the angle between the light-source direction and the surface normal. The light-source direction in terms of both horizontal angle, declination or azimuth, and vertical angle, inclination or elevation, is specified in the calculation. The surface normal, in terms of both declination and inclination, is calculated directly from the slope of the anomaly surface at each grid point from the immediately adjacent grid values. The results are dependent not only on the anomaly surface, but also the specified position of the fictitious light source. The appropriate position of the light source is determined by evaluating the results of an iterative procedure employing a range of light positions, but the optimum inclination is generally of the order of $30° \pm 15°$ depending on the amplitude of the anomalies of interest and the range of amplitudes of the anomaly surface, and the declination is positioned perpendicular to the trend of the anomalies of interest. Presentations are often enhanced by increasing the grid density by interpolation of the original anomaly data set and increasing the vertical exaggeration of the shaded relief intensity. Variations on the shaded relief presentation are used for enhancing specialized features. For example, COOPER (2003) has illustrated a modification to improve mapping of circular gravity and magnetic features of varying radii.

High gradients and areally small anomalies (that is, residual anomalies) are enhanced while regional anomalies are suppressed by shaded relief maps. These maps also are useful to isolate potential errors in the data sets caused by such things as isolated spurious data values and datum level problems at the join between data sets. These can be observed as geologically unrealistic local anomalies and linear trends. Shaded relief maps commonly are presented in gray-shaded images which carry no information on the absolute amplitude of the field. Alternatively, color-shaded relief maps can be prepared by modulating the color contours of the anomaly intensity map with the gray level intensity representing the reflected illumination. The latter do not suppress the absolute amplitude of the field and the regional anomalies.

Examples of various graphical formats for presenting gravity and magnetic anomaly data are shown throughout this book. In addition, numerous illustrations of the use of color graphics for anomaly presentation are given in the supplemental website.

A.7 Key concepts

- Gravity and magnetic survey data are commonly accessible from international, national, state, commercial, and academic data bases in standard formats that facilitate processing the data by digital computers.
- Electronic data analysis is based on analyzing a digital forward model (**AX**) of the data (**B**) using inversion to obtain the unknown coefficients (**X**) from the data and the known coefficients (**A**) of the prescribed model. This fundamental inverse problem (i.e. $\mathbf{AX} = \mathbf{B}$) has the least-squares solution $\mathbf{X} = (\mathbf{A}^t\mathbf{A})^{-1}\mathbf{A}^t\mathbf{B}$ which yields predictions matching the data such that the sum of squared residuals is minimum. Statistical error bars may be ascribed to the solution and its predictions based on the variance of the solution $\sigma^2_{\mathbf{X}} = (\mathbf{A}^t\mathbf{A})^{-1}\sigma^2_{\mathbf{B}}$, where $\sigma^2_{\mathbf{B}}$ is the data variance.
- The inversion solution is never unique, owing to errors in the data, presumed model coefficients, and truncated calculations. In addition, interpretations of gravity and magnetic anomalies are inherently ambiguous (i.e. not unique) because any anomaly can be matched in theory by the effects of an indefinite number of source

distributions. Thus, a sensitivity analysis based on the solution variance is commonly required to establish an "optimal" set or range of anomaly solutions that conform to the inversion errors, as well as any additional constraints on the physical properties of the subsurface.

- The principal format in modern digital data processing is the data grid or array that the fast Fourier transform (FFT) can model with minimal assumptions and maximum computational efficiency and accuracy. The FFT is a spectral model that accurately represents the data as the summation of cosine and sine series over a finite sequence of discrete uniformly spaced wavelengths. This sequence of wavelengths corresponds to an equivalent discrete sequence of frequencies called wavenumbers which are completely fixed by the number of gridded data points. Because of its computational advantages, the FFT dominates modern gravity and magnetic data analysis.
- Data analysis fundamentally interrogates the forward model for the derivative and integral components of the data. These attributes in gravity and magnetic applications are used mainly for anomaly isolation and enhancement, source modeling, and graphical display purposes.
- Isolation and enhancement methods respectively seek to isolate and enhance the anomalies of interest (i.e. the residual anomalies) from the interfering effects of longer-wavelength (i.e. regional) and higher-frequency (i.e. noise) anomalies. The residual–regional separation problem is commonly approached by the use of filters in the spatial and spectral domains.
- Spatial filtering includes the use of smooth contouring and other graphical methods to draw and digitize the regional through a data set. Trend surfaces also are used to identify and extract the regional. In addition, analytical grid or convolution methods are available for isolating and enhancing anomaly derivative and integral properties, but these methods have been largely superseded by the more efficient spectral filtering techniques.
- Spectral filtering involves the use of transfer function coefficients that are multiplied against the wavenumber coefficients from the FFT of the data. Inversely transforming the modified or filtered spectral model efficiently and accurately estimates the spatial attributes of the data. These anomaly attributes commonly include the wavelength and directional properties, as well as the correlative features between two or more anomaly data sets, and anomaly derivatives, integrals, interpolations, and continuations.
- Anomaly source modeling includes the popular forward modeling approach based on trial-and-error inversion. Forward modeling is still widely used, but becomes increasingly unwieldy for problems involving more than a few unknown parameters. For the more computationally extensive problems involving greater numbers of unknowns, inverse modeling usually by linear matrix inversion methods is commonly implemented.
- Graphical displays of the data and data processing results are essential elements of modern data analysis efforts. Graphics are effective for conceptualizing and implementing anomaly analyses and presenting their results.
- Anomaly maps over areas of 6° and larger are commonly presented using the universal transverse Mercator projection, whereas the transverse Mercator projection is used at smaller scales to roughly 0.5°. At even smaller scales, the simple Cartesian (x, y, z)-projection usually suffices for anomaly mapping.
- Electronically generated maps are based on data typically interpolated to grids. Anomaly gridding schemes range from simple linear interpolation, to cubic spline interpolation that preserves the continuity of first and second derivatives at each data point, and kriging and least-squares collocation based on the covariance properties of the data.
- In gridding, the most critical choice is the grid interval which must be at least one-half the wavelength of the smallest anomaly feature of interest to the analysis. At larger grid intervals, the anomaly feature is represented as a longer-wavelength alias or gridding error that corrupts the analysis.
- Effective anomaly graphics list statistical and other attributes that summarize the key properties of the displayed data. These attributes include the range of amplitudes, and amplitude mean, standard deviation, and units, as well as the contour interval and grid interval.
- Removing the mean and dividing by the standard deviation standardizes the data into dimensionless values with zero mean and unit standard deviation. Disparate standardized data thus may be graphically compared using the same plotting parameters (e.g. contour intervals and color scales).

References

Abdelrahman, E.E., El-Araby, T.M., El-Araby, H.M., and Abo-Ezz, E.R. 2001. Three least-squares minimization approaches to depth, shape, and amplitude coefficient determination from gravity data. *Geophysics*, **66**, 1105–1109.

Abdeslem, J.G. 2000. Nonlinear 3-D inversion of gravity data over a sulfide ore body. *Geofis. Int.*, **39**, 179–188.

Acuña, M.H. 2001. The magnetic field of Mars: summary of results from the aerobraking and mapping orbits. *J. Geophys. Res.*, **106**, 23403–23417.

Acuña, M.H. 2002. Space-based magnetics. *Rev. Sci. Instrum.*, **73**, 3717–3736.

Acuña, M.H., Connerney, J.E.P., Ness, N.F. *et al.* 1999. Global distribution of crustal magnetization discovered by the Mars Global Surveyor MAG/ER experiment. *Science*, **284**, 790–793.

Aeromagnetic Standards Committee. 1991. *Guide to Aeromagnetic Specifications and Contracts*. Geological Survey of Canada Open File 2349.

Agar, C.A., and Liard, J.O. 1982. Vertical gravity gradient surveys: field results and interpretations in British Columbia, Canada. *Geophysics*, **47**, 919–925.

Agocs, W.B. 1951. Least-squares residual anomaly determination. *Geophysics*, **16**, 686–696.

Agocs, W.B. 1955. The effect of line spacing illustrated by Marmora Ontario airborne magnetometer control and the determination of optimum spacing. *Geophysics*, **20**, 871–885.

Aiken, C.L.V., Balde, M., Ferguson, J.F. *et al.* 1998. Recent developments in digital gravity data acquisition on land. *The Leading Edge*, **17**, 93–97.

Aina, A. 1986. Reduction to equator, reduction to pole, and orthogonal reduction of magnetic profiles. *Explor. Geophys.*, **17**, 141–145.

Aitken, M.J. 1972. *Physics and Archaeology*. Clarendon Press.

Akaike, H. 1969. Power spectrum estimation through Auto-Regressive model fitting. *Ann. Inst. Stat. Math.*, **21**, 407–419.

Akaike, H. 1974. A new look at the statistical model identification. *IEEE Trans. Automat. Control*, **AC-19**, 716–723.

Algermissen, S.T. 1961. Underground and surface gravity survey, Leadwood, Missouri. *Geophysics*, **27**, 158–168.

Allingham, J.W. 1964. Low-amplitude aeromagnetic anomalies in southeastern Missouri. *Geophysics*, **29**, 537–552.

Alsdorf, D.E., von Frese, R.R.B., Arkani-Hamed, J., and Noltimier, H.C. 1994. Separation of lithospheric, external, and core components of the south polar geomagnetic field at satellite altitudes. *J. Geophys. Res.*, **99**, 4655–4667.

Åm, K. 1972. The arbitrarily magnetized dyke: Interpretation by characteristics. *Geoexplor.*, **10**, 63–90.

Ander, M.E., and Huestis, S.P. 1987. Gravity ideal bodies. *Geophysics*, **52**, 1265–1278.

Ander, M.E., Summers, T., and Gruchalla, M.E. 1999. LaCoste & Romberg gravity meter: System analysis and instrumental errors. *Geophysics*, **64**, 1708–1719.

Andersen, O.L., Knudsen, P., Berry, P.A., Kenyon, S., and Trimmer, R. 2010. Recent developments in high-resolution global altimetric gravity field modeling. *The Leading Edge*, **29**, 540–545.

Anderson, A.J., and Cazenave, A. 1986. *Space Geodesy and Geodynamics*. Academic Press.

Andreasen, G.E., and Zietz, I. 1969. *Magnetic Fields for a 4 × 6 Prismatic Model*. US Geological Survey Professional Paper 666.

Argast, D., Bacchin, M., and Tacey, R. 2009. An extension of the closed-form solution for the gravity curvature (Bullard B) correction in the marine and airborne cases. *Extended Abstracts 2009*. Australian Society of Exploration Geophysics, pp. 1–6.

Arkani-Hamed, J. 1988. Differential reduction-to-pole of regional magnetic anomalies. *Geophysics*, **53**, 1592–1600.

Arkani-Hamed, J. 2007. Differential reduced to pole: Revisited. *Geophysics*, **72**, L1–L11.

Arkani-Hamed, J., and Strangway, D.W. 1986. Magnetic susceptibility anomalies of the lithosphere beneath Europe and the Middle East. *Geophysics*, **51**, 1711–1724.

Asgharzadeh, M.F., von Frese, R.R.B., Kim, H.R., Leftwich, T.E., and Kim, J.W. 2007. Spherical prism gravity effects by Gauss-Legendre quadrature integration. *Geophys. J. Int.*, **169**, 1–11.

Asgharzadeh, M.F., von Frese, R.R.B., and Kim, H.R. 2008. Spherical prism magnetic effects by Gauss–Legendre quadrature integration. *Geophys. J. Int.*, **173**, 315–333.

Athy, L.F. 1930. Density, porosity, and compaction of sedimentary rocks. *Am. Assoc. Petr. Geol. Bull.*, **14**, 1–24.

Bandy, W.L., Gangi, A.F., and Morgan, F.D. 1990. Direct method for determining constant corrections to geophysical survey lines for reducing mis-ties. *Geophysics*, **55**, 885–896.

Baranov, V. 1957. A new method for interpretation of aeromagnetic maps; pseudo-gravimetric anomalies. *Geophysics*, **22**, 359–383.

Baranov, V., and Naudy, H. 1964. Numerical calculation of the formula of reduction to the magnetic pole. *Geophysics*, **29**, 67–79.

Barton, P.J. 1986. The relationship between seismic velocity and density in the continental crust – a useful constraint? *Geophys. J.*, **87**, 195–208.

Bath, G.D. 1962. Magnetic anomalies and magnetization of the Biwabik iron-formation, Mesabi area, Minnesota. *Geophysics*, **28**, 622–650.

Baumann, H., Klingelé, E.E., and Marson, I. 2011. Absolute airborne gravimetry: a feasibility study. *Geophys. Prosp.*, **59**, 1365–1377.

Bazylinski, D.A., and Moskowitz, B.M. 1997. Microbial biomineralization of magnetic iron minerals. *Rev. Mineral.*, **35**, 181–223.

Bean, R.J. 1965. A rapid graphical solution for the aeromagnetic anomaly of the two-dimensional tabular body. *Geophysics*, **31**, 963–970.

Behura, J., Kabir, N., Crider, R., Jilek, P., and Lake, E. 2010. Density extraction from P-wave AVO inversion: Tuscaloosa Trend example. *The Leading Edge*, **29**, 772.

Beiki, M., Bastani, M., and Pedersen, L.B. 2010. Leveling HEM and aeromagnetic data using differential polynomial fitting. *Geophysics*, **75**, L13–L23.

Bell, R.E., and Hansen, R.O. 1998. The rise and fall of oil field technology: The torsion balance gradiometer. *The Leading Edge*, **17**, 81–83.

Bell, R.E., and Watts, A.B. 1986. Evaluation of the BGM-3 sea gravity meter system onboard R.V. *Conrad. Geophysics*, **51**, 1480–1493.

Beltrão, J.F., Silva, J.B.C., and Costa, J.C. 1991. Robust polynomial fitting method for regional gravity estimation. *Geophysics*, **56**, 80–89.

Berckhemer, H., Rauen, A., Winter, H. *et al.* 1997. Petrophysical properties of the 9-km deep crustal section at KTB. *J. Geophys. Res.*, **102**, 18337–18361.

Bevington, P.R. 1969. *Data Reduction and Error Analysis for the Physical Sciences*. McGraw-Hill Book Co.

Bhattacharyya, B.K. 1978. Computer modeling in gravity and magnetic interpretation. *Geophysics*, **43**, 912–929.

Bhattacharyya, B.K., and Chan, K.C. 1977. Reduction of gravity and magnetic data on an arbitrary surface acquired in a region of high topographic relief. *Geophysics*, **43**, 1411–1430.

Bhattacharyya, B.K., and Leu, L.K. 1977. Analysis of magnetic anomalies over Yellowstone national Park: Mapping of Curie-point isothermal surface for geothermal reconnaissance. *J. Geophys. Res.*, **80**, 4461–4464.

Bible, J.L. 1962. Terrain correction tables for gravity. *Geophysics*, **27**, 715–718.

Biegert, E.K., and Millegan, P.S. 1998. Beyond recon: the new world of gravity and magnetics. *The Leading Edge*, **17**, 41.

Billings, S., and Wright, D. 2009. Optimal total-field magnetometer configuration for near-surface applications. *The Leading Edge*, **28**, 522–527.

Birch, F. 1961. The velocity of compressional waves in rocks to 10 kilobars, Part 2. *J. Geophys. Res.*, **66**, 2199–2224.

Blaich, O.A., Faleide, J.I., and Tsikalas, F. 2011. Crustal breakup and continent-ocean transition at South Atlantic conjugate margins. *J. Geophys. Res.*, **116**, 38 pp.

Blake, G.R., and Hartge, K.H. 1986. Bulk density. In *Methods of Soil Analysis, Part I. Physical and Mineralogical Methods*. Agronomy Monograph no. 9, pp. 363–375.

Blakely, R.J. 1995. *Potential Theory in Gravity and Magnetic Applications*. Cambridge University Press.

Blakely, R.J., and Connard, G.G. 1989. Crustal studies using magnetic data. In *Geophysical Framework of the Continental United States*. Geological Society of America Memoir 172, pp. 35–44.

Blakely, R.J., and Grauch, V.J.S. 1983. Magnetic models of crystalline terrane. Accounting for the effect of topography. *Geophysics*, **48**, 1551–1557.

Blakely, R.J., and Hassanzadeh, S. 1981. Estimation of depth to magnetic source using maximum entropy power spectra, with application to the Peru–Chile Trench. In *Nazca Plate; Crustal Formation and Andean Covergence*. Geol. Soc. Am. Memoir 154, pp. 667–682.

Blakely, R.J., and Simpson, R.W. 1986. Locating edges of source bodies from magnetic or gravity anomalies. *Geophysics*, **51**, 1494–1498.

Blakely, R.J., Cox, A., and Iufer, E.J. 1973. Vector magnetic data for detecting short polarity intervals in marine magnetic profiles. *J. Geophys. Res.*, **78**, 6977–6983.

Blakely, R.J., Jachens, R.C., Calzia, J.P., and Langenheim, V.E. 1999. Cenozoic basins of the Death Valley extended terrane as reflected in regional-scale gravity anomalies. In *Cenozoic Basins of the Death Valley Region*. Geol. Soc. Am. Special Paper 333, pp. 1–16.

Blizkovsky, M. 1979. Processing and applications in microgravity surveys. *Geophys. Prosp.*, **27**, 848–861.

Blokh, Y. 1980. Calculation of the magnetic field due to two-dimensional anisotropic objects of arbitrary sections with consideration of demagnetization. *Izv. Earth Phys.*, **16**, 126–130.

Bonvalot, S., Diament, M., and Gabalda, G. 1998. Continuous gravity recording with Scintrex CG-3M meters: a promising tool for monitoring active zones. *Geophys. J. Int.*, **135**, 470–494.

Books, K.G. 1962. Remanent magnetization as a contributor to some aeromagnetic anomalies. *Geophysics*, **27**, 359–375.

Borradaile, G.J., and Henry, B. 1997. Tectonic application of magnetic susceptibility and its anisotropy. *Earth Sci. Rev.*, **42**, 49–93.

Bosum, W., Eberle, D., and Renhli, H.-J. 1988. A gyro-oriented 3-component borehole magnetometer for mineral prospecting with examples of its application. *Geophys. Prosp.*, **36**, 933–961.

Bosum, W., Casten, U., Fieberg, F.C., Heyde, I., and Soffel, H.C. 1997. Three-dimensional interpretation of the DTB gravity and magnetic anomalies. *J. Geophys. Res.*, **102**, 18307–18321.

Bott, M.H.P. 1959. The use of electronic digital computers for the evaluation of gravimetric terrain corrections. *Geophys. Prosp.*, **7**, 46–54.

Bott, M.H.P. 1960. The use of rapid digital computing methods for direct gravity interpretation of sedimentary basins. *Geophys. J. Roy. Astr. Soc.*, **3**, 63–67.

Bott, M.H.P. 1962. A simple criterion for interpreting negative anomalies. *Geophysics*, **27**, 376–381.

Bott, M.H.P., and Ingles, A., 1972. Matrix methods for joint interpretation of two-dimensional gravity and magnetic anomalies with application to the Iceland-Faeroe Ridge. *Geophys. J. Roy. Astr. Soc.*, **30**, 55–67.

Bott, M.H.P., Smith, R.A., and Stacey, R.A. 1966. Estimation of the direction of magnetization of a body causing a magnetic anomaly using a pseudo-gravity transformation. *Geophysics*, **31**, 803–811.

Braile, L.W. 1978. Comparison of four random to grid methods. *Comput. Geosci.*, **4**, 341–349.

Breiner, S. 1973. *Applications Manual for Portable Magnetometers*. Geometrics Co.

Briggs, I.C. 1974. Machine contouring using minimum curvature. *Geophysics*, **39**, 39–48.

Brigham, E.O. 1974. *The Fast Fourier Transform*. Prentice-Hall, Inc.

Brocher, T.M. 2005. Empirical relations between elastic wavespeeds and density in the Earth's crust. *Seism. Soc. Am. Bull.*, **95**, 2081–2091.

Brocher, T.M., Hunter, W.C., and Langenheim, V.E. 1998. Implications of seismic reflection and potential field geophysical data on the structural framework of the Yucca Mountain–Crater Flat region, Nevada. *Geol. Soc. Am. Bull.*, **110**, 947–971.

Brown, A.R., and Lautzenhiser, T.V. 1982. The effects of dipping beds on a borehole gravimeter survey. *Geophysics*, **47**, 25–30.

Brown, J.M., Niebauer, T.M., Richter, B., Klopping, F.J., Valentine, J.G., and Buxton, W.K. 1999. A new miniaturized absolute gravimeter developed for dynamic applications. *EOS Trans. AGU*, **80**, 355.

Browne, B.C. 1937. Second order correction to pendulum measurements at sea. *Mon. Notices Roy. Astr. Soc. Geophys. Suppl.*, **4**, 271–279.

Bruckshaw, J.M., and Kunaratnam, K. 1963. The interpretation of magnetic anomalies due to dykes. *Geophys. Prosp.*, **11**, 509–522.

Burg, J.P. 1975. Maximum entropy spectral analysis. Unpublished Ph.D. dissertation, Stanford University.

Burger, H.R., Sheehan, A.F., and Jones, C.H. 2006. *Introduction to Applied Geophysics: Exploring the Shallow Subsurface*. Norton.

Butler, D.K. 1985. Topographic effects considerations in microgravity surveying. In *Proceedings of International Meeting on Potential Fields in Rugged Topography*. Institut de Geophysique de Universite de Lausanne, Bulletin 7, pp. 34–40.

Butler, D.K. 2003. Implications of magnetic background for unexploded ordnance detection. *J. Appl. Geophys.*, **54**, 111–125.

Butler, D.K. 2005. *Near-surface Geophysics*. Society of Exploration Geophysicists.

Butler, D.K., and Llopis, J.L. 1991. Repeat gravity surveys for anomaly detection in an urban environment. In *61st Annual International Meeting Expanded Abstracts*. Society of Exploration Geophysicists, pp. 534–537.

Cady, J.W. 1980. Calculation of gravity and magnetic anomalies of finite-length right polygonal prisms. *Geophysics*, **45**, 1507–1512.

Campbell, D.L. 1980. Gravity terrain corrections for stations on a uniform slope - a power law approximation. *Geophysics*, **45**, 109–112.

Campbell, W.H. 2003. *Introduction to Geomagnetic Fields*. Cambridge University Press.

Capuano, P., De Luca, G., Di Sena, F., Gasparini, P., and Scarpa, R. 1998. The density of the rock covering Gran Sasso Laboratories in Central Appennines, Italy by underground gravity measurements. *J. Appl. Geophys.*, **39**, 25–33.

Carbone, D., and Rymer, H. 1999. Calibration shifts in a LaCoste-and-Romberg gravimeter: comparison with a Scintrex CG-3M. *Geophys. Prosp.*, **47**, 73–83.

Carlson, R.L., and Herrick, C.N. 1990. Densities and porosities in the oceanic crust and their variations with depth and age. *J. Geophys. Res.*, **95**, 9153–9170.

Carmichael, R.S. 1982. Magnetic properties of rocks and minerals. In *Handbook of Physical Properties of Rocks II*. CRC Press, pp. 229–287.

Carnahan, B., Luther, H.A., and Wilkes, J.O. 1969. *Applied Numerical Methods*. J. Wiley and Sons, Inc.

Chai, Y., and Hinze, W.J. 1988. Gravity inversion of an interface above which the density contrast varies exponentially with depth. *Geophysics*, **53**, 837–845.

Chandler, V.W., and Lively, R.S. 2003. *Rock Properties Database: Density, Magnetic Susceptibility and Natural Remanent Magnetization of Rocks in Minnesota*. Minnesota Geological Survey Report, http://purl.umn.edu/92942.

Chandler, V.W., and Lively, R.S. 2011. *Rock Properties Database: Density, Magnetic Susceptibility and Natural Remanent Magnetization of Rocks in Minnesota: An MGS Rock Properties Database*. Minnesota Geological Survey Online Database, version 2.0.

Chandler, V.W., and Malek, K.C. 1991. Moving window Poisson analysis of gravity and magnetic data from the Penokean orogeny, east-central Minnesota. *Geophysics*, **56**, 123–132.

Chandler, V.W., Koski, J.S., Hinze, W.J., and Braile, L.W. 1981. Analysis of multisource gravity and magnetic anomaly data sets by moving-window application of Poisson's theorem. *Geophysics*, **46**, 30–39.

Chapin, D.A. 1996. A deterministic approach toward isostatic gravity residuals. A case study from South America. *Geophysics*, **61**, 1022–1033.

Chapin, D.A. 1998. Gravity instrumentation: Past, present, and future. *The Leading Edge*, **17**, 100–112.

Chapin, D.A., Crawford, M.F., and Baumeister, M. 1999. A side-by-side test of four land gravity meters. *Geophysics*, **64**, 765–775.

Chapman, M.E., and Bodine, J.H. 1979. Considerations of the indirect effect in marine gravity modeling. *J. Geophys. Res.*, **84**, 3889–3892.

Chappel, B.W., and White, A.J.R. 1974. Two contrasting granite types. *Pacif. Geol.*, **8**, 173–174.

Chen, T., Ferguson, J., Aiken, C., and Brady, J. 2005. Real-time data acquisition and quality control for gravity surveys. *The Leading Edge*, **24**, 702–704.

Christensen, N.I., and Mooney, W.D. 1995. Seismic velocity structure and composition of the continental crust: A global view. *J. Geophys. Res.*, **100**, 9761–9788.

Christensen, N.I., and Salisbury, M.H. 1975. Structure and constitution of the lower oceanic crust. *Rev. Geophys. Space Phys.*, **13**, 57–86.

Claerbout, J.F. 1976. *Fundamentals of Geophysical Data Processing with Applications to Petroleum Prospecting*. Blackwell Scientific Publications.

Clark, D.A. 1983. Comments on magnetic petrophysics. *Bull. Aust. Soc. Explor. Geophys.*, **14**, 40–62.

Clark, D.A. 1999. Magnetic petrology of igneous intrusions: Implications for exploration and magnetic interpretation. *Explor. Geophys.*, **30**, 5–26.

Clark, D.A., and Emerson, D.W. 1991. Notes on rock magnetization characteristics in applied geophysical studies. *Explor. Geophys.*, **22**, 547–555.

Clarke, G.K.C. 1971. Linear filters to suppress terrain effects on geophysical maps. *Geophysics*, **36**, 963–966.

Clement, W.G. 1973. Basic principles of two-dimensional digital filtering. *Geophysics*, **21**, 125–145.

Cogbill, A.H. 1990. Gravity terrain corrections calculated using digital elevation models. *Geophysics*, **55**, 102–106.

Committee, Aeromagnetic Standards. 1991. *Guide to Aeromagnetic Specifications and Contracts*. Geological Survey of Canada Open-File 2349.

Constable, S. 2007. Geomagnetism; in Kono. M. (ed.), Geomagnetism. *Treatise of Geophysics*, **5**, 237–276.

Cook, J.C., and Carts, S.L. 1962. Magnetic effect and properties of typical topsoils. *J. Geophys. Res.*, **67**, 815–828.

Cook, K.L. 1950. Quantitative interpretation of vertical magnetic anomalies over veins. *Geophysics*, **15**, 667–686.

Coons, R.L., Woollard, G.P., and Hershey, G. 1967. Structural significance and analysis of mid-continent gravity high. *Am. Assoc. Petr. Geol. Bull.*, **51**, 2381–2399.

Cooper, G.R.J. 2000. Gridding gravity data using an equivalent layer. *Comput. Geosci.*, **26**, 227–233.

Cooper, G.R.J. 2003. Feature detection using sun shading. *Comput. Geosci.*, **29**, 941–948.

Cooper, G.R.J. 2006. Obtaining dip and susceptibility information from Euler deconvolution using the Hough transform. *Comput. Geosci.*, **32**, 1592–1599.

Cordell, L. 1973. Gravity analysis using an exponential density-depth function – San Jacinto graben. *Geophysics*, **38**, 684–690.

Cordell, L. 1979. Gravimetric expression of graben faulting in Santa Fe Country and the Espanola Basin, New Mexico. In *New Mexico Geological Society Guidebook*. 30th Field Conference, pp. 59–64.

Cordell, L. 1985. Applications and problems of analytical continuation of New Mexico aeromagnetic data between arbitrary surfaces of very high relief. In *Proceedings of International Meeting on Potential Fields in Rugged Topography*. Institut de Geophysique de Universite de Lausanne, Bulletin 7, pp. 96–101.

Cordell, L., and McCafferty, A.E. 1989. A terracing operator for physical property mapping with potential field data. *Geophysics*, **54**, 621–634.

Cordell, L., and Taylor, P.T. 1971. Investigation of magnetization and density of an Atlantic seamount using Poisson's theorem. *Geophysics*, **36**, 919–937.

Cordell, L., Zorin, Y.A., and Keller, G.R. 1991. The decompensative gravity anomaly and deep structure of the region of the Rio Grande Rift. *J. Geophys. Res.*, **96**, 6557–6568.

Cowan, D., and Cooper, G. 2003. Drape-related problems in aeromagnetic surveys: the need for tight-drape surveys. *Explor. Geophys.*, **34**, 87–92.

Cowan, D.R., and Cowan, S. 1993. Separation filtering applied to aeromagnetic data. *Explor. Geophys.*, **24**, 429–436.

Criss, R.E., and Champion, D.E. 1984. Magnetic properties of granitic rocks from the southern half of the Idaho Batholith: Influences of hydrothermal alteration and implications for aeromagnetic interpretation. *J. Geophys. Res.*, **89**, 7061–7076.

Crossley, D. 1994. Exploring Earth's gravity spectrum. *Geotimes*, **39**, 19–21.

Curtis, A. 2004a. Theory of model-based geophysical survey and experimental design. Part 1: linear problems. *The Leading Edge*, **23**, 997–1004.

Curtis, A. 2004b. Theory of model-based geophysical survey and experimental design. Part 2: Nonlinear problems. *The Leading Edge*, **23**, 1112–1117.

Daly, R.A., Manger, B.E., and Clark, S.P. 1966. Density of rocks. In *Handbook of Physical Constants*. Geological Society of American Memoir 97, pp. 19–26.

Dampney, C.N.G. 1969. The equivalent source technique. *Geophysics*, **34**, 39–53.

Dampney, C.N.G. 1977. Gravity interpretation for hydrocarbon exploration - A workshop manual. *Bull. Aust. Soc. Explor. Geophys.*, **8**, 161–180.

Daniels, J.J., and Keys, W.S. 1990. Geophysical well logging for evaluating hazardous waste sites. In *Geotechnical and Environmental Geophysics, I. Review and Tutorial*. Society of Exploration Geophysicists, pp. 263–286.

Dannemiller, N., and Li, Y. 2006. A new method for estimation of magnetization direction. *Geophysics*, **71**, L69–L73.

Davis, J.C. 1986. *Statistics and Data Analysis in Geology*. J. Wiley & Sons.

Davis, T.M. 1971. A filtering technique for interpreting the gravity anomaly generated by a two dimensional fault. *Geophysics*, **36**, 554–570.

de Boer, C.B., Dekkers, M.J., and van Hoof, T.A.M. 2001. Rock-magnetic properties of TRM carrying baked and molten rocks straddling burnt coal seams. *Phys. Earth Planet. Int.*, **126**, 93–108.

De Ritis, R., Ventura, G., Chiappini, M., and von Frese, R.R.B. 2010. Regional magnetic and gravity anomaly correlations of the Southern Tyrrhenian Sea. *Phys. Earth Planet. Int.*, **181**, 27–41.

Dean, W.C. 1958. Frequency analysis for gravity and magnetic interpretation. *Geophysics*, **23**, 97–127.

Debeglia, N., and Corpel, J. 1997. Automatic 3-D interpretation of potential field data using analytic signal derivatives. *Geophysics*, **62**, 87–96.

Debski, W., and Tarantola, A. 1995. Information on elastic parameters obtained from the amplitudes of reflected waves. *Geophysics*, **60**, 1426–1436.

Dehlinger, P. 1978. *Marine Gravity*. Elsevier, London, UK.

Dentith, M. 1995. Textural filtering of aeromagnetic data. *Explor. Geophys.*, **26**, 209–214.

Dentith, M.C., Frankcombe, K.F., and Trench, A. 1994. Geophysical signatures of Western Australian mineral deposits: An overview. *Explor. Geophys.*, **25**, 103–160.

DiFrancesco, D., Kaputa, D., and Meyer, T. 2008. Gravity gradiometer systems: advances and challenges. *Preview*, **136**, 30–36.

Dimri, V.P. 1998. Fractal behavior and detectability limits of geophysical surveys. *Geophysics*, **63**, 1943–1946.

Dobrin, M.B., and Savit, C.H. 1988. *Introduction to Geophysical Prospecting*. McGraw-Hill.

Dods, S.D., Teskey, D.J., and Hood, P.J. 1985. The new series of 1 : 1, 000, 000 scale magnetic anomaly maps of the Geological Survey of Canada: Compilation techniques and interpretation. In *The Utility of Regional Gravity and Magnetic Anomaly Maps*. Society of Exploration Geophysicists, pp. 69–87.

Donovan, T.J., Forgey, R.L., and Roberts, A.A. 1979. Aeromagnetic detection of diagenetic magnetite over oil fields. *Am. Assoc. Petr. Geol. Bull.*, **63**, 245–248.

Donovan, T.J., Hendricks, J.D., Roberts, A.A., and Eliason, P.T. 1984. Low-altitude aeromagnetic reconnaissance for petroleum in the Arctic National Wildlife refuge, Alaska. *Geophysics*, **49**, 1338–1353.

Donovan, T.J., O'Brian, D.O., Bryan, J.G., and Cunningham, K.I. 1986. Near-surface magnetic indicators of buried hydrocarbons, aeromagnetic detection and separation of spurious signals. *Assoc. Petr. Geochem. Explor.*, **2**, 1–20.

Dorman, L.M., and Lewis, B.T.R. 1970. Experimental isostasy 1. Theory of the determination of the Earth's isostatic response to a concentrated load. *J. Geophys. Res.*, **75**, 3357–3365.

Dorman, L.M., and Lewis, B.T.R. 1972. Experimental isostasy, 3, Inversion of the isostatic Green function and lateral density changes. *J. Geophys. Res.*, **77**, 3068–3077.

Dortman, N.B. 1976. *Fiziceskie svoistva gornich porod i polesnich iskopamych*. Nedra, Moskava, Izdat.

Douglas, J.K., and Prahl, S.R. 1972. Extended terrain correction tables for gravity. *Geophysics*, **27**, 377–379.

Dransfield, M., and Zeng, Y. 2009. Airborne gravity gradiometry: Terrain correction and elevation error. *Geophysics*, **74**, 37–42.

Draper, N.R., and Smith, H. 1966. *Applied Regression Analysis*. Blaisdell Publishing.

Drinkwater, M.R., Floberghagen, R., Haagmans, R., Muzi, D., and Popescu, A. 2003. GOCE: ESA's first Earth Explorer Core mission. In *Earth Gravity Field from Space Sensors to Earth Sciences*. Vol. 18 of *Space Sciences Series of ISSI*. Kluwer Academic, pp. 419–432.

Drummond, B.J. 1982. Seismic constraints on the chemical composition of the crust of the Pilbarra Craton, Northwest Australia. *Rev. Brasil. Geosci*, **12**, 113–120.

Dunlop, D.J. 1995. Magnetism in rocks. *J. Geophys. Res.*, **100**, 2161–2174.

Dunlop, D.J., and Özdemir, O. 2007. Magnetizations in rocks and minerals. In *Treatise of Geophysics*. Vol. 5 of *Geomagnetism*. Elsevier, pp. 277–336.

Dziewonski, A.M., and Anderson, D.L. 1981. Preliminary reference earth model. *Phys. Earth Planet. Int.*, **25**, 297–356.

Eaton, G.P., and Watkins, J.S. 1970. The use of seismic refraction and gravity methods in hydrogeological investigations. In *Mining and Groundwater Geophysics/1967, Economic Geology Report 26*. Geological Survey of Canada, pp. 544–568.

Ecker, E., and Mittermayer, E. 1969. Gravity corrections for the influence of the atmosphere. *Bull. Theor. Appl. Geophys.*, **11**, 70–80.

Eckhardt, E.A. 1948. A brief history of the gravity method of prospecting for oil. In *Geophysical Case Histories*. Society of Exploration Geophysicists, pp. 21–34.

Ellwood, B.B., and Burkart, B. 1996. Test of hydrocarbon-induced magnetic patterns in soils: the sanitary landfill as laboratory. In *Hydrocarbon Migration and Its Near-surface Expression*. American Association of Petroleum Geologists Memoir 66, pp. 91–98.

Emerson, D.W. 1990. Notes on mass properties of rocks - density, porosity, permeability. *Explor. Geophys.*, **21**, 209–216.

Emerson, D.W., Clark, D.A., and Saul, S.J. 1985. Magnetic exploration models incorporating remanence, demagnetization and anisotropy: HP 41C handheld computer algorithms. *Explor. Geophys.*, **16**, 1–122.

Engels, M., Barckhausen, U., and Gee, J.S. 2008. A new towed marine vector magnetometer and results from a Central Pacific cruise. *Geophys. J. Int.*, **172**, 115–129.

Eskola, L., and Tervo, T. 1980. Solving the magnetostatic field problem (a case of high susceptibility) by means of the method of subsections. *Geoexplor.*, **18**, 79–95.

Fairhead, J.D., and Odegard, M.E. 2002. Advances in gravity survey resolution. *The Leading Edge*, **21**, 36–37.

Fairhead, J.D., Misener, J.D., Green, C.M., Bainbridge, G., and Reford, S.W. 1997. Large scale compilations of magnetic, gravity, radiometric and electromagnetic data: The new exploration strategy for the 90s. In *Proceedings of Exploration 97, Geophysics and Geochemistry at the Millennium*. Fourth Decennial International Conference on Mineral Exploration, pp. 805–815.

Fairhead, J.D., Green, C.M., and Blitzkow, D. 2003. The use of GPS in gravity surveys. *The Leading Edge*, **22**, 954–959.

Fairhead, J.D., Salem, A., Williams, S., and Samson, E. 2008. Magnetic interpretation made easy: The tilt-depth-dip-Δk method. In *2008 Annual International Meeting Expanded Abstracts*. Society of Exploration Geophysicists, pp. 779–783.

Featherstone, W.E. 1993. GPS coordinate transformations and their use in gravimetry. *Explor. Geophys.*, **24**, 487–492.

Featherstone, W.E. 1995. The Global Positioning System (GPS) and its use in geophysical exploration. *Explor. Geophys.*, **26**, 1–18.

Featherstone, W.E., Dentith, M., and Kirby, J.F. 2000. The determination and application of vector gravity anomalies. *Explor. Geophys.*, **31**, 109–113.

Fedi, M., and Florio, G. 2001. Detection of potential fields source boundaries by enhanced horizontal derivative method. *Geophys. Prosp.*, **49**, 40–58.

Fedi, M., Rapolla, A., and Russo, G. 1999. Upward continuation of scattered potential field data. *Geophysics*, **64**, 443–451.

Fils, M.F., Butt, A.L., and Hawke, P.J. 1998. Mapping the range front with gravity – are the corrections up to it? *Explor. Geophys.*, **29**, 378–383.

Forsberg, R. 1985. Gravity field terrain effect computations by FFT. *Bull. Geod.*, **59**, 342–360.

Forsyth, D.W. 1985. Subsurface loading and estimates of the flexural rigidity of the continental lithosphere. *J. Geophys. Res.*, **90**, 12,623–12,632.

Foster, M.R., and Guinzy, N.J. 1967. The coefficient of coherence: Its estimation and use in geophysical prospecting. *Geophysics*, **32**, 602–616.

Fountain, D.M., and Christensen, N.I. 1989. Composition of the continental crust and upper mantle. In *Geophysical Framework of the Continental United States*. Geological Society of America Memoir 172, pp. 711–742.

Frost, B.R., and Shive, P.N. 1986. Magnetic mineralogy of the lower continental crust. *J. Geophys. Res.*, **91**, 6513–6521.

Fuller, B.D. 1967. Two-dimensional frequency analysis and design of grid operators. In *Mining Geophysics, II*. Society of Exploration Geophysicists, pp. 658–708.

Fuller, M., and Cisowski, S. 1987. Lunar paleomagnetism. In *Geomagnetism*. Academic Press, pp. 307–456.

Gabell, A., Tuckett, H., and Olson, D. 2004. The GT-1A mobile gravimeter. In: *Australian SEG Airborne Gravity Workshop*. Australian Society of Exploration Geophysicists.

Gardner, G.H.F., Gardner, L.W., and Gregory, A.R. 1974. Formation velocity and density – the diagnostic basics for stratigraphic traps. *Geophysics*, **39**, 770–780.

Garland, G.D. 1951. Combined analysis of gravity and magnetic anomalies. *Geophysics*, **16**, 51–62.

Garland, G.D. 1989. *Proceedings of Exploration '87*. Ontario Geological Survey Special Volume 3.

Gay, S.P. 1963. Standard curves for interpretation of magnetic anomalies over long tabular bodies. *Geophysics*, **28**, 161–200.

Gay, S.P. 1986. The effects of cathodically protected pipelines on aeromagnetic surveys. *Geophysics*, **51**, 1671–1684.

Gay, S.P. 2004. Glacial till: A troublesome source of near-surface magnetic anomalies. *The Leading Edge*, **23**, 542–547.

Gehman, C.L., Dennis, L.H., Sanford, W.E., Stednick, J.D., and Beckman, N.A. 2009. Estimating specific yield and storage change in an unconfined aquifer using temporal gravity surveys. *Water Resources Res.*, **45**, 16pp.

Gerald, C.F., and Wheatley, P.O. 1989. *Applied Numerical Analysis*. Addison-Wesley Publ. Co.

Gettings, P., Chapman, D.S., and Allis, R. 2008. Techniques, analysis, and noise in a Salt Lake Valley 4D gravity experiment. *Geophysics*, **73**, WA71–WA87.

Gibb, R.A. 1968. The densities of Precambrian rocks from northern Manitoba. *Can. J. Earth Sci.*, **5**, 433–438.

Gibbs, J.W. 1896. Fourier Series. *Nature*, **59**, 200.

Gibson, R.A., and Millegan, P.S. 1998. *Geologic Applications of Gravity and Magnetics: Case Histories*. Society of Exploration Geophysicists Reference Series No. 8.

Gilbert, R.L.G. 1949. A dynamic gravimeter of novel design. *Proc. Phys. Soc.*, **62**, 445–454.

Girdler, R.W., Taylor, P.T., and Frawley, J.J. 1992. A possible impact origin for the Bangui magnetic anomaly (Central Africa). *Tectonophysics*, **212**, 45–58.

Giret, R., and Naudy, H. 1963. Méthodes actuilles d'interprétations des etudes aéeromagnétiques en recherché pétroliére. In *6th World Petroleum Congress Proceedings*. World Petroleum Congress, pp. 907–914.

Goodkind, J.M. 1999. The superconducting gravimeter. *Rev. Sci. Instrum.*, **70**, 4131–4153.

Götze, H-J., and Lahmeyer, B. 1988. Application of three-dimensional interactive modeling in gravity and magnetics. *Geophysics*, **53**, 1096–1108.

Goyal, H.K., von Frese, R.R.B., and Hinze, W.J. 1990. Statistical prediction of satellite magnetic anomalies. *Geophys. J. Int.*, **102**, 101–111.

Graham, I. 1976. The investigation of the magnetic properties of sediments. In *Geoarchaeology*. Westview Press, pp. 49–63.

Grant, F.S. 1973. Magnetic susceptibility mapping: the first year's experience. In *43rd Ann. Int. Meeting Expanded Abstr.* Society of Exploration Geophysicists.

Grant, F.S. 1984a. Aeromagnetics, geology, and ore environments. I: Magnetite in igneous, sedimentary and metamorphic rocks. An overview. *Geoexplor.*, **23**, 303–333.

Grant, F.S. 1984b. Aeromagnetics, geology, and ore environments: II: Magnetite and ore environments. *Geoexplor.*, **23**, 335–362.

Grant, F.S., and Elsaharty, A.F. 1962. Bouguer gravity corrections using a variable density. *Geophysics*, **27**, 616–626.

Grant, F.S., and Martin, L. 1966. Interpretation of aeromagnetic anomalies by use of characteristic curves. *Geophysics*, **31**, 135–148.

Grant, F.S., and West, G.F. 1965. *Interpretation Theory in Applied Geophysics*. McGraw-Hill.

Grauch, V.J.S. 1987. A new variable-magnetization terrain correction method for aeromagnetic data. *Geophysics*, **52**, 94–107.

Grauch, V.J.S., and Campbell, D.L. 1984. Does draping aeromagnetic data reduce terrain-induced effects? *Geophysics*, **49**, 75–80.

Grauch, V.J.S., and Cordell, L. 1987. Limitations of determining density or magnetic boundaries from the horizontal gradient or pseudogravity data. *Geophysics*, **52**, 94–107.

Grauch, V.J.S., and Hudson, M.R. 2007. Guides to understanding the aeromagnetic expression of faults in sedimentary basins: Lessons learned from the central Rio Grande rift, New Mexico. *Geosphere*, **3**, 396–623.

Grauch, V.J.S., and Hudson, M.R. 2011. Aeromagnetic anomalies over faulted strata. *The Leading Edge*, **30**, 1242–1252.

Grauch, V.J.S., Phillips, J.D., Koning, D.J., Johnson, P.S., and Bankey, V. 2009. *Geophysical Interpretations of the Southern Española Basin, New Mexico, that Contribute to Understanding its Hydrogeologic Framework*. US Geological Survey Professional Paper 1761, 88 pp.

Gravenor, C.P., and Stupavsky, M. 1974. Magnetic susceptibility of the surface tills of Southern Ontario. *Can. J. Earth Sci.*, **11**, 658–663.

Green, A.G. 1983. A comparison of adjustment procedures for leveling aeromagnetic survey data. *Geophysics*, **48**, 745–753.

Green, R. 1960. Remanent magnetization and the interpretation of magnetic anomalies. *Geophys. Prosp.*, **8**, 98–110.

Griffin, W.R. 1949. Residual gravity in theory and practice. *Geophysics*, **14**, 39–56.

Grossling, B.F. 1967. *The Internal Magnetization of Seamounts and its Computer Calculation*. US Geological Survey Professional Paper 554-F.

Grow, J.A., and Bowin, C.O. 1975. Evidence for high-density crust and mantle beneath the Chile trench due to descending lithosphere. *J. Geophys. Res.*, **80**, 1449–1458.

Gubbins, D. 2004. *Time Series Analysis and Inverse Theory for Geophysicists*. Cambridge University Press.

Gubins, A.G. 1997. *Geophysics and Geochemistry at the Millennium: Proceedings of Exploration 97*. 4th Decennial International Conference on Mineral Exploration, Prospectors and Developers Association of Canada.

Gulatee, B.L. 1938. *Magnetic Anomalies*. Survey of India Professional Paper 29.

Gundlach, J.H., and Merkowitz, S.M. 2000. Measurement of Newton's Constant using a torsion balance with angular acceleration feedback. *Phys. Rev. Lett.*, **85**, 28–69.

Gunn, P.J. 1995. An algorithm for reduction to the pole that works at all magnetic latitudes. *Explor. Geophys.*, **26**, 247–254.

Gunn, P.J., and Almond, R. 1997. A method for calculating equivalent layers corresponding to large aeromagnetic and radiometric grids. *Explor. Geophys.*, **28**, 72–79.

Guo, W., Dentith, M.C., Li, Z., and Powell, C.M. 1998. Self demagnetization corrections in magnetic modeling: some examples. *Explor. Geophys.*, **29**, 396–401.

Guo, W., Dentith, M.C., Bird, R.T., and Clark, D.A. 2001. Systematic error analysis of demagnetization and implications for magnetic interpretation. *Geophysics*, **66**, 562–570.

Gupta, V.K., and Fitzpatrick, M.M. 1971. Evaluation of terrain effects in ground magnetic surveys. *Geophysics*, **36**, 582–589.

Gupta, V.K., and Grant, F.S. 1985. Mineral-exploration aspects of gravity and aeromagnetic surveys in the Sudbury-Cobalt area, Ontario. In *The Utility of Regional Gravity and Magnetic Anomaly Maps*. Society of Exploration Geophysicists, pp. 392–412.

Gupta, V.K., and Ramani, N. 1980. Some aspects of regional-residual separation of gravity anomalies in a Precambrian terrane. *Geophysics*, **45**, 1412–1426.

Haalck, F. 1956. A torsion magnetometer for measuring the vertical component of the Earth's magnetic field. *Geophys. Prosp.*, **4**, 424–441.

Hacker, B.R., Abers, G.A., and Peacock, S.M. 2003. Subduction factory 1. Theoretical mineralogy, densities, seismic wave speeds, and H_2O contents. *J. Geophys. Res.*, **108**, 2029.

Hacker, B.R., Abers, G.A., and Peacock, S.M. 2004. Subduction factory 3: An Excel worksheet and macro for calculating the densities, seismic wave speeds, and H_2O contents of minerals and rocks at pressure and temperature. *Geochem., Geophys., Geosyst.*, **5**, Q01005.

Hackney, R.I., and Featherstone, W.E. 2003. Geodetic versus geophysical perspectives of the "gravity anomaly". *Geophys. J. Int.*, **154**, 35–43.

Haggerty, S.E. 1978. Mineralogical constraints on Curie isotherms in deep crustal magnetic anomalies. *Geophys. Res. Lett.*, **5**, 105–108.

Haggerty, S.E. 1979. The aeromagnetic mineralogy of igneous rocks. *Can. J. Earth Sci.*, **16**, 1281–1293.

Hahn, A., and Bosum, W. 1986. *Geomagnetics: Selected Examples and Case Histories*. Gebüder Borntraeger.

Hall, D.H. 1959. Direction of polarization determined from magnetic anomalies. *J. Geophys. Res.*, **64**, 1945–1959.

Hall, D.H., and Hajnal, Z. 1962. The gravimeter in studies of buried valleys. *Geophysics*, **27**, 939–951.

Hamilton, A.C., and Brulé, B.G. 1967. Vibration-induced drift in LaCoste and Romberg geodetic gravimeters. *J. Geophys. Res.*, **72**, 2187–2197.

Hammer, S. 1939. Terrain corrections for gravimeter stations. *Geophysics*, **4**, 184–194.

Hammer, S. 1945. Estimating ore masses in gravity prospecting. *Geophysics*, **10**, 50–62.

Hammer, S. 1950. Density determinations by underground measurements. *Geophysics*, **15**, 637–652.

Hammer, S. 1963. Deep gravity interpretation by stripping. *Geophysics*, **28**, 369–378.

Hammer, S. 1974. Approximation in gravity interpretation calculations. *Geophysics*, **39**, 205–222.

Hammer, S. 1977. Graticule spacing versus depth discrimination in gravity interpretation. *Geophysics*, **42**, 60–65.

Hammer, S. 1982. Critique of terrain corrections for gravimeter stations. *Geophysics*, **47**, 839–840.

Hamoudi, M., Thébault, E., Lesur, V., and Mandea, M. 2007. GeoForschungsZentrum Anomaly Magnetic Map (GAMMA): A candidate model for the World Digital Magnetic Anomaly Map. *Geochem. Geophys. Geosyst.*, **8**.

Haney, M., Johnston, C., Li, Y., and Nabighian, M. 2003. Envelopes of 2D and 3D magnetic data and their relationship to the analytic signal: Preliminary results. In *73rd Annual International Meeting Expanded Abstracts*. Society of Exploration Geophysicists, pp. 592–595.

Haney, M. C., and Li, Y. 2002. Total magnetization direction and dip from multiscale edges. In *72nd Annual International Meeting Expanded Abstracts*. Society of Exploration Geophysicists, pp. 735–738.

Hanna, W.F. 1977. *Weak-field Magnetic Susceptibility Anisotropy and its Dynamic Measurement*. US Geological Survey Bulletin 1418.

Hanna, W.F. 1990. Some historical notes on early magnetic surveying in the U.S. Geological Survey. In *Geological Applications of Modern Aeromagnetic Surveys*. US Geological Survey Bulletin 1924, pp. 63–74.

Hansen, R. O., and deRidder, E. 2006. Linear feature analysis for aeromagnetic data. *Geophysics*, **71**, L61–L67.

Hansen, R. O., and Simmonds, M. 1993. Multiple-source Werner deconvolution. *Geophysics*, **58**, 1792–1800.

Hansen, R.O. 2002. 3D multiple-source Werner deconvolution. In *72nd Annual International Meeting Expanded Abstracts*. Society of Exploration Geophysicists, pp. 802–805.

Hansen, R.O., and Miyazaki, Y. 1984. Continuation of potential fields between arbitrary surfaces. *Geophysics*, **49**, 787–795.

Hansen, R.O., and Suciu, L. 2002. Multiple-source Euler deconvolution. *Geophysics*, **67**, 525–535.

Harbaugh, J.W., and Bonham-Carter, G. 1970. *Computer Simulation in Geology*. John Wiley and Sons.

Hardwick, C.D. 1984a. Important design considerations for inboard airborne magnetic gradiometers. *Geophysics*, **49**, 2004–2012.

Hardwick, C.D. 1984b. Non-oriented cesium sensors for airborne magnetometry and gradiometry. *Geophysics*, **49**, 2024–2031.

Harrison, C.G.A. 1987. The crustal field. In *Geomagnetism, 1*. Academic Press, pp. 513–610.

Hartman, R.R., Tesky, K.J., and Friedberg, J.L. 1971. A system for rapid digital aeromagnetic interpretation. *Geophysics*, **36**, 891–918.

Hartmann, T., and Wenzel, H-G. 1995. The HW95 tidal potential catalogue. *Geophy. Res. Lett.*, **22**, 3353–3556.

Hassan, H.H., and Pierce, J.W. 2005. SAUCE: A new technique to remove cultural noise from HRAM data. *The Leading Edge*, **24**, 246–250.

Hatch, D.M., and Pitts, B. 2010. The De Beers Airship Gravity Project. In *Airborne Gravity 2010 – Abstracts from the ASEG-PESA Airborne Gravity 2010 Workshop*. Geoscience Australia and the Geological Survey of New South Wales, Geoscience Australia Record 2010/23 and GSNSW File GS2010/0457.

Hatch, D.M., Murphy, C., Mumaw, C., and Brewster, J. 2007. Performance of the Air-FTG system aboard an airship platform. *Preview*, **127**, 17–22.

Haxby, W.F., Karner, G.D., LaBrecque, J.L., and Weissel, J.K. 1983. Combined oceanic and continental data sets and their use in tectonic studies. *EOS Trans. Am. Geophys. Union*, **64**, 995–1004.

Hayford, J.F. 1909. *The Figure of the Earth and Isostasy, from Measurements in the United States*. US Coast and Geodetic Survey.

Hearst, J.R. 1977a. Estimation of dip and lateral extent of beds with borehole gravimetry. *Geophysics*, **42**, 990–994.

Hearst, J.R. 1977b. On the range of investigation of a borehole gravimeter. *Soc. Prof. Well Log Analysts*, **42**, 1–11.

Hearst, R.B., and Morris, W.A. 2001. Regional gravity setting of the Sudbury structure. *Geophysics*, **66**, 1680–1690.

Heiland, C.A. 1940. *Geophysical Exploration*. Prentice-Hall.

Heiskanen, W.A., and Moritz, H. 1967. *Physical Geodesy*. Freeman.

Heiskanen, W.A., and Vening Meinesz, F.A. 1958. *The Earth and Its Gravity Field*. McGraw-Hill.

Helbig, K. 1962. Some integrals of magnetic anomalies and their relationship to the parameters of the disturbing body. *Z. Geophys.*, **29**, 83–97.

Henderson, R., Miyazaik, Y., and Wold, R. 1984. Direct indication of hydrocarbons from airborne magnetics. *Explor. Geophys.*, **15**, 213–219.

Henkel, H. 1976. Studies of density and magnetic properties of rocks from northern Sweden. *Pure Appl. Geophys.*, **114**, 235–249.

Henkel, H. 1991. Petrophysical properties (density and magnetization) of rocks from the northern part of the Baltic shield. *Tectonophysics*, **192**, 1–19.

Herring, A.T., and Hall, M.J. 2006. Progress in dynamic gravity since 1984. *The Leading Edge*, **21**, 246–249.

Hildenbrand, T.G., and Ravat, D.N. 1997. Geophysical setting of the Wabash Valley fault system. *Seism. Res. Lett.*, **68**, 567–585.

Hildenbrand, T.G., Berger, B., Jachens, R.C., and Ludington, S. 2000. Regional crustal structure and their relationship to the distribution of ore deposits in the Western United States, based on magnetic and gravity data. *Econ. Geol.*, **95**, 1583–1603.

Hinze, W.J. 1963. *Regional Gravity and Magnetic Anomaly Maps of the Southern Peninsula of Michigan*. Michigan Geological Survey Report of Investigations 1.

Hinze, W.J. 1966. The gravity method in iron ore exploration. In *Mining Geophysics, I*. Society of Exploration Geophysicists, pp. 488–464.

Hinze, W.J. 1985. *The Utility of Regional Gravity and Magnetic Anomaly Maps*. Society of Exploration Geophysicists.

Hinze, W.J. 1990. The role of gravity and magnetic methods in engineering and environmental studies. In *Geotechnical and Environmental Geophysics, 1, Review and Tutorial*. Society of Exploration Geophysicists, pp. 75–126.

Hinze, W.J. 2003. Bouguer reduction density, why 2.67? *Geophysics*, **68**, 1559–1560.

Hinze, W.J., and Braile, L.W. 1988. Geophysical aspects of the craton. In *The Geology of North America Volume D-2*. Geological Society of America, pp. 5–24.

Hinze, W.J., and Merritt, D.W. 1969. Basement rocks of the Southern Peninsula of Michigan. In *Field Excursion Guidebook*. Michigan Basin Geological Society, pp. 28–59.

Hinze, W.J., Bradley, J.W., and Brown, A.R. 1978. Gravimeter survey in the Michigan Basin deep borehole. *J. Geophys. Res.*, **83**, 5864–5868.

Hisarli, M., and Orbay, N. 2002. Determination of crustal density at the atmosphere-crust interface of western Anatolia by using the fractal method. *J. Balkan Geophys. Soc.*, **5**, 3–8.

Hjelt, S.E. 1974. The gravity anomaly of a dipping prism. *Geoexploration*, **12**, 29–39.

Hoerl, A.E., and Kennard, R.W. 1970a. Ridge regression. Applications to nonorthogonal problems. *Technometrics*, **12**, 69–82.

Hoerl, A.E., and Kennard, R.W. 1970b. Ridge regression. Biased estimation for nonorthogonal problems. *Technometrics*, **12**, 55–67.

Homilius, J., and Lorch, S. 1957. Density determination on near-surface layers by gamma absorption. *Geophys. Prosp.*, **5**, 449–468.

Honkasalo, T. 1964. On the tidal gravity correction. *Bull. Theor. Appl. Geophys.*, **6**, 34–36.

Hood, L.L., Zakharian, A., Halekas, J. *et al.* 2001. Initial mapping and interpretation of lunar crustal magnetic anomalies using Lunar Prospector magnetometer data. *J. Geophys. Res.*, **106**, 27825–27839.

Hood, P., and Ward, S.H. 1969. Airborne geophysical methods. *Adv. Geophys.*, **13**, 1–112.

Hood, P.J. 1979. *Proceedings of Exploration 77: Geophysics and Geochemistry in the Search for Metallic Ores*. Geological Survey of Canada, Economic Geology Report 31.

Hood, P.J., Holroyd, M.T., and McGrath, P.H. 1979. Magnetic methods applied to base metal exploration. In *Geophysics and Geochemistry in the Search for Metallic Ores, Proceedings of Exploration 77*. Geological Survey of Canada Economic Geology Report 31, pp. 77–104.

Horn, B.K.P., and Bachman, B.L. 1978. Using synthetic images to register real images with surface models. *Commun. Assoc. Comput. Mach.*, **21**, 914–924.

Howell, L.G., Heintz, K.O., and Barry, A. 1966. The development and use of a high-precision downhole gravity meter. *Geophysics*, **31**, 764–772.

Hsu, S-K. 2002. Imaging magnetic sources using Euler's equation. *Geophys. Prosp.*, **56**, 15–25.

Hsu, S-K., Sibuet, J-C., and Shyu, C-T. 1996. High-resolution detection of geologic boundaries from potential-field anomalies: An enhanced analytic signal technique. *Geophysics*, **61**, 373–386.

Hsu, S-K., Coppens, D., and Shyu, C-T. 1998. Depth to magnetic source using the generalized analytic signal. *Geophysics*, **63**, 1947–1957.

Huang, H. 2008. Airborne geophysical data leveling based on line-to-line correlations. *Geophysics*, **73**, F83–F89.

Huang, H., and Fraser, D.C. 2000. Airborne resistivity and susceptibility mapping in magnetically polarizable areas. *Geophysics*, **65**, 502–511.

Huang, H., and Won, I.J. 2001. Conductivity and susceptibility mapping using broadband electromagnetic sensors. *J. Environ. Eng. Geophys.*, **6**, 31–41.

Huang, J., Vanicek, P., Pagiatakis, S.D., and Brink, W. 2001. Effect of topographical density on geoid in the Canadian Rocky Mountains. *J. Geodesy*, **74**, 805–815.

Hubbert, M.K. 1948. A line integral method of computing gravimetric effects of two-dimensional masses. *Geophysics*, **13**, 215–225.

Hudson, M.R., Grauch, V.J.S., and Minor, S.A. 2008. Rock magnetic characterization of faulted sediments with associated magnetic anomalies in the Albuquerque Basin, Rio Grande rift, New Mexico. *Geol. Soc. Am. Bull.*, **120**, 641–658.

Hughes, D.S., and Pondrom, W.L. 1947. Computation of vertical magnetic anomalies from total magnetic field measurements. *Trans. Am. Geophys. Union.*, **28**, 193–197.

Hunt, C.P., Moskowitz, B.M., and Banerjee, S.K. 1995. Magnetic properties of rocks and minerals. In *Rock Physics and Phase Relations*. Vol. 3 of *A Handbook of Physical Constants*. American Geophysical Union, pp. 189–204.

Hussenoeder, S.A., Tivey, M.A., and Shouten, H. 1995. Direct inversion of potential fields from an uneven track with application to the Mid-Atlantic Ridge. *Geophys. Res. Lett.*, **22**, 3131–3134.

Hutchinson, D.R., Grow, J.A., and Klitgord, K.A. 1983. Crustal structure beneath the southern Appalachians: Nonuniqueness of gravity modeling. *Geology*, **11**, 611–615.

Ibrahim, A., and Hinze, W.J. 1972. Mapping buried bedrock topography with gravity. *Ground Water*, **10**, 18–23.

Ishihara, S. 1979. Lateral variation of magnetic susceptibility of the Japanese granitoids. *J. Geol. Soc. Japan*, **85**, 509–523.

Jackson, D.D. 1972. Interpretation of inaccurate, insufficient and inconsistent data. *Geophys. J. Roy. Astr. Soc.*, **28**, 97–109.

Jacob, T., Bayer, R., Chery, J., and Le Moigne, N. 2010. Time lapse microgravity survey reveals water storage heterogeneity of a karst aquifer. *J. Geophys. Res.*, **115**, 18 pp.

Jacobsen, B.H. 1987. A case for upward continuation as a standard separation filter for potential-field maps. *Geophysics*, **52**, 1138–1148.

Jahren, C.E. 1963. Magnetic susceptibility of bedded iron-formation. *Geophysics*, **29**, 756–766.

Jahren, C.E., and Bath, G.D. 1967. *Rapid estimation of induced and remanent magnetization of rock samples, Nevada Test Site*. US Geological Survey Open-File Report 67-122.

Jakosky, J.J. 1950. *Exploration Geophysics*. Trija Publishing.

Jenkins, G.M., and Watts, D.G. 1968. *Spectral Analysis and its Applications*. Holden-Day.

Johnson, B.D., and van Klinken, G. 1979. Some equivalent bodies and ambiguity in magnetic and gravity interpretation. *Bull. Aust. Soc. Explor. Geophys.*, **10**, 109–110.

Johnson, G.R., and Olhoeft, G.R. 1984. Density of rocks and minerals. In *Handbook of Physical Properties*. Vol. 3. CRC Press, pp. 1–38.

Jones, M.B. 1988. *Correlative Analysis of the Gravity and Magnetic Anomalies of Ohio and Their Geological Significance*. Unpublished MSc. thesis, The Ohio State University.

Joseph, R.I. 1976. Demagnetizing factors in nonellipsoidal samples A review. *Geophysics*, **41**, 1052–1054.

Judd, W.R., and Shakoor, A. 1981. Density. In *Physical Properties of Rocks*. Vol. II-2 of McGraw-Hill/CINDAS Series on Material Properties, pp. 29–43.

Kaban, M.K., and Mooney, W.D. 2001. Density structure of the lithosphere in the southwestern United States and its tectonic significance. *J. Geophys. Res.*, **106**, 721–739.

Kaban, M.K., Schwintzer, P., Artemiva, I.M., and Mooney, W.D. 2003. Density of the continental roots: compositional and thermal contribution. *Earth Planet. Sci. Lett.*, **209**, 53–69.

Kane, M.F. 1962. A comprehensive system of terrain corrections using a digital computer. *Geophysics*, **27**, 455–462.

Kaufman, A.A., and Hansen, R.O. 2008. *Principles of the Gravitational Method*. Elsevier.

Kaufman, A.A., Hansen, R.O., and Kleinberg, R.L.K. 2008. *Principles of the Magnetic Methods in Geophysics*. Elsevier.

Kaufmann, R.D., and Doll, W.E. 1998. Gravity meter comparison and circular error. *J. Environ. Eng. Geophys.*, **2**, 165–171.

Kaula, W.M. 1966. *Theory of Satellite Geodesy*. Blaisdell.

Kaula, W.M. 1967. Geophysical implications of satellite determinations of the Earth's gravitational field. *Space Sci. Rev.*, **7**, 769–794.

Keary, P., and Vine, F.J. 1990. *Global Tectonics*. Blackwell.

Keating, P., and Pilkington, M. 2004. Euler deconvolution of the analytic signal and its application to magnetic interpretation. *Geophys. Prosp.*, **52**, 165–182.

Keller, G.R., Webring, M., Hildenbrand, T.G., Hinze, W.J., Li, X., and Ravat, D. 2006. The quest for the perfect gravity anomaly: Part 2 Mass effects and anomaly inversion. In *Expanded Abstracts*. Society of Exploration Geophysicists, pp. 864–868.

Kellogg, O.D. 1953. *Foundations of Potential Theory*. Dover Publications.

Kelso, P.R., Banerjee, S.K., and Teyssier, C. 1993. Rock magnetic properties of the Arunta Block, Central Australia, and their implications for the interpretation of long-wavelength magnetic anomalies. *J. Geophys. Res.*, **98**, 15987–15999.

Kilty, K.T. 1983. Werner deconvolution of profile potential field data. *Geophysics*, **48**, 234–237.

Kim, H.R., von Frese, R.R.B., Kim, J.W. *et al.* 2002. Ørsted verifies regional magnetic anomalies of the Antarctic lithosphere. *Geophys. Res. Lett.*, **29**, 1–3.

Kim, J.H. 1996. Unpublished Ph. D. dissertation, The Ohio State University.

Kim, J.H., and von Frese, R.R.B. 1993. Enhanced recovery of gravity field from satellite altimetry using geological constraints. *EOS Trans. Am. Geophys. Union*, **74**, 99.

Kim, J.W., Kim, J.H., von Frese, R.R.B., Roman, D.R., and Jezek, K.C. 1998a. Spectral attenuation of track-line noise. *Geophys. Res. Lett.*, **25**, 187–190.

Kim, J.W., Kim, J-H., von Frese, R.R.B., Roman, D.R., and Jezek, K.J. 1998b. Spectral attenuation of track-line noise. *Geophys. Res. Lett.*, **25**, 187–190.

Kim, J.W., von Frese, R.R.B., and Kim, H.R. 2000. Crustal modeling from spectrally correlated free-air and terrain gravity data: A case study of Ohio. *Geophysics*, **65**, 1057–1069.

Kirby, J.F., and Featherstone, W.E. 1999. Terrain correcting Australian gravity observations using the national digital elevation model and fast Fourier transform. *Aust. J. Earth Sci.*, **46**, 555–562.

Klemetti, V., and Keys, D. 1983. Relationships between dry density, moisture content, and decomposition of some New Brunswick peats. In *Symposium on Testing of Peats and Organic Soils*. American Society of Testing and Materials Special Technical Publication 820, pp. 72–82.

Kletetschka, G., Wasilewski, P.J., and Taylor, P.T. 2000a. Hematite vs. magnetite as the signature for planetary magnetic anomalies. *Phys. Earth Planet. Int.*, **119**, 259–267.

Kletetschka, G., Wasilewski, P.J., and Taylor, P.T. 2000b. Mineralogy of the sources for magnetic anomalies on Mars. *Meteor. Planet. Sci.*, **35**, 895–899.

Kletetschka, G., Wasilewski, P.J., and Taylor, P.T. 2000c. Unique thermoremanent magnetization of multidomain size hematite: Implications for magnetic anomalies. *Earth Planet. Sci. Lett.*, **176**, 469–479.

Klontis, A.L., and Young, G.A. 1964. Approximation of residual total magnetic intensity anomalies. *Geophysics*, **29**, 623–627.

Koch, G.S., and Link, R.F. 1971. *Statistical Analysis of Geological Data*. John Wiley and Sons.

Konopliv, A.S., Banerdt, W.B., and Sjogren, W.L. 1999. Venus gravity: 189th degree and order model. *Icarus*, **139**, 3–18.

Konopliv, A.S., Asmar, S.W., Carranza, E., Sjorgen, W.L., and Yuan, D.N. 2001. Recent gravity models as a result of the Lunar Prospector mission. *Icarus*, **150**, 1–18.

Korhonen, J.V., Säävuori, H., Wennerström, M. *et al.* 1993. One hundred seventy eight thousand petrophysical parameter determinations from the regional petrophysical programe. In *Geological Survey of Finland. Special Paper 18*, pp. 137–141.

Kostrov, N.P. 2007. Calculation of magnetic anomalies caused by 2D bodies of arbitrary shape with consideration of demagnetization. *Geophys. Prosp.*, **55**, 91–115.

Kreye, C., Wiedermeier, H., Heyen, R., Stekins-Kobsch, T.H., and Boedecker, G. 2006. Galileo and the Earth's gravity field. In: *Inside GNSS*. www.insidegnss.com/auto/working%20papers-12-06.pdf.

Krutikhovskaya, Z.A., Pashevich, J.K., and Silina, I.M. 1982. *A Magnetic Model and the Structure of the Earth's Crust of the Ukrainian Shield* (in Russian). Naukova Dumka.

Ku, C.C. 1977. A direct computation of gravity and magnetic anomalies caused by 2- and 3-dimensional bodies of arbitrary shape and arbitrary magnetic polarization by equivalent point method and a simplified cubic spline. *Geophysics*, **42**, 610–622.

Ku, C.C., and Sharp, J.A. 1983. Werner deconvolution for automated magnetic interpretation and its refinement using Marquardt's inverse modeling. *Geophysics*, **48**, 754–774.

Ku, C.C., Telford, W.M., and Lim, S.H. 1971. The use of linear filtering in gravity problems. *Geophysics*, **36**, 1174–1203.

Kuhn, M. 2003. Geoid determination with density hypotheses from isostatic models and geological information. *J. Geodesy*, **77**, 50–65.

Kuo, B.Y., and Forsyth, D.W. 1988. Gravity anomalies of the ridge-transform system in the South Atlantic between 34° and 34.5° S: Upwelling centers and variations in crustal thickness. *Marine Geophys. Res.*, **10**, 205–232.

Labo, J. 1986. A practical introduction to borehole geophysics. In *Geophysical Reference Series 2*. Society of Exploration Geophysicists, pp. 1–330.

LaCoste, L. 1983. LaCoste and Romberg straight-line gravity meter. *Geophysics*, **48**, 606–610.

LaCoste, L., Clarkson, N., and Hamilton, G. 1967. LaCoste and Romberg stabilized platform shipboard gravity meter. *Geophysics*, **32**, 99–109.

LaCoste, L.J.B. 1934. A new type long period seismograph. *Physics*, **5**, 178–180.

LaCoste, L.J.B. 1967. Measurements of gravity at sea and in the air. *Rev. Geophys.*, **5**, 477–526.

LaCoste & Romberg, Inc. 2004. *Instruction Manual Model G & D Gravity Meters*. LaCoste & Romberg, Inc.

LaFehr, T.R. 1965. The estimation of the total amount of anomalous mass by Gauss' Theorem. *J. Geophys. Res.*, **70**, 1911–1919.

LaFehr, T.R. 1983. Rock density from borehole gravity surveys. *Geophysics*, **48**, 341–356.

LaFehr, T.R. 1991a. An exact solution for the gravity curvature (Bullard B) correction. *Geophysics*, **56**, 1179–1184.

LaFehr, T.R. 1991b. Standardization in gravity reduction. *Geophysics*, **56**, 1170–1178.

LaFehr, T.R. 1998. On Talwani's "Errors in the total Bouguer reduction". *Geophysics*, **63**, 1131–1136.

LaFehr, T.R., and Nabighian, M.N., 2012. Fundamentals of Gravity Exploration, Society of Exploration Geophysicists.

Lahti, I., and Karinen, T. 2010. Tilt derivative multiscale edges of magnetic data. *The Leading Edge*, **29**, 24–29.

Lam, H-L. 1989. On the prediction of low-frequency geomagnetic pulsations for geophysical prospecting. *Geophysics*, **54**, 635–642.

Lanczos, C. 1961. *Linear Differential Operators*. Van Nostrand-Reinhold.

Lane, R. J. L. 2010. *Airborne Gravity 2010 Abstracts from the ASEG-PESA Airborne Gravity 2010 Workshop*. Geoscience Australia and the Geological Survey of New South Wales, Geoscience Australia Record 2010/23 and GSNSW File GS2010/0457.

Langel, R. A., Sabaka, T.J., Baldwin, R.T., and Conrad, J.A. 1996. The near-Earth magnetic field from magnetosphere and quiet-day ionospheric sources and how it is modeled. *Phys. Earth Planet. Int.*, **98**, 235–267.

Langel, R.A. 1987. The main field. In *Geomagnetism, 1*. Jacobs, J.A. (ed.), pp. 415–459.

Langel, R.A., and Estes, R.H. 1982. A geomagnetic field spectrum. *Geophys. Res. Lett.*, **9**, 250–253.

Langel, R.A., and Hinze, W.J. 1998. *The Magnetic Field of the Earth's Lithosphere*. Cambridge University Press.

LaPointe, P., Chomyn, B.A., Morris, W.A., and Coles, R.A. 1984. Significance of magnetic susceptibility measurements from the Lac Du Bonnet Batholith, Manitoba, Canada. *Geoexplor.*, **22**, 217–229.

Larsson, L.O. 1977. *Statistical Treatment of In-situ Measurements of Magnetic Susceptibility*. Sveriges Geologiska Undersokning, Serie C, Nr 727, Arsbok 71 Nr 2.

Latz, K., Weismiller, R.A., and Van Scoyoc, G.E. 1981. *A Study of the Spectral Reflectance of Selected Eroded Soils of Indiana in Relationbship to their Chemical and Physical Properties*. Laboratory of Applications of Remote Sensing Technical Report. 082181.

Lawson, C.L., and Hanson, R.J. 1974. *Solving Least Squares Problems*. Prentice-Hall, Inc.

Le Borgne, E. 1955. Susceptibilité magnétique anormale du sol superficiel. *Ann. Geophys.*, **11**, 399–419.

Le Borgne, E. 1960. Influence du deu sur les propriets magnétiques du sol et sur celles du schiste et du granite. *Ann. Geophys.*, **16**, 159–196.

Leaman, D.E. 1998. The gravity terrain correction – practical considerations. *Explor. Geophys.*, **29**, 467–471.

Lecoanet, H., Leveque, F., and Segura, S. 1999. Magnetic susceptibility in environmental applications: Comparison of field probes. *Phys. Earth and Plan. Int.*, **115**, 191–204.

Lee, J.B. 2001. FALCON gravity gradiometer technology. *Explor. Geophys.*, **32**, 247–250.

Lee, T.J. 1980. Rapid computation of magnetic anomalies with demagnetization included, for arbitrary shaped bodies. *Geophys. J. Roy. Astr. Soc.*, **60**, 67–75.

Leftwich, T.E., von Frese, R.R.B., Kim, H.R. *et al.* 1999. Crustal analysis of Venus from Magellan satellite observations at Atalanta Planitia, Beta Regio, and Thetis Regio. *J. Geophys. Res.*, **104**, 8441–8462.

Leftwich, T.E., von Frese, R.R.B., Potts, L.V. *et al.* 2005. Crustal modeling of the North Atlantic from spectrally correlated free-air and terrain gravity. *J. Geodynam.*, **40**, 23–50.

Legge, J.A. Jr. 1944. A proposed least-squares method for the determination of the elevation factor. *Geophysics*, **9**, 175–179.

Lemoine, F.G., Kenyon, S.C., Factor, J.K. *et al.* 1998. *The Development of the Joint NASA GSFC and National Imagery and Mapping Agency (NIMA) Geopotential Model EGM96*. NASA/TP-1998-206861. Goddard Space Flight Center.

Lerch, F.J. 1991. Optimum weighting and error calibration for estimation of gravitational parameters. *Bull. Geodynam.*, **65**, 44–52.

Li, X. 2006. Understanding 3D analytic signal amplitude. *Geophysics*, **71**, L13–L16.

Li, X., and Götze, H-J. 2001. Tutorial: Ellipsoid, geoid, gravity, geodesy, and geophysics. *Geophysics*, **66**, 1660–1668.

Li, X., Hildenbrand, T.G., Hinze, W.J., Keller, G.R., Ravat, D., and Webring, M. 2006. The quest for the perfect gravity anomaly: Part 1 new calculation standards. In *Expanded Abstracts*. Society of Exploration Geophysicists, pp. 859–863.

Li, Y., and Oldenburg, D.W. 1996. 3-D inversions of magnetic data. *Geophysics*, **61**, 394–408.

Li, Y., and Oldenburg, D.W. 1998a. 3-D inversion of gravity data. *Geophysics*, **63**, 109–119.

Li, Y., and Oldenburg, D.W. 1998b. Separation of regional and residual magnetic field data. *Geophysics*, **63**, 431–439.

Li, Y., and Oldenburg, D.W. 2000. Joint inversion of surface and three-component borehole magnetic data. *Geophysics*, **65**, 540–552.

Li, Y., Shearer, S.E., Haney, M.M., and Dannemiller, N. 2010. Comprehensive approaches to 3D inversion of magnetic data affected by remanent magnetization. *Geophysics*, **75**, L1–L11.

Li, Y.C., and Sideris, M.G. 1994. Improved gravimetric terrain corrections. *Geophys. J. Int.*, **119**, 740–752.

Lidiak, E.G. 1974. Magnetic characteristics of some Precambrian basement rocks. *J. Geophys.*, **40**, 549–564.

Lilley, F.E.M. 1968. Optimum direction of survey lines. *Geophysics*, **33**, 329–336.

Lilley, F.E.M. 1982. Geomagnetic field fluctuations over Australia in relation to magnetic surveys. *Bull. Aust. Soc. Explor. Geophys.*, **13**, 68–76.

Lilley, F.E.M., and Weitemeyer, K.A. 2004. Apparent aeromagnetic wavelengths of the magnetic signals of ocean swells. *Explor. Geophys.*, **35**, 137–141.

Lilley, F.E.M., Hitchman, A.P., and Wang, L.J. 1999. Time-varying effects in magnetic mapping: Amphidromes, doldrums, and induction hazard. *Geophysics*, **64**, 1720–1729.

Lilley, F.E.M., Hitchman, A.P., Millegan, P.R., and Pedersen, T. 2004. Sea-surface observation of the magnetic signals of ocean swells. *Geophys. J. Int.*, **159**, 565–572.

Lindseth, R.O. 1974. *Recent Advances in Digital Processing of Geophysical Data – A Review*. Society of Exploration Geophysicists Continuing Education Manual.

Lindsley, D.H., Andreasen, G.E., and Balsely, J.R. 1966. Magnetic properties of rocks and minerals. In *Handbook of Physical Constants*. Geological Society America Memoir 97, pp. 544–552.

Lines, L.R., and Levin, F.K. 1988. *Inversion of Geophysical Data*. Society of Exploration Geophysicists Geophysics Reprint Series No. 9.

Lines, L.R., and Treitel, S. 1984. Tutorial – A review of least-squares inversion and its application to geophysical problems. *Geophys. Prosp.*, **32**, 159–186.

Long, L.T., and Kaufmann, R.D., 2013. Acquisition and Analysis of Terrestrial Gravity Data, Cambridge University Press.

Longman, I.M. 1959. Formulas for computing the tidal accelerations due to the moon and the sun. *J. Geophys. Res.*, **64**, 2351–2355.

Longuevergne, L., Boy, J.P., Florsch, N. *et al.* 2009. Local and global hydrologic contributions to gravity variations observed in Strasbourg. *J. Geodynam.*, **48**, 189–194.

Lourenco, J.S., and Morrison, H.F. 1973. Vector magnetic anomalies derived from measurements of a single component of the field. *Geophysics*, **38**, 359–368.

Love, J.J. 2008. Magnetic monitoring of earth and space. *Phys. Today*, **61**, 31–37.

Lu, R.S. 1998. Finite-impulse-response reduction-to-the pole filter. *Geophysics*, **63**, 1958–1964.

Lu, R.S., Mariano, J., and Willen, D.E. 2003. Differential reduction of magnetic anomalies to the pole on a massively parallel computer. *Geophysics*, **68**, 1945–1951.

Ludwig, J.W., Nafe, J.E., and Drake, C.L. 1970. Seismic refraction. In *The Sea, 4*. John Wiley & Sons, pp. 54–84.

Luhmann, J.G. 1986. The solar wind interaction with Venus. *Space Sci. Rev.*, **44**, 241–306.

Luyendyk, A.P.J. 1997. Processing of airborne magnetic data. *AGSO J. Aust. Geol. Geophys.*, **17**, 31–38.

Ma, X.Q., and Watts, D.R. 1994. Terrain correction program for regional gravity surveys. *Comput. Geosci.*, **20**, 961–972.

Machel, H.G. 1996. Magnetic contrasts as a result of hydrocarbon seepage and migration. In *Hydrocarbon Migration and Its Near-surface Expression*. American Association of Petroleum Geologists Memoir 66, pp. 99–109.

Macke, R.J., Kiefer, W.S., Britt, D.T., and Consolmagno, G.J. 2010. Density, porosity and magnetic susceptibility of lunar rocks. In *Proceedings of the 14th Lunar and Planetary Science Conference (Houston, TX)*. NASA Lunar and Planetary Science Institute.

MacLeod, I.C., Jones, K., and Dai, T.F. 1993. 3-D analytic signal in the interpretation of total magnetic field data at low magnetic latitudes. *Explor. Geophys.*, **24**, 679–688.

MacQueen, J.D. 2010. Improved tidal corrections for time-lapse microgravity surveys. *SEG Expanded Abstracts*, **29**, 1142–1145.

MacQueen, J.D., and Harrison, J.C. 1997. Airborne gravity surveys in rugged terrain. In *Expanded Abstracts*. Society of Exploration Geophysicists, pp. 494–497.

Maeda, H., Iyemori, T., Arakai, T., and Kamei, T. 1982. New evidence of meridional current system in the equatorial ionosphere. *Geophys. Res. Lett.*, **9**, 337–340.

Maher, B.A. 1986. Characterization of soils by mineral magnetic measurements. *Phys. Earth Planet. Int.*, **42**, 76–92.

Malin, S. 1987. Introduction to geomagnetism. In *Geomagnetism*, Academic Press, pp. 1–49.

Mallick, K., and Sharma, K.K. 1999. A finite element method for computation of the regional gravity anomaly. *Geophysics*, **64**, 461–469.

Mandea, M., and Purucker, M. 2005. Observing, modeling, and interpreting magnetic field of the solid earth. *Surv. Geophys.*, **26**, 415–459.

Manger, G.E. 1963. *Porosity and Bulk Density of Sedimentary Rocks*. US Geological Survey Bulletin 1144-E.

Mariano, J., and Hinze, W.J. 1993. Modeling complexly magnetized two-dimensional bodies of arbitrary shape. *Geophysics*, **58**, 637–644.

Mariano, J., and Hinze, W.J. 1994. Gravity and magnetic models of the Midcontinent Rift in eastern Lake Superior. *Can. J. Earth Sci.*, **31**, 661–674.

Marquardt, D.W. 1970. Ridge regression, biased linear estimation, and nonlinear estimation. *Technometrics*, **12**, 591–612.

Marsh, B.D. 1971. Aeromagnetic terrain effects. Unpublished M.Sc. thesis, University of Arizona.

Marson, I., and Klingelé, E.E. 1993. Advantages of using the vertical gradient of gravity for 3-D interpretation. *Geophysics*, **58**, 1588–1595.

Marson, J., and Faller, J.E. 1986. g – the acceleration of gravity: its measurement and its importance. *J. Phys. Sci. Instrum.*, **19**, 22–32.

Martelet, G., Sailhac, P., Moreau, F., and Diament, M. 2001. Characterization of geological boundaries using 1-D wavelet transform on gravity data: Theory and application to the Himalayas. *Geophysics*, **66**, 1116–1129.

Martins, C.M., Barbosa, V.C.F., and Silva, J.B.C. 2009. Simultaneous 3D depth-to-basement and density-contrast estimates using gravity data and depth control at few points. *Geophysics*, **75**, 121–128.

Mathews, J.H., and Fink, K.K. 2004. *Numerical Methods Using Matlab*. Prentice-Hall.

Mauring, E., and Kihle, O. 2006. Leveling aerogeophysical data using a moving differential median filter. *Geophysics*, **71**, L5–L11.

Mauring, E., Beard, L.P., Kihle, O., and Smethurst, M.A. 2002. A comparison of aeromagnetic leveling techniques with an introduction to median leveling. *Geophys. Prosp.*, **50**, 43–54.

Maus, S., Barckhausen, U., Bournas, N. *et al.* 2009. EMAG2: A 2-arc min resolution Earth Magnetic Anomaly Grid compiled from satellite, airborne, and marine magnetic measurements. *Geochem. Geophy. Geosyst.*, **10**, 12 pp.

McAdoo, D.C., and Marks, K.M. 1992. Gravity field of the southern ocean from Geosat data. *J. Geophys. Res.*, **96**, 3247–3260.

McCulloh, T.H. 1965. A confirmation by gravity measurement of an underground density profile on core densities. *Geophysics*, **30**, 1108–1132.

McCulloh, T.H. 1966. The promise of precise borehole gravimetry in petroleum exploration and exploitation. In *US Geological Survey Circular 531*. US Geological Survey, pp. 1–12.

McDonald, A.J.W. 1991. The use and abuse of image analysis in geophysical potential field interpretation. *Surv. Geophys.*, **12**, 531–551.

McEnroe, S.A., and Brown, L.L. 2000. A closer look at remanence-dominated aeromagnetic anomalies: Rock magnetic properties and magnetic mineralogy of the Russell Belt microcline-sillmanite gneiss, northwest Adirondack Mountains, New York. *J. Geophys. Res.*, **105**, 16437–16456.

McEnroe, S.A., Robinson, P., Langenhorst, F., Frandsen, C., Terry, M.P., and Ballaran, T.B. 2007. Magnetization of exsolution intergrowths of hematite and ilmenite: Mineral chemistry, phase relations, and magnetic properties of hemo-ilmenite ores with micron to nanometer-scale lamellae from Allard Lake, Quebec. *J. Geophys. Res.*, **112**, 20 pp.

McGinnis, L.D., Kempton, J.P., and Heigold, P.C. 1963. *Relationship of Gravity Anomalies to a Drift-filled Bedrock Valley System in Northern Illinois*. Illinois State Geological Survey Circular 354.

McIntyre, J.I. 1981. Accurate display of fine detail in aeromagnetic data. *Bull. Aust. Soc. Explor. Geophys.*, **12**, 82–88.

Mendonca, C.A., and Silva, J.B.C. 1995. Interpolation of potential field data by equivalent layer and minimum curvature: a comparative analysis. *Geophysics*, **60**, 399–407.

Menichetti, V., and Guillen, A., 1983. Simultaneous interactive magnetic and gravity inversion. *Geophys. Prosp.*, **31**, 929–944.

Menk, F.W. 1993. Low latitude geomagnetic pulsations: a user's guide. *Explor. Geophys.*, **24**, 111–114.

Menke, W. 1989. *Geophysical Data Analysis*. Academic Press.

Merriam, D.F., and Sneath, P.H.A. 1966. Quantitative comparison of contour maps. *J. Geophys. Res.*, **7**, 1105–1115.

Merriam, J.B. 1992. Atmospheric pressure and gravity. *Geophys. J. Int.*, **109**, 488–500.

Merrill, R.T., and Burns, R.E. 1972. A detailed magnetic study of Cobb seamount. *Earth Planet. Sci. Lett.*, **14**, 413–418.

Merrill, R.T., and McElhinny, M.W. 1983. *The Earth's Magnetic Field*. Academic Press.

Meyer, F.D. 1974. Filter techniques in gravity interpretation. *Adv. Geophys.*, **17**, 187–261.

Meyer, J., Hufer, J.H., Siebert, M., and Hahn, A. 1985. On the identification of Magsat anomaly charts as crustal part of internal field. *J. Geophys. Res.*, **90**, 2537–2542.

Meyer de Stadelhofen, C., and Juillard, T. 1987. Truth is an instance of error temporarily appealing. *The Leading Edge*, **6**, 38.

Mikuska, J., Pasteka, R., and Marusiak, I. 2006. Estimation of distant relief effect in gravimetry. *Geophysics*, **71**, 59–69.

Mikuska, J., Marusiak, I., Pasteka, R., Karcol, R., and Beno, J. 2008. The effect of topography in calculating the atmospheric correction in gravimetry. In *Expanded Abstracts*. Society of Exploration Geophysicists, pp. 784–788.

Miller, H.G., and Singh, V.J. 1994. Potential field tilt – A new concept for location of potential field sources. *Appl. Geophys.*, **32**, 213–217.

Milligan, P.R. 1995. Short-period geomagnetic variations recorded concurrently with an aeromagnetic survey across the Bendigo area, Victoria. *Explor. Geophys.*, **26**, 527–534.

Milson, J. 2003. *Field Geophysics*. John Wiley & Sons.

Milson, J.J., and Eriksen, A. 2011. *Field Geophysics*. John Wiley & Sons.

Mims, J.H., and Mataragio, J. 2010. Airborne full tensor gravity. In *Proceedings of SAGEEP Conference*. Society of Exploration Geophysicists, pp. 438–442.

Minty, B.R.S. 1991. Simple micro-levelling for aeromagnetic data. *Explor. Geophys.*, **22**, 591–592.

Mishra, D.C. 2011. *Gravity and Magnetic Methods for Geological Studies*. Taylor and Francis.

Moody, M.V., Chan, H.A., and Paik, H.J. 1986. Superconducting gravity gradiometer for space and terrestrial application. *J. Appl. Phys.*, **60**, 4308–4315.

Moon, C.J., Whateley, M.K.G., and Evans, A.M. 2006. *Introduction to Mineral Exploration*. Blackwell Publishing.

Mooney, H.W., and Bleifuss, R. 1953. Magnetic susceptibility measurements in Minnesota, Part II, analysis of field results. *Geophysics*, **18**, 383–393.

Mooney, W.D., and Kaban, M.K. 2010. The North American upper mantle: Density, composition, and evolution. *J. Geophys. Res.*, **115**, 424–448.

Moreau, F., Gibert, D., Holschneider, M., and Saracco, G. 1999. Identification of sources of potential fields with the continuous wavelet transform: basic theory. *J. Geophys. Res.*, **104**, 5003–5013.

Morelli, C. 1974. *The International Gravity Standardization Net 1971*. International Association of Geodesy Special Publication 4.

Moritz, H. 1980a. *Advanced Physical Geodesy*. H. Wichmann Verlag.

Moritz, H. 1980b. Geodetic Reference System 1980. *Bull. Geod.*, **54**, 395–405.

Morley, L.W. 1970. *Mining and Groundwater Geophysics/1967*. Geological Survey of Canada Economic Geology Report No. 26.

Morris, B., Ugalde, H., and Thomson, V. 2007. Magnetic remanence constraints on magnetic inversion models. *The Leading Edge*, **26**, 960–964.

Morris, W.A., Mueller, E.I., and Parker, C.E. 1995. Borehole magnetics: Navigation, vector components, and magnetostratigraphy. In *65th Annual International Meeting Expanded Abstracts*. Society of Exploration Geophysicists, pp. 495–498.

Mullins, C. 1974. The magnetic properties of the soil and their application to archaeological prospecting. *Archaeo-Physika*, **5**, 143–147.

Mullins, C.E. 1977. Magnetic susceptibility of the soil and its significance in soil science: A review. *J. Soil Sci.*, **28**, 223–246.

Multhauf, R.P., and Good, G. 1987. *A Brief History of Geomagnetism*. Smithsonian Institution Press.

Murthy, I.V.K. 2010. *Gravity and Magnetic Interpretation in Exploration Geophysics*. Geological Society of India, Memoir 40, Bangalore.

Mushayandebvu, M.F., and Davies, J. 2006. Magnetic gradients in sedimentary basins: examples from the western Canada sedimentary basin. *The Leading Edge*, **25**, 69–73.

Mushayandebvu, M.F, van Driel, P., Reid, A.B., and Fairhead, J.D. 2001. Magnetic source parameters of 2D structures using extended Euler deconvolution. *Geophysics*, **66**, 814–823.

Mushayandebvu, M.F, Lesur, V., Reid, A.B., and Fairhead, J.D. 2004. Grid Euler deconvolution with constraints for 2D structures. *Geophysics*, **69**, 489–496.

Muszala, S., Stoffa, P.L., and Lawver, L.A. 2001. An application for removing cultural noise from aeromagnetic data. *Geophysics*, **66**, 213–219.

Nabighian, M.N. 1972. The analytic signal of two-dimensional magnetic bodies with polygonal cross-section. Its properties and use for automated anomaly interpretation. *Geophysics*, **37**, 507–517.

Nabighian, M.N. 1974. Additional comments on the analytic signal of two-dimensional magnetic bodies with polygonal cross-section. *Geophysics*, **39**, 85–92.

Nabighian, M.N. 1984. Toward a three-dimensional automatic interpretation of potential field data via generalized Hilbert transforms Fundamental relations. *Geophysics*, **49**, 780–786.

Nabighian, M.N., and Hansen, R.O. 2001. Unification of Euler and Werner deconvolution in three dimensions via the generalized Hilbert transform. *Geophysics*, **66**, 1805–1810.

Nabighian, M.N., Ander, M.E., Grauch, V.J.S. *et al.* 2005a. Historical development of the gravity method in exploration. *Geophysics*, **70**, 63ND–89ND.

Nabighian, M.N., Ander, M.E., Grauch, V.J.S. *et al.* 2005b. Historical development of the magnetic method in exploration. *Geophysics*, **70**, 33ND–61ND.

NAGDC, North American Gravity Data Committee. 2005. New standards for reducing gravity observations: The revised North American gravity database. *Geophysics*, **70**, 325–332.

Nagy, D. 1966. The gravitational attraction of a right rectangular prism. *Geophysics*, **31**, 362–371.

Naidu, P.S., and Mathew, M.P. 1998. *Analysis of Geophysical Potential Fields*. Elsevier.

Nakatsuka, T., and Okuma, S. 2006. Reduction of magnetic anomaly observations from helicopter surveys at varying elevations. *Explor. Geophys.*, **37**, 121–128.

Namiki, N., Iwata, T., Matsumoto, K. *et al.* 2009. Farside gravity field of the moon from four-way Doppler measurements of SELENE (Kaguya). *Science*, **323**, 900–905.

Naudy, H. 1971. Automatic determination of depths on aeromagnetic profiles. *Geophysics*, **36**, 717–722.

Nelson, J.B. 1994. Leveling total-field aeromagnetic data with measured horizontal gradients. *Geophysics*, **48**, 745–753.

Nettleton, L.L. 1939. Determination of density for reduction of gravimeter observations. *Geophysics*, **4**, 176–183.

Nettleton, L.L. 1940. *Geophysical Prospecting for Oil*. McGraw-Hill.

Nettleton, L.L. 1942. Gravity and magnetic calculations. *Geophysics*, **7**, 293–310.

Nettleton, L.L. 1943. Recent experimental and geophysical evidence of mechanics of salt-dome formation. *Am. Assoc. Petrol. Geol. Bull.*, **27**, 51–63.

Nettleton, L.L. 1954. Regionals, residuals, and structures. *Geophysics*, **19**, 1–22.

Nettleton, L.L. 1976. *Gravity and Magnetics in Oil Prospecting*. McGraw-Hill.

Nettleton, L.L., and Elkins, T.A. 1944. Association of magnetic and density contrasts with igneous rock classifications. *Geophysics*, **9**, 60–78.

Neuendorf, K.K.E., Mehl, Jr., J.P., and Jackson, J.A. 2005. *Glossary of Geology*. American Geological Institute.

Neumeyer, J., Hagedoorn, J., Leitloff, J., and Schmidt, T. 2004. Gravity reduction with three-dimensional atmospheric pressure data for precise ground gravity measurements. *J. Geodynam.*, **38**, 437–450.

Niebauer, T.M. 1988. Correcting gravity measurements for the effects of local air pressure. *J. Geophys. Res.*, **93**, 7989–7991.

Niebauer, T.M. 2007. Absolute gravimeter: instrumentation concepts and implementation. In *Geodesy*. Vol. 3 of *Treatise on Geophysics*. Elsevier, pp. 43–64.

Niebauer, T.M., Sasagawa, G.S., Faller, J.E., Hilt, R., and Klopping, F. 1995. A new generation of absolute gravimeters. *Metrologia*, **32**, 159–180.

Nowell, D.A.G. 1999. Gravity terrain corrections – an overview. *J. Appl. Geophys.*, **42**, 117–134.

O'Brien, D.P. 1972. CompuDepth, a new method for depth-to-basement computation. In *42nd Annual International Meeting Expanded Abstracts*. Society of Exploration Geophysicists.

O'Connell, M.D. 2001. A heuristic method of removing micropulsations from airborne magnetic data. *The Leading Edge*, **20**, 1242–1246.

Odegard, M.E., and Berg, J.W. 1965. Gravity interpretation using the Fourier integral. *Geophysics*, **30**, 424–438.

Okabe, M. 1979. Analytical expressions for gravity anomalies due to homogeneous polyhedral bodies and translations into magnetic anomalies. *Geophysics*, **44**, 730–741.

Oldenburg, D.W. 1974. The inversion and interpretation of gravity anomalies. *Geophysics*, **39**, 526–536.

Olivier, R.J., and Simard, R.G. 1981. Improvement of the conic prism model for terrain correction in rugged topography. *Geophysics*, **46**, 1054–1056.

Olsen, N., Hulot, G., and Sabaka, T.J. 2007. The present field; in Kono, M. (ed.), Geomagnetism. *Treatise Geophys.*, **5**, 33–76.

Olson, P., and Amit, H. 2006. Changes in Earth's dipole. *Naturwissenschaften*, **93**, 519–542.

Onesti, L.A., and Hinze, W.J. 1970. Magnetic observations over eskers in Michigan. *Geol. Soc. Am. Bull.*, **81**, 3453–3456.

O'Reilly, W. 1984. *Rock and Mineral Magnetism*. Blackie and Son.

Oswin, J. 2009. *A Field Guide to Geophysics in Archaeology*. Springer.

Overhauser, A.W. 1953. Paramagnetic relaxation in metals. *Phys. Rev.*, **89**, 689–700.

Packard, M., and Varian, R. 1954. Proton gyromagnetic ratio. *Phys. Rev.*, **93**, 941.

Parasnis, D.S. 1952. A study of rock densities in the English Midlands. In *Geophysical Supplement 6*. Royal Astronomical Society, pp. 252–271.

Parasnis, D.S. 1971. *Physical Property Guide for Rocks and Minerals*. ABEM Geophysical Memorandum 4/71.

Parasnis, D.S. 1986. *Principles of Applied Geophysics*. Chapman and Hall.

Parker, R.L. 1974. Best bounds on density and depth from gravity data. *Geophysics*, **39**, 644–649.

Parker, R.L. 1975. The theory of ideal bodies for gravity interpretation. *Geophys. J. Roy. Astr. Soc.*, **42**, 315–334.

Parker, R.L. 1994. *Geophysical Inverse Theory*. Princeton.

Parker, R.L. 1995. Improved Fourier terrain correction: Part I. *Geophysics*, **60**, 1007–1017.

Parker, R.L. 1996. Improved Fourier terrain correction: Part II. *Geophysics*, **61**, 365–372.

Parkinson, W.D. 1983. *Introduction to Geomagnetism*. Elsevier Science Publishing.

Parkinson, W.D., and Jones, F.W. 1979. The geomagnetic coast effect. *Rev. Geophys. Space Phys.*, **17**, 1999–2015.

Pawlowski, R.S. 1994. Green's equivalent-layer concept in gravity band-pass filter design. *Geophysics*, **59**, 69–76.

Pawlowski, R.S. 1995. Preferential continuation potential-field anomaly enhancement. *Geophysics*, **60**, 390–398.

Pawlowski, R.S. 1998. Gravity gradiometry in resource exploration. *The Leading Edge*, **17**, 51–52.

Pawlowski, R.S., and Hansen, R.O. 1990. Gravity anomaly separation by Wiener filtering. *Geophysics*, **55**, 539–548.

Payne, M.A. 1981. SI and Gaussian CGS units, conversions and equations for use in geomagnetism. *Phys. Earth Planet. Int.*, **26**, 10–16.

Pedersen, L.B., and Rasmussen, R. 1990. The gradient tensor of potential field anomalies: Some implications on data collection and data processing of maps. *Geophysics*, **55**, 1558–1566.

Pedreira, D., Pulgar, J.A., Gallant, J., and Tomé, M. 2007. Three-dimensional gravity and magnetic modeling of crustal indentation and wedging in the western Pyrenees-Cantabrian Mountains. *J. Geophys. Res.*, **112**, 19 pp.

Peters, L.J. 1949. The direct approach to magnetic interpretation and its practical application. *Geophysics*, **14**, 290–320.

Phillips, J.D. 2000. Locating magnetic contacts: A comparison of the horizontal gradient, analytic signal, and local wavenumber methods. Pages 402–405 of: *70th Annual International Meeting Expanded Abstracts*. Society of Exploration Geophysicists.

Phillips, J.D., Saltus, R.W., and Reynolds, R.L. 1998. Sources of magnetic anomalies over a sedimentary basin: Preliminary results for the coastal plain of the Arctic National Wildlife Refuge, Alaska. In *Geologic Applications of Gravity and Magnetics: Case*

Histories. Society of Exploration Geophysicists Geophysical Reference Series, No. 8, pp. 130–134.

Phillips, J.D., Hansen, R.O., and Blakely, R. 2007. The use of curvature in potential-field interpretation. *Explor. Geophys.*, **38**, 111–119.

Pick, M., Picha, J., and Vyskocil, V. 1973. *Theory of the Earth's Gravity Field*. Elsevier.

PicoEnvirotec, Inc. 2009. *PeiComp Program: Program for Calculation of the Coefficients used for Magnetic Compensation, User Manual, Revision 5.1.10*. PicoEnvirotec Inc.

Pignatelli, A., Nicolosi, I., Carluccio, R., Chiappini, M., and von Frese, R. 2011. Graphical interactive generation of gravity and magnetic fields. *Comput. Geosci.*, **37**, 567–572.

Pilkington, M. 1998. Magnetization mapping in rugged terrain. *Explor. Geophys.*, **29**, 560–564.

Pilkington, M., 2006. Joint inversion of gravity and magnetic data for two-layer models. *Geophysics*, **71**, L35–L42.

Pilkington, M., and Keating, P. 2004. Contact mapping from gridded magnetic data. A comparison of techniques. *Explor. Geophys.*, **35**, 306–311.

Pilkington, M., and Todoeschuck, J.P. 1995. Scaling nature of crustal susceptibilities. *Geophys. Res. Lett.*, **22**, 779–782.

Pilkington, M., and Todoeschuck, J.P. 2004. Power-law scaling behavior of crustal density and gravity. *Geophys. Res. Lett.*, **31**, 4.

Pilkington, M., and Urquhart, W.E.S. 1990. Reduction of potential field data to a horizontal plane. *Geophysics*, **55**, 449–455.

Pilkington, M., Miles, W. F., Ross, G. M., and Roest, W. R. 2000. Potential field signatures of buried Precambrian Basin. *Can. J. Earth Sci.*, **37**, 1453–1471.

Plenge, G. 1978. Evaluation of demagnetizing field effect in geomagnetic interpretation. Unpublished M.Sc. thesis, Purdue University.

Plouff, D. 1976. Gravity and magnetic fields of polygonal prisms and application to magnetic terrain corrections. *Geophysics*, **41**, 727–741.

Poisson, S.D. 1826. *Memoire sur la theorie du magnetisme*. Memories de la l'acadamie royale des sciences de l'Institute de France, Paris.

Potts, L.V., and von Frese, R.R.B. 2003. Comprehensive mass modeling of the Moon from spectrally correlated free-air and terrain gravity data. *J. Geophys. Res.*, **108**, 5024–5036.

Press, F. 1970. Regionalized earth models. *J. Geophys. Res.*, **75**, 6575–6578.

Press, W.H., Flannery, B.P., Teukolsky, S.A., and Vetterling, W.T. 2007. *Numerical Recipes: The Art of Computing*. Cambridge University Press.

Prieto, C. 1998. Gravity/magnetic signatures of various geologic models. An exercise in pattern recognition. In *Geologic Applications of Gravity and Magnetics: Case Histories*. Society of Exploration Geophysicists Reference Series No. 8, pp. 20–27.

Pullaiah, G., Irving, E., Buchan, K.L., and Dunlop, D.J. 1975. Magnetization changes caused by burial and uplift. *Earth Planet. Sci. Lett.*, **28**, 133–143.

Purucher, R., and Wonik, T. 1997. Comment on the paper of Taylor and Ravat "An interpretation of the Magsat anomalies of Central Europe". *J. Appl. Geophys.*, **36**, 213–216.

Purucker, M.E. 1990. The computation of vector magnetic anomalies: a comparison of techniques and errors. *Phys. Earth Planet. Int.*, **62**, 231–245.

Ram Babu, H.V., Venkata Raju, D.Ch., and Atchuta Rao, D. 1987. The straight slope rule in gravity interpretation. *J. Assoc. Explor. Geophys.*, **8**, 247–251.

Ramo, A.O., and Bradley, J.W. 1982. Bright spots, milligals, and gammas. *Geophysics*, **47**, 1693–1705.

Ramsey, A.S. 1961. *Theory of Newtonian Attraction*. Cambridge: Cambridge University Press.

Rao, D.B. 1990. Analysis of gravity anomalies of sedimentary basins by an asymmetrical trapezoidal model with quadratic density function. *Geophysics*, **55**, 226–231.

Rappaport, N.J., and Plant, J.J. 1994. 360 degree and order model of Venus topography. *Icarus*, **112**, 27–33.

Rasmussen, R., and Pedersen, L.B. 1979. End corrections in potential field modeling. *Geophys. Prosp.*, **27**, 749–760.

Rauen, A., Soffel, H.C., and Winter, H. 2000. Statistical analysis and origin of the magnetic susceptibility of drill cuttings from the 9.1-km-deep drill hole. *Geophys. J. Int.*, **142**, 83–94.

Ravat, D. 1994. Use of fractal dimension to determine the applicability of Euler's homogeneity equation for finding source locations of gravity and magnetic anomalies. In *Proceedings of the Symposium of the Application of Geophysics to Engineering and Environmental Problems (Boston, MA)*. Environmental and Engineering Geophysical Society, pp. 41–53.

Ravat, D. 1996. Magnetic properties of unrusted steel drums from laboratory and field-magnetic measurements. *Geophysics*, **61**, 1325–1335.

Ravat, D., Lu, Z., and Braile, L.W. 1999. Velocity–density relationships and modeling the lithospheric density variations of the Kenya Rift. *Tectonophysics*, **302**, 225–240.

Ravat, D., Hildenbrand, T.G., and Roest, W. 2003. New way of processing near-surface magnetic data: The utility of the comprehensive model of the magnetic field. *The Leading Edge*, **22**, 584–585.

Ravat, D., Pignatelli, A., Nicolosi, I., and Chiappini, M. 2007. A study of spectral methods of estimating the depth to the bottom of magnetic sources from near surface magnetic anomaly data. *Geophys. J. Int.*, **169**, 421–432.

Ravat, D., Finn, C., Hill, P. *et al.* 2009. *A Preliminary, Full Spectrum, Magnetic Anomaly Grid of the United States with Improved*

Long Wavelengths for Studying Continental Dynamics: A Website for Distribution of Data. US Geological Survey Open-File 2009-1258.

Ravat, D.N., Langel, R.A., Purucker, M., Arkani-Hamid, J., and Alsdorf, D.E. 1995. Global vector and scalar Magsat magnetic anomaly maps. *J. Geophys. Res.*, **100**, 20111–20136.

Reed, J.E. 1980. Enhancement/isolation wavenumber filtering of potential field data. Unpublished M.Sc. thesis, Purdue University.

Reeves, C. 2005. *Aeromagnetic Survey: Principles, Practice, and Interpretation.* Geosoft, Inc.

Reeves, C., and Bullock, S. 2005. *Airborne Exploration: The Foundation of Earth Resources and Environmental Mapping.* Fugro Airborne Surveys (Pty) Ltd. South Africa.

Reeves, C.V. 1993. Limitations imposed by geomagnetic variations on high quality aeromagnetic surveys. *Explor. Geophys.*, **24**, 115–116.

Reford, M.S. 1964. Magnetic anomalies over thin sheets. *Geophysics*, **29**, 532–536.

Reford, M.S. 1979. Problems of magnetic fluctuations in geophysical exploration. In *Solar System Plasma Physics, 3.* North Holland Publishing Co., pp. 356–363.

Reford, M.S. 1984. On "Does draping aeromagnetic data reduce terrain-induced effects?" by V.J.S. Grauch and David L. Campbell (*Geophysics*, **49**, 75–80, January, 1984). *Geophysics*, **49**, 2193–2194.

Reford, M.S. 1986. Magnetic anomalies from the edge, dyke, and thin sheet. In *Interpretation of Gravity and Magnetic Anomalies for Non-specialists.* Notes for Canadian Geophysical Union Short Course, pp. 254–336.

Reford, M.S. and Sumner, J.S. 1964. Aeromagnetics. *Geomagnetics*, **29**, 482–516.

Regan, R.D. 1973. Unpublished Ph.D. dissertation, Michigan State University.

Regan, R.D., and Cain, J.C. 1975. Upward continuation of scattered potential field data. *Geophysics*, **40**, 621–629.

Regan, R.D., and Hinze, W.J. 1976. The effect of finite data length in the spectral analyzing ideal gravity anomalies. *Geophysics*, **41**, 44–55.

Regan, R.D., and Hinze, W.J. 1977. Fourier transforms of finite length theoretical gravity anomalies. *Geophysics*, **42**, 1450–1457.

Regan, R.D., and Hinze, W.J. 1978. Theoretical transforms of the gravity anomalies of two idealized bodies. *Geophysics*, **43**, 631–633.

Regan, R.D., and Rodriguez, P. 1981. An overview of the external magnetic field with regard to magnetic surveys. *Geophys. Surv.*, **4**, 255–296.

Regan, R.D., Cain, J.C., and Davis, W.M. 1975. A global magnetic anomaly map. *J. Geophys. Res.*, **80**, 794–802.

Reid, A. 2001. GXF is here to stay or data exchange standards in gravity and magnetic work. *The Leading Edge*, **20**, 868–869.

Reid, A., FitzGerald, D., and McInerny, P. 2003. Euler deconvolution of gravity data. In *73rd Annual International Meeting Expanded Abstracts.* Society of Exploration Geophysicists, pp. 580–583.

Reid, A.B. 1980. Aeromagnetic survey design. *Geophysics*, **45**, 973–976.

Reid, A.B., Alsop, J.M., Grander, H., Millet, A.J., and Somerton, I.W. 1990. Magnetic interpretation in three dimensions using Euler deconvolution. *Geophysics*, **55**, 80–91.

Reigber, C. 1989. Gravity field recovery from satellite tracking data. In *Theory of Satellite Geodesy and Gravity Field Determination.* Springer Verlag, pp. 197–234.

Resende, M.A., Allan, J.E.M., and Coey, J.M.D. 1986. The magnetic soils of Brazil. *Earth Planet. Sci. Lett.*, **78**, 322–326.

Reynolds, J.M. 2011. *An Introduction to Applied and Environmental Geophysics.* John Wiley & Sons.

Reynolds, R.L. 1977. *Magnetic titanohematite minerals in uranium-bearing sandstone.* US Geological Survey Open-File Rept. 77-335.

Reynolds, R.L., Rosenbaum, J.G., Hudson, M.R., and Fishman, N.S. 1990. Rock magnetism, the distribution of magnetic minerals in the Earth's crust, and aeromagnetic anomalies. In *Geologic Application of Modern Aeromagnetic Surveys.* US Geological Survey Bulletin 1924, pp. 24–45.

Ridgway, J.R., and Zumberge, M.A. 2002. Deep-towed gravity surveys in the southern California Continental Borderland. *Geophysics*, **67**, 777–787.

Rigoti, A., Padilha, A.L., Chamalaun, F.H., and Trivedi, N.B. 2000. Effects of the equatorial electrojet on aeromagnetic data acquisition. *Geophysics*, **65**, 553–558.

Roach, M.J., Leaman, D.E., and Richardson, R.G. 1993. A comparison of regional-residual separation techniques for gravity surveys. *Explor. Geophys.*, **24**, 779–784.

Roberts, A. 2001. Curvature attributes and their application to 3D interpreted horizons. *First Break*, **19**, 85–100.

Roberts, R.L., Hinze, W.J., and Leap, D.I. 1990. Data enhancement procedures on magnetic data from landfill investigations. In *Geotechnical and Environmental Geophysics, II Environmental and Groundwater.* Society of Exploration Geophysicists Investigations in Geophysics, Vol. 5, pp. 261–266.

Robinson, E.A., and Treitel, S. 1980. *Geophysical Signal Analysis.* Prentice-Hall.

Robinson, P.R., Harrison, J., McEnroe, S.A., and Hargraves, R. 2004. Nature and origin of lamellar magnetization in the hematite-ilmenite series. *Am. Mineral.*, **89**, 723–747.

Roest, W.R., and Pilkington, M. 1993. Identifying remanent magnetization effects in magnetic data. *Geophysics*, **58**, 653–659.

Roest, W.R., Verhoef, J., and Pilkington, M. 1992. Magnetic interpretation using the 3-D analytic signal. *Geophysics*, **57**, 116–125.

Romberg, F.E. 1958. Key variables of gravity. *Geophysics*, **23**, 684–700.

Ross, H.E., Blakely, R.J., and Zoback, M.D. 2006. Testing the use of aeromagnetic data for the determination of Curie depth in California. *Geophysics*, **71**, 151–159.

Roth, C.B., Nelson, D.W., and Romkins, M.J.M. 1974. *Prediction of Subsoil Erodibility Using Chemical Mineralogical and Physical Parameters*. US Environmental Protection Agency, Rept. EPA-660/2-74-043.

Roy, A. 1962. Ambiguity in geophysical interpretation. *Geophysics*, **23**, 684–700.

Roy, L., Agarwal, B.N.P., and Shaw, R.K. 2000. A new concept in Euler deconvolution of isolated gravity anomalies. *Geophys. Prosp.*, **48**, 559–575.

Rudman, A.J., and Blakely, R.F. 1965. A geophysical study of basement anomaly in Indiana. *Geophysics*, **30**, 740–761.

Rymer, H. 1989. A contribution to precision microgravity data analysis using LaCoste and Romberg gravity meters. *Geophys. J. Int.*, **97**, 311–322.

Saad, A. 2006. Understanding gravity gradients: a tutorial. *The The Leading Edge*, **25**, 942–949.

Saad, A. H. 1969a. Magnetic properties of ultramafic rocks from Red Mountain, California. *Geophysics*, **34**(6), 974–987.

Saad, A. H. 1969b. Paleomagnetism of Franciscan ultramafic rocks from Red Mountain, California, *J. Geophys. Res.*, **74**, 6567–6578.

Saad, A.H. 1977a. Maximum entropy spectral analysis of magnetic and gravity profiles. *Geophysics*, **42**, 1536–1537.

Saad, A.H. 1977b. *Spectral Analysis of Magnetic and Gravity Profiles: Part I. Theoretical Basis of Spectral Analysis*. Research Report 4226RH001, Gulf Science and Tech. Co.

Saad, A.H. 1978. *Spectral Analysis of Magnetic and Gravity Profiles: Part II. Maximum Entropy Spectral Analysis (MESA) and its Applications*. Research Report 4226RJ002, Gulf Science and Tech. Co.

Saad, A.H. 1991. *Interactive Borehole Gravity Modeling (BHGM) with GAMMA*. ICGC paper, Chevron InterCompany Geophys. Conf.

Saad, A.H. 1992. *GAMMA User's Manual – Interactive Gravity And Magnetic Modeling Applications Program*. Report, Chevron Expl. & Prod. Services Co.

Saad, A.H. 1993. Interactive integrated interpretation of gravity, magnetic, and seismic data tools and examples. In *Offshore Technology Conference, Paper Number OTC# 7079*. Offshore Technology Conference, pp. 35–44.

Saad, A.H. 2009a. Spatial-domain filters for short wavelength sedimentary magnetic anomalies. In *79th Annual International Meeting Expanded Abstracts*. Society of Exploration Geophysicists, pp. 962–966.

Saad, A.H. 2009b. Spatial-domain filters for short wavelength sedimentary magnetic anomalies (Slide Presentation). In *79th Annual International Meeting Expanded Abstracts, 32 slides*. Society of Exploration Geophysicists, pp. 20–27.

Saad, A.H., and Bishop, T.N. 1989. *GAMMA – An Interactive Program for Gravity And Magnetic Modeling Applications in the SEISLINE Environment*. ICGC paper #2, Chevron InterCompany Geophys. Conf.

Sabaka, T.J., Olsen, N., and Purucker, M.E. 2004. Extending comprehensive models for the Earth's magnetic field with Ørsted and CHAMP data. *Geophy. J. Int.*, **159**, 521–547.

Sailhac, P., Galdeano, A., Gibert, D., Moreau, F., and Delor, C. 2000. Identification of sources of potential fields with the continuous wavelet transform: Complex wavelets and application to aeromagnetic profiles in French Guiana. *J. Geophys. Res.*, **105**, 19455–19475.

Sailor, R.V., and Driscoll, M.L. 1993. Noise models for satellite altimeter data. *EOS Trans. Am. Geophys. Un.*, **74**, 99.

Salem, A., Ravat, D., Gamey, T.J., and Ushijima, K. 2002. Analytic signal approach and its applicability in environmental magnetic investigations. *Appl. Geophys.*, **49**, 231–244.

Salem, A., Ravat, D., Smith, R., and Ushijima, K. 2005. Interpretation of magnetic data using an enhanced local wavenumber (ELW) method. *Geophysics*, **70**, 141–151.

Salem, A., Williams, S., Fairhead, J., Ravat, D., and Smith, R. 2007. Tilt-depth method: A simple depth estimation method using first-order magnetic derivatives. *The Leading Edge*, **26**, 1502–1505.

Salem, A., Williams, S., Fairhead, D., Smith, R., and Ravat, D. 2008. Interpretation of magnetic data using tilt-angle derivatives. *Geophysics*, **73**, L1–L10.

Salem, A., Lei, K., Green, C., Fairhead, J.D., and Stanley, G. 2010. Removal of cultural noise from high-resolution aeromagnetic data using a two stage equivalent source approach. *Explor. Geophys.*, **41**, 163–169.

Saltus, R.W. and Blakely, R.J. 2011. Unique geologic insights from "non-unique" gravity and magnetic interpretation. *GSA Today*, **21**, 4–9.

Sambuelli, L., and Strobba, C. 2002. The Buffon's needle problem and the design of a geophysical survey. *Geophys. Prosp.*, **50**, 403–409.

Sander, S., Argyle, M., Elieff, S., Ferguson, S., Lavoie, V., and Sander, L. 2004. The AIRGrav airborne gravity system. In *Proceedings of the ASEG-PESA Symposium (Sydney, Australia)*. Australian Society of Exploration Geophysicists.

Sandwell, D.T. 1992. Antarctica marine gravity field from high-density satellite altimetry. *Geophys. J. Int.*, **109**, 437–448.

Sandwell, D.T., and Smith, W.H.F. 1997. Marine gravity anomalies from Geosat and ERS-1 satellite altimetry. *J. Geophys. Res.*, **102**, 10039–10054.

Sasagawa, G.S., Crawford, W., Eiken, O. *et al.* 2003. A new seafloor gravimeter. *Geophysics*, **68**, 544–553.

Sazhina, N., and Grushinsky, N. 1971. *Gravity Prospecting*. Mir Publishers.

Scharroo, R., and Visser, P. 1998. Precise orbit determination and gravity field improvement for the ERS satellites. *J. Geophys. Res.*, **102**, 8113–8127.

Scheibe, D.M., and Howard, H.W. 1964. *Classical Methods for Reduction of Gravity Observations*. Aeronautical Chart and Information Center Reference Publication No. 12.

Schmidt, P., Clark, D., Leslie, K. *et al.* 2004. GETMAG a SQUID magnetic tensor gradiometer for mineral and oil exploration. *Explor. Geophys.*, **35**, 297–305.

Schmidt, P.W., and Clark, D.A. 2006. The magnetic gradient tensor: its properties and uses in source characterization. *The Leading Edge*, **25**, 75–78.

Schmidt, P.W., McEnroe, S.A., Clark, D.A., and Robinson, P. 2007. Magnetic properties and potential field modeling of the Peculiar Knob metamorphosed iron formation, South Australia: An analog for the source of the intense Martian magnetic anomalies? *J. Geophys. Res.*, **112**, B03102.

Schnetzler, C.C., and Taylor, P.T. 1984. Evaluation of an observational method for estimation of remanent magnetization, with application to satellite data. *Geophysics*, **49**, 282–290.

Schön, J.H. 1996. *Physical Properties of Rocks: Fundamentals and Principles of Petrophysics*. Elsevier.

Schueck, J. 1990. Using a magnetometer for investigating underground coal mine fires, burning coal refuse banks, and for locating AMD sources on surface mines for the interpretation of long-wavelength magnetic anomalies. In *http://wvmdtaskforce.com/proceedings/90/90SCH/90SCH.HTM*. 1990 Mining and Reclamation Conference and Exhibition.

Schultz, A.K. 1989. Monitoring fluid movement with the borehole gravity meter. *Geophysics*, **54**, 1267–1273.

Schwarz, K.P., Sideris, M.G., and Forsberg, R. 1990. The use of FFT techniques in physical geodesy. *Geophys. J. Int.*, **100**, 485–514.

Schwiderski, E. 1980. On charting global ocean tides. *Rev. Geophys. Space Phys.*, **18**, 243–268.

Scollar, I., Tabbagh, A., Hesse, A., and Herzog, I. 1990. *Archaeological Prospecting and Remote Sensing*. Cambridge University Press.

Scott, H.S. 1956. The airborne magnetometer. In *Methods and Case Histories in Mining Geophysics*. 6th Commonwealth Mining and Metallurgical Congress, Ottawa, pp. 26–34.

SEG. 1966. *Geophysical Signatures of Western Australian Mineral Deposits: An Overview*. Society of Exploration Geophysicists, Mining Geophysics: Case Histories I.

Seigel, H.O. 1995. *A Guide to High Precision Land Gravimeter Surveys*. Scintrex Limited.

Shannon, C.E. 1949. Communication in the presence of noise. *Proc. Inst. Radio Engineers*, **37**, 10–21.

Sharma, B., Geldhart, L.P., and Gill, D.E. 1970. Interpretation of gravity anomalies of dike-like bodies by Fourier transformation. *Can. J. Earth Sci.*, **7**, 512–516.

Sharma, P. 1986. *Geophysical Methods in Geology*. Elsevier.

Sharma, P.V. 1966. Rapid computation of magnetic anomalies and demagnetization effects caused by bodies of arbitrary shape. *Pure Appl. Geophys.*, **64**, 89–109.

Sharma, P.V. 1968. Demagnetization effect of a rectangular prism. *Geophysics*, **33**, 132–134.

Sharma, P.V. 1997. *Environmental and Engineering Geophysics*. Cambridge University Press.

Shaw, H. 1932. Interpretation of gravitational anomalies. *Trans. Am. Inst. Mining Metall. Eng.*, **97**, 460–506.

Shearer, S. 2005. Three-dimensional inversion of magnetic data in the presence of remanent magnetization. Unpublished MSc. thesis, Colorado School of Mines.

Sheriff, R.E. 1973. *Encyclopedic Dictionary of Exploration Geophysics*. Society of Exploration Geophysicists.

Sheriff, R.E. 2002. *Encyclopedic Dictionary of Exploration Geophysics* 4th Ed. Anchor Press, Garden City.

Shive, P.N. 1986. Suggestions for the use of SI in magnetism. *EOS Trans. AGU*, **67**, 25.

Shive, P.N., Frost, B.R., and Peretti, A. 1988. The magnetic properties of metaperidotite rocks as a function of metamorphic grade: implications for crustal magnetic anomalies. *J. Geophys. Res.*, **93**, 12187–12195.

Shive, P.N., Blakely, R.J., Frost, B.R., and Fountain, D.M. 1992. Magnetic properties of the lower crust. In *Continental Lower Crust*. Vol. 23 of *Developments in Geotectonics*. Elsevier, pp. 145–177.

Shuey, R.T., and Pasquale, A.S. 1973. End corrections in magnetic profile interpretation. *Geophysics*, **38**, 507–512.

Shurbert, D.H., Keller, G.R., and Friess, J.P. 1976. Remanent magnetization from comparison of gravity and magnetic anomalies. *Geophysics*, **41**, 56–61.

Sideris, M.G. 1985. A Fast Fourier Transform method for computing terrain corrections. *Manusc. Geodaet.*, **10**, 66–73.

Siebert, M., and Meyer, J. 1996. Geomagnetic activity indices. In *The Upper Atmosphere: Data Analysis and Interpretation*. Springer Verlag, pp. 8877–8911.

Siegert, A.J.F. 1942. Determination of the Bouguer correction constant. *Geophysics*, **7**, 29–34.

Sigl, R. 1985. *Introduction to Potential Theory*. Abacus Press.

Silva, J.B.C. 1986. Reduction to the pole as an inverse problem and its application to low-latitude anomalies. *Geophysics*, **51**, 369–382.

Silva, J.B.C., and Barbosa, V.C.F. 2003. 3D Euler deconvolution: theoretical basis for automatically selecting good solutions. *Geophysics*, **68**, 1962–1968.

Silva, J.B.C., Medeiros, W.E., and Barbosa, V.C.F. 2001. Potential-field inversion: Choosing the appropriate technique to solve a geologic problem. *Geophysics*, **66**, 511–520.

Simpson, R.W., and Jachens, R.C. 1989. Gravity methods in regional studies. In *Geophysical Framework of the Continental United States*. Geological Society of America Memoir 172, pp. 34–44.

Simpson, R.W., Jachens, R.C., and Blakely, R.J. 1983. *AIRY-ROOT: A Fortran Program for Calculating the Gravitational Attraction of an Airy Isostatic Root Out to 166.7 km*. US Geological Survey Open-File Report 83–883.

Simpson, R.W., Jachens, R.C., Blakely, R.J. and Saltus, R.W. 1986. A new isostatic residual gravity map of the conterminous United States with a discussion on the significance of isostatic residual anomalies. *J. Geophys. Res.*, **91**, 8348–8372.

Skeels, D.C. 1947. Ambiguity in gravity interpretation. *Geophysics*, **12**, 43–56.

Skeels, D.C. 1963. An approximate solution of the problem of maximum depth in gravity interpretation. *Geophysics*, **28**, 724–735.

Skeels, D.C. 1967. What is residual gravity? *Geophysics*, **32**, 872–876.

Skilbrei, J.R. 1993. The straight-slope method for basement depth determination revisited. *Geophysics*, **58**, 593–595.

Skilbrei, J.R., and Kihle, O. 1999. Display of residual profiles versus gridded image data in aeromagnetic study of sedimentary basins: A case history. *Geophysics*, **64**, 1740–1747.

Slichter, L.B. 1929. Certain aspects of magnetic surveying. *Trans. Am. Inst. Mining Metall. Eng.*, **81**, 315–344.

Slichter, L.B. 1942. Magnetic properties of rocks. In *Handbook of Physical Constants*. Geological Society of American Special Paper 36, pp. 293–297.

Smellie, D.W. 1956. Elementary approximations in aeromagnetic interpretation. *Geophysics*, **21**, 1021–1040.

Smith, A.G., Hurley, A.M., and Briden, J.C. 1981. *Phanerozoic Paleocontinental World Maps*. Cambridge University Press.

Smith, D.E., Zuber, M.T., Neumann, G.A., and Lemoine, F.G. 1997. Topography of the Moon from the Clementine lidar. *J. Geophys. Res.*, **102**, 1591–1611.

Smith, D.E., Zuber, M.T., Solomon, S.C. *et al.* 1999. The global topography of Mars and implications for surface evolution. *Science*, **824**, 1495–1503.

Smith, R. S., Thurston, J., Dai, T-F., and MacLeod, I. N. 1998. ISPI^{TM} the improved source parameter imaging method. *Geophys. Prosp.*, **46**, 141–151.

Smith, R.A. 1959. Some depth formulae for local magnetic and gravity anomalies. *Geophys. Prosp.*, **7**, 55–63.

Smith, R.A. 1960. Some formulae for interpreting local gravity anomalies. *Geophys. Prosp.*, **8**, 607–613.

Smith, R.A., and Bott, M.H.P. 1958. The estimation of limiting depth of gravitating bodies. *Geophys. Prosp.*, **6**, 1–10.

Smith, W.H.F., and Wessel, P. 1990. Gridding with a continuous curvature surface in tension. *Geophysics*, **55**, 293–305.

Smithson, S.B. 1971. Densities of metamorphic rocks. *Geophysics*, **36**, 690–694.

Snyder, D.B., and Carr, W.J. 1984. Interpretation of gravity data in a complex volcano-tectonic setting, southwestern Nevada. *J. Geophys. Res.*, **89**, 10193–10206.

Snyder, J.P. 1987. *Map Projections A Working Manual*. US Geological Survey Prof. Paper 1395.

Sobczak, L.W., Weber, J.R., and Roots, E.F. 1970. Rock densities in the Queen Elizabeth Islands, Northwest Territories. *Geol. Assoc. Can.*, **21**, 5–14.

Sobolev, S.V., and Babeyko, A.Y. 1994. Modeling of mineralogical composition, density and elastic wave velocities in anhydrous magmatic rocks. *Surv. Geophys.*, **15**, 515–544.

Somigliana, C. 1930. Geofisica Sul campo gravitazionale esterno del geoide ellissoidico. *Atti della Accademia Nazionale dei Lincei. Rendiconti. Classe di scienze fisiche, matematiche e naturali*, **6**, 237–243.

Sparlin, M.A., and Lewis, R.D. 1994. Interpretation of the magnetic anomaly over the Omaha Oil Field, Gallatin County, Illinois. *Geophysics*, **59**, 1092–1099.

Speake, C.C., Hammond, G.D., and Trenkel, C. 2001. The torsion balance as a tool for geophysical prospecting. *Geophysics*, **66**, 527–534.

Spector, A., and Grant, F.S. 1970. Statistical models for interpreting aeromagnetic data. *Geophysics*, **35**, 293–302.

Spurgat, M. 1971. Width of two-dimensional anomaly sources from gravity and magnetic second vertical derivatives. Unpublished M.Sc. thesis, Michigan State University.

Stacey, F.D. 1963. The physical theory of rock magnetism. *Adv. Phys.*, **12**, 45–132.

Stacey, F.D., and Banerjee, S.K. 1974. *The Physical Principles of Rock Magnetism*. Elsevier.

Stasinowsky, W. 2010. The advantages of the full tensor over Gz. In *Airborne Gravity 2010 – Abstracts from the ASEG-PESA Airborne Gravity 2010 Workshop*. Geoscience Australia and the Geological Survey of New South Wales, Geoscience Australia Record 2010/23 and GSNSW File GS2010/0457.

Stavrev, P.Y. 1997. Euler deconvolution using differential similarity transforms of gravity or magnetic anomalies. *Geophys. Prosp.*, **45**, 207–246.

Stavrev, P., and Reid, A.B. 2007. Degrees of homogeneity of potential fields and structural indices of Euler deconvolution. *Geophysics*, **72**, L1–L12.

Steenland, N.C. 1963. An evaluation of the Peace River aeromagnetic interpretation. *Geophysics*, **28**, 745–755.

Steenland, N.C. 1968. Structural significance and analysis of Mid-Continent Gravity High: Discussion. *Am. Assoc. Petr. Geol. Bull.*, **52**, 2263–2267.

Steiner, F., and Zilahi-Sebess, L. 1988. *Interpretation of Filtered Gravity Maps*. Akadémiai Kiadó.

Steinhauser, P., Meurers, B., and Reuss, D. 1990. Gravity investigations in mountainous areas. *Explor. Geophys.*, **21**, 161–168.

Stelkens-Kobsch, T.H. 2005. The airborne gravimeter Chekan-A at the Institute of Flight Guidance (IFF). In *Gravity, Geoid, and Space Missions*. International Association of Geodesy Symposia, 129, pp. 113–118.

Stolz, R., Zakosarenko, V., Schulz, M., Chwala, A., Fritzch, L., Meyer, H.-G., and Köstlin, E.O. 2006. Magnetic full-tensor SQUID gradiometer system for geophysical applications. *The Leading Edge*, **25**, 178–180.

Stone, V.C.A., Fairhead, J.D., and Oterdoon, W.H. 2004. Micromagnetic seep detection in the Sudan. *The Leading Edge*, **23**, 734–737.

Strangway, D.W. 1981. Magnetic properties of rocks and minerals. In *Physical Properties of Rocks and Minerals, II-2*. McGraw-Hill/CINDAS Data Series on Material Properties, pp. 331–360.

Stroud, A.H., and Secrest, D. 1966. *Gaussian Quadrature Formulas*. Prentice-Hall.

Studinger, M., Bell, R., and Frearson, N. 2008. Comparison of AIRGrav and GT-1A airborne gravimeters for research applications. *Geophysics*, **73**, 151–161.

Subrahmanyam, C., and Verma, R.K. 1981. Densities and magnetic susceptibilities of Precambrian rocks of different metamorphic grade (Southern Indian Shield). *Geophys. J.*, **49**, 101–107.

Sun, W., Miura, S., Sato, T. *et al.* 2010. Gravity measurements in southeastern Alaska reveal negative gravity rate of change caused by glacial isostatic adjustments. *J. Geophys. Res.*, **115**, 12406–12423.

Swain, C.J. 1976. A Fortran IV program for interpolating irregularly spaced data using the difference equations for minimum curvature. *Comput. Geosci.*, **1**, 231–240.

Swain, C.J. 2000. Reduction-to-the-pole of regional magnetic data with variable field direction, and its stabilization at low inclinations. *Explor. Geophys.*, **31**, 78–83.

Swartz, C.A. 1954. Some geometrical properties of residual maps. *Geophysics*, **19**, 46–70.

Świeczak, M., Kozlovskaya, E., Majdánski, M., and Grad, M. 2009. Interpretation of geoid anomalies in the contact zone between the East European Craton and the Palaeozoic Platform. II: Modelling of density in the lithospheric mantle. *Geophys. J. Int.*, **177**, 334–346.

Syberg, F.J.R. 1972. A Fourier method for the regional-residual problem of potential fields. *Geophys. Prosp.*, **20**, 47–75.

Talwani, M. 1965. Computation with the help of a digital computer of magnetic anomalies caused by bodies of arbitrary shape. *Geophysics*, **30**, 797–817.

Talwani, M. 1998. Errors in the total Bouguer reduction. *Geophysics*, **63**, 1125–1130.

Talwani, M. 2003. The Apollo 17 gravity measurements on the moon. *The Leading Edge*, **27**, 786–789.

Talwani, M., and Ewing, M. 1960. Rapid computation of gravitational attraction of three-dimensional bodies of arbitrary shape. *Geophysics*, **25**, 203–225.

Talwani, M., and Heirtzler, J.R. 1964. Computation of magnetic anomalies caused by two-dimensional structures of arbitrary shape. In *Computers in the Mineral Industries, Part 1*. Stanford University Publication, Geological Sciences, pp. 464–480.

Talwani, M., Worzel, J.L., and Landisman, M. 1959. Rapid gravity computations for two-dimensional bodies with application to the Mendocino submarine fracture zone. *J. Geophys. Res.*, **64**, 49–59.

Tamura, Y. 1987. A harmonic development of the tide-generating potential. *Bull. d'Inf. Marées Terr.*, **99**, 6813–6855.

Tarling, D.H., and Hrouda, F. 1993. *The Magnetic Anisotropy of Rocks*. Chapman and Hall.

Taylor, P.T., and Ravat, D. 1995. An interpretation of the Magsat anomalies of central Europe. *J. Appl. Geophys.*, **34**, 83–91.

Telford, W.M., Geldart, L.P., and Sheriff, R.E. 1990. *Applied Geophysics*. Cambridge University Press, Cambridge.

Tessara, A., Götze, H., Schmidt, S., and Hackney, R. 2006. Three dimensional model of the Nasca plate and the Andean continental margin. *J. Geophys. Res.*, **111**, B09404.

Thompson, D.T. 1982. EULDPH, a new technique for making computer-assisted depth estimates from magnetic data. *Geophysics*, **47**, 31–37.

Thomson, A.W.P., and Lesur, V. 2007. An improved geomagnetic data selection algorithm for global geomagnetic field modeling. *Geophys. J. Int.*, **169**, 951–963.

Thorarinsson, F., and Magnussson, S.G. 1990. Bouguer density determination by fractal analysis. *Geophysics*, **55**, 932–935.

Thurston, J.B., and Brown, R.J. 1992. The filtering characteristics of least-squares polynomial approximation for regional/residual separation. *Can. J. Explor. Geophys.*, **28**, 71–80.

Thurston, J.B., and Smith, R.S. 1997. Automatic conversion of magnetic data to depth, dip, susceptibility contrast using the SPITM method. *Geophysics*, **62**, 807–813.

Thurston, J.B., Guillon, J-C., and Smith, R.S. 1999. Model-independent depth estimation with the SPI™ method. In *69th Annual International Meeting Expanded Abstracts*. Society of Exploration Geophysicists, pp. 403–406.

Thurston, J.B., Smith, R.S., and Guillion, J-C. 2002. A multi-model method for depth estimation from magnetic data. *Geophysics*, **67**, 555–561.

Tite, M.S. 1972. The influence of geology on the magnetic susceptibility of soils on archaeological sites. *Archaeometry*, **13**, 229–236.

Tite, M.S., and Mullins, C. 1971. Enhancement of the magnetic susceptibility of soils on archaeological sites. *Archaeometry*, **14**, 209–219.

Tlas, M., Asfahani, J., and Karmeh, H. 2005. A versatile non-linear inversion to interpret gravity anomaly caused by a simple geometrical structure. *Pure Appl. Geophys.*, **162**, 2557–2571.

Todoeschuck, J.P., Pilkington, M., and Gregotski, M.E. 1994. Using fractal crustal magnetization models in magnetic interpretation. *Geophys. Prosp.*, **42**, 677–692.

Torge, W. 1989. *Gravimetry*. Walter de Gruyter.

Tracey, R. 2006. Accuracy of programs for predicting Earth tides and ocean loading. *Preview*, **122**, 16–23.

Tracey, R., Bacchims, M., and Wynn, P. 2007. A new absolute gravity datum for Australian gravity and new standards for the Australian National Gravity Database. In *Extended Abstracts 2007*. Australian Society of Exploration Geophysicists, 544–568.

Treitel, S., Clement, W.G., and Kaul, R.K. 1971. The spectral determination of depths to buried magnetic basement rocks. *Geophys. J. Roy. Astron. Soc.*, **24**, 415–428.

Tsunakawa, H., Shibuya, H., Takahashi, F. *et al.* 2010. Lunar magnetic field observation and initial global mapping of lunar magnetic anomalies by MAP-LMAG onboard SELENE (Kaguya). *Space Sci. Rev.*, **154**, 219–251.

Turcotte, D.L. 1997. *Fractals and Chaos in Geology and Geophysics*. Cambridge University Press.

Tziavos, I.N., Sideris, M.G., Forsberg, R., and Schwarz, K.P. 1988. The effect of terrain on airborne gravity and gradiometry. *J. Geophys. Res.*, **93**, 9173–9186.

Ugalde, H., and Morris, B. 2008. An assessment of topographic effects on airborne and ground magnetic data. *The Leading Edge*, **27**, 76–79.

Ukawa, M., Nozaki, K., Ueda, H., and Fujita, E. 2010. Calibration shifts in Scintrex CG-3M gravimeters with an application to detection of microgravity changes in Iwo-tou caldera, Japan. *Geophys. Prosp.*, **58**, 1123–1132.

Ulrych, T.J., and Bishop, T.N. 1975. Maximum entropy spectral analysis and autoregressive decomposition. *Rev. Geophys. Space Phys.*, **13**, 183–200.

Uotila, U.A. 1980. Note to users of International Gravity Standardization Net 1971. *J. Geodesy*, **54**, 407–408.

Urquhart, T. 1988. Decorrugation of enhanced magnetic field maps. In *58th Annual International Meeting Expanded Abstracts*. Society of Exploration Geophysicists, pp. 371–372.

Uyeda, S., and Richards, M.L. 1966. Magnetization of four Pacific seamounts near the Japanese Islands. *Bull. Earthquake Res. Inst.*, **44**, 179–213.

Vacquier, V. 1962. A machine method for computing the magnitude and direction of a uniformly magnetized body from its shape and magnetic survey. In *Proc. Benedum Earth Magnetism Symp.* University of Pittsburgh Press, pp. 123–137.

Vacquier, V., Steenland, N.C., Henderson, R.G., and Zietz, I. 1951. *Interpretation of Aeromagnetic Maps*. Geological Society of America Memoir 47.

Vajk, R. 1956. Bouguer corrections with varying surface density. *Geophysics*, **21**, 1004–1020.

Valliant, H.D. 1992. The LaCoste & Romberg air/sea gravity meter: An overview. In *CRC Handbook of Geophysical Exploration*. CRC Press, pp. 142–176.

Van Dam, T.M., and Olivier, F. 1998. Two years of continuous measurements of tidal and nontidal variations of gravity in Boulder, Colorado. *Geophys. Res. Lett.*, **25**, 393–396.

Veitch, R.J. 1980. Averaged demagnetizing factors for collections of not-interacting ellipsoidal particles. *Phys. Earth Planet. Int.*, **25**, 9–12.

Verdun, J., and Klingelé, E.E. 2005. Airborne gravity using a strapped-down LaCoste and Romberg air/sea gravity meter system; a feasibility study. *Geophys. Prosp.*, **53**, 91–101.

Verduzco, B., Fairhead, J.D., Green, C.M., and MacKenzie, C. 2004. New insights into magnetic derivatives for structural mapping. *The Leading Edge*, **23**, 116–119.

Verosub, K.L., and Roberts, A.P. 1995. Environmental magnetism: Past, present, and future. *J. Geophys. Res.*, **100**, 2175–2192.

Vogel, A. 1963. The application of electronic computers to the calculation of effective magnetization. *Geophys. Prosp.*, **11**, 51–58.

von Frese, R.R.B., and Kim, H.R. 2003. Satellite magnetic anomalies for lithospheric exploration. In *Proceedings of the 4th Ørsted International Science Team Conference*. Narayana Press, pp. 115–118.

von Frese, R.R.B., and Noble, V.E. 1984. Magnetometry for archaeological exploration of historical sites. *Historical Archaeol.*, **18**, 38–53.

von Frese, R.R.B., Hinze, W.J., and Braile, L.W. 1981a. Spherical Earth gravity and magnetic anomaly analysis by equivalent point source inversion. *Earth Planet. Sci. Lett.*, **53**, 69–83.

von Frese, R.R.B., Hinze, W.J., Braile, L.W., and Luca, A.J. 1981b. Spherical–Earth gravity and magnetic anomaly modeling by Gauss–Legendre quadrature integration. *Geophys. J.*, **49**, 234–242.

von Frese, R.R.B., Hinze, W.J., and Braile, L.W. 1982. Regional North American gravity and magnetic anomaly correlations. *Geophys. J. Roy. Astr. Soc.*, **69**, 745–761.

von Frese, R.R.B., Hinze, W.J., Olivier, R.J., and Bentley, C.R. 1986. Regional magnetic anomaly constraints on continental breakup. *Geology*, **14**, 68–71.

von Frese, R.R.B., Ravat, D.N., Hinze, W.J., and McGue, C.A. 1988. Improved inversion of geopotential field anomalies for lithospheric investigations. *Geophysics*, **53**, 375–385.

von Frese, R.R.B., Jones, M.B., Kim, J.W., and Kim, J.H. 1997a. Analysis of anomaly correlations. *Geophysics*, **62**, 342–351.

von Frese, R.R.B., Tan, L., Potts, L., Merry, C.J., and Bossler, J.D. 1997b. Lunar crustal analysis of Mare Orientale from topographic and gravity correlations. *J. Geophys. Res.*, **102**, 25657–25675.

von Frese, R.R.B., Jones, M.B., Kim, J.W., and Li, W.S. 1997c. Spectral correlation of magnetic and gravity anomalies of Ohio. *Geophysics*, **62**, 365–380.

von Frese, R.R.B., Tan, L., Kim, J.W., and Bentley, C.R. 1999a. Antarctic crustal modeling from the spectral correlation of free-air gravity anomalies with the terrain. *J. Geophys. Res.*, **104**, 25275–25296.

von Frese, R.R.B., Kim, H.R., Tan, L. *et al.*, 1999b. Satellite magnetic anomalies of the Antarctic crust. *Ann. Geofis.*, **42**, 309–326.

von Frese, R.R.B., Roman, D.R., Kim, J.H., Kim, J.W., and Anderson, A.J. 1999c. Satellite mapping of the Antarctic gravity field. *Ann. Geofis.*, **42**, 293–307.

von Frese, R.R.B., Kim, H.R., Taylor, P.T., and Asgharzadeh, M.F. 2005. Reliability of CHAMP anomaly continuations. In *Earth Observation with CHAMP Results from Three Years in Orbit*. Springer, pp. 287–292.

von Frese, R.R.B., Potts, L.V., Wells, S.B. *et al.* 2009. GRACE gravity evidence for an impact basin in Wilkes Land, Antarctica. *Geochem. Geophys. Geosyst.*, **10**, 14 pp.

Wanliss, J.A., and Antoine, L.A.G. 1995. Geomagnetic micropulsations: Implications for high resolution aeromagnetic surveys. *Explor. Geophys.*, **26**, 535–538.

Ward, S.H. 1990. *Environmental and Groundwater*. Vol. 2 of *Geotechnical and Environmental Geophysics*. Society of Exploration Geophysicists.

Wasilewski, P., and Hood, P. 1991. Magnetic anomalies – land and sea. *Tectonophysics*, **192**, 1–230.

Wasilewski, P.J. 1987. Magnetic properties of mantle xenoliths and the magnetic character of the crust–mantle boundary. In *Mantle Xenoliths*. John Wiley & Sons, pp. 577–588.

Wasilewski, P.J., and Mayhew, M.A. 1982. Crustal xenoliths, magnetic properties and long wavelength anomaly source requirements. *Geophys. Res. Lett.*, **9**, 329–332.

Wasilewski, P.J., Thomas, H.H., and Mayhew, M.A. 1979. The Moho as a magnetic boundary. *Geophys. Res. Lett.*, **6**, 541–544.

Watts, A.B. 2001. *Isostasy and Flexure of the Lithosphere*. Cambridge University Press.

Webring, M.W., Kucks, R.P., and Abraham, J.P. 2004. *Gravity Study of the Guernsey Landfill Site, Guernsey, Wyoming*. US Geological Survey Open-file Report 2004-1383.

Weinstock, H., and Overton, W.C. 1981. Squid applications to geophysics. In *Workshop Proceedings*. Society of Exploration Geophysicists.

Wenzel, H. 1985. *Hochauflösende Kugelfunktionsmodelle für das Gravitationspotential der Erde*. Wissentschaftliche Arbeit University of Hannover Nr. 137.

Werner, S. 1945. *Determination of the Magnetic Susceptibility of Ores and Rocks from Swedish Iron Ore Deposits*. Sveriges Geologiska Undersökning, Årsbok 39, No. 472.

Werner, S. 1953. *Interpretation of Magnetic Anomalies at Sheet-like Bodies*. Sveriges Geologiska Undersökning, ser. C, Årsbok 43, No. 6.

Wiener, N. 1930. Generalized harmonic analysis. *Acta Math.*, **55**, 117–258.

Wiener, N. 1949. *Extrapolation, Interpolation, and Smoothing of Stationary Time Series*. John Wiley and Sons.

Wiggins, R.A. 1972. The general linear inverse problem: Implication of surface waves and free oscillations for Earth structure. *Rev. Geophys. Space Phys.*, **10**, 251–285.

Witten, A.J. 2008. *Handbook of Geophysics and Archaeology*. Equinox.

Woelk, T.S., and Hinze, W.J. 1991. Model of midcontinent rift system in eastern Kansas. *Geology*, **19**, 277–280.

Won, I.J., and Huang, H. 2004. Magnetometers and electro-magnetometers. *The Leading Edge*, **23**, 448–451.

Woollard, G.P. 1943. A transcontinental gravity and magnetic profile of North America and its relation to geologic structure. *Geol. Soc. Am. Bull.*, **54**, 747–790.

Woollard, G.P. 1962. *The Relationship of Gravity Anomalies to Surface Elevation, Crustal Structure and Geology*. University Wisconsin, Geophysical and Polar Research Center, Vol. 62-9.

Woollard, G.P. 1966. Regional isostatic relations in the United States. In *The Earth Beneath the Continents*. American Geophysical Union Monograph 10, pp. 557–594.

Woolridge, A. 2004. Review of modern magnetic gradiometer surveys. In *74th Annual International Meeting Expanded Abstracts*. Society of Exploration Geophysicists, pp. 802–805.

Wu, W., Gu, G., and Liang, M. 2009. Magnetic 3D fast forward modeling with varying terrain. In *79th Annual International Meeting Expanded Abstracts*. Society of Exploration Geophysicists, pp. 497–501.

Xinzhu, L., and Hinze, W.J. 1983. *An Overview of the Utility of Isolation and Enhancement Techniques in Gravity and Magnetic Analysis*. Purdue Unviersity Special Report.

Yarger, H.L., Robertson, R.R., and Wentland, R.L. 1978. Diurnal drift removal from aeromagnetic data using least squares. *Geophysics*, **46**, 1148–1156.

Yunsheng, S., Strangway, D.W., and Urquhart, W.E.S. 1985. Geologic interpretation of a high-resolution aeromagnetic survey in the Amos-Barraute area of Quebec. In *The Utility of Regional Gravity and Magnetic Anomaly Maps*. Society of Exploration Geophysicists, pp. 413–425.

Zeyen, H., and Poous, J. 1993. 3-D joint inversion of magnetic and gravimetric data with a priori information, *Geophys. J. Int.*, **112**, 244–256.

Zhang, J., Zhong, B., Zhou, X., and Dai, Y. 2001. Gravity anomalies of 2-D bodies with variable density contrast. *Geophysics*, **66**, 809–813.

Zhdanov, M.S. 2004. *Geophysical Inverse Theory and Regularization Problems*. Elsevier.

Zhdanov, M.S., Ellis, R., and Mukherjee, S. 2004. Three-dimensional regularized focusing inversion of gravity gradient tensor component data. *Geophysics*, **69**, 925–937.

Zhou, X. 2010. Analytic solution of the gravity anomaly of irregular 2D masses with density contrast varying as a 2D polynomial function. *Geophysics*, **75**, 111–119.

Zhou, X., Zhong, B., and Li, X. 1990. Gravimetric terrain corrections by triangular element method. *Geophysics*, **55**, 232–238.

Zietz, I., and Andreasen, G.E. 1967. Remanent magnetization and aeromagnetic interpretation. *Mining Geophys. Soc. Explor. Geophys.*, **2**, 569–590.

Zietz, I., and Henderson, R.G. 1956. A preliminary report on model studies of magnetic anomalies of three-dimensional bodies. *Geophysics*, **21**, 791–814.

Zimmerman, J.E., and Campbell, W.H. 1975. Tests of cryogenic squid for geomagnetic field measurements. *Geophysics*, **40**, 269–284.

Zumberge, M.A., Ridgway, J.R., and Hildebrand, J.A. 1997. A towed marine gravity meter for near-bottom surveys. *Geophysics*, **62**, 1386–1393.

Zumberge, M.A., Alnes, H., Eiken, O., Sasagawa, G., and Stenvold, T. 2008. Precision of seafloor gravity and pressure measurements for reservoir monitoring. *Geophysics*, **73**, WA133–WA141.

Zurflueh, E.G. 1967. Applications of two-dimensional wavelength filtering. *Geophysics*, **32**, 1015–1035.

Index

accuracy
  alkali-vapor magnetometers, 279
  flux-gate magnetometers, 280
  GPS horizontal positioning, 106
  GPS vertical positioning, 107
  gravity measurements, 89
  magnetic gradiometers, 283
  magnetic measurements, 277
  Overhauser magnetometer, 279
  proton-precession magnetometer, 278
  satellite gravity, 115
  SQUID magnetometers, 282
Adams–Williamson equation, 67
admittance function, 455
AirSea System II gravimeter, 100
Airy–Heiskanen theory, 138
aeromagnetic standards, 287
ambiguity, 11
  gravity interpretation, 176
  gravity source, 60–61
  magnetic source, 250
ambiguity in magnetic interpretation, 339
analytic signal, 409
anhysteretic remanent magnetization, 256
anomaly, 3
anomaly analysis, 444–463
  inversion, 461
  isolation and enhancement methods, 444–460
    analytical grid methods, 447
    correlation filters, 453–455
    derivative and integral filters, 455–457
    directional filters, 450–453
    finite element analysis, 447
    geological methods, 445
    graphical methods, 445
    interpolation and continuation filters, 457–460
    residual and regional, 444
    spatial filtering, 445–447
    spectral filters, 448–460
    trend surface analysis, 446
    wavelength filters, 448–450
anomaly source modeling, 460–463
  forward modeling, 461
  inverse modeling, 463
Apollo missions, 119, 230
applications of gravity and magnetic methods, 415–421
  energy resource applications, 417–418
  general view, 415–416
  lithospheric investigations, 420–421
  mineral resource exploration, 418–420
  near-surface studies, 416–417
Archimedes' principle, 76
asthenosphere, 3

band-pass filter, 450
Bangui anomaly, 294
Basin and Range Province, 72
BGM-3 marine gravimeter, 100
Birch's law, 83
Bouguer correction, 133
boxcar filter, 448
British Geological Survey, 304
Browne correction, 143
Bruns' formula, 107, 116
Bullard-B correction, 140

Cartesian to spherical coordinate transformation, 57
centrifugal force, 24, 101, 130, 142
  radius of gyration, 24
CHAMP mission, 226, 230, 291, 295
Chekan-AM gravimeter, 100
chemical remanent magnetization, 256
Cholesky factorization, 427
Clementine mission, 119, 230
cogeoid, 141
coherency, 428, 454
combined magnetic and gravity potentials, 250
compensated density log, 81
Comprehensive Magnetic Field model, 304
Compton scattering, 81
continuation
  differential, 170, 469

downward, 168, 457
interval, 170, 460
upward, 167, 457
convolution, 443–444
core of the Earth, 2
outer, 223
Coriolis acceleration
Eötvös effect, 101, 142
correlation coefficient, 428
correlation filters, 453–455
Cosmos 49 mission, 230
Coulomb's law, 3
Cramer's rule, 427, 429
cross-correlation, 409
cross-coupling effects, 99
crust, 2
Curie point isotherm, 355
Curie temperature, 223, 255, 256, 303, 371
Curie's law, 255

data bases and standards, 423–424
gravity data exchange standards, 424
data correlation
local favorability indices, 473–474
data graphics, 463–475
data presentation, 474–475
shaded relief maps, 475
graphical parameters and practice, 472–474
standardized and normalized data, 472–473
gridding, 464–472
cubic spline interpolation, 466–468
equivalent source interpolation, 468–471
linear interpolation, 465–466
polynomial interpolation, 471
statistical interpolation, 471–472
map projections and transformations, 463–464
International Terrestrial Reference Frame, 464
Mercator projection, 464
universal transverse Mercator map projection, 464
World Geodetic System-84, 464
metadata, 424
data systems processing, 422–475
key concepts, 475–476
decibel, 437
Definitive Geomagnetic Reference Field, 303
deflection of the vertical, 23
demagnetization, 260, 405–406
demagnetization factor, 260, 405
self, 405
shape, 405
density, 3, 21, 64–65
asthenosphere, 71
bulk density, 65, 66, 76
continental crust, 69–70
crystalline rock, 69–72
felsic, 69
mafic, 69
Earth's interior, 66
Adams–Williamson method, 67
grain density, 65
igneous and metamorphic rock minerals, 68
key concepts, 86–87
lithostatic pressure effects, 73–74
magma, 74
metamorphic rock, 72
Moho contrast, 71
natural density, 65, 66
oceanic crust, 70
ore minerals and metals, 69
overview, 64
rock, 68
saturated bulk density, 65
sedimentary rock, 72–73
variation with depth, 73
sedimentary rock minerals, 68
specific gravity, 65
subduction zones, 71
surface density, 43
tabulations, 85–86
temperature effects, 74
terrestrial waters, 69
true density, 65, 66
upper mantle, 71
vadose zone, 65
volcanic rock, 71
density measurements, 75–85
correlative property, 81–85
density log, 81–82
seismic velocity, 82–85
gravity observations, 77–81
subsurface, 78–81
surface, 77–78
triplet method, 78
laboratory, 76
overview, 75–76
depositional remanent magnetization, 256
derivative and integral filters, 455–457
design matrix, 424
diamagnetism, 254
differential reduction to the pole, 294, 331
digital elevation models, 112
directional filters, 450–453
diurnal variations in the geomagnetic field, 309

Eötvös effect, 109, 142–143
Eötvös units, 102
Earth Gravitational Model EGM2008, 24, 118
Earth Magnetic Anomaly Grid 2, 296

Earth's shape
  flattening, 24
  oblate spheroid, 23, 24
edge effects, 465
electromagnetic distance measurement, 105
electron density, 82
environmental magnetism, 272
equivalent point source, 57
equivalent source, 409, 468
error statistics, 428
  analysis of variance, 428
  coherency, 428
  correlation coefficient, 428
  Fisher distribution, 428
errors
  stacking, 14
  outliers, 13
  propagation, 13
  random, 13, 461
  standard deviation, 13
  systematic, 12, 461
  variance, 428
ERS-1 mission, 114
ERS-2 mission, 114
Euler's equation, 191
extraneous magnetic variations, 302–314
  spatial variations, 302–307
    elevation effects, 304–305
    global, 302–304
    surface and near-surface effects, 306–307
    terrain effects, 305–306
  temporal variations, 308–314
    A_p index, 312
    a_p index, 312
    characteristics, 310
    decorrugation, 314
    disturbed variations, 310
    equatorial electrojets, 310
    internal, 311
    ionosphere, 309
    K_p index, 312
    K-index, 311
    leveling of aeromagnetic data, 313
    magnetic indices, 311
    magnetosphere, 309, 311
    micropulsations, 310, 313
    origin, 310
    quiet variations, 310
    removal, 312–314
    secular variation, 313
    tidal effects, 310
    World Data Centers, 312

ferrimagnetism, 254, 255
ferrite, 255
ferromagnetism, 254, 255
Fisher distribution, 428
forward model, 10, 424
  spectral, 425
forward modeling strategies, 208–211
foundations of geophysical methods, 5, 7
Fourier transforms of idealized sources, 207
free-air effect, 131
FTG gravity gradiometer, 103

GAMMA algorithms, 44–243
gamma–gamma log, 81
Gauss' divergence theorem, 59, 61
Gauss' law, 59, 61, 198, 250, 316
Gauss–Jordan elimination, 427
Gauss–Legendre quadrature, 53, 55, 56, 57, 58, 207, 246, 403
Gaussian elimination, 427
Geodetic Reference System-80, 25, 464
geoid, 23
geoidal height, 141
geomagnetic dynamo, 217
geomagnetic field, 219–227
  CMF, 226
  DGRF, 223
  field-aligned current, 295
  IGRF, 223
  magnetopause current, 295
  overview, 216
  polarity, 221
  ring current, 295
  spatial variations
    overview, 219–223
  tail current, 295
  temporal variations
    daily, 225
    diurnal change, 226
    micropulsations, 226
    overview, 223–227
    secular change, 223, 225
    tidal effects, 225, 226
    westward drift, 223
geophysical data, 11–14
  documentation, 12
    metadata, 12
    quality assurance, 12
  errors, 12–14
geophysical practices, 7–11
  data acquisition phase, 9
  data processing phase, 9–10
  interpretation phase, 10–11
  planning phase, 8–9
Geosat mission, 114, 117
GNSS positioning, 106
GOCE mission, 31, 115, 117

GPS positioning, 102, 105, 107, 143
differential, 106
GPS postioning, 464
GRACE mission, 31, 114, 117
gravimeter drift, 110
microseismic effects, 112
gravitational constant, 21, 25
gravitational field
acceleration, 22
force, 21
gradient of potential, 39
logarithmic potential, 43
mass, 21
potential, 21, 38
tensor, 39, 40, 41, 43
vector, 40, 41
gravitational force, 21–22
gravity
measurement
overview, 25–27
potential theory, 38–63
key concepts, 62–63
spatial variations, 23–25
temporal variations, 25
gravity anomalies, 143–155
Bouguer gravity anomaly equations, 147–151
energy, 181
Faye, 144
filtered, 155
fundamental elements, 143–145
geological, 153–155
decompensative, 154
isostatic, 153
mantle, 154
Helmart, 144
isolation and enhancement methods, 155–173
analytical grid methods, 161, 447
continuation filters, 167–171
correlation filters, 163–165
derivative and integral filters, 165–167
equivalent source methods, 171–172
fundamental principles, 157
geological methods, 159, 445
graphical methods, 159–161, 445
matched filters, 162
pseudomagnetic filters, 171
residual and regional separation, 155
spectral filtering methods, 161–171
spectral filters, 448
trend surface analysis, 446
wavelength filters, 161–162, 448
planetary, 145
Bouguer anomaly limitations, 151–152
complete Bouguer, 147–153
error assessment, 152
Faye, 146
free air, 145–147
simple Bouguer, 147–153
gravity corrections
atmospheric, 126
atmospheric mass effect, 131
North America, 132
combined height and mass correction, 133
curvature effect (Bullard-B), 140–141
equation, 141
North America, 140
vertical gradient, 141
datums, 129–130
Geodetic Reference System-80, 129
geographic and vertical, 129
GEOID03, 129
gravity, 129
gravity datum adjustment, 130
International Terrestrial Reference Frame, 129
World Geodetic System-84, 129
Eötvös effect, 142–143
Browne correction, 143
Earth tides, 25, 127–128
geological effect, 138
gravity stripping, 138
height effect, 131
Faye's correction, 131
vertical gradient, 132
indirect effect, 141
marine surveys, 141
North America, 142
instrumental drift, 124–126
isostatic effect, 138–140
Airy–Heiskanen theory, 138, 153
Pratt–Hayford theory, 138
latitude variation, 130–131
gradient, 131
mass effect, 133
Bouguer correction, 133
Bullard-B correction, 133
curvature correction, 133
density considerations, 133
marine surveys, 134
spherical cap, 133
terrain effect, 134–138
1 in 20 rule, 112
automated methods, 136
critical concerns, 136
Hammer's method, 135
spectral analysis methods, 136
typical values, 134
gravity data acquisition, 88–120
key concepts, 120–121
overview, 88

gravity data exchange standards, 424
  ASEG-GDF2 (Draft 4), 424
gravity data processing, 122–173
  key concepts, 173–174
gravity data sets
  principal facts, 423
gravity field of Earth, 21–27
gravity interpretation, 175–211
  ambiguity, 176
  density contrast and mass, 198
  formation boundaries, 196
  graphical depth estimation techniques, 189–191
    half-width method, 190
    Smith rules, 191
    straight-slope method, 190
  gridded anomaly modeling, 206–208
  ideal body depths, 194
  idealized body interpretation, 205–206
  key concepts, 211–212
  key geological variables, 181–183
    source density contrast, 181
    source depth, 181
    source geometry, 181
    source isolation, 182
  key geophysical variables, 183–189
    anomaly amplitude, 183
    anomaly eccentricity, 187
    anomaly perceptibility, 187
    anomaly shape, 187
    anomaly sharpness, 185
  modeling anomaly sources, 200
    forward modeling, 202
    inverse modeling, 202–205
  modeling strategies, 208–211
    constructing the model, 209–210
    matching the observed and modeled gravity anomalies, 210–211
    selecting the anomaly, 208
  parameters, 181–189
  process, 178–181
  pseudomagnetic anomalies, 199–200
  resolution, 182
  semi-automated depth estimation, 191–193
    Euler deconvolution, 191–192
    statistical spectral techniques, 192
    Werner deconvolution, 192
  simplified techniques, 189
    depth estimation, 189–193
  source depth extent estimation, 193
  two- versus three-dimensional, 176–178
gravity measurements
  absolute gravity, 103–105
    free-fall instruments, 104
    pendulum, 103
  accuracy, 89
  Askania gravimeter, 100
  astaticization, 91
  base stations, 113
  calibration, 93
  calibration error, 96
  Chekan-AM gravimeter, 100
  drift, 110
    vibration effects, 112
  Eötvös torsion balance, 102
  error budgets, 98
  GMN-K gravimeter, 100
  gradiometry, 102–103
    Falcon technology, 103
    full tensor gradiometer, 102
  gravimeter error sources, 95–98
  land surface, 91–98
  moving platform, 98–102
    cross-coupling effects, 99
    electromagnetic accelerometers, 100
    gyrostabilized platform, 31, 99
  overview, 89–91
  position control, 105–107
    horizontal, 105
    vertical, 106
  procedures, 111–113
  resolution, 89
  satellite, 31, 113–120
    active, 115–117
    gradiometer, 115
    gravity mapping progress, 117–120
    overview, 113–114
    passive, 114–115
  superconducting gravimeter, 90
  tare, 96, 97, 98, 124
  underwater, 91, 102
  vibrating spring gravimeter, 98
  zero-length spring, 91–95
  zero-length spring principle, 91
    metal, 93–94
    quartz, 94–95
gravity method
  key concepts, 36–37
gravity modeling
  Bouguer slab, 52
  disk, 53
  end correction, 51
  equivalent source, 57
  extended body, 41–48
  general source, 53
  generic 2D source, 53–54
  generic 3D source, 54–55
  horizontal cylinder, 51
  idealized sources, 48–53
  least-squares 2D source, 56–57
  least-squares 3D source, 55–56
    rectangular prism, 57
  least-squares modeling accuracy, 57

least-squares modeling in spherical coordinates, 57–59
point mass, 39–41
spherical shell, 48
total mass, 59–60
vertical cylinder, 53
gravity stripping, 138
gravity survey design, 107
airborne surveys, 109
land surface surveys, 107–109
marine surveys, 109
gravity survey procedures, 110
gravimeter considerations, 110–111
measurement procedures, 111–113
gravity units, 22–23
cm/s^2, 22
Eötvös unit, 102
gal, 22, 29
gravity unit (g.u.), 22
m/s^2, 22
microgal, 22
milligal, 22
N/kg, 22
gravity variations, 123–143
temporal, 123–128
atmospheric, 126
atmospheric pressure, 124
Earth tides, 127–128
inherent instrumental drift, 123, 124
subsurface mass, 128
tare, 124
temperature, 124
Green's functions, 41, 47
Green's theorem, 53
Grid eXchange Format GXF-Rev 3, 424
gridding, 464–472
cubic spline interpolation, 466–468
equivalent source interpolation, 468–471
linear interpolation, 465
minimum curvature, 468
polynomial interpolation, 471
statistical interpolation, 471–472
GRS67, 131
GRS80, 130, 133
GRS84, 130

Hilbert transform, 382
history of the gravity method, 27–32
history of the magnetic method, 229–231
Hough transform, 367

implementing the gravity method, 32–36
data acquisition phase, 34
data processing phase, 35
interpretation phase, 35
planning phase, 32
reporting phase, 36
implementing the magnetic method, 231–233
data acquisition phase, 231
data processing phase, 232
interpretation phase, 232
planning phase, 231
reporting phase, 232
indirect effect, 142
International Association of Geomagnetism and Aeronomy, 303
International Geomagnetic Reference Field, 303
International Gravity Formula 1980, 130, 131
International Terrestrial Reference Frame, 464
interpolation
linear, 465
cubic spline, 466–468
equivalent source, 468
polynomial, 471
statistical, 471–472
interpolation and continuation filters, 457–460
inverse Fourier transform, 431
inverse modeling, 10
inversion, 10, 429
equivalent point source, 248
linear, 425
non-linear, 425
ionosphere, 225
isostasy, 29, 119, 138

Kaguya satellite, 119
Kenya rift, 71
Koenigsberger ratio Q, 256, 393, 407, 409
kriging, 471

LaCoste & Rombera type gravimeters, 92, 93, 96
calibration shifts, 96
LaCoste & Romberg sea gravimeter, 99
LaCoste & Romberg S-gravimeter, 100
LaCoste & Romberg System 6 gravity, 100
modifications, 100
Laplace's equation, 39, 61, 118, 176, 237, 250, 318, 448, 455, 456, 457
least squares collocation, 471
least-squares modeling accuracy, 57
least-squares modeling in spherical coordinates, 57–59
least-squares solution, 427
Legendre polynomials, 53, 55, 56
lithosphere, 2
local favorability indices, 473–474
logarithmic potential, 43
Lunar Apollo subsatellite missions, 296
Lunar gravity, 119
Lunar Prospector mission, 119, 230

Magellan mission, 119
magnetic airborne surveys
  instrumentation, 290–291
    magnetic compensation, 290
  pre-survey airborne tests, 291
    cloverleaf test, 291
    figure of merit, 291
    lag test, 291
magnetic anomalies, 219
  data repositories, 229
  isolation and enhancement methods, 314–336
    analytical grid methods, 318, 447
    continuation filters, 326–328
    correlation filters, 321–323
    derivative and integral filters, 323–326
    directional filters, 320
    equivalent source methods, 333–335
    fundamental principles, 314–316
    geological methods, 445
    geological, graphical, and visual inspection methods, 316–318
    graphical methods, 445
    pseudogravity, 457
    pseudogravity filters, 332–333
    reduction-to-pole filters, 328–332
    residual and regional, 314
    shaded relief maps, 317
    spectral filtering methods, 318–333
    spectral filters, 448
    stacked profiles, 317
    terrace maps, 323
    textural attributes, 318
    trend surface analysis, 446
    wavelength filters, 318–320, 448
magnetic anomaly, 302, 303
magnetic anomaly interpretation, 338–412
  ambiguity, 339
  demi-pentes depth determination method, 362
  depth determination
    analytical signal (total gradient) method, 382–383
    CompuDepth, 387
    depth estimations corrections, 387
    Euler deconvolution, 364–368
    source parameter imaging (local wavenumber technique), 383–385
    statistical spectral techniques, 368–374
    tilt angle (tilt derivative) method, 385–387
    wavelet transform method, 387
    Werner deconvolution, 374–382
  depth determination methods, 355–387
    graphical techniques, 357–363
    Peters' half-slope method, 360–362
    selecting anomalies, 357
    semi-automatic approaches, 363–387
    Sokolov's method, 362
    straight-slope method, 359–360
  formation boundaries, 393–394
  half-width depth rules, 357
  key concepts, 411–412
  key geological variables, 342–349
    geometry, 342
    source depth, 342
    source isolation, 348
    source location in geomagnetic field, 348
    source magnetic polarization contrast, 342
  key geophysical variables, 349–355
    amplitude, 349
    perceptibility, 354
    shape, 351
    sharpness, 352
  maps, 388–394
    apparent magnetic susceptibility mapping, 389
    electromagnetic apparent magnetic susceptibility, 389
    formation boundaries, 393–394
    pseudogravity anomalies, 393
    reduced-to-equator, 392
    reduced-to-pole magnetic anomalies, 391–393
    terracing, 388
  maximum straight slope depth rule, 359
  modeling anomaly sources, 394–411
    demagnetization effects, 405–406
    forward modeling, 395
    gridded anomaly modeling, 403
    idealized body interpretation, 396
    identifying and characterizing remanence, 407–409
    infinite depth extent prism, 402
    infinite depth extent vertical cylinder, 402
    interpretation with remanence, 409–411
    inverse modeling, 395
    matching observed and modeled magnetic anomalies, 404
    modeling strategies, 404–405
    other idealized sources, 398–403
    remanent magnetization effects, 406–407
    two-dimensional dike, 396–398
    vertical fault, 400
  overview, 338
  Peters' half-slope depth determination method, 360–362
  resolution, 348, 355
  shift of anomaly peak, 342
  Sokolov's depth determination method, 362
  straight slope depth rules, 359
  techniques, 355–394
  the process, 340–342
  two- versus three-dimensional, 340
magnetic charts, 223
magnetic data acquisition, 276–297
  key concepts, 297–299
magnetic data processing, 300–336

introduction, 301–302
key concepts, 336–337
overview, 300
magnetic domains, 255
magnetic effects
of a point dipole, 236–238
of an extended body, 238–243
magnetic field
Gauss A position, 222
Gauss B position, 222
gradient of potential, 39
lines of force, 220
measurement, 227–229
point dipole, 236–237
point dipole potential, 236
potential, 218
tensor components, 237
vector components, 238
magnetic field reversal, 303
magnetic force, 217
Coulomb's law, 217
magnetic moment, 218
magnetization, 216
poles, 216
susceptibility, 217
magnetic instrumentation, 277–283
drillhole measurements, 284
flux-gate, 277, 279–282
gradiometers, 282–283
resonance magnetometers, 277, 278–279
alkali-vapor, 277, 278–279
gyromagnetic ratio of protons, 278
Larmor precession frequency, 278
optical pumping/optical monitoring, 279
Overhauser, 277, 279
proton-precession, 277, 278
Zeeman levels, 279
SQUID, 277, 282, 283
Josephson junction, 282
variometers, 277
vector versus scalar measurements, 283
magnetic interpretation
ambiguity, 250
magnetic jerks, 303
magnetic measurements
dip needle, 230
electron reflectometers, 296
flux-gate magnetometer, 230
Hotchkiss superdip, 230
inherent error, 228
overview, 227–229
resonance magnetometer, 230
magnetic method, 215–233
key concepts, 233–234
potential fields, 235
reduced-to-pole, 243
role, 215
magnetic minerals
iron sulfides, 268
oxides, 257
magnetic modeling
2D generic source, 244–245
2D least-squares, 247–248
3D generic source, 245
3D least-squares, 246–247
rectangular prism, 247
extended source-2D, 238–243
field components, 242
limited length, 242
idealized sources, 243–244
least-squares accuracy, 248
least-squares modeling in spherical coordinates, 248–250
point dipole, 236–238
total magnetic moment, 250
magnetic polarization, 5
magnetic potential theory, 235–251
key concepts, 251
point dipole, 236
magnetic properties, 252–274
anhysteretic remanent magnetization, 256
antiferromagnetism, 255
coercive force, 254
Curie temperature, 256
depositional remanent magnetization, 256
diamagnetism, 254
minerals, 255
origin, 255
ferrimagnetism, 254, 255
ferrite, 255
origin, 255
ferromagnetism, 254, 255
Curie temperature, 255
exchange forces, 255
magnetic domains, 255
origin, 255
fired clay, 271
hysteresis, 254
hysteretic behavior, 254
igneous rocks, 263–264
anorthosite, 267
granite, 264
ilmenohematite, 256
iron objects, 271
Koenigsberger ratio, 220, 256, 259, 264
lithospheric rocks, 265–267
lower crust, 265
Mars, 267
Moon, 267
maghemite, 269
magnetization, 252

magnetic properties (*cont.*)
induced, 252
paleomagnetism, 253
remanent, 252
magnetotactic microorganisms, 269
measurement, 272–273
inductance bridge, 273
magnetometer, 273
spinner magnetometer, 273
metamorphic rocks, 264–265
iron formations, 265
Moho, 266
natural remanent magnetization, 256
oceanic crust, 264
origin, 253
paramagnetism, 254, 255
origin, 255
partial thermoremanent magnetization, 256
relaxation time, 257
remanent magnetization, 256–257
reversed thermoremanent magnetization, 256
rocks and soils, 261–272
saturation magnetization, 254
sedimentary rocks, 267–268
coal, 268
serpentinization, 265
soils, 268–271
glacial, 268
laterites, 269
sanitary landfills, 270
topsoils, 269–270
summary, 271
susceptibility, 254, 259–261
apparent susceptibility, 260
bulk susceptibility, 259
crystalline susceptibility, 260
demagnetization, 260
effective susceptibility, 260
intrinsic susceptibility, 259, 260
magnetocrystalline susceptibility, 260
mass susceptibility, 259
shape anisotropy, 260
specific susceptibility, 259
susceptibility anisotropy ratio, 261
true susceptibility, 260
volume susceptibility, 259
weak-field susceptibility, 259
tabulations, 273–274
thermal remanent magnetization, 256
thermoremanent magnetization, 256
titanohematite series
hematite, 267
viscous remanent magnetization, 256, 260
magnetic relaxation time, 257
magnetic remanence, 256
magnetic space surveys, 291–297
core, lithospheric, and external fields, 292
magnetic mapping progress, 294–297
magnetic storms, 226
magnetic survey
decorrugation, 314
microleveling, 314
magnetic survey design
airborne surveys, 287–291
aircraft, 287
altitude, 289
flight specifications, 287–290
flight track orientation, 287
flight track sampling interval, 288
flight track spacing, 288
objectives, 287
tie-lines, 290
land surface surveys, 284–286
station spacing, 285
temporal variation control, 285–286
marine surveys, 286
magnetic susceptibility, 5, 217
rocks
tabulation, 261
susceptibility anisotropy, 260
tensor components, 260
magnetic terrain effects, 305–306
magnetic units, 218
amperes/meter, 218
conversion of CGS to SI units, 219
conversion of parameters, 220
gamma, 218
gauss, 218
magnetic field strength, 218
magnetic susceptibility, 219
maxwell, 221
nanotesla, 218
oersted, 218
tesla, 218
weber, 218
magnetite, 253
magnetization, 216
magnetization of Earth materials, 252–274
key concepts, 274–275
magnetization of seamounts, 409
magnetosphere, 219
Magsat mission, 226, 230, 294, 324, 392, 472
mantle, 2
map projections, 463
map transformations, 463
Mariner-4 mission, 296
Mariner-9 mission, 119
Mars Global Surveyor, 296
Mars Global Surveyor mission, 119, 230
Mars Orbiter Laser Altimeter mission, 119

Martian gravity, 119
  Goddard Mars Model 2B, 119
Martian magnetic field, 219, 296
mascons, 119
matrix inversion, 426–428
matrix properties, 425
  Cholesky factorization, 427
  Gauss–Jordan elimination, 427
  Gaussian elimination, 427
  identity matrix, 425
  inverse matrix, 425
  transpose, 425
maximum entropy, 440
maximum entropy spectral analysis, 440–443
Mercator projection, 464
metadata, 424
Michigan basin, 73, 355
mineral magnetism, 257–258
  ulvöspinel, 257
  fluids and gases, 258
  grain shape anisotropy, 257
  hematite
    parasitic ferromagnetism, 257
  internal demagnetization, 257
  iron sulfides, 258, 265
    greigite, 258
    monoclinic pyrrhotite, 258, 267
    pyrite, 258
  magnetic domains, 260
  magnetite, 257
    volume susceptibility, 261
  pyrrhotite series, 260
  titanohematite series, 257, 260, 267
    gamma-hematite, 258
    hematite, 257
    ilmenite, 257
    maghemite, 258
  titanomagnetite series, 257
  titanomagnetites, 257
minimum curvature gridding, 468
minimum phase filter, 448

natural remanent magnetization, 256
navigation control, 105–107
Nettleton density profile method, 77
Newton's universal law of gravitation, 3, 21
NGDC720 geomagnetic model, 296
NGDC720 World Magnetic Anomaly Map, 304
normal remanent magnetization, 267
North American Gravity Data Committee, 129, 144
numerical transforms, 433–440
  fast Fourier transform, 438
  Gibbs' error, 438–439, 444
  numerical errors, 438–440
  Shannon's sampling theorem, 439
  wavenumber aliasing, 439–440
  wavenumber leakage, 439
  wavenumber resolution, 440
  window carpentry, 439
Nyquist
  frequency, 8, 435
  sampling theorem, 108
  wavelength, 8, 435, 472
  wavenumber, 8

Okina sub-satellite, 119
Ørsted mission, 226, 230, 291, 295
orthometric height, 141
Ouna sub-satellite, 119
Overhauser effect, 279

paramagnetism, 254, 255
partial thermoremanent magnetization, 256
periodogram, 431
PicoEnvirotec, Inc., 290
Pilbara Craton, 70
Pioneer Venus mission, 119
Pioneer Venus orbiter mission, 297
plate tectonics, 6, 231
POGO missions, 226, 230, 294
Poisson's equation, 61, 176
Poisson's theorem, 61, 167, 171, 172, 199, 243, 250, 318, 322, 328, 335, 350, 391, 393, 409
potential field methods, 3
Pratt–Hayford theory, 138
pseudoanomalies, 61–62
pseudogravity, 409
pycnometer, 76

Q, Koenigsberger ratio, 256, 259

radius of gyration, 24, 130
remanent magnetization, 256
resolution
  gravity anomalies, 182
  gravity measurements, 89
  magnetic anomalies, 348
reversed thermoremanent magnetization, 256
ridge regression, 429

Satellite magnetic mapping progress, 294–297
Scintrex gravimeter, 20
seafloor spreading, 6, 231
secular variation of geomagnetic field, 303, 308
SELENE mission, 119
sensitivity analysis, 24, 429–430
  Cholesky factorization, 430
  damping factor, 429
  error variance, 429
  Marquardt parameter, 429

sensitivity analysis (*cont.*)
  ridge regression, 429
  singular value decomposition, 430
shaded relief maps, 475
SI units, 11
  base, 12
  derived, 12
  supplementary, 12
solar flares, 310
solar storms, 225
solar wind, 219
solid-Earth geophysics, 2
spectral analysis, 430–443
  admittance function, 455
  Akaike information criterion, 443
  analytical transforms, 432–433
  complex conjugate, 431
  complex number, 431
  Dirichlet conditions, 430
  discrete Fourier transforms, 433
  Euler's formula, 431, 433
  fast Fourier transform, 438
  final prediction error, 443
  Fourier analysis, 431
  Fourier series, 434
  Fourier synthesis, 431
  Fourier transform, 431, 444
  Gibbs' error, 448, 449, 451
  Hamming function, 448
  inverse Fourier transform, 431
  Levinson recursion, 442
  maximum entropy, 440
  maximum entropy spectral analysis, 440–443
  numerical transforms, 433–440
  Parseval's theorem, 436
  prediction filter, 441
  wavenumber analysis, 434
    wavenumber aliasing, 439–440
    wavenumber leakage, 439
    wavenumber resolution, 440
  Wiener filter, 445, 453
spherical shell, 48
Sputnik 3 mission, 230, 294
stacking errors, 14
Stasinowsky, W., 103
stationarity, 472
strike-pass filter, 450
strike-sensitive filter, 451
structural index, 193

tachymetry, 105
TAGS Air III, 100
  tare, 96, 97, 98, 124
thermal remanent magnetization, 256
temporal magnetic variations, 223–227, 308–314
tensors, 39–43, 47, 48, 102, 237, 239
terrain correction, 134, 138
thermo-viscous remanent magnetization, 266
TOPEX/Poseidon mission, 114
torsion balance, 102
total magnetic moment, 250

US National Geophysical Data Center, 304
units, 12
universal transverse Mercator map projection, 464

vector components, 41, 42, 46, 237, 239
vector versus scalar magnetic measurements, 283
Vening–Meinesz theory, 139
Venusian magnetic field, 297
viscous remanent magnetization, 256, 265
  magnetic viscosity, 270

wavelength filters, 448–450
westerly drift of geomagnetic field, 303
World Digital Magnetic Anomaly Map, 304
World Geodetic System-84, 464
World Magnetic Model, 304

zero-length spring, 26, 30
  principle, 91

Printed in the United States
By Bookmasters